The Geometrid Moths of Europe
Volume 2

Axel Hausmann

Dedicated to
Prof. Dr. Josef FITTKAU,
friend and mentor,
former director of the
Zoologische Staatssammlung München,
for his 75th birthday,

and to
Claude HERBULOT, Paris,
kind and amicable
'grand master' of Geometridology,
for his 96th birthday.

The Geometrid Moths of Europe

Volume 2

Axel Hausmann

Sterrhinae

Apollo Books, Stenstrup
2004

Editor: Axel Hausmann, Zoologische Staatssammlung München.

Colour plates: David Wilson, Blythburgh, United Kingdom.

Printed by: Vinderup Bogtrykkeri A/S.

This publication should be cited as:
Hausmann, A.: Sterrhinae. – *In* A. Hausmann (ed.): The Geometrid Moths of Europe 2: 1-600.

Published by:
Apollo Books
Kirkeby Sand 19
DK-5771 Stenstrup
Denmark
apollobooks@vip.cybercity.dk
www.apollobooks.com

ISBN 87-88757-37-4 (Vol. 2)
ISBN 87-88757-54-4 (Vols 1-6)

Author's address:
Münchhausenstr. 21,
D-81247 München
Germany
Phone: +49-89-8107-158
Fax: +49-89-8107-300
E-mail: Axel.Hausmann@zsm.mwn.de

Front cover: *Idaea muricata* (Hufn.) (Photo Michael Leipnitz)

Text-figs (photos) by Axel Hausmann, Sven Erlacher, Irina Fritz, Robert Trusch.

Text-figs (SEM photos) by Axel Hausmann, Sonja Knölke, Anna Hell.

Text-figs (drawings) by Axel Hausmann.

Genitalia drawings by Axel Hausmann, Irina Fritz, Sven Erlacher, Edgar Albert.

The scientific studies on the subfamily Sterrhinae, subject of the present Volume, have been supported by the German Ministry of Science, BMBF (Project BIOLOG, 'inventory of the Geometrid moth species of Europe', INGE, 01LC9904/1).

Contents

Preface

Worldwide, about 2,800 Sterrhinae species have been described, so far (SCOBLE et al. 1995; HAUSMANN 2001), and the number of undescribed species may be estimated to be similar, possibly. Suitability of Geometridae and Sterrhinae for research on biodiversity, ecosystems, and environmental changes has been highlighted by many authors, e.g. HOLLOWAY (1984), PEARSON (1994), INTACHAT & HOLLOWAY (2000), HAUSMANN (2001), BREHM (2002), SIHVONEN (2003), LAŠTUVKA & ŠEFROVÁ (2004).

In contradiction to their numerical and ecological importance, Sterrhinae are little 'appreciated' by collectors and poorly studied by amateur and professional scientists. This is due to the fact that Sterrhinae are often small sized, and that species identification and taxonomy are as difficult, as in the Eupitheciini, treated in Vol. 4 (MIRONOV 2003), compare preface and introduction to that volume. The genera *Idaea* (680 species worldwide) and *Scopula* (700 species) are, together with *Eupithecia* (1,300 species), among the most species rich genera in Lepidoptera, but are also among the worst known groups, representing therefore, one of the largest gaps in knowledge.

Distribution patterns of European Sterrhinae are more focussed on the Mediterranean subregion than those of any other Geometridae subfamily: The Sterrhinae fauna of the Iberian peninsula, for example, comprises 137 species (=70% of the 196 European Sterrhinae species), the area of central Europe, though similar in size, is inhabited by 80 species only (41%). In contrast to that pattern, the central European Larentiinae fauna covers 65% of the European species, but the fauna of the Iberian peninsula only 60%. As the southern European fauna is less explored than the other regions, there are limited data available for many Sterrhinae species, so far.

This book provides an updated, valuable, modern tool for species identification. We hope, that it will stimulate the re-examination of all existing collections, in order to improve quality of identification. Parallel to this volume, digital illustrations and additional data are published on the internet, such as the updated European checklist (www.faunaeur.org), the site of the Geometridae research initiative Forum Herbulot (www. herbulot.de) providing a list of family-group names, photographs of type specimens, and additional photographs to the GME volumes (e.g. specimens in nature, larvae), and the precursor of a 'Global Information System on Geometridae' (www.zsm.mwn.de/INGE/ wil.htm; www.biologie.uni-ulm.de/systax/).

It has to be considered a hopeful sign, that there is increasing governmental support for biodiversity research, at European level (support of the German Ministry of Science to this volume; EU support for the updated internet checklist of Lepidoptera, 'Fauna of Europe'), and at global level (GBIF-D programme for internet-based type specimen inventories, e.g. in Geometridae, 'GlobInG', financed by the German Ministry of Science). This reflects increasing sensibility and awareness about our responsibility to protect creation appropriately. But, how can we protect creatures from extinction, if our knowledge about them is scarce? Let us hope, that exploration of global species inventories is no longer obstructed by the nonsense of national and international legislation, which forbids collecting and thus promotes ignorance about nature and ecosystems!

Munich, 10.5.2004 — Axel Hausmann *(Editor-in-Chief)*

Abstract

Volume 2 of the book series 'The Geometrid Moths of Europe' presents a modern revision of the European taxa of the subfamily Sterrhinae, which is represented in Europe by 196 species belonging to six tribes and 18 genera. In Europe, the extremely species-rich genus *Idaea* is subdivided into 23 species-groups, according to external and internal morphological characters. Similarly, the genus *Scopula* is subdivided into seven species-groups.

The species are presented with their full name and reference to the original description, fully referenced junior synonyms, unavailable names (including misidentifications), diagnosis, distribution, phenology, biology including host-plants, habitat (with vertical distribution), parasitoids and differential features from similar species. Special emphasis is given to the descriptions of genitalic features. For each species line drawings of male and female genitalia, colour photographs of adults, and distribution maps for Europe and adjacent countries are provided.

Three new species and three new subspecies are described: *Idaea leipnitzi* sp. n. from Croatia, *Idaea admiranda* sp. n. from Turkey, *Idaea dromikos* sp. n. from eastern Pyrenees, *Idaea bigladiata alberti* subsp. n. from Algeria, *Oar pratana baezi* subsp. n. from Canary islands, and *Glossotrophia alba milleri* subsp. n. from Corsica.

The following 102 taxonomic changes are formally introduced:

One generic name is downgraded to synonymy: *Apostates* (WARREN, 1897), syn. n. of *Rhodostrophia*. One name is raised to genus rank: *Holarctias* (PROUT, 1913), stat. n., from synonymy of *Scopula*.

Six new combinations are proposed: *Idaea mareotica fuscularia* (TURATI, 1929), comb. n., previously subordinated under *I. calunetaria* (with ab. *saturataria* D. LUCAS, 1944 as new synonym). *Brachyglossina hispanaria* (PÜNGELER, 1913); comb. n., *Brachyglossina improbata* (STAUDINGER, 1898), comb. n.; *Brachyglossina brevitarsata* (HAUSMANN, 1991), comb. n.; *Brachyglossina consociata* (STAUDINGER, 1900), comb. n., all transferred from *Idaea*. *Holarctias rufinaria* (STAUDINGER, 1861), comb. n., transferred from *Scopula*. *Pseudocinglis moralesi* (RUNGS, 1945), comb. n., transferred from *Scopula*.

12 taxa are raised to species rank: *Cleta transiens* (PROUT, 1913), stat. n. (from subsp. of *C. ramosaria*); *Idaea mustelata* (GUMPPENBERG, 1892), stat. n. (from subsp. of *Idaea rusticata*); *Idaea albarracina* (REISSER, 1933), stat. n. (from subsp. of *I. incisaria*); *Idaea simplicior* (PROUT, 1934), stat. n. (from subsp. of *I. emarginata*); *Oar reaumuraria* (MILLIÈRE, 1864), stat. n. (from synonymy of *O. pratana*); *Scopula ochraceata* (STAUDINGER, 1901), stat. n. (from subsp. of *S. rubiginata*); *Scopula turbulentaria* (STAUDINGER, 1870), stat. n. (from subsp. of *S. turbidaria*); *Scopula divisaria* (CHRISTOPH, 1893), stat. n. (from synonymy of *S. latelineata*); *Glossotrophia sacraria* (BANG-HAAS, 1910), stat. n. (from subspecies of *G. confinaria*); *Holarctias rufinaria* (STAUDINGER, 1861), stat. n. (from subspecies of *H. sentinaria*); *Rhodostrophia discopunctata* AMSEL, 1935, stat. n. (from synonymy of *R. tabidaria*, replacing misidentified *tabidaria*, e.g. in MÜLLER 1996); *Cyclophora lennigiaria* (FUCHS, 1883), stat. n. (from subsp. of *C. albiocellaria*).

13 taxa are downgraded from species to subspecific rank: *Idaea belemiata helianthemata* (MILLIÈRE, 1870), stat. n.; *Idaea obsoletaria algeriensis* (BETHUNE-BAKER, 1888), stat. n.; *Idaea joannisiata ibericata* (WEHRLI, 1927), stat. n.; *Idaea camparia sodaliaria* (HERRICH-SCHÄFFER, 1852), stat. n.; *Idaea fractilineata subrufaria* (STAUDINGER, 1899), stat. n.; *Idaea tineata praetineata* HAUSMANN, 1994, stat. n.; *Scopula decorata congruata* (ZELLER, 1847), stat. n.; *Scopula submutata nivellearia* (OBERTHÜR, 1922), stat. n.; *Scopula decolor flaccata* (STAUDINGER, 1898), stat. n.; *Glossotrophia diffinaria asiatica* (BRANDT, 1938), stat. n.; *Glossotrophia sacraria semitata* (PROUT, 1913), stat. n.; *Rhodostrophia tabidaria cypriaria* REBEL, 1916, stat. n.; *Casilda antophilaria rosearia* (TREITSCHKE, 1828), stat. n..

69 taxa are downgraded to synonymy (names are listed in the last previously valid combination, in systematic order): *Idaea aureolaria transsylvanaria* (DANNEHL, 1927) (syn. n. to *I. aureolaria*); *Idaea turatii* (SOHN-RETHEL, 1929) and *Idaea taurica* (BANG-HAAS, 1907) (both syn. n. to *I. consanguinaria*); *Idaea sericeata subrecta* (PROUT, 1935) and *Idaea sericeata wehrlii* (PROUT, 1934) (both syn. n. to *I. sericeata calvaria*); *Idaea sericeata altivolaria* (BUBACEK, 1923) (syn. n. to *I. sericeata*); *Idaea ochrata cantiata* (PROUT, 1913) (syn. n. to *I. ochrata*); *Idaea rusticata atrosignaria* LEMPKE, 1967 (syn. n. to *I. rusticata*); *Idaea couloniata* (BALESTRE, 1907) (syn. n. to *I. laevigata*); *Idaea efflorata eriopodata* (GRASLIN, 1863) (syn. n. to *I. efflorata*); *Idaea typicata desertata* (DANNEHL, 1933) (syn. n. to *I. typicata*); *Idaea alyssumata genilaria* (WEHRLI, 1929) (syn. n. to *I. alyssumata*); *Idaea maxima* (WEHRLI, 1934) (syn. n. to *I. calunetaria*); *Idaea calunetaria lhommei* (HERBULOT, 1963) (syn. n. to *I. calunetaria valesiaria*); *Idaea euphorbiata* (BALESTRE, 1906) (syn. n. to *I. belemiata*); *Idaea substraminata* (PROUT, 1913) (syn. n. to *I. belemiata*); *Idaea cavenacata* (CHRÉTIEN, 1929) (syn. n. to *I. belemiata helianthemata*); *Idaea obsoletaria distans* HAUSMANN, 1991 (syn. n. to *I. obsoletaria*); *Idaea dilutaria praeustaria* (LA HARPE, 1853) and *Idaea dilutaria illyrica* (REISSER, 1962) (both syn. n. to *I. dilutaria*); *Idaea politaria abmarginata* (HAUSMANN, 1995) (syn. n. to *I. politaria*); *Idaea longaria faroulti* (ROTHSCHILD, 1914) (syn. n. to *I. longaria*); *Idaea nexata cirtanaria* (H. LUCAS, 1849) and *Idaea nexata marambaudista* (D. LUCAS, 1924) (both syn. n. to *I. nexata*); *Idaea fatimata* (STAUDINGER, 1895) (syn. n. to *Idaea manicaria*); *Idaea seriata paleacata* (GUENÉE, 1858) (syn. n. to *I. seriata canteneraria*); *Idaea camparia cossurata* (MILLIÈRE, 1875) (syn. n. to *I. seriata*); *Idaea zernyi* D. LUCAS, 1934 (syn. n. to *I. subsericeata*); *Idaea bustilloi* (AGENJO, 1967) (syn. n. to *I. mancipiata*); *Scopula arcuaria* (HÜBNER, 1799) (syn. n. of *Idaea trigeminata*); *Idaea cervantaria depressaria* (STAUDINGER, 1892) and *Idaea cervantaria montana* (WEHRLI, 1927) (both syn. n. to *I. cervantaria*); *Idaea contiguaria aurata* (MÜLLER, 1936), *Idaea contiguaria austriae* (MÜLLER, 1936), *Idaea contiguaria hercyniae* (MÜLLER, 1936), *Idaea contiguaria obscura* (FUCHS, 1875), and *Idaea contiguaria pallidaria* (FUCHS, 1904) (the last six taxa syn. n. to *I. contiguaria*); *Idaea infirmaria aquitanaria* (CONSTANT, 1865) (syn. n. to *I. infirmaria*); *Idaea aversata griseocorsa* (SCHAWERDA, 1929) (syn. n. to *I. aversata*); *Idaea degeneraria alticolaria* (SCHAWERDA, 1933) (syn. n. to *I. degeneraria*); *Idaea straminata minuta* (HEYDEMANN, 1934) (syn. n. to *I. straminata*); *Idaea argilata* (GUENÉE, 1858) (syn. n. to *Idaea deversaria*); *Scopula tessellaria proutiana* SHELJUZHKO, 1955 (syn. n. to *S. tessellaria*); *Scopula corrivalaria limburgensis* PROUT, 1935 (syn. n. to *S. corrivalaria*); *Scopula decorata drenowskii* STERNECK, 1941 (syn. n. to *S. decorata*); *Scopula nivellearia pseudhonestata* (WEHRLI, 1926) (syn. n. to *S. submutata nivellearia*); *Scopula submutata mudrica* KOUTSAFTIKIS, 1973 (syn. n. to *S. submutata*);

Scopula 'flaccata' languidata (PROUT, 1913) (syn. n. to *S. decolor*); *Scopula pseudoafghana* EBERT, 1965 (syn. n. to *S. beckeraria*); *Scopula marginepunctata insubrica* (VORBRODT, 1930), *Scopula marginepunctata subatrata* (WAGNER, 1919) and *Scopula marginepunctata britonaria* (OBERTHÜR, 1917) (the last three syn. n. to *S. marginepunctata*); *Scopula fulminataria* (TURATI, 1927) (syn. n. to *S. luridata*); *Scopula frigidaria schoeyeni* (SPARRE-SCHNEIDER, 1883) (syn. n. to *S. frigidaria*); *Scopula emutaria subroseata* (HAWORTH, 1809) (syn. n. to *S. emutaria*); *Idaea medioumbraria* (TURATI, 1930), *Scopula minorata ochroleucaria* (HERRICH-SCHÄFFER, 1847), *Scopula inustaria* (HERRICH-SCHÄFFER, 1847), *Scopula colonaria* (HERRICH-SCHÄFFER, 1852) and *Scopula accessaria* (HERRICH-SCHÄFFER, 1852) (the last five syn. n. to *S. minorata*); *Glossotrophia alba zahmi* HAUSMANN, 1993 (syn. n. to *G. alba*); *Glossotrophia alba capriata* HAUSMANN, 1993 (syn. n. to *Glossotrophia alba brunellii*); *Glossotrophia annae* MENTZER, 1990 (syn. n. to *G. rufomixtaria*); *Rhodostrophia vibicaria minuta* HEYDEMANN, 1933 (syn. n. to *R. vibicaria*); *Rhodostrophia oxyntis* PROUT, 1935 (syn. n. to *R. xesta* PROUT, 1935); *Cyclophora quercimontaria nigrosparsaria* (HEYDEMANN, 1938) and *Cyclophora quercimontaria privataria* (HEYDEMANN, 1938) (both syn. n. to *C. quercimontaria*); *Cyclophora delaeveri* (BERGER, 1949) (syn. n. to *C. punctaria*); *Casilda intermediaria* (TURATI, 1930) (syn. n. to *C. consecraria*).

The following 48 name bearing type specimens are selected:

Neotypes (see www.herbulot.de and nomenclatural chapters in the text) are designated for seven valid species names: *Idaea mediaria* (HÜBNER, 1819), *Idaea rufaria* (HÜBNER, 1799), *Idaea circuitaria* (HÜBNER, 1819), *Oar reaumuraria* (MILLIÈRE, 1864), *Oar pratana* (FABRICIUS, 1794), *Scopula turbidaria* (HÜBNER, 1819), and *Glossotrophia asellaria* (HERRICH-SCHÄFFER, 1847).

Lectotypes (see www.herbulot.de and nomenclatural chapters in the text) are designated for 28 valid species and subspecies (systematic order): *Cleta transiens* PROUT, 1913, *Cleta perpusillaria* (EVERSMANN, 1847), *Idaea lambessata* (OBERTHÜR, 1887), *Idaea ossiculata* (LEDERER, 1871), *Idaea ochrata albida* (ZERNY, 1936), *Idaea nevadata* (WEHRLI, 1934), *Idaea completa* (STAUDINGER, 1892), *Idaea incisaria praecisa* (REISSER, 1933), *Idaea albitorquata* (PÜNGELER, 1909), *Idaea mancipiata* (STAUDINGER, 1871), *Idaea rubraria* (STAUDINGER, 1901), *Idaea degeneraria erschoffi* (CHRISTOPH, 1872), *Cinglis humifusaria* (EVERSMANN, 1837), *Scopula virgulata rossica* (DJAKONOV, 1926), *Scopula orientalis* (ALPHÉRAKY, 1876; *Scopula subtilata* (CHRISTOPH, 1867), *Scopula submutata nivellearia* (OBERTHÜR, 1922), *Scopula submutata taurilibanotica* (WEHRLI, 1932), *Scopula vigilata* (SOHN-RETHEL, 1929), *Scopula ochraceata* (STAUDINGER, 1901), *Scopula beckeraria rebeli* (PROUT, 1913), *Scopula arenosaria* (STAUDINGER, 1879), *Scopula albiceraria vitellinaria* (EVERSMANN, 1851); *Glossotrophia rufomixtaria* (GRASLIN, 1863), *Rhodostrophia vibicaria strigata* (STAUDINGER, 1871), *Rhodostrophia discopunctata* AMSEL, 1935, *Timandra rectistrigaria* (EVERSMANN, 1831), *Cyclophora albipunctata griseolata* (STAUDINGER, 1897).

Further lectotypes are designated for 13 synonyms (alphabetical order): *asiaeminoris* (AMSEL, 1935) (*Cyclophora puppillaria* syn.), *depressaria* (STAUDINGER, 1892) (*Idaea cervantaria* syn.), *europaea* WEHRLI, 1934 (*Idaea camparia* syn.), *falsaria* GUENÉE, 1858 (*Glossotrophia rufomixtaria* syn.), *fatimata* STAUDINGER, 1895 (*Idaea manicaria* syn.), *granadaria* STAUDINGER, 1901 (*Idaea lambessata* syn.), *insularis* WEHRLI, 1926 (*Glosso-*

trophia asellaria syn.), *leukiberica* WEHRLI, 1927 (*Scopula decorata congruata* syn.), *minor* ALPHÉRAKY, 1892 (*Rhodostrophia jacularia* syn.), *montana* WEHRLI, 1927 (*Idaea cervantaria* syn.), *reynaldiata* ROUAST, 1876 (*Idaea alyssumata* syn.), *romani* WEHRLI, 1929 (*Idaea alyssumata reynaldiata* syn.), *violacearia* STAUDINGER, 1901 (*Idaea obsoletaria lilaceola* syn.).

Comparison with the most recent check list of European Geometridae (MÜLLER 1996) reveals 20 new Sterrhinae species: One species, *Ochodontia adustaria* (FISCHER DE WALDHEIM, 1840), transferred from Larentiinae to Sterrhinae. Six species described after MÜLLER (1996): *Idaea leipnitzi* sp. n.; *Idaea admiranda* sp. n.; *Idaea acutipennis* HAUSMANN & HONEY, 2004; *Idaea dromikos* sp. n.; *Scopula scalercii* HAUSMANN, 2003; *Cyclophora serveti* REDONDO & GASTON, 1999. Five species recorded for Europe in previous articles but not in MÜLLER (1996): *Idaea nevadata* (WEHRLI, 1934) (not synonym of *Idaea figuraria*), *Idaea completa* (STAUDINGER, 1892), *Idaea troglodytaria* (HEYDENREICH, 1851), *Scopula divisaria* (CHRISTOPH, 1893) and *Glossotrophia mentzeri* (HAUSMANN, 1993). Eight species result from species 'splitting' rather than from real new entries: *Idaea mutilata* (STAUDINGER, 1876) (sister species of *I. litigiosaria*); *Idaea mustelata* (GUMPPENBERG, 1892) (sister species of *I. rusticata*); *Idaea albarracina* (REISSER, 1933) (sister species of *I. incisaria*); *Idaea simplicior* (PROUT, 1934) (sister species of *I. emarginata*); *Cinglis andalusiaria* WAGNER, 1935 (sister species of *C. humifusaria*); *Scopula turbulentaria* (STAUDINGER, 1870) (sister species of *S. turbidaria*); *Rhodostrophia discopunctata* AMSEL, 1935 replacing misidentified *R. tabidaria* on the European fauna list (with *tabidaria*, however, belonging to the European fauna too); *Cyclophora lennigiaria* (FUCHS, 1883) (sister species of *C. albiocellaria*).

With the above mentioned synonymisations 10 'species' have been removed from the fauna of Europe (MÜLLER 1996). Three other taxa are erroneously mentioned in MÜLLER (1996): *Idaea minuscula* ('RIBBE, 1912') (nomen nudum), *Glossotrophia diffinaria* PROUT, 1913 (misidentified, Turkish species), and *Scopula remotata* (GUENÉE, 1858) (misidentified, Indian species).

Hence there are 20 additional species for the fauna of Europe, which are, however, partially balanced by the 13 species which are deleted from the list or downgraded to infraspecific categories. Thus the species number of European Sterrhinae increases from 189 (MÜLLER 1996) to 196 in GME2.

For some further names, erroneously considered valid species of the European fauna, and for additional species with weakly confirmed occurrence in Europe, see introduction to the checklist.

Acknowledgements

The GME book series is founded upon the knowledge of numerous background contributors, often 'amateurs', in a number of countries. Their assistance is greatly appreciated even if only those, who were directly involved in the contents of this Volume 2 can be named here. A number of colleagues in Lepidopterology granted valuable help in the preparation, or were involved in the editorial process, of Volume 2. I owe them sincere thanks for constructive advice and useful comments. Especially I enjoyed the kind cooperation of:

Dr. Leif AARVIK, Oslo (NO): Provision of faunistic data (revision of maps for Norway).
Dr. David ABRAHAM, Lund (S): Provision of literature.
Ulrich AISTLEITNER, Feldkirch (A): Provision of faunistic data and literature.
Edgar ALBERT, Munich (D): Help in genitalia drawings, and dissections.
Dr. Elena M. ANTONOVA, Moskow (RUS): Provision of literature and faunistic data (Russia).
Dr. Giorgio BALDIZZONE, Asti (I): Provision of faunistic data (Italy).
Dr. Graziano BASSI, Torino (I): Provision of faunistic data (Italy).
Dr. Evgenj BELJAEV, Wladivostok (RUS): Provision of literature.
Dr. Edgardo BERTACCINI, Forlí (I): Provision of faunistic data (Italy).
Stoyan BESHKOV, Sofia (BG): Provision of faunistic data (revision of maps for Bulgaria).
Rolf BLAESIUS, Eppelheim (D): Donation of material, provision of faunistic and biological data.
Ulf BUCHSBAUM, Munich (D): Help in computer questions, dissections and other technical support.
BUNDESMINISTERIUM FÜR BILDUNG UND FORSCHUNG, Bonn (D): Funding of the project 'Inventur der Geometriden-Arten Europas [INGE]' in the BIOLOG-program as scientific basis for this publication.
Dr. Jaroslaw BUSZKO, Torun (PL): Provision of many faunistic data (revision of the maps for Poland), advice in zoogeography.
'BUTTERFLY CONSERVATION',with funding from 'English Nature' for giving very informative and detailed informations on the results of their studies on some threatened English Sterrhinae species.
John CHAINEY, London (GB): Type investigations at the BMNH.
Martin CORLEY, Oxfordshire (GB): Provision of many faunistic data (e.g. Portugal), loan of material. Linguistic editor, constructive revision of the manuscript.
Otmar CZADEK, Tenerife (E): Provision of biological information.
Dr. Jorge DANTART, Barcelona (E): Provision of faunistic data (Spain) and literature.
Dr. Leonardo DAPPORTO, Florence (I): Provision of faunistic data (Tuscany, Italy). Kind support during the author's visit.
Dr. Philippe DARGE, Clenay (F): Provision of literature, kind support during the author's visit.
Dr. Eteri DIDMANIDZE, Tbillissi (Georgia): Provision of faunistic data (Georgia) and literature, loan of specimens. The correspondence was kindly translated by her grand-daughter Mary Odischaria-Munz (D).
Dr. Klaus DOBAT, Tübingen (D): Identification of host-plants.
Dr. Feza DOGANLAR, Adana (TR): Provision of faunistic data (Turkey) and literature, loan of specimens.
Thomas DRECHSEL, Neubrandenburg (D): Provision of faunistic data and host-plant records.
Günter EBERT, Karlsruhe (D): Kind support during the author's visit to the SMNK, loan of material.
Dr. Ulf EITSCHBERGER, Marktleuthen (D): Loan of about 200,000 geometrid moths from his collection.
Gernot EMBACHER, Salzburg (A): Provision of faunistic data (Austria, Greek islands), and literature.
Sven ERLACHER, Munich (D): Extensive help in digital photography, line drawings, preparation of maps, accompanying molecular research, management of the BMBF project.
Siegfried ERLEBACH, Innsbruck (A): Communication of many host-plant relationships.
Andrès EXPÓSITO HERMOSA, Madrid (E): Provision of faunistic data (Spain), taxonomic discussions.
Dr. Claudio FLAMIGNI, Bologna (I): Provision of faunistic data (Italy) Kind support during the author's visit.
FREUNDE DER ZOOLOGISCHEN STAATSSAMMLUNG MÜNCHEN, e.V. (association): Funding of students, guest scientists and visits to museums.
Egbert FRIEDRICH, Jena (D): Loan of material, provision of faunistic data.
Irina FRITZ, Munich (D): Extensive help in digital photography, line drawings, and data base management.
Javier GASTON, Bilbao (E): Provision of faunistic data (Spain), and literature, taxonomic discussions.
Dr. Jörg GELBRECHT, Berlin (D): Provision of many faunistic data (e.g. Turkey, revision of maps for eastern Ger-

many), of host-plant records, and of literature, loan of material, taxonomic discussions. Kind support during the author's visit.
Mauro GIANTI, Cuneo (I): Provision of faunistic data (Italy), taxonomic discussions.
Barry GOATER, Hampshire (GB): Provision of many faunistic data (France, Balearics). Kind support during the author's visit.
Dr. Stanislav GOMBOC, Ljubljana (SLO): Provision of faunistic data (Slovenja), donation of material.
Heinz HABELER, Graz (A): Provision of faunistic data (Croatia, Greece), loan of material.
Penny HALE (E): Provision of faunistic data (southern Spain).
Norman HALL, Reading, Berks (GB): Provision of detailed information on host-plants and differential diagnoses of many species (e.g. *I. saleri*), loan of material.
Prof. Dr. Gerhard HASZPRUNAR, München (D): Support and encouragement of research on the GME series as the director of the ZSM.
Dr. Christoph HÄUSER, Stuttgart (D): Kind support during the author's visit at the SMNS, loan of material. Multiple assistance in the management of the accompaining BMBF project.
Anna HELL, Berlin (D): SEM photography (antennae, legs)..
Dr. Claude HERBULOT, Paris (F): Kind support during the author's visits, release of his entire collection (with many types), provision of literature and faunistic data, taxonomic discussions.
Dr. Jeremy HOLLOWAY, London (GB): Kind support during the author's visits at the BMNH.
Dr. Joachim HOLSTEIN, Stuttgart (D): Multiple assistance in the management of the accompaining BMBF project.
Dr. Peter HUEMER, Innsbruck (A): Kind support during the author's visit at the TLMF, provision of faunistic data. Mediation of data to host-plant relationships.
M. HONEY, London (GB): Kind support during the author's visit at the BMNH, loan of material.
Dr. Hiroshi INOUE, Iruma City (JP): Donation of literature.
Dr. Isabel IZQUIERDO, Madrid (E): Kind support during the visit of R. TRUSCH, loan of type material.
Marius JUNKER, MUNICH (D): Help in literature research, preparation of the maps and dissections.
Dr. Lauri KAILA, Helsinki (FI): Provision of faunistic data (revision of maps for Finland), donation of material.
Jari KAITILA (FI): Provision of faunistic data (e.g. Bulgaria).
Ole KARSHOLT, Copenhagen (DK): Kind support during the author's visit to the ZMUC, loan of material, valuable help in the accompanying 'Fauna of Europe' project.
Gareth KING, MADRID (E): Provision of many host-plant data (from Spain), parasitoids, and literature.
Sonja KNÖLKE, Munich (D): Dissections (about 1,500), SEM photography (antennae).
Detlev KOLLIGS (D): Provision of faunistic data (Germany).
Igor KOSTJUK, Kiyev (UKR): Provision of many literatur data (e.g. all records from Ukraine, revision of maps for eastern Europe), provision of literature, loan of material, taxonomic discussions.
Werner KRAUS, Kaiserslautern (D): Provision of literature and faunistic data (e.g. Spain), loan of material.
Karl KUCHLER †, Munich (D): Loan and donation of material.
Dr. Anatolij KULAK, Minsk (BY): Provision of literature and faunistic data (Byelorussia).
Jaakko KULLBERG, Helsinki (FI): Provision of faunistic data (polar Urals).
Dr. Gerardo LAMAS, Lima (Peru): Discussion of nomenclatural problems.
Dr. Zdeněk LAŠTUVKA, Brno (CZ): Provision of faunistic data (Czech Republic).
Michael LEIPNITZ, Stuttgart (D): Provision of photographs, faunistic data and many host-plant records, donation of material.
Dr. Hans LÖBEL, Sondershausen (D): Donation of material.
Dr. Martin LÖDL, Vienna (A): Kind support during the author's visit to the NHMW, loan of material (also types), research in the collection (*I. allongata*).
Dr. Hans MALICKY, Lunz (A): Donation of extensive material from Mediterranean regions, e.g. Crete.
Dr. Adam MALKIEWICZ, Wroclaw (PL): Provision of faunistic data (revision of maps for Poland).
Risto MARTIKAINEN, Tampere (FI): Donation of literature.
Geoff MARTIN, London (GB): Valuable information about type-specimens deposited in BMNH.
Dr. Robert MAZEL, Perpignan (F): Provision of literature and faunistic data.
Dr. Wolfram MEY, Berlin (D): Kind support during the author's visit to the MNHU, loan of material (also types).
Dr. Mark MEYER, Luxemburg (LUX): revision of maps for Luxemburg and provision of other faunistic data.
Michael MILLER, Munich (D): Dissections (about 2,000), accompanying molecular research.
Dr. Vladimir MIRONOV, St. Petersburg (RUS): Provision of many faunistic data (mainly Russia), loan of material (also types), taxonomic discussions.
Dr. Alessandro MINELLI, Padova (I): Discussion of nomenclatural problems.

Dr. Bernd MÜLLER, Berlin (D): Provision of many host-plant records, loan of material and donation of collection of pupae, kind support during the author's visit.
Jose Manuel NOVOA, Madrid (E): Provision of faunistic data (Spain: Extremadura), and literature.
Kari NUPPONEN, Espoo (FI): Provision of faunistic data (southern Urals and eastern steppes).
Dr. Zuhal OKYAR, Edirne (TR): Provision of literature and faunistic data (European Turkey), taxonomic discussion.
Dr. Paolo PARENZAN, Palermo (I): Kind support during the author's visit, provision of faunistic data, and literature.
Dr. Mark PARSONS, London (GB): Provision of faunistic data and personal comments on English species.
Dr. Linda PITKIN, London (GB): Kind support during the author's visits to the BMNH.
Norbert PÖLL, Bad Ischl (A): Provision of many faunistic data (mainly Austria and Czech Republic), loan of material.
Willi and Jurate DE PRINS, Antwerpen (B): revision of maps (Belgium) and provision of other faunistic data.
Dr. Valter RAINERI, Genova (I): Provision of literature and faunistic data.
Victor REDONDO, Zaragoza (E): Provision of literature and faunistic data, taxonomic discussions.
Jim REID, Hertfordshire (GB): Provision of faunistic data (revision of the maps for Great Britain).
Dr. Lazi RESER, Luzern (CH): Provision of faunistic data, and literature, discussion of taxonomic problems.
Alex RIEMIS, Turnhout (B): Provision of faunistic data (e.g. Turkey).
Luis ROMERA, Madrid (E): Provision of host-plant data (from Spain), and loan of specimens.
Dr. Laszló RONKAY, Budapest (H): Provision of faunistic data.
Dr. Nils RYRHOLM (S): Provision of literature and faunistic data (revision of maps for Sweden).
Paul SAMMUT, Rabat (Malta): Provision of literature, and faunistic data, donation of material from Malta.
Dr. Klaus SATTLER, London (GB): Scientific training, comments, and advice.
Dr. Stefano SCALERCIO, Donnici inf. (I): Kind support during the author's visit, provision of many faunistic data (Italy), loan and donation of material.
Paul SCHAIDER, Igoumenitsa (GR): Provision of many faunistic data, donation of material (Spain, Greece).
Willibald SCHMITZ and J. VIEHMANN, Bergisch Gladbach (D): Provision of faunistic data.
Dr. Malcolm SCOBLE, London (GB): Kind support during the author's visits to the BMNH.
Dr. Andreas SEGERER, Munich (D): Kind help in computer questions (digital photography, network).
Thomas SOBCZYK, Hoyerswerda (D): Biological information (*Scopula decorata*).
Dr. Pasi SIHVONEN, Helsinki (FIN): Taxonomic discussion, donation of material and literature.
Dr. Yuri SINEV, St. Petersburg (RUS): Loan of material (also types).
Bernard SKINNER, South Croydon (GB): Provision of faunistic data (revision of the maps for Great Britain and Ireland), and of host-plant records.
Peder SKOU, Stenstrup (DK): Idea and first moves to initiate the GME-series. Revision of the manuscript with valuable comments and advice, provision of literature, faunistic data (e.g. Spain, France, Denmark) and host-plant records, kind support during the author's visits, taxonomic discussions.
Bjarne SKULE, Rødovre (DK): Provision of faunistic data, transport of material for colour plate photography.
Manfred SOMMERER, Munich (D): Provision of faunistic data, loan of material. Multiple advice on many details.
Dr. Mihai STANESCU, Bucharest (RO): Provision of literature and faunistic data (revision of maps for Romania).
Dr. Dieter STÜNING, Bonn (D): Kind support during the author's visit at the ZFMK, information on type-material of ZFMK, loan of material (also types).
Dr. Gerhard TARMANN, Innsbruck (A): Kind support during the author's visit to the TLMF.
Dr. Valentin TIKHONOV, Pjatigorsk (RU): Loan of material.
Dr. Robert TRUSCH: Extensive help in digital photography, preparation of maps, type research in SMNK MNCN, ZFMK and BMNH, management of the BMBF project.
Francesc VALLHONRAT, Barcelona (E): Provision of faunistic data (Spain: Catalonia).
Dr. Jaan VIIDALEPP, Tartu (EE): Provision of literature, loan of material, taxonomic discussions.
Hartmut WEGNER, Adendorf (D): Provision of faunistic data and many host-plant records, donation of material.
Dr. Norbert ZAHM, Schmelz-Hüttersdorf (D): Provision of faunistic data (Germany, central Italy).
Dr. Alberto ZILLI, Roma (I): Kind support during the author's visit to the MZUR. Provision of faunistic data (Italy) and research in the collection.
David WILSON (GB): Photography of colour plates, shipment of specimens.
ZOÖLOGISCH MUSEUM AMSTERDAM (ZMAN) and EIS-Workgroup Lepidoptera Faunistics (organised by Rob DE VOS): Provision of many faunistic data (revision of maps for the Netherlands).
Dr. Vadim ZOLOTUHIN, Uljanovsk (RU): Provision of faunistic data (Russia).

Finally - but not least - I am grateful to my wife Silvia and my children, who encouraged and supported me, though suffering from the long periods of the preparation of this book.

Introduction

Methodical notes

More than 300,000 **collection specimens** of western Palaearctic Sterrhinae have been examined, about 6,000 dissected and their genitalia analysed. The author has studied the Sterrhinae material of the public collections in London, Munich, Berlin, Vienna, Bonn, Copenhagen, Madrid, Rome, Stuttgart and Karsruhe, and many private collections as well. Additional work has been done on extensive loans from other museums, e.g. from St. Petersburg and Kiev.

All **measurements** (wingspan, length of tarsus, length of antennal cilia or branches) are original data of the author, based on examination of extensive museum material (ZSM). Wingspan measurements are taken exclusively from specimens collected in nature, not from reared specimens, which are often smaller. In the genus *Idaea*, rearing may even lead to larger size.

Antennal structure was usually studied in a few females and 5-20 males per species under the stereo microscope. Further measurements have been taken from scanning electronic exposures by the author and by Sonja Knölke (2002; thesis at the ZSM). Measurements refer to medial part of the antenna in lateral view. Width of flagellum usually measures 0.06-0.14 mm, but flagellum may be thicker in the *aversata* species-group of *Idaea*. Antennae are often slightly flattened dorsoventrally, thus the width of flagellum may measure only 0.04-0.11 mm in ventral view. As a consequence, often, antennal cilia appear to be longer by 20-30% in ventral view. Antennae with intersegmental incisions are measured in the central part of the antennomere, not at the incision. Length of cilia (=sensilla trichoidea) or branches refer to the longest ones, usually found in the medial part of the antenna.

For detailed examination of genitalia and external morphology specimens from all parts of the distribution area have been selected. This is not repeated in the text, unless differences have been found.

Statement of absence of pencil from male hindlegs refers, in accordance to Sterneck (1940) to the lack of a hair brush arising from base of tibia. Nevertheless there may be bushy hairscales inserted all over the length of tibia.

Phenology is given mainly on the base of collection data, literature data are used only as additional source. Original data from four large electronic databases have been consulted, too, including about 15,000 Sterrhinae from southern Italy, 4,000 from southern Germany, and 6,000 from Israel (all original data, ZSM). It is important to note, that the flight period of many species can vary greatly. Therefore single outlying records are usually omitted.

Larval host-plants and **habitats** are taken from literature, the author's own original data, and personal communications. Citation of the outstanding book of Ebert (2001) as secondary literature source has, sometimes, been given preference to primary sources, in order to facilitate data tracing, as many articles are published in obscure journals. Plant families are added, in parenthesis, to the host-plant species, but are omitted in the lists of host-plants from rearings. Vertical distribution refers mainly to the analysis of collection label data, with literature data used as additional source.

For **zoogeographical terms** see volume 1. The used terms refer to actual distribution patterns rather than to glacial refugia (compare proceedings 'zoogeography workshop', Munich 2002, www.herbulot.de).

Distribution maps show the hypothetical resident distribution area (grey), combined with localities from which specimens have been examined (black dots ●; see vol. 1, chapter 3, zoogeography). Old records are marked with a ring (○), when the species has probably disappeared from that locality. Records from outside the residence area (due to migration or artificial transport) are marked with an asterisk (*). Doubtful records are indicated with a question mark. Studied data have been transcribed to more than 50,000 dots on the distribution maps. On the maps, the recent and modern faunistic inventories of Finland (HULDÉN et al. 2000) and south-western Germany (EBERT 2001) are included, too; reliability of these data is ascertained by the accurate data control management of the authors. For supply of further controlled data see acknowledgements, their reliability has been checked. The exact position of the dots on the maps and the whereabouts of the controlled collection specimens can be traced with the original contributor's lists stored at the ZSM.

Additional photographs to this volume will be published in the internet-based 'global information system Geometridae' (GBIF-D: www.zsm.mwn.de/INGE/wil.htm; www.biologie.uni-ulm.de/systax). These sites provide important biodiversity information, e.g. digital images of the neo- and lectotypes (with labels) designated in this volume. Life photographs of adults and immature stages of European Geometridae in nature are available under www.herbulot.de, such as a list of family-group names in Geometridae.

Nomenclatural notes

Though comprehensive information has already been published in SCOBLE (1999), detailed nomenclatural notes to all published names in European Geometridae are included also in this book series for the following reasons: (1) About 1,700 original descriptions (=98% of all existing names) have been checked again. In many cases information on original descriptions has been improved and partially corrected. (2) Much type material has been traced and examined (name bearing types for more than 500 available names), and relevant information is presented. (3) Taxonomic status of all taxa has been critically revised.

Information on type material is presented in the nomenclatural chapters as specified in the original description; in parentheses additional information is given from evidence in the typeholding Natural History Museum.

In cases of misidentification the true identity of the used name is indicated in parentheses, unless referring to a valid European species name of the same genus, which is easy to trace in this volume.

Additional taxa are nomenclaturally available, and are junior synonyms, unless nomenclatural status or taxonomic rank is specified, e.g. subspecific rank, emendation, misidentification or homonymy. Similarly, types have not been traced, unless type deposition is stated.

As recommended by the Code, ICZN 1999, square brackets are not used to indicate years or authors of original publications which have been investigated by secondary sources, subsequently. Names are presented in the original spelling, except for initial upper case of an epithet, which is changed (emended) to lower case. The diacritic marks ö, ä and ü are transcribed to the letters ae, oe and ue, hyphens within the epithet are omitted. As already stated in Vol. 1 gender agreement is ignored in the GME series, detailed discussion and justification is provided by SOMMERER (2002) and the resolution of the SEL-congress 2002. Abbreviated generic names are completed. Such abbreviations as 'v.' or 'aberr.' are unified to 'var.' and 'ab.' respectively. For the type locality both the original description and the type label are considered. When not congruent both are given. Formally, the type locality is presented in accordance with the actual country and region names. Old original locality names are mentioned in parentheses, when strongly differing from those in current use. For exact spellings of originally abbreviated generic names, of type labels, and of type locality from the original description see exact data presented by the GBIF-D project (www.biologie.uni-ulm.de/systax). For detailed information on the family-group names see the Forum Herbulot world list of family-group names under www.herbulot.de.

Some nomenclatural problems, which are, partly, but not explicitly ruled by the Code (ICZN 1999), have been discussed with some well-experienced colleagues. According to their kind advice the following guidelines are used for those questions:

Non binominal names, which later on are used binominally: types are preferably those of the second author, see article 72.7, Code (ICZN) (MINELLI 2002, pers. comm.).

Infrasubspecific names, which have been later raised to available rank: material of both authors may be considered syntypes, the question is not explicitly ruled by the Code (ICZN), see articles 72.7, 45.(5).1, Code (ICZN) (MINELLI 2002, LAMAS 2002, pers. comm.). As long as the name is treated at infrasubspecific rank, it is just 'original material' but not type material s.str.

Names introduced in synonymy, later raised to available rank: types are preferably those of the first author; if no material mentioned there, even material of the second author may be considered, see article 72.4.3, Code (ICZN).

Replacement names for homonyms: types are those of the first author, see article 72.7, Code (ICZN) (MINELLI 2002, pers. comm.)

Emended names: types are, analogously to replacement names, those established by the first, original author.

'Replacement names' for misidentifications are not replacement names, but new descriptions, if they fulfill the criteria for descriptions (description or indication). Hence, types (and thus the locus typicus) are those of the second author (LAMAS 2002, pers. comm.).

Validity of emendations: Historically many Geometridae species names ending in -*aria* were changed to -*ata* (and vice-versa), referring to the old convention of naming species with filiform antennae with -*ata* and those with bipectinate antennae with -*aria*. Such modified names are available as unjustified emendations, when the intention to change the name is explicitly stated (e.g. in BOISDUVAL, 1840: 177; TREITSCHKE 1827: V; DUPONCHEL 1830: 102; HERRICH-SCHÄFFER 1846: 1-4). When, e.g. in long name lists or catalogues, the author has changed all species names according to antennal structure, but without mentioning that reason, they must be regarded as incorrect subsequent spellings, according to article 33.3 Code (ICZN). Even if the original spelling is mentioned in brackets or if there are accompanying descriptions of characters, or illustrations, such names remain incorrect subsequent spellings, e.g. HÜBNER's (1799) '*Geometra aversaria* L.'; the original spelling had been *aversata* LINNAEUS, 1758 (MINELLI 2003, LAMAS 2003, pers. comm.).

LERAUT (1997) proposes replacement of four species-group names in Sterrhinae, which have been commonly used over a very long period, by supposedly senior synonyms. This practice is not adopted here, for reasons see under the nomenclatural chapters of the species *Idaea macilentaria* (no. 25), *Idaea rubraria* (no. 107), *Idaea deversaria* (no. 111), and *Scopula caricaria* (no. 121). Similarly, usage of the name *Scopula subpunctaria* is given preference, despite the existence of a senior synonym (no. 150).

Abbreviations

BMNH	The British Museum (Natural History), London, United Kingdom
CNCI	Canadian National Collection of Insects, Ottawa, Ontario Canada
DBUV	Departament de Biologia Animal, Universitat de Valencia, Spain
DEI	Deutsches Entomologisches Institut, Eberswalde, Germany
EIHU	Entomological Institute, Hokkaido University, Sapporo, Japan
IRSN	Institut Royal des Sciences Naturelles, Brussels, Belgium
ITZA	Instituut voor Taxonomisch Zoölogie, Universiteit van Amsterdam, The Netherlands
IZBE	Institute of Zoology and Botany of the Academy of Sciences, Tartu, Estonia
MCZ	Museum of Comparative Zoology, Harvard University, Cambridge, Massachusetts, U.S.A
MNCN	Museo Nacional de Ciencias Naturales, Madrid, Spain
MNHN	Musée National d'Histoire Naturelle, Paris, France
MNHU	Museum für Naturkunde, Humboldt-Universität zu Berlin, Germany
MRSN	Museo Regionale di Scienze Naturali, Torino, Italy
MZUR	Museo di Zoologia dell'Università 'La Sapienza', Rome, Italy
NHMW	Naturhistorisches Museum, Wien, Austria
NHRS	Naturhistoriska Riksmuseet, Stockholm, Sweden
NSMW	Naturwissenschaftliche Sammlung, Museum Wiesbaden, Germany
OÖLL	Oberösterreichisches Landesmuseum, Linz, Austria
SMF	Senckenberg Museum, Frankfurt a.M., Germany
SMNS	Staatliches Museum für Naturkunde, Stuttgart, Germany
SMTD	Staatliches Museum für Tierkunde Dresden
TMB	Termeszettudomanyi Muzeum Allattara (Hungarian Natural History Museum), Budapest, Hungary.
UZIU	Universitets Zoologiska Institut, Uppsala, Sweden
ZFMK	Zoologisches Forschungsinstitut und Museum A. Koenig, Bonn, Germany
ZISP	Zoological Institute, Russian Academy of Sciences, Saint-Petersburg, Russia
ZMH	Zoological Museum of the University, Helsinki, Finland
ZMKU	Zoological Museum, National Shevchenko University, Kiev, Ukraine
ZMUC	Zoological Museum of the University, Copenhagen, Denmark
ZSM	Zoologische Staatssammlung München, Germany

coll.	collection
prov.	province (dept. = département in France)
pers. comm.	personal communication
var. cit.	various citations, used only for host-plant records: the complete list of references is substituted by 'var. cit.', when at least four identical references for the host-plant were found. These indications usually refer to the well known identification literature listed in volume 1, but additionally to LHOMME 1935; SVENSSON (1993), WARING & TOWNSEND (2003), and EBERT (2001).

Systematic account

Subfamilia **Sterrhinae** MEYRICK, 1892

Seven tribes with 114 valid genera and about 2,800 described species (SCOBLE et al. 1995; SCOBLE 1999), more than half of them belonging to the mega-diverse genera *Idaea* and *Scopula*. Distributed worldwide except Greenland (WOLFF 1964), Iceland, New Zealand and some smaller islands of the Pacific. Subfamily reputedly monophyletic according to STERNECK (1941: 262), but monophyly questioned by various authors, e.g. HOLLOWAY et al. (2001), HAUSMANN (2001), SIHVONEN (2003), and SIHVONEN & KAILA (2004). Possible apomorphies for the subfamily Sterrhinae: length and longitudinal orientation of the forewing vein connecting areole with origin of M1, presence of dark discal spots, which may have pale centres (HOLLOWAY 1997; HOLLOWAY et al. 2001; SIHVONEN & KAILA 2004). Monophyly is furthermore supported by the presence of only one pheromone type, olefinic acetate (ABRAHAM 2002) and by preliminary molecular analysis (28S D1, 28S D2, ND1: ABRAHAM et al. 2001). Important revisionary papers and treatments at subfamily scale: STERNECK (1940; 1941), COVELL (1983), HOLLOWAY (1997), HOLLOWAY et al. (2001), SIHVONEN (2003), and SIHVONEN & KAILA (2004).

Sister group relationship between Sterrhinae and Larentiinae stated by many authors. Weak relationship stressed by HOLLOWAY (1997) basing mainly on hammer-headed ansa in the tympanal organs and on occurrence of secondary sexual features on ♂ sternum A2. Other features common to both subfamilies: absence of tympanic lacinia (COOK & SCOBLE 1992), ventral position of the euphragma opening (FÄNGER 1999), and occurrence of areoles. Pupal morphology confirms close relationship between Sterrhinae and Larentiinae, too (PATOČKA 2003). Within Geometridae, Sterrhinae are the only subfamily with exclusive use of olefinic acetate pheromones, reminiscent of Drepanidae, possibly indicating basal position of Sterrhinae in phylogeny of Geometridae (ABRAHAM et al. 2002b). First results from sequencing mitochondrial ND1 and 28S ribosomal RNA (ABRAHAM et al. 2001), and 28 S D2 (YOUNG 2003) support basal position of both Sterrhinae and Larentiinae in the phylogeny of Geometridae with Larentiinae at the most basal position, and Sterrhinae as sister group to the rest of Geometridae except Larentiinae. In the three mentioned studies, however, the 'Timandrini lineage' (see below) was not examined.

The Sterrhinae are, currently, subdivided into seven tribes worldwide: Sterrhini, Scopulini, Rhodostrophiini, Cyllopodini (neotropical), Timandrini, Cosymbiini, Rhodometrini. All recognized tribes are supposed to be monophyletic except Cosymbiini and Rhodostrophiini (SIHVONEN & KAILA 2004). The Afrotropical 'Aletini' and the Old World 'Problepsini' are considered to fall within the concept of Scopulini (see SIHVONEN & KAILA 2004). Rhodometrini and Rhodostrophiini do not reach Australia. Phylogenetic separation of tribes Sterrhini and Scopulini is suggested for the Tertiary by STERNECK (1940/1941: 201), i.e. younger than 70 my. Worldwide distribution and establishment of the species-rich genera *Cyclophora*, *Scopula* and *Idaea* suggest, that these genera were well established before the disjunction of the Gondwanaland shelves by continental drift, and glaciation of the Antarctic, about 30 my ago, and that tribal separations in Sterrhinae took place long before.

Two main lineages within recent Sterrhinae have been recognized, i.e. the 'Timandrini lineage' (Timandrini, Cosymbiini, Rhodometrini) and the 'Scopulini lineage' including the other four tribes (Sterrhini, Scopulini, Rhodostrophiini and Cyllopodini) (HAUSMANN 1993a; HOLLOWAY 1997; 2001; SIHVONEN 2001; SIHVONEN & KAILA 2004).

The 'Timandrini lineage' is supported by absence of ventral setae on antennal flagellum (autapomorphy) and flat signum sclerite with medial ridge (autapomorphy, in some taxa secondarily lost, e.g. *Traminda*), larval (SINGH 1953) and pupal (NAKAMURA 1994) characters. Male antennae of most species in this lineage bipectinate, with rami arising from ventral part of flagellum. Absence of gnathos (with exceptions) and hindtibial scent pencils in 'Timandrini lineage' reminiscent of Larentiinae (SIHVONEN & KAILA 2004), even the anastomosis of hindwing cell with Sc+R1 in Rhodometrini and habitus approach typical characters of certain Larentiinae (e.g. genus *Lythria*). Preliminary data sets from own molecular research (mitochondrial genes ND1, COI, and 16S rRNA) support monophyly of 'Timandrini lineage' (MILLER et al. 2001; HAUSMANN et al. in prep.). As the basal resolution of the tree is much better than in the 'Scopulini lineage' in the COI studies, the 'Timandrini lineage' may be younger, phylogenetically.

The 'Scopulini lineage' is supported by (1) presence of hindtibial scent pencils, that may be secondarily reduced, (2) presence of membranous pouch between male sternum A7 and A8, often associated with strongly sclerotized sternum A8 (with exceptions), and (3) absence of male medial hindtibial spurs', with exceptions in Rhodostrophiini (HOLLOWAY 1997; SIHVONEN & KAILA 2004). Sternum A2 of most Scopulini species and of many Sterrhini is characterized by a central sclerotized setose pouch and posterolateral processes. Within this lineage basal position of Rhodostrophiini and Cyllopodini has been suggested (SIHVONEN & KAILA 2004), but there are also arguments for basal position of Sterrhini (NAKAMURA 1994; HOLLOWAY 1997; SIHVONEN & KAILA 2004).

Some small genera of both Sterrhini and Scopulini (*Emmiltis, Cleta, Anthometra, Cinglis, Oar*) are characterized by bipectinate antennae, small eyes, hairy thorax and legs, simple ♂ sternum A2, and, usually, by diurnal activity. Possibly, they are phylogenetically 'old' descendants from a common ancestor. Future studies on the phylogeny of Sterrhinae should focus on these taxa.

NOMENCLATURAL NOTE: The name Sterrhinae, which has been in continuous use for more than 70 years (see e.g. PROUT 1930c: 48) is given preference over some senior synonyms in favour of stability of nomencature (FLETCHER 1979; HOLLOWAY 1993: 4; MÜLLER 1996; LERAUT 1997; SCOBLE 1999; SIHVONEN & KAILA 2004). The name Scopulinae has been regarded as valid in SEVEN (1991), but the 'Scopulites' in the original sense (DUPONCHEL, 1845) include only Pyralidae, the type genus is originally misidentified, and the case needs clarification by the Commission (ICZN). As for the 17 other names of the family-group, see HOLLOWAY (1997) and www.herbulot.de.

The English vernacular name 'waves' corresponds well to their German equivalent 'Wellenspanner', as well as to the old German expression 'Bogenstriemigte Spanner' and the Latin 'Phalaenae Geometrae Arcuatostriatae', both used by DENIS & SCHIFFERMÜLLER (1775). The name refers to the transverse lines, present on fore- and hindwing of the vast majority of species.

DIAGNOSIS: Usually small, tender-winged, slender-bodied, many species with great infraspecific variation in size, not rarely with sexual dimorphism regarding wing shape, pattern and coloration. Subsequent generations often much smaller, specimens from rearings sometimes larger than in nature. Coloration whitish or sand coloured with many undulate transverse lines, but usually lacking the broad medial fascia of the Larentiinae. Discal spots or rings usually present, and placed in the pale medial area of wing. Coloration cryptic in most species, but aposematic e.g. in Neotropical Cyllopodini and Afrotropical genera of the *Aletis* group (Scopulini), correlated in both cases with diurnal activity. Pattern of hindwing usually similar to that of forewing, and fringe concolorous. This is not explicitly mentioned in the species descriptions of this book (unless different from that character state) to avoid repetition. Pattern of wing underside usually much less distinct than that of upperside, usually not diagnostic, its pattern is discussed only exceptionally, when bearing important differential features. Frons flat and smooth, with just a few exceptions. Abdominal crests absent. Proboscis usually present, in some species and genera absent. Palpi usually short, longer in some Cosymbiini. Antennae of ♂ ciliate or bipectinate, ♀ antennae filiform with a very few exceptions. Structure of hindtibia diverse in both sexes, in ♂ usually diagnostic, often without spurs. Foretibia without terminal claw. Posterolateral processes on male sternum A2 usually absent, but in many Sterrhini and Scopulini developed ('reversal': SIHVONEN & KAILA 2004). Tympanal organs usually large, ansa narrow at base, dilated medially and terminally hammer-headed or bent (COOK & SCOBLE 1992).

Venation (Text-figs 9-10, 13, 67-69, 149, 152, 156-158, 190, 196, 197, 199, 212, 218, 232, 236-237): Wings and cell sometimes elongate. Forewing with areole (accessory cell). In a very few cases the areole is secondarily broken up and lost, e.g. in some small *Cleta, Idaea* and *Brachyglossina* species (*C. perpusillaria* (EVERSMANN, 1847), *I. microptera* species-group, *B. tibbuana* HERBULOT, 1965 and others). Double areole occurs in Rhodostrophiini, and in the genus *Ochodontia*. Forewing with all radial veins present, R2-R5 usually stalked. Arrangement and origin of radial veins R1-R5 revealing large variation between tribes, genera, and species, often with infraspecific variability. Connecting vein between areole and origin of M1 long, in longitudinal position. On both wings all medial veins present and tubular, M2 arising from cell midway between M1 and M3. On hindwing Sc+R1 usually appressed or fused to Rs at one point at the base of the cell. In Rhodometrini and a few *Idaea* species, e.g. *I. volloni* (LUCAS & JOANNIS, 1907), *I. lilliputaria* (WARREN, 1902), Sc+R1 and Rs anastomosed (PROUT 1930c: 48), these veins are separated in Neotropical Cyllopodini, a cross vein occurs in Afrotropical genera of the *Aletis* group (Scopulini). Rs and M1 usually stalked, M3 and CuA1 separate, rarely connate (e.g. in Timandrini) or even shortly stalked (e.g. in genus *Traminda*). One anal vein present. Frenulum well developed, a single, stout, long bristle in ♂, a brush of bristles in ♀.

MALE AND FEMALE GENITALIA: Though ♂ genitalia were comprehensively analysed by STERNECK (1940; 1941), ♀ genitalia of many Sterrhinae species have, for a long period, not been figured, except for the small number of species in PIERCE (1914). The situation has improved in the last years with some articles with regional focus (e.g. FLAMIGNI & BASTIA 1998; MAZEL 1995; REDONDO & GASTON 1999; 2003; STEKOLNIKOV 1996; DOGANLAR 2003). SIHVONEN (2001 and 2003) proposes everting of vesica as standard procedure in Sterrhinae. Though additional features may be revealed by that technique, it is not used in this study in order to allow direct comparison to 'traditional' figures and slides of non-everted aedeagi.

PHENOLOGY: All European Sterrhinae species overwinter as larva, except for the Cosymbiini, which hibernate as pupa. With regard to this uniformity, larval phenology is not explicitly remarked for each species in this volume, to avoid repetition.

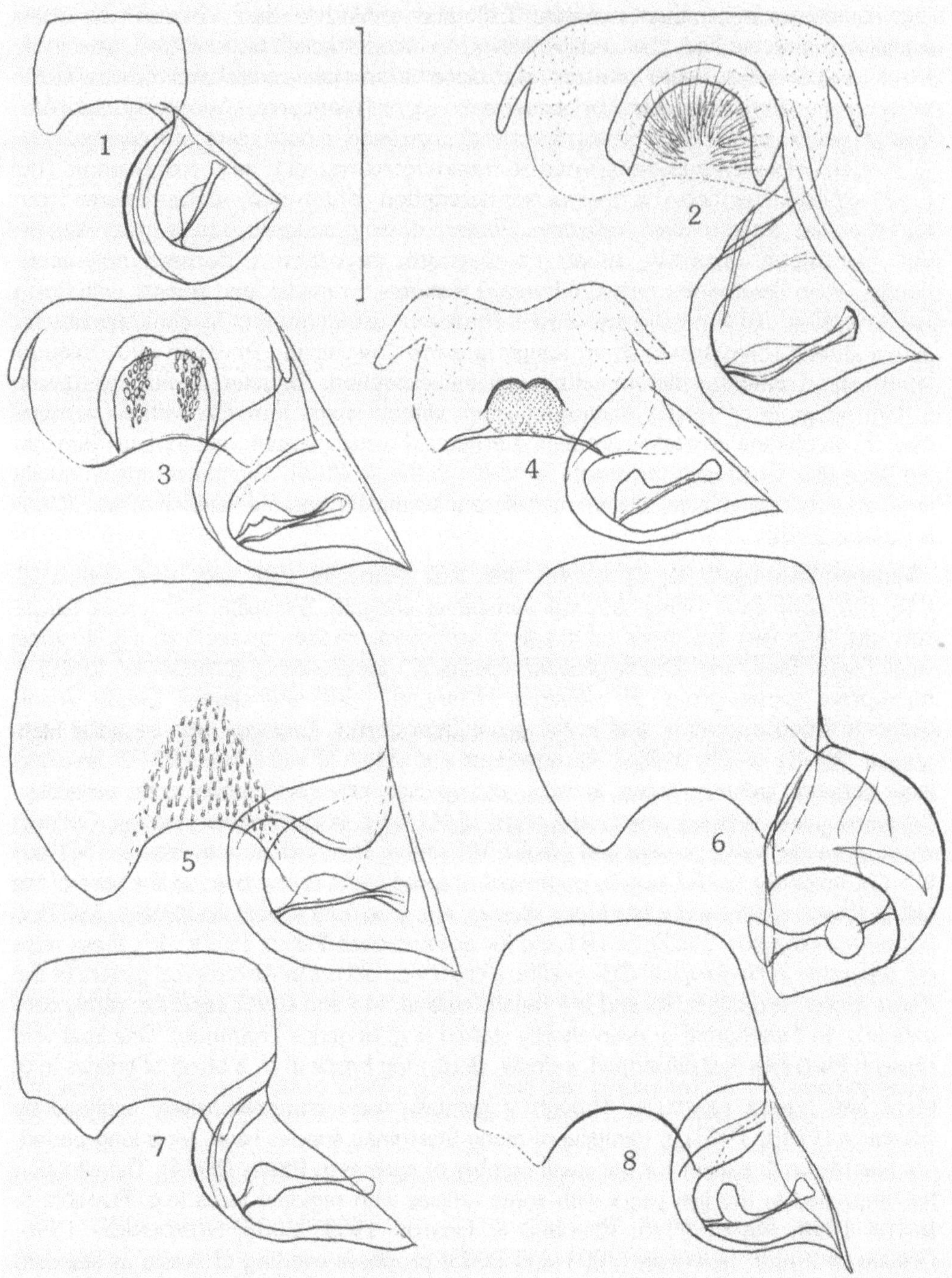

Text-figs 1-8: ♂ sternum A2 and tympanal organ of European Sterrhinae (scale bar = 1 mm). Text-fig. 1: *Emmiltis pygmaearia* (HÜBNER, 1909). Text-fig. 2: *Idaea aversata* (LINNAEUS, 1758). Text-fig. 3: *Scopula ornata* (SCOPOLI, 1763). Text-fig. 4: *Glossotrophia confinaria* (HERRICH-SCHÄFFER, 1847). Text-fig. 5: *Rhodostrophia calabra* (PETAGNA, 1787). Text-fig. 6: *Timandra comae* SCHMIDT, 1931. Text-fig. 7: *Cyclophora albipunctata* (HUFNAGEL, 1767). Text-fig. 8: *Rhodometra sacraria* (LINNAEUS, 1767).

BIOLOGY: Adults often fly at dusk and usually are readily attracted to light, but there are several diurnal species in Europe. Neotropical Cyllopodini (COVELL 1983) and Afrotropical genera of the *Aletis* group (Scopulini) are mainly diurnal, but also attracted to light. Diurnal activity in Cyllopodini is correlated with mimicry of Dioptidae and Riodinidae (COVELL 1983), in *Aletis* with mimicry of nymphalid *Euphaedra* butterflies (STAUDE & CURLE 1997). Some species are readily attracted to bait and flowers. Adult Sterrhinae sucking blood, sweat and tears of Mammalia recorded from south-eastern Asia (BÄNZIGER & FLETCHER 1985). Resting position usually planiform, forewing costa spread at an angle of 90° to the body. Tectiform resting position in Rhodometrini. Egg deposition single, in short rows or small groups, often glued to fine hairs, narrow rootlets or leaves of grass. The latter behaviour recorded for *I. contiguaria* in nature, for many other species in captivity. Larvae often polyphagous on low, herbaceous plants, with certain species facultatively monophagous (EBERT 2001). Often feeding on withered, mouldy, decaying, or dry leaves (mainly in the tribus Sterrhini). Only the larvae of the Cosymbiini arboreal. Some species are recorded as pests, e.g. on tobacco (*Idaea seriata; Scopula turbulentaria, S. minorata*: SANNINO & BALBIANI 1994; SANNINO & ESPINOSA 1999; SIHVONEN & KAILA 2004), corn, tomato (*Scopula luridata*: EL-SAWAF et al. 1968), groundnut (SATPATHI 1995), or in herbaria, insect collections and on stored food, tea etc. (*Idaea inquinata*). Feeding observations in nature rare, mainly those of the Sterrhini, due to the hidden resting and feeding places between fallen leaves near the ground. As a consequence, very few parasitoids are known, e.g. none in the large genus *Idaea*. Pupation usually on the ground between loosely spun leaves and detritus, very rarely in the ground or on the host-plant (e.g. Cosymbiini).

HABITAT: Sterrhinae are typically xerothermophilous, and, thus, less represented at higher latitudes and altitudes (HAUSMANN 2001; BREHM 2002), most species being inhabitants of open landscapes. In contrast to the other subfamilies of European Geometridae, Sterrhinae are almost absent from the northernmost regions, and no Sterrhinae species is mentioned in the article on boreal Geometridae by KRAMPL (1992), though *S. frigidaria* (MÖSCHLER, 1860) belongs to this category. The large genera *Idaea* and *Scopula* have many species in dry steppes, semi-deserts, deserts (oases) and subtropical transitional zones (savannahs), but are less represented in tropical rain forests.

IMMATURE STAGES: Eggs oval, often with red spots, flat type, length 0.50-0.75 mm (TONGE 1932). Larval prolegs with crochets interrupted centrally, seta CD2 on anal proleg below level of seta SV2 (SINGH 1953). Larvae of most species slender, usually resting at an angle of 45° from surface (SUGI 1987; EBERT 2001; SIHVONEN 2003). Cosymbiini pupae with girdle, in genus *Glossotrophia* s.str. pupa with free proboscis. Valuable morphological studies, though of regional focus, are published for eggs (TONGE 1932; SALKELD 1983; YOUNG 2003), larvae (SINGH 1953; MCGUFFIN 1967; YOUNG 2003), and for pupae (PATOČKA 1994a; 1994b; 2003; NAKAMURA 1994). Immature stages of European Sterrhinae extensively discussed also in CHRÉTIEN (1928).

Tribus **Sterrhini** MEYRICK, 1892

(=Idaeini BUTLER, 1881)

More than 800 species (SCOBLE 1999) in 19 genera, 7 further genera assigned tentatively (SIHVONEN & KAILA 2004). Worldwide distribution. Monophyly suggested by SIHVONEN &

KAILA (2004) on the basis of four synapomorphies: (1) uncus present, (2) socii absent (3) undivided valvae and (4) absence of coremata. Sister group relationship to Scopulini stressed by the same authors. This view is supported by similar arrangement of cilia (sensilla trichoidea) on ♂ antennae of many species (KNÖLKE 2002). Sometimes these cilia are grouped into four tufts and arise from short branches, reminiscent of quadripectinate character state in Rhodostrophiini. Absence of special structures on ♂ sternum A2 possibly a plesiomorphic character state, see genera *Anthometra, Emmiltis, Cleta* (with exceptions) and supposedly basal species-groups of *Idaea*. Further common features (plesiomorphic?) of the genera *Anthometra, Emmiltis,* and *Cleta*: ♂ tympanal organs larger than those of ♀, small eyes at large distance apart, bipectinate ♂ antennae, diurnal activity.
For validity of tribal name see introduction to subfamily, HOLLOWAY (1993: 4) and the Forum Herbulot world list of family-group names (www.herbulot.de).

DIAGNOSIS: External features see subfamily. On average much smaller than all the other Sterrhinae tribes. Forewings often more elongate than in Scopulini. Sternum A2 of many species (genus *Idaea*) with central sclerotized setose pouch and posterolateral processes of sternum A2. Hindtarsus of ♂ often shortened, ♀ hindtibia always with 2 spurs. Usually filiform antennae in both sexes, in ♂ ciliate, in a very few species bipectinate.

Venation (Text-figs 9-10, 13, 67-69, 149, 152): Forewing with a single areole, in the African genus *Discomiosis* two (JANSE 1934: 220). For rare cases of secondary loss of areole see diagnosis of subfamily. On forewing R1 often arising from areole, connate with R2-R5, sometimes shortly stalked with R2-R5. R2-R4 and R5 stalked. Sc+R1 of hindwing appressed to Rs at one point at the base of the cell (exceptions under genus *Idaea*). Rs and M1 stalked to 1/4-3/4 distance between apex of cell and termen (differential feature to Scopulini). M3 and CuA1 separate.

MALE GENITALIA: Uncus simple, sclerotized. Socii absent. Gnathos usually present. Valva not divided, usually long and slender. Aedeagus usually broad, with cornuti. Sternum A8 without posterior or lateral projections, except *Idaea sericeata*, *I. allardiata*. In genus *Limeria* sternum A8 triangular. Posterior margin of sternum A8 usually slightly bilobed, shallowly concave between 'lobes', sometimes rounded.

FEMALE GENITALIA: Ovipositor with additional bulbous and setose ventral projections. Corpus bursae often spinose. Ductus bursae usually broad and sclerotized.

REMARKS: Central setose pouch on ♂ sternum A2 probably interacting with shortened hindtarsus or hindtibial pencils, dispersing scales in courtship (HOLLOWAY et al. 2001).

Anthometra BOISDUVAL, 1840

TYPE SPECIES: *Anthometra plumularia* BOISDUVAL, 1840. Monotypic. Synonymy with *Idaea* suggested by STERNECK (1940: 169), but generic rank justified by bipectinate antennae, habitus, and shape of valva.

DIAGNOSIS: Forewings elongate. Antennae long bipectinate in ♂, filiform in ♀. Fringe scales very long. Venation (Text-fig. 1): Cells of both wings long. On forewing areole large. Discocellular veins straight. Frenulum strong in both sexes, in ♀ as brush of setae. Palpi short. Proboscis well developed. Eyes comparatively small, distance between eyes larger than diameter of eye. Hindtibia of both sexes with two terminal spurs, tarsus not

Text-fig. 9: Venation of *Anthometra plumularia* BOISDUVAL, 1840.

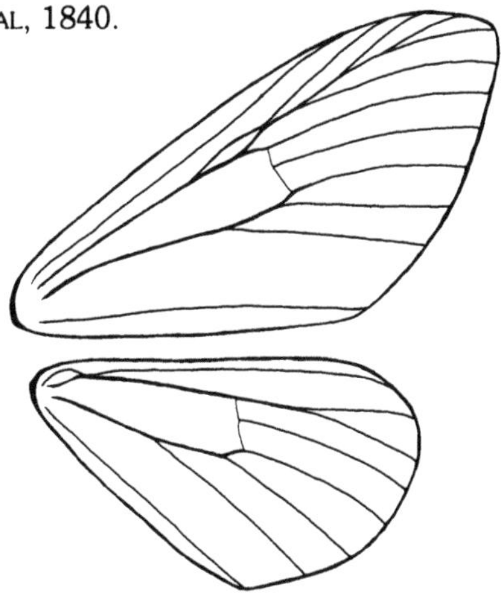

shortened. Tympanal organs in ♂ larger than in ♀. Sternum A2 between tympanal organs without sclerotized, setose patch, rectangular in both sexes.

MALE GENITALIA: Uncus narrow at tip. Gnathos developed, but weak. Valva beak-shaped, with long subterminal process. Aedeagus very broad, with one stout cornutus.

FEMALE GENITALIA: Apophyses anteriores very short. Ductus bursae sclerotized, short and broad. Corpus bursae with two rows of small spinules at the posterior opening and one lateral spinose crest.

BIOLOGY: Adults diurnal.

IMMATURE STAGES: Eggs comparatively large, 1.5 times those of *Idaea* (PROUT 1935a; REISSER 1937). Larva with small, round head, body flat, last segments dilated, with whitish hairs (PROUT 1935a; REISSER 1937). Pupal cremaster often broad, with four pairs of curved setae, the terminal pair not enlarged as in Scopulini (PATOČKA 2003).

1. *Anthometra plumularia* BOISDUVAL, 1840

Anthometra plumularia BOISDUVAL, 1840: Genera Index meth. eur. Lepid.: 231 (southern Spain), referring to RAMBUR'S 'Fauna Andalusiae' published 26 years later. Syntype(s) (ZFMK: 1♂ ex coll. BOISDUVAL, without abdomen, examined).

Anthometra concolorаria LEDERER, 1853: Verh. zool.-bot. Ver. Wien 3: 179 [sep. p. 15] (southern Spain). No separate types. Unnecessary replacement name for *plumularia*.

Fidonia psychinaria ROSENHAUER, 1856: Die Thiere Andalusiens: 379 (Spain, Andalusia: Malaga). Syntype(s).

Unavailable names (infrasubspecific): *fusca* DUFRANE, (1931); *unicolorana* DUFRANE, (1931). – (incorrect subsequent spelling): *plumbaria*: PROUT (1935).

DIAGNOSIS: Wingspan ♂ 14-19, ♀ 13-17 mm. Forewings elongate, larger than hindwings, *Eupithecia*-like. Ground colour brown, variable, sometimes very dark, sometimes comparatively pale. Postmedial line on all wings indistinct, often absent. Rarely also with an indistinct medial line. Fringe concolorous. Discal spots absent. Underside similar. Frons and vertex brown, collar and scales below the eyes paler. Length of palpi 1.0-1.2 times diameter of eye (♂♀). Antennal branches of ♂ 10-13 times width of flagellum. Further features under generic diagnosis.

MALE AND FEMALE GENITALIA: See under generic diagnosis.

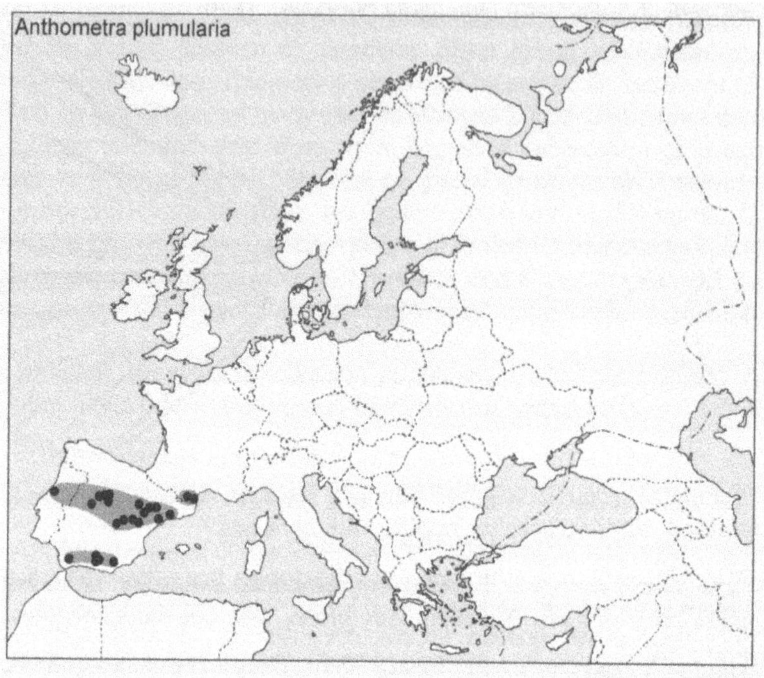

DISTRIBUTION: East-Mediterranean. Restricted to the Iberian peninsula including French Pyrenees.

PHENOLOGY: Univoltine: early July to early August, at lower altitude in eastern Pyrenees (MAZEL & PESLIER 1997) and Portugal (Corley pers. comm.) from early June.

BIOLOGY: Larva in nature possibly oligophagous on Fabaceae. Reputedly on *Genista scorpius* (Fabaceae) (PROUT 1913; LHOMME 1935; subsequently var. cit.). Reared on petals of various plant species, e.g. *Rosa* (Rosaceae) (REISSER 1937; PROUT 1938a: 224), and on leaves and petals of *Trifolium, Medicago* (Fabaceae), *Taraxacum officinale* and *Lactuca sativa* (Asteraceae) (LHOMME 1935). Adults diurnal, ♂ actively flying during daytime, preferably in the morning sun, ♀ inactive. Nectar feeding recorded on *Scabiosa* (REISSER 1937). ♀ depositing a small number of eggs, possibly 20 only (REISSER 1937), indicating K-strategy of the species. ♀ ratio in collections low, about 30%.

HABITAT: Montane species, inhabiting scattered scrubland with *Genista* in valleys and rocky slopes with sparse vegetation (PROUT 1913; REISSER 1937; PÖLL pers. comm.), often on limestone. From 1,400 m up to 2,000 m above sea-level, exceptionally down to 500 m (MAZEL & PESLIER 1997).

SIMILAR SPECIES: No similar species in Europe.

Emmiltis HÜBNER, 1825

TYPE SPECIES: *Geometra pygmaearia* HÜBNER, 1809. Monotypic. Synonymy with *Idaea* suggested by STERNECK (1940: 169), but generic rank justified by bipectinate antennae, and shape of uncus and gnathos.

Text-fig. 10: Venation of *Emmiltis pygmaearia* (HÜBNER, 1809).

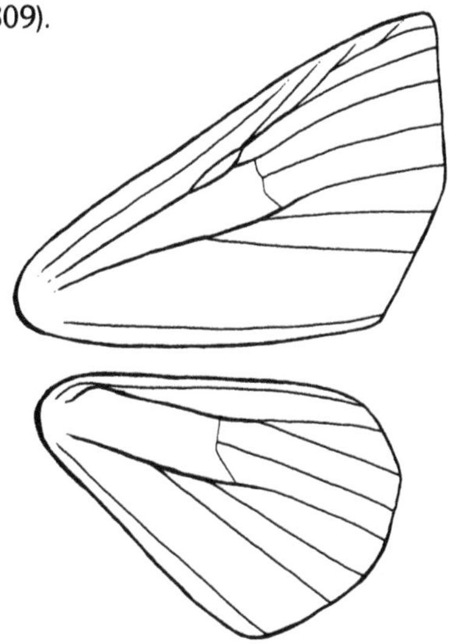

DIAGNOSIS: Small sized. Forewing costa convex in ♂, in ♀ rather straight and apex pointed. Termen angled at R3. Hindwing elongate. Frenulum strong in both sexes, in ♀ as brush of setae. Venation of forewing: R2-R5 on comparatively long stalk. Sc and R1-R3 weaker towards costa (10). Eyes comparatively small, rather distant from each other. Antennae bipectinate in ♂, filiform in ♀. Palpi short with prominent hairs below. Proboscis weak, very short. Hindtibia of both sexes with two spurs, tarsus not shortened. Tympanal organs in ♂ larger than in ♀. Sternum A2 between tympanal organs without sclerotized, setose patch, rectangular in both sexes (Text-fig. 1).

MALE GENITALIA: Uncus triangular, gnathos absent. Base of valva broad, terminally slender. Valva closely connected with saccus, in preparations best maintained in original position. Aedeagus slender, tapering, without cornuti.

FEMALE GENITALIA: Apophyses anteriores short. Sterigma without reniform sclerite. Ductus and corpus bursae very small, membranous, without sclerotizations, in preparations easily lost.

BIOLOGY: Adults diurnal.

IMMATURE STAGES: Larva short, of medium width, tapering towards head, with lateral keel-shaped ridge, reminiscent of larvae of *Idaea* (REBEL 1910b; PROUT 1913)

2. Emmiltis pygmaearia (HÜBNER, 1809)

Geometra pygmaearia HÜBNER, 1809: Samml. Eur. Schmett. 5, Geometrae (1): pl. 65, figs 335, 336 (Europe). Syntype(s) lost.

Phalaena minutaria FABRICIUS, 1781: Spec. Ins. 2: 253 (Italy). Syntype(s).

Emmiltis parvularia HÜBNER, 1825: Verz. bekannter Schmett.: 309 (Europe). No separate types. Unnecessary replacement name.

Acidalia pygmaearia sirentina DANNEHL, 1925: Ent. Z. Frankf. a.M. 39: 12 (Italy, Abruzzi: Mte. Sirente). Lectotype ♂ (ZSM, examined).

Acidalia pygmaearia teriolensis DANNEHL, 1925: Ent. Z. Frankf. a.M. 39: 12 (Italy, South Tyrol, Alto Adige: Terlan). Lectotype ♂ (ZSM, examined).

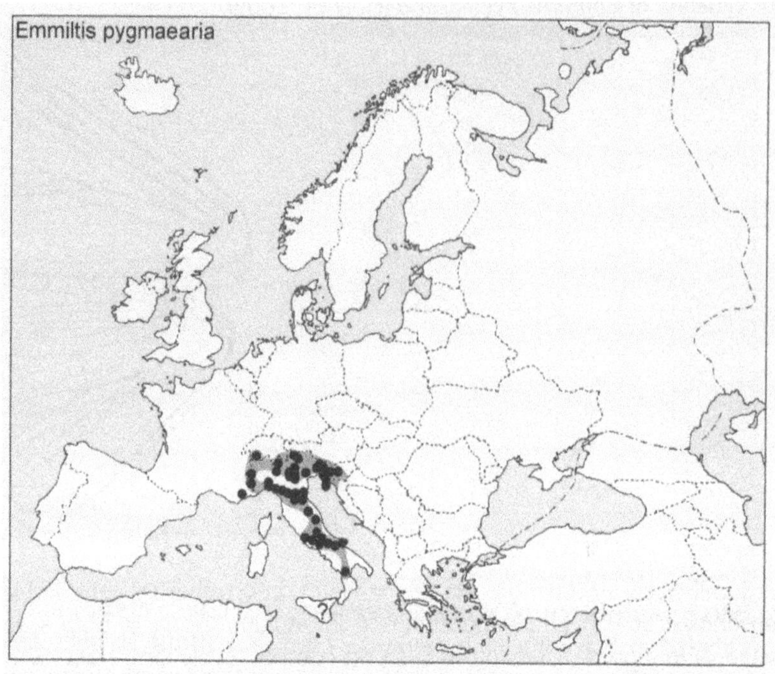

Unavailable names (infrasubspecific): *vannaria*: DANNEHL (1933). – (incorrect subsequent spellings): *pymaearia*: PROUT (1913); *pymearia*: PROUT (1934: 294).

DIAGNOSIS: Wingspan ♂ 9-12, ♀ 11-13 mm. Females usually paler than males. Wings sand coloured, with brown transverse lines rather straight, the subterminal and medial one very broad. Small punctiform discal spots sometimes visible. Brown terminal spots at the base of the fringe in fresh specimens. Frons and vertex pale brown. Palpi bushy scaled, length 1-1.3 times diameter of eye (♂♀). Antennal branches of ♂ 4-5 times width of flagellum (Text-fig. 11). Further features under generic diagnosis.

MALE AND FEMALE GENITALIA: See under generic diagnosis.

DISTRIBUTION: (South-)European. South-eastern France, southern Switzerland, northern Italy, Apennines down to southern Italy (PARENZAN 1994; SCALERCIO pers. comm.), Slovenja, Croatia. Reputedly in Sicily (1♂ in ZISP), possibly mislabelled. In some regions decreasing or extinct (e.g. Brenta region, northern Italy: HELLMANN 1987), threatened.

PHENOLOGY: Bivoltine: mid-April to early September with short gap in July or, sometimes, with overlapping generations. On mountains univoltine, from mid-May to late July (FORSTER & WOHLFAHRT 1981).

BIOLOGY: Larva probably oligophagous, with focus on Fabaceae and Caryophyllaceae, preferring dry leaves (var. cit., e.g. REBEL 1910b; PROUT 1913; LHOMME 1935). Found on *Lotus corniculatus* (Fabaceae) (KITSCHELT 1925). Recorded on *Ononis spinosa* (Fabaceae) (DANNEHL 1927), flowers and leaves of *Stellaria holostea, S. media* and *Cerastium tomentosum* (Caryophyllaceae) (VORBRODT 1930). Reared on withered leaves of *Taraxacum officinale* (Asteraceae) (ERLEBACH pers. comm.). Adults diurnal, flying in sunshine until noon, rarely in the evening. ♀ ratio in collections low, about 20-30%.

HABITAT: Xerothermophilous (FORSTER & WOHLFAHRT 1981). On dry, shortly growing grassland, embankments, usually on stony, low-nutrient soils. The species is highly sensitive to anthropogeous alterations (ORTNER, PÖLL, pers. comm.), and thus, a valuable indicator for environmental changes. Usually from 500 m up to 1,400 m above sea-level, sometimes recorded also in the lowlands, e.g. near Mantova (Po plain) and in Slovenja.

SIMILAR SPECIES: Both allopatric *Cleta ramosaria* and *C. perpusillaria* differ in lacking the transverse line between medial line and terminal area.

REMARKS: No separation of subspecies justified, as there are no constant infraspecific differential features.

Cleta DUPONCHEL, 1845

TYPE SPECIES: *Geometra vittaria* HÜBNER, 1813, junior synonym of *Phalaena ramosaria* VILLERS, 1789. Five species in the western Palaearctic from Spain to central Asia. Incorrect subsequent spelling: *Cletha* (AMSEL 1933). Coloration and genitalia of *Cleta filacearia* (HERRICH-SCHÄFFER, 1847) suggest its isolated (basal?) position within the genus. According to PROUT (1913) the genus is closely related to *Emmiltis*, but it differs in hairy ♂ hindlegs without spurs, in bushy palpi, and in often opened areole of forewing.

DIAGNOSIS: Forewing costa convex in ♂, in ♀ rather straight and apex pointed. Hindwing

Text-figs 11-12: ♂ antennal structure of European Sterrhini species (SEM). Scale bar = 0.05 mm. Text-fig. 11: *Emmiltis pygmaearia* (HÜBNER, 1809). Text-fig. 12: *Cleta ramosaria* (VILLERS, 1789).

Text-fig. 13: Venation of *Cleta ramosaria* (VILLERS, 1789).

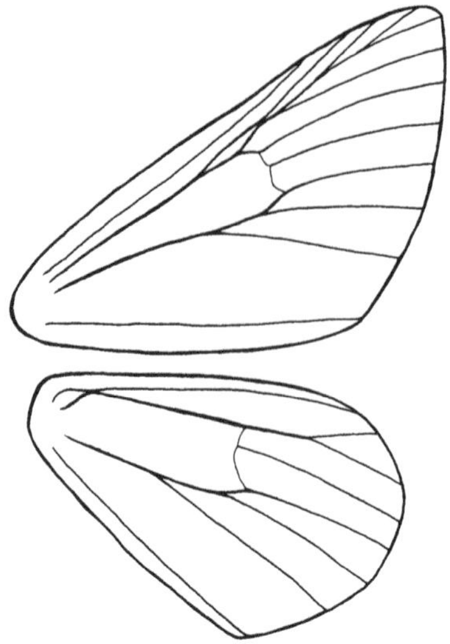

rounded. Frenulum strong in both sexes, in ♀ as brush of setae. Venation (Text-fig. 13): Forewing areole often present, but sometimes broken up as individual aberration, *C. perpusillaria* always without areole. R2-R5 of forewing shortly stalked. Rs and M1 of hindwing on long common stalk, distally strongly diverging. Eyes comparatively small, rather distant from each other. Antennae bipectinate in ♂, filiform in ♀. Palpi long and bushy scaled. Proboscis developed, well sclerotized, but short (1-1.5 mm) in all species. Trochanter of foretibia and hindlegs with long hairy scales in ♂, in ♀ smoothly scaled. Hindtibia of ♂ without, ♀ with two spurs, tarsus not shortened. Tympanal organs in ♂ larger than in ♀; exception: *C. filacearia*. Structure of sternum A2 between tympanal organs diverse, diagnostic in male.

MALE GENITALIA: Uncus slender; exception: *C. filacearia*. Gnathos well developed. Valva slender from base to tip. Aedeagus with one cornutus.

FEMALE GENITALIA: Apophyses anteriores short. Ductus bursae sclerotized, corpus bursae membranous. Shape of ductus and corpus bursae diverse and diagnostic within genus.

BIOLOGY: Adults diurnal.

IMMATURE STAGES: Larva short (REBEL 1903), tapering towards head, corrugated, with intersegmental incisions (PROUT 1913).

3. *Cleta ramosaria* (VILLERS, 1789)

Phalaena Geometra ramosaria VILLERS, 1789: Linn. ent. 2: 329, pl. 6, fig. 8 (southern France; locus typicus possibly erroneous, see PROUT 1934). Syntype(s).
Geometra vittaria HÜBNER, 1813: Samml. Eur. Schmett. 5, Geometrae (1): pl. 83, fig. 429 (Europe). Syntype(s) lost.
Unavailable names (misidentification): *perpusillaria*: sensu RÖSSLER (1877) nec EVERSMANN, 1847.

DIAGNOSIS: Wingspan ♂ 10-13, ♀ 12-14 mm. Females usually paler than males. Wings sand coloured. Dark brown transverse lines fine and sharp. Terminal area dark brown with pale ochreous spots in the apex and the tornus. Discal spots small, punctiform. Underside similar. Fringe concolorous. Frons dark brown with some ochreous, hair-like

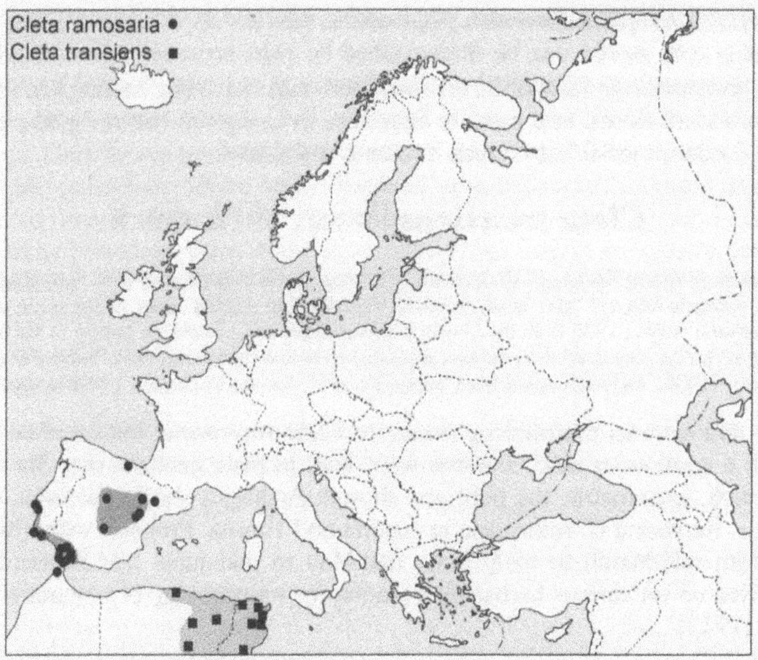

scales. Base of eyes with white scales. Palpi very bushy scaled, length 1.5-2 times diameter of eye (♂♀). Antennal branches of ♂ 3-4 times width of flagellum (Text-fig. 12). Sternum A2 between tympanal organs with V-shaped sclerite in male. Further features under generic diagnosis.

Male genitalia: Valvae straight, slender, symmetrical. Aedeagus very large, long, with one stout, long cornutus. In *Cleta cinneretharia* Amsel, 1933 (Levant) aedeagus broad, straight, with one stout cornutus, bent by 90°.

Female genitalia: Ductus bursae broad and short, anteriorly with semicircular lobe. Two rows of small spinules at the posterior opening of the globular corpus bursae.

Distribution: West-Mediterranean. Local populations in northern Spain (Bilbao), widely distributed in southern Spain and southern Portugal. Erroneously mentioned for Sardinia (Prout 1913), absence from Tuscany confirmed by Flamigni et al. (2001). Presence of the species in southern France 200 years ago not completely excluded (see original description). – Outside Europe in the northern parts of Morocco (Rungs 1981). In Algeria and Tunisia replaced by the sister species *C. transiens* Prout, 1913. Erroneously recorded for Turkey (Riemis 1994).

Phenology: Univoltine: mid-April to mid-June. In Morocco bivoltine, from February to May; July to September (Rungs 1981). Larvae growing very slowly, a biennial life cycle suggested from rearings (King pers. comm.).

Biology: Larva reared on fairly withered leaves of *Rumex* (Polygonaceae) and *Galium* (Rubiaceae) (King & Romera 2004). Adults diurnal (Ribbe 1912, Redondo & Gaston 1999). ♀ ratio in collections low, about 10%.

Habitat: Xerothermophilous, inhabiting hot and dry mainly coastal habitats. From 0 m up to 600 m above sea-level.

SIMILAR SPECIES: Allopatric *Emmiltis pygmaearia* (see no. 2). Allopatric *Cleta transiens*, though being very similar, can be distinguished by paler terminal area and genitalia (see below). *C. perpusillaria* (allopatric) smaller, forewings narrower, medial line straighter, ♂ antennal branches closely appressed to flagellum. In *C. cinneretharia* AMSEL, 1933, from Israel and Jordan, medial line convex, crossing discal spot.

Cleta transiens PROUT, 1913, stat. n.

Cleta ramosaria transiens PROUT, 1913: in SEITZ, Macrolep. 4: 88 (Algeria, Palestine). Lectotype ♂, (BMNH, examined, formerly with red 'type' label), herewith designated to stabilize usage of the name with regard to *C. cinneretharia* AMSEL, 1935 from the Levant (paralectotypes of *C. transiens* belong to the latter species). Introduced as 'forma', but later on in the text supposed to be valid at species rank. Treated at available rank also in PROUT (1934). Herewith raised from subspecies of *C. ramosaria* (SCOBLE 1999) to species rank.

In habitus and external morphology similar to *Cleta ramosaria*, but terminal area much paler, with a continuous sand coloured wavy line. In male genitalia (fig. 3b) valvae with terminal hook, asymmetric, the right one shorter. Aedeagus shorter, broader, with shorter cornutus. Replacing *C. ramosaria* in Algeria and Tunisia. Probably with trivoltine phenology, from mid-March to early May; mid-May to mid-June; mid-September to late October. Reared on various herbaceous plants, for morphology of immature stages see CHRÉTIEN (1917).

4. *Cleta perpusillaria* (EVERSMANN, 1847)

Acidalia perpusillaria EVERSMANN, 1847: Bull. Soc. imp. Moscou 20 (2(3)): 82, pl. 6, fig. 7 (south-eastern European Russia: Krasnoarmeysk, 'Sarepta'). Lectotype ♂ (ZISP, examined), herewith designated to stabilize usage of the name with regard to some closely related taxa; 15 ♂ paralectotypes at ZISP.

Unavailable names (incorrect subsequent spellings): *pepussilaria*: PROUT (1934: 296); *pepussillaria*: FREYER

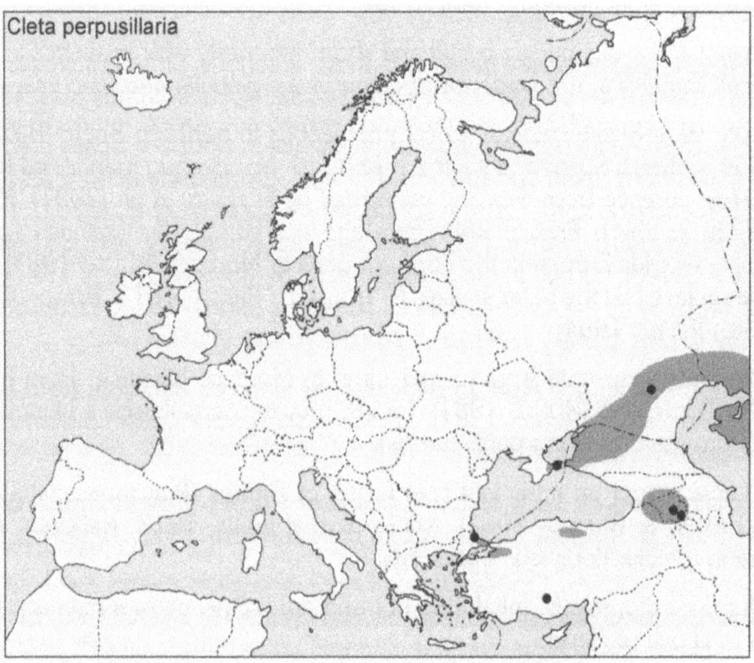

(1847: 59); *perpusilaria*: WARDIKJAN (1985); *perpussilaria*: PROUT (1934: 296); *perpussillaria*: FREYER (1847: pl. 516, fig. 5).

DIAGNOSIS: Wingspan ♂♀ 10-12 mm. Forewing narrower than in *ramosaria*. Venation without forewing areole. Wings sand coloured, females less marked than males, with orange tinge. Dark brown transverse lines fine and sharp. Medial line straight, often crossing the small, punctiform discal spot. Terminal area brown with almost uninterrupted sand coloured wavy line. Underside similar. Fringe concolorous with ground colour. Frons grey with some brownish scales. Palpi less bushy scaled than in congeners, length 1.5-2 times diameter of eye (♂♀). Antennal branches of ♂ 3-5 times width of flagellum, pectination closely appressed to flagellum. Sternum A2 with round, setose, horseshoe-shaped sclerotization, reminiscent of that of many *Idaea* species-groups. Further features under generic diagnosis.

MALE GENITALIA: Valvae straight, very slender, symmetrical. Aedeagus slender, bent at an angle of nearly 90°, with very weak, slender cornutus.

FEMALE GENITALIA: Ductus bursae longer and narrower than in *C. ramosaria*. Corpus bursae elongate.

DISTRIBUTION: (East-)European-Westasiatic. In Europe restricted to south-eastern European Russia, Crimea (KOSTJUK 2004), and European Turkey (OKYAR & AKTAÇ 1999). Records for Iberian peninsula (VIVES MORENO 1994; MÜLLER 1996) erroneous, based on nomenclatural confusion with '*perpusillata*' FERNANDEZ, 1931, a junior synonym of *Perizoma flavosparsata* (WAGNER, 1926) (Larentiinae; AGENJO 1954; MIRONOV 2003). – Outside Europe in Caucasus, Transcaucasus (WARDIKJAN 1985), Turkmeniya and Transcaspia (VIIDALEPP et al. 1992; VIIDALEPP 1996). Isolated populations in Turkey (ZSM; STAUDINGER 1879, RIEMIS 1994). In Jakutia replaced by the sister species *Cleta jacutica* VIIDALEPP, 1976, in Tuva, Altai, Sajan and Mongolia by other undescribed species (compare BIDZILYA et al. 2002).

PHENOLOGY: Uni- or bivoltine: late May to late June. Specimens from early August mentioned in ANIKIN et al. (2000), possibly belonging to a second generation.

BIOLOGY: Larval host-plants unknown. Adults diurnal. ♀ ratio in collections low, about 10%.

HABITAT: Xerothermophilous. Dry lowland steppes, on sandy and chalky soils (Russia: ANIKIN et al. 2000), and scattered oak forest (European Turkey; OKYAR pers. comm.). From 0 m up to 500 m above sea-level (limited data available).

SIMILAR SPECIES: Allopatric *Emmiltis pygmaearia* (see no. 2), *Cleta ramosaria* (3). In *Cleta cinneretharia* AMSEL, 1933 from Israel and Jordan the medial line is convex and crosses the discal spot.

REMARKS: Structure of ♂ sternum A2 merits further attention, with respect to infra- and intergeneric relationships, and to improve reconstruction of Sterrhinae phylogeny. *C. cinneretharia* from the Levant structurally rather different (see HAUSMANN 1991d).

5. *Cleta filacearia* (HERRICH-SCHÄFFER, 1847)

Acidalia filacearia HERRICH-SCHÄFFER, 1847: Syst. Bearb. Schmett. Eur. 3 (24): 27; (1848: 3 (35): pl. 64, figs 392, 393: non binominal) ('southern Europe' (PROUT 1934); '[Austria:] Schneeberg' (HERRICH-SCHÄFFER 1847); 'surely not Austria' (BOHATSCH 1885)). Syntypes ♂♀.

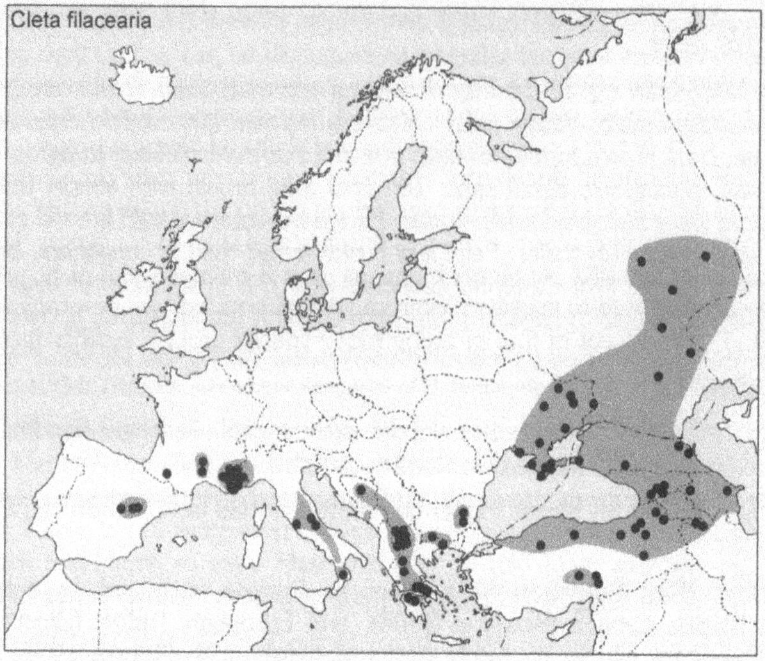

Unavailable names (misidentification): *flaveolaria*: sensu TREITSCHKE (1835) nec HÜBNER, 1809 (identity: *Idaea flaveolaria*).

DIAGNOSIS: Wingspan ♂♀ 14-17 mm, exceptionally 12 mm only. Venation: Forewing areole not rarely broken up distally. Ground colour yellow. Females slightly paler than males. Dark grey transverse lines fine on all the wings. Postmedial line comparatively sharp, medial line diffuse. Terminal line sharp, dark grey. Fringe dark yellowish grey. Discal spots small, punctiform. Fringe ochreous. Underside ochreous with broad dark brown transverse lines. Frons ochreous. Palpi very bushy scaled, length 1.5-2 times diameter of eye (♂♀). Antennal branches of ♂ 2.5-3 times width of flagellum. Sternum A2 without additional sclerites in both sexes. No sexual dimorphism in tympanal organs. Further features under generic diagnosis.

MALE GENITALIA: Uncus broad, slightly concave at tip. Valva comparatively broad, straight. Aedeagus long, straight, with long and slender cornutus.

FEMALE GENITALIA: Ductus bursae very long and slender, anteriorly dilated, with many furrows over the whole length. Corpus bursae globular, membranous.

DISTRIBUTION: Mediterranean-Turanian. From central Spain, and southern France to the Urals. Absent from all Mediterranean islands, from northern Italy and from the northern Balkan peninsula. Records for Catalonia probably misidentified (DANTART 2000). – Outside Europe in northern Turkey, Caucasus, Transcaucasus, in the east to central Asian mountains (VIIDALEPP 1988; 1996).

PHENOLOGY: Usually univoltine: late May to early July. In the southern parts of the distribution area emerging from late April, perhaps with partial second brood. In the mountains from early June to late July.

BIOLOGY: Natural larval host-plants unknown. Reared on withered leaves of *Lactuca sati-*

va (Asteraceae) (REBEL 1903; 1910). Adults diurnal, actively flying in the sunshine, not attracted to light. ♀ ratio in collections low, about 20-30%.

HABITAT: Montane. In southern Italy on flower-rich, wet meadows. In Russia narrow gorges and ditches with grass steppe vegetation (MIRONOV pers. comm.), steppes of *Stipa* (ANIKIN et al. 2000). Usually from 800 m up to 2,000 m above sea-level, sometimes down to 400 m, in the Caucasus and Turkey up to 2,500 m.

SIMILAR SPECIES: *Idaea aureolaria, I. luteolaria* and *I. flaveolaria* can easily be distinguished by their filiform ♂ antenna. ♀ of *I. flaveolaria* with medial lines usually more vague, fringe darker. Ground colour of ♀ *I. luteolaria* with more orange tinge, transverse lines brownish, more vague and convex, terminal line absent. Both sexes of *I. aureolaria* differ in three well marked, blackish, sharp transverse lines on the forewings.

REMARKS: Larva strongly different from those of the *aureolaria* species-group of *Idaea* (REBEL 1903).

Text-figs 14-15: Diagnostic characters (indicated) of European *Cleta* and *Idaea* species. Text-fig. 14: *C. filacearia* (HERRICH-SCHÄFFER, 1847). Text-fig. 15: *I. flaveolaria* (HÜBNER, 1809).

Idaea TREITSCHKE, 1825

(=*Sterrha* HÜBNER, 1825; =*Ptychopoda* CURTIS, 1826; =*Acidalia*: sensu auct. nec HÜBNER 1819)

TYPE SPECIES: *Phalaena aversata* LINNAEUS, 1758. Frequently misspelled, e.g. *Idoea* (COTES & SWINHOE 1888), *Ideaa* (ROMERO et al. 2001), *Ideae* (MENTZER 1990), *Idea* (BERTOLONI 1849). More than fifty generic synonyms, mainly based on type species from tropical countries (FLETCHER & NYE 1979; NIELSEN et al. 1996; HOLLOWAY 1997; SCOBLE 1999). The synonymous generic name *Sterrha* was published just 2 months later than *Idaea*. *Acidalia* TREITSCHKE, 1825 refers to *Scopula*, and is a junior homonym of *Acidalia* HÜBNER 1819, the latter referring to Nymphalidae. The genus *Idaea* includes about 680 described species worldwide with distribution centres in the Mediterranean, African savannahs, and the eremic belt of dry regions and semideserts in western Palaearctic Asia. The current system of the 107 European species is still largely based on that of STERNECK (1940), who subdivided the genus according to length of aedeagus and number of cornuti. The results are not always satisfying, as STERNECK (1940) himself states. Short aedeagus and single cornutus are suggested to be plesiomorphic STERNECK (1940: 11), though in some cases there are 'character conflicts', e.g. the longer aedeagi of the yellow species in the supposedly basal *aureolaria* species-group. Similarly, the presence of ♂ hindtibial spurs and long tarsus are regarded as plesiomorphic. The absence of the central sclerite and setose patch from ♂ sternum A2 in many species-groups may be plesiomorphic too, as it is typical for species-groups with other, supposedly plesiomorphic features (♂ hindtibial spurs, short aedeagus, single cornutus in ♂ genitalia). Today, 60 years after the excellent morphologcal study of STERNECK (1940) there is no comprehensive, improved alternative to his system. In some cases the grouping or the position of single species is changed in this book, mainly when there is strong evidence from ♀ genitalia, which had been ignored by STERNECK. Monophyly of the genus was suggested from molecular analysis of a small data set (ABRAHAM et al. 2001). First results from own molecular studies (sequence analysis of mitochondrial ND1 and COI genes) confirm the relationships in some of STERNECK's groups, e.g. the *aversata* species-group (MILLER et al. 2001; HAUSMANN et al. in prep.). Similar confirmations result from SEM-based analysis of ♂ antennae (own studies; KNÖLKE 2002), and, in most cases, from comparative analysis of ♀ genitalia and structure of ♂ sternum A2. Nevertheless the genus urgently needs comprehensive revision. Phylogenetic analysis should lead to an overall subdivision into species-groups considering the whole world's fauna. Promising perspectives for an improved systematic arrangement arise from additional inclusion of the six character sets molecular analysis, ♂ antennal fine structure, ♂ sternum A2, ♀ genitalia, egg-sculpturation, and larval morphology. Detailed analysis on global scale is in preparation by the author. Major morphological treatments: STERNECK (1940), OZORSKI (1948), BLESZYNSKI (1960), WARDIKJAN (1970), FLAMIGNI & BASTIA (1998).

DIAGNOSIS: Hindwing pattern similar to that of forewing (unless stated otherwise), but antemedial line always absent and medial line at rather proximal position. Forewing comparatively narrow. Hindwing not angled or tailed at M3; exception: *Idaea emarginata* and relatives and some Neotropical species. Venation (Text-figs 67-69): Forewing with single areole usually present, but absent in a few small-sized species, e.g. the eremic *microptera* species-group. Origin of R1 often with infraspecific variation, usually connate or shortly stalked with R2-R5, sometimes arising from common stalk near origin of R5. On hindwing Sc and Rs+M1 anastomosing in one point at base of cell; exception: *I. volloni* (LUCAS & JOANNIS, 1907) from North Africa and Arabia, and some Afrotropical species, which show long fusions, reminiscent of Larentiinae. Rs and M1 stalked, length of stalk often with large infraspecific variation, usually 1/3 to 3/4 distance between apex of cell and termen. Eyes larger than in *Anthometra*, *Emmiltis* and *Cleta*. Frons flat, smooth. Palpi comparatively slender, not bushy, length usually about 0.8-1.2 times diameter of eye, but e.g. in *I. longaria* and some species of the *seriata* species-group shorter. Proboscis usually present, with phylogenetic tendency (underlying synapomorphy) to become weak or absent. Antennae of ♂ (Text-figs 35-66, 143-147) ciliate-setose or ciliate-fasciculate, exceptionally with very short pectinations, e.g. in *I. serpentata* or *I. macilentaria*. Arrangement and length of cilia (sensilla trichoidea) often diagnostic, length varying from about 0.02 to 0.25 mm. For measurements see introduction to this Volume. Hindtibia of ♂ (Text-figs 16-34) usually without spurs, sometimes dilated, often with tarsus shortened. Female always with two terminal spurs on hindtibia and filiform antenna,

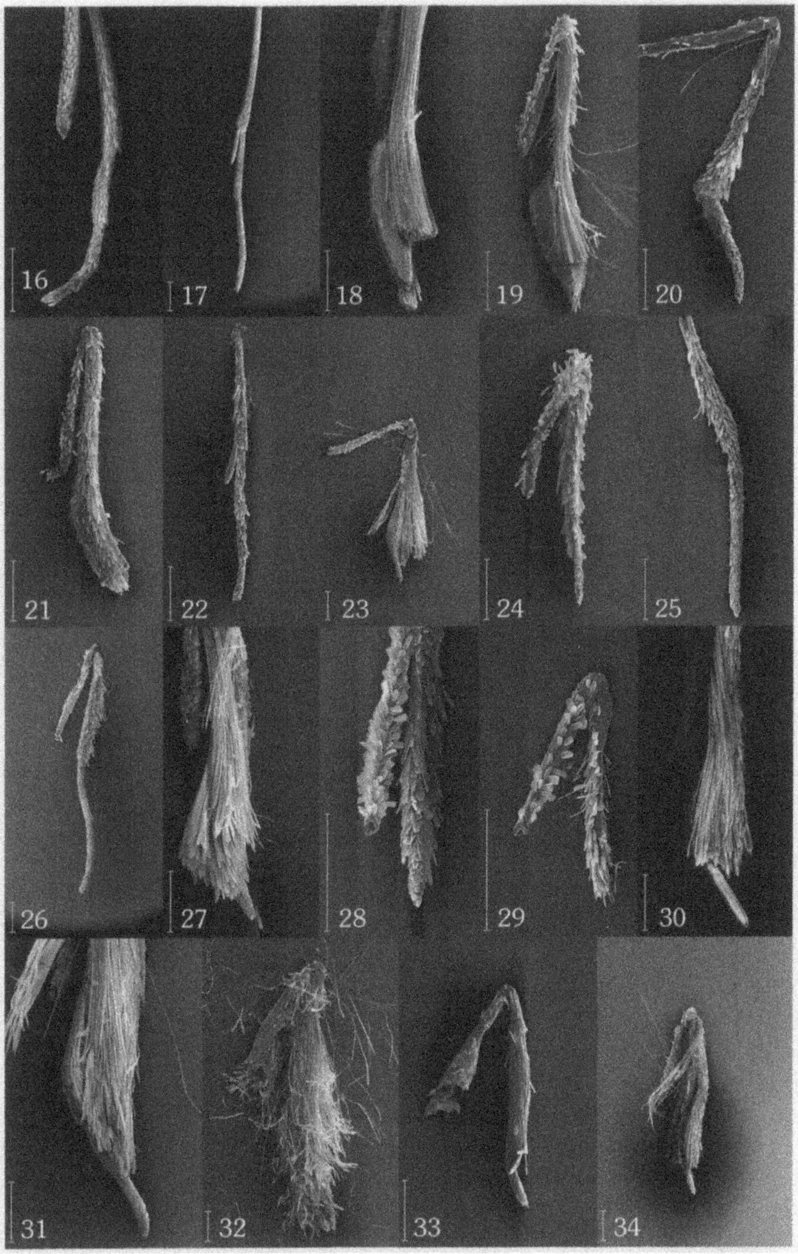

Text-figs 16-34: ♂ hindlegs of European *Idaea* species (SEM). Scale bar = 0.5 mm. Text-fig. 16: *I. serpentata* (HUFNAGEL, 1767). Text-fig. 17: *I. ochrata* (SCOPOLI, 1763). Text-fig. 18: *I. filicata* (HÜBNER, 1799). Text-fig. 19: *I. troglodytaria* (HEYDENREICH, 1851). Text-fig. 20: *I. laevigata* (SCOPOLI, 1763). Text-fig. 21: *I. efflorata* ZELLER, 1849. Text-fig. 22: *I. attenuaria* (RAMBUR, 1833). Text-fig. 23: *I. albarracina* (REISSER, 1933). Text-fig. 24: *I. obsoletaria* (RAMBUR, 1833). Text-fig. 25: *I. dilutaria* (HÜBNER, 1799). Text-fig. 26: *I. bigladiata* HERBULOT, 1975. Text-fig. 27: *I. consolidata* (LEDERER, 1853). Text-fig. 28: *I. carvalhoi* HERBULOT, 1979. Text-fig. 29: *I. saleri* DOMÍNGUEZ & BAIXERAS, 1992. Text-fig. 30: *I. camparia* (HERRICH-SCHÄFFER, 1852). Text-fig. 31: *I. trigeminata* (HAWORTH, 1809). Text-fig. 32: *I. biselata* (HUFNAGEL, 1767). Text-fig. 33: *I. exilaria* (GUENÉE, 1858). Text-fig. 34: *I. eugeniata* (DARDOIN & MILLIÈRE, 1870).

Text-figs 35-50: ♂ antennal structure of European *Idaea* species (SEM). Scale bar = 0.02 mm. Text-fig. 35: *I. serpentata* (HUFNAGEL, 1767). Text-fig. 36: *I. luteolaria* (CONSTANT, 1863). Text-fig. 37: *I. aureolaria* (DENIS & SCHIFFERMÜLLER, 1775). Text-fig. 38: *I. sardoniata* (HOMBERG, 1911). Text-fig. 39: *I. rufaria* (HÜBNER, 1799). Text-fig. 40: *I. consanguinaria* (LEDERER, 1853). Text-fig. 41: *I. sericeata* (HÜBNER, 1813). Text-fig. 42: *I. macilentaria* (HERRICH-SCHÄFFER, 1847). Text-fig. 43: *I. ochrata* (SCOPOLI, 1763). Text-fig. 44: *I. filicata* (HÜBNER, 1799). Text-fig. 45: *I. circuitaria* (HÜBNER, 1819). Text-fig. 46: *I. calunetaria* (STAUDINGER, 1859). Text-fig. 47: *I. alyssumata* (MILLIÈRE, 1871). Text-fig. 48: *I. belemiata* (MILLIÈRE, 1868). Text-fig. 49: *I. obliquaria* (TURATI, 1913). Text-fig. 50: *I. inquinata* (SCOPOLI, 1763).

Text-figs 51-66: ♂ antennal structure of European *Idaea* species (SEM). Scale bar = 0.02 mm. Text-fig. 51: *I. humiliata* (HUFNAGEL, 1767). Text-fig. 52: *I. longaria* (HERRICH-SCHÄFFER, 1852). Text-fig. 53: *I. nexata* (HÜBNER, 1813). Text-fig. 54: *I. descitaria* (CHRISTOPH, 1893). Text-fig. 55: *I. consolidata* (LEDERER, 1853). Text-fig. 56: *I. seriata* (SCHRANK, 1802). Text-fig. 57: *I. camparia* (HERRICH-SCHÄFFER, 1852). Text-fig. 58: *I. subsericeata* (HAWORTH, 1809). Text-fig. 59: *I. pallidata* (DENIS & SCHIFFERMÜLLER, 1775). Text-fig. 60: *I. mancipiata* (STAUDINGER, 1871). Text-fig. 61: *I. trigeminata* (HAWORTH, 1809). Text-fig. 62: *I. biselata* (HUFNAGEL, 1767). Text-fig. 63: *I. exilaria* (GUENÉE, 1858). Text-fig. 64: *I. cervantaria* (MILLIÈRE, 1869). Text-fig. 65: *I. tineata* (THIERRY-MIEG, 1911). Text-fig. 66: *I. metohiensis* (REBEL, 1900).

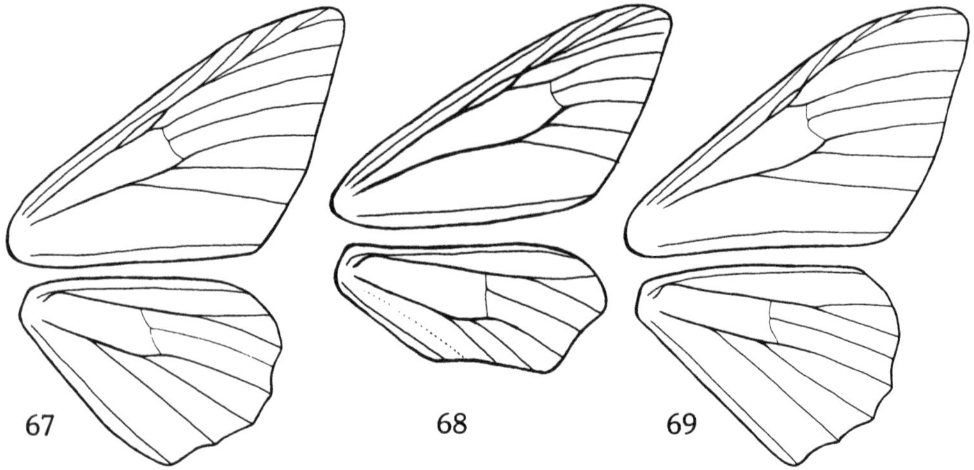

Text-figs 67-69: Venation of *Idaea* species. Text-fig. 67: *I. albitorquata* (PÜNGELER, 1913). Text-fig. 68: *I. saleri* DOMÍNGUEZ & BAIXERAS, 1992. Text-fig. 69: *I. aversata* (LINNAEUS, 1758).

with very sparse cilia. Tympanal organ without sexual dimorphism (see *Cleta, Anthometra* and *Emmiltis* and Text-figs 1-2). Many species-groups with horseshoe-shaped sclerite at ♂ sternum A2, with one single setose patch in its centre. Further features under diagnoses of tribe and subfamily.

MALE GENITALIA: Uncus usually slender and sclerotized, digitiform or triangular, sometimes truncate. Gnathos present, with modifications in the *aversata* species-group. Valva usually long and slender, strongly linked with each other at base (causing problems in spreading of the genitalia on permanent slides). Shape of 'juxta' (='anellus'; 'fultura inferior') sometimes diagnostic. Aedeagus usually with cornuti, their number and shape, and further vesica structures diagnostic in most species. Shape of sternum A8 rarely with diagnostic features, without posterior or lateral projections, sometimes stronger sclerotized, often shallowly concave at posterior end; exception: *I. sericeata, I. allardiata*.

FEMALE GENITALIA: Papillae anales with small additional, ventral, setose projection. Ductus bursae usually well sclerotized, rather long. Corpus bursae anteriorly membranous, posteriorly often sclerotized, with spines or spinules.

PHENOLOGY: Some species plurivoltine, others univoltine with extremely slowly growing larvae, feeding on substrate up to 12 months. All *Idaea* species overwinter as larva.

BIOLOGY: Larva usually polyphagous, clearly preferring dry, withered, mouldy or decaying parts of the host-plant. Usually not producing silky spinnings, and comparatively indifferent to light (compare chapter 'immature stages' of *Brachyglossina*). Feeding observations in nature extremely rare, due to the hidden resting and feeding places between fallen leaves near the ground. Detritus-feeding characteristic also for Nearctic and Neotropical species (COVELL 1983). Seed-feeding on dipterocarps recorded for a Bornean species (HOLLOWAY 1997), the Mediterranean *I. infirmaria* was found in capsules of *Cistus*. Some species have been recorded as pests on tobacco (*I. seriata*) and dried herbs and plant specimens in herbaria (*I. inquinata*), similar observations are published for the Nearctic *I. bonifata* (HULST 1887) (COVELL 1983). Facilitated egg deposition of ♀ is achieved by offering hairs or very thin rootlets (LEIPNITZ pers. comm.). Rearing of larva usually easy on withered leaves of *Taraxacum officinale* and *Polygonum*, young larvae preferring withered or dry petals (e.g. *Taraxacum officinale, Rosa, Rubus, Aster, Achillea millefolium, Campanula, Hieraceum, Lathyrus, Vicia, Chrysanthemum, Polygonum*; MÜLLER pers. comm.). Larvae need dry twigs or wood-wool for resting (MÜLLER pers. comm.).

Adults usually stenotopic, fixed to their habitats, not moving far from native areas. Often readily attracted to light, some species diurnal, mainly those belonging, presumably, to phylogenetically basal groups. Some species readily attracted to pheromones of Sesiidae, e.g. *I. nexata* (Tunisia) or *I. inquinata* (Germany, Spain, Morocco: BLAESIUS pers. comm.). Olefinic acetate identified as sex pheromone component in *I. biselata, I. aversata* and *I. straminata* (ZHU et al. 1996), but surely used by all members of the genus (ABRAHAM et al. 2002b). Hybridization not yet recorded for *Idaea* species, but see remarks on *I. humiliata*. Extensive study of ecology of many species in RYRHOLM (1989).

IMMATURE STAGES: Eggs often with polygonal reticulate sculpturation, species of the *rusticata* species-group and some others with distinct longitudinal, and shallower transverse ridges (see Text-figs 107-108 in Vol. 1). Larva usually strongly furrowed transversely, with lateral keel-shaped ridge, tapering towards head, shorter than those of *Scopula*. Within the genus *Idaea*, there is a gradient from the character state of short and thick larvae with prominent lateral ridge (Text-fig. 70) towards comparative long and slender larvae with shallow lateral ridge (Text-fig. 71, compare photographs in www.herbulot.de). Larval morphology needs more detailed studies in order to get additional character sets for reconstruction of phylogeny. Pupa with six very thin hook-like bristles at cremaster.

REMARKS: Species-group names are introduced according to the recommendations of the Code, as they are easier to remember than the numbering of STERNECK. In the choice of appropriate species-group names well known, widespread and morphologically typical species are preferred.

The ***aureolaria*** species-group

STERNECK's (1940) 2nd and 3rd group of *Idaea*. According to STERNECK (1940: 105) phylogenetically descending from the *rufaria* species-group. Heterogenous group with male

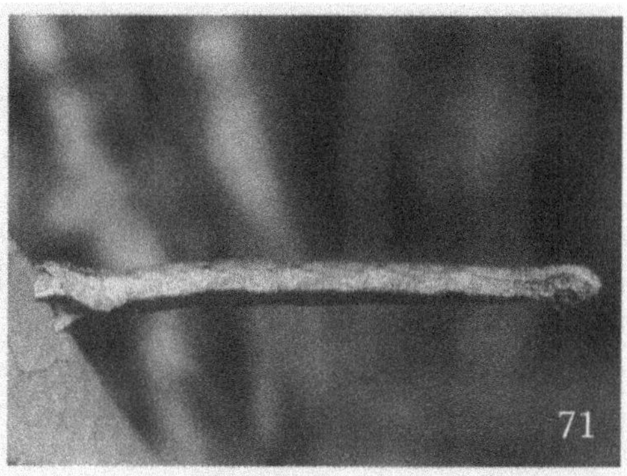

Text-figs 70-71: Larvae of *Idaea* species. Text-fig. 70: Short and thick larva of *I. rhodogrammaria* (PÜNGELER, 1913). Text-fig. 71: Long and slender larva of *I. camparia* (HERRICH-SCHÄFFER, 1852) (Photos: LEIPNITZ).

hindspurs present (exception: *I. serpentata*), ciliate-fasciculate ♂ antennae with tendency to short pectination, reduction or loss of proboscis, small eyes, distant from each other, with whitish scales below eyes (most species), yellow coloration, sexual dimorphism in habitus, and small tympanum (some species). Most of these characters reminiscent of *Cleta*, thus placed at the head of the genus. Sternum A2 of ♂ without central sclerite and setose patch. In ♂ genitalia with modifications of uncus and gnathos, juxta often modified and diagnostic. Female genitalia of some species with short apophyses anteriores, reminiscent of *Cleta* and *Emmiltis*. Egg sculpturation polygonal reticulate. Larvae slender in most species.

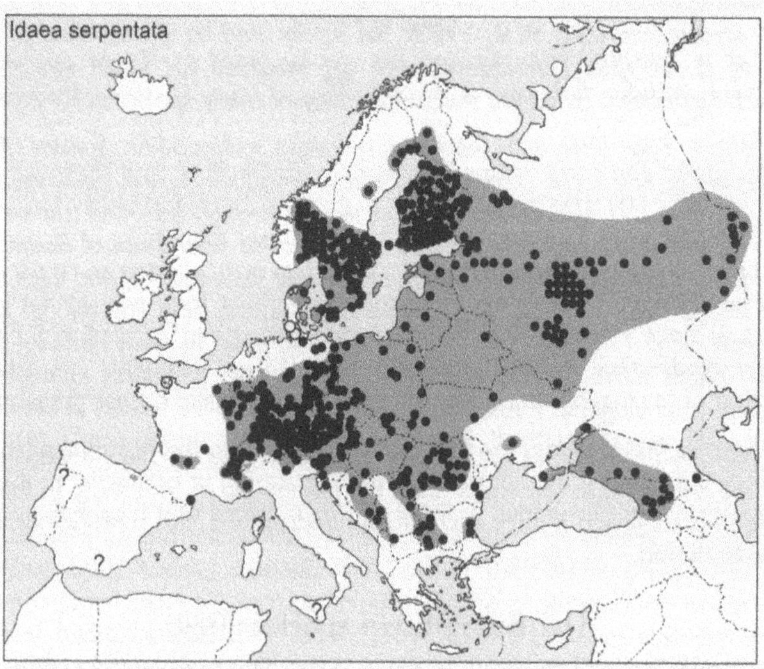

6. *Idaea serpentata* (HUFNAGEL, 1767)

Phalaena serpentata HUFNAGEL, 1767: Berlin. Mag. 4 (6): 613 (Germany: Berlin). Syntype(s) lost.
Phalaena trifasciata SCOPOLI, 1772: Ann. Hist.-Nat. 5: 116 (eastern Germany: Chemnitz). Syntype(s) lost.
Geometra similata THUNBERG, 1784: Diss. ent. sistens Ins. Suecica 1: 10 (Sweden). Lectotype ♂ (UZIU).
Acidalia perochraria FISCHER VON RÖSLERSTAMM, 1836: Abbild. Schmett.kd. 5/6: 46 (central Europe). Syntype(s). Introduced to 'replace' misidentified *ochreata* BORKHAUSEN, 1794 nec SCOPOLI, 1763.
Ptychopoda serpentinata JOHN & KRIZU, 1932: Latv. kuk. noteic. Taurini: 122. Syntype(s).
Acidalia dohlmanni HEDEMANN, 1881: Horae Soc. ent. ross. 16: [257] (eastern Russia: Amur, Chingan Mts.). Syntypes 2♂. Valid at subspecific rank (SCOBLE 1999), in VIIDALEPP (1996) as separate species.
Unavailable names (misidentifications): *exilaria*: sensu SAND (1879) nec GUENÉE, 1858; *ochrata*: sensu TUTT (1897) nec SCOPOLI, 1763; *ochreata*: sensu BORKHAUSEN (1794) nec *ochrata* SCOPOLI, 1763; *rabiginata*: sensu CROMBRUGGHE (1911) nec *robiginata* STAUDINGER, 1863. – (infrasubspecific): *anastomosaria*: GALVAGNI (1927); *bilineata*: REISSER (1961); *demarginata*: REISSER (1961); *flava*: OSTHELDER (1929); *fuscomixata*: PROUT (1913); *griseata*: FUCHS (1904); *nigrolineata*: REISSER (1961); *uniformis*: KAUTZ (1931). – (incorrect subsequent spellings): *fuscomixtata*: PROUT (1914); *limitata* (for *similata*): BORKHAUSEN (1794); *perocharia*: TOMIC et al. (2002); *perochrearia*: HERRICH-SCHÄFFER (1847); *perochriaria*: RÖSSLER (1866); *serpendata*: WOLFSBERGER (1966); *serpentaria*: PASSOS DE CARVALHO (1984); *similiata*: PRÜFFER (1923); *simulata*: HOFFMANN (1915).

DIAGNOSIS: Wingspan ♂ 16-22, ♀ 15-19 mm. Female with forewing apex pointed. Ground colour light brown with orange tinge. Three transverse lines indistinct, greyish brown, undulate. In cold, wet localities with greyish coloration, transverse lines broad, black (BERGMANN 1955). Yellowish wavy line undulate, sometimes broken into spots, proximally darkly bordered. Discal spots very small, on forewing often absent. Medial line proximal to (vestigial) discal spot, or crossing it. Fringe light brown, darker than ground colour. Fringe dots and terminal line absent. Underside similar, with dark brown transverse lines distinct and discal spot present on forewing. Frons and collar dark brown, palpi and vertex sand coloured. Proboscis vestigial, sometimes present, but short. Eyes small, distant from each other. Antennae of ♂ sub-quadripectinate, cilia in tufts on short branches, branches 1.0-1.5 times width of flagellum, cilia about 1.5 times (Text-fig. 35). Hindtibia of ♂ without spurs, tarsus not shortened (Text-fig. 16).

MALE GENITALIA: Uncus rounded. Gnathos absent. Valva with spinose digitiform process at 1/2. Aedeagus broad at base, with two long and stout cornuti, one terminally curved by 90°.

FEMALE GENITALIA: Apophyses anteriores short. Ductus bursae short, membranous. Corpus bursae large, pyriform or oval, anteriorly smoothly sclerotized.

DISTRIBUTION: West-Palaearctic. Widely distributed, but often local, in temperate Europe from France, including the Pyrenees to the Urals. Absent from large parts of northern and southern Europe. Single old records from British Isles, extinct since 1909, on Channel Islands since 1941. In the Iberian peninsula replaced by the sister species *Idaea luteolaria*, but reputedly recorded in northern Portugal (SILVA CRUZ & WATTISON 1935; PASSOS DE CARVALHO 1984; MÜLLER 1996), northern Spain (GÓMEZ DE AIZPÚRUA 1988), in ZSM two specimens from Andalusia (mislabelled?). Reputedly on Corsica (RUNGS 1988; MÜLLER 1996), and Sicily (MÜLLER 1996), but no vouchers traced. Absent from Italy south of Alps (FLAMIGNI & BASTIA 1998), though recorded in literature from various regions (MARIANI 1941; PARENZAN 1994). Abundance decreasing in many regions due to habitat losses and alteration (ERLACHER & FRIEDRICH 1994; EBERT 2001). – Outside Europe in north-easternmost Turkey (RIEMIS 1994), Caucasus, Transcaucasus and central Asian mountains, in the east to Mongolia (VIIDALEPP 1975; VOJNITS 1977) and central Siberia (Jakutia). Eastern Palaearctic populations (*dohlmanni*) considered valid at subspecific (SCOBLE 1999) or at species rank (VIIDALEPP 1996; KOSTJUK & GOLOVUSHKIN 2003).

PHENOLOGY: Usually univoltine: mid-June to mid-August, exceptionally somewhat earlier or later. Under favourable conditions, e.g. south of Alps sometimes bivoltine: mid-May to early September, often with a short gap late June or early July. Protandrous, males emerging one week before females (PROUT 1913).

BIOLOGY: Larva oligo- or polyphagous on withered leaves of herbaceous plants and grasses (var. cit.). Found on *Thymus serpyllum* (Lamiaceae) (Germany: BOLDT 1929). Recorded on *Rumex acetosella* (Polygonaceae), *Galium* (Rubiaceae), *Festuca pratensis* (Poaceae) (Germany: BERGMANN 1955), and *Festuca ovina* (Poaceae) (France: LHOMME 1935). Reared on *Rumex, Lactuca sativa, Taraxacum officinale, Sonchus* and *Galium* (PROUT 1913; LHOMME 1935; BERGMANN 1955; SEPPÄNEN 1970; SKINNER 1984). Adults mainly diurnal, actively flying in sunshine, sometimes attracted also to light. ♀ with reduced flight activity, thus ♀ ratio low in collections, about 10-20%. After deposition of eggs females get more active. Nectar feeding recorded on *Achillea millefolium* (EBERT 2001).

Text-figs 72-74: Diagnostic characters (indicated) of European *Idaea* species. Text-fig. 72: *I. serpentata* (HUFNAGEL, 1767). Text-fig. 73: *I. rufaria* (HÜBNER, 1799). Text-fig. 74: *I. ochrata* (SCOPOLI, 1763).

HABITAT: Xerothermophilous in central Europe. Open habitats on low-nutrient soils, often on limestone: heathlands, uncultivated grassland (Mesobrometum), scattered scrubland, embankments. Sometimes also on wet meadows and moors (KIEFER 1941; SKOU 1986). From 0 m up to 1,100 m above sea-level, up to 1,700 m in southern Alps and Pyrenees (WOLFSBERGER 1971; MAZEL & PESLIER 1997).

SIMILAR SPECIES: Both *I. ochrata*, and *I. luteolaria* with two ♂ hindtibial spurs. *I. luteolaria* differs in more rounded wings, ground colour more yellowish rather than brown, transverse lines straighter, discal spots absent. *I. ochrata* is larger, broader winged, ♂ without antennal branches, wing pattern sharper, postmedial line fine, medial line distal to vestigial 'discal spot', usually differing also in having fringe dots (FRIEDEMANN 1937; EBERT 2001), but that character does not allow secure identification, as fringe dots are sometimes absent.

7. *Idaea luteolaria* (CONSTANT, 1863)

Acidalia luteolaria CONSTANT, 1863: Annls Soc. ent. Fr. (4) 3 (1): 73, pl. 2, figs 1a-c (Spain, Castilia: San Ildefonso near Segovia; France, eastern Pyrennees: Mont-Louis). Syntypes ♂♀ (BMNH: 1♂ examined, labelled as 'lectotype L.B.P[ROUT] sel.'; MNHU: 6♂1♀, from Castilia, examined).

DIAGNOSIS: Wingspan ♂ 15-21, ♀ 15-17 mm. ♀ with wings less rounded. Ground colour dirty yellow, in ♀ paler. Transverse lines rather straight, slightly curved. Postmedial line usually distinct, medial line less, antemedial line often absent. Discal spots absent. Fringe dark grey. Underside with very distinct, dark brown transverse lines. Frons dark brown, palpi, vertex and collar paler brown. Proboscis short. Eyes small, distant from each other. Antennae of ♂ dentate, cilia in four tufts per segment, dentations 1.0 times width of flagellum, cilia 1.5-2.3 times (Text-fig. 36). Hindtibia of ♂ with two spurs, tarsus not shortened. Tympanal organs small.

MALE GENITALIA: Uncus sub-triangular, posteriorly rounded. Gnathos broad, but weak. Valva with small dorsal and larger ventral spine at tip. Aedeagus very large, with one slen-

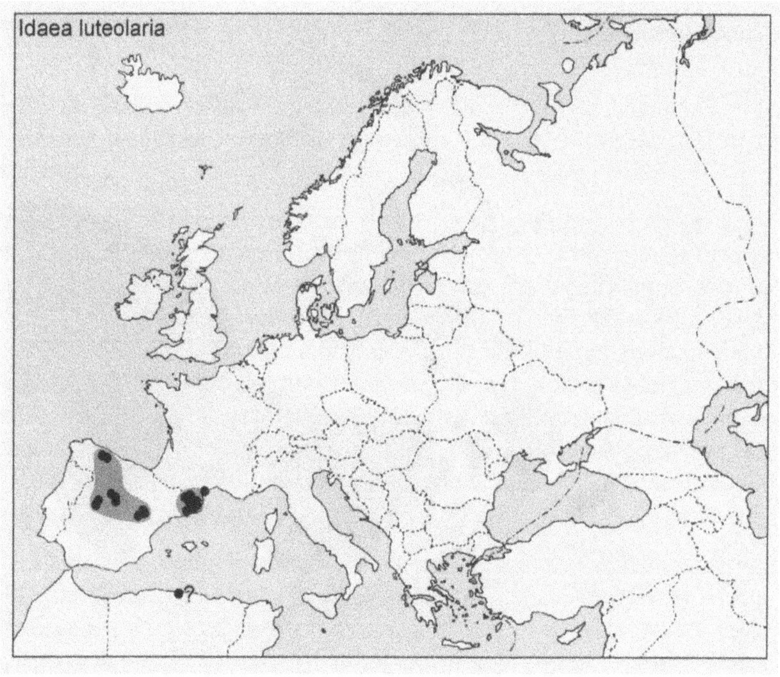

der, double sigmoid, extremely long cornutus (3 mm, i.e. 0.75 times length of abdomen).

FEMALE GENITALIA: Apophyses anteriores short. Ductus bursae dilated at antrum. Semicircular lamella antevaginalis well sclerotized. Corpus bursae elongate pyriform, laterally corrugated longitudinally.

DISTRIBUTION: West-Mediterranean. Mountains of northern and central Spain, Pyrenees and Andorra (DANTART & ROCHE 1992). – Outside Europe reputedly in Algeria (1♂ examined in ZSM), possibly mislabelled, awaiting confirmation.

PHENOLOGY: Univoltine: mid-June to early August. Literature records from early September (YLLA et al. 1990) await confirmation.

BIOLOGY: Larva probably polyphagous on withered leaves of herbaceous plants. In rearings preferring *Achillea millefolium* (Asteraceae) (France: CHRÉTIEN, 1928; LHOMME 1935), and, reputedly, on petals of *Geranium*, and leaves of *Rumex, Leontodon* (MILLIÈRE 1885: 114; cited in PROUT 1913), but misidentification of these larvae suggested by CHRÉTIEN (1928). Adults diurnal (KRAUS 1999). ♀ ratio in collections low, about 10-15%, probably due to reduced flight activity.

HABITAT: Montane. Limited data available on habitat preference. From 1,100 m up to 2,200 m above sea-level.

SIMILAR SPECIES: *Cleta filacearia* (see no. 5), *Idaea serpentata* (6). *I. flaveolaria* with paler ground colour, transverse lines more distinct on upperside and underside, fringe more contrasted, ♂ without hindtibial spurs. North African *I. numidaria* (H. LUCAS, 1849) with ground colour deeper yellow, transverse lines finer, palpi longer.

REMARKS: For rearing and morphology of immature stages see CHRÉTIEN (1928).

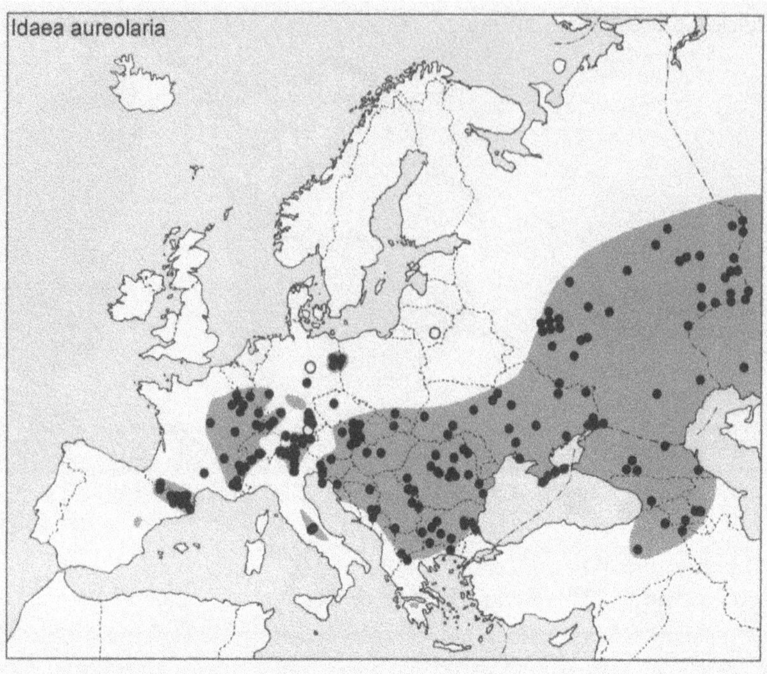

8. Idaea aureolaria (DENIS & SCHIFFERMÜLLER, 1775)

Geometra aureolaria DENIS & SCHIFFERMÜLLER, 1775: Ank. syst. Werkes Schmett. Wienergegend: 102 (Austria: Vienna distr.). Syntype(s) lost.

‡ *Phalaena trilineata* SCOPOLI, 1772: Ann. Hist.-Nat. 5: 115 (north-western Slovenja: Carniolia). Syntype(s) lost. Junior primary homonym nec HUFNAGEL, 1767 (identity: *Idaea aversata*).

‡ *Phalaena bicincta* FOURCROY, 1785: Entom. Parisiensis 2: 271 (France: Paris), Junior primary homonym according to LERAUT (1997): 'nec LINNAEUS, 1768'.

Phalaena Geometra trilinearia HÜBNER, 1787: Beitr. Gesch. Schmett. 1 (2): 23, pl. 2 (3) fig. T (Austria: Vienna distr.). Syntype(s) ♂, lost.

Phalaena Geometra bicinctaria VILLERS, 1789: Linn. ent. 2: 325 (France, Bresse near Lyon ('Bressia prope Lugdunum')). Syntype(s). Probably emendation of *bicincta*.

Acidalia trilineata transsylvanaria DANNEHL, 1927: Ent. Z. Frankf. a.M. 40 (17): 405 (Romania, Transsylvania: Brasov ('Siebenbürgen, Kronstadt'). Lectotype ♂ (ZSM, examined). Downgraded from subspecific rank (SCOBLE 1999) to synonymy (syn. n.).

Unavailable names (infrasubspecific): *anastomosaria*: REISSER (1961); *brunnea*: REISSER (1949); *latistrigata*: VORBRODT & MÜLLER-RUTZ (1917); *nigrocostata*: HIRSCHKE (1910); *pseudolutearia*: OSTHELDER (1929). – (incorrect subsequent spelling): *aureaolaria*: YLLA (1988).

DIAGNOSIS: Wingspan ♂ 16-20, ♀ 14-19 mm. ♀ with apex more pointed. Ground colour deep yellow. Postmedial and medial lines distinct (less in ♀), dark grey, straight, antemedial line slightly curved. Discal spots absent. Fringe dark grey. Populations from southern Europe across south-western Asia to central Asian mountains with transverse lines and fringe blackish, with clinal transitions to typical forms. Frons dark brown with single ochreous scales, palpi grey or ochreous, darker at tip. Eyes normal. Proboscis short. Antennae of ♂ dentate, ciliate-fasciculate, cilia 1.5-2 times width of flagellum (Text-fig. 37). Hindtibia of ♂ with two spurs, tarsus not shortened. Tympanal organs large.

MALE GENITALIA: Uncus sub-triangular, rounded. Gnathos broad, spatulate. Two long, slender processi projecting dorsad from base of juxta. Valva dilated at 1/2. Aedeagus without cornuti.

FEMALE GENITALIA: Ductus bursae very short and broad. Posterior 1/3 – 1/2 of corpus bursae broad, tubular, well sclerotized with numerous spinules. Anterior part of corpus bursae membranous.

DISTRIBUTION: West-Palaearctic. In the north from northern France and northern Germany to the Urals. In southern Europe from northern Spain across France to southern Balkan peninsula, with isolated populations in the Apennines, and Peloponnes (THURNER 1967). Erroneously recorded from southern Italy (MARINI & RUSSO 1980; PARENZAN 1994; rectified by FLAMIGNI & BASTIA 2003a). In large parts of the distribution area split into very isolated populations, sometimes common at localities of occurrence. In many parts of central Europe strongly threatened, populations decreasing, e.g. extinct in prov. Pfalz (Germany: FORSTER & WOHLFAHRT 1981). – Outside Europe from Turkey, Caucasus, Transcaucasus to the central Asian mountains, Mongolia (VIIDALEPP 1975; 1996), Jakutia and Transbaikalia (e.g. KOSTJUK 1992; KOSTJUK & GOLOVUSHKIN 2003).

PHENOLOGY: Univoltine: early June to late July, under favorable conditions emerging from late May, in cold years until early August (BERGMANN 1955). Reputedly with partial second generation early August to early September (FORSTER & WOHLFAHRT 1981).

BIOLOGY: Larva polyphagous on withered leaves of herbaceous plants, with focus on Fabaceae. Recorded on *Vicia dumetorum* (Fabaceae) (DENIS & SCHIFFERMÜLLER, 1775; subsequently var. cit.), *Rumex* (Polygonaceae), *Lactuca* (Asteraceae), *Cytisus oromediterraneus (Genista purgans), Genista cinerea, Coronilla varia*, and *Onobrychis arenaria* (Fabaceae) (var. cit.; most derived from REBEL 1910b and VORBRODT 1914), at least partially based on rearings. Reared on *Vicia cracca* (ERLEBACH pers. comm.), *Coronilla varia* and *Onobrychis* (HAUSMANN & MILLER 2000). Imagines easily flushed up during daytime, almost diurnal, but also readily attracted to light. ♀ ratio in collections low, about 20-30%, probably due to reduced flight activity. Stenotopic (EBERT 2001), K-strategy.

HABITAT: Open habitats on dry, rocky or stony ground, low-nutrient soils on limestone and sand; heathlands, fallows, sun-exposed slopes, scrubby places, sometimes also on wet grassland (REISSER 1949). From 0 m up to 700 m above sea-level, up to 1,500 m in central Alps, preferably from 800 m up 2,000 m in the south of the distribution area, up to 2,400 m in Turkey, Caucasus and central Asian mountains.

SIMILAR SPECIES: Yellower than the other yellowish Sterrhinae *I. flaveolaria, I. luteolaria* and *Cleta filacearia*, feature especially important for identification of ♀. The last three species with transverse lines absent or indistinct, less straight. *C. filacearia* with ♂ antennae bipectinate.

9. *Idaea flaveolaria* (HÜBNER, 1809)

Geometra flaveolaria HÜBNER, 1809: Samml. Eur. Schmett. 5, Geometrae (1): pl. 66, fig. 341 (Europe). Syntype(s) lost.

‡ *Phalaena brunnearia* FABRICIUS, 1794: Ent. Syst. 3 (2): 162 (Italy). Syntype(s) lost. Junior primary homonym nec VILLERS, 1789 (identity: *Selidosema brunnearia*, Ennominae).

Unavailable names (misidentification): *filacearia*: sensu LERAUT (2003: 380) nec HERRICH-SCHÄFFER, 1847

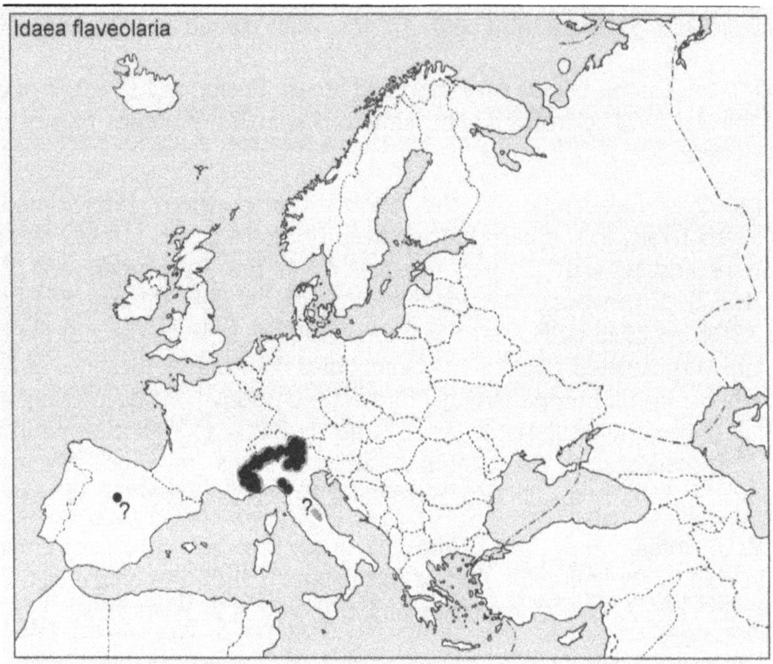

(identity: *Cleta filacearia*); *trifasciata*: sensu WERNEBURG (1864) nec SCOPOLI, 1772 (identity: *Idaea serpentata*). – (infrasubspecific): *nigrolineata*: PROUT (1913).

DIAGNOSIS: Wingspan ♂ 14-18, ♀ 12-14 mm. ♀ with particularly short and slender wings. Ground colour in ♂ yellowish orange, in ♀ light dirty yellow. Postmedial and medial lines light grey, indistinct, nearly invisible (better marked in ♀), antemedial line absent. Discal spots absent. Fringe blackish grey. Transverse lines and discal spots well marked on underside. Frons, vertex and palpi dark brown. Eyes small, distant from each other, with whitish scales below eyes. Proboscis very short (vestigial). Antennae of ♂ dentate, ciliate-fasciculate, dentations about 1.0 times, cilia about 2.0 times width of flagellum. Hindtibia of ♂ without spurs, tarsus not shortened. Tympanal organs small.

MALE GENITALIA: Uncus bifid. Gnathos absent. Tegumen with attached elongate, terminally spinose sclerites and a long caudad projection from base of that sclerite. Valva slender, spatulate. Aedeagus slender, without cornuti.

FEMALE GENITALIA: Apophyses anteriores short. Lamella antevaginalis triangular, cup-shaped. Ductus bursae comparatively slender, laterally attached to corpus bursae. Corpus bursae oval, covered all over with numerous (micro-)spinules.

DISTRIBUTION: European (montane). Widely distributed in the Alps, mainly in the south-western parts. Some local populations in the Apennines (FRILLI & PIZZAGHI 1975; TEOBALDELLI 1976), awaiting confirmation. Reputedly in Spain (VIVES MORENO 1994; MÜLLER 1996), but records for Catalonia erroneous (DANTART 2000), or all data from Spain misidentified (REDONDO & GASTON 1999), one examined specimen from central Spain (ZSM) possibly mislabelled. Erroneously recorded for Slovenja (CARNELUTTI 1992a), data referring to *I. aureolaria* (GOMBOC pers. comm.). Multiple records from Ukraine, Russia, Transcaucasus, Turkey, Iran (e.g. OBRAZTSOV 1930; ANTONOVA 1985; RIEMIS

1994; VIIDALEPP 1996; ANIKIN et al. 2000; DIDMANIDZE 2002) misidentified, referring to *Cleta filacearia*.

PHENOLOGY: Univoltine: late June to mid-August.

BIOLOGY: Larva probably polyphagous on withered leaves of herbaceous plants (REBEL 1910b; LHOMME 1935; FORSTER & WOHLFAHRT 1981). No detailed information available on larval host-plants. ♀ ratio in collections low, about 5%, due to reduced flight activity (beginning wing reduction). Stenotopic, K-strategy.

HABITAT: Xeromontane. On hot, dry southern slopes from 900 m up to 2,400 m above sea-level.

SIMILAR SPECIES (see Text-fig. 15): *Cleta filacearia* (see no. 5), *Idaea luteolaria* (7), *I. aureolaria* (8). Ground colour with more orange tinge than in those species. Individual forms of *I. flaveolaria* with sharp transverse lines (f. '*nigrolineata*') sometimes similar to *I. aureolaria*.

REMARKS: Systematic position debatable. The question needs further study.

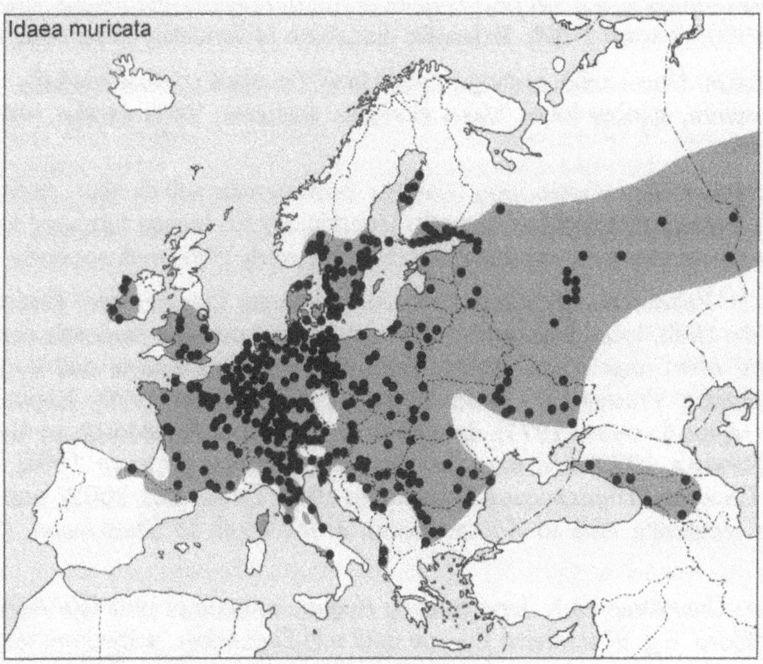

10. Idaea muricata (HUFNAGEL, 1767)

Phalaena muricata HUFNAGEL, 1767: Berlin. Mag. 4 (6): 606 (Germany: Berlin). Syntype(s) lost.

Pyralis auroralis DENIS & SCHIFFERMÜLLER, 1775: Ank. syst. Werkes Schmett. Wienergegend: 124 (Austria: Vienna distr.). Syntype(s) lost.

† *Phalaena variegata* FABRICIUS, 1776: Genera Insect.: 291 (Germany: Hamburg). Junior primary homonym nec SCOPOLI, 1763 (identity: *Scopula incanata*). Syntypes (ZMUC, two syntypes examined, one labelled as 'type', without abdomen, the other only with thorax and 2 wings).

Phalaena (Geometra) aureopolita GOEZE, 1783: Mentioned in LERAUT (1997). Original description not traced.

Phalaena Geometra sanguinaria HÜBNER, 1787: Beitr. Gesch. Schmett. 1 (2): 21, pl. (2) 3, fig. S (Germany, Bavaria: Augsburg distr.). Syntypes ♂♀, lost.

‡ *Phalaena Geometra auroraria* BORKHAUSEN, 1794: Naturg. eur. Schmett. 5: 477 (Europe, probably Germany). Syntypes ♂♀. Junior primary homonym nec HÜBNER, 1787 (identity: *Isturgia limbaria*, Ennominae).

Ptychopoda muricata var. *minor* STERNECK, 1927: Dt. ent. Z. Iris 41: 167 (China: Kwan, Peking). Syntypes 1♂3♀. Valid at subspecific rank (SCOBLE 1999).

Sterrha muricata proutiana BRYK, 1942: Dt. ent. Z. Iris 56: 64 (Kurile islands: Kunashiri, Tomari). Holotype ♂ (NHRS). Valid at subspecific rank (SCOBLE 1999).

Unavailable names (infrasubspecific): *atrorubra*: WISSELINGH (1967); *flava*: MICHAEL (1963); *lutescens*: PROUT (1913); *maidorni*: HANNEMANN (1917); *rubrior*: HOFFMEYER (1952); *totarubra*: LAMBILLON (1905). – (incorrect subsequent spellings): *totaumbra*: KUJAU (1918); *totorubra*: VORBRODT (1930).

DIAGNOSIS: Wingspan ♂♀ 14-19 mm. Ground colour purple. Medial area of forewing with two yellow patches, a third one at base of wing. Hindwing with large yellow discal spot. Postmedial line bluish violet. Terminal area and fringe yellow. Frons black, palpi dark grey, vertex whitish, collar red. Eyes of normal size. Proboscis short. Antennae of ♂ dentate, ciliate-fasciculate, cilia long, 1.5-2.2 times width of flagellum. Hindtibia of ♂ without spurs, tarsus not shortened. Tympanal organs large.

Variability: Yellow areas of forewing may be enlarged and fused (e.g. 'f. *lutescens*'), sometimes, however, reduced (e.g. f. '*totarubra*'). The latter form in some regions common, or even predominating, e.g. on peat-bogs of northern England and Ireland, (PROUT 1913; HUGGINS 1950; SKINNER 1984). Extensive discussion of variability in MICHAEL (1963).

MALE GENITALIA: Uncus bifid, ending in two claws. Gnathos spinose medially. Juxta with paired, posterior, spinose lobes. Valva elongate, spatulate. Valva slender, with one very small cornutus.

FEMALE GENITALIA: Apophyses long. Lamella postvaginalis sub-circular. Antrum heavily sclerotized, posteriorly bisinuous, laterally tapering. ductus bursae furrowed longitudinally. Corpus bursae elongate, spinulose laterally, anteriorly with small appendix.

DISTRIBUTION: Palaearctic. Distributed all over temperate Europe, from British Isles and France to the Urals, local, but sometimes common. In Apennine peninsula some isolated populations, often near coast. Absent from northern Scandinavia and large parts of southern Europe. Present on Corsica (RUNGS 1988, MÜLLER 1996). Reputedly in the Astrachan region (LVOVSKY 1971), data await confirmation. Areal losses by bog drainage (Finland: MIKKOLA 1979) and intensivization of agriculture (HUEMER 1994). – Outside Europe in Caucasus, Transcaucasus (WARDIKJAN 1985; DIDMANIDZE 2002), and widely distributed in temperate Asia to Korea, Japan, in the south to Szetschwan, China (ssp. *minor*).

PHENOLOGY: Univoltine: early June to early August, sometimes until late August. Under good conditions, e.g. in southern Europe until mid-September, sometimes with a gap in late July, possibly indicating a partial second generation, emerging from early August.

BIOLOGY: Larva polyphagous on herbaceous plants, feeding exclusively on withered and dry leaves (BERGMANN 1955; GAUCKLER 1924). Found on *Potentilla palustris* (*Comarum palustre*) (Rosaceae) (FRIEDRICH 1976). Recorded on *Galium palustre* (Rubiaceae), *Rubus, Potentilla palustre* (Rosaceae), *Polygonum* (Polygonaceae), *Plantago* (Plantaginaceae), *Euphorbia* (Euphorbiaceae), *Anemone* (Ranunculaceae), *Pimpinella saxifraga* (Apiaceae), *Calluna vulgaris* (Ericaceae) (var. cit., e.g. REBEL 1910b; LHOMME 1935; ALLAN 1949; BERGMANN 1955; FORSTER & WOHLFAHRT 1981; HUEMER 1994), reputedly on *Festuca* (Poaceae) (SEPPÄNEN 1954; 1970). Reared on withered or dry leaves of *Anemone nemorosa, Rubus idaeus, R. fruticosus, Potentilla reptans,*

Plantago lanceolata, *P. major*, *Lactuca sativa*, *Taraxacum officinale*, *Polygonum aviculare* (BUCKLER 1897; UFFELN 1913; GAUCKLER 1924; SCHNEIDER 1939; ALLAN 1949; BERGMANN 1955, SEPPÄNEN 1970; FRIEDRICH 1976; SKINNER 1984; EBERT 2001; ERLEBACH, MÜLLER pers. comm.). Extensive description of rearings in SCHULTZ (1930). Adults almost diurnal, sometimes flying actively during daytime, easily flushed up from vegetation, active at dusk and at sunrise (SKINNER 1984), sometimes attracted to light. ♀ ratio in collections low, 20-30%. Nectar feeding recorded on *Molinia* (URBAHN & URBAHN 1939; BERGMANN 1955).

HABITAT: Silvicolous, hygrophilous, sometimes, however, also at xerothermic sites: Dry grass- and scrubby places on low-nutrient, sandy soils, damp heathland, forest glades, fringes of coniferous or mixed forests, most common at borders of bogs, mosses, marshes and wet meadows. Usually from 0 m up to 700 m above sea-level, exceptionally up to 1,500 m in southern Alps.

SIMILAR SPECIES: No similar species in Europe.

REMARKS: The type of 'f. *maidorni*' was examined (ZFMK) and belongs to *I. muricata*, and not to *I. dimidiata* as suggested by MICHAEL (1963). For larval morphology see BUCKLER (1897).

The *rufaria* species-group

STERNECK's (1940) 1st group of *Idaea*. Sister-group relationship to *aureolaria* species-group (STERNECK 1940: 105). Homogenous group, with terminal pair of ♂ hindtibial spurs ('plesiomorphic': STERNECK 1940: 104); exception: *I. consanguiberica*. Hindtibia not dilated, without pencil, tarsus not shortened. Proboscis well developed. Antennae of ♂ ciliate-setose or ciliate-fasciculate. Sternum A2 of ♂ without central sclerite and setose patch. Male genitalia with one stout cornutus in the aedeagus ('plesiomorphic': STERNECK 1940: 104), valva linguiform. Female genitalia with appendix bursae; exceptions: *I. ochrata*; *I. consanguiberica*. Usually with univoltine phenology. Egg sculpturation polygonal reticulate; exception: *I. ochrata* with longitudinal ridges. *I. consanguiberica* possibly needs to be removed from this species-group.

11. Idaea determinata (STAUDINGER, 1876)

Acidalia determinata STAUDINGER, 1876: Stettin. ent. Ztg. 37: 141 (Sicily: Valdesi near Palermo). Syntypes 1♂1♀ (MNHU 1♂1♀, examined).
Acidalia fasciata STAUDINGER, 1892: Dt. ent. Z. Iris 5 (1): 145 (Turkey, northern Taurus: Hadjin; southern Antitaurus: near Maras). Syntypes 2♂1♀ (MNHU 2♂1♀, examined).
Acidalia geministrigata FUCHS, 1901: Stettin. ent. Ztg. 62: 122 (Sicily). Holotype.
Unavailable names (infrasubspecific): *kammeli*: STAUDER (1916).

DIAGNOSIS: Wingspan ♂♀ 20-23 mm, in Turkey up to 25 mm. Wings sand coloured. Transverse lines brown. Antemedial line straight, near costa inwardly projecting. Medial line convex, very close to postmedial line, with space between them sometimes filled with dark brown ('f. *kammeli*', about 10-20% of the populations). Postmedial line crenulate. Terminal area light brown with sand coloured spots more or less fused to a wavy line. Discal spots black, punctiform. Terminal line present, fringe dots black, conspicuous. Frons blackish brown, vertex sand coloured. Antennae of ♂ densely ciliate, cilia 0.6-0.8 times width of flagellum. Hindtibia of ♂ with two spurs, tarsus not shortened.

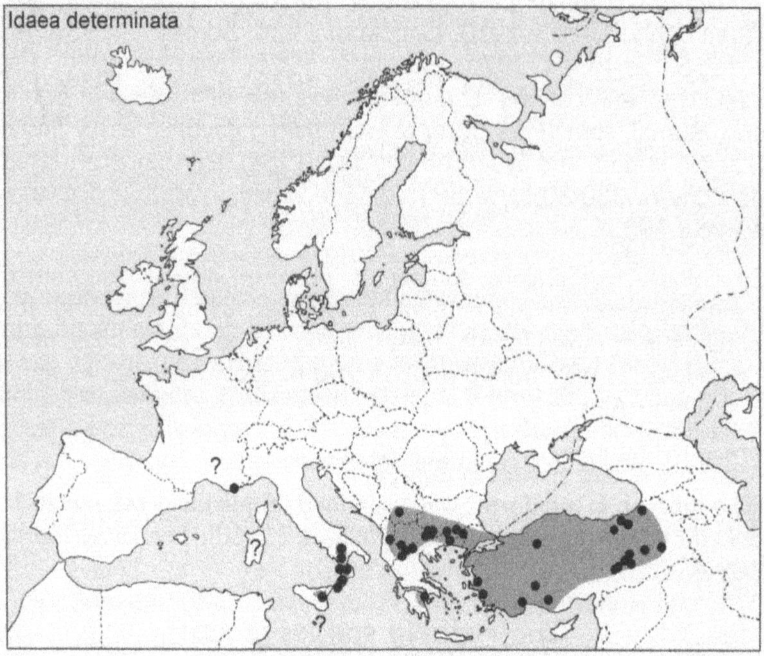

MALE GENITALIA: Uncus terminally tapered. Valva very long, spatulate. Aedeagus very large (nearly 3 mm) and broad. Vesica strongly folded, with one large, stout, curved cornutus, 0.5 times length of aedeagus.

FEMALE GENITALIA: Ductus bursae very short. Corpus bursae approximetely lemon-shaped, strongly sclerotized, with long and slender 'appendix bursae', both filled with many spines, the latter anteriorly membranous.

DISTRIBUTION: (East-)Mediterranean. Very isolated populations in southern France, southern Italy (PARENZAN et al. 1998b), Sicily, and southern Balkan peninsula. Occurrence on Mt. Ventoux, southern France (LUQUET 2000), on Sardinia and on Malta (MÜLLER 1996) awaits confirmation, the last questioned by SAMMUT (2000). Absent from the other Mediterranean islands. – Outside Europe in Turkey.

PHENOLOGY: Univoltine: early June to late July, in Turkey emerging from late May. Records from October (PROUT 1913) probably misidentified.

BIOLOGY: Larva probably polyphagous on withered leaves of herbaceous plants. Reared on petals of various flowers and, later on, on withered leaves of *Lactuca sativa* (HERBULOT 1950b), on petals of *Achillea millefolium*, only larger larvae accepting withered leaves of *Taraxacum officinale* (MÜLLER pers. comm.). Adults easily flushed up from ferns during daytime (STAUDER 1924), readily attracted to light.

HABITAT: Collected in dry, open habitats between aromatic flowers (PROUT 1913), in forest of *Castanea* with undergrowing ferns (STAUDER 1924), and in pine forest (Sicily: SCHWINGENSCHUSS 1942). From 500 m up to 1,200 m above sea-level, in Turkey montane, up to 2,400 m.

SIMILAR SPECIES: *I. ossiculata*, *I. consanguinaria* and *I. deversaria* with antemedial line less distinct, postmedial line less dentate, wavy line in terminal area less distinct, fringe dots usually less distinct or absent. *I. ossiculata* with narrower forewings and longer ♂ antennal cilia. *I. deversaria* without ♂ hindtibial spurs.

REMARKS: Female genitalia suggest closer relationship to *I. consanguinaria*.

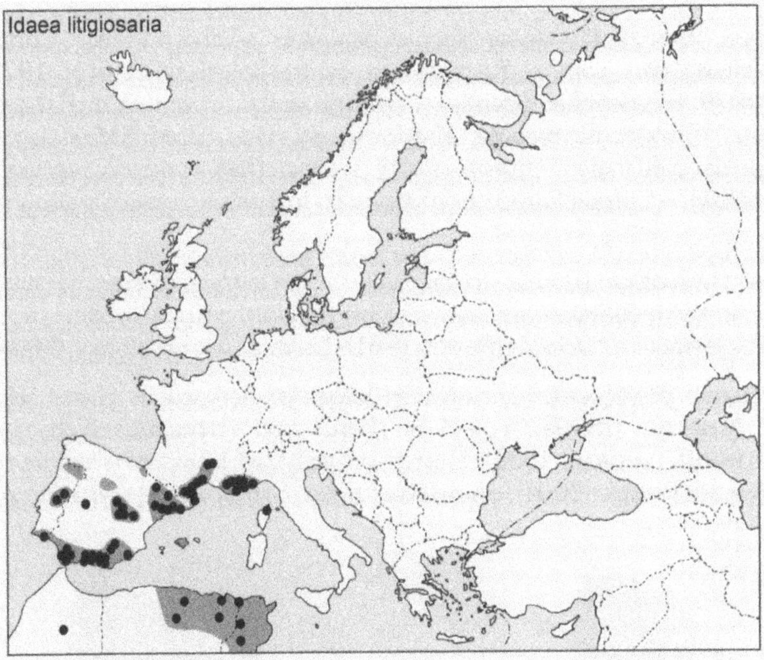

12. *Idaea litigiosaria* (BOISDUVAL, 1840)

Acidalia litigiosaria BOISDUVAL, 1840: Genera Index meth. eur. Lepid.: 226 (southern France). Syntypes (MNHU 2, examined, labelled as 'Orig[inals]' locality not stated, type status uncertain).

Acidalia morosaria HERRICH-SCHÄFFER, 1852: Syst. Bearb. Schmett. Eur. 6 (55): 68 (Europe?). Syntypes ♂♀. Referring to manuscript name of Lederer.

Acidalia agraria JOANNIS, 1891: Bull. Soc. ent. Fr. 1891: 79 (eastern Algeria: Philippeville). Syntypes (MNHN).

Acidalia dissidiata GUENÉE, 1858: in BOISDUVAL & GUENÉE, Hist. nat. Insectes (Spec. gén. Lépid.) 9: 497 (Spain: Andalusia; France: Basses Alpes). Syntypes 2♂ ('coll. BELLIER', in MNHU 1♂ 'type' from Ronda, southern Spain).

Unavailable names (misidentification): *subsericeata*: sensu ROTHSCHILD (1914) nec HAWORTH, 1809. – (incorrect subsequent spelling): *litigiosata*: LEDERER (1861).

DIAGNOSIS: Wingspan ♂♀ 17-24 mm. Variable in size and intensity of wing pattern. Ground colour 'dirty white' with slight sand coloured tinge. Antemedial line of forewing indistinct, sometimes lacking. Medial line absent. Postmedial line grey, undulate, usually vague. Terminal area grey, with whitish wavy line. Wing pattern less distinct in southern Spanish populations. Termen with very small, black, punctiform dots. Discal spots black, punctiform. Underside nearly without pattern, forewing usually darker. Frons blackish brown, vertex whitish, collar sand coloured. Antennae of ♂ densely ciliate, sub-fasciculate, cilia 0.9-1.1 times width of flagellum. Hindtibia of ♂ with two spurs, tarsus not shortened.

MALE GENITALIA: Uncus comparatively broad. Valva spatulate, with two lateral spines at tip. Aedeagus with one stout, slightly curved cornutus, half length of aedeagus. In North African populations cornutus not curved and longer (0.6 times length of aedeagus).

FEMALE GENITALIA: Ductus bursae comparatively long, posteriorly with mushroom-shaped sclerite. Posterior half of corpus bursae strongly sclerotized with many spines, broadly connected to membranous anterior half. Appendix bursae small, closely attached to corpus bursae.

DISTRIBUTION: West-Mediterranean. Iberian peninsula, southern France, Corsica, Liguria, Mallorca (HONEY pers. comm.). Records from central Italy (RACHELI et al. 1978; PROLA & RACHELI 1979) misidentified (identity: *I. rufaria* and *I. straminata*; RAINERI det., ZILLI pers. comm.). Erroneously recorded also for Croatia (1♀ in ZSM/EMEM, leg. MÜCK, mislabelled; similar case under *I. alyssumata*). In southern Italy and Sicily replaced by *I. mutilata* (see below). – Outside Europe from Morocco to Tunisia, reputedly also in Tripolitania (Libya).

PHENOLOGY: Univoltine: late May to early July, in the lowlands sometimes emerging from early May, in the mountains until early August. In North Africa bivoltine, from March to June and September to October (RUNGS 1981). Larva slowly developing (MAIDORN 1914).

BIOLOGY: Larva polyphagous on withered leaves of herbaceous plants, with focus on Fabaceae (CHRÉTIEN 1899a). Found on *Lotus* and *Astragalus* (Fabaceae) (France: CHRÉTIEN 1899a; LHOMME 1935). Reared on withered leaves of *Thymus, Lavandula, Rumex* (KING & ROMERA 2004), on petals of *Rosa, Lathyrus* and *Antirrhinum*, adult lar-

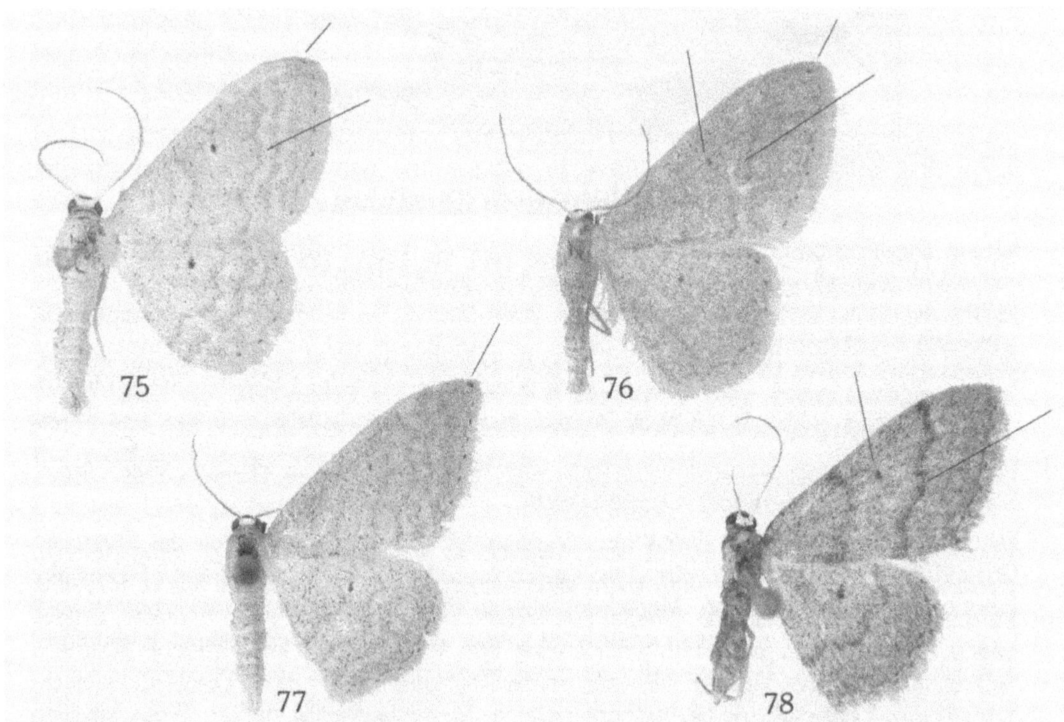

Text-figs 75-78: Diagnostic characters (indicated) of European *Idaea* species. Text-fig. 75: *I. litigiosaria* (BOISDUVAL, 1840). Text-fig. 76: *I. lusohispanica* HERBULOT, 1991. Text-fig. 77: *I. sardoniata* (HOMBERG, 1912). Text-fig. 78: *I. korbi* (PÜNGELER, 1916).

vae on withered or dry leaves of *Lactuca sativa* (MAIDORN 1914). Adults readily attracted to light and bait.

HABITAT: Xerothermophilous. Limited information available about habitat preference. From 0 m up to 2,000 m above sea-level.

SIMILAR SPECIES: Allopatric sister species *I. mutilata* with postmedial line of hindwings usually closer to discal spot, ♂ genitalia with shorter cornutus, ♀ genitalia with sclerotized part of corpus bursae anteriorly narrow. *I. lusohispanica* with ground colour more grey, transverse lines straighter, discal spots smaller, and wavy line less distinct. Pale individual forms of *I. sardoniata* sometimes rather similar, but on average smaller, with terminal dot of forewing apex enlarged, forewing apex pointed, ♂ antennal cilia longer. *I. subsericeata* on average much smaller, transverse lines straighter, discal spots indistinct or absent, ♂ antennal cilia longer, without ♂ hindtibial spurs, tarsus shortened.

HABITAT: For morphology of immature stages see CHRÉTIEN (1899a) and MAIDORN (1914).

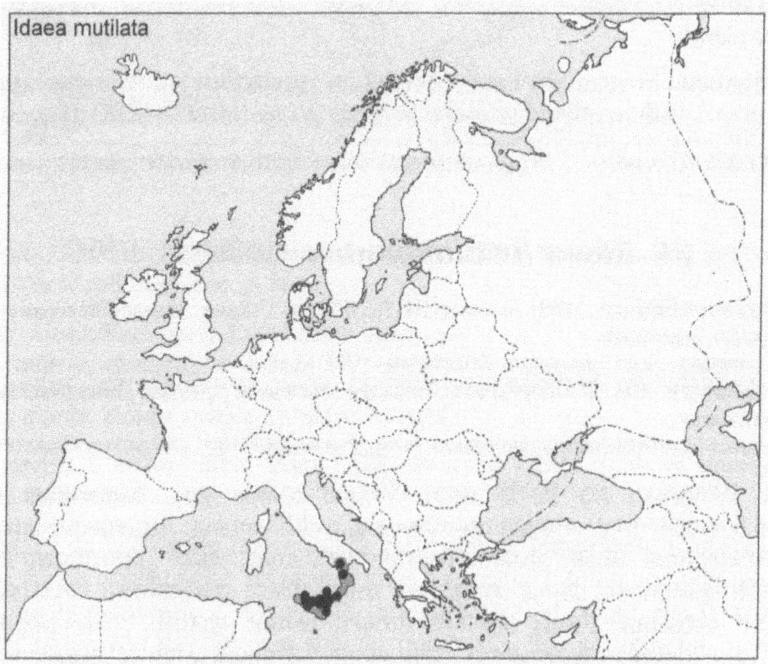

13. *Idaea mutilata* (STAUDINGER, 1876)

Acidalia mutilata STAUDINGER, 1876: Stettin. ent. Ztg. 37: 140 (Sicily: Valdesi near Palermo). Syntypes 1♂2♀ (MNHU 1♂2♀, examined).

‡? *Idaea nitidulata* HERRICH-SCHÄFFER, 1839: in PANZER, Faunae Insect. Germ. initia (Dtl. Insect.) (166): 95 (Type locality not stated, Sicily?). Syntype(s). Listed in PROUT (1934d: 'incertae sedis') and in SCOBLE (1999) as species; PROUT (1935a: 51) suggests possible conspecifity with *determinata* or *mutilata/litigiosaria*; HERRICH-SCHÄFFER (1852: 68) states synonymy of *litigiosaria*; here regarded as nomen dubium. The interpretation awaits confirmation.

‡? *Acidalia nitidularia* HERRICH-SCHÄFFER, 1852: Syst. Bearb. Schmett. Eur. 6 (55): 68. Referring to vol. 3: 20, thus binominal. Emendation, nomen dubium with *nitidulata* (see above).

Unavailable names (misidentification): *litigiosaria*: sensu STAUDINGER (1871), PARENZAN (1994), RAINERI & ZANGHERI (1995, partim), and MÜLLER (1996) nec BOISDUVAL, 1840.

DIAGNOSIS: Wingspan ♂♀ 20-23 mm. Coloration and pattern of wings as in *I. litigiosaria*, but postmedial line of hindwings distinct, usually closer to discal spot, antennae of ♂ densely ciliate, sub-fasciculate, cilia 1.0-1.3 times width of flagellum.

MALE GENITALIA: Uncus slightly broader than in *I. litigiosaria*. Valva spatulate, with two lateral spines at tip more distant from each other than in *I. litigiosaria*. Aedeagus terminally slender, with one slightly curved cornutus, only 0.3-0.35 times length of aedeagus.

FEMALE GENITALIA: Ductus bursae curved, posteriorly with circular sclerite. Posterior half of corpus bursae strongly sclerotized with many spines, narrowly connected to small, membranous anterior part. Appendix bursae small, projecting from corpus bursae.

DISTRIBUTION: Central-Mediterranean. Endemic to Sicily and southern Italy (Calabria; Basilicata).

PHENOLOGY: Univoltine: late May to early July. Doubtful records from mid-March (PARENZAN 1994).

BIOLOGY: Larval host-plants unknown, probably polyphagous on withered leaves of herbaceous plants.

HABITAT: Montane. In southern Italy collected on precipitous rocky slopes, usually from 1,000 m up to 1,900 m above sea-level. In Sicily in oak forest at 600 m above sea-level.

SIMILAR SPECIES: Allopatric *I. litigiosaria* and some further similar species see under no. 12.

14. *Idaea lusohispanica* HERBULOT, 1991

Idaea lusohispanica HERBULOT, 1991: Alexanor 16 (7): 387, fig. 1 (Spain, Aragon: Bronchales). Holotype ♀ (ZSM/HERBULOT, examined).

‡ *Acidalia lambessata* [syn.] *granadaria* STAUDINGER, 1901 (partim): in STAUDINGER & REBEL: Cat. Lepid. palaearct. Faunengeb. (Edn 3) 1: 266 (Spain, Andalusia; Mauritania). Syntypes. Unavailable, because introduced in synonymy.

Unavailable names (misidentification): *lambessata*: sensu STAUDINGER (1901, partim) nec OBERTHÜR, 1887.

DIAGNOSIS: Wingspan ♂♀ 19-22 mm. Ground colour grey, sometimes dark grey. Postmedial line dark grey, straight on forewing, width varying. Antemedial line and wavy line of terminal area usually absent. Termen with small, black, punctiform dots, visible only in fresh specimens. Discal spots very small, black, punctiform. Underside similar, with darker forewings. Frons blackish brown, vertex whitish, collar slightly darker. Antennae of ♂ densely ciliate-setose, cilia about 1.0 times width of flagellum. Hindtibia of ♂ with two spurs, tarsus not shortened.

MALE GENITALIA: Uncus subterminally dilated, tapering to a very short tip. Valva spatulate, with two lateral spines at tip, terminal margin oblique between spines. Aedeagus with one stout cornutus, 0.5 times length of aedeagus, with small lateral sclerite at its base.

FEMALE GENITALIA: Antrum dilated, rhombic. Ductus bursae with lateral spinose patch leading to appendix bursae. The latter membranous, slightly sclerotized only at base. Corpus bursae strongly sclerotized in the posterior half, strongly furrowed with spines at the transition to the membranous anterior half of corpus bursae.

DISTRIBUTION: Endemic to Spain and Portugal. – Outside Europe, from Morocco to Tunisia, replaced by the sister species *I. lambessata*. *I. lambessata* reputedly collected in

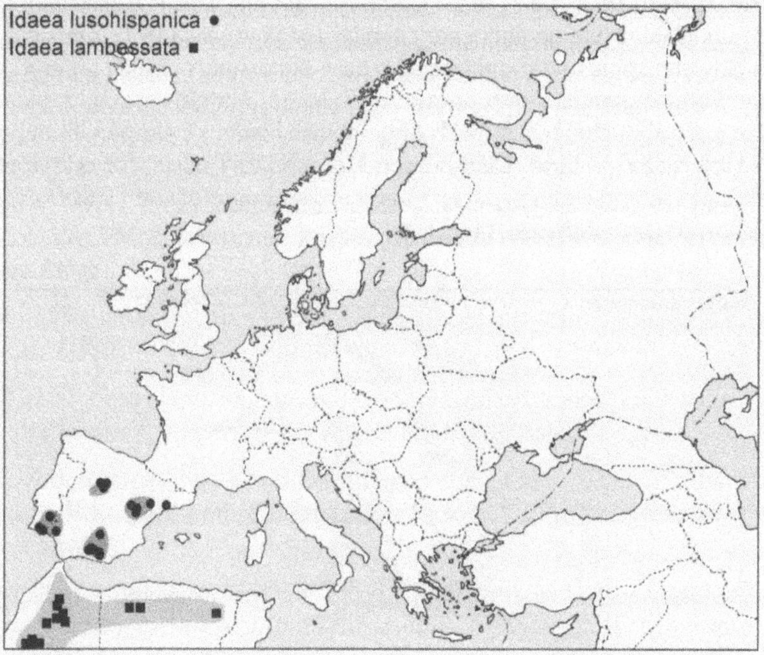

central Spain (1♀, Sierra de Gredos, leg. DÜRCK, ZSM; dissected), but probably mislabelled, as DÜRCK collected also in Morocco.

PHENOLOGY: Univoltine: early June to late July.

BIOLOGY: Larval host-plants unknown, probably polyphagous on withered leaves of herbaceous plants. Adults readily attracted to light.

HABITAT: Usually montane, often from 1,000 m up to 1,800 m above sea-level, sometimes down to sea level.

SIMILAR SPECIES: *I. litigiosaria* (see no. 12). Allopatric *I. lambessata* (OBERTHÜR, 1887) from North Africa usually paler, transverse lines less distinct, often undulate, secure identification only by dissection. *I. korbi* with brownish ground colour, much darker collar, larger discal spots, and distinct, less straight postmedial line of forewing. *I. sardoniata* with postmedial line less distinct, forewing apex pointed, with terminal dot of forewing apex enlarged, collar darker, ♂ antennal cilia longer.

Idaea lambessata (OBERTHÜR, 1887)

Acidalia lambessata OBERTHÜR, 1887: Bull. Soc. ent. Fr. 1887: 67 (eastern Algeria: Lambèze). Lectotype ♂ (ZFMK, examined), herewith designated to stabilize usage of the name with regard to the European sister species; at ZFMK 2♂ paralectotypes.

‡ *Acidalia lambessata* [syn.] *granadaria* STAUDINGER, 1901 (partim): in STAUDINGER & REBEL: Cat. Lepid. palaearct. Faunengeb. (Edn 3) 1: 266 (Spain, Andalusia; Mauritania). Syntypes. Unavailable, because introduced in synonymy.

Unavailable names (misidentification): *subsericeata*: sensu ROTHSCHILD (1914), partim?, nec HAWORTH, 1809. – (infrasubspecific): *nigrolineata*: REISSER (1961).

Usually paler than the European sister species *I. lusohispanica*, transverse lines less distinct, often undulate. Male genitalia very similar too, but valva in *I. lambessata* with distal margin between spines often rectangular, base of cornutus usually tapered, without lateral sclerite. Female genitalia with antrum oval, lateral margins strongly folded, granulated. Anterior part of ductus bursae with large spines leading to appendix bursae. The latter strongly sclerotized at base. Distribution: Morocco to Tunisia, for one doubtful record for central Spain see preceding species. Vertical distribution: From 1,600 m up to 2,500 m. Description of larva in REISSER (1933a).

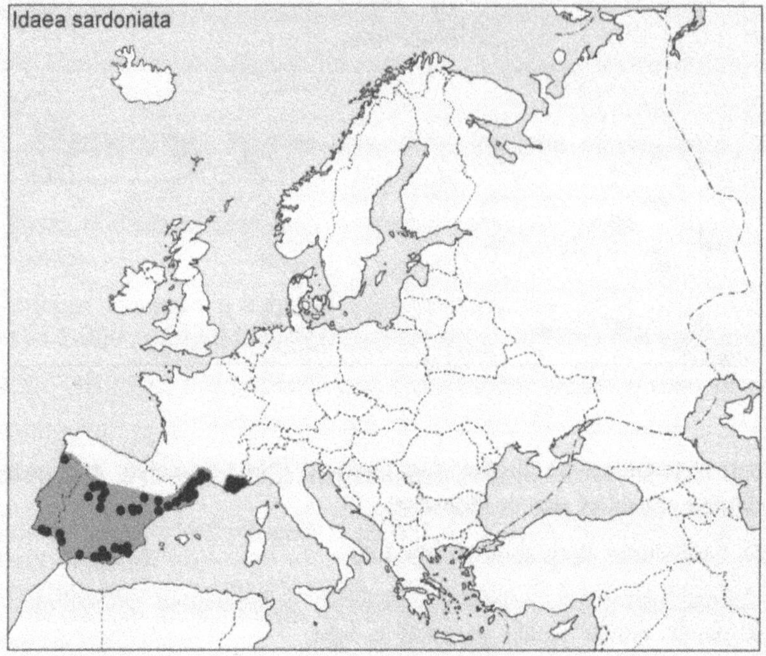

15. Idaea sardoniata (HOMBERG, 1912)

Acidalia sardoniata HOMBERG, 1912: Bull. Soc. ent. Fr. 1912 (12): 266 (France, eastern Pyrenees: Vernet-les-Bains). Lectotype ♂ (MNHN), at ZSM/HERBULOT 2♀ paralectotypes from type locality, 3♂18♀ further paralectotypes, not traced.

Ptychopoda concordaria PÜNGELER, 1913: in PROUT in SEITZ, Macrolep. 4: 93 (pl. 3, fig. f: 1912: non binominal) (south-eastern Spain: Sierra Espuña near Murcia; northern Spain: Pyrenees). Syntypes 5♂5♀ (MNHU/PÜNGELER: 8♂♀ syntypes, examined, all from Sierra Espuña, one labelled as 'type').

Unavailable names (misidentification): *ruficostata*: sensu WEISS (1920), nec ZELLER, 1849. – (infrasubspecific): *plumbaginata*: HOMBERG (1912).

DIAGNOSIS: Wingspan ♂♀ 16-21 mm. Forewing apex pointed. Ground colour variable, usually light brown, sometimes dark brown or orange. Transverse lines brown, often indistinct, slightly undulate. Terminal area darker, divided by wavy line, usually giving the impression of five parallel transverse lines on forewing. Termen with very small, black, punctiform dots, the last one in the forewing apex enlarged. Discal spots small, black, punctiform. Underside similar, with forewings darker. Frons blackish brown, vertex white, collar light brown. Palpi slender, length 1.2-1.3 times diameter of eye. Antennae of ♂ ciliate-fasciculate, cilia 1.3-1.6 times width of flagellum (Text-fig. 38). Hindtibia of ♂ with two spurs, tarsus not shortened.

MALE GENITALIA: Uncus tapered, triangular. Valva with large costal spine slightly subterminal, ventral spine at tip very small. Aedeagus broad, with one broad cornutus, 0.5 times length of aedeagus.

FEMALE GENITALIA: Ductus bursae broad, without modifications at antrum. Corpus bursae completely sclerotized, with large field of small spines. Appendix bursae large, with stout lateral spines in the basal half.

DISTRIBUTION: West-Mediterranean. Iberian peninsula, southern France, north-east to dept. Basses Alpes.

PHENOLOGY: Univoltine: usually mid-June to mid-July, sometimes emerging from late-May (CORLEY et al. 2000).

BIOLOGY: Larval host-plants unknown, probably polyphagous on herbaceous plants. Reared on dry leaves (LHOMME 1935). Adults readily attracted to light.

HABITAT: Xerothermophilous. In southern Spain collected in aisles of pine forest rich in low vegetation (PÖLL pers. comm.). From 0 m up to 1,200 m, exceptionally up to 1,800 m above sea-level.

SIMILAR SPECIES: *I. litigiosaria* (see no. 12), *I. lusohispanica* (14), *I. squalidaria* (112). *I. obsoletaria*, *I. distinctaria* and *I. predotaria* differ in absence of ♂ hindspurs, forewing apex less pointed and terminal dot of forewing apex not enlarged, males can easily be identified by brushing tip of abdomen to remove scales and reveal shape of valva.

REMARKS: Female genitalia suggest closer relationship to *I. ossiculata*.

16. Idaea korbi (PÜNGELER, 1917)

Ptychopoda korbi PÜNGELER, 1917: Mitt. Münch. ent. Ges. 8 (1/4): 20 (Spain, Castilia: Chilleron near Cuenca). Holotype ♀ (MNHU, examined, genitalia figured in AGENJO 1976).
Acidalia korbiae ARNOLD, 1921: Mitt. Münch. ent. Ges. 11 (9): 48. Unjustified emendation.

DIAGNOSIS: Wingspan ♂♀ 18-21 mm. Ground colour greyish brown. Transverse lines dark grey or dark brown. Antemedial line of forewing convex. Postmedial line slightly undulate. Wavy line of terminal area nearly invisible. Termen with very small, black, punctiform dots. Discal spots black, punctiform. Frons blackish brown, vertex white, collar brown. Antennae of ♂ densely ciliate, cilia about 1.2 times width of flagellum. Hindtibia of ♂ with two spurs, tarsus not shortened.

MALE GENITALIA: Uncus broad, terminally sub-rectangular. Valva spatulate, with two lateral spines at tip, arising close together, ventral spine much longer. Aedeagus with one stout, slightly curved cornutus, 0.4 times length of aedeagus.

FEMALE GENITALIA: Ductus bursae posteriorly with circular sclerite, curved towards corpus bursae, subapically connected. Antrum not dilated. Corpus bursae sclerotized, with large field of long spines. Anteriorly with long and broad membranous process, leading to ductus seminalis, with lateral spines towards corpus bursae.

DISTRIBUTION: Endemic to some mountainous regions of Spain: Catalonia (DANTART 2000), southern parts of prov. Aragon and north-eastern prov. Castilla-La Mancha, prov. Segovia.

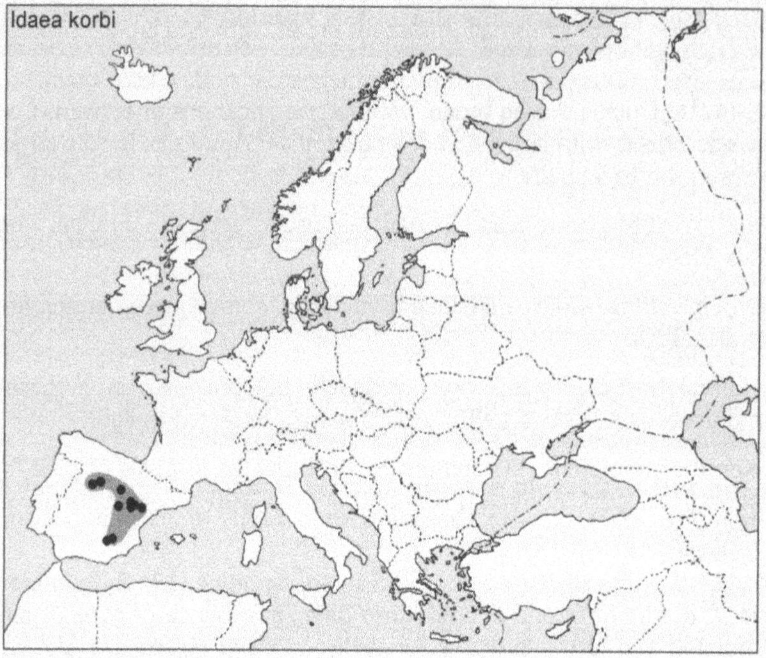

PHENOLOGY: Probably univoltine: most data from mid-June to early August. Other data sets suggest bivoltinism: late May to early June; late July to early August (KRAUS 1999).

BIOLOGY: Larval host-plants unknown, probably polyphagous on withered leaves of herbaceous plants. Adults attracted to light.

HABITAT: Montane. Collected in scattered forest of *Pinus sylvestris* and *P. nigra* (PROUT 1935a; AGENJO 1976; REDONDO & GASTON 1999). From 750 m up to 1,500 m above sea-level.

SIMILAR SPECIES: Dark forms of *I. lusohispanica* (see no. 14).

REMARKS: Taxonomy and distribution: AGENJO (1976). Female genitalia suggest closer relationship to *I. litigiosaria*.

17. *Idaea mediaria* (HÜBNER, 1819)

Geometra mediaria HÜBNER, 1819: Samml. Eur. Schmett. 5, Geometrae (1): pl. 97, fig. 502 (France, Var: St. Tropez). Types of HÜBNER lost. Neotype ♂ (ZSM/HERBULOT, examined: Var, St. Tropez 27.VI.48, leg. HERBULOT), herewith designated to stabilize usage of the name with regard to the sister species, described below.

Ptychopoda mediaria flavidior ROTHSCHILD, 1914: Novit. zool. 21: 343 (central Algeria: Guelt-es-Stel). Syntypes 19♂♀ (BMNH, examined). Valid at subspecific rank (SCOBLE 1999).

Unavailable names (incorrect subsequent spelling): *mediara*: LERAUT (1980).

DIAGNOSIS: Wingspan ♂♀ 14-19 mm. Forewings slender. Wings sand coloured. Antemedial line very indistinct. Terminal area divided by wavy line, forming with postmedial line three fine, straight, greyish brown lines, very close to forewing termen. Medial line fine, straight, oblique to postmedial line. Hindwing with medial fascia often distal to

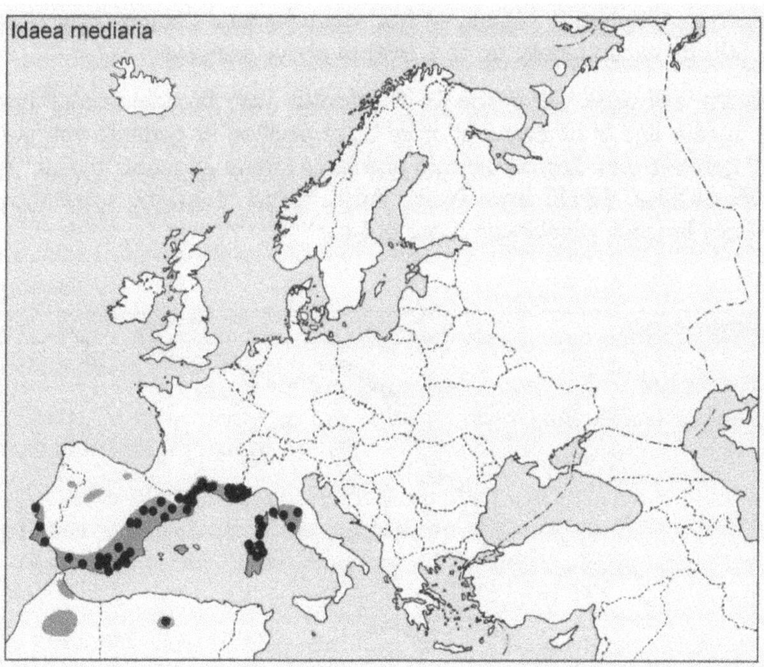

discal spot. Black terminal dots fine. Discal spots black, small. Frons blackish brown, vertex whitish, collar brown. Palpi black, comparatively small. Antennae of ♂ with deep intersegmental incisions, densely ciliate-setose, cilia 0.8-0.9 times width of flagellum. Hindtibia of ♂ with two spurs, tarsus not shortened.

MALE GENITALIA: Uncus slender, tapered. Valva spatulate, terminally rounded, with one lateral spine dorsally. Aedeagus with one bi-sinuous cornutus, 0.4 times length of aedeagus.

FEMALE GENITALIA: Ductus bursae straight, antrum dilated to semicircular sclerite. Corpus bursae globular, completely filled with small spines ('lamina dentata'). Appendix bursae small, caudad directed.

DISTRIBUTION: West-Mediterranean. From Iberian peninsula to north-western and central Italy, present on Mallorca (HONEY pers. comm.), Corsica and Sardinia. – Outside Europe in Morocco and Algeria (subsp. *flavidior*). Replaced by an allopatric sister species, *I. leipnitzi* sp. n., in Sicily, southern Italy and Croatia. Erroneously mentioned for Turkey (RIEMIS 1994).

PHENOLOGY: Univoltine: early June to late July, exceptionally until mid-August. Larva growing very slowly.

BIOLOGY: Larva polyphagous on withered leaves of herbaceous plants (CHRÉTIEN 1926). Reputedly found on *Ulex europaeus* (GRASLIN 1863). Recorded on *Euphorbia spinosa* (Euphorbiaceae), feeding on leaves and petals (MILLIÈRE 1867, data probably from observation in nature; subsequently var. cit.), and on *Dorycnium* (Fabaceae) (CHRÉTIEN 1926). Reared on *Polygonum* and *Taraxacum officinale* (LEIPNITZ pers. comm.). Adults readily attracted to light.

HABITAT: Xerothermophilous. Usually in hot, open lowland habitats near coast. From 0 m up to 1,300 m, exceptionally up to 1,900 m above sea-level.

SIMILAR SPECIES: Allopatric sister species *I. leipnitzi* with broader medial line, terminal dots larger, medial line of hindwing at more basal position, ♂ genitalia with long, straight cornutus, ♀ genitalia with narrow antrum and large lateral appendix bursae. Allopatric *I. renataria* (OBERTHÜR, 1878) from North Africa larger, forewing apex more pointed, transverse lines broader, medial line more oblique.

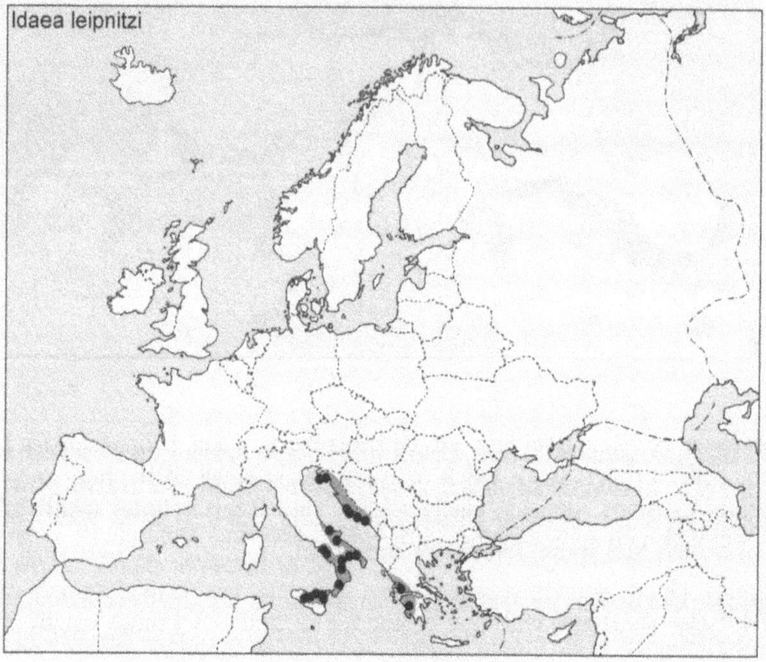

18. *Idaea leipnitzi* sp. n.

Unavailable names (misidentification): *mediaria*: sensu STAUDINGER (1901, partim) nec HÜBNER, 1819.

HOLOTYPE: ♂, 'Dalmatia, Gravosa-Zaton [by Dubrovnik], 25.6.1937, leg. FABIGAN', ZSM.

PARATYPES: 8♂, id., 26.6.1937; 2♂, id., 26.7.1937; 2♂, 'Dalmatia, Gravosa' [Dubrovnik], M7', coll. P. SKOU; 1♂, id., NHMW; 2♂26♀, Croatia, Hvar, 4.-15.7.1983, leg. L. CERVINKA, coll. N. PÖLL; 2♀, id., ZSM; 1♂, 'Dalmatien, Hvar, 1.-17.VI.1962, leg. LÖBERBAUER', ZSM/EMEM; 1♂1♀, [Croatia] ,Jugoslavien/Istrien, Premantura, 3.VII.-23.VII.1982, coll. M. LEIPNITZ'; 1♂3♀, id., XII.1987, e.o., coll. LOSER; 1♂, [Croatia] ,HRV, Insel Krk, Umg. Ponat, Hrusta, 5.VII.1993, leg. et. coll. H. HABELER'; 1♂, [Croatia] ,HRV, Insel Krk, Umg. Ponat, Insel Krk, Jagdhaus, 8.VII.1993, leg. H. HABELER', ZSM; 1♀, [Croatia] ,Dalmatien, YU, Lavanda, 30m, 26.VI.1979, coll. H. HABELER'; 1♂, 'Dalmatia, Gravosa' [Dubrovnik], 1923, leg. F. WAGNER, ZSM; 1♂, id., 1924; 1♀, 'Dalmatien, Siano, 7.-17.VII.1976, leg. ROESLER', ZSM/EMEM; 1♂, 'Dalmatia, Cuziste, VI.1891', ZSM; 1♂, Dalmatia, coll. PÜNGELER in MNHU; 1♀, 'Griechenland, [Peloponnes], Olympia, 100m, 10.X.1975, leg. LUKASCH', ZFMK; 1♀, 'Griechenland, Etolia, Lagune bei Messolonghi, 20.21.VI.1996, leg. et coll. T. GRÜNEWALD.

Further material examined: 1♂, southern Italy, Puglia, Gargano near Foggia, 70m, 18.VI.1992, leg. O. CZADEK, ZSM; 1♂, Puglia, Matera, 6.VI.1991, leg. et coll. P. PARENZAN; 1♂, Basilicata, Ponte Cagnolino, 18.IX.1992, leg. et coll. P. PARENZAN; 1♂, Calabria, Laghi La Vota (CZ), 10 m, 7.VI.1997, leg. et coll. S. SCALERCIO; 1♀, Sicily, Madonie, Castelbuono, Mt. Miccio, 860 m, 10.VI.2001, leg. et coll. N. PÖLL; 1♀, Sicily, Mercuore, Mistretta, leg. J. KLIMESCH, ZSM.

DIAGNOSIS: Wingspan ♂♀ 15-19 mm. Forewings slender. Ground colour glossy, sand coloured. Antemedial line indistinct. Terminal area divided by wavy line, forming with postmedial line three fine, straight, greyish brown lines, very close to forewing termen. Medial line broad, straight, oblique to postmedial line. Hindwing with medial fascia usually proximal to discal spot or crossing discal spot. Termen with distinct black dots. Discal spots very small or absent. Palpi black brown, comparatively small. Frons black brown, vertex whitish, collar brown. Antennae of ♂ with deep intersegmental incisions, densely ciliate-setose, cilia 0.85-1.0 times width of flagellum. Hindtibia of ♂ with two spurs, tarsus not shortened.

MALE GENITALIA: As in *I. mediaria*, but aedeagus longer and cornutus simply curved, longer, 0.7 times lenth of aedeagus.

FEMALE GENITALIA: As in *I. mediaria*, but antrum only slightly dilated. Appendix bursae elongate, orad directed.

DISTRIBUTION: Central-Mediterranean. From Sicily (compare IPPOLITO & PARENZAN 1998), and southern Italy to Greece and Dalmatia (western Croatia). In the north up to Istria. – Outside Europe old and doubtful records from north-western Turkey (STAUDINGER 1879; RIEMIS 1994).

PHENOLOGY: Univoltine: late May to late July; exceptionally in autumn (PARENZAN et al. 1998b: mid-September), probably belonging to a partial second brood.

BIOLOGY: Larva probably polyphagous, or restricted to certain (halophilous?) plants in coastal dunes. Reared on *Euphorbia*, but probably accepting also other plant species (Croatia: LEIPNITZ pers. comm.). Adults readily attracted to light.

HABITAT: Xerothermophilous. Usually in the lowlands near coast, in Greece collected at sea-level in dunes with *Tamarix* (GRÜNEWALD pers. comm.). Usually from 0 m up to 200 m above sea-level, on Sicily collected at 900 m.

SIMILAR SPECIES: Allopatric sister species *I. mediaria* (see no. 17). *I. bengasiaria* (TURATI, 1926), larger, hindwing broader, medial line of hindwing ending more distally at the inner margin, postmedial line of hindwing straight, double.

REMARKS: Larva (see www.herbulot.de) different in habitus from that of *I. mediaria* (LEIPNITZ pers. comm.).

19. Idaea rufaria (HÜBNER, 1799)

Geometra rufaria HÜBNER, 1799: Samml. Eur. Schmett. 5, Geometrae (1): pl. 21, fig. 112 (Europe). HÜBNER's Syntype(s) lost. Neotype ♂ ([Germany], 'Bavaria, Kelheim, 8.VII.1921, leg. L. OSTHELDER', ZSM), herewith designated to stabilize usage of the name with regard to the closely related *I. admiranda* sp. n..

Acidalia rufaria ochridana SILBERNAGEL, 1944: Z. Wien. ent. Ges. 29: 154 (Macedonia: Ochrid, Petrina Planina). Syntypes ♂♀ (SMNK/REISSER: 3♂ syntypes).

Sterrha rufaria imami WILTSHIRE, 1966: Z. Wien. ent. Ges. 51 (9-11): 116 (Afghanistan: Paghman Mts.). Holotype ♂ (NHMW/VARTIAN). Valid at subspecific rank (SCOBLE 1999).

Unavailable names (misidentification): *luteolaria*: sensu RIESEN (1897) nec CONSTANT, 1863. – (infrasubspecific): *abnobaria*: REUTTI (1898); *anastomosaria*: REISSER (1961); *bilineata*: REISSER (1961); *demarginata*: REISSER (1961); *nigrocinctaria*: FUCHS (1904); *nigrolineata*: REISSER (1961); *purpurea*: REISSER (1930); *rufobrunaria*: HÖRHAMMER (1933). – (incorrect subsequent spelling): *rufobrunnaria*: BERGMANN (1955).

DIAGNOSIS: Wingspan ♂ 22-27, ♀ 19-24 mm. ♀ with forewing apex more pointed. Wings

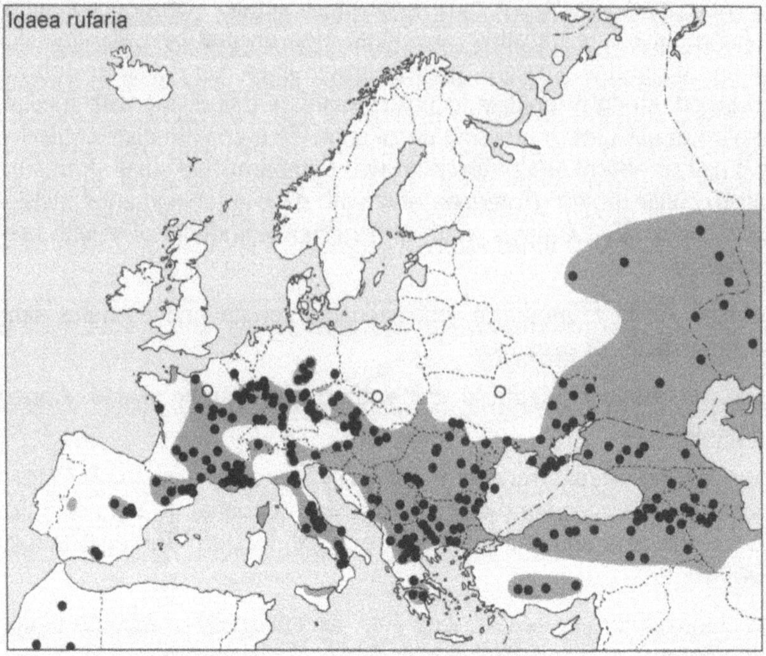

sand coloured, ochreous or light brown in ♂, light brown in ♀, sometimes with orange tinge. In southern and eastern Europe often paler. Transverse lines distinct, postmedial line straight, often darker than the other lines, especially in ♀. Antemedial line of forewing usually well developed. Blackish discal spots sharp and distinct, especially on hindwing. Termen without dots. Fringe concolorous with ground colour. Frons dark brown, vertex whitish to sand coloured, collar brown. Antennae of ♂ densely ciliate-setose, cilia 0.7-0.9 times width of flagellum (Text-fig. 39). Hindtibia of ♂ with two spurs, tarsus not shortened.

MALE GENITALIA: Uncus slender and tapering in terminal half, broad at base. Valva with one terminal spine on dorsal side. Very rarely with a small second spine, ventrally. Aedeagus with stout cornutus, about 0.6 times length of aedeagus, with sclerotized basal sheath, covering about 1/3 of cornutus. Sheath in Turkmeniyan populations larger, covering 1/2 of cornutus, terminal margin oblique.

FEMALE GENITALIA: Ductus bursae of medium length, comparatively broad. Corpus bursae elongate, sclerotized, longitudinally corrugated, with two large, lateral, spinulose patches. Corpus bursae of Turkmeniyan populations less elongate with fused, broad patches of spinules. Membranous appendix bursae arising from anterior end of corpus bursae.

DISTRIBUTION: (Sub-)Mediterranean-Turanian. All over southern Europe except Portugal, warmer parts of western, central and eastern Europe. North of Alps in isolated populations. Absent from all Mediterranean islands except Corsica, and Sicily.– Outside Europe in Morocco, Turkey across Caucasus, Transcaucasus, Turkmeniya, in the east to Afghanistan and central Asian mountains (subsp. *imami*), reputedly also in Mongolia (VOJNITS 1976). Doubtfully recorded for Cyprus.

PHENOLOGY: Univoltine: late June to mid-August, exceptionally somewhat earlier or later.

BIOLOGY: Larva probably polyphagous on withered leaves of herbaceous plants. Recorded on *Stellaria media* (Caryophyllaceae) (REBEL 1910b; LHOMME 1935: BERGMANN 1955; FORSTER & WOHLFAHRT 1981). Reared on *Cerastium brachypetalum* (EBERT 2001; SCHNEIDER 1939), and on withered leaves of *Taraxacum officinale, Vicia cracca* (ERLEBACH, MÜLLER pers. comm.). Imagines diurnal and nocturnal, readily attracted to light.

HABITAT: Xerothermophilous species of open habitats, often on dry, low-nutrient grassland, sun-exposed slopes, preferring sandy soils and limestone. In the north from 0 m up to 600 m above sea-level, up to 1,600 m in southern Alps, up to 2,000 m in the Mediterranean area and in the Asian part of the distribution.

SIMILAR SPECIES (Text-figs 72-74): *I. ochrata* with orange tinge of ground colour, postmedial line less distinct, discal spots absent or very indistinct, fringe usually darker, fringe dots often present, vertex darker, ♂ antennal cilia longer. *I. admiranda* sp. n. with red brown frons, ♂ antennal cilia longer, forewing discal spot usually larger, ground colour paler, all transverse lines similarly marked, postmedial line not more distinct, secure identification by dissection. *I. consanguinaria* differing in shiny sand coloured wings, black fringe dots, and undulate, less contrasted postmedial line. *I. macilentaria* without reddish tinge, forewing discal spot absent, hindwing stronger concave between M1 and M3, ♂ antennal cilia much longer.

REMARKS: Populations from Balkan peninsula ('*ochridana*') without significant and constant differences from nominate subspecies.

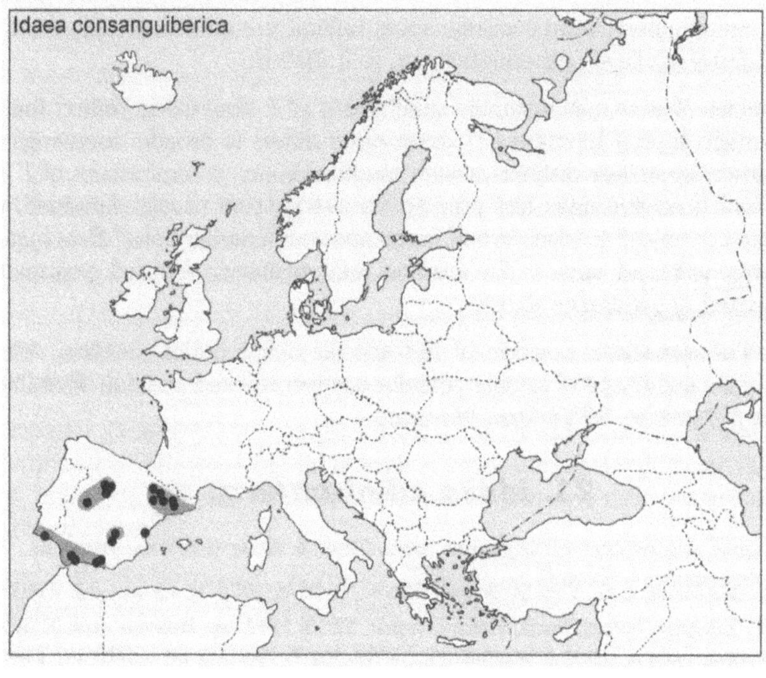

20. Idaea consanguiberica REZBANYAI-RESER & EXPÓSITO, 1992

Idaea consanguiberica REZBANYAI-RESER & EXPÓSITO, 1992: Ent. Z. Frankf. a.M. 102 (15): 275, figs 1,3,4 (Spain, prov. Madrid: Villar del Olmo). Holotype ♂ (MNCN, examined), paratypes 1♂2♀ (MNCN, ZFMK, examined).

DIAGNOSIS: Wingspan ♂♀ 17-21 mm, exceptionally 14 mm only, ♀ with forewing apex pointed. Ground colour straw yellow, sometimes with slight rosy tinge. ♀ suffused with many brown scales, with violet tinge. Postmedial line double and distinct in ♂, indistinct in ♀. In ♂ often also the medial line well developed. Discal spots distinct. Fringe concolorous. Proboscis and palpi short. Frons blackish brown, often with copper coloured tinge, vertex whitish, collar sand coloured to light brown. Antennae of ♂ with deep intersegmental incisions, ciliate-fasciculate, cilia 1.2-1.6 times width of flagellum. Hindtibia of ♂ slender, without spurs. Tarsus 0.65-0.8 times length of tibia.

MALE GENITALIA: Uncus triangular. Saccus sub-triangular, very large. Valva dilated at 1/2, round at tip, with spine on ventral side. Aedeagus long, with long and stout cornutus, about 0.9 times length of aedeagus. Vesica with large granulated patch.

FEMALE GENITALIA: Antrum dilated. Corpus bursae very long and slender, longitudinally corrugated, spinulose nearly all over, anterior fifth membranous. Ductus seminalis arising from posterior end of corpus bursae.

DISTRIBUTION: Endemic to Iberian peninsula from Catalonia (DANTART 2000) to southern Portugal. Very local populations, thus threatened, though sometimes not rare (REDONDO & GASTON 1999).

PHENOLOGY: Univoltine: late July to early September.

BIOLOGY: Larval host-plants unknown, probably polyphagous on low plants. Adults readily attracted to light.

HABITAT: Limited information available about habitat preference. Usually from 100 m up to 700 m above sea-level, exceptionally up to 1,100 m.

SIMILAR SPECIES: The ♂ may resemble small forms of *I. deversaria fallax*, the latter however with much shorter hindtarsus. *I. sardoniata* differs in broader forewings with pointed apex, discal spots less distinct, ♂ with hindibial spurs. ♀ reminiscent of *I. infirmaria*, the latter with forewing apex less pointed, forewing costa usually darkened, clear white vertex more contrasted to black brown frons and much darker collar. *Brachyglossina hispanaria* often with dark spots in the terminal area, proboscis reduced, postmedial line less straight, dotted, ♂ hindtarsus shorter.

REMARKS: The systematic position of this species needs critical revision. Absence of ♂ hindtibial spurs and shape of female genitalia suggest isolated position. Possibly related to *I. debiliata* (STERNECK, 1933) from Morocco.

21. *Idaea admiranda* sp. n.

Unavailable name (misidentification): *litigiosaria*: sensu OKYAR & AKTAÇ (1999) nec BOISDUVAL, 1840.

HOLOTYPE: ♀, 'SO.Türkei, Urfa, 8 km W Siverek, 700 m, 9.-10.VI.1974, lg. GROSS', coll. EMEM in ZSM.

PARATYPES: ♂, 'European Turkey, Tracia, Malkara Region, 22.VII.1991, leg. BESHKOV', coll. P. SKOU, Stenstrup; 2♂1♀, 'European Turkey, Tekirdag, Seymen, 14.7.1983, leg. K. KUCHLER jun.', ZSM; 1♀, 'Türkiye, Erzurum, Palandöken Dagi, 2300 m, 25.7.1992, leg. H. FALKNER', ZSM; 1♂, 'E. Caucasus, Daghestan, Derb[ent], V.[18]78', ZISP; 1♂, [European Turkey], 'Edirne- Centrum, 41 m, 20.VI.1991', leg. et coll. Z. OKYAR; 1♂, [European Turkey], 'Büyükdöllük, 130 m, 31.VII.1991', leg. et coll. Z. OKYAR; 1♂1♀, [European Turkey], 'Kirklareli-Lüleburgaz, 60 m, 12.VIII.1991', leg. et coll. Z. OKYAR; 3♂, [European Turkey], 'Tekirdag-Malkara, Doluköy, 200 m, 26.VII.1992', leg. et coll. Z. OKYAR; 1♂, [TURKEY], ‚Adana-Pozanti, 37.28.35N 34.54.17E, Alpu, 1120m, 17.V.2001', leg. F. DOGANLAR, ZSM.

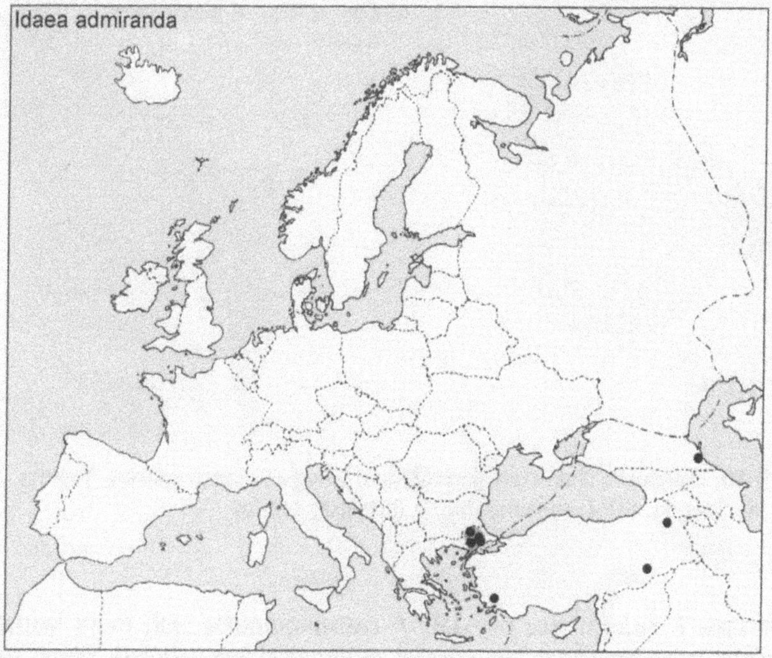

DIAGNOSIS: Wingspan ♂♀ 21-23 mm. Wings sand coloured, slightly glossy, in fresh specimens with warm light brown tinge. Transverse lines brown. Antemedial, medial and postmedial lines distinct, slightly convex, comparatively straight, parallel to each other. Terminal area darker, divided by sharp wavy line, thus appearing as two additional brown transverse lines. Terminal line weakly marked, fringe dots absent. Underside weakly marked. Frons red brown, vertex whitish, collar light brown. Antennae of ♂ densely ciliate-setose, cilia 1.0-1.1 times width of flagellum. Hindtibia of ♂ with two spurs, tarsus not shortened.

MALE GENITALIA: Uncus stout, at base with wing-shaped dilatation. Saccus rounded. Valva extremely long, with two spines at tip. Aedeagus very large, cornutus stout, but comparatively short, 0.35 times length of aedeagus, with lateral sclerite at base.

FEMALE GENITALIA: Ductus bursae comparatively long, broad over the whole length, at antrum tapering, and with small furcate sclerite. Corpus bursae sub-globular, filled all over with spines, anteriorly with edged sclerite. At the anterior end with broad membranous process, without spinules.

DISTRIBUTION: East-Mediterranean. European Turkey. – Outside Europe from western Turkey to eastern Caucasus. Populations very isolated. Rare.

PHENOLOGY: Univoltine: late June to mid-August, in the south emerging from mid-May.

BIOLOGY: Larval host-plants unknown. Adults collected at light.

HABITAT: Limited information available about habitat preference. Collected in 'blended wheat land' (OKYAR pers. comm.). In European Turkey from 0 m up to 300 m above sea-level, in eastern Turkey up to 2,300 m.

Text-figs 79-80: Diagnostic characters (indicated) of European *Idaea* species. Text-fig. 79: *I. admiranda*, sp. n.. Text-fig. 80: *I. consanguinaria* (LEDERER, 1853).

SIMILAR SPECIES: *I. rufaria* (see no. 19). *I. consanguinaria* with more leather-coloured ground colour, transverse lines less parallel, postmedial line undulate, discal spots usually smaller, terminal dots often present, frons without red tinge. Identification of ♂ is easy by brushing abdomen to remove scales and reveal shape of tip of valva. *I. ochrata* without discal spots, darker coloured, with more orange or brownish tinge, frons darker.

REMARKS: The name *Acidalia taurica* BANG-HAAS, 1907, is not available for the new species, types have been examined (see under *I. consanguinaria*). *I. admiranda* may possibly prove to be more closely related to *I. rufaria* than to *I. consanguinaria*, when analysed in more detail.

22. *Idaea consanguinaria* (LEDERER, 1853)

Acidalia consanguinaria LEDERER, 1853: Verh. zool.-bot. Ver. Wien 3: 256, misprinted to 556 (Croatia, Dalmatia: Split ('Spalato'); north-western Turkey: Bursa ('Brussa')). Syntypes.
Acidalia consanguinata BELLIER, 1861: Annls Soc. ent. Fr. (3) 8 (4): 701. Emendation.
Acidalia consecrata STAUDINGER, 1897: Dt. ent. Z. Iris 10 (1): pl. 4, fig. 29; 10 (2) (1898): 305 (Israel: Jerusalem). Syntypes 1♂2♀ (MNHU, examined). Valid at subspecific rank.
Acidalia faillata FUCHS, 1901: Stettin. ent. Ztg. 62: 120 (Sicily: Castelbuono). Syntype(s) (MNHU/PÜNGELER: 1♀ syntype examined, labelled as 'type').
Acidalia taurica BANG-HAAS, 1907: Dt. ent. Z. Iris 20: 78, pl. 3, fig. 5 (Turkey, Taurus: Eibes). Syntypes 2♂ (MNHU: 2♂, examined, abdomen brushed). Herewith downgraded from species rank (RIEMIS 1994; SCOBLE 1999) to synonymy (syn. n.).
Ptychopoda turatii SOHN-RETHEL, 1929: Dt. ent. Z. Iris 43 (1): 15, pl. 2, fig. 21 (Italy, Abruzzi: Gran Sasso, Mte. Autore). Syntypes 1♂1♀. Herewith downgraded from species rank (SCOBLE 1999) to synonymy (syn. n.).
Unavailable names names (misidentification): *determinata*: sensu DOGANLAR (2003) nec STAUDINGER, 1876. – (infrasubspecific): *demarginata*: REISSER (1961). – (incorrect subsequent spellings): *consaguinearia*: ANIKIN et al. (2000); *consanguinearia*: VIIDALEPP (1976); *consanquinaria*: REBEL (1902).

DIAGNOSIS: Wingspan variable, ♂♀ 19-25 mm. Wings sand coloured, sometimes with slight light brown-violet tinge, mainly in females. Transverse lines brown. Slightly curved medial line usually close to undulate postmedial line, distal to black discal spots. Terminal

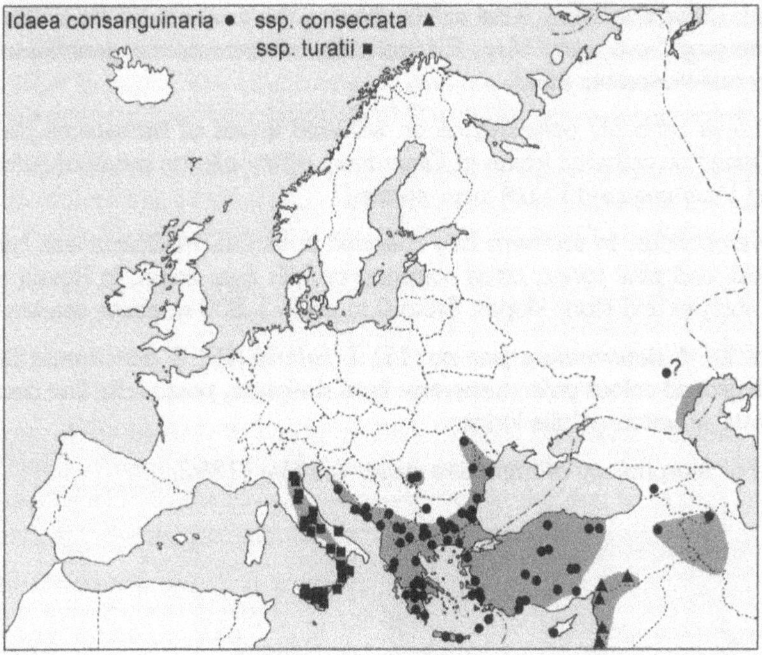

area darker, with distinct wavy line of ground colour. Fringe dots usually present, small, slightly larger in forewing apex. Wing pattern reduced on underside, forewing darker than hindwing. Frons blackish. Antennae of ♂ densely ciliate-setose, cilia 0.5-0.75 times width of flagellum (Text-fig. 40). Hindtibia of ♂ with two spurs, tarsus not shortened.

Subsp. ***turatii*** from Italy on average larger, wingspan 21-26 mm, antennal cilia shorter, only 0.4-0.5 times width of flagellum.

Subsp. ***consecrata*** from Turkey and the Levant with much longer antennal cilia, 0.8-1.2 times width of flagellum. Area between medial line and postmedial line filled with reddish brown in 30-40% of the specimens in Levantine populations, about 10% in southernmost Turkey and northern Lebanon.

MALE GENITALIA: Uncus triangular. Valva terminally rounded, without spines. Aedeagus very broad at base, terminally slender. Cornutus stout, 0.6-0.7 times length of aedeagus.

FEMALE GENITALIA: Ductus bursae widening towards corpus bursae. Corpus bursae globular, filled all over with spines, except for a small lateral area. At anterior end with broad membranous process, filled with spinules, leading to ductus seminalis. In subsp. *turatii* from Italy anterior membranous process almost without spinules.

DISTRIBUTION: East-Mediterranean, subsp. *turatii* Central-Mediterranean. From Sicily, central and southern Italy (ssp. *turatii*) across Balkan peninsula to Crete and Rhodos island. Isolated occurrence in southern European Russia (Astrachan: VIIDALEPP 1996; ANIKIN et al. 2000; Sarepta: ZISP). – Outside Europe in Turkey, Cyprus, Transcaucasus and north-western Iran (e.g. WILTSHIRE 1945). In the Levant replaced by subsp. *consecrata*.

PHENOLOGY: Univoltine: early June to late July, in the lowlands of the south somewhat earlier, emerging from early May. Exceptionally an incomplete generation from late October to mid-November (Italy).

BIOLOGY: Larva probably polyphagous on withered leaves of herbaceous plants (REBEL 1910b). Reared on withered leaves of *Taraxacum officinale*, on petals of *Achillea millefolium* and *Vicia cracca* (MÜLLER pers. comm.).

HABITAT: Xerophilous. In southern Italy collected in various Mediterranean habitats, e.g. scattered oak and pine forest, often common on hills near coast. In Russia and Turkey inhabiting steppes and stony slopes. From 0 m up to 1,800 m above sea-level.

SIMILAR SPECIES: *I. determinata* (see no. 11), *I. rufaria* (19), *I. admiranda* (21). *I. ossiculata* with ground colour paler, transverse lines straighter, postmedial line dentate rather than undulate, ♂ antennal cilia longer.

REMARKS: For morphology of immature stages see MAY (1892).

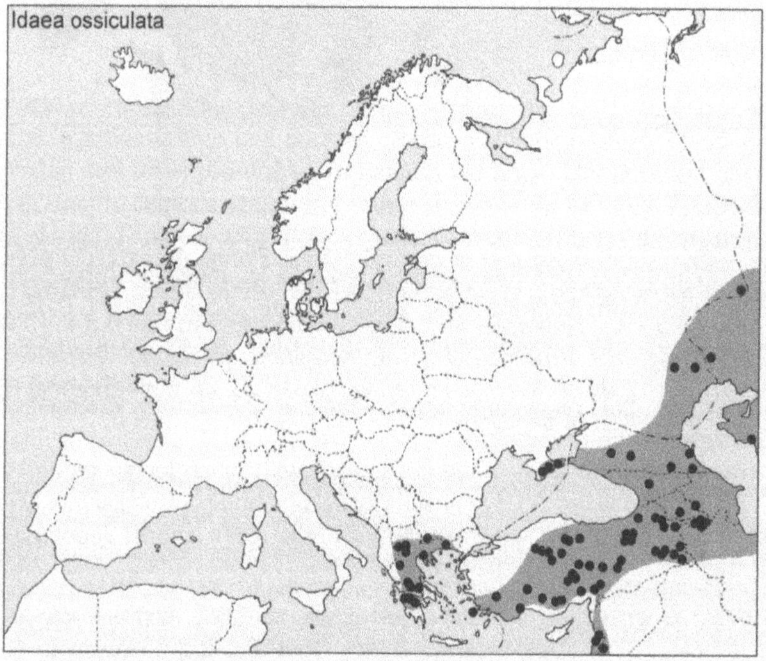

23. *Idaea ossiculata* (LEDERER, 1870)

Acidalia ossiculata LEDERER, 1870: Horae Soc. ent. ross. 8: 18, pl. 2, fig. 5 (northern Iran: Hadschyabad near Astrabad). Lectotype ♂ (MNHU, examined), herewith designated to fix type locality and to stabilize usage of the name. At MNHU three further specimens from Hadschyabad, all labelled as 'Orig[inals]'. In original description 'Turkey, Taurus: Külek' mentioned as second type locality. At MNHU some specimens from Taurus ex coll. LEDERER without type labels, (paralecto-)type status uncertain.

Unavailable names (incorrect subsequent spellings): *assiculata*: VIIDALEPP et al. (1992); *ossicalata*: GUMPPENBERG (1892); *ossicularia*: STEKOLNIKOV (1996:9).

DIAGNOSIS: Wingspan ♂ 20-23, ♀ 15-21 mm. Forewing slender, apex pointed in ♀. Wings sand coloured, rarely melanistic. Wing pattern variable. Distal to the small, black discal spot four greyish transverse lines, terminal field divided by distinct wavy line. Transverse lines not curved, but undulate. Small black fringe dots usually present. Frons blackish, vertex white, collar brown. Palpi sand coloured. Antennae of ♂ densely ciliate-setose, cilia 0.9-1.1 times width of flagellum. Hindtibia of ♂ with two spurs, tarsus longer than tibia.

MALE GENITALIA: Uncus slender, with small terminal spinule. Valva terminally rounded, with tapered subterminal process at 3/4 of costa, at 4/5 in populations of central Asian mountains. Aedeagus with stout cornutus, 0.7-0.8 times length of aedeagus.

FEMALE GENITALIA: Ductus bursae with sclerotized edge towards corpus bursae. Corpus bursae oval, widely filled with spines, except for the most posterior part and the anterior third. Appendix bursae arising between ductus and corpus bursae, with lateral row of spinules.

DISTRIBUTION: (East-)Mediterranean-Turanian. Balkan peninsula, Crimea, southern European Russia. Absent from Greek islands except Kalymnos. – Outside Europe from Turkey across Levant, Transcaucasus, northern Iran, in the east to central Asian mountains.

PHENOLOGY: Univoltine: mid-June to late July, exceptionally in August at high altitude in Turkey.

BIOLOGY: Larva probably polyphagous on withered leaves of herbaceous plants. Reared on *Taraxacum officinale* (MÜLLER pers. comm.). Adults very easily flushed up from vegetation during daytime, but also readily attracted to light.

HABITAT: Xerophilous, often common on rocky slopes, and other dry mountain habitats. From 400 m up to 1,800 m above sea-level, in Turkey usually from 1,000 up to 2,200 m, in northern Iran up to 4,000 m.

SIMILAR SPECIES: *I. determinata* (see no. 11), *I. consanguinaria* (22). *I. metohiensis* with transverse lines less brownish, costal spots present on forewing, postmedial line more dentate and convex, ♂ antennal cilia longer, ♂ hindtibia without spurs. *I. ossiculata* sometimes confused with *I. subsericeata*, and *I. sylvestraria*, but the last two species smaller, forewing apex less pointed, ♂ hindtibial spurs absent. *I. subsericeata* with discal spots weak or absent. *I. sylvestraria* with more grey suffusion.

24. *Idaea sericeata* (HÜBNER, 1813)

Geometra sericeata HÜBNER, 1813: Samml. Eur. Schmett. 5, Geometrae (1): pl. 78, fig. 404 (Europe, Austria or N. Italy). Syntype(s) lost.

‡ *Sterrha sericearia* HÜBNER, 1826: Verz. bekannter Schmett.: 309. Incorrect subsequent spelling.

Acidalia sericeata var. *altivolaria* BUBACEK, 1923: Verh. zool.-bot. Ges. Wien 72: (30) (Corsica: Col de Vergio). Syntype(s). Herewith downgraded from subspecific rank (SCOBLE 1999) to synonymy (syn. n.).

Ptychopoda sericeata var. *calvaria* WEHRLI, 1927: Dt. ent. Z. Iris 41: 59 (Spain, Andalusia: Sierra Nevada). Syntypes 20♂10♀ (ZFMK, examined). The taxon *calvaria* LA HARPE, 1860 does not enter in homonymy being an incorrect subsequent spelling of '*calcearia*' (*Idaea seriata* syn.). Valid at subspecific rank.

Sterrha sericeata var. *wehrlii* PROUT, 1934: Lepid. Cat. 63: 313 (Spain, Andalusia: Sierra Nevada). No separate types. Unnecessary replacement name for *calvaria*.

Sterrha sericeata subrecta PROUT, 1935: in SEITZ, Macrolep. 4, Suppl.: 52, pl. 5, fig. g (Morocco: Great Atlas). Syntype(s) (BMNH). Herewith downgraded from subspecific rank (SCOBLE 1999) to synonymy (of subsp. *calvaria*) (syn. n.).

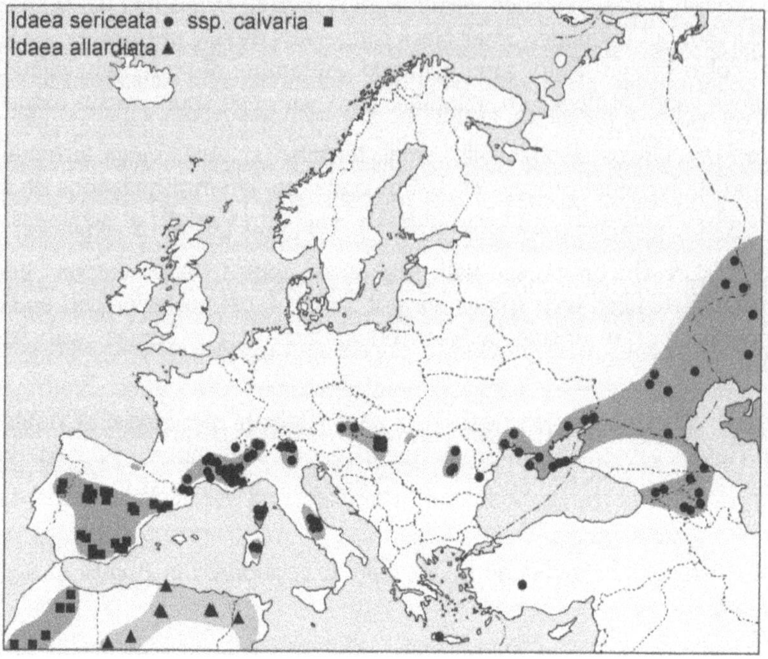

Unavailable names (infrasubspecific): *albomarginata*: ZERNY (1927); *anastomosaria*: REISSER (1961); *extrema*: BUBACEK (1923); *mediofasciata*: HOMBERG (1935); *solida*: PROUT (1935). – (incorrect subsequent spellings): *sericata*: EVERSMANN (1837); *sericaeta*: WARDIKJAN (1970); *sericetaa*: STEKOLNIKOV (1996).

DIAGNOSIS: Wingspan ♂ 19-23, ♀ 15-20 mm. Ground colour white. Forewing with transverse lines developed as broad light brown, undulate fasciae. Medial fascia usually as double line, including discal spot. Termen and base of forewing light brown. Hindwing white with two indistinct light brown fasciae. Underside with darker markings and discal spots. Frons light brown, around proboscis white, vertex with small transverse stripe of white scales, collar brown. Antennae of ♂ densely ciliate-setose, cilia 0.8-1.1 times width of flagellum (Text-fig. 41), in Corsican populations 1.0-1.2 times. Hindtibia of ♂ with two spurs, tarsus not shortened.

Subsp. ***calvaria*** on average larger, ♂ up to 26, ♀ to 22 mm, white pattern slightly extended, hindwing often with more diffuse markings. In Morocco transverse lines slightly straighter, mainly the olive postmedial fascia (PROUT 1935a).

MALE GENITALIA: Uncus slender, tapered. Valva spatulate, terminally rounded. Aedeagus short, broad, with stout cornutus basally bent and inserted on basal sclerite. Populations from Morocco with cornutus more slender and straighter. Vesica weakly granulated, with elongate sclerite. Sternum A8 posteriorly with bifurcate process. The latter slightly broader in Spanish, shorter in Corsican and more slender in Moroccan populations.

FEMALE GENITALIA: Ductus bursae short. Posterior part of corpus bursae sclerotized, corrugated. Corpus bursae with two large, sub-circular lateral patches of spinules, the larger one reaching ductus bursae, the smaller one isolated on the opposite side. Patches smaller in Moroccan populations.

DISTRIBUTION: Mediterranean-Turanian. From Iberian peninsula (subsp. *calvaria*) in local populations all over southern Europe, in the north up to south-eastern Austria and Hungary, in the east to southern Urals. Absent from most Mediterranean islands, but common on Corsica, and Sardinia. With isolated occurrence on Crete (LEIPNITZ pers. comm., examined). Erroneously mentioned for Luxemburg (MÜLLER 1996). – Outside Europe in Morocco (subsp. *calvaria*), isolated occurrence in central Turkey, in the east from Transcaucasus to central Asian mountains. From Algeria to Libya replaced by allopatric sister species *I. allardiata* (MABILLE, 1869).

PHENOLOGY: Univoltine: early June to mid-July, single specimens flying until early August, sometimes emerging from late May.

BIOLOGY: Larva polyphagous, found on various herbaceous plants, preferring fresh, withered or dry leaves of Fabaceae (France: CHRÉTIEN 1899b). Reared on *Plantago, Polygonum aviculare* (France: CHRÉTIEN 1899b; LHOMME 1935), and on petals of various plants, e.g. *Rubus fruticosus* (MÜLLER pers. comm.). Adults very easily flushed up from vegetation (*Juniperus*) during daytime (REISSER 1926; ZERNY 1936), but also readily attracted to light. ♀ ratio at light low, about 20-30%.

HABITAT: Xerothermophilous, preferring dry steppes and other dry open habitats. Usually from 700 m up to 1,800 m above sea-level, sometimes down to sea-level, in the south even higher, e.g. in Morocco up to 3,000 m.

SIMILAR SPECIES: No similar species in Europe. Allopatric North African *I. allardiata* larger, transverse lines straighter, less olive coloured, the two white distal fasciae broader.

REMARKS: Populations from Corsica ('*altivolaria* BUBACEK, 1923') match the populations from the rest of Europe well, transverse lines slightly darker, but not meriting a separate name at subspecific level. Some details in male genitalia, e.g. granulated vesica with elongate sclerite, base of cornutus, reminiscent of certain species of the *obsoletaria* species-group. For morphology of immature stages see CHRÉTIEN (1899b).

25. Idaea macilentaria (HERRICH-SCHÄFFER, 1847)

Acidalia macilentaria HERRICH-SCHÄFFER, 1847: Syst. Bearb. Schmett. Eur. 3 (24): 20. First available name after *macilentaria* BOISDUVAL, 1840 (nomen nudum) and HERRICH-SCHÄFFER, 1846 (pl. 38, figs 230, 231: non binominal) (Spain, Andalusia). Syntype(s) ♀ (BMNH, 1♀ examined). STAUDINGER (1871) was the first to recognize synonymy of *macilentaria* and *antiquaria*, giving precedence to *macilentaria*, thus being the first reviser in the sense of the Code, Art. 24, ICZN.

Acidalia antiquaria HERRICH-SCHÄFFER, 1847: Syst. Bearb. Schmett. Eur. 3 (24): 20; (1848): pl. 56, figs 340, 341 (non binominal) ('Corsica'?). Holotype ♂ (BMNH: 1♂, examined, labelled as 'paratype' from western Germany, 'Wiesbaden'). Valid ('page priority') according to KOCAK (1983), MÜLLER (1996), and SCOBLE (1999).

Unavailable names (misidentification): *sylvestraria*: sensu DUPONCHEL (1830; pl. 177, fig. 8 nec 7) nec HÜBNER, 1799. – (infrasubspecific): *albomarginata*: REISSER (1961); *pulchraria*: FUCHS (1904). – (incorrect subsequent spellings): *antiquuara*: RÖSSLER (1855); *silvestraria*: TOMIC et al. (2002).

DIAGNOSIS: Wingspan ♂ 20-25, ♀ 19-23 mm. Hindwing margin concave between M1 and M3. Wings pale sand coloured. Forewings with three, brown transverse lines. Terminal area dark with broad, distinct, sand coloured wavy line. Costa of forewing usually brownish towards base. Discal spot very small on hindwing, usually absent from forewing. Pattern sharper on underside. Frons brown, vertex and collar light brown. Antennae of

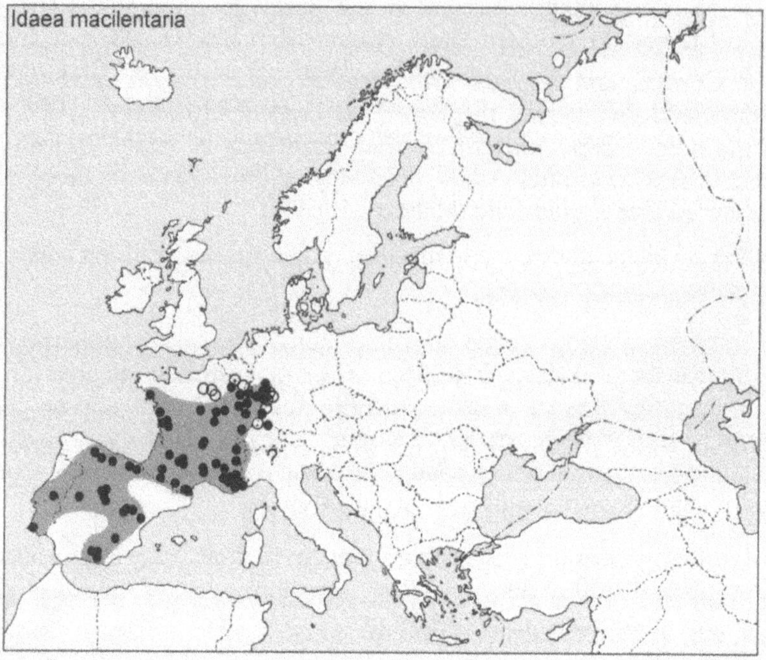

♂ strongly dentate (sub-quadripectinate), with 'branches' about 1.0 times width of flagellum, ciliate-fasciculate, with 4 tufts on each antennomere, cilia 1.5-2 times width of flagellum (Text-fig. 42). Hindtibia of ♂ with two spurs, tarsus not shortened.

MALE GENITALIA: Uncus short and broad. Valva rectangular at tip, with two spines, the costal one longer. Saccus large, rounded. Cornutus comparatively slender, 0.5-0.7 times length of aedeagus. Aedeagus and cornutus curved, but straighter in central European populations.

FEMALE GENITALIA: Antrum broad, bilobed, posterior excavation rectangular. Antrum in central European populations heart-shaped. Ductus bursae comparatively long and broad. Corpus bursae oval, posterior half sclerotized, with many spinules, anterior half membranous.

DISTRIBUTION: (West-)European. From southern Spain across France to Belgium and south-western Germany, and north-western Italy (HELLMANN et al. 1999). Usually not abundant. Recorded for southern Switzerland (VORBRODT 1930), but doubtful (REZBANYAI-RESER 1998). Reputedly in northern Italy, prov. South Tyrol and Friuli (KITSCHELT 1925; DANNEHL 1927; KUSDAS & THURNER 1955), data doubtful, no vouchers traced. Records from southern Italy (ZANGHERI 1963; PARENZAN 1994) and Balkan peninsula (e.g. Slovenja: CARNELUTTI 1992a; Romania: STANESCU 1997b; Serbia: TOMIC et al. 2002; Macedonia: PINKER 1968; MÜLLER 1996; Bulgaria: NESTOROVA 1998) shown to be based on erroneous bibliographic citations and misidentifications, mainly of *I. rufaria* (original examinations; GOMBOC, STANESCU, BESHKOV pers. comm.). Populations often very isolated, though sometimes common at sites of occurrence. In all German localities extinct or on the verge of extinction (KRAUS 1993; EBERT 2001). The xerothermic habitats merit special protection. – Outside Europe erroneously recorded for Caucasus, Transcaucasus and Turkmeniya (VIIDALEPP 1996).

PHENOLOGY: Univoltine: late May to early July, at higher elevations until late July, in the lowlands of the south earlier, from mid-April.

BIOLOGY: Larva probably polyphagous on withered leaves of herbaceous plants. Reputedly on *Salvia* (Lamiaceae), *Achillea millefolium*, *Centaurea* (Asteraceae) and *Scabiosa* (Dipsacaceae) (WILDE 1861; LHOMME 1935; FORSTER & WOHLFAHRT 1981; KOCH 1984), placed in doubt by EBERT (2001). Recorded furthermore on *Stellaria media* (Caryophyllaceae), *Plantago* (Plantaginaceae), *Rumex* (Polygonaceae), *Sonchus* (Asteraceae) (REUTTI 1898; LHOMME 1935; with subsequent citations). Reared on *Rumex* (GÓMEZ DE AIZPÚRUA 1987). Adults attracted to light. Nectar feeding recorded on *Senecio jacobaea* (EBERT 2001).

HABITAT: Xerothermophilous, inhabiting open habitats on low-nutrient soils, e.g. Mesobrometum, dry and stony sites, southern slopes (PROUT 1913, EBERT 2001), rarely also in wet meadows (FORSTER & WOHLFAHRT 1981). In the north from 0 m up to 400 m above sea-level, in the south up to 1,100 m.

SIMILAR SPECIES: Pale forms of *I. rufaria* (see no. 19). *Scopula ternata* differs in absent discal spot of hindwing, forewing costa not brownish, postmedial line more distal, frons darker, vertex paler, ♂ hindtarsus longer.

REMARKS: Closely related to the yellowish *I. numidaria* (LUCAS, 1849) from North Africa. For larval morphology and rearing see GÓMEZ DE AIZPÚRUA (1987).

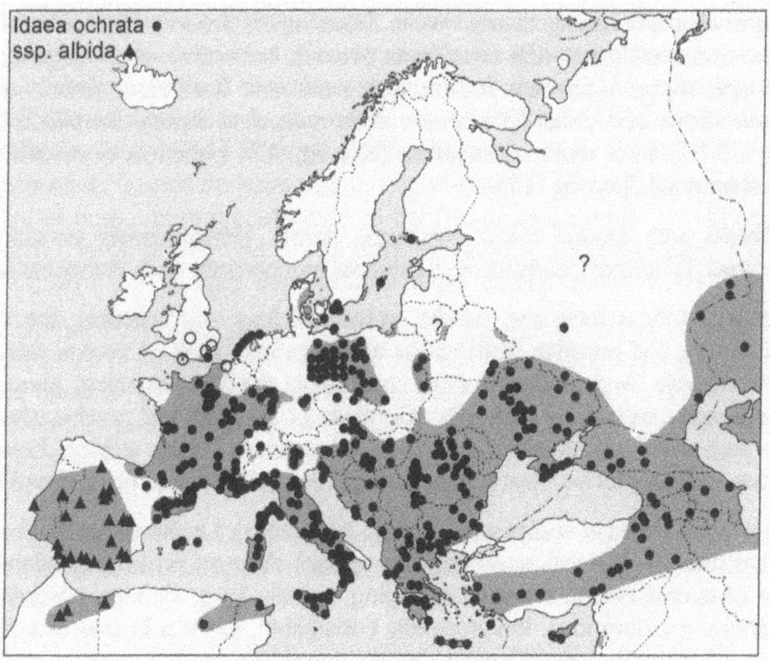

26. *Idaea ochrata* (SCOPOLI, 1763)

Phalaena ochrata SCOPOLI, 1763: Ent. Carniolica: 227, fig. 567 (north-western Slovenja: Carniolia 'Mts.'). Syntype(s) lost.

‡ *Geometra ochreata* DENIS & SCHIFFERMÜLLER, 1775: Ank. syst. Werkes Schmett. Wienergegend: 110. Incorrect subsequent spelling.

Phalaena corrigata FABRICIUS, 1781: Spec. Ins. 2: 265 (Italy). Syntype(s).

‡ *Geometra pallidaria*: sensu HÜBNER, 1799: Samml. Eur. Schmett. 5, Geometrae (1): pl. 18, fig. 96 (Europe). Misidentification nec *pallidata* DENIS & SCHIFFERMÜLLER, 1775 (HÜBNER's figure disputable).

‡ *Phalaena ochrearia* SCHRANK, 1802: Fauna boica 2: 33. Emendation, but junior primary homonym nec ROSSI, 1794 (identity: *Aspilates ochrearia*, Ennominae).

Acidalia ochrearia var. *sicula* ZELLER, 1847: Isis, 1847 (7): 498 (Sicily: Syracus). Syntypes ♂♀ (BMNH: ♂♀ syntypes, examined).

Acidalia accretata FUCHS 1901: Stettin. ent. Ztg. 62: 119 (Sicily). Holotype ♂.

Sterrha ochrata albida ZERNY, 1936: Mem. Soc. Sci. nat. Maroc 42: 73. First available name after *albida* RIBBE, 1912: Lep. Faun. Andal. (Dt. ent. Z. Iris 23): 302 (infrasubspecific: 'such pale forms could be called *albida*') (Spain, Andalusia: Sierra de Alfacar near Granada; ZERNY'S material from Morocco). Lectotype ♂ of RIBBE (MNHU, examined), herewith designated to fix type locality and to stabilize usage of the name. Valid at subspecific rank.

Ptychopoda ochrata 'f.' *cantiata* PROUT, 1913: in SEITZ, Macrolep. 4: 91 (Great Britain: Kent). Lectotype ♂ (BMNH, examined), selected by COCKAYNE (1953). Though introduced as 'form', PROUT (1913) calls it a 'local race' in the text. In PROUT (1934) formally at available rank ('var.'). Herewith downgraded from subspecific rank (SCOBLE 1999; LERAUT 1997; SKINNER 1984) to synonymy (syn. n.).

Unavailable names (misidentification): *perochraria*: sensu DOUBLEDAY (1849) nec FISCHER VON RÖSLERSTAMM, 1836 (identity: *Idaea serpentata*). – (infrasubspecific): *anastomosaria*: REISSER (1961); *bilineata*: REISSER (1961); *demarginata*: REISSER (1961); *flavescens*: GALVAGNI (1926); *impunctaria*: WEBER; *major*: DUFRANE (1931); *nigrolineata*: REISSER (1961). – (incorrect subsequent spellings): *ochraria*: FISCHER VON RÖSLERSTAMM (1836); *perocharia*: STAINTON (1855).

DIAGNOSIS: Wingspan ♂ 19-23, ♀ 18-22 mm, exceptionally 17 mm only. In Sicily ('*sicula*') usually 21-25 mm, rarely under 20 mm. Ground colour light brown with orange tinge. Wing pattern quite variable. Transverse lines distinct, postmedial line sharp, medial line diffuse. Light wavy line usually clearly visible. Discal spots absent or rather indistinct. Fine terminal line and small fringe dots sometimes present, but not always. Fringe usually darker than ground colour. Underside similar, with transverse lines more distinct and darker. Frons brown, vertex and collar light brown. Antennae of ♂ slightly dentate, ciliate-fasciculate, cilia 1.3-1.7 times width of flagellum (Text-fig. 43). Hindtibia of ♂ with two spurs, tarsus not shortened (Text-fig. 17).

Subsp. ***albida*** with ground colour very pale brown. Fringe usually concolorous with transverse lines. Wingspan, pattern, and external morphology as in nominate subspecies.

MALE GENITALIA: Uncus long and slender, setose. Valvae asymmetrical, the right being somewhat shorter and broader. Both valvae with terminal hook on ventral side. Shape of valvae quite variable. Aedeagus with stout cornutus at the distal opening, about 0.7 times length of aedeagus, with basal 'umbrella'. Genitalia of Tunisian and north-eastern Spanish specimens well matching those of nominate subspecies. Subsp. *albida*, however, with longer uncus, valvae shorter, nearly symmetrical, ventral terminal hook broad at base.

FEMALE GENITALIA: Papillae anales with 'double ovipositor', i.e. an additional basal pair of setal patches (modified lateral projection of papillae). Antrum posteriorly bilobed. Length and shape of ductus bursae somewhat varying, usually long, slender, posteriorly sclerotized, anteriorly membranous, longitudinally corrugated. Corpus bursae usually sub-globular. Subsp. *albida* with corpus bursae slightly elongate.

DISTRIBUTION: (Sub-)Mediterranean-Turanian. All over southern Europe, warmer parts of western, central and eastern Europe, also in a small area in south-eastern England (SKINNER 1984) and Denmark (KRISTENSEN 1966). In Sweden resident since 1996 (RYRHOLM pers. comm.). North of Alps in very isolated populations, but with increasing abundance and areal expansion in eastern Germany (GELBRECHT pers. comm.). Strongly threatened in southern Germany, due to habitat losses and disadvantageous mowing regimes (EBERT 2001; WOLF & HACKER 2004), in England by heavy grazing in winter (PARSONS et al. 2001). Records for central European Russia (ANTONOVA & KLEPIKOV 2000) and Lithuania (IVINSKIS 1993) await confirmation. Recorded for Sicily, Corsica, Crete, and Rhodos. Records for Malta (VALLETTA 1973; MÜLLER 1996) placed in doubt by SAMMUT (2000). In central and southern parts of the Iberian peninsula replaced by subsp. *albida*. – Outside Europe in Morocco, and western Algeria (subsp. *albida*), eastern Algeria, Tunisia, Turkey, Cyprus, across Caucasus, Transcaucasus, northern Iran, in the east to northern Afghanistan (VIIDALEPP et al. 1992; VIIDALEPP 1996), occurrence in Kazakhstan doubtful (VIIDALEPP 1988; 1996).

PHENOLOGY: Univoltine: mid-June to mid-August, sometimes one week earlier or later. In Sicily, Crete, Rhodos, southern Turkey and the Levant from early May to late June, subsp. *albida* from late April to mid-July. Larva growing slowly.

BIOLOGY: Larva polyphagous on withered leaves of herbaceous plants and grasses (var. cit.). Found in close vicinity of *Vicia tetrasperma* (Fabaceae), in rearings readily accepting that species, and also a comparatively wide set of other plants (England: PARSONS et al. 2001; CLANCY 2003; WARING & TOWNSEND 2003). Recorded on various species of Compositae, Rubiaceae, Brassicaceae (KRISTENSEN 1966), *Honkenya peploides, Stellaria media* (Caryophyllaceae), *Potentilla* (Rosaceae), *Myosotis* (Boraginaceae), and *Festuca* (Poaceae) (var. cit., e.g. DENIS & SCHIFFERMÜLLER, 1775; REUTTI 1898; REBEL 1910b; LHOMME 1935; ALLEN 1949; PORTER 1997), preferring withered petals (PROUT 1913; ALLAN 1949; SKINNER 1984). Reared on withered leaves and petals of *Taraxacum officinale, Polygonum aviculare, Crepis capillaris, Picris hieracioides, Leontodon autumnalis, Tussilago farfara, Solidago virgaurea, Ononis repens, Galium, Lotus corniculatus, Stellaria media,* and other plant species (TUGWELL 1880; BUCKLER 1897; ALLEN 1949; SKINNER 1984; PORTER 1997; REDONDO et al. 2001; KING & ROMERA 2004). Adults easily flushed up from vegetation, sometimes actively flying during daytime, main activity at dusk, sparingly attracted to light in central Europe, readily in southern Europe. Nectar feeding recorded on *Origanum vulgare, Reseda lutea*, and *Scabiosa columbaria* (EBERT 2001). Eggs deposited by flying females (EBERT 2001).

HABITAT: Xerothermophilous, preferring open dry grassland, sun-exposed slopes, often on vegetated sandy or limestone ground, but also wet heathland, bogs, and fens. In southern England often on early succession golf-courses (CLANCY 2003), on Sicily associated with 'Xerobromion' plant associations (BELLA et al. 1999). In the north from 0 m up to 500 m above sea-level, up to 1,500 m in southern Alps and in the Mediterranean area, exceptionally up to 2,000 m. In Turkey from 1,500 m up to 2,200 m.

SIMILAR SPECIES (see Text-fig. 74): *I. serpentata* (see no. 6), *I. rufaria* (19), *I. admiranda* (21). Identification of ♂ is easy by brushing tip of abdomen to remove scales and reveal shape of valva.

REMARKS: Wing colour, wing pattern, external morphology and genitalia of populations from Sicily ('*sicula*') and England ('*cantiata*') without significant and constant differences from nominate subspecies, therefore not validated at subspecific level here. Distribution and morphology discussed in KRISTENSEN (1966). Subsp. *albida* may be considered a separate species (RESER pers. comm.), but populations from possible transitional zones in northern Spain need further investigation. For rearing and morphology of immature stages see BUCKLER (1897) and TUGWELL (1880).

STERNECK's (1940) 6th and 7th group of *Idaea* (no. 27-36; no. 37-40). Reminiscent of *inquinata* species-group in ♀ genitalia (LENZ & HAUSMANN 1992). Extralimital *I. allongata* (STAUDINGER, 1898), with subsp. *sublongaria* (STAUDINGER, 1899) transferred to *incisaria* species-group in HAUSMANN (1991d: 123). Most species of the former 6th group with proximal half of forewing dark. Areole occasionally open at distal end (*I. filicata, I. intermedia*; compare *Cleta*). R1 not anastomosing with R2-R5. Proboscis present. Antennae of ♂ ciliate-setose. Some species with ♂ hindtibial spurs. Sternum A2 of ♂ without central sclerite and setose patch. Genitalic features suggest close relationship between both STERNECK's groups: subapical costal spines and terminal ventral spines often present on valva, uncus truncate, the single cornutus dilated at base, ♀ genitalia with conical dorsal appendix bursae, anterior part of corpus bursae membranous. Particular egg sculpturation (with distinct longitudinal, and weaker transverse ridges) probably synapomorphic for this group. Larva short and thick. Important revision of the *rusticata* species-group: REDONDO & GASTON (2003).

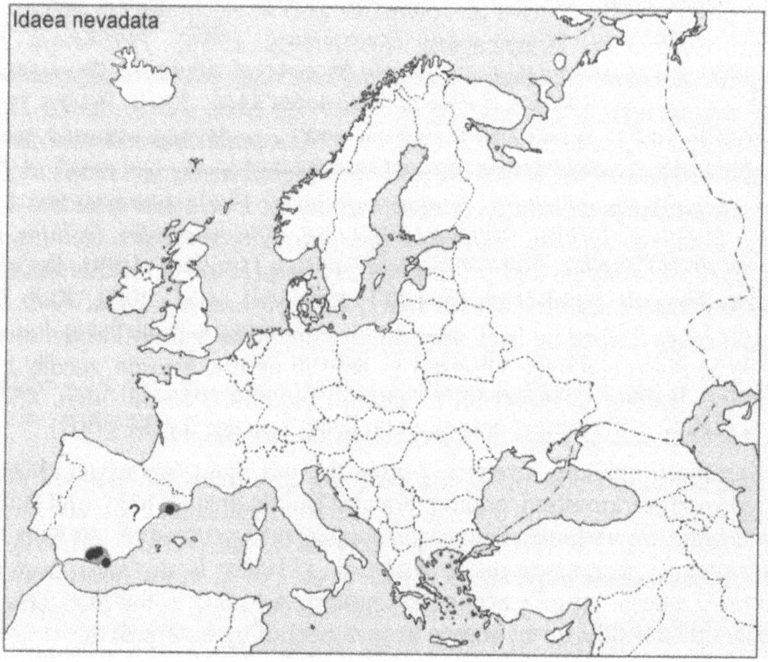

27. *Idaea nevadata* (WEHRLI, 1926)

Ptychopoda nevadata WEHRLI, 1926: Int. Ent. Zeits. 19 (41): 320 (Spain, Andalusia: Muleyhacen, Val de Inferno). Lectotype ♂ (ZFMK, examined), herewith designated to stabilize usage of the name with regard to the similar, possibly sympatric *I. figuraria*; at ZFMK 1♂ paralectotype, at BMNH 2♂ paralectotypes.

Unavailable names (misidentification): *completa*: sensu STAUDINGER (1901, partim) nec STAUDINGER, 1892; ?*figuraria*: sensu ZERNY (1927) nec BANG-HAAS, 1907 (identity: *Idaea figuraria*).

DIAGNOSIS: Wingspan ♂ 13-16, ♀ 12-14 mm. Forewing apex pointed. Ground colour whitish, with sand coloured tinge. Medial line of forewing S-shaped, crossing discal spot, strongly projecting towards base below, often anastomosing with antemedial line.

Proximal half of forewing grey brown. Terminal area with grey brown shadows and distinct, pale wavy line between. Black discal spots distinct, larger on forewing. Frons and palpi dark brown, vertex and collar sand coloured. Antennae of ♂ densely ciliate-setose, cilia 0.8-1.0 times width of flagellum. Hindtibia of ♂ not dilated, with two long spurs. Tarsus not shortened.

MALE GENITALIA: Uncus broad, truncate dorsally. Juxta rhombic. Valva long, curved, at tip rounded, with two small spines. Aedeagus comparatively broad and short (about 0.9 mm), with one long cornutus, 0.8-0.9 times length of aedeagus.

FEMALE GENITALIA: Antrum slightly dilated, rounded. Ductus bursae of medium size. Corpus bursae divided, posterior part sclerotized, elongate, laterally bulging and with long spines, towards ductus bursae a patch of small spinules. Anterior part oval, membranous.

DISTRIBUTION: West-Mediterranean. Very local in southern Spain, possibly endemic. Occurrence in north-eastern Spain (Aragon: REDONDO & GASTON 2003) apparently confirmed by one ♀ from Tarragona (ZSM, 10.VI.1962, leg. SATTLER, dissected). – All examined records from outside Europe (Morocco, Middle Atlas) refer to *I. figuraria*, or *I. completa* (see 'similar species').

PHENOLOGY: Univoltine: late June to late July.

BIOLOGY: Larval host-plants unknown. Adults readily attracted to light.

HABITAT: Montane, associated with scrubby places with *Hormathophylla spinosa* (*Alyssum spinosum*) briars (REDONDO & GASTON 2003), collected also in aisles of pine forest rich in low vegetation (PÖLL pers. comm.). From 1,500 m up to 2,900 m above sea-level.

SIMILAR SPECIES: *I. figuraria* slightly larger, forewings broader, apex less pointed, dark medial fascia of forewing more contrasted, distal border less parallel to postmedial line, the latter darker, more dentate or zigzagging, ♂ antennal cilia shorter, with differences in genitalia of both sexes. Populations of *I. completa* from Anti-Atlas, Morocco, strongly resemble *I. nevadata*, vertex and collar darker, secure identification by genitalia.

REMARKS: Taxonomy and differential features from *I. figuraria* discussed in WEHRLI (1931a) and REDONDO & GASTON (2003).

28. *Idaea figuraria* (BANG-HAAS, 1907)

Acidalia figuraria BANG-HAAS, 1907: Dt. ent. Z. Iris 20: 80, pl. 3, fig. 21 (western Algeria: southern Oran). Syntypes 3♀, not traced in BMNH, MNHU, ZFMK and SMTD. Examination of one ♀ syntype see WEHRLI (1931a).

Ptychopoda completa var. *subcompleta* FERNÁNDEZ, 1931: Eos 7 (2): 216, pl. 2, fig. 9 (Spain: Salamanca). Original types lost. Neotype ♂ (MNCN, figured in REDONDO & GASTON 2003).

Acidalia salmanticensis MENDES, 1935: Broteria (Zool.) 31: 69. First available name after *salmaticensis*: MENDES, 1918 (nomen nudum) (Spain: Salamanca, Mural de Rio). Syntype(s) (MNCN: 1♂, examined, with label 'typus').

Unavailable names (misidentification): *nevadata*: sensu AGENJO (1954) nec WEHRLI, 1926.

DIAGNOSIS: Wingspan ♂♀ 14-19 mm. In wing pattern much resembling *I. nevadata* (see above), but more contrasted, transverse lines sharper, area between postmedial line and medial line dilating towards costa and inner termen (WEHRLI 1931a). Frons and palpi dark brown. Palpi comparatively long, about 1.0 times diameter of eye. Antennae of ♂ ciliate-

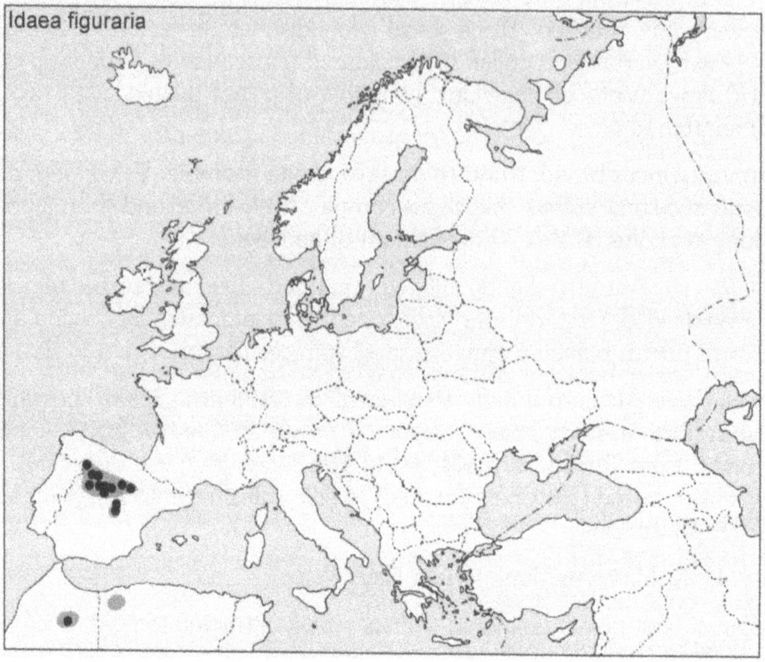

setose, cilia very short, 0.3-0.4 times width of flagellum. Hindtibia of ♂ not dilated, with one spur in one examined ♂, probably due to accidental loss of the other spur. Tarsus not shortened.

MALE GENITALIA: Similar to those of *I. nevadata*, but aedeagus on average longer (0.9-1.1 mm), narrower, cornutus straighter. Valva at tip obliquely sub-truncate. Genitalia of one examined ♂ from Morocco corresponding to these differential characters.

FEMALE GENITALIA: Antrum larger than in *I. nevadata*, irregularly bordered. Spinules near ductus bursae absent. Corpus bursae slender, anteriorly sclerotized to an elongate, narrow tube. Membranous part of corpus bursae less elongate than in *I. nevadata*.

DISTRIBUTION: West-Mediterranean. Rare and local in central Spain (Srra. Guadarrama; prov. Segovia; prov. Salamanca). – Outside Europe in Morocco and western Algeria.

PHENOLOGY: Univoltine: late April to mid-June (REDONDO & GASTON 2003), exceptionally until early July.

BIOLOGY: Larval host-plants unknown. Adults readily attracted to light.

HABITAT: Montane. Limited information available about habitat preference. From 600 m up to 1,200 m above sea-level.

SIMILAR SPECIES: *I. nevadata* (see no. 27). Extralimital *I. erlacheri* (= *I. figuraria* sensu REDONDO & GASTON 2003) with much straighter medial line of forewing, terminal area with indistinct markings.

REMARKS: Conspecifity of above presented ♂ (Morocco) with typical *I. figuraria* is suggested here on the base of the corresponding original description and the short distance (about 400 km) to the type locality in western Algeria, belonging to the same mountainous complex. '*I. figuraria*' sensu REDONDO & GASTON (2003), from Tunisian lowlands, however, is likely to be misidentified and synonym of *I. erlacheri* HAUSMANN, 2002. Conspecifity between populations in Morocco and central Spain was confirmed by examination of genitalia. The question of the identity of *I. figuraria* awaits, however, final confirmation by dissection of type material or by designation of an appropriate neotype. Anyway BANG-HAAS's original illustration and the remarks in WEHRLI (1931a) strongly support our interpretation of the species' identity.

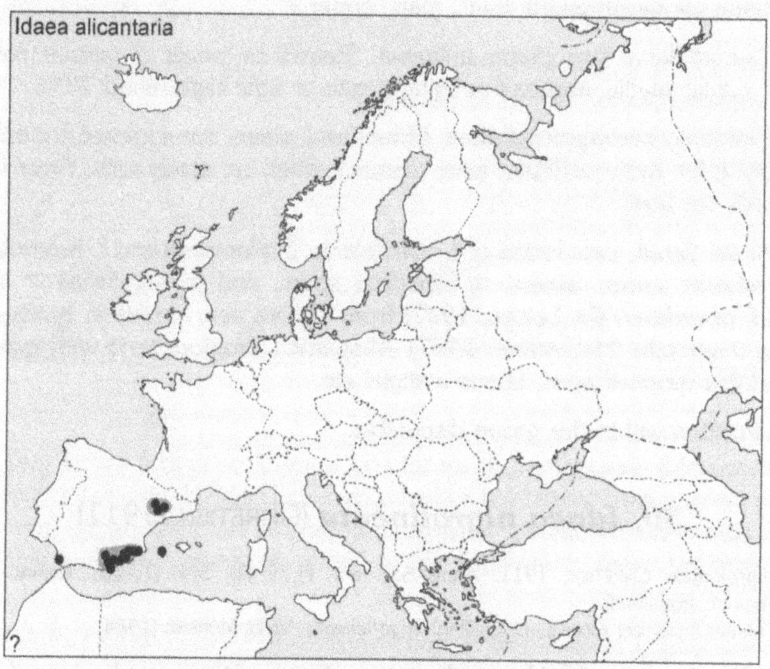

29. *Idaea alicantaria* (REISSER, 1963)

Sterrha alicantaria REISSER, 1963: Z. Wien. ent. Ges. 48 (3): 45, pl. 13, figs 2, 4 (south-eastern Spain: Alicante). Holotype ♀ (SMNK, 8 types examined by R. TRUSCH; 1♂ paratype in MNCN, examined).

DIAGNOSIS: Wingspan ♂♀ 12-16 mm. Forewings broad. Forewing sand coloured, hindwing paler. ♂ paler than ♀. Transverse lines rather straight, indistinct on forewing, better marked, but fine, on hindwing. Black discal spots distinct, very small on hindwing. Terminal dots absent. Fringe concolorous. Frons and palpi dark brown. Vertex and collar sand coloured to light brown. Antennae of ♂ with deep intersegmental incisions, densely ciliate-setose, cilia 1.0-1.35 times width of flagellum. Hindtibia of ♂ not dilated, with two long spurs. Tarsus not shortened.

MALE GENITALIA: Uncus sub-triangular. Juxta posteriorly elongate. Saccus long and slender. Costa of valva with stout tooth at 1/2 and narrow subterminal spine. Aedeagus with one narrow cornutus, vesica granulated beside cornutus.

FEMALE GENITALIA: Antrum with small, rough, digitiform process. Ductus bursae narrow, obliquely and subterminally leading into corpus bursae. Corpus bursae oval, membranous, with oval spinose sclerotization covering 2/3.

DISTRIBUTION: West-Mediterranean. Sometimes common in south-eastern Spain, between prov. Alicante and Almeria. Local records from Ibiza island (HONEY pers. comm.), prov. Huesca and Zaragoza in the north-east (HALL pers. comm.; REDONDO & GASTON 1999). – Outside Europe replaced by the sister species *I. oberthueri* (D. LUCAS, 1917) in Tunisia. Two examined ♀♀ from western Morocco, Agadir (ZSM) differ slightly in genitalia from *I. alicantaria*, possibly belonging to *I. oberthueri*.

PHENOLOGY: Univoltine, with partial second generation: early May to mid-June, at Huesca until early July; mid-September to late September. Larva growing extremely slow, requiring 12 months for development (HALL pers. comm.).

BIOLOGY: Natural larval host-plants unknown. Reared on petals of various plant species and *Erica*. Adults readily attracted to light. ♀ ratio at light high, about 70%.

HABITAT: Extremely xerothermophilous. Hot coastal plains, sun-exposed foothills, dry valleys with *Atriplex halimus* (HALL pers. comm.), often on sandy soils. From 0 m up to 400 m above sea-level.

SIMILAR SPECIES: Small, pale forms of *I. obsoletaria, I. elongaria* and *I. belemiata*, all differing in whitish vertex, absent ♂ hindtibial spurs, and in shortened ♂ hindtarsus. Allopatric *I. oberthueri* (D. LUCAS, 1917) from Tunisia very similar in habitus, but very different in ♂ genitalia (HAUSMANN, 1997). Allopatric *I. troglodytaria* with forewing slender, ♂ hindtibia spoon-shaped, tarsus vestigial etc.

REMARKS: Position within this group disputable.

30. Idaea nigrolineata (CHRÉTIEN, 1911)

?*Acidalia nigrolineata* CHRÉTIEN, 1911: Annls Soc. ent. Fr.79 (4): 508 (Tunisia: Gafsa). Holotype ♂ (ZSM/HERBULOT, examined).
Unavailable names (incorrect subsequent spelling): *nigolieneata*: VIVES MORENO (1984).

DIAGNOSIS: Wingspan ♂♀ 12-15 mm. Forewings slender. Wings sand coloured. Ante- and postmedial lines black brown, well marked, medial area dark. Postmedial line with two dentations projecting towards termen. Discal spots absent. Terminal line present, uninterrupted. Fringe chequered dark brown and whitish. Frons dark brown. Antennae of ♂ densely ciliate-setose, cilia 0.9-1.1 times width of flagellum. Hindtibia of ♂ not dilated, with two spurs. Tarsus not shortened.

MALE GENITALIA: Uncus slender, tapered. Valva subterminally slightly dilated. Aedeagus large, terminally tapering, with one very short, narrow cornutus and slightly granulated vesica.

FEMALE GENITALIA: Apophyses short. Antrum slightly dilated. Ductus bursae narrow, long, at junction with corpus bursae spinose. Corpus bursae curved, U-shaped, strongly spinose laterally, with anterior, membranous appendix.

DISTRIBUTION: West-Mediterranean. Southern and north-eastern Spain (DANTART 1988; SKOU, SKULE pers. comm.). – Outside Europe from Morocco to Tunisia. Rare all over the area of distribution, in small, isolated populations.

PHENOLOGY: Univoltine: mid-June to late July, in the Moroccan mountains to early August. Larva fully developed in June (BLAESIUS pers. comm.).

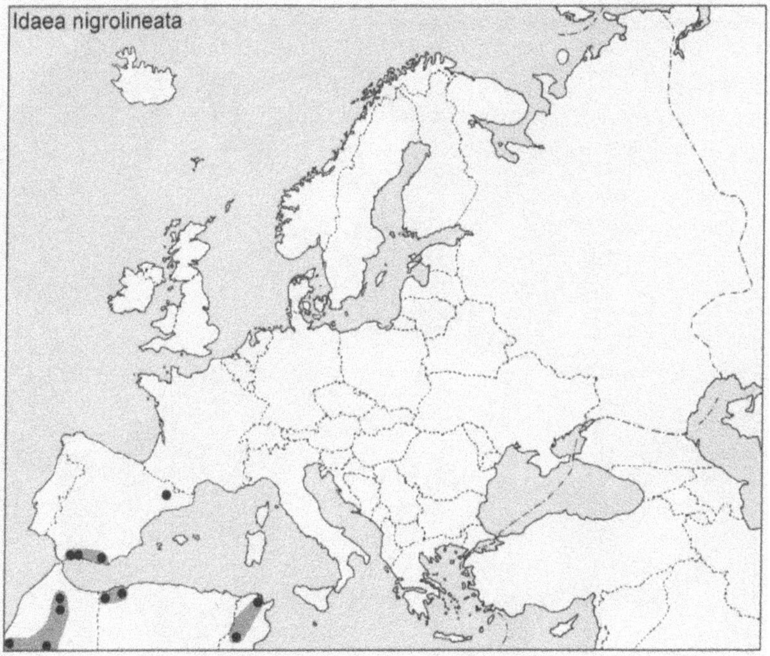

BIOLOGY: Two larvae found on *Viscum cruciatum* (Loranthaceae) (Morocco: BLAESIUS pers. comm.), it remains unclear, however, whether the larvae were feeding on the epiphytic *Viscum*, or, perhaps, on its host, a *Crataegus* species. As only twigs of *Viscum* were collected and the larvae were reared to the adult stage, this is probably a natural host-plant.

HABITAT: Xerothermophilous. Limited information available about habitat preference. In Spain restricted to the lowlands from 0 m up to 700 m above sea-level, in Morocco up to 1,900 m.

SIMILAR SPECIES: From *I. nevadata* and *I. figuraria* easily distinguished by the two dentations of the black postmedial line.

REMARKS: Unique features in habitus, external morphology and genitalia suggest an isolated systematic position.

31. Idaea completa (STAUDINGER, 1892)

Acidalia completa STAUDINGER, 1892: Dt. ent. Z. Iris 5 (1): 152 (eastern Algeria: Constantine). Lectotype ♂ (MNHU, examined), herewith designated to fix type locality and to stabilize usage of the name with regard to the southern Tunisian sister species; at MNHU 2♂1♀ paralectotypes from Tunisia, Tunis.

Idaea completa apostolica HAUSMANN, 1994: Z. Arb.Gem. österr. Ent. 46 (3-4): 83, figs 1,10,11 (Cyprus: Paphos). Holotype ♂ (ZSM, examined). Valid at subspecific rank.

DIAGNOSIS: Wingspan ♂♀ 10-16 mm. Wings sand coloured, often with slight orange or violet tinge. Proximal half of forewing brown, up to medial line. The latter surrounding discal spot. Basal area of hindwing brown. Brown postmedial lines indistinct on all wings. Wavy line indistinct. Palpi and frons brown or red brown, vertex and collar light brown.

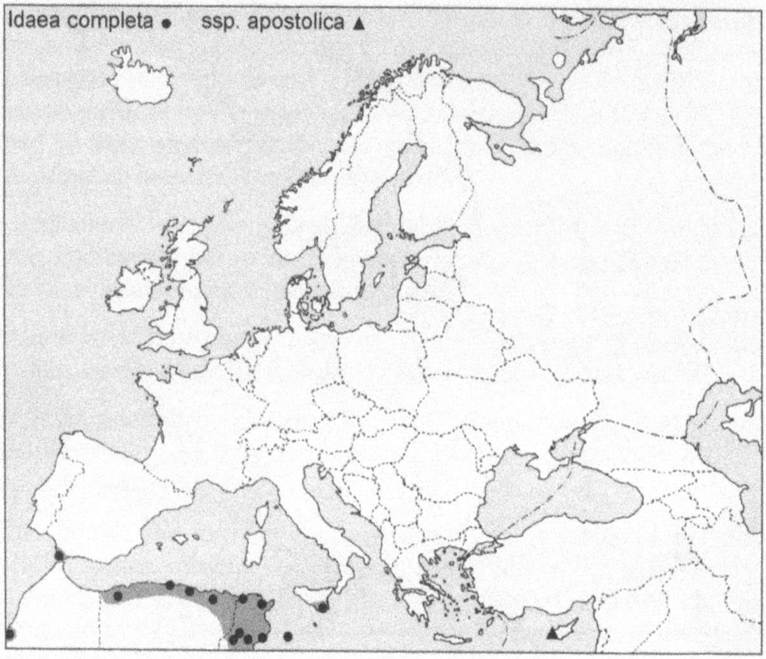

Antennae of ♂ densely ciliate-setose, cilia 1.0-1.2 times width of flagellum. Hindtibia of ♂ not dilated, with two long spurs. Hindtarsus long, 1.2-1.4 times length of tibia.

Variability: Extralimital mountain populations from Morocco (Anti-Atlas) with grey instead of brown markings, terminal area darker, thus resembling *I. nevadata*.

MALE GENITALIA: Uncus basally broad, terminally digitiform. Saccus short. Valva long, with small terminal spine at costa. Aedeagus with narrow, curved cornutus, 0.25-0.35 times length of aedeagus. Vesica granulated.

FEMALE GENITALIA: Apophyses short. Antrum corrugated, with deep posterior excision. Ductus bursae long. Corpus bursae globular or sub-globular, in the medial part and laterally sclerotized and spinose, posteriorly with slender, conical appendix bursae.

DISTRIBUTION: Mediterranean. Southernmost Spain, Lampedusa island (AISTLEITNER & AISTLEITNER 2001) and Sicily (PARENZAN et al. 1998a). – Outside Europe from Morocco to Libya (TURATI 1924; 1934), on Cyprus replaced by subsp. *apostolica*. In southern Tunisia a closely related sister species, *I. erlacheri* HAUSMANN, 2003. In Europe very rare and local, in North Africa sometimes common.

PHENOLOGY: Bi-, tri- or plurivoltine, from late April to early October, usually common in May/June and September. Under good conditions from late February to mid-November. Larval development needs about two months (HOMBERG 1911).

BIOLOGY: Larva probably polyphagous on withered leaves of herbaceous plants (CHRÉTIEN 1917; PROUT 1935a: 225). Reared on *Polygonum aviculare* (PROUT 1913, data derived from HOMBERG 1911; perhaps referring to *I. intermedia*: PROUT 1935a: 225). Adults readily attracted to light, ♀ ratio at light high, about 60-70%.

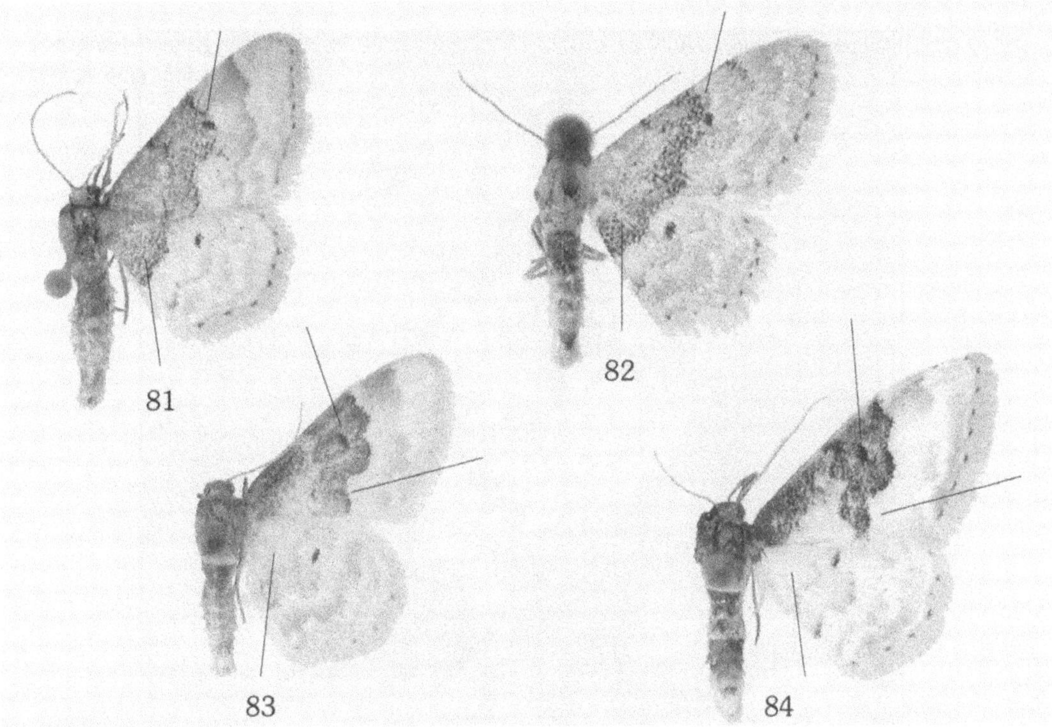

Text-figs 81-84: Diagnostic characters (indicated) of European *Idaea* species. Text-fig. 81: *I. completa* (STAUDINGER, 1892). Text-fig. 82: *I. intermedia* (STAUDINGER, 1879). Text-fig. 83: *I. rusticata* (DENIS & SCHIFFERMÜLLER, 1775). Text-fig. 84: *I. mustelata* (GUMPPENBERG, 1892).

HABITAT: Xerothermophilous lowland species, inhabiting hot, coastal habitats. In Sicily recorded in 'humid areas' (PARENZAN et al. 1998a). From 0 m up to about 200 m above sea-level. In south-western Morocco (Anti-Atlas) up to 1,500 m.

SIMILAR SPECIES: *I. nevadata* (see no. 27), *I. intermedia* (32), *I. rusticata* (33), *I. mustelata* (34), *I. filicata* (35). In this group *I. completa* is the only species with both a dark base of hindwing and discal spots of forewing included in the dark medial area. *I. erlacheri* from southern Tunisia, with forewings narrower, antennal cilia much longer, with strong markings in the distal half of wings and much larger cornutus in ♂ genitalia (HAUSMANN 2003).

REMARKS: For morphology of immature stages see CHRÉTIEN (1917).

32. Idaea intermedia (STAUDINGER, 1879)

Acidalia intermedia STAUDINGER, 1879: Horae Soc. ent. ross. 14 (4): 436 (Turkey, Pontus: Kerasdere near Amasia). Syntypes 2♂2♀ (MNHU: 2♂2♀, examined).

DIAGNOSIS: Wingspan ♂♀ 9-14 mm, small specimens mainly in summer and autumn generations. Wings sand coloured, sometimes with slight grey tinge. Medial area between antemedial line and medial line black brown. Medial line running proximally to discal spot or crossing it. Basal area of hindwing black brown. Postmedial line dark grey, very fine, well marked towards costa of forewing. Palpi and frons blackish, vertex grey or pale sand

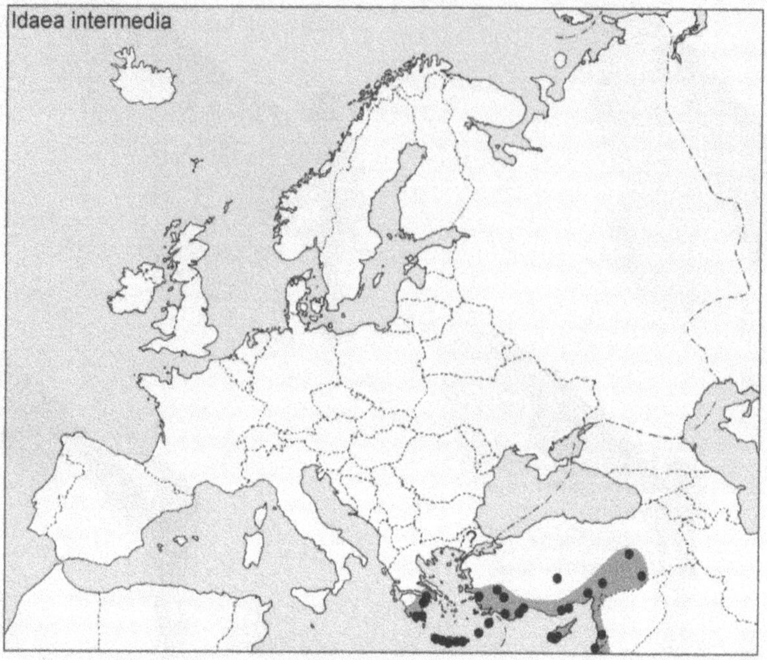

coloured. Antennae of ♂ ciliate-setose, cilia short, 0.6-0.8 times width of flagellum. Hindtibia of ♂ not dilated, with two long spurs. Tarsus not shortened.

MALE GENITALIA: Size variable with size of moth. Uncus broad at base, terminally sub-digitiform. Valva with small terminal spine on costa. Aedeagus short, with narrow, medially slightly dilated cornutus, 0.3-0.4 times length of aedeagus. Vesica slightly granulated, with an elongate field of small sclerites in tile-like arrangement.

FEMALE GENITALIA: Apophyses short. Antrum not dilated, with posterior excision. Ductus bursae of medium length. Corpus bursae oval, with characteristically shaped spinose patches, posteriorly with triangular, conical appendix bursae.

DISTRIBUTION: East-Mediterranean. Southern Greece including the southern Greek islands, and Crete. Reputedly in European Turkey (MÜLLER 1996). Erroneously recorded for Spain (VIVES MORENO 1994; MÜLLER 1996). – Outside Europe in the Levant from western Turkey to Israel, present on Cyprus. One record from central Anatolia, Tuz Gölü. Usually abundant.

PHENOLOGY: Plurivoltine. Almost all months of the year, with a gap from mid-January to mid-April, most abundant from May to July.

BIOLOGY: Larval host-plants unknown. Adults readily attracted to light.

HABITAT: Xerothermophilous lowland species, most abundant in hot, coastal habitats. From 0 m up to 900 m above sea-level, exceptionally up to 1,300 m.

SIMILAR SPECIES: *I. completa* (see no. 31), *I. rusticata* (33), *I. mustelata* (34), *I. filicata* (35). In this group *I. intermedia* is the only species with the feature combination 'discal spot of forewing excluded from dark medial area or crossed by medial line', 'base of forewing bright' and 'base of hindwing dark'. *I. filicata* with spoon-shaped hindtibia of ♂, tarsus vestigial.

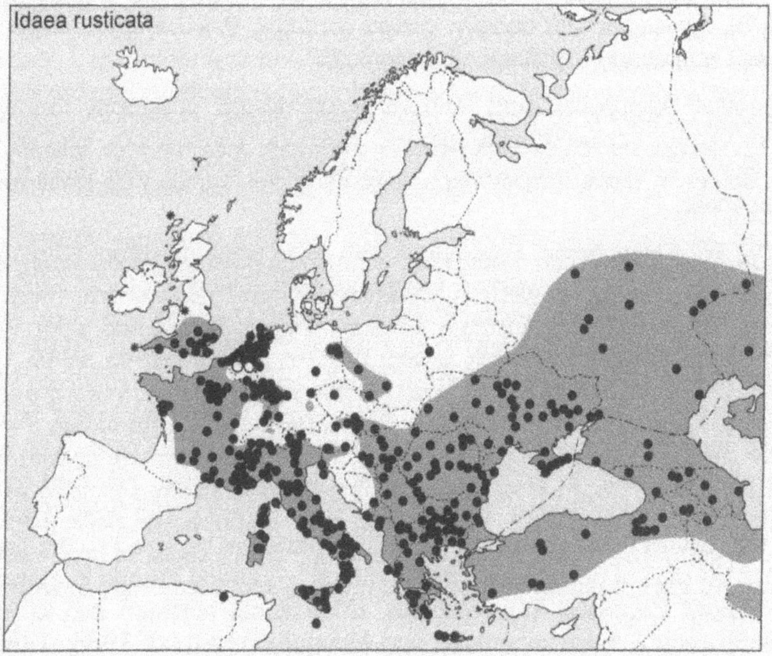

33. *Idaea rusticata* (DENIS & SCHIFFERMÜLLER, 1775)

Geometra rusticata DENIS & SCHIFFERMÜLLER, 1775: Ank. syst. Werkes Schmett. Wienergegend: 113 (Austria: Vienna distr.). Syntype(s) lost.

? *Phalaena humeralis* FOURCROY, 1785: Entom. Parisiensis 2: 287 (France: Paris). Syntype(s). Synonymy awaits confirmation.

? *Phalaena Geometra humerata* VILLERS, 1789: Linn. ent. 2: 380. Emendation of *humeralis*. Synonymy awaits confirmation.

Acidalia rusticaria HERRICH-SCHÄFFER, 1847: Syst. Bearb. Schmett. Eur. 3 (24): 21. Emendation.

Acidalia vulpinaria HERRICH-SCHÄFFER, 1852: Syst. Bearb. Schmett. Eur. 6 (55): 65 (Austria). Syntypes ♂♀. Referring to Vol. 3, p. 14, thus binominal. First available name after *vulpinaria* HERRICH-SCHÄFFER, 1851 (pl. 77, figs 473, 474: non binominal). Junior synonym. Listed as valid species name by LERAUT (1997). Hindtibial spurs mentioned in original description.

Idaea vulpinaria atrosignaria LEMPKE, 1967: Tijdschr. Ent. 110 (19): 259 (Netherlands, Gelderland prov.: Apeldoorn). Holotype ♂ (ITZA). Herewith downgraded from subspecific rank (LERAUT 1997; SKINNER 1984) to synonymy.

Unavailable names (infrasubspecific): *albomarginata*: LEMPKE (1949); *aestiva*: DANNEHL (1933); *demarginata*: REISSER (1961); *subvulpinaria*: OBRAZTSOV.

DIAGNOSIS: Wingspan ♂♀ 14-20 mm, in autumn smaller, sometimes 11 mm only. Wings whitish. Medial area of forewing between ante- and postmedial lines dark brown, sometimes with red tinge, including discal spot. Reddish forms not rare on Sardinia, Balkan peninsula and in Turkey. Black discal spots distinct. Base of hindwing and usually also the medial area concolorous with ground colour. Forewing costa with black suffusion at base. Wavy line of ground colour bordered by grey shadows at both sides. Frons blackish, vertex and collar light brown. Antennae of ♂ densely ciliate-setose, cilia 0.8-1.1 times width of flagellum. Hindtibia of ♂ not dilated, with two spurs ('subsp. *vulpinaria*') or without spurs ('subsp. *rusticata*'), for distribution and constancy of this character see sections on distribution and remarks. Tarsus not shortened.

MALE GENITALIA: Uncus slender. Valva long with small terminal and a larger subterminal spine (at 4/5). Aedeagus with narrow, curved cornutus, 0.4-0.5 times length of aedeagus. In 'subsp. *vulpinaria*' cornutus often sigmoid.

FEMALE GENITALIA: Apophyses short. Antrum slightly dilated, posteriorly concave. Ductus bursae short. Corpus bursae strongly spinose, anteriorly membranous, laterally with spinose bulge, spines in radial arrangement, towards ductus bursae with small round membranous appendix.

DISTRIBUTION: West-Palaearctic. Populations with spurs present ('subsp. *vulpinaria*') from Great Britain and France to western Germany, Switzerland, Corsica, Sardinia, Sicily, Malta across Italy to southern Austria (Carinthia), Istria, Dalmatia and Corfu island. In all these populations spurs occasionally absent (WEHRLI 1924; LEMPKE 1949; FORSTER & WOHLFAHRT 1981). Populations with spurs usually absent ('subsp. *rusticata*') from eastern Germany, Czech Republic, northern and eastern Austria (including Styria) to the Urals, in the north to Estonia, in the south to Macedonia (here not rarely with spur anomalies), Greece and Crete. In England expanding towards north, recorded as 'vagrants' even on islands St. Kilda (PROUT 1935a; REID pers. comm.), and Scilly (HICKS & HALE 1998). In the eastern Pyrenees and in the Iberian peninsula replaced by the sister species *I. mustelata*. At sites of occurrence often common. – Outside Europe in eastern Algeria, and from Turkey, Caucasus, Transcaucasus, Israel across northern Iran to Afghanistan (EBERT 1965), central Asian mountains and Mongolia (VIIDALEPP 1988; 1996; VOJNITS 1976). In Kirgisian and Turkmeniyan populations with spur anomalies. In western Algeria and Morocco replaced by *I. mustelata*.

PHENOLOGY: Usually univoltine: mid-June to mid-August, sometimes emerging from late May. Under good conditions, e.g. in southern Europe sometimes with second generation from mid-July to early October, overlapping with first generation.

BIOLOGY: Larva polyphagous. Found on fallen leaves of *Parthenocissus* (Vitaceae) (Germany: EBERT 2001), *Alyssum saxatile* (Brassicaceae) (Great Britain: WEST 1974), below *Rubus, Prunus spinosa* (Rosaceae) and fruit trees (BERGMANN 1955). *Berteroa incana* (Brassicaceae) suggested as possible host-plant in south-western Germany (EBERT 2001). Recorded also on *Hedera helix* (Araliaceae), *Clematis vitalba* (Ranunculaceae), *Ulmus carpinifolia* (Ulmaceae), and *Senecio* (Asteraceae) (ALLAN 1949; SKINNER 1984; PORTER 1997; WARING & TOWNSEND 2003), preferring withered or mouldy leaves (var. cit.), reputedly also on moss (var. cit.). Hatching of adult from pupa out of moss observed in nature (Piemonte: BASSI pers. comm.). Reared on *Parietaria, Senecio vulgaris, Syringa vulgaris, Hedera helix, Trifolium, Taraxacum officinale, Polygonum aviculare* (original data; BUCKLER 1897; ALLAN 1949; SKINNER 1984; PORTER 1997), and on petals of *Pimpinella anisum* (original data). Larva very inactive. Well fed overwintering larvae developing to extremely large imagines (PROUT 1913). Adults active at dusk (original data from Italy), and readily attracted to light and bait. In Germany active late in the night, never flushed up during daytime (RENNWALD pers. comm.). During daytime resting on piles of compost (EBERT 2001), on fences and on upper side of leaves (SKINNER 1984). Possible occasional migration or high dispersal activities discussed in EBERT (2001).

HABITAT: Xerothermophilous, in scrub downland, chalk embankments (SKINNER 1984), often on sun-exposed slopes with bushes or hedges, on dry grassland, and in gardens. In the north from 0 m up to 400 m above sea-level, exceptionally higher, from 0 m up to 1,200 m in southern Alps and southern Europe, usually from 700 m up to 2,000 m in Turkey and central Asia.

SIMILAR SPECIES: *I. completa* (see no. 31), *I. intermedia* (32), *I. mustelata* (34), *I. filicata* (35). In this group *I. rusticata* is the only species with feature combination 'discal spot of forewing clearly included in the dark medial area' and 'base of hindwing whitish'. *I. filicata* differs furthermore in spoon-shaped hindtibia of ♂, tarsus vestigial. Sister species *I. mustelata* with medial area strongly tapering towards inner margin of forewing.

REMARKS: Formerly separated into two species *vulpinaria* and *rusticata* according to presence or absence of male hindtibial spurs (e.g. URBAHN 1973; FAZEKAS 1979; FORSTER & WOHLFAHRT 1981; VOJNITS 1984; VIVES MORENO 1994; LERAUT 1997). This character, however, does not justify species separation, because specimens with spurs present on one leg and absent on the other one, specimens with one spur on both legs, specimens with extremely short spurs and spurs present and absent mixed up in the same population have been repeatedly found (WEHRLI 1924; PROUT 1935a: 225; LEMPKE 1949; FORSTER & WOHLFAHRT 1981; own observations). Variability of spur number in rearings has been observed repeatedly: among descendants from the eggs of one single female there were males with spurs absent, present and with one single spur (Macedonia, leg. REISSER, ZSM), similar observations are mentioned by PROUT (1913: 131). Compare remarks to 'distribution'. For rearings and morphology of immature stages see BUCKLER (1897).

34. Idaea mustelata (GUMPPENBERG, 1892), stat. n.

Dosithea mustelata GUMPPENBERG, 1892: Nova Acta Acad. Caesar. Leop. Carol. 58 (4): 259, 266. First available name after *mustelata* RAMBUR, 1866 (non binominal) (Spain: Andalusia). Syntype(s) of RAMBUR not traced (see HERBULOT 1987). Herewith raised from subspecies of *Idaea rusticata* (e.g. SCOBLE 1999) to species rank (stat. n.).

‡ *Sterrha rusticata* [syn.] *algirica* PROUT, 1935: in SEITZ, Macrolep 4, Suppl.: 54 (Morocco, Algeria). Manuscript name of BETHUNE-BAKER, unavailable as introduced in synonymy.

Unavailable names (misidentification): ?*figuraria*: sensu ZERNY (1927) nec BANG-HAAS, 1907 (identity: *Idaea figuraria*); *rusticata*: sensu STAUDINGER (1901) nec DENIS & SCHIFFERMÜLLER, 1775.

DIAGNOSIS: Smaller than *I. rusticata*, wingspan ♂♀ 12-16 mm, exceptionally up to 18 mm. Wings usually more slender. Wing colour and pattern similar to *I. rusticata*, but medial area tapering towards inner margin of forewing, often not reaching it. Frons blackish, vertex and collar light brown. Antennae of ♂ densely ciliate-setose, cilia 0.8-1.1 times width of flagellum. Spurs absent from ♂ hindtibia of all examined specimens, though PROUT (1935a) mentions occasional occurrence of spurs.

MALE GENITALIA: Reminiscent of *I. rusticata*, but cornutus straight, 0.5-0.7 times length of aedeagus.

FEMALE GENITALIA: Reminiscent of *I. rusticata*, but posterior excavation of antrum shallower. Spinose patches in different arrangement, e.g. lateral circle of large spines absent. Appendix bursae straight, not looped.

DISTRIBUTION: West-Mediterranean. Eastern Pyrenees and Iberian peninsula. – Outside Europe in Morocco and western Algeria (Oran).

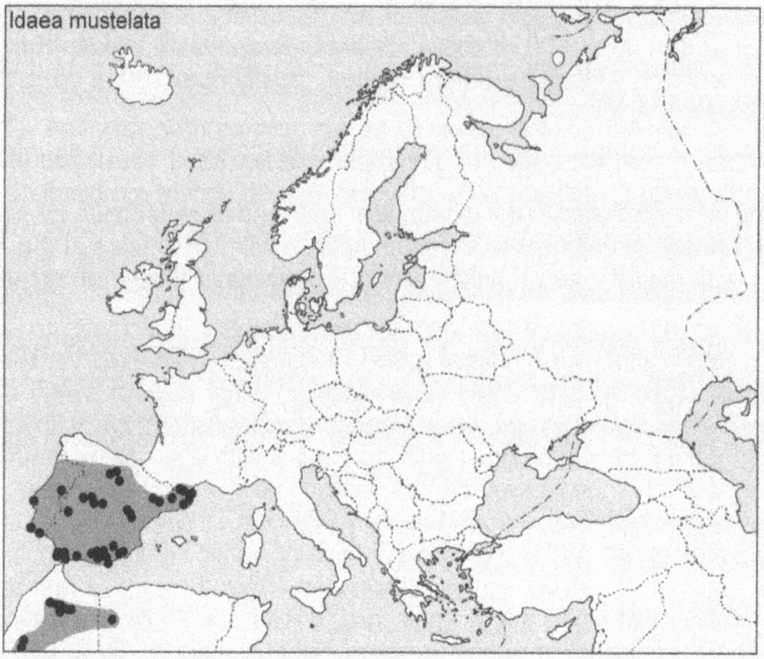

PHENOLOGY: Univoltine: early July to early August. In the southern lowlands bivoltine from April to October (CORLEY pers. comm.).

BIOLOGY: Larval host-plants unknown. Adults attracted to light.

HABITAT: Limited information available about habitat preference. Usually from 1,000 m up to 2,000 m above sea-level, in the north often, in the south sometimes down to sea-level. In Morocco abundant between 1,700 and 2,700 m.

SIMILAR SPECIES: See *I. rusticata* (no. 33).

REMARKS: Positive correlation of occurrence of ♂ hindtibial spurs and reddish coloration (PROUT 1935a) not confirmed by own studies.

35. *Idaea filicata* (HÜBNER, 1799)

Geometra filicata HÜBNER, 1799: Samml. Eur. Schmett. 5, Geometrae (1): pl. 46, fig. 238 (southern Europe). Syntype(s) lost.

Dosithea filicaria DUPONCHEL, 1830: in GODART & DUPONCHEL: Hist. Nat. Lép. Pap. Fr. 8 (1): 49, pl. 173, fig. 6. Emendation.

Unavailable names (infrasubspecific): *angeliata*: SCHULTZ (1931); *autumnalis*: SCHWINGENSCHUSS (1926); *demarginata*: REISSER (1961); *extincta*: WAGNER (1922); *purificata*: DANNEHL (1925); *somnambula*: DANNEHL (1925). – (incorrect subsequent spelling): *silicata*: MANN (1854).

DIAGNOSIS: Wingspan variable ♂♀ 12-21 mm, in the autumn generation not exceeding 17 mm. Ground colour glossy creamy beige. Basal half of forewing dark brown, exceptionally light brown, touching discal spot. Variable in extension of dark proximal area, sometimes dissolved into some very small dots ('f. *extincta*'). Postmedial line marked at costa and inner margin, absent ('f. *purificata*') or dissolved into small dots between. Terminal

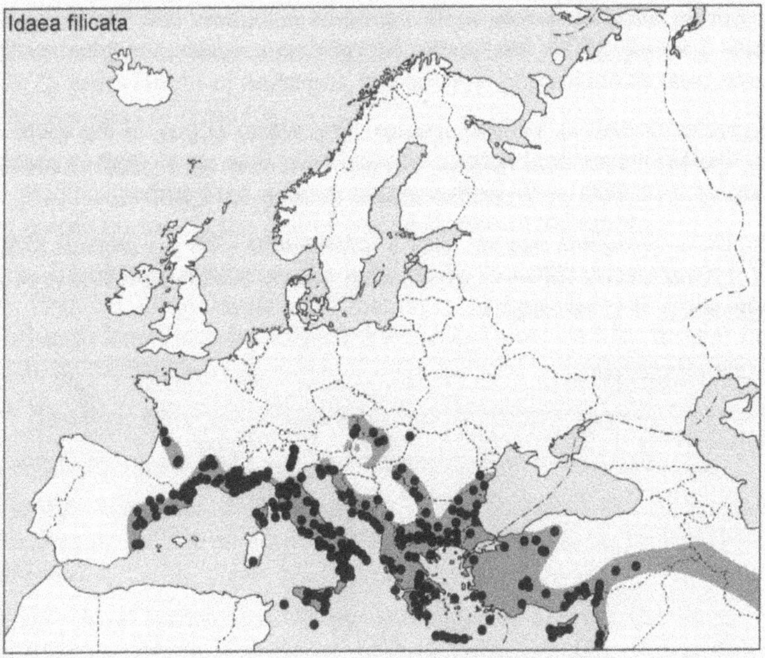

area with grey shadow. Discal spots black, very distinct. Terminal line absent. Fringe concolorous with ground colour. Base of hindwing usually not dark. Frons blackish, vertex and collar light brown. Antennae of ♂ ciliate-setose, cilia 0.8-1.0 times width of flagellum (Text-fig. 44). Hindtibia of ♂ very long, terminally dilated, spoon-shaped. Inner side of hindtibia with pencil of about 0.7-0.9 times length of tibia, tarsus vestigial (Text-fig. 18).

MALE GENITALIA: Uncus truncate. Gnathos broad. Valva with small subapical spine on costa and small terminal spine ventrally. Aedeagus tapering towards tip. At base of aedeagus one small cornutus, about 0.2 times length of aedeagus, swollen and corrugated at base.

FEMALE GENITALIA: Apophyses very short. Antrum slender. Ductus bursae very long, slender, longitudinally corrugated, with spinules. Corpus bursae globular, bent towards anal end, with large area of spinules.

DISTRIBUTION: (South-)European-Westasiatic. Southern Europe from eastern Spain and southern France to Balkan peninsula. With northern extension of distribution area into the Danube river plain up to eastern Austria, and southern Czech Republic. On nearly all Mediterranean islands, but absent from Balearics. In the Mediterranean area often very abundant. – Outside Europe from Turkey and Cyprus to the Levant, northern Iraq, western and southern Iran.

PHENOLOGY: Bi- or trivoltine: early May to late September, exceptionally to late October, often with a gap in July. In the south emerging from mid-April.

BIOLOGY: Larva probably polyphagous on withered leaves of herbaceous plants. Reputedly on petals of *Dianthus*, *Veronica*, and other low herbs (var. cit., e.g. REBEL 1910b; PROUT 1913; LHOMME 1935; FORSTER & WOHLFAHRT 1981). Reared on

Veronica, and on withered leaves of *Polygonum aviculare*, and *Taraxacum officinale* (original data; LHOMME 1935; ERLEBACH, MÜLLER pers. comm.). Adults readily attracted to light.

HABITAT: Xerothermophilous lowland species. Dry, warm slopes, in the south ubiquitous, common in the Mediterranean maquis. Usually from 0 m up to 800 m above sea-level, exceptionally up to 1,400 m, in south-western Asia up to 2,400 m.

SIMILAR SPECIES: *I. completa* (see no. 31), *I. intermedia* (32), *I. rusticata* (33), *I. mustelata* (34). *I. filicata* clearly differs in glossy wing colour, double subapical spot at forewing costa, modification of ♂ hindleg etc.

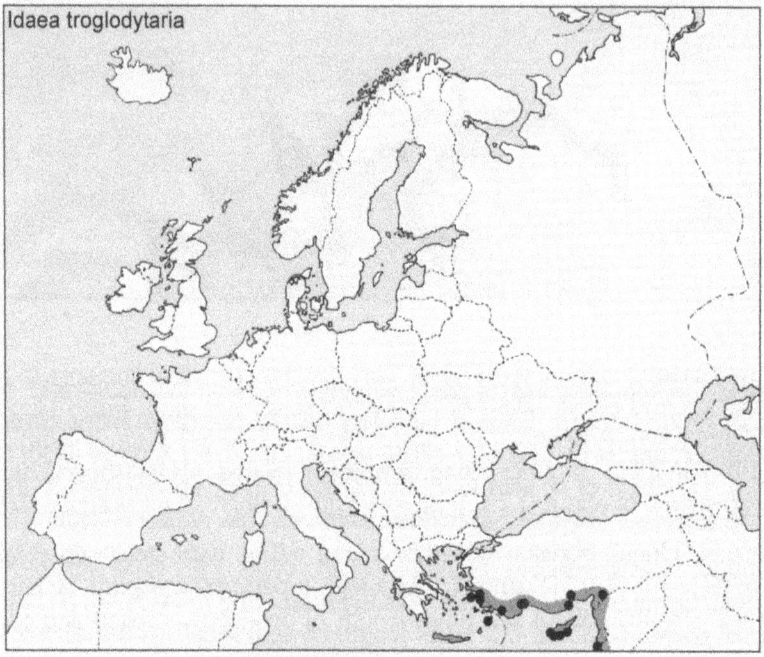

36. *Idaea troglodytaria* (HEYDENREICH, 1851)

Idaea troglodytaria HEYDENREICH, 1851: Lepid Eur. cat. meth. (Edn 3): 60. First available name after *troglodytaria* HERRICH-SCHÄFFER, 1848 (pl. 70, fig. 434: non binominal) (type locality not stated; 'Crete' according to STAUDINGER 1870: 148). Syntype 1♀ (of HERRICH-SCHÄFFER) according to STAUDINGER (1870).

Sterrha albonitens STERNECK, 1939: Z. Wien. Ent. Ver. 24 (12): 165 (Israel: En Kerem, Jerusalem). Holotype ♂ (MNHU/PÜNGELER, examined). First available name after *albonitens* in AMSEL (1935) and PROUT (1935) (nomina nuda).

Unavailable names (misidentification): *monadaria*: sensu PROUT (1913, partim) nec GUENÉE, 1858 (identity: *Idaea elongaria*).

DIAGNOSIS: Wingspan of spring generation ♂♀ 13-17 mm, autumn generation 8-12 mm. Forewing slender. Ground colour glossy white with slight grey or cream-coloured tinge. Wing pattern almost lacking. Discal spots very small. Sometimes with very fine postmedial line. Terminal line absent. Fringe concolorous. Frons grey or grey brown, vertex cream-coloured, collar light brown. Antennae of ♂ ciliate-setose, cilia 1.0-1.2 times width of flagellum. Hindtibia of ♂ spoon-shaped as in *I. filicata*. Inner side of hindtibia with pencil of about 0.7-0.9 times length of tibia, tarsus vestigial (Text-fig. 19).

MALE GENITALIA: Uncus sub-triangular. Gnathos comparatively narrow. Valva subterminally narrow, terminally head-shaped with two lateral spines. Aedeagus with one long cornutus, 0.6-0.75 times length of aedeagus, swollen and corrugated at base. Vesica granulated, mainly near base of cornutus.

FEMALE GENITALIA: Apophyses very short. Antrum slightly dilated. Ductus bursae long, slender, longitudinally corrugated, with spinules, laterally inserted and continuously dilating towards corpus bursae. Distal half of corpus bursae membranous.

DISTRIBUTION: East-Mediterranean. Rhodos island. Records from Crete (STAUDINGER 1870: 148; PROUT 1935a), await confirmation. – Outside Europe in the Levant from southern Turkey to Israel, also on Cyprus. Erroneously recorded for the Far East of Asia (PROUT 1913).

PHENOLOGY: Bivoltine: mid-April to late May; mid-September to early November.

BIOLOGY: Larval host-plants unknown. Adults readily attracted to light.

HABITAT: Xerothermophilous lowland species, most abundant in hot, coastal habitats. Usually from 0 m up to 200 m above sea-level, exceptionally up to 1,100 m, southern Turkey, and Israel.

SIMILAR SPECIES: Small forms of *I. distinctaria* differ in broader forewings, frons dark brown, ♂ hindtarsus present. Pale forms of *I. palaestinensis*, *I. elongaria* usually with postmedial line sharp and distinct, frons dark brown, vertex whitish, ♂ hindtarsus present. Allopatric *I. alicantaria* (see no. 29).

REMARKS: CULOT's figure of '*troglodytaria*' from Mesopotamia misidentified, representing a species of the *obsoletaria* species-group (PROUT 1935a).

37. *Idaea laevigata* (SCOPOLI, 1763)

Phalaena laevigata SCOPOLI, 1763: Ent. Carniolica: 230, fig. 575 (north-western Slovenja: Carniolia). Syntype(s) lost.

‡ *Geometra laevigaria* HÜBNER, 1799: Samml. Eur. Schmett. 5, Geometrae (1): pl. 14, fig. 74. Incorrect subsequent spelling.

Geometra renularia HÜBNER, 1809: Samml. Eur. Schmett. 5, Geometrae (1): pl. 64, fig. 331 (Europe). Syntype(s) lost.

Idaea bellata FREYER, 1840: Neuere Beitr. Schmettkde. 4 (54): 55, pl. 323, fig. 4 (locality not stated, possibly Ljubljana ('Laybach')). Syntype(s).

Acidalia bellaria DUPONCHEL, 1845: Cat. Méth. Lép. Eur.: 375. Emendation.

Dosithea lividellaria PEYERIMHOFF, 1862: Bull. Soc. Hist. nat. Colmar 2: 109 (France: Alsace). Syntype(s).

‡ *Anagoge renulata* GUMPPENBERG, 1892: Nova Acta Acad. Caesar. Leop. Carol. 58 (4): 270, 271. First binominal name after *renulata* RAMBUR, 1866 (non binominal). Incorrect subsequent spelling.

Acidalia couloniata BALESTRE, 1907: Bull. Soc. ent. Fr. 1907 (2): 24 (France, Alpes maritimes: Annonciata hill, near Mentone). Lectotype ♂ (ZSM/HERBULOT, examined), selected by HERBULOT (1953). Herewith downgraded from species (SCOBLE 1999; MÜLLER 1996) to synonymy (syn. n.), previously stated in HERBULOT (1953).

‡ *Ptychopoda (Acidalia) laevigata* var. *roseata* WAGNER, 1919: Zeits. Österr. Ent.-Ver. 4 (2): 11 (Romania: Cornul, jüd. Prahova). Syntypes ♂♀. Junior objective homonym of *roseata* TURATI, 1913 (identity: *I. dimidiata* syn.).

Ptychopoda laevigata var. *storae* ANDREAS, 1928: Int. Ent. Zeits. 22 (27): 251 (Algeria: Stora). Syntypes ♂♀ (MNHU/PÜNGELER 2 'paratypes' from 'Philippeville' examined).

Unavailable names (infrasubspecific): *fuscovittata*: DANNEHL (1927); *johnsoni*: HARRISON (1928); *parvipunctata*: ANDREAS (1928); *suffusa*: HARRISON (1929); *totanigra*: ANDREAS (1928). – (incorrect subsequent spellings): *laeivigata*: BUCHSBAUM (1998); *laeuigata*: FABRICIUS (1787); *renalata*: VIVES MORENO (1984).

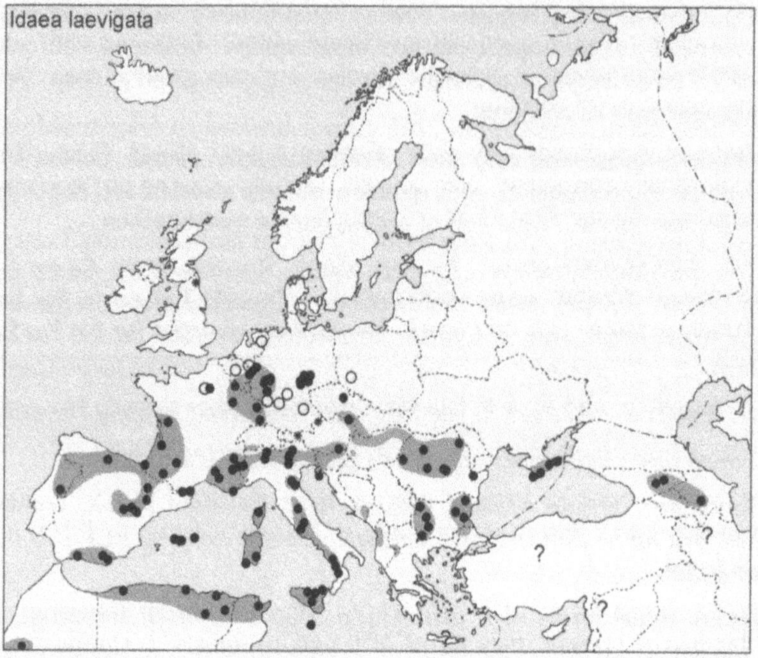

DIAGNOSIS: Wingspan ♂♀ 13-17 mm, reared specimens up to 19 mm. Wings sand coloured with light brown or reddish suffusion, terminal area darker. Variable, mainly in rearings with tendency to melanism and loss of wing pattern, e.g. in North Africa. Ante- and postmedial lines sharp, black brown. Medial line diffuse towards costa, strongly marked on hindwing and towards inner margin of forewing, often anastomosing with antemedial line. Discal spots small. Terminal dots black, distinct. Underside nearly without markings. Frons blackish, vertex sand coloured, collar brown. Antennae of ♂ very shortly and densely ciliate, cilia 0.35-0.6 times width of flagellum. Hindtibia of ♂ not dilated, with pencil. Tarsus 0.5-0.65 times length of tibia (Text-fig. 20).

MALE GENITALIA: Uncus terminally slender, sub-truncate. Gnathos broad, truncate. Juxta anteriorly concave. Valva curved, ventrally with one small spinule at tip. Aedeagus with one stout, terminally furrowed cornutus, 0.65-0.7 times length of aedeagus. Base of cornutus furrowed, laterally widening.

FEMALE GENITALIA: Ductus bursae with a long row of large spines, somewhat dilated at posterior margin. Corpus bursae filled with large spines in the posterior 2/3. Laterally a small membranous sac, posteriorly a conical appendix bursae leading to ductus seminalis.

DISTRIBUTION: Sub-Mediterranean. Widespread but usually rare and local in southern Europe from Spain across Italy to Balkan peninsula and Crimea, and in warmer regions of western and central Europe. Present on Sicily, Malta, Sardinia and Corsica. Reputedly also on Crete (REBEL 1916b; MÜLLER 1996), awaiting confirmation. Isolated record in northern England probably due to artificial transport by man (PROUT 1935a; SKINNER 1984). One old record from south-western Poland (BUSZKO 2000). Abundance decreasing in Thuringia (ERLACHER & FRIEDRICH 1994), extinct in Bavaria (WOLF & HACKER 2004). – Outside Europe from Morocco to Tunisia. Reputedly from Turkey and Syria across Transcaucasus to northern Iran (STAUDINGER 1879; PROUT 1913; RIEMIS 1994; VIIDALEPP 1996; DIDMANIDZE 2002), but no vouchers traced from south-west Asia.

PHENOLOGY: Usually bivoltine: late May to mid-September, with overlapping generations, or with short gap late July to early August. Under bad conditions univoltine early June to early August (BERGMANN 1955; FORSTER & WOHLFAHRT 1981; EBERT 2001). In North Africa trivoltine from early April to late September.

BIOLOGY: Larva polyphagous on dry plants, withered leaves and mouldy plant debris (var. cit.), also moss (SKINNER 1984, VORBRODT 1914). Reputedly on '*Cypressus*' ('*I. couloniata*': PROUT 1913; LHOMME 1935). Probably introduced in England with fibres of coconuts (HARRISON 1927; PROUT 1935a). Reared on withered leaves of *Taraxacum officinale, Salix* (SCHNEIDER 1939; EBERT 2001; SKINNER 1984), *Medicago sativa, Pisum sativum* (RUNGS 1981), on mosses and on petals of *Galium, Chrysanthemum, Gypsophila* (MILLIÈRE 1865; subsequently var. cit., e.g. LHOMME 1935). The larva is more easily found than the adults. The latter repelled by light according to EBERT (2001), but also rarely attracted to light (original data). Dispersal high, not stenotopic, r-strategy (BERGMANN 1955).

HABITAT: Xerothermophilous lowland species. 'Domestic', usually found in houses, gardens, parks, and around piles of compost. Usually from 0 m up to 500 m above sea-level, up to 1,000 m in the Alps, up to 700 m in southern Europe.

SIMILAR SPECIES: Specimens lacking spots of medial line at inner margin may be confused with *I. efflorata*, the latter with straighter postmedial line, forewing with dark brown subapical spot at termen, longer ♂ antennal cilia, only the first tarsomere of ♂ hindleg developed.

38. *Idaea efflorata* ZELLER, 1849

Idaea efflorata ZELLER, 1849: Stettin. ent. Ztg. 10: 218, index (Italy, Tuscany: Artignano; Posignano; Riparbella). Syntypes ♂♂ (BMNH, 1♂, examined). Generic combination indicated in the index.

‡ *Idaea effloraria* HEYDENREICH, 1851: Lepid Eur. cat. meth. (Edn 3): 60. Incorrect subsequent spelling.

Acidalia extarsaria HERRICH-SCHÄFFER, 1852: Syst. Bearb. Schmett. Eur. 6 (55): 67 (Europe). Syntype(s). First available name after *extarsaria* HERRICH-SCHÄFFER, 1848 (pl. 71, fig. 435: non binominal) and HEYDENREICH, 1851 (nomen nudum).

Acidalia eriopodata GRASLIN, 1863: Annls Soc. ent. Fr. (4) 3 (2): 324, pl. 8, fig. 12, 12bis (France, eastern Pyrenees: near Collioure). Syntype(s) ♂ (ZFMK?). Herewith downgraded from subspecific rank (LERAUT 1997; SCOBLE 1999) to synonymy (syn. n.).

Acidalia atromarginata MABILLE, 1869: Annls Soc. ent. Fr. (4) 9 (1): 61, pl. 2, fig. 3 (Corsica: by Bastia). Lectotype ♂ (ZSM/HERBULOT, examined).

Acidalia inesata MILLIÈRE, 1868: Iconogr. Descr. Chen. Lép. inéd. 2 (22): 482, pl. 100, figs 3-5 (Spain, Catalonia: Barcelona region). Syntypes 2♂3♀, probably lost.

Unavailable names (misidentification): *infirmata*: sensu HERRICH-SCHÄFFER (1861) nec *infirmaria* RAMBUR, 1833. – (incorrect subsequent spellings): *atiomarginata*: MABILLE (1869); *extersaria*: KOCAK (1994).

DIAGNOSIS: Wingspan ♂♀ 15-17 mm, in the second generation exceptionally 11 mm only. Wings sand coloured. Antemedial line usually absent except its costal spot. Postmedial line distinct, often dotted. Terminal area with slight light brown suffusion, in the 'f. *eriopodata*' dark brown. The latter form represented in most populations at a frequency of about 10-20%, in the Iberian peninsula commoner. Wavy line diffuse. Forewing apex concolorous with ground colour, with dark brown subapical spot at termen. Hindwings usually dark at base. Frons black brown, vertex sand coloured, collar slightly darker.

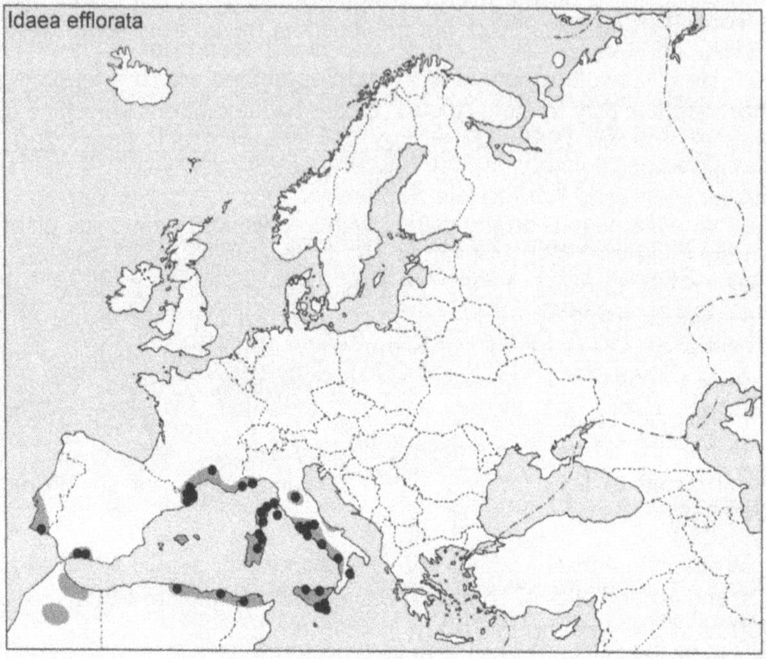

Antennae of ♂ densely ciliate-setose, cilia 0.5-0.7 times width of flagellum. Hindtibia of ♂ not dilated, with long pencil, covering 2/3 of tarsus. First segment of tarsus dilated, flat spade-shaped, 0.65-0.7 times length of tibia, the other tarsomeres vestigial (Text-fig. 21).

MALE GENITALIA: Uncus and gnathos slender. Juxta covered with spinules. Valva terminally tapering to a ventral spinule. Aedeagus with one long and slender cornutus, 0.8 times length of aedeagus, basally truncate. Vesica strongly folded beside basal half of cornutus.

FEMALE GENITALIA: Antrum small, round. Ductus bursae slender, strongly sclerotized. corpus bursae a long tube, longitudinally corrugated, laterally spinulose, anteriorly with membranous sac, posteriorly with large lateral sac completely filled with spinules.

DISTRIBUTION: West-Mediterranean. Spain, Mallorca, Menorca (BONNES & SCHMITZ 1993; HONEY pers. comm.), southern France, Corsica, Sardinia, Sicily, Italy. Occurrence in Portugal (VIVES MORENO 1994; MÜLLER 1996) awaits confirmation, in MNHU one specimen from Portugal, 'Cormora' (=Camara?). – Outside Europe from Morocco to Tunisia. Erroneously mentioned for Turkey (RIEMIS 1994), and Israel (AMSEL 1933).

PHENOLOGY: Uni- or bivoltine: mid-May to mid-July; in southern Italy single specimens from mid-August to late August (partial second generation?).

BIOLOGY: Larval host-plants unknown. Adults, apparently, not optimally attracted to light. Adults beaten from *Buxus* shrubs (CALBERLA 1890).

HABITAT: Xerothermophilous lowland species, often found in buildings. Limited information available about habitat preference. From 0 m up to 1,000 m above sea-level.

SIMILAR SPECIES: *I. laevigata* (see no. 37). Ochreous forms of *I. infirmaria* differ in lacking dark subapical spot of forewing termen, forewing costa dark at base, vertex white, ♂

antennal cilia and hindtarsus shorter. *I. attenuaria*, and *I. incalcarata* with sharp wavy line, wings elongate, costal spot of medial line distinct. Small and pale autumnal forms of *I. efflorata* may be confused with *I. seriata*, the latter usually with ground colour more whitish, subapical spot at forewing termen absent, ♂ antennal cilia much longer.

REMARKS: Lateral, spinulose sac in ♀ genitalia, revealing close relationship to *I. laevigata*.

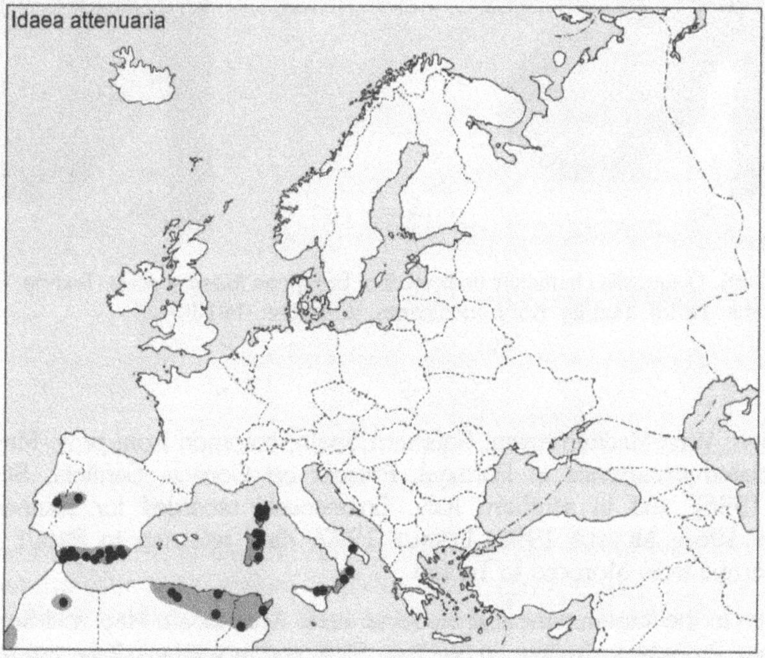

39. Idaea attenuaria (RAMBUR, 1833)

Dosithoea attenuaria RAMBUR, 1833: Annls Soc. ent. Fr. 2 (1): 37, pl. 2, fig. 19 (Corsica). Lectotype ♂ (ZSM/HERBULOT, examined).
Unavailable names (incorrect subsequent spelling): *attenuata*: FUCHS (1904).

DIAGNOSIS: Wingspan ♂♀ first generation 14-17 mm, second generation usually 11-13 mm. Forewings slender, especially in second generation. Apex pointed. Forewings light brown, hindwings sand coloured with grey suffusion. Transverse lines of forewing more or less indistinct, but three blackish costal spots well marked. Diffuse medial line often visible. Terminal area of forewing darker brown with distinct, sharp, straight, whitish line. Forewing apex concolorous with ground colour, obliquely, sharply separated from dark terminal area. Discal spots present. Frons black brown, vertex sand coloured, collar slightly darker. Antennae of ♂ densely ciliate-setose, longest cilia 0.6-0.8 times width of flagellum. Hindtibia of ♂ with one single spur, not dilated. Tarsus not shortened (Text-fig. 22).

MALE GENITALIA: Uncus sub-triangular. Gnathos slender. Valva with subapical spine at costa and a very small subapical spinule ventrally. Aedeagus short, with one very broad, straight cornutus, about 0.75 times length of aedeagus.

FEMALE GENITALIA: Ductus bursae very short and broad. Corpus bursae spinulose, subquadrate, on anterior margin with small membranous process.

Text-figs 85-86: Diagnostic characters (indicated) of European *Idaea* species. Text-fig. 85: *I. attenuaria* (RAMBUR, 1833). Text-fig. 86: *I. incalcarata* (CHRÉTIEN, 1913).

DISTRIBUTION: West-Mediterranean. Southern Spain, common from prov. Murcia to the south. Isolated occurrence in Portugal. Present on Corsica, Sardinia, Sicily, Malta (MÜLLER 1996), and in southern Italy. Erroneously recorded for southern France (HERBULOT 1963; MÜLLER 1996; LERAUT 1997; data referring to PROUT 1935a). – Outside Europe from Morocco to Tunisia.

PHENOLOGY: In the Iberian peninsula bivoltine: early April to late May; mid-September to late October. Probably univoltine on Sardinia, Sicily and in southern Italy: late May to mid-June, exceptionally until early July.

BIOLOGY: Larval host-plants unknown. Adults attracted to light. ♀ ratio at light from about 20% increasing to 80% during flight period (protandry). Resting position with spread wings and hindwing under forewings as in *Eupithecia*.

HABITAT: Xerothermophilous lowland species. Most abundant in hot coastal habitats from 0 m up to 800 m above sea-level.

SIMILAR SPECIES: *I. incalcarata* with broader wings, apex more rounded, terminal area darker, transverse lines better marked, and hindtibial spur of ♂ lacking. *I. efflorata* (see no. 38). *I. deitanaria* darker, base of wing dark, transverse lines of forewing more distinct, ♂ hindtibia with two spurs.

40. *Idaea incalcarata* (CHRÉTIEN, 1913)

Acidalia incalcarata CHRÉTIEN, 1913: Bull. Soc. ent. Fr. 1913 (3): 101 (France, Basses Alpes: Digne). Lectotype ♂ (ZSM/HERBULOT, examined).

‡ *Acidalia disjunctaria* STAUDINGER, 1870: Berl. ent. Z. 14 (1/2): 124 (Spain, Catalonia: Barcelona). Holotype ♂ (MNHU, examined). Junior primary homonym nec WALKER, 1861 (identity: *Cerynea disjunctaria*, Noctuidae, from Borneo).

Unavailable names (misidentification): *attenuaria*: sensu WEISS (1920) nec RAMBUR, 1833.

DIAGNOSIS: Wingspan ♂♀ 14-16 mm; second generation smaller, 11-14 mm. Forewings

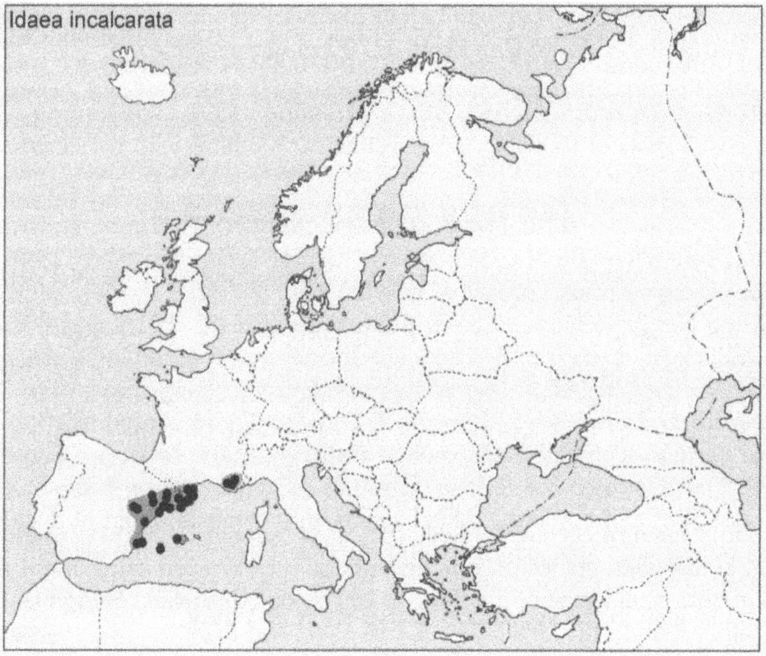

broader than those of *I. attenuaria*. Wings sand coloured, forewings with light brown tinge. Forewing with three distinct, blackish costal spots. Postmedial line often well marked, medial line usually absent. Terminal area of all wings darker brown with sharp, slightly undulate wavy line of ground colour. Discal spots present. Frons black brown, vertex sand coloured, collar slightly darker. Antennae of ♂ densely ciliate-setose, cilia 0.7-0.9 times width of flagellum. Hindtibia of ♂ very slender, without spurs. Tarsus not shortened.

MALE GENITALIA: Uncus sub-triangular, rounded. Gnathos comparatively broad. Valva tapering to a ventral spine at tip. Aedeagus with one long, distally sigmoid cornutus, about 1.0 times length of aedeagus.

FEMALE GENITALIA: Ductus bursae very short and broad, towards corpus bursae with triangular sclerite. Corpus bursae sub-triangular, spinulose posteriorly, anteriorly some large spines and a digitiform membranous process.

DISTRIBUTION: West-Mediterranean, endemic to southern France, north-eastern and eastern Spain, common from prov. Teruel to the north, rare in prov. Alicante. Present on the Balearics: Formentera, Mallorca, and Dragonera (HONEY pers. comm.).

PHENOLOGY: Trivoltine: late April to early June; mid-July to early September; mid-September to mid-October. In the mountains bivoltine with first generation until early July.

BIOLOGY: Larva probably polyphagous on withered leaves of herbaceous plants (var. cit., e.g. PROUT 1913; LHOMME 1935). No detailed information on larval host-plants traced. Adults readily attracted to light. ♀ ratio at light strongly increasing during flight period (protandry) up to about 80%.

HABITAT: Xerothermophilous, but with higher tolerance of cold temperatures than the sister species *I. attenuaria*. From 0 m up to 1,700 m above sea-level.

SIMILAR SPECIES: *I. efflorata* (see no. 38), *I. attenuaria* (39). *I. deitanaria* with base of wing dark, costal spots of forewing more distinct, ♂ hindtibia with two spurs.

The ***typicata*** species-group

STERNECK's (1940) 8th group of *Idaea*. A very 'natural' group (STERNECK 1940: 127), but *I. moniliata* phylogenetically somewhat isolated from the other species. STERNECK's 19th group with the North African species *I. ludovicaria* (CULOT, 1918) and *I. unicalcarata* (PROUT, 1922) clearly belong to the *typicata* species-group according to morphology of male and female genitalia. Common features of the *typicata* species-group: ♂ antennae ciliate-setose. Proboscis present. Sternum A2 of ♂ without central sclerite and setose patch. Male genitalia with patch of spinules or sclerites in the aedeagus. Female genitalia with anterior part membranous, elastic. Genitalia of both sexes reminiscent of *rusticata* species-group: Shape of cornutus, folded vesica, in ♀ genitalia origin of ductus seminalis and corpus bursae with spinose crests and ribbons. Coloration reminiscent of *Scopula marginepunctata* or *Glossotrophia* (PROUT 1913), due to similar resting places on rocks and walls rather than in the vegetation. Larva short and thick.

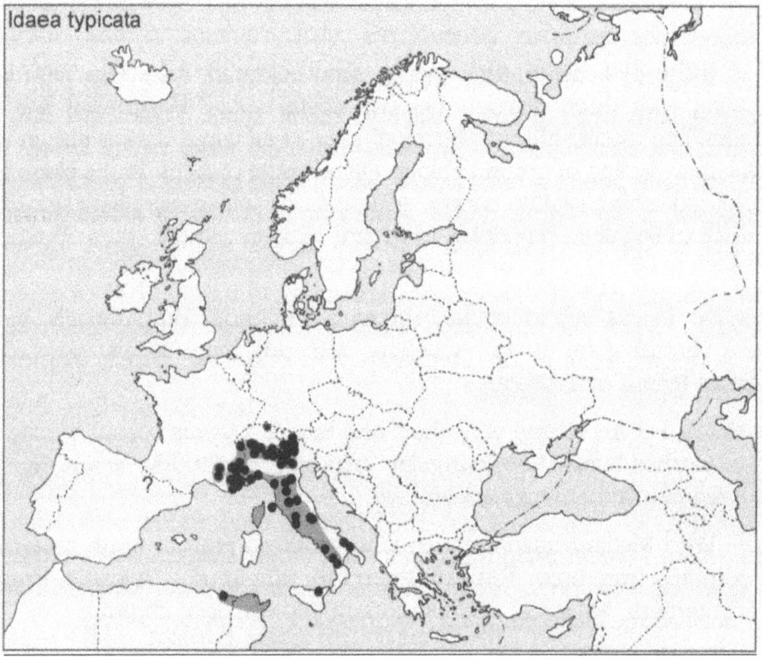

41. *Idaea typicata* (GUENÉE, 1858)

Acidalia typicata GUENÉE, 1858: in BOISDUVAL & GUENÉE, Hist. nat. Insectes (Spec. gén. Lépid.) 9: 464 (France: Basses Alpes). Holotype ♀ (ZFMK: 1♀ ex coll. GUENÉE, examined, without abdomen).

Acidalia asellaria ab. *ruminata* MILLIÈRE, 1885: Annls Soc. ent. Fr. (6) 5: 115, pl. 2, figs. 4, 5 (south-eastern France: Cannes).

Acidalia asellaria var. *hornigaria* STAUDINGER, 1901: in STAUDINGER & REBEL: Cat. Lepid. palaearct.

Faunengeb. (Edn 3) 1: 267 (Italy, South Tyrol: Bolzano). Syntypes ♂♀ (MNHU: 4♂♀, examined; others in BMNH). Valid at subspecific rank.

Acidalia asellaria f. *desertata* DANNEHL, 1933: Ent. Z. Frankf. a.M. 47 (4): 33 (Italy, Lazio: Tivoli). Lectotype ♀ (ZSM, examined). Introduced as 'form' (infrasubspecific), raised to subspecific rank by SCOBLE (1999), herewith downgraded to synonymy (syn. n.), because of the absence of constant differential features from the nominate subspecies.

Unavailable names (misidentification): *asellaria*: sensu STAUDINGER (1871) nec HERRICH-SCHÄFFER, 1847 (identity: *Glossotrophia asellaria*); *dyraria*: sensu TRUSCH (2001) nec ZERNY, 1934 (identity: *Idaea dyraria*, North Africa); *reynaldiata*: sensu PROUT (1913) nec ROUAST, 1876 (identity: *Idaea alyssumata*). – (infrasubspecific): *ciliipunctata*: WEHRLI (1925). – (incorrect subsequent spellings): *assellaria*: HOFMANN (1893); *hornigiaria*: WEHRLI (1925); *horningaria*: VORBRODT (1930); *ciliipunctaria*: VORBRODT (1930).

DIAGNOSIS: Wingspan ♂♀ 16-22 mm (both generations). Wings pale sand coloured. Pattern variable. Transverse lines often diffuse, grey brown, postmedial line usually dentate. Light wavy line zigzagging. Discal spots distinct. Terminal line concolorous with ground colour. Terminal dots elongate, black. Frons black brown, vertex sand coloured, collar brown. Antennae of ♂ densely ciliate-setose, cilia 0.7-0.9 times width of flagellum. Hindtibia of ♂ slender, with two spurs. Tarsus long, 1.2-1.3 times length of tibia.

Subsp. ***hornigaria*** with stronger dark suffusion of wings, transverse lines distinct. Medial line often anastomosing with antemedial line. Wavy line between dark shadows. Terminal dots distinct, sometimes additional fringe dots. External morphology as in nominate subspecies.

MALE GENITALIA: Uncus bottle-shaped. Gnathos often button-like dilated. Valva broad, round at tip, usually with 4-6 spinules. Aedeagus with short cornutus, about 0.15-0.2 times length of aedeagus, sometimes hidden by the large patch of spinules, the latter with basal blisters.

FEMALE GENITALIA: Ductus bursae of medium length, straight. Corpus bursae incised to a lateral sinus with a sub-concentric patch of many very large spines. In the posterior part of the corpus bursae numerous smaller spines, anterior sac of corpus bursae membranous, elastic.

DISTRIBUTION: (South-)European. Nominate subspecies from south-eastern France and northern Italy across Apennine peninsula to southern Italy, with isolated occurrence in northern Sicily. Reputedly also on Corsica (MÜLLER 1996), no vouchers traced. Subsp. *hornigaria* from southern Switzerland to South Tyrolia, single specimens similar to subsp. *hornigaria* occurring also in France (Ht. Alpes, Alpes Maritimes) as individual form. Single records for northern Switzerland (MALICKY 1966), eastern France (Bourgogne: DUTREIX & MOREL 2000), and northern Tyrol (ERLEBACH et al. 1997). Occurrence in easternmost Alps doubtful, examined specimens from Slovenja (CARNELUTTI 1992a) misidentifications of *I. seriata* and *I. camparia* (GOMBOC pers. comm.). Reputedly in the Iberian peninsula, but considered doubtful by VIVES MORENO (1994), dissected males of *I. typicata* from eastern and central Pyrenees (PROUT 1935a; 1♂ in ZSM; REDONDO & GASTON 1999) possibly mislabelled (REDONDO, GASTON pers. comm.). Specimens with genitalia reminiscent of *I. typicata* occur in prov. Burgos, but without ♂ hindtibial spurs (EXPOSITO HERMOSA pers. comm. 2004). – Outside Europe in eastern Algeria and Tunisia. In Morocco replaced by a sister species, *I. dyraria* (ZERNY, 1934).

PHENOLOGY: Uni- or bivoltine: In the lowlands from early June to mid-September, generations sometimes overlapping in the second half of July. On mountains univoltine early June to mid-July.

Text-figs 87-91: Diagnostic characters (indicated) of European *Idaea* species. Text-fig. 87: *I. typicata* (GUENÉE, 1858). Text-fig. 88: *I. alyssumata* (HIMMIGHOFFEN & MILLIÈRE, 1871). Text-fig. 89: *I. urcitana* (AGENJO, 1952). Text-fig. 90: *I. joannisiata* (HOMBERG, 1911). Text-fig. 91: *I. cervantaria* (MILLIÈRE, 1869).

BIOLOGY: Larva probably polyphagous on herbaceous plants (REBEL 1910b), preferring dry leaves (var. cit., e.g. PROUT 1913; LHOMME 1935). Reared on withered leaves of *Taraxacum officinale* (ERLEBACH et al. 1997; TRUSCH 2001), and on petals of various plant species (MÜLLER pers. comm.). Larva inert (PÜNGELER 1892). Adults readily attracted to light and bait, usually resting on rocks during daytime.

HABITAT: Montane. Usually on precipitous rocky slopes, gorges, stony grassland, absent from extended lowland plains. From 0 m up to 1,800 m above sea-level.

SIMILAR SPECIES: *I. alyssumata* on average smaller, usually with warmer brown tinge, without ♂ hindtibial spurs, at the contact zones females have to be dissected. Allopatric *I. dyraria* (ZERNY, 1934) from Morocco and *I. urcitana* with longer ♂ antennal cilia and different genitalia. Species of *consolidata* species-group differ in absence of ♂ hindtibial spurs, less grey suffusion of wings, postmedial line of forewing less angled at radial veins, fringe chequered. *Scopula marginepunctata* usually larger, ♂ without, ♀ with four hindtibial spurs, ♂ antennal cilia longer. *Glossotrophia confinaria* usually larger, with uninterrupted, black terminal line surrounding forewing apex, proboscis extremely long, ♂ with one hindtibial spur, antennal cilia longer.

REMARKS: For morphology of immature stages see PÜNGELER (1892).

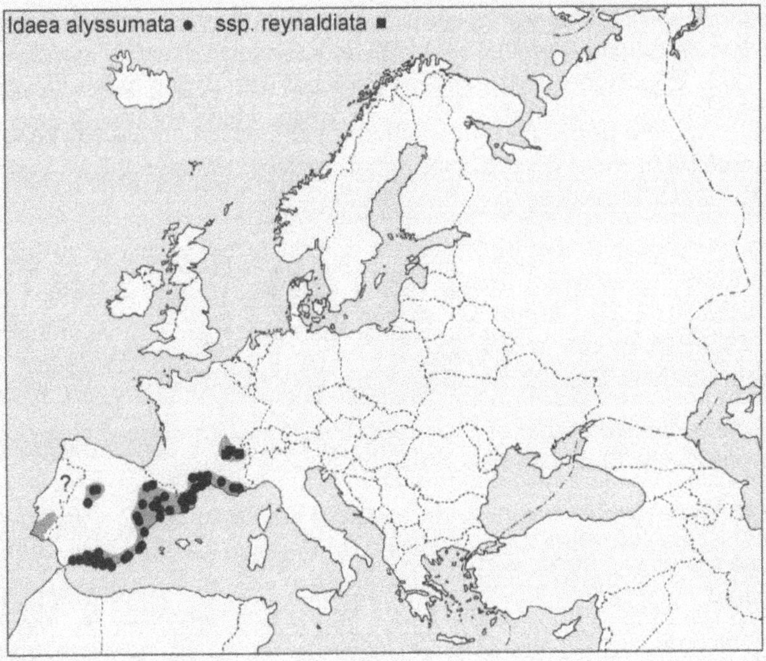

42. *Idaea alyssumata* (HIMMIGHOFFEN & MILLIÈRE, 1871)

Acidalia alyssumata HIMMIGHOFFEN & MILLIÈRE, 1871: in MILLIÈRE: Iconogr. Descr. Chen. Lép. inéd. 3 (28): 197, pl. 121, figs 6-11 (Spain, Catalonia: Barcelona). Syntypes at least 6♂4♀, probably lost.

Ptychopoda alyssata ZERNY, 1927: Dt. ent. Z. Iris 41 (2): 118. Emendation.

Acidalia reynaldiata ROUAST, 1876: Pet. Nouv. Ent. 2: 1 (France: Pierre-Bénite, near Lyon). Lectotype ♂ (NHMW, dissected, with label 'type MILLIÈRE', mentioned in ZERNY 1934) herewith designated to stabilize usage of the name with regard to the sister species; paralectotypes at NHMW 1♂, BMNH; ZSM/HERBULOT 1♂. Validity of subspecific rank (SCOBLE 1999) requires confirmation. Type locality lies outside the actually known distribution area of *I. alyssumata*. Erroneously retained synonym of *Glossotrophia asellaria* by LERAUT (1997).

Psychopoda (sic!) *alyssumata romani* WEHRLI, 1929: Bull. mens. Soc. linn. Lyon 8 (10): 72 (France, by Lyon: St. Didier au Mont d'Or). Lectotype ♂ (ZFMK, examined, previously labelled as 'typ', 11.VII.1927), herewith designated to stabilize usage of the name with regard to closely related, possibly sympatric sister species; at ZFMK 1♂2♀ paralectotypes. Junior synonym of subsp. *reynaldiata*.

Psychopoda (sic!) *alyssumata genilaria* WEHRLI, 1929: Bull. mens. Soc. linn. Lyon 8 (10): 73 (Spain, Andalusia: Sierra Nevada). Syntype(s) (ZFMK: 1♂ with 'type' label, examined). Herewith downgraded from subspecific rank (SCOBLE 1999) to synonymy (syn. n.).

Unavailable names (misidentification): *asellaria*: sensu PROUT (1904) nec HERRICH-SCHÄFFER, 1847 (identity: *Glossotrophia asellaria*). – (infrasubspecific): *anastomosaria*: REISSER (1961). – (incorrect subsequent spellings): *abyssumata*: PROUT (1913: pl. 3f); *allyssumata*: WEHRLI (1929); *reynoldiata*: PROUT (1935: pl. 5i).

DIAGNOSIS: Wingspan ♂♀ 17-21 mm; second generation smaller, 14-17 mm. Wings sand coloured, sometimes with slight orange tinge. Transverse lines marked more or less distinctly, grey brown, postmedial line usually dentate. Discal spots usually present. Black, elongate terminal dots lying in a light terminal line concolorous with ground colour. Frons black brown, vertex sand coloured, collar brown. Antennae of ♂ densely ciliate-setose, cilia 0.7-0.8 times width of flagellum (Text-fig. 47). Hindtibia of ♂ slender, without spurs. Tarsus very long, 1.2-1.5 times length of tibia.

Variability: Populations of central Spain with forewings slender, transverse lines distinct, postmedial line zigzagging, terminal area with dark shadows, possibly meriting a name at subspecific rank. Populations of the eastern Pyrenees with strong dark suffusion, with clinous transitions to the populations of southern Spain, which are usually paler.

Subsp. ***reynaldiata*** more greyish, with rough pattern, transverse lines better marked, possibly with clinous transitions to nominate subspecies.

MALE GENITALIA: Similar to that of *I. typicata*, but uncus bottle-shaped or sub-triangular, gnathos narrower, tip of valva usually with 4 spinules. Length of valva 1.5-1.7 mm. Length of aedeagus 1.25-1.4 mm, i.e. shorter than in *I. typicata*, without larger cornutus, patch of spinules smaller, usually 20-35 spinules. For occasional presence of the larger cornutus in northern Spanish populations see under 'distribution' of *I. typicata*.

FEMALE GENITALIA: Similar to that of *I. typicata*, but lateral, sub-concentric group of spines smaller, corpus bursae here with ribbon-shaped spinose sclerite.

DISTRIBUTION: West-Mediterranean. From southern France to Spain, with isolated occurrences in the mountains north-west of Madrid, in Portugal (CORLEY et al. 2000), and on Menorca (HONEY pers. comm.). Subsp. *reynaldiata* in eastern France near Lyon, in prov. Savoie (GOATER pers. comm.) and prov. Bourgogne (DUTREIX & MOREL 2000). Erroneously recorded for Croatia: ♂ in ZSM/EMEM, leg. MÜCK, mislabelled (similar case under *I. litigiosaria*). Contact zone with *I. typicata* in south-eastern France, e.g. with *I. typicata* on Mt. Ventoux, and *I. alyssumata* in the nearby lowlands.

PHENOLOGY: Bi- or trivoltine: late May to mid-July; late July to mid-September. Single specimens in October perhaps belonging to partial third brood.

BIOLOGY: Larva polyphagous on fresh and withered leaves of herbaceous plants (CHRÉTIEN 1928). *Centaurea aspera* (Asteraceae) suggested as natural host-plant (MILLIÈRE 1871; PROUT 1913). Reared on petals of *Centaurea, Scrophularia, Anthemis* and *Alyssum maritimum* (MILLIÈRE 1871; subsequently var. cit.).

HABITAT: Often collected on rocky slopes, and in stony grassland with abundance of low vegetation and flowers. Usually from 500 m up to 1,800 m above sea-level, but not rarely down to sea-level, in southern Spain up to 2,500 m.

SIMILAR SPECIES: *I. urcitana* extremely similar to small forms of *I. alyssumata* (mainly in the second generation). *I. urcitana* with much longer ♂ antennal cilia, slightly longer ♂ hindtarsus, in ♂ genitalia valva and aedeagus shorter, with 10-12 cornuti only. *I. cervantaria* usually well differing in distinct costal spot of antemedial line, ♂ antennal cilia much longer; correct identification of ♀ sometimes needs dissection. Both allopatric *I. typicata*, and *I. dyraria* (ZERNY, 1934) from Morocco with minor differences in habitus, but with ♂ hindtibial spurs present and constant differential features in genitalia (e.g. one large cornutus, ♀ bursa without medial ribbon-shaped sclerite). Central Spanish populations of *I. alyssumata* much resembling *I. contiguaria*, the latter with medial line of hindwing proximally from discal spot rather than distally, ♂ with longer antennal cilia and shorter hindtarsus. *I. joannisiata* differs in much more glossy wings, pattern more diffuse, ♂ hindtarsus shortened. *Glossotrophia asellaria* with uninterrupted, black terminal line surrounding forewing apex, ♂ hindtibia with one spur.

REMARKS: Due to the great similarity to many other species *I. alyssumata* has very often been misidentified. Therefore, collections and literature are full of errors and many publications need careful verification. For morphology of immature stages see CHRÉTIEN (1928).

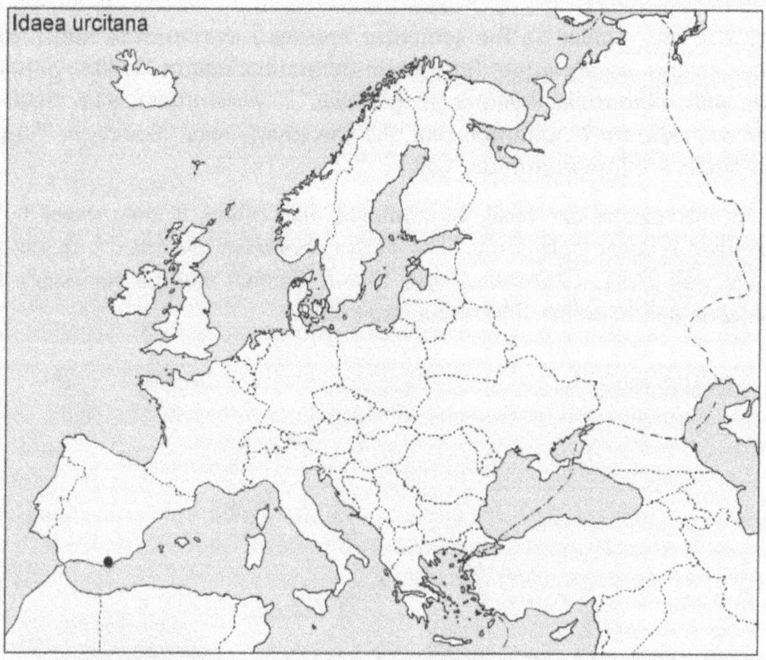

43. *Idaea urcitana* (AGENJO, 1952)

Sterrha urcitana AGENJO, 1952: Fáunula Lepid. Almeriense: 183, pl. 6, fig. 22 (Spain, Andalusia: Almeria). Holotype ♂ (MNCN, examined).

DIAGNOSIS: Wingspan ♂ 15-17 mm, thus at least in first generation smaller than *I. alyssumata*. Very similar to *I. alyssumata* in colour and pattern of wings, ground colour usually paler and forewing costal spots well marked. Discal spots well contrasted from ground colour. Frons black brown, vertex sand coloured, collar brown. Antennae of ♂ densely ciliate-setose, cilia 1.2-1.5 times width of flagellum. Hindtibia of ♂ slender, without spurs. Tarsus very long, 1.3-1.8 times length of tibia.

MALE GENITALIA (n=3): Uncus digitiform. Gnathos of medium width, rounded. Valva comparatively short (1.25-1.5 mm) round at tip, with one to three comparatively long spinules. Aedeagus very short (1.0-1.15 mm), with 10-13 spinules only, some of them very small, cornutus absent.

FEMALE GENITALIA (n=1): Ductus bursae comparatively short. Corpus bursae laterally concave, with sub-concentric group of many very large spines. In posterior part of corpus bursae numerous smaller spines, anterior sac of corpus bursae membranous.

DISTRIBUTION: Endemic to south-eastern Spain, prov. Almeria. Rare. Considering the restricted area of distribution of this (and many other species) its habitats merit particular and severe protection at national and European levels.

PHENOLOGY: Limited data available from late May to mid-June; early September.

BIOLOGY: Immature stages unknown. Adults collected at light.

HABITAT: Xerothermophilous, on hot and dry coastal lowlands. From 0 m up to 200 m above sea-level.

SIMILAR SPECIES: Very similar to the sympatric species *I. cervantaria* and *I. alyssumata*. Both the last species usually larger (important differential feature for first generation), tarsus shorter, with differential features in genitalia. *I. cervantaria* with much longer, *I. alyssumata* with shorter ♂ antennal cilia. All specimens need dissection. Further similar species see under *I. alyssumata* (no. 42).

REMARKS: As this species can easily be confused with others, it may reveal to be present but misidentified in some collections. The author, however, did trace it in just four collections (MCZN, coll. PÖLL, GRÜNEWALD and SKOU). Further study is necessary to clear up distribution area and to refine differential analysis.

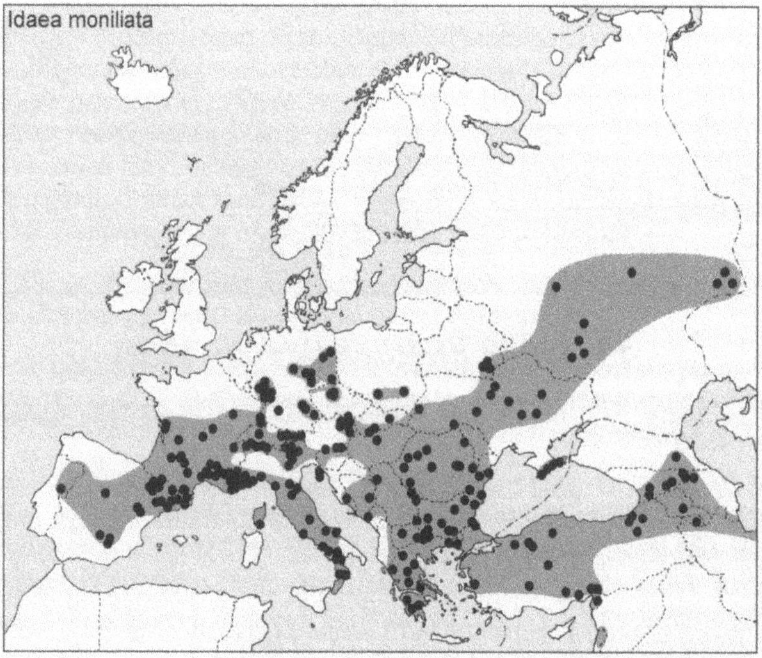

44. *Idaea moniliata* (DENIS & SCHIFFERMÜLLER, 1775)

Geometra moniliata DENIS & SCHIFFERMÜLLER, 1775: Ank. syst. Werkes Schmett. Wienergegend: 117 (Austria: Vienna distr.). Syntype(s) lost.

‡? *Phalaena Geometra pentalineata* VILLERS, 1789: Linn. ent. 2: 383 (southern France). Syntype(s). Synonymy (KOCAK 1983) to be confirmed.

Phalaena omicata FABRICIUS, 1794: Ent. Syst. 3 (2): 199 (Italy). Syntype(s) lost.

‡ *Geometra moniliaria* HÜBNER, 1799: Samml. Eur. Schmett. 5, Geometrae (1): pl. 12, fig. 59. Incorrect subsequent spelling.

Phalaena (Geometra) pluripunctata SCHARFENBERG, 1805: Vollst. naturgesch. d. schädl. Forstinsekten 3: 726.

Unavailable names (incorrect subsequent spellings): *monileata*: WALKER (1863); *moliniata*: RAMANN (1876); *pentalienata*: SCOBLE (1999).

DIAGNOSIS: Wingspan ♂♀ 16-20 mm, in southern Europe slightly smaller. Wings light brown, transverse lines dark brown, sharp and fine. Discal spots elongate. Terminal area with 6-7 whitish spots on all wings. Fringe chequered. Frons dark brown, vertex sand coloured, collar slightly darker. Antennae of ♂ densely ciliate-setose, cilia 0.9-1.05 times width of flagellum, in southern Europe slightly shorter, 0.7-0.9 times. Hindtibia of ♂ slender, with two spurs. Tarsus not shortened.

MALE GENITALIA: Uncus and gnathos sub-triangular. Saccus a short round projection. Valva broad in the basal half, costa sigmoid. Aedeagus broad, with large patch of scale-shaped micro-sclerites, probably homologous to spinules of preceding species. Vesica terminally sclerotized to a small spine.

FEMALE GENITALIA: Lamellae ante- and postvaginalis present as small round and semicircular sclerites. Ductus bursae very short. Antrum bilobed, bent orad. Corpus bursae oval, with two large, oval, strongly spinose patches. Corpus bursae posteriorly with comparatively large appendix bursae, anteriorly membranous and elastic as in the preceding species.

DISTRIBUTION: Eurasiatic. Southern and central Europe from central Spain and France to southern Urals, in the north to eastern Germany, Poland, southern Byelorussia and Moscow district. Absent from all the Mediterranean islands, except Corsica. In most parts of the distribution area rare and local, in some regions (e.g. southern Germany: EBERT 2001; WOLF & HACKER 2004) strongly threatened. – Outside Europe from Turkey and Lebanon (WILTSHIRE 1939a) across Caucasus and Transcaucasus, northern Iran to Turkmeniya and central Asian mountains.

PHENOLOGY: Univoltine: late June to mid-August. Under good conditions, e.g. in southern Europe emerging from late May, sometimes with a partial second brood in September.

BIOLOGY: Larva probably polyphagous on withered leaves of herbaceous plants. Recorded on *Vicia* (Fabaceae), *Viola* (Violaceae), *Myosotis* (Boraginaceae), *Taraxacum officinale, Leontodon* (Asteraceae) (var. cit.), and *Borago* (Boraginaceae) (LHOMME 1935), probably referring to rearings. Reared on petals of various plant species, and on young leaves of *Taraxacum officinale* (ROUGEMONT 1903; MÜLLER pers. comm.). Adults readily attracted to light and bait.

HABITAT: Xerothermophilous. At extremely xerothermic sites, preferring contact zones between dry grassland and hedges, e.g. of *Prunus spinosa* (GELBRECHT et al. 1995), dry open habitats, such as steppes, heathland, fallows, hot rocky slopes, scattered, sun-exposed scrubby places. Mainly on sandy soils, but also on limestone. In the north from 0 m up to 600 m above sea-level, up to 1,500 m in southern Alps and southern Europe, up to 2,200 m in western and central Asia.

SIMILAR SPECIES: No similar species in Europe.

The *spissilimbaria* species-group

The species was unknown to STERNECK (1940). It had been placed, tentatively, in the *typicata* species-group by LERAUT (1980; 1997) and MÜLLER (1996). Phylogenetic relationships and position within genus remain unclear due to a number of unique characters in genitalia of both sexes. Revision: DUFAY (1983).

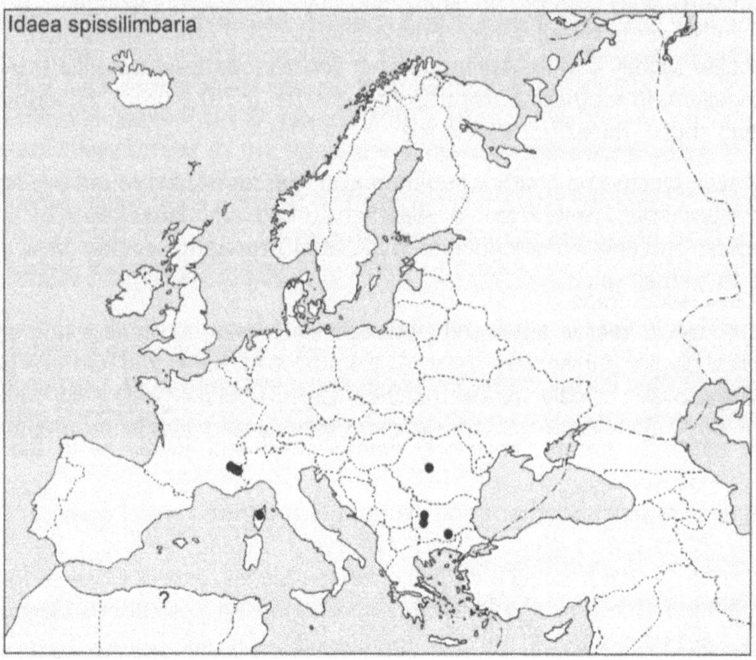

45. *Idaea spissilimbaria* (MABILLE, 1888)

Acidalia spissilimbaria MABILLE, 1888: Bull. Soc. ent. Fr. 1888: 59 (Algeria). Holotype ♂ (ZSM/HERBULOT, examined). Type locality doubtful according to HERBULOT (1968a).

DIAGNOSIS: Wingspan ♂♀ 15-18 mm. Wings glossy, dirty sand coloured. Base of wings darker. Medial line present, somewhat diffuse. Postmedial line undulate. Terminal area broad, dark brown, lighter towards termen. Discal spots present, small. Terminal dots nearly absent. Palpi and frons dark brown, vertex sand coloured. Palpi comparatively long, 1.2-1.3 times diameter of eye. Antennae of ♂ densely ciliate-setose, cilia 1.1-1.2 times width of flagellum. Hindtibia of ♂ without spurs. Tarsus about 1.0 times length of tibia.

MALE GENITALIA: Uncus stout, rounded at tip. Saccus V-shaped. Valva with broad, more strongly sclerotized costa to 2/3 of length, at tip long, slender, digitiform, membranous. Aedeagus with U-shaped cornutus.

FEMALE GENITALIA: Ductus bursae broad. Corpus bursae elongate oval, with oval central patch of spinules. Laterally towards ductus bursae with a long row of larger spines.

DISTRIBUTION: (South-)European. Very rare and local in southern France (Ht. Alpes: DUFAY 1983), Corsica (GRANVILLE 1968; HERBULOT 1968a), Romania (Cibin Mts.: STANESCU & HAUSMANN 2002) and Bulgaria (Sofia and Rodope Mts., GANEV 1983; BESHKOV pers. comm.). – Outside Europe reputedly in Algeria (type locality), doubtful according to HERBULOT (1968a). So far, nine specimens only are known!

PHENOLOGY: Univoltine. Limited data available from mid-July to mid-August.

BIOLOGY: Larval host-plants unknown. Adults collected at light.

HABITAT: Montane. Limited information available about habitat preference. In Bulgaria collected in a village with gardens, grassland, and surrounding mixed *Carpinus* forest on calcareous rocks (BESHKOV pers. comm.). From 900 m up to 1,300 m above sea-level.

SIMILAR SPECIES: *I. subsaturata* smaller, with postmedial line straighter, terminal area with paler shadows, mainly on hindwing, ♂ antennal cilia longer.

The ***circuitaria*** species-group

STERNECK's (1940) 9th group of *Idaea*. 'Very isolated' (STERNECK 1940: 127). Sister group relationship to *incisaria* species-group supported by ♂ genitalia with posterior hooks of juxta, cornutus on a basal sclerite; ♀ genitalia with lateral sclerites of ductus bursae, spinose crests between ductus and corpus bursae. Elastic anterior part of ♀ corpus bursae reminiscent of *typicata* species-group. Without central sclerite and setose patch on ♂ sternum A2. Proboscis present. Antennae of ♂ ciliate-setose. Egg sculpturation shallow, polygonal reticulate. Larvae slender. Revision: HAUSMANN (1994a).

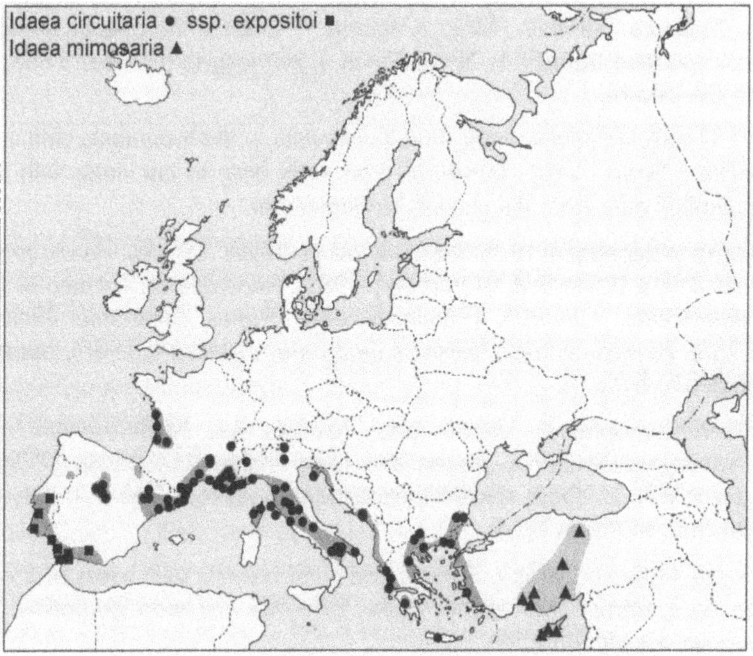

46. *Idaea circuitaria* (HÜBNER, 1819)

Geometra circuitaria HÜBNER, 1819: Samml. Eur. Schmett. 5, Geometrae (1): pl. 97, fig. 499 (Italy, Abruzzi: Mt. Sirente). Syntype(s) of HÜBNER lost. Neotype ♂ (ZSM, examined: 'Mt. Sirente, 5.VIII.[19]26, leg. DANNEHL), herewith designated to fix type locality and to stabilize usage of the name with regard to the sister species *I. rainerii*.

Acidalia circuitata BELLIER, 1860: Annls Soc. ent. Fr. (3) 8 (4): 701. Emendation, referring to *I. circuitaria* sensu HÜBNER, 1919, but partim also to material from Sicily (*I. rainerii*).

Acidalia circuitaria var. *chimaeraria* MILLIÈRE, 1869: Iconogr. Descr. Chen. Lép. inéd. 3 (24): 47, pl. 105, figs. 23-27 (France: Corsica; Lozère; Bordeaux region). Syntypes, probably lost.

Idaea circuitaria expositoi HAUSMANN, 1994: Zeitschr. Arb.Gem. österr. Ent. 46 (1/2): 48, figs 2,7,13 (Spain, Andalusia: Chiclana). Holotype ♂ (ZSM, examined). Valid at subspecific rank.

Unavailable names (infrasubspecific): *expandata*: DANNEHL (1933).

DIAGNOSIS: Wingspan ♂♀ 13-19 mm. Wings light sand coloured. Ante- and postmedial lines sharp, dark brown. Medial line broad, but diffuse. Basal area of forewing light brown. Terminal area of all wings light brown, with sand coloured line before termen. Fringe light brown. Discal spots absent. Frons light brown, vertex and collar sand coloured. Antennae of ♂ densely ciliate-setose, cilia 0.6-0.7 times width of flagellum (Text-fig. 45). Hindtibia of ♂ slender, rough scaled. Tarsus 0.65-0.7 times length of tibia.

MALE GENITALIA: Uncus slender. Saccus strongly developed. Juxta with paired stout posterior hooks. Valva spatulate. Aedeagus with one very short cornutus on a circular sclerite. Cornutus shorter in subsp. *expositoi*.

FEMALE GENITALIA: Ductus bursae short, laterally with paired semicircular sclerites. Corpus bursae pear-shaped, laterally sclerotized in the posterior part. Paired spinose crests towards ductus bursae small.

DISTRIBUTION: (South-)European. Isolated populations in southern Europe from Portugal and central Spain (subsp. *expositoi*) to Greece, usually rare. Also in western France (prov. Vendée) and on Corsica. Replaced by the allopatric sister species *I. rainerii* in south-eastern Spain, Mallorca, Sardinia, Sicily, Calabria. – Outside Europe in western Turkey (Izmir). Sister species *I. rainerii* in North Africa, *I. mimosaria* (GUENÉE, 1858) in Turkey, Cyprus and the Levant.

PHENOLOGY: Usually univoltine: early June to late July, in the mountains until mid-August. In south-western Spain (subsp. *expositoi*) from early May to late June, with partial second brood, limited data from late August to September.

BIOLOGY: Larva polyphagous on herbaceous plants (REBEL 1910b), clearly preferring dry leaves (PROUT 1913; FORSTER & WOHLFAHRT 1981). Recorded on *Astragalus* (Fabaceae), *Osyris* (Santalaceae), *Clematis* (Ranunculaceae), *Rubus* (Rosaceae) (REBEL 1910b; LHOMME 1935). Reared on fresh leaves of *Dorycnium* (LHOMME 1935), rearing difficult. Adults attracted to light.

HABITAT: Xerothermophilous. Various open habitats, e.g. Mediterranean maquis, dry grassland, sun-exposed scrubby places, and rocky slopes (HAUSMANN 1994a). Usually from 0 m up to 800 m above sea-level, exceptionally up to 1,200 m in the Apennines, up to 1,500 m in southern Spain.

SIMILAR SPECIES: Both allopatric *I. rainerii* and *I. mimosaria* with areas of ground colour more extended, *I. rainerii* with slightly darker markings and more contrasted. Dissection allows unambiguous identification.

47. *Idaea rainerii* HAUSMANN, 1994

Idaea rainerii HAUSMANN, 1994: Zeitschr. Arb.Gem. österr. Ent. 46 (1/2): 50, fig. 3,4,8,11,12 (Sicily: Palermo). Holotype ♂ (ZSM, examined).

Unavailable names (misidentification): *circuitaria*: sensu OBERTHÜR (1923) nec HÜBNER, 1819; *circuitata*: sensu BELLIER (1860; partim) nec *circuitaria* HÜBNER, 1819; *fatimata*: sensu SAMMUT (2000) nec STAUDINGER, 1895 (identity: *Idaea manicaria*); *mimosaria*: sensu TURATI (1913) nec HERRICH-SCHÄFFER, 1852.

DIAGNOSIS: Wingspan ♂♀ 15-18 mm. Wing colour and pattern very similar to that of *I. circuitaria*, but markings slightly darker and more contrasted. Light areas more extended, especially in ♀. Antennae of ♂ densely ciliate-setose, cilia 0.6-0.7 times width of flagellum. Hindtibia of ♂ slender, rough scaled. Tarsus long, 0.75-0.85 times length of tibia.

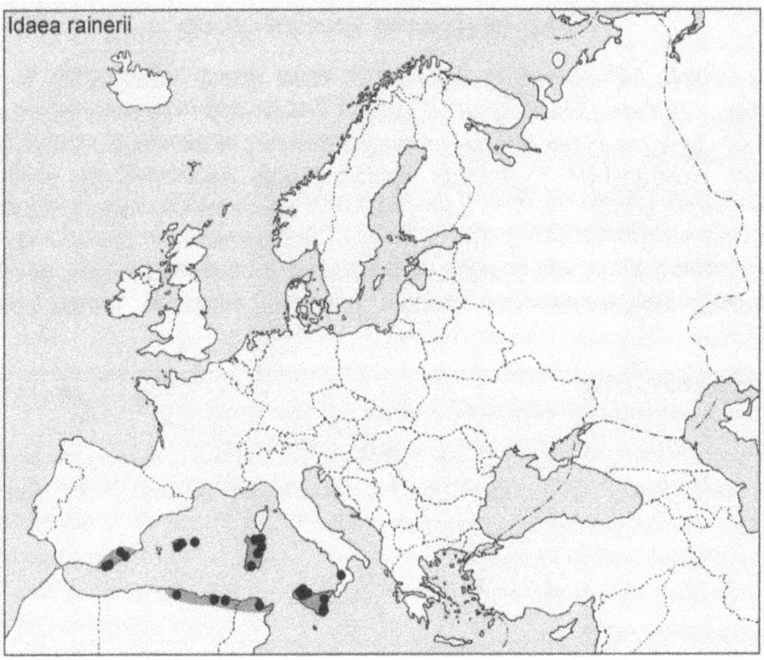

MALE GENITALIA: Similar to those of *I. circuitaria*, but posterior hooks of juxta much longer, valva broader and curved, aedeagus with one stout cornutus, 0.3-0.35 times length of aedeagus.

FEMALE GENITALIA: Similar to those of *I. circuitaria*, but laterally sclerites of ductus bursae circular, corpus bursae elongate oval, without lateral sclerotization, paired spinose crests towards ductus bursae large.

DISTRIBUTION: West-Mediterranean (Circum-Tyrrhenian). South-eastern Spain (prov. Almeria), Balearics (Mallorca, Menorca; dissected: HONEY pers. comm.), Sardinia, Sicily, southern Italy (Calabria: PARENZAN et al. 1998b). – Outside Europe in northern Algeria and northern Tunisia. Replaced by two allopatric sister species in large parts of southern Europe and south-western Asia (see under distribution of *I. circuitaria*). Identity of the record for Libya, Cyrenaica (TURATI 1924) remains unclear.

PHENOLOGY: Uni- or bivoltine: late May to mid-July. Partial second brood from late July to late August indicated by literature data (Menorca: DANTART et al. 1993; Sardinia: HARTIG 1951; Sicily: GRILLO & PARENZAN 1995).

BIOLOGY: Larval host-plants unknown. Adults attracted to light.

HABITAT: In southern Italy collected at sea-level in hot wetland, possibly halophilous. In other regions known from various xerothermic Mediterranean habitats. From 0 m up to 300 m above sea-level, exceptionally up to 800 m (GRILLO & PARENZAN 1995).

SIMILAR SPECIES: *I. circuitaria* (see no. 46).

The *incisaria* species-group

STERNECK's (1940) 12th group of *Idaea*. For sister group relationship to *circuitaria* species-group, see there. Elastic anterior part of ♀ corpus bursae reminiscent of *typicata* species-group. Spinose crests and membranous anterior appendix of corpus bursae may also indicate relationships to *seriata* species-group, supported by similar habitus. Horseshoe-shaped sclerite on ♂ sternum A2 with setose patch distinct, another feature shared with the *seriata* species-group. *I. tineata* transferred to *infirmaria* species-group (HAUSMANN 1994b). Proboscis present. Antennae of ♂ ciliate-fasciculate, or ciliate-setose (*I. calunetaria*). Egg sculpturation shallow, polygonal reticulate. Larvae comparatively long.

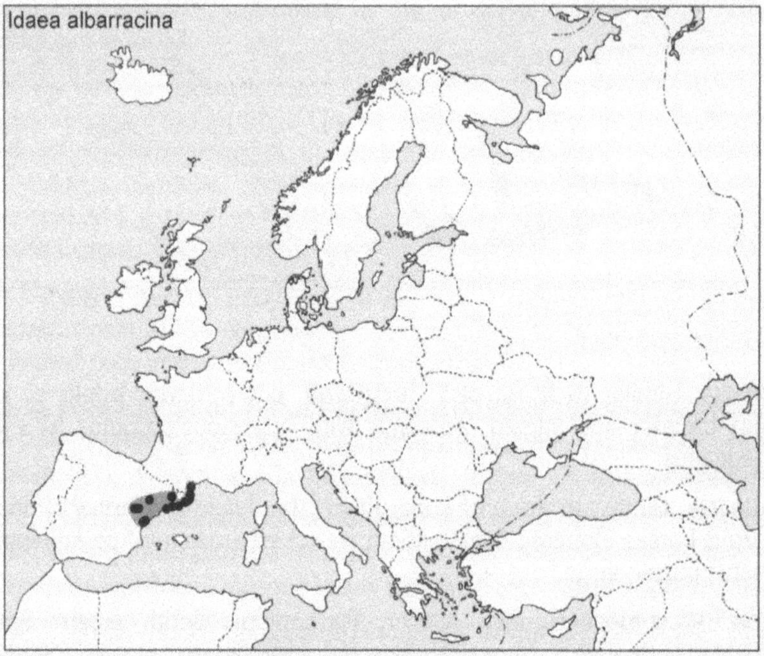

48. *Idaea albarracina* (REISSER, 1933), stat. n.

Ptychopoda incisaria albarracina REISSER, 1933: Eos 9: 250 (Spain, Aragon: Albarracin). Syntypes (SMNK, 1 syntype examined by R. TRUSCH; others at NHMW). Herewith raised from subspecies of *I. incisaria* (e.g. SCOBLE 1999) to species rank (stat. n.).

DIAGNOSIS: On average smaller than the following species, wingspan ♂♀ 14-21 mm; second generation 14-17 mm. Very similar to *I. incisaria* (see below), but wings with stronger grey suffusion, ground colour without yellowish tinge, terminal area slightly darker, antemedial line diffuse or absent, well marked at costa only. External structure without constant differences from *I. incisaria*. Hindtibia and hindtarsus see Text-fig. 23.

MALE GENITALIA: Uncus sub-triangular, Saccus strongly developed. Juxta with digitiform, paired posterior process. Valva long and slender, curved and tapering to an elongate tip. Aedeagus short and broad, with cornutus strongly curved (U-shaped), arising from basal sclerite. At tip of aedeagus a second stout cornutus, triangular, with semicircular base.

FEMALE GENITALIA: Ductus bursae broad, at antrum bilobed, hammer-shaped. Ductus bur-

Text-figs 92-94: Diagnostic characters (indicated) of European *Idaea* species. Text-fig. 92: *I. albarracina* (REISSER, 1933). Text-fig. 93: *I. incisaria* (STAUDINGER, 1892). Text-fig. 94: *I. calunetaria* (STAUDINGER, 1859).

sae at junction with corpus bursae more strongly sclerotized, with two distinct spinulose crests. Corpus bursae membranous, slender, curved.

DISTRIBUTION: Endemic to eastern Spain and eastern French Pyrenees (ZSM; DUFAY 1981; MAZEL 1995). In central and southern Spain, Portugal, Algeria and Tunisia replaced by the sister species *I. incisaria*.

PHENOLOGY: Bivoltine: early May to early July; early August to early October. First generation in the mountains later, until mid-July, in the lowlands sometimes emerging from early April (MAZEL & PESLIER 1997). Rearing requires about two months in summer (HALL pers. comm.).

BIOLOGY: Larval host-plants unknown. Adults readily attracted to light.

HABITAT: No detailed data available for habitat preference. From 0 m up to 1,000 m above sea-level.

SIMILAR SPECIES: Sister species *I. incisaria* usually differs in slight sand coloured tinge, terminal area darker, antemedial line not rarely present. Genitalia allow unambiguous identification. *I. seriata*, and *I. minuscularia* with costal spot of medial line less distinct, terminal area less darkened, more confluent, collar dark instead of white, hindwing margin less concave between medial veins, ♂ hindtarsus longer. *I. calunetaria, I. mareotica* see under *I. incisaria*.

REMARKS: Taxonomy and distribution discussed by REISSER (1933a; 1934) and LAEVER (1960).

49. *Idaea incisaria* (STAUDINGER, 1892)

Acidalia incisaria STAUDINGER, 1892: Dt. ent. Z. Iris 5 (1): 147, pl. 2, fig. 6 (Portugal: Algarve). Syntypes 1♂2♀ (MNHU, examined).

Sterrha incisaria incisarioides WEHRLI, 1934: Ent. Z. Frankf. a.M. 47 (19): 156, 160 figs. 9, 10, 14 (Algeria,

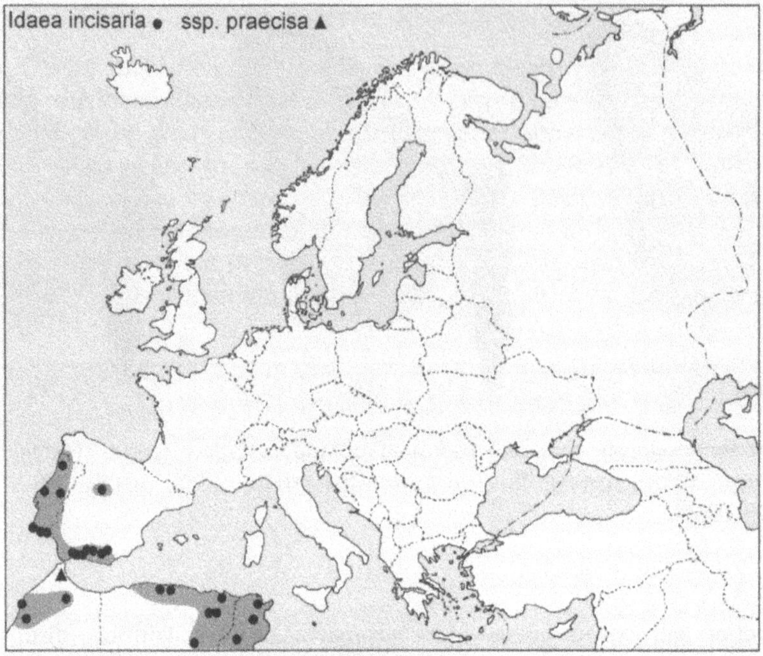

Algier: Hammam Righa). Syntypes 17♂♀ (ZFMK, examined). In Scoble (1999) as subspecies to '*I. maxima*', which is recognized here as synonym of *I. calunetaria*. The question awaits dissection of the types.

Ptychopoda incisaria pulverulenta Reisser, 1933: Eos 9: 251 (eastern Algeria: Philippeville). Syntypes 3 (SMNK, examined by R. Trusch). Valid at subspecific rank.

Ptychopoda incisaria praecisa Reisser, 1933: Eos 9: 248 (Morocco, Rif-Mts.: by Xauen). Lectotype ♂ (ZSM, 'VI.31. leg. Duerck'), herewith designated to stabilize usage of the name with regard to the similar, possibly sympatric sister species; about 90 paralectotypes from type locality, many examined at ZSM, SMNK and NHMV. Valid at subspecific rank, possibly a separate species.

Unavailable names (infrasubspecific): *centropunctata*: Andreas (1928). – (incorrect subsequent spelling): *incisariae*: Wehrli (1934).

Diagnosis: Wingspan ♂♀ 17-21 mm. Hindwing termen concave between veins. Wings dirty white, often with slight sand coloured tinge. Terminal area darker. Postmedial line dark grey, sharp, dentate. Antemedial line comparatively well marked. Discal spots present. Frons brown, vertex whitish to sand coloured, collar slightly darker. Antennae of ♂ ciliate-fasciculate, cilia 1.0-1.2 times width of flagellum. Hindtibia of ♂ dilated, with large pencil slightly shorter than tibia. Tibia slightly covering first segment of tarsus. Tarsus very short, 0.20-0.25 times length of tibia.

Subsp. ***praecisa*** smaller, wingspan ♂♀ 14-19 mm, wings clear white, markings sharp, also the lunular spots in terminal area. Antemedial and medial line usually present. Frons dark brown, vertex white. External structure corresponding to that of *I. incisaria*, but ♂ antennal setae up to 1.5 times width of flagellum.

Male genitalia: Uncus sub-triangular, Saccus strongly developed. Juxta with digitiform, paired posterior processes. Valva long and slender, curved and tapering to a short tip. Aedeagus slightly longer than in *I. albarracina*, cornutus in nominate subspecies usually comparatively short, only slightly curved, basal sclerite small. Second cornutus at tip of aedeagus small, often sub-triangular. Subsp. *pulverulenta* with cornutus much longer and more curved, second cornutus usually vestigial. Subsp. *praecisa* with aedeagus narrower, cornutus stout, basally not dilated. Short spine at tip of aedeagus triangular.

FEMALE GENITALIA: Ductus bursae broad, at antrum with paired round lateral lobes. Corpus bursae with lateral patch of cornuti. Subsp. *praecisa* without constant differences.

DISTRIBUTION: West-Mediterranean. Central and southern Spain and Portugal. In eastern Spain replaced by the sister species *I. albarracina*. – Outside Europe from Morocco to Tunisia (subsp. *pulverulenta*), and western Libya (TURATI 1930a). Subsp. *praecisa* restricted to Rif mountains, Morocco.

PHENOLOGY: Bi- or trivoltine: mid-May to late June; single specimens in July; early September to early October. In the south emerging from late April and flying until early November (CORLEY et al. 2000). One record from Tunisia late February.

BIOLOGY: Larva polyphagous on herbaceous plants, accepting both fresh and dry leaves (PROUT 1913). Reared on withered petals of *Polygonum aviculare, Rumex, Atriplex halimus, Rubia peregrina* and *Galium* (KING & ROMERA 2004).

HABITAT: No detailed data available for habitat preference. From 0 m up to 1,000 m above sea-level.

SIMILAR SPECIES: Sister species *I. albarracina* (see no. 48). *I. seriata*, and *I. minuscularia* see under *I. albarracina*. *I. calunetaria* with postmedial line strongly projecting towards discal spot, ♂ antennal cilia shorter. *I. mareotica* from North Africa may easily be confused with *I. incisaria*, but the first differs in the hindwing termen, which is not concave between veins, forewing narrower, postmedial lines straighter, ♂ genitalia with straight cornutus and straight tapering valva, ♀ genitalia with large, spinose corpus bursae. Allopatric *I. vesubiata* with round hindwing margin, transverse lines broader, discal spots larger, ♂ antennal cilia shorter, ♂ hindtarsus longer. *I. ibizaria* with forewings elongate, ground colour deeper sand coloured, antemedial line with distinct costal spot, ♂ hindtarsus slightly longer.

REMARKS: The taxon *praecisa* may also be considered valid at species rank, further study is required to clear up this question. From southern Spain, prov. Ronda, some specimens have been examined, which resemble '*praecisa*' in habitus. Taxonomic discussion in REISSER (1933a; 1934).

Idaea textaria (LEDERER, 1861)

Acidalia textaria LEDERER, 1861: Wien. ent. Monatschr. 5 (5): 153, pl. 2, fig. 3 (southern Turkey: Antiochia). Holotype ♀.

DIAGNOSIS: Wingspan ♂♀ 16-21 mm. Forewing termen slightly concave subapically. Hindwing termen arched between veins. Ground colour clear white. Medial and antemedial lines indistinct. Postmedial line dentate, fine. Medial and postmedial lines with costal spots. Terminal area grey, with broad white wavy line. Forewing apex white. Discal spots and elongate terminal streaks present. Frons black brown, vertex and collar white. Antennae of ♂ strongly dentate, ciliate-fasciculate, cilia 1.0-1.2 times width of flagellum. Hindtibia of ♂ dilated, with pencil. Tibia slightly covering first segment of tarsus. Tarsus about 0.3 times length of tibia.

MALE GENITALIA (gen.fig. 49d): Uncus short, slender. Saccus round. Valva long and slender, tapering at tip. Aedeagus short and broad, with one very large, stout cornutus, 0.9 times length of aedeagus.

FEMALE GENITALIA (gen.fig. 49b): Ductus bursae slightly curved, comparatively long. Corpus bursae membranous, with scattered spines anteriorly and laterally.

DISTRIBUTION: East-Mediterranean. Possibly distributed on some eastern Greek islands, as it has been recorded on the Turkish coast just a few kilometres from Samos. – Outside Europe in south-western Turkey, Cyprus, and Levant. In Turkmeniya replaced by an allopatric sister species (*I. laszloi* HAUSMANN, 1999). Georgian (DIDMANIDZE 2002) and Armenian (WARDIKJAN 1985) populations perhaps belonging to *I. laszloi*, too (HAUSMANN & LASZLO 1999).

PHENOLOGY: Bivoltine: late April to mid-June; mid-September to mid-October.

HABITAT: No detailed data available for habitat preference. Usually from 0 m up to 900 m above sea-level, exceptionally up to 1,600 m (DOGANLAR 2003).

SIMILAR SPECIES: Ground colour clearer white than in all similar species, pattern sharp and fine, collar white.

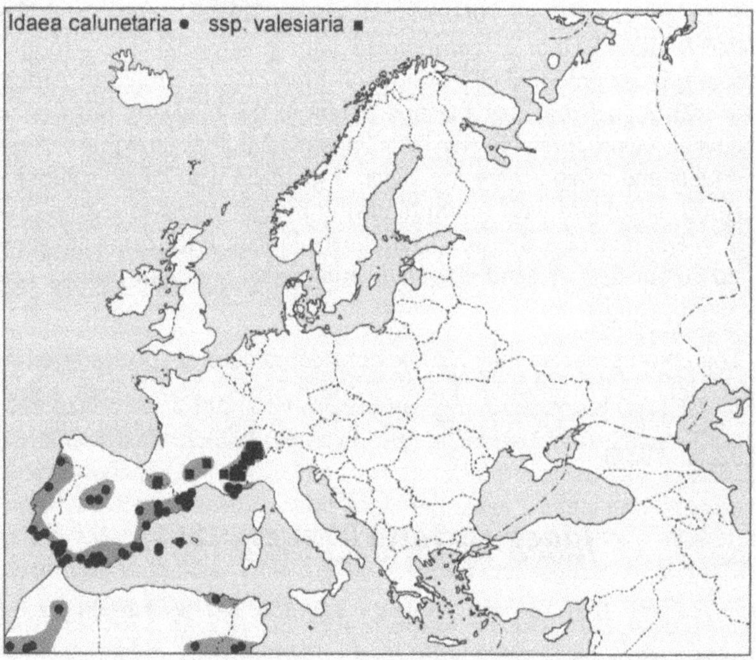

50. ***Idaea calunetaria*** (STAUDINGER, 1859)

Acidalia calunetaria STAUDINGER, 1859: Stettin. ent. Ztg. 20 (7-9): 217 (Spain, Andalusia: Chiclana). Syntypes, at least 5♂♀ (MNHU: 4♂♀, examined).

Acidalia dorycniata BELLIER, 1862: Annls Soc. ent. Fr. (4) 2: 128, pl. 1, fig. 4, 4a (Spain, Catalonia: Barcelona). Syntypes 1♂1♀ (ZFMK: 1♂1♀, examined, ♂ without abdomen). Regarded as valid species by PROUT (1935a). Downgraded from subspecific rank (LERAUT 1997) to synonymy (SCOBLE 1999).

‡ *callunata* RAMBUR, 1866: Cat. syst. Lépid. Andal. 2: pl. 18, fig. 6 (Spain, Andalusia). Syntype(s). Non binominal. Name subsequently mentioned by various authors in synonymy, thus unavailable too.

Acidalia calunetaria var. *valesiaria* PÜNGELER, 1888: Stettin. ent. Ztg. 49: 350 (Switzerland, Valais: Zermatt). Syntype(s) (MNHU;, 8♂♀ syntypes examined). Valid at subspecific rank.

Ptychopoda calunetaria episticta WEHRLI, 1930: Int. Ent. Zeits. 23 (37/38): 436, pl. 2, figs 9,10 (Algeria: Algiers; Morocco: Chabat el Hamma). Syntypes 46♂1♀ (BMNH). Valid at subspecific rank according to SCOBLE (1999), validity questionable.

Ptychopoda baeticaria ZERNY, 1930: Verh. zool.-bot. Ges. Wien 80 (1/2): 21 (Spain: Andalusia). Syntypes, at least 4 (NHMW: 3♂1♀, examined).

Sterrha maxima WEHRLI, 1934: Ent. Z. Frankf. a.M. 47 (19): 160, figs 2, 6, 13 (Morocco: Mrassine). Syntypes 2 (ZFMK). Herewith downgraded from species (SCOBLE 1999) to synonymy (syn. n.), because of the absence of differential features in habitus and ♂ genitalia (WEHRLI 1934: fig. 13).

Sterrha calunetaria lhommei HERBULOT, 1963: Alexanor 3 (1): 19. First available name after *lhommei* HERBULOT, 1962 (Alexanor 2 (5): 154, nomen nudum) (south-eastern France: Le Rozier). Holotype ♂ (ZSM/HERBULOT, examined). Herewith downgraded from subspecies (SCOBLE 1999) to synonymy (of subsp. *valesiaria*) (syn. n.).

Unavailable names (infrasubspecific): *albomarginata*: REISSER (1961). – (incorrect subsequent spellings): *boeticaria*: RUNGS (1981); *callunetaria*: FUCHS (1904); *calunata*: LERAUT (1997).

DIAGNOSIS: Wingspan of nominate subspecies ♂♀ 14-19 mm, in southern Spain exceptionally up to 22 mm. Forewings elongate. Ground colour light grey, terminal area darker. Postmedial line dark grey, sharp, dentate, with strongly prominent tooth towards discal spot, often M-shaped, in southern Spain less prominent and usually dotted. Antemedial line oblique, running parallel to costa for nearly 1/2 of forewing. Discal spots present. Medial and postmedial lines of hindwing straight. Frons black brown, vertex white, collar brown, in southern Spain sand coloured. Antennae of ♂ ciliate-setose, cilia short, 0.6-0.9 times width of flagellum (Text-fig. 46). Hindtibia of ♂ strongly dilated, with large pencil, slightly shorter than tibia. Tibia slightly covering first segment of tarsus. Tarsus very short, 0.2-0.3 times length of tibia.

Subsp. ***valesiaria*** larger, wingspan ♂♀ 17-22 mm. Wing pattern less contrasted, coloration paler, sand coloured.

MALE GENITALIA: Uncus long and very slender, tapered. Valva slightly dilated and round at tip. Aedeagus with one stout, curved cornutus of 0.4-0.5 length of aedeagus. At base of cornutus a large, well developed spinulose vesica patch. Subsp. *valesiaria* with narrower and longer cornutus, about 0.55-0.60 length of aedeagus.

FEMALE GENITALIA: Ductus bursae long, towards corpus bursae with spinose crest. Ductus inserted on corpus bursae sub-rectangularly. Corpus bursae with long spines in the posterior part and patches of smaller spines in the rest except the most anterior part. Subsp. *valesiaria* without constant differences.

DISTRIBUTION: West-Mediterranean. From southern Switzerland, north-western Italy (subsp. *valesiaria*) to southern Spain and southern Portugal. Present on Balearics (e.g. DANTART et al. 1993). Populations of the mountains of south-eastern France, of the Massif Central and of the High Pyrenees correspond well to subsp. *valesiaria* in genitalia, habitus and phenology. Lowland populations of southern France usually well resembling nominate subspecies in habitus correlated with bivoltine phenology. Erroneously recorded from Sicily (MARIANI 1939; MÜLLER 1996), compare RAINERI (1985). – Outside Europe from Morocco to Algeria ('subsp. *episticta*') and southern Tunisia. All examined specimens from northern Tunisia shown to be misidentified (see remarks).

PHENOLOGY: In the lowlands bi- or trivoltine: late April to early June; limited data available from July; early September to late October. In Andalusia sometimes emerging from early April. In North Africa plurivoltine from March to December. Rearing requires 2-3 months in summer (HALL pers. comm.). Subsp. *valesiaria*, however, with univoltine phenology: mid-July to mid-August. At medium elevations of south-eastern France emerging from mid-June and with transitions to bivoltine phenology. Populations of High Pyrenees and mountains in Morocco univoltine (as subsp. *valesiaria*) from early July to late August.

BIOLOGY: Larva polyphagous on withered leaves of herbaceous plants (REBEL 1910b; FORSTER & WOHLFAHRT 1981), preferring dry leaves according to PROUT (1913). Found on *Dorycnium* (Fabaceae) (BELLIER 1862; subsequently var. cit.). Adults fly rapidly, readily attracted to light.

HABITAT: Nominate subspecies recorded from rocky slopes and scattered coniferous forest with undergrowing *Calluna* (STAUDINGER 1859), subsp. *valesiaria* from warm mountain slopes (FORSTER & WOHLFAHRT 1981). From 0 m up to 1,100 m above sea-level, populations of central Pyrenees and subsp. *valesiaria* higher, usually from 600 m up to 1,700 m, in Morocco up to 2,400 m.

SIMILAR SPECIES: *I. albarracina* (see no. 48), *I. incisaria* (49), *I. seriata* (74), *I. minuscularia* (75). *I. incisaria* differs in postmedial line not or little projecting towards discal spot and shorter ♂ antennal cilia. In southern Spain correct identification based on habitus sometimes difficult. *I. mareotica* (DRAUDT, 1912) from North Africa can be distinguished from *I. calunetaria* by shorter ♂ antennal cilia of the first, shorter ♂ hindtarsus and by dissection: ♂ genitalia with straight cornutus and tapering valva, ♀ genitalia with short ductus bursae. Allopatric *I. obliquaria* with much longer ♂ antennal cilia, wings with darker suffusion, costal spots of forewing and discal spots more distinct, postmedial line less projecting towards discal spot.

REMARKS: This taxon has been subdivided into two different species in PROUT (1935a) and FORSTER & WOHLFAHRT (1981) according to size, coloration of collar, and shape of postmedial line. There are, however, no structural features correlated to such differences. Populations from High Pyrenees and subsp. *valesiaria* merit further study with regard to their taxonomic status. Coloration and pattern of the larva of subsp. *valesiaria* very different from that of nominate subspecies (LEIPNITZ pers. comm.). The Tunisian type specimen of the infrasubspecific taxon 'ab. *saturataria* D. LUCAS, 1944', proved to be conspecific to *I. mareotica* (DRAUDT, 1912) after dissection (ZSM/HERBULOT). The taxon '*fuscularia* TURATI, 1930', from western Libya, which is quite similar according to the original description, may refer to *I. mareotica*, too.

The ***obsoletaria*** species-group

STERNECK's (1940) 13th group of *Idaea*. Homogenous, monophyletic group (STERNECK 1940: 137), except for the phylogenetically isolated *I. obliquaria*. Though the strongly granulated vesica in ♂ genitalia may justify placing *I. obliquaria* here, provisionally, female genitalia, and unusual shape of juxta (♂ genitalia) do not match well the characteristics of this group. *I. biselata* is transferred to the *subsericeata* species-group. Proboscis present in the *obsoletaria* species-group, weaker in *I. belemiata*. Palpi about 0.8-1.0 times diameter of eye, shorter in *I. belemiata*. Antennae of ♂ ciliate-setose with comparatively short setae. Horseshoe-shaped sclerite on ♂ sternum A2 with setose patch distinct, in *I. belemiata* and *I. obsoletaria* without setae, in the former narrowly sclerotized, in the latter weak. Male genitalia: Saccus tapered, often with small keel. Aedeagus with weak sclerite beside cornutus, vesica strongly granulated, granules serially arranged, terminal part with longitudinal striations, or reticulate. Female genitalia: Corpus bursae posteriorly strongly sclerotized, anteriorly membranous. Ductus bursae reduced to a short chitinous clasp, antrum less sclerotized, dilated. Egg sculpturation with distinct longitudinal, and weaker transverse ridges. Larvae usually short, thick, transversely furrowed, skin granulated, larvae of *I. elongaria* and *I. palaestinensis*, however, narrower. Revisions: HAUSMANN (1991c; 1992).

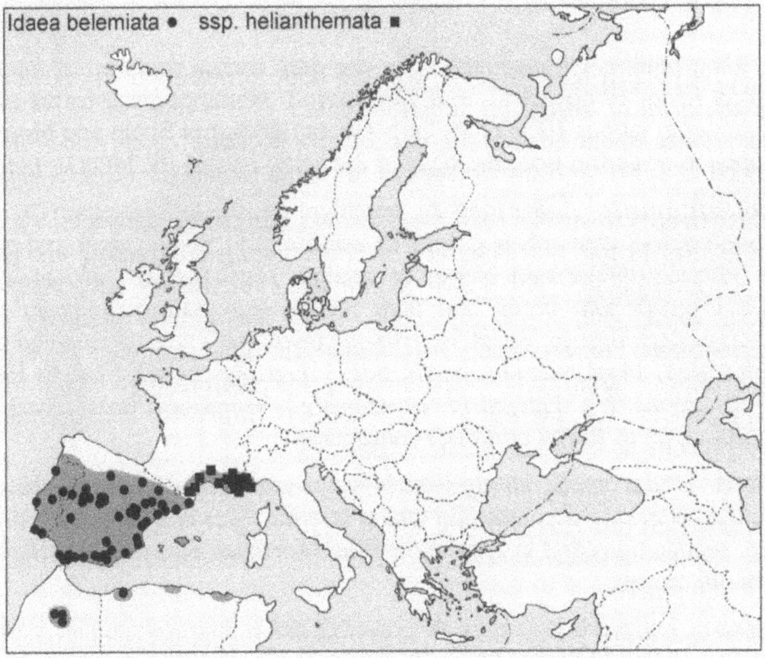

51. Idaea belemiata (MILLIÈRE, 1868)

Acidalia belemiata MILLIÈRE, 1868: Iconogr. Descr. Chen. Lép. inéd. 2 (22): 486, pl. 100, figs. 8, 9 (Spain, Catalonia: Barcelona region). Syntypes 1♂2♀, probably lost.

Acidalia helianthemata MILLIÈRE, 1870: Iconogr. Descr. Chen. Lép. inéd. 3 (26): 134, pl. 113, figs 6-9 (southeastern France: Cannes). Syntypes ♀, probably lost (ZFMK: 1♂ ex coll. GUENÉE, probably no type). Herewith downgraded from species (SCOBLE 1999) to subspecific rank, previously stated by HERBULOT (1946).

Acidalia euphorbiata BALESTRE, 1906: Bull. Soc. ent. Fr. 1906 (7): 80 (France, Alpes maritimes: Mt. Pacanaglia, near Nice). Syntypes ♂♀ (ZSM/HERBULOT: 2♂ from type locality ex coll. BALESTRE, possibly syntypes). Herewith downgraded from species rank (VIVES MORENO 1994; MÜLLER 1996; SCOBLE 1999) to synonymy, previously stated by LERAUT (1997).

Acidalia oberthuriata BALESTRE, 1907: Bull. Soc. ent. Fr. 1907: 23 (France, Alpes maritimes: Mt. Pacanaglia, near Nice). Syntypes ♂♀ (ZSM/HERBULOT: 1♀, examined).

Ptychopoda substraminata PROUT, 1913: in SEITZ, Macrolep. 4: 122, pl. 7, fig. d (Spain, prov. Salamanca: La Granja). Holotype ♂ (BMNH, examined; MNHU 2♂ paratypes). Herewith downgraded from species (SCOBLE 1999) to synonymy (syn. n.). Individual form.

Acidalia cavenacata CHRÉTIEN, 1929: Amat. Papillons 4: 165 (France, Hérault: St. Pons). Lectotype ♂ (ZSM/HERBULOT, examined). Herewith downgraded from species (SCOBLE 1999) to synonymy (syn. n.), previously stated by HERBULOT (1962c).

Unavailable names (misidentification): *obsoletaria*: sensu MILLIÈRE (1867) nec RAMBUR, 1833. – (infrasubspecific): *depravata*: PROUT (1935). – (incorrect subsequent spellings): *cavenata*: CHRÉTIEN (on type label); *helionthemata*: OZORSKI (1948); *oberthueriata*: LERAUT (1980); *oberthüriata*: OBERTHÜR (1916).

DIAGNOSIS: Wingspan ♂♀ 15-19 mm, in Morocco 14-17 mm only. Forewing slender, hindwing slightly concave between M1-M3. Wings light ochreous or sand coloured. Transverse lines often indistinct, sometimes fine and sharp. Medial line rarely visible. Discal spots and fringe dots present, sometimes very small. Frons brown, vertex whitish, collar brown. Proboscis present, but shortened. Palpi slender and very short, about 0.5 times diameter of eye. Antennae of ♂ with intersegmental incisions, ciliate-setose, cilia 0.8-1.05 times width of flagellum (Text-fig. 48). Hindtibia of ♂ slender, short, slightly exceeding 1 mm, with weak pencil covering half of tarsus. Tarsus 0.3-0.5 times length of tibia.

Variability: Wing pattern fairly variable, e.g. the dark tornus spots on all wings present (nominotypical form) or absent ('f. *substraminata*'). Nominotypical forms usually very rare, but commoner (about 10 % of the populations) in central Spain and Morocco. Both forms obtained in a rearing from the eggs of the same female (B. MÜLLER pers. comm.).

Subsp. ***helianthemata*** smaller than the nominate subspecies, wingspan ♂♀ usually 14-17 mm, sometimes 11 mm only (e.g. type of '*couloniata*'). Wing colour and pattern fairly variable. Ground colour light brown to orange. Dark tornus spots always absent. Nominotypical forms with broad and dark medial line getting gradually commoner towards south-eastern France. Small, less orange forms with strongly marked postmedial line mainly in Alpes Maritimes ('f. *oberthuriata*'). Unicolorous grey brown forms dominant in some regions, e.g. Hérault ('f. *cavenacata*'). Transverse lines, discal spots and external morphology as in the nominate subspecies.

MALE GENITALIA: Uncus comparatively slender. Valva spatulate, dilated and rounded at tip. Aedeagus short, with one cornutus and lateral sclerite. Cornutus about 0.6 times length of aedeagus, slightly narrower in specimens from Morocco. Subsp. *helianthemata* without constant differences.

FEMALE GENITALIA: Corpus bursae oval. At posterior end numerous spinules, and two anteriorly projecting chitinous crests, one with one single spine, the other with 5-10 smaller spines. Anterior border of sclerotized part of corpus bursae with a few small spines. Corpus bursae more slender and posterior crests larger in specimens from Morocco. Subsp. *helianthemata* without constant differences.

DISTRIBUTION: West-Mediterranean. Nominate subspecies in the Iberian peninsula, subsp. *helianthemata* in southern France between eastern Pyrenees and Provence (prov. Var), recently recorded in north-western Italy (Liguria: BERTACCINI & CAMPRI 2003). Present on Mallorca (HONEY pers. comm.). Erroneously recorded for Sicily (MARIANI 1941; MÜLLER 1996), and for Crete (1♀, Rethimnon, e.o., dissected, coll. KRAUS, Kaiserslautern), the latter very probably mislabelled or confused in parallel rearings. – Outside Europe in Morocco and Algeria (nominate subspecies).

PHENOLOGY: Univoltine: late June to mid-August, two specimens from prov. Hérault from early June. Larva growing very slowly, requiring 10-12 months for development (PROUT 1913).

BIOLOGY: Larva mono- or oligophagous, possibly with preference for *Euphorbia* species, reputedly polyphagous on withered or dry leaves and petals of herbaceous plants (PROUT 1913, MILLIÈRE 1870). Found on *Euphorbia spinosa* (Euphorbiaceae) (HOMBERG in LHOMME 1935), and on detritus under *E. spinosa*, in rearings feeding on withered and dry leaves of this plant ('*I. euphorbiata*' and '*I. oberthuriata*': BALESTRE 1906; 1907; subsequently var. cit.). Reared also on *Tuberaria lignosa* ('*Helianthemum tuberaria*':

Text-figs 95-98: Diagnostic characters (indicated) of European *Idaea* species. Text-fig. 95: *I. belemiata* (MILLIÈRE, 1868). Text-fig. 96: *I. elongaria* (RAMBUR, 1833). Text-fig. 97: *I. palaestinensis* (STERNECK, 1933). Text-fig. 98: *I. obsoletaria* (RAMBUR, 1833).

MILLIÈRE 1870; subsequently var. cit.), and on fresh and withered leaves of various herbaceous plants, mainly Fabaceae ('*I. cavenacata*': CHRÉTIEN 1929). Larva may live for months without feeding (MILLIÈRE 1867: 233). Adults readily attracted to light. ♀ ratio at light high, about 60-80%.

HABITAT: Xerothermophilous. On dry steppes, warm glades, and in scattered xerothermic forest. From 0 m up to 2,000 m above sea-level, in Morocco from 1,500 m up 2,200 m. Subsp. *helianthemata* from 0 m up to 1,200 m above sea-level, but commoner in the hot lowlands.

SIMILAR SPECIES: Unicolorous forms of *I. belemiata* very similar to *I. obsoletaria*, the latter differing in less pointed forewing apex, in dark grey suffusion (valid only for Spanish populations), hindwings not concave between M1-M3, ♂ antennal cilia shorter, proboscis longer. Males may be identified by brushing of the abdomen to remove scales and reveal shape of tip of valva, which is less dilated in *I. obsoletaria*; females sometimes need dissection. *I. obsoletaria* flies two weeks earlier, and according to PROUT (1913) 'never' at the same localities. *I. alicantaria* (see no. 29). *I. distinctaria* and *I. predotaria* larger in the first generation, coloration more glossy, forewing costa paler, ♂ hindtibia with large pencil, dilated. *I. infirmaria* with forewing costa and fringe usually darker, coloration often with red or dark grey suffusion, ♂ antennal cilia much shorter. Identification of ♀ sometimes requires dissection. *I. dimidiata* without tornus spots on hindwing, ♂ antennal cilia and ♂ hindtarsus longer.

REMARKS: Description of immature stages in CHÉTIEN (1929).

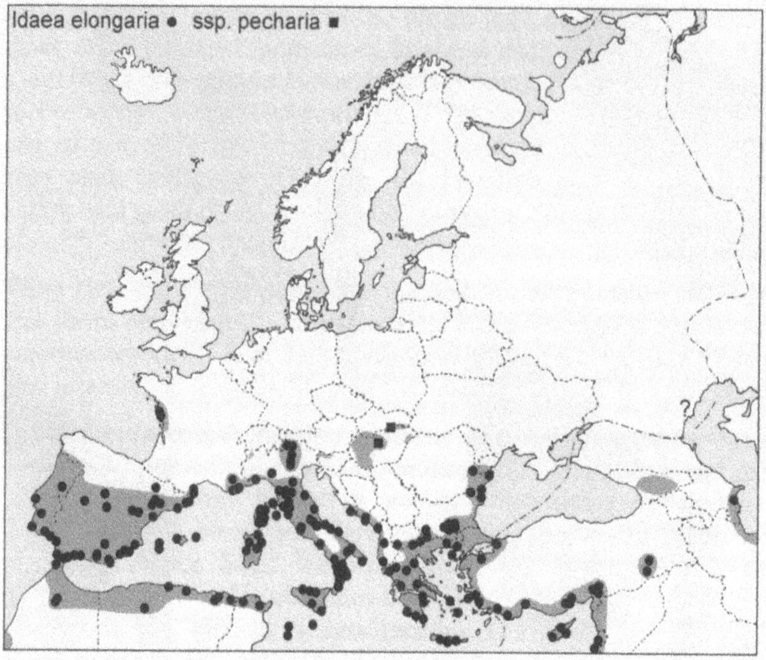

52. *Idaea elongaria* (RAMBUR, 1833)

Acidalia elongaria RAMBUR, 1833: Annls Soc. ent. Fr. 2 (1): 38, pl. 2, fig. 20 (Corsica: Ajaccio). Syntype 1♂ (MNHU: 1♂, examined, labelled '*elongaria* type' and 'orig[inal]').

Idaea aridata ZELLER, 1847: Isis 1847: 513 (Sicily: Messina, Siracusa; Italy: Rome). Syntypes. Junior synonym.

‡ *Idaea aridaria* HEYDENREICH, 1851: Lepid Eur. cat. meth. (Edn 3): 60. Incorrect subsequent spelling.

Acidalia monadaria GUENÉE, 1858: in BOISDUVAL & GUENÉE, Hist. nat. Insectes (Spec. gén. Lépid.) 9: 475 (southern Turkey: Tarsus). Holotype ♂ (ZFMK, examined).

Acidalia pecharia STAUDINGER, 1863: Stettin. ent. Ztg. 24 (7-9): 266 (Hungary: Ofen). Syntypes 2♂1♀ (MNHU: 5♂♀, examined, all labelled 'orig[inals]'). Valid at subspecific rank (in RAINERI 1996 retained valid at species rank).

Acidalia (Cnidia) infermata GUMPPENBERG, 1892: Nova Acta Acad. Caesar. Leop. Carol. 58 (4): 223, 232. First available name after *infermata* RAMBUR, 1866 (non binominal) and *infermata* STAUDINGER, 1871 (introduced in synonymy) (Spain: Andalusia). Syntype(s) of RAMBUR (ZSM/HERBULOT: 1♂, ex coll. MABILLE, possibly a syntype; 'not traced' according to HERBULOT (1987)).

Acidalia zephyrata MILLIÈRE, 1872: Iconogr. Descr. Chen. Lép. inéd. 3 (30): 268, pl. 132, figs 1-8 (south-eastern France: Cannet Valley near Cannes). Syntypes ♂♀ (MNHU: 2♂ ex coll. MILLIÈRE, examined).

Ptychopoda marialudovicata D. LUCAS, 1942: Bull. Soc. ent. Fr. 47 (8): 122 (Tunisia: Soudia). Holotype ♂ (ZSM/HERBULOT, examined).

Unavailable names (infrasubspecific): *abundata*: DANNEHL (1933); *albomarginata*: REISSER (1961); *bilineata*: REISSER (1961); *nigrolineata*: REISSER (1961). – (nomen nudum): *confusaria*: SNELLEN (1862).

DIAGNOSIS: Wingspan ♂♀ variable, first generation 15-20 mm, second generation 12-18 mm. Quite variable in size, ground colour and intensity of wing pattern. Forewing often slender and apex pointed. Ground colour dirty white or light sand coloured. Wing pattern variable. Transverse lines fine, straight, parallel to each other, oblique. Postmedial line slightly dentate. Light wavy line distinct on all wings. Discal spots present. Frons brown, vertex whitish, collar sand coloured or light brown. Antennae of ♂ ciliate-setose, 0.6-0.85 times width of flagellum. Hindtibia of ♂ slightly dilated, with long pencil. Tarsus 0.2-0.4 times length of tibia.

Subsp. ***pecharia*** with wingspan ♂♀ 16-18 mm. Ground colour dark grey with slight brown tinge, darker towards termen. Wing pattern indistinct, fine and sharp postmedial line usually visible. Wavy line indistinct. Discal spots present. External morphology as in the nominate subspecies.

MALE GENITALIA: Uncus bottle-shaped. Valva bent in the last fourth, slightly dilated subterminally. Aedeagus short and broad, with one cornutus, 0.45-0.6 times length of aedeagus, sometimes up to 0.7 times. Vesica strongly corrugated at tip, condensed to a small globular structure beside cornutus. Reminiscent of genitalia of *I. belemiata*, the latter however with valva dilated terminally rather than subterminally. Subsp. *pecharia* with cornutus on average slightly longer, 0.6-0.7 times length of aedeagus.

FEMALE GENITALIA: Antrum sub-rectangular. Corpus bursae elongate. Sclerotized part with chitinous, spinose crest at posterior end, numerous lateral spinules and some spinules at anterior margin. Subsp. *pecharia* similar, but antrum shorter, sclerotized ribbon of ductus bursae curved, corpus bursae comparatively broad, and spinulose patch of corpus bursae smaller, more at the posterior end rather than laterally. Female genitalia figured in WARDIKJAN (1985) under '*I. elongaria*' are misidentified (identity: *I. persidis* (WILTSHIRE, 1966)).

DISTRIBUTION: Mediterranean-Turanian. From Iberian peninsula across southern France, Italy to Balkan peninsula. Present on all Mediterranean islands. In western and northern Hungary, and southern Slovakia replaced by subsp. *pecharia*. Reputedly in southern European Russia (LVOVSKY 1971; VIIDALEPP 1996), but all examined specimens proved to be misidentified (mainly *I. descitaria*). In the north of the distribution local, in the south sometimes ubiquitous and common. – Outside Europe in North Africa from Morocco to Egypt (WILTSHIRE 1949a), Levant, Turkey, Transcaucasus (examined specimens from Azerbaijan and 'Georgia'), across northern Iran to Turkmeniya (VIIDALEPP 1996; HAUSMANN & LASZLO 1999), in the east possibly to Afghanistan and central Asian mountains (see remarks). The taxon '*pecharia*' is erroneously recorded for southern European Russia, Transcaspia and central Asian mountains (PROUT 1913; VIIDALEPP 1988; 1996), usually confused with *I. descitaria*.

PHENOLOGY: Usually bivoltine: early May to late June, exceptionally somewhat later; late July to mid-September. In Spain and Cyprus under favourable conditions a (partial) third brood from early October to early November. In North Africa tri- or plurivoltine from March to November. Subsp. *pecharia* univoltine (KRAMPL 1984), early August to late August.

BIOLOGY: Larva polyphagous on herbaceous plants (MILLIÈRE 1872; REBEL 1910b; FORSTER & WOHLFAHRT 1981). Rearing slow with dry plant pieces and petals (LEIPNITZ pers. comm.). Subsp. *pecharia* reared on withered leaves of *Taraxacum officinale* (MÜLLER pers. comm.). Adults readily attracted to light.

HABITAT: Xerothermophilous, most common in Mediterranean lowlands near the coasts, Mediterranean maquis, dry and stony grassland, rocky slopes. From 0 m up to 1,200 m above sea-level, exceptionally up to 2,200 m (Greece, Turkmeniya), on Crete montane, from 1,100-1,500 m. Subsp. *pecharia* from 100 m up to 200 m.

SIMILAR SPECIES: *I. alicantaria* (see no. 29), *I. troglodytaria* (36), *I. dimidiata* (84). Pale forms of *I. descitaria* (e.g. from Astrachan desert) rather similar, transverse lines more diffuse, ♂ antennal cilia and ♂ hindtarsus longer. Confusions with *I. seriata* mainly occur in the small second generation, *I. seriata* with ground colour usually more whitish, transverse lines less straight, wavy line in terminal area less distinct, collar brown, ♂ antennal cilia and ♂ hindtarsus longer. *I. palaestinensis* with wings deeper sand coloured, transverse lines on veins with sharp black markings, postmedial line of forewing usually more convex, that of hindwing double S-shaped, collar usually paler, ♂ hindtarsus longer, sometimes requiring dissection for correct identification.

REMARKS: *I. effeminata* (STAUDINGER, 1892), described from Afghanistan very similar in genitalia (see VIIDALEPP 1988), further study is needed to analyse relationships and possible synonymy of both taxa. *Idaea substriata* (TURATI, 1930) from Libya, Cyrenaica, may prove to be synonym, too. Genitalia of both sexes reveal sister species relationship to *I. peluraria* (REISSER, 1939) from south-western Asia. The taxon *pecharia* is suggested to be valid at species rank by GELBRECHT (pers. comm.), and FORSTER & WOHLFAHRT (1981).

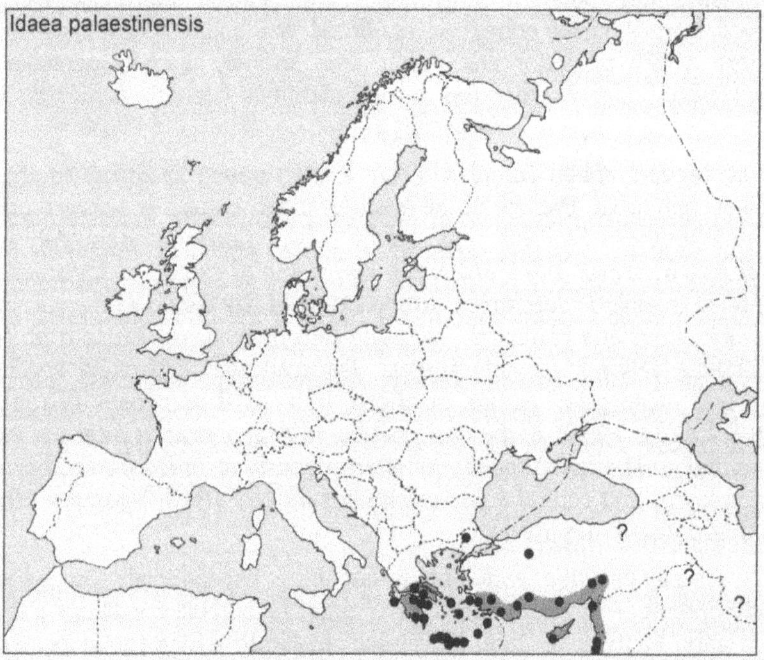

53. *Idaea palaestinensis* (STERNECK, 1933)

Ptychopoda palaestinensis STERNECK, 1933: Zeits. österr. Ent. Ver. 18 (8/9): 65 (Israel: Haifa, Jerusalem, Jericho). Syntypes 5♂17♀ (MNHU: 1♂1♀, examined)

‡ *Ptychopoda elongaria antitaurica* WEHRLI, 1932: Mitt. Münch. ent. Ges. 22 (1): 4 (northern Lebanon: Bscharre). Lectotype ♂ (ZFMK, examined), designated in HAUSMANN (1995). Junior primary homonym nec WEHRLI, 1931 (identity: *Idaea dimidiata antitaurica*).

Sterrha elongaria var. *seitunensis* PROUT, 1934: Lepid. Cat. 63: 389 (northern Lebanon: Bscharre). No separate types. Replacement name for *antitaurica* WEHRLI, 1932.

Unavailable names (misidentification): *obsoletaria*: sensu PROUT (1921) nec RAMBUR, 1833. – (incorrect subsequent spelling): *zeitunensis*: WILTSHIRE (1957).

DIAGNOSIS: Wingspan ♂♀ 14-18 mm. Wings dull sand coloured to light brown. Postmedial and antemedial lines sharp, fine, with small black dots on veins. Small, black terminal dots distinct. Discal spots fine, exceptionally absent. Palpi and frons blackish brown, vertex whitish, collar pale sand coloured. Antennae of ♂ with intersegmental incisions, ciliate-setose, cilia 0.6-0.9 times width of flagellum. Hindtibia of ♂ slender, with long pencil. Tarsus comparatively long, 0.4-0.6 times length of tibia.

MALE GENITALIA: Uncus bottle-shaped. Valva bent in the distal third, dilated subterminally. Aedeagus long, with one long and very narrow cornutus, 0.55-0.65 times length of aedeagus. Vesica with longitudinal striations at tip.

FEMALE GENITALIA: Antrum cup-shaped widening. Sclerotized part of corpus bursae very long and slender (slightly shorter in north-western part of distribution), numerous spinules at posterior end and some spines at anterior margin.

DISTRIBUTION: East-Mediterranean. In Europe restricted to European Turkey (OKYAR & AKTAÇ 1999), Greece and most of its islands, including Crete and Rhodos. – Outside Europe on Cyprus and from south-western Turkey to the Levant. Reputely in northern half of Iraq (WILTSHIRE 1957) and south-western Iran (WILTSHIRE 1966), awaiting confirmation.

PHENOLOGY: Bi- or trivoltine: late May to late June; late July to late August; single records from October (REISSER 1962; LEIPNITZ pers. comm.). Abundance higher in the first generation. In the south of the distribution (Israel) probably trivoltine from late April to mid-June; late August to mid-November.

BIOLOGY: Larval host-plants unknown. Adults readily attracted to light.

HABITAT: Xerothermophilous. Limited information available about habitat preference, collected in various types of dry grassland. From 0 m up to 600 m above sea-level, in Turkey exceptionally up to 1,000 m.

SIMILAR SPECIES: *I. troglodytaria* (see no. 36), *I. elongaria* (52), *I. dimidiata* (84). Extraterritorial *I. peluraria* (REISSER, 1939) from southern Turkey to Iran with more orange tinge, discal spots and terminal dots absent, transverse lines more diffuse. Extraterritorial *I. persidis* (WILTSHIRE, 1966) from southern Turkey (♂♀, prov. Elazig, Kale and Hazar Lake, new for the fauna of Turkey) to Iran very similar, slightly paler, for secure identification dissection is needed. *I. obsoletaria* usually smaller and darker, collar darker, postmedial line with black markings weak or absent, hindtibia without pencil. *Scopula minorata* usually with postmedial lines of both wings more dentate, ♂ antennal cilia much longer, ♂ hindtarsus longer, ♀ hindtibia with four spurs.

54. *Idaea obsoletaria* (RAMBUR, 1833)

Acidalia obsoletaria RAMBUR, 1833: Annls Soc. ent. Fr. 2 (1): 39, pl. 2, fig. 21 (Corsica: Ajaccio). Syntype(s) (MNHU 1♂, examined, with labels 'orig[inal]' and 'Granada', probably by error).

‡ *rufillaria* HERRICH-SCHÄFFER, 1845: Syst. Bearb. Schmett. Eur. 3: pl. 14, figs 84, 85; text (1846): 16 (non binominal); on p. 17 (1847) clearly quoted as misspelling.

Acidalia rufularia HERRICH-SCHÄFFER, 1847: Syst. Bearb. Schmett. Eur. 3 (24): 17 (Austria). Syntype(s).

Acidalia obsoletata BELLIER, 1861: Annls Soc. ent. Fr. (3) 8 (4): 702. Emendation.

Acidalia algeriensis BETHUNE-BAKER, 1888: Trans. ent. Soc. Lond. 1888 (1): 118 (Algeria: Sebdou). Holotype ♂ (BMNH). Herewith downgraded from species (SCOBLE 1999) to subspecific rank (stat. n.), as there are only minor differences from *I. obsoletaria dierli* in ♂♀ genitalia.

Idaea obsoletaria distans HAUSMANN, 1991: Nachr.Bl. Bayer. Ent. 40 (2): 37, fig. 3 (Italy, South Tyrol: Terlan), Holotype ♂ (ZSM, examined). Herewith downgraded from subspecies (SCOBLE 1999; HAUSMANN 1991c) to synonymy (syn. n.), previously proposed by LERAUT (1997).

Idaea obsoletaria dionigii HAUSMANN, 1991: Nachr.Bl. bayer. Ent. 40 (2): 37, fig. 4,5,13 (Italy, Calabria: Aieta). Holotype ♂ (ZSM, examined). Valid at subspecific rank.

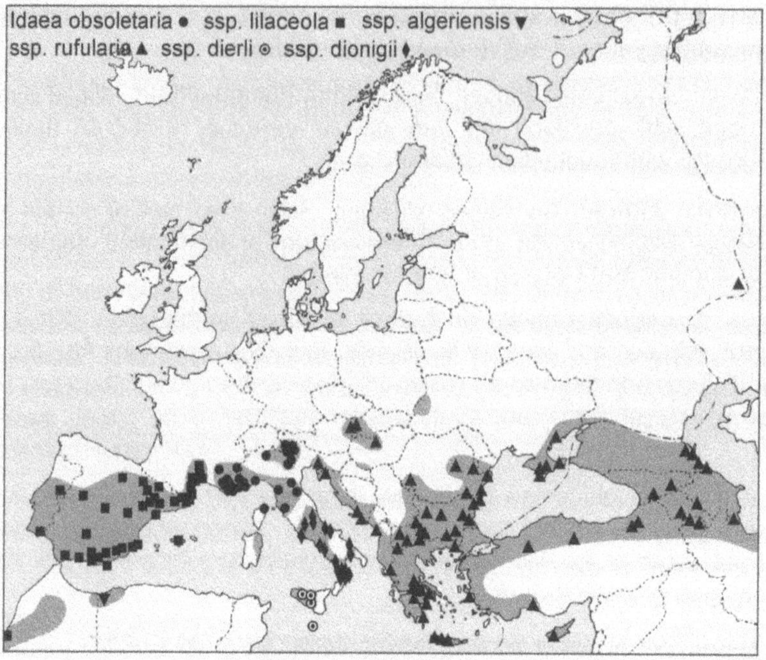

Idaea obsoletaria dierli HAUSMANN, 1991: Nachr.Bl. Bayer. Ent. 40 (2): 39, fig. 9 (Sicily: Madonie Mts.). Holotype ♂ (ZSM, examined). Valid at subspecific rank.

‡ *Idaea obsoletaria violacearia* HAUSMANN, 1991: Nachr.Bl. bayer. Ent. 40 (2): 37, fig. 6. First available name after *violacearia* STAUDINGER, 1901 (infrasubspecific) (Spain: Barcelona). Lectotype ♂ of STAUDINGER (MNHU, examined), herewith designated to fix type locality and to stabilize usage of the name; at MNHU 1♂3♀ paralectotypes from type locality, Balearics (Mallorca) and from Greece (Attika). Junior subjective homonym nec WALKER, 1861 (N. America; identity: *Idaea violacearia*).

Idaea obsoletaria lilaceola HAUSMANN, 2003: Ent. Zeits. Stuttgart 113 (11): 320 (Spain: Barcelona). No separate types. Replacement name for *violacearia* HAUSMANN, 1991. Valid at subspecific rank.

Unavailable names (misidentification): *distinctaria*: sensu STAUDINGER (1871) nec BOISDUVAL, 1840. – (nomen nudum): *tenellaria*: BOISDUVAL (1840).

DIAGNOSIS: Wingspan ♂♀ 13-17 mm (m=15.3; n=53). Hindwing margin round. Ground colour light brown. Forewing costa concolorous with ground colour. Transverse lines indistinct or very thin, straight or slightly undulate. Discal spots and terminal dots black, very small. Frons brown, vertex whitish, collar sand coloured to light brown. Antennae of ♂ ciliate-setose, cilia 0.6-0.85 times width of flagellum. Hindtibia of ♂ slender, short, slightly exceeding 1 mm, without pencil. Tarsus 0.4-0.6 times length of tibia (Text-fig. 24).

Subsp. ***lilaceola*** smaller than nominate subspecies, wingspan ♂♀ 12-16 mm, exceptionally 17 mm (m=14.2; n=61). Ground colour greyish brown with slight violet tinge. Transverse lines indistinct, sometimes absent. External morphology as in nominate subspecies, but tarsus shorter, 0.3-0.4 times length of tibia.

Subsp. ***dionigii*** larger than nominate subspecies, wingspan ♂♀ 14-19 mm (m=16.7; n=62). In wing coloration, wing pattern, and external morphology not constantly differing from nominate subspecies.

Subsp. ***dierli*** larger than nominate subspecies, wingspan ♂♀ 14-19 mm (m=16.5;

n=17). Wings sand coloured with orange tinge. Transverse lines indistinct or absent. Discal spots small. Terminal dots very small or absent. External morphology as in nominate subspecies, but ♂ antennal cilia slightly longer, 0.7-0.9 times width of flagellum, ♂ hindtarsus shorter, 0.3-0.5 times length of tibia.

Subsp. ***rufularia*** similar to nominate subspecies in size. Ground colour light brown. Transverse lines, mainly postmedial line, better marked than in the other subspecies. External morphology see nominate subspecies, but ♂ antennal cilia slightly longer, up to 1.0 times width of flagellum.

MALE GENITALIA: Uncus digitiform. Saccus tapered. Valva simple, spatulate, slightly curved. Aedeagus short, with one slender, lance-shaped cornutus, 0.4-0.55 times length of aedeagus, up to 0.6 times in subsp. *dionigii* (with clinous transitions towards north). Length of cornutus about 5 times width. Subsp. *lilaceola* with cornutus short and weakly sclerotized, only 0.3-0.4 times length of aedeagus. Length of cornutus about 4 times width. Subsp. *dierli* with aedeagus and cornutus very short and broad, cornutus 0.3-0.35 times length of aedeagus. Length of cornutus about 3 times width. Subsp. *rufularia* with cornutus shorter and broader than in nominate subspecies, cornutus 0.3-0.35 times length of aedeagus. Length of cornutus about 3-4 times width.

FEMALE GENITALIA: Antrum dilated. Corpus bursae broad, oval, posterior half sclerotized, strongly spinose. Outlet of ductus seminalis near ductus bursae. Without constant differential features between the subspecies.

DISTRIBUTION: Mediterranean-Turanian. Subsp. *lilaceola* in southwestern France, in the Iberian peninsula and Mallorca, the border towards nominate subspecies running quite sharply between the French provinces Hérault and Vaucluse. Nominate subspecies from south-eastern France, southern Switzerland (REZBANYAI-RESER 1985; 1988a), northern Italy to Corsica and Sardinia (MÜLLER 1996), subsp. *dionigii* in central and southern Italy, subsp. *dierli* on Sicily, Malta, reputedly on Lampedusa (ROMANO & ROMANO 1995), and subsp. *rufularia* from south-eastern Austria, southern Czech Republic, and southern Slovakia (KRAMPL 1984) across Balkan peninsula to Crimea, with isolated occurrence in southern Urals. – Outside Europe local populations in western Morocco (subsp. *lilaceola*), and northern Algeria (subsp. *algeriensis* BETHUNE BAKER, 1888). Reputedly in western Libya (TURATI 1930a). Widely distributed from Turkey and Cyprus (HAUSMANN 1995) across Transcaucasus and northern Iran to northern Afghanistan and central Asian mountains (subsp. *rufularia*).

PHENOLOGY: Univoltine: late June to early August, single specimens until early September. In south-western Morocco emerging from early June.

BIOLOGY: Larva polyphagous on herbaceous plants, inert (REBEL 1910b; PROUT 1913, FORSTER & WOHLFAHRT 1981), growing very slowly. Young larvae reared on petals of various plants, e.g. *Pimpinella anisum*, and Cruciferae, later on accepting also withered leaves of *Taraxacum officinale* (original data), furthermore reared on *Tuberaria guttata* (LHOMME 1935). Adults readily attracted to light and bait. ♀ ratio at light high, about 60-90%. Nectar feeding recorded on *Scabiosa* (DANNEHL 1927).

HABITAT: Xerothermophilous. Sometimes common in hot open lowland habitats, such as Mediterranean maquis, dry steppes, rocky slopes or scattered oak forest. From 0 m up to 700 m above sea-level, exceptionally up to 1,300 m (Alpes Maritimes, Apennines, Greece, Crete). Subsp. *lilaceola* in southern Spain montane, abundant from 1,000 up to

1,500 m, exceptionally in the lowlands. In Afghanistan and central Asian mountains from 800 up to 2,000 m.

SIMILAR SPECIES: *I. sardoniata* (see no. 15), *I. alicantaria* (29), *I. belemiata* (51), *I. palaestinensis* (53), *I. fractilineata* (88). *I. distinctaria* and *I. predotaria* differ in less marked transverse lines, more glossy coloration, collar and vertex darker, first generation larger. Extraterritorial *I. peluraria* (REISSER, 1939) from southern Turkey to Iran without discal spots and terminal dots, transverse lines more diffuse. *I. infirmaria* with forewing costa and fringe usually darker, coloration often with red or dark grey suffusion, ♂ antennal cilia shorter. *Scopula minorata* with postmedial lines of both wings more dentate, ♂ antennal cilia much longer, ♂ hindtarsus longer, ♀ hindtibia with four spurs.

REMARKS: Taxonomy and distribution discussed in HAUSMANN (1991c).

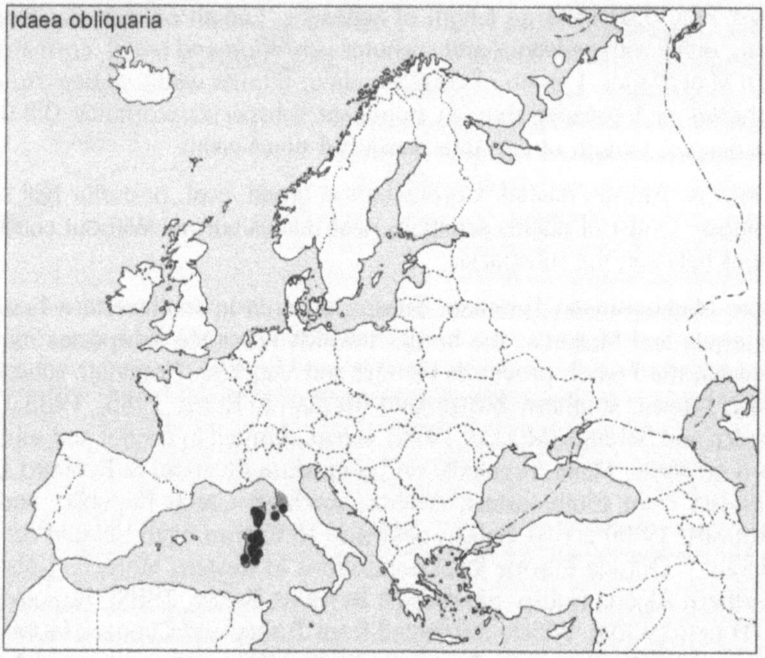

55. *Idaea obliquaria* (TURATI, 1913)

Acidalia obliquaria TURATI, 1913: Entom. Rec. J. Var. 25: 17 (Sardinia: Monte Chiesa). Syntypes 9♂1♀ (MNHU: 1♂1♀ with 'cotype' labels, examined; unnecessary and invalid neotype designation in RAINERI (1996, fig. 19): MSNT: ♂, Aritzo).

Ptychopoda albitorquata f. *napoleon* PROUT, 1913: in SEITZ, Macrolep. 4: 110, pl. 5, fig.b (Corsica: Bastelica). Holotype ♀ (BMNH). Junior synonym. Available despite the use of the term 'forma', because Prout quotes 'probably specific rank' in the same sentence.

Unavailable names (infrasubspecific): *anastomosaria*: REISSER (1961).

DIAGNOSIS: Wingspan ♂♀ 15-18 mm. Ground colour grey, terminal area with dark grey shadows. Ante- and postmedial lines distinct, dark grey, dentate. Discal spots distinct. Black terminal line interrupted at veins. Frons blackish brown, vertex and collar whitish to sand coloured. Antennae of ♂ ciliate-fasciculate, cilia 1.5-1.7 times width of flagellum (Text-fig. 49). Hindtibia of ♂ long (about 2.5 mm), somewhat dilated, with long pencil. Tarsus 0.25-0.35 times length of tibia.

MALE GENITALIA: Uncus sub-triangular, slightly truncate at tip. Juxta with strongly sclerotized, paired posterior thorns. Valva simple. Aedeagus short and broad, with one short triangular cornutus of 0.4-0.5 times length of aedeagus. Vesica strongly granulated.

FEMALE GENITALIA: Without ductus bursae and antrum, connection towards ostium membranous. Corpus bursae elongate, posteriorly with two smooth lateral sclerites, anteriorly membranous.

DISTRIBUTION: Tyrrhenian. Endemic to Corsica and Sardinia, with isolated occurrence in western Tuscany (REZBANYAI-RESER 1997). Erroneously recorded for Sicily (PROUT 1935a), and Israel (AMSEL 1933; 1935).

PHENOLOGY: Usually univoltine: late June to mid-August, exceptionally until mid-September, possibly referring to a partial second brood. On Sardinia emerging from early June or late May.

BIOLOGY: Larva probably on herbaceous plants, preferring dry leaves (PROUT 1935a). Reared on dry and fresh leaves of *Taraxacum officinale*, withered leaves of *Lactuca sativa*, and dry petals of *Rosa* (REISSER 1929; 1933). Adults readily attracted to light.

HABITAT: Montane. Limited information available about habitat preference. From 800 m up to 2,000 m above sea-level.

SIMILAR SPECIES (see Text-fig. 105): Allopatric *I. calunetaria* (see no. 50). *I. seriata* differs in much longer ♂ hindtarsus, with less dark suffusion. Allopatric *I. camparia* with broader forewings, transverse lines less oblique and less sharp, ♂ hindtarsus longer.

REMARKS: Systematic position unclear (STERNECK 1940: 139; see remarks to the species-group). PROUT (1935a) suggests a position in the *incisaria* species-group. Larva very similar to that of *I. seriata* (LEIPNITZ pers. comm.). For morphology of immature stages see REISSER (1929).

The *inquinata* species-group

STERNECK's (1940) 14th group of *Idaea*. STERNECK (1940: 140) suggests alternative position after the *rusticata* species-group, basing on ♂ genitalia and external morphology. Such a transfer would be supported by the lack of central sclerite and setose patch on ♂ sternum A2, but contradicted by the hexagonal reticulate egg sculpturation. Palpi short, 0.7-0.8 times diameter of eye. Proboscis present. Antennae of ♂ ciliate-setose, with short cilia. Hindtarsus of ♂ long. Male genitalia: saccus without keel, valva with lateral spinules at tip, aedeagus with 2-4 cornuti. Female genitalia: corpus bursae with numerous spinose patches. This species-group may be subdivided into the first two species, with a number of related species in the Middle East, characterized by fringe dots rather than terminal dots, by a pencil at the ♂ hindtibia and by the granulated vesica in ♂ genitalia, and the remaining species without ♂ hindtibial pencil, but with strong vesica folds at the base of cornuti in ♂ genitalia. Larva short, thick, furrowed, skin granulated.

56. *Idaea inquinata* (SCOPOLI, 1763)

Phalaena inquinata SCOPOLI, 1763: Ent. Carniolica: 230, fig. 576 (north-western Slovenja: Carniolia). Syntype(s) lost.

Phalaena herbariata FABRICIUS, 1798: Syst. Ent. (Suppl.): 457 (type locality not stated, 'in herbariis'). Syntype(s) lost.

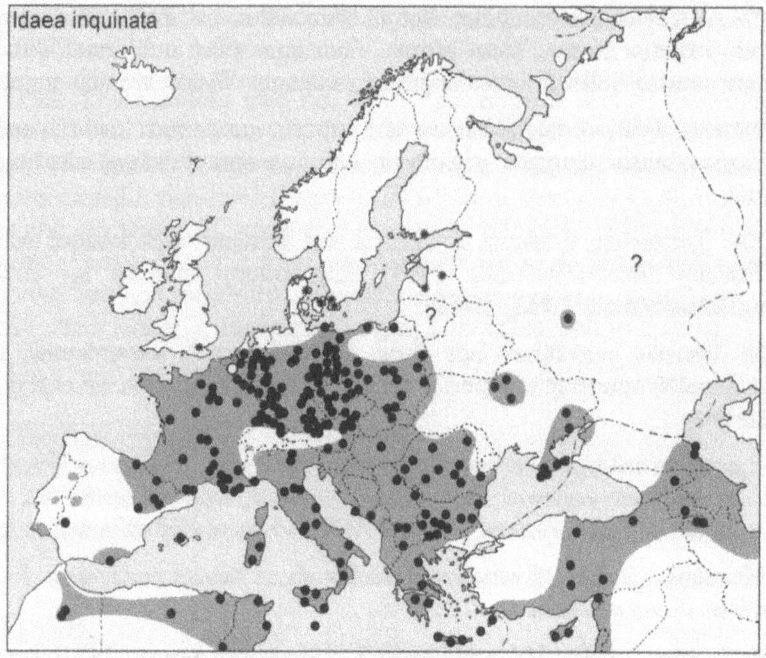

‡ *Geometra pusillaria* HÜBNER, 1799: Samml. Eur. Schmett. 5, Geometrae (1): pl. 19, fig. 99 (Europe). Syntype(s) lost. HÜBNER (1825: 312) refers to 'Schiff.', thus misidentification nec *pusillata* DENIS & SCHIFFERMÜLLER, 1775 (identity: *Eupithecia pusillata*, Larentiinae, see Vol. 4).

Acidalia microsaria BOISDUVAL, 1840: Genera Index meth. eur. Lepid.: 222 (Europe). No separate types. Unnecessary replacement name for *pusillaria*.

Acidalia incomptaria BOISDUVAL, 1840: Genera Index meth. eur. Lepid.: 222 (southern France), referring to RAMBUR'S 'Fauna Andalusiae'. Syntype(s) not traced in ZFMK.

‡ *Ptychopoda pusillata* HEINEMANN, 1859: Schmett. Deutschl. 1: 722. Incorrect subsequent spelling for misidentified *pusillaria*: sensu HÜBNER (see above).

‡ *Acidalia subherbariata* STAUDINGER, 1897: Dt. ent. Z. Iris 10 (1): pl. 4, fig. 31 (Israel: Jordan Valley; Jerusalem). Syntypes 5♂2♀ (MNHU, examined). Junior primary homonym nec RÖSSLER, 1877 (identity: *Idaea subsaturata*).

Acidalia herbariata var. ? *adherbariata* STAUDINGER, 1898: Dt. ent. Z. Iris 10 (2): 306 (Israel: Jordan Valley; Jerusalem). No separate types. Replacement name for *subherbariata*. Valid at subspecific rank.

Unavailable names (misidentification): *circuitaria*: sensu STAINTON (1855) nec HÜBNER, 1819. – (infrasubspecific): *aestiva*: FUCHS (1898); *bilineata*: REISSER (1961); *demarginata*: REISSER (1961); *mediofasciata*: BUBACEK (1923); *teutoburgensis*: SCHULTZ. – (nomen nudum): *musaica*: ANDRES-SEITZ in WILTSHIRE (1949). – (incorrect subsequent spellings): *herbariat*: ANDRES & SEITZ (1924); *inguinata*: FORSTER & WOHLFAHRT (1981: pl. 2).

DIAGNOSIS: Wingspan ♂♀ 14-18 mm, exceptionally 19 mm, in second generation sometimes 11 mm only. Forewing broad, apex round. Wings sand coloured, terminal area with brown diffuse spots. Second generation often darker. Ante- and postmedial lines distinct, dark brown, close to each other. Medial line diffuse, usually anastomosing with antemedial line at inner margin. Discal spots and fringe dots distinct. Frons blackish brown, vertex and palpi sand coloured, collar slightly darker. Antennae of ♂ ciliate-setose, cilia short, 0.5-0.65 times width of flagellum (Text-fig. 50), in populations from Rhodos, Cyprus, Israel and Jordan (subsp. *adherbariata*) 0.8-0.9 times. Hindtibia of ♂ slender, with pencil. Tarsus 0.4-0.5 times length of tibia, in one ♂ examined from Tunisia only 0.25 times.

MALE GENITALIA: Uncus slender. Gnathos broad. Valva slightly bent at 1/2, with one dorsal and one ventral spinule at tip. Aedeagus short, with two short, narrow cornuti, one with base curved at an angle of 180°. Populations from southern Turkey with aedeagus slightly longer and broader.

FEMALE GENITALIA: Antrum strongly sclerotized, widely dilated, cup-shaped. Ductus bursae short. Corpus bursae with spinose crest at posterior end, and two large spinose patches, laterally and centrally. Outlet of ductus seminalis laterally from membranous anterior part of corpus bursae.

DISTRIBUTION: West-Palaearctic. Widely distributed in Europe, in the north up to England, northern Germany and southern Baltic states (VIIDALEPP 1996), in the east to Crimea and western European Russia. Frequent artificial transport with dried herbs and foliage, vegetable packing material, e.g. to the Netherlands or Great Britain (PROUT 1935a; SKINNER 1984). The same is suggested for the isolated records in Finland, Denmark (SKOU 1986), and Sweden (RYRHOLM pers. comm.). Recorded from all larger Mediterranean islands including Malta and Rhodos. Reputedly in central European Russia (LASTUKHIN 2001), data await confirmation. Abundance decreasing in Thuringia (ERLACHER & FRIEDRICH 1994). Usually rare all over Europe, occasionally common in haystacks. – Outside Europe from Morocco (RUNGS 1981; HAUSMANN 1998) to Tunisia, from Turkey and Cyprus across Caucasus, Transcaucasus and northern Iran to central Asian mountains (VIIDALEPP 1988; 1996). From Jordan and Israel to Egypt and eastern Libya (Cyrenaica) replaced by subsp. *adherbariata*.

PHENOLOGY: Usually bivoltine: late April to mid-July; late July to mid-September, generations sometimes overlapping, first generation usually commoner. Under favourable conditions plurivoltine, and recorded in all months of the year, mainly when developing in buildings. Larva rapidly growing.

BIOLOGY: Larva polyphagous, exclusively feeding on dried and withered herbaceous plants, preferring dry petals (var. cit.). Found on dried flowers (Great Britain: PORTER 1997; GERMANY: MILLER pers. comm.), and in dry corn-cobs (Morocco: RUNGS 1981). Larva often found as pest in herbaria (first mentioned by FABRICIUS, 1798: 457; Morocco: RUNGS 1981). Recorded on hay and straw (var. cit.), on dried flowers in floral decoration, especially *Nigella orientalis* (SKINNER 1984), straw decoration (AISTLEITNER & AISTLEITNER 2000a), on dry leaves of *Malva sylvestris* and *Sanicula* (SORHAGEN 1881; PROUT 1913), on peppermint and balm tea (SCHNEIDER 1939; EBERT 2001), and other stored dry food, feeding also on seeds (var. cit.). Exceptionally pest in insect collections (FRIEDRICH 1976; EBERT 2001). Reputedly on moss (FORSTER & WOHLFAHRT 1981; BERGMANN 1955). Reared on withered leaves and petals of *Taraxacum officinale* (original data; MÜLLER pers. comm.) and *Mentha aquatica* (FIUMI & CAMPORESI 1988). Adults often found during daytime when resting on walls. Activity mainly diurnal, but photophobic (EBERT 2001). Adults readily attracted to pheromones of Sesiidae (Germany, Spain, Morocco: BLAESIUS pers. comm.), also to light and bait, easily collected with dry petals of *Sambucus nigra* (BERGMANN 1955).

HABITAT: Domestic, often found in human buildings and food stores, in nature usually when resting on walls and fences. From 0 m up to 700 m above sea-level, up to 1,000 m in Alps and central Asian mountains, one Spanish record from 1,700 m.

SIMILAR SPECIES: *I. blaesii* differs in more diffuse wing pattern, medial area broader, discal spots more distinct, ♂ antennal cilia and ♂ hindtarsus shorter. *I. holliata* and *I affinitata*

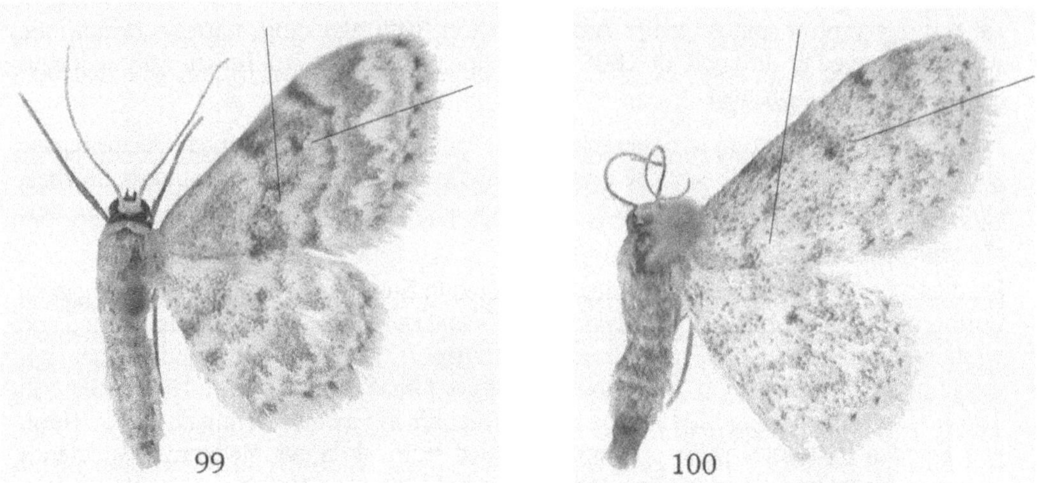

Text-figs 99-100: Diagnostic characters (indicated) of European *Idaea* species. Text-fig. 99: *I. inquinata* (SCOPOLI, 1763). Text-fig. 100: *I. blaesii* LENZ & HAUSMANN, 1992.

from the Levant differ in smaller size, shorter ♂ antennal cilia, and genitalia: ♂ with stouter cornuti, in *I. affinitata* one cornutus strongly curved; ♀ with less sclerotized, narrower antrum.

REMARKS: Some closely related species from the Levant urgently await examination of types, e.g. *I. saida* (WILTSHIRE, 1968), *I. affinitata* (BANG-HAAS, 1907), *I. holliata* (HOMBERG, 1909), and *I. banghaasi* (PROUT, 1934). From male and female genitalia at least three species can be distinguished from *I. inquinata* (subsp. *adherbariata*). So far, the identification of only one, *I. holliata*, could be ascertained by examination of paratypes.

57. ***Idaea blaesii*** LENZ & HAUSMANN, 1992

Idaea blaesii LENZ & HAUSMANN, 1992: Mitt. Münch. ent. Ges. 82: 47, figs. 1-10 (Portugal, Algarve: by Vilamoura). Holotype ♂ (ZSM, examined).

Unavailable names (misidentification): *biselata*: sensu MONTEIRO & PASSOS DE CARVALHO (1984) nec HUFNAGEL, 1767.

DIAGNOSIS: Wingspan ♂ 13-16, ♀ 14-18 mm. Wings sand coloured. Ante- and postmedial lines distinct, dark brown, distant from each other. Medial line diffuse, never anastomosing with antemedial line. Discal spots and fringe dots distinct. Frons black brown, vertex pale sand coloured, collar light brown. Antennae of ♂ ciliate-setose, cilia short, 0.3-0.45 times width of flagellum. Hindtibia of ♂ slender, with large pencil. Tarsus 0.35-0.45 times length of tibia, first segment dilated.

MALE GENITALIA: Uncus slender. Gnathos broad, sub-rectangular. Valva slightly curved, with one dorsal and one ventral spinule at tip. Aedeagus short, with four narrow cornuti, the central one nearly half length of aedeagus, two of the distal cornuti connected at their base. Granulated vesica conglomeration strongly developed.

FEMALE GENITALIA: Antrum not dilated. Ductus bursae broad, of medium length. Corpus bursae with long curved spinose crest in the central part, and two lateral spinose patches. Outlet of ductus seminalis laterally from membranous anterior part of corpus bursae.

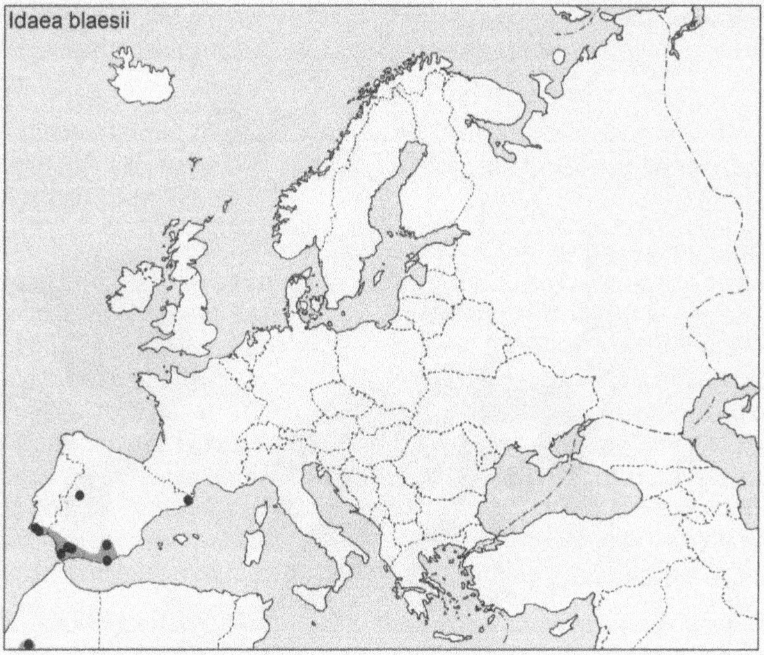

DISTRIBUTION: West-Mediterranean. In Europe restricted to southern Portugal and southern Spain, with additional, isolated records from Extremadura (NOVOA pers. comm.) and north-eastern Spain (ZSM). – Outside Europe in Morocco, Great Atlas (HAUSMANN 1994e).

PHENOLOGY: Bivoltine: late May to late September, generations overlapping.

BIOLOGY: Larva probably polyphagous. Reared on withered leaves of various plant species, preferring *Taraxacum officinale*, young larvae reared on petals of various plant species (LENZ & HAUSMANN 1992).

HABITAT: Xerothermophilous. Collected on dry, scattered scrubland. From 0 m up to 1,000 m above sea-level, up to 1,900 m in Morocco.

SIMILAR SPECIES: *I. inquinata*, and some other allopatric species (see no. 56).

58. *Idaea dilutaria* (HÜBNER, 1799)

Geometra dilutaria HÜBNER, 1799: Samml. Eur. Schmett. 5, Geometrae (1): pl. 19, fig. 100 (Europe). Syntype(s) lost. Figure unambiguous.

Acidalia interjectaria BOISDUVAL, 1840: Genera Index meth. eur. Lepid.: 224 (Europe). Unnecessary replacement name for *dilutaria*. No separate types, BOISDUVAL's material from southern Spain and Montpellier (ZFMK: 1♂ ex coll. BOISDUVAL, examined, identity: *I. dilutaria*). Erroneously listed as synonym of *I. fuscovenosa* by LERAUT (1997) and FORSTER & WOHLFAHRT (1981).

Acidalia holosericata DUPONCHEL, 1842: in GODART & DUPONCHEL, Hist. nat. Lépid. Pap. Fr. (Suppl.) 4: 109, pl. 59, fig. 7 (Austria: Vienna). Syntype(s).

Sterrha stramentata EVERSMANN, 1842: Bull. Soc. Nat. Moscou 15 (3): 556 (Russia: foothills of the Urals). Syntype(s) (ZISP).

Acidalia holosericaria BRUAND, 1846: Mém. Soc. Emul. Doubs (2) 2 (4/4): 139. Emendation (SCOBLE 1999).

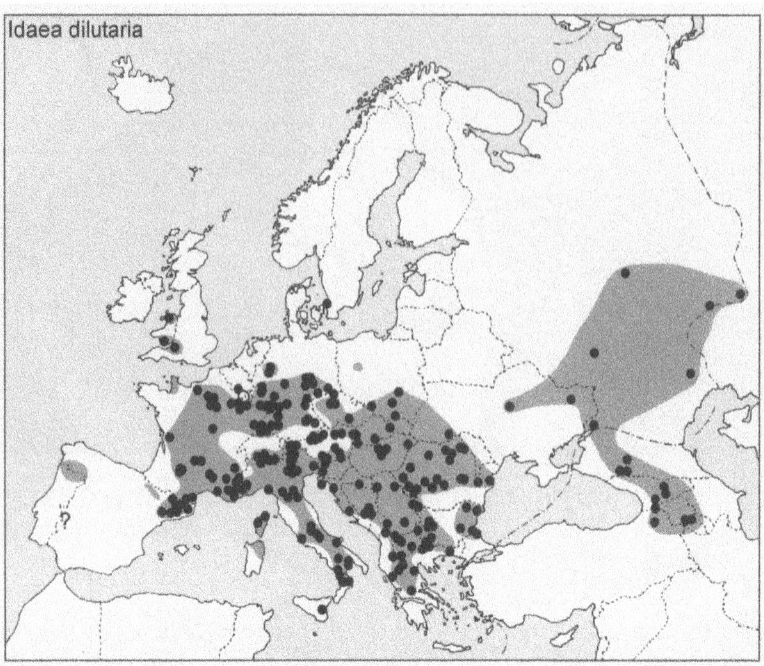

Acidalia holosericearia HERRICH-SCHÄFFER, 1847: Syst. Bearb. Schmett. Eur. 3 (24): 19. First available name after *holosericearia* HERRICH-SCHÄFFER, 1845 (pl. 14, figs 80, 81: non binominal). Emendation.

Acidalia holosericearia var.? *praeustaria* LA HARPE, 1853: Neue Denkschr. Allg. schweiz. Ges. ges. Naturw. 13: 23 (type locality not stated; probably Croatia, Dalmatia: Fiume/Spalato). Syntype(s) (MNHU: 1♂ with label '*praeustaria* Mann' examined, possibly a syntype; further syntypes from NHMW examined by REISSER (1962)). Herewith downgraded from subspecific rank (SCOBLE 1999) to synonymy (infrasubspecific form) (syn. n.), previously stated by LERAUT (1997).

Ptychopoda praeustata HERRICH-SCHÄFFER, 1861: Neue Schmett. Eur. 3: 28 [24], fig. 141. Emendation.

‡ *Sterrha barbagiaria* STERNECK, 1940: Z. Wien. Ent. Ver. 25: 101, 140 (Sardinia). Syntype(s). Nomen dubium. Synonymy tentative.

Sterrha dilutaria illyrica REISSER, 1962: Z. Wien. ent. Ges. 47 (1): 23 (Slovenja: Cerovlje, by Trieste). Holotype ♀ (SMNK, examined by R. TRUSCH). Herewith downgraded from subspecific rank (SCOBLE 1999) to synonymy (syn. n.).

Unavailable names (infrasubspecific): *anastomosaria*: PREISSECKER (1922); *nigrolineata*: REISSER (1961); *subfasciata*: PROUT (1913). – (incorrect subsequent spellings): *holocerisearia*: REINHARD et al. (1999); *holoceriseata*: REINHARD et al. (1999); *holosericeata*: STAINTON (1858); *holosericiaria*: FREYER (1846; non binominal); *holosericiata*: HEINEMANN (1859); *interiectaria*: SPEISER (1903); *dilectaria* (for *dilutaria*): GAEDE (1929); *dilutatria*: BARANIAK & SOSINSKI (2000); *zolosricata* (for *holosericata*): WAGNER (1922).

DIAGNOSIS: Wingspan ♂♀ 14-18 mm. Hindwing margin round. Wings sand coloured, glossy, forewing costa concolorous. Transverse lines and terminal area brownish grey, the latter with wavy line concolorous with ground colour. Terminal area sometimes very dark ('f. *praeustaria*'). Medial line often crossing the vestigial 'discal spot'. Discal spots and terminal dots usually absent, or very small. Frons dark brown, vertex and palpi sand coloured, collar slightly darker. Antennae of ♂ with intersegmental incisions, ciliate-setose, cilia 0.8-1.05 times width of flagellum. Hindtibia of ♂ slender, without pencil. Tarsus very long, 0.75-0.9 times length of tibia (Text-fig. 25).

MALE GENITALIA: Uncus digitiform or sub-triangular. Gnathos rounded. Valva slightly curved, with dorsal spinule at tip very small, or absent; ventral spinule slightly larger.

Aedeagus elongate, broad at base, with one narrow, needle-shaped and one broad lance-shaped cornutus, the latter slightly narrowing towards base, arising from swollen, corrugated vesica.

FEMALE GENITALIA: Antrum rhombic, not dilated. Ductus bursae long and slender. Corpus bursae posteriorly with large oval sclerite. Central part of corpus bursae curved, spinulose, anterior part membranous.

DISTRIBUTION: Euro-Caucasian. Widely distributed in southern Europe from northern half of Iberian peninsula to Balkan peninsula. Widespread in the western and southern half of central Europe, with a very isolated, small resident population in southern Sweden (RYRHOLM 1989). For England recorded from three localities only. Present on Corsica, Sardinia and Sicily. Reputedly in Portugal (VIVES MORENO 1994; MÜLLER 1996), and Extremadura (BLAZQUEZ et al. 2000), data await confirmation. Abundance decreasing in Thuringia (ERLACHER & FRIEDRICH 1994). – Outside Europe from Caucasus and Transcaucasus to north-easternmost Turkey. Erroneously recorded for southern Turkey (DOGANLAR 2003). Reputedly in Morocco, Algeria, all examined specimens however misidentified, identity often small *I. bigladiata*, or *I. terminolineata* (ROTHSCHILD, 1933), see remarks to distribution of *I. lutulentaria*.

PHENOLOGY: Univoltine: late June to early August, on mountains one week later, under good conditions one week earlier. In southern Europe regularly emerging from early June.

BIOLOGY: Larva probably polyphagous on herbaceous plants, on dry, withered or mouldy leaves (var. cit.). Found on fallen leaves of young shrubs of *Quercus* (Fagaceae) (Germany: BOLDT 1929). Recorded on *Helianthemum nummularium* (Cistaceae) (var. cit.), suggested not to be the only host-plant (PARSONS et al. 2001). Reputedly on mosses under hedges and on twigs (REBEL 1910b; VORBRODT 1914; BERGMANN 1955). Reared also on *Plantago lanceolata, Taraxacum officinale* (withered leaves), *Polygonum aviculare, Thymus praecox* (*T. polytrichus*) and moss (ALLAN 1949; BERGMANN 1955; SKINNER 1984; PORTER 1997; PARSONS et al. 2001; ERLEBACH, MÜLLER pers. comm.). Larva stripping stems of host-plant, and then feeding on the partially withering leaves (PROUT 1905; 1913). Larva very inactive, slowly growing. Adults often resting in hedges (BERGMANN 1955; EBERT 2001), easily flushed up from there during daytime, but also readily attracted to light. In northernmost population stenotopic, dispersal low, flight distances 20-130 m in mark-recapture studies (Sweden: RYRHOLM 1989).

HABITAT: Xerothermophilous. In the north usually restricted to the most xerothermic sites in the lowlands, rocky hot slopes and limestone downland, hedges, scrubby places and sun-exposed aisles in open forests. In southern Europe more widespread in open habitats. Usually from 0 m to 1,000 m above sea-level, from the lowlands up to 2,000 m in southern Alps, southern Europe, Morocco and Turkey (xeromontane habitats).

SIMILAR SPECIES: Different from similar species (*I. fuscovenosa, I. bigladiata, I. humiliata*) in glossy coloration, forewing costa concolorous, terminal line and fringe dots absent. *I. fuscovenosa* with paler ground colour, wings broader, discal spots distinct. *I. humiliata* usually with discal spots more distinct. *I. humiliata* and *I. bigladiata* with longer ♂ hindtarsus and shorter ♂ antennal cilia. Male of *I. pallidata* larger, wings broader, coloration less glossy, with more yellow tinge.

REMARKS: Monography: RYRHOLM (1989).

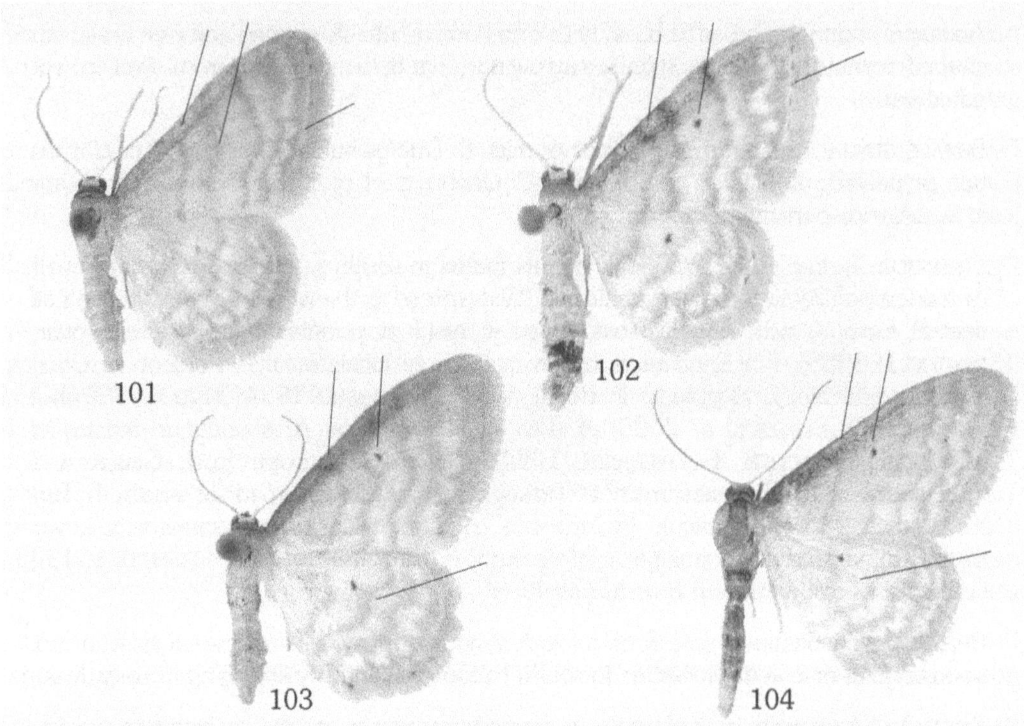

Text-figs 101-104: Diagnostic characters (indicated) of European *Idaea* species. Text-fig. 101: *I. dilutaria* (HÜBNER, 1799). Text-fig. 102: *I. fuscovenosa* (GOEZE, 1781). Text-fig. 103: *I. humiliata* (HUFNAGEL, 1767). Text-fig. 104: *I. bigladiata* HERBULOT, 1975.

59. *Idaea fuscovenosa* (GOEZE, 1781)

Phalaena Geometra fuscovenosa GOEZE, 1781: Entom. Beytr. 3 (3): 428 (France: Paris). Syntype(s).

? *Phalaena plebeia* FOURCROY, 1785: Entom. Parisiensis 2: 285 (France: Paris). Syntype(s). No homonymy of *Phalaena plebeja* LINNAEUS, 1761, as spelled with 'j' (identity: *Archiearis parthenias,* Archiearinae). Synonymy with *I. humiliata* not excluded.

Ptychopoda fuscovenosa var. *corsula* SCHAWERDA, 1929: Zeits. österr. Ent. Ver. 14 (3): 29 (Corsica: Monte d'Oro; Evisa). Syntypes 20. Valid at subspecific rank.

Unavailable names (misidentifications): *dilutaria*: sensu LEDERER (1853) and DOGANLAR (2003) nec HÜBNER, 1799; *holosericata*: sensu ZAPATER & KORB (1892) nec DUPONCHEL, 1842 (identity: *Idaea dilutaria*); *humiliata*: sensu TUTT (1889) nec HUFNAGEL, 1767; *interjectaria*: sensu HERRICH-SCHÄFFER (1847) nec BOISDUVAL, 1840 (identity: *Idaea dilutaria*); *osseata*: sensu HAWORTH (1809) nec DENIS & SCHIFFERMÜLLER, 1775 (identity: *Idaea humiliata*). – (infrasubspecific): *subochreata*: WESTWOOD (1844); *anastomosaria*: PREISSECKER (1922). – (incorrect subsequent spellings): *fusciovenosa*: PROUT (1913); *fusconervosa*: UHERKOVICH (1976); *fuscovenosata*: WARDIKJAN (1985:101); *interiectaria*: SPULER (1910); *osseatae*: HERRICH-SCHÄFFER (1840).

DIAGNOSIS: Wingspan ♂♀ 14-19 mm. Forewing broad. Hindwing margin slightly concave between medial veins. Wings whitish or pale sand coloured. Forewing costa concolorous or, often, with grey brown suffusion at base. Transverse lines brown, dilated and darkened to three spots at forewing costa. Medial line usually proximal to discal spot. Blackish discal spots and terminal dots distinct. Frons black brown, vertex whitish, collar sand coloured. Antennae of ♂ ciliate-setose, cilia 0.7-0.9 times width of flagellum. Hindtibia of ♂ slender, without pencil. Tarsus nearly as long as tibia.

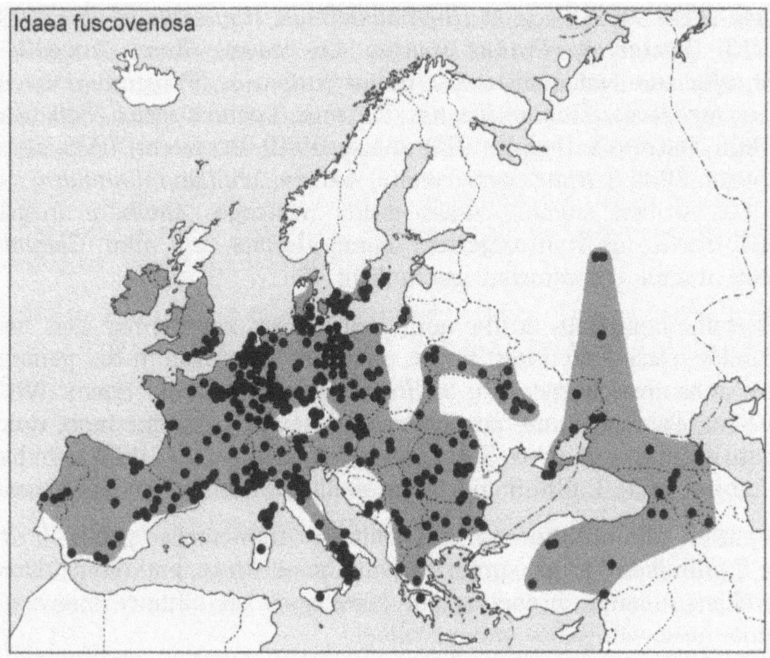

Subsp. ***corsula*** smaller, wingspan 13-15 mm, ♂ antennal cilia shorter, 0.6-0.65 times width of flagellum.

MALE GENITALIA: Uncus short. Gnathos comparatively narrow, usually tapering towards tip. Valva bent at 1/2, with 0-2 small dorsal spinules, and one slightly larger ventral spinule at tip. Aedeagus with two narrow, slightly curved cornuti, the smaller one at base dilated, arising from swollen, corrugated vesica.

FEMALE GENITALIA: Antrum concave, not dilated. Ductus bursae long and very slender. Corpus bursae pyriform, laterally sclerotized, longitudinally corrugated here. Large spinulose patch close to membranous part of corpus bursae.

DISTRIBUTION: Euro-Caucasian. Widely distributed in southern Europe from Iberian peninsula to Crimea and European Russia, and in favourable localities in western and central Europe, in the north up to Denmark, southern Sweden and Baltic states (ŠULCS et al. 1981). Recorded from Corsica (subsp. *corsula*), Sardinia (LEIPNITZ pers. comm.) and Sicily. Deleted from the fauna of southern Portugal (CORLEY et al. 2000). Usually rare, often in isolated populations. – Outside Europe from Turkey to Caucasus and Transcaucasus (WARDIKJAN 1970; 1985; VIIDALEPP 1996). All examined specimens from North Africa revealed to be misidentified, referring to *I. bigladiata*, or *I. terminolineata* (ROTHSCHILD, 1933), see remarks to *I. lutulentaria* (no. 61).

PHENOLOGY: Univoltine: mid-June to late July, sometimes until mid-August. Larva slowly growing.

BIOLOGY: Larva polyphagous. Reputedly found on shrubs of *Quercus* (Fagaceae) (BERGMANN 1955). Recorded to feed on many different low plants and shrubs, preferring fallen leaves both dry and withering (var. cit.). Reputedly on mosses of hedges or trees (var. cit., see EBERT 2001), e.g. on *Rhytidiadelphus triquetrus* in winter (POPE 1908; PROUT 1913). Reared on *Prunus spinosa* (dry leaves), *Rosa, Anagallis arvensis, Taraxacum officinale* (withered leaves), *Rubus fruticosus, Polygonum aviculare* (withered or decaying leaves), *Galium, Rumex, Thymus, Lactuca sativa, Salix* (var. cit., e.g. MILLIÈRE 1865; URBAHN & URBAHN 1939; ALLAN 1949; BERGMANN 1955; SKINNER 1984; KING & ROMERA 2004; LEIPNITZ pers. comm.), and on *Achillea millefolium*, feeding also in winter (MÜLLER pers. comm.). Adults resting in hedges, shrubs or in grass (PROUT 1913), easily flushed up from vegetation during daytime, e.g. from *Clematis* (REISSER 1949). Active at dusk, but attracted also to light.

HABITAT: Xerothermophilous. In the north restricted to xerothermic sites, such as sun-exposed scrubby places, dry forest fringes, grassland with much moss, gardens, often in uncultivated areas near hedges, e.g. of *Rosa rugosa* and *Salix repens* (WEGNER pers. comm.), where fallen leaves are not removed. In southern Europe more widespread. In the north usually from 0 m up to 500 m above sea-level, up to 1,000 m in the Alps and southern Europe, up to 2,000 m in Corsica, southernmost Europe and Caucasus.

SIMILAR SPECIES: *I. dilutaria* (see no. 58). Different from similar species (*I. dilutaria, I. bigladiata, I. humiliata*) in pale ground colour, broad wings, and distinct discal spots. *I. humiliata* differs in darker ground colour, discal spots less distinct, forewing costa reddish. *I. lutulentaria* with yellow ground colour.

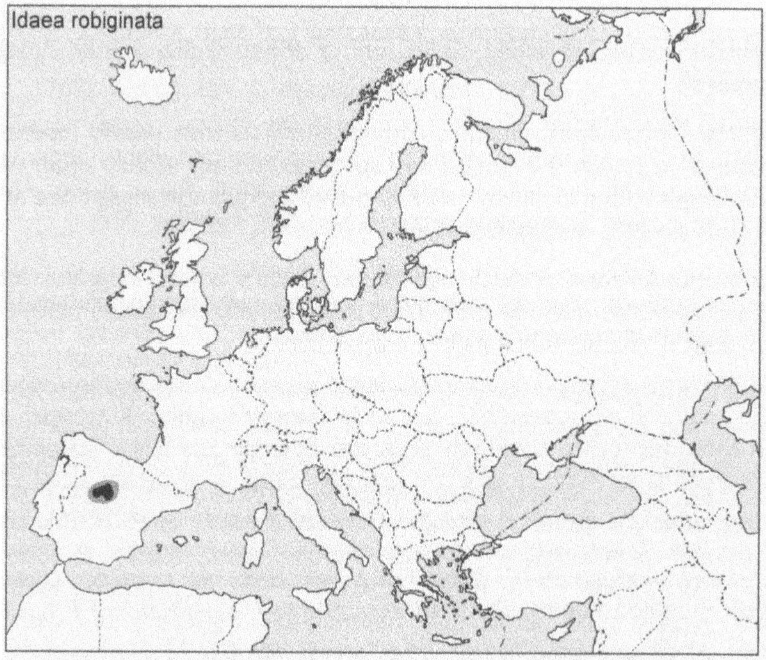

60. Idaea robiginata (STAUDINGER, 1863)

Acidalia robiginata STAUDINGER, 1863: Stettin. ent. Ztg. 24 (7-9): 267 (Spain, Castilia: San Ildefonso). Syntypes 5♂1♀ (MNHU: 5♂1♀, examined).

Acidalia rubiginaria FUCHS, 1904: in SPULER, Schmett. Eur. 2: 17 (Spain, Castilia: San Ildefonso). No separate types. Unnecessary replacement name for *robiginata*.

DIAGNOSIS: Wingspan ♂♀ 16-20 mm. Wings sand coloured with orange tinge. Transverse lines dark brown, fine and sharp, slightly undulate. Discal spots present, that of forewing usually crossed by medial line. Fringe darker than ground colour. Frons light brown, palpi, vertex and collar slightly paler. Antennae of ♂ ciliate-setose, cilia 0.65-0.8 times width of flagellum. Hindtibia of ♂ slender, without pencil. Tarsus long, 0.6-0.75 times length of tibia.

MALE GENITALIA: Uncus and gnathos short, strongly truncate. Valva broad in basal half, tapering distally, at tip with 0-2 spinules. Aedeagus with two stout cornuti, 0.5-0.6 times length of aedeagus. Vesica not corrugated.

FEMALE GENITALIA: Antrum not dilated, terminally convex. Ductus bursae long and very slender. Corpus bursae curved, bean-shaped, laterally sclerotized and spinulose, longitudinally corrugated here.

DISTRIBUTION: West-Mediterranean. Endemic to central Spain (prov. Segovia and Madrid). Reputedly in Portugal (SILVA CRUZ & WATTISON 1935; VIVES MORENO 1994; MÜLLER 1996), awaiting confirmation. One specimen from 'Portugal' also in coll. HERBULOT (ZSM; 'leg. MONTEIRO'). Rare species with very isolated populations.

PHENOLOGY: Univoltine: late June to late July. One specimen in early September (EXPÓSITO HERMOSA pers. comm.). Larva in rearings slowly feeding in winter, pupation in April (PROUT 1913).

BIOLOGY: No secure data on larval host-plants. Larvae reared on petals of *Geranium*, and leaves of *Rumex, Leontodon* (MILLIÈRE 1885: 114), if CHRÉTIEN's (1928) suggestion of confusion between *I. luteolaria* and *I. robiginata* in MILLIÈRE's rearings is correct. Adults active in the evening or at dusk, attracted also to light.

HABITAT: Montane, inhabiting scrubby places, and shaded areas with *Pinus sylvestris* or *Fagus sylvatica* (EXPÓSITO HERMOSA pers. comm.). From 1,400 m up to 1,900 m above sea-level.

SIMILAR SPECIES: No similar species in Europe.

REMARKS: Genitalia and habitus suggesting isolated position of this species, female genitalia with certain similarity to *Idaea fuscovenosa*.

61. Idaea lutulentaria (STAUDINGER, 1892)

Acidalia lutulentaria STAUDINGER, 1892: Dt. ent. Z. Iris 5 (1): 153 (Spain, Andalusia: Sierra de Alfacar, near Granada, Chiclana; Portugal: Algarve). Syntypes 17 ♂♀ (MNHU: 10♂♀; examined).

Acidalia lutulentaria var. *aurata* MENDES, 1912: Broteria (Zool.) 10 (3): 175, pl. 3, figs. 18-20 (S. Fiel, Portugal). Syntypes (BMNH: 4♂♀, examined).

Unavailable names (nomen nudum): ?*zerlina*: MENDES (1903). – (incorrect subsequent spelling): *lutulentata*: PROUT (1913).

DIAGNOSIS: Wingspan ♂♀ 14-18 mm. Ground colour pale yellow, forewing costa darkened at base, black brown. In Portugal sometimes larger and deeper yellow ('f. *aurata*'). Transverse lines light brown. Discal spots black, sharp, distinct. Terminal line absent. Fringe concolorous. Frons brown, vertex dirty white, collar dark yellow. Antennae of ♂ with long antennomeres and intersegmental incisions, ciliate-setose, cilia 0.55-0.8 times width of flagellum. Hindtibia of ♂ slender, without pencil. Tarsus very long, 0.9-1.0 times length of tibia.

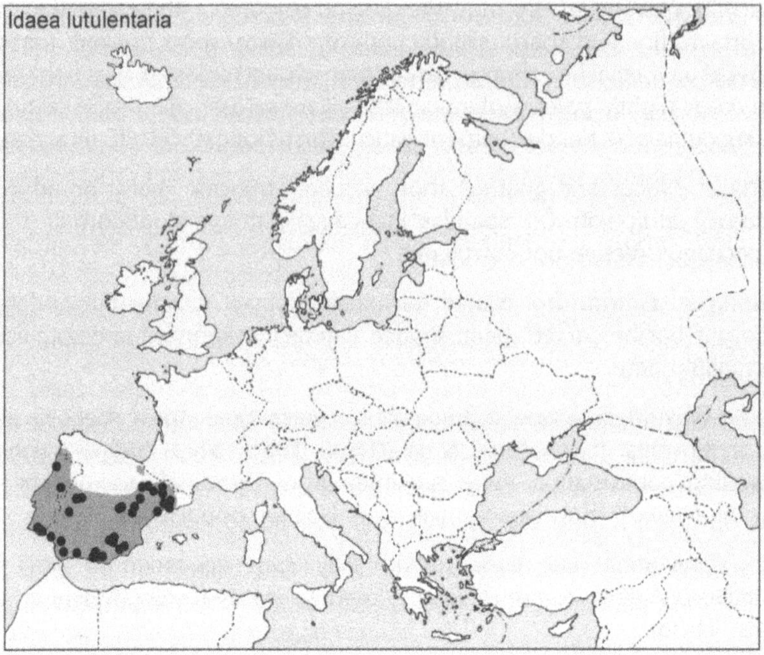

MALE GENITALIA: Uncus short digitiform. Lateral edges of juxta spinulose. Saccus shallow. Valva with two spinules at tip. Aedeagus with three stout cornuti of nearly equal length, at base with swollen, corrugated vesica. Terminal cornutus curved. Sometimes two other, very small, needle-shaped cornuti.

FEMALE GENITALIA: Antrum strongly dilated. Ductus bursae short, with membranous appendix. Corpus bursae slightly curved, longitudinally corrugated and spinulose nearly all over.

DISTRIBUTION: West-Mediterranean. Endemic to Portugal and Spain. – In Morocco replaced by the allopatric sister species *I. terminolineata* (ROTHSCHILD, 1933) (HAUSMANN 2003a).

PHENOLOGY: Univoltine: early June to mid-July, rarely until early August, e.g. on mountains. Larva slowly growing.

BIOLOGY: Larva probably polyphagous on withered leaves of herbaceous plants. Reared on dry or moulding leaves of *Sarothamnus scoparius* (MENDES 1912; PROUT 1913). Adults nocturnal, attracted to light.

HABITAT: Montane. Limited information available about habitat preference, collected in aisles of pine forest rich in low vegetation (PÖLL pers. comm.). In Portugal mainly in *Quercus suber* woodland (CORLEY pers. comm.). From 700 m up to 2,000 m above sea-level, in Portugal at lower altitudes, exceptionally down to sea-level.

SIMILAR SPECIES: *I. fuscovenosa* (see no. 59).

REMARKS: Sister species *I. terminolineata* reminiscent of *I. fuscovenosa* in habitus, see HAUSMANN (2003a).

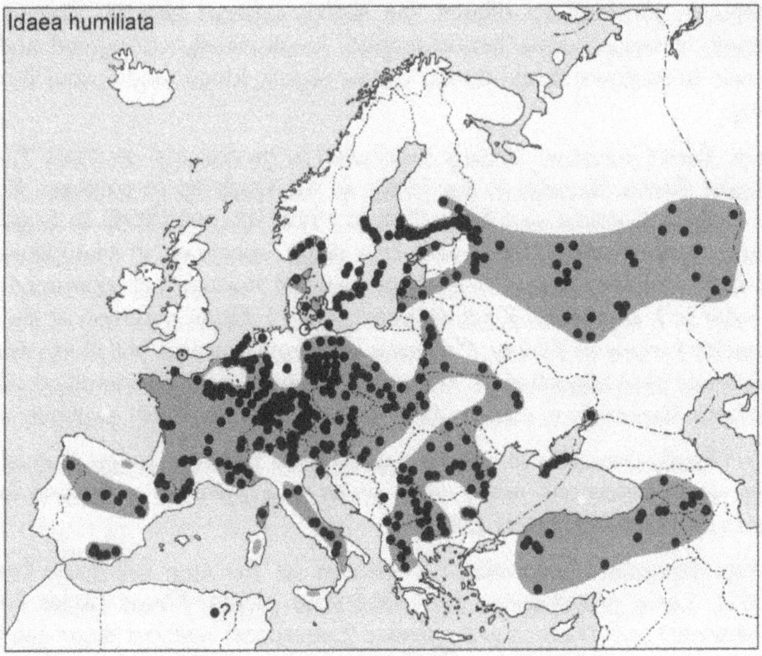

62. *Idaea humiliata* (HUFNAGEL, 1767)

Phalaena humiliata HUFNAGEL, 1767: Berlin. Mag. 4 (6): 614 (Germany: Berlin). Syntype(s) lost.

Geometra osseata DENIS & SCHIFFERMÜLLER, 1775: Ank. syst. Werkes Schmett. Wienergegend: 110 (Austria: Vienna distr.). Syntype(s) lost. Valid at subspecific rank for France in LERAUT (1997), but with clinous transitions to nominate subspecies. Junior synonym in SCOBLE (1999).

Geometra ossearia HÜBNER, 1799: Samml. Eur. Schmett. 5, Geometrae (1): pl. 19, fig. 102 (Europe). Syntype(s) lost. Figure unambiguous.

Unavailable names (misidentification): ?*spataceata*: sensu PANZER (1804) nec SCOPOLI, 1763 (identity: ?*Idaea deversaria*). – (infrasubspecific): *anastomosaria*: REISSER (1961); *impuncta*: LEMPKE (1949); *nigrolineata*: REISSER (1961); *suspiciata*: KRULIKOWSKY (1908; also nomen nudum). – (incorrect subsequent spellings): *bisseata* (for *osseata*): FREYER (1846); *hulimiata*: HUEMER & TARMANN (1993); *humuliata*: KARSHOLT & NIELSEN (1976); *odeata* (for *osseata*): JUNG (1782); *osearia*: TOMIC et al. (2002).

DIAGNOSIS: Wingspan ♂♀ 14-19 mm, ♀ on average slightly smaller, with narrower forewings. Pale sand coloured, often with light brown or reddish brown tinge, forewing costa darkened red brown. Intensity of reddish costal streak increasing with humidity and low temperatures (BERGMANN 1955). Transverse lines grey brown, postmedial line often darker than the others. Medial line usually crossing discal spot or running proximally. Discal spots small, black, sharp, rarely absent. Terminal line fine. Frons brown, vertex pale sand coloured, collar darker. Antennae of ♂ densely ciliate-setose, cilia 0.5-0.65 times width of flagellum (Text-fig. 51). Hindtibia of ♂ slender, without pencil. Tarsus very long, 1.0-1.1 times length of tibia.

MALE GENITALIA: Uncus short digitiform to sub-triangular. Juxta slightly spinulose laterally. Saccus V-shaped. Valva with two spinules at tip. Aedeagus with 3-4 cornuti: one large cornutus, 0.4-0.65 times length of aedeagus, with swollen, corrugated vesica at base. The other cornuti considerably smaller.

FEMALE GENITALIA: Antrum not dilated, but slightly tapered laterally. Ductus bursae of medium length. Corpus bursae lemon-shaped, longitudinally corrugated and spinulose nearly all over. In southern Spain ductus bursae slightly longer and corpus bursae anteriorly tapering.

DISTRIBUTION: Euro-Caucasian. Widely distributed in central and southern Europe from southern Spain (Sierra Nevada) to the Urals, in the north up to southern Scandinavia. Present on Corsica, Sardinia, and Sicily (TURATI 1913; MÜLLER 1996). In England, on the Isle of Wight, extinct since 1931, with one single specimen in Hampshire in 1954 (SKINNER 1984). Probably absent from large parts of Portugal, all examined specimens proved to refer to *I. bigladiata* (CORLEY pers. comm.). Often common at sites of occurrence. – Outside Europe in Turkey, Caucasus and Transcaucasus. All North African specimens, which had been identified as '*humiliata*', shown to be misidentified, identity usually *I. bigladiata*. Reputedly in eastern Algeria (1♀, leg. KORB, ZSM), probably mislabelled.

PHENOLOGY: Usually univoltine in one long generation: late May to late August, mainly in July, under good conditions rarely with single specimens of a second brood mid-September to mid-October (EBERT 2001).

BIOLOGY: Egg deposition unspecific, i.e. females let the eggs fall down from grasses (EBERT 2001). Larva polyphagous on herbaceous plants. Found under *Hippocrepis comosa* (Fabaceae) and *Thymus pulegioides* (Lamiaceae), without direct feeding record (EBERT 2001). Recorded as preferring *Ononis*, especially *O. spinosa* and *O. repens* (Fabaceae) (var. cit., see EBERT 2001), *O. arvensis* (Great Britain: ALLAN 1949; Russia: ANIKIN et al. 2000), also on fallen leaves (BERGMANN 1955) and on *Potentilla* (Rosaceae) (BOLDT 1929). Reared on *Taraxacum officinale, Polygonum aviculare, Salvia pratensis, Rumex, Veronica,* and on *Achillea millefolium* (var. cit., e.g. MILLIÈRE 1865; SCHNEIDER 1939; ALLAN 1949; SKINNER 1984; EBERT 2001; ERLEBACH pers. comm.). Adults easily flushed up from grass during daytime (afternoon or evening), actively flying at sunset and after sunrise (SKINNER 1984), suboptimally attracted to light.

HABITAT: North of Alps xerothermophilous. Open dry grassland, sun-exposed slopes, gardens, embankments, dry salt marshes and calcareous or sandy dunes, mainly where *Ononis repens* occurs. In the south more restricted to temperate habitats in the mountains. In the north from 0 m up to 700 m above sea-level, up to 1,500 m in southern Alps, sometimes up to 2,000 m in southern Europe and Turkey. In southernmost Europe montane, rarely below 1,000 m.

SIMILAR SPECIES: Different from similar species (*I. dilutaria, I. fuscovenosa*, see no. 58, 59) in darker ground colour, small, fine discal spots, and reddish forewing costa. *I. bigladiata* on average larger, paler, ♂ hindtarsus slightly longer, secure identification often requires dissection, see also remarks.

REMARKS: Specimens with intermediate habitus and genitalia (e.g. ♂ with three cornuti) in the Sierra Nevada region, southern Spain, possibly indicating local gene flow between *I. humiliata* and *I. bigladiata*. Further study is needed to clear up distribution borders. For morphology of immature stages see ABBOTT (1898).

63. *Idaea bigladiata* HERBULOT, 1975

Idaea bigladiata HERBULOT, 1975: Alexanor 9 (4): 192 (Algeria: Tlemcen). Holotype ♂ (ZSM/HERBULOT, examined).

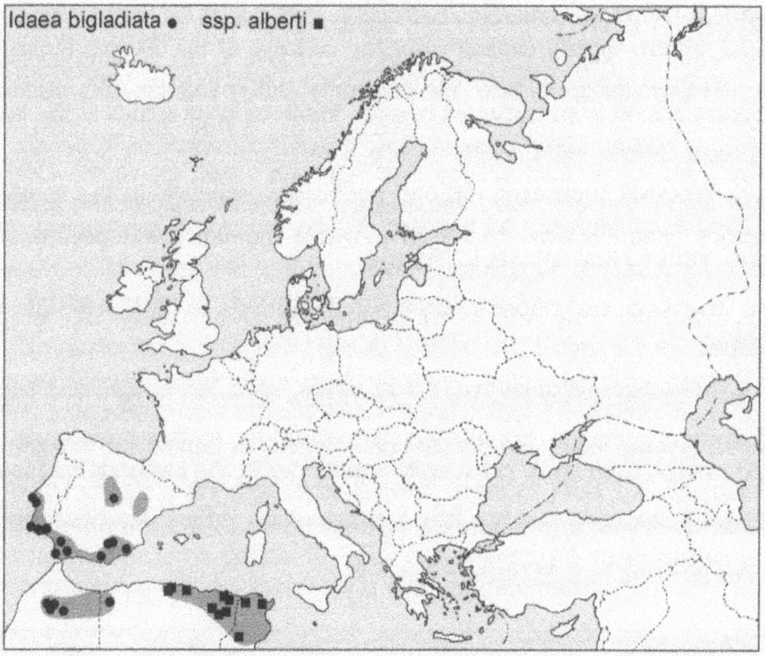

Idaea bigladiata alberti subsp. n.: HOLOTYPE: ♂, 'Tunisia, Ain Draham, 20 Oct. 1913', ZSM/HERBULOT. – PARATYPES: 2♂, id.; 2♂, [Algeria], 'Le Tarf, Mai 05; 18-6-04', ZSM/HERBULOT; 1♂, 'Algeria, Tala Rana, 1909', ZSM/HERBULOT; 2♂1♀, [Algeria], 'Fort National, 23.-26-6-37', ZSM/HERBULOT; 1♀, 'Tunis, Tunisia', ZSM/HERBULOT; 1♀, [Algeria], 'Le Tarf, 28-6-06', ZSM; 1♀, 'Tunisie, Soudia, 23-5-41', ZSM; 3♂3♀, 'Tunisie, Tunis, V.06; IV.12; 2.V.11; 3.VI.11', ZSM; 1♀, 'Tunisia, Oasis Gafsa', ZSM. – Description below. Further material from prov. Algier. The new subspecies is dedicated to Edgar ALBERT, Munich, in gratitude for his diligence in the preparation of the genitalia drawings.

Unavailable names (misidentification): *fuscovenosa*: sensu MONTEIRO & PASSOS DE CARVALHO (1984) nec GOEZE, 1781; *humiliata*: sensu STAUDINGER (1901; partim) nec HUFNAGEL, 1767. – (incorrect subsequent spelling): *biglandiata*: AGENJO (1976).

DIAGNOSIS: Larger than preceding species, wingspan ♂♀ 17-24 mm. Wings pale sand coloured, forewing costa usually slightly darkened brown at base. Wing pattern variable, transverse and wavy lines grey brown, usually distinct. Medial line distal to discal spot, rarely crossing. Discal spots black, distinct. Frons brown, vertex pale sand coloured, collar darker. Antennae of ♂ ciliate-setose, cilia 0.4-0.6 times width of flagellum. Hindtibia of ♂ slender, without pencil. Tarsus extremely long, about 1.2 times length of tibia (Text-fig. 26).

Subsp. ***alberti*** subsp. n. smaller, 17-20 mm, exceptionally 15 mm only. Without other constant differential features from nominate subspecies, but for genitalic differences see below. Types under nomenclatural paragraphs.

MALE GENITALIA: Uncus triangular, sometimes sub-truncate. Lateral edges of juxta smooth. Saccus V-shaped, shallow in subsp. *alberti*. Valva very long, with two spinules at tip, shorter in subsp. *alberti*. Aedeagus with two stout, lance-shaped cornuti, both 0.5-0.6 times length of aedeagus, one with swollen, corrugated vesica at base. Aedeagus shorter and cornuti broader in subsp. *alberti*.

FEMALE GENITALIA: Antrum not dilated, but slightly concave at tip, and tapered laterally. In subsp. *alberti* antrum slightly dilated, distinctly concave at tip. Ductus bursae comparatively long, shorter in subsp. *alberti*. Corpus bursae sub-triangular, with distinct spinulose appendix bursae arising from posterior margin. Patch of large spines at the latero-anterior part of corpus bursae, here, subterminally, a small membranous appendix.

DISTRIBUTION: West-Mediterranean. Iberian peninsula, especially in the southern half. – Outside Europe from Morocco to western Algeria (nominate subspecies), replaced by subsp. *alberti* from central Algeria to Tunisia.

PHENOLOGY: Trivoltine: early April to early June; mid-June to early July; late September to late October.

BIOLOGY: Larval host-plants unknown. Adults readily attracted to light and bait.

HABITAT: Xerothermophilous. No detailed data traced for habitat preference. From 0 m up to 2,000 m above sea-level, but usually commoner in the lowlands and foothills.

SIMILAR SPECIES: *I. dilutaria* (see no. 58), *I. fuscovenosa* (59), *I. humiliata* (62).

REMARKS: See remarks to *I. humiliata* (no. 62).

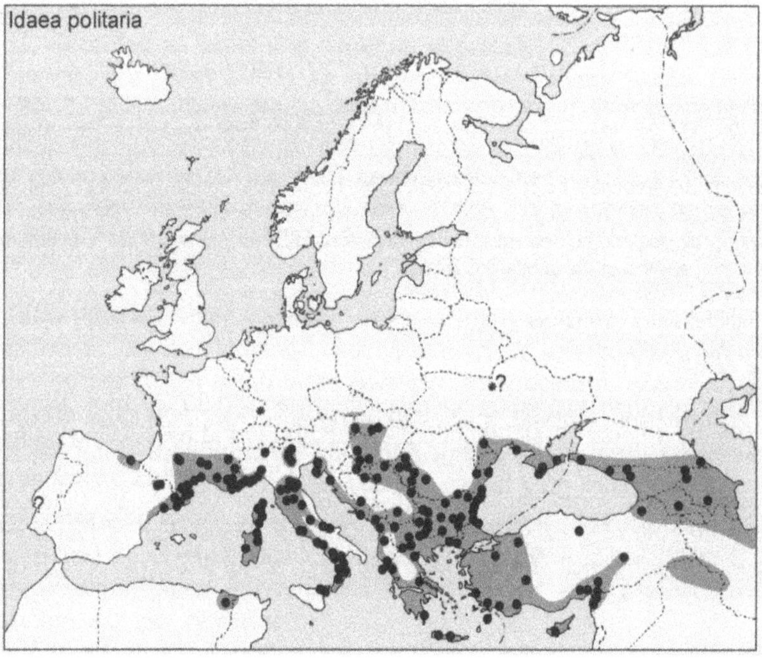

64. *Idaea politaria* (HÜBNER, 1799)

Geometra politaria HÜBNER, 1799: Samml. Eur. Schmett. 5, Geometrae (1): pl. 12, fig. 61 (southern Europe). Syntype(s) lost.

‡ *Phalaena Geometra politata* HÜBNER, 1793: Samml. auserl. Vög. und Schmett.: [5], pl. 1 (southern Europe). Syntype(s) lost. Junior primary homonym nec FABRICIUS, 1781, the latter being an emendation of CRAMER's *politia* (identity: *Nepheloleuca politia*, Ennominae, Neotropis).

Ptychopoda mundipennis WARREN, 1898: Novit. zool. 5: 21 ('Madagascar': incorrect locality). Holotype ♂ (BMNH).

Idaea politata abmarginata HAUSMANN, 1995: Mitt. Münch. ent. Ges. 85: 83. First available name after *abmarginata* BOHATSCH, 1885 (Wien. ent. Ztg. 4: 145: infrasubspecific) (Croatia, Slavonia: Lipik). Syntype(s), of BOHATSCH (ZFMK: 1♀ examined; ZSM: material from Cyprus mentioned in HAUSMANN 1995). Herewith downgraded from subspecific rank (HAUSMANN 1995; SCOBLE 1999) to synonymy (syn. n.).

DIAGNOSIS: Wingspan ♂♀ 14-18 mm. Wings pale sand coloured, glossy, forewing costa darkened black brown at base. Terminal area concolorous or darker. Postmedial line brown, distinct. Medial line indistinct. Discal spots black, distinct. Frons brown, vertex pale sand coloured, collar darker. Antennae of ♂ with intersegmental incisions, ciliate-setose, cilia 0.6-0.75 times width of flagellum. Hindtibia of ♂ slender, with elongate lateral scales. Tarsus very long, 0.80-0.9 times length of tibia.

Variability: Nominate form with dark, glossy grey brown terminal area, usually predominating in populations from Portugal to Italy except north-easternmost provinces. Forms with terminal area nearly concolorous with ground colour ('f. *abmarginata*') predominating in south-eastern Alps and in the eastern Mediterranean with varying frequency.

MALE GENITALIA: Uncus short digitiform. Lateral edges of juxta slightly spinulose. Valva slightly curved, with two, rarely three, spinules at tip. Aedeagus with one narrow, curved terminal cornutus, one large central (0.4-0.5 times length of aedeagus) and 4-7 smaller cornuti. One examined ♂ from Turkmeniya without differences.

FEMALE GENITALIA: Antrum dilated. Ductus bursae of medium length, bulgily dilated, with very small spinulose patch. Corpus bursae pyriform, longitudinally corrugated, and spinulose in the posterior, lateral half.

DISTRIBUTION: Mediterranean (European-Westasiatic). From northernmost Iberian peninsula to Balkan peninsula and Crimea. In the north up to southern Switzerland (REZBANYAI-RESER 1985), Czech Republic and Slovakia (HRUBY 1964; LAŠTUVKA 1998), and eastern Austria (1♂, Oberweiden, ZSM; not mentioned in HUEMER & TARMANN 1993). One specimen found in south-western Germany, not resident (EBERT 2001). Reputedly in Portugal (e.g. SILVA CRUZ & WATTISON 1935; VIVES MORENO 1994; MÜLLER 1996), and northern Ukraine ('Kiev': leg. SHELJUZHKO, ZMKU; KOSTJUK 1986), data await confirmation. Not yet recorded from Malta and Crete. – Outside Europe from Turkey, Cyprus, Caucasus and Transcaucasus to northern Iran and Turkmeniya (KUZNETZOV 1960; dissected material at ZSM). One specimen from eastern Algeria (ZSM). In central Asian mountains replaced by the sister species *Idaea subpolitata* MIRONOV, 1986. Description of another sister species from Israel in preparation by the author.

PHENOLOGY: Univoltine: early June to late July, exceptionally single specimens of a partial second brood until mid-September. In Turkmeniya emerging from late May.

BIOLOGY: Polyphagous on herbaceous plants (REBEL 1910b; FORSTER & WOHLFAHRT 1981). Reared on dry petals and dry leaves of various plant species (LEIPNITZ pers. comm.). Adults readily attracted to light.

HABITAT: Xerothermophilous lowland species, typically occurring in hot wetland near coast, along rivers, and other Mediterranean habitats. From 0 m up to 900 m above sea-level. In the south-eastern part of the distribution exceptionally up to 1,800 m.

SIMILAR SPECIES: Nominate form without similar species in Europe. Forms with bright terminal area somewhat reminiscent of *I. mancipiata* and *I. biselata*, both species however without glossy coloration, forewing apex not darkened, *I. biselata* with dark ♂ hindtibial pencil.

The **longaria** species-group

STERNECK's (1940) 15th group of *Idaea*. Monotypic in STERNECK (1940: 141), but the Canarian sister species *I. abnorma* (PINKER, 1960), and, possibly, *I. atlantica* (STAINTON, 1859) from Madeira belong here. Relationships to the other species-groups unclear. Genitalia may indicate certain relations to the *nexata* species-group. Sternum A2 of ♂ with very narrow, semicircular, sclerotized ridge, but setose patch absent, these features reminiscent of the *nexata* species-group, too. Proboscis weak and short. Palpi very short, only 0.5 times diameter of eye. Egg sculpturation polygonal reticulate (GLASER 1958). Larva of the long and slender type.

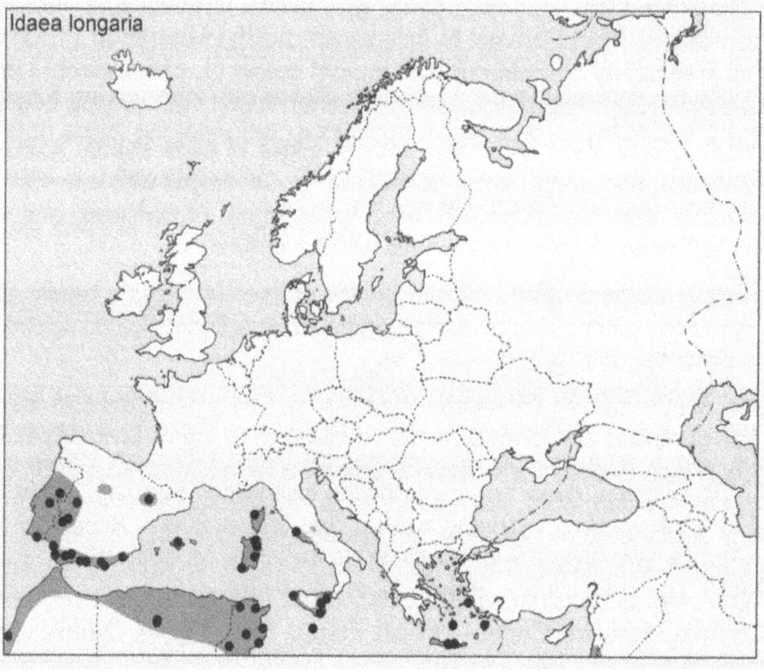

65. *Idaea longaria* (HERRICH-SCHÄFFER, 1852)

Acidalia longaria HERRICH-SCHÄFFER, 1852: Syst. Bearb. Schmett. Eur. 6 (55): 66 (Spain, Andalusia: Ronda). Syntypes, at least 1♂1♀ (MNHU 1♂1♀, examined). Referring to vol. 3: 16, hence binominal.

Ptychopoda faroulti ROTHSCHILD, 1914: Novit. zool. 21: 344 (central Algeria: Guelt-es-Stel). Holotype ♀ (BMNH). Herewith downgraded from subspecific rank (SCOBLE 1999; TRUSCH & MÜLLER 2001) to synonymy (syn. n.).

Unavailable names (misidentification): *sublongaria*: sensu STERNECK (1940) nec STAUDINGER, 1899 (identity: *Idaea allongata sublongaria*, Levant). – (non binominal): *prolongata* RAMBUR, 1866. – (incorrect subsequent spelling): *faroutti*: RAINERI (1996).

DIAGNOSIS: Wingspan ♂♀ 17-22 mm, in autumn generation sometimes 14 mm only. Forewing longer than hindwing, reminiscent of *Eupithecia*. ♀ with forewing narrower, apex pointed. Hindwing termen slightly concave between M1-M3. Fairly variable in coloration and wing pattern. Ground colour white or dirty white in female, light or dark grey in male. Populations of Iberian peninsula with particularly dark males. Transverse lines oblique on forewing. Postmedial line dentate, close to termen also on hindwing. Discal spots present. Frons dark brown, vertex and collar dirty white. Antennae of ♂ ciliate, sub-fasciculate, cilia 1.25-1.6 times width of flagellum (Text-fig. 52), in the Iberian peninsula up to 2 times. Hindtibia of ♂ slender, with elongate lateral scales. Tarsus 0.3-0.45 times length of tibia.

MALE GENITALIA: Uncus slender. Saccus very broad. Valva very slender, tapering to a small hook. Aedeagus with 6-15 cornuti, usually about 10, basal and distal cornuti smaller than the central.

FEMALE GENITALIA: Antrum dilated, cup-shaped. Ductus bursae of medium length, distinctly curved. At the junction with corpus bursae a small curved patch of micro-spinules, and a larger, ribbon-shaped patch of spinules. Corpus bursae membranous, bilobed at both sides of the spinose ribbon.

DISTRIBUTION: Mediterranean. Southern Iberian peninsula, and most Mediterranean islands. Isolated occurrences in north-eastern Spain and central Italy (Lazio: REZBANYAI-RESER 1997; RACHELI et al. 1978). – Outside Europe from the Canary island Fuerteventura across whole of North Africa to the Levant. Reputedly in Turkey (RIEMIS 1994), data await confirmation. On western Canary islands replaced by the sister species *I. abnorma*.

PHENOLOGY: Bi- or trivoltine: common from early March to early May; late September to early December. Specimens collected from late May to mid-July either referring to late emerged representatives of first generation or to a summer generation. Autumn (winter) generation in North Africa sometimes until late January, in Corsica and Sardinia and Greek islands emerging from early September.

BIOLOGY: Larva probably polyphagous on herbaceous plants, preferring dry leaves (PROUT 1913). Rearing easy on withered leaves of *Polygonum aviculare*, and *Lactuca sativa* (KALCHBERG 1895; GLASER 1958; KING & ROMERA 2004), and on petals of *Taraxacum officinale, Hieracium,* and *Campanula* (TRUSCH & MÜLLER 2001; MÜLLER pers. comm.). Adults readily attracted to light. Resting position with spread wings and hindwings under the forewings as in *Eupithecia*.

HABITAT: Xerothermophilous. Usually restricted to hot open habitats in the lowlands near the coast, often on dry, sandy or stony soils, rarely up to 600 m above sea-level, in Algeria and the Levant up to about 1,000 m.

SIMILAR SPECIES: *I. calunetaria* (see no. 50), *I. seriata* and *I. minuscularia* with shorter and broader wings, postmedial line of hindwing more proximal and less parallel to termen. *I. calunetaria* with shorter ♂ antennal cilia and shorter ♂ hindtarsus. *I. seriata* and *I. minuscularia* with longer ♂ hindtarsus. Dark males of *I. longaria* may be confused with *I. carvalhoi*, the latter with forewings narrower, ground colour darker, ♂ hindtarsus slightly shorter. Allopatric *I. abnorma* from Canary islands smaller, genitalia of both sexes smaller, 0.6 times those of *I. longaria*, antrum only slightly dilated, spinose patch of corpus bursae very small. Small pale forms of the extralimital *I. allongata* (STAUDINGER, 1898) from the Levant with yellowish or sand coloured tinge, postmedial line more diffuse, less close to termen.

REMARKS: Relationships between *I. longaria* and *I. abnorma* merit further attention and taxonomic revision, see also differential diagnosis in PINKER (1960). For rearing and morphology of immature stages see GLASER (1958).

The ***nexata*** species-group

STERNECK's (1940) 16th group of *Idaea*. Phylogenetic relationships between the included species questionable (STERNECK 1940: 141). Extralimital *I. subpurpurata* from the Middle East transferred to the *gracilipennis* species-group (HAUSMANN 1991d: 127). Sternum A2 of ♂ with very narrow, semicircular, sclerotized ridge, but setose patch absent. Proboscis short. Frons slightly prominent near palpi. *I. manicaria* should, perhaps, be transferred to *Brachyglossina* (see remarks to *I. manicaria*). Possible relationships also to the *fractilineata* species-group (see there), and to the genus *Cleta* (see remarks to *I. nexata*).

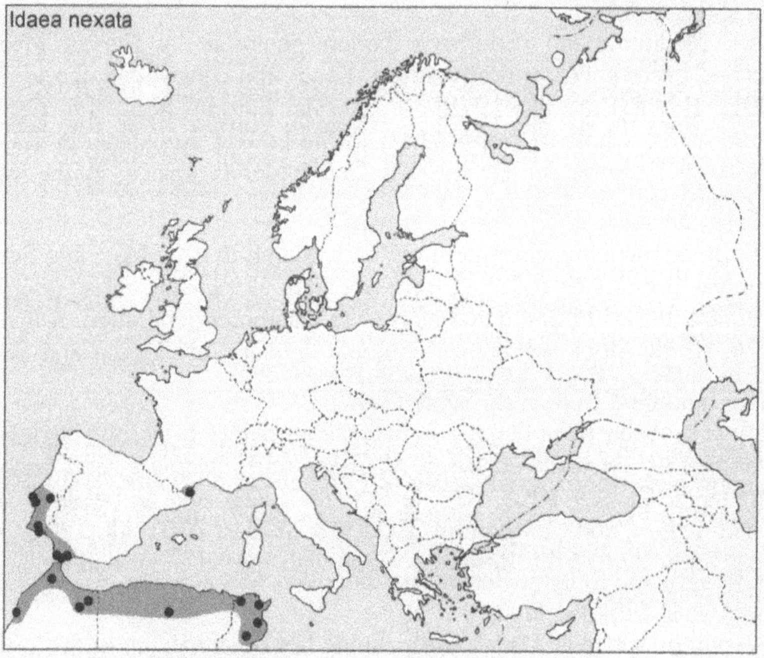

66. Idaea nexata (HÜBNER, 1813)

Geometra nexata HÜBNER 1813: Samml. Eur. Schmett. 5, Geometrae (1): pl. 78, fig. 401; pl. 82, fig. 421 (Europe). Syntype(s) lost.

‡ *Emmiltis nexaria* HÜBNER, 1825: Verz. bekannter Schmett.: 309. Incorrect subsequent spelling.

Acidalia cirtanaria H. LUCAS, 1849: Explor. Sci. Alg. Anim. Artic. 3: 395, pl. 4, fig. 3 (eastern Algeria: Constantine, gardens of Salah'-Beh). Syntype(s) (MNHN). Herewith downgraded from subspecific rank (SCOBLE 1999) to synonymy (syn. n.).

Acidalia nexata var. *marambaudista* D. LUCAS, 1924: Bull. Soc. ent. Fr. 1924 (5/6): 78 (Morocco: Taourirt region). Lectotype ♂ (ZSM/HERBULOT, examined). Herewith downgraded from subspecific rank (SCOBLE 1999) to synonymy, previously stated in HERBULOT (1984a).

Unavailable names (incorrect subsequent spelling): *nexataria*: VIVES MORENO (1994).

DIAGNOSIS: Wingspan ♂♀ 9-12 mm. Ground colour white. Terminal area of forewing dark brown, divided by white wavy line. Medial and antemedial lines dark brown, often anastomosing. Discal spots very fine. Hindwing white with just a few dark brown markings, mainly towards termen. Frons and vertex brown. Eyes small, distant from each other. Proboscis stout, but short. Antennae of ♂ ciliate-fasciculate, cilia 1.35-1.55 times width of flagellum (Text-fig. 53). Hindtibia of ♂ slender, without pencil. Tarsus 0.4-0.6 times length of tibia. Trochanter of foreleg and hindfemur with long hairy scales in ♂, in ♀ smoothly scaled.

MALE GENITALIA: Uncus digitiform. Gnathos beak-shaped. Valva terminally tapering. Aedeagus broad and short, usually with 8 cornuti: two long, curved, basal cornuti, and about 6 terminal cornuti of medium length.

FEMALE GENITALIA: Antrum not dilated, but laterally with tapered process. Ductus bursae of medium length, with three or four spinules near corpus bursae. Corpus bursae membranous, irregularly shaped, at both sides of the junction with ductus bursae.

DISTRIBUTION: West-Mediterranean. Southern Iberian peninsula and isolated occurrence in eastern French Pyrenees (GIBEAUX 1993). – Outside Europe from Morocco to Tunisia, and Libya (Cyrenaica: TURATI & ZANON 1922). Reputedly in Israel (AMSEL 1933; MNHU/PÜNGELER: Jerusalem, Haifa), but no records since then.

PHENOLOGY: Tri- or plurivoltine: early March to early July; mid-October to late October, but recorded also in December (PROUT 1913) and early February (RIBBE 1912).

BIOLOGY: Larva posssibly mono- or oligophagous, preferring Scrophulariaceae. Found on *Linaria gharbensis* (Scrophulariaceae) (Morocco: RUNGS 1981). Reared on petals of *Chaenorhinum origanifolium* (*Linaria origanifolia*) (MILLIÈRE 1864). Adults diurnal, also attracted to light, readily attracted to pheromones of Sesiidae (Tunisia: BLAESIUS pers. comm.). ♀ ratio low in collections.

HABITAT: Xerothermophilous lowland species, in Europe restricted to coastal dunes (GELBRECHT pers. comm.). From 0 m up to 500 m above sea-level, in central Algeria up to 1,000 m.

SIMILAR SPECIES: No similar species in Europe.

REMARKS: Habitus, size and position of eyes, and hairy scales on ♂ legs reminiscent of genus *Cleta*. Phylogenetic relationships cannot be excluded, the question merits further attention.

67. Idaea manicaria (HERRICH-SCHÄFFER, 1852)

Acidalia manicaria HERRICH-SCHÄFFER, 1852: Syst. Bearb. Schmett. Eur. 6 (55): 67. First available name after *manicaria* HERRICH-SCHÄFFER, 1851 (pl. 77, fig. 471: non binominal) (Spain, Andalusia: Ronda). Holotype ♂ (MNHU, examined). Referring to vol. 3: 17, thus binominal.

Acidalia volitaria JOANNIS, 1891: Bull. Soc. ent. Fr. 1891: 79 (eastern Algeria: Philippeville). Syntypes 5 (MNHN).

Acidalia fatimata STAUDINGER, 1895: Dt. ent. Z. Iris 7 (2): 288, pl. 9, fig. 18 (Spain, Granada). Lectotype ♂ (MNHU, examined). Herewith designated to fix type locality and to stabilize usage of name with regard to some very similar species; at MNHU paralectotypes from Algeria: Prov. Biskra (2♀), Tenied el Had, 100km SW Algier (1♀), Mauritania (1♀, coll. PÜNGELER). Herewith downgraded from species rank (MÜLLER 1996; SCOBLE 1999) to synonymy (syn. n.).

Unavailable names (misidentification): *attenuaria*: sensu SAMMUT (2000) nec RAMBUR, 1833.

DIAGNOSIS: Wingspan ♂♀ 16-19 mm in spring generation, 11-15 mm in summer and autumn generations, exceptionally 9 mm only. Ground colour light ochreous with slight orange tinge. Terminal field of forewing light brown, often paler towards termen. Transverse lines distinct, brown, not waved. Discal spots brown. Frons dark brown, vertex light brown. Proboscis weak and short, sometimes absent. Antennae of ♂ ciliate-fas-

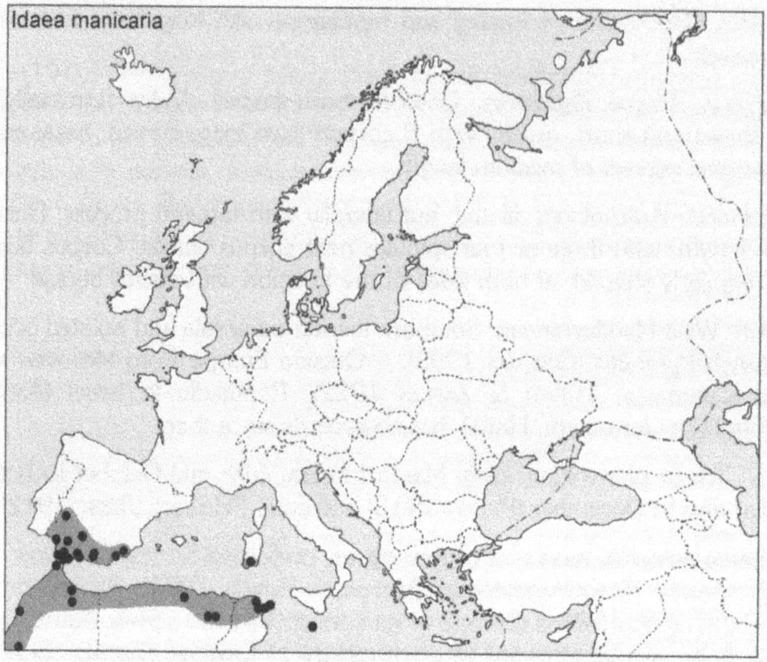

ciculate, cilia long, 1.6-1.8 times width of flagellum. Hindtibia of ♂ slender, without pencil. Tarsus not shortened, 1.0-1.2 times length of tibia.

MALE GENITALIA: Uncus slender. Juxta anteriorly bilobed. Valva subterminally dilated and with a small subterminal spinule on ventral side. Aedeagus comparatively slender, with three small cornuti in the central part and about six terminal cornuti.

FEMALE GENITALIA: Antrum and ductus bursae reduced to a laterally tapering transverse sclerite. Corpus bursae membranous, spinulose, slightly corrugated, reniform, asymmetrically connected with ductus bursae.

DISTRIBUTION: West-Mediterranean. Southern Spain, Sardinia (RAINERI 1991), and Malta (VALLETTA 1973; SAMMUT 2000, examined; MÜLLER 1996: '*fatimata*'). Erroneous records for Catalonia ('*fatimata*') corrected by DANTART (2000). – Outside Europe from Morocco to Tunisia.

PHENOLOGY: Bi- or trivoltine, from mid-March to late May; mid-July to early October. First generation in the mountains later, until late June.

BIOLOGY: Larva probably polyphagous on herbaceous plants. Reared on strongly withered, moist plant detritus, not accepting dry plants (Spain: LEIPNITZ pers. comm.). Larvae extremely photophobic, night-active, building loose silky caves, pulling pieces of moist plant detritus into the cave (LEIPNITZ pers. comm.). Adults attracted to light, ♀ ratio at light very high, about 90%.

HABITAT: Thermophilous. Often on hot localities near coast. From 0 m up to 1,300 m above sea-level.

SIMILAR SPECIES: *I. exilaria* darker, with red brown coloration, hindwing without discal spot, ♂ hindtarsus much shorter. *I. subsaturata* with wings more rounded, medial and antemedial lines more diffuse, postmedial line more convex, proboscis well developed.

REMARKS: ♂ and ♀ genitalia features, reduced proboscis, coloration and larval behaviour reminiscent of *Brachyglossina*, such as the *B. culoti* species-group from North Africa and the *B. sonyae* species-group from Arabia. Thus, transfer to *Brachyglossina* could be necessary.

The *consolidata* species-group

STERNECK's (1940) 18th group of *Idaea*. Structurally homogenous group, monophyletic (STERNECK 1940: 141). Close relationship to the *seriata* species-group stated by MENTZER (1990a). With phylogenetic tendency to dilatation of ♂ hindtibia and tarsomeres. Some extralimital species without proboscis. Hindtarsus of ♂ strongly shortened. Sternum A2 of ♂ with distinct horseshoe-shaped sclerite and setose patch; very weakly developed, however, in *I. descitaria*. Male genitalia large, uncus slightly hook-shaped at tip (in lateral view), saccus often enlarged, aedeagus long, vesica strongly folded and slightly granulated, with one cornutus (secondarily lost in *I. consolidata*). In ♀ genitalia ductus bursae long, corresponding to long aedeagus of ♂, corpus bursae strongly spinulose, ductus seminalis arising posteriorly. Larva long, dark.

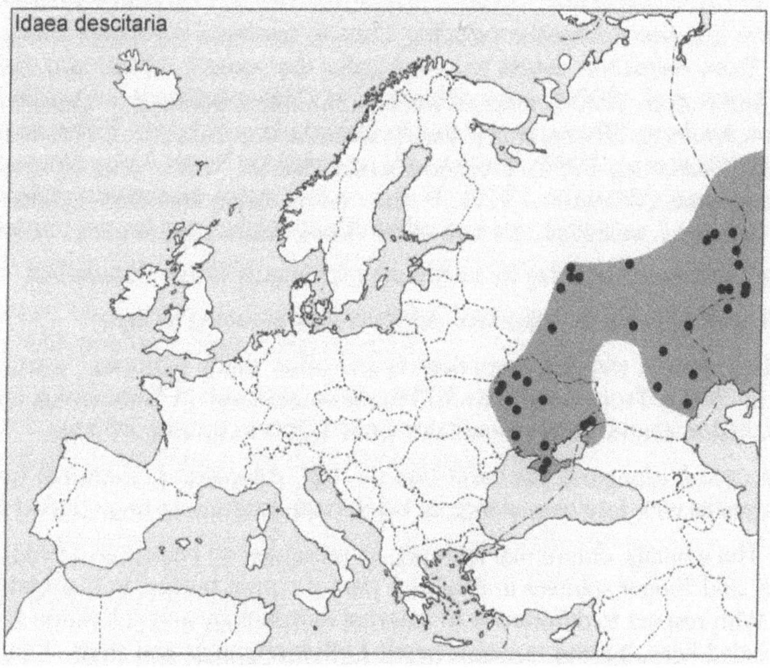

68. *Idaea descitaria* (CHRISTOPH, 1893)

Acidalia descitaria CHRISTOPH, 1893: Dt. ent. Z. Iris 6: 94 (south-eastern European Russia: Sarepta; Urals: Guberli; Taganrog; central Asia: Ashkhabad, Kuldscha). Syntype(s) (ZISP), lectotype should be selected from the Urals.

Acidalia velitchkovskyi REBEL, 1907: Verh. zool.-bot. Ges. Wien 57 (2/3): (37), fig. (south-western European Russia: Walouiki). Syntypes 2♀ (NHMW: 2♀ exmained).

‡? *Sterrha elongaria* [var.?] *favillata* PROUT, 1935: in SEITZ, Macrolep. 4, Suppl.: 63. First name after *favilla-*

ta PROUT, 1934 (unavailable, introduced in synonymy at 4th position) (south-eastern European Russia: Krasnoarmeysk, 'Sarepta'). Holotype (BMNH). Probably a junior synonym of *descitaria*. Introduced as 'provisional name'.
Unavailable names (misidentification): *effeminata*: sensu ANIKIN et al. (2000) nec STAUDINGER, 1892 (identity: *Idaea elongaria* syn.?); *elongaria*: sensu PROUT (1935, partim) nec RAMBUR, 1833; *pecharia*: sensu PROUT (1913, partim) nec STAUDINGER, 1863 (identity: *Idaea elongaria pecharia*). – (incorrect subsequent spelling): *velitschkovskyi*: PROUT (1913).

DIAGNOSIS: Wingspan ♂♀ 16-21 mm. Ground colour grey brown. Postmedial line dark grey strongly dentate, sharp. Discal spots present. Some examined specimens from Astrachan desert smaller (12-16 mm), ground colour paler, sand coloured. Frons and collar dark brown, vertex whitish. Antennae of ♂ ciliate-fasciculate, cilia 1.1-1.3 times width of flagellum (Text-fig. 54), in the sister species *I. soldaitisi* HERBULOT, 1994 1.3-1.5 times. Hindtibia of ♂ slender, with large pencil. Tarsus 0.35-0.5 times length of tibia, in the sister species *I. soldaitisi* 0.6-0.7 times.

MALE GENITALIA: Uncus slender triangular. Saccus distinctly projecting, digitiform. Valva simple. Aedeagus long and slender, with one stout cornutus, about 0.6 times length of aedeagus.

FEMALE GENITALIA: Antrum slightly dilated, posteriorly truncate. Ductus bursae of medium length and width. Corpus bursae '*Eupithecia*-like', globular, with posterior spinulose crest and large lateral spinulose patch. Appendix bursae from posterior part of corpus bursae.

DISTRIBUTION: Eurasiatic. Ukraine including Crimea, southern European Russia to southern Urals. From Astrachan desert reported under the name *I. effeminata* (STAUDINGER, 1892) by ANIKIN et al. (2000), original material in ZISP (examined). – Outside Europe in Turkmeniya, southern Siberia, in the east to Dahuria (e.g. KOSTJUK 1992; KHOMENKO et al. 1994; KOSTJUK et al. 1994). Erroneously recorded for North Africa (WILTSHIRE 1983) and Transcaucasus (VIIDALEPP 1976). In the central Asian mountains replaced by the allopatric vicariant *I. soldaitisi*, for two other closely related species see remarks.

PHENOLOGY: Bivoltine: mid-May to early July; mid-August to late September.

BIOLOGY: Larval host-plants unknown. Adults readily attracted to light.

HABITAT: Dry lowland steppes, semi-deserts and other sandy biotopes, 'warm clearings' (ANIKIN et al. 2000). From 0 m up to 300 m above sea-level, in Turkmeniya up to 2,300 m. *I. soldaitisi* in central Asian mountains from 1,200 m up to 2,700 m.

SIMILAR SPECIES: *I. elongaria pecharia* (see no. 52). Allopatric *I. soldaitisi* from central Asian mountains with forewing elongate, colour with less brown tinge, tarsus longer.

REMARKS: The genitalic differential features, as presented by HERBULOT (1994), i.e. 'larger gnathos' and 'longer spinules in posterior part of corpus bursae' in *I. soldaitisi* are not constant. With respect to differences in external morphology and coloration species rank is not excluded here, but the question needs further attention and study. *I. curtopedata* (EBERT, 1965) from Afghanistan may be conspecific, too. For close relations to *I. conioptera* (HAMPSON, 1903) see PROUT (1935a).

69. Idaea vesubiata (MILLIÈRE, 1873)

Acidalia vesubiata MILLIÈRE, 1873: Revue Mag. Zool. (3) 1: 6 (south-eastern France: Saint Martin, Vésubie). Syntypes, perhaps lost (MNHU, 2 'cotypes' examined, but type status uncertain).
Unavailable names (infrasubspecific): *demarginata*: REISSER (1961); *tripartita*: WEHRLI (1924).

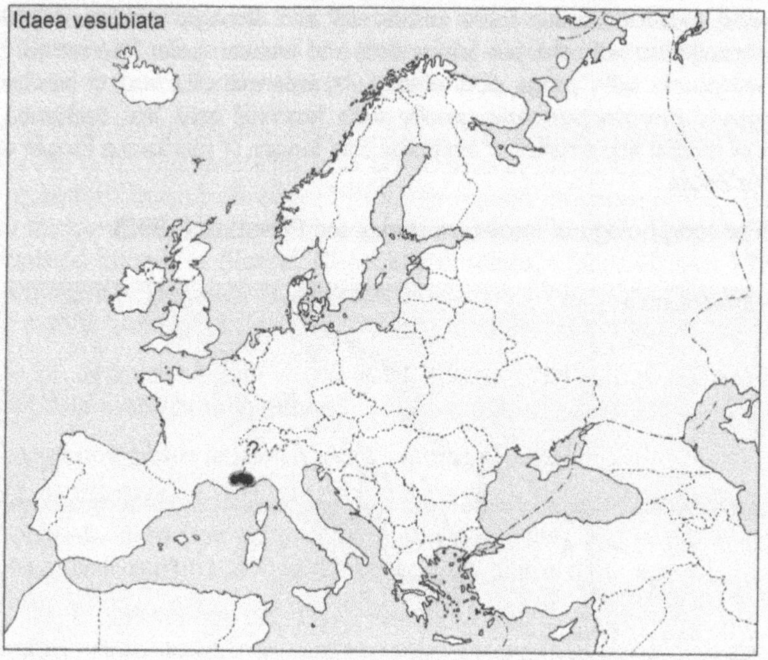

DIAGNOSIS: Wingspan ♂♀ 15-20 mm. Ground colour clear white to dirty white, terminal area darker, mainly in males, sometimes with slight blue tinge ('f. *tripartita*'). Postmedial line dark grey, dentate, sharp. Medial and antemedial lines present, sharp. Discal spots distinct. Terminal dots elongate. Frons black brown, vertex whitish, collar brown. Antennae of ♂ ciliate-setose, cilia 0.7-0.8 times width of flagellum. Hindtibia of ♂ not dilated, with pencil. Tarsus slender, 0.25-0.35 times length of tibia.

MALE GENITALIA: Uncus slender. Saccus triangular. Valva tapering at tip. Aedeagus long and slender, with one very small, distal cornutus.

FEMALE GENITALIA: Ductus bursae long, S-shaped, with lateral spinose crest in the lower half. Corpus bursae sac-shaped, densely spinulose all over. Appendix bursae arising from posterior part of corpus bursae, slender, membranous, looped.

DISTRIBUTION: Endemic to south-eastern France (Provence; Alpes Maritimes; Basses Alpes), and north-western Italy (Piedmont). Reputedly also in southern Switzerland (Valesia) (RAPPAZ 1979; MÜLLER 1996), awaiting confirmation. Distribution area very restricted, rare and local, threatened.

PHENOLOGY: Bivoltine: late May to early June; mid-July to mid-August.

BIOLOGY: No detailed information on larval host-plants traced. Reared on dry and withered leaves of various herbaceous plants (PÜNGELER 1892; LHOMME 1935). Adults resting on rocks during daytime (PÜNGELER 1892; PROUT 1913), attracted to light.

HABITAT: Possibly montane. Limited information available about habitat preference. From 500 m up to 1,100 m above sea-level.

SIMILAR SPECIES: *I. typicata* (see no. 41), *I. albarracina* (48), *I. incisaria* (49). *I. libycata* sand coloured, postmedial line more proximally and strongly convex, ♂ antennal cilia longer. *I. consolidata* with sharper fringe dots and weaker, paler interneural dots at termen. *I. contiguaria* with fringe dots absent, ♂ antennal cilia and ♂ hindtarsus much longer. *Scopula marginepunctata* usually with terminal area less darkened, forewing costal spot of medial line smaller, ♂ antennal cilia longer, ♂ hindtarsus longer and ♀ hindleg with four spurs.

REMARKS: For morphology of immature stages see PÜNGELER (1892).

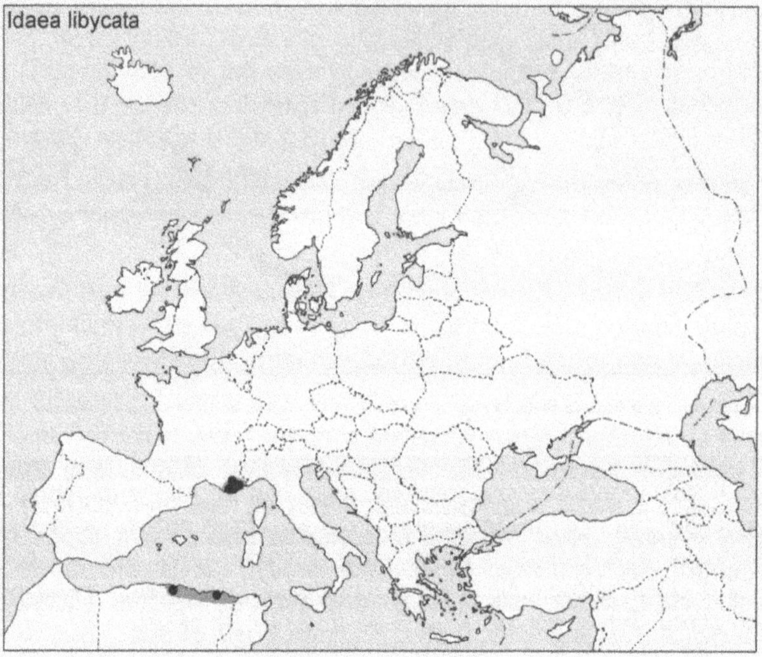

70. Idaea libycata (BARTEL, 1906)

Acidalia libycata BARTEL, 1906: Ent. Zeits. (Int. ent. Ver. Guben) 20 (1): 2 (Algeria: Algier). Syntypes ♂♀ (MNHU: 4 with 'cotype' labels, examined; others in IRSN).

Unavailable names (infrasubspecific): *anastomosaria*: REISSER (1961); *dimeglionaria*: ANDREAS (1929); *obscura*: BRETTSCHNEIDER; *rusicadaria*: ANDREAS (1929); *sriginaria*: ANDREAS (1929). – (incorrect subsequent spelling): *lybicata*: PROUT (1913).

DIAGNOSIS: Wingspan ♂♀ 18-21 mm. Forewing broad with costa convex. Coloration and wing pattern fairly variable. Ground colour light sand coloured, often with strong brown suffusion. Postmedial line dark brown, dentate, rather broad. Medial and antemedial lines usually indistinct, never sharp. Discal spots distinct. Terminal area broad, with brown 'clouds'. Fringe chequered, with distinct punctiform terminal dots at base. Frons black brown, vertex sand coloured, collar brown. Antennae of ♂ with deep and large intersegmental incisions, ciliate-fasciculate, cilia 1.0-1.5 times width of flagellum. Hindtibia of ♂ terminally dilated, with large pencil. Tarsus slightly dilated, 0.3-0.4 times length of tibia.

MALE GENITALIA: Uncus slender triangular. Saccus broad, distinctly projecting. Valva rounded at tip. Aedeagus long, with one stout, sometimes curved cornutus, about 0.4 times length of aedeagus.

FEMALE GENITALIA: Antrum dilated. Ductus bursae long, straight, without spines. Corpus bursae with large spinose patch and short membranous anterior end. Appendix bursae broad, membranous, with ductus seminalis at posterior end.

DISTRIBUTION: West-Mediterranean, disjunct. In Europe restricted to southern France: various localities in dept. Var, and with local occurrence in the Hautes Alpes. Erroneously recorded for eastern Spain in RAINERI (1996), the figured specimen of the TURATI collection (fig. 3) misidentified, identity probably *I. joannisiata* (see also REDONDO & GASTON 1999). – Outside Europe in Algeria. Reputedly in western Libya (TURATI 1930a). Rare and local, threatened. Last date of a European specimen collected in nature and examined by the author from 1972.

PHENOLOGY: Bi- or trivoltine: mid-May to early August; mid-October. For autumn generation, limited data available from nature.

BIOLOGY: Larval host-plants unknown. Adults attracted to light.

HABITAT: Xerothermophilous lowland species. Limited information available about habitat preference. Usually from 0 m up to 200 m above sea-level, but in the French Hautes Alpes one male collected at 1,200 m (BLAESIUS pers. comm.).

SIMILAR SPECIES: *I. typicata* (see no. 41), *I. vesubiata* (69). *I. consolidata* with postmedial line straighter, closer to termen, less brownish coloured. *Scopula marginepunctata* with postmedial line straighter, ♂ antennal cilia and ♂ hindtarsus longer and ♀ hindleg with four spurs.

REMARKS: For larval morphology see HOMBERG (1910b).

71. Idaea consolidata (LEDERER, 1853)

Acidalia consolidata LEDERER, 1853: Verh. zool.-bot. Ver. Wien 3: 257, (173) (north-western Turkey: Bursa ('Brussa')). Syntypes, at least 2 (MNHU: 2, examined).
Unavailable names (infrasubspecific): *autumnalis*: SCHWINGENSCHUSS (1926); *nigrolineata*: REISSER (1961).

DIAGNOSIS: Wingspan ♂♀ 19-22 mm, autumn generation on average smaller. Ground colour dirty white, more or less suffused with grey brown scales. Postmedial line sharp, dentate. Medial and antemedial lines indistinct, but well marked at their costal spots. Discal spots distinct. Fringe appearing chequered by terminal dots at base of fringe. Frons black brown or black, vertex sand coloured, collar brown. Antennae of ♂ ciliate, sub-fasciculate, cilia 0.9-1.2 times width of flagellum (Text-fig. 55). Hindtibia of ♂ dilated, with large pencil. Tarsus broad, 0.3-0.4 times length of tibia (Text-fig. 27).

MALE GENITALIA: Uncus slender triangular. Saccus large and broad, subterminally laced. Valva obliquely truncate at tip. Aedeagus very long, without cornutus.

FEMALE GENITALIA: Ductus bursae very long and slender, anteriorly with spinulose crest. Lateral half of corpus bursae densely spinose. Membranous half separated from appendix bursae by spinose loop. The latter leading to the looped ductus seminalis.

DISTRIBUTION: East-Mediterranean. Very isolated populations in central and southern Italy, north-western Slovenja close to the Austrian border, more widespread in the Balkan peninsula, on some Greek islands, e.g. Naxos, Crete, Rhodos. Recorded for Sicily and, reputedly, for Sardinia (RAGUSA 1889; TURATI 1911; RAINERI & ZANGHERI 1995; MÜLLER

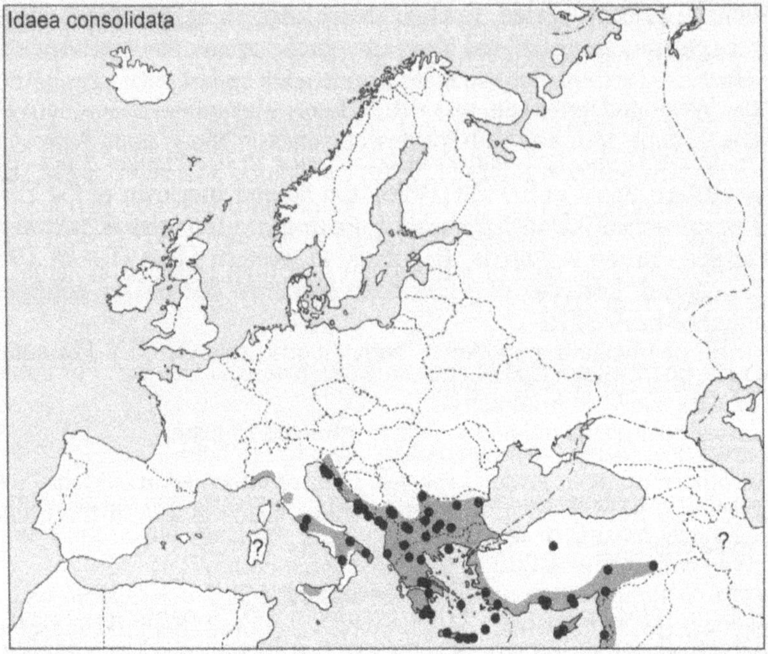

1996), data await confirmation. Erroneously recorded for Spain and Portugal (e.g. PROUT 1913; VIVES MORENO 1994; MÜLLER 1996). – Outside Europe in southern Turkey, with isolated occurrence near Ankara, Transcaucasus, Cyprus, in the south to Israel and Jordan.

PHENOLOGY: Bivoltine: late May to early July; mid-August to early October. In the south-eastern parts of the distribution emerging from late April.

BIOLOGY: Larva polyphagous on herbaceous plants (REBEL 1910b). Reared on *Taraxacum officinale* (LEIPNITZ pers. comm.). Adults readily attracted to light.

HABITAT: Xerothermophilous. Limited information available about habitat preference. In Europe usually from 0 m up to 700 m above sea-level, on Crete and in south-western Asia most abundant from 500 m up 1,400 m.

SIMILAR SPECIES: *I. typicata* (see no. 41), *I. vesubiata* (69), *I. libycata* (70). *I. metohiensis* with pointed forewing apex, medial line of forewing without costal spot, ♂ antennal cilia and ♂ hindtarsus longer. Allopatric *I. joannisiata* with more glossy coloration, wings narrower, wing pattern more diffuse, discal spots smaller, ♂ antennal cilia shorter. Allopatric *I. incisaria* with hindwing margin concave beteen veins, wing pattern sharper, with elongate terminal dashes rather than punctiform fringe dots, ♂ hindtarsus shorter. *I. contiguaria* with more distinct costal spots, medial line zigzagging, postmedial line close to termen, fringe dots absent, ♂ antennal cilia and ♂ hindtarsus longer. *I. camparia* on average smaller, ground colour paler, with elongate terminal dashes rather than punctiform fringe dots, with longer ♂ antennal cilia and narrower ♂ hindtarsus. *Scopula marginepunctata* with forewing costal spot of medial line less distinct, ♂ antennal cilia and ♂ hindtarsus longer and ♀ hindleg with four spurs. *Glossotrophia* species without fringe dots, but with uninterrupted terminal line surrounding forewing apex, transverse line more diffuse, proboscis very long, ♂ hindtibia with one spur in European species.

REMARKS: For morphology of immature stages see MAY (1892).

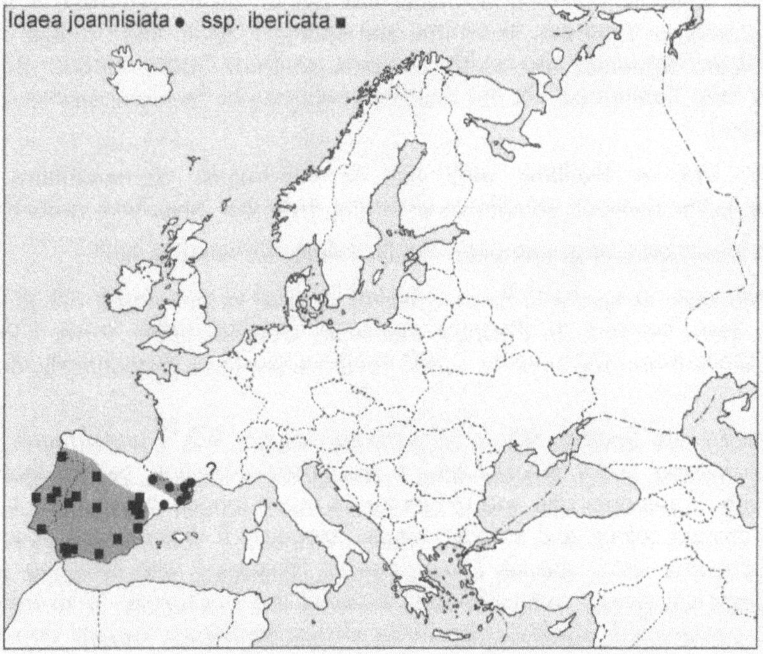

72. *Idaea joannisiata* (HOMBERG, 1911)

Acidalia joannisiata HOMBERG, 1911: Bull. Soc. ent. Fr. 1911 (15): 306 (France, eastern Pyrenees: Vernet-les-Bains). Lectotype ♂ (MNHN); 3♂2♀ paralectotypes (ZSM/HERBULOT, 1♂ paralectotype examined).

Ptychopoda consolidata f. an var. aut spec.? *ibericata* WEHRLI, 1927: Int. ent. Zeit. 20 (37/38): 340 (Spain, Andalusia: Sierra Nevada). Syntypes 2♂ (ZFMK, examined). Herewith downgraded from species (VIVES MORENO 1994; SCOBLE 1999) to subspecific rank (stat. n.).

Unavailable names (misidentification): *alyssumata*: sensu PROUT (1904) nec MILLIÈRE, 1871; *camparia*: sensu BELLIER (1858) nec HERRICH-SCHÄFFER, 1852; *campata*: sensu RAMBUR (1866; non binominal) nec *camparia* HERRICH-SCHÄFFER, 1852; *consolidata*: sensu PROUT (1913; partim: 106) nec LEDERER, 1853. – (incorrect subsequent spellings): *campariata* (for *campata* sensu RAMBUR): VIVES MORENO (1994); *jonnisiata*: PROUT (1935: 56).

DIAGNOSIS: Wingspan ♂♀ 16-22 mm. Ground colour dirty white, glossy, with light brown tinge. Transverse lines undulate, sometimes broad, but never sharp. Postmedial line with black dots on the veins. Discal spots small. Frons brown, vertex sand coloured, collar brown. Antennae of ♂ ciliate-setose, cilia 0.7-0.85 times width of flagellum. Hindtibia of ♂ dilated, with pencil. Tarsus broad, 0.3-0.35 times length of tibia.

Subsp. ***ibericata*** usually with stronger brown suffusion, transverse lines more indistinct.

MALE GENITALIA: Uncus triangular. Saccus very large, broad, distinctly projecting. Length from base of saccus to tip of valva about 2.4 mm. Valva slender, tapered at tip. Aedeagus very long, straight, with one centrally positioned, thin, straight cornutus, 0.15-0.2 times length of aedeagus, cornutus slightly dilated at base.

FEMALE GENITALIA: Ductus bursae very long and slender, anterior third bent back by 180°. This anterior part usually laterally spinose, sometimes with a very few spinules only. Corpus bursae with a large, densely spinose patch anteriorly, with membranous sac, posteriorly tapered towards ductus seminalis.

DISTRIBUTION: Endemic to Iberian peninsula and French eastern Pyrenees. Nominate subspecies restricted to Pyrenees, in central and southern Spain and Portugal replaced by subsp. *ibericata*. Reputedly also on Mt. Ventoux, southern France (LUQUET 2000). Local, and usually rare, threatened. On the Balearics replaced by the sister species *I. acutipennis* (see below).

PHENOLOGY: Uni- or bivoltine: early July to mid-August, on mountains until mid-September, in the lowlands sometimes emerging from late May, here probably bivoltine.

BIOLOGY: Larval host-plants unknown. Adults readily attracted to light.

HABITAT: Montane. In southern Spain collected in aisles of pine forest rich in low vegetation (PÖLL pers. comm.). In Portugal mainly in *Quercus suber* forest (CORLEY pers. comm.). Usually from 500 m up to 1,800 m above sea-level, exceptionally down to sea-level.

SIMILAR SPECIES (see Text-fig. 90): *I. alyssumata* (see no. 42). *I. alyssumata, I. consolidata, I. cervantaria*, and *I. ibizaria* differ in less glossy coloration, pattern sharper. *I. cervantaria* with ♂ antennal cilia and ♂ hindtarsus much longer. Allopatric *I. consolidata* with paler ground colour, and broader wings. Allopatric *I. ibizaria* with postmedial line dentate. Allopatric sister species *I. acutipennis* (Balearics) with forewing apex more pointed, wings less glossy, pattern finer, postmedial line much closer to termen, ♂ antennae more segmentated, sub-fasciculate, cilia longer, hindtarsus shorter. Correct identification sometimes difficult on the base of habitus features, mainly in worn females, easy however by dissection or brushing of the abdomen.

REMARKS: Discussion of taxonomy, biology and distribution in DANTART (1997).

73. *Idaea acutipennis* HAUSMANN & HONEY, 2004

Idaea acutipennis HAUSMANN & HONEY, 2004: Ent. Zeits. Stuttgart 114 (1): 12, figs. 1-2 (Balearics: Dragonera). Holotype ♂ (BMNH, examined).

DIAGNOSIS: Wingspan ♂♀ 18 mm. Forewing apex pointed. Ground colour dirty white, less glossy than in *I. joannisiata*. Transverse lines indistinct, finer than in *I. joannisiata*. Postmedial line fine, sub-dentate, parallel to termen, closer to termen than in *I. joannisiata*. Discal spots small. Frons brown, vertex sand coloured. Antennae of ♂ with small intersegmental incisions, sub-fasciculate, cilia rather long, 1.2-1.35 times times width of flagellum. Hindtibia of ♂ dilated, with pencil. Tarsus broad, but short, 0.2-0.25 times length of tibia only.

MALE GENITALIA: As described for *I. joannisiata* but genitalia smaller (length from base of saccus to tip of valva 1.8 mm), saccus smaller, valva broader and sub-truncate at tip, aedeagus slightly curved, cornutus slender at base, slightly longer than in *I. joannisiata*, about 0.25 times length of aedeagus, positioned at base of aedeagus.

FEMALE GENITALIA: unknown.

DISTRIBUTION: Endemic to Dragonera (close to Mallorca island).

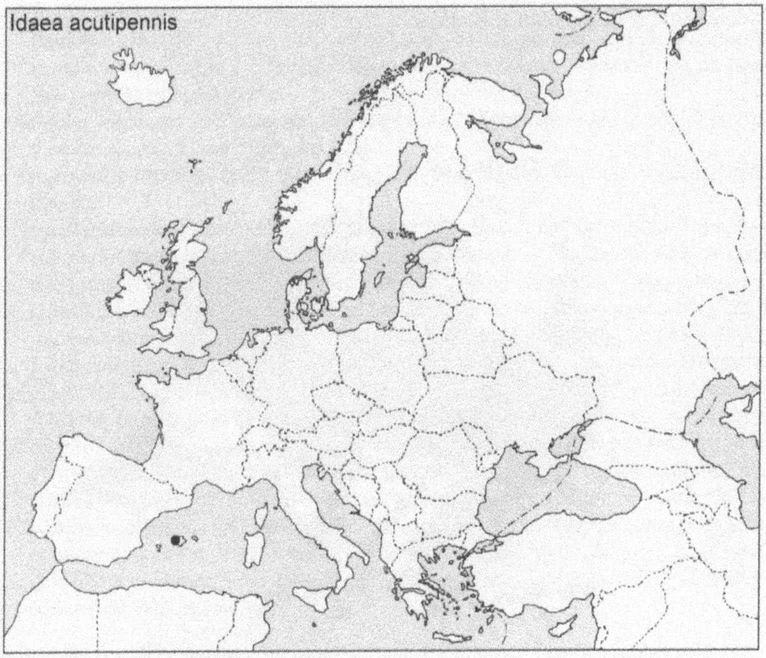

PHENOLOGY: Limited data available from early April.

BIOLOGY: Larval host-plants unknown. Adults collected at light.

HABITAT: Hot coastal habitats at sea-level. Possibly halophilous.

SIMILAR SPECIES: Allopatric sister species *I. joannisiata* (see no. 72). *I. ibizaria* is very similar, brushing of ♂ abdomen or female dissection is necessary for identification. Confusions are possible also with *I. minuscularia*, the latter, however, with forewing apex less pointed, ground colour more whitish, ♂ with longer hindtarsus.

The ***seriata*** species-group

STERNECK's (1940) 20th group of *Idaea*. Sensu stricto a homogenous group, with doubtful systematic relationships to the species pair *I. carvalhoi* and *I. fathmaria* (STERNECK 1940: 153). In forewing venation *I. carvalhoi* differs from the other species in R1 arising subapically from cell, separately from R2-R5. Palpi comparatively small, 0.5-1.0 times diameter of eye. Proboscis usually developed, vestigial in *I. carvalhoi* and *I. fathmaria*. Sternum A2 of ♂ with distinct horseshoe-shaped sclerite and setose patch, in *I. carvalhoi* and *I. fathmaria* setose patch absent, sclerotized ridge narrow, semicircular, reminiscent of *I. longaria*. Male genitalia: Aedeagus with very long and slender cornutus. Female genitalia: species-group sensu stricto with corpus bursae asymmetrical, laterally dilated, and narrow anterior appendix bursae. Egg sculpturation hexagonal reticulate. Larva long and slender in the species-group sensu stricto, short and thick in *I. carvalhoi* and *I. fathmaria*. The *seriata* species-group has been revised by MENTZER (1990a), who states close relationship to the *consolidata* species-group. *I. saleri* is transferred to the *infirmaria* species-group.

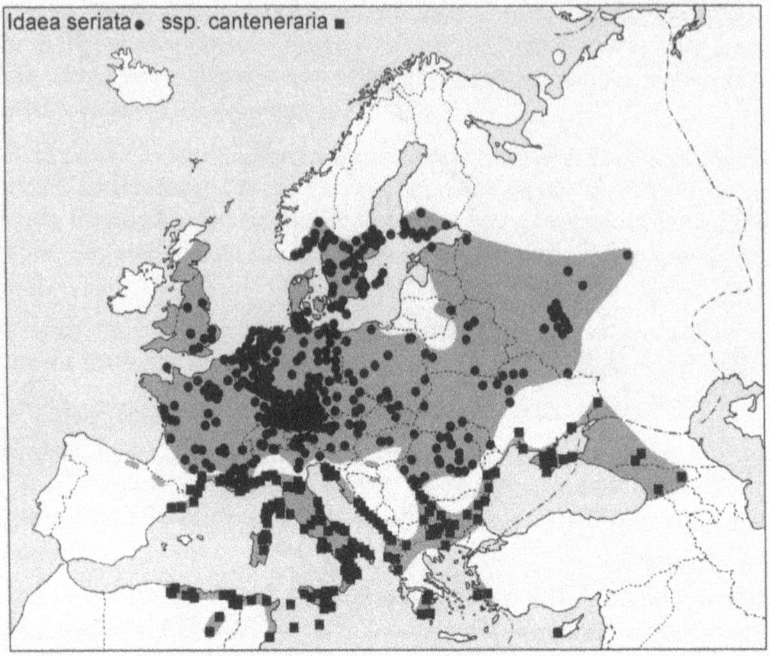

74. *Idaea seriata* (SCHRANK, 1802)

Phalaena seriata SCHRANK, 1802: Fauna boica 2 (2): 57 (Germany, Bavaria: Ingolstadt). Syntype(s).

‡? *Phalaena albulata* HUFNAGEL 1767: Berlin. Mag. 4 (6): 616 (Germany: Berlin). Syntype(s) lost. Nomen dubium, as the ambiguous description may be attributed also to other species.

‡ *Geometra incanata*: sensu DENIS & SCHIFFERMÜLLER, 1775: Ank. syst. Werkes Schmett. Wienergegend: 117 (Austria: Vienna distr.). Syntype(s) lost. Misidentification nec LINNAEUS, 1758 (identity: *Scopula incanata*).

‡ *Phalaena Geometra moniliata*: sensu BORKHAUSEN, 1794: Naturg. eur. Schmett. 5: 520. Misidentification nec DENIS & SCHIFFERMÜLLER, 1775.

‡ *Geometra virgularia*: sensu HÜBNER, 1799: Samml. Eur. Schmett. 5, Geometrae (1): pl. 19, fig. 104. Misidentification (and emendation) nec *virgulata* DENIS & SCHIFFERMÜLLER, 1775 (identity: *Scopula virgulata*) (see HÜBNER 1825).

‡ *Geometra incanaria*: sensu HÜBNER 1799: Samml. Eur. Schmett. 5, Geometrae (1): pl. 20, fig. 106. Misidentification (and incorrect subsequent spelling) nec *incanata* LINNAEUS, 1758 (identity: *Scopula incanata*) (see HÜBNER 1825).

Acidalia canteneraria BOISDUVAL, 1840: Genera Index meth. eur. Lepid.: 222 (France: Provence). Syntype(s) (ZFMK, 1♂ ex coll. BOISDUVAL, examined). Valid at subspecific rank.

Idaea incanata australis ZELLER, 1847: Isis 1847 (6): 511 (Sicily: Messina, Catania, Syracus; Italy, Campania: Napols). Syntypes, not traced in BMNH. Junior synonym (subsp. *canteneraria*).

Idaea calcearia ZELLER, 1849: Stettin. ent. Ztg. 10 (7): 217, index (Italy, Tuscany: Livorno, Antignano, Posignano). Syntypes. Generic combination indicated in the index. Junior synonym (subsp. *canteneraria*).

Acidalia paleacata GUENÉE, 1858: in BOISDUVAL & GUENÉE, Hist. nat. Insectes (Spec. gén. Lépid.) 9: 478 (France: Iles d'Hyères). Holotype ♂ (ZFMK, examined). Herewith downgraded from subspecific rank (SCOBLE 1999) to synonymy (of subsp. *canteneraria*) (syn. n.).

Dosithea incanaria var. *cubicularia* PEYERIMHOFF, 1862: Bull. Soc. Hist. nat. Colmar 2: 156 (France: Alsace). Syntypes.

Geometra vicinata WERNEBURG, 1864: Beitr. Schmettk. 1: 447, 536; 2: 198. Syntype(s). Introduced to replace the misidentified *incanata* sensu DENIS & SCHIFFERMÜLLER, 1775.

‡ *bischoffaria* LA HARPE, 1864: Neue Denkschr. Allg. schweiz. Ges. ges. Naturw. 20: 4. Non binominal. Misidentification nec GEYER (1838) (identity: *Pasiphila rectangulata*, Larentiinae). Frequently used name for a melanistic form.

Eupithecia cossurata MILLIÈRE, 1875: in RAGUSA, Boll. Soc. ent. ital. 7 (4): 254 (southern Italy: Pantelleria island). Syntype(s) ♂, probably lost. So far, retained as valid subspecific name for '*I. camparia*' (e.g. SCOBLE 1999), the latter species, however, not occurring on Pantelleria island. See remarks to *I. camparia*.

Acidalia incanaria var. *obscura* MILLIÈRE, 1882: Ann. Soc. Linn. Lyon 29: 16, pl. 3, fig. 3 (France: Doubs, Lyon). Syntypes.

‡ *rupicolaria* MILLIÈRE, 1882: Ann. Soc. Linn. Lyon 29: 16. Non binominal and introduced in synonymy.

Unavailable names (misidentifications): *albitorquata*: sensu PARENZAN (1988) nec PÜNGELER, 1909; *camparia*: sensu MANN (1853) nec HERRICH-SCHÄFFER, 1852; *consolidata*: sensu TURATI (1911) nec LEDERER, 1853; *dilutaria*: sensu GEYER (1838) nec HÜBNER, 1799; *immutata*: sensu GRAAF (1856) nec LINNAEUS, 1758 (identity: *Scopula immutata*); *laevigaria*: sensu VER HUELL (1836) nec *laevigata* SCOPOLI, 1763; *virgulata*: sensu HAWORTH (1809) nec DENIS & SCHIFFERMÜLLER, 1775 (identity: *Scopula virgulata*). – (homonym): ?*virgata*: sensu HAWORTH (1802) nec FOURCROY, 1785 (identity unclear, referring to a 'green moth'). – (infrasubspecific): *anastomosaria*: REISSER (1961); *fuscomarginata*: LEMPKE (1949); *grisescens*: LA HARPE (1864); *hauderi*: KAUTZ (1913); *lutescens*: LEMPKE (1949); *mediofasciata*: LEMPKE (1949); *undulata*: OSTHELDER (1829). – (incorrect subsequent spellings): *afra* (? for *bischoffaria*): BETHUNE-BAKER (1885); *bischoffiaria*: FUCHS (1884); *bishoffaria*: RAINERI (1996); *calceata*: BELLIER (1860); *calvaria* (for *calcearia*): LA HARPE (1860); *cantenaria*: RAINERI (1996); *cantenerata*: BELLIER (1860); *conteneraria* ROCCI (1914); *curbicularia*: STANESCU (1997); *faria* (for *bischoffaria*): SJÖSTEDT (1928); *incanataria*: TREITSCHKE (1825); *miniscularia*: DANNEHL (1927; also misidentification); *paleaceata* BERCE (1873).

DIAGNOSIS: Wingspan variable, ♂♀ 13-20 mm. Fairly variable in coloration and pattern of wings. Ground colour dirty white, with tendency to melanism. Ante- and postmedial lines dark grey, sharp, the latter marked black on the veins, sometimes dissolved into dots. Medial line usually diffuse, often dentate. Discal spots usually distinct. Terminal line fine, fringe dots present. Underside with forewing darker than hindwing. Frons and collar brown, vertex dirty white, collar dark sand coloured or brown. Antennae of ♂ with intersegmental incisions, ciliate-fasciculate, cilia long, 1.2-1.8 times width of flagellum (Text-fig. 56). Hindtibia of ♂ slender, slightly dilated distally, with weak pencil. Tarsus 0.5-0.7 times length of tibia.

Variability: In nominate subspecies comparatively high tendency to melanism. Inheritance of black forms follows MENDEL's rules according to PROUT (1913), but this hypothesis is contradicted by BERGMANN (1955). Development of pale forms favoured by increasing temperature and dryness (BERGMANN 1955). As other individual aberrations are shown to have genetical reasons which follow MENDEL's rules, but can be induced also by high temperature (KÜHN & ENGELHARDT 1940), this may be valid also for melanistic or pale forms.

Subsp. ***canteneraria*** with ground colour clear white, on Corsica and Sardinia dirty white. Wing pattern as in the nominate subspecies, but postmedial line always dotted, medial line usually absent. Discal spots distinct. External morphology as in the nominate subspecies. On Corsica, Sardinia and Pantelleria island forms with strongly contrasted wing pattern reminiscent of the nominate subspecies.

MALE GENITALIA: Uncus very slender. Saccus large. Valva round at tip. Aedeagus with one very long and narrow cornutus, about 0.8 times length of aedeagus (1.3-1.5 mm). Vesica not granulated. Subsp. *canteneraria* without constant differences from nominate subspecies, but valva often slightly tapering. In

Algeria and Tunisia vesica sometimes slightly granulated (see gen.fig. 74b) and aedeagus shorter (1.1-1.2 mm), slightly reminiscent of *I. minuscularia*.

FEMALE GENITALIA: Ductus bursae broad, medium length. Corpus bursae globular or slightly elongate, anteriorly with long membranous appendix, laterally with large, semicircular, spinulose crest and some additional spinules at the posterior part.

DISTRIBUTION: Euro-Caucasian. Widely distributed in western, central and northern Europe, in the north up to Denmark and southern Scandinavia, in the east to central European Russia. Replaced by subsp. *canteneraria* from north-eastern Spain, across central and eastern Mediterranean, including all the islands except Crete, in the east to Crimea. In the western Mediterranean and on the Balearics replaced by the sister species *I. minuscularia*. – Outside Europe in eastern Algeria and Tunisia, Turkey (ZSM: 'Kania', locality not traced), Cyprus (HAUSMANN 1995), Caucasus (VIIDALEPP 1996) and north-western Transcaucasus (Georgia: DIDMANIDZE 2002). In Morocco and western Algeria replaced by *I. minuscularia*.

PHENOLOGY: Nominate subspecies usually bivoltine: early May to late September, generations often overlapping. Exceptionally with partial third brood, from early October to mid-November. Single specimens in all months of the year after development in houses (EBERT 2001). In the north of the distribution univoltine, from mid-June to mid-August (FORSTER & WOHLFAHRT 1981; SKINNER 1984; HULDEN et al. 2000). Southern subsp. *canteneraria* in the coastal lowlands plurivoltine over the whole year. At higher elevations trivoltine: early May to early July; mid-August to late September; early October to mid-November. Growth of larvae asynchronous (BERGMANN 1955).

BIOLOGY: Larva polyphagous. Preferring dry and withered leaves of herbaceous plants (var. cit.), or mouldy plant debris (REBEL 1910b). Found in fruit capsules of *Silene vulgaris* (Caryophyllaceae) (Germany: EBERT 2001), and near ground on withered leaves of *Poa annua* (Poaceae), and *Taraxacum officinale* (Asteraceae), in rearings feeding on these two species (Germany: WEGNER pers. comm.). Larva also found in mouldy plant debris in human buildings, such as haystacks, barns, sheds, factories, piles of compost (BERGMANN 1955), near balcony plants and straw (EBERT 2001), as pest in food stores with dry plants (BERGMANN 1955) and tobacco (Sicily: PARENZAN pers. comm.), and on dried insects in collections (BERGMANN 1955). Larva beaten from *Hedera helix* (Araliaceae) (Great Britain: SKINNER 1984; PORTER 1997; WARING & TOWNSEND 2003). Recorded on *Taraxacum officinale* (Asteraceae), *Prunus, Malus* (Rosaceae), *Urtica, Viburnum, Lonicera caprifolium* (Caprifoliaceae), *Urtica* (Urticaceae), *Plantago* (Plantaginaceae), *Trifolium* (Fabaceae), Poaceae, leaves of deciduous trees (var. cit., e.g. BERGMANN 1955; SKOU 1986; EBERT 2001), reputedly also on moss and lichens (URBAHN & URBAHN 1939; FORSTER & WOHLFAHRT 1981). Reared on withered leaves of *Taraxacum officinale, Polygonum aviculare, Geum urbanum, Rumex* and *Brassica oleracea* (original data; PROUT 1913; BERGMANN 1955; FRIEDRICH 1976; PORTER 1997; TRUSCH & MÜLLER 2001; ERLEBACH, LEIPNITZ pers. comm.). Adults usually found during daytime when resting on white walls. Main activity at dawn (PROUT 1913; BERGMANN 1955), also, sparingly, coming to light and bait, in southern Europe readily attracted to light. Nectar feeding recorded on *Buddleja* and *Lonicera caprifolium* (EBERT 2001).

HABITAT: Domestic in central Europe, usually found near human buildings, rare outside towns, villages and farms. In southern Europe ubiquitous and often very abundant in the

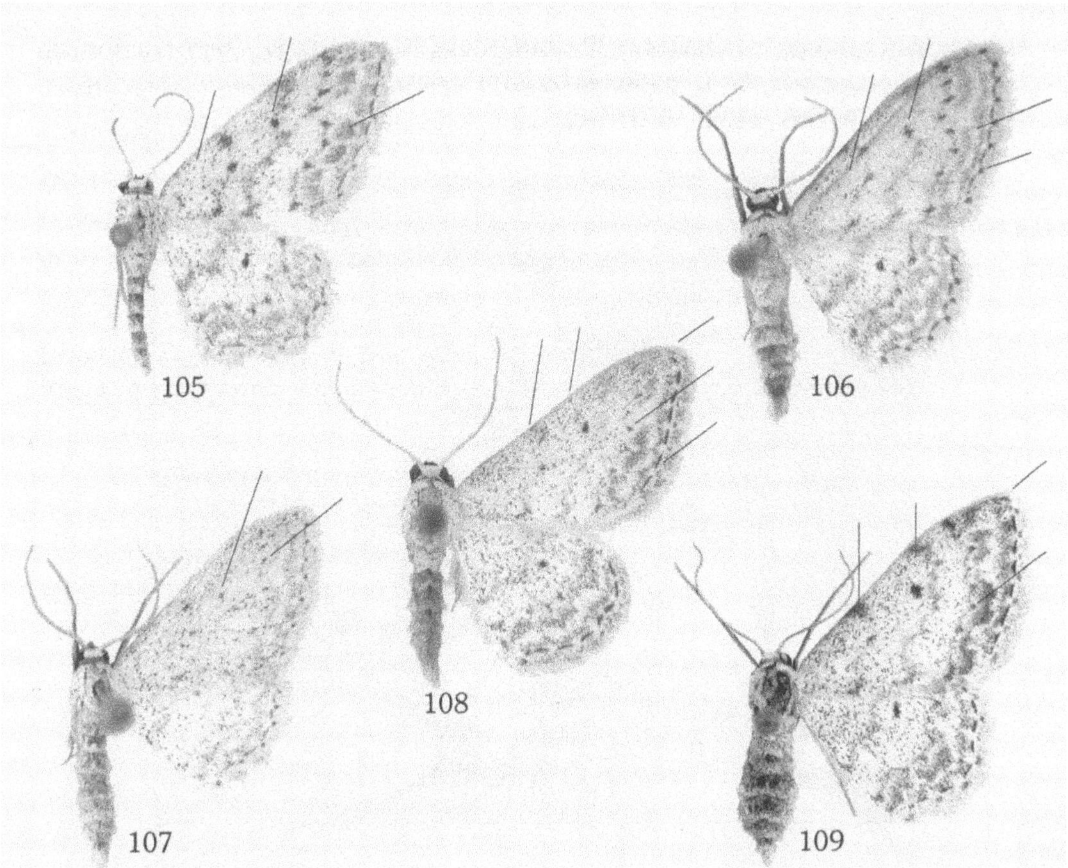

Text-figs 105-109: Diagnostic characters (indicated) of European *Idaea* species. Text-fig. 105: *I. obliquaria* (TURATI, 1913). Text-fig. 106: *I. seriata* (SCHRANK, 1802). Text-fig. 107: *I. minuscularia* (RIBBE, 1912). Text-fig. 108: *I. albitorquata* (PÜNGELER, 1908). Text-fig. 109: *I. camparia* (HERRICH-SCHÄFFER, 1852).

Mediterranean maquis. Usually from 0 m up to 500 m above sea-level, in south-eastern France higher. Subsp. *canteneraria* up to 1,000 m above sea-level. Exceptionally up to 1,800 m in xeromontane localities in southern Alps, Apennines, Sicily and Corsica.

SIMILAR SPECIES: *I. efflorata* (see no. 38), *I. albarracina* (48), *I. incisaria* (49), *I. calunetaria* (50), *I. elongaria* (52), *I. obliquaria* (55), *I. longaria* (65). *I. albitorquata* differs in postmedial line usually more dentate, terminal line dissolved to elongate streaks, fringe dots weak or absent, collar much paler, ♂ hindtibia dilated, ♂ hindtarsus shorter. *I. camparia* with distinct costal spots, terminal shadows distinct, separate and not confluent, fringe dots weak or absent, ♂ hindtarsus shorter. Sister species *I. minuscularia* on average smaller, dots of postmedial line and discal spots smaller, differences in genitalia see under *I. minuscularia*.

REMARKS: Subspecies *canteneraria* usually well distinguishable from nominate subspecies in habitus. On the base of theses habitus features, often with sharp geographical borders resulting, e.g. between lowlands of southern France, Basses Alpes and Hautes Alpes. For relationships to the West-Mediterranean sister species see remarks to *I. minuscularia*.

Hybrids from captivity experiments are mentioned in PROUT (1938a: 227), i.e. between *I. seriata canteneraria* ♂ x *I. inquinata* ♀ and *I. seriata canteneraria* ♂ x *I. contiguaria* ♀, referring to pers. comm. of THURAU.

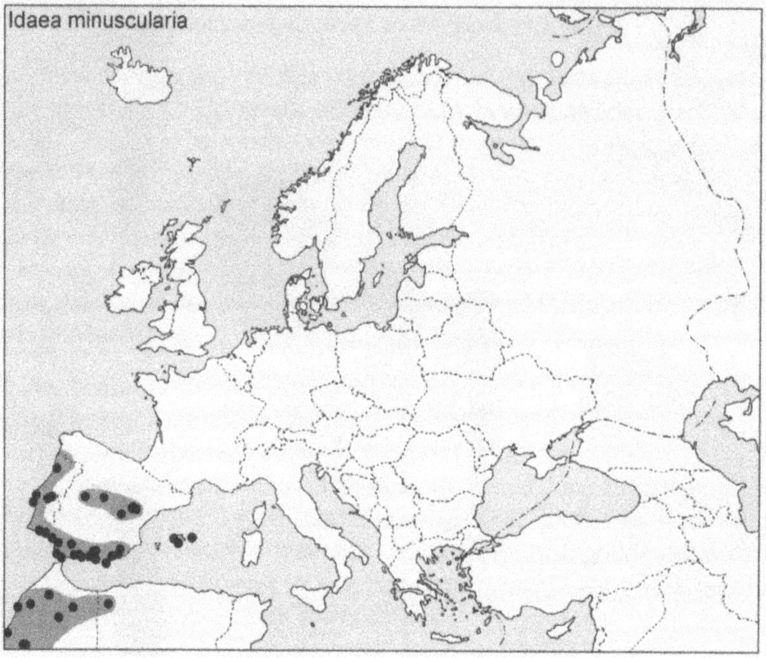

75. *Idaea minuscularia* (RIBBE, 1912)

Acidalia virgularia var. *minuscularia* RIBBE, 1912: Dt. ent. Z. Iris 23 (Lep. Faun. Andal.): 305 (Spain, southern Andalusia; southern Prov. Murcia; Morocco: 'Mauretania'). Syntypes ♂♀ (MNHU?). Lectotype should be selected from southern Spain.

Sterrha herbuloti AGENJO, 1952: Faunula Lepid. Almeriense: 187, pl. 6, figs 19-20, gen.fig. 7 (Spain, Andalusia: Almeria). Holotype ♂ (MNCN, examined).

Unavailable names (misidentification): *canteneraria*: sensu REBEL (1910) nec BOISDUVAL, 1840 (identity: *Idaea seriata*); *seriata*: sensu MONTEIRO & PASSOS DE CARVALHO (1984). – (incorrect subsequent spellings): *miniscularia*: DANNEHL (1927); *minuscula*: VIVES MORENO (1984).

DIAGNOSIS: Wingspan ♂♀ 12-19 mm, in winter and springtime exceptionally up to 23 mm (HALE pers. comm.). Ground colour clear white, sometimes dirty white, terminal area often grey. Wing pattern very similar to that of *I. seriata canteneraria*, dots of postmedial line and discal spots smaller. Frons and collar brown, vertex whitish. Antennae of ♂ with intersegmental incisions, ciliate-fasciculate, cilia long, 1.3-1.7 times width of flagellum. Hindtibia of ♂ slender, with weak pencil. Tarsus 0.5-0.7 times length of tibia.

MALE GENITALIA: As in *I. seriata*, but aedeagus shorter (1.0-1.1 mm). Vesica strongly granulated, with short spinules beside cornutus.

FEMALE GENITALIA: As in *I. seriata*, but corpus bursae usually more elongate. Lateral, semicircular spinulose crest of corpus bursae slightly smaller. Anteriorly membranous appendix much shorter and narrower.

DISTRIBUTION: West-Mediterranean. Iberian peninsula and Balearics. Erroneously recorded for southern Italy (PARENZAN 1994; MÜLLER 1996). – Outside Europe in Morocco and western Algeria. Erroneously recorded for the Lebanon (RAINERI 1996).

PHENOLOGY: Plurivoltine, in all months of the year.

BIOLOGY: Larval host-plants unknown. Adults readily attracted to light.

HABITAT: Xerothermophilous species of hot lowland habitats, usually near coast. From 0 m up to 500 m above sea-level, exceptionally up to 1,200 m.

SIMILAR SPECIES: *I. longaria* (see no. 65), *I. acutipennis* (73), *I. seriata* (74).

REMARKS: Downgrading of this taxon to subspecies of *I. seriata* may be discussed. Breeding experiments and molecular analysis could help to clarify this question.

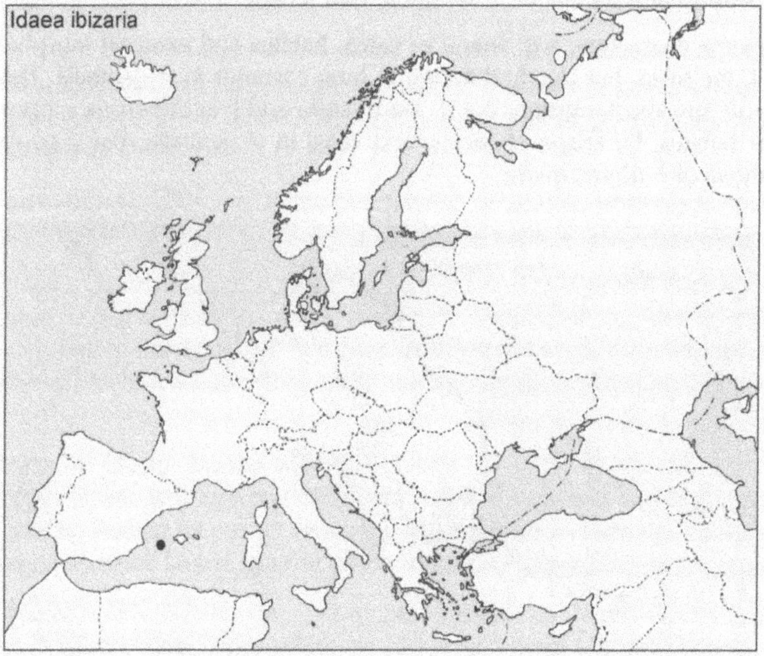

76. Idaea ibizaria MENTZER, 1980

Idaea ibizaria MENTZER, 1980: Entomol. scand. 11 (2): 206, figs 1-2, 5-6 (Balearics: Ibiza). Holotype ♂ (NHRS/MENTZER) (coll. SKOU: 1♂ paratype, examined).

DIAGNOSIS: Wingspan ♂♀ 19-22 mm. Forewing elongate. Wings sand coloured. Transverse lines brown, distinct at costa, mainly antemedial line. Medial line diffuse. Postmedial line strongly dentate. Discal spots small. Frons brown, vertex sand coloured. Antennae of ♂ ciliate-fasciculate, cilia 1.1-1.2 times width of flagellum. Hindtibia of ♂ curved and dilated in the distal half, with large pencil. Tarsus 0.25-0.4 times length of tibia.

MALE GENITALIA: Uncus short. Saccus large, broad. Valva long, terminally tapering. Aedeagus with one very long and narrow cornutus of about 0.8-0.9 times length of aedeagus. Vesica not granulated.

FEMALE GENITALIA: Ductus bursae long, with continuous transition to slender, laterally spinulose corpus bursae, anterior end broadly digitiform. Appendix bursae large, globular, posteriorly leading to ductus seminalis. Compare description in MENTZER (1990a).

DISTRIBUTION: Endemic to Ibiza and Formentera islands (MENTZER 1980; DANTART et al. 1993). Until now, only 6♂3♀ are known (NHRS, coll. SKOU, DANTART and VALLHONRAT).

PHENOLOGY: Bi- or trivoltine: Limited data available from early May, early July to late July, late September, late October.

BIOLOGY: Larval host-plants unknown. Adults collected at light.

HABITAT: Xerothermophilous. Dunes in the coastal lowlands, at sea-level.

SIMILAR SPECIES: *I. incisaria* (see no. 49), *I. acutipennis* (73). Allopatric *I. albitorquata* with more whitish ground colour, ♂ antennal cilia longer.

REMARKS: Some characters, e.g. shape of valva, habitus and external morphology reminiscent of *I. incisaria*, but clearly differing in long cornutus in ♂ genitalia. Relationships to the *consolidata* species-group, e.g. *I. joannisiata* and *I. acutipennis* supported by the very similar habitus, by shape of saccus and valva in ♂ genitalia, but ♀ genitalia better matching those of *I. albitorquata*.

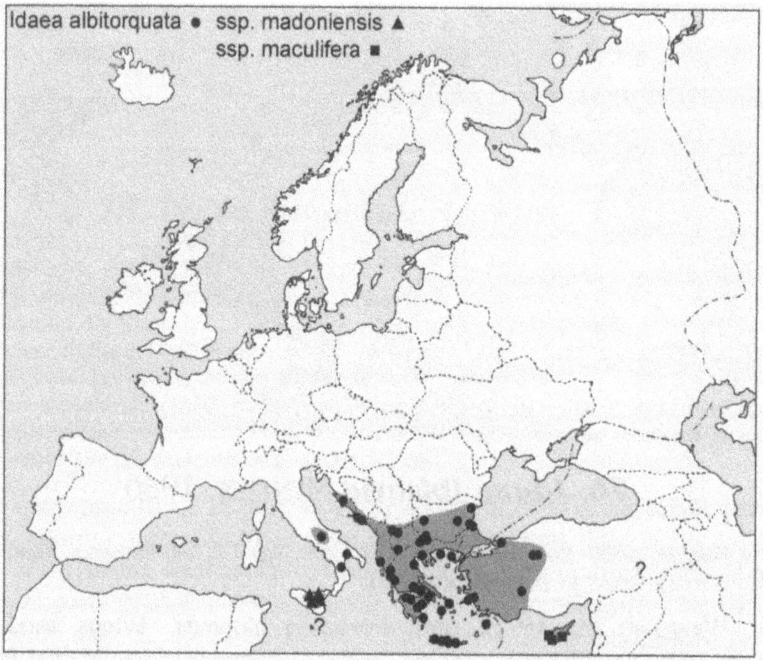

77. *Idaea albitorquata* (PÜNGELER, 1909)

Acidalia albitorquata PÜNGELER, 1909: Dt. ent. Z. Iris 21 (4): 293, pl. 4, fig. 10 (Greece, Peloponnes: Tiryns). Lectotype ♂ (MNHU, examined, originally labelled as 'type'), herewith designated to stabilize usage of the name and to avoid nomenclatural confusion between subspecies; 20♂♀ paralectotypes from type locality, Peloponnes, Kalavrita; Turkey: Amasia, Aidin; Sicily: Ficuzza (MNHU 7; ZSM 1; MSNT 1).

Idaea albitorquata madoniensis HAUSMANN, 1993: Naturalista sicil. (4) 17 (1-2): 87 (Sicily: Madonie). Holotype ♂ (ZSM, examined). Valid at subspecific rank.

Idaea albitorquata maculifera HAUSMANN, 1994: Z. Arb.Gem. österr. Ent. 46 (3-4): 86, figs 2,12 (Cyprus: Paphos). Holotype ♀ (ZSM, examined). Valid at subspecific rank.

Unavailable names (misidentification): *australis*: sensu WEISS (1920) nec ZELLER, 1847 (identity: *Idaea seriata*); *boeklini*: sensu MENTZER (1990: partim) nec MENTZER, 1990 (Holotype; identity: *Idaea camparia*). – (infrasubspecific): *albomarginata*: REISSER (1961); *anastomosaria*: REISSER (1961); *diffluata*: REISSER (1961); *nigrolineata*: REISSER (1961). – (incorrect subsequent spelling): *ablitorquata*: PROUT (1913).

DIAGNOSIS: Wingspan ♂♀ 16-22 mm, in autumn generation smaller, sometimes 14 mm only. Forewing usually with rather straight costa and termen. Forewing venation with R1 long stalked. Ground colour clear white. Transverse lines diffuse, except dentate postmedial line. Shadows in terminal area usually weak or absent. Discal spots present. Terminal line dissolved into elongate streaks, fringe dots absent or very weak. Frons dark brown or blackish, vertex white, collar almost concolorous with vertex. Antennae of ♂ with intersegmental incisions, ciliate-fasciculate, cilia long, 1.2-1.7 times width of flagellum. Hindtibia of ♂ somewhat dilated, with large pencil. Tarsus 0.35-0.4 times length of tibia.

Subsp. ***maculifera*** with dirty white ground colour, shadows in terminal area distinct, ♂ antennal cilia longer, 1.5-1.8 times width of flagellum, ♂ hindtarsus slightly shorter, about 0.3 times length of tibia.

Subsp. ***madoniensis*** with medial line of hindwing proximally from discal spot, ♂ antennal cilia comparatively long, 1.5-1.8 times width of flagellum.

MALE GENITALIA: Uncus triangular. Saccus large. Valva round at tip. Aedeagus with one long cornutus, 0.8-0.9 times length of aedeagus, sigmoid at tip. Vesica with compact spinose patch beside cornutus. No constant differences between the subspecies, except smaller spinose patch in subsp. *maculifera* and slightly larger juxta in subsp. *madoniensis*.

FEMALE GENITALIA: Ductus bursae with continuous transition to slender, spinose corpus bursae. Corpus bursae with narrow, digitiform anterior appendix and large, membranous lateral appendix, leading to ductus seminalis. No constant differences between the subspecies except separate lateral spinose patch in corpus bursae of subsp. *maculifera*.

DISTRIBUTION: East-Mediterranean. Nominate subspecies from western Croatia to Peloponnes and European Turkey. Also on Kyklades islands (Tinos), Crete and Rhodos. In Sicily and Apennines replaced by subsp. *madoniensis*. Erroneously recorded from one locality in Spain, due to misidentification or erroneous labelling (Aragon: Albarracin; examined from various collections, all labelled 'e.l. coll. SOFFNER' and 'e.o. leg. NOACK'). All other records (e.g. PROUT 1935a; VIVES MORENO 1994; etc.) supposed to refer to such specimens, too. Reputedly on Malta (PROUT 1935a), possibly misidentified. – Outside Europe in Turkey and on Cyprus (subsp. *maculifera*).

PHENOLOGY: Bivoltine: early May to early July; late August to early October. Under favourable conditions (e.g. in subsp. *maculifera*) trivoltine from early May to early November.

BIOLOGY: Larval host-plants unknown. Adults readily attracted to light.

HABITAT: Xerothermophilous. Limited information available about habitat preference. From 0 m up to 800 m above sea-level, in Turkey up to 1,000 m. Subsp. *madoniensis* montane, from 1,000 m up 1,200 m.

SIMILAR SPECIES: *I. seriata* (see no. 74), *I. ibizaria* (76). *I. camparia* differs in a number of characters, none of them allowing secure identification in all cases: usually with rounder forewing, coloration with slight sand coloured tinge, postmedial line more dissolved, costal spots of transverse lines more distinct, terminal shadows often darker. Identification of ♂ is facilitated by brushing tip of abdomen to remove scales and reveal shape of valva.

REMARKS: Turkish populations with transitional characters between subsp. *maculifera* and nominate subspecies (HAUSMANN 1994c).

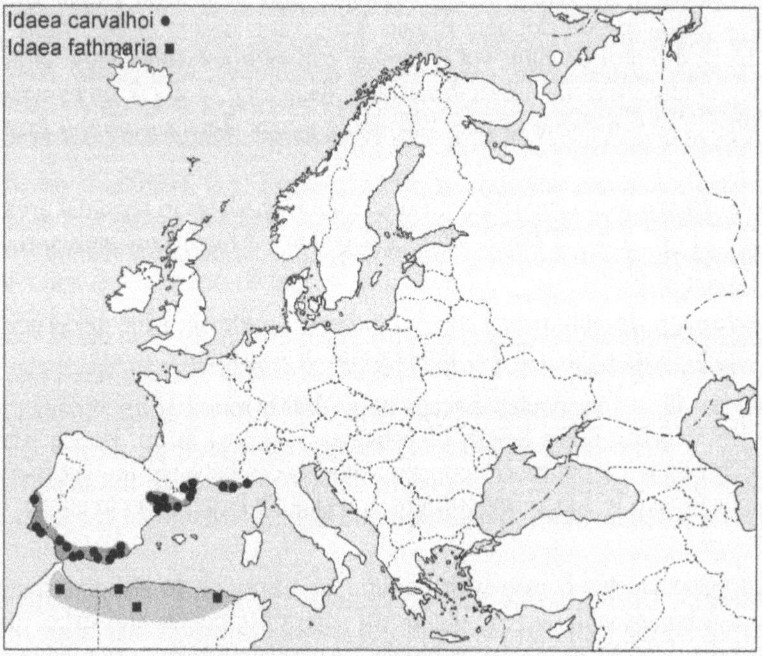

78. *Idaea carvalhoi* HERBULOT, 1979

Idaea carvalhoi HERBULOT, 1979: Alexanor 11 (4): 178 (France, Var: les Arcs, Baume). Holotype ♀ (ZSM/HERBULOT, examined).
Unavailable names (misidentification): *fathmaria*: sensu SCHMIDLIN (1964, partim) nec OBERTHÜR, 1876.

DIAGNOSIS: Wingspan ♂♀ 14-17 mm, exceptionally 13 mm only. Wings very slender and elongate. Forewing venation: R1 arising subapically from cell, separately from R2-R5. Ground colour dark grey brown. Transverse lines of forewing dark grey, oblique, postmedial line straight, antemedial line angled. Dark terminal line almost uninterrupted. Discal spots usually absent or indistinct. Frons and collar dark brown, vertex white. Abdomen unicolorous, dark. Proboscis present, short or vestigial. Palpi short, about 0.5 times diameter of eye. Antennae of ♂ with deep intersegmental incisions, ciliate-fasciculate, cilia long, 1.5-1.8 times width of flagellum, decreasing to about 1.0 times in the distal half of the antenna. Hindtibia of ♂ slender, without pencil. Tarsus 0.2-0.3 times length of tibia (Text-fig. 28). Length of hindtibia and hindtarsus about 2 mm.

MALE GENITALIA: Uncus broad, round at tip. Valva round at tip. Aedeagus with one comparatively stout cornutus, 0.7-0.85 times length of aedeagus. Sister species *I. fathmaria* (OBERTHÜR, 1876) with longer aedeagus, cornutus very thin, as long as the aedeagus.

FEMALE GENITALIA: Ductus bursae short. Corpus bursae elongate, width varying, strongly corrugated longitudinally, with small, globular, membranous appendix at the anterior end. Sister species *I. fathmaria* with anterior part of corpus bursae extremely long and narrow.

DISTRIBUTION: West-Mediterranean. Southern France and Iberian peninsula, recently recorded in north-western Italy (Liguria: BERTACCINI & CAMPRI 2003). – Outside Europe replaced by the sister species *I. fathmaria* (OBERTHÜR, 1876) from Morocco to eastern Algeria. The latter erroneously recorded for Turkey (RIEMIS 1994; 1996).

PHENOLOGY: Usually univoltine: most data from late May to late July, but in the south emerging from late April (KRAUS 1999). Larva slowly developing (HALL pers. comm.).

BIOLOGY: No detailed information for larval host-plants traced. Sister species *I. fathmaria* on fresh or withered leaves of herbaceous plants (PROUT 1913). Main activity of adults immediately after dusk, readily attracted to light. ♀ ratio at light very high, 80-95%, the same in the North African sister species (REISSER 1933). Resting position *Eupithecia*-like with hindwing not touching the abdomen.

HABITAT: Xerothermophilous. Most common in hot lowland habitats near the coast with Mediterranean vegetation (HERBULOT 1947c; BERTACCINI & CAMPRI 2003). From 0 m up to about 500 m above sea-level, in Andalusia exceptionally up to 1,700 m.

SIMILAR SPECIES (see Text-fig. 115): *I. longaria* (see no. 65). Allopatric sister species *I. fathmaria* without differences in habitus and external structure, distinguishable only by dissection. *I. saleri* smaller, with shorter wings, vertex dark, with a white dorsal line on the abdomen of both sexes, ♂ hindleg short. Additionally, when resting, *I. carvalhoi* draws its forewings more forewards than does *I. saleri*.

REMARKS: Genitalically examined specimens from many localities along the southern Spanish coast revealed to be all *I. carvalhoi*, whilst from various 'nearby' localities in Morocco (e.g. Xauen, Rif-mountains) only the sister species *I. fathmaria* could be found. VIVES MORENO (1994; note 89) retains occurrence of the latter in Europe as possible. Immature stages of *I. fathmaria* described by REISSER (1933a).

The ***camparia*** species-group

STERNECK's (1940) 22nd group of *Idaea*. Monotypic species-group. '*I. fatimata*' sensu STERNECK (1940) was misidentified, refers to *I. erubescens* HAUSMANN, 2003, and must be transferred to the *fractilineata* species-group, as sister species of *I. exilaria*. Male genitalia with comparatively short valva, aedeagus strongly curved. Lateral dilatation of corpus bursae and spinose crests in ♀ genitalia suggest relationship to the *seriata* species-group. Proboscis present. Sternum A2 of ♂ with weakly sclerotized semicircular sclerite and setose patch. Egg sculpturation hexagonal reticulate. Larva long and slender (Text-fig. 71), reminiscent of core groups of both the *seriata* and the *subsericeata* species-groups.

79. Idaea camparia (HERRICH-SCHÄFFER, 1852)

Acidalia camparia HERRICH-SCHÄFFER, 1852: Syst. Bearb. Schmett. Eur. 6 (55): 65. First available name after *camparia* HERRICH-SCHÄFFER, 1851 (pl. 76, fig. 465: non binominal) (western Turkey: Smyrna). Syntypes 1♂1♀ (from coll. LEDERER (HOMBERG 1911)). Referring to vol. 3: 15, thus binominal.

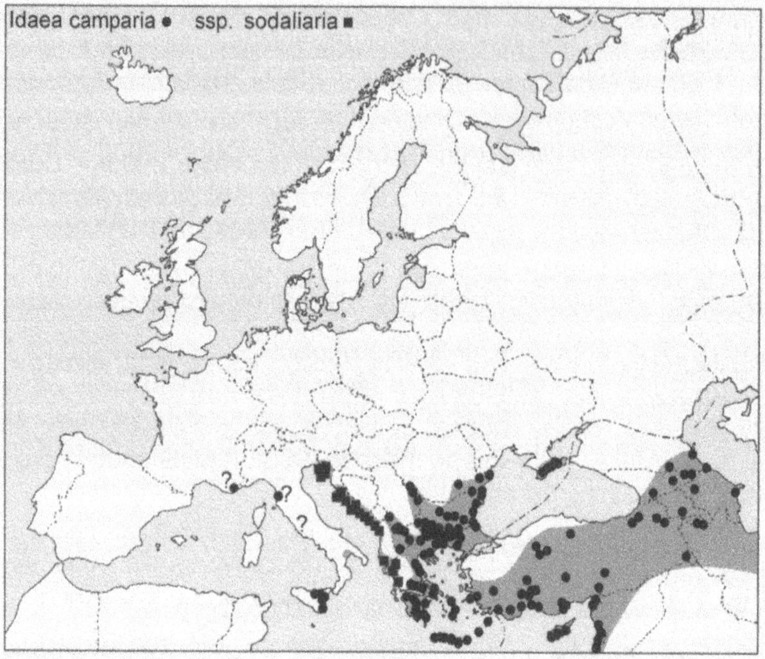

Acidalia sodaliaria HERRICH-SCHÄFFER, 1852: Syst. Bearb. Schmett. Eur. 6 (55): 65. First available name after *sodaliara* HERRICH-SCHÄFFER, 1851 (pl. 76, fig. 466: non binominal) (Croatia: Rijeka ('Fiume')). Syntypes 1♂1♀. Referring to vol. 3: 15, thus binominal. 'Page priority' would give precedence to *sodaliaria* over *camparia*. As the Code does not recognize 'page priority' and in favour of stability, this is not accepted here. Herewith downgraded again from species (SCOBLE 1999) to subspecific rank, previously stated e.g. by LERAUT (1997).

Ptychopoda camparia europaea WEHRLI, 1934: Mitt. Münch. ent. Ges. 24 (1): 9 (Bulgaria: Sistov). Lectotype ♂ (ZFMK, examined), herewith designated to fix type locality and to stabilize usage of the name; at ZFMK 14 paralectotypes (collected before 1934): 8♂3♀ Balkans, 2♂2♀ Crimea. Synoynmy: MENTZER (1990a).

Idaea boeklini MENTZER, 1990: Nota lepid. 12 (4): 300, figs 9-10,17,22 (Sicily: Taormina). Holotype ♀ (NHRS/MENTZER; ZSM: paratypes, examined). Discussion of synonymy see RAINERI (1992a).

Unavailable names (misidentification): *bischoffaria* (partim): sensu REBEL (1901) nec LA HARPE, 1864 (identity: *Idaea seriata*); *cossurata*: sensu PROUT (1913) nec MILLIÈRE, 1875 (identity: *Idaea seriata*). – (infrasubspecific): *albomarginata*: REISSER (1961); *nigrolineata*: REISSER (1961); *uniformis*: REISSER (1961). – (incorrect subsequent spellings): *campata*: BELLIER (1861); *europea*: GANEV (1983); *sodalaria*: PROUT (1913: 110).

DIAGNOSIS: Wingspan ♂ 15-19, ♀ 18-21 mm, in subsequent generations sometimes 13 mm only in both sexes. Forewing broad. Ground colour dirty white with bone coloured tinge. Populations from Balkan peninsula with paler ground colour and stronger marked pattern ('*europaea*'), but with clinous transitions to nominotypical forms from Turkey. Transverse lines dilated to dark spots at the forewing costa. Postmedial line dentate, sometimes also medial line. Terminal area with three groups of dark spots. Terminal line interrupted at the veins. Fringe dots weak or absent. Discal spots distinct. Underside with forewing darker and postmedial line distinct, close to termen. Frons black brown, vertex white, collar sand coloured or light brown. Antennae of ♂ dentate, with deep intersegmental incisions, ciliate-fasciculate, cilia long, 1.3-1.6 times width of flagellum (Text-fig. 57). Hindtibia of ♂ long, more than 2 mm, slightly dilated, standing over half of first tarsomere, with long pencil from base of tibia. Tarsus 0.4-0.45 times length of tibia (Text-fig. 30).

Variability: Populations of eastern Sicily ('*cossurata* auct.'; see remarks) with strong grey brown suffusion of ground colour, spots in terminal area large and strongly confluent, both wings concolorous on the underside. In southern Sicily, however, there are rather pale populations, similar to the nominate subspecies. Transitions to the nominate forms recorded also from rearings (SCHWINGENSCHUSS 1942). Specimens from Crete, Rhodos and central Anatolia often reminiscent of eastern Sicilian populations in habitus.

Subsp. ***sodaliaria*** with ground colour clear glossy white, pattern reduced, spots in terminal area indistinct.

MALE GENITALIA: Uncus and gnathos digitiform. Saccus large and broad. Valva broad, with one small, ventral, subapical spinule. Aedeagus long, strongly curved, with small, stout cornutus, about 0.25 times length of aedeagus. No constant differences between the subspecies, cornutus slightly broader in Asian populations.

FEMALE GENITALIA: Ductus bursae hourglass-shaped (helically twisted). Corpus bursae globular, membranous, laterally joining ductus bursae, with two semicircular sclerites one of them strongly spinulose, the other less. No constant differences between the subspecies, but antrum less dilated in the west of the distribution and second sclerite more spinulose in the Levant.

DISTRIBUTION: (East-)Mediterranean-Turanian. Balkan peninsula including all Greek islands, in the east to Crimea. In the western Balkan peninsula (north-westernmost Greece to Trieste) replaced by subsp. *sodaliaria*. Isolated occurrence on Sicily and southern Italy (PARENZAN 1994). To be deleted from the fauna of Pantelleria island (see remarks). Erroneously recorded from Spain (e.g. VIVES MORENO 1994; MÜLLER 1996) usually referring to misidentified *I. joannisiata*. RAINERI's (1992) dissected ♀ from 'Andalusia' supposedly mislabelled, such as two 'Spanish' specimens (ZSM, leg KORB). Occurrence in eastern France indicated by 1♂1♀ at ZSM (e.o., Alpes Maritimes, Fayence; R. BLAESIUS), but requiring further confirmation. Reputedly in north-western Italy, Livorno (MNHU/PÜNGELER), central Italy (TURATI 1914; MARIANI 1941; PROLA & RACHELI 1979), awaiting verification. – Outside Europe in Turkey, Cyprus, Levant, Transcaucasus, Iraq, Iran, Turkmeniya. Doubtful records for Libya, Cyrenaica (PROUT 1913; TURATI 1934; PROUT 1935a; VIIDALEPP et al. 1992) await confirmation. Erroneously recorded for Morocco (RUNGS 1981).

PHENOLOGY: Usually trivoltine: early May to late October, without gaps between generations. In the south (e.g. Levant, Crete, Cyprus) emerging from early April. On mountains and in the north probably bivoltine, from May to September.

BIOLOGY: Natural host-plants unknown. Reared on withered, not too dry, leaves of various herbaceous plants (LEIPNITZ pers. comm.), on *Hedera helix* and *Lactuca sativa* (MAY 1892). Adults readily attracted to light, in late autumn also to bait (Greece: WEGNER pers. com.).

HABITAT: Xerothermophilous. Inhabiting dry, stony grassland, and scrubby places on sun-exposed slopes. Usually from 0 m up to 800, sometimes up to 1,200 m above sea-level, in eastern Turkey and Lebanon up to 2,000, in Iran up to 2,600 m.

SIMILAR SPECIES (see Text-fig. 109): *I. obliquaria* (see no. 55), *I. consolidata* (71), *I. seriata* (74). *I. albitorquata* differs in a number of characters, none of them allowing secure identification in all cases: usually forewing less round, coloration clearer white, postmedial line distinct, dentate and straighter, costal spots of transverse lines more diffuse, discal spots smaller, terminal shadows weaker. Identification of ♂ is facilitated by brushing tip of abdomen to remove scales and reveal shape of valva. *I. contiguaria* with darker costal spots of forewing, medial line proximally from discal spot, ♂ hindtarsus longer. *Scopula marginepunctata* larger, ♂ hindtarsus longer, ♀ hindtibia with four spurs.

REMARKS: Some examined specimens from Pantelleria island (leg. AISTLEITNER), reminiscent of *I. camparia* from eastern Sicily, and well matching the original description of *cossurata* MILLIÈRE, 1875, proved to belong to *I. seriata* (compare colour plates, fig. 74n). For morphology of immature stages see MAY (1892) and REBEL (1903).

The *subsericeata* species-group

STERNECK's (1940) 10th (*I. mancipiata, I. sylvestraria*), 21st (*I. dimidiata, I. subsaturata*), 24th (*I. subsericeata, I. pallidata*) group of *Idaea*. *I. biselata,* and *I. trigeminata* are tranferred here from 13th and 28th group, respectively. Male genitalia of both the previous 21th and 24th group characterize two 'very natural' sub-groups (STERNECK 1940: 153; 157), the former with asymmetrical valvae, the latter with apical hooks on the aedeagus. The latter mentioned character links the *subsericeata* sub-group with *I. mancipiata*. Similar shape of cornutus links the *dimidiata* sub-group with *I. trigeminata*. Female genitalia: reminiscent of those of *Eupithecia*, corpus bursae usually covered all over with micro-spinules; exceptions: *I. subsaturata, I. pallidata*. Hindwing usually rounded, but distinctly concave between M1 and M3 (*I. dimidiata, I. subsaturata*). Proboscis present. Antennae of ♂ usually ciliate-fasciculate (exceptions: *I. biselata* and *I. trigeminata*), often with curved cilia, e.g. in *I. pallidata* and *I. dimidiata*, linking STERNECK's 21st and 24th group. Central sclerite and setose patch absent from ♂ sternum A2 (species no. 80-83), or setose patch present (no. 84-85); *I. biselata* with large and distinct horseshoe-shaped sclerite and patch of long setae, in *I. trigeminata* similar, but more weakly sclerotized. Egg sculpturation hexagonal reticulate. Larva usually slender, but short and thick in *I. biselata* and *I. trigeminata*. Larva of *I. dimidiata* resting with last segment not attached to the substrate (BOLDT 1929; EBERT 2001: fig.).

80. Idaea subsericeata (HAWORTH, 1809)

Phalaena subsericeata HAWORTH, 1809: Lepid. Brit. (2): 352 (United Kingdom). Syntype(s) (BMNH: 1♂ examined, with red 'type' label).

Acidalia perfluaria BOISDUVAL, 1840: Genera Index meth. eur. Lepid.: 226 (southern France: Montpellier). Syntypes ♂♀ (ZFMK: 1♂1♀ ex coll. BOISDUVAL, examined; BMNH).

Acidalia pinguedinata ZELLER, 1847: Isis 1847 (7): 521 (Sicily: Cisterne, near Messsina). Syntypes 1♂1♀ (BMNH).

Idaea asbestaria ZELLER, 1849: Stettin. ent. Ztg. 10 (7): 216, index (Italy, Tuscany: Antignano, Posignano). Syntypes ♂♀ (BMNH). Generic combination indicated in the index.

Cabera puraria HEYDENREICH (1851): Lepid. Eur. cat. meth. (Edn 3): 54 (type locality not stated). Syntype(s).

Acidalia pinguedinaria HERRICH-SCHÄFFER, 1852: Syst. Bearb. Schmett. Eur. 6 (55): 66. First available name after *pinguedinaria* HERRICH-SCHÄFFER, 1848 (pl. 66, figs 409, 410: non binominal) (Sicily: Messina). Syntypes 1♂1♀. Referring to vol. 3: 16, hence binominal.

Acidalia oloraria RÖSSLER, 1857: Jahrb. nass. Ver. Naturk. 12: 393 (western Germany: Nassau). Syntype(s).

Acidalia mancuniata KNAGGS, 1865: Entom. mon. Mag. 2: 130 (United Kingdom: Manchester). Syntypes, at least 6.

Acidalia veterata GREGSON, 1866: in ANONYMOUS: 'Entom. mon. Mag. 3: 158' (reference according to SCOBLE 1999, but not found there).

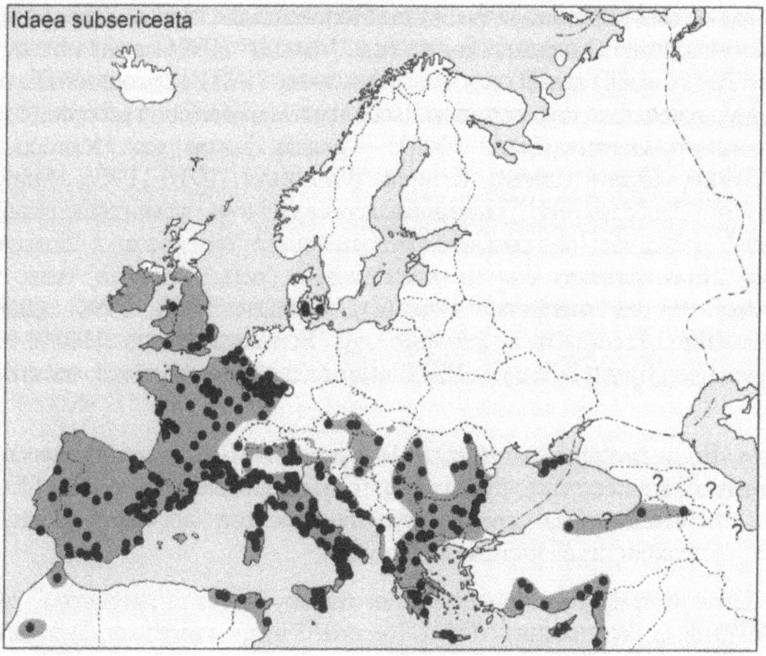

Acidalia diaphanaria BANG-HAAS, 1910: Dt. ent. Z. Iris 24: 43, pl. 3, fig. 16 (Tunisia: Tunis, Ain Draham). Syntypes ♂♂ (MNHU/PÜNGELER, examined).

Ptychopoda zernyi D. LUCAS, 1934: Bull. Soc. Ent. Fr. 38: 312 (Morocco: Khenifra). Holotype ♀ (ZSM/HERBULOT, examined). Herewith downgraded from species rank (SCOBLE 1999) to synonymy, previously stated by HERBULOT (1984a).

Unavailable names (misidentification): *pallidata*: sensu FORSTER & WOHLFAHRT (1981: figs 9) nec DENIS & SCHIFFERMÜLLER, 1775. – (infrasubspecific): *albomarginata*: REISSER (1961); *impuncta*: LEMPKE (1949); *obscura*: REBEL (1910). – (incorrect subsequent spellings): *olivaria* (for *oloraria*): WALKER (1863); *olorata*: RÖSSLER (1866); *subcericeata*: CALLE et al. (2000: pl. 3); *subscriceata*: PROUT (1934); *subsericata*: STAUDINGER (1861); *subsericearia*: DOUBLEDAY (1849); *subsericiata*: DIDMANIDZE (2002).

DIAGNOSIS: Wingspan ♂♀ 18-22 mm, in subsequent generations 13-20 mm. Ground colour white, slightly glossy. Transverse lines light grey, straight. Distances between transverse lines variable. Distal to postmedial line two whitish (wavy) lines. Discal spots and fringe dots very small (mainly on forewing) or absent. Frons dark brown, vertex white, collar ochreous. Antennae of ♂ ciliate, sub-fasciculate, cilia long, 1.25-1.7 times width of flagellum (Text-fig. 58), in the eastern Mediterranean area sometimes up to 2.0 times. Hindtibia of ♂ broad, with large whitish pencil, tibia at distal end prolonged and covering half of first tarsomere. Tarsus short, 0.25-0.35 times length of tibia.

MALE GENITALIA: Uncus slender. Gnathos sub-rectangular. Saccus a small, narrow process. Valva normal, round at tip. Aedeagus slender, with a small, very fine hook at tip.

FEMALE GENITALIA: Ductus bursae slender, with antrum slightly dilated and membranous appendix at anterior end. Corpus bursae pyriform, densely spinulose nearly all over, at posterior end only laterally spinulose.

DISTRIBUTION: West-Palaearctic. Widely distributed in southern Europe, more local in western and central Europe, in the north to Great Britain, Ireland, Denmark and Czech

Republic, in the east to Crimea. Present on Mediterranean islands, except Malta. Isolated occurrence in southern European Russia (e.g. VIIDALEPP 1996) awaits further confirmation. In the north usually rare (FORSTER & WOHLFAHRT 1981), in southern Europe, e.g. in southern Italy sometimes very common. Recent areal expansions recorded for Rhine valley, south-western Germany (EBERT 2001). – Outside Europe from Morocco to Tunisia, southern Turkey, Cyprus, Levant, Armenia (WARDIKJAN 1970; 1985). Many examined specimens from Caucasus and Transcaucasus revealed to be misidentified (mainly *I. mancipiata* and *I. pallidata*), the species is perhaps absent from some of those areas, such as, possibly, from northern Iran to south-western parts of central Asian mountains. Citations from the last mentioned areas (e.g. VIIDALEPP 1988; 1996) usually referring back to unverified records from literature, e.g. SCHWINGENSCHUSS (1939). Records for Libya, Cyrenaica (TURATI & ZANON 1922) refer to the sister species *I. maurusia* (TURATI, 1924).

PHENOLOGY: Bi- or trivoltine: late April to late June; mid-July to mid-September, with single specimens of an incomplete third brood, from mid-September to early October. In the lowlands of the Mediterranean area tri- or plurivoltine from late March to late October. In the Levant plurivoltine in all months of the year.

BIOLOGY: Larva polyphagous on withered or mouldy leaves of herbaceous plants (REBEL 1910b; FORSTER & WOHLFAHRT 1981). Recorded (resp. reared) on *Anagallis arvensis* (Plumbaginaceae), *Stellaria media* (Caryophyllaceae), *Plantago* (Plantaginaceae), *Polygonum aviculare, Rumex* (Polygonaceae), *Prunus domestica* (Rosaceae), *Solidago virgaurea, Lactuca sativa,* and on withered leaves of *Taraxacum officinale* (Asteraceae) (var. cit., e.g. LHOMME 1935; ALLAN 1949; SKOU 1986; SKINNER 1984; MÜLLER pers. comm.). Reared on withered leaves of *Hieracium, Taraxacum officinale, Vicia cracca, Galium, Campanula, Fagus* and on petals of *Pimpinella anisum* (original data; HAUSMANN & MILLER 2000). Main activity of adults at dusk (SKINNER 1984), but readily attracted to light and bait.

HABITAT: Xerothermophilous. In the north restricted to warm slopes, sun-exposed forest fringes, rides and glades, e.g. of oak forest, hedgerows, grassland, and dry salt marshes

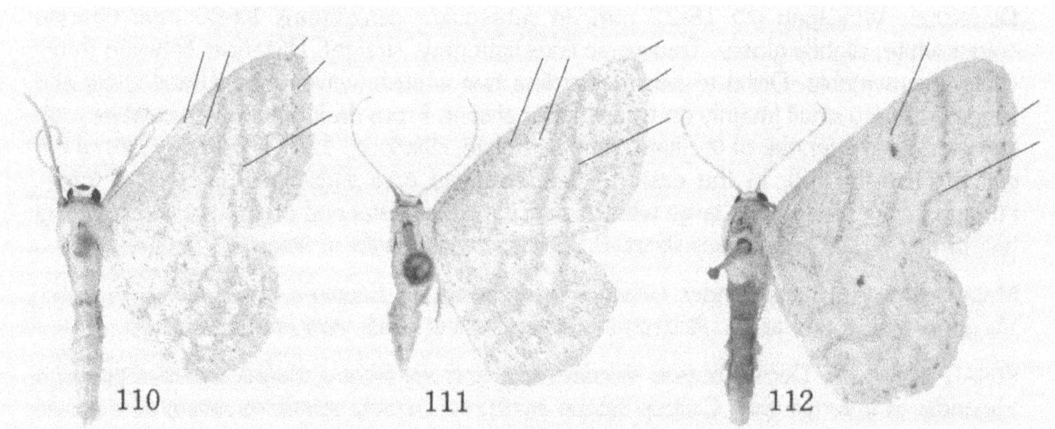

Text-figs 110-112: Diagnostic characters (indicated) of European *Idaea* species. Text-fig. 110: *I. subsericeata* (HAWORTH, 1809). Text-fig. 111: *I. pallidata* (DENIS & SCHIFFERMÜLLER, 1775), female. Text-fig. 112: *I. mancipiata* (STAUDINGER, 1871).

near coast (SKOU 1986, SKINNER 1984; WARING & TOWNSEND 2003), on sand, chalk and limestone. In southern Europe almost ubiquitous and common in the Mediterranean maquis, uncultivated grassland, scattered oak forest etc.. From 0 m up to 600 m above sea-level, in the south up to 1,000 m, exceptionally higher, e.g. in Turkey up to 1,900 m.

SIMILAR SPECIES: *I. litigiosaria* (see no. 12), *I. mutilata* (13), *I. ossiculata* (23). ♀ of *I. pallidata* differs in less glossy white coloration, more suffused with black scales, transverse lines more diffuse and yellowish, fringe dots and discal spots absent, forewing apex less rounded. *I. mancipiata* with distinct discal spots.

REMARKS: Large pencil on the hindtibia reminiscent of equivalents in *I. trigeminata* and *I. biselata*, perhaps indicating certain relationships. For misidentified genitalia in literature see rectification in REZBANYAI-RESER (1988b). For rearings and morphology of immature stages see BUCKLER (1897).

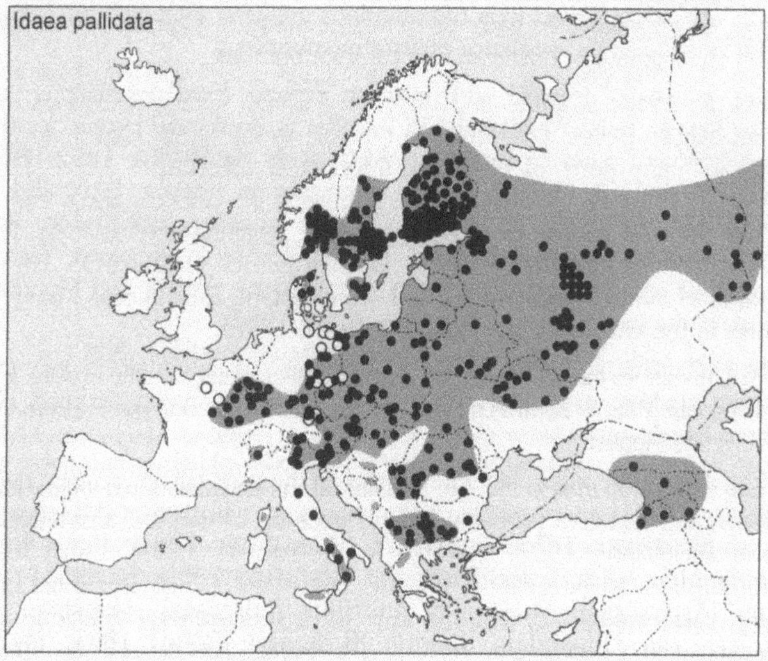

81. Idaea pallidata (DENIS & SCHIFFERMÜLLER, 1775)

Geometra pallidata DENIS & SCHIFFERMÜLLER, 1775: Ank. syst. Werkes Schmett. Wienergegend: 110 (Austria: Vienna distr.). Syntype(s) lost. Erroneously regarded as homonym of *Phalaena pallidata* HUFNAGEL, 1767 by MENTZER (1984), see remarks to validity of SCHIFFERMÜLLER names in HAUSMANN (2001:77).
Acidalia pallidaria TREITSCHKE, 1825: Schmett. Eur. 5 (2): 439. Emendation. See *Geometra pallidaria*: sensu HÜBNER (1799) (identity: *I. ochrata*).
Acidalia byssinata TREITSCHKE, 1828 (♀): Schmett. Eur. 6 (2): 36 (Austria: Vienna; Styria). Syntypes ♀.
Acidalia byssinaria BOISDUVAL, 1840: Genera Index meth. eur. Lepid.: 225. Emendation.
Idaea pallidata similiformis VASILENKO, 1990: in: ZOLOTARENKO [Arthropods and helminths]. – Fauna Sibiri 12: 108, fig. (Russia: Novosibirsk region). Holotype ♂. Valid at subspecific rank (SCOBLE 1999).
Unavailable names (misidentification): *subsericeata*: sensu FORSTER & WOHLFAHRT (1981: fig. 8) nec HAWORTH, 1809. – (infrasubspecific): *anastomosaria*: REISSER (1961); *centribitangens*: SILBERNAGEL (1943); *demarginata*: REISSER (1961). – (incorrect subsequent spellings): *bissinata*: FREYER (1829: 37); *paludata*: MANN (1855).

DIAGNOSIS: Strongly sexually dimorphic. Wingspan ♂ 18-22, ♀ 13-19 mm. Wings sand coloured in ♂. ♀ white, with slight black suffusion, mainly towards costa. Transverse lines pale brown, slightly waved, broad in ♂, finer in ♀. Distally from postmedial line two pale lines, concolorous with ground colour. Discal spots and fringe dots absent. Palpi pale sand coloured. Frons dark brown, vertex white (♀) or sand coloured (♂), collar darker, ochreous. Antennae of ♂ ciliate-fasciculate, cilia curved and long, 1.7-2.2 times width of flagellum (Text-fig. 59). Hindtibia of ♂ slender, without pencil, but with elongate lateral scales. Tarsus short, 0.25-0.35 times length of tibia.

MALE GENITALIA: Uncus slender digitiform. Gnathos slightly broader. Saccus strongly sclerotized, narrow, distinctly projecting. Valva long, with triangular dorsal process at 2/3. Aedeagus slender, with weakly sclerotized needle-shaped cornutus in the centre, and a stout hook at the tip.

FEMALE GENITALIA: Antrum laterally tapering, posteriorly concave. Ductus bursae comparatively broad, anteriorly dilated with membranous margins. Corpus bursae oval, densely covered with small spinules, posterior quarter membranous.

DISTRIBUTION: Eurasiatic. Central and northern Europe, local populations in southern Europe from Italy to Balkan peninsula, in the east to the Urals. Usually local and rare. Areal losses in many parts of Germany (MÜLLER & GELBRECHT 1992; ERLACHER & FRIEDRICH 1994; WOLF & HACKER 2004). Reputedly on Corsica, Sicily and in Greece (RUNGS 1988; MÜLLER 1996), no vouchers traced, awaiting confirmation. Records for Catalonia misidentified (DANTART 2000). – Outside Europe in Caucasus, Transcaucasus, northern parts of central Asian mountains (Kirghizstan), Siberia and Mongolia (subsp. *similiformis*), in the east to Amur region (VIIDALEPP 1996).

PHENOLOGY: Univoltine: early May to late June, females exceptionally in July, in the north emerging from late May, in southern Europe with (partial) second generation, limited data available from late August.

BIOLOGY: Egg deposition unspecific, i.e. females let the eggs fall down when flying around grasses (SKALA 1910). Larva probably polyphagous on withered or dry leaves of herbaceous plants (BERGMANN 1955; FORSTER & WOHLFAHRT 1981). Larva found under *Achillea millefolium* (Asteraceae) (Germany: BERGMANN 1955). Recorded on *Achillea millefolium* (Asteraceae), *Vaccinium myrtillus* (Ericaceae), *Valeriana officinalis* (Valerianaceae), and *Filipendula ulmaria* (Rosaceae) (STRØM 1891; PROUT 1913; LHOMME 1935). Reared on *Achillea millefolium, Lactuca sativa, Hieracium pilosella* and *Taraxacum officinale* (HABICH 1892; KVALHEIM 1944; SEPPÄNEN 1954; SEPPÄNEN 1970). Adults usually recorded during daytime or at dusk, very easily flushed up from vegetation, ♂♂ readily attracted to light. ♀-ratio in collections low, 25-35%.

HABITAT: On sun-exposed, low-nutrient grassland in aisles, glades and fringes of forests, preferring dry places. In southern Italy collected on flower-rich, wet grassland on sun-exposed mountain slopes. On limestone (EBERT 2001), sandy and silicious soils (BERGMANN 1955). From 0 m up to 1,000 m above sea-level, south of the Alps montane, from 1,200 m up 1,900 m.

SIMILAR SPECIES: *I. dilutaria* (♂, see no. 58), *I. subsericeata* (♀, 80). *I. sylvestraria* with distinct discal spots. ♂ of *I. pallidata* reminiscent of *I. nitidata*, the latter larger, wings more rounded, transverse lines more undulate, ♂ antennal cilia much shorter, ♂ hindtarsus longer.

REMARKS: For misidentified genitalia in literature see rectification in REZBANYAI-RESER (1988b). For morphology of immature stages see HABICH (1892) and MAY (1893).

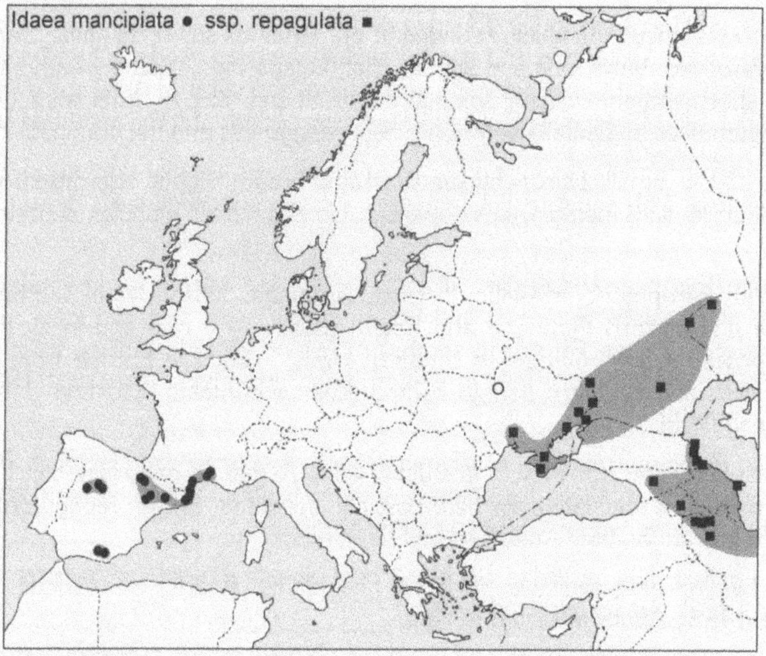

82. *Idaea mancipiata* (STAUDINGER, 1871)

Acidalia mancipiata STAUDINGER, 1871: in STAUDINGER & WOCKE, Cat. Lepid. eur. Faunengeb. (Edn 2) 1: 147 (Spain, Andalusia: Granada). Lectotype ♂ (MNHU, examined), herewith designated to stabilize usage of the name with regard to misidentified paralectotype; at MNHU 5 paralectotypes, 4 from type locality, one from Castilia, the latter misidentified (identity: *I. litigiosaria*).

Sterrha mancipiata var. *repagulata* PROUT, 1934: Lep. Cat.: 369. First available name after *repagulata* PROUT, 1913 (in SEITZ, Macrolep. 4: 117, pl. 7, fig. d: infrasubspecific, introduced as 'local form') (south-eastern European Russia: Krasnoarmeysk, 'Sarepta'). Syntypes 1♂2♀ (BMNH/ZELLER: 1♂ examined, with 'holotype' label).

Sterrha bustilloi AGENJO, 1967: Eos 42 (3/4): 299, figs. 1-6 (Spain, prov. Madrid: Vaciamadrid). Holotype ♂ (MNCN, examined). Herewith downgraded from species rank (VIVES MORENO 1994; MÜLLER 1996; SCOBLE 1999) to synonymy (syn. n.).

Unavailable names (misidentification): *minuscula* (misspelling): sensu VIVES MORENO (1984) nec RIBBE, 1912 (identity: *Idaea minuscularia*). – (incorrect subsequent spelling): *repagulate*: DIDMANIDZE (2002).

DIAGNOSIS: Wingspan ♂♀ 15-20 mm, second generation 14-16 mm. Broad winged, hindwing margin rounded. Wings pale sand coloured, with light brown suffusion, more strongly suffused in central Spanish populations ('*bustilloi*'), and, sometimes, in southern Spain. Transverse lines fine, slightly convex, brown in the nominate subspecies. Terminal area almost concolorous with ground colour. Discal spots black, sharp, distinct. Frons dark brown, vertex whitish or sand coloured, collar usually darker. Antennae of ♂ ciliate, subfasciculate, cilia very long, 2.0-2.6 times width of flagellum (Text-fig. 60). Hindtibia of ♂ slender, curved, with whitish pencil, not covering tarsus. Tarsus shortened, 0.2-0.35 times length of tibia.

Subsp. ***repagulata*** with ground colour glossy white, with slight sand coloured tinge, transverse lines grey. Hindtarsus of ♂ in populations from Transcaucasus to central Asian mountains as in the nominate subspecies, but in two examined ♂♂ from Uralsk longer, 0.4-0.45 times length of tibia.

MALE GENITALIA: Uncus triangular, rounded at tip. Gnathos sub-rectangular. Saccus large, strongly sclerotized. Valva long and slender, slightly tapering. Aedeagus without cornutus, but with a short, straight external spine at the distal end. Size of distal spine varying. No constant differences in genitalia between subsp. *repagulata* and the nominate subspecies.

FEMALE GENITALIA: Small. Ductus bursae short and broad, slightly sclerotized, posteriorly slightly spinulose, with membranous appendix. Corpus bursae globular, densely spinulose all over.

DISTRIBUTION: European-Westasiatic, disjunct. Nominate subspecies in southern France (Camargue and eastern Pyrenees) and Spain, usually very rare and local, threatened. Subsp. *repagulata* from Crimea to southern Urals. – Outside Europe from Caucasus, Transcaucasus, and northern Iran to central Asian mountains (VIIDALEPP 1988; 1996) (subsp. *repagulata*).

PHENOLOGY: Bivoltine: mid-May to late June; early August to early October. First generation on mountains until late July, here possibly univoltine. Subsp. *repagulata* bivoltine: late May to late June, early August to mid-September.

BIOLOGY: Limited data available on larval host-plants. Reared on 'various low herbs' (WILTSHIRE 1952). Adults attracted to light.

HABITAT: In north-eastern Spain and southern France in salt marshes and dry, sandy steppe habitats, usually near coast, in northern Iran in oasis gardens (WILTSHIRE 1952). Usually from 0 m up to 300 m above sea-level, in central and southern Spain sometimes up to 1,500 m, subsp. *repagulata* from 0 m up to 1,500 m.

SIMILAR SPECIES: *I. belemiata* (unicolorous forms, see no. 51), *I. politaria* (pale forms, 64), *I. subsericeata* (80). *I. sylvestraria* with darker suffusion of wings, fringe dots more distinct, ♂ antennal cilia shorter, ♂ hindtarsus longer. *I. subsaturata* usually smaller, vertex darker, ♂ hindtarsus longer. *I. biselata* with darker terminal area, ♂ hindtibia with large, dark pencil, identification of ♀ sometimes requiring dissection in the small area of sympatrical occurrence (north-eastern Spain, southern France). *Scopula minorata* with postmedial line more dentate, ♂ hindtarsus much longer, ♀ hindtibia with four spurs rather than two. *I. m. repagulata* similar to *S. immutata*, the latter larger, usually with more yellow suffusion, mainly in the males, ♂ hindtarsus much longer, ♀ hindtibia with four spurs.

83. Idaea sylvestraria (HÜBNER, 1799)

Geometra sylvestraria HÜBNER, 1799: Samml. Eur. Schmett. 5, Geometrae (1): pl. 18, fig. 94 (Europe). Syntype(s) lost.

‡ *Acidalia marginepunctata* STEPHENS, 1831: Illust. Br. Ent. (Haust.) 3: 310 (southern England, New Forest: Brockenhurst). Syntype(s). Junior subjective homonym nec GOEZE, 1781 (identity: *Scopula marginepunctata*).

‡ *Acidalia straminata* TREITSCHKE, 1835: Schmett. Eur. 10 (2): 205 (Germany: Franconia, Saxonia). Junior primary homonym nec BORKHAUSEN, 1794 (identity: *Idaea straminata*).

‡ *Idaea straminaria* HERRICH-SCHÄFFER, 1839: in PANZER, Faunae Insect. Germ. initia (Dtl. Insect.) (166): 93. Incorrect subsequent spelling (Emendation according to SCOBLE 1999).

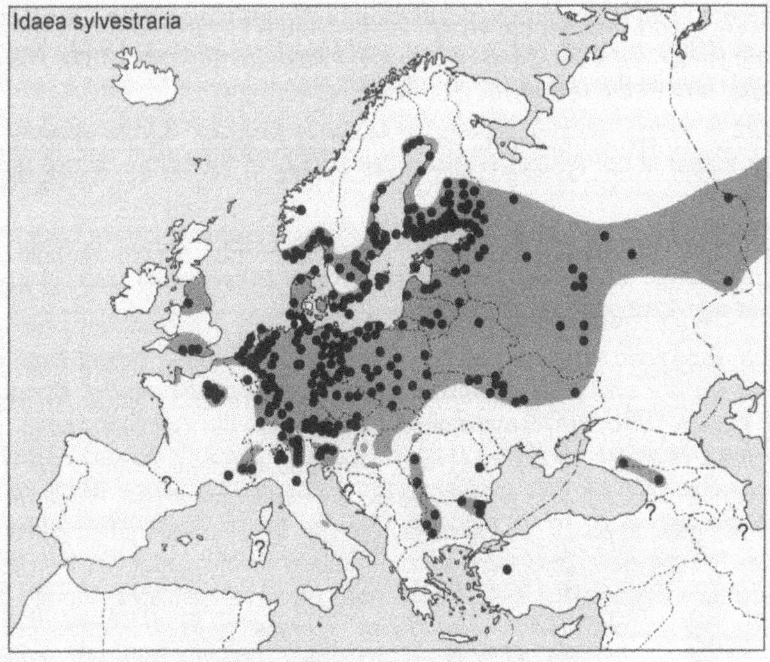

Acidalia grammicaria BOISDUVAL, 1840: Genera Index meth. eur. Lepid.: 225 (Europe). No separate types. Unnecessary replacement name for *sylvestraria*.
Acidalia circellata GUENÉE, 1858: in BOISDUVAL & GUENÉE, Hist. nat. Insectes (Spec. gén. Lépid.) 9: 482 (Great Britain: Manchester). Holotype ♂ ('coll. DOUBLEDAY'; ZFMK: 1♂ ex coll. GUENÉE, 'Angleterre', possibly the holotype). Junior synonym (individual aberration).
Acidalia folognearia STAUDINGER, 1863: Stettin. ent. Ztg. 24 (7-9): 264 (Belgium: Brussels). Syntypes 2♂1♀ (MNHU: 2♂, examined). Junior synonym (f. *circellata*).
Acidalia graciliata MANN, 1867: Verh. zool.-bot. Ges. Wien 17: 841 (Italy, South Tyrol: by Bozen). Holotype ♂.
Acidalia shiskensis MATSUMURA, 1925: J. Coll. Agric. Hokkaido Imp. Univ. 15: 158, fig. 7 (Russia, Sakhalin: Shiska). Lectotype ♂ (EIHU: selected and examined by SATO 2001).
‡ *Acidalia straminata minuta* HEYDEMANN, 1933: Schr. Naturw. Ver. Schl.-Holst. 20 (1): 164 (northern Germany: Amrum island; Rendsburg). Syntypes lost. Preoccupied, junior subjective homonym of *Idaea minuta* (SCHAUS, 1901) from N. America. Validity at subspecific rank refused here.
Unavailable names (misidentification): *obsoletaria*: sensu WESTWOOD (1854) nec RAMBUR, 1933. – (infrasubspecific): *clausa*: LEMPKE (1949); *demarginata*: REISSER (1961); *virgata*: COCKAYNE (1950). – (incorrect subsequent spellings): *circellaria*: MORRIS (1861); *circellatta*: MILLIÈRE (1869: 51); *marginepunctaria*: DOUBLEDAY (1849); *silvestraria*: DIDMANIDZE (2002); *sylverstraria*: FIUMI & CAMPORESI (1988); *sylvestaria*: ANIKIN et al. (2000); *tolognearia*: SNELLEN (1873).

DIAGNOSIS: Wingspan ♂♀ 17-20 mm, exceptionally 15 mm only. Ground colour dirty white, with brown suffusion. Transverse lines usually broad, only slightly darker than ground colour. Distal to postmedial line two pale wavy lines, concolorous with ground colour. Discal spots fine, black, usually lighly encircled. Fringe dots black, sharp, distinct. Underside with darker markings. Frons dark brown, vertex white, collar sand coloured. Antennae of ♂ ciliate, sub-fasciculate, cilia 1.1-1.3 times width of flagellum. Hindtibia of ♂ slender, with elongate lateral scales, pencil absent. Tarsus 0.45-0.6 times length of tibia.

Variability: Transverse lines sometimes sharper, dark brown, with postmedial line often dotted, not rarely with stronger brown suffusion of wings ('f. *circellata*'). Such forms common in Great Britain (mosses of Lancashire and Cheshire: SKINNER 1984), and European Atlantic coast, rare in the rest of the distribution area.

MALE GENITALIA: Uncus slender digitiform. Gnathos broader. Saccus shallow, rounded. Valva broad, round at tip. Aedeagus with weak sclerite in the centre, vesica slightly granulated.

FEMALE GENITALIA: Ductus bursae slender, towards corpus bursae laterally spinulose. Corpus bursae large, oval or lemon-shaped, densely spinulose all over, at anterior end with globular membranous appendix.

DISTRIBUTION: Eurasiatic. Western, central and northern Europe, absent from Ireland. In the south to central and south-eastern France, northern and central Apennines (e.g. FLAMIGNI & BASTIA 1998; 2003) and Balkan peninsula, in the east to the Urals. Reputedly in Spain (VIVES MORENO 1994; MÜLLER 1996), and Sardinia (HARTIG 1951; MÜLLER 1996), no vouchers traced, data possibly based on misidentifications. Some records from central Italy (RACHELI et al. 1978; PROLA & RACHELI 1979) misidentified (identity: *I. consanguinaria*; RAINERI det., ZILLI pers. comm.). Absent from Catalonia (DANTART 2000) and southern Italy (PARENZAN 1994). Populations often very isolated, though sometimes common at sites of occurrence. Abundance decreasing in Thuringia (ERLACHER & FRIEDRICH 1994). – Outside Europe in Caucasus (ZISP), northern Transcaucasus (VIIDALEPP 1996; DIDMANIDZE 2002), in the east across northern central Asian mountains and Siberia to Far East of Asia. Reputedly in 'Transcaspia' (PROUT 1913) and Turkey (RIEMIS 1994; VIIDALEPP 1996), awaiting confirmation.

PHENOLOGY: Usually univoltine (e.g. in wetland habitats and in the north): mid-June to mid-August. Under good conditions bivoltine, from early June to early September, generations overlapping. Single specimens emerging from early May and flying until early October.

BIOLOGY: Oligophagous (3) on *Calluna vulgaris* (Ericaceae) and some other herbaceous plants of low-nutrient soils. Egg positioning preferably on stems and twigs of *Calluna vulgaris*, in captivity, and in nature (PROUT 1938a: 227; URBAHN & URBAHN 1939). Larva found on *Calluna vulgaris* (Germany: WEGNER pers. comm.), and under *Artemisia campestris* (Asteraceae) (Poland: URBAHN & URBAHN 1939). Larva recorded also on *Thymus serpyllum* (Lamiaceae), *Genista tinctoria* (Fabaceae), *Chenopodium* (Chenopodiaceae) (var. cit.), on *Vaccinium oxycoccos* (Ericaceae) (MILLIÈRE 1869), reputedly also on lichenes, and on *Vaccinium myrtillus* (BERGMANN 1955). Reared on withered leaves of *Taraxacum officinale* (also petals), *Thymus praecox, Chenopodium album, Polygonum aviculare, Plantago*, and *Rubus fruticosus* (var. cit., e.g. ALLAN 1949; SKINNER 1984; PORTER 1997; MÜLLER pers. comm.). Main activity of adults at dusk, ♂♂ readily attracted to light. During daytime resting among *Calluna* (PROUT 1913) or *Thymus* (original data; EBERT 2001). Stenotopic, K-strategy, single populations occupying small territories. Both ecotypes (see habitat) valuable as environmental indicators (EBERT 2001).

HABITAT: Open habitats on low-nutrient soils, usually at localities with abundance of *Calluna* or *Thymus*. Occurring in two ecotypes, a xerothermophilous and a hygrophilous type, which may show geographically disjunct distributions, e.g. in south-western Germany (EBERT 2001). Xerothermophilous form inhabiting dry open habitats, such as heathland, fallows, dry forest fringes, often on sandy soils, dunes, or beach fringes. Hygrophilous form in wet open localities, e.g. on moors or bogs, *Sphagnum* bogs in humid forests mentioned by ANIKIN et al. (2000). In Bolognese (northern Italy) restricted to stony and rocky riverside habitats (FLAMIGNI & BASTIA 1998). From 0 m up to 500 m above sea-level, the hygrophilous type up to 700 m (EBERT 2001), up to 1,300 m in southern Alps.

SIMILAR SPECIES: *I. ossiculata* (see no. 23), *I. pallidata* (81), *I. mancipiata* (82).

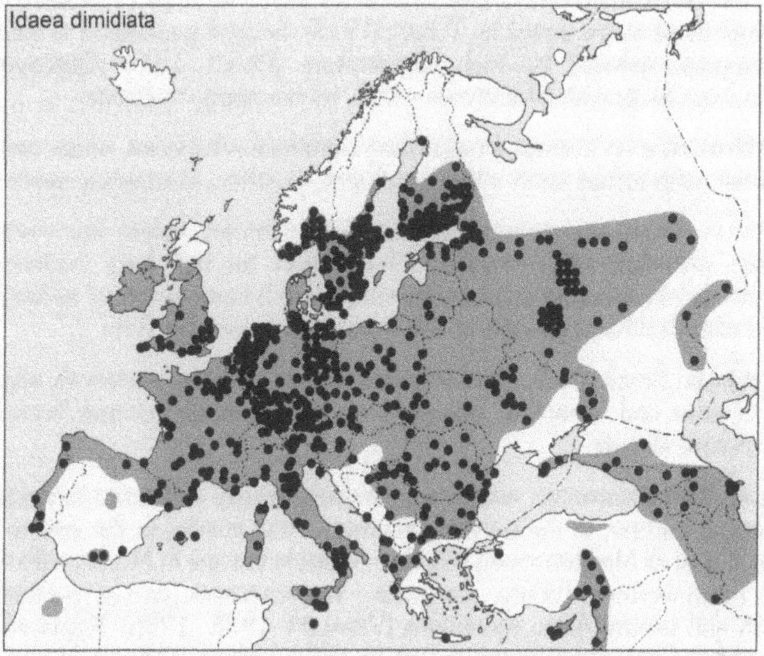

84. *Idaea dimidiata* (HUFNAGEL, 1767)

Phalaena dimidiata HUFNAGEL, 1767: Berlin. Mag. 4 (6): 602 (Germany: Berlin). Syntype(s) lost.

Geometra scutulata (DENIS & SCHIFFERMÜLLER, 1775): Ank. syst. Werkes Schmett. Wienergegend: 117 (Austria: Vienna distr.). Syntype(s) lost.

‡ *Phalaena scutata* FABRICIUS, 1787: Mant. Ins. 2: 211. Incorrect subsequent spelling of *scutulata*.

‡ *Geometra scutularia* HÜBNER, 1799: Samml. Eur. Schmett. 5, Geometrae (1): pl. 14, fig. 72 (Europe). Syntype(s) lost. Incorrect subsequent spelling.

Acidalia dimidiata roseata TURATI, 1913: Atti Soc. ital. Sci. nat. 51: 321, pl. 9, fig. 25 (Sardinia: Aritzo). Syntypes 3♂1♀. Though calling it 'form' in the text, TURATI understands the taxon at subspecific rank.

Ptychopoda dimidiata antitaurica WEHRLI, 1931: Mitt. Münch. ent. Ges. 21 (2): 41 (southern Turkey: Maras; Maydan Ikbis ('Akbès')). Syntypes 3♂2♀ (ZFMK, examined). Valid at subspecific rank.

Ptychopoda dimidiata tauricola WEHRLI, 1934: Mitt. Münch. ent. Ges. 24 (1): 8. In ZFMK, one ♀ is labelled as 'type' of *tauricola* (southern Turkey: Maras). Unnecessary replacement name for *antitaurica* WEHRLI, 1931, which is not a junior homonym of *antitaurica* WEHRLI, 1932 (see *I. palaestinensis*).

Unavailable names (misidentification): *lividata*: sensu HAWORTH (1809) nec CLERCK, 1759 (identity: *Idaea aversata*); *subsaturata*: sensu WILTSHIRE (1949) and PROUT (1938, partim: 225) nec GUENÉE, 1858. – (infrasubspecific): *delictata*: PROUT (1913); *fuscomarginata*: LEMPKE (1949); *lutescens*: LEMPKE (1949); *mediofasciata*: LEMPKE (1949); *nigrociliata*: CHALMERS-HUNT (1961); *subfuscaria*: SCHAWERDA (1912); *suffusa*: LEMPKE (1949). – (nomen nudum): ?*orientis*: PROUT (1938: 225; type from Tel Aviv in MNHU; without abdomen). – (incorrect subsequent spellings): *antiaurica*: COVELL in HODGES (1983); *dimideata*: PROUT (1913: 103).

DIAGNOSIS: Wingspan ♂♀ 15-20 mm. Ground colour whitish with more or less strong sand coloured tinge. Terminal area of forewing with 3-4 distinct, dark brown spots close to tornus. Medial and antemedial lines indistinct or absent. Postmedial line with black markings on veins or dotted, with distinct costal spot. Fringe dots and discal spots sharp, black. Frons brown, vertex whitish, collar ochreous. Antennae of ♂ with intersegmental incisions, ciliate-fasciculate, cilia very long, 2.2-2.4 times (maximum) width of flagellum. Hindtibia of ♂ slender, with large whitish pencil, covering 1/3 of tarsus. Tarsus long, 0.9-1.0 times length of tibia.

Variability: On Sardinia ground colour sometimes with rosy tinge ('f. *roseata*'), recessive inheritance of these forms stated by TURATI (1913). Second generation in rearings more reddish coloured, induced by high temperature (PROUT 1913; BERGMANN 1955). Populations from Sicily with pale brown wings, tornus spots confluent.

Subsp. ***antitaurica*** on average smaller than nominate subspecies, wings narrower, coloration darker, with tornus spots either confluent, or, often, completely absent.

MALE GENITALIA: Uncus and gnathos slender. Saccus broad. Valvae asymmetric, distally hook-shaped, with digitiform processes, the left short, the right long. Aedeagus sigmoid with cornutus distally tapering to a long needle, 0.4-0.5 times length of aedeagus. Subsp. *antitaurica* with digitiform process of left valva shorter (gen.fig. 84b).

FEMALE GENITALIA: Small. Ductus bursae membranous, slender, posteriorly slightly spinulose. At posterior end a narrow, curved, sclerotized process. Corpus bursae globular, densely spinulose all over.

DISTRIBUTION: West-Palaearctic, and North America. Widely distributed in western, southern and central Europe, in the north to southern Scandinavia, in the east to the Urals. Present on almost all Mediterranean islands. – Outside Europe in Morocco (RUNGS 1981), and from north-western Turkey, Caucasus, Transcaucasus, across northern Iran to Afghanistan and central Asian mountains (VIIDALEPP 1988; 1996). Replaced by subsp. *antitaurica* from Cyprus and southern Turkey, to the Levant. Widespread also in Canada and northernmost parts of United States (MCGUFFIN 1967; COVELL 1983; HANDFIELD et al. 1997).

PHENOLOGY: Usually bivoltine late May to late September, with single specimens of a partial third generation in October. Under bad conditions, i.e. at higher elevations north of Alps, in the northern parts of the distribution and in Caucasus, univoltine, from mid-June to mid-August. In hot localities in southern Europe and in the Levant trivoltine from May to October or plurivoltine in all months of the year.

BIOLOGY: Larva polyphagous. Reputedly found on *Prunus spinosa* (Rosaceae) (Germany: SCHNEIDER 1939; EBERT 2001) and under *Salix* (Salicaceae), and *Alnus* (Betulaceae) (BOLDT 1929). Reported as preferring withered, decaying, or mouldy leaves of low plants (var. cit.), e.g. *Plantago* (Plantaginaceae), *Filipendula ulmaria* (Rosaceae), *Anthriscus sylvestris, A. caucalis, Pimpinella saxifraga* (Apiaceae), *Galium mollugo* (Rubiaceae) (var. cit., e.g. ALLAN 1949; BERGMANN 1955; SKINNER 1984). Reared on withered leaves of *Taraxacum officinale, Polygonum aviculare, Rumex* (SKINNER 1984; SEPPÄNEN 1970; PORTER 1997; ERLEBACH pers. comm.), *Lactuca sativa*, in captivity also carnivorous (BERGMANN 1955; FRIEDRICH 1976). Adults hiding in hedges, shrubs and other vegetation, not escaping far when disturbed (PROUT 1913), partially diurnal (BERGMANN

1955), but also readily attracted to light and bait. Nectar feeding recorded on *Molinia* (URBAHN & URBAHN 1939; BERGMANN 1955) and *Cirsium* (WARING & TOWNSEND 2003). Stenotopic, K-strategy (BERGMANN 1955).

HABITAT: In central Europe silvicolous, meso- to hygrophilous. Inhabiting damp woodland, marshes, ditches, sun-exposed scrubby places, and forests with much undergrowing vegetation rich in fallen, decaying leaves, hedges, gardens, riverside vegetation, and around lakes. In the east more xerophilous, often on dry steppe habitats. North of Alps restricted to the lowlands from 0 m to 700 m above sea-level, from 0 m to 1,200 m in the Alps and in southern Europe, up to 2,000 m from Turkey to Afghanistan.

SIMILAR SPECIES: *I. belemiata* (see no. 51). Forms without tornus spots (subsp. *antitaurica*) may resemble *I. palaestinensis*, and *I. elongaria*, both with hindwings more rounded, ♂ antennal cilia much shorter.

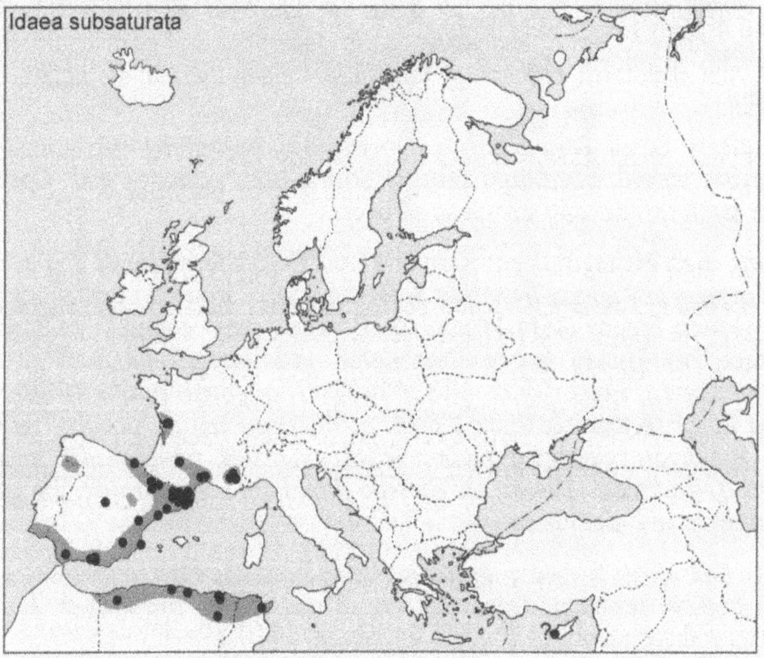

85. *Idaea subsaturata* (GUENÉE, 1858)

Acidalia subsaturata GUENÉE, 1858: in BOISDUVAL & GUENÉE, Hist. nat. Insectes (Spec. gén. Lépid.) 10: 542 (southern France: near Cette). Holotype ♂ (ZFMK, ex coll. GUENÉE, examined).

Acidalia miserata STAUDINGER, 1859: Stettin. ent. Ztg. 20 (7-9): 217 (Spain, Andalusia: Granada). Lectotype ♂ (MNHU, examined), selected and genitalia figured by AGENJO (1967).

Acidalia subherbariata RÖSSLER, 1877: Stettin. ent. Ztg. 38: 366 (Spain, Basque country: Bilbao). Lectotype ♂ (MNCN, examined).

Acidalia subsaturata var. *lecerfiata* HOMBERG, 1910: Bull. Soc. ent. Fr. 1910: 18 (Algeria: by Algiers). Lectotype ♂ (MNHN). Valid at subspecific rank.

†? *Acidalia subsaturata* var. *holli* CULOT, 1917: Noct. & Géom. Eur. 3: 33, pl. 4, fig. 75 (Algeria: Hussein-Dey). Holotype ♂. In the text non binominal and with doubts about validity; on plate as 'var.'.

Unavailable names (infrasubspecific): *brunneofasciata*: ANDREAS (1910); ?*orientis*: PROUT (1935) (see remarks under 'distribution'; in AMSEL (1935) as nomen nudum).

DIAGNOSIS: Wingspan ♂♀ 14-15 mm, in the second generation 12-13 mm, in North Africa up to 19 mm. Wings sand coloured to light brown, in European populations often with dark suffusion. Postmedial line dark brown, slightly curved. In the terminal area, close to the postmedial line about six dark brown spots, also near apex, sometimes confluent. Medial line of hindwing usually distinct. Discal spots small and indistinct. Fringe dots small, sharp, black. Frons brown, vertex and collar ochreous. Antennae of ♂ dentate, ciliate-fasciculate, cilia long and curved, 1.8-2.1 times width of flagellum. Hindtibia of ♂ slender, with large whitish pencil, covering 1/3 of tarsus. Tarsus long, about 1.0 times length of tibia.

Subsp. ***lecerfiata*** with paler ground colour and usually paler brown markings, spots in terminal area confluent or indistinct, medial line of hindwing absent, discal spots distinct on pale background, ♂ antennal cilia slightly longer.

MALE GENITALIA: Uncus sub-triangular, slightly truncate. Saccus broad. Valvae asymmetric, with distal, digitiform processes, the left short, the right long. Aedeagus with slender cornutus, 0.6-0.7 times length of aedeagus. Subsp. *lecerfiata* and populations from southern Spain with digitiform process of left valva longer than in nominate subspecies (gen.fig. 85b).

FEMALE GENITALIA: Small. Ductus bursae membranous, posteriorly with spinulose ribbon. With a narrow, curved, sclerotized process arising from posterior end. Corpus bursae elongate or pyriform, densely spinulose all over.

DISTRIBUTION: West-Mediterranean. Rare and local in southern France and in the Iberian peninsula, present in Andorra (DANTART & ROCHE 1992). Isolated populations in western and south-eastern France (ZSM/HERBULOT; BLAESIUS pers. comm.). – Outside Europe from Morocco to Tunisia (subsp. *lecerfiata*), and western Libya (TURATI 1930a). Reputedly in western Israel (AMSEL 1933; 1935; 'f. *orientis*': PROUT 1935a; STERNECK 1939), and Egypt (ANDRES & SEITZ 1924), no vouchers traced, possibly referring to *I. dimidiata*. Reputedly collected in Cyprus (ZSM, examined; mislabelled?), mentioned for Cyprus also by WILTSHIRE (1949b; as '*Sterrha dimidiata subsaturata*'), the latter record possibly referring to *I. dimidiata*.

PHENOLOGY: Uni- bi- or trivoltine: In the lowlands from late May to mid-October, on the mountains possibly univoltine, most data from mid-July to mid-August. Larva rapidly growing, larval development in 4 weeks (HOMBERG 1910a).

BIOLOGY: Larva probably polyphagous on withered leaves of herbaceous plants. Reared on *Lactuca sativa* (ANDREAS 1910), *Polygonum aviculare*, petals of *Taraxacum officinale* and various other flowers and plants (HOMBERG 1910a; CHRÉTIEN 1926). Adults attracted to light.

HABITAT: Xerothermophilous. Limited information available about habitat preference. From 0 m up to 2,000 m above sea-level.

SIMILAR SPECIES: *I. spissilimbaria* (see no. 45), *I. manicaria* (67), *I. mancipiata* (82). *I. biselata* larger, ground colour and terminal area paler, ♂ hindtibia with large, dark pencil.

REMARKS: Erroneously retained synonym of *dimidiata* by STERNECK (1940). For morphology of immature stages see CHRÉTIEN (1926).

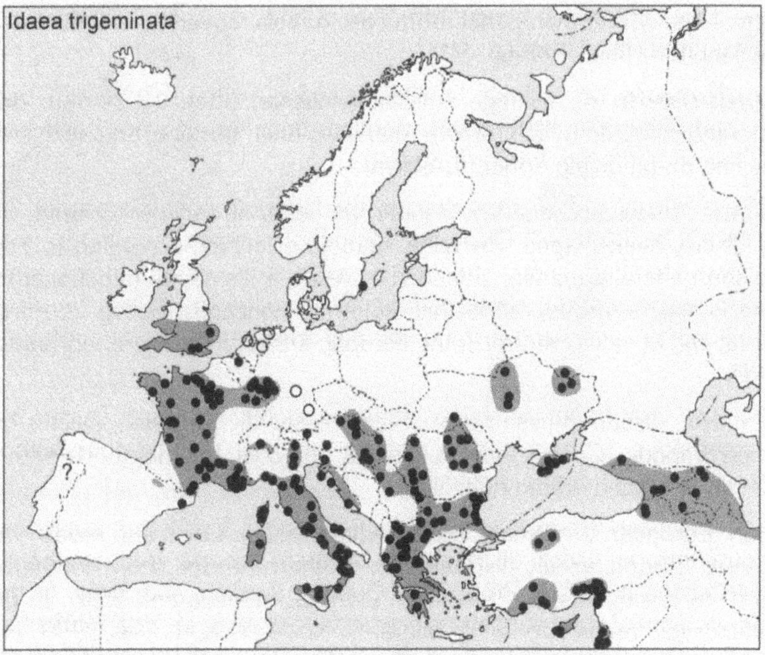

86. Idaea trigeminata (HAWORTH, 1809)

Phalaena trigeminata HAWORTH, 1809: Lepid. Brit. (2): 354 (United Kingdom: Kent). Syntype(s).

‡? *Geometra arcuaria* HÜBNER, 1799: Samml. Eur. Schmett. 5, Geometrae (1): pl. 26, fig. 137 (western Germany, Rhine valley: Jülich). Syntype(s) lost. Nomen dubium. HÜBNER's figure slightly reminiscent of *I. trigeminata*, thus placed here tentatively as possible senior synonym. In SCOBLE (1999) valid at species rank, subordinated under *Scopula*. In any case I propose to suppress this name in order to promote nomenclatural stability.

‡? *Phalaena cordigera* FOURCROY, 1785: Entom. Parisiensis 2: 288 (France, Paris). Nomen dubium, synonymy awaits confirmation.

‡ *Geometra scutularia*: sensu HÜBNER (1799): Samml. Eur. Schmett. 5, Geometrae (1): pl. 14, fig. 73. Misidentification nec HÜBNER, 1799: pl. 14, fig. 72 (identity: *Idaea dimidiata*).

Idaea reversata TREITSCHKE, 1828: Schmett. Eur. 6 (2): 286 (Hungary). Syntype(s) (MNHU 1 'Orig[inal]' from Greece, Parnass, see STAUDINGER (1870): Horae Soc. ent. ross. 7: 145,: pl. 2 fig. 2).

Ptychopoda trigeminata tenuirussata ZERNY, 1933: Dt. ent. Z. Iris 47 (2): 91 (northern Lebanon: Bscharre). Syntypes 3♂ (NHMW). Valid at subspecific rank.

Unavailable names (misidentification): *bisetaria*: sensu DUPONCHEL (1830) nec *biselata* HUFNAGEL, 1767; *bisetata*: sensu STEPHENS (1831) nec ROTTEMBURG, 1777 (identity: *Idaea biselata*); *scutulata* sensu HEINEMANN (1859) nec DENIS & SCHIFFERMÜLLER, 1775 (identity: *Idaea dimidiata*). – (infrasubspecific): *demarginata*: REISSER (1961); *nigrolineata*: REISSER (1961). – (incorrect subsequent spellings): *trigeminaa*: STEKOLNIKOV (1996); *trigeminaria*: MORRIS (1861).

DIAGNOSIS: Wingspan ♂♀ 16-22 mm, in subsequent generations sometimes 14 mm only. Wings broad, rounded. Ground colour yellowish white. Medial and antemedial lines indistinct or absent. Medial line proximal to discal spot, sometimes touching. Forewing costa dark brown in the basal third. Postmedial line of forewing dark brown, slightly dentate. Terminal area with distinct dark brown confluent spots close to postmedial line. Hindwing with much paler markings. Discal spots present. Frons dark brown, vertex whitish, collar darker. Antennae of ♂ with intersegmental incisions, ciliate-setose, cilia 1.1-1.5 times width of flagellum (Text-fig. 61). Hindtibia of ♂ slender, with one whitish pencil close to tibia and one large dark brown pencil from base of tibia, covering 1/2 tarsus. Tarsus 0.5-0.65 times length of tibia (Text-fig. 31).

Subsp. ***tenuirussata*** on average smaller, wingspan often 12-14 mm only, ground colour dark sand coloured to light brown, markings finer, terminal area with smaller spots, postmedial line on hindwing dotted or absent.

MALE GENITALIA: Uncus and gnathos slender, the latter strongly sclerotized. Saccus shallow. Valva distally beak-shaped. Cornutus sigmoid or curved according to position, with sclerotized short lateral appendix, attached to vesica at its centre (functional morphology see *infirmaria* species-group), about half length of aedeagus. Subsp. *tenuirussata* with broader distal end of valva, ventral spine abruptly arising from there, cornutus straighter, juxta rounder.

FEMALE GENITALIA: Ostium bursae round. Sterigma slightly furrowed. Ductus bursae long, slender, membranous. Corpus bursae globular or oval, membranous without spines. Genitalia easily damaged during dissection.

DISTRIBUTION: European-Westasiatic (Sub-Mediterranean). Local and usually rare in western and central Europe, widely distributed in southern Europe, from southern France to westernmost European Russia. Present on Corsica, Sardinia and Sicily. In the north to the Netherlands (old records, possibly migrants: KUCHLEIN & DE VOS 1999; DE VOS pers. comm.), western Germany and southern Czech Republic, with very isolated resident population on Öland island (Sweden: RYRHOLM pers. comm.). Very local in north-eastern Spain (DANTART & REQUENA 1992; VIVES MORENO 1994; MÜLLER 1996; REDONDO & GASTON 1999). Extinct in Bavaria (WOLF & HACKER 2004). Reputedly, occasional anthropogenous transport to central Europe is caused by the trade of plants (BERGMANN 1955). Collections and literature, however, are full of wrong data, due to confusion with *I. biselata*. – Outside Europe in Turkey, Cyprus, Levant, Transcaucasus northern Iran and Turkmeniya (e.g. WILTSHIRE 1945; VIIDALEPP 1996; DIDMANIDZE 1978; 2002). Replaced by subsp. *tenuirussata* in western Lebanon, Amanus mountains, Cyprus and Israel (see PROUT 1935a).

PHENOLOGY: In temperate climates univoltine: early June to late July. Under good conditions with partial second brood late July to mid-September. In the lowlands of southern Europe regularly bivoltine, mid-May to early July; late July to mid-September; exceptionally emerging from late April. In the Levant trivoltine: mid-March to early May; early June to early July; September.

BIOLOGY: Larva probably polyphagous: Recorded on *Betula* (Betulaceae), *Acer campestre* (Aceraceae), *Taraxacum officinale* (Asteraceae), *Polygonum aviculare* (Polygonaceae), and *Hedera helix* (Araliaceae) (LHOMME 1935; ALLAN 1949; SKOU 1986; WARING & TOWNSEND 2003). Reared on *Hedera helix, Taraxacum officinale, Lactuca sativa, Polygonum aviculare, Plantago* and other herbaceous plants (BUCKLER 1897; RUHLAND 1912; ALLAN 1949; BERGMANN 1955, SKINNER 1984; PORTER 1997). Rearing difficult on withered petals and leaves (LEIPNITZ pers. comm.). Main activity of adults at dusk, readily attracted also to light. During daytime often resting on upperside of leaves or on fences.

HABITAT: Thermophilous, in western and central Europe, usually restricted to the hottest sites, e.g. near rocks, heathland, oak and hazel forest (SKOU 1986), woodland rides and edges, gardens, hedges on chalk downland (WARING & TOWNSEND 2003). In southern Europe with wider spectrum of habitats, e.g. in the Mediterranean maquis, scattered oak

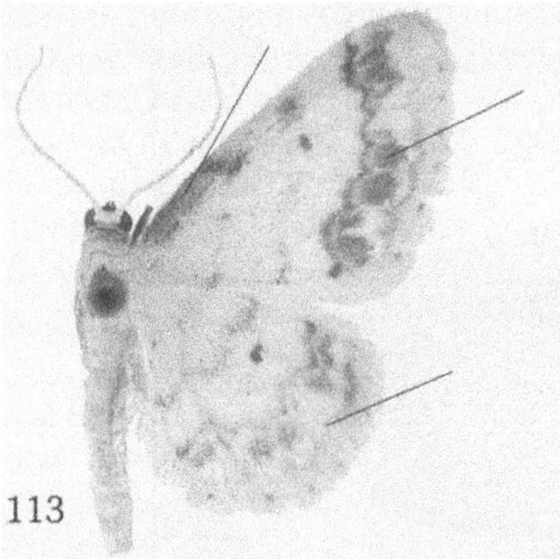

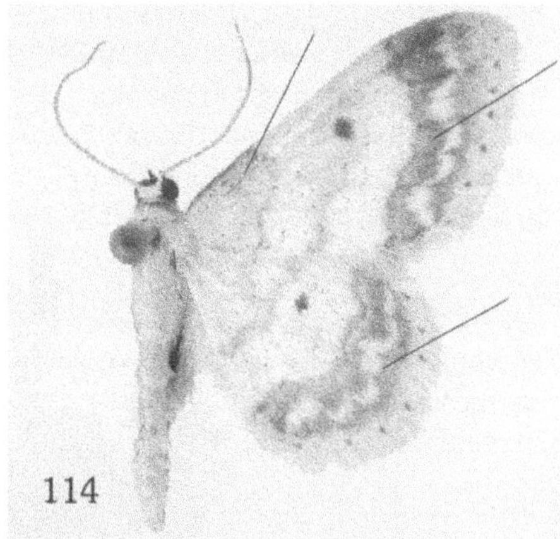

Text-figs 113-114: Diagnostic characters (indicated) of European *Idaea* species. Text-fig. Text-fig. 113: *I. trigeminata* (HAWORTH, 1809). Text-fig. 114: *I. biselata* (HUFNAGEL, 1767).

woodland, forest fringes. Usually restricted to the lowlands from 0 m up to 700 m above sea-level, in southernmost Europe up to 1,400 m, in Turkey up to 1,900 m.

SIMILAR SPECIES: *I. biselata* with terminal area usually paler, markings more confluent, hindwing pattern similar to that of forewing, forewing costa pale at base, postmedial line usually light brown, medial line distal to discal spot, ♂ antennal cilia shorter, ♂ hindtibia with dark pencil.

REMARKS: According to STERNECK (1940: 162) in the *infirmaria* species-group (28th group). Relationship to *I. biselata* supported by hindtibial pencil. Relationship to *I. subsaturata* and *I. dimidiata* supported by genitalia (possibly homologous terminal hook of valva and shape of cornutus, ♀ with long, membranous ductus bursae). For rearings and morphology of immature stages see BUCKLER (1897).

87. Idaea biselata (HUFNAGEL, 1767)

Phalaena biselata HUFNAGEL, 1767: Berlin. Mag. 4 (6): 618 (Germany: Berlin). Syntype(s) lost.

Geometra fimbriata DENIS & SCHIFFERMÜLLER, 1775: Ank. syst. Werkes Schmett. Wienergegend: 117 (Germany: Saxonia). Syntype(s) lost.

‡ *Phalaena bisetata* ROTTEMBURG, 1777: Naturf. 11: 89. Incorrect subsequent spelling.

Ptychopoda fimbriolata STEPHENS, 1831: Illust. Br. Ent. (Haust.) 3: 306 (Great Britain: London). Syntype(s).

‡ *Acidalia bisetaria* BOISDUVAL, 1840: Genera Index meth. eur. Lepid.: 222. Emendation of *bisetata*. 10 years earlier erroneously used for *I. trigeminata* by DUPONCHEL (see there).

Acidalia plumipedata WALKER, 1861: List. Spec. lepid. Insects Colln Br. Mus. 23: 797 ('country unknown'). Holotype ♂ (BMNH).

Acidalia bisetata var. *extincta* STAUDINGER, 1897: Dt. ent. Z. Iris 10 (1): 15 (eastern Russia, Amur region: Onon, Vladivostok a.s.o.). Syntypes, at least 12♂♀ (MNHU, examined). Valid at subspecific rank.

Ptychopoda shimizuensis MATSUMURA, 1925: J. Coll. Agric. Hokkaido Imp. Univ. 15 (3): 159, pl. 11, fig. 5 (Russia: Sakhalin: Shimizu, Kawakami). Syntypes 2♂1♀ (EIHU, examined by SATO 2001).

Unavailable names (misidentifications): *cinereata*: sensu STEPHENS (1831; misspelled) nec *cinerata* FABRICIUS, 1781 (identity: *Scopula decorata*); *decoraria*: sensu STEPHENS (1829) nec *decorata* DENIS & SCHIFFERMÜLLER, 1775 (identity: *Scopula decorata*); *dilutata*: sensu HAWORTH (1809) nec *dilutaria* HÜBNER, 1799; *dilutaria*: sensu STEPHENS (1829) nec HÜBNER, 1799; *reversaria*: sensu DUPONCHEL (1830) nec *reversata* TREITSCHKE,

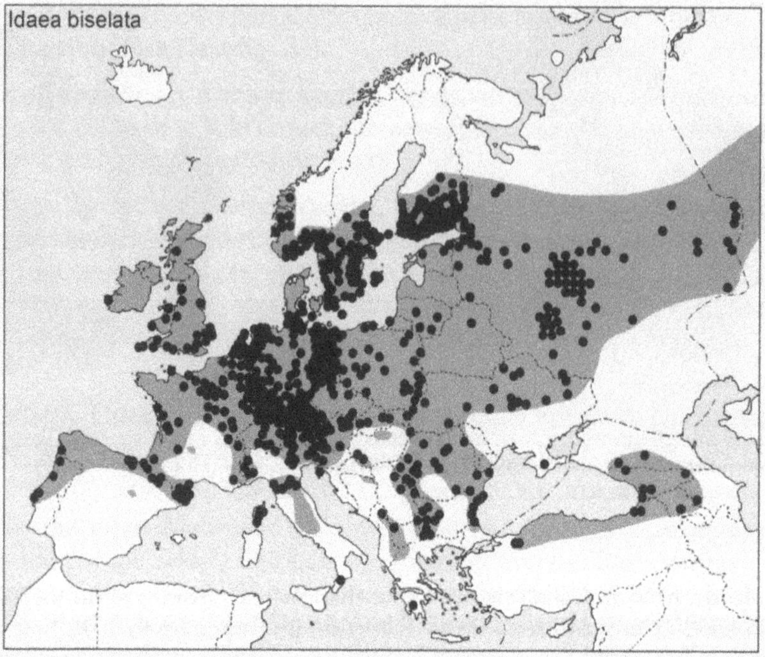

1828 (identity: *Idaea trigeminata*); *scutularia*: sensu DUPONCHEL (1830) nec *scutulata* DENIS & SCHIFFERMÜLLER, 1775 (identity: *Idaea dimidiata*). – (infrasubspecific): *griseata*: PREISSECKER (1922); *infuscata*: PROUT (1913); *schaefferaria*: FUCHS (1904). – (incorrect subsequent spellings): *schefferaria*: HEINRICH (1916); *schiminzuensis*: DIDMANIDZE (2002).

DIAGNOSIS: Wingspan ♂♀ 16-21 mm, second generation 15-17 mm. ♂ with very long abdomen. Wings pale sand coloured. Medial and antemedial lines usually indistinct. Postmedial line brown, convex. Discal spots present. Terminal area light brown to dark brown, with wavy line of ground colour. Frons brown, vertex whitish, collar sand coloured. Palpi brown. Antennae of ♂ with intersegmental incisions, densely ciliate-setose, cilia 0.8-1.0 times width of flagellum (Text-fig. 62). Hindtibia of ♂ slender, with large whitish pencil, mainly from distal part of tibia and large dark brown pencil from base of tibia, completely covering tarsus (Text-fig. 32), yet described by HUFNAGEL (1767: 626). Tarsus 0.4-0.5 times length of tibia.

Variability: Darkening of terminal area induced by humidity (BERGMANN 1955), such forms ('f. *fimbriolata*') rare in continental Europe, commoner in Great Britain. Specimens from northern Iberian peninsula with terminal area only slightly darker than ground colour, markings grey brown rather than brown.

Subsp. ***extincta*** smaller, wingspan 15-19 mm, all markings less distinct.

MALE GENITALIA: Uncus and gnathos digitiform. Saccus with strongly sclerotized, boat-shaped sclerite, laterally tapering. Valva slightly dilated at tip, with 2-4 ventral, subapical spinules. Aedeagus broad, with stout, sigmoid cornutus and vesica sclerotized at tip.

FEMALE GENITALIA: Antrum slightly tapering laterally. Ductus bursae broad, membranous, anteriorly with globular, membranous appendix. Corpus bursae usually elongate, membranous.

DISTRIBUTION: Eurasiatic. Widely distributed from western Europe to the Urals, except the extreme north, in the south to Portugal, Corsica (BRUSSEAUX 1994; ZSM, dissected), northern Apennines (Tuscany) and northern Greece. Isolated occurrences in southern Calabria (SCALERCIO & INFUSINO 2003; examined) and Peloponnes. Deleted from the fauna of southern Portugal (CORLEY et al. 2000). – Outside Europe in the Caucasus, Transcaucasus, northern Turkey, Siberia, northern Kazakhstan, in the east to Far East of Asia (Korea, Ussuri region: subsp. *extincta*).

PHENOLOGY: North of Alps univoltine: mid-June to mid-August, exceptionally up to ten days later or earlier. Under good conditions a partial second brood from early August to mid-September, south of Alps from mid-September to early October. Growth of larvae slow, asynchronous (BERGMANN 1955; EBERT 2001). In southern part of West-Siberian plain reputedly overwintering as pupa (VASILENKO 1994).

BIOLOGY: Larva polyphagous on herbaceous plants (var. cit.), mainly withered or dry leaves (FORSTER & WOHLFAHRT 1981), also Poaceae and fallen leaves of *Tilia* (Tiliaceae), *Corylus* (Corylaceae), and *Alnus* (Betulaceae) (BERGMANN 1955), reputedly *Leontodon* (Asteraceae) (Finland: VALLE, 1946). Reared on *Taraxacum officinale, Polygonum aviculare, Rubus fruticosus, Plantago, Salvia pratensis, Trifolium, Rumex, Campanula, Lonicera, Tilia*, and Poaceae, preferring withered leaves (var. cit., e.g. SCHNEIDER 1939; ALLAN 1949; BERGMANN 1955; SEPPÄNEN 1970; FRIEDRICH 1976; SKINNER 1984; EBERT 2001). Larva inert (PROUT 1913). Adults active from early dusk on, readily attracted to light and bait. Movement restricted to short distances (HAUSMANN 1990a). Nectar feeding recorded on *Molinia, Cirsium, Eupatorium cannabinum, Eleocharis palustris* and *Sambucus ebulus* (URBAHN & URBAHN 1939; BERGMANN 1955; EBERT 2001; WARING & TOWNSEND 2003). Olefinic acetate identified as sex pheromone component (ZHU et al. 1996).

HABITAT: Meso- to hygrophilous, silvicolous. Inhabiting a wide range of different habitats. In central Europe sometimes extremely common in deciduous forests with rich understorey, dense scrubby places, hedgerows on wet soils, and forest fringes. South of Alps montane, restricted to the *Fagus* belt. Rarely leaving the habitat into open land (HAUSMANN 1990a; 1990b). Usually from 0 m up to 800 m above sea-level, sometimes up to 1,200 m, regularly up to 1,700 m in the Alps. South of Alps from 600 up to 1,500 m, only exceptionally lower (FLAMIGNI & BASTIA 1998).

SIMILAR SPECIES: *I. politaria* (pale forms, see no. 64), *I. mancipiata* (82), *I. subsaturata* (85), *I. trigeminata* (86).

REMARKS: *I. biselata* was tentatively placed by STERNECK (1940: 139) in the *obsoletaria* species-group (13th group) and he suggested that the presence of many unique characters indicate a phylogenetically young species. Relationships to the *subsericeata* species-group (mainly *I. trigeminata*) supported by habitus, sigmoid cornutus which is laterally linked with a sclerite and membranous corpus bursae. These features suggested to be homologous to equivalent structures in *I. trigeminata*. Laterally tapering antrum and ductus bursae with globular membranous appendix reminiscent of equivalents in *I. subsericeata* and *I. pallidata*.

The *fractilineata* species-group

STERNECK's (1940) 23rd group of *Idaea*, combined with *I. exilaria* and its two North African sister species *I. truschi* HAUSMANN, 2003, and *I. erubescens* HAUSMANN, 2003, transferred here from STERNECK's 31st group. The last three species perhaps with certain phylogenetic relations to *Brachyglossina* and/or the *nexata* species-group. STERNECK (1940: 157) includes in the *fractilineata* species-group also the North African *I. merklaria* (OBERTHÜR, 1885), the latter making, however, this group 'structurally heterogeneous'. Palpi comparatively short. Eyes small. Proboscis weak or absent (*I. exilaria*). Sternum A2 of ♂ with very narrow, semicircular, sclerotized ridge, but setose patch absent, reminiscent of equivalents in the *longaria* and the *nexata* species-groups. Eggs at poles with polygonal, laterally with elongate oval sculpturation (*I. fractilineata*). Larva of *I. fractilineata* with skin strongly furrowed, granulated, that of *I. exilaria* extremely short and thick.

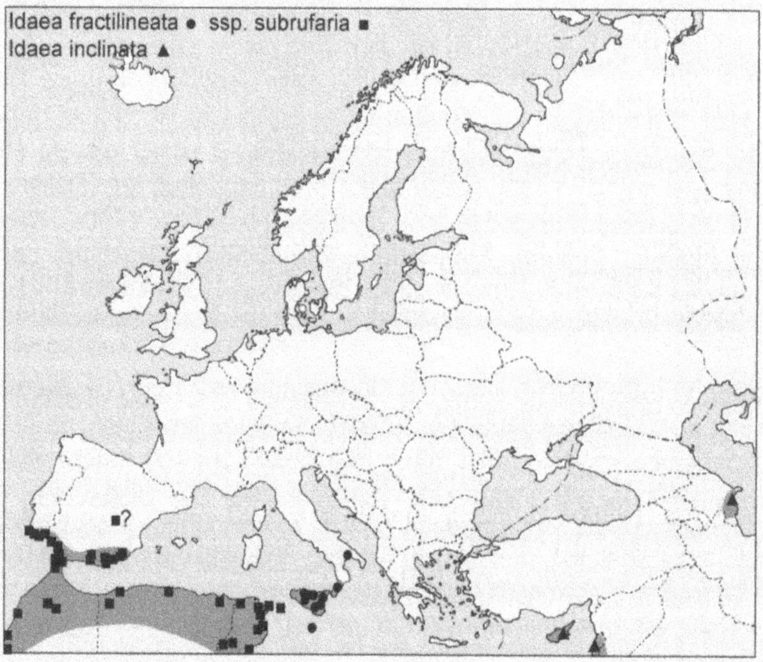

88. *Idaea fractilineata* (ZELLER, 1847)

Acidalia fractilineata ZELLER, 1847: Isis 1847 (7): 518 (Sicily, Mt. Etna: Catania, Vieranni). Syntypes 2♂ (BMNH: 1♂, with red 'type' label; MNHU: 1♂, examined).

Acidalia fractilinearia HERRICH-SCHÄFFER, 1852: Syst. Bearb. Schmett. Eur. 6 (55): 66. First binominal name after *fractilinearia* HERRICH-SCHÄFFER, 1848 (pl. 66, figs 405, 406: non binominal). Emendation. Referring to vol. 3: 16, thus binominal.

Acidalia subrufaria STAUDINGER, 1900: Dt. ent. Z. Iris 12 (2): 386 (Algeria: Tenet Haad). Syntypes 4♂ (MNHU 3♂ syntypes examined). Herewith downgraded from species (SCOBLE 1999) to subspecific rank (stat. n.).

Acidalia subrufaria var. *fusaria* CHRÉTIEN, 1917: Annls Soc. ent. Fr. 85 (3): 393 (Tunisia: Gafsa). Lectotype ♂ (ZSM/HERBULOT, examined), selected by HERBULOT (1982).

Ptychopoda parallelolineatella D. LUCAS, 1948: Bull. Soc. ent. Fr. 53 (5/6): 88 (Tunisia: Metlaoui). Syntype(s) (ZSM/HERBULOT: 1♀, examined). In HERBULOT (1984a) as 'holotype'.

Unavailable names (incorrect subsequent spelling): *fuscaria*: PROUT (1934).

DIAGNOSIS: Wingspan ♂♀ 14-18 mm. Hindwing strongly concave between M1-M3. ♀ with forewing apex pointed. Fairly variable in colour and pattern of wings. Wings sand coloured, sometimes with strong red brown or orange suffusion. Red brown transverse lines close to each other, delimiting the narrow medial area. Broad terminal area red brown, with wavy line concolorous with ground colour. Terminal line and fringe dots absent. Discal spots very small. Forelegs, palpi, frons, vertex and collar red brown. Antennae of ♂ with deep intersegmental incisions, ciliate-fasciculate, cilia long, 1.7-1.9 times width of flagellum. Hindtibia of ♂ slender, with weak pencil. Tarsus 0.4-0.5 times length of tibia.

Subsp. ***subrufaria*** smaller, wingspan ♂♀ 12-16 mm, ♀ exceptionally 17 mm, in subsequent generations sometimes 8 mm only. Wing coloration and pattern still more variable than in the nominate subspecies, terminal area not rarely concolorous with ground colour. Transverse lines sometimes very indistinct. Discal spots often absent. ♂ antennal cilia 1.4-1.8 times width of flagellum. Hindtarsus 0.3-0.45 times length of tibia.

MALE GENITALIA: Small. Uncus short and broad, sub-truncate. Saccus V-shaped. Valva with subapical ventral spine. Aedeagus with one stout cornutus, swollen at base, 0.4-0.5 times length of aedeagus, and with a minute second cornutus. Subsp. *subrufaria* with valva usually narrower, cornutus narrower, rarely swollen at base, longer, 0.5-0.6 times length of aedeagus. *I. inclinata* with aedeagus and cornutus much shorter and valva with terminal spine.

FEMALE GENITALIA: Small. Apophyses anteriores very short and broad. Ductus bursae short, weakly sclerotized, asymetrically sub-triangular. Corpus bursae oval, membranous, laterally extended beside ductus bursae. Female genitalia easily damaged during dissection. Subsp. *subrufaria* without constant differences from nominate subspecies. *I. inclinata* with ductus bursae much longer.

DISTRIBUTION: West-Mediterranean. Nominate subspecies on Sicily and, in very isolated populations in southern Italy (PARENZAN & HAUSMANN 1992). On Malta, Lampedusa (ROMANO & ROMANO 1995), Pantelleria (AISTLEITNER & AISTLEITNER 2001) and Iberian peninsula replaced by the subsp. *subrufaria*. Reputedly collected in Cuenca, central Spain (ZSM, examined), possibly mislabelled. – Outside Europe from Morocco to Tunisia (subsp. *subrufaria*). In the eastern Mediterranean, from southern Turkey and Cyprus to Egypt (dissected) and eastern Libya (dissected; TURATI & ZANON 1922; TURATI 1924; PROUT 1935a) replaced by the sister species *I. inclinata* (LEDERER, 1855).

PHENOLOGY: Nominate subspecies on mountains univoltine: late June to mid-July. Subsp. *subrufaria* plurivoltine: mid-April to mid-October. Sister species *I. inclinata* from the Levant in all months of the year.

BIOLOGY: Larva preferring fresh leaves of herbaceous plants, sometimes with cannibalism (CHRÉTIEN 1917; PROUT 1935a). Reared on withered leaves of *Taraxacum officinale*, and on petals of Asteraceae and *Campanula* (TRUSCH & MÜLLER 2001). Adults readily attracted to light.

HABITAT: Xerothermophilous, often near coast, but also with xeromontane populations. Collected in forest of *Pinus* (Sicily: SCHWINGENSCHUSS 1942). From 0 m up to 300 m above sea-level in Calabria, from 1,000 m up to 1,800 m in Sicily. Subsp. *subrufaria* usually from 0 m up to 500 m, up to 1,500 m in Morocco.

SIMILAR SPECIES: Usually well characterized by the shape of hindwing and the warm brown coloration. Extreme forms need dissection for correct identification. Some forms may be confused with *I. obsoletaria*, others with the North African *I. lobaria* (CHRÉTIEN, 1909), both the last species however differ in the whitish vertex, *I. obsoletaria* in much shorter ♂ antennal cilia. Allopatric sister species *I. inclinata* with terminal area often dark brown, hindwing paler at base, discal spots often absent, antennal cilia slightly shorter, hindtibia shorter (< 1 mm), hindtarsus about 0.3 times length of tibia only, genitalia different (see above).

REMARKS: Species rank postulated for *subrufaria* by AGENJO (1952), RUNGS (1981), VIVES MORENO (1994) and SCOBLE (1999).

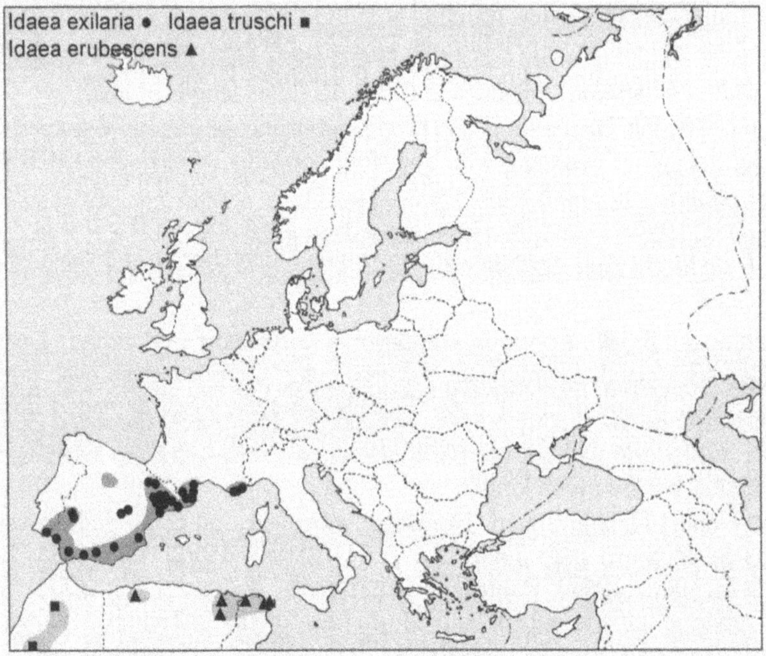

89. *Idaea exilaria* (GUENÉE, 1858)

Acidalia exilaria GUENÉE, 1858: in BOISDUVAL & GUENÉE, Hist. nat. Insectes (Spec. gén. Lépid.) 9: 449 (southern France). Syntypes 1♂2♀ (ZFMK, examined).

Acidalia transmutata STAUDINGER, 1871: in STAUDINGER & WOCKE, Cat. Lepid. eur. Faunengeb. (Edn 2) 1: 146. First available name after *transmutata* RAMBUR, 1866 (non binominal) (Spain, Andalusia). Lectotype ♂ (of RAMBUR: ZSM/HERBULOT, examined).

Acidalia esterelata MILLIÈRE, 1879: Naturaliste 1 (18): 138 (France, Alpes Maritimes: near Troyas). Syntypes ♂♀, probably lost.

Unavailable names (misidentification): *filacearia*: sensu PROUT (1913, pl. 4a) nec HERRICH-SCHÄFFER, 1847 (identity: *Cleta filacearia*); *purpureomarginata*: sensu VIVES MORENO (1994) nec BOHATSCH, 1879 (identity: *Idaea purpureomarginata*, Middle East). – (infrasubspecific): *gynochromaria*: HOMBERG (1911). – (incorrect subsequent spellings): *exiliaria*: ABOS CASTEL (1990); *transmutaria*: WALKER (1890).

DIAGNOSIS: Wingspan ♂ 13-17, ♀ 16-20 mm. Sexually dichroic. ♀ pale brown with red tinge, without any wing pattern or with indistinct transverse lines. Ground colour of ♂ orange to red brown. Transverse lines fine and sharp, deep red brown, medial line and antemedial line sometimes absent. Distal to postmedial line usually a broad, deep red brown fascia. Discal spots and fringe dots absent. Coloration of ♂ exceptionally similar to that of ♀ ('f. *gynochromaria*'). Palpi, frons, vertex and collar unicolorous pale brown. Proboscis absent, exceptionally visible as a very short rudiment. Antennae of ♂ with deep intersegmental incisions, ciliate-fasciculate, cilia long, 1.2-1.45 times width of flagellum (Text-fig. 63). Hindtibia of ♂ slender, without pencil, but distally with elongate scales. Tarsus very short, 0.15-0.25 times length of tibia (Text-fig. 33), exceptionally longer, in one specimen from northern Spain 0.65 times.

MALE GENITALIA: Uncus and saccus digitiform. Gnathos broad. Valva broad, with tapered, subapical process on ventral margin. Aedeagus short, with one very stout, slightly curved cornutus at tip, and two smaller cornuti beside.

FEMALE GENITALIA: Ductus bursae broad, well sclerotized, with lateral, sclerotized appendix at junction with corpus bursae. Corpus bursae membranous, posteriorly spinulose.

DISTRIBUTION: West-Mediterranean. From south-eastern France to Iberian peninsula. – Outside Europe replaced by the sister species *I. truschi* HAUSMANN, 2003 in Morocco and by *I. erubescens* HAUSMANN, 2003 in Algeria and Tunisia. Erroneously recorded for Israel (AMSEL 1933).

PHENOLOGY: Bivoltine: early May to mid-September. North African sister species bivoltine, emerging from late March (*I. truschi*) or trivoltine, with second generation in July, third generation mid-October to late October (*I. erubescens,* limited data available).

BIOLOGY: Larva reputedly polyphagous on flowers (PROUT 1913). Reared on *Cistus, Anemone, Rosmarinus* and *Astrantia major* (LHOMME 1935). Adults diurnal, main activity during daytime and in the evening (RIBBE 1912). Also comes to light.

HABITAT: Xerothermophilous lowland species. Limited information available about habitat preference. From 0 m up to 1,000 m above sea-level. Sister species *I. truschi* in Morocco up to 2,300 m.

SIMILAR SPECIES: *I. manicaria* (see no. 67). Allopatric sister species *I. erubescens* HAUSMANN, 2003 and *I. truschi* HAUSMANN, 2003 with hindtarsus 0.25-0.4 times length of tibia, ground colour paler and considerable genitalic differences (*I. truschi* with hook-shaped spine at tip and needle-shaped cornutus in the slender aedeagus, saccus broad, *I. erubescens* with subapical costal spinule on valva and one broad cornutus, saccus curved, ductus bursae narrower and longer, corpus bursae with a very few spinules). Identification of ♂ is possible by brushing tip of abdomen to remove scales and reveal shape of valva (slender in both *I. erubescens* and *I. truschi*).

REMARKS: STERNECK (1940) figures under '*I. fatimaria*' the genitalia of *I. erubescens*. '*I. fatimaria*' has been reported from Europe, based on the southern Spanish syntype (now paralectotype) and Maltese records (VALLETTA 1993; MÜLLER 1996). The taxon is synonym of *I. manicaria* (see nr. 67).

The *cervantaria* species-group

STERNECK's (1940) 25th and 26th group of *Idaea*, including also *I. decidua* (WARREN, 1900) from central Asia. Relationships between the first two species ascertained (STERNECK 1940: 158). *I. rupicolaria* and *I. contiguaria* with similar features in ♂♀ genitalia, e.g. shape of valva, saccus and corpus bursae, supporting fusion of both STERNECK's groups. Palpi comparatively long, 1.0-1.3 times diameter of eye. Proboscis present. Sternum A2 of ♂ with horseshoe-shaped sclerite and patch of long setae. Male genitalia: Valva with costal process, aedeagus narrow, cornutus absent. ♂ and ♀ genitalia with many unusual characters for *Idaea*, thus phylogenetically isolated position is suggested. Egg sculpturation hexagonal reticulate, in *I. contiguaria*, possibly, with distinct longitudinal, and weaker transverse ridges (CHRÉTIEN 1928). Larva comparatively long and slender.

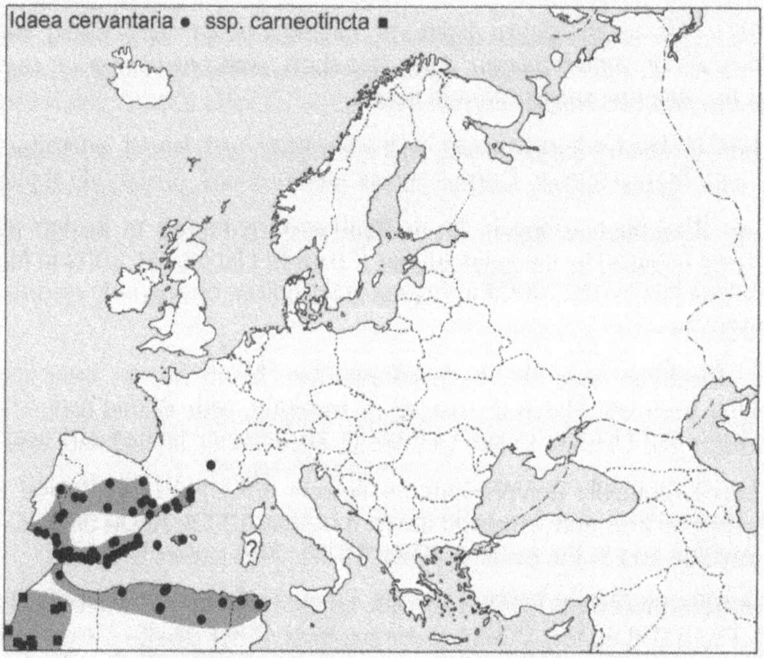

90. Idaea cervantaria (MILLIÈRE, 1869)

Acidalia cervantaria MILLIÈRE, 1869: Iconogr. Descr. Chen. Lép. inéd. 3 (23): 21, pl. 103, figs 1-3 (Spain, Catalonia: Barcelona region). Syntypes 2♂, probably lost.

Acidalia cervantaria var. *depressaria* STAUDINGER, 1892: Dt. ent. Z. Iris 5 (2): 289 (southern Spain: Granada). Lectotype ♂ (MNHU, examined), herewith designated to fix type locality and to stabilize usage of the name; at MNHU 4♂4♀ paralectotypes from southern Spain; Algeria: prov. Constantine, near Lambèze; Tunisia: Tunis. Herewith downgraded from subspecific rank (SCOBLE 1999) to synonymy (syn. n.).

Ptychopoda cervantaria var. *montana* WEHRLI, 1930: Int. Ent. Z. 23 (37/38): 431. First available name after *montana* WEHRLI, 1927 (infrasubspecific) (Spain, Andalusia: Sierra Nevada). Lectotype ♂ (ZFMK, examined), herewith designated to stabilize usage of the name; at ZFMK 47 paralectotypes. Herewith downgraded from subspecific rank (SCOBLE 1999) to synonymy (syn. n.).

Sterrha cervantaria carneotincta ZERNY, 1936: Mém. Soc. Sci. nat. Phys. Maroc 42: 77 (Morocco: Tachdirt). Syntypes (NHMW, ZFMK 3♂ syntypes examined). Valid at subspecific rank.

Unavailable names (misidentification): *mauretanica*: sensu WEHRLI (1931; misspelled) nec *mauritanica* BETHUNE-BAKER, 1885 (identity: *Brachyglossina mauritanica*). – (infrasubspecific): *fusculata*: REISSER (1927).

DIAGNOSIS: Wingspan ♂♀ 15-19 mm, in rearings up to 22 mm, in subsequent generations sometimes 13 mm only. Wings sand coloured, with slight dark suffusion. Postmedial line dentate, sometimes dotted. Antemedial line with large dark costal spot. Terminal line distinct, dissolved into streaks. Fringe dots absent. Discal spots present. Frons dark brown, vertex and collar sand coloured. Antennae of ♂ dentate with deep intersegmental incisions, ciliate-fasciculate, cilia very long, 1.7-2.2 times width of flagellum (Text-fig. 64).

Hindtibia of ♂ slender, with long white pencil. Tarsus long, 0.55-0.65 times length of tibia, on Mallorca slightly shorter, in North Africa sometimes up to 0.8 times.

Variability: Populations of eastern Pyrenees and mountains of the southern Iberian peninsula ('f. *montana*'), with stronger dark suffusion, but with clinous transitions to nominate forms. Similarly the pale lowland forms with diffuse pattern ('f. *depressaria*') occur mixed up with nominotypical forms.

Subsp. ***carneotincta*** usually with rosy tinge.

MALE GENITALIA: Uncus digitiform or bottle-shaped. Saccus large, digitiform. Juxta large, elongate posteriorly. Valva broad, at tip with digitiform costal process and larger ventral process. Aedeagus with one slightly sclerotized needle-shaped sclerite at tip. Valva slightly broader in populations from southern Spain and North Africa.

FEMALE GENITALIA: Very small. Ductus bursae sclerotized, in the anterior quarter with a strongly sclerotized transverse ridge. Corpus bursae very small, membranous. Lateral membrane also beside ductus bursae.

DISTRIBUTION: West-Mediterranean. Southernmost France and Iberian peninsula, with isolated occurrence in French dept. Ardèche (ZSM/HERBULOT). Present on Mallorca and Ibiza. In the Spanish lowlands near coasts very common. Reputedly collected in 'Sardinia' (1♀, ZSM/HERBULOT, dissected), probably mislabelled. – Outside Europe from Morocco to Tunisia. In south-western Morocco and Great Atlas replaced by the subsp. *carneotincta*. Erroneously mentioned for Turkey (PFEIFFER 1927; RIEMIS 1994).

PHENOLOGY: In the Iberian peninsula usually trivoltine: early April to early November. Under warm conditions, e.g. in North Africa plurivoltine in all months of the year.

BIOLOGY: Larva polyphagous on withered leaves of herbaceous plants (LHOMME 1935). Reared on petals of *Lobularia maritima* (MILLIÈRE 1871; subsequently var. cit.), and on petals of Asteraceae (TRUSCH & MÜLLER 2001). Adults readily attracted to light.

HABITAT: Xerothermophilous species of open habitats. Limited information available about habitat preference. Adults often resting on house walls. Usually from 0 m up to about 1,000 m above sea-level, in southern Spain up to 2,000 m ('*montana*'), in Morocco up to 3,000 m.

SIMILAR SPECIES (see Text-fig. 91): *I. alyssumata* (see no. 42), *I. urcitana* (43), *I. joannisiata* (72). Mountain forms ('f. *montana*') may be confused with *I. contiguaria*, the latter with less dark suffusion, medial line of forewing proximally from discal spot, ♂ antennal cilia shorter. *I. rupicolaria* with shorter ♂ antennal cilia, forewings narrower, apex pointed.

REMARKS: Taxonomy discussed in WEHRLI (1931a). For morphology of immature stages see CHRÉTIEN (1928).

91. *Idaea contiguaria* (HÜBNER, 1799)

Geometra contiguaria HÜBNER, 1799: Samml. Eur. Schmett. 5, Geometrae (1): pl. 20, fig. 105 (Europe, probably northern Germany (see TREITSCHKE, 1835: 225)). Syntype(s) lost. Junior primary homonym nec BORKHAUSEN, 1794 (identity unknown, BORKHAUSEN knew only the larva), the latter to be rejected according to MÜLLER (1996), LERAUT (1997), SCOBLE (1999), a practice accepted here in favour of nomenclatural stability.

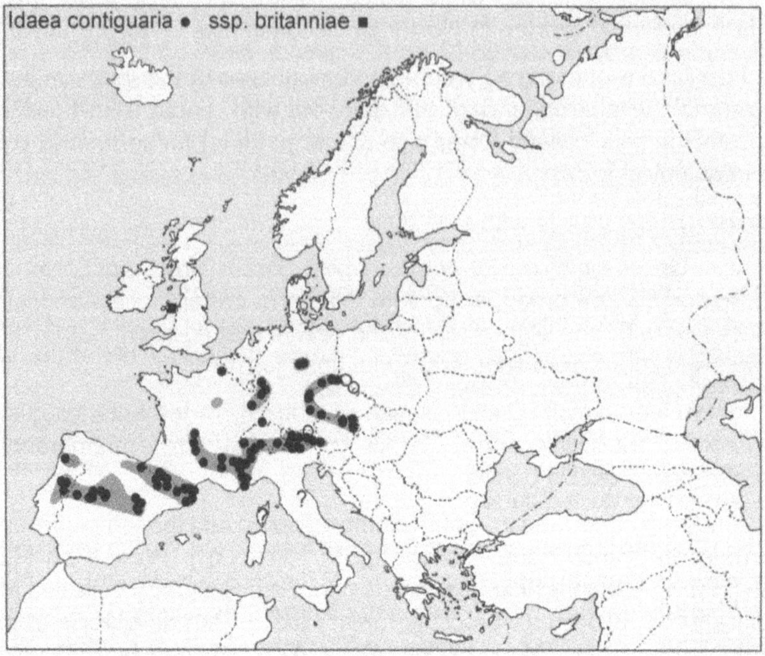

Idaea eburnata WOCKE, 1850: Zt. Ent. 4 (4, Lepidoptera): 48, pl. 4, fig. 13 (Poland, Silesia: Breslau). Syntype(s) ♂ (MNHU: 1♂, examined, labelled as 'type').

Idaea contiguata TREITSCHKE, 1835: Schmett. Eur. 10 (2): 224. Emendation.

Acidalia contiguaria var. *obscura* FUCHS, 1875: Stettin. ent. Ztg. 36 (4-6): 227 (western Germany: Rheingau). Syntypes 13♂♀ (NHMW: 1 syntype examined; ?NSMW). Herewith downgraded from subspecific rank (SCOBLE 1999) to synonymy (syn. n.).

Acidalia contiguaria var. *pallidaria* FUCHS, 1904: in SPULER, Schmett. Eur. 2: 9 (Spain). Syntype(s) (?NSMW). Herewith downgraded from subspecific rank (SCOBLE 1999) to synonymy (syn.n.).

Acidalia contiguaria hercyniae MÜLLER, 1936: Mitt. Münch. ent. Ges. 26 (1): 24, pl. 1, figs E1,E2 (northern Germany: Harz). Syntype(s) (OÖLL). Herewith downgraded from subspecific rank (SCOBLE 1999) to synonymy (syn.n.).

Acidalia contiguaria austriae MÜLLER, 1936: Mitt. Münch. ent. Ges. 26 (1): 21, pl. 1 fig B5 (Austria). Syntypes (MNHU: 2 syntypes examined; OÖLL). Herewith downgraded from subspecific rank (SCOBLE 1999) to synonymy (syn.n.).

Acidalia contiguaria britanniae MÜLLER, 1936: Mitt. Münch. ent. Ges. 26 (1): 31 (United Kingdom, England: Wales, Penmaenmawr). Syntype(s) (OÖLL). Valid at subspecific rank (SCOBLE 1999; SKINNER 1984).

‡ *Idaea contiguaria aurata* SCOBLE, 1999: Geometrid Moths of the World 2: 490. First available name after *aurata* MÜLLER, 1936 (Mitt. Münch. ent. Ges. 26 (1): 10, pl. 1, figs C1,C2: infrasubspecific) (Austrian Alps). Syntype(s) of MÜLLER (OÖLL). Herewith downgraded from subspecific rank (SCOBLE 1999) to synonymy (syn.n.). Junior subjective homonym of *aurata* MENDES, 1912 (*Idaea lutulentaria* syn.).

Unavailable names (infrasubspecific): *anastomosaria*: MÜLLER (1936); *bellieri*: PROUT (1938: 225); *conspersa*: MÜLLER (1936); *dirutaria*: FUCHS (1904); *domestica*: KLIMESCH (1933); *fuscalata*: FUCHS (1884); *griseospersa*: MÜLLER (1936); *grisescens*: CULOT (1917); *melaina*: MÜLLER (1936); *nigra*: MÜLLER (1936); *nigrescens*: MÜLLER (1936); *tenuis*: MÜLLER (1936). – (incorrect subsequent spellings): *brittaniae*: SOUTH (1961); *eburneata*: ZERNY (1927); *fusculata*: PROUT (1913); *herciniae*: FORSTER & WOHLFAHRT (1981).

DIAGNOSIS: Wingspan ♂♀ 16-21 mm, in rearings up to 24 mm, in the autumn generation sometimes 13 mm only. Fairly variable in coloration and pattern of wings. Wings whitish to pale sand coloured. Strong tendency to melanism, mainly in rearings. Markings very well contrasted and comparatively sharp. Transverse lines dentate or zigzagging, with large costal spots. Medial line of forewing proximally surrounding discal spot. Terminal area with three groups of dark spots. Terminal line distinct, dissolved into streaks. Fringe dots absent. Discal spots present. Frons dark brown, vertex and collar sand coloured. Antennae of ♂ sub-dentate, ciliate-fasciculate, cilia long, 1.4-1.6 times width of flagellum. Hindtibia of ♂ slender, with long pencil. Tarsus long, 0.55-0.7 times length of tibia.

Variability: Populations of Iberian peninsula ('*pallidaria*') with ground colour usually lighter, with clinous transitions to nominate forms of central Europe. According to MÜLLER (1936) and PROUT (1938a: 226) populations from Great Britain to Rhine valley with dominant nigrism, in all other European populations with recessive melanism. For results from experimental rearing see also KLIMESCH (1933).

Subsp. ***britanniae*** occurring in a pale form on the coast, and a darker, more robust form in mountainous localities (MÜLLER 1936). Not examined. Not downgraded here due to the isolated occurrence, but subspecific status disputable.

MALE GENITALIA: Uncus digitiform or bottle-shaped. Saccus large, V-shaped. Valva broad at base, with digitiform costal process at 1/2, at tip obliquely truncate. Aedeagus slender terminally with two spinulose crests, populations of southern France and Iberian peninsula with smaller spinules.

FEMALE GENITALIA: Very small. Ductus bursae broad, fused with corpus bursae to an oval, sclerotized complex.

DISTRIBUTION: (West-)European. Very local and rare in Great Britain (uplands of north-western Wales: subsp. *britanniae*) and in central Europe, in the north to central Germany (Harz mountains), and south-western Poland. Widely distributed in mountainous regions from Austria to Portugal. Reputedly in central and southern Italy (TEOBALDELLI 1976; PROLA & RACHELI 1979, PARENZAN 1994), probably misidentified, possibly referring to *I. consolidata* (FLAMIGNI pers. comm.). Doubtful records also from Corsica (RUNGS 1988; MÜLLER 1996), awaiting confirmation. All records from Balkan peninsula (former Yugoslavia: MÜLLER 1996; Romania: STANESCU 1997b; Serbia: TOMIC et al. 2002; Macedonia: PINKER 1968; Bulgaria: NESTOROVA 1998) are considered erroneous, as all examined specimens were misidentified. Records for Romania could not be confirmed, too (STANESCU pers. comm.). Reputedly anthropogenous transport through horticultural trade (BERGMANN 1955). Strongly threatened in most parts of central and western Europe, with recent areal losses (e.g. Bavaria: WOLF & HACKER 2004). – Outside Europe reputedly in Morocco (RUNGS 1981), probably misidentified.

PHENOLOGY: Usually univoltine, mid-June to late July, in the mountains to mid-August. In central Europe exceptionally, in southern Europe more frequently with partial second generation, from early September to early October. Growth of larvae asynchronous (EBERT 2001).

BIOLOGY: Larva polyphagous, possibly with focus on Crassulaceae and Ericaceae. Found on *Calluna vulgaris* (Ericaceae), *Umbilicus rupestris* (Crassulaceae), and *Empetrum nigrum* (Empetraceae) (Great Britain: PROUT 1913; MÜLLER 1936; FRIEDRICH 1976; SKINNER 1984; subsequently var. cit.), *Erica cinerea* (Ericaceae) (Great Britain: FRIEDRICH 1976), and *Sedum album* (Crassulaceae) (Germany: MÜLLER 1936; Pyrenees: CHRÉTIEN 1928). Recorded as preferring *Sedum album*, *S. telephium* (Crassulaceae), *Plantago* (Plantaginaceae) and *Polygonum* (Polygonaceae) (var. cit., e.g. MILLIÈRE 1873; REBEL 1910b; LHOMME 1935; BERGMANN 1955), also fallen leaves of shrubs (BERGMANN 1955). Reared on withered leaves of *Rosa, Minuartia, Lactuca sativa*, *Stellaria, Erica, Galium mollugo*, *Polygonum aviculare*, and *Taraxacum officinale* (BUCKLER 1897; REBEL 1910b; KLIMESCH 1933; LHOMME 1935; ALLAN 1949; BERGMANN 1955; FRIEDRICH 1976; SKINNER 1984; CARTER & HARGREAVES 1987; PORTER 1997). Main activity of adults at dusk, also readily attracted to light. During daytime inert (PROUT 1913), resting on rocks and walls rather than in vegetation (var. cit.). Nectar feeding recorded on *Sedum* (FUCHS 1875). Mating in the morning (EBERT 2001). Stenotopic, K-strategy.

HABITAT: Xeromontane. According to BUSZKO (2000) also at moist localities. Often on precipitous rocky slopes, gorges, heather-clad hillsides, and in scattered mountain *Larix* forest with undergrowing *Calluna*. From 700 m up to 2,200 m above sea-level, sometimes down to 200 m, in the Iberian peninsula rarely below 1,000 m.

SIMILAR SPECIES: *I. alyssumata* (see no. 42), *I. vesubiata* (69), *I. consolidata* (71), *I. camparia* (79), *I. cervantaria* (90).

REMARKS: Nomenclature, zoogeography and inheritance of aberrations extensively discussed in MÜLLER (1936).

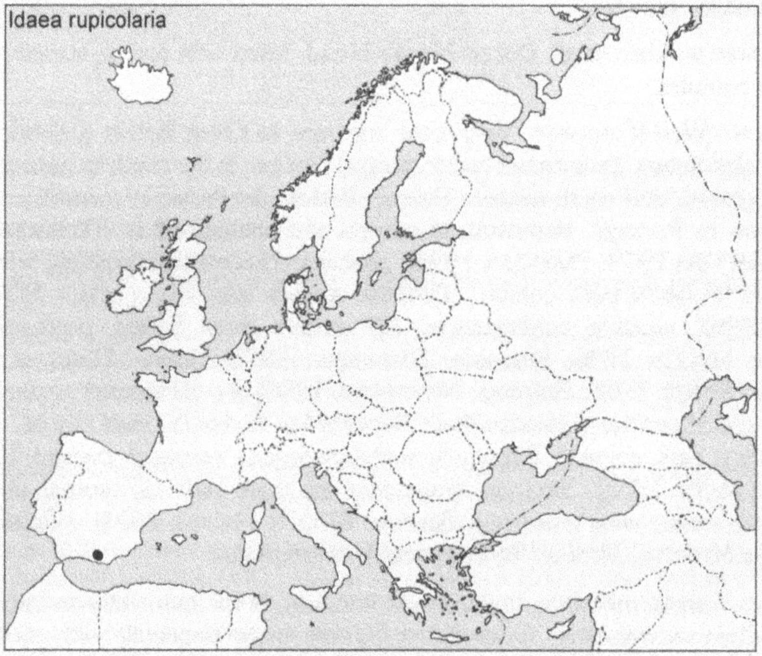

92. Idaea rupicolaria (REISSER, 1927)

Ptychopoda rupicolaria REISSER, 1927: Zeits. österr. Ent. Ver. 12 (11): 106, pl. 4, figs 5-8 (Spain, Andalusia: Sierra Nevada, Puerta de Lobo). Syntypes 5♂28♀ (SMNK: 13♂♀, ZSM: 1♀, NHMW: 5♂♀, examined). The taxon *rupicolaria* MILLIÈRE, 1882 is not a senior homonym as it is unavailable, see under *I. seriata*.

Sterrha reisseri PROUT, 1934: Lepid. Cat. 63: 445 (Spain, Andalusia: Sierra Nevada, Puerta de Lobo). No separate types. Unnecessary replacement name.

Unavailable names (infrasubspecific): *fuscalata*: REISSER (1927). – (incorrect subsequent spelling): *fusculata*: BANG-HAAS (1928).

DIAGNOSIS: Wingspan ♂♀ 19-23 mm. Forewing very slender, apex pointed. Ground colour grey with strong dark suffusion. Transverse lines usually indistinct. Wavy line broad, light grey, often dissolved into a row of light spots on both wings. Discal spots absent. Terminal line present, dissolved into streaks alternating with small light spots. Frons and collar blackish, vertex sand coloured. Antennae of ♂ ciliate-fasciculate, flagellum narrow, cilia long, 1.55-1.75 times width of flagellum. Hindtibia of ♂ slender, with long but weak pencil. Tarsus long, 0.65-0.7 times length of tibia.

MALE GENITALIA: Uncus slender, digitiform or bottle-shaped. Saccus large, digitiform. Valva broad at base, projecting at 1/2 of costa. Aedeagus long and slender, terminally with two long, very narrow, curved sclerites.

FEMALE GENITALIA: Very small. Ductus bursae sclerotized, antrum laterally tapering, posteriorly concave. Corpus bursae oval, membranous.

DISTRIBUTION: Endemic to southern Spain. Restricted to a few localities in the Sierra Nevada, Granada province. Rare and local. Vulnerable on account of limited distribution and habitats.

PHENOLOGY: Univoltine: mid-July to late August.

BIOLOGY: Larval host-plants unknown. Adults attracted to light.

HABITAT: High-montane. Limited information available about habitat preference. Usually from 2,000 m up to 2,900 m above sea-level, sometimes down to 1,600 m.

SIMILAR SPECIES: *I. cervantaria* (see no. 90), *I. contiguaria* (91). *I. rupicolaria* is well characterized by elongate wings and pointed forewing apex.

The ***deitanaria*** species-group

Preliminarily, a separate group is established for this species, as there are no relationships to STERNECK's 19th group, as suggested by REISSER & WEISERT (1977). STERNECK's 19th group has to be included in the *typicata* species-group concept (see there). Shape of valva and paired spinose crest of aedeagus reminiscent of the *cervantaria* species-group, but contradicted by the strikingly different female genitalia. Confusingly, ♂ genitalia, e.g. cornutus and lateral spines of aedeagus, with very similar equivalents in *Cleta cinneretharia* AMSEL, 1935 from the Middle East. Habitus similar to that of *I. attenuaria*, thus possible relations to the *rusticata* species-group may be discussed, too. This position would be supported by some supposedly plesiomorphic features, e.g. long ♂ hindtarsus, presence of ♂ hindtibial spurs and ♂ sternum A2 without central sclerite and setose patch. Egg sculpturation hexagonal reticulate. Larva of the short, thick type.

93. Idaea deitanaria (REISSER & WEISERT, 1977)

Sterrha deitanaria REISSER & WEISERT, 1977: Z. Arb.Gem. österr. Ent. 28 (4): 106, pl. 1, figs 1,2, figs 1-4, (Spain, Andalusia: Otivar, Sierra del Chaparral, prov. Granada). Holotype ♂ (coll. WEISERT, Vienna)

DIAGNOSIS: Wingspan ♂♀ 12-16 mm, in subsequent generations sometimes 8 mm only. Forewings slender, apex pointed. Ground colour brown or dark brown, terminal area darker. Not rarely melanistic with nearly the whole wings black. Transverse lines distinct, blackish brown. Medial line straight, rather broad. Discal spots usually small, indistinct. Frons black, vertex light brown or greyish, collar slightly darker. Antennae of ♂ densely ciliate-setose, cilia short, 0.5-0.7 times width of flagellum. Hindtibia of ♂ slender, with two spurs. Tarsus not shortened, as long as the tibia.

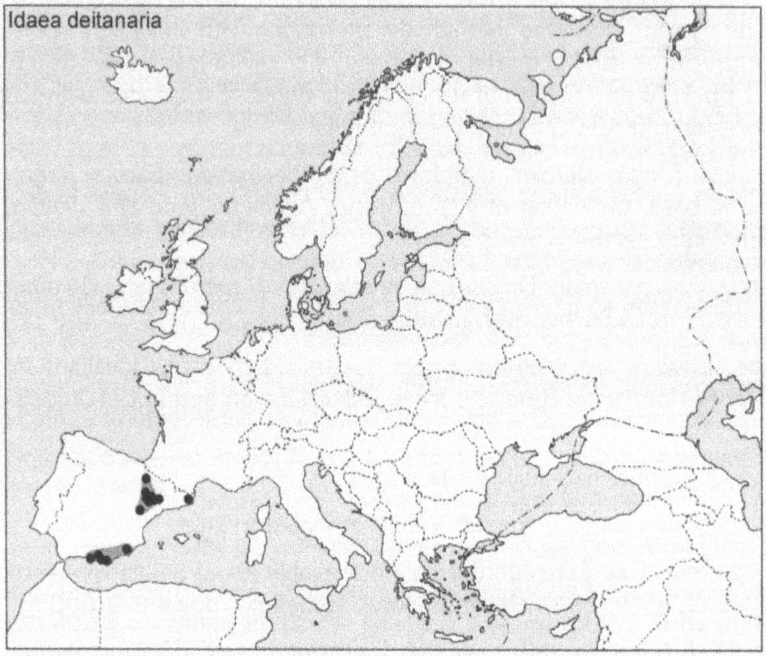

MALE GENITALIA: Uncus bottle-shaped, terminally rounded. Saccus short. Valva broad, with dorsal tooth at 1/2 and small terminal spinule. Aedeagus slender, with two terminal crests, each bearing about four spines.

FEMALE GENITALIA: Apophyses anteriores comparatively long. Ductus bursae short, narrow, antrum semimembranous, strongly dilated. Corpus bursae very slender, laterally spinulose, anteriorly with membranous sac.

DISTRIBUTION: West-Mediterranean. Eastern Spain, from Catalonia to Andalusia.

PHENOLOGY: Usually bivoltine: mid-April to mid-June; early September to late October. At higher elevations, e.g. in Sierra Nevada, possibly univoltine, from early July to early August. Larval development requires six weeks in rearings (REISSER & WEISERT 1977).

BIOLOGY: Larva reared on *Taraxacum officinale* (REISSER & WEISERT 1977). Adults readily attracted to light.

HABITAT: Xerothermophilous. Sometimes extremely common in hot lowland localities, usually near coast. Preferring open habitats with halophilous and gypsophilous plant associations (DANTART 1997). Usually from 150 m up 500 m above sea-level, exceptionally higher. One record from Sierra Nevada, at 2,000 m (coll. GRÜNEWALD).

SIMILAR SPECIES: *I. attenuaria* (see no. 39), *I. incalcarata* (40). *I. saleri* with dark brown frons, vertex grey, ♂ antennal cilia much longer, ♂ hindtibia without spurs, tarsus shortened. Small, dark forms of *I. infirmaria* differ in the white vertex, ♂ antennal cilia shorter, ♂ hindtibia without spurs, tarsus shortened.

REMARKS: Taxonomy, biology and distribution discussed in DANTART (1997). Immature stages figured in the original description.

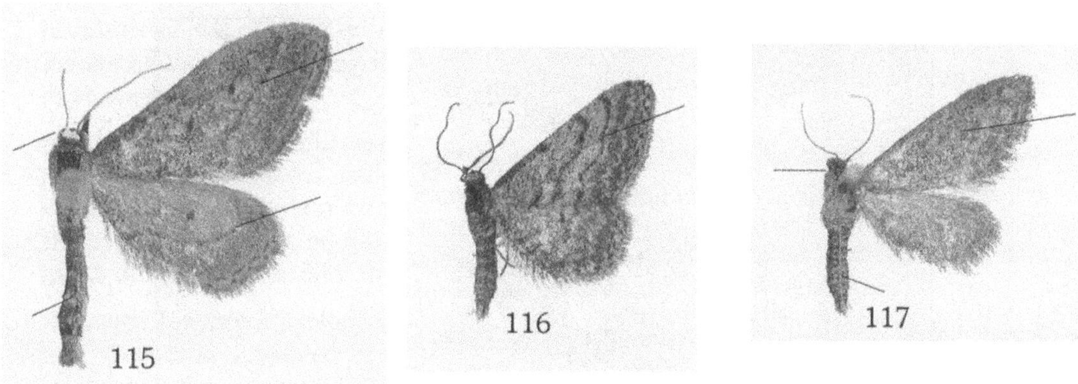

Text-figs 115-117: Diagnostic characters (indicated) of European *Idaea* species. Text-fig. 115: *I. carvalhoi* HERBULOT, 1979. Text-fig. 116: *I. deitanaria* (REISSER & WEISERT, 1977). Text-fig. 117: *I. saleri* DOMÍNGUEZ & BAIXERAS, 1992.

The ***infirmaria*** species-group

STERNECK's (1940) 4th, 5th, and 28th group of *Idaea*, with two additional species, *I. tineata* and *I. saleri* included in this group. *I. trigeminata* is transferred from the 28th group to the *subsericeata* species-group (compare similar suggestions in STERNECK 1940: 162). Relations between *I. lobaria* (CHRÉTIEN, 1909), *I. tineata*, *I. infirmaria* and *I. rhodogrammaria* outlined in PROUT (1935a) and HAUSMANN (1994b: 68), supported e.g. by small male genitalia with setose base of uncus, slender and spatulate valva, by absent cornutus in some species (no. 94, 95), truncate or sub-bilobed uncus of the species no. 94-96, by trapeziform ♀ tergum A8, and by similarities in habitus. *I. saleri* may, preliminarily, be placed here based on similar habitus, short palpi, and similar ♂ genitalia. Hindwing in various species concave between veins M1 and M3, reminiscent of the *fractilineata* species-group. North African *I. lobaria* even similar in habitus to *I. fractilineata*. Certain relationships of the *infirmaria* species-group with the *obsoletaria* species-group seem to be supported by habitus, single cornutus in ♂ genitalia, sclerotization and position of spines in ♀ genitalia. Wings usually with indistinct transverse lines on dark ground colour. Proboscis present, but short. Palpi very short and weak. Sternum A2 of ♂ with horseshoe-shaped sclerite, setose patch weak or absent. Male genitalia small, uncus usually truncate. Cornutus absent in some species, in others present, tapering at both ends and linked with vesica in the central part, with anchor-mechanism in copula (STERNECK 1940: 161). Similar structure in *I. trigeminata*, suggested to be analogous rather than homologous by STERNECK (1940: 162). Egg sculpturation polygonal reticulate. Larvae short and thick (Text-fig. 70).

94. *Idaea saleri* DOMÌNGUEZ & BAIXERAS, 1992

Idaea saleri DOMÌNGUEZ & BAIXERAS, 1992: Nota lepid. 15 (2): 103, figs 1-5 (Spain, Valencia: El Saler). Holotype ♂ (DBUV) (MNCN: 1♂ paratype, examined).

DIAGNOSIS: Wingspan ♂♀ 11-14 mm. Wings slender and elongate. Cells very long. Hindwing slightly concave between M1 and M3, angled between M3 and CuA1 (Text-fig. 68). Ground colour dark grey brown. Transverse lines of forewing dark grey, indistinct,

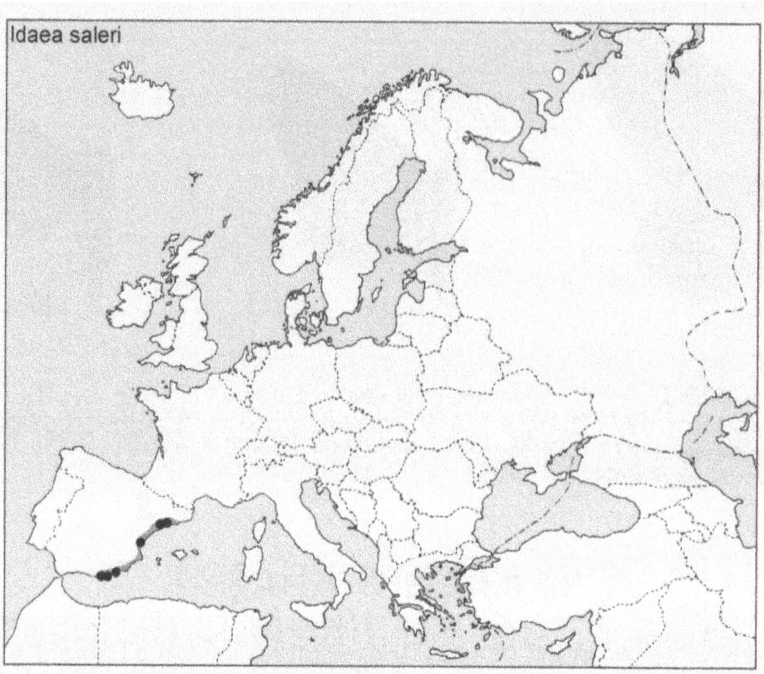

postmedial line oblique, straight. Hindwings without markings. Terminal line absent. Discal spots absent. Frons and collar dark brown, vertex grey. Abdomen with pale dorsal line in both sexes bracketed by two rows of angular spots or dashes, dorsal line very rarely absent. Proboscis stout, but short. Palpi short, about 0.5 times diameter of eye. Antennae of ♂ ciliate-fasciculate, cilia long, about 1.5-2.1 times width of flagellum. Hindtibia of ♂ slender, with weak pencil, together with the short tarsus not longer than 1.2 mm (Text-fig. 29).

MALE GENITALIA: Saccus shallow, round. Valva slender, round at tip. Aedeagus with one comparatively stout cornutus, 0.3-0.35 times length of aedeagus. The similar North African species *I. scabraria* (CHRÉTIEN, 1909) clearly belonging to the *microptera* species-group, with valva extremely narrow, distally tapering, cornutus very long and thin.

FEMALE GENITALIA: Antrum dilated. Anterior part of ductus bursae with numerous transparent spinules. Corpus bursae globular, membranous, without spinules. *I. scabraria* with corpus bursae very long and slender, strongly spinulose.

DISTRIBUTION: Endemic to easternmost Spanish coastal regions between prov. Tarragona and Almeria.

PHENOLOGY: Bi- or trivoltine: late April to late September, single specimens until mid-October. Rearing requires about three months in summer (HALL pers. comm.).

BIOLOGY: Larva reared on petals of *Helichrysum* as long as possible, also feeding on petals of *Cistus, Helianthemum, Geranium* etc. In later instars *Erica* accepted as additional host-plant, but rearing exclusively on *Erica* not successful, and a broad set of other plant species was also refused (HALL pers. comm.). Larvae in captivity growing very slowly, perhaps due to the offer of sub-optimal host-plant species. Adults readily attracted to light. Resting position *Eupithecia*-like, hindwings not touching abdomen.

HABITAT: In hot coastal lowland habitats, such as sandy dunes, or saltmarshes with *Tamarix* (HALL pers. comm.; DANTART 1997), possibly halophilous. From 0 m up to 300 m above sea-level.

SIMILAR SPECIES: *I. carvalhoi* (see no. 78), *I. deitanaria* (93). In the eremic belt from North Africa to south-western Asia there are many similar, allopatric species from the *microptera* species-group (STERNECK's 11th group), e.g. *I. scabraria* from Algeria with similar white markings on the abdomen, but different genitalia (see above). Small, dark forms of *I. infirmaria* differ in white vertex, ♂ antennal cilia much shorter.

REMARKS: Taxonomy, biology and distribution discussed in DANTART (1997). Relationships to *I. descitaria* suggested in DOMÌNGUEZ & BAIXERAS (1992). Certain relations to the *microptera* species-group (STERNECK's 11th group) with some very similar species are not excluded, too.

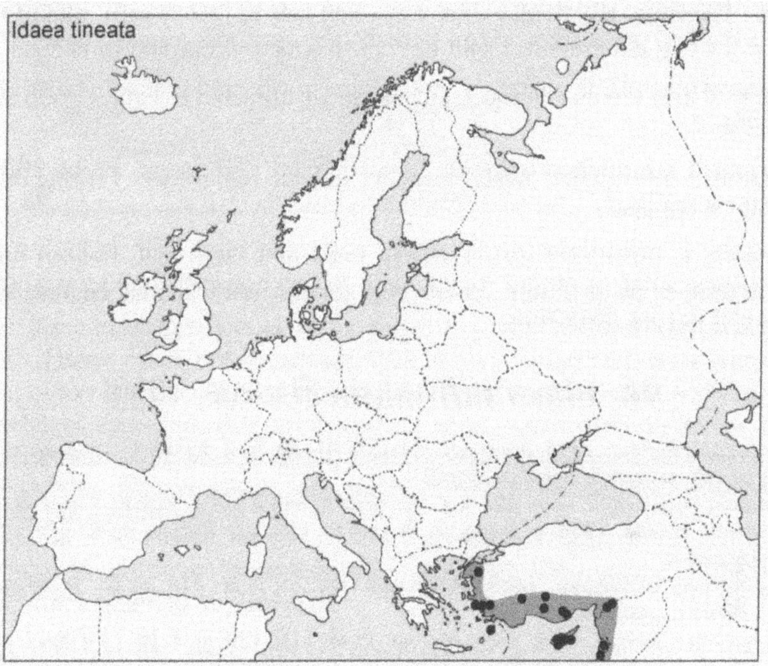

95. *Idaea tineata* (THIERRY-MIEG, 1911)

Acidalia tineata THIERRY-MIEG, 1911: Annls Soc. ent. Belg. 54: 468 (Turkey, Amanus: Maydan Ikbis ('Akbès')). Syntypes 5♀.

Ptychopoda antennata WEHRLI, 1931: Mitt. Münch. ent. Ges. 21 (2): 41 (Turkey, Amanus: Maydan Ikbis ('Akbès')). Holotype ♂ (ZFMK, examined).

Idaea praetineata HAUSMANN, 1994: Bonn. zool. Beitr. 45 (1): 68, figs 2, 3, 11, 13, 13a (northern Jordan: Wadi Sir near Amman). Holotype ♂ (ZSM, examined). Herewith downgraded from species (HAUSMANN 1994; SCOBLE 1999) to subspecific rank (stat. n.), as there are transitional populations in the Lebanon.

Unavailable names (incorrect subsequent spelling): *lineata*: PROUT (1934).

DIAGNOSIS: Wingspan ♂ 11-14, ♀ 12-16 mm. Forewing slender. Hindwing rounded. Ground colour dirty white in ♂, dark grey in ♀. Blackish postmedial lines almost dissolved into spots, usually distinct in ♂. In ♀ transverse lines indistinct or absent. Small punctiform discal spots present. Frons blackish, vertex white, collar ochreous. Palpi black, narrow and short, 0.6-0.7 times diameter of eye. Proboscis short, weak. Antennae of ♂ thick, ciliate, cilia short, 0.3-0.5 times width of flagellum (Text-fig. 65). Hindtibia of ♂ not dilated, with pencil covering 1/2 of tarsus. Tarsus shortened, 0.35-0.45 length of tibia.

MALE GENITALIA: Uncus broadly rounded, sometimes slightly bilobed. Valva spatulate. Juxta posteriorly with paired tapering hooks. Aedeagus without cornuti.

FEMALE GENITALIA: Lamella postvaginalis semicircular. Antrum swollen, laterally sclerotized. Corpus bursae pyriform, corrugated anteriorly, with longitudinally folded, conical appendix bursae laterally.

DISTRIBUTION: East-Mediterranean. In Europe recorded only from islands Samos (e.g. REBEL 1934), and Rhodos. – Outside Europe in southern Turkey, Cyprus, Syria, Lebanon, Jordan and Israel. Very abundant at the sites of occurrence.

PHENOLOGY: Trivoltine: late April to late June; mid-July to mid-August; mid-October, very abundant in the first generation. Flight periods may vary according to altitude.

BIOLOGY: Larval host-plants unknown. Adults readily attracted to light, ♀ ratio at light very high, 80-95%.

HABITAT: Limited information available about habitat preference. From 200 m up to 1,500 m above sea-level.

SIMILAR SPECIES: *I. infirmaria* usually larger, paler and often with reddish tinge, costal spots of forewing small or absent. Small, dark forms, however may be very similar to *I. infirmaria* and require dissection.

96. Idaea infirmaria (RAMBUR, 1833)

Dosithoea infirmaria RAMBUR, 1833: Annls Soc. ent. Fr. 2 (1) : 36, pl. 2, fig. 18 (Corsica: Ajaccio). Lectotype ♂ (ZSM/HERBULOT, examined, abdomen absent).

Acidalia carnearia MANN, 1855: Verh. zool.-bot. Ver. Wien 5: 544 (Corsica: Ajaccio, Corte). Syntypes 2♂.

Acidalia ledererata GUENÉE, 1858: in BOISDUVAL & GUENÉE, Hist. nat. Insectes (Spec. gén. Lépid.) 9: 475 (Corsica). Holotype ♂.

Acidalia nigrobarbata STAUDINGER, 1859: Stettin. ent. Ztg. 20 (7-9): 216 (Spain, Andalusia: Chiclana). Holotype ♂ (MNHU, examined).

Acidalia aquitanaria CONSTANT, 1865: Annls Soc. ent. Fr. (4) 5 (2): 195, pl. 7, fig. 11 (France, prov. Landes). Syntypes (BMNH: 1♂ examined, labelled as 'lectotype'). Downgraded from subspecific rank (LERAUT 1997; SCOBLE 1999) to synonymy (syn. n.).

Idaea infirmaria mitescens SCOBLE, 1999: Geometrid Moths of the World (2): 497. First available name after *mitescens* PROUT, 1935: in SEITZ, Macrolep. 4, Suppl.: 60, pl. 6, fig. h (infrasubspecific) (Algeria: Bône). Syntype(s). Valid at subspecific rank.

Unavailable names (misidentification): *attenuata*: sensu RAMBUR (1866; non binominal) nec *attenuaria* RAMBUR, 1833; *euphorbiata*: sensu MENTZER (1990) nec BALESTRE, 1906 (identity: *Idaea belemiata*). – (infrasubspecific): *anastomosaria*: REISSER (1961). – (nomen nudum): *crudata*: PROUT (1934); *lucia*: SCHAWERDA (1931). – (incorrect subsequent spellings): *acquitanaria*: TURATI (1913); *aquinataria*: VALLETTA (1973); *attentuata*: SCOBLE (1999); *infirmar*: SAMMUT (1983); *infirmata*: PROUT (1935); *infirmatia*: DE PRINS (1998); *iniformata* (for *infirmaria*): LEDERER (1861).

DIAGNOSIS: Wingspan ♂♀ 13-18 mm, exceptionally 10 mm only. Extremely variable in coloration and pattern of wings. Ground colour brownish grey, often mixed with red, mainly in ♀, sometimes also pale ochreous, mainly in ♂. Percentage of specimens with brown suffusion higher in Spain. Rosy or red coloured forms ('f. *carnearia*') common in Italy, south-eastern France and Morocco, but occurring, less abundantly, also elsewhere.

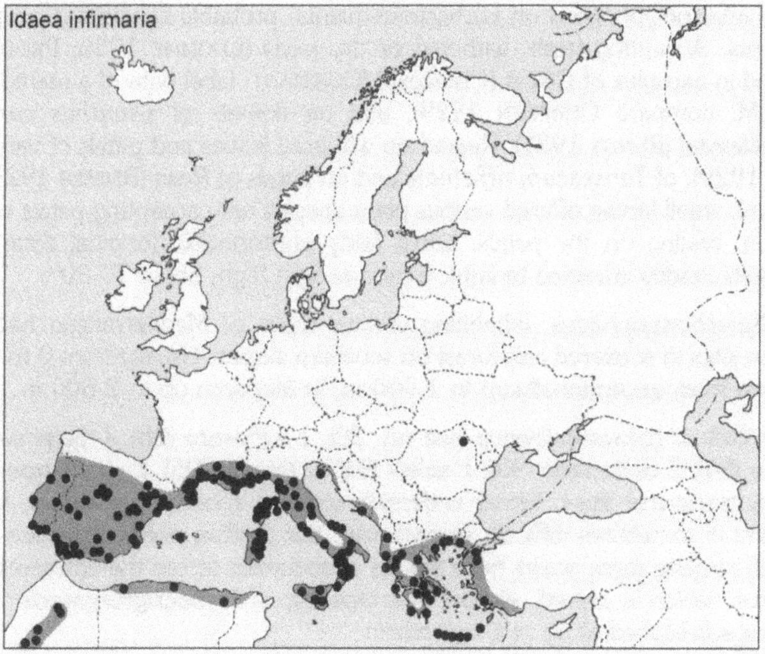

Postmedial line dark brown, exceptionally red or violet, fine, slightly dentate. Antemedial and medial lines indistinct. Discal spots present. Costa of forewing usually blackish towards base. Fringe often, but not always, dark brown, darker than ground colour. Frons dark brown, vertex white, collar red brown. Palpi dark brown, narrow and short, 0.5-0.7 times diameter of eye. Proboscis comparatively short, weak. Antennomeres of ♂ short and broad, with intersegmental incisions, ciliate-setose, cilia short, 0.4-0.6 times width of flagellum. Hindtibia of ♂ not dilated, with pencil covering 1/2 of tarsus. Tarsus shortened, 0.45-0.5 times length of tibia.

MALE GENITALIA: Uncus broad, posteriorly slightly bilobed. Valva spatulate. Aedeagus with one long cornutus tapering at both ends.

FEMALE GENITALIA: Ductus bursae short. Corpus bursae elongate pyriform, posterior 2/3 well sclerotized and corrugated, anterior third membranous. Between both parts a ring of small sclerites. Membranous part of corpus bursae larger and broader from Italy to the east.

DISTRIBUTION: Mediterranean. From Iberian peninsula across France, Italy to Greece. Isolated occurrence in eastern Romania (NEMES 1970). Present on all the major Mediterranean islands, also Malta (VALLETTA 1973; MÜLLER 1996). Often very abundant, populations more isolated towards north. One single record from Belgium (DE PRINS 1998: 157) probably due to anthropogenous transport. – Outside Europe from Morocco to Tunisia (subsp. *mitescens*), and western Turkey. Absent from Cyprus.

PHENOLOGY: Usually univoltine: mid-June to late July. Sometimes emerging from late May. At higher altitudes until early August. Under good conditions with a partial second generation from early September to mid-October. Larva extremely inert, developing very slowly (original data; REISSER 1929), in rearings however, exceptionally, emerging after just three months (HALL pers. comm.).

BIOLOGY: Larva polyphagous on herbaceous plants, probably specialised on petals and fruit capsules. Accepting fresh, withered or dry parts (LHOMME 1935; PROUT 1935a). Larva found in capsules of *Cistus* (Cistaceae) (CONSTANT: label data of a reared specimen at the ZSM; compare CHRÉTIEN 1929), and on flowers of *Dianthus caryophyllus* (Caryophyllaceae) (RUNGS 1981). Reared on withered leaves and petals of various plants (CHRÉTIEN 1929), of *Taraxacum officinale* and on petals of *Rosa* (REISSER 1929; LEIPNITZ pers. comm), small larvae offered various plant species only accepting petals of *Achillea millefolium*, resting on the petals. Larva fairly photophobic (original data; CHRÉTIEN 1929). Adults readily attracted to light, ♀ ratio at light high, about 70-80%.

HABITAT: Xerothermophilous. Inhabiting various types of Mediterranean habitats, e.g. maquis, hot sites in scattered oak forest on southern slopes. Usually from 0 m up to 800 m above sea-level, exceptionally up to 1,900 m, in Morocco up to 2,600 m.

SIMILAR SPECIES: *I. consanguiberica* (see no. 20), *I. efflorata* (38), *I. belemiata* (51), *I. obsoletaria* (54), *I. deitanaria* (93), *I. saleri* (94), *I. tineata* (95). *I. rhodogrammaria* differs in rosy transverse lines, lighter underside and less shortened ♂ tarsus, but correct identification is sometimes difficult. In particular, the ♀ often needs dissection. In ♂ it is sufficient to remove some scales from the tip of abdomen to see the differential features in the uncus, which is curved, slender and tapering in *I. rhodogrammaria*, somewhat broader and sub-bilobed at tip in *I. infirmaria*.

REMARKS: Relationship to *I. tineata* is highlighted by nearly all external characters and the sexual dimorphism, with ♂♂ often paler. Description of immature stages in CHRÉTIEN (1929) and REISSER (1929).

97. *Idaea rhodogrammaria* (PÜNGELER, 1913)

Ptychopoda rhodogrammaria PÜNGELER, 1913: in PROUT, in SEITZ, Macrolep. 4: 119, pl. 3, fig. h (Spain, prov. Murcia: Sierra d'Espuña). Syntypes '3♂♂' (MNHU 1♂ 'type', 2♂ 'cotypes'; ZSM 1♂ 'cotype', examined).

DIAGNOSIS: Wingspan ♂♀ 12-16 mm. Wings sand coloured with yellow or orange tinge. ♀ usually with dark brown suffusion. Medial and postmedial lines, terminal shadow and

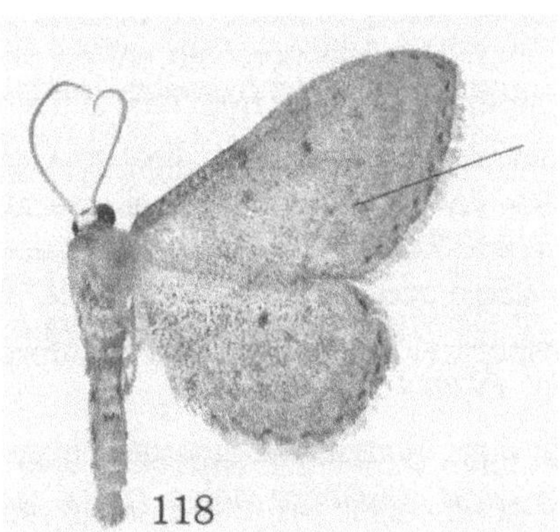

Text-figs 118-119: Diagnostic characters (indicated) of European *Idaea* species. Text-fig. 118: *I. infirmaria* (RAMBUR, 1833). Text-fig. 119: *I. rhodogrammaria* (PÜNGELER, 1913).

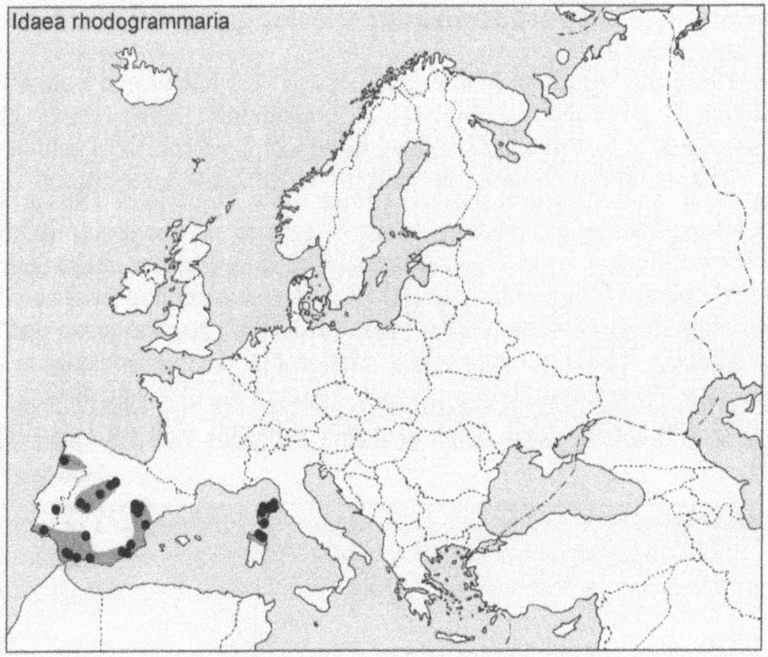

forewing costa pink or violet. Punctiform discal spots black, distinct. Fringe concolorous with ground colour. Underside whitish with transverse lines indistinctly marked brown. Frons dark brown, vertex white, collar brown. Palpi dark brown, narrow and short, 0.5-0.7 times diameter of eye. Proboscis comparatively short, weak. Antennomeres of ♂ short and broad, with intersegmental incisions, ciliate-setose, cilia short, 0.35-0.6 times width of flagellum. Hindtibia of ♂ not dilated, with pencil covering 1/4-1/3 of tarsus. Tarsus 0.5-0.6 length of tibia.

MALE GENITALIA: Uncus slender, tapering. Valva spatulate. Aedeagus broad, curved, with one stout, strongly curved cornutus tapering at both ends.

FEMALE GENITALIA: Antrum strongly sclerotized, U-shaped. Ductus bursae short. Corpus bursae oval, posterior half well sclerotized and corrugated, anterior half membranous. Between both halves a ring of spines.

DISTRIBUTION: West-Mediterranean, disjunct. Spain, Portugal, Corsica, Sardinia. Reputedly in France (LERAUT 1997), awaiting verification. The only record for Catalonia misidentified (DANTART 2000).

PHENOLOGY: Usually univoltine: mid-June to mid-August, sometimes until mid-September.

BIOLOGY: No detailed information on larval host-plants traced. Reared on dry petals and leaves (LEIPNITZ pers. comm.). Growth of larvae extremely slow. Adults readily attracted to light.

HABITAT: Xerothermophilous. Limited information available about habitat preference. From 0 m up to 900 m above sea-level, in central Spain up to 1,500 m, in Morocco up to 1,700 m.

SIMILAR SPECIES: *I. infirmaria* (see no. 96).

The *ostrinaria* species-group

STERNECK's (1940) 32nd group of *Idaea*. Three species are transferred from this group to *Brachyglossina*: *B. improbata*, comb. n., *B. brevitarsata*, comb. n., *B. consociata*, comb. n. (see remarks to *Brachyglossina*). Position of *I. metohiensis* remains tentative and questionable. *I. ostrinaria* with relationships to STERNECK's 29th group and parts of the 16th group (*I. subpurpurata*) (PROUT 1938a: 228; HAUSMANN 1991d; 1994), but genitalia indicating also certain relations to the *eugeniata* species-group. Reminiscent of many *Brachyglossina* species in ♂ (small terminal cornuti) and ♀ genitalia (looped ductus bursae, central spines of corpus bursae and laterodorsal inseration of ductus seminalis). Morphology of neotropical *Idaea rufulata* (WARREN, 1900) species-group and nearctic *I. demissaria* (HÜBNER, 1831) not supporting relationship, despite similarity in habitus. In the *ostrinaria* species-group proboscis present. Sternum A2 of ♂ with distinct horseshoe-shaped sclerite and setose patch. Male genitalia: aedeagus with 25-100 small cornuti. Larva short.

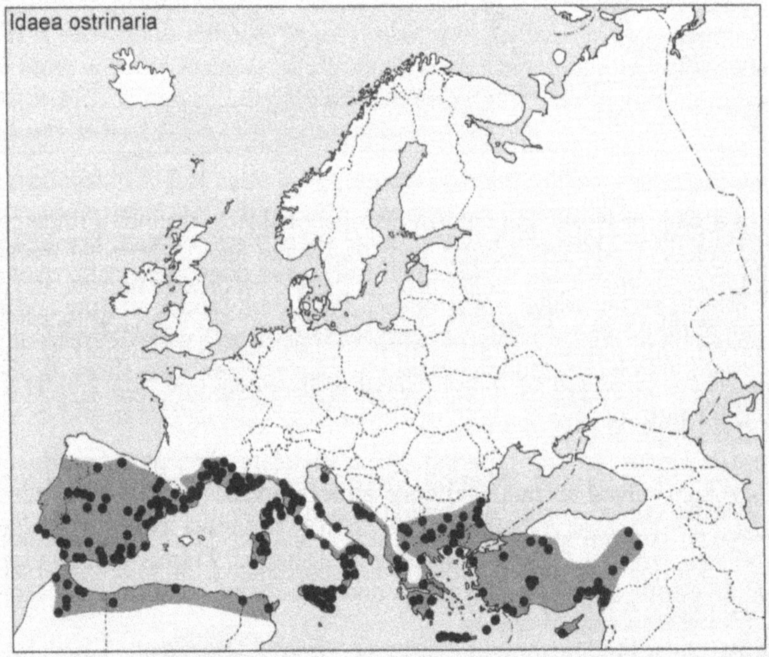

98. Idaea ostrinaria (HÜBNER, 1813)

Geometra ostrinaria HÜBNER, 1813: Samml. Eur. Schmett. 5, Geometrae (1): pl. 83, fig. 430 (Europe). Syntype(s) lost.

Sterrha ostrinaria demarginata SCHWINGENSCHUSS, 1938: Ent. Rdsch. 55 (45): 532 (Turkey: central Anatolia). Syntypes. Valid at subspecific rank (SCOBLE 1999), but possibly with clinous transitions to nominate subspecies.

Unavailable names (infrasubspecific): *eucrines*: PROUT (1935); *lutea*: PROUT (1935); *oenoparia*: PÜNGELER (1913); *purpuraria*: TURATI (1913). – (incorrect subsequent spellings): *oenopararia*: SEITZ (1912: pl. 3h); *ostrina*: HÜBNER (1825); *ostrinata*: BELLIER (1861).

DIAGNOSIS: Wingspan ♂♀ 14-20 mm. Forewing apex pointed. Wings sand coloured, sometimes with olive tinge. On the forewing, base of costa and terminal area purple, but apical area concolorous with ground colour. Postmedial line wavy, deep purple. Hindwing with reduced pattern, often deeper purple towards termen. Discal spots, terminal line and fringe dots absent. Underside pale straw coloured, without red. Frons dark brown, vertex white, collar sand coloured. Antennae of ♂ ciliate-setose, cilia 1.0-1.1 times width of flagellum. Hindtibia of ♂ with long, but weak pencil. Tarsus 0.35-0.5 times length of tibia, in Morocco extremely variable, 0.35-0.8 times, in Turkey 0.4-0.6 times.

Variability: Forms with strong purple suffusion dominant in parts of Sardinia, occasionally in Corsica, Spain and elsewhere ('f. *purpuraria*'; 'f. *oenoparia*';), more frequent in females than in males, often emerging from the same rearings together with normal moths (LÖBEL pers. comm.). In North Africa sometimes with yellow ground colour ('f. *lutea*').

MALE GENITALIA: Uncus narrow. Saccus comparatively broad, often V-shaped. Valva long and slender. Aedeagus very large, long, with terminal patch of about 25-50 cornuti, distal cornuti smaller.

FEMALE GENITALIA: Ductus bursae sclerotized, long and slender, looped towards corpus bursae. Inner margin of loop sclerotized. Corpus bursae large, pyriform, in the centre with numerous short, but stout, star-shaped spines. Anterior part of corpus bursae membranous.

DISTRIBUTION: Mediterranean. From Iberian peninsula across Italy to Balkan peninsula, most common near coasts. Present on Elba, Corsica, Sardinia, Sicily, Greek islands. – Outside Europe from Morocco to Tunisia, in Turkey, and on Cyprus ('subsp. *demarginata*').

PHENOLOGY: Univoltine: mid-May to early July, exceptionally until late July, in the lowlands of the south sometimes emerging from early April. Second brood very rare in nature, limited data available from late September to mid-October. Larva slowly developing (WAGNER 1926).

BIOLOGY: Larva polyphagous on herbaceous plants (REBEL 1910b). Found on withered leaves of *Thymus* under plants of *Thymus* (Lamiaceae) (Spain: KING & ROMERA 2004). Recorded on *Heliotropium europaeum* (Boraginaceae), *Campanula* (Campanulaceae), and *Calluna vulgaris* (Ericaceae) (France: LHOMME 1935). Reared on various species of Caryophyllaceae, Convolvulaceae, Campanulaceae, and Ericaceae, first feeding on pollen and stamens, then on petals, finally on leaves (MILLIÈRE 1868; RIBBE 1912), reared also on withered leaves of *Thymus, Campanula, Polygonum aviculare*, and *Taraxacum officinale* (LHOMME 1935; TORMO & MUÑOZ 1995; HAUSMANN & MILLER 2000; KING & ROMERA 2004; MÜLLER pers. comm.), reputedly on *Pteridium aquilinum* (DOGANLAR 2003).

HABITAT: Xerothermophilous. Preferring hot, coastal lowland habitats. Usually from 0 m up to 800 m above sea-level, exceptionally up to 1,300 m. In southern Spain, southern Turkey and Morocco from 0 m up to 1,700 m.

SIMILAR SPECIES: No similar species in Europe.

REMARKS: For morphology of immature stages see MAY (1893).

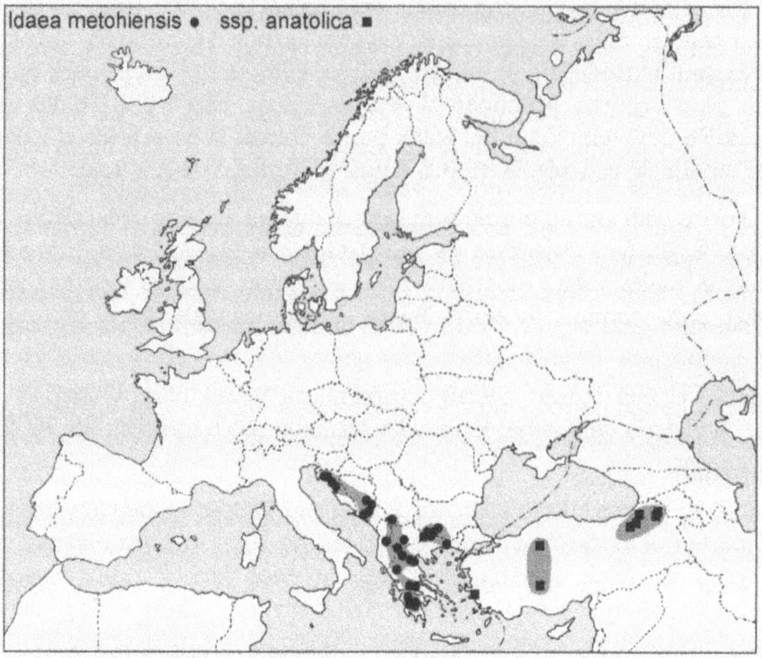

99. *Idaea metohiensis* (REBEL, 1900)

Acidalia metohiensis REBEL, 1900: Verh. zool.-bot. Ges. Wien 50 (8): 451 (Herzegowenia: Gacko ('Metohia')). Syntypes 2♂ (NHMW).

Sterrha metohiensis anatolica SCHWINGENSCHUSS, 1938: Ent. Rdsch. 55 (59): 700 (Turkey: central Anatolia). Holotype ♂. Valid at subspecific rank.

Unavailable names (infrasubspecific): *albomarginata*: REISSER (1961); *antiopa*: REISSER (1934). – (incorrect original spelling): *metochiensis*: TULESCHKOV (1951).

DIAGNOSIS: Wingspan ♂♀ 18-25 mm. Forewing slender, apex pointed. Ground colour dirty white, in Dalmatia lighter, in Turkey (subsp. *anatolica*) with stronger dark suffusion. Medial line sometimes indistinct. Postmedial line convex, strongly dentate on all wings. Terminal area darker beside pale wavy line. Discal spots present. Fringe dots distinct, sometimes also with terminal line. Frons and collar dark brown, vertex white. Antennae of ♂ ciliate-fasciculate, cilia long, 1.5-2.0 times width of flagellum (Text-fig. 66). Hindtibia of ♂ slender, without pencil. Tarsus long, 0.6-0.75 times length of tibia.

MALE GENITALIA: Uncus slender digitiform. Gnathos comparatively broad. Saccus broadly rounded. Valva broad, in Croatia slightly narrower. Aedeagus comparatively short and broad with two patches of very small spinules.

FEMALE GENITALIA: Ductus bursae of medium length, towards corpus bursae with oblique row of spinules. Corpus bursae globular, membranous, with long spines, usually arranged in the form of two stars.

DISTRIBUTION: East-Mediterranean. From western Croatia to Greece. – Outside Europe local populations in Turkey (subsp. *anatolica*).

PHENOLOGY: Univoltine: mid-June to late July, sometimes until early August. Data from early May (PROUT 1935a) await confirmation.

BIOLOGY: Natural host-plants of larva unknown. Reared on *Polygonum aviculare* (HAUSMANN & MILLER 2000). Adults readily attracted to light, resting on rocks during daytime (PROUT 1913).

HABITAT: Montane. Rocky slopes and stony grassland. Usually from 1,400 m up to 2,500 m above sea-level, here sometimes common, exceptionally down to 200 m.

SIMILAR SPECIES: *I. ossiculata* (see no. 23), *I. consolidata* (71). *S. marginepunctata* with terminal dots rather than fringe dots, postmedial line S-shaped at forewing costa, ♂ hindtarsus longer, ♀ hindtibia with four spurs.

REMARKS: Habitus and antennal cilia do not support position in the *ostrinaria* species-group as suggested by STERNECK (1940).

The *eugeniata* species-group

STERNECK's (1940) 30th group of *Idaea*. Relationship to the 32nd group supported by 'spiral position' of cornuti (STERNECK 1940: 164). Habitus, ♂ genitalia and posterior loop of corpus bursae (♀ genitalia) well matching equivalents in the *aversata* species-group. Habitus and ♀ genitalia also reminiscent of some members of the *rusticata* species-group, e.g. *I. troglodytaria*. Palpi comparatively short. Proboscis present. Sternum A2 of ♂ with distinct horseshoe-shaped sclerite and setose patch. Egg sculpturation hexagonal reticulate. Larva comparatively slender. Revisions: REZBANYAI-RESER (1987); RAINERI (1989).

100. Idaea eugeniata (DARDOIN & MILLIÈRE, 1870)

Acidalia eugeniata DARDOIN & MILLIÈRE, 1870: Iconogr. Descr. Chen. Lép. inéd. 3 (25): 118, pl. 111, figs 4-6 (southern France: near Marseille). Syntypes 2♂2♀ (coll. DARDOIN).
Acidalia seeboldiata RÖSSLER, 1877: Stettin. ent. Ztg. 38 (7-9): 366 (Spain, Basque country: Bilbao). Syntypes 2♂1♀.
Acidalia algeriaca CULOT, 1917: Noct. et Géom. Eur. 3: 54, pl. 7, fig. 139 (Algeria). Syntype(s). Valid at subspecific rank (SCOBLE 1999).
Unavailable names (infrasubspecific): *jacobsi*: PROUT (1913); *pseudodegeneraria*: WEHRLI (1926). – (incorrect original spelling): *speudodegeneraria*: WEHRLI (1926).

DIAGNOSIS: Wingspan ♂♀ 20-25 mm, in subsequent generations sometimes 15 mm only. Forewing broad. Ground colour variable, usually light brown with orange tinge, sometimes greyish. Postmedial line convex, dotted. Medial and antemedial line sometimes absent, medial line, however, not rarely broad, sometimes fusing with antemedial line to a broad dark fascia ('f. *jacobsi*'). Discal spots and fringe dots small. Frons dark brown, collar sand coloured, vertex slightly paler. Antennae of ♂ ciliate-setose, cilia short, 0.65-0.75 times width of flagellum. Hindtibia of ♂ dilated, exceeding first tarsomere, with large pencil. Tarsus 0.3-0.4 times length of tibia (Text-fig. 34).

Subsp. ***algeriaca*** larger, coloration usually more brownish rather than reddish or grey. The form '*jacobsi*' commoner in North Africa than in Europe.

MALE GENITALIA: Uncus triangular. Gnathos linguiform. Saccus V-shaped, of medium size. Valva medially narrowing, dilated at tip. Aedeagus very large, usually with seven stout cornuti of increasing size towards tip.

FEMALE GENITALIA: Ductus bursae narrow, of medium length. Corpus bursae oval, elongate, spinulose, posteriorly looped.

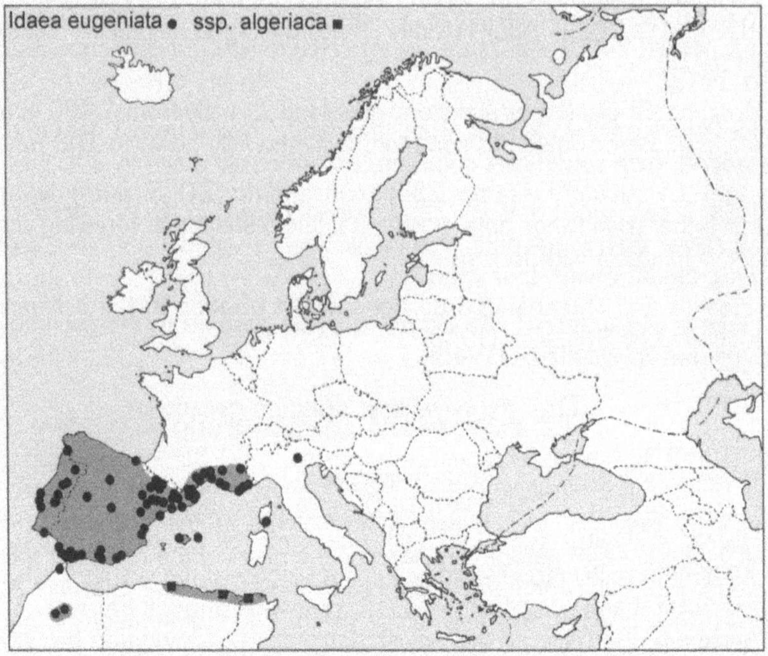

DISTRIBUTION: West-Mediterranean. From south-eastern France to Iberian peninsula. Present on Mallorca and Menorca (VALLHONRAT & PETITPIERRE 1985). One single specimen collected in South Tyrolia (HAUSMANN 1993d), another in south-eastern Corsica (BRUSSEAUX 1995; ZSM/HERBULOT, dissected). – Outside Europe in isolated populations from Morocco to Tunisia (subsp. *algeriaca*).

PHENOLOGY: Bi- or trivoltine. In the south of the distribution trivoltine from early March to late October, in southern Spain exceptionally emerging from early February (HALE pers. comm.). In the north bivoltine, late May to mid-July; early September to late October. Rearing requires about two months in summer (HALL pers. comm.).

BIOLOGY: Larva probably polyphagous on withered leaves of herbaceous plants. Reared on *Geranium*, *Cymbalaria muralis* (LHOMME 1935), *Stellaria media, Dactylis glomerata, Poa pratensis, Taraxacum officinale, Sonchus oleraceus, S. arvensis* (SOFFNER 1955). Adults readily attracted to light. Mating late in the night (SOFFNER 1955).

HABITAT: Xerothermophilous. Inhabiting various types of open habitats, such as dry, stony grassland, fallows, sun-exposed scrubby places. From 0 m up to 1,200 m above sea-level.

SIMILAR SPECIES: *I. distinctaria* and *I. predotaria* differ by less intensive coloration, postmedial line not dentate or dotted, fringe dots usually weaker or absent. Identification of ♂ is easy by brushing tip of abdomen to remove scales and reveal shape of valva. *Brachyglossina hispanaria* often with dark spots in the terminal area, the row of postmedial dots more undulate, proboscis reduced, ♂ antennal cilia longer. *Scopula rubellata* usually paler, postmedial line dentate and zigzagging, ♂ hindtibia with two spurs, tarsus not shortened, ♀ with four hindtibial spurs. *S. minorata* with much longer ♂ antennal cilia, ♂ hindtarsus longer, ♀ hindtibia with 4 spurs.

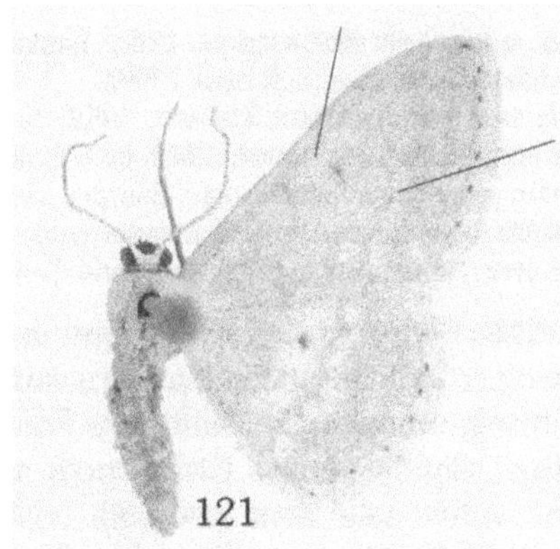

Text-figs 120-121: Diagnostic characters (indicated) of European *Idaea* species. Text-fig. 120: *I. eugeniata* (MILLIÈRE, 1870). Text-fig. 121: *I. predotaria* (HARTIG, 1951).

REMARKS: Brown and red forms, and forms with and without dark medial fascia can be reared from the eggs of the same female.

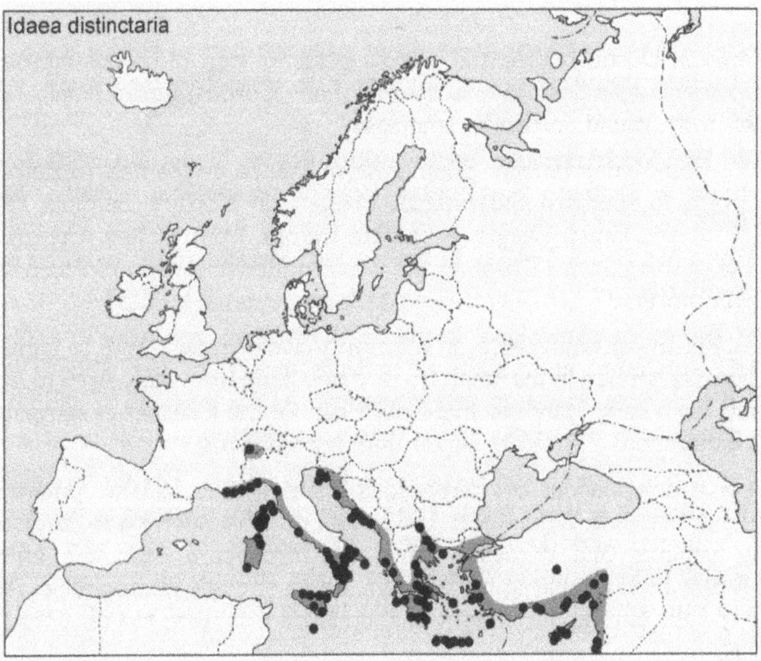

101. Idaea distinctaria (BOISDUVAL, 1840)

Acidalia distinctaria BOISDUVAL, 1840: Genera Index meth. eur. Lepid.: 224 (France, Provence), referring to RAMBUR'S 'Fauna Andalusiae' published 26 years later. Syntype(s) (ZFMK: 1♂ ex coll. BOISDUVAL, examined, probably a unique type specimen: see PROUT 1935a). Valid name according to RAINERI (1989).

Idaea ruficostata ZELLER, 1849: Stettin. ent. Ztg. 10 (7): 215, index. First available name after *ruficostata* ZELLER, 1847 (non binominal) (Italy, Tuscany: Montenero, Antignano, Posignano; Lazio: Rome; Sicily). Syntypes ♂♀ (BMNH). Original material from ZELLER (1847) was from Turkey: Tlos. Generic combination indicated in the index.

‡ *Idaea ruficostaria* HEYDENREICH, 1851: Lepid Eur. cat. meth. (Edn 3): 59. Incorrect subsequent spelling (emendation according to SCOBLE 1999).

Acidalia incarnaria HERRICH-SCHÄFFER, 1852: Syst. Bearb. Schmett. Eur. 6 (55): 66. First available name after *incarnaria* HERRICH-SCHÄFFER, 1848 (pl. 71, fig. 436: non binominal) and *incarnaria* ZELLER, 1849 (introduced in synonymy of *ruficostata*) (Europe). Syntype(s).

Unavailable names (infrasubspecific): *autumnalis*: SCHWINGENSCHUSS (1926); *fumata*: REZBANYAI-RESER (1987). – (incorrect subsequent spelling): *ruficosta*: CARNELUTTI (1992).

DIAGNOSIS: Wingspan ♂♀ 16-22 mm, in subsequent generations sometimes 12 mm only. Forewing broad. Ground colour very variable, glossy grey with brown or violet tinge, often with strong orange or red suffusion. Forewing costa yellowish or orange. Transverse lines indistinct, slightly convex. Discal spots small. Fringe dots very small or absent. Frons dark brown, vertex pale sand coloured, collar ochreous. Antennae of ♂ ciliate-setose, cilia short, 0.65-0.8 times width of flagellum, up to 1.0 times in the Levant. Hindtibia of ♂ dilated, exceeding first tarsomere, with large pencil. Tarsus short, 0.25-0.4 times length of tibia.

MALE GENITALIA: Uncus digitiform, comparatively narrow. Gnathos longer than uncus. Saccus V-shaped. Valva medially dilated. Aedeagus very large, with two long, but comparatively thin cornuti in the centre, 1-2 small cornuti distally and 3-5 smaller basal cornuti. Populations of Rhodos, Cyprus and the Levant usually with reduced number of basal cornuti.

FEMALE GENITALIA: Ductus bursae and narrow posterior part of corpus bursae fused to a long stalk, laterally spinulose in the anterior half. Corpus bursae small, rectangularly looped posteriorly, round anteriorly, spinulose.

DISTRIBUTION: East-Mediterranean. South-eastern France across Italy to Balkan peninsula. In the north to southern Switzerland. Present on Corsica, Sardinia, Sicily, Malta (SAMMUT 2000) and Greek islands. – Outside Europe from Turkey, Cyprus across the Levant to Egypt (WILTSHIRE 1949a). In the western Mediterranean replaced by the sister species *I. predotaria*.

PHENOLOGY: Bi-, tri- or plurivoltine. In the north bivoltine, late April to early July; early August to late September. In the south tri- or plurivoltine from early April to late October, in the Levant from early March to mid-November. At the localities of sympatrical occurrence both generations 2-4 weeks earlier than *I. predotaria*.

BIOLOGY: Larva polyphagous on herbaceous plants (REBEL 1910b), preferring flowers (PROUT 1913; FORSTER & WOHLFAHRT 1981; data possibly referring to the sister species). Reared on withered and decaying leaves of *Taraxacum officinale* (original data; REZBANYAI-RESER 1987), refusing fresh leaves. Larva strongly photophobic. Adults readily attracted to light. Strong fluctuations in abundance from year to year (PROUT 1913 for both species).

HABITAT: Xerothermophilous lowland species. Almost ubiquitous in Mediterranean habitats, such as maquis, stony, dry grassland, scattered oak forest, *Ficus-Olea* plantations, riverside vegetation and others. From 0 m up to 800 m above sea-level, up to 1,200 m in Sicily and Turkey.

SIMILAR SPECIES: *I. sardoniata* (see no. 15), *I. troglodytaria* (36), *I. belemiata* (51), *I. obsoletaria* (54), *I. eugeniata* (100). Sister species *I. predotaria* on average slightly smaller and more greyish, but identification usually impossible without examination of genitalia; identification of ♂ is easy by brushing tip of abdomen to remove scales and reveal shape of valva. Unicolorous, small forms of *I. degeneraria* usually differ in more distinct terminal line, ♂ hindtarsus shorter.

REMARKS: For possible glacial speciation mechanisms see REZBANYAI-RESER (1987). Larvae of *I. distinctaria* from Sardinia strikingly different from those of Balkan peninsula (LEIPNITZ pers. comm., adults dissected by the author).

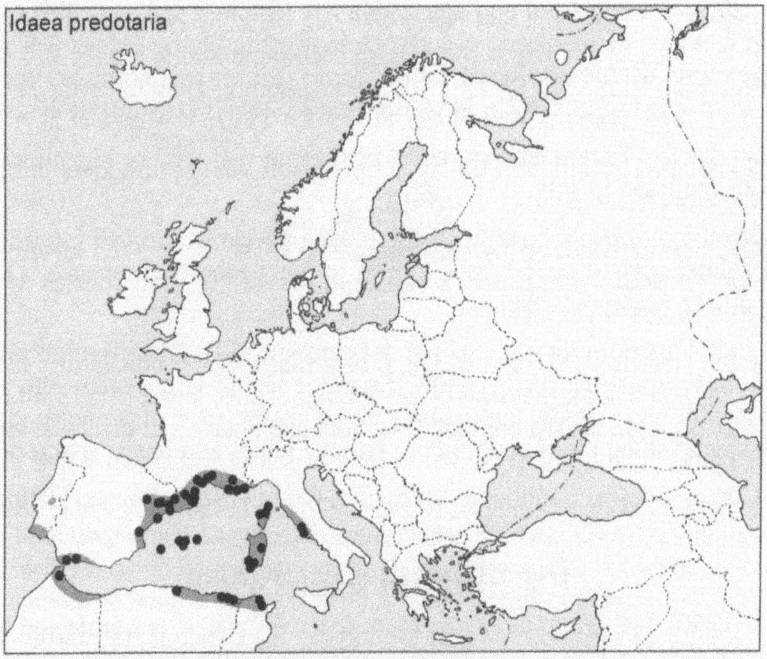

102. Idaea predotaria (HARTIG, 1951)

Sterrha incarnaria predotaria HARTIG, 1951: in HARTIG & AMSEL, Fragm. Ent. Roma 1 (1): 38 (Sardinia: Teulada ('Teudala')). Lectotype ♂ (MZUR).

Idaea griseanova REZBANYAI-RESER, 1987: Mitt. ent. Ges. Basel 37 (4): (141) 154, figs. 1a-1c, 4a-4c, pl. 5, 6, 9, 10 (Algeria: Bône). Holotype ♂ (ZFMK, examined).

Unavailable names (misidentification): *incarnaria*: sensu GUENÉE (1858) nec HERRICH-SCHÄFFER, 1852 (identity: *Idaea distinctaria*); *ruficostata*: sensu RIBBE (1912) nec ZELLER, 1849 (identity: *Idaea distinctaria*). – (infrasubspecific): *grisea*: THIERRY-MIEG (1889).

DIAGNOSIS: Wingspan ♂♀ 16-19 mm, in subsequent generations sometimes 12 mm only. Habitus as in *I. distinctaria*, but ground colour grey with slight violet tinge, less variable, usually less glossy and very rarely with orange or red tinge. Forewing costa usually sand coloured. Frons dark brown, vertex pale sand coloured, collar ochreous. Antennae of ♂ ciliate-setose, cilia 0.75-0.9 times width of flagellum. Hindtibia of ♂ dilated, exceeding first tarsomere, with large pencil. Tarsus short, 0.3-0.4 times length of tibia.

MALE GENITALIA: Uncus sub-triangular. Gnathos lingulate. Saccus broad. Valva spatulate, costa parallel to ventral margin. Aedeagus very large, with two stout distal cornuti, and 4-5 smaller, but stout basal cornuti.

FEMALE GENITALIA: Ductus bursae comparatively short and broad. Corpus bursae reniform, elongate, spinulose, posteriorly looped.

DISTRIBUTION: West-Mediterranean. From north-western Italy across southern France to eastern and southern Iberian peninsula. Present on Corsica, Sardinia, Balearics. Isolated occurrence in central Italy (Lazio: RAINERI 1989; REZBANYAI-RESER 1997). – Outside Europe from Morocco to Tunisia. In the eastern Mediterranean replaced by the sister species *I. distinctaria*. More rare and local than *I. distinctaria*.

PHENOLOGY: Usually bivoltine, mid-May to late June; mid-August to early October, in the south emerging from mid-April and flying until late October, collected at light also early January. At localities of sympatric occurrence both generations usually 2-4 weeks later than *I. distinctaria* (REZBANYAI-RESER 1997). Some authors mention early second brood specimens from July (MAZEL 1995; MAZEL & PESLIER 1997; DANTART et al. 1993).

BIOLOGY: No detailed information on larval host-plants with secure identification traced. Adults readily attracted to light.

HABITAT: Xerothermophilous species of the coastal lowlands. Limited information available about habitat preference, probably similar to that of the sister species. Usually from 0 m up to 300 m above sea-level.

SIMILAR SPECIES: *I. sardoniata* (see no. 15), *I. belemiata* (51), *I. obsoletaria* (54), *I. eugeniata* (100). Sister species *I. distinctaria* on average slightly larger often with reddish suffusion, but identification usually impossible without examination of genitalia; identification of ♂ is easy by brushing tip of abdomen to remove scales and reveal shape of valva.

REMARKS: Comprehensive treatments: REZBANYAI-RESER (1987); RAINERI (1989).

The *nitidata* species-group

STERNECK's (1940) 31st group of *Idaea*. Monotypic. *I. exilaria* is transferred to the *fractilineata* species-group, some extralimital species (*vindicata*, *culoti*, *paroranaria*) to the genus *Brachyglossina* (see there). Some genitalic characters of *I. nitidata* (shape of juxta, valva, saccus, paired terminal cornuti) reminiscent of *cervantaria* species-group. Habitus, shape of juxta and valva similar to *I. deversaria*. Proboscis present. Antennomeres dentate, flagellum very broad in lateral view. Forewing with one radial vein (R2) lacking, R1 stalked with R3-R5, sometimes connate with R5. Sternum A2 of ♂ with distinct horseshoe-shaped sclerite and setose patch. Larva short, thick, with smooth skin.

103. Idaea nitidata (HERRICH-SCHÄFFER, 1861)

Acidalia nitidata HERRICH-SCHÄFFER, 1861: Neu. Schmett. Eur. 3: 28 [24], fig. 142 (Romania: Mehadia). Syntypes, at least 3♂1♀ (MNHU: 3♂1♀, examined).

‡ *Acidalia tectaria* LEECH 1897: Ann. Mag. nat. Hist. (6) 20: 102 (central China: Chang-yang; Korea: Gensan). Syntypes 12 (BMNH). Junior primary homonym nec WALKER, 1866 (identity: *Scopula lactaria* WALKER, 1861).

‡ *Acidalia nitidulata* STAUDINGER, 1897: Dt. ent. Z. Iris 10: 16 (eastern Russia: Amur region). Introduced in synonymy, junior subjective homonym nec HERRICH-SCHÄFFER, 1839 (identity unclear, but congeneric).

Unavailable names (infrasubspecific): *anastomosaria*: REISSER (1961); *uniformis*: REISSER (1961).

DIAGNOSIS: Wingspan ♂♀ 22-25 mm, in northern Italy and Far East of Asia sometimes 20 mm only. Wings strongly rounded. Ground colour glossy sand coloured with yellow tinge. All transverse lines dark sand coloured, undulate. Discal spots and fringe dots absent. Frons dark brown, vertex whitish to sand coloured, collar slightly darker. An-

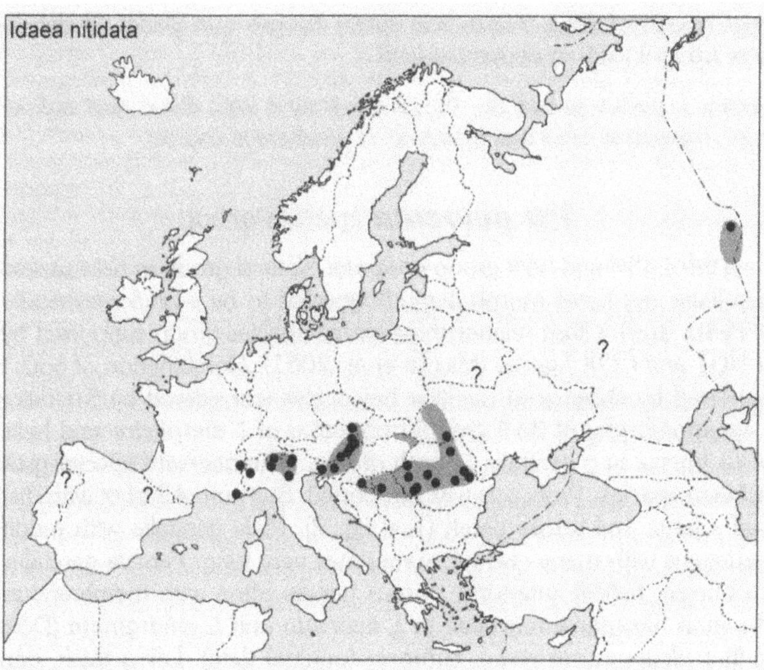

tennae of ♂ with very broad flagellum in lateral view, dentate, ciliate-setose, cilia 0.5-0.6 times width of flagellum in lateral view, up to 1.0 in dorsal view. Hindtibia of ♂ stout but not dilated, with large pencil. Tarsus 0.45-0.5 times length of tibia.

MALE GENITALIA: Uncus sub-triangular. Gnathos broad. Saccus large, V-shaped. Juxta narrow, posteriorly elongate. Valva broad, with short digitiform subapical process on costa. Aedeagus long, with central narrow cornutus and two small hook-shaped cornuti at tip.

FEMALE GENITALIA: Ductus bursae divided into two dilated, sclerotized bulbs. Corpus bursae posteriorly with long row of spinules and lateral membranous appendix, leading to ductus seminalis. Anteriorly globular, spinulose.

DISTRIBUTION: Eurasiatic, disjunct. Rare and local from southern Alps across south-eastern Austria, to northern Balkan peninsula. Very isolated populations in south-western Alps and in central Urals. Data from Byelorussia and southern European Russia (MERZHEEVSAYA et al. 1976; VIIDALEPP 1996) await confirmation. Reputedly in the northern Iberian peninsula (e.g. VIVES MORENO 1994, MÜLLER 1996), data probably based on confusion with *I. deversaria*. – Outside Europe in western Siberia and Sajan mountains, Far East of Asia (Dahuria, Amur and Ussuri regions, Korea). Records from Turkey (STAUDINGER 1859; RIEMIS 1994; VIIDALEPP 1996) and Transcaucasus (PROUT 1913) probably misidentified.

PHENOLOGY: Univoltine: mid-June to late August.

BIOLOGY: Larva polyphagous on herbaceous plants (PROUT 1913, FORSTER & WOHLFAHRT 1981). Reared on *Lactuca sativa* (PROUT 1913; VORBRODT 1930), on petals of various plant species and on leaves of *Taraxacum officinale* (MÜLLER pers. comm.). Adults easily flushed up from vegetation (shrubs and grass) during daytime (HERBULOT 1945), also at light (e.g. KITSCHELT 1925). ♀ inert, ♀ ratio in collections low, about 10-20%.

HABITAT: Xerothermophilous. Warm and sunny fringes and glades of deciduous forest. From 200 m up to 1,000 m above sea-level.

SIMILAR SPECIES: *I. pallidata* (see no. 81). *I. deversaria* with discal spot and terminal dots more distinct, transverse lines less undulate, ♂ hindtarsus shorter.

The *aversata* species-group

STERNECK's (1940) 33rd and 34th group of *Idaea*. Natural group in habitus, external morphology, genitalia and larval morphology. Suggested to be a phylogenetically old group (STERNECK 1940: 168). Close relationships within species-group supported by molecular analysis of ND1 and COX I genes (MILLER et al. 2001). Combination of both STERNECK's groups supported by absence of peculiar hair-scales (considered by STERNECK the most important synapomorphy of 33rd group) from valva of *I. simplicior* and helical anterior part of ductus bursae in ♀ genitalia of both groups. The *aversata* species-group includes the largest *Idaea* species. Proboscis well developed. Sternum A2 of ♂ with distinct horseshoe-shaped sclerite and setose patch (Text-fig. 2). Male genitalia with modifications of gnathos, aedeagus with many cornuti, sometimes very long. Female genitalia with elongate ductus bursae, helical anteriorly, corpus bursae often with membranous appendix. Egg sculpturation hexagonal reticulate in *I. aversata* and *I. emarginata* (DÖRING 1955), granulated in *I. degeneraria* and *I. rubraria* (original data). Larva thick, comparatively long, that of *I. emarginata* shorter.

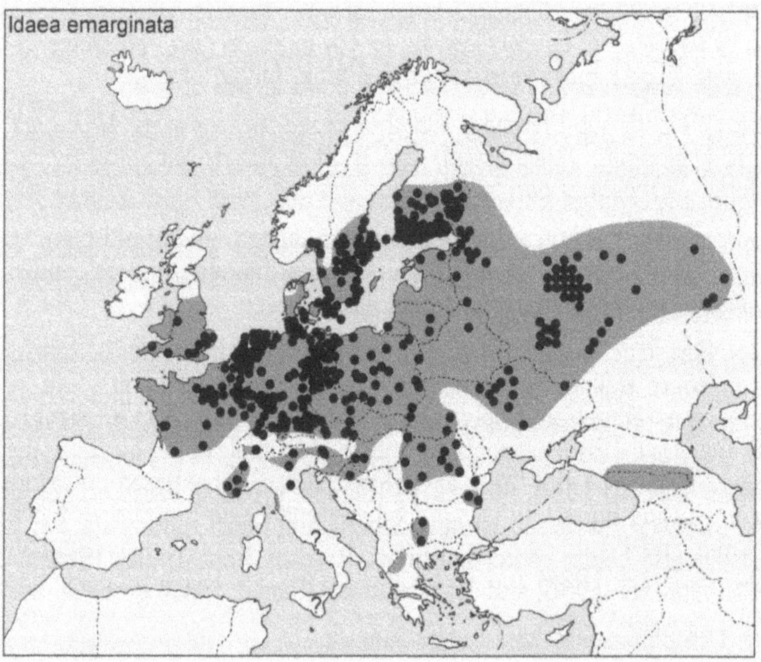

104. *Idaea emarginata* (LINNAEUS, 1758)

Phalaena Geometra emarginata LINNAEUS, 1758: Syst. Nat. (Ed. 10) 1: 524 (Europe). Syntype(s) (LSL). As 'habitat' (type locality?) Stockholm is stated in LINNAEUS 1761.

Phalaena erosata HUFNAGEL, 1767: Berlin. Mag. 4 (5): 526 (Germany: Berlin). Syntype(s) lost.

Phalaena demandata FABRICIUS, 1775: Syst. Ent.: 630 (Germany: Leipzig ('Lipsiae')). Syntype(s).

‡ *Phalaena marginata* FABRICIUS, 1787: Mant. Ins. 2: 200. Incorrect subsequent spelling (for *emarginata*) and junior primary homonym nec LINNAEUS, 1758 (identity: *Lomaspilis marginata*, Ennominae).

Phalaena Geometra aresiaria VILLERS, 1789: Linn. ent. 2: 303 (France, Bresse near Lyon ('Bressia')). Holotype with antennae accidently broken.

‡ *Geometra emarginaria* HÜBNER, 1799: Samml. Eur. Schmett. 5, Geometrae (1): pl. 20, fig 107. Incorrect subsequent spelling.

Phalaena Geometra demandaria ESPER, 1801: Die Schmett. in Abbild. 5 (4): 104, pl. 20, figs 4-6. Emendation.

Phalaena rumigerata DONOVAN, 1810: Nat. Hist. Brit. Ins. 14: 67, pl. 493, fig. 2 (United Kingdom). Syntype(s).

‡ *Phalaena quadripunctata* DONOVAN, 1810: Nat. Hist. Brit. Ins. 14: 68, pl. 493, fig. 3 (United Kingdom). Syntype(s). Junior primary homonym nec GOEZE, 1781 (identity: *Aspilates gilvaria*, Ennominae).

Acidalia emarginata var. *mosquensis* HEYNE, 1899: Soc. Ent. 14: 105. (Russia: by Moscow), Syntype(s).

Unavailable names (misidentification): ?*dimidiata*: sensu HAWORTH (1809) nec HUFNAGEL, 1767. – (infrasubspecific): *brunnescens*: LEMPKE (1949); *obsoleta*: LEMPKE (1949); *pallida*: LEMPKE (1949). – (incorrect subsequent spelling): *damandataria*: ESPER (1801: pl. 20).

DIAGNOSIS: Wingspan ♂♀ 19-23 mm. ♀ on average slightly smaller, but more robust-bodied. Forewing termen distinctly concave below apex, somewhat deeper in ♀. Hindwing termen arched, with prominent teeth at M3 and CuA1. Wings sand coloured to light brown, in ♀ usually light brown with orange tinge, with slight tendency to melanism. Darker forms induced by humidity and cold temperatures (BERGMANN 1955). Ante- and postmedial lines brown, fine, regularly convex. Medial line usually absent in ♂, dark brown, broad but diffuse in ♀. Discal spots distinct. Terminal line brown, uninterrupted. Fringe slightly chequered. Frons brown, vertex whitish, collar slightly darker. Antennae of ♂ not dentate, ciliate-setose, cilia 0.9-1.2 times width of flagellum (Text-fig. 143). Hindtibia of ♂ comparatively slender, with large pencil. Tarsus 0.5-0.6 times length of tibia.

MALE GENITALIA: Uncus digitiform. Gnathos broad. Saccus large, prominent. Valva slightly sigmoid, strongly covered with hair-scales in the basal half. Aedeagus broad, with 8-10 small cornuti.

FEMALE GENITALIA: Apophyses comparatively long. Ductus bursae broad, antrum dilated, laterally tapering. Corpus bursae as long as ductus bursae, globular, scattered spinulose, laterally bulbed at junction with ductus bursae.

DISTRIBUTION: European-Westasiatic. Widely distributed in temperate Europe, from England to the Urals, in the north to southern Scandinavia. Isolated occurrences in Balkan peninsula, and in the Po plain (Modenese: FLAMIGNI & BASTIA 1998). Records from central Italy and Sicily (CURÒ 1878, RAINERI & ZANGHERI 1995; MÜLLER 1996) probably misidentified, for data from central Italy see PROLA & RACHELI (1979). In the Iberian peninsula replaced by the two sister species *I. dromikos* and *I. simplicior*. Usually rare, in northern Germany commoner. – Outside Europe in Caucasus, Turkmeniya, western Siberia and Altai mountains (VIIDALEPP 1996). In Morocco replaced by *I. simplicior*.

PHENOLOGY: Usually univoltine: late June to mid-August. Exceptionally emerging from mid-May. Under good conditions single specimens of a partial second generation until late September. Larvae slowly growing (BUCKLER 1897).

BIOLOGY: Larva polyphagous on various low plants, possibly preferring *Convolvulus* (Convolvulaceae) (var. cit.) or fallen leaves of *Betula* (Betulaceae) (BOLDT 1929). Recorded also on leaves of deciduous shrubs, such as *Alnus* (Betulaceae), *Corylus* (Corylaceae),

Rubus (Rosaceae), and on low plants such as *Galium* (Rubiaceae), *Plantago* (Plantaginaceae), *Sedum* (Crassulaceae), *Cytisus scoparius* (Fabaceae), *Polygonum, Rumex acetosella* (Polygonaceae) (var. cit., e.g. LHOMME 1935; ALLAN 1949; BERGMANN 1955; FORSTER & WOHLFAHRT 1981; SKINNER 1984; SKOU 1986; EBERT 2001), old, fallen leaves and grass (BERGMANN 1955). Reared on withered leaves of *Taraxacum officinale, Polygonum aviculare, Rumex, Senecio vulgaris*, and *Lactuca sativa* (BUCKLER 1897; BERGMANN 1955; SKINNER 1984; PORTER 1997; MÜLLER pers. comm.). Adults (mainly ♂♂) easily flushed up from grass in the morning (BERGMANN 1955), actively flying at late dusk and in the early night, occasionally attracted to light and bait. Protandrous, females emerging two weeks after males (BERGMANN 1955).

HABITAT: Meso- to hygrophilous, silvicolous. Preferring warm sites in wetlands (FORSTER & WOHLFAHRT 1981; BERGMANN 1955, SKOU 1986), wet meadows in scrubby places, fenland, fringes and aisles of deciduous damp woodland with much undergrowing vegetation, rich in fallen leaves. Most frequently observed in small birch forests at the border of degenerated bogs with abundance of *Molinia caerulea* (WEGNER pers. comm.). Exceptionally on dry sandy soils (EBERT 2001). From 0 m up to 600 m above sea-level, exceptionally higher, e.g. up to 1,300 m in southern Alps (Piemonte: FLAMIGNI 1999).

SIMILAR SPECIES: Wing shape unique. Allopatric sister species *I. simplicior* with wings more rounded, ground colour slightly paler, fringe unchequered, tarsus slightly shorter. Allopatric sister species *I. dromikos* somewhat darker, with orange tinge, transverse lines angled at costa, fringe unchequered, ♂ antennae dentate. Correct identification of ♂ is easy by brushing of tip of abdomen to remove scales and reveal shape of valva.

REMARKS: For rearings and morphology of immature stages see BUCKLER (1897).

105. *Idaea dromikos* sp. n.

Unavailable names (misidentification): *emarginata*: sensu LHOMME 1935 (partim) nec LINNAEUS, 1758.

HOLOTYPE: ♂, [southern France, Aude], 'Carcass[on], coll. P. MABILLE', ZSM/HERBULOT.

PARATYPES: 1♂, 'Pyr. or., Vernet les Bains, 1.-17.VIII.1951, leg. GROSS', EMEM in ZSM. 1♂, 'Vernet-les-Bains, Pyrénées Or.les, Süd-Frankreich' 'VB 12.VII.[18]94' 'ex coll. L. VIARD an R. OBERTHÜR 1922', ZFMK; 2♂1♀ 'Pyren. orient., Sorède, e.l. 3.II., 2.III., 15.III., 10.VII.[19]09, M. SIELMANN, Arnstadt', ZFMK; 1♀, id., ZSM/HERBULOT; 4♂5♀, southern Spain 'Malaga, Algatocin, 30 km SW Ronda, 865 m, 28.VII.1998, e.o. 1.X.1998', leg. et coll. N. PÖLL; 1♂, id., ZSM/HERBULOT; 2♂5♀, eastern Spain 'Teruel, Olalla, 1,100 m, 30.VII.2000', leg. et coll. V. REDONDO; 2♂3♀, id., leg. et coll. J. GASTÓN; 1♂, central Spain, 'Avila, Puerto Castilla, 1,185 m, 10.VIII.2001', leg. et coll. J. GASTÓN; 1♂, northern Spain, 'Burgos, Obarenes, 715 m, 18.VII.2003', leg. et coll. J. GASTÓN; 1♂, northern Spain, 'Burgos, Llorengoz, 850 m, 7.VIII.2003', leg. et coll. J. GASTÓN; 1♀, northern Spain 'Zaragoza, Monasterio de Veruela, 750 m, 9.VIII.2000', leg. et coll. V. REDONDO; 1♂, northern Spain 'Zaragoza, Sierra de Moncayo, 1,100 m, 9.VIII.2000', leg. et coll. V. REDONDO; 1♂, northern Spain 'la Rioja, Villoslada, Sierra Cebollera, 1,000 m, 16.VII.2001', leg. J.L. SONCHEZ, coll. V. REDONDO; 1♀, ,Spain, prov. Teruel, 5km NW of Montalban, 950m, 17.VII.2003, BJARNE SKULE leg.', coll. P. SKOU.

ETYMOLOGY: The name is drawn from Greek language, it means 'fast running', and is dedicated to the grand master of geometridology, Claude HERBULOT, Paris, who was once French university champion in 110 m hurdle race.

DIAGNOSIS: Wingspan ♂♀ 19-22 mm. Forewing termen concave below apex. Hindwing termen arched, with prominent teeth at M3 and CuA1. Ground colour light brown. Ante- and postmedial lines brown, sometimes indistinct, angled at costa. Medial fascia diffuse. Discal spots distinct. Terminal line brown, uninterrupted. Fringe unchequered. Frons red

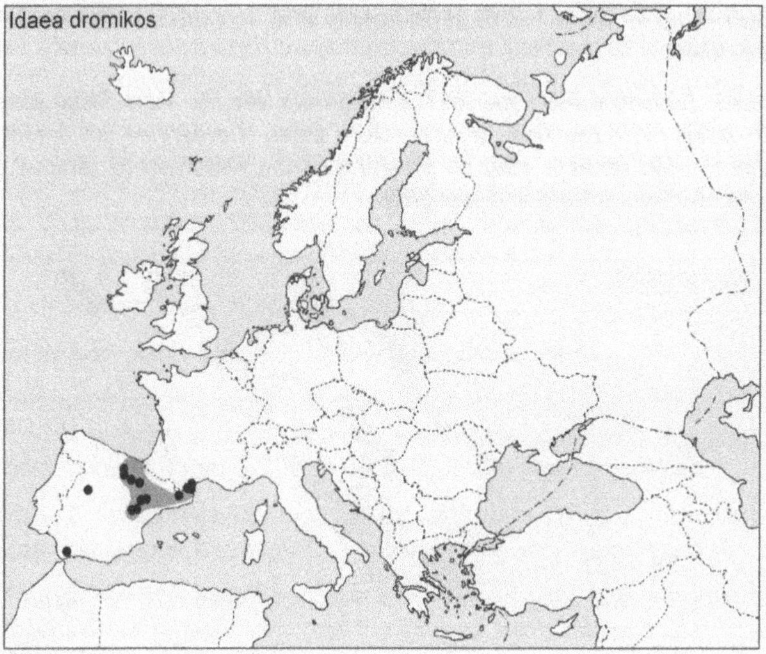

brown, vertex sand coloured, collar slightly darker. Antennae of ♂ with broad flagellum in lateral view, dentate, with deep intersegmental incisions, ciliate-setose, cilia 0.8-1.0 times width of flagellum. Hindtibia of ♂ comparatively slender, with large pencil. Tarsus 0.45-0.55 times length of tibia.

Male genitalia: Uncus triangular, broad at base, tapered at tip. Gnathos cleft into two long and stout spines. Saccus slender, prominent. Valva long, medially narrow, broad and with costal spine at tip. Basal hair-scales absent. Aedeagus long and slender, with 5-8 cornuti.

Female genitalia: Apophyses comparatively long. Ductus bursae broad, antrum much dilated, laterally rounded. Corpus bursae as long as ductus bursae, elongate, strongly spinose, laterally bulbed at junction with ductus bursae.

Distribution: West-Mediterranean (allopatric vicariant of *I. emarginata*). Endemic to southernmost France (prov. Aude; eastern Pyrenees) and northern half of Spain, with isolated occurrence in southern Spain, prov. Malaga (Pöll pers. comm.; Redondo et al. in prep.). Records for Portugal (‚*I. emarginata*': e.g. Vives Moreno 1994; Müller 1996) refer to the next species (Corley pers. comm.).

Phenology: Univoltine: mid-July to mid-August (original data, Pöll pers. comm., Redondo et al. in prep.).

Biology: Larva possibly polyphagous. Reared on dry leaves of *Taraxacum officinale* (Pöll pers. comm.). Adults collected at light.

Habitat: Limited data available on habitat preference. In southern Spain near cultivated areas (Pöll pers. comm.). Collected in lowland, hills and lower mountainous regions, usually from 500 m up to about 1,200 m (REDONDO et al. in prep.), in the north of the distribution also lower.

SIMILAR SPECIES: Allopatric sister species *I. emarginata* (see no. 104). Sister species *I. simplicior* with wings more rounded, ground colour paler, ♂ antennae not dentate. Correct identification of both sexes is easy by brushing of the abdomen to remove scales and reveal shapes of valva, antrum and sterigma.

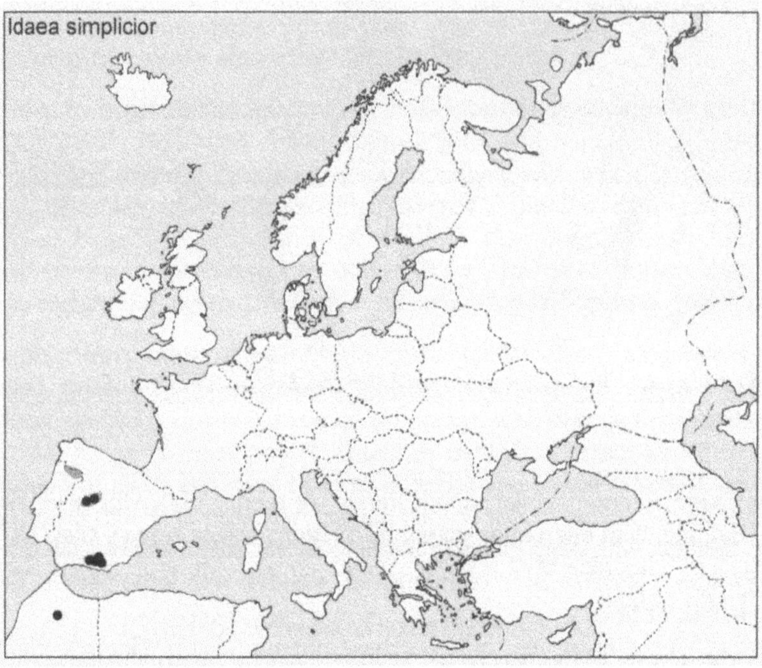

106. Idaea simplicior (PROUT, 1934), stat. n.

‡ *Ptychopoda emarginata* f. *simplicior* WEHRLI, 1927: Int. Ent. Zeits. 20 (37): 342 (Spain, Andalusia: Sierra Nevada). Holotype ♂ (ZFMK, examined). Infrasubspecific (only 'var.' means subspecific rank in WEHRLI's terminology).

Sterrha emarginata var. *simplicior* PROUT, 1934: Lepid. Cat. 63: 418. Raised by PROUT (1934) to available rank. No separate types. Herewith raised from subspecific rank (PROUT 1934) and synonymy (VIVES MORENO 1994) of *I. emarginata* to species rank (stat. n.).

Unavailable names (misidentification): *emarginata*: sensu ZERNY (1936) nec LINNAEUS, 1758.

DIAGNOSIS: Wingspan ♂♀ 19-22 mm. Wing pattern and external morphology as in *I. emarginata*, but wing margins rounded, forewing straight or slightly concave below apex, coloration slightly paler than in *I. emarginata*, medial fascia often orange rather than grey, fringe unchequered. Antennae of ♂ not dentate, ciliate-setose, cilia 0.9-1.1 times width of flagellum. Tarsus slightly shorter than in *I. emarginata*, 0.45-0.55 times length of tibia.

MALE GENITALIA: Uncus narrow, digitiform. Gnathos narrow, spatulate. Saccus developed as narrow clasp, anteriorly shallow. Juxta with two small spinulose crests, reminiscent of equivalents in *I. degeneraria*. Valva slender, broad at base, without hair-scales in the basal half. Aedeagus long and slender, with 8-12 small cornuti.

FEMALE GENITALIA: Apophyses shorter than in *I. emarginata* and *I. dromikos*. Ductus bursae narrower than equivalent in *I. emarginata*, antrum less tapered laterally. Corpus bursae half length of ductus bursae, globular, slightly spinulose, laterally bulbed at junction with ductus bursae.

DISTRIBUTION: West-Mediterranean (allopatric vicariant of *I. emarginata*). Higher mountains of central and southern Spain (ZSM, examined). Records for Portugal (‚*I. emarginata*': e.g. VIVES MORENO 1994; MÜLLER 1996) refer to this species (CORLEY pers. comm.). – Outside Europe in Morocco (Rif and Middle Atlas).

PHENOLOGY: Univoltine: mid-July to mid-August.

BIOLOGY: Larval host-plants unknown. Adults attracted to light.

HABITAT: Montane. Inhabiting rocky and stony southern slopes of the Sierra Nevada, with abundance of low-growing aromatic flowers, and in pine forest with rich ground vegetation (PÖLL pers. comm.). From 1,400 m up to 2,000 m above sea-level.

SIMILAR SPECIES: *I. emarginata* (see no. 104), *I. dromikos* (105). *I. rubraria* with round hindwing, forewing termen less angled, ♂ antennal cilia and ♂ hindtarsus shorter.

REMARKS: Though already stated by STERNECK (1940: 166), the genitalic differences remained unconsidered in taxonomy and the species status unrecognized. Usually cited as synonym (e.g. VIVES MORENO 1994).

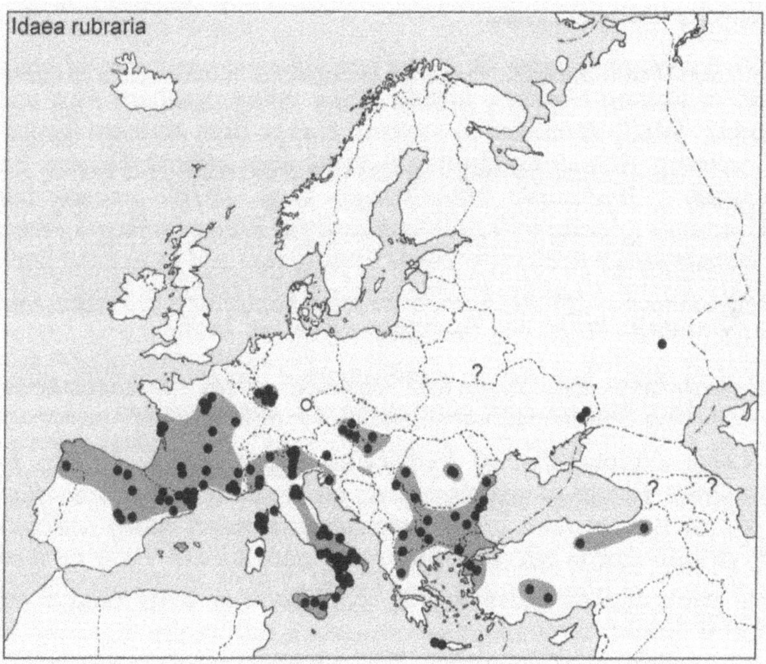

107. *Idaea rubraria* (STAUDINGER, 1901)

Acidalia rubraria STAUDINGER, 1901: in STAUDINGER & REBEL: Cat. Lepid. palaearct. Faunengeb. (Edn 3) 1: 272. First available name after *rubraria* STAUDINGER, 1871 (in STAUDINGER & WOCKE, Cat. Lepid. eur. Faunengeb. (Edn 2) 1: 149: infrasubspecific) (Croatia: Split, 'Spal[ato]'). Lectotype ♀ (MNHU, examined; with 'gen.prp. Urbahn 1/30'), herewith designated to stabilize usage of the name and to fix type locality; at MNHU 1♀ paralectotype from Nassau, western Germany. For validity see under *A. bilinearia*.

Acidalia bilinearia A. FUCHS, 1889: Jb. Nassau. Ver. Naturk. 42: 211. First available name after *bilinearia* A. FUCHS, 1878 (Stettin. ent. Ztg. 39: 331: infrasubspecific) (western Germany: Rheingau). Syntypes. Valid species name according to LERAUT (1997), SCOBLE (1999) and EBERT (2001), rejected here, as the name had been forgotten and never used for about 100 years. Even the son of the original author F. FUCHS (1904: 18) accepted *rubraria* as valid name.

Unavailable names (misidentification): *degeneraria*: sensu HERRICH-SCHÄFFER (1847) nec HÜBNER, 1799. – (infrasubspecific): *ritteraria*: ANDREAS (1929); *therinaria*: FUCHS (1904).

DIAGNOSIS: Wingspan ♂♀ 21-26 mm. Wings sand coloured, in ♀ often with slight reddish or violet tinge. Ante- and postmedial lines fine, light brown. Medial line usually broad, but diffuse, red brown or dark brown. Postmedial line distinctly angled below costa. Discal spots large. Terminal line fine, brown, uninterrupted. Fringe dots absent. Frons dark brown, vertex pale ochreous, collar slightly darker. Antennae of ♂ with small intersegmental incisions, ciliate-setose, cilia short 0.65-0.9 times width of flagellum. Hindtibia of ♂ large, dilated, with large pencil from base of tibia, tibial scales covering half of tarsus. Tarsus 0.30-0.35 times length of tibia.

MALE GENITALIA: Uncus digitiform, truncate. Gnathos rounded, usually narrow at base. Valva comparatively short and broad, with small spinulose crest at tip, tapered dorsally. Aedeagus with 5-7 small and two slightly larger, terminal cornuti.

FEMALE GENITALIA: Ductus bursae narrow, curved, at antrum somewhat dilated asymmetrically. Corpus bursae elongate, strongly spinulose, helical towards ductus bursae, with lateral sac-shaped or helical appendix.

DISTRIBUTION: Sub-Mediterranean. Scattered populations in warm sites of central Europe, in the north to western Germany (middle Rhine valley), southern Alps and southern Czech Republic. Widely distributed in southern Europe from northern Iberian peninsula to Balkan peninsula. Present on Mallorca (GOATER pers. comm.), Corsica, Sardinia and Sicily. Reputedly in Byelorussia (MERZHEEVSAYA et al. 1976), probably misidentified. Isolated occurrences in eastern Ukraine and southern European Russia (Volsk/Saratov). Extinct in Bavaria (WOLF & HACKER 2004). Usually rare and in isolated populations, in southern Italy common. – Outside Europe scattered populations in Turkey, reputedly also in Georgia (DIDMANIDZE 2002) and Azerbaijan (VIIDALEPP 1996).

PHENOLOGY: Univoltine: late July to mid-September, single specimens emerging from mid-June, thus flying between both generations of *I. degeneraria*.

BIOLOGY: Larva polyphagous on herbaceous plants (REBEL 1910b; FORSTER & WOHLFAHRT 1981). Reared on withered leaves and petals of *Taraxacum officinale* (original data; MÜLLER pers. comm.). In the first three larval stages clearly preferring petals of *Pimpinella anisum* though offered a wide set of plant species. Later on readily accepting withered leaves of *Rubus idaeus*, and *Convolvulus arvensis* (original data). Adults readily attracted to light.

HABITAT: In central Europe xerothermophilous at the warmest localities. In southern Europe preferring temperate microclimates in valleys and on mountains. From 0 m up to 1,000 m above sea-level, in southern Italy and Turkey up to 1,700 m.

SIMILAR SPECIES: *I. simplicior* (see no. 106). *I. degeneraria* with smaller discal spots, postmedial line more diffuse, usually less angled below costa, often with broad medial fascia,

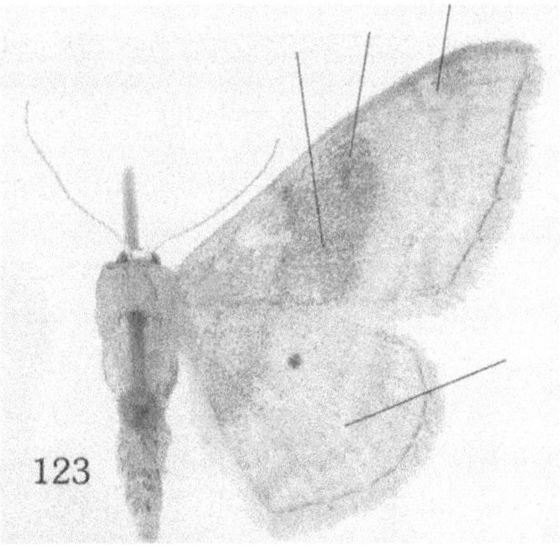

Text-figs 122-123: Diagnostic characters (indicated) of European *Idaea* species. Text-fig. 122: *I. rubraria* (STAUDINGER, 1901). Text-fig. 123: *I. degeneraria* (HÜBNER, 1799).

♂ hindtarsus shorter. Furthermore on average slightly smaller, less robust, coloration often with greenish tinge. Correct identification is easy in both sexes by brushing of the abdomen to remove scales and reveal shape of valva, uncus and antrum (Text-figs 131, 137). *I. aversata* on average larger and with less reddish tinge, medial line sharper, fringe dots present, discal spots smaller.

REMARKS: Description of immature stages in FUCHS (1889) and MAY (1893).

108. Idaea aversata (LINNAEUS, 1758)

Phalaena Geometra aversata LINNAEUS, 1758: Syst. Nat. (Ed. 10) 1: 526 (Finland). Syntype(s) (LSL).
Phalaena Geometra remutata LINNAEUS, 1758: Syst. Nat. (Ed. 10) 1: 528 (Europe). Syntype(s) (LSL).
Phalaena lividata CLERCK, 1759: Icones Ins. Rar. 1, pl. 5, fig. 10 (Sweden). Syntype(s) probably lost.
Phalaena trilineata HUFNAGEL, 1767: Berlin. Mag. 4 (6): 606 (Germany: Berlin). Syntype(s) lost. (=f. *remutata*).
Phalaena grisata FABRICIUS, 1776: Genera Insect.: 288 (northern Germany; Hamburg). Syntype(s).
‡ *Phalaena murinata* FABRICIUS, 1794: Ent. Syst. 3 (2): 198 (England). Syntype(s) lost. (=f. *remutata*). Junior primary homonym nec SCOPOLI, 1763 (identity: *Minoa murinata*, Larentiinae).
‡ *Geometra aversaria* HÜBNER, 1799: Samml. Eur. Schmett. 5, Geometrae (1): pl. 11, fig. 56. (Europe). Incorrect subsequent spelling (see introduction).
‡ *Phalaena fuliginata* HAWORTH, 1809: Lepid. Brit. (2): 350 (England). Syntype(s). Junior primary homonym nec HUFNAGEL, 1767 (identity: *Triphosa dubitata*, Larentiinae).
Idaea aversaria var. *latifasciaria* HEYDENREICH, 1851: Lepid Eur. cat. meth. (Edn 3): 60. Introduced to replace '*remutata* sensu BORKHAUSEN nec *remutaria* HÜBNER'.
Sterrha aversata indeviata PROUT, 1929: Novit. zool. 35 (2): 142 (Algeria: Hammam Rirha; Tunisia: Tunis, Ain Draham). Syntypes, at least 3♂. Valid at subspecific rank (SCOBLE 1999).
Ptychopoda aversata var. *griseocorsa* SCHAWERDA, 1929: Zeits. österr. Ent.Ver. 14 (3): 29 (Corsica: Col de Bavella, Col de Vergio, Col de Vizzavona). Syntypes. Herewith downgraded from subspecific rank (SCOBLE 1999) to synonymy (syn. n.).
Sterrha aversata japonica INOUE, 1955: Tinea 2 (1/2): 75, pl. 6, fig. 2 (Japan: Hokkaido, Kushiro, Kawaya). Holotype ♂ (BMNH/INOUE). Valid at subspecific rank (SCOBLE 1999).
Unavailable names (misidentification): ?*barbalis*: sensu GUÉRIN (1818) nec CLERCK, 1759 (identity: *Pechipogo strigilata*, Noctuidae). – (infrasubspecific): *atrata*: FUCHS (1904); *amoenata*: FUCHS (1901); *approximata*:

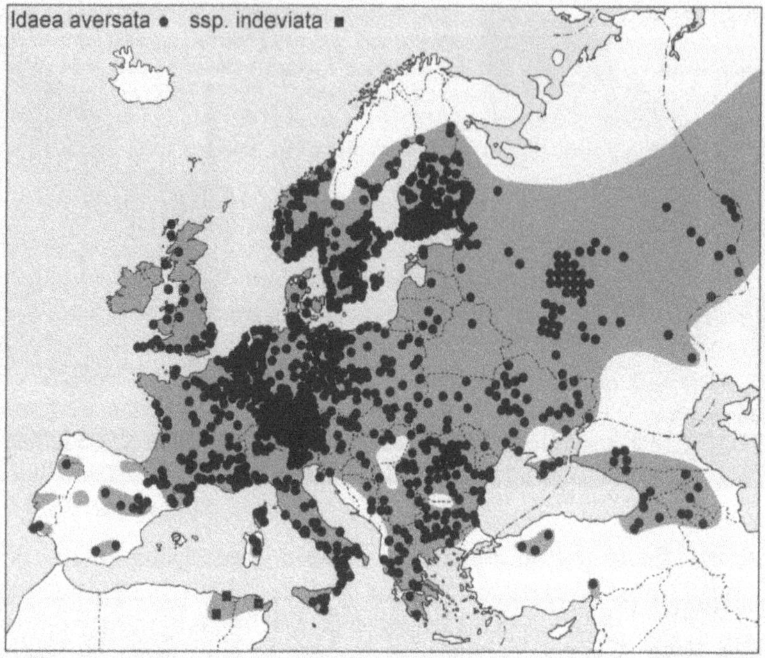

COCKAYNE (1952); *aurata*: FUCHS (1900); *auratadiluta*: HANNEMANN (1917); *aureospoliata*: BOLDT (1925); *demarginata*: REISSER (1961); *diluta*: HANNEMANN (1917); *dilutata*: PREISSECKER (1923); *effuscata*: GALVAGNI (1903); *fuchsi*: DUFRANE (1931); *impuncta*: LEMPKE (1949); *mediofasciata*: SCHWINGENSCHUSS (1953); *nigrolineata*: REISSER (1961); *pseudaurata*: SCHAWERDA (1933); *quadrilineata*: COCKAYNE (1952); *spoliata*: STAUDINGER (1870); *suaveolaria*: FUCHS (1902) (replacement name for *amoenata*); *suffumata*: LAMBILLON (1909); *suffusa*: LEMPKE (1949); *tenuifasciata*: LEMPKE (1949); *unilineata*: LEMPKE (1949). – (nomen nudum): *bifasciata*: HEYDENREICH (1846). – (incorrect subsequent spellings): *adversaria*: GLASER (1857); *adversata*: SCHRECKENSTEIN (1800); *auersata*: FABRICIUS (1787); *aurodiluta*: PROUT (1935); *latefasciata*: WEHRLI in VORBRODT & MÜLLER-RUTZ (1913); *latefesciata*: WEHRLI (1917); *spoiliata*: ROBSON (1902).

DIAGNOSIS: Wingspan ♂♀ 23-31 mm, in the second generation sometimes 20 mm only. Wings sand coloured, sometimes with orange tinge ('f. *aurata*'; 'f. *aureospoliata*'), with slight tendency to melanism (URBAHN 1972). Dichroic with medial area between medial and postmedial line dark brown or concolorous with ground colour ('f. *remutata*'), the latter usually much commoner. Transverse lines sharp, dark brown. Medial line usually distal to discal spot, rarely crossing it. Postmedial line usually distinctly angled below costa. Terminal area of ground colour. Discal spots small. Small fringe dots and terminal line present. Frons dark brown, vertex whitish, collar slightly darker. Antennae of ♂ ciliate-setose, cilia 0.9-1.2 times width of flagellum (Text-fig. 144). Hindtibia of ♂ large, dilated, with large pencil from base of tibia, tibial scales covering tarsus. Tarsus 0.35-0.45 times length of tibia (Text-fig. 128).

Variability: Forms without dark fascia ('f. *remutata*'), recessively inherited (BERGMANN 1938; 1955), but usually common at a frequency of 60-80%, in Great Britain 50%, nominate form slightly commoner in cold and mountainous sites (BERGMANN 1955). For breeding experiments see PROUT (1938a: 229) and ADKIN (1903). Coloration with orange tinge recessively inherited (BERGMANN 1955). For results from experimental rearing see BERGMANN (1938).

Corsican populations ('*griseocorsa*') on average smaller and with slightly stronger dark suffusion, nominate form very rare, but with clinous transitions towards normal habitus in Corsican lowlands.

MALE GENITALIA: Uncus curved, broad. Gnathos divided into paired, strongly sclerotized, round, spinulose processes. Saccus shallow. Valva tapered, with small spinulose crest at tip. Aedeagus with 5-6 comparatively stout cornuti. Sister species *I. gelbrechti* with valva bent at tip, not spinulose, gnathos processi very large.

FEMALE GENITALIA: Lamella postvaginalis with semicircular sclerotization. Ductus bursae narrow, straight or slightly curved, at antrum dilated. Corpus bursae oval, strongly spinulose, looped at junction with ductus bursae. *I. gelbrechti* with antrum much more dilated.

DISTRIBUTION: Palaearctic. Widely distributed and common all over Europe, in the north to central Scandinavia, in the east to the Urals. Present on Corsica, Sardinia and Sicily. Reputedly also on Mallorca awaiting confirmation. Usually common. – Outside Europe from Algeria to Tunisia (subsp. *indeviata*), from northern Turkey, Caucasus, Transcaucasus, across northern Iran to western Siberia and central Asian mountains, in the east to Japan (subsp. *japonica*). In the south a very isolated occurrence in southernmost Turkey (DOGANLAR 2003). In Morocco (Middle Atlas) replaced by *I. gelbrechti* HAUSMANN, 2003.

PHENOLOGY: Uni- bi- or trivoltine. Univoltine in the north and on the mountains of central Europe, usually above 600 m above sea-level,: mid-June to mid-August. North of the Alps under good conditions, in the south regularly bivoltine, from mid-May to late September, generations overlapping or with a gap in the second half of July, second generation often less abundant. Exceptionally with a partial third generation, until late October. Subsp. *indeviata* from mid-July to early August (PROUT 1935a). Growth of larvae asynchronous (PROUT 1935a; BERGMANN 1955).

BIOLOGY: Larva polyphagous on withered leaves of low plants, shrubs and deciduous trees (var. cit.). Found on *Vaccinium myrtillus*, *Calluna vulgaris* (Ericaceae), and *Alnus* (Betulaceae) (Germany: BERGMANN 1955), under deciduous shrubs on fallen, mouldy leaves (BOLDT 1929), and near ground on withered leaves of *Poa annua* (Poaceae), and *Taraxacum officinale* (Asteraceae), in rearing feeding on these two plant species (Germany: WEGNER pers. comm.). Recorded also on *Plantago* (Plantaginaceae), *Melampyrum* (Scrophulariaceae), *Clematis* (Ranunculaceae), *Genista, Cytisus scoparius* (Fabaceae), *Stellaria media* (Caryophyllaceae), *Galium* (Rubiaceae), *Myosotis* (Boraginaceae), *Primula* (Primulaceae), *Corylus* (Corylaceae), *Geum urbanum*, *G. rivale*, *Poterium* (Rosaceae), *Ribes* (Saxifragaceae), *Polygonum aviculare*, *Rumex* (Polygonaceae), *Taraxacum officinale* (Asteraceae) (var. cit., e.g. FREYER 1827; KOCH 1856; WILDE 1861; LHOMME 1935; ALLAN 1949; BERGMANN 1955; SKINNER 1984; SKOU 1986; EBERT 2001). Reared on catkins of *Salix babylonica* (ERLEBACH pers. comm.), on *Taraxacum officinale, Polygonum aviculare, Salvia pratensis, Plantago, Lonicera tatarica*, *Alnus, Galium, Stellaria, Crataegus, Prunus spinosa, Betula,* and *Campanula medium* (original data; ALLAN 1949; BERGMANN 1955; SEPPÄNEN 1970; FRIEDRICH 1976; MÜLLER, ERLEBACH pers. comm.), usually on withered leaves. In rearings larvae survive some months without feeding (PROUT 1913). Adults readily attracted to light and bait. Nectar feeding recorded on *Buddleja*, on grasses, on *Eleocharis palustris* (EBERT 2001) and on *Cirsium* (WARING & TOWNSEND 2003). Olefinic acetate identified as sex pheromone component (ZHU et al. 1996).

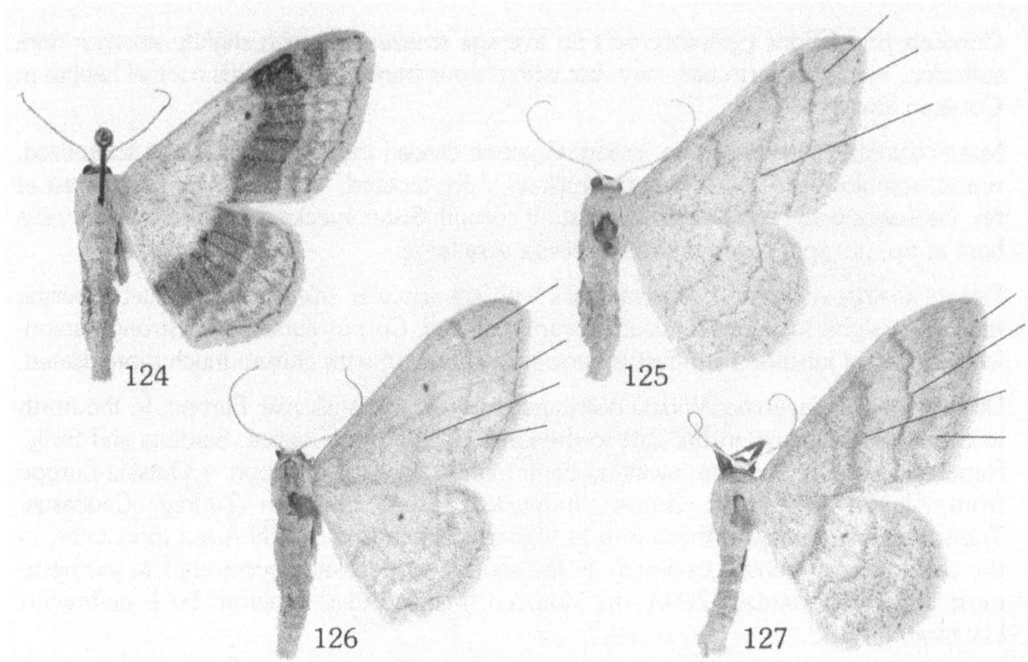

Text-figs 124-127: Diagnostic characters (indicated) of European *Idaea* species. Text-fig. 124: *I. aversata* (LINNAEUS, 1758), nominate form. Text-fig. 125: *I. aversata* (LINNAEUS, 1758), 'f. *remutata*'. Text-fig. 126: *I. straminata* (BORKHAUSEN, 1794). Text-fig. 127: *I. deversaria* (HERRICH-SCHÄFFER, 1847).

HABITAT: Mesophilous, silvicolous. In temperate parts of Europe almost ubiquitous. Most common near hedges, forest edges, forests with abundant understorey, gardens and parks, where fallen leaves are not removed. Less frequent in open land, here almost exclusively ♂♂ (HAUSMANN 1990a; 1990b). From 0 m up to 1,500 m above sea-level, up to 1,700 m in southern Europe, up to 2,200 m in Turkey and central Asian mountains.

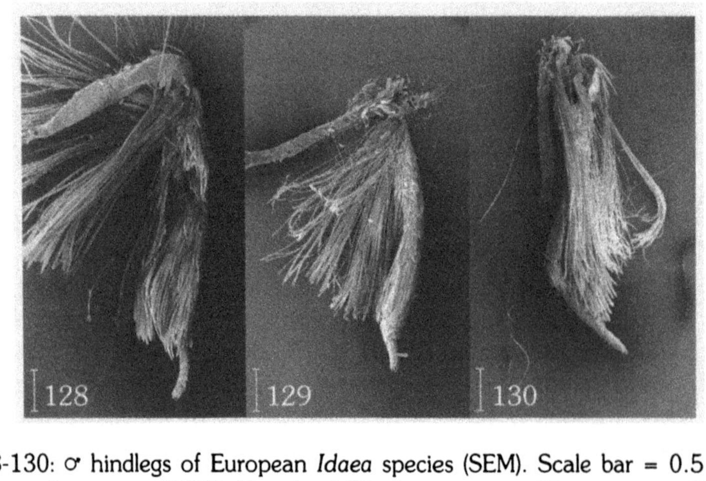

Text-figs 128-130: ♂ hindlegs of European *Idaea* species (SEM). Scale bar = 0.5 mm. Text-fig. 128: *I. aversata* (LINNAEUS, 1758). Text-fig. 129: *I. straminata* (BORKHAUSEN, 1794). Text-fig. 130: *I. deversaria* (HERRICH-SCHÄFFER, 1847).

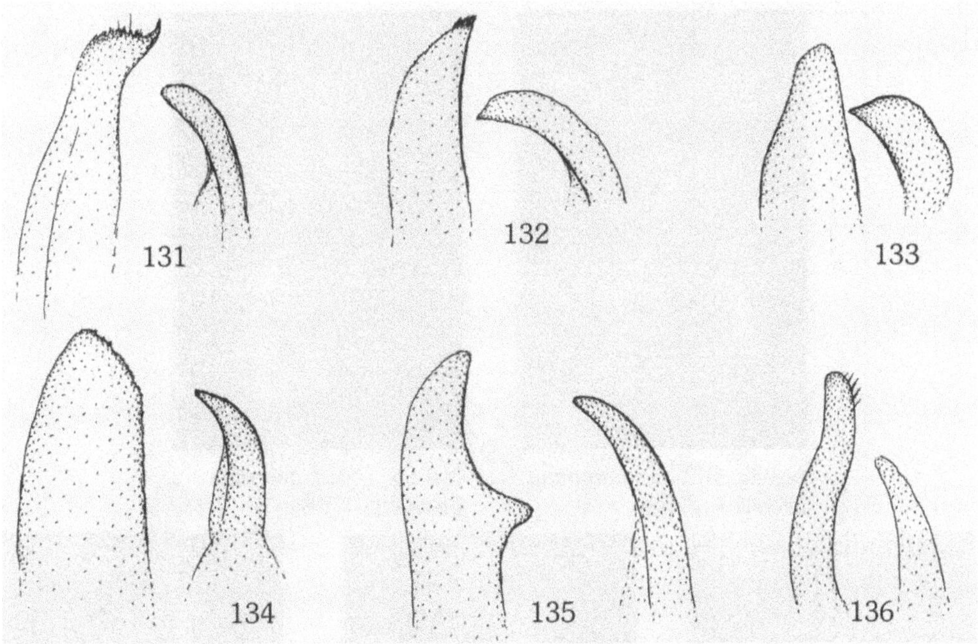

Text-figs 131-136: ♂ abdomina of European *Idaea* species after brushing to remove scales: tip of valva and tip of uncus, in lateral view. Text-fig. 131: *I. rubraria* (STAUDINGER, 1901). Text-fig. 132: *I. aversata* (LINNAEUS, 1758). Text-fig. 133: *I. degeneraria* (HÜBNER, 1799). Text-fig. 134: *I. straminata* (BORKHAUSEN, 1794). Text-fig. 135: *I. deversaria* (HERRICH-SCHÄFFER, 1847). Text-fig. 136: *I. squalidaria* (STAUDINGER, 1882).

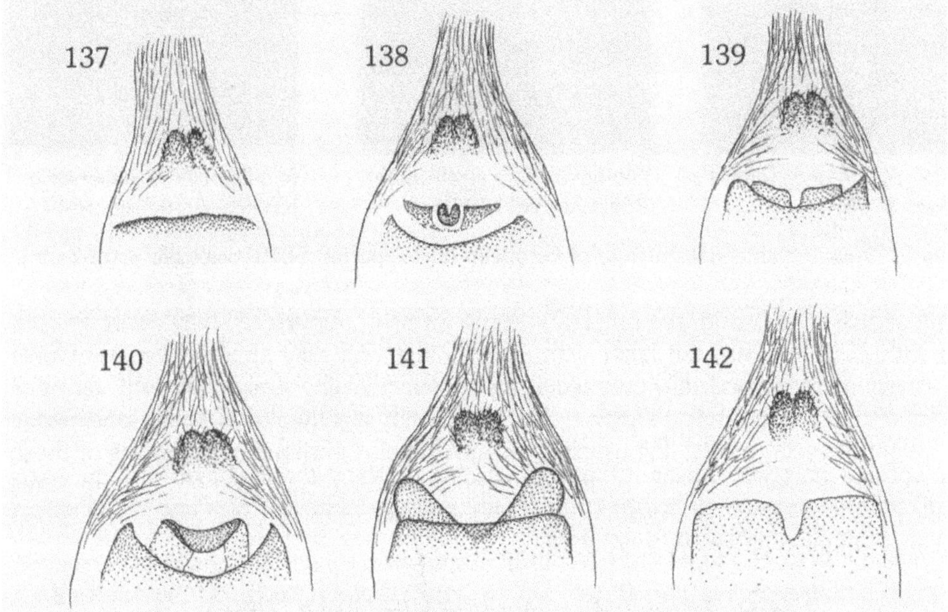

Text-figs 137-142: ♀ abdomina of European *Idaea* species after brushing to remove scales, ventral view. Text-fig. 137: *I. rubraria* (STAUDINGER, 1901). Text-fig. 138: *I. aversata* (LINNAEUS, 1758). Text-fig. 139: *I. degeneraria* (HÜBNER, 1799). Text-fig. 140: *I. straminata* (BORKHAUSEN, 1794). Text-fig. 141: *I. deversaria* (HERRICH-SCHÄFFER, 1847). Text-fig. 142: *I. squalidaria* (STAUDINGER, 1882).

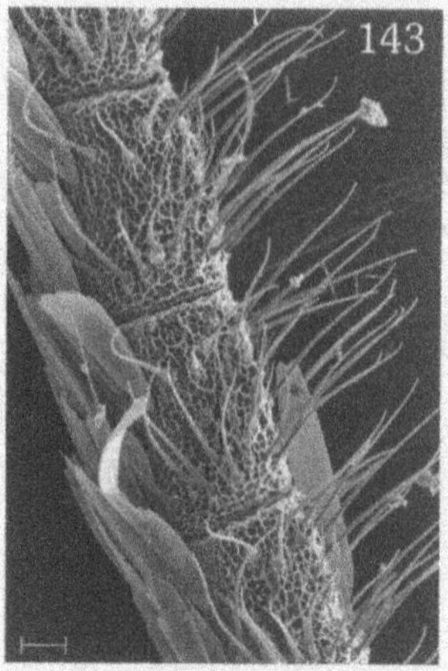

Text-fig. 143: *I. emarginata* (LINNAEUS, 1758).

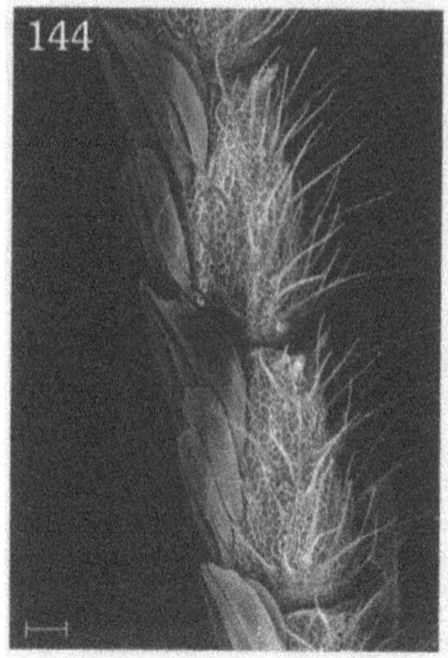

Text-fig. 144: *I. aversata* (LINNAEUS, 1758).

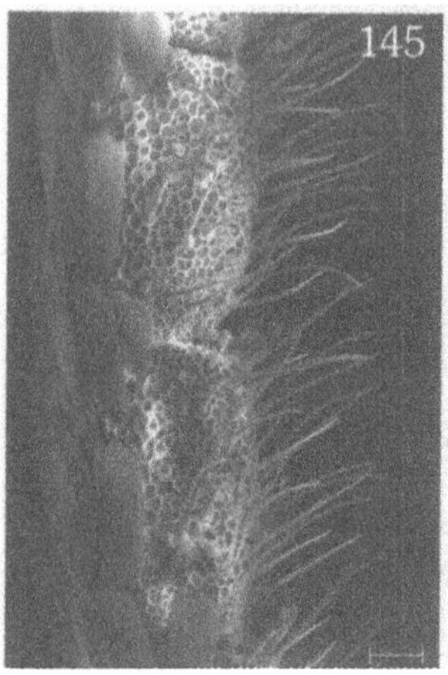

Text-fig. 145: *I. degeneraria* (HÜBNER, 1799).

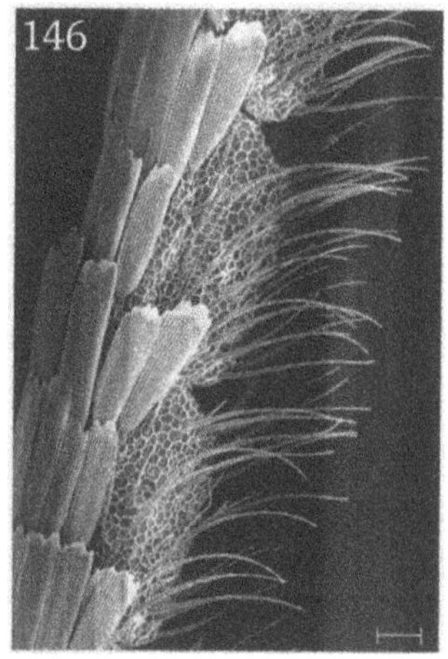

Text-fig. 146: *I. straminata* (BORKHAUSEN, 1794).

Text-fig. 147: *I. deversaria* (HERRICH-SCHÄFFER, 1847).

Text-figs 143-147: ♂ antennal structure of European *Idaea* species (SEM). Scale bar = 0.02 mm.

SIMILAR SPECIES: *I. rubraria* (see no. 107). Some forms ('f. *remutata*') may easily be confused with *I. straminata*, the latter with glossy wings and silky appearance, transverse lines weakly marked, especially the medial line, postmedial line straight at costa, terminal line less distinct, fringe dots absent. *I. deversaria* with slightly glossy wings, postmedial line straight at costa, medial line usually more proximal. Certain extreme forms of these three species, however, difficult to identify without brushing the abdomen. In both sexes identification is possible by brushing tip of abdomen to remove scales and reveal shape of valva or sterigma (Text-figs 132, 138).

109. *Idaea degeneraria* (HÜBNER, 1799)

Geometra degeneraria HÜBNER, 1799: Samml. Eur. Schmett. 5, Geometrae (1): pl. 11, fig. 57 (Europe). Syntype(s) lost.
Idaea degenerata TREITSCHKE, 1828: Schmett. Eur. 6 (2): 267. Emendation.

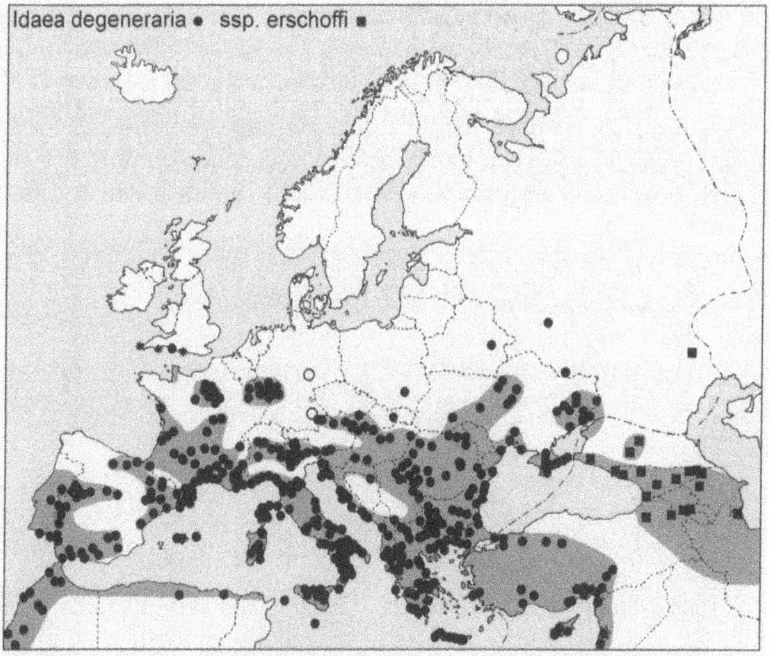

Acidalia degeneraria var. *meridiaria* MILLIÈRE, 1868: Iconogr. Descr. Chen. Lép. inéd. 2 (22): 491, 498, pl. 100, figs 12-15 (southern France: Montpellier, Marseille, Amélie-les-Bains). Syntypes, probably lost.

Acidalia erschoffi CHRISTOPH, 1872: Horae Soc. ent. ross. 9: 5, pl. 1, fig. 2 (northern Iran: Gjas). Lectotype ♀ (ZISP, examined), herewith designated to stabilize usage of the name; at ZISP 2 further ♂ paralectotypes. Valid at subspecific rank.

‡ *Acidalia degeneraria* f. *floridaria* TURATI, 1913 (31.III.): Atti Soc. ital. Sci. nat. 51 (3/4): 275, 325, pl. 9, fig. 23. First binominal citation after *floridata* PÜNGELER, 1912 (in PROUT in SEITZ: fig. 3k; non binominal; p. 135 published 5.V.1913) (Sardinia, Corsica). Syntypes at least 64♂♀ (MNHU/PÜNGELER 1♂ 'type', 36♂♀ 'cotypes; ZFMK 27♂♀ 'cotypes', examined). Infrasubspecific, never raised to subspecific (available) rank.

Ptychopoda degeneraria var. *alticolaria* SCHAWERDA, 1933 (15.V.): Zeits. österr. Ent.Ver. 18 (5), pl. 8, fig. 11 (Corsica: Vizzavona, Ponte de Fulminato; Evisa). Syntypes ♂. Herewith downgraded from subspecific rank (SCOBLE 1999) to synonymy (syn. n.).

‡ *Ptychopoda degeneraria* var. *altivolaria* SCHAWERDA, 1933 (15.VI.): Zeits. österr. Ent.Ver. 18 (6): 47. Incorrect subsequent spelling, and junior secondary homonym nec BUBACEK, 1923 (identity: *I. sericeata*).

Unavailable names (misidentification): *rubraria*: sensu GUMPPENBERG (1890) nec STAUDINGER, 1871. – (infrasubspecific): *aestiva*: FUCHS (1899); *affumigata*: DANNEHL (1925); *depravata*: STAUDINGER (1901); *vernalis*: FUCHS (1904). – (incorrect subsequent spellings): *affumicata*: DANNEHL (1927); *degenararia*: HUEMER & TARMANN (1993); *degenaria*: DOMÍNGUEZ & BAIXERAS (1989); *erschoffiaria*: PROUT in SEITZ (1913:pl.3i); *ershoffiaria*: DIDMANIDZE (2002).

DIAGNOSIS: Wingspan ♂♀ 21-26 mm, in subsequent generations sometimes 17 mm only. Ground colour usually ochreous or light brown. Wing pattern variable. Transverse lines usually more indistinct than in *I. rubraria*. Postmedial line angled below costa. Area between medial and antemedial lines usually filled with dark brown. Base of hindwing usually dark brown. Discal spots small. Terminal line fine, brown, uninterrupted. Fringe dots absent. Frons dark brown, vertex whitish to pale ochreous, collar ochreous. Antennae of ♂ ciliate-setose, cilia short, 0.5-0.8 times width of flagellum (Text-fig. 145). Hindtibia of ♂ dilated, with large pencil from base of tibia, tibial scales covering half of tarsus. Tarsus very short, 0.15-0.25 times length of tibia.

Variability: In the mountains of Corsica usually smaller, straw yellow with dark suffusion and indistinct pattern ('f. *alticolaria*'), with clinous transitions to the nominotypical forms flying in the lowlands of Corsica. Ground colour often with red suffusion ('f. *floridaria*') in North Africa, Corsica, Sardinia and Tuscany. Heritage of these red forms following MENDEL's rules (PROUT 1913). Forms with medial fascia concolorous to ground colour ('f. *depravata*') not rare in southern Europe. Populations of North Africa with particularly dark medial fascia.

Subsp. ***erschoffi*** less variable, usually with coloration more contrasted, ground colour pale ochreous, less greyish, often with very dark medial fascia, forewing termen more concave below apex.

MALE GENITALIA: Uncus very broad, beak-shaped. Gnathos reduced to a narrow clasp with medial excision. Saccus prominent. Valva short and broad, round at tip. Aedeagus without cornuti.

FEMALE GENITALIA: Ductus bursae narrow, straight, at antrum slightly dilated, posteriorly bilobed. Corpus bursae pyriform or elongate oval, strongly spinulose, laterally slightly bulbed at junction with ductus bursae, appendix bursae arising from posterior end.

DISTRIBUTION: (Sub-)Mediterranean-Turanian. Widely distributed in southern Europe from Iberian peninsula to Ukraine and scattered populations in southern European Russia. Present on all Mediterranean islands. Rare, and in isolated populations in the warmest localities of western and central Europe from southern England to southern Byelorussia. In many of the English localities suggested 'vagrant', e.g. on Scilly islands (SKINNER 1984; HICKS & HALE 1998). – Outside Europe from Morocco to Algeria, Turkey, Cyprus, Levant. Replaced by subsp. *erschoffi* from Caucasus, Transcaucasus, across northern Iraq, Iran and Turkmeniya to Afghanistan and central Asian mountains.

PHENOLOGY: Bi- or trivoltine. In the north and on the mountains bivoltine: late April to late June; early August to late September. Single specimens of first generation until July, mainly in the mountains. In North Africa, Levant and lowlands of southern Europe regularly trivoltine from late March to late October, usually with a gap from early July to mid-August. Rearing requires 2-3 months in summer (HALL pers. comm.).

BIOLOGY: Larva polyphagous on herbaceous plants, preferring dry and withered parts (var. cit.). Recorded on many plant species, such as *Achillea millefolium* (Asteraceae), *Scabiosa* (Dipsacaceae), *Convolvulus* (Convolvulaceae), *Frangula alnus* (Rhamnaceae), *Polygonum* (Polygonaceae), *Rubus* (Rosaceae), *Stachys*, e.g. *S. officinalis* (Lamiaceae), *Cerastium* (Caryophyllaceae), and *Veronica* (Scrophulariaceae) (var. cit., e.g. LHOMME 1935; ALLAN 1949; BERGMANN 1955; EBERT 2001), probably all referring to rearings. Reared on withered leaves of *Taraxacum officinale, Polygonum aviculare, Achillea millefolium, Convolvulus, Clematis, Trifolium, Rubus fruticosus, Stellaria media, Lonicera periclymenum, Veronica polita, Plantago lanceolata, Rumex, Salix, Silene*, and on petals of *Medicago* and *Chenopodium album* (original data; BUCKLER 1897; ALLAN 1949; SKINNER 1984; KRAUS 1993; PORTER 1997; HAUSMANN & MILLER 2000; KING & ROMERA 2004; MÜLLER pers. comm.). Adults readily attracted to light and bait. Nectar feeding recorded on *Mentha* (KOCH 1856).

HABITAT: Xerothermophilous in central Europe, inhabiting open, dry grassland, steppes, heathland, preferably on southern slopes. In southern Europe ubiquitous, e.g. very common in the lowland along the coasts, in the Mediterranean maquis, *Olea-Ficus* plantations, and scattered oak forest. From 0 m to 500 m above sea-level in the north. Up to about 1,200 m, exceptionally up to 1,700 m in southern Europe, but with focus on the lowlands. In Morocco, Turkey and central Asia up to 2,000 m.

SIMILAR SPECIES: *I. sardoniata* (see no. 15), *I. distinctaria* (101), *I. predotaria* (102), *I. rubraria* (107). Correct identification is easy in both sexes by brushing the abdomen to remove scales and reveal shape of valva, uncus or sterigma (Text-figs 133, 139).

REMARKS: Rearings and immature stages described in BUCKLER (1897).

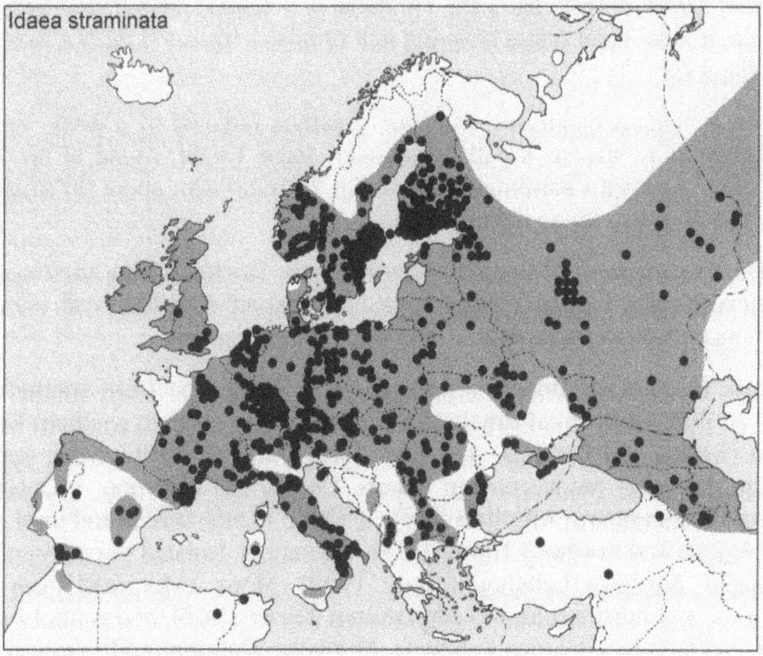

110. Idaea straminata (BORKHAUSEN, 1794)

Phalaena Geometra straminata BORKHAUSEN, 1794: Naturg. eur. Schmett. 5: 568 (Europe, probably Germany). Syntypes ♂♀.
Phalaena inornata HAWORTH, 1809: Lepid. Brit. (2): 349 (United Kingdom). Syntype(s).
Idaea suffusata TREITSCHKE, 1828: Schmett. Eur. 6 (2): 272 (Hungary). Syntypes ♂♀.
Acidalia suffusaria BOISDUVAL, 1840: Genera Index meth. eur. Lepid.: 226. Emendation.
Acidalia inornaria DOUBLEDAY, 1849: Syn. List. Brit. Lep.: 20. Emendation (SCOBLE 1999).
Acidalia agrostemmata GUENÉE, 1858: in BOISDUVAL & GUENÉE, Hist. nat. Insectes (Spec. gén. Lépid.) 9: 512 (central France: Chateaudun). Syntypes 2♀ (ZFMK: 2♀ ex coll. GUENÉE, examined, without abdomen).
Ptychopoda karafutonis MATSUMURA, 1925: J. Coll. Agric. Hokkaido Imp. Univ. 15 (3): 159, pl. 11, fig. 4 (Russia, Sakhalin: Ichinosawa, Toyohara, Kawakami, Kiminai, Sakayehama, Nyiwo, Rikovskoje). Syntypes 8♂9♀ (EIHU: examined by SATO 2001). Valid at subspecific rank (EBERT 2001; SATO 2001), in SCOBLE (1999) as separate species.
Ptychopoda inornata sibirica DJAKONOV, 1926: Ezheg. Gosud. Muz. N.M. Mart'yanova 4 (1): 21 (central Russia, Siberia: Minussinsk, Altai Mts.; Ongubai; Karakol River; Altaiskoe). Syntype(s). Valid at subspecific rank (SCOBLE 1999; EBERT 2001).
Acidalia inornata minuta HEYDEMANN, 1934: Schr. Naturw. Ver. Schlesw.-Holst. 20 (2): 164 (northern Germany: Amrum island). Syntype(s). Herewith downgraded from subspecific rank (SCOBLE 1999) to synonymy (syn. n.).
Unavailable names (misidentification): *aversata*: sensu GRAAF (1851) nec LINNAEUS, 1758; *deversaria*: sensu

BLESZYNSKI (1960: fig. 409) and KOCH (1984: 736) nec HERRICH-SCHÄFFER, 1847. – (infrasubspecific): *approximata*: COCKAYNE (1952); *impuncta*: LEMPKE (1949); *nigrolineata*: REISSER (1961); *rufescens*: COCKAYNE (1952). – (incorrect subsequent spelling): *agrostemnata*: CULOT (1917).

DIAGNOSIS: Wingspan ♂♀ 24-28 mm, second generation smaller, sometimes 20 mm only. Small island forms ('*minuta*') recorded from northern Germany by HEYDEMANN (1934). Wings unicolorous glossy sand coloured, fresh specimens with olive tinge. Transverse lines usually indistinct, slightly convex. Discal spots present. Terminal line fine, usually interrupted at veins. Fringe dots absent. Frons dark brown, vertex whitish, collar ochreous. Antennae of ♂ with intersegmental incisions, densely ciliate-setose, cilia 0.85-1.1 times width of flagellum (Text-fig. 146). Hindtibia of ♂ slightly dilated, with very large pencil from base of tibia, tibial scales covering half of tarsus. Tarsus 0.35-0.4 times length of tibia (Text-fig. 129).

MALE GENITALIA: Uncus digitiform, truncate. Gnathos reduced to a weak, narrow clasp with medial excision. Saccus broad, prominent. Valva broad, round at tip, subapically dilated at costa. Aedeagus extremely long (about 20 mm!) with about 30 small cornuti in distal part and one very long chitinous ridge.

FEMALE GENITALIA: Antrum U-shaped, dorsally open. Ductus bursae very long, laterally spinulose at both sides, looped. Corpus bursae sac-shaped, spinulose, with extremely long appendix bursae, laterally spinulose at one side, spirally curved.

DISTRIBUTION: Palaearctic. Widely distributed all over Europe from southern Portugal (CORLEY et al. 2000) and Great Britain to the Urals. In the north to southern Scandinavia. Present on Corsica and Sardinia, data from Sicily (MÜLLER 1996), await confirmation. Absent from the other Mediterranean islands. Usually not common. – Outside Europe recorded for Rif mountains, Morocco (REISSER 1933; RUNGS 1981) and local occurrence in eastern Algeria and northern Tunisia (ZSM, examined). Isolated populations in Turkey, and, reputedly, northern Lebanon (ZERNY 1933). More widespread from Caucasus, Transcaucasus, and northern Iran to Afghanistan (EBERT 1965), and central Asian mountains. Replaced by subsp. *sibirica* in Siberia, Mongolia, Dahuria, north-eastern China and Korea, by subsp. *karafutonis* on Sakhalin and Kuriles (VIIDALEPP 1996; SATO 2001; BELJAEV 2001).

PHENOLOGY: Uni- or bivoltine: In the north and on southern mountains univoltine mid-June to mid-August. Under good conditions and in the south bivoltine mid-May to late July; early August to early October, with transitional populations along geographical and altitudinal gradients. In the southernmost part of the distribution area sometimes until late October, second generation better separated from the first.

BIOLOGY: Larva polyphagous on withered leaves of herbaceous plants (var. cit.). Found near ground on withered leaves of *Taraxacum officinale* (Asteraceae) (Germany: WEGNER pers. comm.), under shrubs of *Quercus* (Fagaceae) (Germany: BOLDT 1929), and in capsules of '*Agrostemma dioica*' (Caryophyllaceae; =*Melandrium rubrum*?) (France: GUENÉE, 1858: as '*agrostemmata*'), reputedly found on *Rumex* (Polygonaceae) (Poland: URBAHN & URBAHN 1939). Recorded on *Lysimachia vulgaris* (Primulaceae), *Rumex acetosa* (Polygonaceae), *Sonchus* (Asteraceae), *Stellaria* (Caryophyllaceae), *Viola* (Violaceae), *Galium* (Rubiaceae), *Vaccinium myrtillus* (Ericaceae), *Ononis repens* (Fabaceae), and *Rubus* (Rosaceae) (var. cit., e.g. ALLAN 1949; BERGMANN 1955; SKOU 1986). Reared on withered leaves of *Lactuca sativa, Taraxacum officinale, Polygonum aviculare, Rosa, Salix, Prunus spinosa,* and *Plantago* (URBAHN 1935; SCHNEIDER 1939; ALLAN 1949; BERGMANN 1955; SKINNER 1984; PORTER 1997; EBERT 2001; WEGNER, MÜLLER pers. comm.). Main activity of adults at dusk, also attracted to light and bait. Collected in a 22 m high canopy of *Quercus* (KÖPPEL 1997). Olefinic acetate identified as sex pheromone component (ZHU et al. 1996).

HABITAT: Xerothermophilous, often on sand or limestone. Inhabiting various habitats, e.g. sunny slopes, glades in thermophilous pine-oak and oak forest, dry shrub-land, forest fringes, heathland, peat bogs, and, in the south, Mediterranean maquis. North of Alps usually restricted to lowlands under 500 m above sea-level, in Alps and southern Europe from 0 m up to 1,800 m, from Turkey to central Asian mountains up to 2,000 m.

SIMILAR SPECIES: *I. distinctaria* (see no. 101), *I. aversata* (108), *I. squalidaria* (112). *I. deversaria* subsp. *fallax* can usually be distinguished by its fringe dots and by the more convex and undulate transverse lines. The other populations of *I. deversaria* with less glossy coloration than in *I. straminata*, medial line more distinct, postmedial line usually undulate, and terminal area slightly darker with wavy line present. Correct identification is easy in both sexes by brushing of the abdomen to remove scales and reveal shape of valva in ♂, antrum and sterigma in ♀ (Text-figs 129, 134). First generation in central Europe 10-20 days earlier than that of *I. deversaria*. Specimens collected after early August usually belong to *I. straminata*.

REMARKS: For misidentified genitalia in literature see rectification in REZBANYAI-RESER (1989).

111. Idaea deversaria (HERRICH-SCHÄFFER, 1847)

Acidalia deversaria HERRICH-SCHÄFFER, 1847: Syst. Bearb. Schmett. Eur. 3 (24): 17, (non binominal on pl. 50, figs. 305-308) (western Germany: Frankfurt a.M.). Syntypes. Some years ago, the senior name '*maritimaria*' was discovered (see below). As the name '*deversaria*' has been in continuous use for about 150 years, it is preferable not to reject it in favour of (doubtful) senior synonyms.

‡? *Phalaena spataceata* SCOPOLI, 1763: Ent. Carn.: 228, fig. 570 (Slovenja: Ljubljana ('Labacum')). Syntype(s) lost. Nomen dubium, description and illustration not sufficient for unambiguous identification, latin description, however, excluding identity as *Scopula floslactata* (see such attribution in SCOBLE 1999).

‡? *Acidalia maritimaria* BRUAND, 1846: Mém. Soc. Emul. Doubs (2) 2 (3/4): 140 (France: Doubs). Syntype(s), neither traced nor examined. Perhaps senior synonym. After having been forgotten and never used for about 150 years, the name was first re-discovered by LERAUT (1997). It has been accepted, subsequently, e.g. by SCOBLE (1999) and EBERT (2001).

Ptychopoda diffluata HERRICH-SCHÄFFER, 1861: Neue Schmett. Eur. 3: 28 [24], fig. 138 (Romania: Mehadia). Syntypes, at least 3 (MNHU: 3, examined).

‡ *Acidalia inornata* [var.] *maritimata* GUENÉE, 1858: in BOISDUVAL & GUENÉE, Hist. nat. Insectes (Spec. gén. Lépid.) 9: 511 (western France: Bourgogne). Syntypes 1♂1♀ (ZFMK: 1♂1♀, ex coll. GUENÉE, examined, synonymy of *deversaria* ascertained). Incorrect subsequent spelling?

Acidalia argilata GUENÉE, 1858: in BOISDUVAL & GUENÉE, Hist. nat. Insectes (Spec. gén. Lépid.) 9: 469 (southern France: Lozère). Holotype, ♀ (ZFMK, examined). Herewith downgraded from species (SCOBLE 1999) to synonymy (syn. n.), in LERAUT (1997) erroneously as synonym of *I. pallidata*.

‡ *subversaria* LA HARPE, 1864: Neue Denkschr. Allg. schweiz. Ges. ges. Naturw. 20 (2): 5 (Switzerland: Schüpfen). Holotype ♂. Non binominal.

‡ *Acidalia deversata* ZELLER, 1868: Verh. zool.-bot. Ges. Wien 18: 585. First binominal name after *deversata* ZELLER, 1853 (non binominal). Incorrect subsequent spelling (emendation according to SCOBLE 1999).

Acidalia hyalinata CHRISTOPH, 1885: in ROMANOFF, Mém. Lépid. 2: 89, pl. 5, fig. 3 (Russia: Ordoubad; Borjom).

Acidalia italicata MILLIÈRE, 1885: Annls Soc. Ent.Fr. (6) 5: 113, pl. 2, fig. 1 (Italy, Acqui Terme).

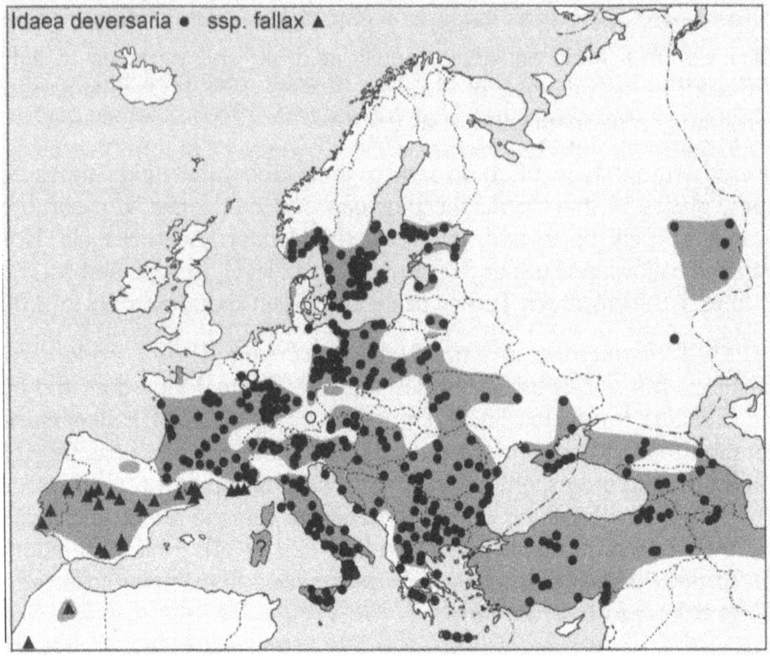

Idaea deversaria fallax HAUSMANN, 2003: Ent. Zeits. Stuttgart 113 (11): 323, fig. 7 (Spain, Aragon: Bronchales). Holotype ♂ (ZSM, examined). Valid at subspecific rank.

Unavailable names (misidentification): *inornata*: sensu BLESZYNSKI (1960: fig. 408) and KOCH (1984: 736) nec HAWORTH, 1809 (identity: *Idaea straminata*); *suffusata*: sensu GUENÉE (1858: 512) nec TREITSCHKE, 1828 (identity: *Idaea straminata*). – (infrasubspecific): *anastomosaria*: PREISSECKER (1922); *demarginata*: REISSER (1961); *ferenigra*: HOMBERG; *habichi*: SCHAWERDA (1908); *laureata*: FUCHS (1901); *nigrolineata*: REISSER (1961); *uniformis*: REISSER (1961). – (incorrect subsequent spellings): *argillata*: STAUDINGER (1861); *italica*: MILLIÈRE (1885:pl. 2); *sufsusata*: GUENÉE (1858: 511).

DIAGNOSIS: Wingspan ♂♀ 22-28 mm. Wings straw coloured, often with slight yellowish or orange tinge. Terminal area slightly darker with wavy line present. Transverse lines brown, well contrasted to ground colour, postmedial line slightly undulate. Discal spots, terminal line and fringe dots present; the last exceptionally lacking. Frons dark brown, vertex whitish, collar ochreous. Antennae of ♂ with intersegmental incisions, densely ciliate-setose, cilia 0.9-1.1 times width of flagellum (Text-fig. 147). Hindtibia of ♂ slightly dilated, with very large pencil from base of tibia, tibial scales covering half of tarsus. Tarsus 0.25-0.35 times length of tibia (Text-fig. 130).

Variability: Forms with dark brown terminal area ('f. *diffluata*') up to 50% of certain populations, e.g. in the Balkan peninsula. In Crimea and in the Caucasus on average smaller, sometimes 20 mm only, usually with postmedial line darker and broader. Populations from Caucasus to northern Iran with slightly paler ground colour, with clinous transitions to typical European forms, not meriting a separate subspecific name ('*hyalinata*').

Subsp. ***fallax*** smaller, sometimes 20 mm only, ground colour darker (light brown), with very indistinct pattern little contrasting with ground colour, terminal area concolorous with ground colour. Coastal populations of Iberian peninsula somewhat better matching these differential features than inland populations.

MALE GENITALIA: Uncus and gnathos very narrow. Valva apically tapering with large subapical tooth on costa. Juxta with paired spinulose crests. Aedeagus very long (8-10 mm), looped at base. Distally with one comparatively large cornutus, medially usually with 6-7 smaller cornuti, not rarely with 2-3 spines at each cornutus. Subsp. *fallax* usually with 4-5 smaller cornuti and costal tooth of valva in slightly more distal position.

FEMALE GENITALIA: Antrum bilobed. Ductus bursae long, helical. Corpus bursae globular, distal 2/3 spinulose. Appendix bursae long, weakly sclerotized, laterally spinulose at one side, usually curved spirally.

DISTRIBUTION: West-Palaearctic. Widely distributed all over Europe from western France to the Urals. In the north to southern Scandinavia. Present on Sicily and Crete. Data from Corsica and Sardinia (RUNGS 1988; MÜLLER 1996) await confirmation. In the Iberian peninsula and along the coast of south-eastern France (dept. Var) replaced by subsp. *fallax*. – Outside Europe in Morocco (subsp. *fallax*), from Turkey across Caucasus, Transcaucasus, northern Iran to central Asian mountains (VIIDALEPP 1988; 1996; VIIDALEPP et al. 1992).

PHENOLOGY: Usually univoltine: mid-June to late July, exceptionally in August, on mountains two weeks later. In the lowlands of the Mediterranean area emerging from late May. In the north exceptionally, in the south not rarely specimens of a partial second brood from mid-August to late October.

BIOLOGY: Larva polyphagous on herbaceous plants and shrubs, preferring withered and fallen leaves, accepting also moss (BERGMANN 1955). Found on *Sedum reflexum* (Crassulaceae) (RÖSSLER 1881) and under *Prunus spinosa* (Rosaceae), *Quercus* (Fagaceae), *Tilia* (Tiliaceae), and *Populus* (Salicaceae) (BERGMANN 1955), reputedly also on *Lysimachia* (Primulaceae), *Ononis* (Fabaceae) and *Rumex* (Polygonaceae) (SEPPÄNEN 1954). Reared on withered leaves of *Taraxacum officinale*, *Plantago* and leaves of various deciduous shrubs and trees, in the winter on *Hylocomium* (VORBRODT 1930; URBAHN 1935; BERGMANN 1955; ERLEBACH, MÜLLER pers. comm.). Adults readily attracted to light and bait.

HABITAT: Xerothermophilous. North of Alps restricted to warm sites in rocky valleys and on southern slopes, sun-exposed glades in deciduous forests, usually with oak, e.g. pine-oak forests, in the north near coasts. In southern Europe almost ubiquitous, common in the Mediterranean maquis and in scattered oak forest. North of Alps usually restricted to lowlands under 500 m above sea-level, from 0 m up to 1,800 m in southern Alps and southern Europe. In the Iberian peninsula and in Morocco usually from 1,000 m up to 2,500 m, rarely down to 200 m. In the Asian part of the distribution from 800 m up to 2,700 m.

SIMILAR SPECIES: *I. determinata* (see no. 11), *I. sardoniata* (15), *I. consanguiberica* (20), *I. nitidata* (103), *I. aversata* (108), *I. straminata* (110), *I. squalidaria* (112). Subsp. *fallax* very similar to syntopic *I. straminata*, but fringe dots usually distinct, postmedial line usually more convex and slightly undulate. Correct identification is easy in both sexes by brushing of the abdomen to remove scales and reveal shape of valva in ♂, antrum and sterigma in ♀ (Text-figs 135, 141). *Scopula ternata* and *S. floslactata* with broader forewings, ♀ with four hindtibial spurs. *S. ternata* with darker suffusion, transverse lines more diffuse, and with ♂ hindtibial spurs. *S. floslactata* with postmedial line dentate.

REMARKS: For misidentified genitalia in literature see rectification in REZBANYAI-RESER (1989). Holotype of *argilata* with glossy ground colour, discal spots distinct, fringe dots absent, underside glossy, without pattern, discal spots absent, frons brown, tongue strongly developed, postmedial line not indented or bent. Synonymy to *I. deversaria* verified by brushing of the abdomen. This name may prove to be a senior synonym of *fallax*.

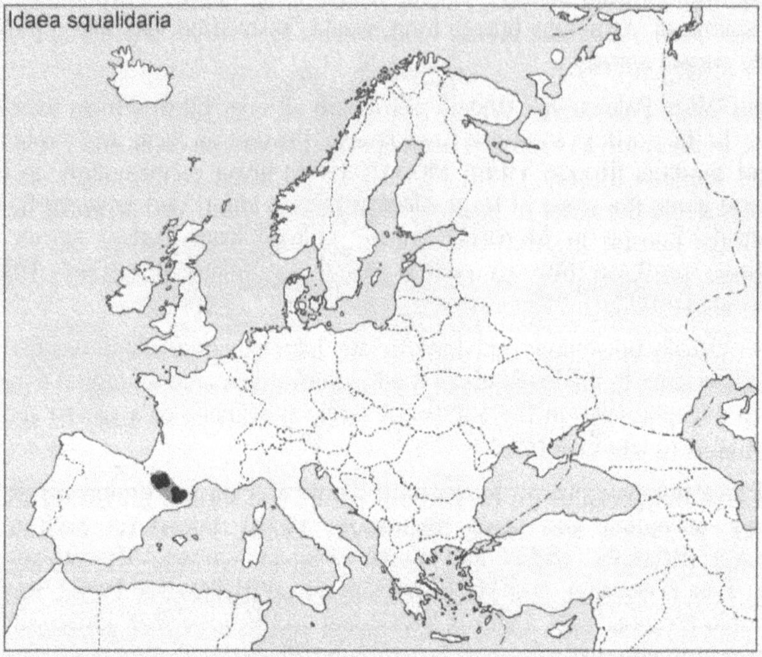

112. *Idaea squalidaria* (STAUDINGER, 1882)

Acidalia squalidaria STAUDINGER, 1882: Stettin. ent. Ztg. 43 (7-9): 405 (northern Spain, southern Pyrenees: Panticosa). Syntypes 4♂1♀ (MNHU: 2♂, MNHU/PÜNGELER: 2♂1♀, examined).

DIAGNOSIS: Wingspan ♂ 22-25 mm, ♀ smaller, 17-20 mm. Forewings slender, apex pointed, termen straight. Wings glossy, pale straw coloured, with dense grey brown suffusion. Transverse lines oblique, often indistinct, slightly darker than ground colour, better marked in ♀. Postmedial line usually 'double', i.e. with parallel dark shadow distally. Discal spots and fringe dots absent. Terminal line grey brown, very thin. Underside of forewing darker than that of hindwing. Palpi dirty white, frons blackish brown, vertex whitish, collar ochreous. Antennae of ♂ ciliate, sub-fasciculate, cilia 1.1-1.4 times width of flagellum. Flagellum with very small intersegmental incisions, antennomeres without projections or crests. Hindtibia of ♂ slender, with weak pencil, tibial scales not covering tarsus. Tarsus 0.4-0.5 times length of tibia.

MALE GENITALIA: Uncus narrow, digitiform. Saccus small, rounded. Juxta with long bilobed posterior projection. Valva long and slender, rounded and stoutly setose at tip. Aedeagus slender, straight, length about 2 mm, with one terminal cornutus and four smaller, curved cornuti. Lateral chitinous ridge possibly homologous with that in *I. straminata*.

FEMALE GENITALIA: Ductus bursae long, straight, helical at junction with corpus bursae. Posterior third of ductus bursae dilated, bilobed at antrum. Corpus bursae small, globular, with densely spinulose appendix bursae.

DISTRIBUTION: West-Mediterranean. Endemic to Pyrenees of both southern France (LHOMME 1935; MAZEL & PESLIER 1997; LUTRAN et al. 1998) and northern Spain, present also in Andorra (DANTART & ROCHE 1992).

PHENOLOGY: Univoltine: late June to early August.

BIOLOGY: Larval host-plants unknown. Adults readily attracted to light.

HABITAT: High montane, preferring glades in subalpine coniferous forest, but also on meadows (DANTART 1997). From 1,200 m up to 2,400 m above sea-level, most common at about 2,000 m (DANTART 1997).

SIMILAR SPECIES: Confusions may occur with certain forms of *I. straminata, I. deversaria* or *I. sardoniata*, but these species with discal spots present, forewings broader, apex less pointed, the last two species with fringe dots. Hindtibia of ♂ of *I. sardoniata* with two spurs, ♂ hindtarsus shorter in *I. straminata* and *I. deversaria*. *I. subsericeata* with less dark suffusion and forewing apex less pointed.

REMARKS: Taxonomy, distribution and biology discussed in DANTART (1997).

Brachyglossina WAGNER, 1914

TYPE SPECIES: *Brachyglossina acidaliaria* WAGNER, 1914 (Tunisia). 24 species in the western Palaearctic. Revision: PROUT (1938a: 229-233). Validity of generic rank refused by STERNECK (1940), but re-established by WILTSHIRE (1966; 1990) and SCOBLE (1999), based on the authority of FLETCHER. STERNECK (1940) included many of the species in question in his 31st and 32nd species-groups of *Idaea*. Four species are transferred here from *Idaea* to *Brachyglossina*: *B. hispanaria* (PÜNGELER, 1913), comb. n.; *B. improbata* (STAUDINGER, 1898), comb. n.; *B. brevitarsata* (HAUSMANN, 1991), comb. n.; *B. consociata* (STAUDINGER, 1900), comb. n.. Tibial pencil and antennal characters reveal isolated position of *B. hispanaria* from the North African species-group around the generic type species and the most common species *B. culoti* (WEHRLI, 1930). Habitus and genitalia of both sexes of *B. hispanaria* support very close relationships to the Levantine species-group including *B. consociata*, *B. brevitarsata* and *B. improbata*. Possible relations to the *ostrinaria* and the *fractilineata* species-groups of *Idaea* see above. Synapomorphies: Coloration and behaviour of larvae (see 'immature stages'). The North African species of the genus urgently need comprehensive revision.

DIAGNOSIS: Most features as in *Idaea*. Wings usually rounded, sand coloured, ochreous or brownish. Wing pattern quite uniform, often with dotted or dentate postmedial line and darkened terminal area. Venation (Text-fig. 149) as in *Idaea*, but R1 of forewing more frequently stalked, rarely arising from areole, areole absent in some smaller species (e.g. *B. tibbuana* HERBULOT, 1965) or with infraspecific variation being sometimes open at the apex of the areole (e.g. *B. chaspia* BRANDT, 1938, *B. brevitarsata*). Proboscis absent or reduced, short rudiments of proboscis occurring also in several North African species, with infraspecific variability. Palpi short, usually 0.6-0.8 times diameter of eye. Antennae of ♂ ciliate-fasciculate, cilia very long (exception: *B. hispanaria*), flagellum with prominent broad teeth, which are shallower in only a very few species (Text-fig. 148). Sternum A2 of ♂ with distinct horseshoe-shaped sclerite and setose patch. Hindtibia of ♂ usually very slender, smooth scaled and without spurs or pencil; exception: *B. hispanaria*. Hindtarsus of ♂ shortened.

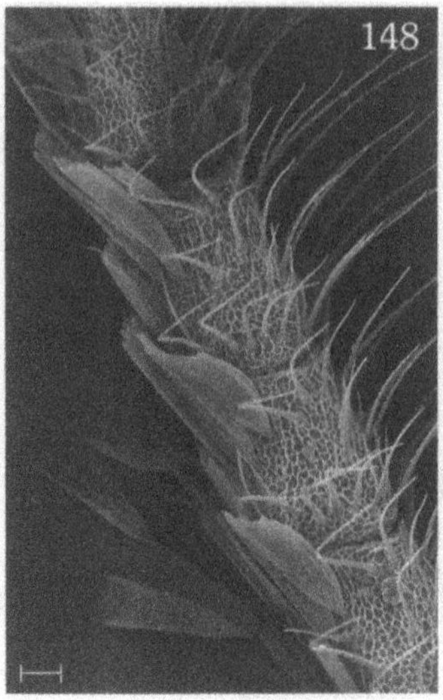

Text-fig. 148: ♂ antennal structure of *Brachyglossina hispanaria* (PÜNGELER, 1913) (SEM). Scale bar = 0.02 mm.

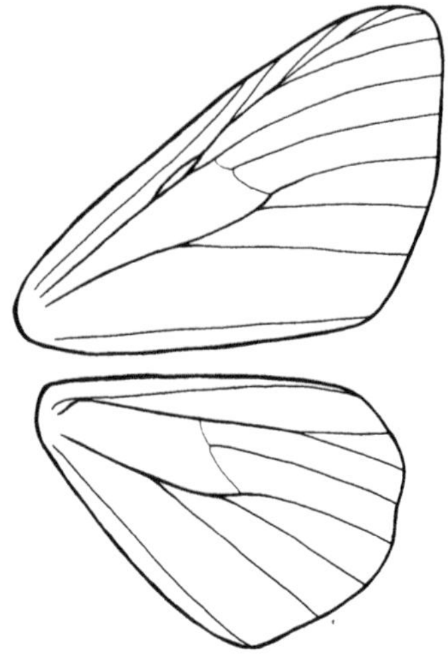

Text-fig. 149: Venation of *Brachyglossina hispanaria* (PÜNGELER, 1913).

MALE GENITALIA: Uncus often curved at tip. Juxta often with long posterior appendages. Saccus strongly sclerotized, prominent in most species. Valva usually long and slender, often with subapical tooth or lobe. Aedeagus with 3-25 cornuti (exception: *B. paroranaria* with one cornutus), typically at tip of aedeagus. Vesica strongly folded. Sternum A8 of ♂ posteriorly usually with two shallow, sclerotized lobes.

FEMALE GENITALIA: Antrum usually well sclerotized and dilated. Anterior part of ductus bursae and latero-posterior part of corpus bursae furrowed. Corpus bursae spinulose laterally.

IMMATURE STAGES: Larvae usually chequered, with dark and pale transverse rings, in *B. hispanaria* more brownish (Text-figs 150-151). In contrast to *Idaea* larvae, living in loosely spun silky 'caves', resting head-down, pulling plant material into the cave, feeding on very moist, mouldy plant material, strongly photophobic, immediately hiding when exposed to light (GELBRECHT, LEIPNITZ, MÜLLER, pers. comm.).

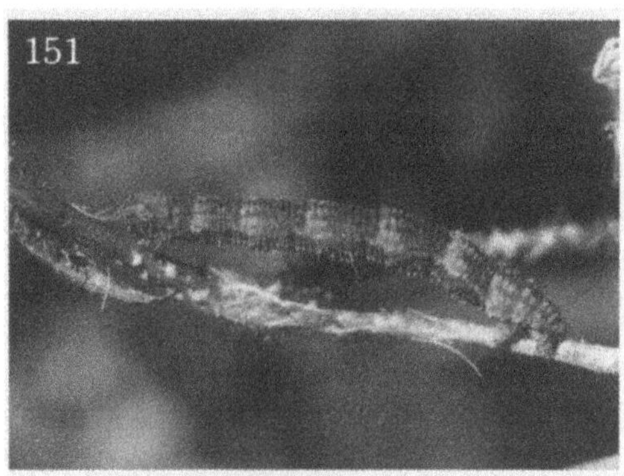

Text-figs 150-151: Larvae of *Brachyglossina* species. Text-fig. 150: Larva of *B. tantalidis* (TURATI, 1924) from Tunisia, with typical transverse ring pattern. Text-fig. 151: Larva of *B. hispanaria* (PÜNGELER, 1913) (Photos: LEIPNITZ).

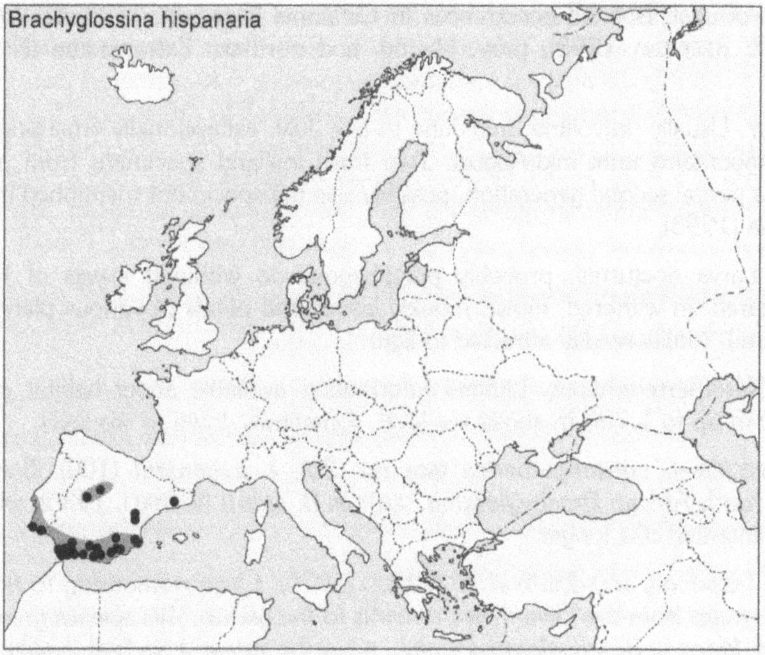

113. Brachyglossina hispanaria (PÜNGELER, 1913), comb. n.

Ptychopoda hispanaria PÜNGELER, 1913 (10.IV.): in PROUT: in SEITZ, Macrolep. 4: 128, pl. 3, fig. i (1912, non binominal) (Spain, prov. Murcia: Sierra d'Espuña). Syntypes 4♂2♀ (MNHU/PÜNGELER, 4♂2♀ syntypes, labelled as 'type' and 5 'cotypes'; all labelled as '*hispunaria*' by PÜNGELER). Manuscript name of JACOBS, in original description misspelled by PÜNGELER. Nevertheless valid according to the Code, as there is no evidence of error from the original description itself.

‡ *Acidalia hispunaria* JACOBS, 1913 (September): Ent. Mo. Mag. 49: 302 and PROUT, 1913 (15. December): in SEITZ, Macrolep. 4: 93: Nomina nuda.

DIAGNOSIS: Wingspan ♂♀ 19-23 mm, sometimes 17 mm only. Wings rounded. Wings light brown or sand coloured with orange tinge. Terminal area usually with some dark brown spots, sometimes confluent. Rarely also basal area dark brown. Postmedial line fine, often dotted. Discal spots present. Fringe dots distinct. Frons dark brown, vertex sand coloured, collar light brown. Proboscis weak, short, length about 1 mm. Antennae of ♂ with small intersegmental incisions, antennomeres without projections, ciliate, subfasciculate, cilia 1.3-1.5 times width of flagellum (Text-fig. 148). Hindtibia of ♂ stout but not dilated, with long pencil, tibial scales not covering tarsus. Length of tarsus variable, 0.3-0.5 times length of tibia.

MALE GENITALIA: Uncus slender digitiform. Gnathos rectangular. Saccus narrow, prominent. Juxta without dorsal projections. Valva slender. Aedeagus long and slender, with about 20 small cornuti.

FEMALE GENITALIA: Antrum strongly dilated, semicircular. Ductus bursae with narrow lateral sclerotization. Corpus bursae elongate pyriform, latero-posteriorly furrowed, latero-anteriorly spinulose. At anterior end of corpus bursae without spinules, bent caudad towards ductus seminalis.

DISTRIBUTION: West-Mediterranean. Endemic to the Iberian peninsula: southern Spain and southern Portugal. Isolated occurrences in Catalonia (DANTART 1991; DANTART 2000; DANTART & REQUENA 1993), prov. Madrid, and northern Extremadura (NOVOA pers. comm.).

PHENOLOGY: Usually univoltine: mid-June to late July, exceptionally emerging from late May, on mountains until mid-August. Two fresh lowland specimens from mid-August belong to a partial second generation, possibly also the specimens mentioned by DANTART & REQUENA (1993).

BIOLOGY: Larva nocturnal, probably polyphagous on withered leaves of herbaceous plants. Reared on withered, moist, mouldy leaves and petals of various plants (LEIPNITZ pers. comm.). Adults readily attracted to light.

HABITAT: Xerothermophilous. Limited information available about habitat preference. From 500 m up to 1,700 m above sea-level, sometimes down to sea-level.

SIMILAR SPECIES: *I. consanguiberica* (see no. 20), *I. eugeniata* (100). Some similar allopatric North African *Brachyglossina*, such as *B. culoti* (WEHRLI, 1930), without proboscis, ♂ antennal cilia longer.

REMARKS: Taxonomy and distribution: AGENJO (1976). Close relationship to *B. improbata* species-group from the Levant, see remarks to the genus. This species-group is transferred from *Idaea* to *Brachyglossina* (comb. n.) as the larvae show both typical behaviour and habitus for *Brachyglossina* (LEIPNITZ, GELBRECHT pers. comm.).

Limeria STAUDINGER, 1892

TYPE SPECIES: *Limeria macraria* STAUDINGER, 1892. Monotypic, with one, western Palaearctic species. Systematic position within Sterrhini unclear.

DIAGNOSIS: Wings extremely slender, mainly in ♀. Forewing apex pointed. Hindwing elongate with sigmoid waved termen. Venation (Text-fig. 152): R1 of forewing arising from apex of areole. Forewing cell very long. Hindwing: Sc+R1 and Rs approximate over 2/3

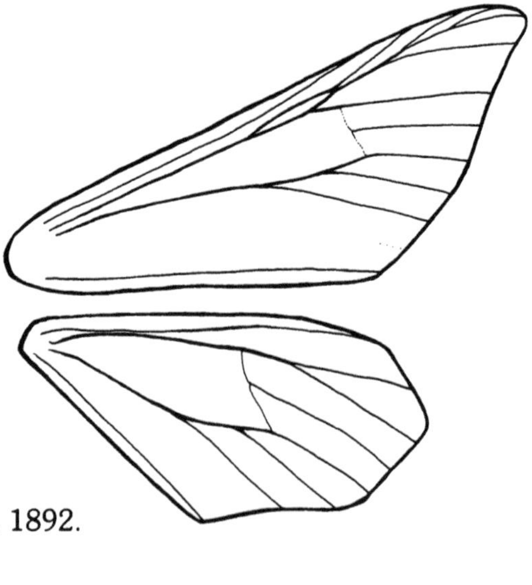

Text-fig. 152:
Venation of *Limeria macraria* STAUDINGER, 1892.

of cell, discocellulars oblique. Eyes small, diameter of eye equalling distance between eyes. Proboscis vestigial or very short. Palpi short, stout and truncate, about 0.5-0.7 times diameter of eye. Antennae of ♂ sub-bipectinate (ciliate-fasciculate), cilia on short branches of about 1.0 times width of flagellum. Cilia curved, very long, 2.7-3.0 times width of flagellum. Flagellum very thin. Legs and abdomen (♂) very long in proportion to the thorax. Hindtibia of both sexes extremely slender, without pencil, with two short spurs of unequal length. Hindtibial spurs of ♂ very short and weak, sometimes absent. Hindtarsus of ♂ long, about 1.1 times length of tibia.

MALE GENITALIA: Very small. Uncus and saccus narrow digitiform. Gnathos large, arrowhead-shaped. Valva at tip trilobed. Aedeagus short, without cornuti. Sternum A8 of ♂ triangular.

FEMALE GENITALIA: Very small. Apophyses short. Ductus bursae sclerotized, short and broad. Corpus bursae very small, membranous.

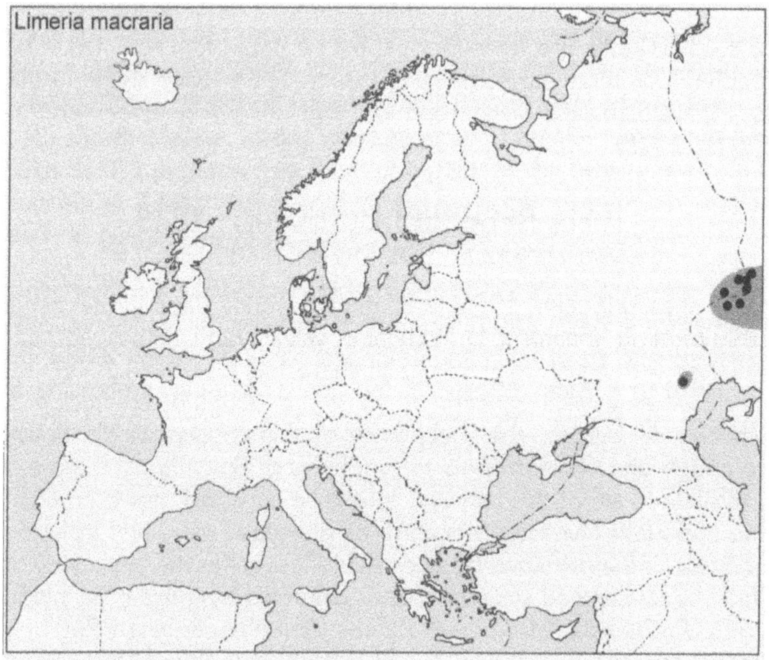

114. *Limeria macraria* STAUDINGER, 1892

Limeria macraria STAUDINGER, 1892: Dt. ent. Z. Iris 5: 195, pl. 2, fig. 14 (south-eastern European Russia: Krasnoarmeysk nr. Volgograd ('Sarepta')). Syntypes 4♂2♀.
Unavailable names (incorrect subsequent spelling): *marsaria*: RIEMIS (1998).

DIAGNOSIS: Wingspan ♂♀ 13-18 mm, females on average slightly smaller than males. Ground colour dirty white to sand coloured, with grey and brown suffusion, hindwing paler than forewing. Medial and antemedial line absent. Postmedial line close to termen on all wings, on forewing oblique, on hindwing strongly convex. Discal spots present, that of forewing close to apex. Terminal line and fringe dark brown. Frons brown, vertex sand coloured. Further features under generic diagnosis.

MALE AND FEMALE GENITALIA: see under generic diagnosis.

DISTRIBUTION: Turanian. In Europe restricted to Volga plain and southern Ural region. Very rare and local. – Outside Europe in southern Siberia (northern Kazakhstan), in the east to western slopes of Altai mountains, in the south to Turkmeniya ('Aschchabad'). Records of from North Africa (e.g. TURATI 1930a) are misidentified and refer to *Idaea volloni* (LUCAS & JOANNIS, 1907). The record for Turkey (VIIDALEPP 1996; RIEMIS 1998) probably based on misidentification, too. Worldwide very few specimens are known of this species (e.g. BMNH, ZSM, MNHU, coll. VIIDALEPP, NUPPONEN).

PHENOLOGY: Univoltine, with very short flight period from late July to early August, sometimes from mid-July.

BIOLOGY: Immature stages unknown. Adults nocturnal, never found by day or by netting, preferably flying on the hottest nights, when temperature does not fall below 30°C (NUPPONEN pers. comm.).

HABITAT: Dry steppes (NUPPONEN pers. comm.). From 0 m up to 600 m above sea-level.

SIMILAR SPECIES: No similar species in Europe, but perhaps often assumed to belong to the 'Microlepidoptera', as there are some species with similar wing shape in the Pyralidae, subfamilies Pyraustinae and Acentropinae, e.g. *Acentria ephemerella* (DENIS & SCHIFFERMÜLLER, 1775).

Tribus **Scopulini** DUPONCHEL, 1845

(=Acidalites DUPONCHEL, 1845; =Problepsini WILTSHIRE, 1990)

More than 900 species (SCOBLE 1999) in 25 genera worldwide, 5 further genera assigned tentatively (SIHVONEN & KAILA 2004). Number of valid generic names reduced to seven in a cladistic analysis with equal weighting of all characters (SIHVONEN 2003), many genera would be included in a wide concept of the genus *Scopula*, embracing 85% of the Scopulini species. In Europe, the acceptance of this concept would downgrade *Oar, Cinglis, Holarctias* and *Glossotrophia* to synonyms of *Scopula*. At the present stage of research, I hesitate to adopt this concept. It may be accepted, however, after consideration of other important character sets, such as molecular data, and inclusion of further species worldwide, if the proposed changes still reflect better the phylogeny of Scopulini. The tribe may be understood sensu stricto or delimited by a wider definition, including e.g. *Somatina, Problepsis, Dithalama, Zythos* (SIHVONEN & KAILA 2004). I prefer the latter, traditional concept, in accordance with HOLLOWAY (1997), HOLLOWAY et al. (2001), and STERNECK (1941: 207), who state synonymy of Problepsini to Scopulini. Monophyly of the tribe supposed to be indicated by modifications of ♂ sternum A8 and A2 (HOLLOWAY 1997; SIHVONEN & KAILA 2004), but status as syn- or autapomorphisms not tested analytically. Absence of ♂ metathoracic medial ridge in epinotum and ovate signum, divided into flat, separate spinules, directed away from longitudinal axis (with secondary losses) support monophyly, too (HOLLOWAY et al. 2001; SIHVONEN 2003; SIHVONEN & KAILA 2004). Sternum A8 typically modified, with posterior sclerite ('mappa') and paired lateral appendages ('cerata'), rarely reduced secondarily ('reversals'). Absence of cerata in the Indo-Pacific genera *Dithalma* and *Zythos* suggested to indicate plesiomorphic character state (SIHVONEN & KAILA 2004). Absence of modifications of ♂ sternum A2 in some genera, such as *Oar, Cinglis, Holarctias*, possibly plesiomorphic, too. Traditional view of Sterrhini as sister group of Scopulini confirmed by recent analysis (SIHVONEN 2003; SIHVONEN & KAILA 2004). Revisions at tribe level: STERNECK (1941), HOLLOWAY (1997), HOLLOWAY et al. (2001), SIHVONEN (2003); SIHVONEN & KAILA (2004).

NOMENCLATURAL NOTE: The family-group name Scopulini is usually accepted for the tribe but not for the subfamily, despite priority. For justification see introduction to subfamily and the Forum Herbulot world list of family-group names in Geometridae www.herbulot.de. The type genus was originally misidentified, as DUPONCHEL includes only Pyralidae in his 'Scopulites'. The case needs to be ruled by the Commission.

DIAGNOSIS: Medium sized, on average larger than Sterrhini species. Proboscis present, vestigial in *Stigma*, hypertrophic in nominate subgenus of *Glossotrophia*. Antennae usually filiform in both sexes, in ♂ usually ciliate-fasciculate, e.g. in the genus *Scopula*; rarely bipectinate, e.g. in the genera *Oar, Cinglis*, and most species of *Problepsis*. ♂ hindlegs often with scent hair pencils. Hindtibia of ♀ with 4 spurs; exception: *Glossotrophia*. Sternum A2 of ♂ often with horseshoe-shaped sclerite, with setose patch in its center, the latter usually paired (Text-figs 3-4). Posterior margin of tergum A8 usually slightly concave.

Venation (Text-figs 156-158, 190, 196-197): Forewing with one, single areole; exception: many species of the extraterritorial genus *Somatina* with double areole, reminiscent of venation of Rhodostrophiini. Most genera with forewing R1 arising free from apex of areole and R5 shortly stalked, arising close to apex of areole; exceptions: *Oar, Cinglis*. Hindwing with Sc+R1 appressed to Rs at one point at the base of the cell. Rs and M1 separate, connate or shortly stalked (differential feature from the Sterrhini), with distinct stalk, however, in *Holarctias*. M3 and CuA1 separate.

MALE GENITALIA: Uncus reduced or absent, usually replaced by paired setose socii. Gnathos absent. Valva cleft into two processes ('valva' and 'fibula'), often both reduced in length. Shape and sclerotization of fibula often diagnostic. Juxta (in Scopulini usually

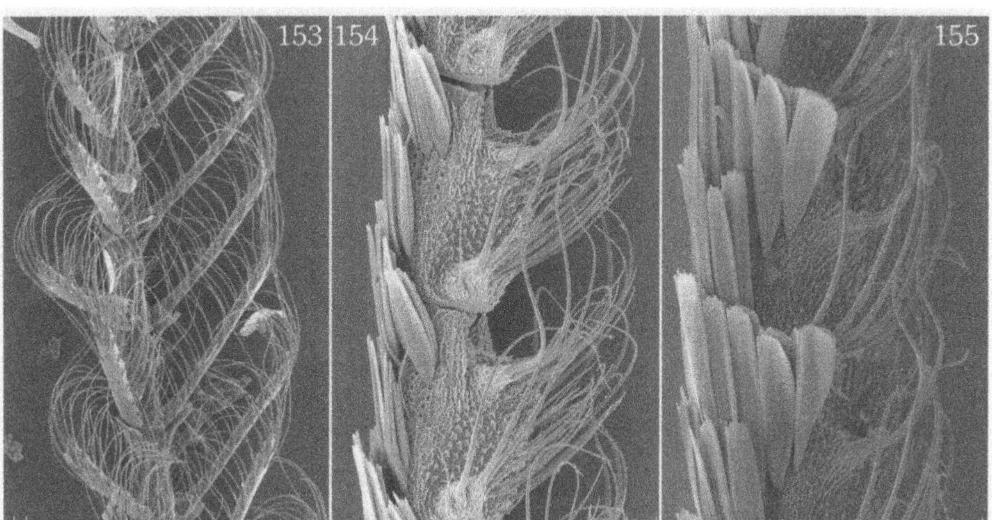

Text-figs 153-155: ♂ antennal structure of European Scopulini species (SEM). Scale bar = 0.02 mm. Text-fig. 153: *Oar reaumuraria* (MILLIÈRE, 1864). Text-fig. 154: *Scopula ornata* (SCOPOLI, 1763). Text-fig. 155: *Glossotrophia confinaria* (HERRICH-SCHÄFFER, 1847).

called 'anellus') with lateral lobes or processes, sometimes diagnostic. Aedeagus without vesica glomerates. Sternum A8 typically with posterior sclerite ('mappa') and paired lateral appendages ('cerata'), usually diagnostic for species, shape of its base sometimes diagnostic for genera or species-groups.

FEMALE GENITALIA: Ovipositor longitudinally corrugated. Shape of colliculum, antrum or sterigma (lamella antevaginalis) often diagnostic. Ductus bursae long and slender, membranous in most species. Corpus bursae usually oval, membranous. Signum very often an ovate patch of small, flat, separate spinules.

BIOLOGY: Larvae of most species polyphagous on herbaceous plants. Some species recorded as pests, e.g. on tobacco, corn, tomato, or groundnut, see introduction to subfamily. Adults usually nocturnal and readily attracted to light, but diurnal e.g. in the Mediterranean genus *Oar* or in the African *Aletis/Cartaletis* group. For lacriphagy of some species see genus *Scopula*. See also introduction to subfamily.

IMMATURE STAGES: Larva usually long and slender (Text-fig. 159). Pupal cremaster with four pairs of setae, the terminal pair usually enlarged, stout, diverging (PATOČKA 2003).

REMARKS: Central setose pouch on ♂ sternum A2 probably interacting with shortened hindtarsus or hindtibial pencils, dispersing scales in courtship (HOLLOWAY et al. 2001).

Oar PROUT, 1913

TYPE SPECIES: *Phalaena pratana* FABRICIUS, 1794. Two species in the western Palaearctic region. According to SIHVONEN (2003) synonym of *Scopula*, but venation, structure of ♂ sternum A2 and A8, shape of aedeagus and presence of cornuti provisionally justify generic rank. Small and distant eyes, long palpi, bipectinate ♂ antennae and hairy trochanter of foreleg reminiscent of genus *Cleta*, Sterrhini. Close relationship to basal Sterrhini also suggested by PROUT (1913), see remarks to the subfamily. Particular origin of forewing vein R5 resembling that of Rhodostrophiini, where R5 delimits the second areole.

DIAGNOSIS: Wings short and broad. Hindwing termen round. Venation (Text-fig. 156): Forewing with areole at basal position, below cell apex, R1 and R2-R4 connate or rarely shortly stalked, R5 stalked with M1, arising connate or slightly separate from areole, but never connate or stalked with R2-R4 from apex of areole. Hindwing with Rs and M1 shortly stalked, M3 and CuA1 separate. Proboscis very short, vestigial. Eyes small, quite distant from each other. Frons roughly scaled. Palpi long, extremely bushy, with prominent hair-like scales in ♂, less bushy in ♀, 2-3 times diameter of eye in both sexes. Antennae of ♂ (Text-fig. 153) bipectinate, branches densely ciliate, flagellum dentate in lateral view. Antennae of ♀ filiform. Trochanter of foreleg with long, hairy scales. Hindtibia of ♂ with two spurs, hindtarsus not shortened. Hindtibia of ♀ with 4 spurs. Central sclerite and setose patch absent from ♂ sternum A2.

MALE GENITALIA: Socii weakly sclerotized, linked by a membrane. Anellus with bifurcate ventral sclerite, the latter with long and narrow lateral processes. Aedeagus with lateral projections and patches of small cornuti at tip. Shape of sternum A8 diagnostic.

FEMALE GENITALIA: Apophyses very long and thin. Sclerotization of antrum diagnostic. Corpus bursae oval, posterior part corrugated and sclerotized. Signum absent.

Text-fig. 156: Venation of *Oar pratana* (FABRICIUS, 1794).

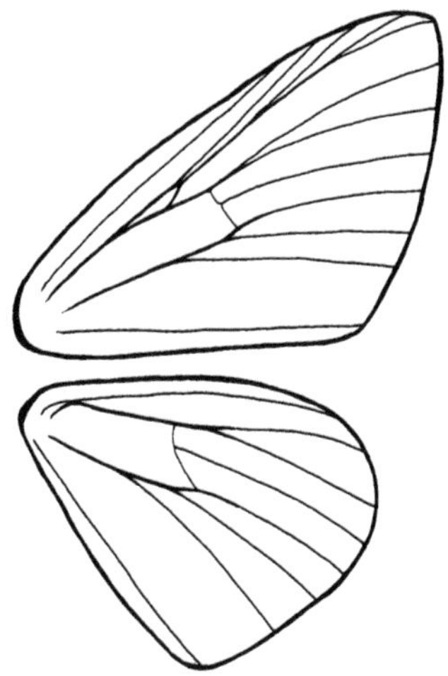

IMMATURE STAGES: Egg sculpturation polygonal reticulate. Larva with last three segments dilated, tapering towards head. Segment A6 laterally with a dark spot over the stigma. Pupa short, abdomen slender.

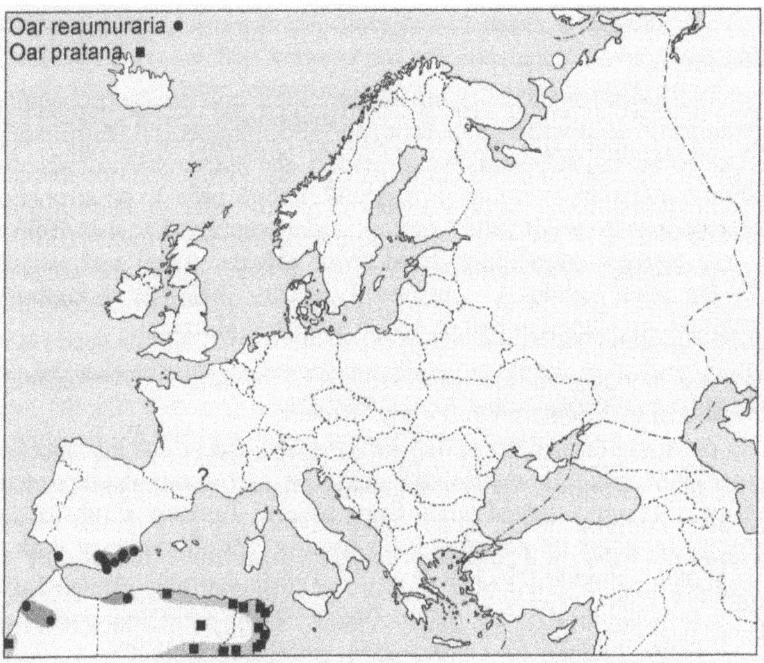

115. *Oar reaumuraria* (MILLIÈRE, 1864), stat. n.

Cleta reaumuraria MILLIÈRE, 1864: Iconogr. Descr. Chen. Lép. inéd. 2 (11): 2, pl. 51, figs 1-2 ('France, by Montpellier; Spain or Algeria'; neotype from Spain: Alicante). Original type locality 'Algier' according to BOHATSCH (1885), 'probably Algeria' according to STAUDINGER (1898: 315:). No subsequent records from France in 150 years of intensive collecting, the ♀ types may be mislabelled or refer to temporary colonization. Original syntypes: 2♀ (southern France) and 2♂ ('coll. GUENÉE', Spain), not traced in ZFMK, BMNH, MNHN. Therefore, the following neotype is designated to fix type locality and to stabilize usage of the name: ♂,

'Hispania mer. or., Alicante, 15.VI.1960, leg. K. SATTLER', ZSM, examined. Herewith raised from synonymy (SCOBLE 1999) of *Oar pratana* to species rank (stat.n.).

Fidonia megearia OBERTHÜR, 1881: Ét. Ent. 6: 84, pl. 3, fig. 8 (western Algeria: Oran). Syntypes 3♀ (ZFMK, 2♂ ex coll. OBERTHÜR, one with 'type' label, examined).

Oar pratana occidens PROUT, 1935: in SEITZ, Macrolep. 4, Suppl.: 49, pl. 5, fig. g (western Algeria, Oran: Perrégaux; Taourirt). Syntype(s) (BMNH: 1♂ examined, with red 'type' label).

Unavailable names (misidentification): *pratana*: sensu STAUDINGER (1901) nec FABRICIUS, 1794 (identity: *Oar pratana*). – (incorrect subsequent spelling): *megiaria*: BETHUNE-BAKER (1894).

DIAGNOSIS: Wingspan ♂ 17-19, ♀ 15-16 mm, in the second generation sometimes 14 mm only. Wings sand coloured, usually with strong brown suffusion, ♀ lighter, with less brown suffusion. Antemedial line distinct, dark brown. Medial line angled twice. Postmedial line slightly dentate, distally lightly bordered. Terminal area and area between medial and postmedial lines chocolate brown. Hindwing brown, darker towards termen, medial fascia less distinct than on the forewing. Discal spots present. Fringe slightly chequered. Frons greyish, mixed with dark brown scales. Longest ♂ antennal branches 4.5-6 times width of flagellum (Text-fig. 153). Further features under generic diagnosis.

MALE GENITALIA: Socii prominent. Valva with dorsal process. Sclerotized ridge of anellus with comparatively short lateral processes, rounded at tip. Aedeagus broad at base, with tapered lateral process at 1/2, and two patches of very small spinules at tip. Sternum A8 without spines posteriorly, but with paired, very small, slightly spinulose lobi.

FEMALE GENITALIA: Antrum and lamella antevaginalis fused to a circular sclerite, slightly notched posteriorly.

DISTRIBUTION: West-Mediterranean. South-eastern Spain and isolated populations in prov. Cadiz. Occurrence in southern France (LHOMME 1935; HERBULOT 1963; LERAUT 1980; 1997; MÜLLER 1996) urgently needs confirmation, the author was not able to trace any reliably labelled material from France. Possibly all records refer to an erroneous interpretation of the type locality of MILLIÈRE'S *reaumuraria* (compare LHOMME 1935). – Outside Europe in very isolated populations in Morocco (Middle Atlas) and western Algeria. Replaced by the sister species *O. pratana* (FABRICIUS, 1794) from central Algeria to Arabian peninsula, on Canarian islands and in southern Morocco.

PHENOLOGY: In the Iberian peninsula usually trivoltine: late March to late July in two overlapping generations; mid-September to early October.

BIOLOGY: No detailed information on larval host-plants traced for this species, compare the sister species *O. pratana*. Adults (♂♂) actively flying during daytime, from 8-10 o' clock (SKOU pers. comm.), netted when flying around *Tamarix* shrubs (BLAESIUS pers. comm.), in large numbers on *Ziziphus lotus* (AGENJO 1952), active at dusk around the supposed host-plant *Suaeda* (HACKER & WOLF 1983), sparingly attracted also to light (KRAUS 1999). Stenotopic (AGENJO 1952; KRAUS 1999), patrolling a small territory of about 5 m^2, possibly attracted by ♀ (SKOU pers. comm.), K-strategy.

HABITAT: Hot lowland habitats, usually on sandy soils near coast. Probably halophilous (KRAUS 1999; see *O. pratana*). From 0 m up to 300 m above sea-level, exceptionally up to 700 m, up to 1,500 m in Morocco.

SIMILAR SPECIES: Allopatric sister species *O. pratana* very similar in habitus, correct identification requires dissection in ♀, in ♂ brushing of the abdomen is sufficient to remove scales and reveal shape of sternum A8.

REMARKS: Wing pattern fairly constant, higher variability in *O. pratana*.

Oar pratana (FABRICIUS, 1794)

Phalaena pratana FABRICIUS, 1794: Ent. Syst. 3 (2): 161 (North Africa, 'Barbaria'; neotype from Algeria: Bône). Original syntype(s) not traced in ZMUC, MNHN. Neotype, ♂, [eastern Algeria], 'Bône, 22.V., L. LHOMME', ZSM/HERBULOT, examined; herewith designated to stabilize usage of the name, with regard to the sister species *O. reaumuraria*.

Coremia oppressa WALKER, 1870: Entom. 5: 154 (Egypt: Hor Tamarib, Suqkin). Syntype(s) ♂. Possibly valid at subspecific rank (SCOBLE 1999).

Cidaria ectypata MABILLE, 1888: Bull. Soc. ent. Fr. 1888: 58 (Tunisia: Gabès). Syntype(s) (ZSM/HERBULOT: 1♀ examined, other 3♂ with possible type status).

Fidonia megiaria var. *obscuraria* BETHUNE-BAKER, 1894: Trans. ent. Soc. London 1894 (1): 45, pl. 1, fig. 12 (Egypt: Alexandria). Syntype(s). Synonym to subsp. *oppressa*.

Emmiltis nigrescens HAMPSON, 1896: in WALSINGHAM & HAMPSON, Proc. zool. Soc. London 1896 (1): 267, pl. 10, fig. 2 (Yemen: Aden). Syntype(s) ♂, (BMNH). Synonym to subsp. *oppressa*.

Fidonia? pratana var. *mortuaria* STAUDINGER, 1898: Dt. ent. Z. Iris 10 (2): 314 (Israel: Jordan Valley near Dead Sea). Syntypes '♀♀' (MNHU, 4♂1♀, with labels 'orig.', examined?). Valid at subspecific rank (SCOBLE 1999).

Oar pratana baezi subsp. n.: HOLOTYPE: ♂, 'Gran Canaria, Maspalomas, 28.V.1965, leg. J. KLIMESCH', ZSM. – PARATYPES: 7♂5♀, id.; 19♂, 'Gran Canaria, Maspalomas, XI.[19]58, IV-V.[19]61, II.[19]62, XI.[19]63, leg. PINKER', ZSM; 6♂, id., EMEM in ZSM; 2♂, 'Gran Canaria, Maspalomas, 16.IV.2000', leg. et coll. M. BAEZ. The name is dedicated to Dr. Marcos BAEZ, Tenerife, for his merits in the lepidopterous exploration of the Canary islands.

Unavailable names (infrasubspecific): *lactea*: D. LUCAS (1948); *pallens*: D. LUCAS (1948). – (incorrect subsequent spelling): *pratanaria*: TURTON (1802).

DIAGNOSIS: North African populations in habitus and external morphology as described for *O. reaumuraria*. As coloration features are not constant in 'subsp.' *oppressa* (usually blackish, but mixed with pale forms) and subsp *mortuaria* (usually greyish, but mixed with light and blackish forms), their validity needs confirmation by further study. PROUT (1938a: 224) states high individual variability, geographical subdivisions difficult to define.

Subsp. ***baezi*** subsp. n. smaller, wingspan ♂♀ 13-17 mm, markings darker, black brown, medial fascia straighter, less zigzagging, discal spot usually touching medial line. Types see above (nomenclatural paragraph).

MALE GENITALIA: Socii shorter, lateral processes of anellus longer, aedeagus cross-shaped, terminal spinules beside the large cornutus larger than in *O. reaumuraria*, posterior margin of sternum A8 with 3-8 long spines. Subsp. *baezi* with much smaller genitalia, cornuti smaller, sternum A8 with two short posterior processes only.

FEMALE GENITALIA: Antrum with distinct posterior notch, corpus bursae with broad sclerite posteriorly.

DISTRIBUTION: Extralimital: From central Algeria to Egypt, from southern Levant ('subsp. *mortuaria*') across Arabian peninsula to Yemen and Sudan ('subsp. *oppressa*'). On Canary islands and southern Morocco (Agadir region) replaced by subsp. *baezi* subsp. n..

PHENOLOGY: Plurivoltine, in all months of the year.

BIOLOGY: Possibly monophagous on *Suaeda* (Chenopodiaceae). Larva reared on a halophilous plant, probably '*Salicaria*' (=*Lythrum salicaria*? Lythraceae) (STAUDINGER 1898), on *Suaeda vermiculata* (Algeria: PROUT 1913) on *S. vera* (*S. fruticosa*) (ANDRES & SEITZ 1924), and on an unidentified *Suaeda* species (Israel: HALPERIN & SAUTER 1992). Halophilous plants, such as *Suaeda vera, Salsola* and *Arthrocnemum indicum* suspected as host-plants by RUNGS (1981). Adults diurnal, very easily flushed up from vegetation, rarely attracted to light, too. ♀ ratio in collections low, about 10-15%.

HABITAT: Xerothermophilous, inhabiting dry, sandy open habitats with sparse vegetation, such as dunes, deserts and oases. Halophilous (STAUDINGER 1898).

Cinglis GUENÉE, 1858

TYPE SPECIES: *Fidonia humifusaria* EVERSMANN, 1837. Two species in the western Palaearctic region. According to SIHVONEN (2003) synonym of *Scopula*, but venation, antennal pectination, structure of ♂ sternum A2 and A8, reduction of socii in ♂ genitalia and sclerotization of ♀ corpus bursae provisionally justify generic rank. Closely related to the genus *Pseudocinglis*, the latter distributed in western Asia and North Africa (Morocco: *Pseudocinglis moralesi* (RUNGS, 1945), comb.n.: HAUSMANN 2004). Small and distant eyes, long palpi, and bipectinate ♂ antennae reminiscent of *Oar*, but also of some presumably basal Sterrhini, e.g. *Cleta*.

DIAGNOSIS: Chequered wing pattern and chequered fringe diagnostic. Hindwing margin arched between veins. Venation (Text-fig. 157): Forewing with large areole close to cell apex and exceeding cell apex, R1 and R2-R4 distinctly stalked, R5 separate or connate. Hindwing with Rs and M1 shortly stalked. Discocellulars oblique. Cells of all wings short. Eyes small, distant from each other. Proboscis present, but very short. Palpi long and bushy. Antennae of ♂ bipectinate to tip (unknown for Palaearctic *Scopula*), branches densely ciliate. Antennae of ♀ filiform. Hindtibia of ♂ stout, with two spurs, hindtarsus not shortened. Hindtibia of ♀ with 4 spurs. Sternum A2 of ♂ not modified, i.e. without central sclerite and setose patch.

MALE GENITALIA: Socii connected by a sclerotized membrane. Fibula broad. Anellus with short and narrow lateral processes. Aedeagus with one single cornutus. Sternum A8 with anterior margin concave, left ceras vestigial, right ceras long. Not polymorphic (n=8).

FEMALE GENITALIA: Apophyses anteriores stout. Sterigma with circular sclerite around ostium bursae. Corpus bursae without signum, but with lateral sclerite.

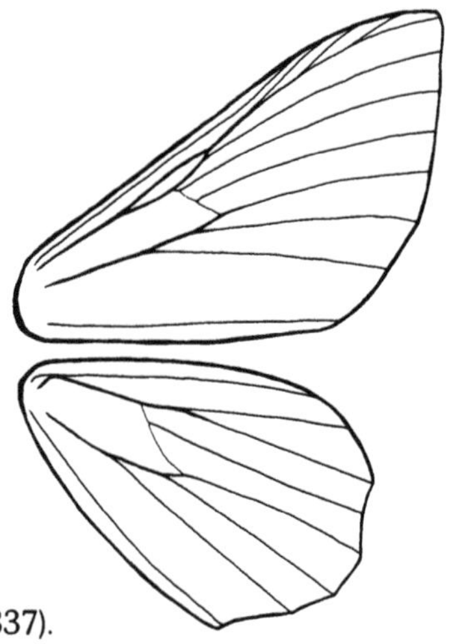

Text-fig. 157: Venation of *Cinglis humifusaria* (EVERSMANN, 1837).

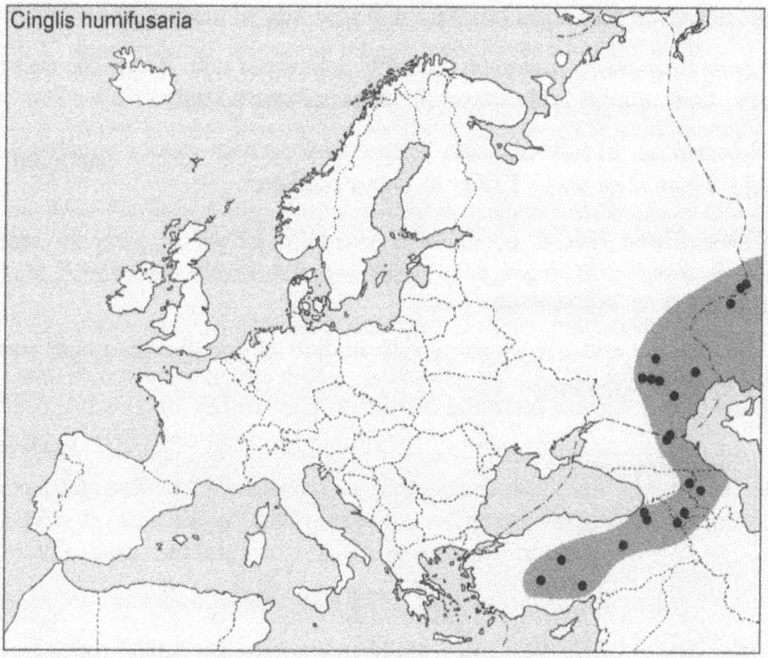

116. Cinglis humifusaria (EVERSMANN, 1837)

Fidonia humifusaria EVERSMANN, 1837: Bull. Soc. imp. Moscou 10 (6): 57 (southern European Russia: Lower Volga). Lectotype ♂ (ZISP, examined), herewith designated to stabilize usage of the name with regard to the Iberian sister species.

Unavailable names (incorrect subsequent spelling): *humifuscaria*: EVERSMANN (1844); *humifusiaria*: BOLLAND (1975); *humisfuscaria*: EXPOSITO (1978).

DIAGNOSIS: Wingspan ♂♀ 18-23 mm, in the second generation sometimes 16 mm only. Ground colour white. Transverse lines and veins broadly dark brown marked, forming a reticulate pattern. Postmedial line double. Discal spots large. Terminal line dark brown, fine, uninterrupted. Fringe strongly chequered. Frons grey brown. Eyes encircled with white scales resembling spectacles. Abdomen dark, with white segmental transverse rings. Length of palpi 1.3-1.8 times diameter of eye (♂♀). Longest ♂ antennal branches 1.5-2 times width of flagellum. Further features under generic diagnosis.

MALE GENITALIA: Aedeagus with two straight spinulose crests. Sternum A8 deeply concave anteriorly, left ceras vestigial, very short, right ceras tapering at tip. Further features under generic diagnosis.

FEMALE GENITALIA: Postvaginal membrane strongly corrugated. Sterigma with small circular sclerite around ostium bursae. Ductus bursae comparatively long and slender. Lateral sclerite of corpus bursae bipartite. Female genitalia figured in WARDIKJAN (1985) under '*C. humifusaria*' are misidentified (identity: *Tephrina* spec., Ennominae). Further features under generic diagnosis.

DISTRIBUTION: (East-)European-Westasiatic. Widely distributed in southern European Russia. – Outside Europe from central Turkey to Caucasus, Transcaucasus and from Turkmeniya to central Asian mountains. In the Iberian peninsula and in North Africa, perhaps also in the Levant, replaced by the sister species *C. andalusiaria*.

PHENOLOGY: Bivoltine: late April to mid-June; mid-July to early August.

BIOLOGY: Larval host-plants unknown, possibly associated with *Artemisia* (see *C. andalusiaria*). Adults both diurnal and nocturnal, attracted also to light.

HABITAT: Halophilous. In habitats with 'Artemisia-Poaceae' steppe vegetation (MIRONOV pers. comm.). From 0 m up to 1,000 m above sea-level.

SIMILAR SPECIES: Sister species *C. andalusiaria* with longer ♂ antennal branches and longer palpi. *S. tessellaria* larger, postmedial line not double, abdominal segments with indistinct pale rings or unicolorous.

REMARKS: Taxonomic and geographical delimitation of the species needs further study, see remarks to *C. andalusiaria*.

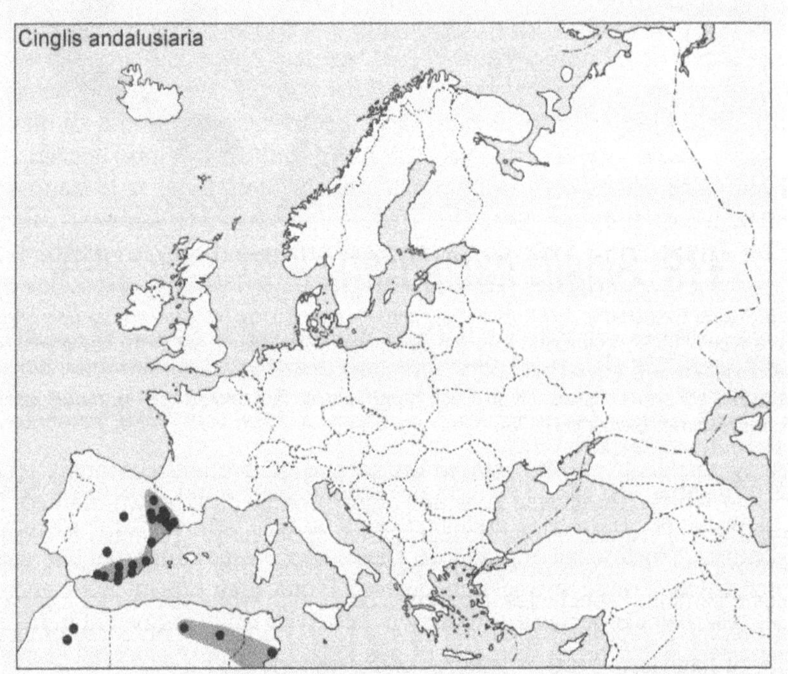

117. Cinglis andalusiaria WAGNER, 1935

Cinglis andalusiaria WAGNER, 1935: Zeits. österr. Ent. Ver. 20 (3): 17 (Spain, Andalusia Lanjaron). Syntypes ♂♀ (NHMW: 1♂1♀, examined, labelled as 'type', other 1♂1♀ from Algezeares, labelled as 'paratype').
Unavailable names (misidentification): *humifusaria*: sensu STAUDINGER (1871) nec EVERSMANN, 1837. – (infrasubspecific): *reducta*: THIERRY-MIEG (1915).

DIAGNOSIS: Wingspan ♂♀ 18-22, second generation 16-18 mm. Without constant differential features from the sister species *C. humifusaria* in habitus and external morphology, except longer ♂ antennal branches and longer palpi: length of antennal branches about 2.2-2.5 times width of flagellum, length of palpi 1.7-2.0 times diameter of eye in ♂, in ♀ slightly longer. None of the differential characters 'larger size, body more robust, coloration more chocolate brown, white pattern reduced, discal spots less distinct, frons not white' (WAGNER 1935; PROUT 1935a) revealed to be significant and constant.

MALE GENITALIA: Aedeagus with spinulose crests vestigial. Sternum A8 less concave anteriorly than in *C. humifusaria*, left ceras longer, right ceras rounded at tip.

FEMALE GENITALIA: Postvaginal membrane slightly furrowed. Sterigma with large circular sclerite around ostium bursae. Ductus bursae short and broad. Lateral sclerite of corpus bursae not bipartite.

DISTRIBUTION: West-Mediterranean. From northern to southern Spain. – Outside Europe in very scattered populations from Morocco to Libya. For one ♂ from Jordan see remarks.

PHENOLOGY: Bivoltine: early May to late June; late July to mid-September.

BIOLOGY: Larva recorded on *Artemisia herba-alba* (Asteraceae) (Spain: REDONDO & GASTON 1999). Adults both diurnal and nocturnal, attracted to light. Easily flushed up during daytime from *Artemisia herba-alba* (PROUT 1913).

HABITAT: Dry steppes, adults associated with occurrence of *Artemisia herba-alba* (BOLLAND 1975). From 0 m up to 500 m above sea-level, exceptionally up to 800 m. In Morocco, Middle Atlas, collected at 1,700 m above sea-level.

SIMILAR SPECIES: *C. humifusaria* (see no. 115). *S. tessellaria*, see under *C. humifusaria*.

REMARKS: Following recent publications (e.g. SCOBLE 1999; REDONDO & GASTON 1999; DANTART 2000; TRUSCH & MÜLLER 2001; REDONDO et al. 2001) species rank is accepted here, preliminarily. However, on the base of the currently available, limited information on North African and Levantine populations, clinous transitions, as suggested by PROUT (1935a), cannot be excluded. More detailed studies are needed. Genitalia of one examined single ♂ from Jordan match better those of *C. andalusiaria*.

Scopula SCHRANK, 1802

(=*Acidalia*: sensu TREISCHKE, 1825 in OCHSENHEIMER, nec HÜBNER, 1819)

TYPE SPECIES: *Phalaena paludata* LINNAEUS, 1767 (in SCHRANK as '*Pyralis paludalis*'), (junior synonym of *S. ornata*). In the traditional, strict concept including 700 species worldwide, some 800 species in the wider definition proposed by SIHVONEN (2003), see introduction to tribe. *Acidalia* TREITSCHKE, 1825, which had been used as valid genus name for many years, is a junior synonym of *Scopula* and a junior homonym of *Acidalia* HÜBNER, 1819, which refers to Nymphalidae. The excellent study of STERNECK (1940/1941) is used as basis for the systematic arrangement in this book. Some of STERNECK's subgroups are suggested to be monophyletic (STERNECK 1940/1941: 205). Monophyly is strongly supported for the *ornata* species-group, by several characters. For the need of large scale revision worldwide see remarks to genus *Idaea*. Particular care is required with regard to the polymorphism of the genitalia, and to the frequent spur anomalies (see below). Important revisionary work has been started by SIHVONEN (2003) and SIHVONEN & KAILA (2004). Morphology of Armenian species discussed in WARDIKJAN (1970).

DIAGNOSIS: Forewing comparatively broad. Usually with dirty white or sand coloured wings, discal spot black and distinct. Underside of forewing often darker than that of hindwing. Less robust and less hairy than other Scopulini, such as *Somatina* and *Problepsis*. Venation (Text-fig. 158) with low infraspecific variation. Forewing with single areole, not exceeding cell apex. R1 arising from apex of areole, R2-R4 and R5 distinctly stalked; exception: the *immistaria* species-group. Hindwing with Rs and M1 usually slightly separate, but sometimes connate or shortly stalked. Palpi short, smooth scaled. Antennae of both sexes filiform, in ♂ usually ciliate-fasciculate, exceptionally with very short branches (extralimital species), those of ♀ almost without ciliation; exception: *S. submutata* and *S. vigilata*. Hindtibia of ♂ usually without spurs, often with pencil, rarely with 1-2 spurs. Infraspecific variation of ♂ hindtibial spur number observed in many species, e.g. *S. immorata, S. decorata, S. decolor, S. rubellata, S. ternata*. Hindtarsus of ♂ long or slightly shortened. Hindtibia of ♀ always with 4 spurs. Sternum A2 of ♂ with central sclerite and paired setose patch (Text-fig. 3). Further features under diagnosis of tribe.

MALE GENITALIA: Genital capsule ovate, saccus broad. Socii setose, usually long. Fibula and valva sometimes asymmetric. Sternum A8 with mappa, usually with cerata, in a very few species reduced (reversal). Cerata usually setose at tip, often asymmetrical. Further features under diagnosis of tribe. Shape and sclerotization of fibula and valva often diagnostic for species-groups. Shape of socii and sternum A8 usually diagnostic for species. Identification often possible by removing scales from abdomen, but high caution is necessary, as the cerata are easily broken in this procedure. Many species with polymorphic length of cerata. Despite these polymorphic patterns differential features in shape and length of cerata may be fairly stable e.g. in the sister species *S. turbidaria, S. turbulentaria* and *S. ochraceata*. Similarly, geographical borders between morphologically different taxa are sometimes surprisingly sharp, e.g. between *S. beckeraria beckeraria* and *S. b. rebeli*.

FEMALE GENITALIA: Lamella antevaginalis usually sclerotized strongly, often semicircular. Signum often developed as ovate patch of small, flat, separate spinule-bearing sclerites, pointing away from longitudinal axis. Though rather different from the signum of most other Geometridae, such patches are called 'signum', here.

BIOLOGY: Larva rapidly hatching from egg, usually after just 3-7 days. Larvae forming questionmark-shape or crook when irritated, e.g. *S. immorata, S. immutata, S. ternata*. Larvae of some other species hanging straight down from host-plant, and swinging when disturbed, e.g. *S. imitaria, S. umbelaria, S. nigropunctata*. Rearings are usually

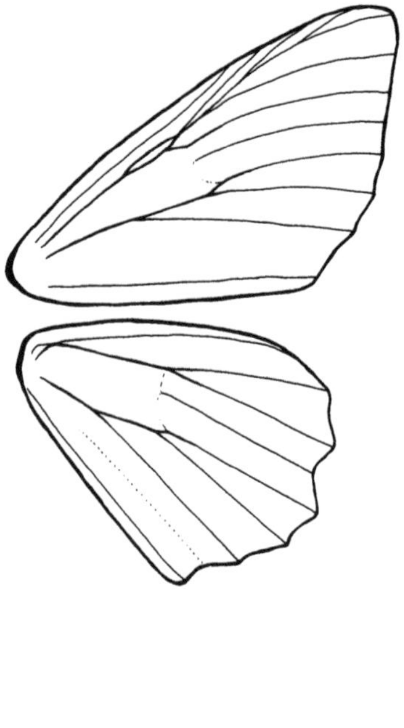

Text-fig. 158: Venation of *Scopula ornata* (SCOPOLI, 1763).

Text-fig. 159: Long and slender larva of *Scopula rubellata* (STAUDINGER, 1871) (Photo: LEIPNITZ).

easy on many plant species, similarly to those of *Idaea*, but there are also some specialists, e.g. the *ornata* species-group (MÜLLER pers. comm.).

IMMATURE STAGES: Eggs with strong longitudinal and weaker transverse ridges. Reddish spots appearing some days after being deposited. Larva long and very slender (Text-fig. 159), often with furrowed, keel-shaped lateral ridge. Pupal cremaster with four pairs of setae, terminal pair diverging, longer and stouter than the other setae (HOLLOWAY 1997; PATOČKA 2003; SIHVONEN 2003).

REMARKS: For transfer of '*Scopula' moralesi* (RUNGS, 1945) to the genus *Pseudocinglis* HAUSMANN, 1994, see under *Cinglis* (dissected ♂ holotype from Morocco examined in MNCN).

Subgenus ***Scopula*** SCHRANK, 1802

(= *Eucidalia* FLETCHER, 1979, replacing *Eucidalia* STERNECK, 1941: nom. invalidatum)

TYPE SPECIES: *Phalaena paludata* LINNAEUS, 1767 (in SCHRANK as '*Pyralis paludalis*') (junior synonym of *S. ornata*).

MALE GENITALIA: Socii often broad, shorter than their distance apart. Fibula with normal sclerotization, at most dark brown terminally. Shape of fibula diagnostic for species-groups. Genitalia not rarely asymmetric. Aedeagus usually broad, often with cornutus. Cerata often broad and stout.

FEMALE GENITALIA: Antrum and lamella antevaginalis often broad and large, the latter sometimes asymmetric.

The ***immorata*** species-group

STERNECK's 1st group of *Scopula*. Heterogeneous group (STERNECK 1941: 42). Possible relationship of *S. immorata* and *S. tessellaria* to *Cinglis* supported by habitus and ♂♀ genitalic details (♂: shape of cornutus; ♀: ring-shaped antrum). Palpi comparatively long and bushy in *S. immorata* and *S. tessellaria*, shorter in the other species. Proboscis well developed. Antennae of ♂ dentate, with deep intersegmental incisions, ciliate-fasciculate. Hindtibia of ♂ without spurs, tarsus shortened in many species. Setose patch on ♂ sternum A2 with short setae, in *S. nigropunctata* and *S. nemoraria* central sclerite exceptionally large. Male genitalia: Anellus with long lateral processi. Aedeagus short and thick, plesiomorphic according to STERNECK (1941: 42). Sternum A8 apparently without polymorphisms, cerata strongly spinulose.

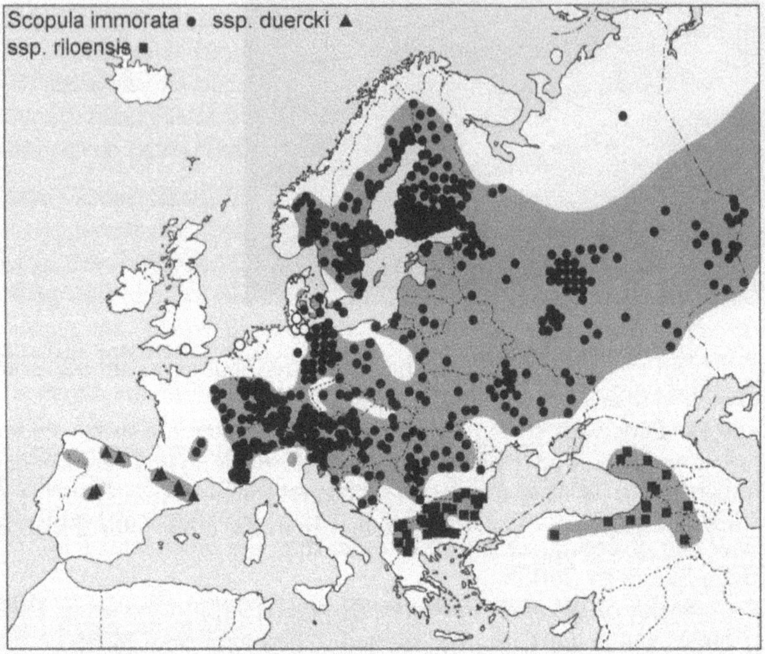

118. Scopula immorata (LINNAEUS, 1758)

Phalaena Geometra immorata LINNAEUS, 1758: Syst. Nat. (Ed. 10) 1: 528 (Europe). Syntype(s) (LSL).

Phalaena contaminata SCOPOLI, 1763: Ent. Carniolica: 225, fig. 559 (north-western Slovenja: Carniolia 'Mts.'). Syntype(s) lost.

Phalaena graminata HUFNAGEL, 1767: Berlin. Mag. 4 (6): 614 (Germany: Berlin). Syntype(s) lost.

‡ *Phalaena festucaria* BRAHM, 1791: Insektenkal. Samml. Oek. 2: 321 (western Germany: near Mainz). Syntype(s). Junior primary homonym nec HÜBNER, 1790 (identity: *Macaria artesiaria*, Ennominae)

‡ *Phalaena fuscata* FABRICIUS, 1794: Ent. Syst. 3 (2): 200 (Denmark: Seelandia). Syntype 1♂ (ZMUC, examined). Junior primary homonym nec HUFNAGEL, 1767 (identity: *M. murinata*, Larentiinae).

‡ *Geometra immoraria* HÜBNER, 1799: Samml. Eur. Schmett. 5, Geometrae (1): pl. 25, fig. 133. Incorrect subsequent spelling.

‡ *Acidalia serenata* TURATI, 1905: Naturalista sicil. 18: 44, pl. 8, fig. 3 (type locality not stated). Syntype(s).

Acidalia immorata L. var. *riloensis* ZÜLLICH, 1936: Z. Wien. ent. Ges. 21: 55 (Bulgaria: Rilo Mts.). Valid at subspecific rank.

Scopula immorata duercki SHELJUZHKO, 1955: Mitt. Münch. ent. Ges. 44/45: 289 (Spain, Castilia: Sierra de Gredos). Holotype ♂ (ZSM, examined). Valid at subspecific rank.

Unavailable names (infrasubspecific): *albofasciata*: REBEL (1914); *albomarginata*: HABICH (1903); *anastomosaria*: REISSER (1961); *bistrigata*: GALVAGNI (1914); *fuscomarginata*: HÖFER (1922); *luctuata*: REBEL (1914); *pallidior*: SKALA (1913); *porosa*: KRULIKOVSKY (1910); *unicolorata*: SCHNEIDER (1932); *unistrigata*: GALVAGNI (1914). – (incorrect subsequent spelling): *contaminara*: TOMIC et al. (2002); *imoraria*: TOMIC et al. (2002); *palliodor*: ZECEVIC & RADOVANOVIC (1974); *unicoloraria*: PROUT (1935).

DIAGNOSIS: Wingspan ♂ 23-27, ♀ 22-26 mm, second generation 19-25 mm. Ground colour grey, with broad, diffuse, undulate, brown transverse lines. ♀ paler, with ground colour more extended between the well marked transverse lines. Terminal area brown, with the whitish wavy line often dissolved into dots in ♂, confluent in ♀. Discal spots absent, on hindwing sometimes small and inconspicuous. Fringe slightly chequered. Palpi, frons, vertex and collar with mixed whitish and blackish scales. Palpi 1.2-1.5 times diameter of eye, exceptionally up to nearly 2 times. Antennae of ♂ with deep intersegmental incisions, ciliate-fasciculate, cilia curved, 1.5-1.9 times width of flagellum, exceptionally up to 2.2 times. Hindtibia of ♂ without spurs, with hairy scales, but without pencil. Exceptionally one or two vestigial spurs present.

Variability: With slight tendency to melanism (SCHNEIDER 1932; SVENDSEN & HAURITZ 1991), low temperature inducing dark grey forms, higher temperatures forms with yellowish tinge (BERGMANN 1955). Populations from the Ukraine, Ural, Altai and eastern Asia with ochreous or light brown markings, much lighter than in typical central European populations.

All the univoltine, montane populations from Spain (subsp. ***duercki***), Balkan peninsula, Pontus, Caucasus (subsp. ***riloensis***), and western High Alps, larger, ♂ 25-30, ♀ 24-26 mm. With stronger brown suffusion, wavy line often better marked, transverse lines sometimes distinct (chiefly subsp. *riloensis*). Subsp. *duercki* with palpi up to nearly 2 times diameter of eye.

MALE GENITALIA: Fibula tapering, curved. Lateral processi of anellus long and slender. Aedeagus short and broad, with broad terminal cornutus, round lateroterminal lobe with 1-3 small spinules. Sternum A8 with short left and broad and long right ceras. Not polymorphic (n=13).

FEMALE GENITALIA: Lamella antevaginalis rounded or slightly notched. Semicircular clasp around ostium bursae. Ductus bursae broad, sclerotized, obliquely joining corpus bursae. Signum laterally spinulose, drop-shaped or sub-triangular.

DISTRIBUTION: Euro-Siberian. Widely distributed in temperate areas of Europe, in the east to the Urals, isolated populations in Spain (subsp. *duercki*) and in the Balkan peninsula (subsp. *riloensis*). Absent from Italy south of Alps (FLAMIGNI et al. 2001), and from Mediterranean islands. Erroneously recorded from Sicily (e.g. MÜLLER 1996), data referring to *S. tessellaria*. In the Netherlands only one old record from 1901 (KUCHLEIN & DE VOS 1999). Extinct in Great Britain (East Sussex) since 1961 (SKINNER 1984; PORTER 1997; WARING & TOWNSEND 2003). – Outside Europe in northern Turkey, Caucasus, Transcaucasus, northern Iran. All these populations may be assigned to subsp. *riloensis*, provisionally. In the east across Siberia to Mongolia and western Amur region (VOJNITS 1976; VIIDALEPP 1996).

PHENOLOGY: Usually bivoltine: mid-May to early July; late July to late August. Under good conditions emerging two weeks earlier. On mountains, e.g. subsp. *duercki* from Spain, subsp. *riloensis* from Balkan peninsula or in the western High Alps, univoltine, from mid-June to late July, with transitional populations along altitudinal gradients. Second generation incomplete or absent also in the north of the distribution.

BIOLOGY: Larva polyphagous on withered leaves of herbaceous plants. Egg deposition observed on *Brachypodium pinnatum* (Poaceae) (EBERT 2001). Larva found on *Thymus pulegioides*, *Origanum vulgare* (Lamiaceae) (Germany: PROUT 1935a, EBERT 2001; SCHNEIDER 1939; LEIPNITZ pers. comm.), *Artemisia* (Asteraceae), *Calluna vulgaris* (Ericaceae) (Poland: URBAHN & URBAHN 1939), *Hieracium* (Asteraceae), *Lychnis viscaria*

(Caryophyllaceae), *Plantago major* (Plantaginaceae) (Finland: Reuter 1890; Seppänen 1954). Recorded furthermore on *Lonicera caprifolium* (Caprifoliaceae), *Polygonum* (Polygonaceae), *Erica* (Ericaceae), *Thymus serpyllum* (Lamiaceae), *Plantago* (Plantaginaceae), *Achillea millefolium*, *Taraxacum officinale* (Asteraceae) (var. cit., e.g. Denis & Schiffermüller, 1775; Lhomme 1935; Valle 1946; Allan 1949; Bergmann 1955), data perhaps partially referring to rearings. Reared on *Achillea millefolium*, *Salvia pratensis*, *Lactuca sativa*, *Lonicera tatarica*, *Taraxacum officinale*, *Galium*, *Polygonum aviculare* and *Plantago lanceolata* (var. cit., e.g. Woodbridge 1906; Hoffmann & Klos 1917; Schneider 1939; Allan 1949; Bergmann 1955; Seppänen 1970; Ebert 2001; Sommerer pers. comm.). Adults very easily flushed up from vegetation during daytime, almost diurnal, main activity at early dusk and after sunrise (Rennwald pers. comm.), but also attracted to light. Nectar feeding rarely observed, once at *Thymus pulegioides* (Ebert 2001).

Parasitoids: Diptera, Tachinidae: *Paratryphera barbatula* (Herting 1960; 1984).

Habitat: Thermophilous species of open habitats, in both dry or slightly wet localities, preferring uncultivated open land, grassland rich in flowers, heathland, steppes, forest fringes, glades and aisles. On low-nutrient sandy or silicious soils. In central and northern Europe usually from 0 m up to 1,200 m above sea-level, in the Alps, in southern Europe and Turkey up to 2,400 m and almost absent from the lowlands. In eastern Turkey and northern Iran up to 3,000 m.

Similar species: Confusions may occur with *S. tessellaria*, especially in females, *S. tessellaria* on average larger, markings and veins darker brown, well contrasted on whitish ground colour, fringe distinctly chequered, discal spot of hindwing longer. In ♂, identification is possible by brushing of the abdomen to remove scales and reveal shape of sternum A8. *S. immorata riloensis* from Caucasus, eastern Pontus and northern Iran particularly similar to *S. tessellaria*.

Remarks: Rearings and immature stages described in Woodbridge (1906).

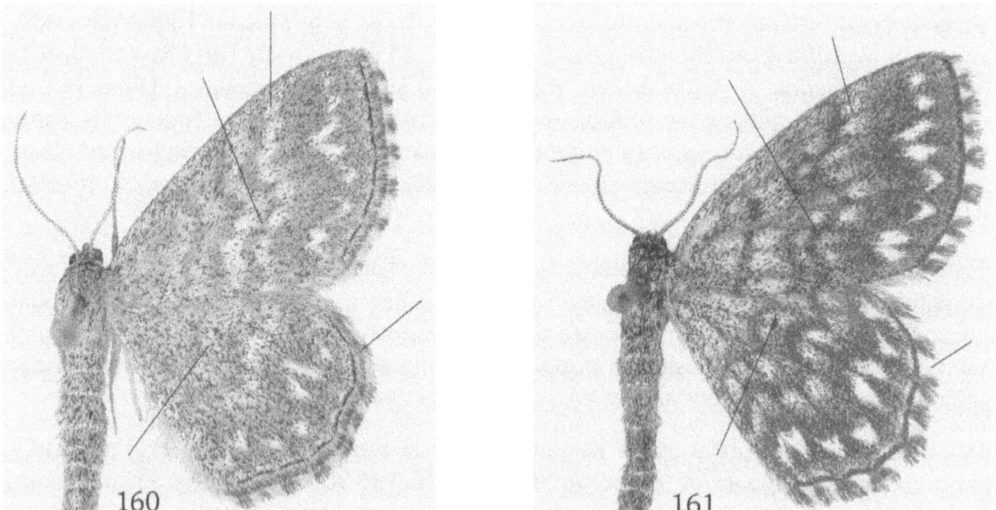

Text-figs 160-161: Diagnostic characters (indicated) of European *Scopula* species. Text-fig. 160: *S. immorata* (Linnaeus, 1758). Text-fig. 161: *S. tessellaria* (Boisduval, 1840).

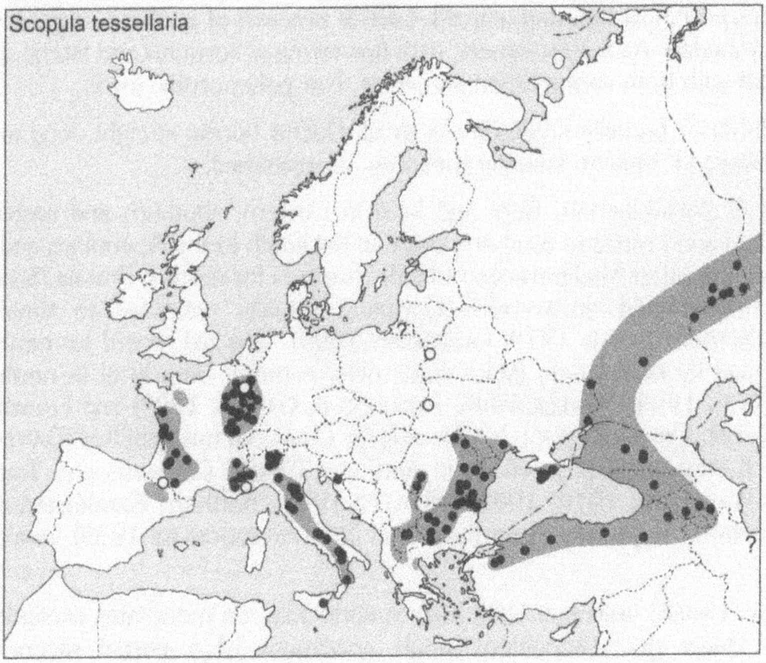

119. Scopula tessellaria (BOISDUVAL, 1840)

Strenia tessellaria BOISDUVAL, 1840: Genera Index meth. eur. Lepid.: 228 (northern Italy). Syntype(s) (ZFMK: 1♂ ex coll. BOISDUVAL, examined, without abdomen).

‡? *Phalaena pellicea* FOURCROY, 1785: Entom. Parisiensis 2: 315 (France: Paris). (Senior) synonymy possible (LERAUT 1997). If verified, validity must be rejected as nomen oblitum (not used for 220 years), in order to stabilize nomenclature.

Eupisteria pulverulentaria SÉLYS-LONGCHAMPS, 1844: Bull. Soc. ent. Fr. (2) 2: 13 (southern Italy: Caserta). Syntype(s).

Acidalia tabianaria TURATI, 1905: Naturalista sicil. 18: 43, pl. 8, figs. 5-14 (Italy, Parma Prov.: Tabiano; Salso Maggiore). Syntypes 4♂♀ (MSNT: 2♂1♀, examined by RAINERI (1996, gen.figs. 7-8), one syntype lost).

Scopula tessellaria proutiana SHELJUZHKO, 1955: Mitt. Münch. ent. Ges. 44/45: 289 (Kirghizstan, Semiretshje: river Maloje 120 km S Aksu). Holotype ♂ (ZSM, examined). Herewith downgraded from subspecific rank (VIIDALEPP 1996; SCOBLE 1999) to synonymy (syn. n.), without constant and significant differences from nominate subspecies.

Unavailable names (infrasubspecific): *meissli*: SCHAWERDA (1913). – (incorrect subsequent spellings): *tabiascaria*: PROUT (1913:3k); *tersellaria*: EXPOSITO (1978); *tesellaria*: DIDMANIDZE (2002); *tesselaria*: BRUAND (1842); *tessellata*: CULOT (1917: fig. 195).

DIAGNOSIS: Wingspan ♂ 25-29, ♀ 23-28 mm. Ground colour whitish, with distinct, undulate, dark brown transverse lines. Veins with dark scales forming a reticulate wing pattern. Markings lighter in south-western France. Coloration of ♀ lighter, due to the more extended ground colour between the transverse lines. Terminal area dark brown, with the whitish wavy line usually dissolved into dots. Discal spots usually present on hindwing. Fringe distinctly chequered. Palpi, frons, vertex and collar dark brown with single whitish scales. Eyes encircled with white scales resembling spectacles (compare *Cinglis*). Palpi about 1.3-1.7 times diameter of eye. Antennae of ♂ with deep intersegmental incisions, ciliate-fasciculate, cilia curved, 1.7-2.2 times width of flagellum. Hindtibia of ♂ without spurs, with hairy scales, but without pencil.

MALE GENITALIA: Fibula tapering, curved. Lateral processi of anellus comparatively short, broad and rounded. Aedeagus slender, with fine terminal cornutus and lateral spine at tip. Sternum A8 with both cerata broad and short. Not polymorphic (n=9).

FEMALE GENITALIA: Lamella antevaginalis small. Ductus bursae straight, long and slender, slightly sclerotized. Signum laterally spinulose, drop-shaped.

DISTRIBUTION: Euro-Siberian. Rare and local in western, southern and eastern Europe from Belgium and France to southern Urals, in the south to the Apennines and Sicily, but absent from the other Mediterranean islands. Records for eastern Prussia (SPEISER 1903) probably misidentified. In western Germany actually restricted to three localities (Saarland: SCHMIDT-KOEHL 1979; GELBRECHT 1999). One old record for eastern Poland not confirmed by recent data (MALKIEWICZ pers. comm.). Very local in northern Spain (VIVES MORENO 1994; MÜLLER 1996; REDONDO & GASTON 1999) and French Pyrenees (RONDOU 1934; ORHANT 1996), but records for Catalonia misidentified (DANTART 2000). – Outside Europe local populations in northern Turkey, Caucasus and Transcaucasus (Armenia: WARDIKYAN 1970; 1985; VIIDALEPP 1996), northern Kazakhstan and central Asian mountains. Reputedly in northern Iran (SCHWINGENSCHUSS 1939), awaiting confirmation.

PHENOLOGY: Usually univoltine: late May to early July, on mountains exceptionally until late July. Under good conditions single specimens of a partial second brood in September.

BIOLOGY: Larva reputedly on *Medicago sativa* (Fabaceae), *Erica* (Ericaceae) (PROUT 1913; LHOMME 1935; FORSTER & WOHLFAHRT 1981) and *Pulsatilla vulgaris* (Ranunculaceae) (WARLET 1968; SCHMIDT-KOEHL 1979). Reared on *Origanum vulgare*, not accepting other plant species (GELBRECHT, LEIPNITZ pers. comm.), but also on leaves of *Taraxacum officinale* (MÜLLER pers. comm.), and *Medicago* (FIUMI & CAMPORESI 1988). Adults almost diurnal, very easily flushed up from vegetation during daytime, sometimes also attracted to light. ♀ ratio in collections low, 20-30%.

HABITAT: Usually montane. Inhabiting uncultivated open habitats on low-nutrient soils, both on dry and slightly wet ground. In Russia on chalk steppes (ANIKIN et al. 2000). Usually from 1,000 m up to 1,800 m above sea-level, also in the western Alps, but in the two distribution areas in western France and from northern France to Germany in the lowlands from 0 m up to 500 m. On central Asian mountains up to 3,000 m.

SIMILAR SPECIES: *C. humifusaria* (see no. 116), *C. andalusiaria* (117). *S. immorata* (chiefly the ♀, 118).

120. *Scopula corrivalaria* (KRETSCHMAR, 1862)

Acidalia corrivalaria KRETSCHMAR, 1862: Berl. Ent. Zeits. 6: 136, pl. 1, figs 6a, 6b (Germany: Berlin). Syntypes ♂♀.

Scopula corrivalaria eccletica PROUT, 1935: in SEITZ, Macrolep. 4, Suppl.: 43, pl. 5, fig. b (eastern Russia: Ussuri). Syntype(s) (BMNH). Valid at subspecific rank (SCOBLE 1999).

Scopula corrivalaria limburgensis SCOBLE, 1999: Geometrid moths of the world 2: 845. First available name after *limburgensis* PROUT, 1935: in SEITZ, Macrolep. 4, Suppl.: 43 (infrasubspecific) (Netherlands: Limburg). Syntype(s) (BMNH). Herewith downgraded from subspecific rank (SCOBLE 1999) to synonymy (syn. n.).

Unavailable names (misidentification): *sylvestraria* (var.): sensu HERRICH-SCHÄFFER (1845, pl. 18, figs. 106, 107: non binominal) nec HÜBNER, 1799 (identity: *Idaea sylvestraria*). – (incorrect subsequent spelling): *corrivaleria*: ŠULCS & VIIDALEPP (1972); *corrivallaria*: PROUT (1913).

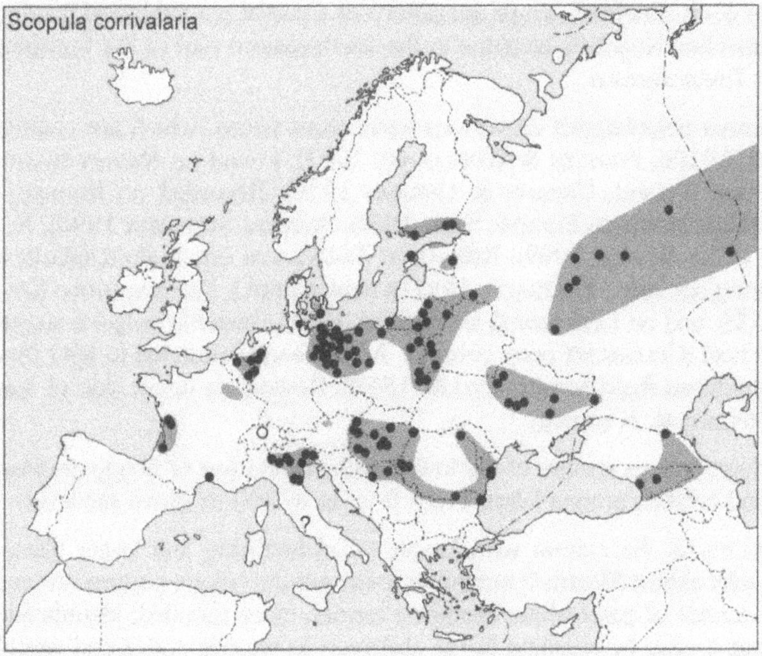

DIAGNOSIS: Wingspan ♂♀ 19-24 mm, second generation 17-20 mm. Larger in the west, smaller in the east of the distribution. Hindwing termen slightly angled at M3. Wings cream coloured with slight grey tinge. Medial line distinct, close to the strongly dentate postmedial line. Terminal area with dark bordered wavy line. Terminal dots and discal spots black, distinct. Medial line of hindwing usually crossing discal spot or distal to it. Wing pattern on underside very distinct. Frons and palpi blackish, vertex pale sand coloured, collar ochreous to light brown. Antennae of ♂ with broad flagellum and deep intersegmental incisions, ciliate-fasciculate, cilia long, 2.0-2.4 times width of flagellum. Hindtibia of ♂ stout, with pencil, tarsus shortened, 0.5-0.65 times length of tibia.

MALE GENITALIA: Socii short, broad and rounded. Fibula and valva sclerotized, tapering. Aedeagus curved, without cornutus. Sternum A8 with both cerata short, strongly spinose, mappa rounded, of variable extension. Not polymorphic (n=6).

FEMALE GENITALIA: Ductus bursae straight, with lateral sclerotization. Corpus bursae oval. Spinulose patch of signum elongate.

DISTRIBUTION: Palaearctic (disjunct). Widespread, but always local and rare in the lowlands of western and central Europe, from France to central Urals, in the north to Denmark, Baltic states and southern Finland, in the south to northern Apennines (FLAMIGNI et al. 2001) and Bulgaria (BESHKOV & ABADJIEV 2000). Absent from most parts of southern Europe. Records from central Italy (RACHELI et al. 1978; PROLA & RACHELI 1979) erroneous (identity: *Scopula minorata*; RAINERI det., ZILLI pers. comm.). Doubtful records from Astrachan region (LVOVSKY 1971). Strongly threatened by destruction of wetland habitats. – Outside Europe in eastern Caucasus and Transcaucasus (ZISP; VIIDALEPP 1996; DIDMANIDZE 2002), subsp. *eccletica* in Transbaikalia, Far East of Asia and Japan.

PHENOLOGY: Usually univoltine: mid-June to mid-July, sometimes two weeks earlier or later. Under good conditions single specimens of a partial second brood from mid-August to mid-September. Regularly bivoltine in the south-eastern part of the European distribution and in Transcaucasus.

BIOLOGY: Larva polyphagous on various herbaceous plants, which are characteristic for bogs (REBEL 1910b; FORSTER & WOHLFAHRT 1981). Found on *Rumex hydrolapathum* (Polygonaceae) (Poland: URBAHN & URBAHN 1939). Recorded on *Rumex*, mainly *R. hydrolapathum* (northern Europe: SKOU 1986; Sweden: SVENSSON 1993), *R. aquaticus* (SEPPÄNEN 1970; SCOBLE 1999). Reared on *Taraxacum officinale* (URBAHN & URBAHN 1939: 'rearing not easy'; ERLEBACH, MÜLLER pers. comm.), *Lactuca sativa* (ZELLER 1869; REBEL 1910b), and on *Lysimachia thyrsiflora*, but *Lysimachia vulgaris* suggested to be the natural host (GELBRECHT pers. comm.). Adults readily attracted to light (SKOU 1986), but ♀ with reduced flight activity (ZELLER 1869). Resting on upper side of leaves during daytime. Stenotopic, K-strategy.

HABITAT: Hygrophilous species of the lowland plains. On wet or boggy grassland, marshland, wetland habitats around lakes. From 0 m up to 300 m above sea-level.

SIMILAR SPECIES: *S. flaccidaria* with longer tailed hindwing and larger discal spots. *S. immutata* with almost filiform ♂ antennae, paler ground colour, pattern on underside less distinct, underside of palpi white, hindwing termen more rounded; identification of ♂ of *S. immutata* is easy by brushing tip of abdomen to remove scales and reveal shape of the long right ceras. Small *S. corrivalaria* may even resemble *S. minorata*, the latter with rounded hindwing, discal spots smaller, ♂ hindtarsus longer.

REMARKS: No material of the east Asian populations could be examined, but habitus and ♂ sternum A8 are figured in ESAKI et al. (1957). *S. contramutata* is not related to this species, but to *S. immutata* (subgen. *Calothysanis*). For morphology of immature stages see ZELLER (1869) and RÜCKER (1921).

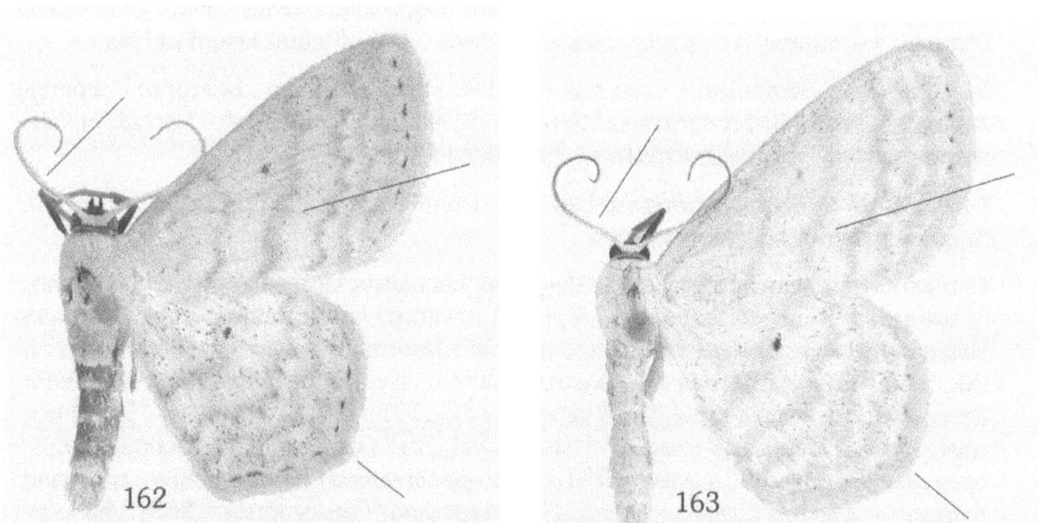

Text-figs 162-163: Diagnostic characters (indicated) of European *Scopula* species. Text-fig. 162: *S. corrivalaria* (KRETSCHMAR, 1862). Text-fig. 163: *S. immutata* (LINNAEUS, 1758).

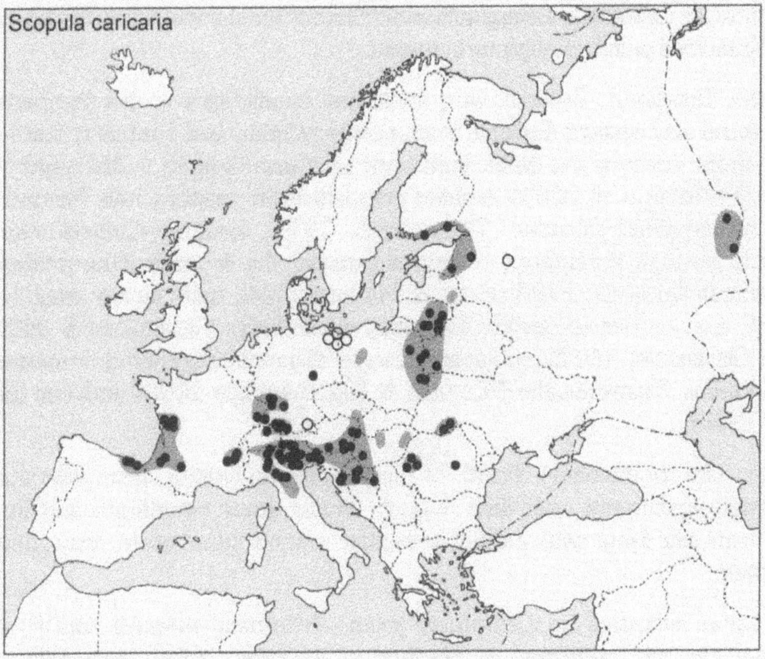

121. Scopula caricaria (REUTTI, 1853)

Acidalia caricaria REUTTI, 1853: Beitr. rhein. Naturg. 3 (Lep. Baden): 14, 113 (Germany: Baden). Lectotype ♂ (SMNK, figured in EBERT 2001). Some years ago, the senior name '*virginalis*' was discovered (see below). As the name '*caricaria*' has been in continuous use for about 150 years, it is preferable not to reject it in favour of the senior synonym.

Phalaena virginalis FOURCROY, 1785: Entom. Parisiensis 2: 277 (France: Paris). Syntype(s). FOURCROY's very short description characterizes well this species. Probably a senior synonym. After having been forgotten and never used for more than 200 years, the name was first re-discovered by LERAUT (1997), and used, subsequently, e.g. by SCOBLE (1999) and EBERT (2001).

Phalaena Geometra immaculataria VILLERS, 1789: Linn. ent. 2: 324 (central France, by Lyon ('Lugdunum')). Syntype(s). Senior synonym, but name forgotten for more than 200 years and never used.

‡ *Acidalia phlearia* REUTTI, 1853: Beitr. rhein. Naturg. 3 (Lep. Baden): 113. Introduced in synonymy of *caricaria*. Unavailable, though listed in SCOBLE (1999)

Unavailable names (infrasubspecific): *anonyma*: SCHAWERDA (1920); *nigrocingulata*: HARTIG (1924). – (incorrect subsequent spelling): *caracaria*: KITSCHELT (1925).

DIAGNOSIS: Wingspan ♂ 19-25, ♀ 18-23 mm. Wings rounded. Ground colour clear white, glossy. Forewing with four straight, yellowish grey (rarely brown) transverse lines of slightly variable width. Small sharp black discal spots usually present on the hindwing and absent from the forewing. Without terminal dots. Fringe concolorous. Frons dark brown, vertex white, collar sand coloured. Palpi slender, brown, underside white. Antennae of ♂ with deep intersegmental incisions, ciliate-fasciculate, cilia curved, 1.4-1.9 times width of flagellum. Hindtibia of ♂ broad, without spurs. Hindtarsus short, 0.5-0.6 times length of tibia.

MALE GENITALIA: Socii short, far apart. Fibulae symmetric, elongate, curved. Anellus asymmetric, right lateral process much longer. Aedeagus broad at base. Vesica with two sclerites at an angle of 90°. Sternum A8 with both cerata of medium length, the right slightly shorter. Not polymorphic (n=4).

FEMALE GENITALIA: Lamella antevaginalis oval. Ductus bursae short and broad. Corpus bursae oval. Spinulose patch of signum elongate.

DISTRIBUTION: Eurasiatic, disjunct. Very local and usually rare in the temperate zone of western, central and eastern Europe, from northern Spain and France to northern Balkan peninsula, in the north to the Baltic states and southern Finland, in the south to northern Apennines (FLAMIGNI et al. 2001). Isolated populations in central Urals. Records from central Italy misidentified (Marche: TEOBALDELLI 1976; identity: *Cabera exanthemata*, Ennominae). Strongly threatened in central Europe, due to destruction and alteration of wetland habitats (HUEMER 1996; WOLF & HACKER 2004), with severe areal losses in the last century, e.g. extinct in eastern Germany since 1960 (GELBRECHT & MÜLLER 1987; MÜLLER & GELBRECHT 1992). – Outside Europe disjunctly distributed in western Siberia, Sajan mountains, Transbaikalia (KOSTJUK & GOLOVUSHKIN 2003) and Far East of Asia (VIIDALEPP 1996).

PHENOLOGY: Uni- or bivoltine. North of Alps usually univoltine, from late June to early August, single specimens until late August. Under good conditions and in the south emerging from late May, with partial or regular second generation, from mid-August to early October.

BIOLOGY: Larva recorded on *Centaurea jacea, Artemisia vulgaris* and *A. campestris* (Asteraceae) (var. cit., referring to rearings of MILLIÈRE 1866), reputedly also *Carex* (Cyperaceae), and *Genista* (Fabaceae) (France: LHOMME 1935). Reared also on *Lactuca sativa* (URBAHN & URBAHN 1939). Adults easily flushed up from vegetation during daytime, and readily attracted to light. Main activity from early dusk (RENNWALD pers. comm.). Stenotopic, K-strategy.

HABITAT: Hygrophilous. Preferring open habitats, such as wet grassland rich in *Carex* (REUTTI 1853), Molinietum plant associations (HUEMER 1996), peat bogs, river valleys, often at fringes of damp woodland. From 0 m up to 600 m above sea-level, exceptionally up to 1,000 m.

SIMILAR SPECIES: Confusions occur mainly with the other two glossy white species *S. nemoraria* and *S. subpunctaria*. *S. nemoraria* larger, frons white, discal spots absent from all wings, terminal dots present on forewing, terminal line absent, postmedial line strongly undulate. *S. subpunctaria* larger, discal spots present on all wings, terminal dots present, terminal line with interruptions, postmedial line slightly undulate. Small and pale forms of *S. immutata* may be similar, too, but wings with more grey or yellowish suffusion, postmedial line more distinct, closer to termen, terminal line absent.

122. *Scopula nemoraria* (HÜBNER, 1799)

Geometra nemoraria HÜBNER, 1799: Samml. Eur. Schmett. 5, Geometrae (1): pl. 17, fig. 89 (Europe). Syntype(s) lost.

‡ *Acidalia nemoraria* [syn.] *aliata* HERRICH-SCHÄFFER, 1847: Syst. Bearb. Schmett. Eur. 3 (24): 22 (Austria: Vienna). Syntype(s). Introduced in synonymy of *nemoraria*.

Unavailable names (incorrect subsequent spellings): *memoraria*: HOFFMEYER (1966: 43); *nemoria*: PROUT (1913).

DIAGNOSIS: Wingspan ♂♀ 26-30, in Far East of Asia smaller, 23-27 mm. Wings broad and rounded. Ground colour clear white, glossy. Transverse lines yellowish grey, strongly

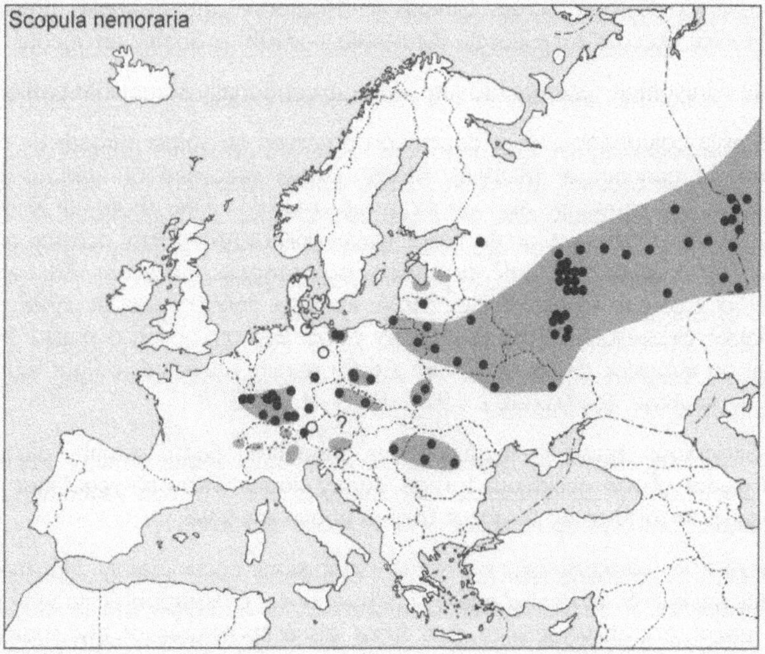

undulate, medial line strongly concave at cell, sometimes with cut off yellowish grey ring at position of discal spot. Discal spots absent. Terminal dots distinct towards apex of all wings. Fringe concolorous. Frons, vertex and collar clear white. Palpi grey brown. Antennae of ♂ with deep intersegmental incisions, ciliate-fasciculate, cilia curved, 1.8-2.1 times width of flagellum. Hindtibia of ♂ stout, without spurs. Hindtarsus short, 0.5-0.6 times length of tibia.

MALE GENITALIA: Socii short, triangular, close to each other. Fibulae symmetric, elongate, curved. Anellus with lateral processes long and slender. Aedeagus broad at base. Vesica with two sclerites, the distal large, slightly pointed. Sternum A8 anteriorly slightly concave. Left ceras long, the right very short, in Far East of Asia slightly longer. Not polymorphic (n=5).

FEMALE GENITALIA: Lamella antevaginalis posteriorly concave, anteriorly with two round sclerites. Ductus bursae very short. Corpus bursae oval. Spinulose patch of signum elongate.

DISTRIBUTION: Eurasiatic. Temperate zone of central and eastern Europe from north-western France (LHOMME 1935; HERBULOT 1985) to the Urals, in the south to northern Italy (WOLFSBERGER 1966; RAINERI & ZANGHERI 1995), Hungary and Romania, in the north to southern Baltic states and St. Petersburg region. In Austria and Slovenja reputedly with wide distribution (CARNELUTTI 1992; HUEMER & TARMANN 1993), but only two specimens from Innsbruck traced (ZSM). Records for Catalonia (DANTART 2000), and central Italy (Marche: SPADA 1893) misidentified. In many parts of central Europe strongly decreasing and threatened, e.g. on the verge of extinction in many German regions (GELBRECHT & MÜLLER 1987; KRAUS 1993; ERLACHER & FRIEDRICH 1994; EBERT 2001; WOLF & HACKER 2004). Populations often very isolated, though sometimes common at sites of occurrence. – Outside Europe across Siberia to Primorye and Sakhalin (PROUT 1935a; VIIDALEPP 1996). Record for Georgia (DIDMANIDZE 2002) probably erroneous.

PHENOLOGY: Univoltine: late May to late June, exceptionally somewhat earlier or later.

BIOLOGY: Larva polyphagous. Egg deposition observed on spider threads on *Hypericum perforatum* (Hypericaceae) (RESSLER 1928). Larva recorded on various herbaceous plants, shrubs, but reputedly also on *Populus tremula, Tilia* (FORSTER & WOHLFAHRT 1981, BERGMANN 1955), and on dry fallen leaves (BERGMANN 1955). Reared on *Plantago lanceolata, Taraxacum officinale, Impatiens noli-tangere, Lonicera, Tilia* and on withered leaves of *Lactuca sativa* (VOGEL 1924; AICHELE 1927; RESSLER 1928; URBAHN & URBAHN 1939; SCHNEIDER 1939; BERGMANN 1955; FRIEDRICH 1976; EBERT 2001), rearing difficult on *Brassica* (RESSLER 1928). Adults readily attracted to light, but not rarely found during daytime, too (AICHELE 1927; RESSLER 1928).

HABITAT: Silvicolous, meso- to hygrophilous. Inhabiting damp scrubby places, fringes, aisles and glades of wet deciduous forests, sunny slopes with rich vegetation. Restricted to the lowlands from 0 m up to about 600 m above sea-level.

SIMILAR SPECIES: *S. caricaria* (see no. 121). Confusions occur mainly with the other two glossy white species *S. caricaria* and *S. subpunctaria*. *S. subpunctaria* with less rounded wings, straighter transverse lines, dark frons, discal spots present, and shorter ♂ antennal cilia. *S. floslactata* with similar transverse lines but dark frons and ground colour yellowish. *Cabera exanthemata* (SCOPOLI, 1763) (Ennominae) without terminal dots below forewing apex, postmedial lines concave, but not undulate, ♂ antennae long bipectinate.

REMARKS: Rearings and immature stages described in AICHELE (1927) and RESSLER (1928).

123. Scopula umbelaria (HÜBNER, 1813)

Geometra umbelaria HÜBNER, 1813: Samml. Eur. Schmett. 5, Geometrae (1): pl. 85, figs 437, 438 (Europe). Syntype(s) lost.

‡ *Phalaena Geometra sylvestrata* BORKHAUSEN, 1794: Naturg. eur. Schmett. 5: 516 (Europe). Syntype(s). Junior primary homonym nec HÜBNER, 1786 (identity: *Bapta temerata*, Ennominae).

Acidalia compararia HERRICH-SCHÄFFER, 1847: Syst. Bearb. Schmett. Eur. 3 (24): 26 (pl. 49, figs 299-302: non binominal) (Austria). Syntypes ♂♀.

Acidalia majoraria LEECH, 1897: Ann. Mag. nat. Hist. (6) 20: 96 (Japan: Oiwake). Syntypes 7♂2♀ (BMNH). Valid at subspecific rank (SCOBLE 1999).

Scopula umbelaria graeseri PROUT, 1935: in SEITZ, Macrolep. 4, Suppl.: 44, pl. 5, fig. c (eastern Russia, Ussuri: Amurland). Valid at subspecific rank (SCOBLE 1999; EBERT 2001).

Unavailable names (misidentification): *strigilaria*: sensu RAMANN (1876) nec HÜBNER, 1799 (identity: *Scopula nigropunctata*). – (infrasubspecific): *anastomosaria*: REISSER (1961). – (incorrect subsequent spellings): *conpararia*: PROUT (1913); *umbellaria*: HÜBNER (1825); *umbellata*: GARTNER (1866).

DIAGNOSIS: Wingspan ♂ 29-34, ♀ 26-31 mm. Forewing apex slightly pointed, hindwing termen slightly angled at M3. Ground colour light grey with sand coloured tinge, irrorated with grey scales. Transverse lines greyish ocheous to light brown, fine, straight. Small black discal spots on hindwing often, on forewing sometimes present. Terminal line usually present, fine, light grey, not rarely with small black fringe dots. Frons blackish brown, vertex white, collar ochreous. Palpi slender, brownish. Antennae of ♂ with deep intersegmental incisions, ciliate-fasciculate, cilia curved, 1.7-2.1 times width of flagellum. Hindtibia of ♂ stout, without spurs. Hindtarsus short, 0.5-0.6 times length of tibia.

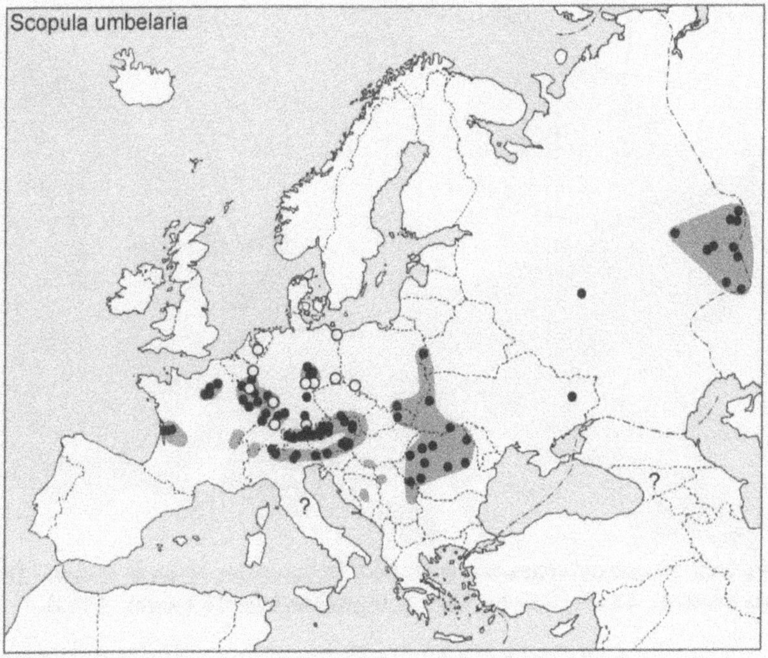

Subsp. ***graeseri*** on average smaller, wingspan 26-32 mm, paler.

Subsp. ***majoraria*** larger, wingspan up to 36 mm, transverse lines indistinct, terminal dots absent.

MALE GENITALIA: Fibulae asymmetric, the left very long and narrow, the right furcate. Anellus asymmetric with the right process longer. Aedeagus broad. Vesica with 3-4 terminal sclerites, one of them long, curved and tapered. Sternum A8 with both cerata long and narrow, no polymorphism found (n=5). One examined ♂ of subsp. *majoraria* (Japan) with symmetrical cerata of medium size, as long as the mappa. In one ♂ from eastern Russia (Primorye) left ceras of medium size, the right long, possibly due to polymorphism.

FEMALE GENITALIA: Lamella antevaginalis asymmetric with large lateral sclerite. Ductus bursae short. Corpus bursae elongate. Signum elongate, with numerous spines, filling almost the whole bursa.

DISTRIBUTION: Palaearctic. Local and usually rare in the temperate area of western, central and eastern Europe, from France to the Urals, in the south to northern Italy and Romania, in the north to Poland. Reputedly in central Italy (Marche: TEOBALDELLI 1976; PROLA & RACHELI 1979), data await confirmation. For erroneous records from Iberian peninsula see DANTART (2000). In the Netherlands extinct since 1906 (DE VOS pers. comm.). Recent areal losses in many German regions (MÜLLER & GELBRECHT 1992; KRAUS 1993; ERLACHER & FRIEDRICH 1994; WOLF & HACKER 2004). – Outside Europe in western Siberia, Altai, Sajan, and northern parts of central Asian mountains. Reputedly in Georgia (DIDMANIDZE 2002). In the east from Mongolia and central Siberia to Far East of Asia (subsp. *graeseri*) and Japan (subsp. *majoraria*). In China replaced by the sister species *S. szechuanensis* (PROUT, 1913). *S. quinquelinearia* from North America, similar in habitus, but not related structurally.

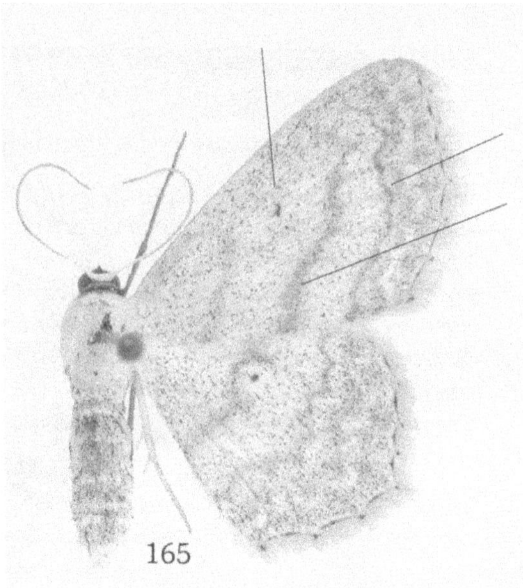

Text-figs 164-165: Diagnostic characters (indicated) of European *Scopula* species. Text-fig. 164: *S. umbelaria* (HÜBNER, 1813). Text-fig. 165: *S. nigropunctata* (HUFNAGEL, 1767).

PHENOLOGY: Univoltine: mid-May to late June, exceptionally somewhat earlier or later, in France regularly emerging from early May.

BIOLOGY: Larva polyphagous on herbaceous plants of many different families, possibly preferring *Vincetoxicum* (Asclepiadaceae). Found on *Vincetoxicum hirundinaria* (Germany: LEIPNITZ pers. comm.). Reputedly found on *Prunus spinosa* (Rosaceae) (SCHNEIDER 1939; BERGMANN 1955), but data doubtful according to EBERT (2001). Recorded on *Vincetoxicum hirundinaria* (Asclepiadaceae) (var. cit., e.g. SPULER 1910; REBEL 1910b; LHOMME 1935; BERGMANN 1955; EBERT 2001), *Polygonum* (Polygonaceae), *Clematis* (Ranunculaceae), *Solidago virgaurea, Artemisia, Achillea millefolium* (Asteraceae), *Plantago* (Plantaginaceae), *Stellaria media* (Caryophyllaceae), *Vicia, Cytisus purgans* (Fabaceae) (VORBRODT 1930; LHOMME 1935; BERGMANN 1955). Reared on *Vincetoxicum hirundinaria, Solidago, Lysimachia*, *Vicia cracca, Clematis vitalba*, *Taraxacum officinale*, and *Ligustrum vulgare* (URBAHN 1965; FRIEDRICH 1976; SCHADEWALD in EBERT 2001; MÜLLER, LEIPNITZ, ERLEBACH pers. comm.), *Ligustrum* suggested as natural host-plant by EBERT (2001). Larva in spring only drinking water, not feeding on grass as suggested in literature (MÜLLER pers. comm.). Adults attracted to light and flowers.

HABITAT: Xerothermophilous (FORSTER & WOHLFAHRT 1981; EBERT 2001), often on sand or limestone. Scattered warm deciduous woods, southern slopes, sites with vast occurrence of *Vincetoxicum* (BERGMANN 1955), forest clearings, railroad embankments and fallows (BUSZKO 2000), damp forest mentioned by SCHULTZ (1930). Usually from 0 m up to 800 m above sea-level, up to 1,200 m in southern Alps.

SIMILAR SPECIES: From similar species usually clearly differing in size. *S. nigropunctata* smaller, ground colour and transverse lines darker, discal spot of forewing present, ♂ hind-tarsus shorter, usually flying one month later. *S. incanata* with distinct discal spots of all wings.

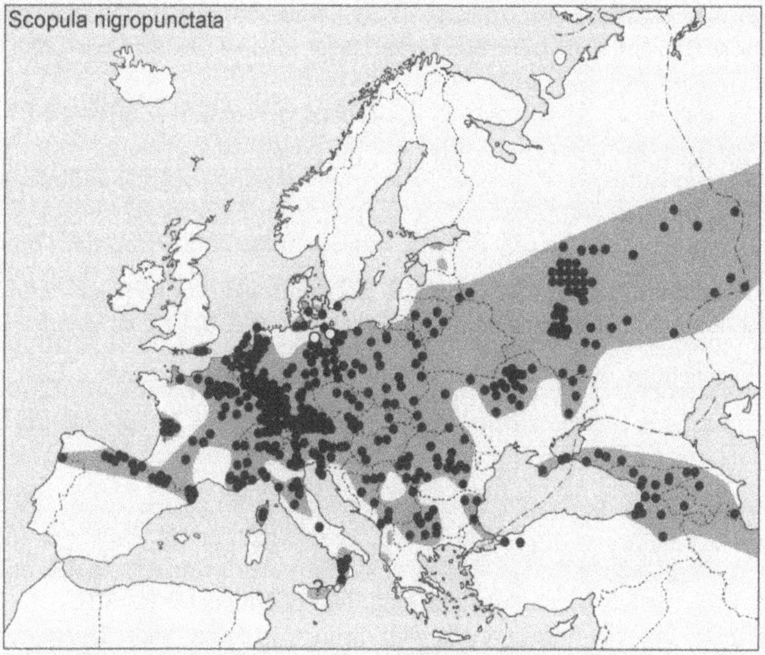

124. Scopula nigropunctata (HUFNAGEL, 1767)

Phalaena nigropunctata HUFNAGEL, 1767: Berlin. Mag. 4 (5): 526 (Germany: Berlin). Syntype(s) lost.

‡ *Phalaena repandata*: sensu SCOPOLI, 1763: Ent. Carniolica: 215, fig. 531 (north-western Slovenja: Carniolia). Misidentification nec LINNAEUS, 1758, referring to Syst. Nat., p. 524 (identity: *Alcis repandata*, Ennominae). Synonymy uncertain, SCOPOLI's description and illustration reminiscent of *S. subpunctaria*.

‡ *Geometra strigilata*: sensu DENIS & SCHIFFERMÜLLER, 1775: Ank. syst. Werkes Schmett. Wienergegend: 117 (Austria: Vienna distr.). Misidentification nec LINNAEUS, 1758 (identity: *Polypogon strigilata*, Noctuidae).

Phalaena tristriaria FABRICIUS, 1794: Ent. Syst. 3 (2): 130 (Italy). Syntypes lost.

Phalaena Geometra nemorata BORKHAUSEN, 1794: Naturg. eur. Schmett. 5: 518 (Europe, probably Germany). Syntypes.

‡ *Geometra strigilaria* HÜBNER, 1799: Samml. Eur. Schmett. 5, Geometrae (1): pl. 20, fig. 109. Incorrect subsequent spelling of the misidentified *strigilata* sensu DENIS & SCHIFFERMÜLLER, 1775. HÜBNER's figure ambiguous, reminiscent of *S. incanata*, too.

Phalaena inspersata SCHRANK, 1802: Fauna boica 2 (2): 57 (Germany, Bavaria: Ingolstadt). Syntype(s).

Calothysanis exemptaria HÜBNER, 1823: Verz. bekannter Schmett.: 301 (Austria: Vienna distr.). No separate types. Unnecessary replacement name for *strigilaria*.

‡ *Acidalia prataria* BOISDUVAL, 1840: Genera Index meth. eur. Lepid.: 227. Unnecessary replacement name for misidentified '*strigilaria* HÜBNER' 'because of the similarly spelled name *strigillaria*'.

Acidalia prataria var. *catenaria* BRUAND, 1846: Mem. Soc. Doubs (2) 2 (3/4): 122 (France: Doubs, Besancon).

Acidalia subcandidata WALKER, 1862: List. Spec. lepid. Insects Colln Br. Mus. 26: 1607 (eastern China: Shanghai). Holotype ♂ (BMNH). Valid at subspecific rank (SCOBLE 1999; EBERT 2001).

Craspedia imbella WARREN, 1901: Novit. zool. 8: 22 (Japan). Syntypes 3♀ (BMNH). Valid at subspecific rank (INOUE 1958; SCOBLE 1999).

Scopula nigropunctata chosensis BRYK, 1949: Ark. Zool. 41A (1): 164. (Korea: Shuotsu). Holotype ♂ (NHRS). Valid at subspecific rank (SCOBLE 1999).

Scopula nigropunctata subimbella INOUE, 1958: Tinea 4 (2): 243 (Japan: Hokkaido, Kushiro, Shibecha). Holotype ♂ (BMNH). Valid at subspecific rank (INOUE 1958; SCOBLE 1999).

Unavailable names (misidentifications): *incanata*: sensu ZELLER (1853) nec LINNAEUS, 1758; *variegata*: sensu STEPHENS (1853) nec SCOPOLI, 1763 (identity: *Scopula incanata*). – (infrasubspecific): *anastomosaria*: PREISSECKER (1926); *basinuda*: LEMPKE (1949); *crassestrigata*: LEMPKE (1949); *demarginata*: REISSER (1961);

nigrisignata: PROUT (1934); *ochrea*: STERNECK (1927); *rubella*: STERNECK (1927); *strigularia*: KITSCHELT (1933); *subalbulata*: STERNECK (1927); *variegata*: STERNECK (1927). – (incorrect subsequent spellings): *rependata*: WERNEBURG (1864); *strigillaria*: SCHRECKENSTEIN (1800); *subcanditata*: STERNECK (1927).

DIAGNOSIS: Wingspan ♂♀ 25-29 mm, in the second generation sometimes 21 mm only. Forewing broad, hindwing termen angled at M3. Wings sand coloured with more or less strong grey brown suffusion. Postmedial line brown, distinct, slightly dentate, medial line broad, oblique, on hindwing proximal, on forewing far distal to discal spot. Discal spots distinct. Terminal line dark, usually distinct, without fringe dots. Frons dark brown, vertex whitish, collar ochreous. Palpi slender, brownish. Antennae of ♂ with deep intersegmental incisions, ciliate-fasciculate, cilia curved, 1.5-2 times width of flagellum. Hindtibia of ♂ dilated, without spurs. Hindtarsus comparatively short, 0.3-0.4 times length of tibia.

Variability: Populations of Pyrenee mountains usually more contrasted. East Asian and Japanese populations (subsp. *subcandidata*, *chosensis*, *imbella*, *subimbella*) with smaller discal spots, medial line less distinct, in China hindwing termen less angled at M3.

MALE GENITALIA: Socii very short, rounded, connected by a convex sclerite. Fibulae asymmetric, the left elongate, the right bilobed. Valvae strongly curved. Aedeagus short. Vesica with a large, terminally rounded sclerite. Sternum A8 anteriorly broad and straight. Cerata very thin, the left short, the right long. Not polymorphic (n=13).

FEMALE GENITALIA: Lamella antevaginalis with distinct, irregular folds. Ductus bursae very short. Corpus bursae elongate oval. Signum large, with numerous spines, covering large parts of the surface.

DISTRIBUTION: Palaearctic. Widely distributed in Europe, from Portugal to the Urals, in the north to southern Scandinavia and Baltic states. Absent from Ireland, southern half of Iberian peninsula, southern Greece and from Mediterranean islands except Corsica. Reputedly in Sicily, based on old and doubtful citations (PARENZAN et al. 1998b). In Great Britain resident in the south-easternmost part and suspected immigrant at various other localities (SKINNER 1984). Areal regression in the Baltic states suspected by ŠULCS & VIIDALEPP (1972). – Outside Europe in northern and eastern Turkey, Caucasus, Transcaucasus, northern Iran, in the east across Mongolia and Siberia to Far East of Asia (subsp. *subcandidata*; subsp. *chosensis*) and Japan (subsp. *imbella*; subsp. *subimbella*)

PHENOLOGY: In Europe usually univoltine: mid-June to early August, but single specimens emerging from late May and flying until mid-August. In the lowlands south of the Alps and in Asia in two, not always sharply separated generations from early May to early October, in China emerging from late March.

BIOLOGY: Larva polyphagous, probably preferring *Clematis vitalba* (Ranunculaceae) and, perhaps, *Lonicera periclymenum* (Caprifoliaceae) (Germany: EBERT 2001; northern Europe: SKOU 1986; Great Britain: CARTER & HARGREAVES 1987; WARING & TOWNSEND 2003). Found on *Lonicera periclymenum* (Caprifoliaceae) (Poland: URBAHN & URBAHN 1939), *Stachys sylvatica* (Lamiaceae) (UFFELN 1908), *Clematis recta* (Ranunculaceae), *Soldanella alpina* (Primulaceae), *Ligustrum vulgare* (Oleaceae), near *Rubus caesius* (Rosaceae), *Acer platanoides* (Aceraceae), *Origanum vulgare* (Lamiaceae), *Linaria vulgaris,* and *Veronica chamaedrys* (Scrophulariaceae) with signs of feeding on the plants (Germany: EBERT 2001). Recorded on *Vicia* (Fabaceae), *Viola* (Violaceae), *Veronica* (Scrophulariaceae), *Stachys recta, S. sylvatica, Origanum vulgare* (Lamiaeae), *Vaccinium myrtillus* (Ericaceae), *Plantago* (Plantaginaceae), *Lonicera* (Caprifoliaceae),

Corylus (Corylaceae), *Alnus* (Betulaceae), *Clematis* (Ranunculaceae), also on fallen leaves (var. cit., e.g. REBEL 1910b; CHRÉTIEN 1928; ALLAN 1949; BERGMANN 1955; SVENSSON 1993; EBERT 2001). Reared furthermore on *Taraxacum officinale, Polygonum aviculare, Campanula medium, Vincetoxicum hirundinaria* (var. cit., e.g. REISSER 1949; ALLAN 1949; BERGMANN 1955; SKINNER 1984). Main activity of adults at dusk, but also attracted to light and bait. Nectar feeding recorded on *Clematis vitalba* (EBERT 2001), adults beaten from this plant (WARING & TOWNSEND 2003). Trivial movement restricted to short distances (HAUSMANN 1990a).

HABITAT: Silvicolous, meso- to hygrophilous. Mainly in deciduous or mixed woodland, forest fringes, scrubby places, preferably near *Clematis*, also anthropogenous habitats such as embankments or gardens, almost absent from open habitats (HAUSMANN 1990a). Usually from 0 m up to 1,000 m above sea-level, up to 1,800 m in southern Alps and in southern Europe, up to 2,200 m in Turkey and Transcaucasus.

SIMILAR SPECIES: *S. umbelaria* (see no. 123). *S. flaccidaria* smaller, with less dark suffusion, postmedial line dotted, discal spots larger, hindwing tailed. Faded specimens may resemble also *S. floslactata* or *S. ternata*, but *S. nigropunctata* differs in angled hindwing termen, discal spot of forewing sharp, medial line of forewing thinner, more convex.

REMARKS: In central and eastern Asia there are some further species closely related to *S. nigropunctata*. In Far East of Asia with a sympatric, very similar species of the subgenus *Calothysanis* with hindwing more rounded, unnamed according to STERNECK (1941: 48). For rearings and morphology of immature stages see BUCKLER (1897).

The *virgulata* species-group

STERNECK's 2nd group of *Scopula*. Monotypic. *S. parallelaria* is not related to this group (PROUT 1935a), though STERNECK (1941) lists it as synonym of *S. virgulata*, probably based on misidentified material. Different from all other groups in double fibula and in the aedeagus with two cornuti. Proboscis developed. Palpi slender. Hindtibia of ♂ slightly dilated, with pencil, without spurs. Hindtarsus not shortened. Central sclerite on ♂ sternum A2 comparatively small.

125. *Scopula virgulata* (DENIS & SCHIFFERMÜLLER, 1775)

Geometra virgulata DENIS & SCHIFFERMÜLLER, 1775: Ank. syst. Werkes Schmett. Wienergegend: 117 (Austria: Vienna distr.). Syntype(s) lost.

Geometra strigaria HÜBNER, 1799: Samml. Eur. Schmett. 5, Geometrae (1): pl. 18, fig. 95 (Europe). Syntype(s) lost.

‡ *Leptomeris virgularia* HÜBNER, 1825: Verz. bekannter Schmett.: 311. Incorrect subsequent spelling. Partly misidentified, see synonymy of *I. seriata*.

Leptomeris sulcaria HÜBNER, 1825: Verz. bekannter Schmett.: 310 (Europe). No separate types. Unnecessary replacement name for *strigaria*.

Acidalia substrigaria STAUDINGER, 1900: Dt. ent. Z. Iris 12 (2): 391 (central Russia, southern Yakutia: Witim; Altai: Ongadai). Syntypes 1♂1♀ (MNHU, examined). Valid at subspecific rank (SCOBLE 1999; EBERT 2001).

Acidalia virgulata rossica DJAKONOV, 1926: Ezheg. Gosud. Muz. N.M. Mart'yanova 4 (1): 16 (western Russia: St. Petersburg, Lachta). Lectotype ♂ (ZISP, examined), herewith designated to stabilize usage of the name. Valid at subspecific rank.

‡ *Acidalia strigaria* f. *insubrica* VORBRODT, 1930: Mitt. Schw. Ent. Ges. 14 (7): 325 (Switzerland, Tessin).

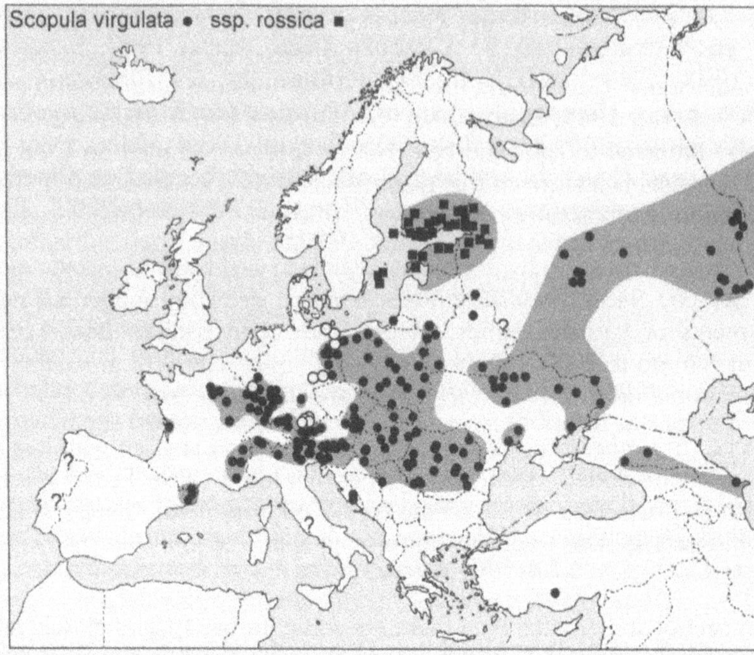

Formally introduced as 'form', but subspecific rank ('race') outlined in the text. Junior primary homonym nec VORBRODT, 1930 (identity: *Scopula marginepunctata*).

Scopula virgulata var. *albicans* PROUT, 1934: Lepid. Cat. 61: 191. First available name after *albicans* PROUT, 1913 (infrasubspecific) (Japan: Oiwake). Syntypes 1♂1♀ (BMNH). Valid at subspecific rank (SCOBLE 1999).

Scopula virgulata subtilis PROUT, 1935: in Seitz, Macrolep. 4, Suppl.: 44 (Korea). Holotype. Valid at subspecific rank (SCOBLE 1999; EBERT 2001).

Scopula virgularia ticinensis FORSTER, 1973: Nachr.Bl. bayer. Ent. 22 (1): 15 (Switzerland, Tessin). No separate types. Replacement name for *insubrica*. Though subspecific rank stated in FORSTER & WOHLFAHRT (1981), habitus features show clinous transitions to nominate subspecies, hence synonym.

Unavailable names (misidentification): *adjunctaria*: sensu PEYERIMHOFF (1863) nec BOISDUVAL, 1840 (identity: *Scopula incanata*). – (infrasubspecific): *anastomosaria*: PREISSECKER (1926); *mus*: KAUCKI (1922); *nigerrima*: REBEL (1913); *nigrolineata*: REISSER (1961); *rehfousiana*: CULOT (1917). – (incorrect subsequent spellings): *frigaria* (for *strigaria*): WALKER (1861); *rehfousaria*: CULOT (1918); *rehfousiaria*: CULOT (1919); *tizinensis*: FORSTER & WOHLFAHRT (1981); *trigaria*: CULOT (1917).

DIAGNOSIS: Wingspan ♂ 19-24, ♀ 19-22 mm, in the second generation sometimes 16 mm only. Forewing broad and rounded in ♂, more elongate with apex pointed in ♀, hindwing rounded. Ground colour light brown. Transverse lines fine, straight, parallel, grey brown. Discal spot absent on forewing, usually present on hindwing. Terminal line fine, dark, usually present, without fringe dots. Frons black brown, vertex whitish with ochreous tinge, collar ochreous. Palpi brown, underside whitish. Antennae of ♂ with deep intersegmental incisions, ciliate-fasciculate, cilia curved, 1.4-1.7 times width of flagellum. Further features under diagnosis of species-group.

Subsp. ***rossica*** on average slightly smaller, with much darker suffusion of wings and more diffuse pattern, discal spots usually absent.

Subsp. ***substrigaria*** from the Altai similar to subsp. *rossica*, but with coloration slightly more ochreous, larger. Subsp. ***subtilis*** with paler, slightly ochreous wings, transverse lines fine, discal spot of hindwing distinct. Subsp. ***albicans*** with whitish ground colour.

MALE GENITALIA: Socii broad and short. Fibula with two separate, curved projections. Aedoeagus with one large, curved cornutus and a second, smaller cornutus at tip. Sternum A8 anteriorly rounded. Cerata with stout terminal spines, the left short, broad, strongly spinose, the right long, strongly curved. Not polymorphic (n=9).

FEMALE GENITALIA: Apophyses anteriores short. Lamella antevaginalis with circular folds, anteriorly sclerotized strongly. Corpus bursae drop-shaped, distinctly sclerotized posteriorly. Signum absent.

DISTRIBUTION: Palaearctic. Nominate subspecies widely distributed in central and eastern Europe, in the east to the Urals, in the south to northern Italy, Croatia and Romania, with isolated occurrences all over France and eastern Pyrenees. Reputedly in Portugal (e.g. SILVA CRUZ 1967; MÜLLER 1996), and in central Italy (PROLA & RACHELI 1979), data await confirmation, no vouchers traced. Erroneously recorded for central Spain (prov. Madrid: ROMERO et al. 2001; examined, identity: *S. turbidaria*). Subsp. *rossica* around the northern parts of the Baltic Sea, resident also on the Swedish island Gotland (RYRHOLM pers. comm.). Areal losses from bog drainage reported from Finland (MIKKOLA 1979). Extinct in Thuringia (ERLACHER & FRIEDRICH 1994). Strongly threatened in many parts of central Europe (see EBERT 2001; WOLF & HACKER 2004), mainly by destruction of dry habitats on low-nutrient soils and fertilizing. Appropriate habitats merit protection (GELBRECHT et al. 1995). – Outside Europe in the Caucasus and Azerbaijan, in the east across Siberia, Altai and Mongolia (subsp. *substrigaria*) to Far East of Asia (subsp. *subtilis*) and Japan (subsp. *albicans*). Very isolated occurrence in southern Turkey (Konya: NHMW/VARTIAN).

PHENOLOGY: Usually bivoltine: mid-May to to mid-September, often with a gap between the generations in mid-July or late July. In the north emerging from early June, in the south exceptionally a partial third generation late September and early October. Under bad conditions, e.g. at higher elevation or in the north, univoltine from mid-June to late July, with transitions along altitudinal and latitudinal gradients.

BIOLOGY: Females observed sitting on *Bromus erectus* and dropping down eggs without gluing them onto the plant (EBERT 2001). Larva polyphagous on herbaceous plants (REBEL 1910b), possibly preferring short species of Poaceae (FORSTER & WOHLFAHRT 1981, BERGMANN 1955), *Carex* (Cyperaceae) and *Inula* (Asteraceae) (northern Europe: SKOU 1986; SVENSSON 1993). Reputedly also on *Betula alba* (TREITSCHKE 1828). Reared on *Vaccinium myrtillus* (Finland: SEPPÄNEN 1970), *Poa* (ERLEBACH pers. comm.), *Lactuca sativa* (withered leaves), *Potentilla, Taraxacum officinale*, young or withered leaves of *Lonicera xylosteum* and other shrubs (STANGE 1881; LHOMME 1935; URBAHN & URBAHN 1939; BERGMANN 1955; EBERT 2001; MÜLLER pers. comm.). Adults diurnal, but also attracted to light. Stenotopic, K-strategy, local populations with small extension of area (EBERT 2001).

HABITAT: Xerothermophilous, preferring sandy soils or limestone. Usually on subcontinental, dry, sandy steppe habitats (basiphilic xerothermic meadows), often on embankments (GELBRECHT et al. 1995) and restricted to sun-exposed localities, protected from wind, such as southern slopes, and warm scrubby places. Sometimes also in wet open habitats (FORSTER & WOHLFAHRT 1981, BERGMANN 1955; SKOU 1986). In the foothills of the Alps and in northern Europe in dry sites on peat bogs. From 0 m up to 1,000 m above sea-level, up to 1,300 m in southern Alps, reputedly on 2,200 m in Romania (DIÓSZEGHY 1935).

SIMILAR SPECIES: Subsp. *rossica* sometimes quite similar to *S. frigidaria*, the latter with transverse lines more distant from forewing termen, slightly convex and less oblique (usually better visible on the underside of wings), forewing apex less pointed, terminal line absent, frons white towards proboscis. Allopatric *S. turbidaria* and *S. turbulentaria* with postmedial line usually slightly convex and undulate, discal spot present also on forewing.

REMARKS: Description of immature stages in STANGE (1881).

The *ornata* species-group

STERNECK's 3rd group of *Scopula*, a 'very natural group' (STERNECK 1941: 49). Well characterized by the thin black terminal line of forewing surrounding the apex. Palpi comparatively stout, 1.2-1.3 times diameter of eye. Proboscis well developed. Hindtibia with infraspecific spur anomalies. Antennae of ♀ densely ciliate. Central sclerite on ♂ sternum A2 with paired, distinct patches of large, broad setae (Text-fig. 3). Male genitalia: Socii large and broad, crossed. Valva usually bent, larger than fibula, often spinose at tip. Aedeagus often curved. Some species with polymorphisms in the length of cerata. Female genitalia with short ductus bursae.

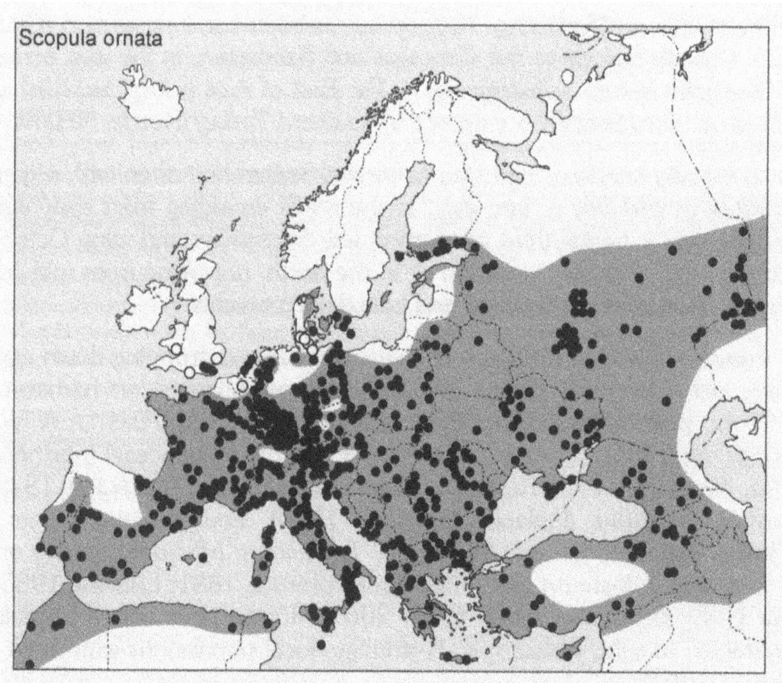

126. *Scopula ornata* (SCOPOLI, 1763)

Phalaena ornata SCOPOLI, 1763: Ent. Carniolica 9: 219, fig. 545 (north-western Slovenja: Carniolia 'Mts.'). Syntype(s) lost.

Phalaena instilata HUFNAGEL, 1767: Berlin. Mag. 4 (5): 526 (Germany: Berlin). Syntype(s) lost.

Phalaena Geometra paludata LINNAEUS, 1767: Syst. Nat. (Ed. 12) 1 (2): 873 (Portugal ('Lusitania')). Syntype(s) (LSL).

‡ *Phalaena nivearia* FABRICIUS, 1775: Syst. Ent.: 620 (England). Syntype(s). Synonymy according to LERAUT (1997), but description not unambiguously referring to *S. ornata*, e.g. 'pectinate antennae' contradicting, nomen dubium (*Cabera*?).

‡ *Phalaena institata* ROTTEMBURG, 1777: Naturf. Halle 11: 74 (Europe). Syntype(s). Incorrect subsequent spelling.

Phalaena Geometra interrupta GOEZE, 1781: Ent. Beytr. 3 (3): 427 (France: Paris). Syntype(s).

Phalaena intersecta FOURCROY, 1785: Entom. Parisiensis 2: 283 (France: Paris). Syntype(s).

‡ *Geometra ornataria* HÜBNER, 1799: Samml. Eur. Schmett. 5, Geometrae (1): pl. 14, fig. 70. Incorrect subsequent spelling.

Pyralis paludalis SCHRANK, 1802: Fauna Boica 2 (2): 62. Emendation of *paludata*.

Acidalia ornata subornata PROUT, 1913: in SEITZ, Macrolep. 4: 79 (Japan: Oiwake, Yokohama). Syntype(s) (BMNH). Valid at subspecific rank (SCOBLE 1999).

Scopula ornata enzela PROUT, 1935: in SEITZ, Macrolep. 4, Suppl.: 46, pl. 5, fig. e (Iran, southern Caspian Sea: Enzeli). Valid at subspecific rank.

Scopula cinis INOUE, 1946: Bull. Lepidopt. Soc. Japan 1 (1): 4 (Japan: Nagano Prefecture, Yunomata). Holotype ♂ (BMNH/INOUE). Junior synonym of subsp. *subornata*.

Unavailable names (infrasubspecific): *albomarginata*: REISSER (1961); *defecta*: STAUDER (1924); *obsoleta*: LEMPKE (1949); *paucisignata*: KRAUSSE (1912). – (incorrect subsequent spellings): *instillata*: ZELLER (1844); *ornate*: DIDMANIDZE (2001); *ornatoria*: BLANCHARD (1840); *ornatura*: TOMIC et al. (2002); *palutata*: REINHARD et al. (1999).

DIAGNOSIS: Wingspan ♂♀ 20-26 mm, second generation slightly smaller, sometimes 18 mm only. Hindwing termen concave between M1 and M3. Ground colour clear white. Antemedial line dissolved into 2-4 small dots. Medial line very indistinct or absent. Postmedial line slightly dentate, distally bordered by brown double spots in tornus and between radials of both wings. Brown spots of hindwing tornus with black proximal border. Rest of terminal area greyish with white wavy line. Forewing apex white. Discal spot usually very indistinct on forewing, present on hindwing. Fringe whitish or light grey, slightly chequered. Frons black, vertex and underside of palpi whitish, collar brownish. Antennae of ♂ with deep intersegmental incisions, ciliate-fasciculate, cilia long and curved, 1.8-2.3 times width of flagellum (Text-fig. 154). Hindtibia stout, with pencil, tarsus 0.6-0.8 times length of tibia.

Variability: Dark, well contrasted wing pattern favoured by cold, wet conditions; high temperature and dryness inducing whitish, weakly marked specimens (BERGMANN 1955). Such forms with reduced markings ('f. *paucisignata*') commoner in Sardinia, Caucasus, Ukraine and Russia, but in all European populations mixed with nominotypical forms. Iranian subsp. *enzela* with constantly reduced pattern.

MALE GENITALIA: Socii comparatively slender. Valva slender, smooth. Fibula tapering. Aedeagus slender, without cornuti. Sternum A8 with two long, narrow cerata. West-Mediterranean populations including North Africa, Sardinia, Sicily and southern Italy with base of sternum A8 developed as round, bulbed sclerite. Without polymorphism of cerata length (n=20).

FEMALE GENITALIA: Lamella antevaginalis with anterior margin smooth, rounded and with V-shaped central fold. Posterior margin of corpus bursae sclerotized, triangular. Signum narrow, elongate, length varying, with numerous small spinules.

DISTRIBUTION: Palaearctic. Widely distributed all over Europe, from Portugal to the Urals. Present on all the Mediterranean islands. Absent from central and northern Scandinavia, and from large parts of the British Isles, but in England often occurring as 'vagrants' without being resident (REID pers. comm.). In southern Sweden resident (RYRHOLM pers. comm.). Areal losses in eastern Germany (MÜLLER & GELBRECHT 1992; ERLACHER & FRIEDRICH 1994). – Outside Europe from Morocco to Tunisia, Cyprus, Turkey, Caucasus, Transcaucasus, across northern Iran (subsp. *enzela*) to Afghanistan and central Asian mountains, Siberia, in the east to Dahuria and Amur region (KHOMENKO et al. 1994; KOSTJUK et al. 1994; VIIDALEPP 1996), Korea and Japan (subsp. *subornata*). In Pakistan replaced by the sister species *S. kashmirensis* (MOORE, 1888).

PHENOLOGY: Usually bi- or trivoltine: early May to early July; mid-July to late August, exceptionally until mid-September. Under good conditions emerging from late April. Under bad conditions, e.g. on mountains, with first generation emerging later, or even univoltine, early June to late July. In the lowlands of the Mediterranean, North Africa and Turkey trivoltine from April to early November.

BIOLOGY: Larva oligophagous (1-2) on herbaceous plants of low-nutrient soils, preferring Lamiaceae. Eggs and larva found on *Thymus pulegioides, T. serpyllum* (Lamiaceae) (Germany: UFFELN 1908; SCHNEIDER 1939; EBERT 2001; LEIPNITZ pers. comm.), *Origanum vulgare* (Lamiaceae) (Germany: UFFELN 1908; LEIPNITZ pers. comm.), and *Mentha suaveolens* (Lamiaceae) (Morocco: RUNGS 1981). Egg positioning also on *Veronica serpyllifolia* (Scrophulariaceae), *Achillea millefolium* (Asteraceae), *Chamaespartium sagittale* (Fabaceae) and grasses (Poaceae) (EBERT 2001), but without recorded feeding on these plant species. Frequently recorded on *Thymus and Origanum* (Lamiaceae) (var. cit.), *Mentha* (Lamiaceae) (REBEL 1910b; LHOMME 1935; NORDSTRÖM et al. 1941; PORTER 1997), reputedly also on *Veronica* (Scrophulariaceae), *Achillea millefolium* (Asteraceae), *Taraxacum officinale* (Asteraceae) and *Rumex* (Polygonaceae) (VORBRODT 1914), the last three refused in feeding experiments, accepting exclusively *Thymus pulegioides* (EBERT 2001). Reared on *Achillea millefolium, Mentha, Origanum vulgare, Thymus serpyllum*, and *Mentha aquatica* (PROUT 1913; SCHNEIDER 1939; ALLAN 1949; EBERT 2001; ERLEBACH, MÜLLER pers. comm.). Adults easily flushed up from vegetation during daytime, main flight activity at dusk, attracted to light in the early evening, occasionally at bait. Nectar-feeding recorded on *Echium vulgare, Thymus pulegioides, Calluna vulgaris, Dactylorhiza maculata, Lotus corniculatus* (EBERT 2001), *Anchusa azurea* (Turkey: KOCAK 1990a), and *Ranunculus* (ESCHE 1996).

HABITAT: In central Europe xerothermophilous, preferring uncultivated, dry open habitats on low-nutrient ground, often on hills and sunny slopes, frequent in Thymo-Festucetum ovinae (WEGNER pers. comm.), with abundance of *Thymus* and *Origanum*. Usually on calcareous and silicious soils. From 0 m up to 1,600 m above sea-level, in Morocco, Turkey and central Asian mountains up to 2,000 m.

SIMILAR SPECIES: Confusions occur with *S. decorata*, mainly with *S. d. congruata*. *S. decorata* with hindwing termen less concave between M1 and M3, apex with grey spots, transverse lines more distinct, medial line of forewing with a dark costal spot, postmedial line more dentate, the last two spots in the hindwing tornus grey, concolorous with the others, discal spot of forewing usually more distinct, fringe dark grey. Furthermore ♂ antennal cilia shorter, ♂ hindtarsus longer. In both sexes identification is possible by brushing tip of abdomen to remove scales and reveal diagnostic characters (cerata, lamella antevaginalis). *S. orientalis* larger, postmedial line more dentate, terminal area darker, less confluent.

REMARKS: *S. kashmirensis* (MOORE, 1888) is considered another subspecies of *S. ornata* by STERNECK (1941: 50), reputedly without constant structural differences. Retained as valid at species rank by SCOBLE (1999).

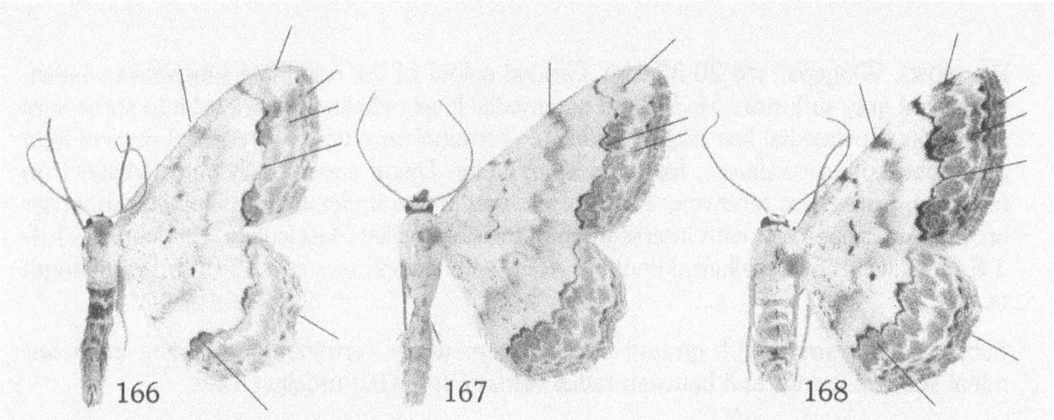

Text-figs 166-168: Diagnostic characters (indicated) of European *Scopula* species. Text-fig. 166: *S. ornata* (SCOPOLI, 1763). Text-fig. 167: *S. orientalis* (ALPHÉRAKY, 1876). Text-fig. 168: *S. decorata* (DENIS & SCHIFFERMÜLLER, 1775).

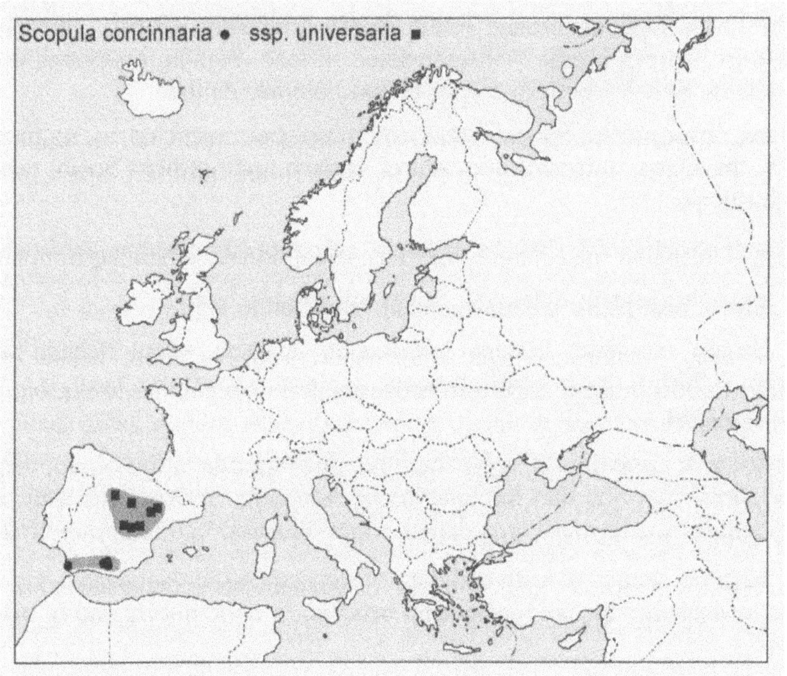

127. *Scopula concinnaria* (DUPONCHEL, 1842)

Dosithea concinnaria DUPONCHEL, 1842: in GODART & DUPONCHEL, Hist. nat. Lépid. Pap. Fr. (Suppl.) 4: 19, pl. 52, fig. 2 (Spain, Andalusia). Syntype(s) ♀.

Acidalia concinnata GUENÉE, 1858: in BOISDUVAL & GUENÉE, Hist. nat. Insectes (Spec. gén. Lépid.) 9: 486 (Spain, Andalusia). No separate types. Unnecessary replacement name for *concinnaria*.

Acidalia (Idaea) concinnaria var. *hesperidata* GUMPPENBERG, 1890: Nova Acta Acad. Caesar. Leop. Carol. 54 (5): 491. First available name after *hesperidata* RAMBUR, 1866 (non binominal) and *hesperidata* STAUDINGER, 1871 (in synonymy) (Spain: Andalusia). Syntypes, at least 3♂3♀, of RAMBUR (ZSM/HERBULOT: 3♂3♀, examined).

Acidalia concinnaria universaria ZERNY, 1927: Eos Madr. 3 (3): 399 (Spain, Aragon: Albarracin). Syntypes. (NHMW: 1♀, examined). Valid at subspecific rank (SCOBLE 1999, confirmed).

Unavailable names (incorrect subsequent spelling): *concinuaria*: PROUT (1913).

DIAGNOSIS: Wingspan ♂♀ 26-32 mm. Ground colour of the nominate subspecies whitish, with slight grey suffusion. Medial and antemedial lines indistinct, dissolved into some very small dots. Postmedial line slightly undulate. Terminal area with two parallel rows of light grey spots, often confluent, forewing apex white. Discal spot usually very indistinct on forewing, present on hindwing. Frons black, vertex and underside of palpi whitish, collar brown. Antennae of ♂ with intersegmental incisions, ciliate-fasciculate, cilia curved, 1.4-1.8 times width of flagellum. Hindtibia stout, with pencil, tarsus 0.85-0.95 times length of tibia.

Subsp. ***universaria*** with ground colour clearer white, terminal area darker, inner terminal spots of tornus and between radial veins of forewing usually brown.

MALE GENITALIA: Socii large and broad. Valva spoon-shaped, smooth. Lateral processes of anellus distinct, V-shaped. Aedeagus without cornuti. Sternum A8 with two long and stout cerata. Not polymorphic (n=4), but asymmetrically modified cerata (to half length) mentioned by STERNECK (1941: 22; 49).

FEMALE GENITALIA: Sclerites around ostium bursae irregularly shaped. Lamella antevaginalis reniform. Corpus bursae with granulated surface. Signum developed as elongate, bipartite sclerite of half length of corpus bursae, laterally dentate.

DISTRIBUTION: Endemic to Spain. Nominate subspecies restricted to southern Spain, replaced by the subsp. *universaria* in central, eastern and northern Spain, possibly with clinous transitions.

PHENOLOGY: Univoltine: mid-June to mid-July, exceptionally emerging from late May.

BIOLOGY: Larval host-plants unknown. Adults attracted to light.

HABITAT: Usually montane. Limited information available about habitat preference. Usually from 1,000 m up to 1,500 m above sea-level, sometimes lower, e.g. to 200 m in Huesca province.

SIMILAR SPECIES: *S. concinnaria* larger than the other species of this group, wing pattern reduced to discal spot and thin terminal fascia. Allopatric *S. orientalis* with postmedial line more dentate and terminal area darker, more dissolved into two rows of dark spots.

REMARKS: Habitus, shapes of juxta and base of sternum A8 in ♂ genitalia, and signum in ♀ genitalia reveal sister species relationship between *S. concinnaria* and *S. orientalis*.

128. Scopula orientalis (ALPHÉRAKY, 1876)

Acidalia decorata var. *orientalis* ALPHÉRAKY, 1876: Trudy russk. Ent. Obshch. 8: 197 (south-western European Russia: Taganrog). Lectotype ♂ (ZISP, examined), herewith designated to stabilize usage of the name.

Unavailable names (misidentification): *concinnaria*: sensu STERNECK (1941, partim) nec DUPONCHEL, 1842. – (infrasubspecific): *magna*: PROUT (1913). – (incorrect subsequent spelling): *orientata*: WARDIKJAN (1970).

DIAGNOSIS: Wingspan ♂♀ 24-29 mm, in Turkey sometimes 22 mm only. Hindwing termen slightly concave between M1 and M3. Ground colour white. Antemedial and medial lines fine, the latter dentate. Postmedial line dentate. Terminal area with two rows of grey spots, with brownish double spots close to postmedial line in the tornus and between the radial veins of both wings. Forewing apex usually white. Discal spot grey on forewing,

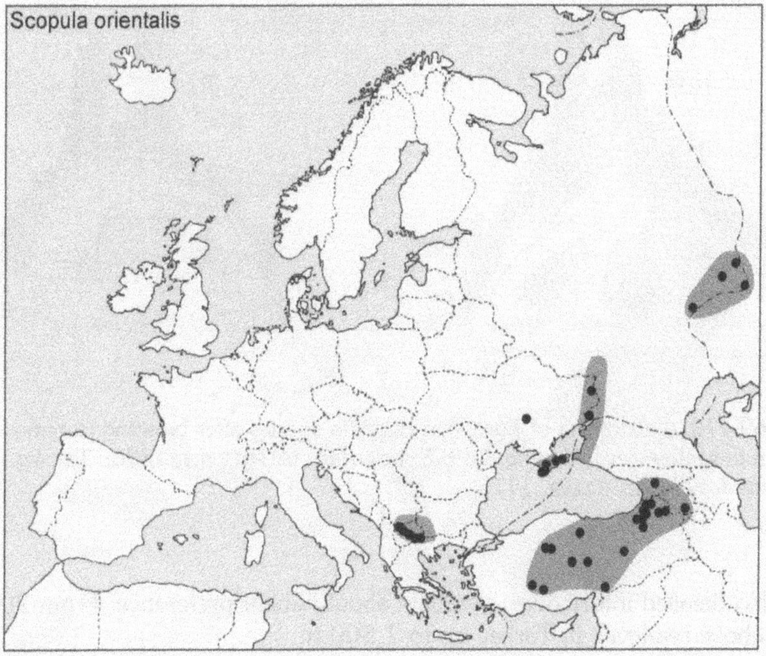

black on hindwing. Fringe light grey. Frons blackish brown, vertex whitish, collar ochreous. Antennae of ♂ with small intersegmental incisions, ciliate-fasciculate, cilia 1.4-1.7 times width of flagellum. Hindtibia stout, but not dilated, with pencil, tarsus 0.8-0.85 times length of tibia.

MALE GENITALIA: Similar to those of *S. concinnaria*. Socii broad, rounded. Valva broad, rounded. Fibula short, slender. Aedeagus without cornutus. Sternum A8 with mappa posteriorly concave, base shallow, rounded, both cerata stout, long and strongly curved, in the lectotype and one further ♂ from the Urals slender with weak spines at tip (gen.fig. 128b), in all the other examined ♂♂ (n=12; also 3♂♂ from Urals) dilated at tip, strongly spinose. As no other differential features are found, these different cerata forms are regarded to be due to polymorphism.

FEMALE GENITALIA: Antrum sclerotized to a slender tube with a laterally bulgy sclerite. Sterigma sclerotized to a broad ring. Ductus bursae short. Corpus bursae laterally with granulated surface. Signum long and slender, fused to a single sclerite, posteriorly joining ductus bursae.

DISTRIBUTION: (East-)European-Westasiatic. Very isolated populations in Macedonia, Bulgaria, Ukraine, southern European Russia, southern Urals. – Outside Europe in Turkey, Caucasus, Transcaucasus and northern Iran. Recorded by VIIDALEPP (1996) from central Asian mountains.

PHENOLOGY: Usually univoltine: mid-May to early July, in the mountains two weeks later. One specimen from early September (Armenia: ZSM) belonging to a (partial?) second generation.

BIOLOGY: Larval host-plants unknown. Adults attracted to light.

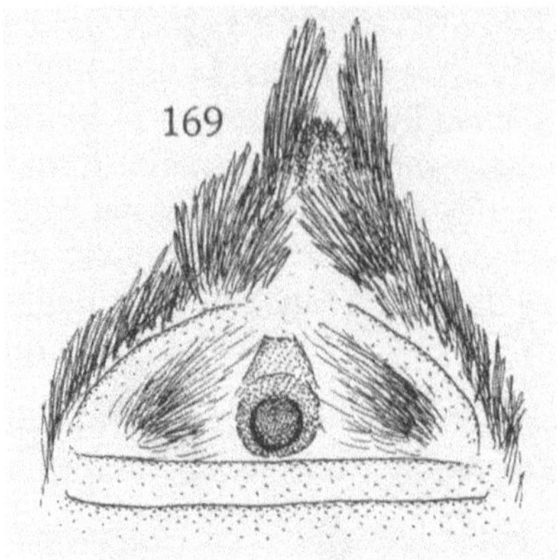

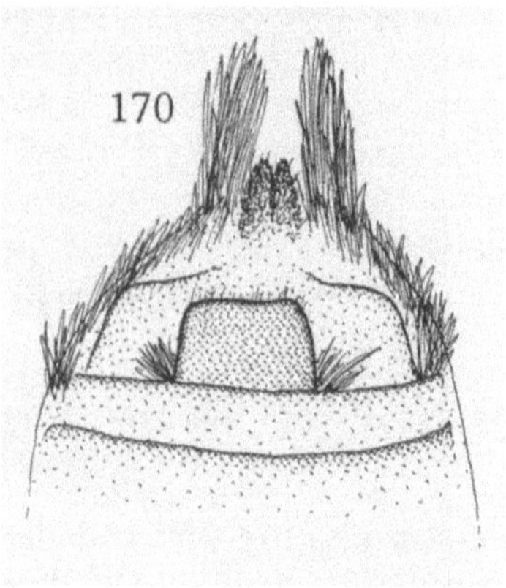

Text-figs 169-170: ♀ abdomina of European *Scopula* species after brushing to remove scales (tip of abdomen in ventral view). Text-fig. 169: *S. orientalis* (ALPHÉRAKY, 1876). Text-fig. 170: *S. decorata* (DENIS & SCHIFFERMÜLLER, 1775).

HABITAT: No detailed information available about habitat preference. From 200 m up to 1,100 m above sea-level, in Turkey up to 1,800 m.

SIMILAR SPECIES: *S. ornata* (see no. 126), allopatric *S. concinnaria* (127). *S. decorata* smaller, forewing apex and costal spots darker, forewing with distal row of terminal spots confluent, hindwing without brownish spots, shorter ♂ antennal ciliation. Transcaucasian *S. decorata* may, however, be very similar to *S. orientalis*. In both sexes identification is possible by brushing tip of abdomen to remove scales and reveal diagnostic characters (cerata, lamella antevaginalis; see genitalia figures and Text-figs 169-170).

REMARKS: Taxonomy discussed in OBRAZTSOV (1931). For sister species relationship to *S. concinnaria* see remarks to no. 127.

129. Scopula decorata (DENIS & SCHIFFERMÜLLER, 1775)

Geometra decorata DENIS & SCHIFFERMÜLLER, 1775: Ank. syst. Werkes Schmett. Wienergegend: 117 (Austria: Vienna distr.). Syntype(s) lost.

‡? *Phalaena cinerata* FABRICIUS, 1781: Spec. Ins. 2: 265 (Germany). Syntype(s). Synonymy according to FABRICIUS (1787). Junior primary homonym nec FABRICIUS, 1781 (p. 255) (identity unknown).

Geometra violata THUNBERG, 1784: Diss. ent. sistens Ins. Suecica 1: 14, pl. 3, fig. 9 (Sweden: Uppland, Uppsala). Lectotype ♂ (UZIU). Valid at subspecific rank.

Phalaena Geometra caerulata GMELIN, 1790: in LINNAEUS: Syst. Nat. (edn 13) 1 (5): 2490 (Sweden). Syntype(s).

‡ *Geometra decoraria* HÜBNER, 1799: Samml. Eur. Schmett. 5, Geometrae (1): pl. 14, fig. 71. Incorrect subsequent spelling.

Idaea decorata congruata ZELLER, 1847: Isis 1847 (7): 508 (Sicily: Syracus). Syntypes (BMNH: 2♂1♀, MNHU: 1♂, examined). Downgraded from species (MÜLLER 1996; SCOBLE 1999) to subspecific rank (stat. n.).

‡ *Idaea congruaria* HEYDENREICH, 1851: Lepid Eur. cat. meth. (Edn 3): 60. First binominal name after *congruaria* HERRICH-SCHÄFFER, 1948 (pl. 66, fig 404: non binominal). Incorrect subsequent spelling.

‡ *Acidalia ornata* var.? *aequata* STAUDINGER, 1879: Horae Soc. ent. ross. 14 (4): 440 (northern Turkey, Pontus: Amasia). Holotype ♂ (MNHU, examined). Infrasubspecific, as 'preliminary considered as aberration' by STAUDINGER.

Acidalia violata var. *armeniaca* THIERRY-MIEG, 1916: Misc. ent. 23: 52 (Armenia: Kirghiz). Syntype(s) ♂ (MNHN). Valid at subspecific rank (SCOBLE 1999).

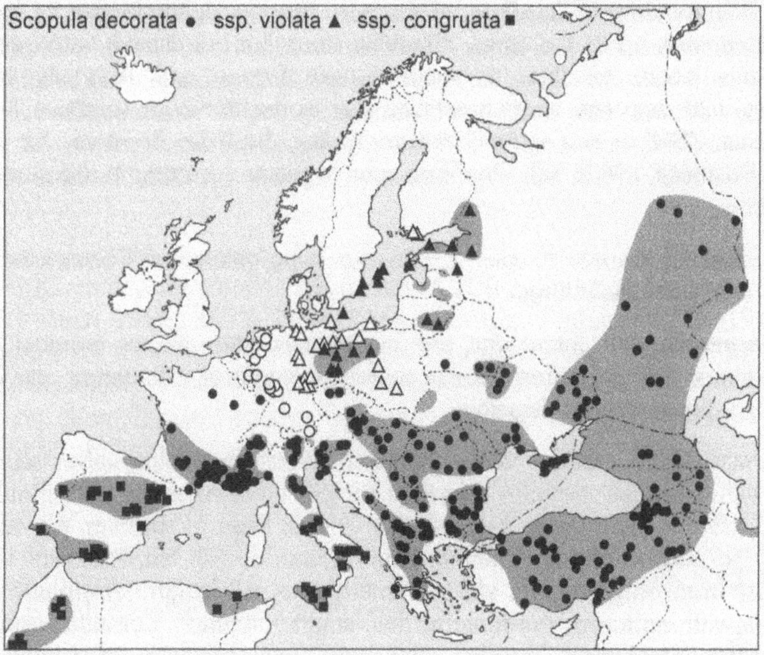

Acidalia decorata var. *leukiberica* WEHRLI, 1927: Dt. ent. Z. Iris 41 (1): 57 (Spain, Castilia: Escorial). Lectotype ♀ (ZFMK, examined), herewith designated to fix type locality and to stabilize usage of the name; at ZFMK 5 paralectotypes from Catalonia, Andalusia, southern France and Sicily; further paralectotypes from Algiers and Oran, not traced. Junior synonym (subsp. *congruata*).

† *Acidalia decorata* var. *rebeli* DRENOWSKI, 1930: Mitt. Bulg. ent. Ges. 5: 113 (Bulgaria: Alibotusch-Bereng). Junior primary homonym nec PROUT, 1913 (identity: *Scopula beckeraria*).

Scopula decorata eurhythma PROUT, 1935: in SEITZ, Macrolep. 4, Suppl.: 47 (China, Shantung Promontory: Wei-hai-wei). Holotype ♂ (BMNH). Valid at subspecific rank (SCOBLE 1999).

Scopula decorata f. *drenowskii* STERNECK, 1941: Z. Wien Ent. Ver. 26 (2): 53 (Bulgaria: Alibotusch). Syntype(s). Though formally introduced as a 'form', STERNECK (1941: 54) calls it 'a separate species, hence not understood at infrasubspecific rank. Herewith downgraded from subspecific rank (SCOBLE 1999) to synonymy (syn. n.).

Scopula decorata przewalskii VIIDALEPP, 1975: Nasekom. Mongol. 3 (6): 446, figs 2,3 (Mongolia, central Aimak: 125 km S Ulan Bator). Holotype ♂. Valid at subspecific rank (SCOBLE 1999, confirmed).

Unavailable names (misidentification): *orientalis*: sensu STERNECK (1941) nec ALPHÉRAKY, 1876; *ornata*: sensu ZETTERSTEDT (1839) and ROTHSCHILD (1914) nec SCOPOLI, 1763; *ornataria*: sensu ESPER (1806: pl. 51, fig. 2) nec *ornata* SCOPOLI, 1763; *subtilata*: sensu KOUTSAFTIKIS (1973) and STEKOLNIKOV (1996: gen.fig. 14-9) nec CHRISTOPH, 1867. – (infrasubspecific): *ablutata*: DANNEHL (1927); *albomarginata*: REISSER (1961); *limnica*: KOUTSAFTIKIS (1973); *pumilio*: ROTHSCHILD (1914); *radiata*: SCHWINGENSCHUSS (1962). – (incorrect subsequent spellings): *drenovskii*: PINKER (1968); *leuciberica*: REDONDO & GASTON (1999); *oblutata*: PROUT (1935); *violeta*: TER HAAR (1904).

DIAGNOSIS: Wingspan ♂♀ 22-27 mm, in subsequent generations sometimes 17 mm only. Hindwing slightly concave between M1 and M3. Ground colour creamy white. Antemedial and medial lines usually present, the latter sometimes dentate, with costal spots distinct. Postmedial line dentate, distally with grey spots, brownish double spots only on the forewing. Near termen very small spots, usually strongly confluent. Terminal area darker, on hindwing almost concolorous with ground colour. Forewing apex with spots. Discal spot usually grey on forewing, black on hindwing. Fringe dark grey. Frons blackish brown, vertex whitish, collar brown. Antennae of ♂ almost filiform, ciliate-fasciculate, cilia 1.0-1.3 times width of flagellum, in southern Europe slightly longer, in Asian populations exceptionally up to 1.6 times. Hindtibia stout, but not dilated, with pencil, tarsus 0.8-0.9 times length of tibia, in south-eastern Europe and Asia slightly shorter. Occasionally with one very short hindtibial spur (some ♂♂ from southern France, 1♂ from Armenia, ZSM) or two spurs (1♂ from Turkey, 1♂ from Armenia; 1♂ under '*orientalis*' in STERNECK 1941: 52). One long spur regularly occurring in the populations of northern Iran.

Subsp. ***violata*** on average smaller, 19-26 mm, wing pattern well contrasted, terminal area dark, with slight violet tinge.

Subsp. ***congruata*** with antemedial and medial lines often absent, terminal area pale, hindwings almost without pattern, collar ochreous, length of ♂ antennal cilia exceptionally up to 1.6 times width of flagellum.

MALE GENITALIA: Socii crossed. Valva digitiform. Fibula exceeding valva, setose at tip. Aedeagus with one small cornutus. Sternum A8 with left ceras short, with spinulose brush on an oval sclerite. Right ceras long, slightly curved. Base of sternum A8 semicircular, projecting. No polymorphism of cerata length found (n=34), but STERNECK (1941: 53) mentions 1♂ from Bulgaria with left arm enlarged to half length ('*drenowskii*') and 1♂ from Siberia with right arm shortened to half length ('*violata*'). Cornutus usually slightly larger in eastern populations. In subsp. *congruata* base of sternum A8 shallow, left ceras comparatively short and broad, mappa usually more asymmetric than in the nominate subspecies. Subsp. *przewalskii* with shorter right ceras and spines of left ceras on the inner side.

FEMALE GENITALIA: Lamella antevaginalis developed as a flat, slightly asymmetric, sub-rectangular sclerite. Antrum (colliculum) small, elongate. Ductus bursae of medium length. Corpus bursae oval or globular, medially with granulated surface. In subsp. *congruata* lamella antevaginalis more asymmetric and very broad, corpus bursae oval, more or less elongate.

DISTRIBUTION: Palaearctic. Widely distributed in Europe, from France to the Urals, usually rare and local. Around the Baltic Sea replaced by the subsp. *violata*. In the Iberian peninsula, Sicily and in southern Italy replaced by subsp. *congruata*. Absent from all Mediterranean islands except Sicily. On Corsica and Sardinia replaced by the sister species *S. honestata*. In the northern half of Europe with severe areal losses, e.g. extinct in the Netherlands since 1942 (KUCHLEIN & DE VOS 1999), from most regions of Germany, actually only in a very few localities in north-eastern Germany (MÜLLER & GELBRECHT 1992; ERLACHER & FRIEDRICH 1994; GELBRECHT 1999; EBERT 2001; WOLF & HACKER 2004; SOBCZYK & GELBRECHT 2004), in Poland since 1980 (BUSZKO, MALKIEWICZ pers. comm.), and in Finland (HULDEN et al. 2000). The species is highly sensitive to overshading vegetation, such as shrubs or even grass (GELBRECHT et al. 1995). It is vulnerable to habitat alteration resulting from the increase of nutrients and air-borne nitrates (EBERT 2001). Special habitat programs are required to protect this species. – Outside Europe from Morocco to Tunisia (subsp. *congruata*), Turkey, Caucasus, Transcaucasus, northern Iran, in the east to central Asian mountains, Tadjikistan and Afghanistan, with isolated occurrence in southern Iran (prov. Fars, coll. SOMMERER). Replaced by subsp. *eurhythma* in western China, and by subsp. *przewalskii* VIIDALEPP, 1975 from Sajan mountains to northern Mongolia and Dahuria (KHOMENKO et al. 1994; VIIDALEPP 1996).

PHENOLOGY: Usually bivoltine: early May to early September. Under favourable conditions south of the Alps and in Asia trivoltine from mid-April to mid-October. Under bad conditions, e.g. in southern Scandinavia (subsp. *violata*) and on mountains above ca. 1,000 m, univoltine from late June to early August, sometimes with a partial second generation in September (e.g. in Armenia). Subsp. *congruata* univoltine on mountains above ca. 1,200 m, bivoltine around 1,000 m, or trivoltine in the lowlands. In Morocco emerging from late March.

BIOLOGY: Larva possibly monophagous (3) on *Thymus*. Found on *Thymus serpyllum* (Lamiaceae) (Germany: BERGMANN 1955; EBERT 2001; SOBCZYK & GELBRECHT 2004; Finland: SEPPÄNEN 1954; 1970; MÜLLER pers. comm.; Portugal: RIBBE 1912), sitting freely exposed on dry parts of the host-plant during daytime (MÜLLER pers. comm.; SOBCZYK & GELBRECHT 2004). Frequently recorded on *T. serpyllum* (var. cit.) as well as, reputedly, on other *Thymus* species (var. cit., e.g. MILLIÈRE 1867; BERGMANN 1955; SCOBLE 1999), preferring petals and young leaves (SKOU 1986). Reputedly on *Melissa officinalis* (Lamiaceae), and *Clinopodium vulgare* (Lamiaceae), as well (France: LHOMME 1935). Reared on *Origanum vulgare* (SOBCZYK & GELBRECHT 2004). Adults easily flushed up from vegetation during daytime, rarely with active flight during daytime, often resting on sandy soil, readily attracted to light. Imagines flying around *Thymus* (PROUT 1913). r-strategy, with high turnover from year to year.

HABITAT: Xerothermophilous. Sun-exposed, dry localities with abundance of *Thymus serpyllum*, and almost free from other vegetation, on extremely low-nutrient soils (SOBCZYK & GELBRECHT 2004). Also in former gravel- or sand-pits (GELBRECHT et al. 1995), dry heathland with short grass and scattered occurrence of *Pinus* (GAEDIKE & HEINICKE 1999), sandy dunes near coast, fallows, rocky steppes. Preferring sandy and calcareous soils. North of Alps restricted to the hot lowlands from 0 m up to about 300 m above sea-level, up to 1,600 m in southern Alps, and Balkan peninsula, sometimes up to 3,100 m in Turkey, northern Iran and central Asian mountains. Subsp. *congruata* from 0 m up to 1,900 m above sea-level, below 1,000 m usually rare, in Morocco up to 2,300 m.

SIMILAR SPECIES: *S. ornata* (see no. 126), *S. orientalis* (128). In both sexes identification is possible by brushing tip of abdomen to remove scales and reveal diagnostic characters (cerata, lamella antevaginalis; see genitalia figures and Text-figs 169-170). *S. honestata* with creamy white ground colour, large discal spots, and distinct antemedial line. *S. subtilata* differs in the broader medial line ending in a large brownish costal spot, wavy line broad, zigzagging, brown markings in the terminal area darker, proximally with broader black border, discal spots distinct.

REMARKS: The taxa *violata* and *congruata* are considered valid at subspecific rank (see HAUSMANN 1993e), despite the existence of transitional forms to the nominate subspecies. Clear delimitation of the distribution areas still await further research. Populations from south-western France to central Spain with transitional feature combinations between *decorata* and *congruata*. Populations of Germany and Czech Republic with transitional feature combinations between *decorata* and *violata*.

130. Scopula honestata (MABILLE, 1869)

Acidalia honestata MABILLE, 1869: Annls Soc. ent. Fr. (4) 9 (1): 62, pl. 2, fig. 8 (Corsica: Bastia). Lectotype ♂ (ZSM/HERBULOT, examined).
Unavailable names (infrasubspecific): *cyanolata*: SCHAWERDA (1928).

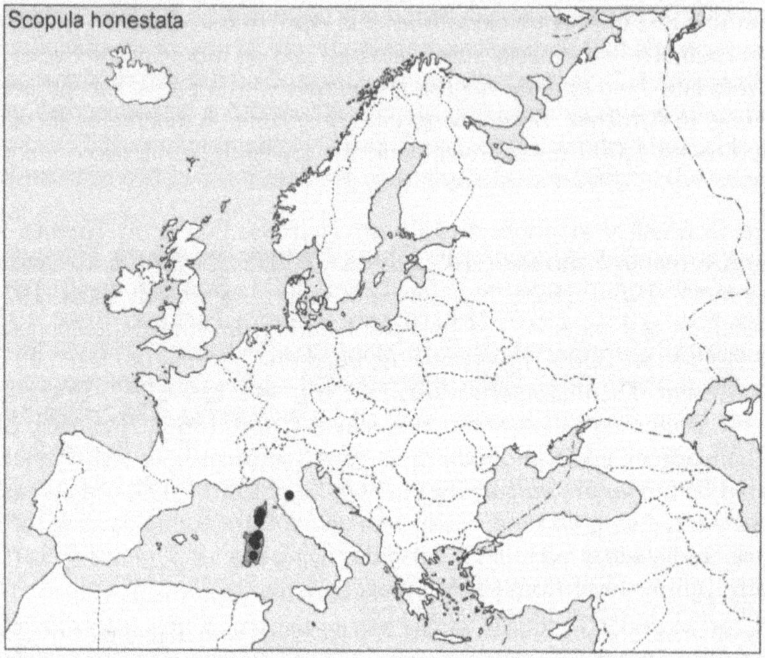

DIAGNOSIS: Wingspan ♂♀ 21-26 mm. Ground colour creamy white with slight, warm brown tinge. Transverse lines of forewing with distinct costal spots. Antemedial line black, distinct. Postmedial line strongly zigzagging on all wings. Terminal area brightly coloured, with bluish grey spots and warm brown terminal fascia and double spots on forewing. Discal spot of forewing black, distinct. Frons blackish brown, vertex whitish, collar light brown. Antennae of ♂ with small intersegmental incisions, ciliate-fasciculate, cilia 1.4-1.6 times width of flagellum. Hindtibia stout, but not dilated, with pencil, tarsus 0.8-0.9 times length of tibia.

MALE GENITALIA: Very similar to those of *S. decorata* but base of sternum A8 projecting as round sclerite, right ceras comparatively short and stout in populations from Corsica, slightly longer in Sardinia, curved at tip, mappa rounded. Left ceras on average longer than in *S. decorata*. Not polymorphic (n=6).

FEMALE GENITALIA: Without constant differential features from *S. decorata*.

DISTRIBUTION: Tyrrhenian. Endemic to Corsica, and Sardinia. One specimen recorded in prov. Firenze, Tuscany (ZSM, examined).

PHENOLOGY: Univoltine: late June to late July, single specimens until late August, mainly at higher elevation.

BIOLOGY: Larval host-plants unknown. Main flight activity of adults late in the night (REISSER 1926).

HABITAT: Xeromontane. No detailed information available about habitat preference. From about 800 m up to 1,800 m above sea-level.

SIMILAR SPECIES: The creamy white ground colour is unique within the group, discal spots large, antemedial line distinct, postmedial line of forewing strongly zigzagging. *S. decorata* (see no. 129).

REMARKS: Valid at species rank, see HAUSMANN (1993c).

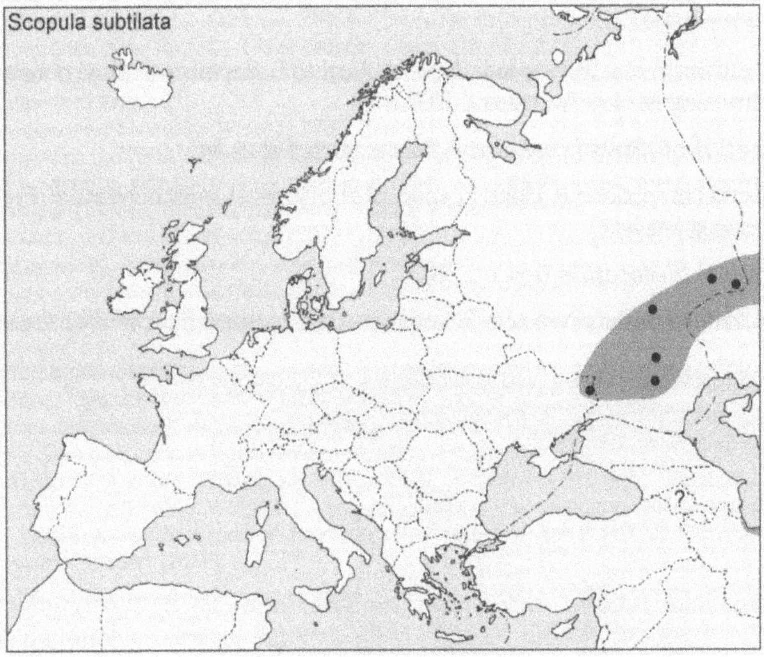

131. *Scopula subtilata* (CHRISTOPH, 1867)

Acidalia subtilata CHRISTOPH, 1867: Stettin. ent. Ztg. 28 (4-6): 236 (south-eastern European Russia: Krasnoarmeysk, 'Sarepta'). Lectotype ♂ (ZISP, examined), herewith designated to stabilize usage of the name with regard to the very similar other species of this group.
Unavailable names (incorrect subsequent spelling): *subtilala*: PROUT (1913).

DIAGNOSIS: Wingspan ♂ 23-27 mm, ♀ 22-25 mm. Hindwing termen dentate, mainly at M3. Ground colour white. Transverse lines of forewing with distinct costal spots, chiefly that of the medial line. Postmedial line strongly zigzagging on forewing. Terminal area of forewing with both brown double spots large and confluent. Terminal area of hindwing with indistinct pattern except for the brown tornus spot. Discal spots dark grey. Fringe dark brown. Underside with large black hindwing discal spots. Palpi, frons and collar brown, vertex white. Antennae of ♂ with deep intersegmental incisions, ciliate-fasciculate, cilia long, about twice width of flagellum. Hindtibia slender, with two spurs, tarsus longer than tibia. STERNECK (1941: 54) and VASILENKO (2001) mention one ♂ hindtibial spur only, due to accidental loss, individual aberration or error.

MALE GENITALIA: Socii tapering, crossed. Valva broad, sub-rectangular. Aedeagus with one comparatively long, straight cornutus. Sternum A8 with both cerata of medium length, stout, with spinulose crest at tip. Not polymorphic (n=8).

FEMALE GENITALIA: Lamella antevaginalis developed as flat, heraldic emblem-shaped sclerite. Antrum (colliculum) small. Ductus bursae short. Corpus bursae globular.

DISTRIBUTION: (East-)European-Westasiatic. Southern Urals, southern European Russia, eastern Ukraine (KOSTJUK 2004). Erroneously mentioned for Greece (KOUTSAFTIKIS 1973; MÜLLER 1996). – Outside Europe in north-eastern Kazakhstan (specimens in ZISP examined by MIRONOV, pers. comm.), Siberia (VASILENKO 2001), northern Iran (ZISP), and Turkmeniya (VIIDALEPP et al. 1992; VIIDALEPP 1996). Reputedly also in Transcaucasus (Armenia: PROUT 1913).

PHENOLOGY: Bivoltine: late May to late June; August to September (limited data available), in Siberia from early May (VASILENKO 2001).

BIOLOGY: Larval host-plants unknown. Adults attracted to light.

HABITAT: Xerothermophilous lowland species of dry short steppe habitats. From 0 m up to 300 m above sea-level.

SIMILAR SPECIES: *S. decorata* (see no. 128).

REMARKS: Unjustified designation of a 'holotype' in the redescription of VASILENKO (2001).

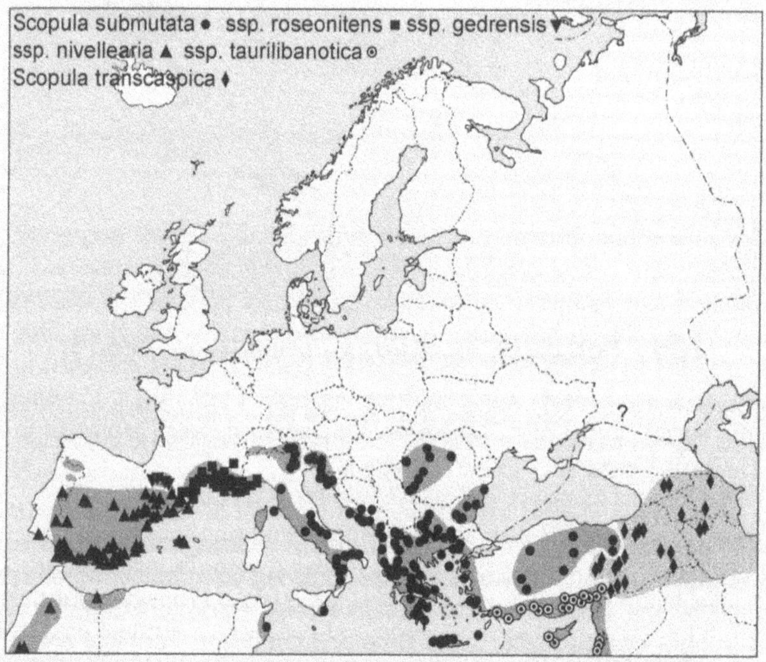

132. Scopula submutata (TREITSCHKE, 1828)

Idaea submutata TREITSCHKE, 1828: Schmett. Eur. 6 (2): 277 (Croatia). Syntype(s).

Acidalia gianellaria TURATI, 1905: Naturalista sicil. 18 (2-3): 44, pl. 8, figs 15-17 (Italy: Val d'Aosta, Donnaz). Syntypes ♂♀ (MSNT: 1 specimen ex coll. TURATI, examined by RAINERI 1996).

Acidalia submutaria BOISDUVAL, 1840: Genera Index meth. eur. Lepid.: 221. Emendation.

‡ *Acidalia submutataria* HERRICH-SCHÄFFER, 1847: Syst. Bearb. Schmett. Eur. 3 (24): 25 ('*submutaria*' on pl. 17, fig. 97: non binominal (1845)). Incorrect subsequent spelling (emendation according to SCOBLE 1999).

Acidalia submutata var. *nivellearia* OBERTHÜR, 1922: Études Lépid. comp. 19: 287, pl. 537, fig. 4492 (Morocco: Mrassine). Lectotype ♂ (ZFMK, examined, with 'holotype' label), herewith designated to stabilize usage of the name with regard to closely related taxa; at ZFMK 2 paralectotypes. Herewith downgraded from species (AGENJO 1952; VIVES MORENO 1994; MÜLLER 1996; SCOBLE 1999) to subspecific rank. Previously downgraded in DANTART et al. (1993) and HAUSMANN (1995).

Acidalia submutata var. *pseudhonestata* WEHRLI, 1926: Dt. ent. Z. Iris 40 (2/3): 116 (Spain, Andalusia: Granada). Syntypes 1♂2♀ (ZFMK: 1♂1♀, examined; 'BMNH': SCOBLE 1999); in description one further ♂ from Turkey, Amasia, mentioned. Herewith downgraded from subspecific rank of '*S. nivellearia*' (SCOBLE 1999), of *S. submutata* (REDONDO & GASTON 1999) or from species rank (CALLE et al. 2000) to synonymy (syn. n.).

Acidalia submutata roseonitens WAGNER, 1926: Zeits. österr. Ent.Ver. 11: 82, 85, fig. (southern France: Esterel; Italy, Liguria: Ventimiglia). Syntypes (NHMW, figured in description). Valid at subspecific rank.

Acidalia submutata var. *flava* WAGNER, 1926: Zeitschr. Österr. Ent. Ver. 11 (9): 82, 83, 86, fig. First available name after *flava* KITT, 1925: Zeits. österr. Ent.Ver. 10: 28 (infrasubspecific) (France, Pyrenees: Vernet-les-Bains). Holotype ♂ of KITT.

Acidalia submutata taurilibanotica WEHRLI, 1932: Mitt. Münchn. Ent. Ges. 22 (1) 4 (northern Lebanon: Bscharre). Lectotype ♀ (ZFMK, examined), herewith designated to fix type locality and to stabilize usage of the name with regard to the similar sister species; at ZFMK and BMNH 61 paralectotypes from southern Turkey, Maras and Lebanon, Bscharre and Beirut. Valid at subspecific rank.

Acidalia submutata syrilibanotica WEHRLI, 1933: In ZERNY: Dt. Ent. Z. Iris 47 (2): 86 (northern Lebanon: Bscharre). Syntypes ♂♀. Junior synonym of *taurilibanotica*.

Scopula submutata mudrica KOUTSAFTIKIS, 1973: Annls Mus. Goulandris 1: 214, pl. 31, fig. 17 (Greece, northern Greek islands: Limnos, Moudhro). Holotype ♀. Herewith downgraded from subspecific rank (SCOBLE 1999) to synonymy (syn. n.).

Scopula submutata gedrensis HAUSMANN, 2003: Ent. Zeits. Stuttgart 113 (11): 323, fig. 8 (France, Hautes Pyrenees: Gèdre). Holotype ♂ (ZSM/HERBULOT, examined). Valid at subspecific rank.

Unavailable names (misidentification): *contiguaria*: sensu DUPONCHEL (1830) nec HÜBNER, 1799 (identity: *Idaea contiguaria*). – (infrasubspecific): *albomarginata*: REISSER (1961); *anastomosaria*: REISSER (1961); *cyanata*: SCHAWERDA (1921); *marginata*: PROUT (1913); *nigricans*: THIERRY-MIEG (1916). – (incorrect subsequent spellings): *giannelaria*: FORSTER & WOHLFART (1981: pl.5); *granellaria*: VALLETTA (1973; misidentified); *nivellaria*: ZERNY (1927); *pseudohonestata*: AGENJO (1952); *submatata*: RUNGS (1981); *submitata*: EXPOSITO (1978); *submutala*: MILLIÈRE (1868).

DIAGNOSIS: Wingspan ♂♀ 21-25 mm, subsequent generations smaller, sometimes 17 mm only. Hindwing termen rounded, slightly arched between veins. Coloration and wing pattern with high infrapopulational and geographical variability (see below). Wings whitish or pale sand coloured, with dense dark suffusion. Forewing with three distinct black costal spots. Postmedial line distinct, zigzagging, distally bordered by a grey fascia with often bluish tinge. Distal part of terminal area usually brownish. Discal spot usually present on all wings. Frons and collar blackish brown, vertex whitish. Antennae of ♂ with intersegmental incisions, ciliate-fasciculate, cilia 1.0-1.3 times width of flagellum. Antennae of ♀ ciliate-setose, cilia about 0.3-0.5 times width of flagellum. Hindtibia stout, but not dilated, with pencil, tarsus 0.8-1.0 times length of tibia.

Variability: Populations of north-western Italy and South Tyrolian Porphyr Alps ('*gianellaria*') with strong dark suffusion, similar to that of *Idaea typicata hornigaria* and *Glossotrophia confinaria prouti*, but with transitions to the nominotypical populations from lake Garda region to Trieste. In Macedonia often with warm brownish tinge. East-Mediterranean subsp. *taurilibanotica* with distinct, often bluish terminal shadows. Sister species *S. transcaspica* as described for *S. submutata*, but ground colour often more creamy white, ♂ antennal flagellum nearly filiform, with very shallow intersegmental incisions.

Subsp. ***roseonitens*** usually with strong sand coloured suffusion, pattern not well contrasted, transverse lines often with orange tinge. In the departments Var and Vaucluse, however, reminiscent of subsp. *nivellearia* in habitus. Populations of the eastern Pyrenees well matching subsp. *roseonitens* in habitus.

Subsp. ***gedrensis*** larger, 25-27 mm, with strong dark grey suffusion and slight violet tinge.

Subsp. ***nivellearia*** very variable in habitus, usually with sharp, contrasted pattern, postmedial line dark brown, zigzagging. On high mountains of southern Spain with pale ground colour and reduced, but sharp markings, in the lowlands often with dark suffusion. Populations of Iberian peninsula ('*pseudhonestata*') and North Africa without constant differences in habitus and morphology.

MALE GENITALIA: Socii broad, distinctly projecting. Valva broad, curved with short terminal spines. Aedeagus with one cornutus. Sternum A8 with right ceras long, curved, left ceras very short, broad, spinose. Nominate subspecies with base of sternum A8 biangulate, posterior margin of mappa with deep dorsal notch. West-Mediterranean populations with rounded base of sternum A8 and cornutus usually curved at tip. Subsp. *nivellearia* and subsp. *gedrensis* with posterior margin of mappa slightly concave, or even rounded. Without polymorphism of cerata length (n=37). Sister species *S. transcaspica* with aedeagus and cornutus broad, both cerata long, the left, sometimes, shortened to half length and strongly curved (HAUSMANN 1997; 1999).

FEMALE GENITALIA: Lamella antevaginalis irregularly shaped, often sub-trapezoid and/or with asymmetrically bilobed posterior margin. Corpus bursae with sclerotized, rounded posterior end. Signum absent.

DISTRIBUTION: Mediterranean. Widely distributed from Iberian peninsula and Mallorca (DANTART et al. 1993: subsp. *nivellearia*) across Pyreenees (subsp. *gedrensis*), southern France and Liguria (subsp. *roseonitens*) to Sicily and Balkan peninsula, in the east reputedly in isolated populations in southern European Russia (VIIDALEPP 1996), in the north to southern Switzerland. Present also on Corsica (MÜLLER 1996) and many Greek islands including Crete, not recorded from Malta and Sardinia. – Outside Europe from Morocco to western Algeria (subsp. *nivellearia*), Tunisia and from Turkey and Cyprus to the Levant (subsp. *taurilibanotica*). From eastern Turkey and Transcaucasus to Turkmeniya replaced by the sister species *S. transcaspica* PROUT, 1935 with subsp. *taftanica* BRANDT, 1941 from eastern Iran, Afghanistan and northern Pakistan. All records of '*S. submutata*' from this area (e.g. VIIDALEPP 1988; 1996; WARDIKJAN 1970; 1985; WILTSHIRE 1939a; 1957; 1967) refer to *S. transcaspica*. Erroneously recorded for Mongolia (VOJNITS 1976). Doubtfully recorded from Libya, Cyrenaica (TURATI 1924, '*submutulata*').

PHENOLOGY: Nominate subspecies and subsp. *roseonitens* bivoltine: mid-May to early October, usually with a short gap in July. Under good conditions in the lowlands one or two weeks earlier, on the mountains one week later. In the north of the distribution generations sometimes slightly overlapping. On Crete trivoltine until late October. Subsp. *gedrensis* univoltine from early July to late July with partial second generation in August and September. Subsp. *nivellearia* in the lowlands trivoltine from early April to early June; early June to late July; mid-September to late October. On mountains bivoltine as the nominate subspecies, on high mountains of southern Spain two short, overlapping generations from late June to early August. Subsp. *taurilibanotica* bivoltine in the north and on the mountains, trivoltine in the lowlands and in the Levant, from late March to early June; late June to mid-August; mid-October to early November.

BIOLOGY: Larva probably oligophagous (2) on Lamiaceae and Fabaceae of low-nutrient soils. Found on *Thymus vulgaris* (Lamiaceae) and *Dorycnium* (Fabaceae) (France: MILLIÈRE 1867; subsequently var. cit.). Reared on withered or dry leaves of *Thymus serpyllum* (FRIEDRICH 1976; LEIPNITZ pers. comm.). Adults readily attracted to light.

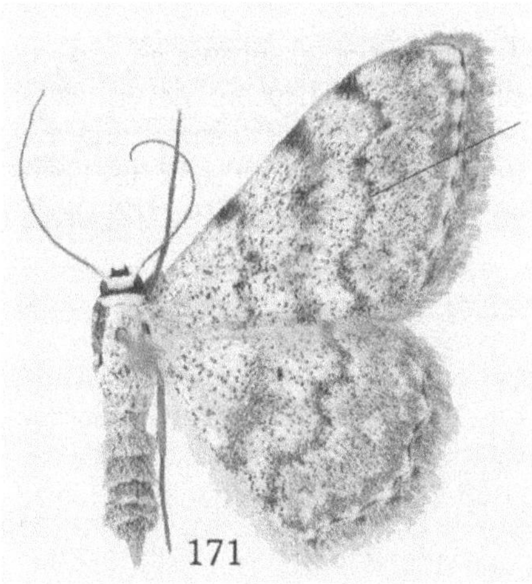

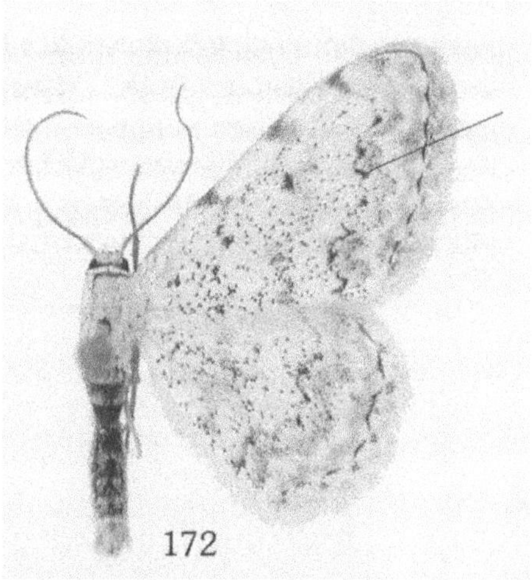

Text-figs 171-172: Diagnostic characters (indicated) of European *Scopula* species. Text-fig. 171: *S. submutata* (Treitschke, 1828). Text-fig. 172: *S. vigilata* (Wagner, 1926).

HABITAT: Xerothermophilous. Sometimes common in dry, open habitats, such as coastal rocky slopes, or Mediterranean maquis, preferably on limestone. From 0 m up to 1,500 m above sea-level, subsp. *gedrensis* from 1,000 m up 1,500 m, up to 2,300 m in southern Spain, Morocco, Turkey and Lebanon.

SIMILAR SPECIES: Coloration of *S. marginepunctata* with slight sand coloured tinge, costal spots indistinct, terminal line not surrounding forewing apex, hindwing termen less concave between veins, ♀ antennae not ciliate. *S. vigilata* smaller, postmedial line usually sharper, ground colour with less dark suffusion, ♂ with one hindtibial spur. *Glossotrophia* species usually with more diffuse wing pattern, European species with longer proboscis, one hindtibial spur in ♂, two spurs in ♀. In both sexes identification is possible by brushing tip of abdomen to remove scales and reveal diagnostic characters (cerata, lamella antevaginalis; see genitalia figures and Text-figs 173-175).

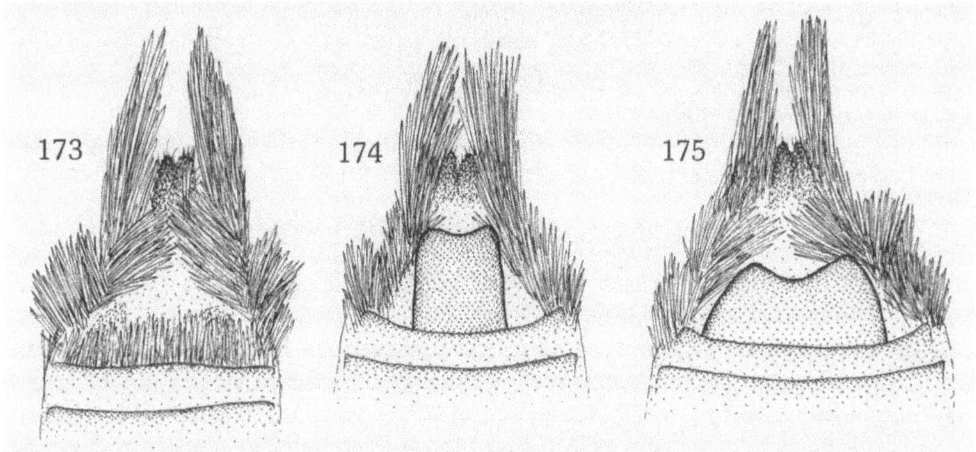

Text-figs 173-175: ♀ abdomina of *Scopula* species after brushing to remove scales (tip of abdomen in ventral view). Text-fig. 173: *S. submutata* (Treitschke, 1828). Text-fig. 174: *S. vigilata* (Wagner, 1926). Text-fig. 175: *S. transcaspica* (Prout, 1935).

REMARKS: With hybridization zones between some of the mentioned subspecies, e.g. in the eastern Pyrenees (subsp. *roseonitens*, subsp. *nivellearia*). They are nevertheless retained as valid here, in order to reflect their high stability of habitus features and the slight differences in genitalia. Validity of the taxon *nivellearia* at species rank has been postulated repeatedly, with 'subsp. *pseudhonestata*' in Spain (AGENJO 1952: 192; VIVES MORENO 1995: 332). See nomenclatural remarks.

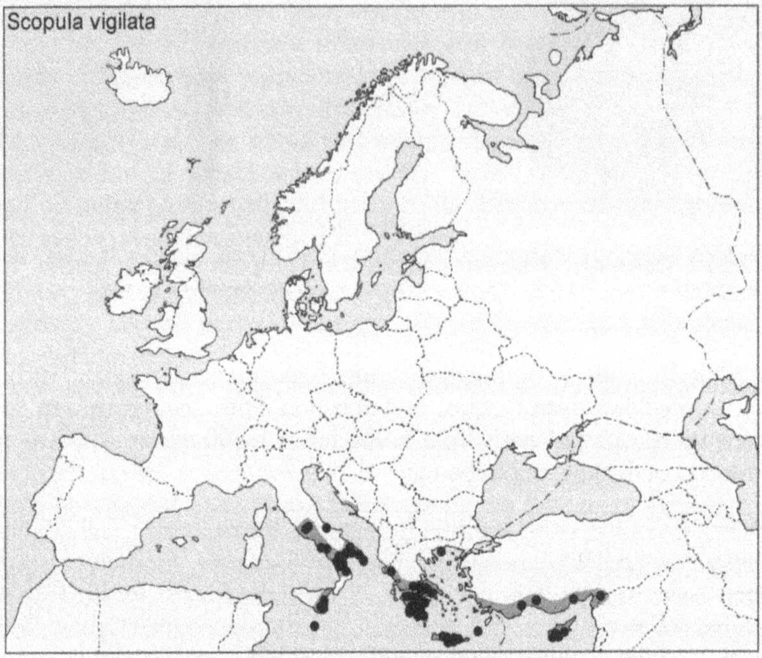

133. *Scopula vigilata* (WAGNER, 1926)

Acidalia submutata var. *vigilata* WAGNER, 1926: Zeitschr. Österr. Ent. Ver. 11 (9): 84, 86, fig.. First available name after *vigilata* PROUT, 1913 (infrasubspecific, introduced in synonymy) (Sicily: Palermo). Lectotype ♂ of PROUT (BMNH: examined, with original red 'type' label ex coll. ZELLER, leg. MANN), herewith designated to fix type locality and to stabilize usage of the name; paralectotype(s) from Capri. WAGNER's material from Sicily: Taormina, central Italy: Tivoli, Syria and Mesopotamia. Species and genus name on p. 81, thus not uninominal. See also HAUSMANN (1995).

Acidalia submutata var. *turatii* WAGNER, 1926: Zeitschr. Österr. Ent. Ver. 11 (9): 82, 85, fig. (Sicily, Mt. Etna: Nicolosi). Syntype(s). Referring to species and genus name on p. 81, thus not uninominal. Dark individual form of lava slopes.

Acidalia submutata var. *submutulata* WAGNER, 1926: Zeitschr. Österr. Ent. Ver. 11 (9): 82, 85, fig. First available name after *submutulata* REBEL, 1902 and GALVAGNI, 1909 (both infrasubspecific) (Greece, Peloponnes: Kalavrita, N of Morea). Syntype(s) (of REBEL, NHMW: 1♀, dissected). New synonymy.

Unavailable names (incorrect subsequent spelling): *submutatula*: RAINERI (1996).

DIAGNOSIS: Wingspan ♂♀ 19-22 mm, subsequent generations 15-20 mm. Hindwing termen rounded. Ground colour creamy, ivory white, slightly suffused with some dark scales, on Mt. Etna, Sicily, and exceptionally elsewhere with strong dark suffusion. Wing pattern sharp. Forewing with three distinct black costal spots. Postmedial line dark brown, zigzagging, not rarely dotted. Terminal area with distinct wavy line, proximally bordered by a grey, distally by a brownish fascia. Discal spots usually distinct, mainly that of forewing. Frons and collar black brown, vertex whitish, palpi light brown. Proboscis comparatively long. Antennae of ♂ with intersegmental incisions, ciliate-fasciculate, cilia 1.2-1.6 times width of flagellum. Antennae of ♀ ciliate-setose, cilia 0.5-0.7 times width of flagellum. Hindtibia not dilated, with one spur, without pencil, tarsus longer than tibia. Exceptionally with an additional proximal pair of ♂ hindtibial spurs (HERBULOT 1950a).

Variability: Generally little varying. On the lava slopes of Mt. Etna, Sicily, with strong dark suffusion ('substrate race'), but with clinous transitions to nominotypical forms in the lowlands of prov. Catania.

MALE GENITALIA: Socii broad, distinctly projecting. Valva broad, curved with short terminal spines. Fibula very small. Aedeagus with one cornutus. Both cerata of sternum A8 medium-sized, stout, curved, strongly spinulose. Not polymorphic (n=5). East Mediterranean populations with aedeagus slightly shorter.

FEMALE GENITALIA: Lamella antevaginalis sub-rectangular, posteriorly rounded or slightly convex. Corpus bursae elongate, surface latero-posteriorly 'rough', i.e. covered with extremely small spinules.

DISTRIBUTION: East-Mediterranean. From central Italy to Sicily and Malta, Greece, Crete, and other Greek islands, with isolated occurrence in western Croatia (GERSTBERGER 1990; MÜLLER 1996). – Outside Europe in Cyprus and southern Turkey, reputedly in Lebanon (SOHN-RETHEL 1929).

PHENOLOGY: Tri- or plurivoltine: late April to late June; mid-July to late October. Generations not always clearly separated.

BIOLOGY: Larva possibly mono- or oligophagous on Lamiaceae. Reared on *Thymus pulegioides* (Lamiaceae) (original data), refusing a wide range of plant species from Asteraceae, Caryophyllaceae, Ranunculaceae, Fabaceae, Scrophulariaceae and Convolvulaceae (original data). Adults readily attracted to light.

HABITAT: Xerothermophilous lowland species of the Mediterranean maquis. Usually from 0 m up to 600 m above sea-level, exceptionally higher, e.g. on Mt. Etna, Sicily (1,000 m) or Parnass Mts., Greece (1,500 m).

SIMILAR SPECIES: *S. submutata* (see no. 132) larger, ♂ without spurs, *Glossotrophia* species with more diffuse wing pattern, longer proboscis, ♀ with two hindtibial spurs. From all other species easily distinguished by the uninterrupted terminal line surrounding the forewing apex.

The ***rubiginata*** species-group

STERNECK's 4th group of *Scopula*, including also the eastern Palaearctic *S. impersonata* (WALKER, 1861) and *S. manifesta* (PROUT, 1911). Forewing short and broad, hindwing rounded. Ground colour dark (reddish, ochreous or brownish). Vertex brownish. Palpi bushy, large, 1.2-1.5 times diameter of eye. Proboscis developed. Central sclerite on ♂ sternum A2 with patch of comparatively short setae. Male genitalia: Socii distant from each other. Valva sclerotized, with terminal spine. Fibula short and broad. Base of sternum A8 concave. Female genitalia: Ductus bursae comparatively long. Corpus bursae globular or oval. Signum divided into numerous small, flat spinule-bearing sclerites pointing away from longitudinal axis.

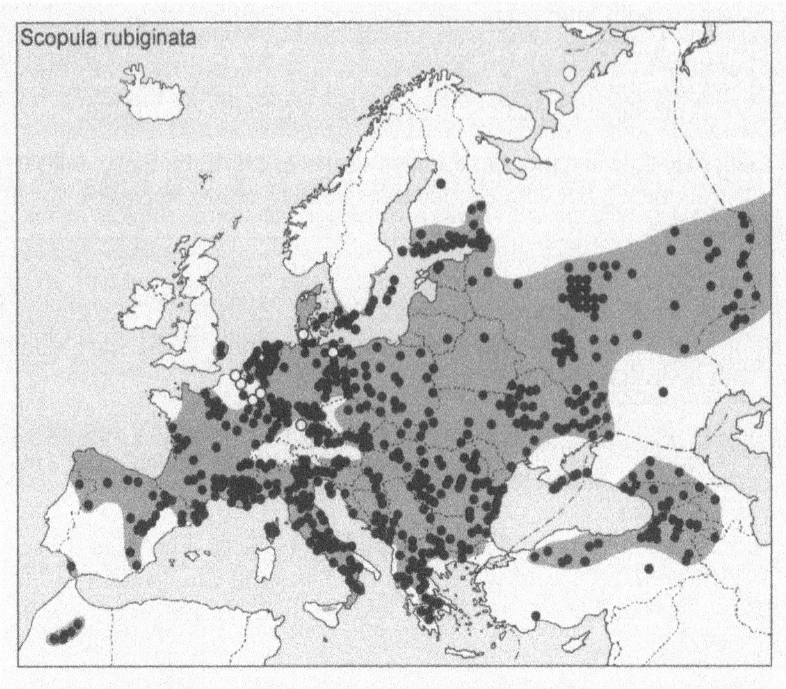

134. Scopula rubiginata (HUFNAGEL, 1767)

Phalaena rubiginata HUFNAGEL, 1767: Berlin. Mag. 4 (6): 610 (Germany: Berlin). Syntype(s) lost.
Geometra rubricata DENIS & SCHIFFERMÜLLER, 1775: Ank. syst. Werkes Schmett. Wienergegend: 110 (Austria: Vienna distr.). Syntype(s) lost.
Geometra vittata THUNBERG, 1784: Diss. ent. sistens Ins. Suecica 1: 8 (Sweden, Uppland: Uppsala). Lectotype ♂ (UZIU)
Phalaena domialla FOURCROY, 1785: Entom. Parisiensis 2: 273 (France: Paris). Syntype(s).
Phalaena Geometra variata VILLERS, 1789: Linn. ent. 2: 382 (France, Bresse near Lyon ('Bressia')). Syntype(s).
‡ *Geometra rubricaria* HÜBNER, 1799: Samml. Eur. Schmett. 5, Geometrae (1): pl. 21, fig. 111. Incorrect subsequent spelling.
Idaea subangularia HERRICH-SCHÄFFER, 1839: in PANZER, Faunae Insect. Germ. initia (Dtl. Insect.) (166): 91 (Germany). Syntype(s).
Unavailable names (infrasubspecific): *bilineata*: REISSER (1961); *bruneomarginata*: SCHAWERDA (1916); *clarirufa*: LEMPKE (1949); *fuliginosa*: STRAND (1917); *obscurata*: SKALA (1913); *pallifasciata*: LEMPKE (1949); *purpureofasciata*: DANNEHL (1927); *rufolineata*: LEMPKE (1949); *scotina*: BUBACEK (1923). – (incorrect subsequent spellings): *brunneomarginata*: PROUT (1935); *domiala*: TOMIC et al. (2002); *rubicata*: REINHARD et al. (1999); *rubigenata*: SANCHEZ & CIFUENTES (1990); *rubignata*: RAINERI (1996); *vitata*: TOMIC et al. (2002).

DIAGNOSIS: Wingspan ♂♀ 16-22 mm, small specimens also in the first generation, but second generation on average smaller and darker. Ground colour variable, brown, light brown or rosy brown. Area between medial and postmedial line usually paler. Transverse lines dark brown, distinct. Wavy line indistinct. Discal spots absent or very small. Fringe dots and usually also the dark terminal line absent. Frons brown, vertex and collar light brown. Antennae of ♂ with intersegmental incisions, ciliate-fasciculate, cilia long, 1.6-1.8 times width of flagellum, up to 2.1 times in southern Europe, up to 2.6 times in Morocco. Hindtibia not dilated, with pencil, tarsus 1.0-1.2 times length of tibia.

Variability: North of Alps and on mountains of southern Europe usually with strong rosy tinge, in the lowlands of southern Europe rarely. Wing pattern and pale medial area more contrasting in southern Europe. In Great Britain smaller and darker, two large rosy and sharply patterned specimens supposed to be transported from France (PROUT 1935a). Low temperatures induce darker coloration (BERGMANN 1955).

MALE GENITALIA: Socii narrow, length corresponding to distance between them. Anellus posteriorly not concave, often truncate. Aedeagus long and narrow, without cornutus or with an indistinct small sclerite. Sternum A8 with both cerata short and narrow, slightly spinulose on the terminal half. Not polymorphic (n=18).

FEMALE GENITALIA: Antrum attached to large triangular sclerite. Shape of signum variable, usually drop-shaped, spinules small.

DISTRIBUTION: Palaearctic. Widely distributed in Europe from Iberian peninsula and southern England to the Urals, in the north up to southern Scandinavia. Resident in eastern England, as suspected immigrant in many other localities (SKINNER 1984; WARING & TOWNSEND 2003; REID pers. comm.). Absent from Ireland and Mediterranean islands, except Corsica and Sicily. Areal losses and decreasing abundance in many regions north of the Alps (ERLACHER & FRIEDRICH 1994; EBERT 2001; WOLF & HACKER 2004), due to alteration of habitats. – Outside Europe in Morocco (Middle Atlas), Turkey, Caucasus, Transcaucasus, Siberia, northern parts of central Asian mountains (VIIDALEPP 1988; 1996), in the east to Mongolia (VIIDALEPP 1975; VOJNITS 1976) and Dahuria (e.g. KOSTJUK 1992; KOSTJUK & GOLOVUSHKIN 2003).

PHENOLOGY: North of the Alps usually bivoltine: mid-May to mid-September, generations slightly overlapping or with a gap in early July. South of the Alps bi- or trivoltine late April to early October, exceptionally earlier or later. Under bad conditions, e.g. in cold years, rarely univoltine, from mid-June to early August.

BIOLOGY: Larva oligophagous on various herbaceous plants (FORSTER & WOHLFAHRT 1981; BERGMANN 1955), possibly preferring Fabaceae (CARTER & HARGREAVES 1987) and Lamiaceae of low-nutrient soils. Egg deposition observed in nature on *Thymus glabrescens* (Lamiaceae) (Italy: KITSCHELT 1925). Larva found on *Calluna vulgaris* (Ericaceae), and *Artemisia campestris* (Asteraceae) (Germany: WEGNER pers. comm.), between *Rumex acetosella* (Polygonaceae) (Germany: EBERT 2001), on *Thymus serpyllum* (Lamiaceae), and *Medicago lupulina* (Fabaceae) (Germany: UFFELN 1908). Repeatedly recorded on *Thymus* and *Calluna* (var. cit.), but also on *Vicia, Lotus, Trifolium, Cytisus scoparius, Medicago* (Fabaceae), *Thalictrum* (Ranunculaceae), *Galium* (Rubiaceae), *Taraxacum officinale* (Asteraceae), *Convolvulus arvensis* (Convolvulaceae) (var. cit., e.g. LHOMME 1935; SEPPÄNEN 1954; BERGMANN 1955). Reared on *Onobrychis, Polygonum aviculare, Lotus corniculatus, Vicia cracca, Medicago lupulina, Thymus praecox, Trifolium, Convolvulus vulgare* and *Taraxacum officinale* (MILLIÈRE 1868; BUCKLER 1897; ALLAN 1949; BERGMANN 1955; FRIEDRICH 1976; SKINNER 1984; KRAUS 1993; PORTER 1997; MÜLLER pers. comm.). Cannibalism reported even when sufficient food available (MÜLLER pers. comm.). Adults easily flushed up from vegetation, sometimes with active flight during daytime, main activity in the late afternoon and at dusk, sometimes also attracted to light and bait. Nectar feeding recorded on *Scrophularia canina, Rhinanthus minor, Reseda lutea, Senecio erucifolius, Thymus pulegioides, Daucus carota* and *Medicago sativa* (EBERT 2001).

HABITAT: North of the Alps inhabiting xerothermic open habitats, e.g. uncultivated, dry

Text-figs 176-179: Diagnostic characters (indicated) of European *Scopula* species. Text-fig. 176: *S. rubiginata* (HUFNAGEL, 1767). Text-fig. 177: *S. ochraceata* (STAUDINGER, 1901). Text-fig. 178: *S. turbidaria* (HÜBNER, 1819). Text-fig. 179: *S. turbulentaria* (STAUDINGER, 1870).

grassland, fallows, heathland, embankments. Here restricted to the lowlands from 0 m up to 500 m above sea-level. Preferring low-nutrient sandy, calcareous or gravelly soils. In southern Alps and southern Europe more ubiquitously distributed, from 0 m up to 1,000 m, exceptionally up to 1,600 m. In Morocco, Turkey and Transcaucasus up to 2,200 m and almost absent from the lowlands.

SIMILAR SPECIES: *S. ochraceata* usually slightly larger, ground colour homogenously ochreous, terminal area not darker, transverse lines finer, discal spots small, but distinct, frons darker, vertex paler. *S. turbidaria* and *S. turbulentaria* with discal spots small, but distinct, ground colour homogenously brown, more or less suffused with grey or dark grey.

REMARKS: For rearings and morphology of immature stages see BUCKLER (1897).

135. *Scopula ochraceata* (STAUDINGER, 1901), stat. n.

Acidalia rubiginata var. (et ab.) *ochraceata* STAUDINGER, 1901: in STAUDINGER & REBEL: Cat. Lepid. palaearct. Faunengeb. (Edn 3) 1: 273 (south-eastern European Russia: Krasnoarmeysk, 'Sarepta'). Lectotype ♂ (MNHU,

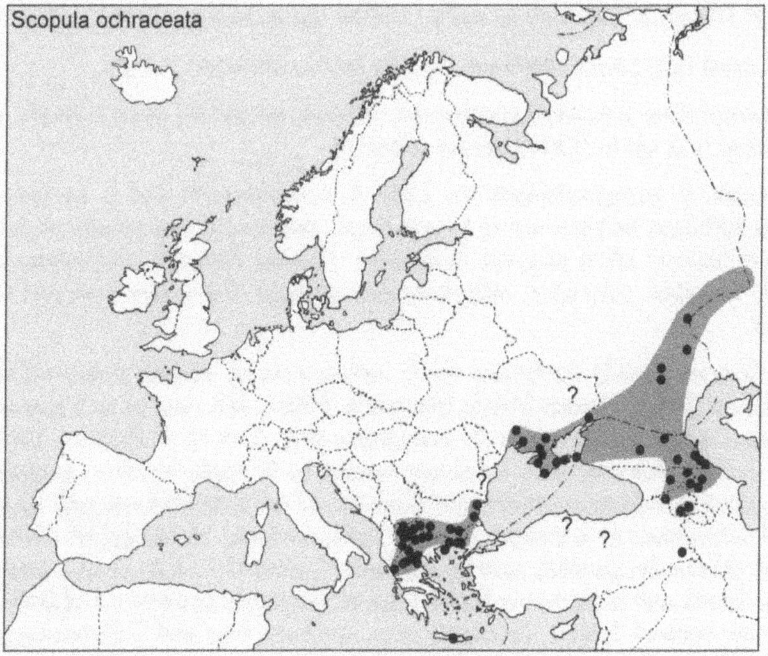

examined), herewith designated to fix type locality and to stabilize usage of the name with regard to the sister species; at MNHU 9 paralectotypes, probably belonging to different species: from type locality, Naryn, 'Dirbent', Achalzich, Marasch/Taurus, AlsöDobas. Herewith raised to species rank (stat. n.), from subsp. of *S. rubiginata* (VIIDALEPP 1996; SCOBLE 1999).

DIAGNOSIS: Wingspan ♂♀ 17-23 mm. Wings light ochreous, variability low. Transverse lines fine, but distinct, dark grey brown. Ground colour not paler between medial and postmedial line. Wavy line indistinct. Discal spots present, very small. Fine dark terminal line usually present. Frons dark or blackish brown, vertex pale ochreous, and collar light brown. Antennae of ♂ with intersegmental incisions, ciliate-fasciculate, cilia long, 1.5-2.0 times width of flagellum. Hindtibia not dilated, with pencil, tarsus 0.9-1.1 times length of tibia.

MALE GENITALIA: Socii short and rounded, half as long as distance between them. Anellus strongly concave posteriorly. Aedeagus broader than that of *S. rubiginata*, with one indistinct small central sclerite. Sternum A8 anteriorly concave, with both cerata stout and broad, slightly exceeding mappa, strongly spinulose along the whole length. Not polymorphic (n=12).

FEMALE GENITALIA: Posterior margin of antrum concave. Signum broad, drop-shaped, with posterior indentation, the spinules comparatively large.

DISTRIBUTION: (East-)European-Westasiatic. Macedonia and northern Greece to southern Urals, with isolated occurrence on Crete. Reputedly in north-eastern Ukraine (ALBERTI & SOFFNER 1962) and, possibly, in western Romania, data await confirmation. – For transitional populations to *S. turbulentaria* in eastern Turkey, Transcaucasus, and Turkmeniya see remarks. In central Asian mountains replaced by the sister species *S. halimodendrata* (ERSCHOV, 1874).

PHENOLOGY: Bivoltine: mid-May to early October, generations overlapping.

BIOLOGY: Larval host-plants unknown. Adults readily attracted to light.

HABITAT: Xerophilous species of continental climates, inhabiting open habitats, mainly dry steppes. From 0 m up to 900 m above sea-level.

SIMILAR SPECIES: *S. rubiginata* (see no. 134). *S. turbulentaria* and *S. turbidaria* usually with darker suffusion and less warm brown tinge, postmedial line usually slightly dentate. Secure identification often requires dissection. Central Asian *S. halimodendrata* with wings more rounded, coloration with more orange tinge, transverse lines and discal spots less distinct.

REMARKS: Species rank is confirmed for *S. ochraceata* by various authors (DANIEL et al. 1951; URBAHN 1966; PINKER 1968; MESZAROS 1985), but refused in VIIDALEPP (1996). Though often subordinated under *S. rubiginata* (e.g. CAPUSE & KOVACS 1987; SCOBLE 1999), *S. ochraceata* seems to be closer related to *S. turbulentaria*, phylogenetically. Nearly allopatric distribution pattern and similarity of genitalic features may suggest downgrading of *ochraceata* to a subspecies of *S. turbulentaria*. However, on Crete and eastern Balkan peninsula, possibly also in northern Turkey (1♂ of *S. ochraceata* examined from prov. Sivas) and in Armenia there is some sympatric occurrence of both. *S. turbulentaria* from eastern Turkey, Transcaucasus, northern Iran and Turkmeniya (see below) in habitus and genitalia approaching features of *S. ochraceata*, possibly as a consequence of hybridization. With further investigations both taxa may prove to be a 'ring species' around the Black Sea, being well separated in the Balkan peninsula. Also the relationships between *S. ochraceata* and *S. halimodendrata* require taxonomic revision with extensive material from western Asia.

136. Scopula turbidaria (HÜBNER, 1819)

Geometra turbidaria HÜBNER, 1819: Samml. Eur. Schmett. 5, Geometrae (1): pl. 97, fig. 501 (Europe). Syntype(s) lost. HÜBNER's figure shows a specimen with unusual, reddish light brown tinge. In MNHU one 'Orig[inal]' from Spain, Andalusia: Granada, possibly the original for the figure of HERRICH-SCHÄFFER, 1851, but coloration quite different from that of HÜBNER's figure. Neotype ♂ (ZSM/HERBULOT, examined; [southern France], 'Vernet-le[s]-Bains, Pyrenées Orientales, Ch. OBERTHÜR, Août 1880, coll. D. LUCAS'), herewith designated to stabilize usage of the name with regard to the sister species *S. turbulentaria*.

Acidalia macraria GUENÉE, 1858: in BOISDUVAL & GUENÉE, Hist. nat. Insectes (Spec. gén. Lépid.) 9: 455 (Spain? Iles d'Hyères?). Holotype ♂ ('coll. BELLIER'; not traced in ZFMK).

‡ *lutosata* RAMBUR, 1866: Cat. syt. Lépid. Andalousie 2: pl. 15, fig. 7 (Spain: Andalusia). Syntype(s) (ZSM/HERBULOT: 1♂ ex coll. RAMBUR, examined; MNHU: 1♀ from Spain, Andalusia: Granada, labelled 'Orig. *lutosaria* Rbr.'). Non binominal, and later on not cited at available rank, e.g. in STAUDINGER, 1871 (in synonymy), or in GUMPPENBERG 1890 (in synonymy).

Unavailable names (misidentification): *rubricaria*: sensu DUPONCHEL (1830; pl. 176, fig. 7 nec 5-6) nec HÜBNER, 1799 (identity: *Scopula rubiginata*); *turbulentaria*: sensu RUNGS (1981) nec STAUDINGER, 1870. – (infrasubspecific): *anomala*: BUBACEK (1926). – (nomen nudum): *lutosaria*: BOISDUVAL (1840). – (incorrect subsequent spelling): *turbidata*: CULOT (1917).

DIAGNOSIS: Wingspan ♂♀ 19-24 mm, in subsequent generations sometimes 16 mm only. Very variable (see HAUSMANN 1991d). Ground colour dark grey with slight brown tinge to pale ochreous in continuous transitions. Usually with stronger brown tinge than in *S. turbulentaria*. Pale forms commoner in subsequent generations. Transverse lines fine, but distinct, dark grey, rather straight and parallel to each other. Postmedial line usually slightly dentate. Pale ochreous wavy line distinct. Discal spots present, very small. Fine dark

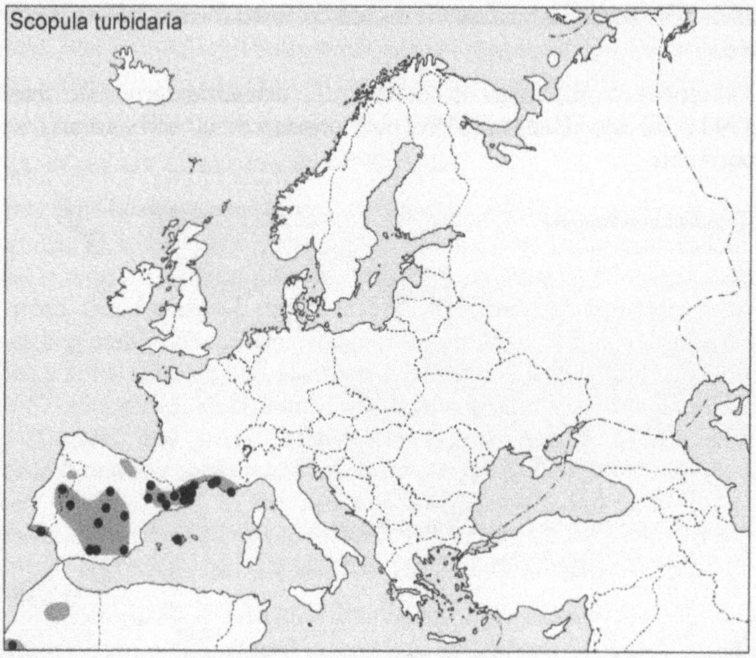

terminal line often dissolved into very small dots, sometimes absent. Frons black brown, vertex pale ochreous, collar darker. Antennae of ♂ with intersegmental incisions, ciliate-fasciculate, cilia 1.6-1.9 times width of flagellum. Hindtibia not dilated, with pencil, tarsus 0.85-1.0 times length of tibia.

MALE GENITALIA: Length of socii less than half the distance between them. Aedeagus with one small, needle-shaped central cornutus. Sternum A8 with concave base. Both cerata long, stout, spinulose only on the terminal third. Not polymorphic (n=7).

FEMALE GENITALIA: Lamella antevaginalis sclerotized to a narrow ring. Antrum with tapering, lateral projections, cup-shaped. Signum drop-shaped.

DISTRIBUTION: West-Mediterranean. Iberian peninsula and southern France. Present also on Mallorca (HONEY pers. comm.). Reputedly in north-western Italy (HELLMANN et al. 1999), data await confirmation. Distribution of the allopatric vicariant sister species (*S. turbulentaria*) see below. – Outside Europe in Morocco.

PHENOLOGY: Bi- or trivoltine: late April to mid-September. In North Africa under favourable conditions plurivoltine from early March to early December.

BIOLOGY: Larval host-plants unknown. Adults attracted to light.

HABITAT: Xerothermophilous lowland species. No detailed information available about habitat preference. From 0 m up to 800 m above sea-level.

SIMILAR SPECIES: *S. virgulata* (see no. 125), *S. rubiginata* (134), *S. ochraceata* (135). Correct identification of. *S. turbidaria* is very difficult due to the great variation. Postmedial line often dentate, ground colour more greyish, wavy line distinct. Small forms may also be confused with *S. minorata*, the latter with longer ♂ antennal cilia and shorter ♂ hindtarsus. For secure identification of sand coloured forms of *S. turbidaria* dissection is needed.

REMARKS: Differences in the length of cerata in *S. turbidaria* sensu lato mentioned by STERNECK (1941: 55) and RAINERI (1996), but considered by both authors to be due to individual aberration.

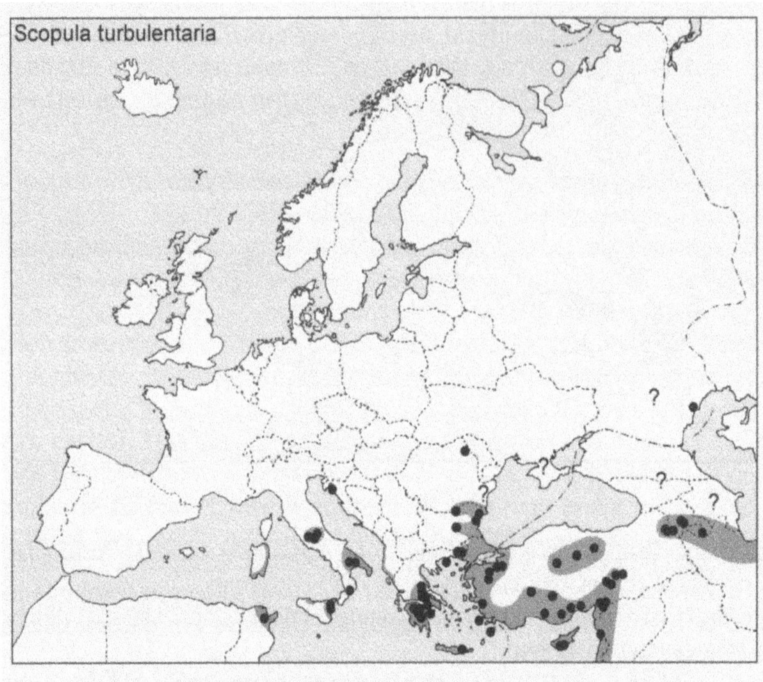

137. Scopula turbulentaria (STAUDINGER, 1870), stat. n.

Acidalia turbidaria var. *turbulentaria* STAUDINGER, 1870: Horae Soc. ent. ross. 7: 151, pl. 2, fig. 3 (Greece: Attika). Syntypes 1♂2♀ (MNHU: 1♂1♀, examined). Herewith raised from subspecies of *S. turbidaria* (SCOBLE 1999) to species rank (stat. n.).

Craspedia collata WARREN, 1901: Novit. zool. 8: 193 (Lebanon: Beirut). Holotype ♂ (BMNH).

Craspedia habenata WARREN, 1901: Novit. zool. 8: 194 (Lebanon: Beirut). Holotype ♂ (BMNH).

Acidalia immutata var. *syriacata* NEUBURGER, 1904: Soc. Ent.. 19 (15): 115 (Lebanon). Syntype(s) (MNHU/PÜNGELER: 1♂ syntype examined). This taxon has erroneously been regarded as synonym of *S. immutata*, repeatedly.

Acidalia turbidaria syriturcica WEHRLI, 1934: Mitt. Münch. ent. Ges. 24 (1): 4 (southern Turkey: Amanus mts. 'YD'; Syria). Syntypes, at least 10♂2♀ (ZFMK: 17♂♀ from Amanus, Marasch and Akbès ex coll. WEHRLI examined, probably all syntypes.

Scopula turbidaria steinbacheri PROUT, 1935: in SEITZ, Macrolep. 4, Suppl.: 36, pl. 6, fig. g (northern Iran: Elburs Mts.: Mazanderan, Sabatku, Darekeroudbar). Syntypes 15 (BMNH). Valid at subspecific rank (SCOBLE 1999).

Unavailable names (misidentification): *ochroleucata*: sensu PROUT (1913) nec *ochroleucaria* HERRICH-SCHÄFFER, 1847. – (infrasubspecific): *nubilata*: THIERRY-MIEG (1915). – (incorrect subsequent spellings): *syritaurica*: PROUT (1935); *turbillentaria*: KOUTSAFTIKIS (1973).

DIAGNOSIS: Wingspan ♂♀ 19-23 mm. Very similar to *S. turbidaria*, but with less brown tinge. No significant and constant differences in external morphology.

Subsp. ***steinbacheri*** with warmer coloration and sharper, more distinct transverse lines, ♂ antennal ciliation about 2.0 times width of flagellum.

MALE GENITALIA: As in *S. turbidaria*, but socii longer, slightly shorter than the distance between them. Sternum A8 with both cerata narrow and short, not exceeding mappa, spinulose along nearly the whole length (compare *S. rubiginata*). Not polymorphic (n= 22).

FEMALE GENITALIA: Very similar to those of *S. turbidaria*, but signum often triangular, posterior margin of antrum often rounded or triangular.

DISTRIBUTION: East-Mediterranean(-Iranian). Sardinia, Sicily, central and southern Italy across southern Greece, Crete and Rhodos, in the east to eastern Bulgaria. Isolated occurrences in western Croatia (Rab island: ZSM), in Albania (URBAHN 1966) and one specimen from the Astrachan desert (ZISP). Reputedly in Romania (e.g. STANESCU 1997b), awaiting confirmation, some examined specimens misidentified (STANESCU pers. comm.). Records for Ukraine and southern European Russia (VIIDALEPP 1996) probably referring to *S. ochraceata*, all examined specimens proved to be misidentified. – Outside Europe in Tunisia, and from Turkey and Cyprus across Levant, and southern Transcaucasus (nominate subspecies) to northern Iran and Turkmeniya (VIIDALEPP et al. 1992; subsp. *steinbacheri*), in the east to central Asian mountains (VIIDALEPP 1988; 1996), in the south an isolated record in Saudi Arabia (WILTSHIRE 1986; 1990). Reputedly also in Georgia, and Azerbaijan (DIDMANIDZE 1978; 2002; VIIDALEPP 1996).

PHENOLOGY: In Europe bi- or trivoltine from early June to early October with peaks in June and September, in the lowlands sometimes emerging from mid-April. In the Levant and other Asian localities under favourable conditions plurivoltine from early March to early December.

BIOLOGY: Larvae found 'as pest' on tobacco in southern Italy (SANNINO & ESPINOSA 1999). Reared on tobacco and *Plantago lanceolata* (SANNINO & ESPINOSA 1999). Pupation in the ground. Adults attracted to light.

HABITAT: Xerothermophilous lowland species, often in dry, open, uncultivated habitats, such as fallows, on stony grassland. From 0 m up to 800 m above sea-level, up to 1,300 m in Turkey and the Levant, up to 2,100 m in northern Iran.

SIMILAR SPECIES: Allopatric sister species *S. turbidaria* usually differing in the brown tinge of the coloration, and in genitalia. Other similar species see *S. turbidaria* (no. 136).

REMARKS: Detailed discussion of this species with description of immature stages in SANNINO & ESPINOSA (1999). All examined '*turbidaria*'-specimens from the Balkan peninsula (except southern Greece) proved to belong to *S. ochraceata*. *S. turbulentaria* and *S. ochraceata* almost allopatric, but possibly with hybridization zone from eastern Turkey to Turkmeniya (see remarks to *S. ochraceata*).

Subgenus ***Calothysanis*** HÜBNER, 1823

(=*Ustocidalia* STERNECK, 1932)

TYPE SPECIES: *Geometra imitaria* HÜBNER, 1799. Structurally very homogenous and uniform in genitalia. The arrangement of species in the *incanata* and *ternata* species-groups tentative (STERNECK 1941: 96). STERNECK (1932) provides a valuable subgeneric diagnosis, the introduced name, however, has to be replaced by reasons of priority.

DIAGNOSIS: Without constant differential features from subgenus *Scopula* in habitus and external morphology. See under generic diagnosis.

MALE GENITALIA: Socii narrow, usually longer than the distance between them. Fibula strongly sclerotized, black at least at tip, in contrast to the subgenus *Scopula* not diagnostic for subgroups. Aedeagus long and slender, without cornutus. Sternum A8 usually with narrow cerata.

FEMALE GENITALIA: Antrum usually small, with hut-shaped sclerite. Lamella antevaginalis often with semicircular sclerotization.

The *incanata* species-group

STERNECK's 6th group of *Scopula*. Heterogenous group, based on the shape of sternum A8, including many species. Proboscis developed. Palpi sometimes slightly elongate, e.g. in *S. incanata* 1.2-1.4 times diameter of eye. Central sclerite on ♂ sternum A2 with patch of comparatively short setae. Male genitalia: Socii long and slender, shorter in *S. decolor*. Valva membranous, elongate. Aedeagus narrow, without cornuti. Base of sternum A8 convex, mappa rounded. Female genitalia usually with large signum. For close relationship to the following group see remarks to the *ternata* species-group. STERNECK (1941) places the species *S. beckeraria* and *S. subrubellata* STERNECK 1941 in the *ternata* species-group, for sister species relationship to *S. rubellata* see HAUSMANN (2003a). Though STERNECK establishes a separate group (5th group) for species with shorter socii in ♂ genitalia, including the European *S. decolor*, the latter species is arranged here in the *incanata* species-group.

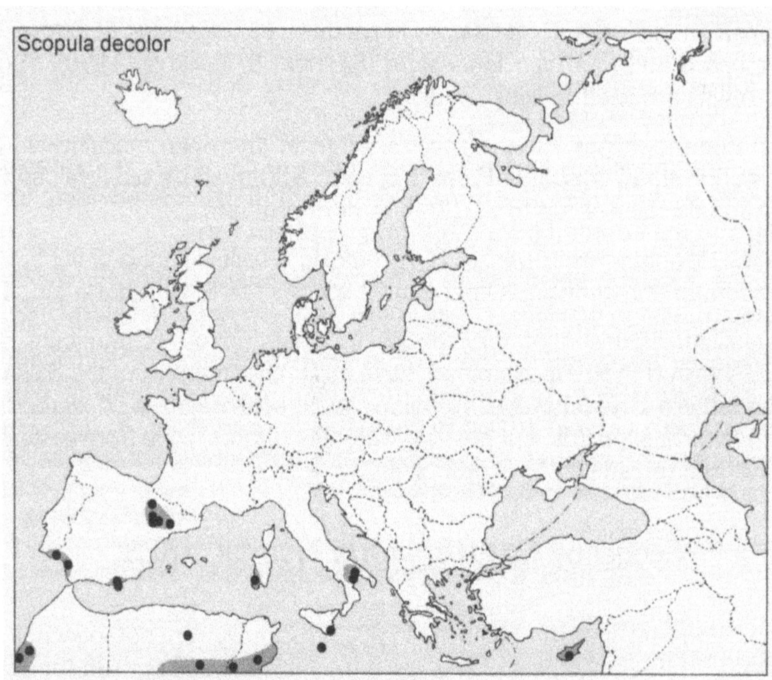

138. *Scopula decolor* (STAUDINGER, 1897)

Acidalina decolor STAUDINGER, 1897: Dt. ent. Z. Iris 10 (1): pl. 4, fig. 5; 10 (2) (1898): 269 (Algeria, Algier prov.: Chellala). Holotype ♂ (MNHU, examined).

Acidalia flaccata STAUDINGER, 1898: Dt. ent. Z. Iris 10 (2): 309, pl. 9, fig. 22 (Palestine: Jordan valley near

Dead Sea, peninsula Messra). Syntypes ♂♀ (MNHU: 6♂♀ syntypes examined). Herewith downgraded from species (SCOBLE 1999) to subspecific rank (stat. n.).

Acidalia flaccata languidata PROUT, 1913: in SEITZ, Macrolep. 4: 76 (Algeria: Biskra). Syntype(s) (BMNH). Herewith downgraded from subspecies of '*S. flaccata*' (SCOBLE 1999) to synonymy (syn. n.).

DIAGNOSIS: Wingspan ♂♀ 24-27 mm in spring, 23-25 mm in subsequent generations. Hindwing termen slightly angled. Ground colour creamy white with slight orange tinge, sometimes with rough dark grey suffusion. Antemedial line often absent or indistinct. Medial and postmedial lines convex, medial area light brown, paler in the distal part. On hindwing medial and postmedial lines getting more distinct towards the inner termen. Discal spots usually present, very small. Frons, vertex and collar whitish. Antennae of ♂ with deep intersegmental incisions, ciliate-fasciculate, cilia 1.4-1.8 times width of flagellum, in North African populations slightly longer. Hindtibia of ♂ stout, with pair of distal spurs, exceptionally with one additional proximal spur (1♂ from Spain), tarsus not shortened.

Subsp. ***flaccata*** similar in habitus, but with longer ♂ antennal cilia, 2.0-2.1 times width of flagellum. Hindtibia of ♂ usually with one single terminal spur, some examined specimens, however, without spur, or with two spurs.

MALE GENITALIA: Socii long, digitiform. Valva long and slender. Fibula long, spine-shaped. Sternum A8 with concave, shallow projection at base. Both cerata short, without spinules. Not polymorphic (n=8). No difference between subspecies, but cerata somewhat longer in populations from Cyprus.

FEMALE GENITALIA: Apophyses anteriores stout, with broad base. Lamella antevaginalis with circular folds. Antrum cup-shaped. Ductus bursae long. Signum variable, usually drop-shaped, and with medial ridge.

DISTRIBUTION: Mediterranean. Very local populations in Portugal, Spain (e.g. LAJONQUIÈRE 1963; RAINERI 1991; DANTART 1991), Malta (VALLETTA 1973; SAMMUT 2000), Sicily (BELLA et al. 1999), Sardinia (RAINERI 1991), and southern Italy (Basilicata: PARENZAN & HAUSMANN 1992; PARENZAN 1994). – Outside Europe from Morocco to Tunisia, and Libya, Cyrenaica (TURATI & ZANON 1922; TURATI 1924), replaced by subsp. *flaccata* on Cyprus (specimens at ZSM) and in Israel.

PHENOLOGY: Plurivoltine. European data from late April to late October, often with a gap from late July to mid-August. On Sicily recorded even in mid-December (BELLA et al. 1999). In North Africa and in the Levant in all months of the year.

BIOLOGY: Larva oligophagous (2) on halophilous Chenopodiaceae, possibly also Caryophyllaceae. Egg positioning recorded on *Atriplex halimus* (Chenopodiaceae), larva reared on that plant and on *Chenopodium album* (Spain: REDONDO et al. 2001). *Atriplex halimus* suggested as natural host-plant also on Cyprus (LEIPNITZ pers. comm.), in rearings accepting other *Atriplex* species, *Chenopodium album* and *Silene vulgaris* (LEIPNITZ pers. comm.). Adults readily attracted to light, but collected also during daytime (LAJONQUIÈRE 1963).

HABITAT: Xerothermophilous, halophilous lowland species. In Europe restricted to hot, sandy steppe habitats, dunes, open habitats with sparse, halophilous vegetation, in North Africa and Middle East inhabiting deserts and oases. From 0 m up to 300 m above sea-level.

SIMILAR SPECIES: Differing from almost all other Sterrhinae in the whitish frons.

REMARKS: Detailed treatments see RAINERI (1991) and DANTART (1991). The differential diagnosis provided by PROUT (1913; 1935) refused by STERNECK (1941).

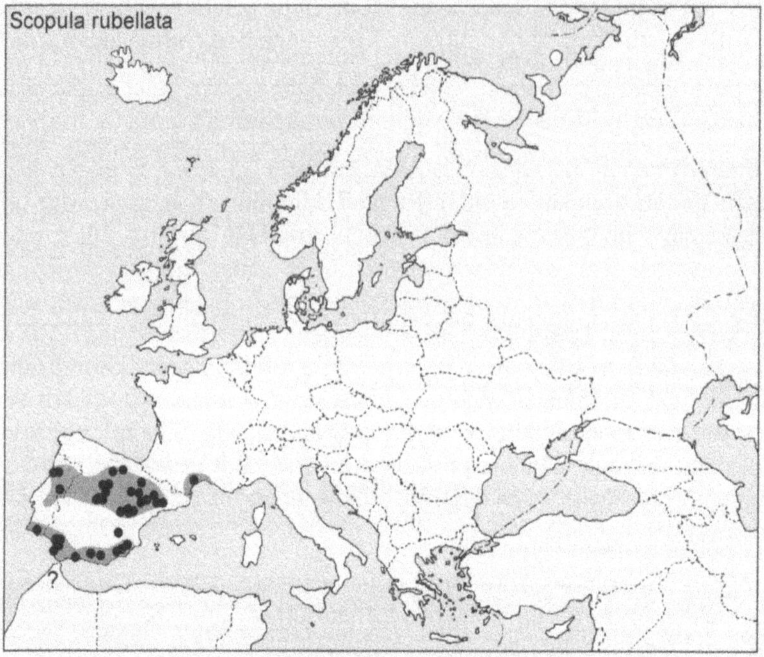

139. *Scopula rubellata* (STAUDINGER, 1871)

Acidalia rubellata STAUDINGER, 1871: in STAUDINGER & WOCKE, Cat. Lepid. eur. Faunengeb. (Edn 2) 1: 146. Introduced in synonymy but available according art. 11.6.1. ICZN. First available name after *rubellaria* BOISDUVAL, 1840 (nomen nudum), and *rubellata* RAMBUR, 1866 (non binominal) (Spain, Andalusia: Granada). Holotype ♂, of RAMBUR (ZSM/HERBULOT, examined).

Unavailable names (misidentification): *beckeraria*: sensu ZAPATER & KORB (1892) nec LEDERER, 1853; *consanguinaria*: sensu MENDES (1903) nec LEDERER, 1853 (identity: *Idaea consanguinaria*).

DIAGNOSIS: Wingspan ♂♀ 19-25 mm. Wings warm sand coloured, often with strong orange tinge. Medial line usually present. Postmedial line fine, strongly dentate. Terminal area with grey shadings. Discal spots and terminal dots usually distinct. Frons dark brown, vertex whitish, collar light brown. Antennae of ♂ filiform, ciliate-setose (sub-fasciculate), cilia 1.3-1.7 times width of flagellum. Hindtibia of ♂ slender, usually with two spurs, not rarely however one of them very short or even absent, tarsus not shortened.

MALE GENITALIA: Socii long and narrow. Fibula narrow, digitiform. Lateral projections of anellus comparatively long and slender. Sternum A8 with both cerata short, not polymorphic (n=4). Cerata slightly spinulose at tip.

FEMALE GENITALIA: Lamella antevaginalis cup-shaped or reniform. Ductus bursae long, widening towards junction with the oval corpus bursae. Signum very large.

DISTRIBUTION: West-Mediterranean. Iberian peninsula, French Pyrenees (HERBULOT 1954), reputedly in the Rhône delta (DUFAY 1972). – Outside Europe reputedly in northern Morocco (RUNGS 1981 as '*Idaea rubellata*'), awaiting confirmation, all examined specimens proved to be misidentified (identity usually *Idaea eugeniata*); North African distribution questioned also by HERBULOT (1954). Erroneously recorded from western Libya (TURATI 1930a), possibly referring to the reddish *Idaea hausmanni* TRUSCH & MÜLLER, 2001.

PHENOLOGY: Bi- or trivoltine: early April to late September, most common in June and August.

BIOLOGY: Larva reared on *Polygonum aviculare* (MENDES 1910) and *Taraxacum officinale* (WEHRLI 1927). Adults attracted to light.

HABITAT: Lowland species of hot and dry localities, inhabiting steppes, and scattered xerophilous forest. From 0 m up to 1,200 m above sea-level.

SIMILAR SPECIES: *Idaea eugeniata* (see no. 100). Allopatric *S. beckeraria* often with slightly glossy coloration of wings, and with light brown tinge, terminal area usually darker, ♂ hindtibia without spurs, tarsus slightly shortened. Allopatric *S. scalercii* with whitish ground colour of wings, ♂ hindtibia without spurs.

REMARKS: 'Eastern populations' (Turkestan) separated by STERNECK at species rank: *S. subrubellata* STERNECK, 1941.

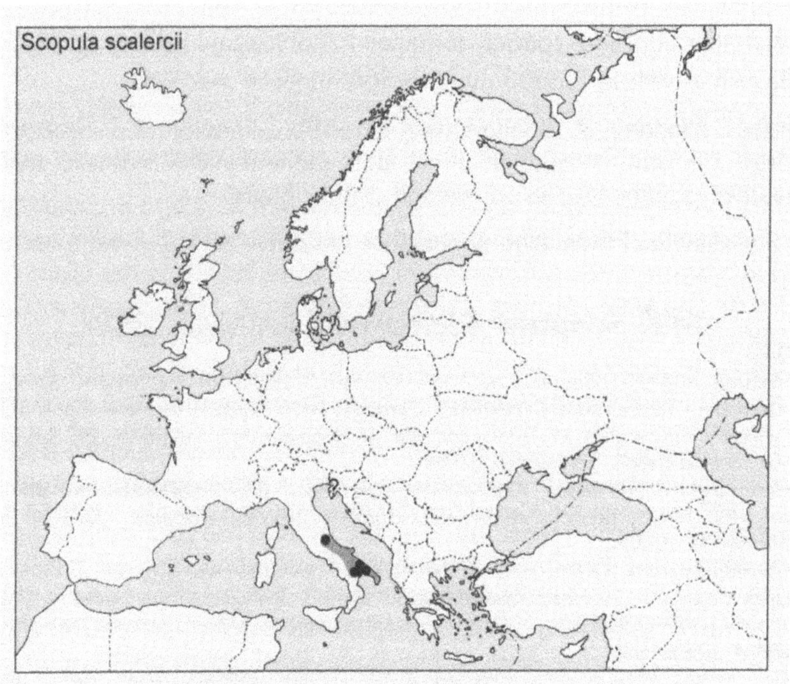

140. Scopula scalercii HAUSMANN, 2003

Scopula scalercii HAUSMANN, 2003: Ent. Zeits. Stuttgart 113 (11): 324, figs 9, 20, 23 (Italy, Puglia: Gravina). Holotype ♀ (ZSM, examined).

Unavailable names (misidentification): *beckeraria*: sensu PARENZAN & HAUSMANN (1992) nec LEDERER, 1853.

DIAGNOSIS: Wingspan ♂♀ 23–26 mm, in the second generation sometimes 20 mm only.

Wings white with slight sand coloured tinge. Medial line diffuse. Postmedial line sharp, dentate, darker at costa. Terminal area with grey shadows. Discal spots usually distinct, in the ♀ holotype smaller. Terminal line dissolved to small but distinct dots. Palpi whitish, upper side darker, frons black brown, vertex whitish to pale sand coloured, collar light brown. Antennae of ♂ filiform, ciliate-fasciculate, cilia 1.4–1.5 times width of flagellum. Hindtibia of ♂ stout, with pencil, without spurs, tarsus 0.8–0.9 times length of tibia.

MALE GENITALIA: Socii long and narrow. Fibula long, narrow, digitiform. Lateral projections of anellus comparatively long and slender. Sternum A8 with very short cerata, not polymorphic (n=5). Cerata sigmoid, spinulose and outwardly curved at tip.

FEMALE GENITALIA: Antrum sclerotized, with globular sclerite; vertical view on ostium bursae resulting from pressure of permanent slide. Sterigma (lamella antevaginalis) with sclerotized crest. Ductus bursae of medium length, subapically joining corpus bursae. The latter oval, with large and comparatively broad patch of about 150 flat spinules.

DISTRIBUTION: Central-Mediterranean. Endemic to central and southern Italy (regions Abruzzi, Basilicata and Puglia). Very local and rare.

PHENOLOGY: Bivoltine: mid-May to late May; mid-August (limited data available). PARENZAN's (1994) statement 'early July to late September' awaits re-examination.

BIOLOGY: Larval host-plants unknown. Adults attracted to light.

HABITAT: Xerothermophilous species, restricted to hot lowland habitats, at sites protected from wind, such as valleys. From 0 m up to 500 m above sea-level.

SIMILAR SPECIES: Allopatric *S. rubellata* (see no. 139). Coloration of *S. beckeraria* often slightly glossy, with light brown tinge, discal spots and terminal dots usually smaller, postmedial line finer, ♂ antennal cilia on average slightly longer.

REMARKS: For identity of Dalmatian populations see remarks to *S. beckeraria*.

141. *Scopula beckeraria* (LEDERER, 1853)

Acidalia beckeraria LEDERER, 1853: Verh. zool.-bot. Ver. Wien 3: 258 (174) (south-eastern European Russia: Sarepta; Azerbaijan, Transcaucasus: Elizabethpol). Syntypes 1♂1♀ (Sarepta), 2 (Elizabethpol). The lectotype should be selected from Sarepta. At MNHU (box Stgr. 6) specimens from Sarepta ex coll. LEDERER, possibly types, but without type labels. No types at NHMW.

Acidalia beckeraria assimilaria PROUT, 1913: in SEITZ, Macrolep. 4: 62. First available name after *assimilaria* STAUDINGER, 1892 (infrasubspecific) (Central Asia, Ferghana: Alai Mts.). Holotype ♀ (MNHU). Valid at subspecific rank (SCOBLE 1999).

Acidalia beckeraria amataria WEHRLI, 1927: in BANG-HAAS, Horae macrolep. Reg. pal. 1: 92, pl. 11, fig. 16 (central Russia: Sajan Mts., Tunkinsk Weissgebirge SW Irkutsk). Syntypes 2♂ (not traced in ZFMK; in ZSM one ♂ with corresponding label data, probably a syntype, examined). Valid at subspecific rank (SCOBLE 1999), possibly valid at species rank.

Acidalia rebeli PROUT, 1913: in SEITZ, Macrolep. 4: 62 (Croatia, Dalmatia, Zadar ('Zara')). Introduced to replace misidentified *agraria* sensu REBEL. Lectotype ♂ (NHMW, examined), herewith designated from the original material of REBEL to stabilize usage of the name with regard to the other subspecies; at NHMW 1♀ paralectotype from Istria, Pola. Valid at subspecific rank (HAUSMANN 2003a).

Scopula pseudoafghana EBERT, 1965: Stuttg. Beitr. Naturk. 142: 8, pl. 1, fig. 9; pl. 2, fig. 9 (central Afghanistan: Bamian). Holotype ♂ (SMNK, examined). Herewith downgraded from species (SCOBLE 1999) to synonymy (syn. n.).

Scopula beckeraria hermonicola HAUSMANN, 1997: Entomofauna 18 (1): 6, fig. 3 (northern Israel: Sede Nehamya). Holotype ♂ (ZSM, examined). Valid at subspecific rank.

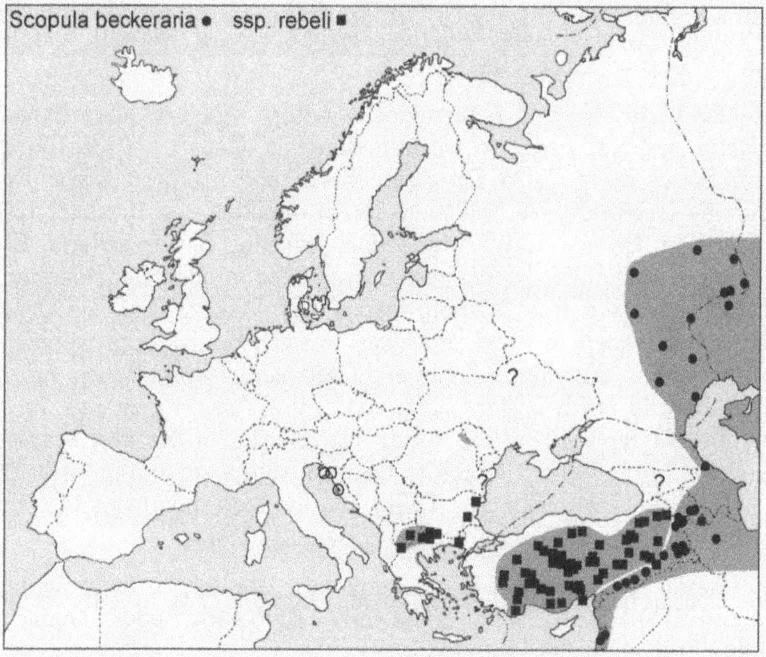

Unavailable names (misidentification): *agraria*: sensu REBEL (1908) nec JOANNIS, 1891 (identity: *Idaea litigiosaria*). – (infrasubspecific): *anastomosaria*: REISSER (1961); *uniformis*: REISSER (1961). – (incorrect subsequent spelling): *beskeraria*: DIDMANIDZE (2001).

DIAGNOSIS: Wingspan ♂♀ 22-26 mm, second generation smaller, sometimes 19 mm only. Up to 28 mm in the Altai Mts.. Wings sand coloured, with warm, light brown tinge, terminal area usually darker. Medial line diffuse. Postmedial line sharp, dentate, at costa usually darker. Discal spots small, terminal dots present, elongate. Palpi pale sand coloured, upper side darker, frons brown or black brown, vertex whitish or pale sand coloured, collar unicolorous light brown. Antennae of ♂ filiform, ciliate-fasciculate, cilia 1.4-1.8 times width of flagellum. Hindtibia of ♂ stout, with pencil, tarsus 0.8-0.9 times length of tibia.

Subsp. ***rebeli*** without constant differential features from the nominate subspecies in habitus and external morphology, antennal cilia usually 1.4-1.5 times width of flagellum.

MALE GENITALIA: Socii long and narrow. Fibula long, narrow, digitiform. Lateral projections of anellus comparatively short and broad. Sternum A8 with both cerata of medium length, slightly curved and narrow in the nominate subspecies. From eastern Afghanistan to Mongolia shortened (subsp. *assimilaria*, subsp. *amataria*). From Levant to Transcaucasus and Turkmeniya slightly shortened, inwardly curved and very stout (subsp. *hermonicola*). In the subsp. *rebeli* cerata very long and thin. Such long cerata usually visible without brushing abdomen. Cerata of '*rebeli*' short according to STERNECK (1941), but no locality mentioned for the examined specimen(s). No true infrapopular polymorphisms found (n=91, all subspecies summarized), even close to the distributional border between nominate subspecies and subsp. *rebeli*. In a very few localities in south-eastern Turkey and northernmost Lebanon, however, both specimens with long or short cerata are found, such as intermediate specimens.

FEMALE GENITALIA: Sterigma with circular furrows. Lamella antevaginalis with a large transverse fold. Antrum sub-rectangular, flattened. Signum broadly fusiform or oval. No constant differences between subspecies.

DISTRIBUTION: Eurasiatic. Isolated populations of subsp. *rebeli* in Macedonia, northern Greece, Bulgaria, and, reputedly, in western Romania (CARADJA 1930; NEMES 1970). Nominate subspecies in southern European Russia and southern Urals. Records for northern Ukraine (ANTONOVA & SMETANIN 1978) doubtful, see KOSTJUK (2004), one examined male from Crimea (IZBE) misidentified, identity: *S. corrivalaria*. Erroneously recorded for Spain, all examined specimens proved to be misidentified, identity: *S. rubellata*. Mislabelled specimens from 'Canary islands' (leg. PINKER, ZSM). – Outside Europe from Turkey (subsp. *rebeli* in most provinces, eastern border approximately running through provinces Kars, Erzurum, Batman and Gaziantep), Levant (subsp. *hermonicola*), Transcaucasus, Iran, to north-western India and central Asian mountains. In the mountains of central Asia replaced by the subsp. *assimilaria*, in the east across southern Siberia, Mongolia, north-western China to Dahuria (subsp. *amataria*; KHOMENKO et al. 1994; KOSTJUK et al. 1994; VIIDALEPP 1996). On the high mountains of southern Iran replaced by the sister species *S. hoerhammeri* BRANDT, 1941.

PHENOLOGY: Usually bivoltine mid-May to early July; late July to early September. On mountains above 2,000 m usually univoltine early July to early August. Under good conditions trivoltine from early April to mid-October.

BIOLOGY: Limited data available on larval host-plants. Reared on leaves of *Taraxacum officinale* with occasional cannibalism (MÜLLER pers. comm.). Adults attracted to light.

HABITAT: Xerothermophilous or xeromontane steppe species of continental climates. In European Russia a lowland species, from 0 m up to 200 m above sea-level, in central Asia from the lowlands up to 3,500 m. Subsp. *rebeli* in the Balkan peninsula restricted to the lowlands, from 0 m up to about 400 m above sea-level, in Turkey common from 1,000 up to 2,000 m, but occasionally also lower and on the summits near 3,000 m.

SIMILAR SPECIES: *S. marginepunctata*, mainly in some central Asian populations very similar, but coloration usually less ochreous and less glossy, with darker suffusion, postmedial line usually less marked at costa, correct identification sometimes requires dissection or brushing of the abdomen to remove scales and reveal shape of sternum A8. Allopatric *S. rubellata* (see. no. 139), and *S. scalercii* (140). Allopatric and extralimital *S. subrubellata* STERNECK, 1941 from Turkestan usually with spurs at the ♂ hindtibia. Extralimital *S. cumulata* (ALPHÉRAKY, 1883) (=*subrubellata*?) from central Asia often very similar, slightly larger, ♂ usually with hindtibial spurs. Extralimital *S. hoerhammeri* BRANDT, 1941 larger, with ♂ hindtibial spurs. Male genitalia of *S. incanata* very similar to *S. beckeraria rebeli*, the former can, however, be easily recognized by the bicolorous collar and absence of brown tinge of the wing colour.

REMARKS: It may be discussed whether *rebeli* should be raised to species rank or not. Further study is required to clarify this. For examination of the lectotype of *S. b. rebeli* from Zadar, western Croatia, see HAUSMANN (2003a). More material is needed to stabilize usage of the names *rebeli* and *scalercii*. One examined ♂ from the Croatian island Krk (coll. HABELER) proved to have short cerata though, in habitus, it matches well the populations from Macedonia. If *scalercii* turns out to be a synonym of *rebeli*, the populations from Macedonia and most parts of Turkey would need a new name.

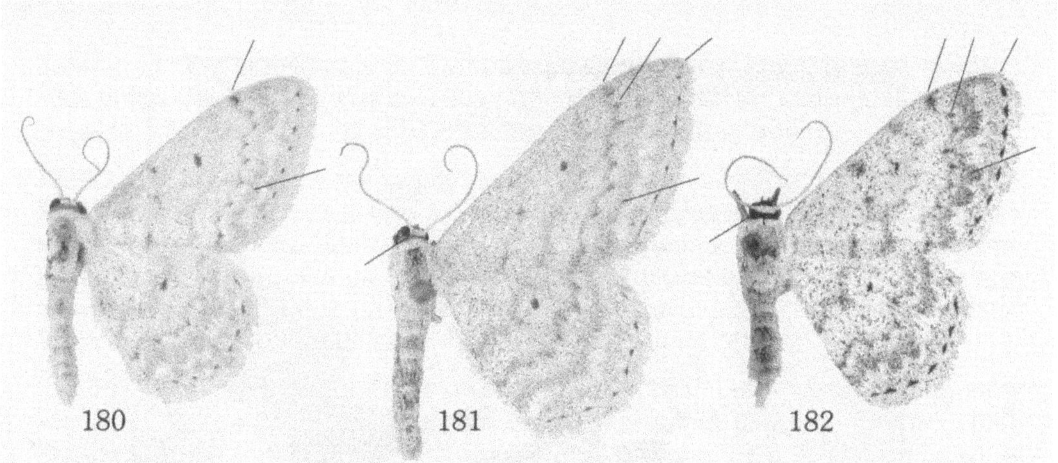

Text-figs 180-182: Diagnostic characters (indicated) of European *Scopula* species. Text-fig. 180: *S. beckeraria* (LEDERER, 1853). Text-fig. 181: *S. incanata* (LINNAEUS, 1758). Text-fig. 182: *S. marginepunctata* (GOEZE, 1781).

142. Scopula incanata (LINNAEUS, 1758)

Phalaena Geometra incanata LINNAEUS, 1758: Syst. Nat. (Ed. 10) 1: 528 (Europe). Syntype(s) (LSL).
Phalaena variegata SCOPOLI, 1763: Ent. Carniolica: 222, fig. 553 (north-western Slovenja: Carniolia). Syntype(s) lost.
Phalaena mediata FABRICIUS, 1776: Genera Insect.: 290 (Germany). Syntype(s) (ZMUC: one ♀ 'Type' without abdomen, examined).
Phalaena Geometra planata VILLERS, 1789: Linn. ent. 2: 386 (southern France). Tentatively synonymized by LERAUT (1997).
Arrhostia incanaria HÜBNER, 1825: Verz. bekannter Schmett.: 312. Emendation. Partly misidentified (see synonymy of *Idaea seriata*).
Idaea mutata TREITSCHKE, 1828: Schmett. Eur. 6 (2): 274, 275 (Austria: Styria; Croatia; Hungary). Syntypes.
Acidalia mutataria DUPONCHEL, 1830: in GODART & DUPONCHEL: Hist. Nat. Lép. Pap. Fr. 8 (1): 102, pl. 177, fig. 4. Emendation.
Acidalia adjunctaria BOISDUVAL, 1840: Genera Index meth. eur. Lepid.: 225 (Italy: Lombardian Alps). Syntype(s) (ZFMK: 1♀ with label 'typicum specimen', examined).
Dosithea demutaria BRUAND, 1846: Mém. Soc. Emul. Doubs (2) 2 (3/4): 138 (France, Doubs: Besançon). Syntype(s).
Acidalia incanata ibericata REISSER, 1935: Ent. Rdsch. 53 (5): 78, pl. 1, figs C3-8 (central Spain: Sierra de Gredos). Syntypes ca. 50♂8♀ (BMNH 3♂, NHMW 6, SMNK 32, ZSM 2♂1♀, MNCN 1♂ syntypes examined). Valid at subspecific rank.
Acidalia incanata albida SILBERNAGEL, 1944: Z. Wien. ent. Ges. 29: 154 (Macedonia: Istok). Syntypes ♂♀. Valid at subspecific rank according to DANIEL et al. (1951).
Scopula incanata rubeni VIIDALEPP, 1979: Uchen. Zap. Tartu gos. Univ. 12 (483): 85 (Russia: Tuva: Kyzyl). Holotype ♂ (IZBE). Valid at subspecific rank (SCOBLE 1999).
Unavailable names (misidentification): *immutata*: sensu PANZER (1804) nec LINNAEUS, 1758. – (infrasubspecific): *anastomosaria*: REISSER (1961); *bilineata*: REISSER (1961); *catenata*: HÖRHAMMER (1933); *nigrolineata*: REISSER (1961); *seminigra*: REBEL (1912). – (non binominal): *mutaria*: HERRICH-SCHÄFFER (1847). – (incorrect subsequent spellings): *adiunctaria*: OSTHELDER (1929); *demutata*: GUENÉE (1858; in synonymy); *incarnata*: CARNELUTTI (1992).

DIAGNOSIS: Wingspan ♂♀ 24-30 mm, second generation 20-25 mm, ♀ on average smaller. Forewing apex slightly pointed. Ground colour light grey, sometimes with darker suffusion. Wing pattern dark grey, variable. Antemedial line often marked with dark dots on the veins. Medial line usually diffuse. Postmedial line dentate, not curved at costa.

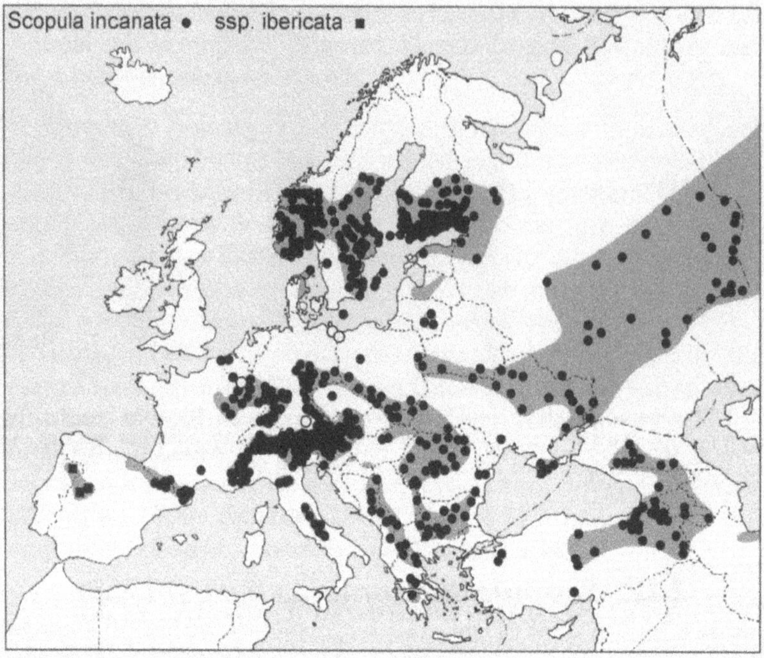

Terminal area usually grey, with whitish wavy line. Black discal spots very distinct on all wings. Terminal dots distinct. Frons blackish, vertex light grey. Collar bicolorous, light brown, towards vertex blackish. Antennae of ♂ with intersegmental incisions, ciliate-fasciculate, cilia 1.1-1.5 times width of flagellum. Hindtibia of ♂ stout, with pencil, tarsus 0.7-0.8 times length of tibia. Hindtibia of one examined ♂ from central Italy with one very short spur (ZSM).

Variability: Dark suffusion induced by cold and wet conditions; dryness and high temperature leading to pale specimens with fine markings (BERGMANN 1955). In Caucasus, central and north-eastern Turkey often with dark suffusion.

Subsp. ***ibericata*** on average larger, with whitish ground colour, also in the terminal area, sharply marked. Populations of central Italy quite similar, in the Balkan peninsula larger, pale forms mixed up with normal specimens.

MALE GENITALIA: Socii long and narrow. Fibula thorn-shaped. Base of sternum A8 convex, both cerata long and narrow, length variable. In 3 of 18 examined ♂♂ cerata shortened (gen.fig. 142b), probably due to polymorphism.

FEMALE GENITALIA: Antrum with pentagonal sclerite. Lamella antevaginalis sclerotized, oval. Ductus bursae long. Corpus bursae oval. Signum large, elongate, usually with posterior indentation.

DISTRIBUTION: Eurasiatic, in Europe with slight disjunction, similar to boreo-montane distribution patterns. Widely distributed in Europe from Iberian peninsula to the Urals. Scattered populations in the northern parts of central Europe, and in southern Europe. Absent from British Isles and Mediterranean islands, but reputedly present on Sicily (MÜLLER 1996). In the western Iberian peninsula replaced by subsp. *ibericata*. – Outside

Europe in Turkey, Caucasus, Transcaucasus, northern Iran, Siberia, northern Kazakhstan, Altai, in the east to Mongolia and Yakutia (subsp. *rubeni*). Reputedly in Nepal at 3,500 m (FLETCHER 1961), possibly misidentified (see remarks).

PHENOLOGY: In central Europe usually bivoltine: early May to mid-September, occasionally until early October, usually with a gap in July, but sometimes with overlapping generations. In the southern Alps under good conditions trivoltine, from late March to mid-November. In northern Europe univoltine, from mid-June to early August. Montane populations, from about 700 m up to 1,200 m univoltine, from late June to early August, in Spain and Turkey two weeks later.

BIOLOGY: Larva polyphagous (FORSTER & WOHLFAHRT 1981). Larva found on *Polygonum aviculare* (Polygonaceae), *Sorbus aucuparia* (Rosaceae) (Finland: REUTER 1890; SEPPÄNEN 1954). Recorded on various herbaceous plants of dry localities, such as *Thymus serpyllum, Origanum vulgare, Teucrium chamaedrys* (Lamiaceae), *Campanula* (Campanulaceae), *Lychnis, Dianthus* (Caryophyllaceae), *Hippocrepis comosa, Cytisus scoparius* (Fabaceae) (var. cit., e.g. SCHMID 1885; REBEL 1910b; LHOMME 1935; VALLE 1941; BERGMANN 1955; EBERT 2001), preferring *Thymus* and *Polygonum* (northern Europe: SKOU 1986; Sweden: SVENSSON 1993). Reared on *Thymus, Origanum vulgare* (LEIPNITZ pers. comm.), *Clematis vitalba, Campanula medium, Lonicera, Rhamnus cathartica* and *Taraxacum officinale* (SCHNEIDER 1939; BERGMANN 1955; EBERT 2001). Adults readily attracted to light and flowers. During daytime resting on rocks (REISSER 1935), walls, or – in tectiform wing position - on twigs of *Berberis* (OSTHELDER 1929).

HABITAT: Xerothermophilous, preferring mountainous habitats (BERGMANN 1955; FORSTER & WOHLFAHRT 1981; EBERT 2001), rocky slopes, dry, stony grassland, quarries, sun-exposed scrubby places, often at sites with abundance of *Thymus*. Usually on calcareous or silicious soils, in Denmark on grey sand dunes (SKOU 1986). From 0 m up to 1,400 m above sea-level, up to 2,000 m in southern Alps. In southern Europe regularly montane, usually from 1,000 m up to 2,300 m, up to 2,700 m in northern Iran. In north-eastern Spain, however, sometimes down to 300 m, correlated with bivoltinism.

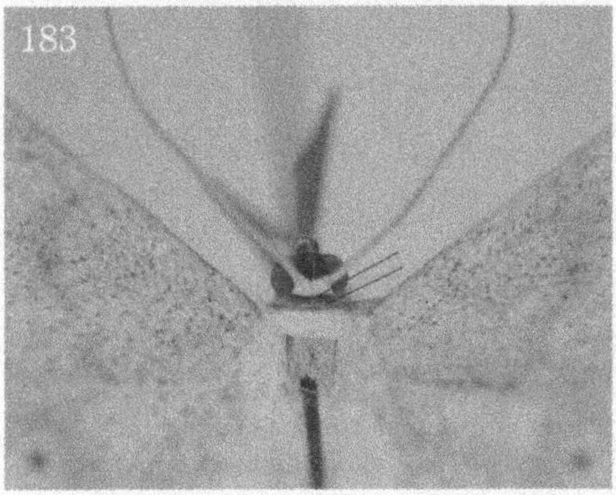

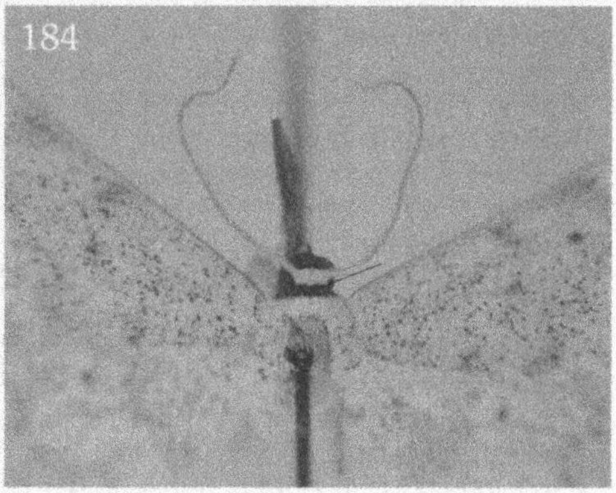

Text-figs 183-184: Diagnostic characters (indicated) of the head of European *Scopula* species. Text-fig. 183: *S. incanata* (LINNAEUS, 1758) with bicolorous collar. Text-fig. 184: *S. marginepunctata* (GOEZE, 1781) with unicolorous, dark collar.

SIMILAR SPECIES: *S. marginepunctata* on average smaller, forewing apex more rounded, collar unicolorous, dark brown, wings with sand coloured tinge and less grey, postmedial line bent at costa. In ♂ genitalia, and sometimes also in habitus reminiscent of *S. beckeraria rebeli* but the latter with ground colour glossy, sand coloured or light brown, collar unicolorous light brown. *S. umbelaria* (see no. 123).

REMARKS: A series from Nepal, identified by D.S. FLETCHER as *S. incanata* (ZSM) with much longer ♂ antennal cilia but minor differences in ♂ genitalia, requires further study.

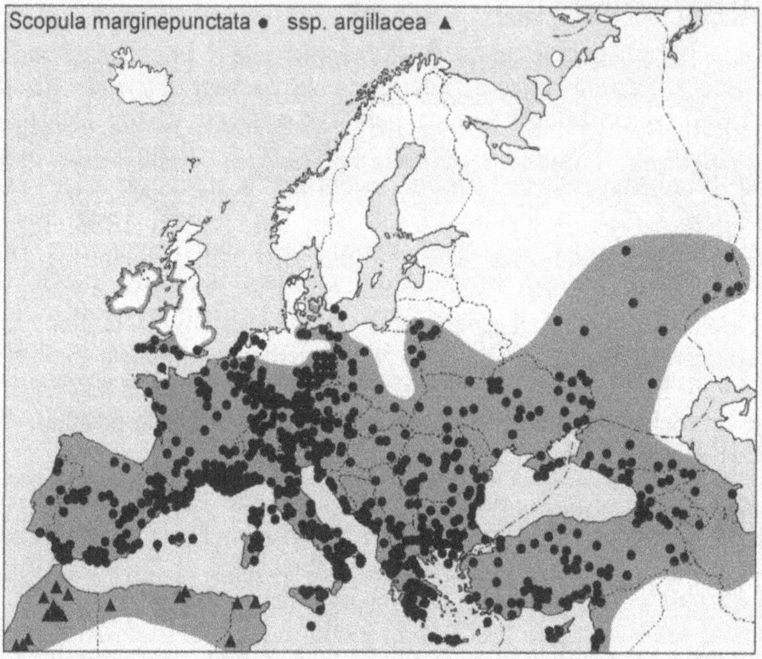

143. *Scopula marginepunctata* (GOEZE, 1781)

Phalaena Geometra marginepunctata GOEZE, 1781: Ent. Beytr. 3 (3): 385 (Europe). Syntype(s).

Phalaena Geometra coniugata BORKHAUSEN, 1794: Naturg. eur. Schmett. 5: 521 (Europe, probably Germany). 2 syntypes.

Phalaena aniculosata RAMBUR, 1829: Annls Sci. Observ. 2: 266, pl. 6, fig. 4 (southern France: Montpellier). Holotype ♀ (ZSM/HERBULOT, examined, without abdomen).

‡ *Acidalia puellaria* BOISDUVAL, 1840: Genera Index meth. eur. Lepid.: 221 (Switzerland, 'Helvetia'). Introduced to replace misidentified '*immutata* [sensu] Fr. pl. 54'.

Acidalia promutata GUENÉE, 1858: in BOISDUVAL & GUENÉE, Hist. nat. Insectes (Spec. gén. Lépid.) 9: 491 (Europe). Syntype(s).

Acidalia apertaria WALKER, 1863: List. Spec. lepid. Insects Colln Br. Mus. 26: 1612 (type locality not stated). Holotype ♂ (BMNH).

Acidalia pastoraria JOANNIS, 1891: Bull. Soc. ent. Fr. 1891: 80 (Turkey: 'Césarée'). Syntypes.

Acidalia marginepunctata var. *madoniata* FUCHS, 1901: Stettin. ent. Ztg. 62: 123 (Sicily). Syntypes 8♂♀.

Acidalia marginepunctata var. *britonaria* OBERTHÜR, 1917: in CULOT: Noct. & Géom. Eur. 3: 72, pl. 10, fig. 207 (France, Bretagne: Cancale). Syntypes 1♂1♀ (ZFMK, examined). Herewith downgraded from subspecific rank (LERAUT 1994; 1997) to synonymy (syn. n.).

Acidalia marginepunctata var. *subatrata* WAGNER, 1919: Zeits. österr. Ent.Ver. 4 (2): 11 (Italy, Friuli: Udine, Cividale distr., Attimis). Syntypes ♂♀ (NHMW: 1♂1♀, examined, labelled as 'cotypes' of '*friulana*', mislabelled with non-existent taxon name, ♂ without abdomen). Herewith downgraded from subspecific rank (SCOBLE 1999) to synonymy (syn. n.).

Acidalia marginepunctata 'f.' *insubrica* VORBRODT, 1930: Mitt. Schweiz. Ent. Ges. 14: 324 (Switzerland: Mixos). Formally introduced as 'form', but subspecific rank outlined in text. Herewith downgraded from subspecific rank (FORSTER & WOHLFAHRT 1981) to synonymy (syn. n.).

Scopula marginepunctata argillacea REISSER, 1933: EOS 9: 240. First available name after *argillacea* PROUT, 1913 (infrasubspecific) (Morocco: Rif mountains). Syntype(s) of REISSER. Valid at subspecific rank.

Scopula marginepunctata terrigena PROUT, 1935: in SEITZ, Macrolep. 4, Suppl.: 39, pl. 4, fig. i (northern Iran, Elburs Mts.). Syntypes 26 (BMNH). Valid at subspecific rank (SCOBLE 1999).

Unavailable names (misidentifications): *confinaria*: sensu LAEVER (1966, fig.) nec HERRICH-SCHÄFFER, 1847 (identity: *Glossotrophia confinaria*); *contiguaria*: sensu STEPHENS (1829) nec HÜBNER, 1799 (identity: *Idaea contiguaria*); *contiguata*: sensu HAWORTH (1809) nec *contiguaria* HÜBNER, 1799; *immutaria*: sensu HÜBNER (1799) nec *immutata* LINNAEUS, 1758; *immutata*: sensu DENIS & SCHIFFERMÜLLER (1775) nec LINNAEUS, 1758; *incanata*: sensu HAWORTH (1809) nec LINNAEUS, 1758; *mutaria*: sensu HEYDENREICH (1851) nec *mutata* TREITSCHKE, 1828 (identity: *Scopula incanata*); *mutata*: sensu METZNER (1849) nec TREITSCHKE, 1828; *mutataria*: sensu GRAAF (1852) nec *mutata* TREITSCHKE, 1828. – (infrasubspecific): *anastomosaria*: REISSER (1961); *basinotata*: COCKAYNE (1950); *bilineata*: REISSER (1961); *demarginata*: REISSER (1961); *diffluata*: REISSER (1961); *marginevirgata*: DANNEHL (1927); *griseofasciata*: TURATI (1915); *mundata*: PROUT (1913); *nigropunctata*: HARTMANN & STERNECK (1915); *orphnaeata*: FUCHS (1904); *promutaria*: MORRIS (1861); *samothracica*: KOUTSAFTIKIS (1973); *zernyi*: SCHAWERDA (1927). – (incorrect subsequent spellings): *aniculosaria*: RAMBUR (on type label); *conjugata*: PROUT, 1934; *marginata*: VORBRODT (1930); *margineopunctata*: DIDMANIDZE (1978); *marginepunctala*: HOFFMEYER (1966); *marginepunctaria*: LATTIN (1951); *marginipunctata*: PIERCE (1914).

DIAGNOSIS: Wingspan ♂♀ 22-25 mm, occasional females up to 27 mm, subsequent generations smaller, sometimes 18 mm only. Ground colour dirty white with slight sand coloured tinge, sometimes with strong dark suffusion. Wing pattern dark grey, very variable. Medial and antemedial lines diffuse or absent. Postmedial line dentate. Terminal area with grey shadings. Discal spots and terminal dots distinct. Frons and collar blackish brown, vertex sand coloured. Antennae of ♂ with intersegmental incisions, ciliate-fasciculate, cilia 1.3-1.7 times width of flagellum. Hindtibia of ♂ stout, with pencil, tarsus 0.75-0.95 times length of tibia.

Variability: Dark forms chiefly occurring in the wet climates of the Atlantic coast ('*britonaria*') and on mountains, e.g. Pyrenees, Alps ('*insubrica*', '*subatrata*') or Turkey. There are, however, clinous transitions from such 'regional forms' to nominotypical populations (compare LERAUT 1994). In North Africa (subsp. *argillacea*) wings sand coloured, wing pattern fine and weak.

Subsp. ***terrigena*** larger, up to 29 mm. Wings with sand coloured or light brown tinge. Medial fascia and terminal area broad and dark brown marked.

MALE GENITALIA: Socii long and narrow. Fibula hook-shaped. Sternum A8 with shallow base, both cerata absent, replaced by weak chitinous folds. Without polymorphism of cerata length (n=29).

FEMALE GENITALIA: Antrum with half-moon-shaped or triangular sclerite. Ductus bursae long. Corpus bursae oval. Signum large, oval, with posterior indentation.

DISTRIBUTION: Eurasiatic. Widely distributed in Europe, from Portugal to the Urals, in the north to southernmost Scandinavia and southern Baltic states, ubiquitous and common in the south. Present on all Mediterranean islands. – Outside Europe from Morocco to Tunisia (subsp. *argillacea*), Turkey, Cyprus, Levant, Caucasus, Transcaucasus (nominate subspecies). Replaced by subsp. *terrigena* from northern Iran to central Asian mountains and Mongolia (VIIDALEPP 1975; 1988; 1996).

PHENOLOGY: Bivoltine: mid-May to late September, exceptionally somewhat earlier or later, the generations often overlapping in July, towards north with a gap between the generations in July. In the Alps, in southern Europe and south-western Asia above ca. 1,400 m univoltine, from mid-June to late August, mainly in July. In the lowlands of southern Europe tri- or plurivoltine mid-March to early November, under good conditions, e.g. on Crete and in the Levant until mid-December.

BIOLOGY: Larva polyphagous on many herbaceous plants of dry localities, often on Lamiaceae. Found on *Gypsophila struthium* (Caryophyllaceae) (Spain: REDONDO et al. 2001), *Thymus serpyllum* (Lamiaceae) (Germany: SCHNEIDER 1939; EBERT 2001), and *Teucrium montanum* (Lamiaceae) (Austria: ERLEBACH pers. comm.). Recorded on *Thymus serpyllum, Origanum vulgare, Teucrium montanum, Glechoma hederacea, Stachys recta* (Lamiaceae), *Artemisia vulgaris, Achillea millefolium* (Asteraceae), *Hippocrepis comosa, Vicia* (Fabaceae), *Gypsophila repens, Stellaria* (Caryophyllaceae), *Plantago* (Plantaginaceae), *Sedum* (Crassulaceae), *Galium* (Rubiaceae), *Filipendula ulmaria, Potentilla reptans* (Rosaceae), *Polygonum aviculare* (Polygonaceae), *Valeriana* (Valerianaceae) (var. cit., e.g. REBEL 1910b; LHOMME 1935; ALLAN 1949; BERGMANN 1955; SKINNER 1984; EBERT 2001), possibly preferring *Sedum*, e.g. *S. album, S. telephium, S. reflexum* (Crassulaceae) (Germany: BERGMANN 1955). Reared on petals of various plants, and on leaves of *Taraxacum officinale, Polygonum aviculare, Salvia pratensis, Lactuca sativa, Achillea millefolium, Stellaria media, Galium mollugo, G. verum, Trifolium, Geranium* and *Vicia* (original data; ERLEBACH, LEIPNITZ, MÜLLER pers. comm.; GUENÉE 1858; URBAHN & URBAHN 1939; SCHNEIDER 1939; ALLAN 1949; BERGMANN 1955; HAUSMANN & MILLER 2000; EBERT 2001). Main activity of adults at dusk and at night, readily attracted to light, bait and flowers, e.g. *Sedum* (FUCHS 1875). Usually resting on rocks during day time.

PARASITOIDS: THOMPSON (1947) mentions *Atractodes gilvipes* (Hymenoptera, Ichneumonidae), but data are very doubtful, as the genus *Atractodes* needs thorough taxonomic revision (ZWAKHALS pers. comm.).

HABITAT: Xerothermophilous, in central Europe local in rocky, stony or sandy localities, open or scrubby sites, often on southern slopes. In southern Europe ubiquitous and common in many different habitats. North of Alps from 0 m up to 800 m above sea-level, up to 1,700 m in southern Alps, up to 2,000 m in southernmost Europe, up to 2,700 m in Morocco, south-western Asia and central Asian mountains.

SIMILAR SPECIES: *S. incanata* (see no. 142), *S. submutata* (132), *Idaea typicata* (41), *I. vesubiata* (69), *I. libycata* (70), *I. consolidata* (71), *I. metohiensis* (99). *S. luridata* on average smaller, wings sand coloured, pattern more diffuse, frons paler, vertex darker, discal spots smaller, medial line zigzagging or dentate. *Glossotrophia* species with more diffuse wing pattern, the European species with longer proboscis, one hindtibial spur in ♂, two spurs in ♀.

144. *Scopula luridata* (ZELLER, 1847)

Idaea luridata ZELLER, 1847: Isis, 1847 (1): 20 (Greece: Rhodos). Syntype(s) ♀ (BMNH: 1♀ examined, with red 'type' label, according to HERRICH-SCHÄFFER the only type specimen).

Acidalia luridaria HERRICH-SCHÄFFER, 1852: Syst. Bearb. Schmett. Eur. 6 (55): 69. First available name after *luridaria* HERRICH-SCHÄFFER, 1848 (pl. 68, fig. 419: non binominal). Emendation.

‡ *Acidalia luridaria* [syn.] *formosaria* HERRICH-SCHÄFFER, 1852: Syst. Bearb. Schmett. Eur. 6 (55): 69 (Crete). Holotype ♂. Unavailable, introduced in synonymy.

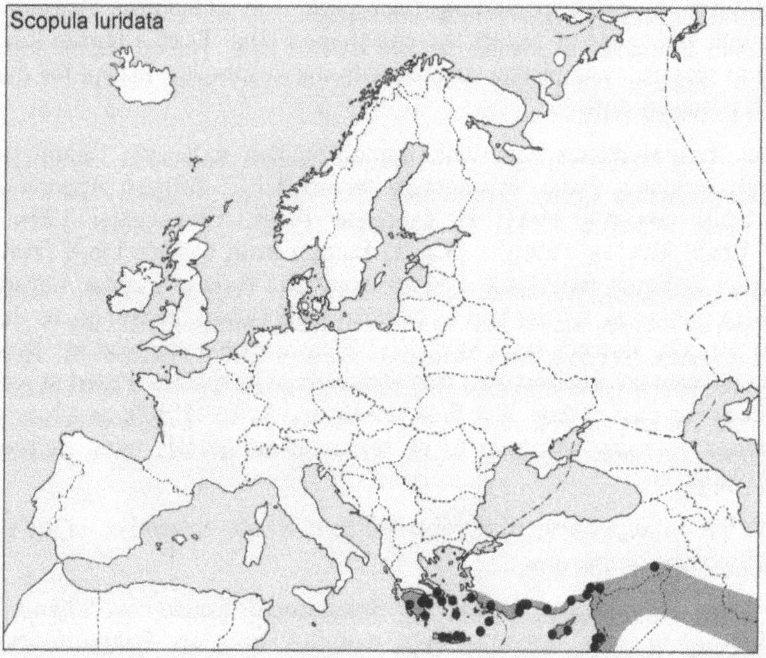

Acidalia coenosaria LEDERER, 1855: Verh. zool.-bot. Ver. Wien 5: 209, pl. 3, fig. 3 (Lebanon: Beirut; Cyprus). Syntypes 5♂♀ (MNHU: 3♂1♀, examined).

‡? *Acidalia coenosaria chinensis* STERNECK, 1931: Dt. ent. Z. Iris 45 (2): 79 (Corea). Syntypes 2♂. Junior primary homonym nec STERNECK, 1927 (identity: *Scopula immutata chinensis*).

? *Scopula luridata* var. *sternecki* PROUT, 1934: Lepid. Cat. 63: 198 (Corea). No separate types. Replacement name for *chinensis* STERNECK, 1931. Valid at subspecific rank (SCOBLE 1999). Conspecifity awaits verification (see PROUT 1935: 39).

Acidalia distracta BUTLER, 1881: Proc. Zool. Soc. London 1881 (3): 616 (Pakistan: Karachi). Syntypes 3 (BMNH). Valid at subspecific rank (SCOBLE 1999).

Acidalia absconditaria BUTLER, 1883: Proc. Zool. Soc. London 1883: 170 (India: Mhow). Tentative synonymy in PROUT (1934).

Acidalia fulminataria TURATI, 1927: Atti Soc. Ital. Sci. Nat. 66 (3/4): 324, pl. 8, fig. 6 (Libya, Cyrenaica: Ain Mara). Holotype ♂. Herewith downgraded from species rank (SCOBLE 1999) to synonymy (syn. n.).

Unavailable names (misidentification): *satsumaria*: sensu STERNECK (1927) nec LEECH, 1897 (identity: *Scopula insolata* BUTLER, 1889). – (incorrect subsequent spelling): *coenosata*: SNELLEN (1862).

DIAGNOSIS: Wingspan ♂♀ 18-24 mm. Ground colour pale sand coloured. Wing pattern yellowish or brownish grey, variable, diffuse. Medial line and postmedial lines dentate. Forewing costal spots and spots in the terminal area distinct. Discal spot on hindwing usually present, on forewing often diffuse. Terminal dots elongate, distinct. Frons brown, often with sand coloured scales towards proboscis, vertex and collar pale sand coloured. Antennae of ♂ with intersegmental incisions, ciliate-fasciculate, cilia long, 1.9-2.3 times width of flagellum. Hindtibia of ♂ with pencil, tarsus 0.9-1.0 times length of tibia.

MALE GENITALIA: Socii long and narrow. Fibula narrow, tapering. Sternum A8 with strongly projecting base, both cerata long and stout, mappa elongate and asymmetrically linked with sternum. No polymorphism of cerata length found (n=23), but type of *similata* LE CERF, 1924 from Ethiopia with left ceras short, possibly a polymorphic form of *S. luridata*.

FEMALE GENITALIA: Sterigma posteriorly sclerotized, sub-rectangular. Antrum elongate, posteriorly with triangular or semicircle, cap-shaped lobe. Ductus bursae long. Corpus bursae oval or globular, completely covered with micro-spinules, except for the right lateral and the posterior part.

DISTRIBUTION: East-Mediterranean, with Saharo-Sindian extension. Southern Greece, Greek islands including Crete. Erroneously recorded for northern Apennines, central Italy, and Sicily (MARIANI 1941; P. ZANGHERI 1969; TEOBALDELLI 1976; FIUMI & CAMPORESI 1988; MÜLLER 1996). – Outside Europe from eastern Libya (TURATI 1927, '*fulminataria*') to Egypt (WILTSHIRE 1949a), southern Turkey, Cyprus, Levant and the whole Arabian peninsula, across Iran to western India (subsp. *distracta*), in the south to Yemen and Somalia. Records from Morocco ('*Glossotrophia coenosaria*': RUNGS 1981) refer to nomenclatural confusion with true *Glossotrophia* species. Listed in some faunistic studies from Far East of Asia, e.g. SHIN (1996) and INOUE (1982), as subsp. *sternecki*, but populations probably will have to be separated at species rank, as proposed by STERNECK (1941).

PHENOLOGY: Tri- or plurivoltine: in Europe late April to late November, in the Levant and Arabia in all months of the year.

BIOLOGY: Larva polyphagous, often on Solanaceae. Found on *Silene 'vulgaris'* (Caryophyllaceae) (Cyprus: LEIPNITZ pers. comm.), and on *Solanum* (Solanaceae) (Lebanon: WILTSHIRE 1939a). Recorded on *Silene* (Caryophyllaceae) (Lebanon: WILTSHIRE 1939a), *Asparagus* (Liliaceae), *Cyphomandra* (Solanaceae), *Opuntia* (Cactaceae), and *Rosa* (Rosaceae) (Israel: HALPERIN & SAUTER 1992), and as pest on seeds of maize, tomato, berseem (clover), sweet potato and many ornamental plants (Egypt: EL-SAWAF et al. 1968). Reared on *Rosa* (WILTSHIRE 1949b). Adults readily attracted to light.

HABITAT: Xerothermophilous. Near coasts and in xeromontane localities. In Europe restricted to the lowlands. In south-western Asia from 0 m up to 2,300 m above sea-level, in the southern Arabian peninsula montane, from 1,800 m up to 3,100 m.

SIMILAR SPECIES: *S. marginepunctata* (see no. 143). *S. submutata* and *Glossotrophia* species with terminal line encircling forewing apex, the latter with longer proboscis, European species with one spur on the ♂ hindtibia, two in the ♀.

REMARKS: Synonymy of the taxon *fulminataria* results from the examination of western Egyptian specimens (ZSM; coll. SOMMERER), well matching TURATI's description and not far from the locus typicus of *fulminataria*. For larval morphology and rearings see EL-SAWAF et al. (1968) and PÜNGELER (1892).

145. Scopula imitaria (HÜBNER, 1799)

Geometra imitaria HÜBNER, 1799: Samml. Eur. Schmett. 5, Geometrae (1): pl. 10, fig. 51 (Europe). Syntype(s) lost.

† *Phalaena fimbriata* FOURCROY, 1785: Entom. Parisiensis 2: 276 (France: Paris). Syntype(s). Junior primary homonym nec SCHREBER, 1759 (identity: *Noctua fimbriata*, Noctuidae).

Acidalia syriacaria CULOT, 1918: Noct. et Géom. Eur. 3: pl. 12, fig. 246 (Syria). Syntype(s) (ZFMK: 4♂ Akbès, 3♀ Beirut, possibly syntypes, examined). In the text (p. 83) non binominal and infrasubspecific, on plate treated at species rank. Valid at subspecific rank.

Unavailable names (misidentification): *tristriaria*: sensu WERNEBURG (1864) nec FABRICIUS, 1794 (identity: *Scopula nigropunctata*). – (infrasubspecific): *aequilineata*: SCHWINGENSCHUSS (1926); *bitinctata*: DANNEHL

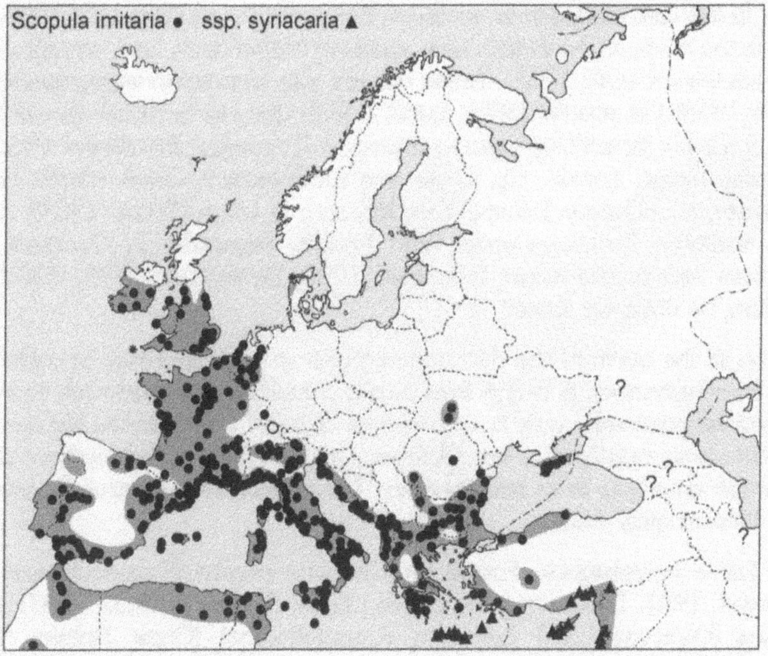

(1933); *brevipennis*: COCKAYNE (1952); *erubescens*: PROUT (1934); *fasciata*: VORBRODT (1930); *kesslitzi*: HIRSCHKE (1911); *pallida*: CHALMERS-HUNT (1961); *rosea*: TURATI (1924); *umbrata* DUFRANE (1931). – (incorrect subsequent spellings): *imitarla*: SCALERCIO & INFUSINO (2003); *imitata*: CURTIS (1826); *immitaria*: HÜBNER (1823); *imitata*: CURTIS (1826).

DIAGNOSIS: Wingspan ♂♀ 22-26 mm, subsequent generations on average smaller, sometimes 18 mm only. Hindwing termen strongly angled at M3. Very variable in coloration and pattern of wings. Wings sand coloured, often with yellowish, orange or light brown tinge. Transverse and terminal lines fine and sharp. Medial line oblique, straight, often with distal dark shadow. Frons and palpi blackish, vertex pale sand coloured, collar light brown. Antennae of ♂ with deep intersegmental incisions, ciliate-fasciculate, cilia, 1.3-1.7 times width of flagellum. Hindtibia of ♂ with pencil, tarsus slightly shortened, 0.65-0.85 times length of tibia.

Coloration of subsp. ***syriacaria*** with strong rosy tinge. Transverse lines often more weakly marked, postmedial line straighter, angle of hindwing termen shallower.

MALE GENITALIA: Socii long, broad at base. Fibula narrow, digitiform. Sternum A8 with variable length and shape of cerata: In subsp. *syriacaria* from the Levant rather straight, in West-Mediterranean populations long with curved right ceras, in central, southern and eastern Mediterranean (including Crete) with strongly curved cerata, often slightly shortened, mainly the right one. With transitions between long and shortened arms, no true polymorphism with discontinuous feature pattern found (n=18).

FEMALE GENITALIA: Lamella antevaginalis strongly sclerotized, oval or round. Antrum with posterior semicircular, cap-shaped lobe. Signum elongate and slender, spinules small, usually fused medially.

DISTRIBUTION: Sub-Mediterranean. Widely distributed in the Mediterranean from Portugal to Crimea. In the east reputedly to southern European Russia (VIIDALEPP 1996, ANIKIN et al. 2000). In the north to the British Isles, southern Netherlands, and, as migrant, to western Denmark (FALCK et al. 1985). Single records also from south-westernmost Germany (HEINEMANN 1859; GELBRECHT 1999; EBERT 2001). One old 'artificial' record for eastern Germany, probably by anthropogenous transport (Thuringia: BERGMANN 1955). Present on all Mediterranean islands. On Crete and south-eastern Greek islands replaced by subsp. *syriacaria*. – Outside Europe from Morocco to Libya (TURATI 1924), subsp. *syriacaria* in southern Turkey, Cyprus, and Levant. Reputedly in Caucasus, northern Transcaucasus and northern Iran (VIIDALEPP 1996; DIDMANIDZE 1978; 2002), awaiting confirmation, no vouchers traced.

PHENOLOGY: In the north of the distribution bivoltine, from mid-May to early July; early August to mid-September. In British Isles usually univoltine, from early July to mid-August, a partial second generation only in warm years (SKINNER 1984). In the Mediterranean tri- or plurivoltine from mid-April to late October. On Crete, and in the southern and eastern Mediterranean emerging from late February. Larval development strongly asynchronous (southern Italy: original data).

BIOLOGY: Larva polyphagous. Found on *Clematis vitalba* (Ranunculaceae) (northern Italy: DANNEHL 1927; ERLEBACH pers. comm.; Great Britain: BUCKLER 1897), *Ligustrum ovalifolium* (Oleaceae), and *Lonicera* (Caprifoliaceae) (Great Britain: WARING & TOWNSEND 2003), *Chaenorhinum origanifolium* (Scrophulariaceae) (Pyrenees: CHRÉTIEN 1928). Recorded on *Galium* (Rubiaceae), *Lotus* (Fabaceae), *Erica, Calluna vulgaris* (Ericaceae), *Artemisia* (Asteraceae), *Lonicera periclymenum* (Caprifoliaceae), *Clematis* (Ranunculaceae), *Ligustrum vulgare* (Oleaceae), *Crataegus, Prunus spinosa,* and *Rubus fruticosus* (Rosaceae), (var. cit., e.g. SEEBOLD 1879; REBEL 1910b; VORBRODT 1914; ALLEN 1949; LHOMME 1935; SKINNER 1984). Reared on leaves of *Taraxacum officinale, Polygonum aviculare, Convolvulus arvensis, Vicia, Stellaria, Rubus fruticosus, Lactuca sativa, Galium, Rumex, Senecio vulgaris, Rosa, Syringa vulgaris, Ligustrum* (original data; BUCKLER 1897; ALLAN 1949; FRIEDRICH 1976; PORTER 1997; HAUSMANN & MILLER 2000; MÜLLER pers. comm.). Main activity of adults at dusk and in the early evening, readily attracted to light in the Mediterranean, sparingly in the British Isles (SKINNER 1984).

HABITAT: Xerothermophilous lowland species, north of Alps local, e.g. on coastal sandhills, and sunny hedgerows. In the south common and ubiquitous, e.g. in the Mediterranean maquis, *Olea-Ficus* plantations, stony grassland, scattered oak forest. Usually from 0 m up to 1,000 m above sea-level, exceptionally up to 1,400 m in southernmost Europe and Morocco.

SIMILAR SPECIES: *Timandra comae* differs in bipectinate ♂ antenna, medial and postmedial lines leading to the apex rather than to the costa, discal spot absent from hindwing.

REMARKS: For rearings and morphology of immature stages see BUCKLER (1897).

146. Scopula immutata (LINNAEUS, 1758)

Phalaena Geometra immutata LINNAEUS, 1758: Syst. Nat. (Ed. 10) 1: 528 (Europe). Syntype(s) (LSL).
‡ *Phalaena Geometra pallidata*: sensu BORKHAUSEN, 1794: Naturg. eur. Schmett. 5: 325. Misidentification nec DENIS & SCHIFFERMÜLLER, 1775.

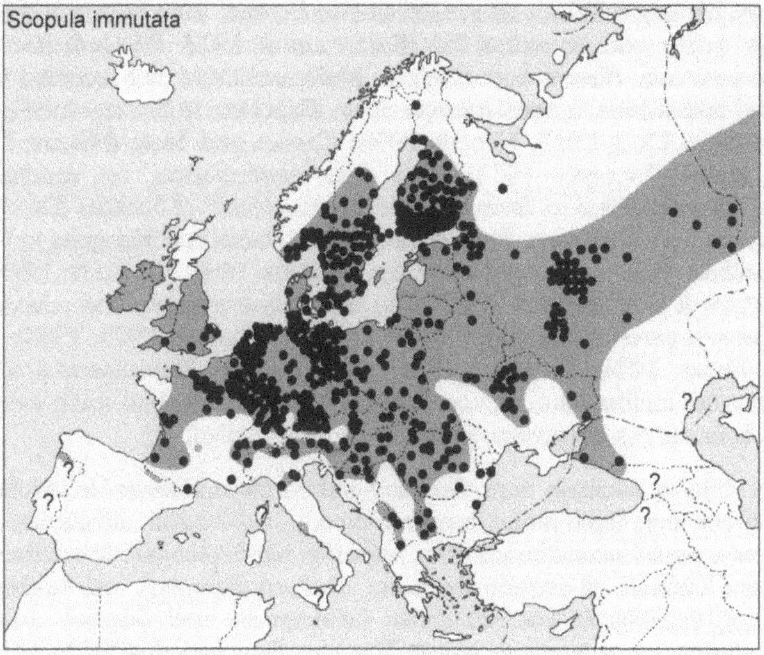

‡ *Arrhostia immutaria* HÜBNER, 1825: Verz. bekannter Schmett.: 311. Incorrect subsequent spelling. Partly misidentified (see synonymy of *Scopula marginepunctata*).
‡ *Acidalia caespitaria* BOISDUVAL, 1840: Genera Index meth. eur. Lepid.: 225. Introduced to replace misidentified *sylvestrata*: sensu FREYER, 1833: 142, pl. 77.
Acidalia myrtillata DADD, 1911: Berl. Ent. Zeits. 55 (3/4): 262 (Germany: Berlin, Spandau forest). Syntype(s) ♂♀.
Unavailable names (misidentifications): *exanthemaria*: sensu HÜBNER (1825) nec *exanthemata* SCOPOLI, 1763 (identity: *Cabera exanthemata*, Ennominae); ?*punctata*: sensu WERNEBURG (1864) nec SCOPOLI, 1763 (identity: *Scopula subpunctaria*); *sylvestraria*: sensu HÜBNER (1799: fig.97) nec HÜBNER, 1799 (fig.94: identity: *Idaea sylvestraria*); *sylvestrata*: sensu TREITSCHKE (1825), and FREYER (1933) nec *sylvestraria* HÜBNER, 1799. – (infrasubspecific): *anastomosaria*: PREISSECKER (1936); *atra*: REBEL (1914); *coarctata*: SCHULTZ (1931); *flavescens*: LEMPKE (1949); *grisescens*: LEMPKE (1949); *juncta*: LEMPKE (1949); *pulverata*: COCKAYNE (1950). – (incorrect subsequent spelling): *immulata*: SAMOUELLE (1819).

DIAGNOSIS: Wingspan ♂♀ 22-25 mm, ♀♀ in northern Europe and second generation in the south smaller, 19-23 mm. Hindwing rounded. Ground colour white with yellow suffusion of variable intensity, mainly in the males. Females more whitish. Dryness and high temperature inducing yellowish forms with distinct markings (BERGMANN 1955). Pattern very variable. Transverse lines yellowish, convex, meeting costa at right angles. Discal spots sharp, black, that of forewing rarely absent. Terminal dots usually absent or very small. Palpi blackish, underside white. Frons black, vertex white, collar ochreous. Antennae of ♂ with intersegmental incisions, ciliate-fasciculate, cilia 1.3-1.7 times width of flagellum. Hindtibia of ♂ stout, with pencil, tarsus shortened, 0.6-0.7 times length of tibia.

MALE GENITALIA: Socii long and narrow. Fibula knife-shaped, laterally more strongly sclerotized. Valva long and slender. Sternum A8 round, with cerata strongly curved, the left short, the right long. Not polymorphic (n=10).

FEMALE GENITALIA: Sterigma poorly sclerotized, with two longitudinal folds. Ductus bursae comparatively short. Signum very slender, spinules small, usually fused posteriorly.

DISTRIBUTION: Eurasiatic. Widely distributed all over Europe, from France and British Isles to the Urals. In the south to central Italy (RACHELI et al. 1978; PROLA & RACHELI 1979) and Balkan peninsula. Absent from Sardinia, Malta and Crete, not recorded in Albania. Very isolated populations in northernmost Spain (REDONDO & GASTON 1999), reputedly in Portugal (SILVA CRUZ 1967; MÜLLER 1996), Corsica and Sicily (MÜLLER 1996), and Astrachan region (LVOVSKY 1971), data await confirmation, no vouchers traced. Abundance decreasing due to destruction or 'improvement' of habitats (EBERT 2001). – Outside Europe from Caucasus, Kazakhstan, across Siberia and Mongolia to Far East of Asia and Sakhalin (subsp. *contramutata*: e.g. KOSTJUK 1992; VIIDALEPP 1996; BELJAEV 2001; KOSTJUK & GOLOVUSHKIN 2003). The taxon *contramutata* was retained as valid at species level in previous literature (STERNECK 1941; VIIDALEPP 1975; 1987). Reputedly in Turkey (RIEMIS 1994; VIIDALEPP 1996), and northern Transcaucasus (DIDMANIDZE 2002), data await confirmation, no vouchers traced, some examined south-western Asian specimens from various collections proved to be misidentified.

PHENOLOGY: Uni- or bivoltine: from Germany and northern Hungary to the Baltic states univoltine in one long flight period from mid-June to mid-August, sometimes with single specimens of a partial second brood from August to mid-September. In southern Europe, France, warm lowlands of western Germany, southern Germany, and Austria regularly bivoltine, from mid-May to late September, exceptionally until October, usually with a short gap between the generations in July. The long flight period is due to asynchronous development of the larvae (BERGMANN 1955).

BIOLOGY: Larva polyphagous. Found on *Primula elatior* (Primulaceae) (Germany: SCHNEIDER 1939; EBERT 2001), *Leontodon* (Asteraceae), *Lamium* (Lamiaceae) (Germany: UFFELN 1908), *Viola canina* (Violaceae), *Plantago major* (Plantaginaceae) (Finland: SEPPÄNEN 1954), *Valeriana officinalis* (Valerianaceae), and *Filipendula ulmaria* (Rosaceae) (Great Britain: PROUT 1913; ALLAN 1949; FRIEDRICH 1976; SKINNER 1984; PORTER 1997; WARING & TOWNSEND 2003; Sweden: SVENSSON 1993). Recorded on various herbaceous plants and grasses, e.g. *Plantago lanceolata* (Plantaginaceae), *Lamium* (Lamiaceae), *Sedum album* (Crassulaceae), *Crepis paludosa*, *Taraxacum officinale*, *Achillea millefolium* (Asteraceae), *Vaccinium myrtillus, Calluna vulgaris* (Ericaceae), *Holcus lanatus*, *Phalaris arundinacea*, and *Poa* (Poaceae), but also feeding on old, fallen leaves (var. cit., e.g. NOLCKEN 1870; REBEL 1910b; NORDSTRÖM et al. 1941; VALLE 1946; BERGMANN 1955), possibly preferring *Filipendula* (SKOU 1986). Reared on *Artemisia campestris, A. vulgaris, Thesium linophyllon* (petals), *Senecio vulgaris, Filipendula vulgaris, Taraxacum officinale, Clematis vitalba, Achillea millefolium, Lysimachia, Polygonum aviculare, Plantago lanceolata, Crataegus, Lactuca sativa*, and *Campanula medium* (MILLIÈRE 1866; ALLAN 1949; FRIEDRICH 1976; SKINNER 1984, BERGMANN 1955; KRAUS 1993; PORTER 1997; ERLEBACH, LEIPNITZ pers. comm.). Pupation in an open web on the host-plant (SKOU 1986). Adults easily flushed up from vegetation, sometimes also actively flying during daytime, active at early dusk, sometimes attracted to light. Nectar feeding recorded on *Solidago gigantea, Lolium perenne, Elymus repens* (EBERT 2001) and *Molinia* (URBAHN & URBAHN 1939).

HABITAT: Meso- to hygrophilous, preferring open wetland habitats as well as forests aisles and glades: bogs, swamps, wet grassland with *Carex* and *Molinia*, riverside vegetation. Mainly on sandy and silicious soils. Usually from 0 m up to 700 m above sea-level, in Alps and southern Europe sometimes up to 900 m, rarely up to 1,850 m (HELLMANN 1987).

SIMILAR SPECIES (see Text-fig. 163): *Idaea mancipiata* (see no. 82), *Scopula corrivalaria* (120), *S. caricaria* (121). *S. immutata* is usually well characterized by the feature combination 'sharp black discal spots on all wings', and 'yellowish suffusion'. *S. subpunctaria* on average larger, ground colour clear white, medial and postmedial lines at forewing costa bent towards base of wing.

REMARKS: For rearings and morphology of immature stages see BUCKLER (1897).

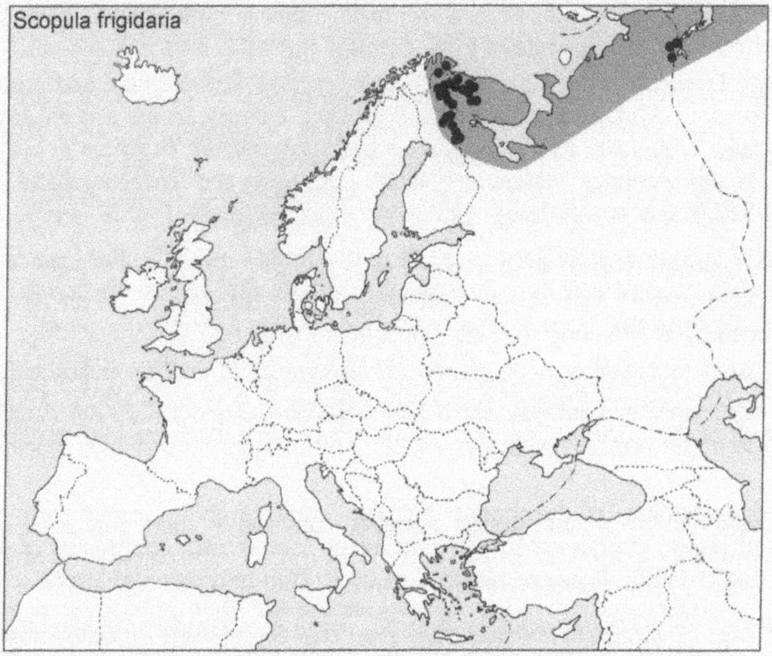

147. Scopula frigidaria (MÖSCHLER, 1860)

Acidalia frigidaria MÖSCHLER, 1860: Wien. Ent. Monats. 4 (12): 373, pl. 10, fig. 1 (eastern Canada: Labrador). Holotype ♂ (MNHU, examined).

Acidalia impauperata WALKER, 1861: List. Spec. lepid. Insects Colln Br. Mus. 22: 721 (Canada: Hudson Bay, Albany River, St. Martin's Falls). Lectotype ♂ (BMNH).

‡ *Acidalia defixaria* WALKER, 1861: List. Spec. lepid. Insects Colln Br. Mus. 23: 796 (type locality not stated). Lectotype ♂ (BMNH). Junior primary homonym nec WALKER, 1861 (22: p. 731; identity: *Scopula defixaria*, Venezuela).

Acidalia? arcticaria WALKER, 1863: List. Spec. lepid. Insects Colln Br. Mus. 26: 1594 (Canada: Arctic region). Holotype ♂ (BMNH).

Acidalia okakaria PACKARD, 1867: Proc. Bost. Soc. nat. Hist. 11: 43 (Canada, Labrador: Okak; Hopedale). Holotype ♂ (MCZ).

Acidalia schoeyeni SPARRE-SCHNEIDER, 1883: Ent. Tidskr. 4 (3/4): 80 (Norway: Varanger Fjord). Herewith downgraded from subspecific rank (VIIDALEPP 1996; SCOBLE 1999) to synonymy (syn. n.).

Unavailable names (misidentification): *spuraria*: sensu PACKARD (1876) nec *spuriaria* CHRISTOPH, 1858 (identity: *Holarctias sentinaria*). – (incorrect subsequent spellings): *schöyenii*: SANDBERG (1885); *schoyeni*: PAGENSTECHER (1902).

DIAGNOSIS: Wingspan ♂♀ 20-23 mm. Ground colour dark grey brown, terminal area often darker. Pattern diffuse. Dark, straight postmedial line usually distinct. Discal spots and terminal dots absent. Palpi slender. Frons brown, towards proboscis white, vertex and collar dirty white with some dark scales. Antennae of ♂ with intersegmental incisions, ciliate-fasciculate, cilia comparatively short, 1.1-1.4 times width of flagellum. Hindtibia of ♂ stout, with pencil, tarsus 0.8-0.9 times length of tibia.

MALE GENITALIA: Socii of medium length. Fibula tapered, but comparatively short. Sternum A8 with convex base, cerata almost without spinules, both cerata short, not polymorphic (n=3). North American populations with fibula more tapered, base of sternum A8 more convex, mappa larger, cerata without any spinules.

FEMALE GENITALIA: Similar to those of *S. imitaria*. Lamella antevaginalis oval, with transverse fold. Antrum with cap-shaped lobe. Signum elongate, fusiform.

DISTRIBUTION: Holarctic. Local and rare in northernmost Scandinavia, and arctic parts of European Russia. – Outside Europe in northernmost Siberia, in the east to Transbaikalia and Kamchatka (VIIDALEPP 1996; BELJAEV & KURANISHI 2000; KOSTJUK & GOLOVUSHKIN 2003). In North America widespread from Labrador and Newfoundland to Alaska (MÖSCHLER 1860; KROGERUS 1954; MCGUFFIN 1967; HANDFIELD et al. 1997).

PHENOLOGY: Univoltine: early June to late July (SVENSSON 1993). In Far East of Asia from mid-July to late August (BELJAEV & KURANISHI 2000; BELJAEV & VASILENKO 2002). In North America from late June to mid-August (MCGUFFIN 1967).

BIOLOGY: Larva recorded and/or reared on *Vaccinium myrtillus* (Ericaceae) (Finland: SEPPÄNEN 1970; northern Europe: SKOU 1986; Sweden: SVENSSON 1993). Adults active in the long dusk of the northern summer 'night'. ♀ ratio in collections low, usually less than 10%.

HABITAT: Open habitats of the tundra, but also 'common in the forest zone' (JALAVA & KULLBERG in press), coniferous forests and mixed forests with abundance of *Vaccinium myrtillus* (SKOU 1986). Lowlands from 0 m up to 300 m above sea-level.

SIMILAR SPECIES: *S. virgulata rossica* (see no. 125). *S. cajanderi* usually larger, ground colour with silver tinge, pattern more distinct, postmedial line undulate or dentate, discal spots distinct. *S. ternata* larger, paler, with two ♂ hindtibial spurs.

REMARKS: Nearctic populations with slight differences in ♂ genitalia, but, across Siberia, with clinous transitions to North European populations. Thus, the taxon *schoeyeni* is downgraded from subspecific rank (see VIIDALEPP 1996) to synonymy.

The ***ternata*** species-group

STERNECK's 7th group of *Scopula*. Heterogenous group, based on the shape of sternum A8, including many species. Central sclerite on ♂ sternum A2 with a patch of comparatively short setae. Male genitalia: Base of sternum A8 convex, mappa heart-shaped, with posterior indentation (STERNECK 1941). This feature, however, is quite variable: Aberrations with shallow incision or even rounded mappa occur in most species. Accurate morphological and molecular analysis will probably reveal that this group is not monophyletic and has to be joined with the *incanata* species-group.

148. *Scopula ternata* SCHRANK, 1802

Scopula ternata SCHRANK, 1802: Fauna boica 2 (2): 57 (Germany, Bavaria: Burghausen). Syntype(s).
Acidalia fumata STEPHENS, 1831: Illust. Br. Ent. (Haust.) 3: 312 (Scotland; England: Cumberland; Westmoreland; Devon, near Ashburton). Syntypes.

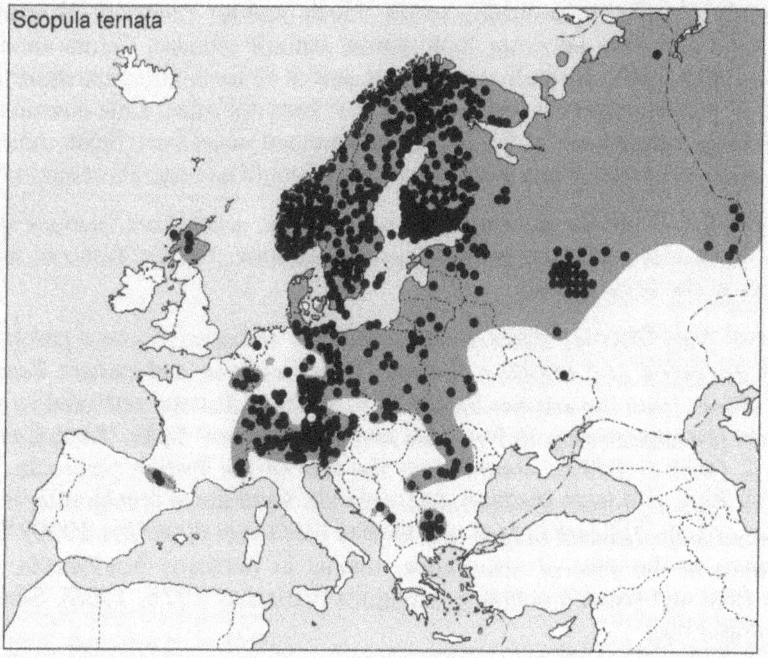

Idaea commutata FREYER, 1832: Neuere Beitr. Schmettkde. 1 (13): 143, pl. 77, fig. 4 (Austia, Styria: 'Steyrische Alpen'). Holotype ♀.
Cabera saltuata SPEYER, 1839: Isis 1839 (2): 125 (Germany, Hessen: Arolsen).
Acidalia commutaria BOISDUVAL, 1840: Genera Index meth. eur. Lepid.: 226. Emendation.
Acidalia nitidaria BOISDUVAL, 1840: Genera Index meth. eur. Lepid.: 226 (Lapponia). Syntype(s) (ZFMK: 1♂ ex coll. BOISDUVAL, examined).
Acidalia gypsaria BOISDUVAL, 1840: Genera Index meth. eur. Lepid.: 226 (northern Germany). Syntype(s) (ZFMK: 1♀ ex coll. BOISDUVAL, examined).
Idaea simplaria FREYER, 1852: Neuere Beitr. Schmettkde. 6 (99): 183, pl. 594, figs 1, 2 (Germany, Bavaria: Füssen). Syntypes ♂♀.
Acidalia duplexaria LA HARPE, 1855: Neue Denkschr. Allg. schweiz. Ges. ges. Naturw. 14 (7): 4. Synonymy according to PROUT (1934) and VIVES MORENO (1994).
Scopula ternata var. *perfumata* PROUT, 1934: Lep. Cat. 61: 179. First available name after *perfumata* REUTER, 1893: Acta. Soc. F.F. Fenn. 9 (6): 62 (infrasubspecific) (northern Finland). Syntype(s).
Scopula aequicerata TRÄFF, 1965: Opusc. ent. 30 (1-2): 131, fig. 2, 4b, 5b, 6b (Sweden: Njuonjevare). Holotype ♂. Synonymy see SVENSSON (1967).
Unavailable names (misidentification): ?*suffusaria*: sensu PEYERIMHOFF (1862) nec BOISDUVAL, 1840 (identity: *Idaea straminata*). – (infrasubspecific): *anastomosaria*: PREISSECKER (1926); *demarginata*: REISSER (1961); *laszloi*: DIÓSZEGHY (1934); *purissima*: DJAKONOV (1926); *resplendaria*: DANNEHL (1925); *unicolor*: BERGMANN (1955). – (nomen nudum): *graminaria*: FISCHER VON RÖSSLERSTAMM (1827). – (incorrect subsequent spellings): *commetata*: LIENIG (1846); *fumaria*: MORRIS (1861); *grawinaria*: PROUT (1913).

DIAGNOSIS: Wingspan ♂ 24-29, ♀ 21-25 mm, in northern Europe and Scotland smaller, ♂ 22-26 mm. Wings pale sand coloured with grey suffusion, in ♂ usually stronger than in ♀. Transverse lines diffuse, postmedial line more distinct, slightly undulate. Discal spots and terminal dots absent. The latter, exceptionally, present as elongate streaks in the ♀. Frons black brown, vertex and collar sand coloured. Antennae of ♂ with intersegmental incisions, ciliate-fasciculate, cilia comparatively short, 1.0-1.3 times width of flagellum. Hindtibia of ♂ with two spurs, tarsus very long, 1.2-1.6 times length of tibia.

MALE GENITALIA: Socii comparatively narrow. Fibula tapered. Sternum A8 round, usually with long right and short left ceras, both almost without spinules. Cerata anomalies at a frequency of about 10% for both cerata long, and 5 % for both cerata short (e.g. in 20 examined ♂♂ from northern Norway, and 20 ♂♂ from the Alps). One aberrant specimen found with three cerata (gen.fig. 148d). Both examined males from Sajan mountains with symmetrical short cerata). For variability of cerata length see also HAUSMANN (1999).

FEMALE GENITALIA: Lamella antevaginalis semicircular, sclerotized, surface granulated, often with transverse fold. Antrum with cap-shaped lobe. Signum fusiform, usually with fused spines at the anterior end.

DISTRIBUTION: Euro-Siberian, distribution reminiscent of boreo-montane patterns. Widely distributed in central and northern Europe, from Scotland and eastern France to the Urals, but absent from the warmer lowlands. In southern Europe restricted to very isolated mountain populations, e.g. in Pyrenees and northernmost Spain (ABOS CASTEL 1994; REDONDO & GASTON 1999). Absent from the rest of the Iberian peninsula, from Italy south of the Alps, and from Mediterranean islands. Occasional immigration is suggested for the Netherlands, resident only at one locality (KUCHLEIN & DE VOS 1999). Sometimes very common at the sites of occurrence, mainly in northern Scandinavia. – Outside Europe in Altai and Sajan mountains, Mongolia (VIIDALEPP 1975; 1996), Siberia, in the east to Jennisej river.

PHENOLOGY: Univoltine: In central Europe from late May to late July, on mountains and in northern Europe emerging from late June, on mountains flying until mid-August. Growth of larvae slow.

BIOLOGY: Larva oligophagous (2), most commonly found on *Vaccinium myrtillus* (Ericaceae) (Germany: SCHNEIDER 1939; SCHULTZ 1953; BERGMANN 1955; EBERT 2001; WEGNER pers. comm.; northern Italy: ERLEBACH pers. comm.; Great Britain: PORTER 1997), and on *V. uliginosum* (Germany: WEGNER pers. comm.; Finland: SEPPÄNEN 1954), usually restricted to the former (SCHULTZ 1953). Repeatedly recorded on *Vaccinium, Calluna vulgaris, Erica* (Ericaceae), and *Lathyrus niger* (Fabaceae) (var. cit., e.g. LHOMME 1935; ALLAN 1949; BERGMANN 1955; SKINNER 1984; SKOU 1986; EBERT 2001), and on *Salix repens* (Salicaceae) (ALLAN 1949). Reared on *Lonicera tatarica, Vaccinium myrtillus, Stellaria, Polygonum aviculare, Fragaria vesca, Lactuca sativa,* and *Taraxacum officinale* (ALLAN 1949; BERGMANN 1955; SEPPÄNEN 1970; ERLEBACH pers. comm.). Pupation in a loose cocoon on the host-plant. Adults very easily flushed up from vegetation during daytime, with active flights in the afternoon or at dusk looking for nectar sources, attracted also to light.

HABITAT: Meso- to hygrophilous, montane in the south of the distribution area. Open habitats with *Vaccinium myrtillus*, such as fringes of peat bogs, heathland and mosses, but usually most frequent in aisles or at fringes of sparse forest of *Pinus sylvestris*. Preferring acid soils (SKOU 1986) on sandy or silicious ground. In northern Europe from 0 m up to 1,000 m above sea-level, in central European lowlands very scattered populations, here montane, usually from 600 m up to 2,500 m, exceptionally lower.

SIMILAR SPECIES: *Idaea macilentaria* (see no. 25), *I. deversaria* (111), *Scopula nigropunctata* (124), *S. frigidaria* (147). *S. floslactata* with less dark suffusion, postmedial line dentate, often with discal spot on hindwing and terminal dots, frons unicolorous brown, collar darker, ♂ hindtibia with pencil and tarsus shortened. *S. cajanderi* with dis-

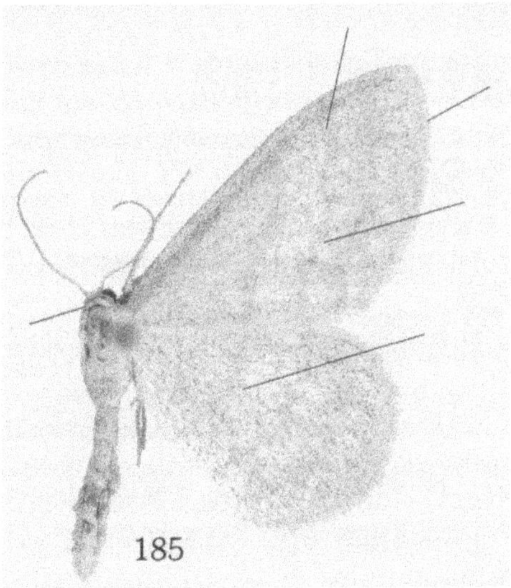

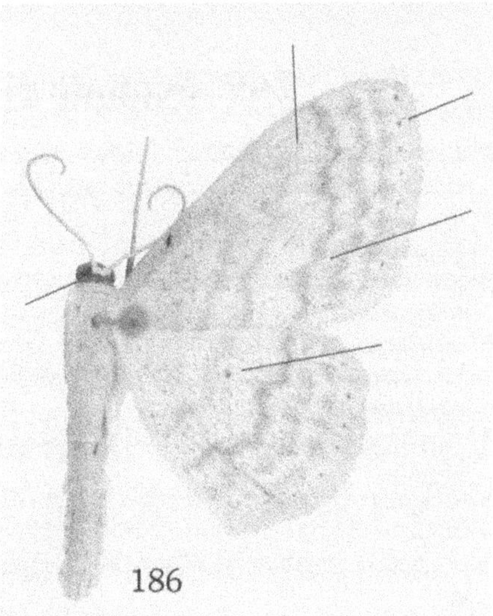

Text-figs 185-186: Diagnostic characters (indicated) of European *Scopula* species. Text-fig. 185: *S. ternata* SCHRANK, 1802. Text-fig. 186: *S. floslactata* (HAWORTH, 1809).

cal spots distinct, postmedial line broader, often dentate, more contrasted with the silvery ground colour. Dark forms of *Cabera exanthemata* (SCOPOLI, 1763) (Ennominae) with whitish frons, ♂ antennae bipectinate.

REMARKS: No constant differences in coloration and pattern of wings between northern and central European populations, no subspecific separation justified. Cerata anomalies led to the description of a separate species (TRÄFF 1965), later synonymized (SVENSON, 1967). Biology discussed in SCHULTZ (1953).

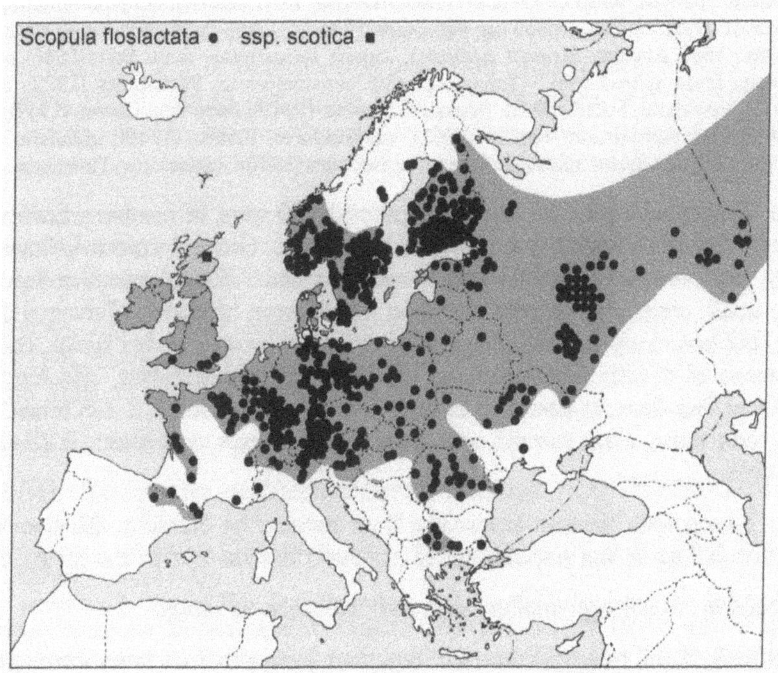

149. *Scopula floslactata* (HAWORTH, 1809)

Phalaena floslactata HAWORTH, 1809: Lepid. Brit. (2): 351 (United Kingdom). Syntype(s). Introduced to replace misidentified *G. remutaria* sensu HÜBNER (pl. 18, fig. 98).

‡? *Phalaena concatenata* HUFNAGEL, 1767: Berlin. Mag. 4 (5): 522 (Germany: Berlin). Syntype(s) lost. Nomen dubium.

‡ *Geometra remutata*: sensu DENIS & SCHIFFERMÜLLER, 1775: Ank. syst. Werkes Schmett. Wienergegend: 117. Misidentification nec LINNAEUS, 1758 (identity: *Idaea aversata*).

‡? *Phalaena centrata* FABRICIUS, 1776: Genera Insect.: 290 (northern Germany; Hamburg). Syntype(s). Synonymy with *remutata*: sensu DENIS & SCHIFFERMÜLLER, 1775 according to FABRICIUS (1794), but better considered a nomen dubium.

‡? *Phalaena Geometra fulvostriata* GOEZE, 1781: Ent. Beytr. 3 (3): 426 (France: Paris). Syntype(s). Nomen dubium.

‡? *Phalaena Geometra brunneata* GOEZE, 1781: Ent. Beytr. 3 (3): 427 (France: Paris). Syntype(s). Nomen dubium.

‡? *Phalaena fulvicans* FOURCROY, 1785: Entom. Parisiensis 2: 274 (France: Paris). Syntype(s). Nomen dubium.

‡? *Phalaena strigata* FOURCROY, 1785: Entom. Parisiensis 2: 283 (France: Paris). Syntype(s). Junior primary homonym nec SCOPOLI, 1763 (identity: *Cabera pusaria*, Ennominae) and nomen dubium.

‡? *Phalaena Geometra dentilinearia* BORKHAUSEN, 1794: Naturg. eur. Schmett. 5: 231 (Europe, probably Germany). Holotype ♂. Nomen dubium.

‡ *Geometra remutaria*: sensu HÜBNER, 1799: Samml. Eur. Schmett. 5, Geometrae (1): pl. 18, fig. 98 (Europe). Misidentification (and incorrect subsequent spelling) nec *remutata* LINNAEUS, 1758 (identity: *Idaea aversata*).

‡? *Phalaena cariata* SCHRANK, 1802: Fauna boica 2 (2): 56 (Germany, Bavaria: Ingolstadt). Syntype(s). Synonymy tentative.

Phalaena lactata HAWORTH, 1809: Lepid. Brit. (2): 351 (United Kingdom). Syntype(s).

Phalaena sublactata HAWORTH, 1809: Lepid. Brit. (2): 351 (United Kingdom). Syntype(s).

Acidalia floslactata claudata PROUT, 1913: in SEITZ, Macrolep. 4: 67 (Japan). Holotype (BMNH). Valid at subspecific rank (SCOBLE 1999).

Scopula lactata scotica COCKAYNE, 1951: Entomologist 84: 154 (Scotland: Morayshire, Forres). Holotype ♂ (BMNH, examined). Valid at subspecific rank (SKINNER 1984).

Unavailable names (misidentifications): *aversata*: sensu SCHWARZ (1793, partim) nec LINNAEUS, 1758 (identity: *Idaea aversata*); ?*fumata*: sensu LEECH (1897) nec STEPHENS, 1831 (identity: *Scopula ternata*); *nemorata*: sensu HAWORTH (1802, nomen nudum) nec *nemoraria* HÜBNER, 1799; ?*pudicaria*: sensu HERZ (1905) nec MOTSCHULSKY, 1866 (identity: *Scopula pudicaria*, Japan); *sylvestraria*: sensu SEPP (1846) nec HÜBNER, 1799 (identity: *Idaea sylvestraria*). – (infrasubspecific): *anastomosaria*: PREISSECKER (1922); *conjunctiva*: PROUT (1913); *exstirpata*: FUCHS (1901); *flavescens*: LEMPKE (1949); *impuncta*: LEMPKE (1949); *nigrolineata*: REISSER (1961); *nigrolineata*: REISSER (1961); *quadripuncta*: LEMPKE (1949); *undularia*: HELLWEGER (1914). – (incorrect subsequent spellings): *remutatia*: NOLCKEN (1870); *sublactatta*: TOMIC et al. (2002).

DIAGNOSIS: Wingspan ♂♀ 22-28 mm, exceptionally 30 mm, in northern Europe smaller, 20-25 mm. Variable in wing pattern and coloration. Ground colour yellowish white. Postmedial line dentate. Terminal area sometimes with 'cloudy' spots or fascia. Discal spots very small, often absent from forewing, rarely from hindwing. Terminal dots usually present, but not rarely lacking. Palpi and frons dark brown, vertex white, collar ochreous. Antennae of ♂ with intersegmental incisions, ciliate-fasciculate, cilia long, 1.5-2.0 times width of flagellum, in northern Europe somewhat shorter, 1.2-1.6 times. Hindtibia of ♂ long, very stout, often curved, with large pencil, tarsus very short, 0.25-0.35 times length of tibia.

Variability: Forms with broader transverse lines induced by humidity (BERGMANN 1955). Wing pattern diffuse in the populations of northern Europe and in the subsp. *claudata*.

Subsp. ***scotica*** on average smaller, with dark suffusion of wings.

MALE GENITALIA: Socii long and narrow. Tegumen between socii sclerotized, prominent, reminiscent of uncus. Fibula long, tapering. Sternum A8 usually with both cerata long, but the left often shortened to length of mappa, e.g. in Austrian populations 45% with shortened left ceras (n=20). Length of left ceras variable in both forms, but transitions discontinuous. For variability of ceras length see also AGENJO (1953).

FEMALE GENITALIA: Lamella antevaginalis weakly sclerotized, semicircular, anteriorly rough. Ductus bursae long, straight. Signum near ductus bursae, very small, of undefined shape, spinules small.

DISTRIBUTION: Eurasiatic. Widely distributed from British Isles to the Urals. Absent from southern Europe except scattered populations in Pyrenees and in the Balkan peninsula. Records from Italy south of the Alps, e.g. MARIANI (1941), erroneous (FLAMIGNI et al. 2001). In Scotland replaced by subsp. *scotica*. Abundance decreasing in Thuringia (ERLACHER & FRIEDRICH 1994). – Outside Europe across Siberia and Mongolia to Korea, Sakhalin (VIIDALEPP 1996) and Japan (subsp. *claudata*; INOUE 1957; SCOBLE 1999).

PHENOLOGY: Univoltine: early or mid-May to late June, exceptionally until mid-July, e.g. on mountains. In the south of the distribution area under good conditions emerging from late April and with partial second generation from late July to late August.

BIOLOGY: Larva polyphagous. Found on *Vaccinium myrtillus* (Ericaceae) (Germany: BERGMANN 1955; WEGNER pers. comm.), *Lonicera xylosteum* (Caprifoliaceae), *Alnus glutinosa* (Betulaceae) (Germany: BERGMANN 1955), *Galium odoratum* (Rubiaceae) (Germany: SCHMID 1885; EBERT 2001), *Rumex acetosa* (Polygonaceae), *Galium verum* (Rubiaceae), *Taraxacum officinale* (Asteraceae) (Finland: SEPPÄNEN 1954). Recorded on *Vaccinium myrtillus* (Ericaceae), *Rumex, Polygonum* (Polygonaceae), *Solidago, Taraxacum officinale* (Asteraceae), *Galium odoratum* (Rubiaceae), *Epilobium* (Onagraceae), *Vicia sepium* (Fabaceae), *Lonicera* (Caprifoliaceae), *Populus, Salix* (Salicaceae), and *Alnus* (Betulaceae) (var. cit., e.g. REBEL 1910b; LHOMME 1935; ALLAN 1949; BERGMANN 1955, SKOU 1986; SVENSSON 1993), reputedly also on dry, fallen leaves (BERGMANN 1955). Reared on *Rosa*, *Alnus*, *Rumex, Polygonum aviculare, Campanula medium* and *Taraxacum officinale* (var. cit., e.g. BERGMANN 1955; SKINNER 1984; PORTER 1997). Adults resting on upper side of leaves (e.g. *Impatiens*, *Prunus padus, P. avium, Atropa belladonna*) during daytime, easily collected by beating from branches of deciduous trees. Main activity at dusk, rarely attracted to light.

HABITAT: Silvicolous, meso- to hygrophilous. Deciduous and mixed forest, fringes, aisles and glades. Preferring acid soils (SKOU 1986). Usually from 0 m up to 800 m above sea-level, exceptionally up to 1,500 m, e.g. in southern Alps.

SIMILAR SPECIES: *Idaea deversaria* (see no. 111), *Scopula nemoraria* (122), *S. nigropunctata* (124), *S. ternata* (148). *Cabera exanthemata* (SCOPOLI, 1763) (Ennominae) with whitish frons, ♂ antennae bipectinate.

150. Scopula subpunctaria (HERRICH-SCHÄFFER, 1847)

Acidalia subpunctaria HERRICH-SCHÄFFER, 1847: Syst. Bearb. Schmett. Eur. 3 (24): 23; 3 (25): pl. 51, figs. 311-313 (Austria). Syntypes ♂♀.

‡ *Phalaena punctata* SCOPOLI, 1763: Ent. Carniolica: 218, fig. 543 (north-western Slovenja: Carniolia). Syntype(s) lost. Junior primary homonym nec CLERCK, 1759 (SCOBLE 1999; identity unknown).

Acidalia punctaria DUPONCHEL, 1830: in GODART & DUPONCHEL: Hist. Nat. Lép. Pap. Fr. 8 (1): 100, pl. 177,

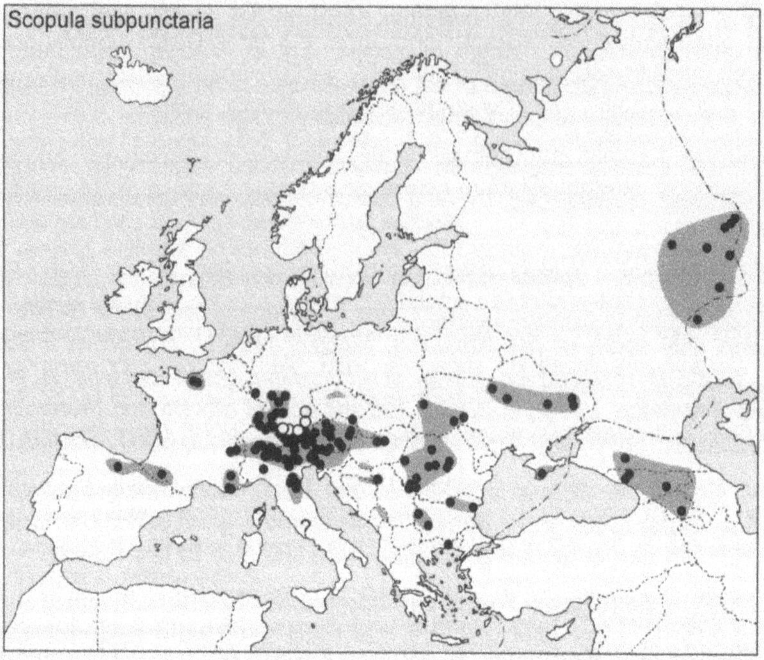

fig. 3. Emendation, compare vol. 7 (2): 102. Senior synonym. Though available, not accepted here as valid name, in order to stabilize usage of the 'traditional' name *subpunctaria*.

Scopula cerusaria LA HARPE, 1850: Bull. Soc. vaud. Sci. Nat. 3 (22): 57 (Austria). No separate types. Unnecessary replacement name for *subpunctaria*.

Acidalia depunctata GUENÉE, 1858: in BOISDUVAL & GUENÉE, Hist. nat. Insectes (Spec. gén. Lépid.) 9: 500 (Austria). No separate types. Unnecessary replacement name for *subpunctaria*.

Unavailable names (misidentification): *nemoraria*: sensu FREYER (1853) nec HÜBNER, 1799. – (infrasubspecific): *anonyma*: GALVAGNI & PREISSECKER (1924); *exstirpata*: PROUT (1913); *isoceras*: PROUT (1935); *nigrolineata*: REISSER (1961); *seminigra*: CZEKELIUS (1935). – (incorrect subsequent spellings): *cerussaria*: SPEYER (1860); *extirpata*: LHOMME (1935).

DIAGNOSIS: Wingspan ♂♀ 23-29 mm, exceptionally 20 mm only in the rare second brood specimens. Ground colour glossy white, very slightly powdered with black scales, mainly towards forewing costa. Transverse lines yellowish grey, of varying intensity, postmedial line slightly undulate, at forewing costa bent towards base of wing. Discal spots and terminal dots usually present on all wings, exceptionally lacking. Palpi brown with white underside. Frons dark brown, vertex and collar white. Antennae of ♂ with intersegmental incisions, ciliate-fasciculate, cilia 1.2-1.5 times width of flagellum. Hindtibia of ♂ long and very stout, with large pencil, tarsus short, 0.2-0.3 times length of tibia.

MALE GENITALIA: Saccus broad and round. Socii very long, terminally curved, flag-shaped. Fibula digitiform, with swollen base. Aedeagus with sclerite at tip. Sternum A8 with both cerata long, symmetrical, or left ceras shortened to half length and strongly curved. On average about 35% with symmetrical cerata (n=63), but with regional differences, e.g. in France and western Alps 25% only (n=20), in northern Austria 40% (n=20) and in northern Serbia 70% (n=10). Dimorphism of sternum A8 first described by PROUT (1935a: 42), see also discussion in HAUSMANN (1999).

FEMALE GENITALIA: Lamella antevaginalis ring-shaped, posteriorly and laterally strongly sclerotized. Antrum dilated, cup-shaped. Ductus bursae short. Corpus bursae large, pyriform, with very large signum, covering about 3/4 of corpus bursae; usually with many spines fused to a medial ridge.

DISTRIBUTION: Palaearctic. Distributed in a narrow belt from northern Spain across southern Alps, Balkan peninsula to southern Urals. North of Alps very local from southern Germany to Slovakia. Reputedly on Corsica (RUNGS 1988; MÜLLER 1996), and in central Italy, Lazio (PROLA & RACHELI 1979), data await confirmation, no vouchers traced. – Outside Europe in easternmost Turkey, Caucasus, Transcaucasus (e.g. WARDIKJAN 1970), and from southern Siberia and Altai to Dahuria (KOSTJUK 1992; VIIDALEPP 1996; KOSTJUK & GOLOVUSHKIN 2003). Recorded also from Far East of Asia and Japan, but for possible confusion with the similar *S. prouti* (DJAKONOV, 1935) see remarks.

PHENOLOGY: Univoltine: mid-June to late July, under bad conditions and on mountains up to two weeks later. Specimens flying in August with very high ♀-ratio, indicating protandry. Under good conditions south of Alps emerging from late May, with a partial second generation, from mid-August to late September. Larvae slowly growing (HABICH 1884).

BIOLOGY: Larva probably polyphagous on herbaceous plants. Found near *Hippocrepis comosa* (Fabaceae) (EBERT 2001). Recorded on *Artemisia campestris*, *A. vulgaris* (Asteraceae), *Hippocrepis comosa, Melilotus officinalis* (Fabaceae) (CHRÉTIEN 1928; LHOMME 1935; FORSTER & WOHLFAHRT 1981; partly referring to rearings of MILLIÈRE 1866). Reared on *Lonicera*, *Taraxacum officinale* and *Lactuca sativa*, accepting fresh, withered and dry leaves (HABICH 1884; GROSS 1887; BERGMANN 1955; FRIEDRICH 1976; MÜLLER, LEIPNITZ pers. comm.). Adults often resting in hedges and shrubs, males readily attracted to light, females preferring bait (DANNEHL 1927). Nectar feeding recorded on *Calamagrostis epigejos* (EBERT 2001).

HABITAT: Xerothermophilous lowland species, usually on low-nutrient, rocky or stony grassland, southern slopes, scrubby places. Often on calcareous ground. Usually from 100 m up to 700 m above sea-level, up to 1,100 m in Alps, in the south exceptionally up to 1,500 m.

SIMILAR SPECIES: *S. immutata* (see no. 146). Often confused with the other two glossy white species *S. caricaria* (121) and *S. nemoraria* (122). *S. subpunctaria* is usually well characterized by the glossy white coloration, fine dark suffusion, black discal spots and terminal dots. The sister species *S. prouti* (DJAKONOV, 1935) differs in round hindwings, longer ♂ hindtarsus, longer ♂ antennal cilia, transverse lines broader and less undulate.

REMARKS: *S. prouti* and *S. subpunctaria* occur sympatrically in Japan and Far East of Asia according to various authors (STERNECK 1941: 94; KOSTJUK 1992; VIIDALEPP 1996; SHIN 1996). ESAKI et al. (1957) list only *S. prouti* for the fauna of Japan, and figure genitalia matching well those of *S. subpunctaria*. Further study is needed to improve knowledge about taxonomy and distribution area. For morphology of immature stages see e.g. HABICH (1884), GROSS (1887) and SCHNEIDER (1924).

151. Scopula flaccidaria (ZELLER, 1852)

Geometra (Acidalia) flaccidaria ZELLER, 1852: Stettin. ent. Ztg. 13 (6): 184 (north-western Turkey: Bursa ('Brussa')). Syntypes (BMNH: 2♂♂, examined).

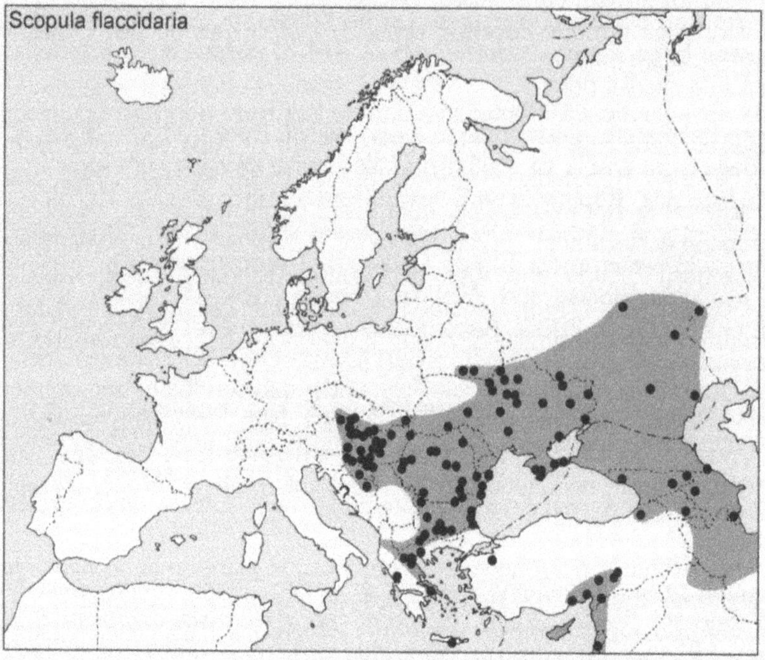

Unavailable names (misidentification): *emutaria*: sensu DOGANLAR (2003) nec HÜBNER, 1809. – (infrasubspecific): *roseata*: AIGNER-ABAFI (1907). – (incorrect subsequent spellings): *flacidaria*: DIDMANIDZE (2001); *rosea*: REBEL (1910).

DIAGNOSIS: Wingspan ♂♀ 20-23 mm. Forewing termen convex, hindwing shortly tailed at M3. Wings pale sand coloured, often with red or light brown tinge. Medial line oblique, dark, broad and diffuse. Postmedial line dentate or dotted. Discal spots black, large, mainly that of hindwing. Terminal streaks fine, elongate. Populations from the locus typicus (Turkey) always very pale, discal spots smaller, pattern more diffuse. Frons dark brown, vertex sand coloured, collar light brown. Antennae of ♂ with deep intersegmental incisions, ciliate-fasciculate, cilia long, 1.7-2.2 times width of flagellum. Hindtibia of ♂ with pencil, tarsus 0.7-0.8 times length of tibia.

MALE GENITALIA: Socii long and narrow. Fibula very long, narrow, digitiform. Sternum A8 with both cerata long, slightly curved, right ceras often strongly curved, with transitions. Not polymorphic (n=17). Sister species *S. albidaria* with shorter fibula, aedeagus broader and curved, both cerata strongly curved.

FEMALE GENITALIA: Lamella antevaginalis ring-shaped, anteriorly granulated. Antrum posteriorly notched, with cap-shaped lobe. Signum broad fusiform or drop-shaped. Sister species *S. albidaria* with antrum cup-shaped, signum very narrow, reduced to a sclerotized ridge.

DISTRIBUTION: (East-)European-Westasiatic. From eastern Austria across Balkan peninsula, southern Byelorussia (KULAK pers. comm.), and Ukraine to southern European Russia. One isolated record from Crete (‚Protoria'; SMNK). Recently recorded for Slovenja (GOMBOC 1996). – Outside Europe in Turkey, Caucasus, Transcaucasus, Cyprus (WILTSHIRE 1948), Israel (HAUSMANN 1997), northern Iraq (WILTSHIRE 1957), northern Iran, and

Turkmeniya (ZSM, dissected; VIIDALEPP et al. 1992). Erroneously recorded from western Libya (TURATI 1930a), possibly a misspelling of '*flaccata*' (=*S. decolor*). Records from central Asian mountains refer to the sister species *S. albidaria* (STAUDINGER 1901). With regard to *S. iranaria* BYTINSKI-SALZ & BRANDT, 1937, see remarks.

PHENOLOGY: Usually bivoltine: mid-May to mid-June; mid-July to late August; specimens from early to late September may belong to a partial third generation. In the Levant regularly trivoltine from April to October.

BIOLOGY: Larva recorded on various herbaceous plants (FORSTER & WOHLFAHRT 1981), such as *Polygonum*, and *Plantago* (REBEL 1910b), data possibly referring to rearings. Reared on leaves of *Taraxacum officinale* (MÜLLER pers. comm.). Adults readily attracted to light.

HABITAT: Xerothermophilous species of dry lowland steppes, preferring continental climates. Recorded also in 'deciduous forests' (ANIKIN et al. 2000). In northern Iraq collected in swampy localities in the mountains (WILTSHIRE 1957). From 0 m up to 300 m above sea-level. In Turkey up to 1,200 m.

SIMILAR SPECIES: *S. corrivalaria* (see no. 120), *S. nigropunctata* (124). *S. emutaria* with ground colour more glossy and rosy, discal spots smaller, hindwing termen angled rather than tailed at M3, medial line broader, closer to postmedial line. Allopatric sister species *S. albidaria* with more whitish ground colour, discal spots and antemedial line more distinct, medial fascia broader, underside of forewing darker.

REMARKS: Taxonomic status of *S. iranaria* BYTINSKI-SALZ & BRANDT, 1937, described from Keredj, Iran, needs further study, some dissected specimens from Transcaucasus and Turkmeniya well matching the genitalia of *S. flaccidaria*.

152. Scopula emutaria (HÜBNER, 1809)

Geometra emutaria HÜBNER, 1809: Samml. Eur. Schmett. 5, Geometrae (1&2): pl. 63, fig. 323 (Germany). Syntype(s) lost.

Phalaena subroseata HAWORTH, 1809: Lepid. Brit. (2): 351 (United Kingdom: Norfolk). Syntype(s). Herewith downgraded from subspecific rank (SCOBLE 1999) to synonymy (syn. n.).

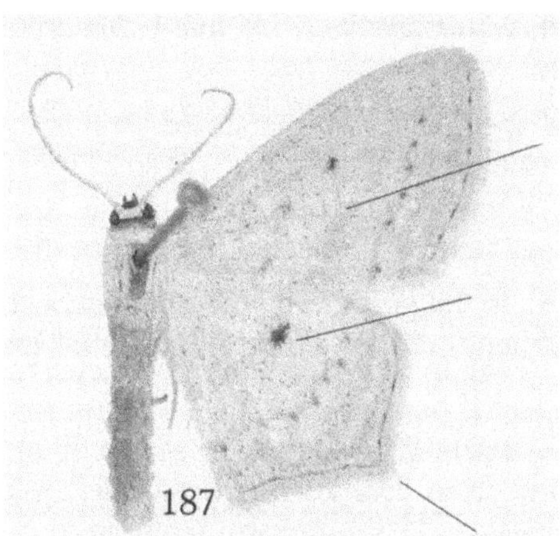

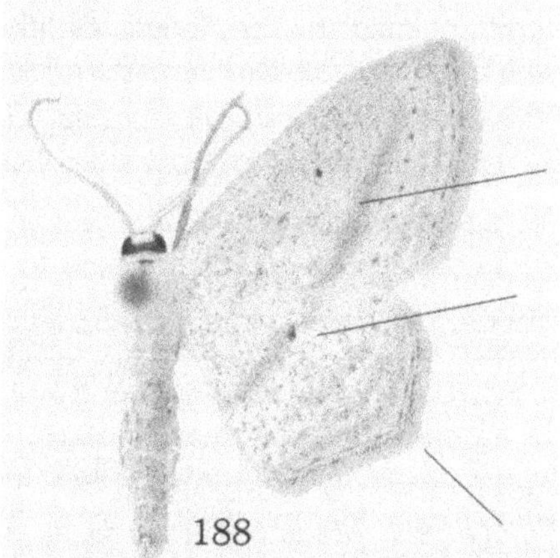

Text-figs 187-188: Diagnostic characters (indicated) of European *Scopula* species. Text-fig. 187: *S. flaccidaria* (ZELLER, 1852). Text-fig. 188: *S. emutaria* (HÜBNER, 1809).

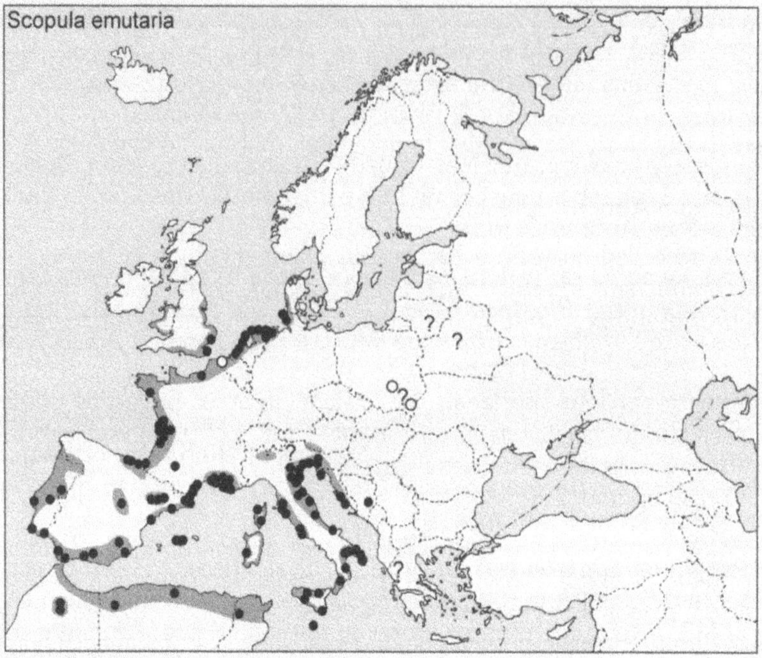

Unavailable names (infrasubspecific): *infrarosea*: AGENJO (1934). – (incorrect subsequent spelling): *emmutaria*: ZECEVIC & RADOVANOVIC (1974).

DIAGNOSIS: Wingspan ♂♀ 21-25 mm, subsequent generations smaller, sometimes 18 mm only. Hindwing termen angled at M3. Wings pale sand coloured, often with rosy or violet tinge, terminal area usually slightly darker. Medial line oblique, dark, broad and diffuse. Postmedial line dotted. Discal spots black, small. Terminal line fine, usually uninterrupted. Frons dark brown, vertex white, collar brown. Antennae of ♂ with deep intersegmental incisions, ciliate-fasciculate, cilia long, 1.7-2.1 times width of flagellum, cilia on short projections. Hindtibia of ♂ stout, with pencil, tarsus 0.8-0.9 times length of tibia.

MALE GENITALIA: Small. Socii long and narrow. Fibula digitiform. Sternum A8 with both cerata of medium size, length slightly varying, with transitions. No true polymorphism with discontinuous feature pattern found (n=5).

FEMALE GENITALIA: Lamella antevaginalis ring-shaped, anteriorly granulated. Antrum dilated, with cap-shaped posterior lobe. Corpus bursae with large oval patch of small spinules.

DISTRIBUTION: (West-)Mediterranean-Atlantic. Widespread, but local along the Mediterranean coasts, from Spain to north-eastern Italy and Dalmatia, present on the islands, local populations near Sarajevo (ZFMK). In the north along the Atlantic coasts of western Europe to Great Britain, northern Germany, and westernmost Denmark (FIBIGER & SVENDSEN 1981; KARSHOLT & STADEL NIELSEN 1998). In Germany actually restricted to the islands Sylt and Amrum (GELBRECHT 1999). Isolated populations in the southernmost Alps (FORSTER & WOHLFAHRT 1981; HUEMER & TARMANN 1993), reputedly also in South Tyrolia near Merano (GRILLO et al. 1996). Absent from Switzerland (REZBANYAI-RESER 1993). Data from eastern Austria (HUEMER & TARMANN 1993) probably referring to *S. flaccidaria*. Reputedly with isolated populations in Lithuania and Byelorussia (IVINSKIS 1993; MERZHEEVSAYA et al. 1976; VIIDALEPP 1996), data await confirmation. Old data from southern Poland, not confirmed by recent records (BUSZKO 2000). Erroneously recorded from Slovakia (HRUBY 1964), central and eastern Balkan peninsula (MÜLLER 1996; NESTOROVA 1998; TOMIC et al. 2002), Crete (REISSER 1974; MÜLLER 1996; specimen in SMNK examined), European Turkey (OKYAR & AKTAÇ 1999), and southern European Russia (ANIKIN et al. 2000). Populations often very isolated, therefore threatened in many regions. – Outside Europe from Morocco to Tunisia. Erroneously recorded from Turkey (RIEMIS 1994; VIIDALEPP 1996; DOGANLAR 2003).

PHENOLOGY: Uni-, bi- or plurivoltine. In southern Europe usually bivoltine: mid-April to late June; early August to late September with single specimens of a partial third generation in October. First generation in the north up to one month later, with or without a partial second generation from late August to early October. In the lowlands of Sicily and North Africa plurivoltine from mid-March to mid-November, but at higher elevations in Sicily and Morocco possibly univoltine, limited data from mid-July to early August.

BIOLOGY: Larva probably oligophagous (2-3) on halophilous plants. Egg deposition recorded on *Hieracium umbellatum* (Asteraceae) (WEGNER 1986). Larva found on *Rumex acetosella* (Polygonaceae) (Germany: WEGNER pers. comm.). Recorded on *Calystegia (Convolvulus) sepium* (Convolvulaceae), *Beta vulgaris maritima* (Chenopodiaceae) (Great Britain: SKINNER 1984; PORTER 1997; SCOBLE 1999), *Armeria (Statice) maritima* and *Limonium vulgare* (Plumbaginaceae) (southern Europe: MILLIÈRE 1873; FORSTER & WOHLFAHRT 1981), the latter suspected as natural host-plant in Great Britain, rearing on *Beta* failed (WARING & TOWNSEND 2003). Further host-plants: *Lonicera*, e.g. *L. periclymenum* (Caprifoliaceae), *Atriplex, Halimione portulacoides* (Chenopodiaceae), *Valerianella* (Valerianaceae), *Polygonum, Rumex* (Polygonaceae) (var. cit., e.g. REBEL 1910b; LHOMME 1935; ALLAN 1949; SKOU 1986). Reared on *Taraxacum officinale, Polygonum aviculare, Trifolium campestre, Medicago lupulina, Lotus corniculatus*, preferring fresh leaves (var. cit., e.g. BUCKLER 1897; SICH 1904; ALLAN 1949; LEIPNITZ pers. comm.), on *Hypochoeris radicata* and *Rumex acetosella* (WEGNER 1986). Main activity of adults at early dusk (WEGNER 1986), frequently found in torchlight, when resting on grass. Also, rather sparingly, attracted to light. Stenotopic (WARNECKE 1936b).

HABITAT: Halophilous. Preferring warm sand dunes, protected from wind (Germany: WEGNER 1986), but recorded also from wet coastal localities, e.g. salt marshes, from inland bogs and open habitats with *Calystegia sepium* (Great Britain: SKINNER 1984; SUTTON & BEAUMONT 1989; Denmark: SKOU 1986; Sicily: BELLA et al. 1999; Mediterranean: PROUT 1935a). Usually from 0 m up to 400 m above sea-level, up to 900 m on Sicily (GRILLO & PARENZAN 1995), up to 1,400 m in Morocco.

SIMILAR SPECIES: *S. flaccidaria* (see no. 151).

REMARKS: Distribution, phenology and biology discussed in FIBIGER (1976). For morphology of immature stages see BUCKLER (1897) and SICH (1904).

153. Scopula minorata (BOISDUVAL, 1833)

Geometra (Idea) (sic!) *minorata* BOISDUVAL, 1833: Nouv. Ann. Mus. Hist. nat. Paris 2: 263 (Faun. Ent. Madag.: 115) (Mauritius). Syntype(s).

Acidalia ochroleucaria HERRICH-SCHÄFFER, 1847: Syst. Bearb. Schmett. Eur. 3 (24): 24 (southern Europe). Syntypes ♂♀. First available name after *ochroleucata* HERRICH-SCHÄFFER, 1844 (pl. 3, figs 19-21: non binom-

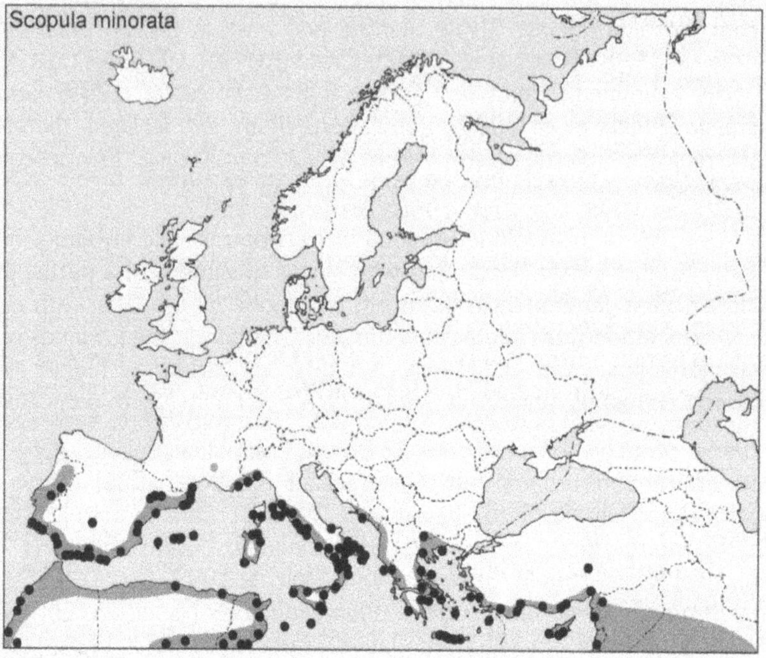

inal). With clinous transitions to nominate subspecies, thus downgraded from subspecific rank (SCOBLE 1999) to synonymy (syn. n.).

Acidalia inustaria HERRICH-SCHÄFFER, 1847: Syst. Bearb. Schmett. Eur. 3 (24): 24 (central Italy). Holotype ♀ (MNHU, examined). First available name after *inustata* HERRICH-SCHÄFFER, 1844 (pl. 3, fig. 15, 16: non binominal). Herewith downgraded from species (e.g. FLAMIGNI et al. 2001; SCOBLE 1999; MÜLLER 1996) to synonymy (syn. n.).

Acidalia colonaria HERRICH-SCHÄFFER, 1852: Syst. Bearb. Schmett. Eur. 6 (55): 67 (locality not stated, [Europe]?). Syntype(s). Referring to vol. 3: 17, thus binominal. First available name after *colonaria* HERRICH-SCHÄFFER, 1851 (pl. 87, fig. 534: non binominal). Herewith downgraded from species (SCOBLE 1999) to synonymy (syn. n.).

Acidalia accessaria HERRICH-SCHÄFFER, 1852: Syst. Bearb. Schmett. Eur. 6 (55): 64 (locality not stated, [Europe], Sicily? (PROUT 1935)). Holotype ♂. Referring to vol. 3: 14, thus binominal. First available name after *accessaria* HERRICH-SCHÄFFER, 1851 (pl. 87, fig. 533: non binominal). Herewith downgraded from species (SCOBLE 1999) to synonymy (syn. n.).

Acidalia mauritiata GUENÉE, 1858: in BOISDUVAL & GUENÉE, Hist. nat. Insectes (Spec. gén. Lépid.) 9: 476 (Mauritius). No separate types. Unnecessary replacement name for *minorata*.

Acidalia luculata GUENÉE, 1858: in BOISDUVAL & GUENÉE, Hist. nat. Insectes (Spec. gén. Lépid.) 9: 472 (Réunion: Bourbon island). Syntypes 1♂1♀ (BMNH).

Acidalia recessaria GUENÉE, 1858: in BOISDUVAL & GUENÉE, Hist. nat. Insectes (Spec. gén. Lépid.) 9: 476 (Europe). No separate types. Unnecessary replacement name for *accessaria*.

Acidalia ochroleucata GUENÉE, 1858: in BOISDUVAL & GUENÉE, Hist. nat. Insectes (Spec. gén. Lépid.) 9: 458. First binominal name in identical spelling after *ochroleucata* HERRICH-SCHÄFFER, 1844 (non binominal) (southern Europe, Cyprus). Syntypes 2♂ ('coll. LEDERER').

Acidalia consentanea WALKER, 1861: List. Spec. lepid. Insects Colln Br. Mus. 22: 745 (South Africa: Cape). Holotype ♂ (BMNH).

? *Acidalia hypochra* MEYRICK, 1888: Proc. Linn. Soc. N.S.W. (2) 2: 848 (Australia: Queensland, Duaringa; New South Wales, Sydney; South Australia, Mt. Lofty). Syntype(s) ♂ (BMNH). Synonymy uncertain.

Cleta ochroleucata var. *inustata* GUMPPENBERG, 1890: Nova Acta Acad. Caesar. Leop. Carol. 54 (5): 532. First binominal name in identical spelling after *inustata* HERRICH-SCHÄFFER, 1844 (non binominal).

Acidalia corcularia REBEL, 1894: in REBEL & ROGENHOFER, Ann. K.K. naturhist. Hofmus. Wien 9 (1): 70 (Canary islands). Syntypes 2♂1♀ (NHMW). Valid at subspecific rank (SCOBLE 1999).

Acidalia holobapharia MABILLE, 1900: Annls Soc. ent. Fr. 68(4): 739 (Madagascar: Diego Suarez). Lectotype ♂ (BMNH).

Emmiltis (Craspedia) mombasae WARREN, 1904: Novit. zool. 11: 467 (Kenya: Mombassa). Syntypes 2♂ (BMNH).

Acidalia tremula BASTELBERGER, 1909: Dt. ent. Zeits. 1909: 319 (Mozambique). Holotype ♀.

Acidalia turbidaria cheimerinaria REBEL, 1928: Verh. zool.-bot. Ges. Wien 78: (30) (Cyprus: Limassol). Syntypes (NHMW: 1♂1♀, examined).

Ptychopoda medioumbraria TURATI, 1930: Boll. Lab. Zool. gen. agr. R. Scuola Agric. Portici 23: 100, 112, pl. A, fig. 5 (Libya, Tripolitania: Sidi Messri). Holotype ♂ (MSNT, examined by RAINERI (1996, gen.fig. 11)). Herewith downgraded from species rank (under *Idaea* in SCOBLE 1999) to synonymy (syn. n.).

Acidalia (Ustocidalia) turbulentaria tripolitana STERNECK, 1933: Zeits. österr. Ent. Ver. 18 (2): 12 (Libya: Tripoli). Holotype ♂ ('coll. KAUTZ').

Scopula ochroleucaria ab. loc. *serrans* PROUT, 1935: in SEITZ, Macrolep. 4, Suppl.: 37, pl. 4, fig. h (Jordan: Ghor el Safieh). Holotype ♀ (BMNH).

Acidalia dresnayi D. LUCAS, 1937: Bull. Soc. Ent. Fr. 42 (16): 232 (Morocco: Rabat). Holotype ♀ (ZSM/HERBULOT, examined).

Unavailable names (misidentification): *elongaria*: sensu PROUT (1903) nec RAMBUR, 1833 (identity: *Idaea elongaria*); *remotata*: sensu OBERTHÜR (1887) nec GUENÉE, 1858 (identity: *Scopula remotata*, India); *turbulentaria*: sensu KAUTZ (1932) and STERNECK (1933) nec STAUDINGER, 1870 (identity: *Scopula turbidaria*). – (infrasubspecific): *anastomosaria*: REISSER (1961); *mortua*: HAUSMANN (1991). – (incorrect subsequent spellings): *acessaria*: LEDERER (1853); *achroleucaria*: PINKER & BACCALADO (1975); *ochroleuca*: STAUDER (1924); *mediomubrata*: RAINERI (1996); *medioumbrata*: RAINERI (1996); *mimorata*: EXPOSITO (1978).

DIAGNOSIS: Wingspan ♂♀ 14-20 mm, exceptionally 12 mm only. One of the smallest *Scopula* species. Wings sand coloured, often with red or light brown tinge. Medial line usually distinct, sometimes dilated. Postmedial line dentate on both wings. Small, black discal spots and terminal dots present. Frons dark or black brown, vertex and collar sand coloured or light brown. Antennae of ♂ with intersegmental incisions, ciliate-fasciculate, cilia long, 1.7-2.1 times width of flagellum. Hindtibia of ♂ stout, with pencil, tarsus 0.7-0.8 times length of tibia.

MALE GENITALIA: Socii of medium length. Fibula narrow, digitiform. Aedeagus with one small needle-shaped cornutus at tip. Sternum A8 with left ceras long, the right short, a very rare pattern in *Scopula*. Some variation in the length of the left ceras, but without true transition to short-armed forms (HAUSMANN 1999). Holotype of *medioumbraria* with medium sized left ceras (RAINERI 1996: fig. 11). 3 % of the dissected specimens (n=118) with left arm nearly as short as the right. In one slide left ceras distinctly furcate. Mappa distinctly notched posteriorly, without exceptions.

FEMALE GENITALIA: Lamella antevaginalis round. Antrum narrow, elongate, with oval sclerite, posteriorly concave. Ductus bursae long and slender, dilated to a flat lateral lobe. Signum oval or fusiform, spinules small and weak, towards ductus bursae with strongly sclerotized 'stalk'.

DISTRIBUTION: Old World (intercontinental distribution). In Europe restricted to the Mediterranean coasts and lowlands from Portugal to Bulgaria. Present on all islands, but absent from the northern Adriatic coast. – Outside Europe from Morocco to the Levant and Turkey, widespread in Arabia and north-eastern Africa, with continuous distribution across tropical Africa, in the south to Madagascar and South Africa (JANSE 1934: 206). Also on Cape Verde islands, Canary islands, and with isolated localities in the Oriental region: A series from eastern Afghanistan and western Pakistan (ZSM) indicates a wider distribution in the east than previously known. One specimen from Sumatra in the ZSM, previously identified as *S. hypochra* by L.B. PROUT, revealed to be conspecific with *S. minorata*.

PHENOLOGY: Plurivoltine. Usually in all moths of the year, but in northernmost populations from April to November. Rearing requires less than two months in summer (HALL pers. comm.).

BIOLOGY: Larva very polyphagous (WILTSHIRE 1990). Found as 'pest' on tobacco in southern Italy (SANNINO & BALBIANI 1984). In a German greenhouse found on *Pachypodium lamerei* (Apocynaceae) (STÜNING pers. comm.). Recorded on *Plumbago* (Plumbaginaceae), *Petunia* (Petuniaceae), *Tamarix indica* (Tamaricaceae), on representatives of the families Acanthaceae, Convolvulaceae, Fabaceae (SCOBLE 1999), and on *Tamarix gallica* (Tamaricaceae) (ROBINSON et al. 2001). Reared on withered leaves of *Taraxacum officinale* and *Galium* (original data; REZBANYAI-RESER 1997). Adults readily attracted to light and bait. r-strategy.

HABITAT: Xerothermophilous lowland species, usually near coast, sometimes very common. From 0 m up to 200 m above sea-level, exceptionally up to 700 m. In Africa and Afghanistan less strongly associated with coastal habitats, in the savannahs up to 1,500 m above sea-level.

SIMILAR SPECIES: *Idaea palaestinensis* (see no. 53), *I. obsoletaria* (54), *I. mancipiata* (82), *I. eugeniata* (100). *S. corrivalaria* (120), *S. turbidaria* (136), *S. turbulentaria* (137).

REMARKS: '*S. inustaria*' was hitherto regarded as a separate species (e.g. PROLA & RACHELI 1979; RAINERI & ZANGHERI 1995; MÜLLER 1996; SCOBLE 1999; FLAMIGNI et al. 2001), but the synonymy was ascertained by examination of types and original material of PROLA & RACHELI (1979) in MZUR. The taxon *ochroleucaria* must be downgraded from subspecies to synonymy, due to the absence of significant and constant differential features from the nominate subspecies in habitus, external morphology and genitalia.
In south-eastern France, Col de la Cayolle, at 2,250 m above sea-level, a male *Scopula* has been collected by J. LENZ (1.VIII.1981), resembling *S. minorata* in habitus, but with stronger dark suffusion. Male genitalia indicate an identity as another (new?) species (gen.fig. 153c), with both cerata short and shallow base of sternum A8, reminiscent of *S. frigidaria*. Antennae similar to those of *S. minorata* but hindtibia without pencil, hindtarsus 1.1 times length of tibia. More material is needed for description.

The *immistaria* species-group

STERNECK's 8th group of *Scopula*. Homogenous group, based on the shape of ♂ sternum A8, including also the phylogenetically isolated *S. superciliata* (PROUT, 1913) from Japan. Species-group well characterized also by ♀ genitalia (see below) and venation: R5 of forewing arising connate with R2-R4 from apex of areole. Palpi bushy, 1.0-1.3 times diameter of eye in all species. In habitus and external morphology, four synapomorphies link all species of this group (also *S. superciliata*), except *S. cajanderi*: yellowish coloration, ♂ antennae with strongly curved cilia. Antennae of ♀ densely ciliate-setose, longest cilia 1.0 times width of flagellum. Proboscis long, 4-5 mm, a convergence with the long proboscis of *Glossotrophia* s.str., which also suck nectar from Caryophyllaceae. Central sclerite on ♂ sternum A2 with patch of comparatively short setae. Male genitalia: Base of sternum A8 concave, mappa rounded, cerata vestigial in four of six species. Not polymorphic in length of cerata. Female genitalia: Signum absent. Relationships between the species *S. latelineata*, *S. arenosaria* and *S. albiceraria* very close, more detailed study is needed to improve knowledge of taxonomy and distribution areas.

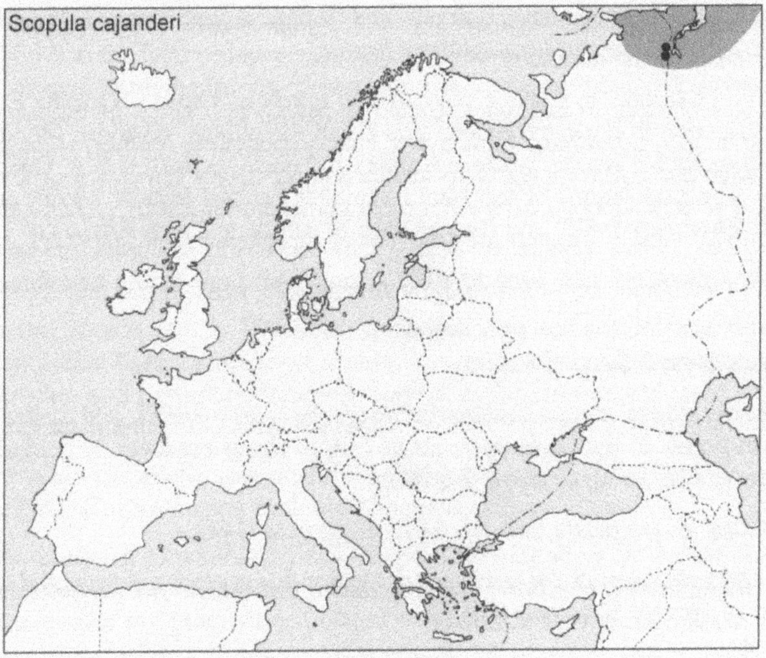

154. Scopula cajanderi (HERZ, 1903)

Acidalia cajanderi HERZ, 1903: Ofvers. finska Vetensk.-Soc. Förh. 45 (15): 13, fig. (central Russia, Yakutia, Lena-region: Vilui-Delta). Lectotype ♂ (ZMH, figured in SIHVONEN 2001).
Acidalia anaitaria HERZ, 1903: Ofvers. finska Vetensk.-Soc. Förh. 45 (15): 14, fig. (central Russia, Yakutia, Lena-region: Vilui-Delta). Lectotype ♀ (ZMH, figured in SIHVONEN 2001; ♀ paralectotype in ZISP, examined).
Scopula achlyoides PROUT, 1935: in SEITZ, Macrolep. 4, Suppl.: 35 (north-western Mongolia: Tannuola: Schawyr). Holotype ♂ (BMNH, figured in SIHVONEN 2001). Synonymy according to SIHVONEN 2001 (in SCOBLE 1999 subspecies of *S. elwesi*).
Scopula elwesi sajanensis PROUT, 1935: in SEITZ, Macrolep. 4, Suppl.: 35, pl. 4, fig. h (central Russia: Sajan Mts., Munko Sardyk). Holotype ♂ (BMNH), SIHVONEN (2001) designates a further 'Lectotype'. Synonymy according to SIHVONEN 2001 (in SCOBLE 1999 subspecies of *S. elwesi*).
Scopula septentrionicola MCDUNNOUGH, 1939: The Canad. Entom. 71: 189 (Canada, Yukon Terr.: Nation Riv). Holotype ♂ (CNCI, figured in SIHVONEN 2001). Valid at subspecific rank.
Unavailable names (incorrect subsequent spelling): *achlyodes*: VIIDALEPP (1996).

DIAGNOSIS: Wingspan ♂ 22-27, ♀ 19-23 mm, in the Urals ♂ 22-23, ♀ 19-20 mm. Sexually dimorphic: Females with forewing apex more pointed, ground colour much lighter and pattern much more contrasted than in males. Ground colour grey, silvery shining, with strong dark suffusion in ♂. All transverse lines and discal spots present. Postmedial line undulate or dentate. Elongate terminal streaks present, distinct in ♀. Palpi bushy. Frons, vertex and collar light brown mixed with a varying number of blackish scales. Antennae of ♂ with intersegmental incisions, ciliate-fasciculate, cilia 1.5-1.65 times width of flagellum, in one examined Nearctic specimen 1.8 times. Hindtibia of ♂ comparatively slender, with two spurs, tarsus longer than tibia.

MALE GENITALIA: Socii long and narrow. Fibula long and tapering. Valva long, sclerotization comparatively strong. Aedeagus large, with elongate sclerite. Sternum A8 concave at base, mappa narrower than base, both cerata short, without spinules, weakly sclerotized. Not polymorphic (n=6).

FEMALE GENITALIA: Lamella antevaginalis and sterigma with two U-shaped sclerites. Corpus bursae oval, membranous, without signum.

DISTRIBUTION: Holarctic. In Europe restricted to northern Urals. – Outside Europe with disjunct distribution in northern Siberia and Sajan mountains, northern Mongolia ('*sajanensis*'; '*achlyoides*'), Alaska, western Canada and north-western U.S.A. (subsp. *septentrionicola*). The populations of the Altai mountains ('*elwesi* PROUT, 1922') are possibly conspecific (SIHVONEN 2001) and may be valid at subspecific rank (VIIDALEPP 1996).

PHENOLOGY: Univoltine: late June to early August with peak in the first three weeks of July.

BIOLOGY: Larval host-plants unknown.

HABITAT: Rocky slopes, steppe habitats on slopes (SIHVONEN 2001), and tundra (JALAVA & KULLBERG in press). In the Urals up to above 500 m above sea-level, in northern Asia up to 2,500 m, in U.S.A. up to about 3,300 m.

SIMILAR SPECIES: *S. frigidaria* (see no. 147), *S. ternata* (148).

REMARKS: Discrimination of different subspecies for this species not accepted in the comprehensive revision of SIHVONEN (2001).

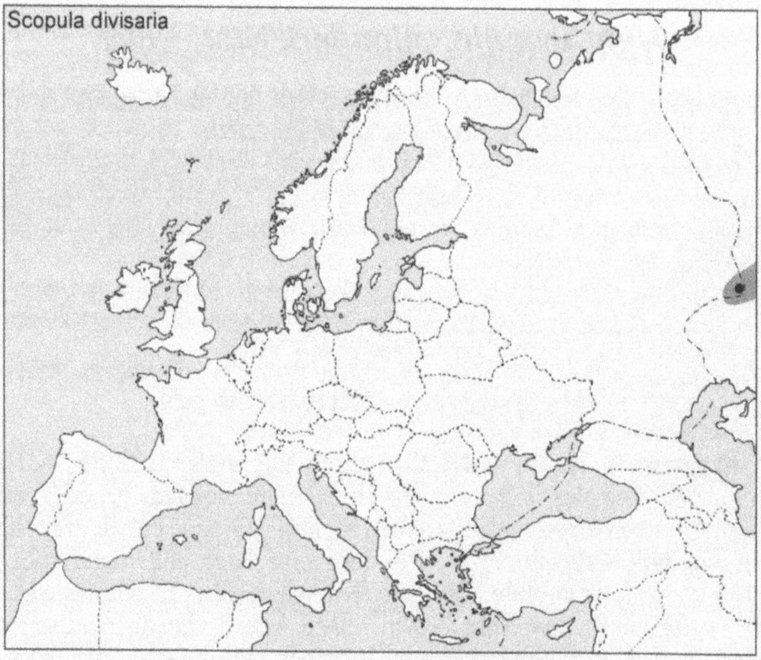

155. *Scopula divisaria* (CHRISTOPH, 1893), stat. n.

Acidalia divisaria CHRISTOPH, 1893: Dt. ent. Z. Iris 6: 94 (Russia, southern Urals: Guberli). Syntypes ♂♀ (ZISP, examined). Latin description 'fascia postica distincta, incrassata, sinuata' unambiguous. Herewith raised from synonymy of *Scopula latelineata* (PROUT 1934) to species rank (stat. n.).

Unavailable names (misidentification): *latelineata*: sensu STAUDINGER (1901, partim) nec GRAESER, 1892 (identity: *S. latelineata*, central Asia).

DIAGNOSIS: Wingspan ♂♀ 22-26 mm. Differing from the sister species *S. latelineata* (GRAESER, 1892) in the light yellow ground colour, hindwings whitish with usually only the medial line present. Antemedial line present on the forewing. Medial line darker, sharper, sigmoid waved. Terminal area usually with indistinct pattern. Terminal dots black, distinct. Discal spots distinct, at least on forewing. Palpi comparatively long, 1.2-1.4 times diameter of eye. Frons flat. Collar slightly darker than vertex. Antennae of ♂ with very small intersegmental incisions, ciliate-fasciculate, cilia strongly curved, 1.4-1.8 times width of flagellum. Hindtibia of ♂ stout, with pencil, tarsus long, 1.0-1.1 times length of tibia.

MALE GENITALIA: Saccus narrow, V-shaped. Aedeagus long and slender. Sternum A8 very shallowly concave at base, cerata short, round. In populations from Kazakhstan, the cerata more elongate and posteriorly tapering. Type of *S. latelineata* with saccus round and broad, base of sternum A8 distinctly concave.

FEMALE GENITALIA: Lamella antevaginalis slender, semicircular. Sclerotization of antrum slender, elongate. Corpus bursae oval, membranous, without signum.

DISTRIBUTION: Endemic to southern Urals, or possibly extending to eastern Kazakhstan (VIIDALEPP et al. 1992 as '*latelineata*'), see remarks. Sister species *S. latelineata* from Alexander mountains to Turkmeniya. Isolated populations, rare in collections.

PHENOLOGY: Univoltine: late May to late June. Doubtfully conspecific populations from eastern Kazakhstan from early May to late May.

BIOLOGY: Larval host-plants unknown. Adults readily attracted to light.

HABITAT: Restricted to lowland steppes and hills up to about 300 m above sea-level. In eastern Kazakhstan up to 1,300 m, but conspecifity doubtful.

SIMILAR SPECIES: Certain forms of *S. arenosaria* similar in habitus, but on average smaller, hindwing less marked, medial line less sigmoid and less distinct, discal spots and terminal dots small or absent, ♂ genitalia different (e.g. cerata and aedeagus shorter). Type specimen of *S. latelineata* from Alexander mountains with ground colour straw yellow, colour of hindwing similar to that of forewing, antemedial line absent, medial line straight, brown postmedial and wavy lines sharp and distinct on all wings, with additional row of sharp spots close to termen; terminal line uninterrupted, without black dots; discal spots extremely small. Frons slightly convex in the type specimen, brown, with white scales towards proboscis, vertex and collar sand coloured. Antennae of ♂ very shallowly dentate. Hindtarsus of ♂ 0.9 times length of tibia.

REMARKS: Doubtfully conspecific populations from mountains and lowlands of eastern Kazakhstan (ZSM) with straighter and more diffuse medial fascia, forewing antemedial line absent and minor differences in ♂ genitalia. Further study is required, which may lead to the description of 1-2 further species (see colour plates and figured genitalia).

156. *Scopula arenosaria* (STAUDINGER, 1879)

Acidalia arenosaria STAUDINGER, 1879: Stettin. ent. Ztg. 40 (7-9): 325 (southern European Russia: Astrachan between Ural and Volga river). Lectotype ♂ (MNHU, examined), herewith designated to stabilize usage of the name; paralectotypes from type locality: MNHU 2♂3♀, MNHU/PÜNGELER 1♂. 'Narün desert' (Kirghizstan) erroneously stated as type locality in the description, compare introduction to STAUDINGER (1879).

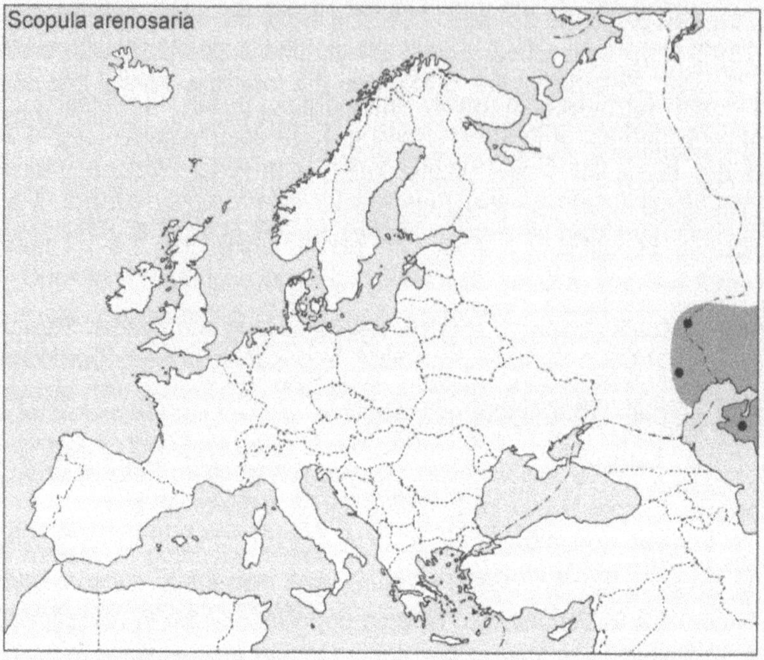

DIAGNOSIS: Wingspan ♂♀ 20-23 mm, in the second generation sometimes 17 mm only. Forewing yellow, hindwing whitish. European populations with medial line of forewing fine or, rarely, broadened. Ante- and postmedial lines absent. Hindwing without or with weak pattern. Populations from Kazakhstan with dark wavy line usually present on all wings, hindwing with indistinct medial line. Small forewing discal spots present. Terminal dots absent in European populations, in Kazakhstan very small. Frons light brown, vertex and collar yellowish. Antennae of ♂ with intersegmental incisions, ciliate-fasciculate, cilia strongly curved, 1.6-2.0 times width of flagellum. Hindtibia of ♂ stout, with pencil, tarsus 1.0 times length of tibia.

MALE GENITALIA: Saccus broader than in *S. divisaria*. Aedeagus shorter and broader. Base of sternum A8 straight or very shallowly concave. Cerata short, round, much smaller than in *S. divisaria*.

FEMALE GENITALIA: Lamella antevaginalis slender with U-shaped sclerite. Antrum slender, elongate. Corpus bursae oval, membranous, without signum.

DISTRIBUTION: Endemic to steppes of south-eastern European Russia and Kazakhstan. In the east probably to the Ili valley and the foothills of the Tian Shan (eastern Kazakhstan). Erroneously recorded from Mongolia (VOJNITS 1976), and Uzbekistan (VIIDALEPP 1996), the latter referring to the erroneous type locality (see nomenclatural paragraph). Populations isolated, rare in collections, possibly threatened.

PHENOLOGY: Probably bivoltine: early May to late June; mid-July to late July (limited data available).

BIOLOGY: Larval host-plants unknown.

HABITAT: Xerothermophilous lowland species, inhabiting dry steppes. From 0 m up to 500 m above sea-level.

SIMILAR SPECIES: *S. divisaria* (see no. 155). *S. immistaria* broader winged, coloration paler sulphury yellow, hindwing concolorous with forewing, pattern sharper, postmedial line dentate, with black markings on the veins, discal spots present on all wings. *S. albiceraria* with pattern much sharper, postmedial line double on all wings, close to termen, antemedial line present.

REMARKS: Further study is needed to clarify the status of the central Asian populations.

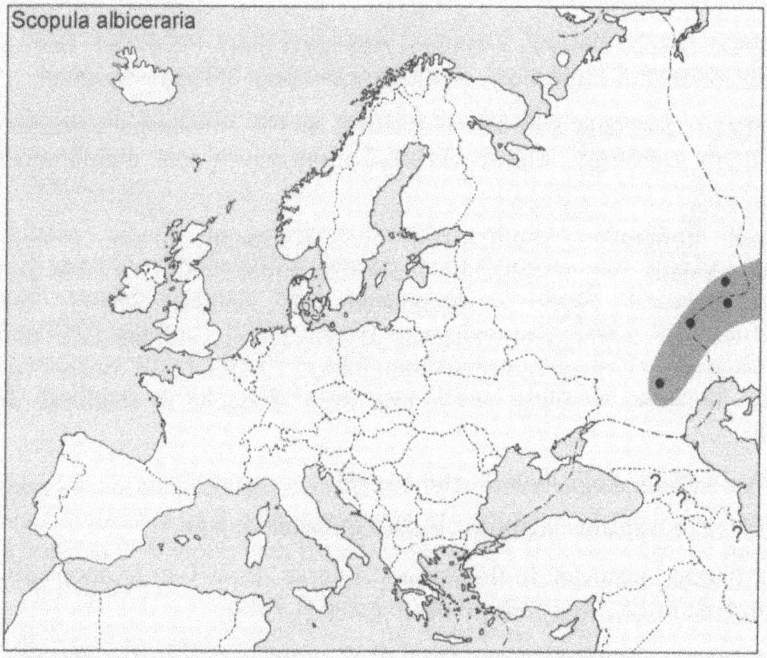

157. *Scopula albiceraria* (HERRICH-SCHÄFFER, 1847)

Acidalia albiceraria HERRICH-SCHÄFFER, 1847: Syst. Bearb. Schmett. Eur. 3 (24): 23. First available name after *albiceraria* HERRICH-SCHÄFFER, 1844 (pl. 3, figs 17, 18: non binominal) (south-eastern European Russia: Krasnoarmeysk ('Sarepta'); erroneous locality 'southern Europe' in original description). Syntypes 2♂.

Acidalia sulphuraria FREYER, 1847 (October): Neuere Beitr. Schmettkde. 6 (86): 59, pl. 516, fig. 6 (south-eastern European Russia: Krasnoarmeysk ('Sarepta')). Syntypes (MNHU: 4 syntypes examined). The MNHU specimens (Sarepta, leg. LEDERER) are regarded as syntypes, based on FREYER's statement, that LEDERER provided the material; similar information is given by HERRICH-SCHÄFFER 1852).

Acidalia vitellinaria EVERSMANN, 1851: Bull. Soc. imp. Moscou 24 (2): 641 (Russia, eastern Siberia: Irkutsk). Lectotype ♂ (ZISP, examined), herewith designated to stabilize usage of the name; at ZISP 1♀ paralectotype from type locality, 1♂ from Mongolia and 1♂ from Kirghizstan, the latter probably belonging to the nominate subspecies. Valid at subspecific rank (see remarks).

Unavailable names (misidentifications): *arenosaria*: sensu VOJNITS (1976) nec STAUDINGER, 1879; *ochroleucaria*: sensu HERRICH-SCHÄFFER (1851, pl. 76, figs 469, 470) nec HERRICH-SCHÄFFER, 1847. – (infrasubspecific): *infuscata*: PROUT (1913). – (introduced in synonymy): *mannerheimiata*: ERSCHOV (1871). – (incorrect subsequent spellings): *albicearia*: HERZ (1904); *mannerheimatia*: SCOBLE (1999); *mannerheimaria*: PROUT (1934).

DIAGNOSIS: Wingspan of nominate subspecies ♂♀ 22-24 mm. Wings pale sand coloured. All transverse lines present, brown, fine, straight, parallel to each other. Medial area between medial and antemedial line often with shaded brown. Hindwing paler, with medial and antemedial lines absent or diffuse. Discal spots very small. Terminal dots small in

European populations. Palpi pale sand coloured. Frons brown, vertex sand coloured, collar sometimes slightly darker. Antennae of ♂ with intersegmental incisions, ciliate-fasciculate, cilia strongly curved, 1.4-1.8 times width of flagellum. Hindtibia of ♂ stout, with pencil, tarsus 0.9-1.0 times length of tibia.

Subsp. ***vitellinaria*** larger, up to 27 mm, more sharply marked, transverse lines broader, postmedial line slightly dentate, discal spots distinct, black, terminal dots distinct and elongate.

MALE GENITALIA: Saccus narrow, V-shaped. Aedeagus short and broad. Base of sternum A8 distinctly concave. Cerata short, tapering posteriorly. Mappa very small.

FEMALE GENITALIA: Sterigma with lateral sclerites, lamella antevaginalis slender, crescent-shaped. Antrum sclerotized, slightly dilated. Corpus bursae oval, membranous, without signum.

DISTRIBUTION: Mongolian. South-easternmost European Russia, southern Urals. Populations isolated, rare in collections, possibly threatened. – Outside Europe from Turkmeniya (VIIDALEPP 1996) to Kazakhstan, and southern Siberia. Reputedly in Caucasus (VIIDALEPP 1996), Transcaucasus (PROUT 1913; Georgia: DIDMANIDZE 2002; Armenia: VIIDALEPP 1974) and northern Iran (PROUT 1913), but no vouchers traced from these areas. Replaced by subsp. *vitellinaria* from Mongolia to southern Yakutia and Amur basin.

PHENOLOGY: Bivoltine: early June to late August.

BIOLOGY: Larval host-plants unknown. Adults attracted to light.

HABITAT: In Europe restricted to dry lowland steppes, from 0 m to about 300 m above sea-level, in Asia usually from 500 m up to about 1,700 m.

SIMILAR SPECIES: *S. arenosaria* (see no. 156). *S. immistaria* broader winged, coloration paler sulphury yellow, postmedial line dentate, not double, usually with black markings on the veins, antemedial line weak or absent, discal spots present on all wings, males can easily be identified by brushing tip of abdomen to remove scales and reveal length of cerata (medium sized).

REMARKS: Taxon *vitellinaria* may be valid at specific rank. Further attention and study is needed to resolve this question.

158. Scopula immistaria (HERRICH-SCHÄFFER, 1852)

Acidalia immistaria HERRICH-SCHÄFFER, 1852: Syst. Bearb. Schmett. Eur. 6 (55): 68. First available name after *immistaria* HERRICH-SCHÄFFER, 1851 (pl. 76, figs 467, 468: non binominal) (Azerbaijan: Kirowabad ('Elisabethpol')). Syntypes, at least 2 (MNHU: 2♂, examined). Referring to vol. 3: 23, thus binominal.
Scopula immistaria beshkovi GELBRECHT & HAUSMANN, 1997: Linzer biol. Beitr. 29 (2): 984, figs 1, 12, 15, 16 (Bulgaria: Pobitide Kamani near Varna). Holotype ♂ (ZSM, examined). Valid at subspecific rank.
Unavailable names (nomen nudum): *elisabetharia*: HEYDENREICH (1851).

DIAGNOSIS: Wingspan ♂♀ 24-27 mm, subsequent generations smaller, sometimes 22 mm only. Ground colour light yellow. Postmedial line dentate. Terminal area usually with indistinct pattern. Discal spots sharp, black. Terminal dots distinct, in central and southern Turkey, however, weak or absent. Frons dark brown, vertex and collar light yellow.

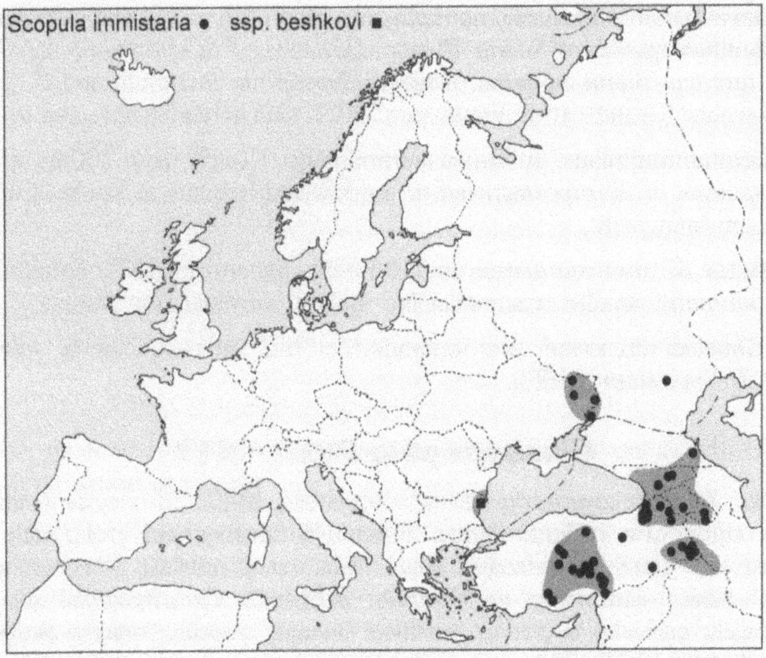

Length of palpi = diameter of eye, in southern and eastern populations of nominate subspecies longer. Antennae of ♂ with intersegmental incisions, cilia 1.5-2.0 times width of flagellum. Hindtibia of ♂ stout, with pencil, tarsus 0.9-1.0 times length of tibia.

Variability: With strong tendency to regional wing patterns, e.g. postmedial line and terminal dots darker in the populations from Transcaucasus. Terminal dots and black markings on the veins weak or completely lacking in the south of the distribution area.

Subsp. ***beshkovi*** smaller, wingspan ♂♀ 18-23 mm. Ground colour whitish, almost without yellow tinge. Postmedial line strongly dentate. Terminal area with two additional straight, dark fasciae. Frons black with some white scales towards proboscis.

MALE GENITALIA: Sternum A8 deeply notched at base. Cerata long, in subsp. *beshkovi* slightly shorter than in the nominate subspecies. Mappa posteriorly slightly concave in subsp. *beshkovi*, round in the nominate subspecies.

FEMALE GENITALIA: Lamella antevaginalis narrow semicircular. Antrum elongate. Ductus bursae dilating towards corpus bursae. Corpus bursae elongate, membranous, without signum. No difference between subspecies.

DISTRIBUTION: (Irano-)Pontic. Nominate subspecies in southern European Russia (1♂, Sarepta, MNHU) and eastern Ukraine (KOSTJUK pers. comm.). In north-eastern Bulgaria replaced by subsp. *beshkovi*. – Outside Europe in Turkey, from Caucasus across Transcaucasus to northern Iraq, northern Iran and Turkmeniya. Isolated occurrence in south-eastern Iran (one ♂ examined from Kerman). Sister species *S. lehmanni* HAUSMANN, 1991 endemic to northern Jordan and western Syria (NHMW).

PHENOLOGY: Bivoltine: mid-May to early September, generations usually overlapping in July. On high mountains univoltine, from late June to early August.

BIOLOGY: Larva in nature probably monophagous (3) on *Silene* (Caryophyllaceae). Found on an unidentified species of *Silene* (Bulgaria: GELBRECHT & HAUSMANN 1997). Reared on *Silene tatarica*, *Silene vulgaris*, petals of *Gypsophila fastigiata* and *G. paniculata* (Caryophyllaceae) (GELBRECHT & HAUSMANN 1997). Adults readily attracted to light.

HABITAT: Xerothermophilous, in Turkey xeromontane. Usually from 700 m up to 2,800 m above sea-level, in Europe restricted to hot lowland habitats at sea-level with sparse vegetation, on sandy soils.

SIMILAR SPECIES: *S. arenosaria* (see no. 156), *S. albiceraria* (157). Extralimital sister species *S. lehmanni* smaller, postmedial line straight, terminal dots absent.

REMARKS: Detailed discussion and separation of the sister species *S. lehmanni* in GELBRECHT & HAUSMANN (1997).

Glossotrophia PROUT, 1913

TYPE SPECIES: *Acidalia confinaria* HERRICH-SCHÄFFER, 1847. 20 species in the western Palaearctic region, one in South Africa. Structurally homogenous group with regard to most features in ♂ genitalia (aedeagus, valva, fibula, uncus, anellus). Morphological analysis reveals close relationship to *Scopula* (subgenus *Calothysanis*) and *Stigma*. Phylogenetically probably a young, modified lineage, possibly making *Scopula* paraphyletic (SIHVONEN 2003). Without large scale revision of all Scopulini genera including molecular data and more tropical representatives, the generic concept should, however, not be changed. Important synapomorphies of *Glossotrophia*: female spur number and shape of fibula in ♂ genitalia. Single ♂ hindspur and serial arrangement of scales in the distal half of wing as very rare or unique features in Geometridae occurring in many species of all five subgenera, supporting monophyly of *Glossotrophia*. Revisions: HAUSMANN (1993b, 1993c, 1994).

DIAGNOSIS: Terminal line uninterrupted, surrounding forewing apex, phylogenetically independent from the very similar, homologous feature in the *ornata* species-group of *Scopula* (STERNECK 1941: 50). Venation (Text-fig. 190) reminiscent of that of most *Scopula*: With one areole, not exceeding cell apex. R1 connate with R2-R5. R5 on rather short stalk with R2-R4. In the hindwing Rs and M1 always separate, M3 and CuA1 separate, too. Proboscis long, very long in the nominotypical subgenus. Palpi 1.0-1.3 times diameter of eye, bushy. Antennae of ♂ ciliate, with long cilia. Male hindtibia usually with one spur. Female hindtibia always with two spurs, but the South African species reputedly without spurs (JANSE 1934: 130; 217). Tarsus long. Central sclerite on ♂ sternum A2 (Text-fig. 4), smaller than in *Scopula*, with double arched posterior sclerotization, setae very short, fused into one single patch; exception: subgenus *Microglossotrophia*, without sclerite on ♂ sternum A2.

MALE GENITALIA: Fibula strongly sclerotized, short and rounded. Inner ring of anellus modified to a conical tubus, similar modifications only in *Stigma* and a very few *Scopula* species. Two subgenera with cerata very short or absent. Polymorphic cerata length only in the nominotypical subgenus (HAUSMANN 1999).

IMMATURE STAGES: Pupa of subgenus *Glossotrophia* with long, looped, external proboscis (Text-fig. 189; PATOČKA 2003). The other subgenera with proboscis closely attached to the pupa. Terminal pair of cremaster setae only slightly larger than the other setae, unusual for Scopulini (PATOČKA 2003).

Text-fig. 189: Pupa of *Glossotrophia confinaria* (HERRICH-SCHÄFFER, 1847) (Photo: LEIPNITZ).

Subgenus ***Glossotrophia*** PROUT, 1913

TYPE SPECIES: *Acidalia confinaria* HERRICH-SCHÄFFER, 1847. Subgenus with nine species in the western Palaearctic region.

DIAGNOSIS: Antennae of ♂ ciliate-fasciculate, cilia arising from crests on the antennomeres. Proboscis hypertrophic, length 10-14 mm. Hindtibia of ♂ with one spur; exception: *G. diffinaria*. Spur anomalies are, however, found in all species (see below). Legs very long. Arrangement of scales usually normal in the distal half of wings, but serial in *G. rufomixtaria* and *G. chalcographata* BRANDT, 1938. Serial arrangement in *G. chalcographata* always, in *G. rufomixtaria* sometimes restricted to forewing. Further features under generic diagnosis.

MALE GENITALIA: Socii long and slender. Fibula broad and short, slightly tapering at tip. Aedeagus long and slender. Base of sternum A8 with distinct projection (synapomorphic with subgenera *Libanonia* and *Bicalcarella*). Polymorphic cerata length in all European species and subspecies, except *G. rufomixtaria*.

FEMALE GENITALIA: Antrum posteriorly dilated, covered with a large sclerite. Lamella antevaginalis strongly sclerotized, corrugated. Signum with comparatively large spines.

BIOLOGY: Larval host-plants almost exclusively Caryophyllaceae. Adults sucking nectar from the same plants (PROUT 1913), often bearing *Silene* pollen grains on the proboscis (HAUSMANN & DÖTTERL 2004).

IMMATURE STAGES: Pupa with long, looped, external proboscis (Text-fig. 189, see also Text-fig. 113, Vol. 1).

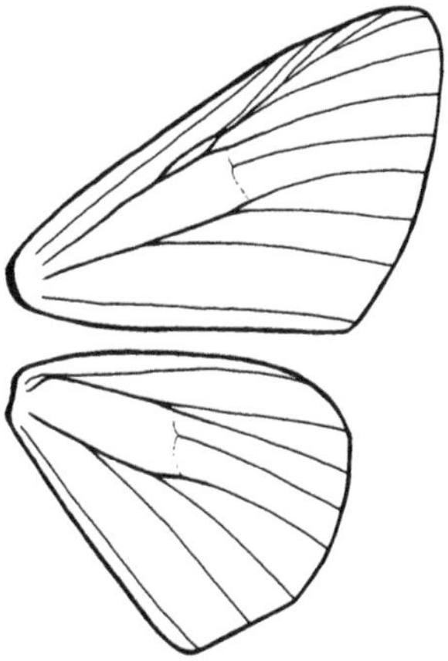

Text-fig. 190:
Venation of *Glossotrophia confinaria* (HERRICH-SCHÄFFER, 1847).

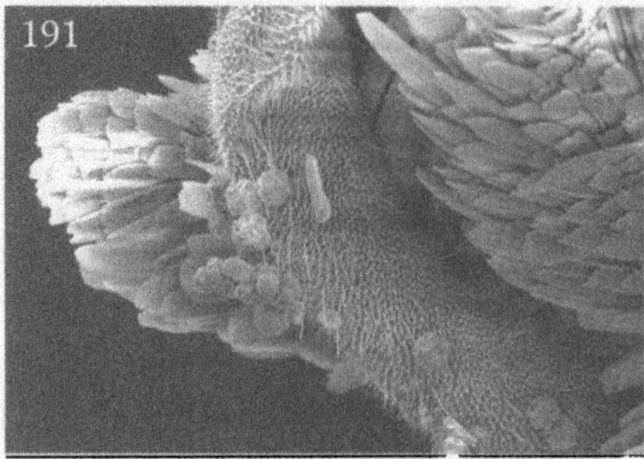

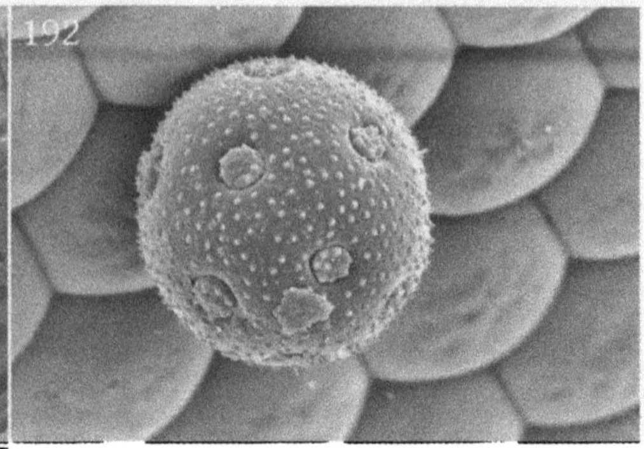
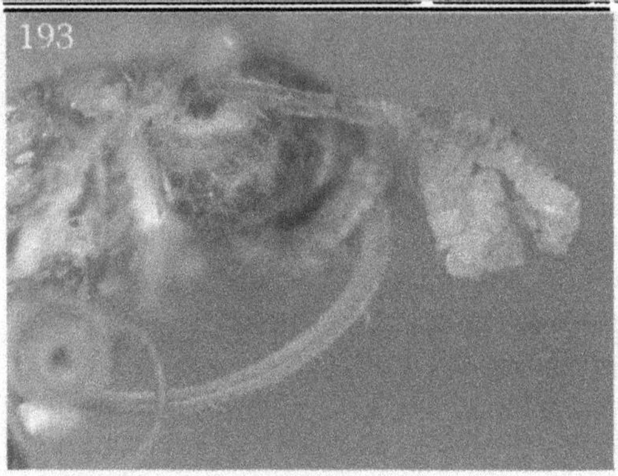

Text-figs 191-193: Pollen grains and pollinium attached to collection specimens of *Glossotrophia confinaria* (HERRICH-SCHÄFFER, 1847) (after HAUSMANN & DÖTTERL 2004). Text-fig. 191 (SEM): pollen grains of the *Silene dioica*-group (Caryophyllaceae), attached to the long proboscis. Text-fig. 192 (SEM): pollen grain of the *Silene dioica*-group (Caryophyllaceae), attached to the compound eye. Text-fig. 193: Pollinium of *Platanthera* (Orchidaceae), attached to the head.

159. Glossotrophia confinaria (HERRICH-SCHÄFFER, 1847)

Acidalia confinaria HERRICH-SCHÄFFER, 1847: Syst. Bearb. Schmett. Eur. 3 (24): 21 (pl. 51, figs 315-317: non binominal) (Hungary). Syntypes ♂♀ (BMNH: 1♂ examined, labelled as 'paratype').

Acidalia corrivularia MILLIÈRE, 1869: Iconogr. Descr. Chen. Lép. inéd. 3 (24): 54 (Europe). Syntypes ♂♀, probably lost.

Glossotrophia confinaria aetnaea PROUT, 1935: in SEITZ, Macrolep. 4, Suppl.: 48 (Sicily: Etna Mt.). Syntypes (BMNH, 1♂ examined, with red 'type' label). Valid at subspecific rank.

Glossotrophia confinaria scoblei HAUSMANN, 1993: Atalanta 24 (1/2): 277, figs 6, 22, 25 (Sicily: Le Madonie). Holotype ♂ (ZSM, examined). Valid at subspecific rank for populations of north-western Sicily.

Glossotrophia confinaria prouti HAUSMANN, 1993: Atalanta 24 (1/2): 270, figs 4, 20, 24 (Italy, South Tyrolia: Terlan). Holotype ♂ (ZSM, examined). Valid at subspecific rank.

Unavailable names (misidentification): *coenosaria*: sensu PROUT (1913) nec LEDERER, 1855 (identity: *Scopula luridata*); *falsaria*: sensu GUENÉE (1858, partim) nec HERRICH-SCHÄFFER, 1852 (identity: *G. rufomixtaria*); *infirmaria*: sensu BOISDUVAL (1840) nec RAMBUR, 1833 (according to GUENÉE 1858; identity: *Idaea infirmaria*); *luridata*: sensu STAUDINGER (1871) nec ZELLER, 1847 (identity: *Scopula luridata*). – (infrasubspecific): *albomarginata*: REISSER (1961); *anastomosaria*: PROUT (1935); *arenacea*: PROUT (1913); *commutata*: DANNEHL (1933); *perfalsaria*: PROUT (1934). – (incorrect subsequent spellings): *aetnea*: RAINERI (1996); *arenaria*: PROUT (1913); *falcaria*: PROUT (1913); *laridata*: PROUT (1935).

DIAGNOSIS: Wingspan ♂ 21-25, ♀ 24-29 mm, subsequent generations on average 2 mm smaller, sometimes 17 mm only. Ground colour grey, in the Balkan peninsula sometimes with yellowish brown tinge. Transverse lines very diffuse, on the forewing with three distinct costal spots. Postmedial line often slightly sharper, dentate. Terminal area usually with dark shadows. Discal spots indistinct. Underside glossy, usually without pattern. Proboscis extremely long. Frons brown, vertex and collar pale sand coloured. Antennae of ♂ with intersegmental incisions, ciliate-fasciculate, cilia long, 1.9-2.4 times width of fla-

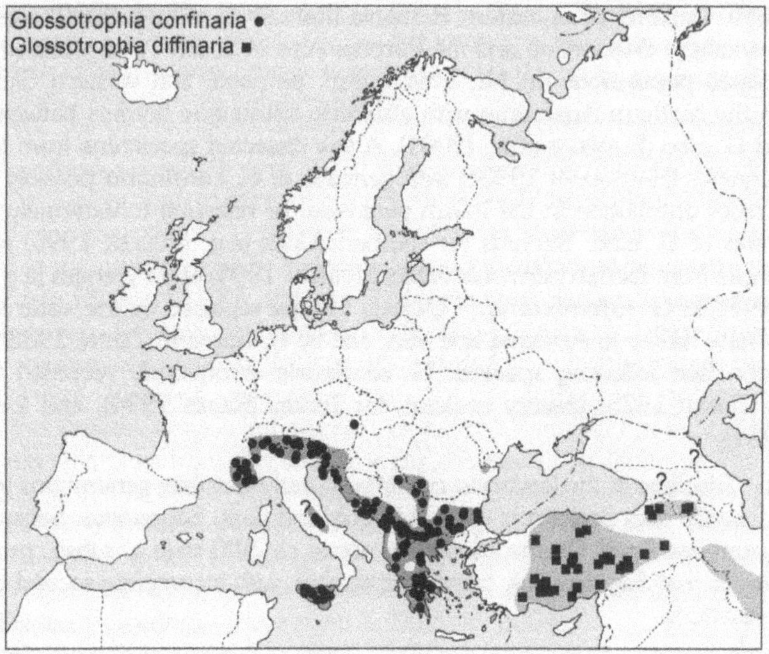

gellum (Text-fig. 155). Hindtibia of ♂ slender, with one spur, tarsus longer than tibia. Reexamination of the southern Greek '*diffinaria*' ♂ (HAUSMANN 1993b) showed that the hindtibial spur is absent only on one side; the same was found in one ♂ from Macedonia (ZSM), and in one ♂ from northern Slovenja (coll. FRIEDRICH) and must be regarded as an individual aberration. All the other examined males from central and southern Greece (n=17) with one hindtibial spur on both legs.

Subsp. ***prouti*** differs in darker ground colour with brown tinge, transverse lines broad and with stronger contrast to the light ground colour.

Subsp. ***aetnaea*** with second generation adults very small, sometimes 14 mm only. Ground colour dark grey.

Coloration of subsp. ***scoblei*** with yellow suffusion.

MALE GENITALIA: See diagnosis of subgenus. Basal projection of sternum A8 long, comparatively slender. Right ceras usually long and slender, the left short. Polymorphic aberrants are found with both cerata long (about 30% in southern Alps, n=45; about 20% in the Balkan peninsula, n=24) and with both arms short (4% in the Balkan peninsula, not yet found in southern Alps). In the subsp. *scoblei* and *aetnaea* symmetrical cerata are commoner (about 25% short and 20% long; n=15). Detailed study of cerata polymorphism see HAUSMANN (1993b; 1993c; 1999).

FEMALE GENITALIA: See diagnosis of subgenus. Sclerite on antrum long, posteriorly distinctly concave.

DISTRIBUTION: East-Mediterranean. From south-eastern France across northern Italy, Balkan peninsula to Bulgaria. In the north an isolated record from the suburbans of

Vienna (ZSM). Occurrence in eastern Romania (RÁKOSY & WIESER 2000) awaits confirmation. In southern Switzerland and the Porphyr Alps of South Tyrol replaced by subsp. *prouti*. Isolated populations on Mt. Etna (subsp. *aetnaea*) and western Sicily (subsp. *scoblei*). In the northern Apennines with allopatric distribution borders between *G. confinaria* and *G. alba* (FLAMIGNI et al. (2001). A few dissected specimens from the central Italian Apennines (HAUSMANN 1993c) with genitalia of *G. confinaria* possibly a relict of a former, wider distribution in the Italian peninsula, or referring to extremely rare genitalic aberrants of *G. alba*. Records for southern Urals (e.g. MÜLLER 1996) refer to *G. sacraria*, those from Iberian peninsula (VIVES MORENO 1994) and Pyrenees (e.g. MAZEL & PESLIER 1997), to *G. rufomixtaria*. – Outside Europe replaced by the sister species *G. diffinaria* from Turkey to Armenia and Iran, and by *G. uberaria* ZERNY 1933 in Cyprus and Lebanon (see following species). *G. confinaria* erroneously recorded for Libya, Cyrenaica (TURATI 1925; identity unclear), for Turkey (RIEMIS 1994), and for Morocco (RUNGS 1981).

PHENOLOGY: Trivoltine in the lowlands: mid-May to early October, generations not sharply separated, usually with peaks mid-June, late July and early September. Subsp. *aetnaea* emerging from late April. On the mountains above ca. 500 m (e.g. subsp. *prouti*) bivoltine late May to mid-July; late July to mid-September, with incomplete second generation at higher elevations.

BIOLOGY: In nature possibly monophagous (3) on *Silene*, or oligophagous on Caryophyllaceae. Egg deposition on flowers of Caryophyllaceae (HAUSMANN & DÖTTERL 2004). Larva found on *Silene vulgaris* (Caryophyllaceae) (Croatia: REBEL 1893), and on rocks near *Silene* (Croatia: WAGNER 1926). Recorded on *Silene vulgaris*, *Dianthus* and *Petrorhagia saxifraga* (Caryophyllaceae) (southern and western Alps: PROUT 1913; DANNEHL 1927; VORBRODT 1930; LHOMME 1935; FORSTER & WOHLFAHRT 1981), reputedly also on *Linaria* and lichens (REBEL 1893; 1910), the last two data doubtful. Reared on *Silene, Dianthus*, and *Thymus* (PÜNGELER 1896; WAGNER 1926; ERLEBACH pers. comm.). Adults readily attracted to light. Nectar feeding recorded with strong focus on Caryophyllaceae, recorded e.g. on *Dianthus, Silene vulgaris, S. dioica* (Caryophyllaceae), and on *Platanthera* (Orchidaceae) (HAUSMANN & DÖTTERL 2004). Females more actively visiting flowers (HAUSMANN & DÖTTERL 2004).

HABITAT: Xerophilous, inhabiting dry open habitats, such as stony grassland on slopes, calcareous rocks, or uncultivated sites with abundant Caryophyllaceae. Usually common in the lowlands from 0 m up to 500 m above sea-level, exceptionally up to 1,500 m in the southern Alps, regularly up to 1,900 m in the Balkan peninsula.

PARASITOIDS: Unidentified parasitoids mentioned by DANNEHL (1927).

SIMILAR SPECIES: *Idaea typicata* (see no. 41), *I. consolidata* (71), *Scopula submutata* (132), *S. marginepunctata* (143), *G. sacraria* (163). *G. asellaria* differs in shorter proboscis, scales serially arranged in distal half of wings. *G. alba* can be distinguished, unambiguously, only by dissection, on average paler; almost no sympatric occurrence with *G. confinaria*.

Glossotrophia diffinaria (PROUT, 1913)

Glossotrophia diffinaria PROUT, 1913: in SEITZ, Macrolep. 4: 83, pl. 4, fig. h (northern Turkey, Pontus: Amasia). Introduced to replace misidentified *luridata* sensu STAUDINGER, 1871. Syntypes at least 4 (MNHU: 4, examined, one labelled as 'type' the others as 'cotype').

Glossotrophia asiatica BRANDT, 1938: Ent. Rdsch. 55 (49): 574, figs 226-231. Syntypes, at least 6♂♀ (Iran: Fort Mian-Kotal). Herewith downgraded from species (SCOBLE 1999) to subspecific rank (stat. n.).

Unavailable names (misidentification): *luridata*: sensu STAUDINGER (1871, partim) nec ZELLER, 1847 (identity: *Scopula luridata*); *semitata*: sensu WARDIKJAN (1985: 44) nec PROUT, 1913. – (infrasubspecific): *ochrearia*: WAGNER (1931). – (incorrect subsequent spelling): *diffinara*: PROUT (1913); *difinaria*: DIDMANIDZE (2001).

Recorded for Rhodos island (BENDER 1963), data await confirmation, no vouchers traced. Approaching Europe also in NW. Turkey, with records from about 200 km east of Istanbul. For the absence of *G. diffinaria* from Greece see diagnosis of *G. confinaria*. Distribution in the east to western Transcaucasus, in the south to northern Iraq and Iran (subsp. *asiatica* BRANDT, 1938, stat. n.). Populations of eastern Caucasus (ZISP) and Transcaucasus (e.g. Georgia: DIDMANIDZE 1978; 2002; VIIDALEPP 1996; '*falsaria*') need taxonomic revision (see remarks), such as 1♀ recorded from north-western Saudi Arabia (WILTSHIRE 1990). Bivoltine phenology, from May to September, second generation sometimes incomplete. Usually in mountainous habitats from 700 m up 1,800 m above sea-line. Main difference from *G. confinaria* the absence of ♂ hindspur. The larger sister species *G. uberaria* ZERNY 1933 from Lebanon, however, with hindspur present. In ♂ genitalia of *G. diffinaria* symmetrical cerata commoner than in *G. confinaria* (25% short and 40% long; n=20; HAUSMANN 1999). Spectrum of nectar plants wider than in *G. confinaria*, recorded on various Caryophyllaceae, but also on *Linum catharticum* and, possibly, on *Polygonum* (HAUSMANN & DÖTTERL 2004).

G. diffinaria is not downgraded to subsp. of *G. confinaria* in the present study as there are constant and clear distributional borders based on the structure of the ♂ hindtibia and on the frequency of symmetrical ♂ genitalia.

The paralectotype of the taxon '*falsaria*' (see nomenclatural paragraphs of *G. rufomixtaria*) in genitalia reminiscent of typical *G. diffinaria*: Sternum A8 with long and slender right ceras, the left short. Hindtibia, however, with two spurs (!) of unequal length. From southern Armenia to northern Iran some examined males with stout, curved cerata of medium length (symmetrical or with left ceras short). All south-western Asiatic populations of this species complex urgently need accurate taxonomic revision on the base of extensive material.

160. Glossotrophia alba HAUSMANN, 1993

Glossotrophia alba HAUSMANN, 1993: Atalanta 24 (1/2): 284, figs 7, 17, 31-35, 41, 41a (Italy: Abruzzi: Mtgna. Grande). Holotype ♂ (ZSM, examined; =lectotype of *romanaria* DANNEHL, 1933).

‡ *Acidalia luridata confinaria romanaria* DANNEHL, 1933: Ent. Z. Frankf. a.M. 47: 81. Infrasubspecific; junior primary homonym nec MILLIÈRE, 1869 (identity: *Glossotrophia asellaria*).

‡ *Glossotrophia confinaria* var. (ab?) *dannehli* PROUT, 1934: Lepid. Cat. 63: 444. 'Replacement name' of the infrasubspecific *romanaria* DANNEHL. Infrasubspecific, see HAUSMANN, 1993.

Glossotrophia alba africana HAUSMANN, 1993: Mitt. Münch. ent. Ges. 83: 81, figs 20, 37 (Tunisia: Tunis). Holotype ♂ (ZSM, examined). Valid at subspecific rank.

Glossotrophia alba brunellii HAUSMANN, 1993: Atalanta 24 (1/2): 287, figs 9, 28, 30 (Italy, Basilicata: Maratea). Holotype ♂ (ZSM, examined). Valid at subspecific rank.

Glossotrophia alba zahmi HAUSMANN, 1993: Atalanta 24 (1/2): 286, figs 8, 27 (Italy: Abruzzi: Maiella, Valle dell'Orfento, La Cesa). Holotype ♂ (ZSM, examined). Herewith downgraded from subspecific rank (SCOBLE 1999; HAUSMANN 1993c) to synonymy (syn. n.).

Glossotrophia alba capriata HAUSMANN, 1993: Atalanta 24 (1/2): 288, figs 10, 29 (Italy, Campania: Capri island). Holotype ♂ (BMNH, examined). Herewith downgraded from subspecific rank (SCOBLE 1999; HAUSMANN 1993c) to synonymy (syn. n.).

Glossotrophia alba milleri subsp. n.: HOLOTYPE: ♂, 'France, Haute Corse, 20 m, Stareso bei Calvi, 23.IX.1997, leg. M. MILLER', ZSM. – PARATYPES: 1♀, id., 20.IX.1997; 1♂, 'Frankreich, Korsika, Mt. Renoso, Ghisoni, leg. N. KEIL', ZSM; 1♀, 'Insel Corse, Castifao, 255 m, 28.VII.1999, leg. MAY', ZSM. The name is

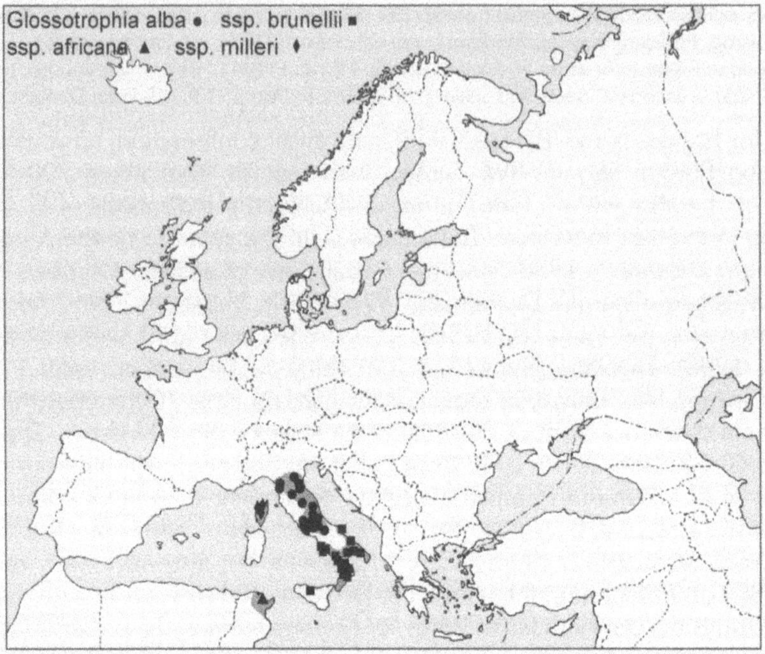

dedicated to Michael A. MILLER, Munich, for his support in preparing this volume.

Unavailable names (misidentification): *Glossotrophia rufomixtaria*: sensu MENTZER (1990) nec GRASLIN, 1863. – (infrasubspecific): *desertata*: DANNEHL (1933); *extenuata*: PROUT (1913). – (incorrect subsequent spelling): *bronellii*: SCOBLE (1999).

DIAGNOSIS: Wingspan ♂ 21-25, ♀ 24-28 mm, subsequent generation(s) on average 2 mm smaller, sometimes 17 mm only. Ground colour dirty white to light grey. Wing pattern as described for *G. confinaria*. Proboscis extremely long. Frons brown, vertex and collar pale sand coloured. Antennae of ♂ with intersegmental incisions, ciliate-fasciculate, cilia long, 1.9-2.4 times width of flagellum, on Corsica up to 2.7 times. Hindtibia of ♂ slender, with one spur, tarsus longer than tibia. One ♂ found with two hindtibial spurs (n=40 examined).

Subsp. ***brunellii*** smaller, ♂ 18-22, ♀ 20-25 mm, in the second generation sometimes 16 mm only. Wings with dark grey suffusion.

Subsp. ***milleri*** subsp. n. with a wingspan of 19-25 mm, all examined specimens from the second generation. Wings with strong dark grey suffusion, mainly in the terminal area (see colour plates fig. 160f). Costal spots of forewing very distinct. Antennal cilia longer than in the nominate subspecies, up to 2.7 times width of flagellum. Types see above (nomenclatural paragraph).

MALE GENITALIA: See diagnosis of subgenus. Aedeagus slightly shorter than in *G. confinaria*. Basal projection of sternum A8 long, comparatively slender. Right ceras usually of medium length and curved, the left short. Polymorphic aberrants are found with both cerata short (about 10% in all subspecies) and with both arms of medium size (about 2%; n=77). Such polymorphic genitalia have been found in the rearing of the eggs of a single ♀. For detailed study of cerata polymorphism see HAUSMANN (1993a; 1993b; 1999).

FEMALE GENITALIA: See diagnosis of subgenus. Sclerite on antrum slightly shorter than in *G. confinaria*, posteriorly less concave.

DISTRIBUTION: Central-Mediterranean. Italy south of Po river plain. In southern Italy and northern Sicily replaced by subsp. *brunellii*, on Corsica by subsp. *milleri*, here restricted to the lowlands of the northern half of the island. On Crete replaced by the sister species *G. mentzeri*. – Outside Europe in northern Tunisia (subsp. *africana*).

PHENOLOGY: Usually bivoltine: early May to early October, generations not sharply separated, with peaks early June and late August. In the lowlands sometimes with a partial third generation emerging from mid-September, on mountains above 1,000 m emerging in June and second generation incomplete. Larvae growing rapidly. Protandrous.

BIOLOGY: Larva possibly monophagous (3) on *Silene*, or oligophagous on Caryophyllaceae (see *G. confinaria*). Reared on fresh, withered and dry petals and leaves of *Silene vulgaris*, and *S. uniflora* (*S. maritima*) (Caryophyllaceae) (southern Italy: original data), suboptimally accepting *Silene dioica* (*Melandrium rubrum*) (Caryophyllaceae), not accepting a lot of other offered plant species (original data). Young larvae feed on the flowers of *Silene* and rest, excellently camouflaged, on the anthers. Skin of larvae with violet tinge when feeding on the calyx, with green tinge when feeding on leaves, and greyish, when feeding on withered or dry leaves. Adults readily attracted to light. Females more actively visiting flowers than males (HAUSMANN & DÖTTERL 2004). Nectar feeding recorded exclusively on Caryophyllaceae, e.g. *Silene vulgaris* and *S. dioica* (HAUSMANN & DÖTTERL 2004). Deposition of up to 80 eggs usually in the first two nights after mating.

HABITAT: Xerothermophilous, inhabiting dry open habitats, such as stony grassland on slopes, calcareous rocks, or other uncultivated sites with Caryophyllaceae. Common near the coasts. Usually in the lowlands from 0 m up to 800 m above sea-level, but at xeromontane sites, i.e. hot and sun-exposed rocks exceptionally up to 1,500 m.

SIMILAR SPECIES: See similar species of *G. confinaria* (no. 159). On Corsica particularly similar to *G. asellaria*, the latter differing in shorter proboscis, and in the serial arrangement of scales in the terminal area of wings.

REMARKS: The unusual coloration on Corsica suggests the populations to be strongly isolated.

161. Glossotrophia mentzeri HAUSMANN, 1993

Glossotrophia mentzeri HAUSMANN, 1993: Mitt. Münch. ent. Ges. 83: 78, figs 18, 19, 34-36, 56, 56a (Crete: Idaeon Antron). Holotype ♂ (ZSM, examined).

DIAGNOSIS: Slightly larger than *G. alba*, wingspan ♂ 22-26, ♀ 25-29 mm, second generation smaller, sometimes 18 mm only. Similar to *G. alba brunellii*, but wings ash grey, with pattern extremely diffuse and fringe often yellowish. Proboscis extremely long. Frons and palpi black brown, darker than in *G. alba*. Antennae of ♂ with intersegmental incisions, ciliate-fasciculate, cilia long, 1.9-2.4 times width of flagellum. Hindtibia of ♂ slender, with one spur, tarsus longer than tibia. One ♂ found with two hindtibial spurs (n=30 examined; HAUSMANN 1993b).

MALE GENITALIA: See diagnosis of subgenus. Aedeagus shorter than in *G. confinaria* and *G. alba*, the latter with exceptions. Basal projection of sternum A8 shorter and broader than in *G. alba*. Right ceras usually of medium size (but on average shorter than in *G. alba*) and strongly curved, the left short. Polymorphic aberrants are found with both cerata short (about 5%) and with both arms of medium length (about 10%; n=42), the latter

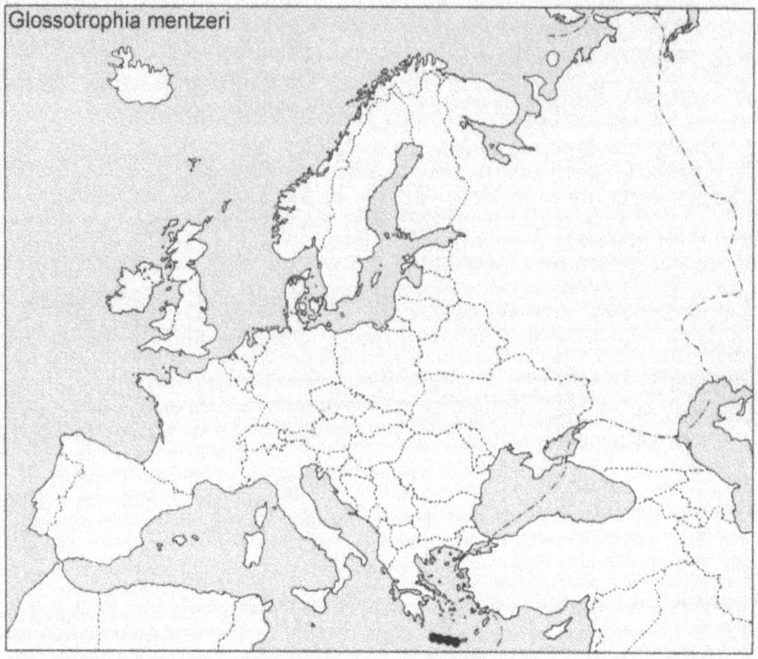

much commoner than in *G. alba*. For detailed study of cerata polymorphism see HAUSMANN (1993b; 1999).

FEMALE GENITALIA: See diagnosis of subgenus. Very similar to those of *G. alba*, sclerite on antrum comparatively short, posteriorly almost not concave or even rounded. Anterior part of lamella antevaginalis with circular sclerotization, reminiscent of that of *G. rufomixtaria*.

DISTRIBUTION: Endemic to Crete.

PHENOLOGY: Usually bivoltine: mid-May to mid-August, generations overlapping. Below 900 m emerging from late April, with partial or full third generation mid-September to early October.

BIOLOGY: Larva possibly monophagous (3) on *Silene*, or oligophagous on Caryophyllaceae (see *G. confinaria*). Reared on *Silene vulgaris* (LEIPNITZ pers. comm.). Adults readily attracted to light.

HABITAT: Montane, on hot rocky slopes and stony grassland rich in flowers. Usually from 500 m up to 1,400 m above sea-level, exceptionally, however, down to sea-level.

SIMILAR SPECIES: *I. consolidata* (see no. 71). Other similar (allopatric) species see under *G. confinaria* (159).

REMARKS: Allopatric vicariant of the closely related *G. alba*, but not downgraded in this study considering the differential features in habitus and genitalia.

162. *Glossotrophia rufomixtaria* (GRASLIN, 1863)

Acidalia rufomixtaria GRASLIN, 1863: Ann. Soc. ent. Fr. (4) 3: 357 (southern France: eastern Pyrenees), referring to the manuscript name '*rufomixtata* RAMBUR' (see below). Lectotype ♀ (ZFMK: ex coll. GRASLIN, exam-

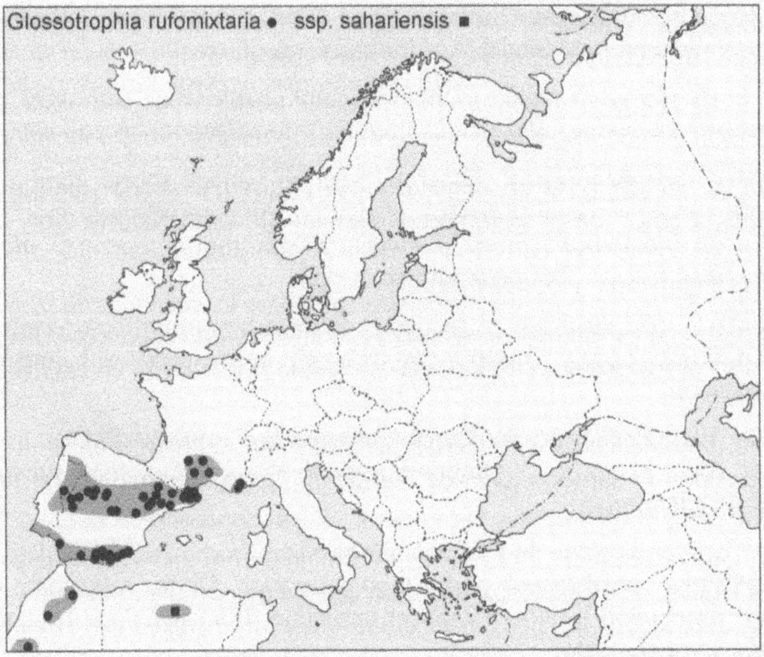

ined), herewith designated to stabilize usage of the name with regard to closely related species; at ZFMK 2♂1♀ paralectotypes from Pyrenees and southern France.

Acidalia rufomixtata STAUDINGER, 1871: in STAUDINGER & WOCKE, Cat. Lepid. eur. Faunengeb. (Edn 2) 1: 151. First available name after *rufomixtata* RAMBUR, 1866 (non binominal) (Spain: Andalusia). Syntype(s) (ex coll. RAMBUR: 1♀ ZSM/HERBULOT, examined, in HERBULOT (1987) as 'holotype'; MNHU/STAUDINGER 1♀ from 'Ronda' labelled as 'orig.', type status of both uncertain).

Glossotrophia annae MENTZER, 1990: Nota lepid. 13 (1): 44, figs 1, 2, 5, 8 (Spain, Aragon, prov. Zaragoza: Botorrita). Holotype ♂ (NHRS/MENTZER). Herewith downgraded from species rank (MÜLLER 1996; SCOBLE 1999, REDONDO & GASTON 1999) to synonymy, previously stated by HAUSMANN (1993c).

Glossotrophia rufomixtaria sahariensis HAUSMANN, 1993: Mitt. Münch. ent. Ges. 83: 86, figs 23, 39 (central Algeria: Guelt-es-Stel). Holotype ♂ (NHMW, examined). Valid at subspecific rank.

Acidalia falsaria GUENÉE, 1858: in BOISDUVAL & GUENÉE, Hist. nat. Insectes (Spec. gén. Lépid.) 9: 489. First available name after *falsaria* HERRICH-SCHÄFFER (1852:fig. 464) (non binominal) and *falsaria* LEDERER (1855) (in synonymy) (Spain, Andalusia: Ronda). Lectotype ♀ (MNHU, examined: 'Ronda'), herewith designated, to stabilize usage of the name. Paralectotypes 1♂ Azerbaijan, Elisabethpol (MNHU, examined); further paralectotypes (e.g. from South Tyrolia) in collection GUENÉE. New synonymy (but compare also MENTZER 1990b). See remarks to *G. diffinaria*.

Unavailable names (misidentifications): *dentatolineata*: sensu STERNECK (1941) nec WEHRLI (1926) (identity: *Glossotrophia asellaria* subsp.); *luridata*: sensu MANN (1855) nec ZELLER (1847) (identity: *Scopula luridata*); *submutata*: sensu MONTEIRO & PASSOS DE CARVALHO (1984) nec TREITSCHKE (1828) (identity: *Scopula submutata*). – (infrasubspecific): *anastomosaria*: REISSER (1961); *perrufa*: WEHRLI (1926). – (incorrect subsequent spellings): *rufomixta*: CULOT (1918); *ruformixtata*: PROUT (1913).

DIAGNOSIS: Wingspan ♂ 21-26, ♀ 23-28 mm, second generation 17-21 mm. Wings broad. Scales serially arranged in terminal area of wings. Fairly variable in coloration and pattern of wings. Ground colour whitish to grey, often with strong sand coloured or orange tinge. Basal half of hindwing usually darkened by many black scales. Transverse lines often diffuse, on the forewing with three black costal spots. Discal spots usually indistinct. Proboscis extremely long. Frons dark brown, vertex and collar pale sand coloured. Antennae of ♂ with intersegmental incisions, ciliate-fasciculate, cilia 1.6-2.0 times width of flagellum. Hindtibia of ♂ slender, usually with one spur, tarsus longer than tibia. Spur anomalies found in several males (with two spurs or without spurs), exceptionally even differing between left and right hindtibia of the same specimen.

Variability: In the Zaragoza region (northern Spain) usually white with weak pattern ('f. *annae*'), but with exceptions and clinous transitions to neighbouring populations.

MALE GENITALIA: See diagnosis of subgenus. Socii on average shorter than in the other species of the subgenus. Basal projection of sternum A8 comparatively short, broad and rounded. Cerata symmetrical, curved, of medium length, shorter than in *G. alba* and usually not exceeding mappa. Not polymorphic (n=20).

FEMALE GENITALIA: See diagnosis of subgenus. Sclerite on antrum comparatively broad, posteriorly distinctly concave. Anterior part of lamella antevaginalis with conspicuous circular sclerotization.

DISTRIBUTION: West-Mediterranean. Iberian peninsula and southern France, in the north-east to dept. Alpes Maritimes. – Outside Europe in Morocco (nominate subspecies) and Algeria (subsp. *sahariensis*).

PHENOLOGY: Uni- or bivoltine: In the lowlands bivoltine, from mid-April to late June; late July to mid-October, possibly with partial third generation. On the mountains, and in the north of the distribution (southern France) univoltine, from mid-June to early August, exceptionally somewhat later.

BIOLOGY: Larva probably oligophagous (2) on various herbaceous plants of low-nutrient ground, with focus on Caryophyllaceae. Found on *Gypsophila struthium hispanica* (Caryophyllaceae) (KING 2000), *Silene* and *Dianthus* (Caryophyllaceae) (CHRÉTIEN 1928). Reputedly also on *Thymus serpyllum* (Lamiaceae) (LHOMME 1935). Reared on *Dianthus pungens* and *Silene nutans* (Caryophyllaceae) (GRASLIN 1863; subsequently var. cit.), young larvae in rearings not accepting *Silene* (KING & ROMERA 2004). Nectar feeding recorded exclusively on Caryophyllaceae, e.g. *Dianthus* and *Silene dioica* (HAUSMANN & DÖTTERL 2004). Females more actively visiting flowers than males (HAUSMANN & DÖTTERL 2004).

PARASITOIDS: Ichneumonidae: spec. indet. (KING 2000). – Braconidae: *Homolobus truncator*.

HABITAT: Xerothermophilous, but preferring shaded northern slopes according to

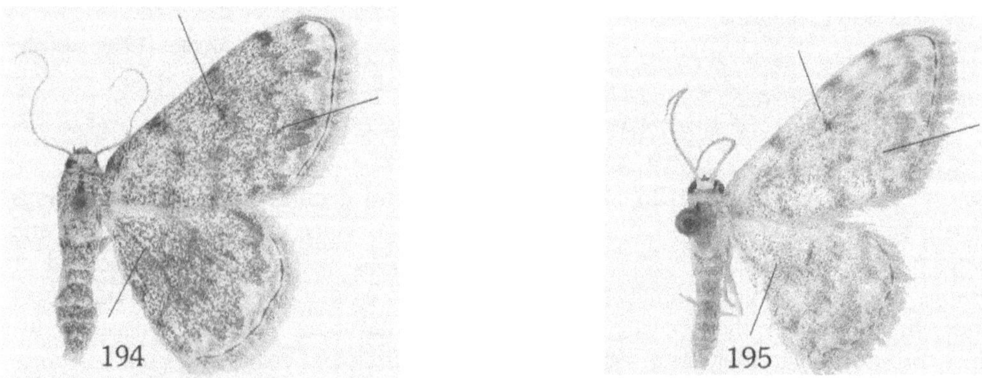

Text-figs 194-195: Diagnostic characters (indicated) of European *Glossotrophia* species. Text-fig. 194: *G. rufomixtaria* (GRASLIN, 1863). Text-fig. 195: *G. asellaria dentatolineata* WEHRLI, 1926.

CHRÉTIEN (1928). On gypsum soils (KING pers. comm.). Usually from 200 m up to 1,300 m above sea-level, in southern Spain and North Africa more montane, from 600 m to 2,100 m, but sometimes down to sea-level also here.

SIMILAR SPECIES: Spanish populations of G. *asellaria* on average smaller, proboscis shorter, basal half of hindwing not blackish, transverse lines and discal spots usually more distinct, antennal cilia longer. G. *rufomixtaria sahariensis* from Algeria very similar to Tunisian G. *alba africana*, the former easily recognized by the serial arrangement of scales.

REMARKS: Serial arrangement of scales and shallower base of sternum A8 indicating certain relations to subgenus *Parenzanella*. Pupa with externally looped proboscis (CHRÉTIEN 1928). For synonymy of the taxon *annae* see nomenclatural paragraphs.

Subgenus ***Libanonia*** HAUSMANN, 1993

TYPE SPECIES: *Glossotrophia semitata* PROUT, 1913 (subspecies of G. *sacraria*). Two species in the western Palaearctic region. Sister group relationship to the nominate subgenus (HAUSMANN 1993b).

DIAGNOSIS: Scales serially arranged on the distal half of forewing. Proboscis long (6-10 mm), but not as hypertrophic as in the nominate subgenus. Antennae of ♂ ciliate, subfasciculate, cilia arising directly from the furrowed surface of antennomere (HAUSMANN 1993b, fig.). ♂ hindibia with one spur, no anomalies found.

MALE GENITALIA: Base of sternum A8 strongly projecting. Both cerata long and narrow, not polymorphic (n=28).

FEMALE GENITALIA: Antrum dilated, concave posteriorly. Lamella antevaginalis strongly sclerotized, corrugated, but smaller than in the nominate subgenus.

163. Glossotrophia sacraria (BANG-HAAS, 1910), stat. n.

Acidalia sacraria BANG-HAAS, 1910: Dt. ent. Z. Iris 24: 42, pl. 3, fig. 18 (Russia: Uralsk). Syntypes ♂♀ (MNHU: 2♂♂, examined, the old glycerine preparation with broken cerata, but clearly conspecific with '*semitata*' as the very long basal projection reveals). Herewith raised from subspecies of G. *confinaria* (SCOBLE 1999) to species rank (stat. n.) and included in the species concept of 'G. *semitata*'.

Glossotrophia romanaria semitata PROUT, 1913: in SEITZ, Macrolep. 4: 83 (Lebanon: Baalbeck). Holotype ♂ (BMNH, examined). Herewith downgraded from species (SCOBLE 1999) to subspecific rank.

Glossotrophia asellaria taurica WEHRLI, 1930: Int. Ent. Z. 23 (37/38): 430 (southern Turkey: Maras). Syntypes, at least 1♂2♀ (ZFMK: 1♀, ex coll. WEHRLI, examined; BMNH). Valid at subspecific rank.

Glossotrophia ariana EBERT, 1965: Stuttg. Beitr. Naturk. 142: 10, pl. 1, fig. 10, pl. 2, fig. 10 (eastern Afghanistan: Sarobi). Holotype ♂ (SMNK, examined). Valid at subspecific rank.

Glossotrophia tangii EBERT, 1965: Stuttg. Beitr. Naturk. 142: 12, pl. 1, fig. 11, pl. 3, fig. 1 (eastern Afghanistan: Sarobi). Holotype ♂ (SMNK, examined). Junior synonym of G. *s. ariana*.

Glossotrophia ghirshmani WILTSHIRE, 1966: Z. Wien. ent. Ges. 51 (9-11): 129, pl. 9, figs 29.30 (eastern Afghanistan: Kabul). Holotype ♂ (BMNH, examined). Junior synonym of G. *s. ariana*.

Glossotrophia semitata fumata HAUSMANN, 1993: Mitt. Münchn. ent. Ges. 83: 88, figs 24, 25, 42, 57, 57a (Cyprus: Platres). Holotype ♂ (BMNH, examined). Valid at subspecific rank.

Unavailable names (misidentification): *asellaria*: sensu PROUT (1913, partim) nec HERRICH-SCHÄFFER, 1847; *romanaria*: sensu PROUT (1913, partim) and WARDIKJAN (1985) nec MILLIÈRE, 1869; *rufomixtata*: sensu PÜNGELER (1909, partim) nec GRASLIN, 1863.

DIAGNOSIS: Wingspan ♂♀ 18-22 mm, subsequent generations smaller, sometimes 14 mm only. Nominate subspecies with white ground colour, without dark suffusion, wing pattern

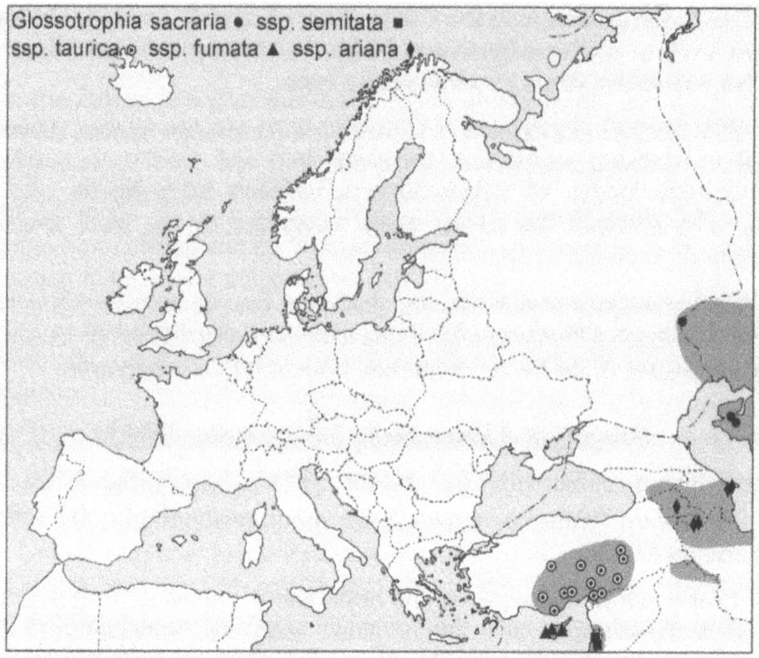

very weak. Wings of the other subspecies usually sand coloured, with dark grey suffusion of varying intensity. Transverse lines on forewing with distinct costal spots, those of post-medial line and terminal shadow adjacent, forming a conspicuous black double spot. Discal spots usually present, but indistinct. Proboscis long (6-10 mm). Frons and palpi sand coloured, with some brown scales dorsally. Antennae of ♂ filiform, cilia long, 2.2-2.9 times width of flagellum. Hindtibia of ♂ slender, with one short spur, tarsus longer than tibia.

MALE AND FEMALE GENITALIA: see diagnosis of subgenus.

DISTRIBUTION: (East-)Mediterranean-Turanian. In Europe restricted to south-easternmost European Russia (nominate subspecies). Reputedly on Rhodos island (TURATI & FIORI 1930; BENDER 1963; MÜLLER 1996), but no vouchers traced. – Outside Europe in the Levant (subsp. *semitata*), Cyprus (subsp. *fumata*), Turkey (subsp. *taurica*), and from Transcaucasus across Iran, Turkmeniya to Afghanistan (subsp. *ariana*), in western Kazakhstan (Ust-Urt plateau, north-eastern shores of Caspian Sea) similar to the nominate subspecies. In central Asian mountains replaced by the sister species *G. rufotinctaria*.

PHENOLOGY: Nominate subspecies tri- or plurivoltine, e.g. in the Levant from mid-February to mid-November, on the mountains bivoltine. For the nominate subspecies limited data available from mid-May.

BIOLOGY: Larval host-plants unknown. Adults readily attracted to light. Nectar feeding recorded on a wide spectrum of plants, such as Caryophyllaceae species, *Syringa, Linum catharticum,* possibly also *Polemonium* and *Polygonum* (HAUSMANN & DÖTTERL 2004).

HABITAT: Nominate subspecies restricted to lowland steppes from 0 m up to 300 m above sea-level, all the other subspecies with considerable vertical distribution, from 0 m up to 2,200 m above sea-level.

SIMILAR SPECIES: *Scopula marginepunctata* (see no. 143), *S. luridata* (144). *G. confinaria* (allopatric) and *G. diffinaria* on average larger, without conspicuous black costal double spot close to forewing apex. *G. diffinaria* without ♂ hindtibial spur. Allopatric *G. asellaria* with longer ♂ antennal cilia and shorter proboscis, scales serially arranged in the terminal area of wings, without conspicuous black costal double spot close to forewing apex. Males of both species can easily be separated by brushing tip of abdomen to remove scales and reveal shape of sternum A8.

REMARKS: Previously, the taxon *sacraria* was treated as subspecies or synonym of *G. confinaria* (PROUT 1913: 82; VIIDALEPP 1976; SCOBLE 1999), or even synonym of *Scopula marginepunctata* (LAEVER 1966).

Subgenus ***Parenzanella*** HAUSMANN, 1993

TYPE SPECIES: *Acidalia asellaria* MILLIÈRE, 1847. Monotypic, restricted to the Mediterranean area.

DIAGNOSIS: Scales in strongly serial arrangement on the distal half of both wings. Antennae of ♂ ciliate-fasciculate, cilia arising from small crests on the antennomeres. Length of proboscis usually 4.5-5.5 mm. ♂ hindibia with one spur. No spur anomalies found except one ♂ from Malta with two hindtibial spurs.

MALE GENITALIA: Socii shorter than in the subgenus *Glossotrophia*, Fibula more rounded, aedeagus shorter. Sternum A8 small, at the base only slightly projecting. Both cerata vestigial, synapomorphic with subgenus *Microglossotrophia*, the latter with several eremic species. Without polymorphism of cerata length.

FEMALE GENITALIA: Antrum posteriorly dilated, cup-shaped. Shape of sclerite on antrum diagnostic at subspecies level. Lamella antevaginalis flat, more weakly sclerotized than in the nominate subgenus. Signum with smaller and weaker spinules.

BIOLOGY: Larvae and adults less strongly associated with Caryophyllaceae than in those of the nominate subgenus (HAUSMANN & DÖTTERL 2004).

IMMATURE STAGES: Pupa with normal adnate proboscis.

164. Glossotrophia asellaria (HERRICH-SCHÄFFER, 1847)

Acidalia asellaria HERRICH-SCHÄFFER, 1847: Syst. Bearb. Schmett. Eur. 3 (24): 18 (pl. 56, figs 342, 343: non binominal) (Corsica). Syntypes 1♂ and at least 1♀, not traced in MNHU and BMNH. Neotype ♂ ('Ile de Corse, Gorges au Lancone, Jun. 1923', ZSM/HERBULOT), herewith designated to stabilize usage of the name with regard to the sympatric occurrence of the similar *G. alba milleri*. Original illustration of HERRICH-SCHÄFFER shows a specimen with pale ground colour and well contrasted transverse lines. A similarly coloured specimen was selected for the neotype, excluding confusion with the darker, sympatrically occurring *G. alba milleri*.

Glossotrophia romanaria var. *philipparia* PROUT, 1913: in SEITZ, Macrolep. 4: 84 (eastern Algeria: Philippeville). Syntypes, at least 4 ♂♀ (ZSM 1♂1♀, MNHU 1♂1♀, examined). Valid at subspecific rank.

Acidalia isabellaria MILLIÈRE, 1868: Iconogr. Descr. Chen. Lép. inéd. 2 (22): 453, pl. 97, figs 3, 4 (Spain, Catalonia: Barcelona, Mt. Putchet). Syntypes 2♂, probably lost. Valid at subspecific rank.

Acidalia romanaria MILLIÈRE, 1869: Iconogr. Descr. Chen. Lép. inéd. 3 (24): 52, pl. 106, figs 4-11 (Italy, Lazio: Rome). Syntypes 15♂18♀ (MNHU: 3♂2♀, examined; MNHN (SCOBLE 1999)). Valid at subspecific rank.

Glossotrophia dentatolineata var. *insularis* WEHRLI, 1926: Dt. ent. Z. Iris 40 (2/3): 120 (Corsica: Col

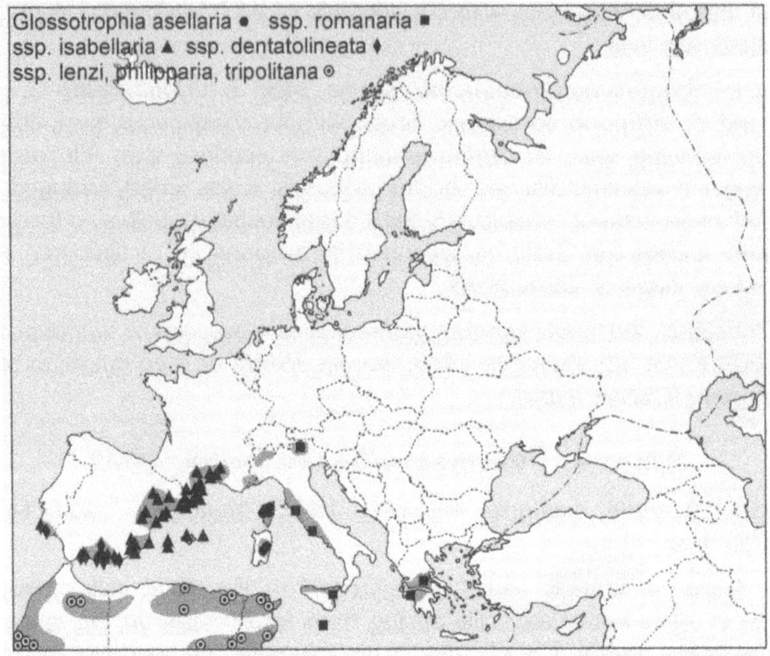

Teglime). Lectotype ♂ (ZFMK, examined), herewith designated to fix type locality and to stabilize usage of the name; at ZFMK 3♂1♀ paralectotypes from Corsica and Sardinia, others in BMNH.

Glossotrophia dentatolineata WEHRLI, 1926: Dt. ent. Z. Iris 40: 118. First available name after *dentatolineata* RAMBUR, 1866 (non binominal) (Spain, Andalusia: by Granada). Syntypes (of WEHRLI: ZFMK 22♂♀, examined; original material of RAMBUR: ZSM/HERBULOT, examined). Nomenclature see HAUSMANN (1993a). Valid at subspecific rank.

Glossotrophia tripolitana TURATI, 1929: Boll. Lab. Zool. Gen. agr. R. Scuola Agric. Portici 23: 100, 110, pl. A, fig. 1 (Libya, Tripolitania: Tripoli, Sidi Messri). Syntypes 2♂1♀. Valid at subspecific rank.

Glossotrophia asellaria gerstbergeri HAUSMANN, 1993: Mitt. Münch. ent. Ges. 83: 98, figs 30, 31, 51, 52, 65 (Canary islands, Gran Canaria: San Bartholomae). Holotype ♂ (ZSM, examined). Valid at subspecific rank.

Glossotrophia asellaria lenzi HAUSMANN, 1993: Mitt. Münch. ent. Ges. 83: 97, figs 28, 29, 48, 49, 59, 59a, 64 (Morocco, Great Atlas: Goundafa). Holotype ♂ (ZSM, examined). Valid at subspecific rank.

Unavailable names (misidentification): ?*confinaria*: sensu BELLIER (1862) nec HERRICH-SCHÄFFER, 1847; *rufomixta*: sensu OBERTHÜR (1922) nec *rufomixtaria* GRASLIN, 1863; *rufomixtaria*: sensu TURATI & FIORI (1930) nec GRASLIN, 1863; *rufomixtata*: sensu PROUT (1913, partim: Tenerife) nec *rufomixtaria* GRASLIN, 1863. – (infrasubspecific): *demarginata*: REISSER (1961). – (incorrect subsequent spellings): *assellaria*: DOMÌNGUEZ & BAIXERAS (1989); *dentatalineata*: REBEL (1930).

DIAGNOSIS: Wingspan of nominate subspecies ♂ 18-21, ♀ 19-23 mm, second generation ♂ 16-18; ♀ 18-19 mm. Wings sand coloured, on Corsica sometimes with strong dark grey suffusion, on Sardinia paler, slightly reminiscent of subsp. *romanaria*. Transverse lines usually distinct. Postmedial line dentate. Discal spots usually distinct. Proboscis of medium length, 4.5-5.5 mm. Palpi and frons brown, vertex creamy white, collar sand coloured to light brown. Antennae of ♂ with shallow intersegmental incisions, ciliate-fasciculate, cilia long, 2.3-2.9 times width of flagellum in the nominate subspecies. Hindtibia of ♂ slender, with one spur, tarsus longer than tibia. Spur anomalies very rare (see subsp. *dentatolineata*).

Subsp. ***romanaria*** similar in size, wingspan ♂ 18-22, ♀ 22-24 mm, second generation ♂ 15-18, ♀ 16-19 mm. Similar to nominate subspecies, but ground colour lighter, slightly glossy. Wing pattern greyish, postmedial line not dentate, slightly undulate, grey shadows in the terminal area distinct. Discal spots often large, but diffuse. Antennal cilia of ♂ 2.2-2.7 times width of flagellum, on Malta slightly longer. Spur anomalies rare, one ♂ from Malta with two hindtibial spurs. Eastern Algerian populations (subsp. *philipparia*) quite similar in habitus.

Subsp. ***isabellaria*** slightly smaller, wingspan ♂ 17-20, ♀ 20-22 mm, subsequent generations ♂ 15-17, ♀ 17-19 mm. Similar to nominate subspecies, but with strong red brown or dark brown suffusion, infrapopular variability high, with single paler specimens mixed in the populations. Transverse lines and discal spots usually indistinct. Postmedial line usually dentate. Antennal cilia of ♂ shorter than in the nominate subspecies, 2.1-2.3 times width of flagellum.

Subsp. ***dentatolineata*** larger, wingspan ♂ 19-23, ♀ 23-24, subsequent generation(s) ♂♀ 16-20 mm. Larger and paler than subsp. *isabellaria*. Similar to nominate subspecies, but with slight grey brown suffusion, wing pattern diffuse. Postmedial line usually dentate. Discal spots usually diffuse. Antennal cilia of ♂ shorter than in the nominate subspecies, 2.2-2.3 times width of flagellum. One aberrant ♂ (prov. Almeria) with two hindtibial spurs.

MALE GENITALIA: Socii longer than the distance between them. Nominate subspecies with cerata rudiments short and rounded. Subsp. *romanaria* with cerata rudiments short, usually tapered, at base with triangular projection. Absence of cerata rudiments in Maltese specimens reminiscent of North African subspecies. In subsp. *isabellaria* socii as long as distance the between them, cerata rudiments large, tapered, mappa very short, base straight edged, with small projection only. Subsp. *dentatolineata* with socii as long as the distance between them, or shorter. Cerata rudiments of variable size, but smaller than in subsp. *isabellaria* and posteriorly rounded. Mappa larger than in subsp. *isabellaria*, base similar. See also diagnosis of subgenus.

FEMALE GENITALIA: Nominate subspecies with sclerite on antrum posteriorly distinctly concave. Subsp. *romanaria* usually with a small medial process in this deep posterior excavation. In subsp. *isabellaria* sclerite on antrum posteriorly rounded, in subsp. *dentatolineata* posteriorly concave. See also diagnosis of subgenus.

DISTRIBUTION: West-Mediterranean. Ring species (see Vol. 1: 72; and SUDHAUS & REHFELD 1992: 53). Nominate subspecies on Corsica and Sardinia. In southern France, northern and eastern Spain replaced by subsp. *isabellaria*, sometimes common on the Mediterranean coast, present on all Balearic islands (coll. SKOU; VALLHONRAT & PETITPIERRE 1985; DANTART et al. 1993). Subsp. *dentatolineata* in southern Spain and southern Portugal. Subsp. *romanaria* in very isolated populations and very rare in the warmest and lowest valleys of southern Alps and Valesia (FORSTER & WOHLFAHRT 1981), central Italy, Sicily, Malta and Greece (Thermopylae; northern Peloponnes). Recorded also from French Alpes Maritimes (GOATER pers. comm., dissected). – Outside Europe on Canary islands (subsp. *gerstbergeri*), in Morocco (subsp. *lenzi*), northern Algeria and northern Tunisia (subsp. *philipparia*). Subsp. *tripolitana* from Saharan areas of eastern Algeria and southern Tunisia to Libya. All records from the eastern Mediterranean to central Asian mountains (e.g. Turkey: RIEMIS 1994; Lebanon: WILTSHIRE 1939a; Transcaucasus: VIIDALEPP 1976) refer to *G. sacraria*, which is similar in habitus, but quite different in structure.

PHENOLOGY: Nominate subspecies usually bivoltine: early June to early July; mid-August to mid-September; one record from mid-November. Subsp. *romanaria* bi- or trivoltine:

Limited data available from early April to early May; mid-June; early August to mid-September. On Peloponnes in early October, probably belonging to a third generation. Subsp. *isabellaria* bi- or trivoltine. In southern France bivoltine early June to mid-July; late August to mid-September. From Pyrenees to central and eastern Spain trivoltine mid-April to mid-October. Subsp. *dentatolineata* in the lowlands bi- or trivoltine: late March to late May; late September to mid-October, so far no data from summer months in the lowlands. Above 500 m from mid-May to late June; early July to early August. North African populations (subsp. *lenzi, philipparia, tripolitana*) plurivoltine in all months of the year.

BIOLOGY: Larva polyphagous, preferring dry leaves of herbaceous plants (FORSTER & WOHLFAHRT 1981). Found on leaves of *Fagonia cretica* (Zygophyllaceae) (Tenerife: CZADEK pers. comm.: subsp. *gerstbergeri;* Tunisia: CHRÉTIEN 1917: subsp. *tripolitana*), *Anarrhinum brevifolium* (Scrophulariaceae), *Salvia aegyptiaca* (Lamiaceae) (Tunisia: CHRÉTIEN 1917), *Gypsophila struthium hispanica* (Caryophyllaceae), and *Antirrhinum majus* (Scrophulariaceae) (central Spain: KING & ROMERA 2004: subsp. *isabellaria*). Recorded on leaves of *Linaria* and *Antirrhinum* (Scrophulariaceae) (MILLIÈRE 1869: subsp. *romanaria*; subsequently var. cit.). Reared on petals of *Lobularia maritima, Centaurea* and other plants (MILLIÈRE 1871; subsequently var. cit.: subsp. *isabellaria*). Adults attracted to light. Nectar feeding recorded on various plants, e.g. Caryophyllaceae, *Syringa* and, possibly, *Polemonium* (HAUSMANN & DÖTTERL 2004).

HABITAT: Xerothermophilous. Preferring habitats on calcareous or sandy ground with sparse vegetation. Nominate subspecies from 400 m up to 1,000 m, usually in the inner parts of the islands. Subsp. *romanaria* near coasts, in the lowlands from 0 m up to 300 m above sea-level. Subsp. *isabellaria* usually from 0 m up to 400 m above sea-level in the north, but up to 1,000 m in Andorra, up to 1,200 m in eastern and southern Spain (subsp. *dentatolineata*), up to 1,500 m in Morocco (subsp. *lenzi*).

SIMILAR SPECIES: *G. confinaria* (see no. 159), *G. alba* (160, most misidentifications with material from Corsica), *G. rufomixtaria* (162), *G. sacraria* (163). *Idaea* species such as *I. typicata, I. alyssumata* or *I. dyraria* without terminal line around the forewing apex and with much shorter ♂ antennal cilia.

REMARKS: Larvae of subsp. *isabellaria* more reddish on the dorsum than those of the nominate subspecies (PROUT 1913).

Holarctias PROUT, 1913, stat. n.

TYPE SPECIES: *Haematopis sentinaria* GEYER, 1837 (Canada, Labrador). Three species, holarctic. By many authors treated as synonym of *Scopula* (HODGES 1983; SCOBLE 1999; SIHVONEN 2003). With regard to differential features in venation (e.g. position of cell, origin of forewing R1 and the distictly stalked Rs and M1 of hindwing), ♂ sternum A2, tympanal cave and genitalia (e.g. broad sclerotized socii of ♂) herewith raised again to generic rank.

DIAGNOSIS: Venation (Text-fig. 196): On forewing one single areole present, close to cell apex, exceeding distal margin of cell. R1 and R2-R5 on a short, common stalk, origins of R1 and R5 close to each other. On hindwing Rs and M1 distinctly stalked, M3 and CuA1 separate. Eyes small. Palpi extremely bushy and long (about twice diameter of eye). Proboscis well developed. Frons convex. Thorax and legs with long hairy scales. Hindtibia with two long spurs in ♂, in ♀ usually with four long spurs, but proximal spurs sometimes weak, reduced to one or absent. Even on the same individual different spur number may occur on right and left hindleg. Tarsus sometimes slightly shortened. Tympanal cave large, sclerotized. Central sclerite and setose patch absent from ♂ sternum A2.

MALE GENITALIA: Genitalia very strongly sclerotized, even the broad socii. Aedeagus furcate at tip. Sternum A8 ovoid with cerata absent or very short. Mappa small.

FEMALE GENITALIA: Apophyses anteriores usually very short and stout. Ductus bursae short and broad, strongly sclerotized. Signum absent.

BIOLOGY: Adults diurnal (BIDZILYA et al. 2002).

165. Holarctias rufinaria (STAUDINGER, 1861), comb. n., stat. n.

Acidalia rufinaria STAUDINGER, 1861: in STAUDINGER & WOCKE: Cat. Lepid. Eur. 1: 64 (Russia: eastern Siberia). No separate types. Replacement name for *Acidalia rufularia*, see below. Herewith raised from subspecies of the nearctic *H. sentinaria* (GEYER, 1837) (SCOBLE 1999) to species rank.

‡ *Acidalia rufularia* EVERSMANN, 1851: Bull. Soc. imp. Moscou 24 (1)(2): 641 (Russia: eastern Siberia). Syntype(s) (ZISP?). Junior primary homonym nec HERRICH-SCHÄFFER, 1847 (identity: *Idaea obsoletaria*).

Acidalia rufociliaria BREMER, 1864: Mém. Acad. Sci. St.-Pet. (7) 8 (10): 78, pl.7, fig. 2 (Russia: eastern Siberia: Dahuria). Syntype(s).

Unavailable names (misidentification): *frigidaria*: sensu PROUT (1913: pl. 4 fig. i) nec MÖSCHLER, 1860. – (incorrect subsequent spellings): *ruficiliaria*: PROUT (1913); *rufilaria*: PROUT (1913).

DIAGNOSIS: Wingspan ♂♀ 21-23 mm. Males on average slightly larger than females. Ground colour reddish brown, darkened towards wing base. Females paler, with orange or ochreous tinge. Dark brown medial and postmedial lines slightly convex, the former broad and diffuse, the latter sharper. Veins slightly darker. Discal spot elongate, brown, indistinct. Terminal line distinct, uninterrupted. Frons and vertex reddish, brownish or ochreous. Antennae with chequered scales, those of ♂ with deep intersegmental incisions, ciliate-fasciculate, tufts of cilia densely appressed, cilia curved, 1.7-2.3 times width of flagellum. Hindtibia of ♂ not dilated, with two long spurs, tarsus not shortened. Hindtibia of ♀ see subgeneric diagnosis. Populations of northern Urals closely matching those of north-eastern Asia.

Sister species *H. rufinularia* (STAUDINGER, 1901) from Altai and Sajan mountains larger, ♂♀ 22-25 mm. Ground colour not darkened towards wing base. Veins with strong dark suffusion. Discal spot distinct.

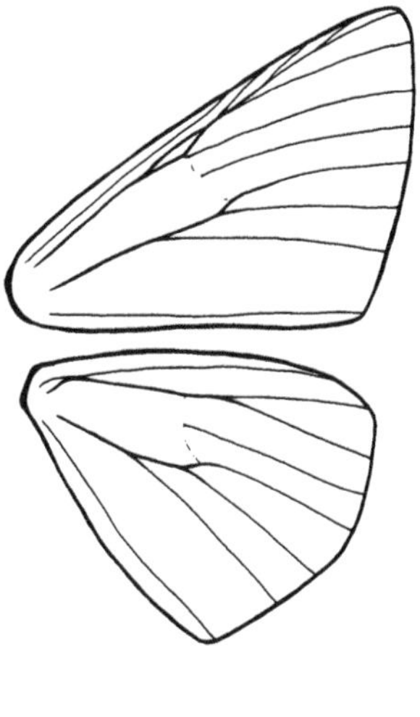

Text-fig. 196: Venation of *Holarctias rufinaria* (STAUDINGER, 1861).

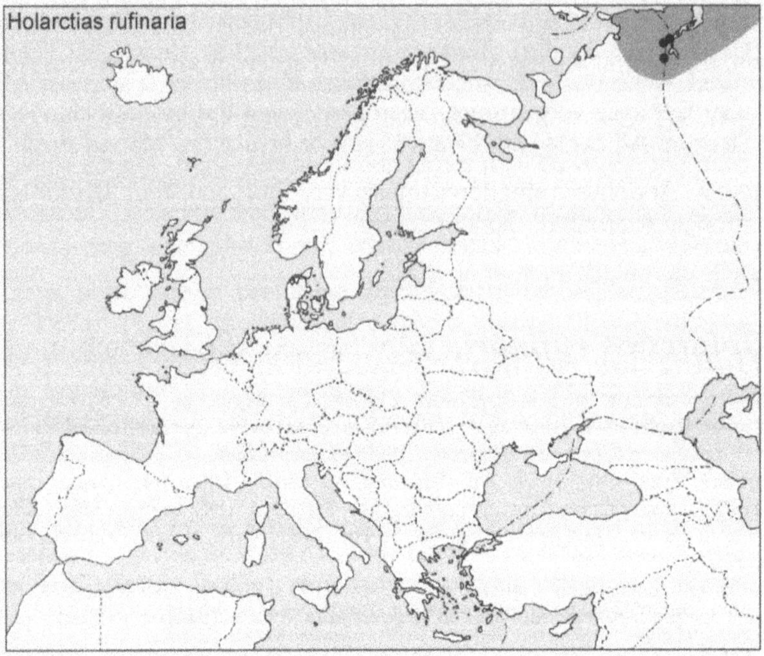

North American sister species *H. sentinaria* (GEYER, 1837) with wingspan about 23 mm (Labrador), 16-27 mm according to MCGUFFIN (1967). Ground colour strongly darkened towards wing base. Veins concolorous. Discal spot indistinct.

MALE GENITALIA: Socii short, broad, sclerotized. Valva and fibula both slightly sclerotized, elongate. Juxta with short lateral appendages. Aedeagus with subterminal cornutus long and slender. Sternum A8 narrow, without cerata. *H. sentinaria* differs in longer, irregularly shaped socii (attached to tegumen and appearing short in MCGUFFIN 1967), longer lateral appendages of juxta, short and broad cornutus, much broader sternum A8. *H. rufinularia* with longer socii, stouter cornutus, sternum A8 with two short, ear-shaped cerata without spinules.

FEMALE GENITALIA: Ductus bursae and antrum sclerotized, very broad. Hat-shaped posterior sclerite at antrum broad, laterally tapering. Ductus bursae laterally bulbed. Corpus bursae comparatively small, membranous, without signum. In *H. rufinularia* and *H. sentinaria* elongate pear-shaped. The latter with posterior sclerite of antrum laterally less elongate.

DISTRIBUTION: Eurasiatic. In Europe restricted to the northern Urals. – Outside Europe in northern Siberia, in the east to Yakutia. Replaced by the sister species *H. rufinularia* (STAUDINGER, 1901) from Altai and Sajan mountains, Kirghizstan (1♂; ZSM/HERBULOT), across northern Mongolia, to Dahuria (KOSTJUK & GOLOVUSHKIN 2003). In North America replaced by the sister species *H. sentinaria* (GEYER, 1837).

PHENOLOGY: Univoltine: Limited data available from early July to mid-July in the tundra, emerging from mid-June in the taiga. Sister species *H. sentinaria* until mid-August.

BIOLOGY: Larval host-plants unknown. Adults diurnal or active at dusk, in the long arctic summer 'night'.

HABITAT: Grassland in sparse taiga forests (MIRONOV pers. comm.), open habitats in the tundra, 'common on meadows in the forest zone' (JALAVA & KULLBERG in press). In the north in the lowlands, limited data available from 0 m up to 500 m above sea-level. Sister species *H. rufinularia* montane, from 1,400 m up to 3,200 m.

SIMILAR SPECIES: No similar species in Europe.

REMARKS: Species rank is postulated for all three species *H. sentinaria* (GEYER, 1837), *H. rufinularia* (STAUDINGER, 1901), and *H. rufinaria*, status formally changed in KOSTJUK & GOLOVUSHKIN (2003), differential diagnosis under 'diagnosis' and 'male genitalia'

Problepsis LEDERER, 1853

TYPE SPECIES: *Caloptera ocellata* FRIVALDSZKY, 1845. 50 species in the Old World, in the south to South Africa and Australia, absent from America. Preliminary results from analysis of complete COI gene (1550 bp) support close relationship to *Scopula* (HAUSMANN et al. in prep.), as also does comprehensive cladistic analysis of morphological characters (SIHVONEN 2003). Genus *Problepsis* roughly divided into two subgroups, the first with bipectinate ♂ antennae and separate socii in ♂ genitalia (with *P. ocellata*), reminiscent of *Scopula*. The second subgroup with dentate antennae and modified socii. Sister group relationship to the genus *Somatina*, the latter with many tropical species, usually characterized by the absence of opalescent scales and by a double areole, but with exceptions from this feature combination in a few species. Improvement of the differential diagnosis between these two genera and large scale revision at species level is urgently needed (in preparation). For relationships to Rhodostrophiini see remarks to that tribe.

DIAGNOSIS: Almost all species with eye spots encircled by large opalescent scales, vertically projecting from surface of wing, sometimes also transverse lines with similar scales. Structure and arrangement of these opalescent scales unique in Geometridae. Venation (Text-fig. 197): Usually with only one single areole. The areole may, exceptionally, be divided in rare aberrants, e.g. in one *P. ocellata* from Crete (n=5 examined). R5 usually on a long common stalk with R2-R4. Hindwing similarly to that of *Scopula* with Rs and M1 separate, M3 and CuA1 separate. Antennae of ♂ bipectinate in the typical group, sometimes with very long branches, however in some species (e.g. *P. phoebaria* ERSCHOV, 1870, *P. discophora* FIXSEN, 1887) dentate, ciliate-fasciculate. Antennae of ♀ ciliate, cilia short. Thorax and femur with long, 'hairy' scales. Frons flat. Palpi usually not

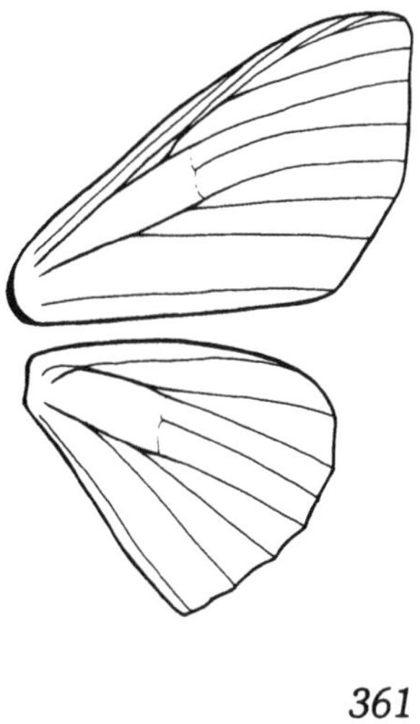

Text-fig. 197: Venation of *Problepsis ocellata* (FRIVALDSZKY, 1845).

exceeding diameter of eye. Proboscis well developed. Hindtibia of ♀ with four spurs. Hindtibia of ♂ often dilated, with large pencil, without spurs. Hindtarsus of ♂ shortened.

MALE GENITALIA: Quite uniform, posteriorly with two membranous, setose socii, which are fused in some species. Inner, ventral margin of tegumen usually spinose. Valva cleft, with two narrow, tapering processes (valva and fibula). Aedeagus usually divided at tip. Sternum A8 long, with broad base and head-shaped posterior dilatation, cerata absent.

FEMALE GENITALIA: Ductus bursae sclerotized, broad. Signum quite similar to that of the genus *Scopula*, i.e. divided into numerous flat sclerites, which are fused into a ridge along the longitudinal axis.

BIOLOGY: Larval host-plants strongly focussed on Oleaceae (REISSER 1957; HOLLOWAY 1997; DUKE & DUKE 1998). Larva not forming crook when disturbed, but resting in straight position (compare genus *Scopula*).

IMMATURE STAGES: Egg sculpturation polygonal reticulate (REISSER 1957). Larva long and slender, with small head (REISSER 1957). Pupal cremaster with one pair of stout terminal setae (REISSER 1957; NAKAMURA 1994). Further features in REISSER (1957).

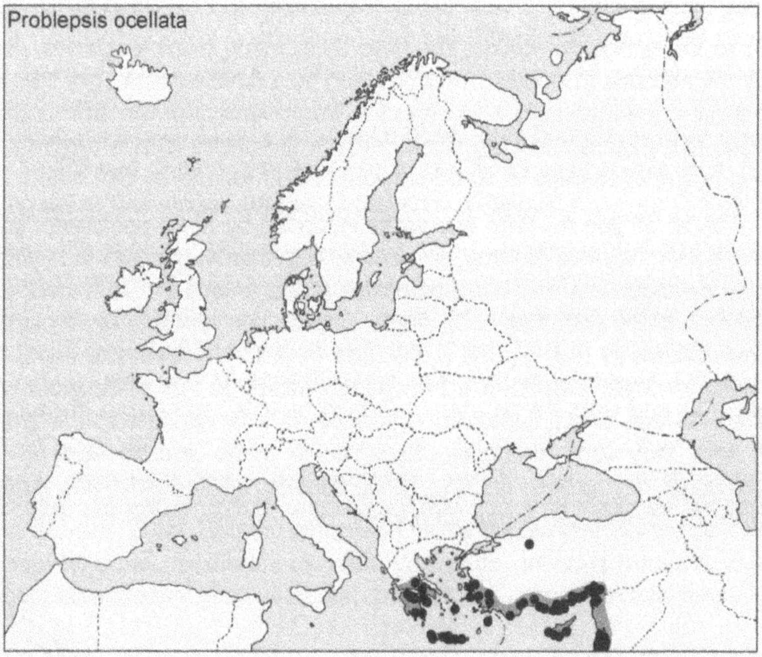

166. ***Problepsis ocellata*** (FRIVALDSZKY, 1845)

Caloptera ocellata FRIVALDSZKY, 1845: Evk. Királ. Magy. Termész. Tars. 1: 185, pl. 3, fig. 1 (Crete; western Turkey: Izmir ('Smyrna')). Syntype(s).
Argyris ommatophoraria GUENÉE, 1858: in BOISDUVAL & GUENÉE, Hist. nat. Insectes (Spec. gén. Lépid.) 10: 13 (Lebanon: Beirut). Syntypes 1♂1♀ ('coll. LEDERER').
Argyria cinerea BUTLER, 1886: Proc. zool. Soc. London 1886 (3): 387, pl. 35, fig. 8 (Pakistan: Campbellpore nr. Rawal Pindee). Syntype(s) (BMNH). Valid at subspecific rank (HAUSMANN 1998; SCOBLE 1999).
Unavailable names (incorrect subsequent spelling): *occellata*: LAMBILLON (1905).

DIAGNOSIS: Wingspan ♂ 28-33, ♀ 29-37 mm, subsequent generations smaller, sometimes

26 mm only. Ground colour whitish with slight grey brown tinge. On both wings the large eye spots surrounded by opalescent scales. Two rows of small spots distal to postmedial line. Palpi, frons and vertex black brown. Antennae of ♂ bipectinate and densely ciliate, longest branches about three times width of flagellum. Hindtibia of ♂ stout, with pencil, tarsus 0.4-0.45 times length of tibia.

MALE GENITALIA: Socii terminally fused, broad, in the subsp. *cinerea* slender and closely attached to each other. Anterior margin of sternum A8 trilobed, 'wing nut'-shaped, in the subsp. *cinerea* narrower.

FEMALE GENITALIA: Ductus bursae sclerotized, comparatively short. Corpus bursae oval. Signum with medial spines fused to a sclerite, posteriorly leading to the curved sclerite near ductus bursae.

DISTRIBUTION: East-Mediterranean. Greece, including Crete, and the other Greek islands. – Outside Europe from western Turkey to Cyprus and the Levant, with one isolated record from northern Turkey (ZSM). Replaced by subsp. *cinerea* in isolated populations on the mountains of Oman (HAUSMANN 1998), eastern Afghanistan and Pakistan, not yet recorded in southern Iran (REISSER 1957).

PHENOLOGY: Bivoltine: mid-May to mid-July; mid-August to late September. First generation under good conditions two weeks earlier, under bad conditions somewhat later, with incomplete second generation. Mountain populations beyond 1,200 m above sea-level (Turkey, Cyprus, Oman) univoltine, from late June to early August. In the lowlands of the Levant tri- or plurivoltine, from mid-March to mid-November.

BIOLOGY: Larva oligophagous (1) on Oleaceae, or monophagous on *Olea*. Recorded on *Olea* (Israel: BODENHEIMER 1930; HALPERIN & SAUTER 1992). Reared on *Olea* and *Ligustrum vulgare* (Oleaceae) (REISSER 1957; MÜLLER, LEIPNITZ pers. comm.). Larvae resemble twigs. Adults readily attracted to light.

HABITAT: Usually on Mediterranean terraces with *Olea* and *Ficus*. In Europe from 0 m up to 1,000 m above sea-level, in Turkey sometimes up to 1,400 m. An isolated mountain population in the Oman (subsp. *cinerea*) at 2,000-2,400 m.

SIMILAR SPECIES: No similar species in Europe.

REMARKS: Monography: REISSER (1957).

Tribus **Rhodostrophiini** PROUT, 1935

More than 180 species (SCOBLE 1999) in 17 genera, 5 further genera assigned tentatively (SIHVONEN & KAILA 2004). Almost worldwide with distribution of main lineages focussed on the Palaearctic and the transitional northern parts of Indo-Pacific regions. Suggested to be a phylogenetically old group (PROUT 1939b: 142). Monophyly supported by sclerotized lobes at the posterior end of sternum A8, by shape of uncus and valva. Relationship to *Problepsis* and *Somatina* suggested by PROUT (1935a) and STERNECK (1941: 155), the latter referring to shape of uncus. Transitional features in *Craspediopsis* and *Pylargosceles*, e.g. presence of cerata (STERNECK 1941: 181). Frequent misspelling: Rhodostrophini.

DIAGNOSIS: Venation with large infraspecific variability with regard to origin of forewing radial veins and shape of areoles. Often with double areole, separating vein between are-

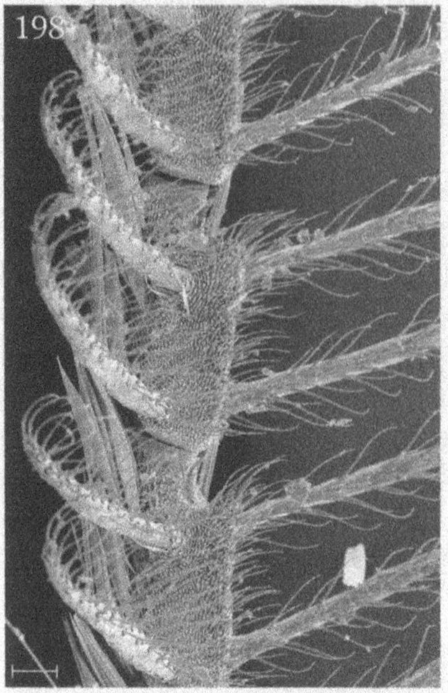

Text-fig. 198: ♂ antennal structure of *Rhodostrophia calabra* (PETAGNA, 1787) (SEM). Scale bar = 0.05 mm.

oles arising from cell. Hindwing with Sc+R1 and Rs appressed or fused at one point. Palpi short or medium-sized. Proboscis well developed. Antennae of ♂ quadripectinate (Text-fig. 198) in the nominotypical genus, or ciliate-fasciculate with four robust fascicles at each antennomere, rarely bipectinate, e.g. in *Pylargosceles* and *Erythrolophus*. Legs usually long and slender. Hindtibia of ♂ with 3-4 spurs, female with 4 spurs. Hindtarsus of ♂ not shortened. Sternum A2 of ♂ without central sclerite, no setose patches except antero-central setose patch in the *calabra* species-group of *Rhodostrophia* (Text-fig. 6).

MALE GENITALIA: Uncus long, setose, slender at base, terminally often bi- or trilobed, sometimes weak. Gnathos usually present, sometimes membranous, but absent in *Craspediopsis*. Shape of distal processes of valva usually diagnostic for species or species-groups. Anellus flat, elongate, often with two longitudinal ridges. Aedeagus usually long and slender; exceptions: *Craspediopsis* and *Dithecodes*. Sternum A8 usually without cerata; exceptions: *Craspediopsis* and *Pylargosceles*.

IMMATURE STAGES: Larva very slender and long (Text-fig. 200). Head flat laterally and orally. Pupa with elongate cremaster, terminal pair of setae (D2) long and stout, three anterior pairs smaller, recurved (PATOČKA 2003), reminiscent of Scopulini.

Rhodostrophia HÜBNER, 1823

(=*Pellonia* DUPONCHEL, 1829)

TYPE SPECIES: *Phalaena calabra* PETAGNA 1786. 72 species, from Europe to Taiwan, distribution focussed on the western and central Asian parts of the Palaearctic region, some species also in the temperate transitional zones of the northern Indopacific region. The genus is absent from Corsica and Sardinia. *Apostates* is downgraded to synonymy, see under *jacularia* species-group. Systematic position and validity of the Indo-Pacific genus *Tanaotrichia* disputable, too. Possible paraphyly of *Rhodostrophia* with regard to the genus *Discomiosis* stressed by SIHVONEN & KAILA (2004). Definition of infrageneric subgroups mainly based on differential characters published in PROUT (1935a) and STERNECK

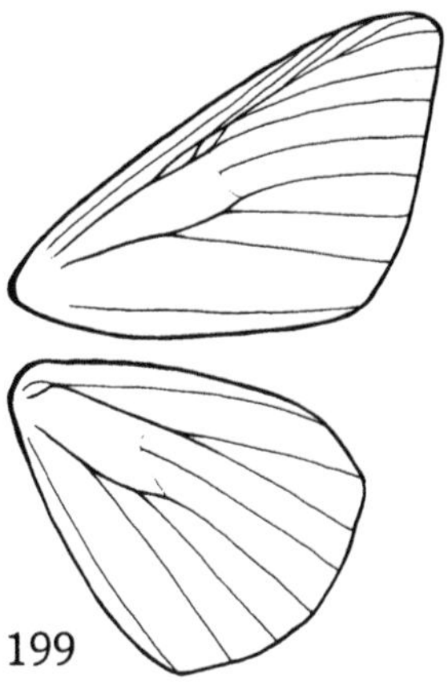

Text-fig. 199: Venation of *Rhodostrophia calabra* (PETAGNA, 1787).

Text-fig. 200: Larva of *Rhodostrophia calabra* (PETAGNA, 1787) (Photo: LEIPNITZ).

(1941). Taxonomic discussion of many species in PETERSEN (1937). Comprehensive revision at species and species-group level is urgently needed.

DIAGNOSIS: Venation (Text-fig. 199): Usually with double areole; exception: *R. solitaria* (CHRISTOPH, 1887). R2-R4 stalked, R5 from apex of second areole. Hindwing with Rs and M1 shortly stalked, M3 and CuA1 usually separate. Antennae of ♂ long quadripectinate, with thin pectinations to tip. Frons slightly convex. Length of palpi equal to or slightly exceeding diameter of eye. Some species (*calabra* species-group; Text-figs 203-207) with peculiar modifications of ♂ hindtibial spurs to 'pseudospurs' of unknown function, which are dilated, densely covered with scales and terminally rounded. Similar structures also in the genus *Tanaotrichia*. Further features under diagnosis of tribe.

MALE GENITALIA: Uncus very long, shape often diagnostic at species level. Gnathos present. Shape of valva diagnostic, mainly the distal parts. Aedeagus long and slender, usually curved or sigmoid, often with diagnostic characters at tip. Sternum A8 bilobed posteriorly, lobes bent down in natural position.

FEMALE GENITALIA: Ovipositor without ventral processes. Ductus bursae sclerotized, posteriorly dilating. Corpus bursae membranous. Signum developed as paired elongate sclerites with longitudinal ridges.

IMMATURE STAGES: Larva extremely long and slender (Text-fig. 200). Pupal characters under diagnosis of tribe.

The *jacularia* species-group

Small group of three species, well characterized by dark and undulate transverse fasciae, structure of ♂ hindtibia with four normal spurs, ♂ genitalia (shape of the flat valva, aedeagus with cornutus) and ♀ genitalia (short stout apophyses anteriores, ovipositor usually spinose). The monotypic genus *Apostates* is downgraded to synonymy of *Rhodostrophia*, as *Rhodostrophia solitaria* (CHRISTOPH, 1887), comb. n. matches well the *jacularia* species-group in wing coloration, spur formula, male and female genitalia.

Forewing venation of *R. solitaria* usually without second areole, but fairly variable, not justifying separation at generic rank. *R. tyugui* VASILENKO, 1998 from Altai Mts., with brachypterous females, running and jumping by day, on the ground, rapidly like a spider (NUPPONEN pers. comm.).

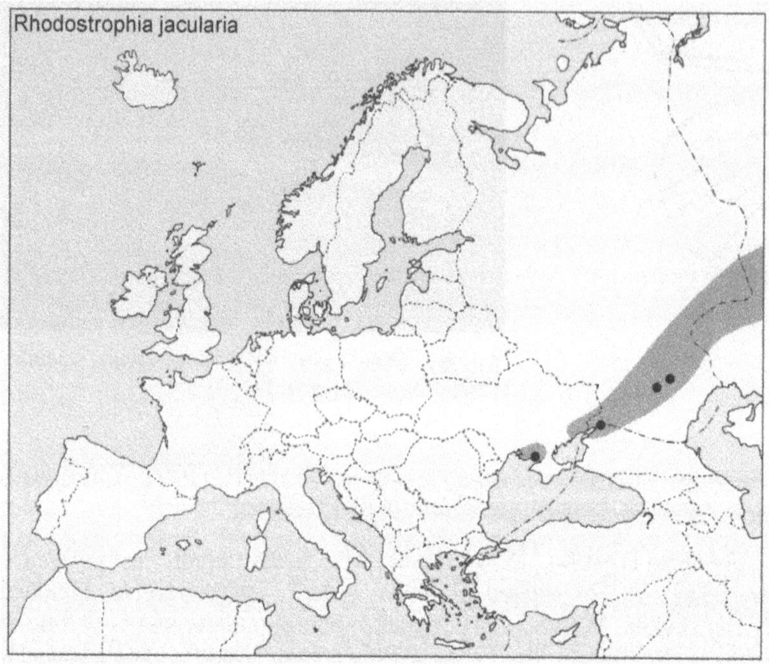

167. *Rhodostrophia jacularia* (HÜBNER, 1813)

Geometra jacularia HÜBNER, 1813: Samml. Eur. Schmett. 5, Geometrae (1): pl. 84, fig. 431 (Europe, probably southern Ural). Syntype(s) lost.

Eusarca jacularia var. *minor* ALPHÉRAKY, 1892: in ROMANOFF, Mém. Lépid. 6: 71 (western China: Ordosse, Tai-tou-hai, near River Oulane-morine). Lectotype ♂ (ZISP, examined), herewith designated to stabilize usage of the name; at ZISP 2♂ paralectotypes. Valid at subspecific rank (SCOBLE 1999), disputable.

Unavailable names (infrasubspecific): *carnosaria*: STAUDINGER (1900). – (incorrect subsequent spellings): *iacularia*: EVERSMANN (1844); *lacularia*: ANIKIN et al. (2000).

DIAGNOSIS: Wingspan ♂ 25-28, ♀ 23-25 mm, in Asian populations up to 32 mm. Ground colour light brown, hindwings paler. Forewings with conspicuous brown, whitish bordered transverse fasciae, postmedial fascia broader and undulate. Discal spot distinct on forewing. Markings more diffuse on hindwing. Frons light brown, vertex sand coloured. Longest branches of ♂ antennae 4-5 times width of flagellum, antennomeres ventrally slightly convex. Hindtibia of ♂ with four normal spurs.

MALE GENITALIA: Uncus very long, slightly dilated and slightly bifid at tip. Valva broad, ventro-terminal margin strongly spinose. Aedeagus with small base, slightly curved, with stout cornutus at tip.

FEMALE GENITALIA: Papillae anales spinose, an almost unique character in Geometridae. Apophyses stout, apophyses anteriores short. Ductus bursae sclerotized, straight. Corpus bursae globular, with large, 'coat-shaped' sclerite.

DISTRIBUTION: Euro-Siberian. Volga plain and southern Urals. In the ZSM one male from southern Ukraine, Skadovsk. – Outside Europe across northern Kazakhstan, southern Siberia (Altai and Sajan Mts.) and Mongolia to Dahuria (e.g. VOJNITS 1976; KOSTJUK et al. 1994; KOSTJUK & GOLOVUSHKIN 2003). Reputedly also in Turkey (RIEMIS 1994; VIIDALEPP 1996), but no vouchers traced. In Altai Mts. a closely related sister species, *R. tyugui* VASILENKO, 1998.

PHENOLOGY: Usually univoltine: late May to late June; one specimen from mid-August belonging to a partial second generation. On higher mountains, e.g. in the Altai and in Mongolia, later, from mid-June to late July.

BIOLOGY: Larval host-plants unknown. ♀ ratio at light very low, about 5-10%, probably due to much lower flight activity than in males. Adults nocturnal (NUPPONEN pers. comm.).

HABITAT: Xerophilous species of continental climates. In Europe restricted to sandy lowland steppes, from 0 m up to about 300 m above sea-level. In the Altai and Mongolia xeromontane, up to 2,700 m.

SIMILAR SPECIES: No similar species in Europe.

REMARKS: Genitalia of both sexes of *R. tyugui* very similar to those of *R. jacularia* from Sarepta (NUPPONEN, SIHVONEN pers. comm.).

The ***vibicaria*** species-group

By some authors suggested to be subordinated under the *calabra* species-group. STERNECK (1941: 194) separates *R. vibicaria* and the Indian *R. vinacearia* (MOORE, 1868) from the *calabra* species-group. *R. acidaria* (STAUDINGER, 1892) from central Asian mountains and *R. stigmatica* (BUTLER, 1889) from India and Nepal also matching this group well in habitus, external and genital morphology. Hindwing termen slightly angled at M3. Antennal flagellum not dentate ventrally (see *calabra* species-group). Hindtibia of ♂ with three normal spurs. Sternum A2 without setae. Tip of valva forked, with long ventral process. Aedeagus without cornutus. Larva extremely long and slender.

168. Rhodostrophia vibicaria (CLERCK, 1759)

Phalaena vibicaria CLERCK, 1759: Icones Ins. Rar. 1: pl. 3, fig. 2 (Sweden: 'Liumkihl'). Lectotype ♂ (LSL). Type locality stated in LINNAEUS (1761).

Phalaena cruentata SCOPOLI, 1763: Ent. Carniolica: 226, fig. 562 (Slovenja: Ljubljana ('Labacum')). Syntype(s) lost.

Phalaena rubrofasciata HUFNAGEL, 1767: Berlin. Mag. 4 (6): 612 (Germany: Berlin). Syntype(s) lost.

Phalaena Geometra rubrociliaria GOEZE, 1781: Ent. Beytr. 3 (3): 387 (Europe). Syntype(s).

Phalaena artriosa FOURCROY, 1785: Entom. Parisiensis 2: 275 (France: Paris). Syntype(s).

‡? *Phalaena Geometra cruentataria* VILLERS, 1789: Linn. ent. 2: 323 (Europe). Probably an incorrect subsequent spelling of *cruentata*.

Pellonia vibicaria var. *unicolorata* STAUDINGER, 1871: in STAUDINGER & WOCKE, Cat. Lepid. eur. Faunengeb. (Edn 2) 1: 154 (central Russia: Altai). Syntypes (MNHU: 1♂1♀, examined). Valid at subspecific rank (SCOBLE 1999).

Pellonia vibicaria var. (et ab.?) *strigata* STAUDINGER, 1871: in STAUDINGER & WOCKE, Cat. Lepid. eur. Faunengeb. (Edn 2) 1: 154 (Spain: St. Ildefonso). Syntypes, at least 3 (MNHU, 1♂2♀ examined). Lectotype ♂ (MNHU, examined: 1/3), herewith designated to stabilize usage of the name and to fix type locality; at MNHU 2♀ paralectotypes from type locality and from Sierra de Alfacar; in original description also northern Iran ('Hyrcania') mentioned. Valid at subspecific rank.

Rhodostrophia vibicaria var. *augustiniaria* FERNANDEZ, 1931: Eos 7: 215, pl. 2, fig. 7 (Spain, Andalusia: Sierra Nevada). Holotype. Junior synonym of subsp. *strigata*.

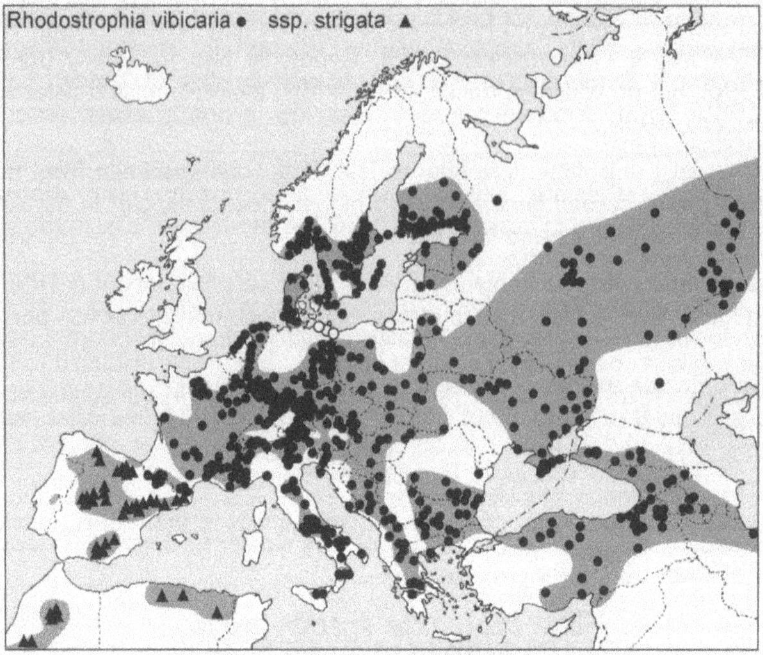

Rhodostrophia vibicaria minuta HEYDEMANN, 1933: Int. ent. Z. 27: 371 (northern Germany: Amrum island). Syntypes 9♂2♀, lost. Herewith downgraded from subspecific rank (FORSTER & WOHLFAHRT 1981) to synonymy, compare WARNECKE (1936).

Unavailable names (misidentification): *calabraria*: sensu LE PELETIER & SERVILLE (1825) nec *calabra* PETAGNA, 1787). – (infrasubspecific): *adulterina*: HEYDEMANN (1933); *fasciata*: REBEL (1910); *intermedia*: KEMPNY (1896); *obsoleta*: LEMPKE (1949); *rectilinearia*: MEVES (1914); *rosans*: PROUT (1935); *roseata*: ERSCHOFF (1888). – (nomen nudum): *fanoeensis*: PROUT (1935). – (incorrect subsequent spellings): *fibicaria*: HEROLD (1844); *rubrifasciata*: PROUT (1934); *rubrociliata*: PROUT (1913).

DIAGNOSIS: Wingspan ♂ 27-33 mm, ♀ on average slightly smaller. In the second generation and in the northernmost parts of the distribution smaller, sometimes 23 mm only. Wings sand coloured with purple transverse lines, thin, but distinct. Medial and postmedial lines of hindwing distant from each other. Small discal spots present on forewing, rarely also on hindwing. Fringe usually purple. Frons purple or red brown, vertex sand coloured. Longest branches of ♂ antennae 5-7 times width of flagellum. Hindtibia of ♂ with three spurs: proximal spur long and slender, two distal spurs of normal length.

Variability: Extent of purple coloration quite variable individually and between populations: In the north of the distribution, the wing area between medial and postmedial lines usually filled with purple. Strong red suffusion is common on montane rocky steppes (BERGMANN 1955). Those forms become rarer towards south and east. Influence of temperature and humidity on the coloration discussed in BERGMANN (1955).

Subsp. ***strigata*** with transverse lines sharp, without reddish tinged fasciae, with very rare exceptions. Fringe usually sand coloured.

MALE GENITALIA: Uncus broad, subapically dilated, convex at tip. Distal margin of valva not spinulose, forked. Shape of the uncus and the digitiform processi of valva comparatively variable. Aedeagus straight.

FEMALE GENITALIA: Apophyses fine, apopyses anteriores comparatively long. Sterigma sclerotized, rectangular. Ductus bursae sclerotized, straight. Corpus bursae with two sclerites, the smaller with spinulose margins.

DISTRIBUTION: Palaearctic. Widely distributed all over Europe, except northernmost Scandinavia, British Isles and nearly all the Mediterranean islands, but present on Sicily. In the east to the Urals. In the Iberian peninsula replaced by subsp. *strigata*. Decreasing abundance and areal losses in south-western Germany (EBERT 2001), often due to intensification of pastures and afforestation (RETZLAFF et al. 1993; EBERT 2001). – Outside Europe from Morocco to Algeria (subsp. *strigata*) and from Turkey across Caucasus, Transcaucasus and northern Iran to central Asian mountains, western and southern Siberia (Altai and Sajan Mts.: subsp. *unicolorata*) and central Mongolia (VIIDALEPP 1975).

PHENOLOGY: Usually univoltine: early June to late July, exceptionally somewhat earlier, on the mountains up to two weeks later. At lower elevation in North Africa (subsp. *strigata*) emerging from mid-May. Rarely small, fresh specimens from late July to mid-September, belonging to a partial second generation. From Istria to eastern Austria and Hungary and along the Danube river to the Black Sea regularly bivoltine. Overwintering as larva, exceptionally as pupa.

BIOLOGY: Larva polyphagous, but with clear focus on Fabaceae. Repeatedly found on *Calluna vulgaris* (Ericaceae), and *Genista pilosa* (Fabaceae) (Finland: SEPPÄNEN 1954; Poland: URBAHN & URBAHN 1939; Germany: VOLLRATH et al. 1965; WEGNER pers. comm.), and on *Cytisus scoparius* (Fabaceae) (Germany: VOLLRATH et al. 1965; EBERT 2001), but also on *Lychnis vulgaris* (Caryophyllaceae) (Finland: SEPPÄNEN 1954), *Sanguisorba minor, Prunus spinosa* (Rosaceae), *Onobrychis viciifolia, Genista tinctoria, Genista pilosa* (Fabaceae), *Taraxacum officinale*, and *Leucanthemum vulgare* (Asteraceae) (Germany: EBERT 2001; ROESSLER 1866; Austria and northern Italy: ERLEBACH pers. comm.). Repeatedly recorded on Fabaceae and Ericaceae, such as *Cytisus scoparius, Genista tinctoria, Calluna vulgaris* (var. cit., e.g. REBEL 1910b; LHOMME 1935; BERGMANN 1955; FORSTER & WOHLFAHRT 1981, SKOU 1986; GÓMEZ DE AIZPÚRUA 1989), *Hippocrepis comosa, Onobrychis sativa, Coronilla, Medicago sativa falcata, Astragalus glycyphyllos, Anthyllis* (Fabaceae), but also on *Prunus spinosa, Rosa* (Rosaceae), *Rumex* (Polygonaceae), *Vaccinium myrtillus* (Ericaceae), *Inula conyza, Cichorium intybus, Leucanthemum vulgare* (Asteraceae), *Acinos arvensis* (Lamiaceae) (REBEL 1910b; LHOMME 1935; SCHNEIDER 1939; NORDSTRÖM et al. 1941; BERGMANN 1955; FORSTER & WOHLFAHRT 1981, SVENSSON 1993; EBERT 2001). Reared on *Laburnum anagyroides* (SCHNEIDER 1929; 1934; 1939; BERGMANN 1955), *Coronilla varia* (KRAUS 1993), *Polygonum, Genista* (ZSM), on *Astragalus danicus* (first instars), later on *Medicago sativa* (MÜLLER pers. comm.). Main activity of adults at dusk, easily flushed up from vegetation during daytime, rarely with diurnal flight activity, but readily attracted to light, too. Nectar feeding recorded on *Echium* (EBERT 2001).

HABITAT: Xerothermophilous, often on sand or limestone, in the south xeromontane. Preferring dry, rocky slopes or steppes with low vegetation, but with abundance of *Cytisus*, fallows, heathland, dams, sun-exposed forest fringes, warm scrubby places or scattered forests with *Quercus*, in the north often on wind protected dunes near coast, also on moors (SKOU 1986). North of Alps usually from 0 m up to 700 m above sea-level, up to 1,500 m in the Alps, exceptionally higher, up to 2,600 m in southern Europe and Asia. In southern Europe and Asia rarely below 1,000 m, from 1,600 m up to 3,000 m in North Africa and in northern Iran.

SIMILAR SPECIES: *R. calabra* differs in the yellow tinge of ground colour, hindwing rounded, antemedial line broader, medial line closer to postmedial line, forming a comparatively narrow postmedial fascia. *R. auctata* (STAUDINGER, 1879) from Turkey and Transcaucasus with strongly pointed forewing apex in both sexes, forewing discal spot more distinct, hindwing paler than forewing, subterminal line closer to termen.

REMARKS: Sister species, *R. serraginaria* KAILA et al., 1996, described from Tian Shan, with small differences, differential diagnosis needs further attention and study.

The ***calabra*** species-group

Small group of western Palaearctic species, very uniform in habitus. Wing coloration usually with red transverse fasciae. Hindwing margin rounded. Antennal flagellum dentate ventrally. Structure of ♂ hindtibia diagnostic, with 3-4 spurs, often modified to 'pseudospurs'. Pseudospurs are present also in *Tanaotrichia*, generic validity and possible subordination under *Rhodostrophia* may be discussed (STERNECK 1941: 197; 261). Male genitalia of the *calabra* species-group: uncus long, usually with subapical dilatation, tip of valva with spinose process, aedeagus without cornutus. Female genitalia with long lateral appendages of lamella postvaginalis connected with latero-anterior margin of tergum A8. Species identification sometimes very difficult in females, as there are only minor differences in habitus and genitalia. Sternum A2 of ♂ with antero-central setose patch (Textfig. 5). Further features under generic diagnosis. For zoogeography see remarks to *R. tabidaria*. Taxonomic discussion of many species in PETERSEN (1937) and AGENJO (1952).

Rhodostrophia auctata (STAUDINGER, 1879)

Pellonia auctata STAUDINGER, 1879: Horae Soc. ent. ross. 14 (4): 441 (northern Turkey: Karaman). Syntypes, at least 7 (MNHU: 6♂1♀, examined).

This species approaches European territory, with records from north-western Turkey, about 100 km east of Istanbul. In the east distributed to Transcaucasus and northern Iran. Wingspan 27-35 mm, second generation smaller, sometimes 25 mm only. Differential features in habitus see under similar species of *R. vibicaria*. Male genitalia (gen.fig. 168b) with uncus very long and slender, valva broad with long digitiform process. Bivoltine phenology from late April to late June and mid-July to mid-September. Montane species, usually collected from 1,000 up to 2,600 m above sea-level.

169. Rhodostrophia pudorata (FABRICIUS, 1794)

Phalaena pudorata FABRICIUS, 1794: Ent. Syst. 3 (2): 179 (North Africa, 'Barbaria'). Syntype(s).

Geometra (Acidalia) sicanaria ZELLER, 1852: Stettin. ent. Ztg. 13 (6): 180 (Sicily: Syracus). Syntypes (BMNH: 1♂ examined, with 'lectotype' label). Valid at subspecific rank.

Pellonia perezaria OBERTHÜR, 1875: An. Soc. esp. Hist. nat. 4: 372, pl. 17, figs 5, 6 (south-eastern Spain: Cartagena). Syntype(s) (ZFMK: 1♀ with 'holotype' label, examined). Valid at subspecific rank.

Rhodostrophia sicanaria quadricalcarata PROUT, 1913: in SEITZ, Macrolep. 4: 38, pl. 5, fig. c (Gibraltar). Holotype ♂ (BMNH).

Unavailable names (misidentification): *calabraria*: sensu GEYER (1828) nec *calabra* PETAGNA, 1787.

DIAGNOSIS: Wingspan ♂ 29-34 mm, ♀ 31-35 mm. Forewing dark sand coloured, often with olive green tinge, hindwing yellow. Basal area of forewing, transverse fascia between medial and postmedial line, and fringe purple; Terminal area of forewing towards fringe

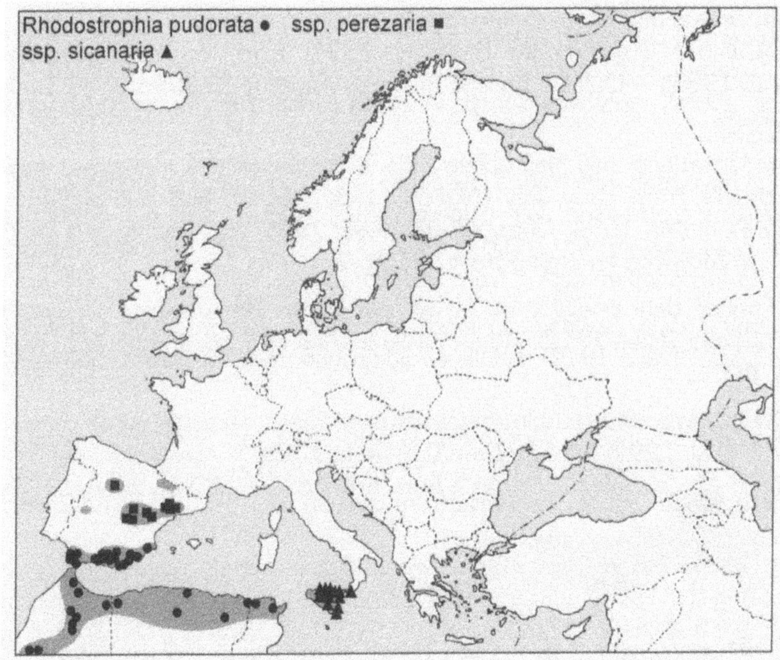

sometimes purple, too. Transverse fascia almost absent from hindwing, but distinct close to the inner termen. Discal spots absent. Frons purple or red brown, vertex sand coloured, often with purple scales. Longest branches of ♂ antennae about 4 times width of flagellum. Nominate subspecies with four normal spurs at the ♂ hindtibia, exceptionally three (1♂, southern Spain, ZSM; 1♂, northern Morocco: STERNECK 1941: 195).

Subsp. ***perezaria*** smaller, wingspan sometimes 27 mm only. Females on average smaller than males. Forewing dark sand coloured, sometimes with olive green tinge, hindwing usually pale yellow towards costa. Transverse fascia usually without red, indistinct on both wings. Two examined specimens from Juneda (Catalonia) however rather similar to the nominate subspecies in habitus. ♂ with four hindtibial spurs, the proximal spurs often of strongly unequal length.

Subsp. ***sicanaria*** as large as the nominate subspecies, females on average larger than males. Brightly coloured with deep purple markings strongly contrasting with the deep yellow hindwing. Fringe deep purple. Transverse fascia of hindwing very distinct close to tornus. ♂ with three hindtibial spurs, without spur anomalies, proximal spur very long and slender (Text-fig. 203).

MALE GENITALIA: Uncus broad, subapically dilated, terminally tapering. Valva very broad, at tip with a reniform, laterally spinose process. Without significant and constant differences between the subspecies.

FEMALE GENITALIA: Ductus bursae sclerotized, triangular. Corpus bursae with two large band-shaped sclerites, tapered anteriorly. Without significant and constant differences between the subspecies, except slightly shorter signum sclerites in subsp. *perezaria*. Only minor differences from other congeners in ♀ genitalia.

DISTRIBUTION: West-Mediterranean. Nominate subspecies in southern Spain, subsp.

perezaria in central and north-eastern Spain, subsp. *sicanaria* on Sicily and in southernmost Italy (Calabria, Aspromonte: PARENZAN 1994). Erroneously recorded for Tuscany (GRILLO et al. 1996). – Outside Europe in North Africa from Morocco to Tunisia (nominate subspecies).

PHENOLOGY: Univoltine: mid-May to late June. In the mountains sometimes until mid-July. Small specimens from early July (Algeria), early August, and mid-September (Sicily: GIULIANO & PARENZAN 1994; GRILLO & PARENZAN 1995) may belong to a partial second generation. In Morocco emerging from mid-April.

BIOLOGY: Limited data available on larval host-plants. Recorded on *Cytisus scoparius* (Fabaceae) (Spain: RIBBE 1912). Adults easily flushed up from shrubs, but readily attracted to light, too.

HABITAT: Xerothermophilous. Hot lowlands near coast, on Sicily most common in oak forest (BELLA et al. 1999). Usually from 0 m up to 1,300 m above sea-level, up to 1,700 m in northern Africa. Subsp. *sicanaria* from 600 up to 1,800 m, exceptionally down to 300 m.

SIMILAR SPECIES: *R. calabra* sometimes rather similar, certain olive forms in the Iberian peninsula being strongly reminiscent of *R. p. perezaria* in habitus. Identification of males easy by number and shape of hindspurs. Secure identification of ♀ usually, but not always, possible by habitus features (*R. pudorata*: hindwing brighter yellow, postmedial fascia indistinct towards forewing). The Iberian subspecies *R. calabra separata* differs furthermore in having discal spots present on upper- and underside of wings, and postmedial fascia usually divided into a double line, the proximal being distinct and sharp on the hindwing. In a few, rare cases doubts about correct identification of a ♀ may remain even after dissection.

REMARKS: *R. p. sicanaria* from Sicily has repeatedly been considered a separate species (e.g. PARENZAN 1994; RAINERI & ZANGHERI 1995), based on the two conspicuous differential features long ♂ proximal spur and bright habitus.

Text-figs 201-202: Diagnostic characters (indicated) of European *Rhodostrophia* species. Text-fig. 201: *R. pudorata perezaria* (OBERTHÜR, 1875). Text-fig. 202: *R. calabra separata* PROUT, 1935.

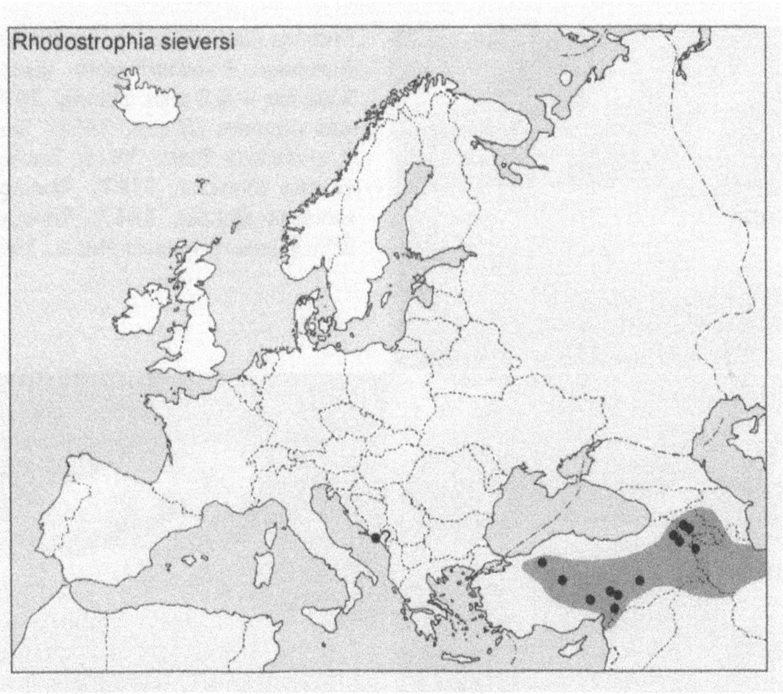

Rhodostrophia sieversi (CHRISTOPH, 1882)

Aspilates sieversi CHRISTOPH, 1882: Horae Soc. ent. ross. 17 (1/2): 115 (Armenia: Ordubad). Holotype ♀ (ZISP).

DIAGNOSIS: Wingspan ♂♀ 29-37 mm. Ground colour of all wings yellow. Basal area of forewing, transverse postmedial fascia and terminal area towards fringe purple. Yellow medial area broad. Postmedial fascia crenulate, mainly on hindwing. Discal spots large. Underside of forewing dark grey in the basal half in both sexes. Frons purple. Longest branches of ♂ antennae about 4 times width of flagellum. ♂ with three normal hindtibial spurs, the proximal rather long.

MALE GENITALIA: Uncus slender to tip. Valva with two terminal projections, the costal broad, round and spinose, the ventral long, digitiform and smooth, both with subapical teeth.

FEMALE GENITALIA: Ductus bursae broad. Paired band-shaped signa narrow and short.

DISTRIBUTION: East-Mediterranean. Reputedly collected in south-western Croatia (1♂ with handwritten label 'Dalm. Ragusa' = Dubrovnik; ZSM). The occurrence in Europe awaits further confirmation. – Outside Europe in Turkey, even close to the Bosporus. Recorded also from Transcaucasus, northern Iraq and northern Iran.

PHENOLOGY: Univoltine: early May to late June. In Armenia bivoltine with a second generation in September.

BIOLOGY: Larval host-plants unknown. Adults attracted to light.

HABITAT: Limited information available about habitat preference. The locality in Croatia in the coastal plain at sea-level. In Turkey and Transcaucasus up to 1,400 m above sea-level.

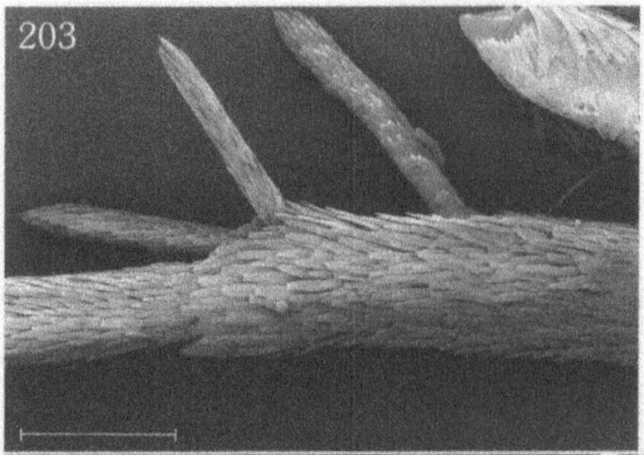

Text-figs 203-207: ♂ hindtibial spurs in European *Rhodostrophia* species (SEM). Scale bar = 0.5 mm. Text-fig. 203: *R. pudorata sicanaria* (ZELLER, 1852). Text-fig. 204: *R. cretacaria* REBEL, 1916. Text-fig. 205: *R. calabra* (PETAGNA, 1787). Text-fig. 206: *R. tabidaria* (ZELLER, 1847), Rhodos. Text-fig. 207: *R. discopunctata* AMSEL, 1935.

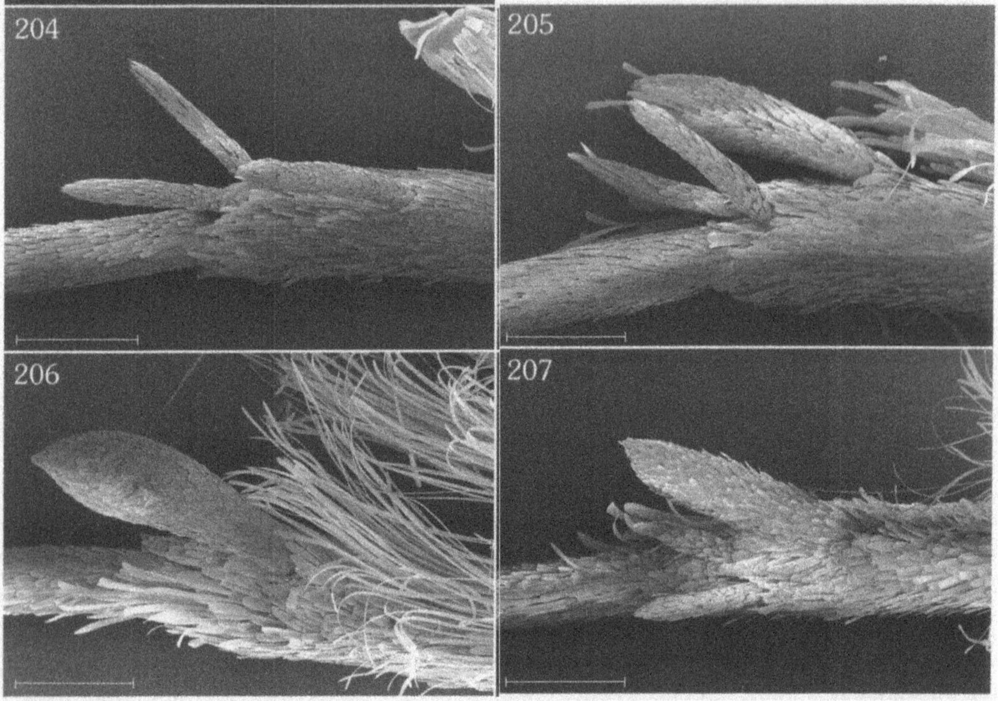

SIMILAR SPECIES: Differing from the other species of the species-group in the yellow ground colour of forewings, postmedial fascia of hindwing crenulate, underside of forewing with large grey patch at the inner temen. Both *R. discopunctata*, and *R. tabidaria* with darker ground colour, medial area of forewing narrower, dark grey patch of forewing underside lacking, postmedial fascia of hindwing straighter, ♂ hindtibia with large proximal pseudospur.

REMARKS: The inclusion of this species in the European fauna needs to be confirmed, see under 'distribution'. Absence of pseudospurs possibly plesiomorphic, but male genitalia seem to indicate relationships to *R. discopunctata*.

170. Rhodostrophia cretacaria REBEL, 1916

Rhodostrophia sicanaria cretacaria REBEL, 1916: Ann. K.K. Naturhist. Hofmus. Wien 30: 135, pl. 4, fig. 1 (Crete: Frankokasteli; Chalepa; Suda Bay; Nerokuri; Mélabes; Asitaes; Harmeni; Kristallenia). Syntypes 9♂8♀ (NHMW).

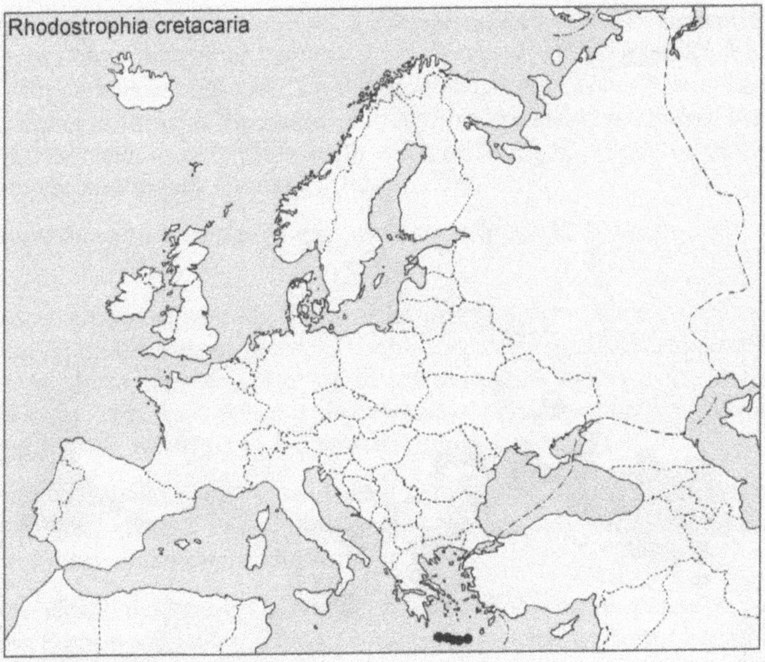

Unavailable names (misidentification): *calabra*: sensu PROUT (1913; partim) nec PETAGNA, 1787. – (incorrect subsequent spellings): *cretacearia*: STERNECK (1941) and MÜLLER (1996).

DIAGNOSIS: Wingspan ♂♀ 27-32 mm. Ground colour dirty olive green, hindwing sand coloured towards costa. Transverse fasciae purple, violet or brownish. Postmedial fascia usually distinctly bordered by a double line. Area between hindwing transverse lines broader than that of forewings. Discal spots distinct on all wings. Frons red brown, with olive green scales. Longest branches of ♂ antennae about four times width of flagellum. ♂ with three hindtibial spurs (Text-fig. 204), the proximal long and slender, but subterminally often slightly dilated and terminally rounded, reminiscent of 'pseudospurs' of the following species.

MALE GENITALIA: Similar to those of *R. calabra*. Uncus comparatively short and broad. Valva with broad, truncate spinose terminal process. Ventral digitiform process rough, surface scale-like, smooth appendix very small. Aedeagus slender.

FEMALE GENITALIA: Corpus bursae pyriform. Paired band-shaped signa short, narrow, laterally dentate.

DISTRIBUTION: Endemic to Crete.

PHENOLOGY: Univoltine: mid-April to mid-June.

BIOLOGY: Larval host-plants unknown. Adults attracted to light.

HABITAT: No detailed information available about habitat preference. From 0 m up to 700 m above sea-level.

SIMILAR SPECIES: No similar species on Crete.

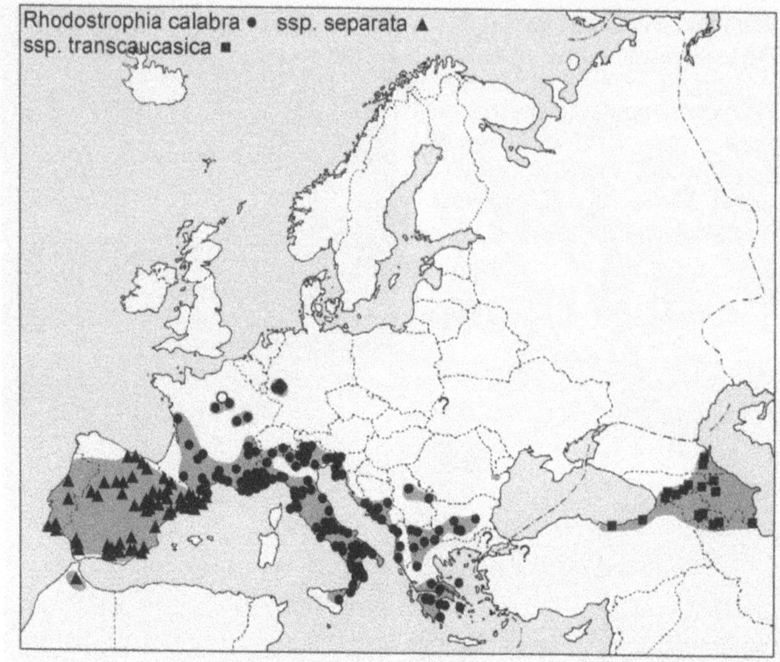

171. Rhodostrophia calabra (PETAGNA, 1787)

Phalaena calabra PETAGNA, 1786: Spec. Ins. ult. Calabr.: 38, pl. 1, fig. 23 (Italy, Calabria: Aspromonte centr. ('Montes Quinquefrondi')). Syntype(s).

‡ *Phalaena trifasciata* CYRILLO, 1787: Entom. Neap. Spec. prim.: 2, pl. 1, fig. 6 (Italy, Campania: Napoli). Syntype(s). Junior primary homonym nec SCOPOLI, 1772 (identity: *Idaea serpentata*). Listed under *R. vibicaria* in LERAUT (1997) under *R. calabra* in SCOBLE (1999).

Phalaena calabraria HÜBNER, 1790: Beitr. Gesch. Schmett. 2 (4): 92, pl. 2, fig. M. Emendation.

Aspilates taeniaria FREYER, 1834: Neuere Beitr. Schmettkde. 2 (22): 59, pl. 131, fig. 3 (Croatia, Dalmatia: Dubrovnik ('Ragusa')). Syntype(s) ♀.

Rhodostrophia calabraria var. *muscosa* BASTELBERGER, 1908: Ent. Zeits. Stuttgart 21 (36): 217 (Germany: Kreuznach). Syntypes 3♂1♀ (SMF).

Rhodostrophia calabra transcaucasica PROUT, 1920: Novit. zool. 27: 268 (Transcaucasus: Borjom). Syntypes 6♀ (BMNH). Valid at subspecific rank (SCOBLE 1999).

Rhodostrophia calabra separata PROUT, 1935: in SEITZ, Macrolep. 4, Suppl.: 23. First available name after *separata* THIERRY-MIEG, 1889 (infrasubspecific) (southern France, Drôme). Syntypes (of PROUT: BMNH). Valid at subspecific rank, see remarks.

Unavailable names (infrasubspecific): *coacta*: PROUT (1935); *iberica*: PETERSEN (1937); *languida*: DANNEHL (1927); *punctaria*: CARADJA (1931); *rubrofasciata*: DANNEHL (1933); *sanguinea*: THIERRY-MIEG (1889); *subseparata*: PROUT (1935); *unilinea*: PROUT (1913); *violettaria*: VORBRODT (1917). – (incorrect subsequent spellings): *callabra*: PROUT (1913); *punctata* PROUT (1934); *trancaucasic*: DIDMANIDZE (2001).

DIAGNOSIS: Wingspan ♂ 28-33, ♀ 28-35 mm. Ground colour ochreous with olive tinge, hindwing lighter towards costa. Postmedial transverse fascia comparatively narrow, usually filled with purple. Discal spots in nominate subspecies absent, with a very few exceptions. Frons red brown. Longest branches of ♂ antennae about four times width of flagellum (Text-fig. 198). Hindtibia of ♂ with three hindtibial spurs, the proximal strongly dilated ('pseudospur'), length 1.15-1.35 mm (Text-fig. 205). Distal spurs slender, comparatively short, usually of unequal length, in one ♂ (Greek island Serifos) reduced to one very short spur. No other spur anomalies found (see *R. discopunctata*).

Subsp. ***separata*** smaller, wingspan ♂♀ 26-33 mm. Ground colour light ochreous, pattern often more brownish rather than red. In the lowlands of the southern Iberian peninsula often with dark violet suffusion, on the mountains paler. Discal spots usually slightly visible, borders of postmedial fascia darker, the fascia appearing as double line. Antennal branches of ♂ sometimes only three times width of flagellum. Hindtibia of ♂ with pseudospur narrower and slightly shorter (1.0-1.2 mm).

Subsp. ***transcaucasica*** similar to the nominate subspecies, but with discal spots present.

MALE GENITALIA: Uncus subapically dilated, dilatation less than half length of uncus. Valva with truncate spinose terminal process. Ventral digitiform processes shorter than in *R. discopunctata*. Juxta narrower. No significant difference between the nominate subspecies and subsp. *separata*. Subsp. *transcaucasica* with somewhat longer uncus, terminal dilatation shorter, restricted to the terminal third of uncus.

FEMALE GENITALIA: Ductus bursae often bulbed anteriorly. Corpus bursae usually elongate oval, comparatively slender. Paired band-shaped signa usually short, especially in subsp. *separata*, in France and western Germany.

DISTRIBUTION: (South-)European-Caucasian, disjunct. From Iberian peninsula (subsp. *separata*) across France and Italy to Balkan peninsula including western Greek islands (e.g. Serifos and Skiros). Present on Sicily. In the north in scattered populations to central France and western Germany. Doubtfully recorded for the Ukraine (ROMANISZYN & SCHILLE 1929; VIIDALEPP 1996). Records for European Turkey (SEVEN 1991) await confirmation. – Outside Europe with isolated occurrences in northern Morocco, northern and central Turkey (e.g. PETERSEN 1937; KOCAK 1990b; 1991; 1993; RIEMIS 1994; DOGANLAR pers. comm., examined), eastern Caucasus and Transcaucasus (subsp. *transcaucasica*). In the eastern Mediterranean replaced by the closely related sister species *R. tabidaria*.

PHENOLOGY: Univoltine: mid-May to late June, exceptionally from early May, in the western Alps and southern European mountains from late May to late July. In the lowlands of the southern Iberian peninsula emerging from mid-April.

BIOLOGY: Larva probably oligophagous, with focus on Fabaceae. Found on *Cytisus scoparius* (Fabaceae) (Germany: FUCHS 1879). Recorded on *Cytisus scoparius, Genista, Dorycnium* (Fabaceae), *Scabiosa* (Dipsacaceae), *Rumex, Polygonum* (Polygonaceae), *Thymus* (Lamiaceae), and *Asperula* (Rubiaceae) (var. cit., e.g. REBEL 1910b; PROUT, 1913; LHOMME 1935; FORSTER & WOHLFAHRT 1981), mainly *G. tinctoria* (LHOMME 1935). Reared on *Cytisus scoparius* (RIBBE 1912; ERLEBACH pers. comm.). Adults nocturnal and diurnal, easily flushed up from vegetation during daytime.

HABITAT: Xerothermophilous. Mediterranean maquis and scattered oak forest, with abundance of *Genista*. From 0 m up to 1,000 m above sea-level, exceptionally up to 1,600 m. Subsp. *separata* from 0 m up to 2,000 m.

SIMILAR SPECIES: *R. vibicaria* (see no. 168), *R. pudorata* (169). *R. discopunctata* differs in large discal spots, females on average larger, postmedial fascia slightly broader, and slightly dentate, forewing basal area often reddish, ♂ hindtibia with shorter pseudospur (usually shorter than twice length of distal spurs). *R. tabidaria* with discal spots, wing coloration with orange tinge.

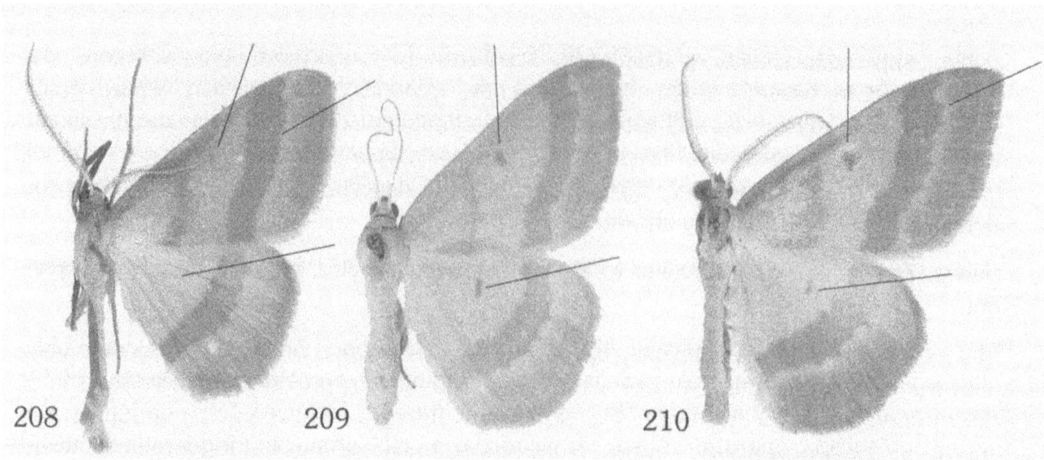

Text-figs 208-210: Diagnostic characters (indicated) of European *Rhodostrophia* species. Text-fig. 208: *R. calabra* (PETAGNA, 1787). Text-fig. 209: *R. tabidaria* (ZELLER, 1847). Text-fig. 210: *R. discopunctata* AMSEL, 1935.

REMARKS: For morphology of immature stages see FUCHS (1879). The populations from Caucasus and Transcaucasus ('*transcaucasica*') need more detailed study to examine possible validity at species rank. Traditionally, the subspecies name *separata* is applied to the populations of the Iberian peninsula despite the southern French type locality. Further study is needed to delimitate distributional borders between the subspecies.

172. Rhodostrophia tabidaria (ZELLER, 1847)

Geometra tabidaria ZELLER, 1847: Isis 1847 (7): 506 (Greece: Rhodos; western Turkish coast). Syntypes 6♂♀ (BMNH, 1♂ examined, with 'lectotype' label).
Rhodostrophia sicanaria cypriaria REBEL, 1916: Jahresb. Wien. Ent. Ver. 26: 107 (Cyprus: by Nicosia). Syntypes 1♂1♀ (NHMW). Herewith downgraded from species (HAUSMANN 1994c; SCOBLE 1999) to subspecific rank (stat. n.).
Rhodostrophia calabra cypria PROUT, 1920: Novit. Zool. 27: 268 (Cyprus: Aghirda). Holotype (BMNH).
Unavailable names (misidentification): *calabraria*: sensu ZELLER (1847) nec PETAGNA, 1787.

DIAGNOSIS: Wingspan ♂ 30-32, ♀ 31-36 mm. Ground colour of forewing olive green or ochre, hindwing yellowish towards costa. All wings with with orange suffusion, sometimes strongly. Antemedial line indistinct or thin. Postmedial transverse fascia comparatively narrow, usually filled with dark violet. Discal spots usually present, sometimes distinct. Frons red brown. Longest branches of ♂ antennae about 4 times width of flagellum. Hindtibia of ♂ as described for *R. calabra* with three hindtibial spurs, pseudospur strongly dilated, length 1.3-1.4 mm on easternmost Greek islands (Text-fig. 206), 1.0-1.1 mm on Cyprus (here slightly narrower). Distal spurs very short, of unequal length.

MALE GENITALIA: Very similar to those of *R. calabra*, see no. 171. Uncus dilatation usually half length of uncus. *R. t. cypriaria* with broader valva and shorter aedeagus.

FEMALE GENITALIA: Without clear differential features from those of *R. calabra*, see no. 171. Paired band-shaped signa as short and narrow as in the western Mediterranean populations of *R. calabra*.

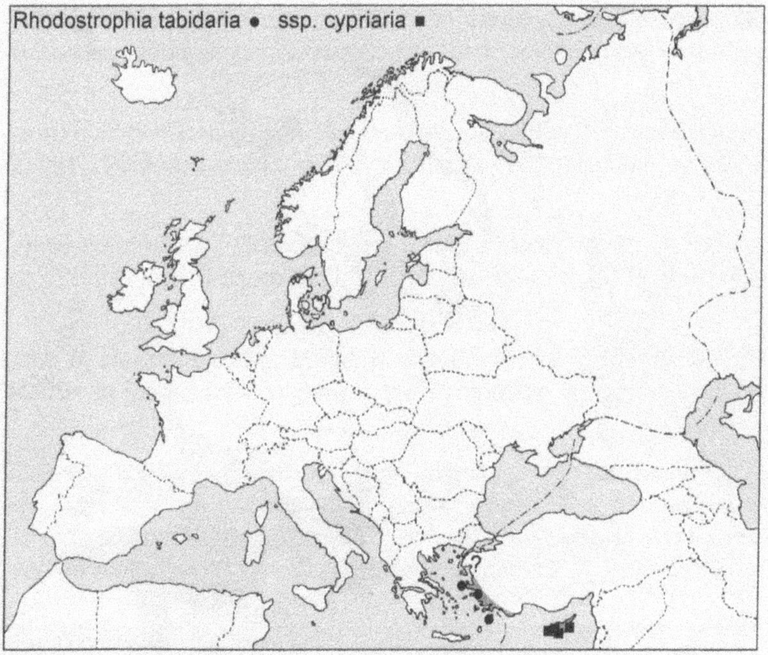

DISTRIBUTION: In Europe restricted to the easternmost Greek islands, e.g. Rhodos, Chios. – Outside Europe in westernmost Turkey and on Cyprus (subsp. *cypriaria*).

PHENOLOGY: Univoltine: early April to late May, thus earlier than the congeners.

BIOLOGY: Larval host-plants unknown. Adults attracted to light.

HABITAT: Xerothermophilous. No detailed information available on habitat preference. From 0 m up to 900 m above sea-level.

SIMILAR SPECIES: *R. calabra* (see no. 171), *R. sieversi* (169b). *R. discopunctata* differs in the larger size of females, discal spots larger, pseudospur of ♂ hindtibia shorter (usually shorter than twice length of distal spurs).

REMARKS: The name *tabidaria* has erroneously been used, for 150 years, for the following species, see under *R. discopunctata*. Differential features from *R. calabra* are weak in genitalia, but considerable in habitus and external morphology. The taxon *cypriaria* is downgraded to subsp. of *tabidaria*. From a zoogeographical point of view the distribution pattern of the closely related species pair *R. calabra* and *R. tabidaria* (with '*cypriaria*') is strongly reminiscent of the pattern in the Geometrinae species *Pseudoterpna coronillaria* (HÜBNER, 1817) (see Vol. 1), with separate taxa in similarly isolated populations on the easternmost Greek islands, westernmost Turkey and on Cyprus. Interestingly both genera are of similar size and feed on Fabaceae.

173. Rhodostrophia discopunctata AMSEL, 1935, stat. n.

Rhodostrophia discopunctata AMSEL, 1935: Veröff. Dt. Kolon. U. Übersee-Mus. Bremen 1 (2): 243 (Israel: Kiriath Anavim near Jerusalem). Lectotype ♂ (coll. PÜNGELER in MNHU, examined, formerly labelled as 'type'), herewith designated to fix type locality and to stabilize usage of the name; at MNHU 5♀ paralecto-

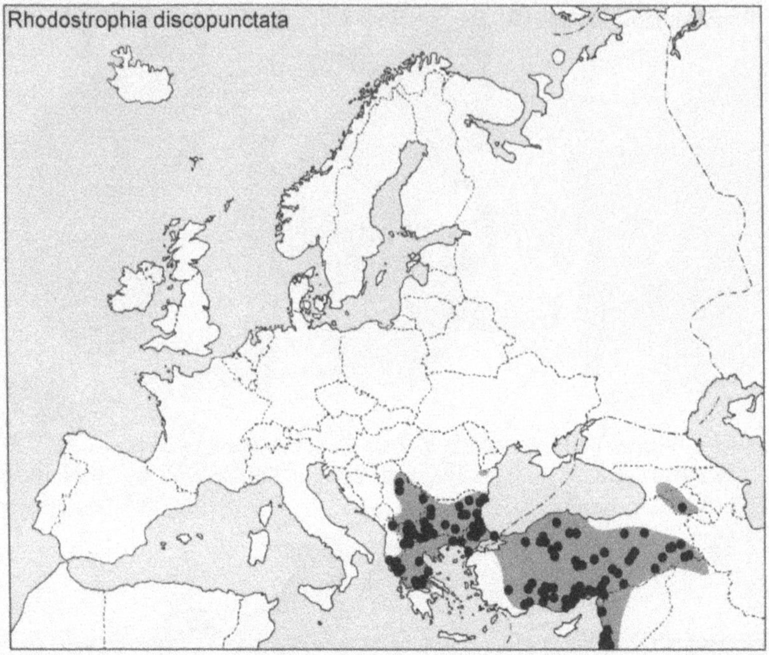

types from the type locality, Dead Sea and Beirut; in description furthermore mentioned: Karmel; Abu Gosh; Ramallah; Wadi el Kelt; Jericho. Herewith raised from synonymy to species rank (stat. n.), replacing misidentified *R. tabidaria* (see below), e.g. in MÜLLER (1996) and VIIDALEPP (1996).

‡ *Pellonia tabidaria*: sensu GUENÉE, 1858: in BOISDUVAL & GUENÉE, Hist. nat. Insectes (Spec. gén. Lépid.) 10 (2): 9. Misidentification nec ZELLER, 1847.

Unavailable names (misidentification): *calabraria*: sensu ZELLER (1847) nec *calabra* PETAGNA, 1787. – (infrasubspecific): *suavis*: CARADJA (1931); *tenuistrigata*: CARADJA (1931).

DIAGNOSIS: Wingspan ♂ 27-34, ♀ 32-37 mm, sexual dimorphism in size greater than in the related species. Ground colour ochreous, often with olive tinge, mainly on forewing. Postmedial transverse fascia often broad, usually filled with purple. Discal spots distinct and large. Frons red brown. Longest branches of ♂ antennae 3-4 times width of flagellum. Hindtibia of ♂ with three hindtibial spurs (Text-fig. 207), the proximal pseudospur dilated, length 1.0-1.2 mm. Distal spurs usually long and slender, usually of about equal length. Exceptionally with four spurs (2♂♂ from Turkey and Israel), the additional proximal spur short, slender or dilated. Hindtibia of ♀ with four spurs, one of the proximal spurs often very short, absent at a frequency of 10-20% (see also STERNECK 1941: 196; PROUT 1935a: 24; GUENÉE 1858: 9).

MALE GENITALIA: Uncus dilatation about half length of uncus. Valva with rounded spinose terminal process. Ventral digitiform processes fused, laterally rough, with scale-like surface, much longer than in *R. calabra*. Juxta broader.

FEMALE GENITALIA: Similar to those of the two preceding species. Ductus bursae anteriorly straight. Corpus bursae large, pyriform. Paired band-shaped signa comparatively long.

DISTRIBUTION: East-Mediterranean. In Europe restricted to Balkan peninsula, but absent from Peloponnes and from all Greek islands. – Outside Europe in Turkey, the Levant, western Transcaucasus (VIIDALEPP 1996; DIDMANIDZE 2002), and northern Iraq (WILTSHIRE 1957). Records for southern European Russia (VIIDALEPP 1996) probably erroneous.

PHENOLOGY: Univoltine: early May to mid-June, in the mountains exceptionally until late July. In the Levant, depending on altitude, bi- or trivoltine from mid-March to mid-November.

BIOLOGY: Larval host-plants unknown. Adults easily flushed up from vegetation, e.g. from *Coronilla* (ZERNY 1933), sometimes with active flights during daytime, readily attracted to light, too.

HABITAT: Xerothermophilous. Grassland and scrubby places with *Coronilla varia libanotica* (ZERNY 1933). From 0 m up to 1,200 m above sea-level, in the Lebanon up to 2,000 m.

SIMILAR SPECIES: *R. sieversi* (see no. 169b), *R. calabra* (171), *R. tabidaria* (172). The last two species differ in longer pseudospur on the ♂ hindtibia (longer than twice length of distal spurs).

REMARKS: This species was, for many years, known under the name '*R. tabidaria*'. Unfortunately, this usage was based on a misidentification: The populations of the eastern Greek islands (with the types of *tabidaria* from Rhodos) and westernmost Turkey have to be separated at species level (see preceding species).

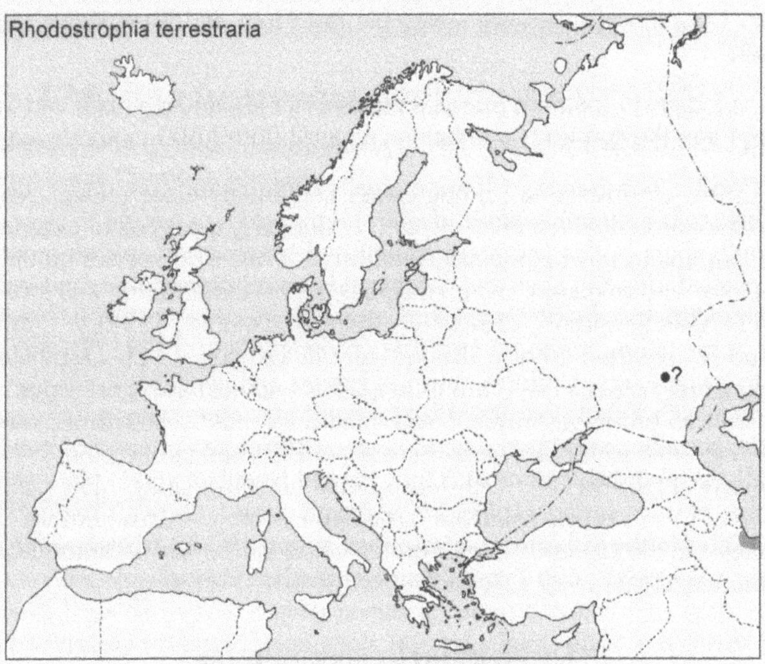

Rhodostrophia terrestraria (LEDERER, 1869)

Eusarca terrestraria LEDERER, 1869: Soc. ent. ross. 6: 90, pl. 5, fig. 9 (Iran: Astrabad). Syntypes 2♂.
Eusarca pellonaria CHRISTOPH, 1885: In ROMANOFF, Mém. Lépid. 2: 126, figs 6a, 6b (Southern Turkmeniya: Achal Tekke region). Syntypes ♂♀. Valid at subspecific rank.

DIAGNOSIS: Wingspan ♂ 32-35, ♀ 30-34 mm. Forewing dark sand coloured, often greyish, hindwing pale sand coloured, with brown tinge towards termen. Forewing with broad, straight, brown postmedial fascia and a similar fascia in the terminal area. Discal spot present only on forewing. Palpi, frons and vertex dark sand coloured. Longest branches of ♂ antennae 4-5 times width of flagellum. Hindtibia of ♂ with four normal spurs, both proximal and distal pair of strongly unequal length.

Variability: Populations from Turkmeniya and Alexander mountains (subsp. *pellonaria*) usually with yellow hindwings, reminiscent of *R. dispar* (STAUDINGER, 1892).

MALE GENITALIA: Valva with two broad, strongly spinulose terminal lobes. Aedeagus strongly curved, with narrow cornutus.

FEMALE GENITALIA: Apophyses long and stout. Ductus bursae curved and swollen, two small band-shaped signa at the junction with corpus bursae. The latter small, narrow sac-shaped.

DISTRIBUTION: Turanian. Reputedly collected in southern European Russia (1♂ from 'Sarepta'; leg. STAUDINGER; ZSM), possibly mislabelled. Not mentioned by ANIKIN et al. (2000). Not traced in various collections in southern European Russia (ZOLOTUHIN pers. comm.). Occurrence in Europe is not excluded, but awaits confirmation. – Outside Europe in northern Iran, Turkmeniya and northern Afghanistan (subsp. *pellonaria*). Erroneously recorded for 'Transcaucasus' (SCOBLE 1999).

PHENOLOGY: Univoltine (limited data available from Iran and Turkmeniya): late July to early August.

HABITAT: The locality in southern European Russia in a lowland river plain. In Turkmeniya and northern Iran the species is montane at about 2,000-2,500 m above sea-level.

REMARKS: Possible occurrence in Europe awaits confirmation (see under 'distribution'). Closely related to the central Asian *R. dispar*. Further study is needed to clear up the systematic relationships in this taxonomically difficult '*terrestraria* species-group', characterized by pointed forewing apex and considerable sexual dimorphism, and including several western Palaearctic species, such as *R. dispar* (STAUDINGER, 1892), *R. xesta* PROUT, 1935 (syn. n.: *R. oxyntis* PROUT, 1935), *R. furialis* BRANDT, 1941. Distributed mainly from Iran to central Asia. Uniform group in habitus and genitalia. Forewings dark sand coloured, often with broad postmedial fascia, hindwings often yellowish. Forewing venation variable, but often with open second areole, or R5 fused in one point only with the other radial veins. Hindtibia of ♂ with four normal hindtibial spurs. Male genitalia with terminal lobes of valva strongly spinulose, aedeagus curved, cornutus narrow. The group is possibly related to the *jacularia* species-group, where similar sexual dimorphisms occur. *R. pleonasma* WILTSHIRE, 1967 may be an intermediate species.

The *badiaria* species-group

Small group of broad winged, sand coloured species, uniform in habitus, and with slight sexual dimorphism. Several extralimital species, from southern Turkey across Iran to central Asia. Hindtibia of ♂ with four normal hindtibial spurs. Sternum A2 of ♂ without setae. Male genitalia with narrow uncus, valva not spinulose at tip, aedeagus curved. Relationship to '*Apostates*', actually placed in the *jacularia* species-group, stated by STERNECK (1941).

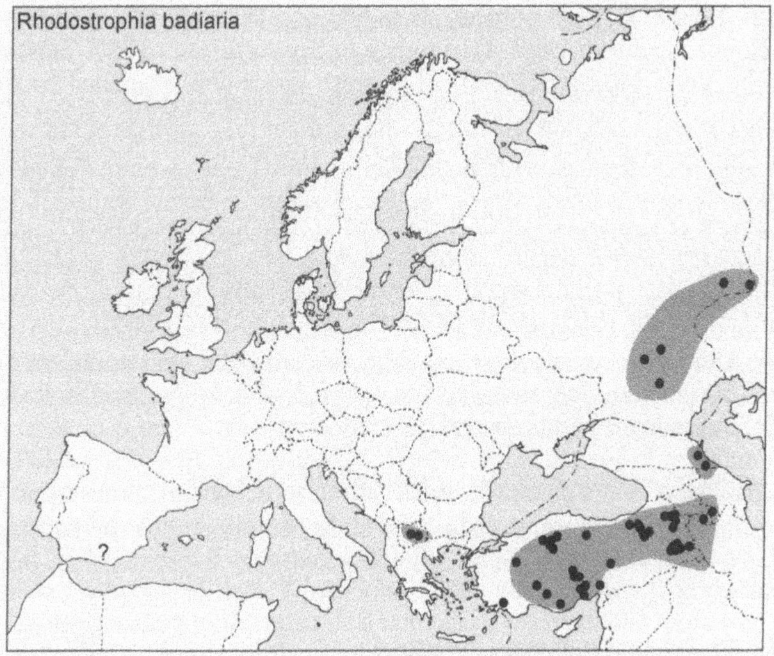

174. *Rhodostrophia badiaria* (FREYER, 1841)

Fidonia badiaria FREYER, 1841: Neuere Beitr. Schmettkde. 4 (59): 115, pl. 354, fig. 2 (south-eastern European Russia: Krasnoarmeysk ('Sarepta')). Syntype(s).

Eusarca telaria HERRICH-SCHÄFFER, 1847: Syst. Bearb. Schmett. Eur. 3 (24): 34 (southern European Russia: Volga river; southern Urals: Orenburg). Syntypes.

‡ *Eusarca telaria* [syn.] *cervinaria* HERRICH-SCHÄFFER, 1847: Syst. Bearb. Schmett. Eur. 3 (24): 34 (probably southern European Russia). Introduced in synonymy.

Fidonia praecanaria EVERSMANN, 1847: Bull. Soc. imp. Nat. Moscou 20 (2): 81, pl. 6, figs 4, 5 (Russia: lower Volga river, Kamyschin; southern Urals, Orenburg). Syntype(s). Introduced to replace misidentified *emucidaria* sensu EVERSMANN (see below).

‡ *pollenaria* FREYER, 1854: Neuere Beitr. Schmettkde. 7 (107): 61 (south-eastern European Russia: Krasnoarmeysk ('Sarepta')). No separate types. Non binominal; unnecessary replacement name for *badiaria*.

Unavailable names (misidentification): *emucidaria*: sensu EVERSMANN (1844) nec HÜBNER, 1813 (identity: *Dyscia fagaria*, Ennominae); *praecisaria*: sensu PROUT (1913, fig. 3d) nec STAUDINGER, 1892 (identity: *Rhodostrophia praecisaria*).

DIAGNOSIS: Wingspan ♂ 28-33 mm, ♀ on average smaller, sometimes 25 mm only. Coloration variable, usually sand coloured, sometimes with grey or ochreous tinge. Ante- and postmedial lines dark grey, slightly undulate, often dissolved into spots or streaks. Medial shadow usually present. Discal spots present. Palpi, frons and vertex pale sand coloured. Longest branches of ♂ antennae about 4 times width of flagellum. Hindtibia of ♂ with four normal spurs, both proximal and distal pair of strongly unequal length.

MALE GENITALIA: Uncus narrow, slightly bifid at tip. Valva bilobed at tip, not spinulose. Aedeagus strongly curved, at tip with one very small, lateral spinule, rarely with two spinules.

FEMALE GENITALIA: Ductus bursae sclerotized, comparatively narrow, closely connected to sterigma. Corpus bursae small, elongate oval, paired band-shaped signa small and narrow, anteriorly with a small spinule.

DISTRIBUTION: Pontic. Isolated populations in Macedonia, more widespread in southern European Russia, southern Urals. Occurrence in Spain (WEHRLI 1927; EXPOSITO 1978) questioned by VIVES MORENO (1994), all Spanish specimens examined by the author, revealed to have been collected by KORB, who is known for having caused various mislabellings. – Outside Europe in Turkey, Caucasus, Transcaucasus. In large parts of northern Iran replaced by the sister species *R. iranica* SCHWINGENSCHUSS, 1939 (WILTSHIRE 1970; DOGANLAR & HAUSMANN 2004). Status of Turkmeniyan populations (VIIDALEPP et al. 1992; VIIDALEPP 1996) require separate study. The very similar species *R. bahara* BRANDT, 1938 distributed from central Turkey and Azerbaijan to northern Iraq and south-western Iran (WILTSHIRE 1957; 1970; DOGANLAR & HAUSMANN, 2004).

PHENOLOGY: Univoltine: usually from mid-May to early July, but dependent on altitude, e.g. above 2,500 m early July to early August. One fresh late specimen, from late August at 2,000 m, indicating a partial second generation.

BIOLOGY: Larval host-plants unknown. Adults attracted to light.

HABITAT: Limited information available about habitat preference. In European Russia inhabiting dry lowland steppes, in Macedonia and Turkey montane, from 900 m up to 3,000 m above sea-level.

SIMILAR SPECIES: No similar species in Europe.

Rhodostrophia vastaria (CHRISTOPH, 1876)

Eusarca vastaria CHRISTOPH, 1876: Horae Soc. ent. ross. 12: 261, pl. 7, figs 31,32 (Russia: steppes by Krasnowodsk). Syntypes ♂♀.

DIAGNOSIS: Wingspan ♂ 25-27 mm. Ground colour of all wings light brown, with slight grey suffusion. ♀ paler (PROUT 1913). Wing pattern very weak, or absent, largely restricted to the dark forewing discal spots and, sometimes, to darker shadows in the terminal area of the forewing. Palpi, frons and vertex concolorous with ground colour. Longest branches of ♂ antennae 4-5 times width of flagellum. Hindtibia of ♂ with four normal spurs.

MALE GENITALIA: Uncus narrow, slightly bifid at tip. Valva broad, slightly bilobed at tip, strongly spinulose. Aedeagus with one long and narrow 'cornutus' at tip, similar to that of *R. jacularia*. Sternum A8 deeply notched posteriorly.

DISTRIBUTION: Turanian. Recorded in south-western Kazakhstan (Ust-Urt Plateau) and along the Caspian shore line in western Turkmeniya. Reputedly in Tian Shan (PROUT 1913), data await confirmation. Possible occurrence in the Astrachan semi-deserts (see remarks).

PHENOLOGY: Univoltine (limited data available): early May to late May.

HABITAT: Lowland steppes from 0 m up about 300 m above sea-level.

REMARKS: This species has been recorded in the steppes of south-western Kazakhstan, not far from Europe. As there is no zoogeographical barrier separating similar habitats in south-easternmost European Russia, occurrence in Europe is possible, too.

Tribus **Timandrini** STEPHENS, 1850

(=†Calothysanini HERBULOT, 1963)

45 species in four genera in the currently accepted concept (HOLLOWAY 1997; SIHVONEN & KAILA 2004), embracing also the Nearctic genus *Haematopis*, and the tropical Old World genera *Synegiodes*, and *Traminda*. Distribution center in the Indo-Pacific region. Inclusion of *Traminda* tentative, mainly based on the absence of a pupa girdle (HOLLOWAY 1997), though venation and striation of valva in ♂ genitalia matches well equivalent features in Cosymbiini. Long medial projection of tympanic ansa in *Traminda* suggesting relationship to Rhodometrini (HOLLOWAY 1997). For sister group relationship between Timandrini, Cosymbiini and Rhodometrini ('Timandrini lineage') see introduction to the subfamily. Preliminary data sets from own molecular research (mitochondrial genes ND1, COI, and 16S rRNA) support monophyly of 'Timandrini lineage' (MILLER et al. 2001; HAUSMANN et al. in prep.). STERNECK (1941: 226) states similarities between Timandrini and genus *Pylargosceles* (Rhodostrophiini). Further study is needed to refine tribal definition and delimitation. Reduction of uncus, shape of sacculus and longitudinal striation of valva in many Timandrini strongly reminiscent of Cosymbiini and some Rhodometrini (*Ochodontia*). In ♀ genitalia the presence of a flat signum sclerite with longitudinal ridge supports these relationships, too. Timandrini usually with straight reddish line on both wings, on forewing ending in the apex. Full number of ♂♀ hindtibial spurs may suggest a basal position in the phylogeny of Sterrhinae. Monophyly of the tribe supported by shape of hindwing termen (tailed or slightly projecting at M3), ♂ sternum A8 with posterior membranous lobes and presence of gnathos (SIHVONEN & KAILA 2004).

DIAGNOSIS: Forewing with oblique transverse lines, pattern more distinct on the underside. Venation: Forewing with one single areole, R1 arising free from areole, proximal to the origin of R5 or connate (in Cosymbiini distally). Hindwing with Sc+R1 and Rs appressed or fused at one point. Rs and CuA1 arising from upper and lower cell apex, with slight variation. Antennae of ♂ bipectinate. Hindtibia without pencil. Ansa of tympanal organ

Text-fig. 211: ♂ antennal structure of *Timandra comae* SCHMIDT, 1931 (SEM). Scale bar = 0.05 mm.

on a long stalk, broad in the central part (Text-fig. 6). Sternum A2 of ♂ sclerotized, but without additional sclerites or setose patches (Text-fig. 6). Antennae of ♂ strongly bipectinate to about 3/4 - 4/5 (Text-fig. 211), branches arising, closely attached, from ventral part of antennomeres. Four hindtibial spurs in both sexes.

MALE GENITALIA: Uncus usually vestigial, but small in *Synegiodes*, larger in *Traminda*, with sclerotized lateral appendages ('socii'). Saccus broad, concave. Valva elongate, divided into a sclerotized costal and a membranous ventral part. In the genus *Traminda*, however, valva broad, Larentiinae-like, with strong longitudinal striation reminiscent of Cosymbiini.

FEMALE GENITALIA: Ovipositor without ventral projections. Signum triangular or rhombic, with longitudinal ridge.

IMMATURE STAGES: Pupa slender cuneiform, with 'edged shoulders', reminiscent of Cosymbiini (and Rhopalocera), but anteriorly tapering. Segments 1 and 9 setose (NAKAMURA 1994), cremaster with four pairs of short setae (PATOČKA 2003). Attached to substrate without silk girdle (compare Cosymbiini).

Timandra DUPONCHEL, 1829

(=†*Calothysanis*: sensu auct. nec HÜBNER, 1823)

TYPE SPECIES: *Timandra griseata* PETERSEN, 1902. 21 species in Palaearctic, Nearctic and Indo-Pacific regions (SCOBLE 1999). Revisions of Palaearctic species: KAILA & ALBRECHT (1994; 1995), SIHVONEN (2001a).

DIAGNOSIS: Forewing apex pointed. Hindwing tailed or angled at M3. Venation (Text-fig. 212): Forewing with one single areole below cell apex. R1 arising terminally or slightly subterminally from areole. R2-R4 and R5 connate from apex of areole or slightly stalked. Hindwing with discocellular vein straight and not oblique. Rs and M1 connate or slightly stalked. M3 and CuA1 connate or slightly separate. Antennae of ♂ long bipectinate, appressed type (Text-fig. 211). Proboscis stout, but comparatively short. Frons convex and smoothly scaled. Palpi about 1.2 times diameter of eye (♂), 1.5 times in ♀, last segment narrow. Full number of hindtibial spurs in both sexes.

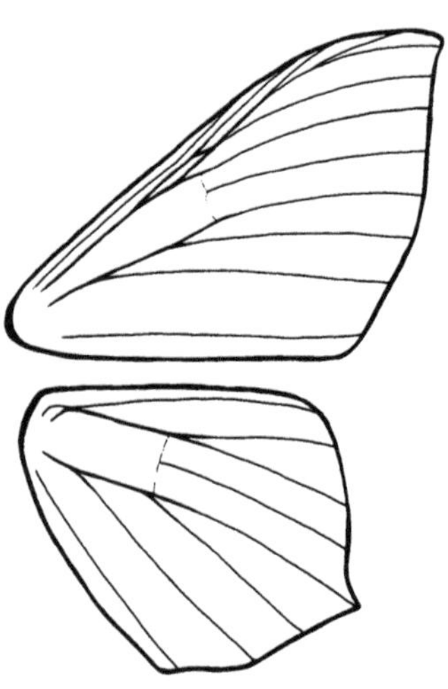

Text-fig. 212: Venation of *Timandra comae* SCHMIDT, 1931.

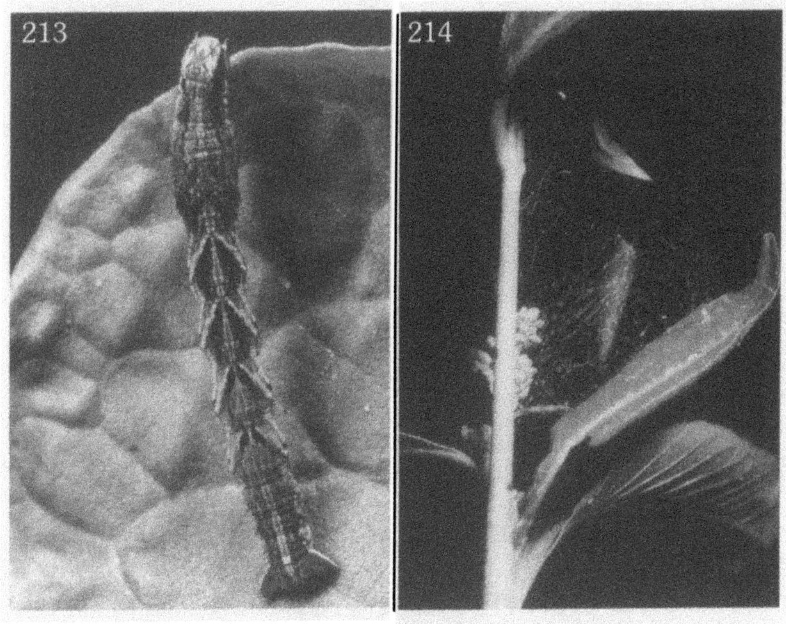

Text-figs 213-214: Immature stages of *Timandra comae* SCHMIDT, 1931. Text-fig. 213: Adult larva (Photo: LEIPNITZ). Text-fig. 214: Pupa.

MALE GENITALIA: Uncus reduced, with sclerotized appendages ('socii'), possibly evolved from modified gnathos. Anellus absent. Saccus broad, concave. Valva with sclerotized costa, often forked at tip. Combination of aedeagus, shape of valva and socii diagnostic at species level. Sternum A8 normal.

FEMALE GENITALIA: Apophyses comparatively long and narrow. Ductus bursae usually sclerotized, corpus bursae with membranous appendix at junction with ductus bursae. Sternum A8 strongly sclerotized, antero-laterally with two setose patches.

BIOLOGY: Larva feeding on Polygonaceae.

IMMATURE STAGES: Larva with small head and first segment narrow, segments T2, T3 and A1 strongly swollen (Text-fig. 213). Pupa hanging free in a loosely woven web of silky threads between leaves, often on the host-plant (Text-fig. 214), or among plant debris near the ground (WARING & TOWNSEND 2003). Further features under diagnosis of tribe.

175. *Timandra comae* SCHMIDT, 1931

Timandra amata f. (spec.n.?) *comae* SCHMIDT, 1931: Int. Ent. Zeits. 25 (6): 57, figs (Spain, Murcia: Alberca). Holotype ♂ (TMB, examined by KAILA & ALBRECHT 1994). The name does not need emendation, see below ('*T. comai* KAILA & ALBRECHT 1994').

‡ *Timandra amata*: sensu STAUDINGER, 1871: in STAUDINGER & WOCKE, Cat. Lepid. eur. Faunengeb. (Edn 2) 1: 153. Misidentification nec LINNAEUS, 1758 (identity: *Cyclophora punctaria*).

‡ *Phalaena vibicaria*: sensu HUFNAGEL, 1767: Berlin. Mag. 4 (5): 514. Misidentification and junior primary homonym nec CLERCK, 1759 (identity: *Rhodostrophia vibicaria*).

‡ *Phalaena angulata* FOURCROY, 1785: Entom. Parisiensis 2: 270 (France: Paris). Syntype(s) (ZMUC (SCOBLE 1999)). Junior primary homonym nec HUFNAGEL, 1767 (identity unknown).

‡ *Geometra amataria*: sensu DENIS & SCHIFFERMÜLLER, 1775: Ank. syst. Werkes Schmett. Wienergegend: 103 (Austria: Vienna distr.). Misidentification nec LINNAEUS, 1761 (identity: *Cyclophora punctaria*).

Calothysanis amataria brykaria NORDSTRÖM, 1943: in BRYK, Folium entom.: Fests. 60. Geb. F. Bryk: 19, fig. (Poland, Silesia: 'Wartha'). Holotype ♂ (NHRS).

Timandra comae palmonii HAUSMANN, 1997: Entomofauna 18 (1): 8, figs 1, 5 (northern Israel: Sede Nehamya). Holotype ♂ (ZSM, examined). Valid at subspecific rank.

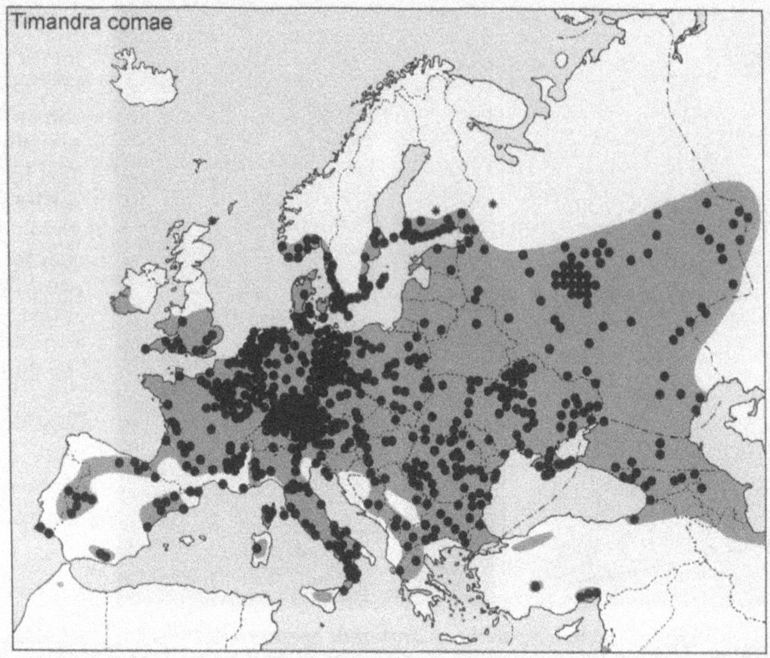

Timandra comai KAILA & ALBRECHT 1994: Ent. Scandinavica 25: 461, 471. Unjustified emendation, as from latinized name 'Coma' results 'comae' in accordance with the example to art. 31.1.1. Code ICZN.

Unavailable names (misidentification): *griseata*: sensu FORSTER & WOHLFAHRT (1981) nec PETERSEN, 1902. – (infrasubspecific): *atropurpurea*: MICHEL (1936); *bipartita*: PROUT (1913); *crassestrigata*: LEMPKE (1949); *deleta*: REBEL (1910); *derufata*: LEMPKE (1949); *effusaria*: KLEMENSIEWICZ (1894); *extrema*: LEMPKE (1949); *impuncta*: LEMPKE (1949); *latistrigata*: REBEL (1915); *nigra*: REBEL (1910); *pulverata*: COCKAYNE (1952); *roseata*: HIRSCHKE (1910); *rufomarginata*: LEMPKE (1949); *serenata*: DANNEHL (1927); *suffumata*: PROUT (1913); *witzi*: SCHNEIDER (1938). – (incorrect subsequent spellings): *amatoria*: FABRICIUS (1781); *amatorio*: DONOVAN (1793); *deleata*: HOFFMEYER (1966);

DIAGNOSIS: Wingspan ♂♀ 25-30 mm, subsequent generations smaller, usually 23-28 mm, exceptionally 19 mm only. Wings sand coloured, with darker suffusion. Reddish medial line distinct on all wings, on forewing anastomosing with postmedial line and leading into apex. Discal spots present on forewing, absent from hindwing. Fringe and terminal area usually with purple or rosy tinge. Frons red brown or dark brown, vertex pale sand coloured, collar darker. Longest branches of ♂ antennae 12-14 times width of flagellum (Text-fig. 211). Hindtibia of ♂ with four spurs of unequal length.

Variability: Reddish terminal coloration broader in the second generation, first and third generation with stronger grey suffusion (BERGMANN 1955; FORSTER & WOHLFAHRT 1981). Not rarely with darkened area between medial and postmedial lines. Rarely melanistic (URBAHN 1972).

MALE GENITALIA: Shape of socii variable, shoe-shaped, with slightly spinulose margin. Costa of valva furcate, membranous tip of valva bilobed, ventral part with a long, narrow, slightly sclerotized process arising from the sub-triangular sacculus. Aedeagus with one short cornutus, cornutus with some spinules at tip.

FEMALE GENITALIA: Ductus bursae broad in the anterior half, in south-eastern Europe often narrower. Membranous, posterior appendix of corpus bursae reputedly at an angle of about 100° to corpus bursae (KAILA & ALBRECHT 1994), but own dissections revealed great variability of this feature. Corpus bursae elongate oval. Signum with longitudinal ridge, slightly sclerotized (sub-triangular). Position of junction between anterior and posterior part of corpus bursae medial (KAILA & ALBRECHT 1994).

DISTRIBUTION: West-Palaearctic. Distributed all over Europe except northern Scandinavia, northern Russia, northern British Isles, and some Mediterranean islands (Balearics, Malta, Greek islands). In the Iberian peninsula rare and local. – Outside Europe in Morocco (POWELL & RUNGS 1941; RUNGS 1981), Turkey, northern Israel (subsp. *palmonii*), Caucasus, Transcaucasus, to Turkmeniya (KUZNETZOV 1960; VIIDALEPP et al. 1992, as '*griseata*'; VIIDALEPP 1996; coll. SOMMERER) and western Siberia. The genus is represented by three other, closely related species *T. recompta, T. paralias* and *T. comptaria* from the central Asian mountains, across Mongolia to Far East of Asia and Japan. The North American *T. amaturaria* (WALKER, 1866) shows close relationships, too.

PHENOLOGY: Often bivoltine: early May to early September, with generations overlapping in late June or July, second generation more abundant than the first. In Finland rare from early June to mid-July; common from mid-July to late September (KAILA & ALBRECHT 1994). North of Alps under favourable conditions with single specimens of partial third and fourth generations until October or November (KAILA & ALBRECHT 1994; EBERT 2001). In southern European lowlands regularly tri- or plurivoltine from early April to mid-November. Growth of larvae very asynchronous within the same broods (SCHNEIDER 1924; EBERT 2001). Larva overwintering (EBERT 2001), exceptionally also the pupa (BERGMANN 1955; FORSTER & WOHLFAHRT 1981; SKOU 1986).

BIOLOGY: Larva oligophagous (1), strongly focussed on *Rumex* and *Polygonum* species (Polygonaceae). Egg deposition observed on *Fallopia (Polygonum) convolvulus* or *F. dumetorum* (Germany: SCHULTZ 1930). Larvae found on *Rumex sanguineus, R. obtusifolius, R. crispus, R. hydrolapathum, R. acetosa, R. acetosella, R. thyrsiflorus, R. rugosus, Polygonum aviculare, P. lapathifolium, Fallopia convolvulus, F. baldschuanica* (Germany: original data; WEGNER pers. comm.; EBERT 2001; Poland: URBAHN & URBAHN 1939). Repeatedly recorded on various *Rumex* and *Polygonum* species (var. cit.), clearly preferring *Rumex* (BERGMANN 1955), feeding also on seeds (PROUT 1913). Reputedly on *Atriplex* (Chenopodiaceae), *Fagopyrum esculentum, Polygonum persicaria* (Polygonaceae), *Calystegia sepium (dumetorum)* (Convolvulaceae), *Taraxacum officinale* (Asteraceae), *Veronica* (Scrophulariaceae), *Stellaria media* (Caryophyllaceae) (var. cit., e.g. REBEL 1910b; LHOMME 1935; SCHNEIDER 1939; 1943; URBAHN & URBAHN 1939; ALLAN 1949; EBERT 2001). Reared on *Polygonum salicifolium* (POWELL & RUNGS 1941) and *Rumex aquaticus* (SCHNEIDER 1924). Main activity of adults at dusk, sometimes with active flight during daytime, readily attracted to light, sometimes to bait. Nectar feeding recorded on *Echium vulgare, Calluna vulgaris* and *Solidago canadensis* (EBERT 2001). r-Strategy (HAUSMANN 1990a).

PARASITOIDS: Diptera, Tachinidae: *Compsilura concinnata* (THOMPSON 1947; HERTING 1960).

HABITAT: In temperate parts of Europe ubiquitous, slightly hygrophilous, preferring uncultivated areas with pioneer vegetation, in southern Europe often in deep valleys and isolated wetlands. Usually inhabiting open habitats, such as uncultivated grassland, fallows, gardens, wet sparse scrubland, moors. From 0 m up to 700 m above sea-level, up to

1,000 m in the Alps, up to 1,600 m in the southern Alps (WOLFSBERGER 1971), but usually not above 1,100 m in southern Europe.

SIMILAR SPECIES: *T. griseata* differs mainly in the darker suffusion of wings, ground colour with less warm, sand coloured tinge, on average slightly larger, forewing discal spots less distinct. As even dissection of genitalia does not provide valuable differential features, phenological and geographical data may help for preliminary identification. Individual forms of *T. comae* with stronger dark suffusion may be extremely similar to *T. griseata*, correct identification of such specimens is almost impossible in northern Europe. In southern Europe, this species is sometimes confused with *Scopula imitaria* (see no. 145).

REMARKS: The species pair *T. comae* and *T. griseata* has recently been separated (KAILA & ALBRECHT 1994; 1995; MÜLLER 1996; SCOBLE 1999; SIHVONEN 2001), the discussion about the validity of species rank is still controversial and merits further attention. In this book they are treated as different species, based on habitus features, which are, to a certain extent, correlated with phenological patterns. First results from molecular studies, using mitochondrial genes ND1 and COI seem to confirm species rank for both taxa (ND1: MILLER et al. 2001; COI: TRUSCH et al. 2002; MILLER et al. in prep.). Species rank is also postulated for *T. recompta* (KAILA & ALBRECHT 1994; SCOBLE 1999; SIHVONEN 2001).

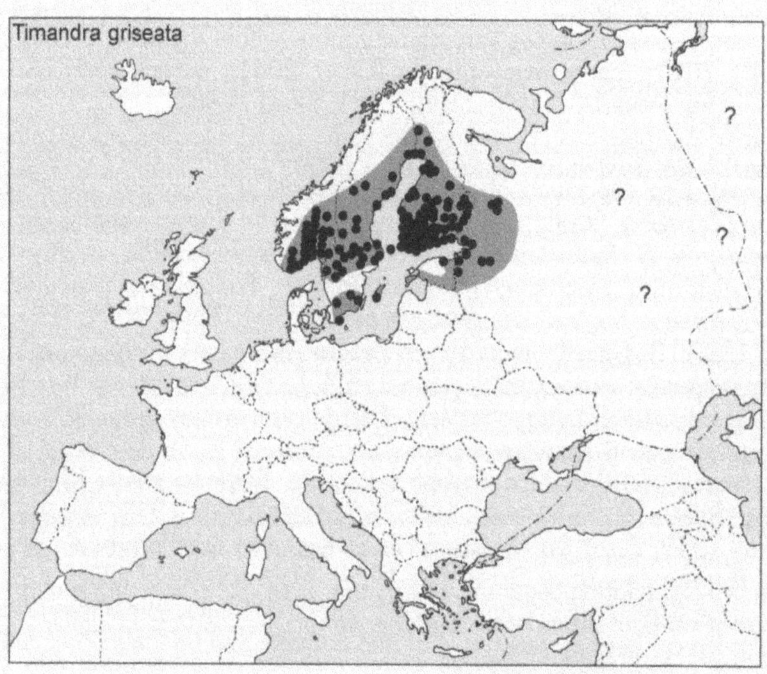

176. *Timandra griseata* PETERSEN, 1902

Timandra amata var. *griseata* PETERSEN, 1902: Lep. Estl.: 116 (Estonia: Lechts). Lectotype ♂ (IZBE, examined by KAILA & ALBRECHT 1994).
Unavailable names (incorrect subsequent spellings): *grisearia*: PROUT (1913).

DIAGNOSIS: Wingspan ♂♀ 23-30 mm. Wings pale sand coloured, with strong grey suffusion. Brown or greyish medial line distinct on all wings, on the forewing anastomosing with postmedial line and leading into the apex. Discal spots weak or absent on forewing, absent from hindwing. Fringe and terminal area usually concolorous, rarely with rosy tinge. Wing pattern not rarely diffuse. Frons dark brown, vertex whitish, collar sand coloured. Longest branches of ♂ antennae 12-14 times width of flagellum. Hindtibia of ♂ with four spurs of unequal length.

MALE GENITALIA: Without significant and constant differential features from *T. comae*. Appendix of sacculus on average slightly broader at base than in *T. comae* (KAILA & ALBRECHT 1994).

FEMALE GENITALIA: Very similar to those of *T. comae*. Ductus bursae usually not dilated in the anterior half. Membranous, posterior appendix of corpus bursae usually arising at a greater angle to corpus bursae (KAILA & ALBRECHT 1994: 120°), but own dissections revealed great variability of this feature in *T. comae*. Position of junction between anterior and posterior part of corpus bursae on the ventral side (KAILA & ALBRECHT 1994).

DISTRIBUTION: (North-)European, boreal. Scandinavia, eastern Estonia, northern European Russia. Possible occurrence in central Urals (see colour plates, fig. 175e) and central European Russia (LASTUKHIN 2001) awaits confirmation. Reputedly, one single female collected in northern Denmark (LARSEN 1995; MÜLLER 1996; KARSHOLT & STADEL NIELSEN 1998), the very late date ('24.IX-15.X.'), however, not typical for that species. All the other records from regions south of Scandinavia (e.g. in VIIDALEPP 1996) refer to misidentified grey forms of the preceding species.

PHENOLOGY: Usually univoltine: late June to early August, single specimens until mid-September (KAILA & ALBRECHT 1994). Sometimes emerging from early June.

BIOLOGY: Limited data available unambiguously referring to this species. Larva found on *Rumex acetosa* (Polygonaceae) (Finland: SEPPÄNEN 1954), probably belonging to this species. Recorded on *Polygonum, Rumex* (Polygonaceae) and *Atriplex* (Chenopodiaceae) (Finland: VALLE 1946; Sweden: SVENSSON 1993). The latter author distinguished between the two species. Adults active at dawn, attracted to light.

HABITAT: Limited information available about habitat preference, but probably similar to the preceding species. From 0 m up to 200 m above sea-level.

SIMILAR SPECIES: *T. comae* (see no. 175). *T. rectistrigaria* differs in the stronger projecting frons, wings with darker suffusion, transverse lines reaching the forewing costa subapically.

REMARKS: For validity of *T. comae* and *T. griseata* at species rank see remarks to the preceding species.

177. Timandra rectistrigaria (EVERSMANN, 1851)

Ennomos rectistrigaria EVERSMANN, 1851: Bull. Soc. imp. Nat. Moscou 24 (2): 637 (Russia: eastern Siberia: Irkutsk). Lectotype ♀ (ZISP, examined), herewith designated to stabilize usage of the name with regard to some similar species.
Timandra puziloi ERSCHOV, 1870: Trudy russk. Ent. Obshch. 4: 197 (eastern Russia: Irkutsk). Syntypes(s) ♂.
Unavailable names (infrasubspecific): *obsoleta*: PROUT (1913).

DIAGNOSIS: Wingspan ♂ 25-28 mm, ♀ 23-27. Hindwing termen with a small and sharp tooth at M3. Wings pale sand coloured, with very strong brown or grey brown suffusion,

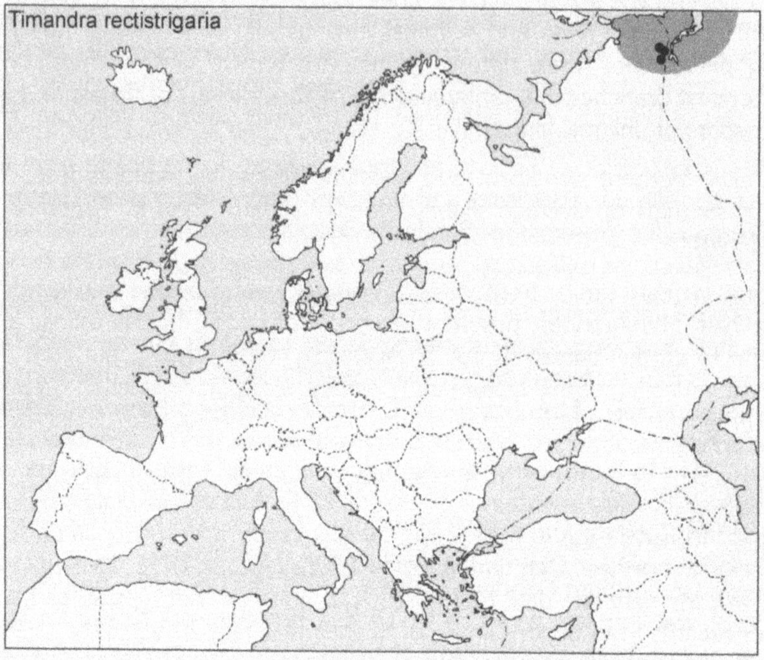

covering the ground colour almost completely. Wing pattern and discal spots very diffuse. Medial and postmedial lines of forewing reaching the costa subapically. Discal spot absent from hindwing, elongate and very diffuse on the forewing. Fringe pale sand coloured. Frons strongly convex, light brown, vertex and collar sand coloured. Longest branches of ♂ antennae 10-12 times width of flagellum. Hindtibia of ♂ with four spurs.

MALE GENITALIA: Strongly sclerotized, ventral positioning hardly achievable on permanent slides. 'Socii' fused, margins spinose. Valva broad, costa sclerotised, ventral margin with broad subterminal projection. Aedeagus narrow, without cornutus.

FEMALE GENITALIA: Ovipositor sclerotized. Apophyses anteriores very short. Antrum T-shaped, with long lateral projections. Corpus bursae elongate, slender.

DISTRIBUTION: Eurasiatic (arctic). Very few records from polar Urals (VIIDALEPP 1996; JALAVA & KULLBERG in press). – Outside Europe from Siberia to Far East of Asia (Amur region and Kamchatka).

PHENOLOGY: Univoltine (limited data available for Europe): early July to mid-July. In Far East of Asia from late June to late July (BELJAEV & VASILENKO 2002).

BIOLOGY: Larval host-plants unknown. Adults diurnal and active in the long dusk of the arctic summer 'night'.

HABITAT: Sandy riverside meadows, tundra, and 'common in the forest zone' (JALAVA & KULLBERG in press). From 200 m up to 500 m above sea-level.

SIMILAR SPECIES: *T. griseata* (see no. 176).

Tribus **Cosymbiini** PROUT, 1911

(=†Ephyrini GUENÉE, 1858; =Zonosomini WHITE, 1876; =Cyclophorini MOORE, 1887)

More than 180 species (SCOBLE 1999) in 11 genera, five further genera assigned tentatively (SIHVONEN & KAILA 2004). Worldwide distribution. For sister group relationship to Cosymbiini and Rhodometrini ('Timandrini lineage') see introduction to the subfamily. Preliminary data sets from own molecular research (mitochondrial genes ND1, COI, and 16S rRNA) support monophyly of the 'Timandrini lineage' (MILLER et al. 2001; HAUSMANN et al. in prep.). Structurally homogenous group, with several common features in habitus, imaginal morphology (♂♀ genitalia, venation, ♂ antennae), morphology of immature stages, and ecology. SIHVONEN & KAILA (2004) mention five autapomorphies supporting monophyly of tribe: Absence of two pores on the anterior part of the vertex, hindwing with cell rings, ♂ without proximal hindtibial spurs (but most tropical taxa with 3-4 spurs), absence of uncus, anteriorly truncate pupa, with girdle. Forewing radial venation unique in the core group, additional comprehensive analysis is needed for the tropical taxa. Reduction of uncus and shape of sacculus strongly reminiscent of Timandrini and genus *Ochodontia* in the Rhodometrini. Presence of a flat signum sclerite with longitudinal ridge in ♀ genitalia supports these relationships, too. Extralimital genus *Pseudosterrha* tentatively placed here (HAUSMANN 1993a), with many features reminiscent of the Timandrini genus *Traminda*, but in wing shape, habitus and hindwing venation with anastomosis of Sc+R1 and Rs similar to the Rhodometrini. The species rich, extralimital 'genus' *Anisodes* from the Indo-Pacific and Neotropical regions recently subordinated under *Cyclophora* in a broader definition (HOLLOWAY 1997; SCOBLE 1999). Further comprehensive revision at global scale is needed. For validity of the name Cosymbiinae, see 'Forum Herbulot world list of family-group names in Geometridae' (www.herbulot.de).

DIAGNOSIS: Wings with cell rings rather than discal spots (with a few exceptions). Venation (Text-fig. 218): Forewing with one small areole, rarely absent, never double, origin of R5 basally from origin of R1, thus R1 stalked with R2-R4. Hindwing Sc+R1 and Rs appressed or fused at one point; exception: longer fusion in *Pseudosterrha*. Palpi often long. Antennae of ♂ usually bipectinate to about 2/3 (Text-fig. 215). Tympanic ansa broad in the central part. Sternum A2 of ♂ sclerotized, but without additional sclerites or setose patches (Text-fig. 7).

MALE GENITALIA: Uncus truncate or divided, weak. Gnathos absent. Saccus broad, often concave. Valva with longitudinal striation and ventral fibula, shape diagnostic.

FEMALE GENITALIA: Ovipositor without ventral projections. Signum developed as lemon-shaped sclerite with longitudinal ridge.

BIOLOGY: Deposition of single eggs often along the edges of leaves. Larvae usually arboreal.

IMMATURE STAGES (Text-figs 216-217): Egg elongate in the core group, in tropical taxa also round, with hexagonal or irregularly polygonal sculpturation, cell walls rounded. Pupa anteriorly truncate, with anchor-shaped setae at cremaster. Attached to substrate supported by a silk girdle reminiscent of Papilionidae, Pieridae, Elachistidae, and Hedylidae, but girdle around tergum A3. Described by HUFNAGEL (1767: 623), see also STARNECKER (2001), and PATOČKA (1994; 2003). Quite similar girdle pupae occur in tropical '*Anisodes*' (COMMON 1990: fig. 112; HOLLOWAY 1997). Further features under generic diagnosis of *Cyclophora*.

Text-fig. 215: ♂ antennal structure of *Cyclophora punctaria* (LINNAEUS, 1758) (SEM). Scale bar = 0.05 mm.

Cyclophora HÜBNER, 1822

(=*Cosymbia* HÜBNER, 1823; =*Ephyra* DUPONCHEL, 1829 nec PÉRON & LESUEUR, 1810; =*Zonosoma* LEDERER, 1853)

TYPE SPECIES: *Phalaena albipunctata* HUFNAGEL, 1767. 274 species worldwide (SCOBLE 1999) in the broad concept of HOLLOWAY (1997), who includes also the Neotropical and Indo-Pacific species of the former genus *Anisodes*. About 60 species in the strict, 'traditional' concept. The diagnosis given below is valid only for *Cyclophora* s. str., which is focussed in its distribution on the Holarctic region. Separation of this core group as 'sub-

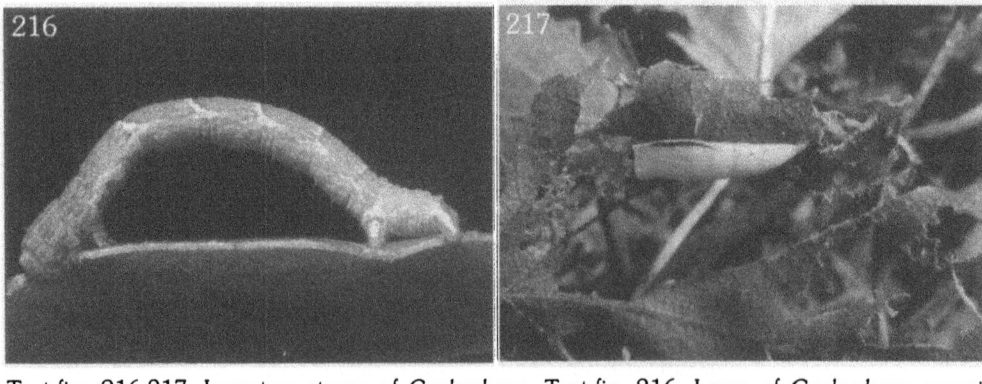

Text-figs 216-217: Immature stages of *Cyclophora*. Text-fig. 216: Larva of *Cyclophora quercimontaria* (BASTELBERGER, 1897) (Photo: NIPPE). Text-fig. 217: Pupa of *Cyclophora albipunctata* (HUFNAGEL, 1767).

genus *Cyclophora*' may be discussed. Major treatments: STERNECK (1941); KLIMASZEWSKI (1960); URBAHN (1971); PATOČKA (1994). For validity of genus name see HERBULOT (1962a).

In Europe three species-groups may be distinguished based on male genitalia: the *pendularia* species-group with long and narrow fibula, the *porata* species-group with dilated fibula and the *punctaria* species-group with large and strongly setose ventral projection of sacculus. The latter structure is probably not homologous to the naked fibula of the *porata* species-group.

Within the genus *Cyclophora*, the species barriers often seem to be low, and hybridizations can easily be obtained in rearings (see e.g. under 'unavailable names' of *C. pendularia*), natural hybrids have been repeatedly reported (THOMANN 1955; see also remarks to *C. quercimontaria* and *C. punctaria*). Hybrids may be intermediate also in genitalia (STERNECK 1941: 219). Small morphological differences between species and frequent hybridizations suggest a comparatively recent phylogenetic radiation of the western Palaearctic species, possibly by glacial and interglacial speciation.

DIAGNOSIS: All wings with a small whitish discal spot, often encircled by a dark ring. Some species with seasonal dichroism. Hindwing rounded, angled or slightly tailed at M3. Venation (Text-fig. 218): Forewing with areole below cell apex. R1 distinctly stalked with R2-R4, R5 arising from areole, a quite rare feature in the Sterrhinae. Hindwing with Rs and M1 connate or very shortly stalked, M3 and CuA1 separate or connate. Discocellular veins of all wings straight. Proboscis well developed, but comparatively short (3-4 mm). Length of palpi 1.1-1.6 times diameter of eye in ♂, in ♀ slightly longer, in ♀ of *C. puppillaria* up to 1.8 times. Third segment of palpi narrow, smoothly scaled. Frons flat. Antennae of ♂ bipectinate, appressed type, last third ciliate (Text-fig. 215). Hindibia of ♂ with two, of ♀ with four spurs, the pairs of strongly unequal length. Extralimital, tropical taxa often with 3-4 male hindtibial spurs, longer palpi and differential features in venation.

MALE GENITALIA: Uncus weakly sclerotized. Saccus broad, strongly concave. Valva spatulate, with strong longitudinal striation in the distal half, and with a long ventral process ('fibula'). Shape of fibula often diagnostic. Aedeagus without cornuti. Sternum A8 normal. Often with minor differences between the species.

FEMALE GENITALIA: Ovipositor lobes large, oval, with a medial sclerotized ridge between. Apophyses posteriores short and thin. Apophyses anteriores absent. Tergum A8 weakly sclerotized. Ductus bursae sclerotized, funnel-shaped. Ductus seminalis arising from corpus bursae close to ductus bursae. Corpus bursae elongate. Signum with long longitudinal ridge, laterally bilobed. Often with minor or without differences between the species. On permanent slides of female genitalia, shape of ductus bursae, corpus bursae and signum depend on the position and require accurate analysis. Shape of corpus bursae and signum rarely diagnostic at species level.

PHENOLOGY: Usually bi-, tri-, or plurivoltine. In contrast to all other European Sterrhinae regularly overwintering as pupa.

BIOLOGY: Deposition of eggs preferably along the edges of leaves. Larvae usually arboreal, feeding on deciduous trees. *C. maderensis* from Madeira and Canary islands has been found on the conifer *Pinus canariensis* (A. BECK pers. comm.). Collecting of larvae in nature easy by beating from trees. Rearing from eggs very easy. Girdled pupa in all

Text-fig. 218: Venation of *Cyclophora albipunctata* (HUFNAGEL, 1767).

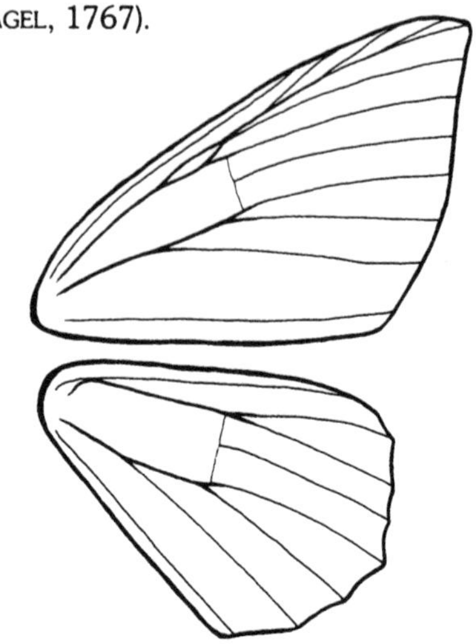

species of the genus attached to leaves or twigs, on a tightly spun web, head upright, falling to the ground with the leaves in autumn. Adults resting on stems or leaves of trees, usually on the host-plant species of the larvae. Readily attracted to light and bait.

IMMATURE STAGES: Egg elongate. Larva of medium length, cylindical, on the last segments slightly broader, head capsule large, slightly bifid dorsally. Ground colour with dichroic variation in many species, green or brown, or polychroic. Di- or polychroism also in the pupae of most species, see remarks under 'biology' of *C. albipunctata*. Further features under diagnosis of tribe.

REMARKS: The genus includes many tropical species, which previously were placed in the genus *Anisodes* (HOLLOWAY 1997; SCOBLE 1999). Many of them differ considerably from the above given genitalic features, e.g. in the presence of cornuti (♂), or in a strongly spinulose corpus bursae (♀). Accurate, comprehensive worldwide revision should reveal and define sub-groupings in this species-rich genus. Some European species are frequently misidentified, e.g. *C. porata, C. ruficiliaria, C. quercimontaria, C. punctaria, C. suppunctaria* and reddish forms of *C. linearia*.

The *pendularia* species-group

Cell rings large and distinct. Fibula long and narrow. Further features under generic diagnosis. Interspecific differences small or absent in ♂ and ♀ genitalia of *C. albipunctata, C. albiocellaria, C. lennigiaria, C. ariadne, C. serveti*, and *C. annularia* (compare STERNECK 1941: 213).

178. *Cyclophora pendularia* (CLERCK, 1759)

Phalaena pendularia CLERCK, 1759: Icon. Insect. rariorum 1: pl. 7, fig. 5 (Sweden). Lectotype (NHRS). For identity of type and validity of the name *pendularia* see NORDSTRÖM (1941: 127) and FLETCHER (1979: 54).
Geometra orbicularia HÜBNER, 1799: Samml. Eur. Schmett. 5, Geometrae (1): pl. 12, fig. 60 (Europe). Syntype(s) lost.
Unavailable names (misidentification): *ianthinarium*: sensu PROUT (1912) nec STICHEL, 1901 (identity: *Cyclophora albipunctata*). – (infrasubspecific): *biobsoleta*: COCKAYNE (1952); *namurcensis*: LAMBILLON

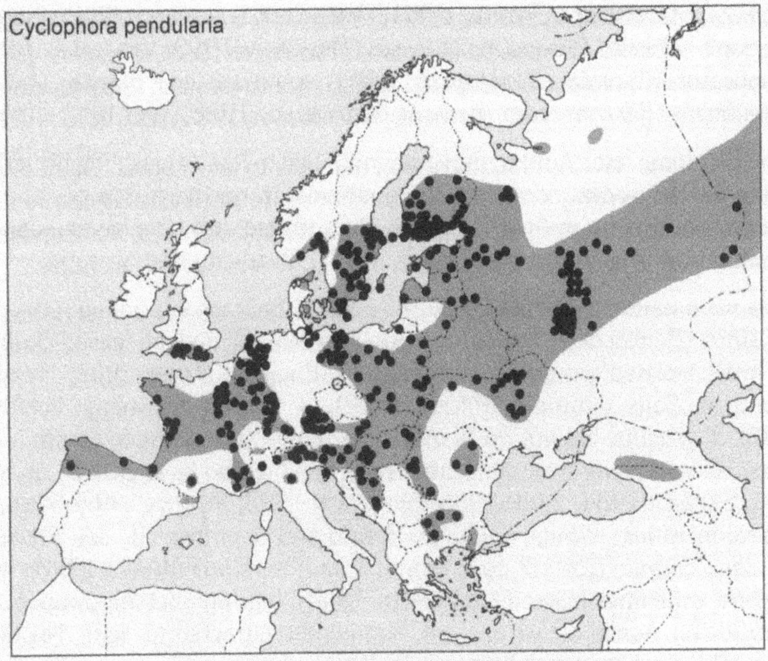

(1905); *roseonigrata*: COCKAYNE (1950). – (hybrids): *brightoni*: TUTT (1907) (*pendularia* ♂ x *linearia* ♀); *headi*: TUTT (1907) (*pendularia* ♂ x *annularia* ♀); *orbialbiocellaria*: HEIN (1936) (*pendularia* ♂ x *albiocellaria* ♀); *orbiculopendula*: TUTT (1907) (*pendularia* ♂ x *albipunctaria* ♀). – (incorrect subsequent spellings): *orbicula*: PROUT (1913).

DIAGNOSIS: Wingspan ♂♀ 20-26 mm, second generation smaller, sometimes 20 mm only. Forewing apex pointed, termen angled or convex, hindwing termen angled or shortly tailed at M3. Coloration and pattern of wings with low variability. Ground colour grey, with strong dark grey and brownish suffusion, with slight purple tinge in fresh specimens. Medial fascia usually darker and strongly dentate, often with red brown tinge. Postmedial line dissolved to a row of blackish dots. Cell rings distinct, often red brown. Terminal dots present. Frons black brown, vertex grey. Abdomen with black dorsal streaks. Longest branches of ♂ antennae about 5 times width of flagellum. Further features under generic diagnosis.

MALE GENITALIA: Fibula weakly sclerotized, slightly curved, thin, tapering, short, free projection shorter than half length of valva.

FEMALE GENITALIA: Sterigma with deep U-shaped invagination. Ductus bursae very short, about 1/4 length of corpus bursae, towards ostium bursae slightly dilated. Corpus bursae rather slender.

DISTRIBUTION: European-Westasiatic (temperate zone). Usually local and rare from southern England and north-western Iberian peninsula to the Urals. In the north to southern Scandinavia, in the south to Bulgaria (BESHKOV pers. comm.) and European Turkey (OKYAR & AKTAÇ 1999). An isolated population in northern Italy, Reggio Emiglia (FLAMIGNI et al. 2001). Absent from large parts of central Europe and Baltic states. Abundance decreasing in Great Britain (SKINNER 1984; PORTER 1997) and Germany (STEINER 2001; WOLF & HACKER 2004), vulnerable to the destruction of riverside forests and urbanization of heathland (STEINER 2001; WARING & TOWNSEND 2003). Habitats meriting protection. – Outside Europe in Caucasus (TIKHONOV 1993; VIIDALEPP 1996), northern Transcaucasus (Georgia: DIDMANIDZE 2002), south-western Siberia, and southern Siberian mountains. Records from Armenia (WARDIKJAN 1985) refer to *C. albipunctata*.

PHENOLOGY: Bivoltine: late April to mid-June; mid-July to late August. Flight period sometimes shifting by 1-2 weeks, according to conditions. In northern Europe, and northern parts of central Europe univoltine, from late June to late July. Larval abundance higher in the second brood than in the first (DAVEY 2001). Overwintering as pupa.

BIOLOGY: Larva in nature monophagous (2) on *Salix* (Salicaceae), or oligophagous (2) on *Salix* and *Alnus* (Betulaceae). Found on various small- or medium-leaved *Salix* species, preferring small, isolated groups of young shrubs (England: DAVEY 2001; PARSONS et al. 2001), found on *Salix viminalis* (PROUT 1913), *S. caprea* (Germany: STEINER 2001; ALLAN 1949), *S. fragilis*, but not on *S. alba* (Germany: WEGNER pers. comm.). Recorded on various species of *Salix* (var. cit.), preferring small leaved species such as *S. cinerea* (var. cit., e.g. SKINNER 1984; SKOU 1986; SVENSSON 1993; PORTER 1997; STEINER 2001), reputedly also on *Alnus*, *Ulmus*, *Quercus, Betula* and *Acer* (var. cit., see STEINER 2001). Reared on *Salix caprea* (LEIPNITZ pers. comm.), and on *Alnus* (KRAUS 1993), rearing on the latter plant sometimes failed (GREMMINGER 1944), but regional importance suggested by STEINER (2001). Adults active at dusk, sometimes attracted to light. Nectar feeding recorded on *Molinia* (URBAHN & URBAHN 1939).

PARASITOIDS: Diptera, Tachinidae: *Ceranthia abdominalis* (HERTING 1960; 1984; ANDERSEN 1996). – Hymenoptera, Braconidae: *Apanteles spurius* (THOMPSON 1945; SHENEFELT 1972). – Hymenoptera, Ichneumonidae: *Limneria* spec.? (URBAHN & URBAHN 1939).

HABITAT: Hygrophilous lowland species, preferring warm riverside hardwood forests (STEINER 2001), damp woodland and marshes (SKOU 1986), sallow scrub on open heathland, damp heathland (SKINNER 1984; WARING & TOWNSEND 2003), especially alder-ash carrs and willow thickets (BUSZKO 2000). Usually from 0 m up to 300 m above sea-level, exceptionally up to 500 m.

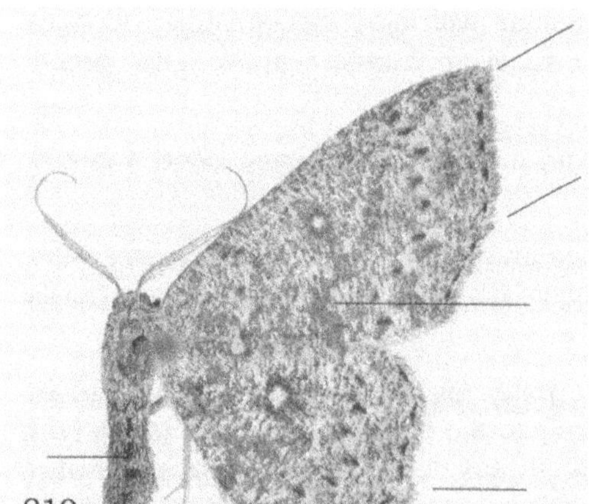

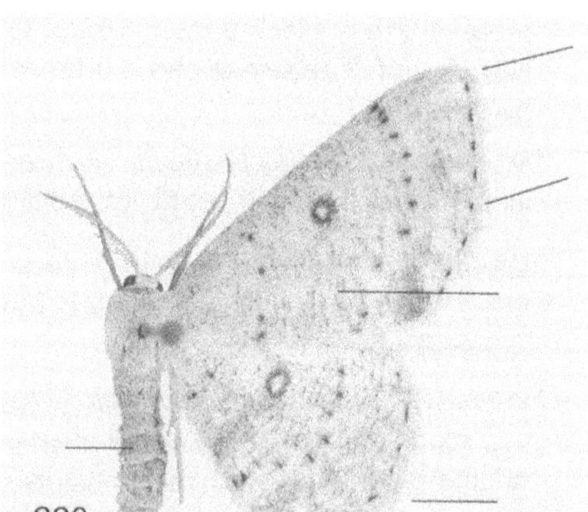

Text-figs 219-220: Diagnostic characters (indicated) of European *Cyclophora* species. Text-fig. 219: *C. pendularia* (CLERCK, 1759). Text-fig. 220: *C. albipunctata* (HUFNAGEL, 1767).

SIMILAR SPECIES: *C. albipunctata* 'f. *decoraria*' and certain melanistic forms ('f. *subroseata*') may be similar to *C. pendularia*, but *C. albipunctata* differs in the unicolorous abdomen without black dots, forewing termen straighter, apex less pointed, hindwing less tailed at M3, medial fascia less distinct, straighter, not dentate, wings usually with much less dark suffusion.

REMARKS: In captivity hybrid copulae can be induced with congeneric species, but usually only a small portion of the eggs are fertile, see under 'unavailable names' and HEIN (1936).

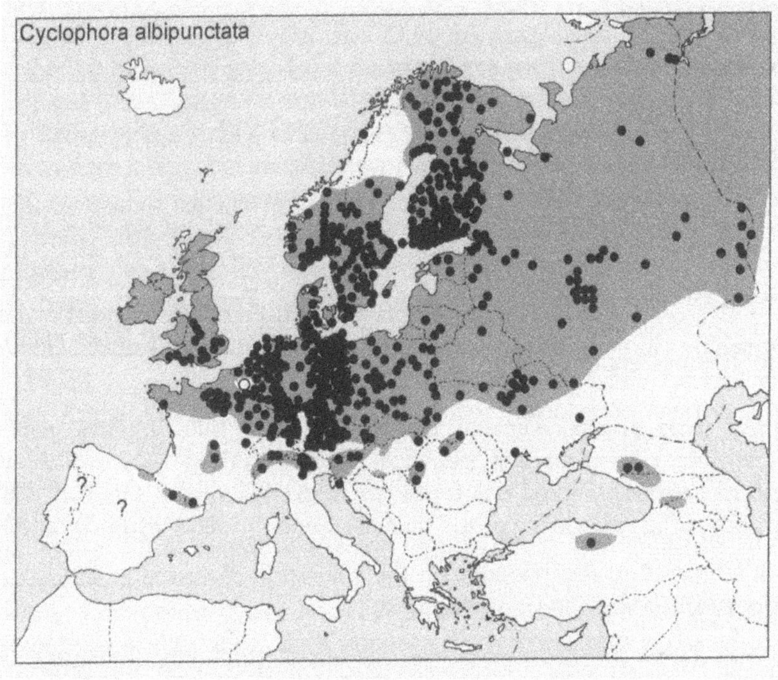

179. Cyclophora albipunctata (HUFNAGEL, 1767)

Phalaena albipunctata HUFNAGEL, 1767: Berlin. Mag. 4 (5): 526 (Germany: Berlin). Syntype(s) lost. For validity of the name *albipunctata* see FLETCHER (1979: 54) and nomenclatural remarks under *C. pendularia*.

Phalaena suspensa RETZIUS, 1783: Gen. et Spec. Ins.: 43 (Europe). Syntype(s).

‡ *Geometra pendularia*: sensu DENIS & SCHIFFERMÜLLER, 1775: Ank. syst. Werkes Schmett. Wienergegend: 107 (and HÜBNER 1799, ESPER 1801 etc.). Misidentification(s) nec CLERCK, 1759. DENIS & SCHIFFERMÜLLER mention *Betula alba* (=*B. pendula*) as host-plant, quite characteristic for this species.

Phalaena Geometra pupillaria BRAHM, 1791: Insektenkal. Samml. Oekon. 2 (1): 385 (western Germany: Mainz). Syntype(s).

‡ *Phalaena pandularia*: sensu FABRICIUS, 1798: Syst. Ent. (Suppl.): 451. Misidentification and incorrect subsequent spelling for *pendularia*.

Phalaena circularia FABRICIUS, 1798: Syst. Ent. (Suppl.): 451 (France). Syntype(s) lost.

Ephyra vusarmana WALKER, 1861: List. Spec. lepid. Insects Colln Br. Mus. 22: 637 ('Ceylon': incorrect locality). Holotype ♀ (BMNH).

Zonosoma ianthinarium STICHEL, 1901: Berl. Ent. Zeits. 1901, S.B. (20) (eastern Germany: Arneburg). Holotype.

Zonosoma pendularia var. *subroseata* WOODFORDE, 1902: Entom. 35: 276 (Great Britain: North Staffordshire). Syntype(s).

Zonosoma pendularia var. *griseolata* STAUDINGER, 1897: Dt. ent. Z. Iris 10 (1): 23 (eastern Russia: Amur). Lectotype ♂ (MNHU, examined), herewith designated to fix type locality and to stabilize usage of the name;

at MNHU 2♂1♀ paralectotypes; in description also northern Finland mentioned. Valid at subspecific rank.

Cosymbia albipunctata coreana BRYK, 1949: Ark. Zool. 41A (1): 160 (Korea: Shuotsu). Holotype ♀. Junior synonym of subsp. *griseolata*.

Cosymbia excellens BRYK, 1942: Dt. ent. Z. Iris 56: 63 (Kurile islands: Kunashiri, Tomari). Holotype ♂ (NHRS). Junior synonym of subsp. *griseolata*.

Unavailable names (misidentification): *amataria*: sensu VALLOT (1802) nec LINNAEUS, 1761 (identity: *Cyclophora punctaria*). – (infrasubspecific): *aestiva*: VORBRODT (1930); *albescens*: LEMPKE (1949); *brunnearia*: LAMBILLON (1905); *decoraria*: NEWMAN (1861); *depulsa*: BASTELBERGER (1908); *flavescens*: PROUT (1913); *foliata*: LEMPKE (1949); *griseata*: LEMPKE (1949); *hatertica*: SCHULTZ (1931); *impictaria*: MEVES (1914); *linearia*: LAMBILLON (1905); *magnocellata*: LEMPKE (1949); *mediofasciata*: LEMPKE (1949); *nigroroseata*: WOOD; *nigrostriata*: LUTZAU (1908); *nigrosubrosea*: BOWMAN; *obsoleta*: LAMBILLON (1905); *orbiculoides*: WOODFORDE (1919); *radiata*: DELAHAYE (1900); *rufescens*: LEMPKE (1949); *striata*: LEMPKE (1949); *subochreata*: WOODFORDE (1910); *unicolaria*: LAMBILLON (1905). – (hybrids): *pendulomontaria*: URBAHN (1939) (*albipunctaria* ♂ x *quercimontaria* ♀); *kessleri*: THOMANN (1955) (*albipunctaria* ('*pendularia*') ♂ x *puppillaria* ♀); *penduloorbicula*: TUTT (1907; *albipunctaria* ♂ x *pendularia* ♀). – (incorrect subsequent spellings): *albopunctata*: HOLLOWAY (1997); *janthinaria*: REBEL (1910); *obsoletaria*: PROUT (1934); *unicoloria*: PROUT (1913).

DIAGNOSIS: Wingspan ♂♀ 23-27 mm, second generation smaller, sometimes 20 mm only. Coloration and pattern of wings variable. Ground colour usually dirty white. Postmedial line dissolved to a row of blackish dots. Terminal area with more or less distinct, confluent shadows. Cell rings distinct. Terminal dots present. Frons brown, vertex dirty white. Abdomen unicolorous, grey. Longest branches of ♂ antennae 5-6 times width of flagellum. Further features under generic diagnosis.

Variability: Cell rings occasionally reduced to spots ('f. *depulsa*'), which may be rosy. Melanistic forms not rare in rearings, but occasionally occurring also in nature, with transitions to normal forms. Ground colour darker in populations of Great Britain. Forms with strong rosy suffusion, mainly at the medial fascia ('f. *decoraria*') common in some regions of Great Britain, but occurring also elsewhere, usually in the second generation. Influence of temperature and humidity on the coloration discussed in BERGMANN (1955). Subsp. ***griseolata*** with sand coloured tinge.

MALE GENITALIA: Fibula strongly curved, well sclerotized, stout, long, free projection longer than half length of valva. Aedeagus comparatively long.

FEMALE GENITALIA: Similar to those of *C. pendularia*. Ductus bursae of medium length, about half length of corpus bursae, towards ostium bursae distinctly dilated.

DISTRIBUTION: Eurasiatic (temperate zone). From British Isles and western France across central and northern Europe to the Urals. In Scandinavia up to the extreme north. In the south isolated populations in southern Alps (e.g. WOLFSBERGER 1966; RAINERI 1991), Krk island (HABELER pers. comm.), northern Romania (STANESCU 1997), Pyrenees (GÓMEZ DE AIZPÚRUA 1988; DANTART & ROCHE 1992; MAZEL & PESLIER 1997; REDONDO & GASTON 1999), reputedly also in northern Portugal (PASSOS DE CARVALHO 1984; VIVES MORENO 1994; MÜLLER 1996). Erroneously mentioned for central Spain (ROMERO et al. 2001; ROMERO pers. comm.). Absent from Italy south of Alps (FLAMIGNI et al. 2001). Abundance decreasing, but reasons unclear (Germany: STEINER 2001). – Outside Europe in northern Turkey, Caucasus (MÜLLER pers. comm.), Transcaucasus (Armenia: WARDIKJAN 1985 as '*pendularia*'; Georgia: VIIDALEPP 1996), western Tian Shan (PROUT 1913), western Siberia to Dahuria, in Far East of Asia and Japan replaced by the subsp. *griseolata*, in North America by the sister species *C. pendulinaria* (GUENÉE, 1858).

PHENOLOGY: Usually bivoltine: mid-April to early June; mid-July to late August. Flight period sometimes shifting by 1-2 weeks, according to conditions, sometimes emerging from mid-March. Specimens in September and October in southern Alps and southern Germany probably belonging to a partial third generation. In northern Europe and northernmost parts of central Europe univoltine, from late May to early July, exceptionally later, e.g. in Scotland from late June to late July. Overwintering as pupa.

BIOLOGY: Larva monophagous (3) on *Betula* (Betulaceae). Found on *Betula pendula, B. pubescens, B. nana*, often on saplings (Germany: STEINER 2001; WEGNER pers. comm.; Austria: SCHWINGENSCHUSS 1953; ERLEBACH pers. comm.; Finland: SEPPÄNEN 1954; Great Britain: BUCKLER 1897; PORTER 1997). Repeatedly recorded on *Betula* (var. cit.), rarely on *Alnus* (Betulaceae), *Fagus* and *Quercus* (Fagaceae) (REBEL, 1910; LHOMME 1935), but the last three may refer to host-plants from rearing. Reared on *Betula, Alnus, Quercus, Ulmus* (var. cit., see ALLAN 1949; BERGMANN 1955; STEINER 2001). Pupation attached to leaves. Pupa green when pupating on living leaves and turning to brown when the leaves are falling, but usually brown when pupating on dry leaves (original data; VOOGD pers. comm.). Adults readily attracted to light, sometimes to bait. During daytime resting on stems of birches, rarely of oaks (WOODFORDE 1902). Nectar feeding recorded on *Cirsium* and *Senecio* (WARING & TOWNSEND 2003).

PARASITOIDS: Diptera, Tachinidae: *Ceranthia abdominalis* (ANDERSEN 1996). – Hymenoptera, Braconidae: *Apanteles juniperatae, Rogas unicolor* (THOMPSON 1945; SHENEFELT 1972).

HABITAT: Hygrophilous, silvicolous. Sometimes common in wet birch forests, forest fringes, moors, marshes, scrubby places, damp heathland, near water courses, and in gardens, usually frequent in pioneer birch woodland and birch succession in degenerated bogs. From 0 m up to 800 m above sea-level, up to 1,500 m in southern Alps and Pyrenees.

SIMILAR SPECIES: *C. pendularia* (see no. 178). Pale forms of *C. porata* always with traces of the orange tinge of coloration, dark clouds in the terminal area not confluent.

REMARKS: For hybrid rearings see URBAHN & URBAHN (1939), and under 'unavailable names'. In captivity hybrid copulae can be induced with congeneric species, but usually only a small portion of the eggs are fertile. For natural hybridization between *C. albipunctata* ('*pendularia*') and *C. puppillaria* see THOMANN (1955).

180. Cyclophora albiocellaria (HÜBNER, 1789)

Phalaena Geometra albiocellaria HÜBNER, 1789: Beitr. Gesch. Schmett. 1 (4): 3, pl. (4) 1, fig. A (Austria: Vienna distr.). Syntypes ♂♀, lost.

Geometra ocellaria HÜBNER, 1799: Samml. Eur. Schmett. 5, Geometrae (1): pl. 13, fig. 64 (Europe). Syntype(s) lost. Unambiguously figured as belonging to the eastern populations with large discal spots.

Ephyra argusaria BOISDUVAL, 1840: Genera Index meth. eur. Lepid.: 220 (Europe; own material without type status from 'Hungary'). No separate types. Unnecessary replacement name for *ocellaria*.

Unavailable (infrasubspecific): *therinata*: BASTELBERGER (1900). – (incorrect subsequent spellings): *albiceolaria*: DIDMANIDZE (2002); *albiocelaria*: PATOČKA (1994); *albiocellara*: GUENÉE (1858); *albiocellata*: TURATI in RAINERI (1996); *albiocelllaria*: GANEV (1984); *elbiocellaria*: GANEV (1984); *terinaria*: FORSTER & WOHLFAHRT (1981); *therinaria*: REBEL (1911); *thermata*: BASTELBERGER (1906).

DIAGNOSIS: Wingspan ♂♀ 20-25 mm, second generation smaller, sometimes 18 mm only. Forewing termen concave below apex. Ground colour pale yellow with slight orange

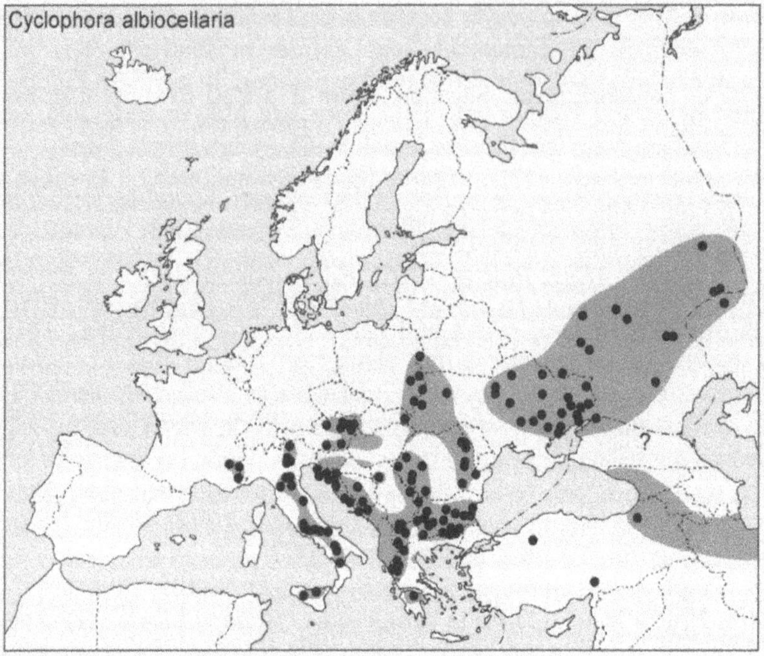

tinge. Postmedial line dissolved to dark brown dots. Medial area towards postmedial line and towards inner termen with strong black brown coloration on all wings, in second generation, however, with reduced black coloration. Cell rings large, round on all wings, with white centre. Terminal dots absent. Fringe darker, ochreous. Frons light brown, vertex concolorous with ground colour. Longest branches of ♂ antennae about 5 times width of flagellum. Further features under generic diagnosis.

Variability: The second generation often lighter, reddish, with less dark suffusion ('f. *therinata*').

MALE GENITALIA: Spinulose process at base of fibula comparatively large. Fibula slightly curved, well sclerotized, long. Aedeagus short. Differences small from *C. ariadne* and *C. serveti*.

FEMALE GENITALIA: Ductus bursae of medium length, about 1/3 length of corpus bursae, laterally dilated in the most posterior part only.

DISTRIBUTION: West-Palaearctic. From southern Alps, south-eastern parts of central Europe and Balkan peninsula to the Urals. In the south to Corsica, central and southern Italy, and Sicily (e.g. SCHWINGENSCHUSS 1942). In the west with local populations in the French dept. Hautes Alpes. From western Germany to Iberian peninsula replaced by the sister species *C. lennigiaria*. In south-eastern France (dept. Var, Basses Alpes, warm valleys of Hautes Alpes) and western Germany, Rhine Valley single specimens with habitus of *albiocellaria* are mixed in typical populations of *C. lennigiaria* (ZSM/HERBULOT), possibly as a result of sympatric occurrence with occasional hybridization (France), or to old 'silent' genes (western Germany), see remarks. HERBULOT (1946) mentions one specimen with transitional characters from dept. Var, south-eastern France, at a locality where no

Acer monspessulanum grows. – Outside Europe from Turkey across Caucasus, Transcaucasus, northern Iran and Turkmeniya to central Asian mountains (WARDIKYAN 1985; VIIDALEPP et al. 1992; VIIDALEPP 1988; 1996; HAUSMANN & LASZLO 1999; DIDMANIDZE 2002).

PHENOLOGY: Usually bivoltine: late April to early June; mid-July to mid-September. Under good conditions trivoltine, from early April to late September. Overwintering as pupa.

BIOLOGY: Larva monophagous (1-2) on *Acer* (Aceraceae). Usually found and recorded on *Acer campestre* (var. cit., e.g. Austria: MAY 1892), preferring hedges (MAY 1892; REBEL 1910b). In rearings also feeding on *A. monspessulanum* (BASTELBERGER 1900), but only accepting *A. campestre* in rearings of GELBRECHT (pers. comm.). Adults attracted to light.

HABITAT: Inhabiting sparse deciduous forest and scrubby places with *Acer*, often on sun-exposed slopes. From 0 m up to 300 m above sea-level, usually from 400 m up to 1,300 m in southern Alps and southern Europe.

SIMILAR SPECIES: Sister species *C. lennigiaria* with much smaller cell rings, often oval, ground colour usually with stronger orange tinge, dark coloration more restricted to a medial fascia rather than extended to the inner termen. Allopatric *C. ariadne* with forewing apex not pointed, hindwing rounded, medial fascia straight, terminal streaks present.

REMARKS: For validity at species rank of *C. lennigiaria* see remarks to the following species. For morphology of immature stages see MAY (1892).

181. Cyclophora lennigiaria (FUCHS, 1883), stat. n.

Zonosoma lennigiaria FUCHS, 1883: Stettin. ent. Ztg. 44 (4-6): 268 (western Germany: Rheingau). Syntype(s) (?NSMW). Herewith raised from subspecies of *C. albiocellaria* (SCOBLE 1999) to species rank (stat. n.).
Ephyra albiocellaria var. *occidentalis* D. LUCAS, 1932: Bull. Soc. ent. Fr. 37: 166 (France, Charente Inférieure: Forêt de Benon). Lectotype ♂ (ZSM/HERBULOT, examined), designated in HERBULOT (1984a).
Cyclophora lennigiaria mauretanica REISSER, 1933: Eos Madr. 9: 258 (Morocco: Rif Mts.). Syntypes 9♂6♀ (NHMW 1♂1♀, SMNK 4♀2♀, 1♂ coll. SCHAWERDA, 3♂3♀ coll. DÜRCK, but not in ZSM). Valid at subspecific rank (SCOBLE 1999).

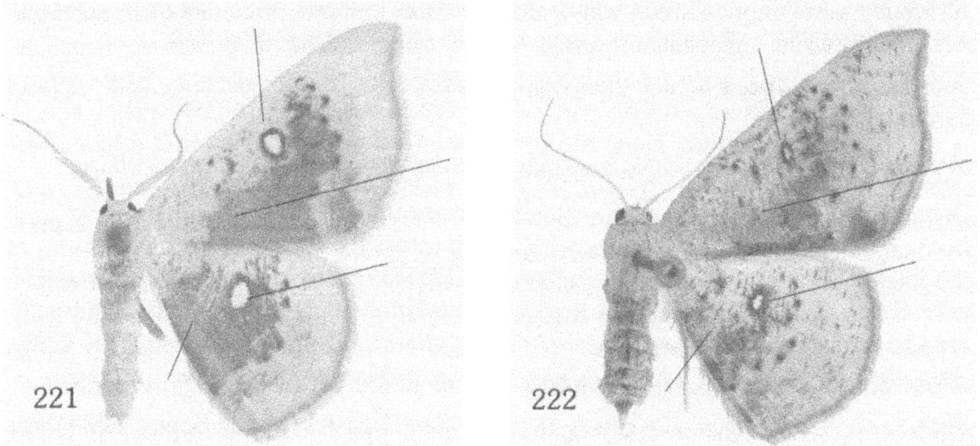

Text-figs 221-222: Diagnostic characters (indicated) of European *Cyclophora* species. Text-fig. 221: *C. albiocellaria* (HÜBNER, 1789). Text-fig. 222: *C. lennigiaria* (FUCHS, 1883).

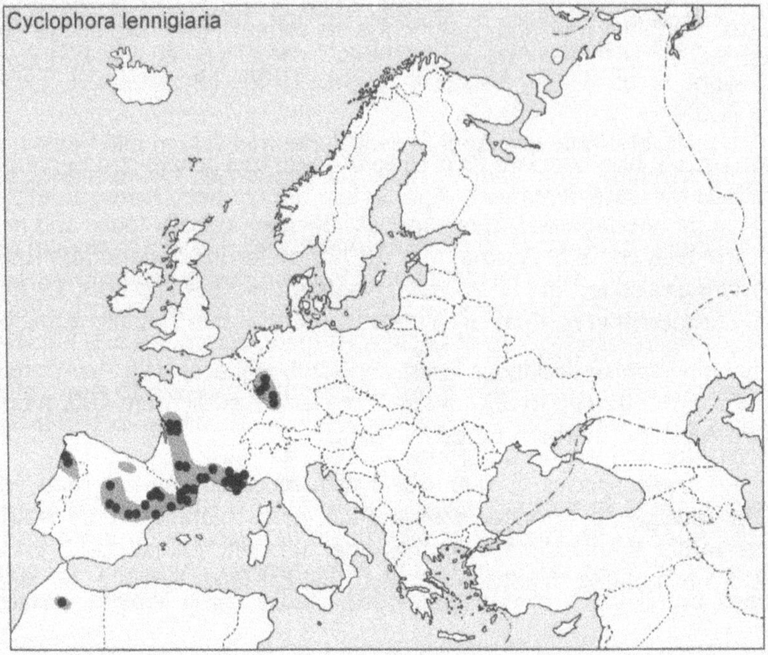

Unavailable names (misidentification): *albiocellaria*: sensu FUCHS (1880) nec HÜBNER, 1789. – (infrasubspecific): *aestiva*: FUCHS (1883); *suppupillaria*: FUCHS (1883). – (incorrect subsequent spellings): *lenigiaria*: BASTELBERGER (1900); *lennigaria*: TURATI in RAINERI (1996); *lenniugiaria*: TOMIC et al. (2002).

DIAGNOSIS: Wingspan ♂♀ 20-25 mm, second generation smaller, sometimes 18 mm only. Forewing termen concave below apex. Ground colour yellowish, usually with strong orange tinge. Postmedial line dissolved to dark brown dots. Medial area towards postmedial line with strong black brown coloration on all wings, in the second generation, however, usually with strongly reduced black coloration. Cell rings small, usually oval on all wings, with small white centre. Terminal dots absent. Fringe darker, ochreous. Frons light brown, vertex concolorous with ground colour. Longest branches of ♂ antennae about 5 times width of flagellum. Further features under generic diagnosis.

Subsp. ***mauretanica*** lighter than nominate subspecies, more yellowish, with reduced dark markings.

MALE AND FEMALE GENITALIA: No constant differential features from *C. albiocellaria*.

DISTRIBUTION: West-Mediterranean (Sub-Mediterranean). Distributed with the host-plant from an isolated occurrence in western Germany across the warmer regions of France to the Iberian peninsula. German populations threatened, possibly on the verge of extinction. For heterogenous habitus in the populations from south-eastern France and western Germany see under 'distribution' of *C. albiocellaria*. Erroneously recorded for Serbia (ZECEVIC & RADOVANOVIC 1974). – Outside Europe in Morocco (subsp. *mauretanica*).

PHENOLOGY: Usually bivoltine: late April to early June; mid-July to mid-September. Under good conditions trivoltine, from early April to late September. Overwintering as pupa.

BIOLOGY: Larva monophagous (1) on *Acer monspessulanum* (Aceraceae), repeatedly found (Germany: FUCHS 1883; BASTELBERGER 1900; WARNECKE 1933) and recorded (var. cit.) on that plant species. Reared on *A. monspessulanum* (GÓMEZ DE AIZPÚRUA 1989; KRAUS 1993), in rearings not accepting other species of *Acer* (BASTELBERGER 1900; PROUT 1913; WARNECKE 1933; GELBRECHT pers. comm.), but possibly sometimes accepting also *A. platanoides* (LHOMME 1935). Adults attracted to light, and visiting flowers of *Calluna vulgaris* and *Leucanthemum vulgare* (FUCHS 1883).

HABITAT: Xerothermophilous. Sun-exposed scrubby places, hedges and forest fringes with the host-plant. From 0 m up to 300 m above sea-level, up to 1,200 m in southern Europe, up to 1,700 m in Morocco.

SIMILAR SPECIES: Sister species *C. albiocellaria* (see no. 180). *C. lennigiaria* may be confused with similar forms of *C. puppillaria*, the latter usually larger, medial area with less dark suffusion, medial line less distinct, the row of postmedial dots less convex, fringe usually concolorous.

REMARKS: Validity of species rank of *C. lennigiaria* is still controversial, accepted by some authors (e.g. BASTELBERGER 1900; WARNECKE 1933; FORSTER & WOHLFAHRT 1981; SCOBLE 1999; DANTART 2000), rejected by others (e.g. STERNECK 1941; HERBULOT 1963; URBAHN 1971; LERAUT 1980; MÜLLER 1996; KRAUS 1999), who usually rank it at subspecific level. As *C. ariadne* is also considered bona species, based only on habitus features, species rank of *C. lennigiaria* is accepted here, provisionally. Additional study is needed, e.g. breeding or hybridization experiments, preferably at the contact zone in south-eastern France. Separation of the species suggested for the last glaciation (WARNECKE 1933). For morphology of immature stages see FUCHS (1883), BASTELBERGER (1900), and GÓMEZ DE AIZPÚRUA (1989).

182. Cyclophora ariadne (REISSER, 1939)

Codonia (Cosymbia) ariadne REISSER, 1939: Z. Wien. Ent. Ver. 24: 169 (Crete: Silva Rouwa). Syntypes at least 19♂♀ (NHMW 1♂1♀, SMNK 16♂♀, ZSM 1♀, examined).

DIAGNOSIS: Wingspan ♂♀ 22-27, second generation smaller, 21-24 mm. Forewing termen not concave below apex, hindwing rounded. Ground colour yellow with red suffusion. Second generation with less dark suffusion and less marked, dotted postmedial line. Medial fascia distinct, reddish, straight, distally touching the cell ring. Cell rings with white centre, usually small, chiefly on hindwing. Small, elongate terminal dots (streaks) present. Fringe darker, reddish. Frons red brown, vertex ochreous. Longest branches of ♂ antennae about 5 times width of flagellum. Further features under generic diagnosis.

MALE GENITALIA: Differences small from *C. albiocellaria*, *C. lennigiaria* and *C. serveti*. Aedeagus longer and narrower.

FEMALE GENITALIA: Ductus bursae straight and narrow near junction with corpus bursae, strongly dilating in the posterior half, comparatively long, about 1.5 mm, i.e. 0.7-1.0 times length of corpus bursae.

DISTRIBUTION: Endemic to the island of Crete. Possibly threatened and meriting special protection.

PHENOLOGY: Uni- or bivoltine: At lower elevations (100 m up 1,100 m above sea-level) limited data from early May to late May; early September to late October. On mountains

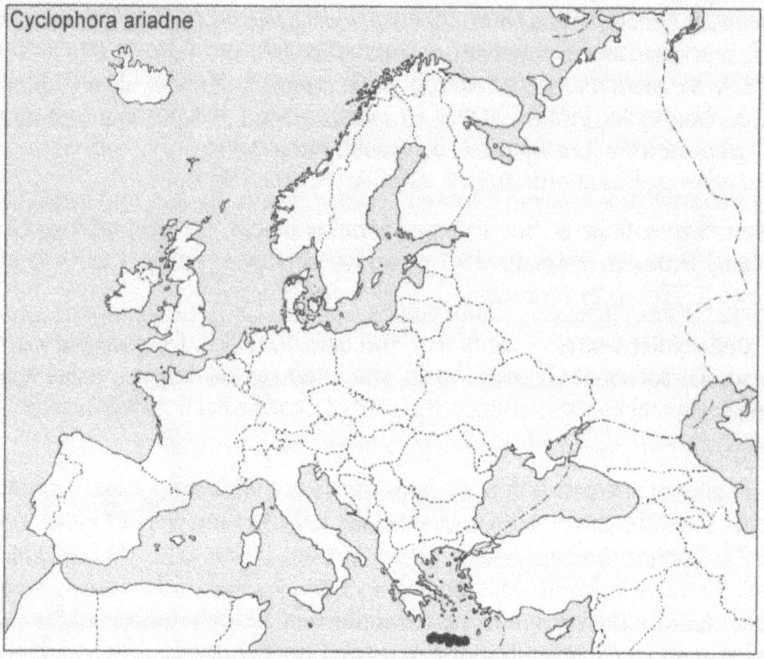

(1,100 m up 1,400 m) possibly univoltine, limited data available from late June to early August.

BIOLOGY: Larva recorded on *Platanus orientalis* (Platanaceae) (REISSER 1939b). Adults attracted to light.

HABITAT: Forests of *Platanus* on the Ida mountains (REISSER 1939b), but in the north collected even close to the coast. From 100 m up to 1,400 m above sea-level.

SIMILAR SPECIES: Allopatric *C. albiocellaria* (see no. 180). *C. ariadne* may be confused with similar forms of the variable *C. puppillaria*, the latter usually larger (second generation), ground colour more reddish, medial line more diffuse, fringe usually concolorous, abdomen with dark spots.

183. Cyclophora serveti REDONDO & GASTON, 1999

Cyclophora serveti REDONDO & GASTON, 1999: Monografias S.E.A. 3: 30, figs 185, 186, 189, pl. 5 (Spain, Zaragoza: Torralba de los Frailes). Holotype ♂ (coll. GASTON, figured in original description; MNCN: 1♂1♀ paratypes, examined).

DIAGNOSIS: Wingspan ♂♀ 21-26 mm. Forewing termen concave below apex, apex subfalcate. Hindwing termen angled at M3. Ground colour red brown, hindwing slightly paler. Medial line distinct, dark grey. Postmedial line dotted, dots very small. Terminal area of forewing with indistinct dark grey shadows. Discal spots developed as extremely small white dots. Fringe reddish. Frons red brown, vertex slightly paler. Longest branches of ♂ antennae 7-8 times width of flagellum. Further features under generic diagnosis.

MALE GENITALIA: Valva spatulate, dilating towards tip. Fibula very narrow, tapered at tip. Base of fibula with spinulose process. Aedeagus short.

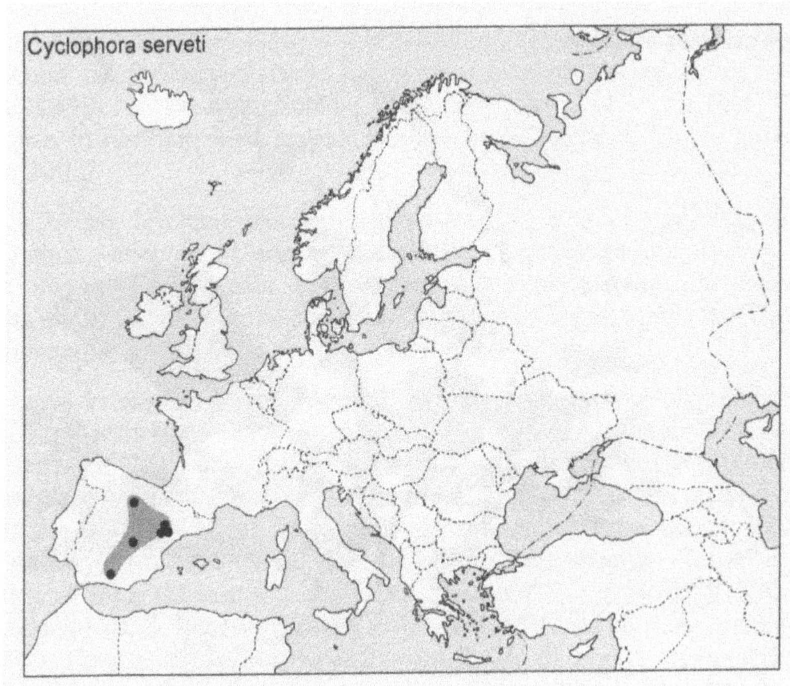

FEMALE GENITALIA: Ductus bursae comparatively short (about 0.9 mm), straight, narrow at junction with corpus bursae, strongly dilating posteriorly.

DISTRIBUTION: Endemic to Iberian peninsula. Known from a very few localities between the provinces Zaragoza in the north-east, Burgos in the north-west and Granada in the south.

PHENOLOGY: Bivoltine: Limited data available: mid-May to mid-June; late July to early September.

BIOLOGY: Larval host-plants unknown. Adults attracted to light.

HABITAT: Mixed open forest of *Quercus* and *Pinus* (REDONDO & GASTON 1999). From 800 m up to 1,300 m above sea-level.

SIMILAR SPECIES: Reminiscent of *C. suppunctaria*, but forewing colour more red brown rather than rosy or light ochreous, hindwing paler, terminal area with grey shadows. *C. serveti* may also be confused with similar forms of *C. hyponoea*, but medial line of the former straighter and more distinct on all wings. *C. punctaria* differs in paler, sand coloured ground colour. *C. porata* with cell rings distinct, medial line usually more diffuse, ground colour paler, sand coloured, hindwing not paler than forewing.

184. Cyclophora annularia (FABRICIUS, 1775)

Phalaena annularia FABRICIUS, 1775: Syst. Ent.: 625 (Germany). Syntype(s).
Phalaena Geometra annulata SCHULZE, 1775: Naturf. Halle 6: 92, pl. 4, fig. 3 (Germany). Syntype(s).
Geometra omicronaria DENIS & SCHIFFERMÜLLER, 1775: Ank. syst. Werkes Schmett. Wienergegend: 107 (Austria: Vienna distr.). Syntype(s) lost.
Phalaena circulifera FOURCROY, 1785: Entom. Parisiensis 2: 285 (France: Paris). Syntype(s).

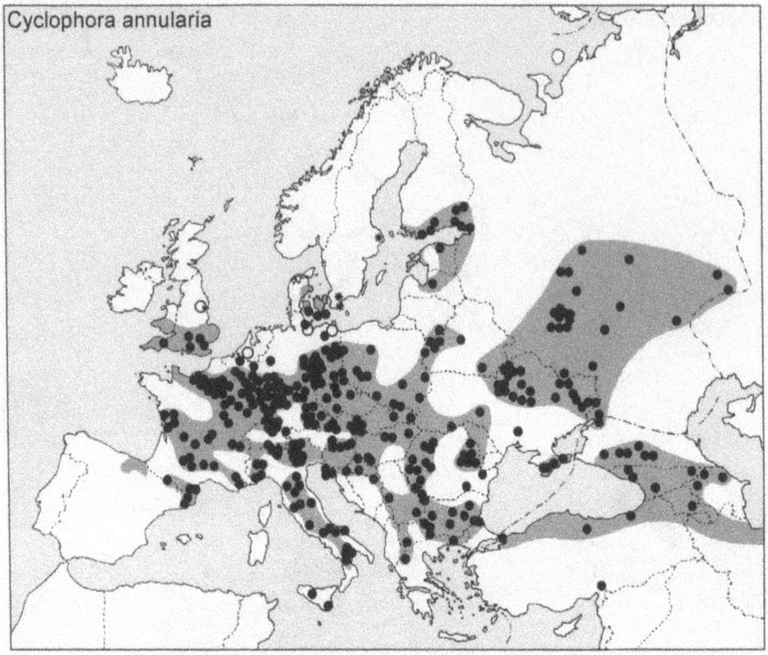

Phalaena denticulata THUNBERG, 1788: D.D. Mus. Nat. Acad. Upsal. 6: 75, fig. (Europe). Lectotype ♀ (UZIU), see KARSHOLT & SCHMIDT NIELSEN (1985).

Unavailable names (misidentification): ?*pendularia*: sensu LANG (1789) nec CLERCK, 1759. – (infrasubspecific): *aestiva*: PROUT (1913); *biobsoleta*: RIDING (1898); *decolorata*: COCKAYNE (1952); *extenuata*: DANNEHL (1933); *fasciata*: LEMPKE (1949); *obsoleta*: TUTT (1902); *sertaria*: DANNEHL (1927); *striata*: COCKAYNE (1952). – (incorrect subsequent spellings): *amicronaria*: TOMIC et al. (2002); *amulata*: SCHMIDT (1880); *annullata*: HORMUZAKI (1899); *annullate*: WOLFSBERGER (1955); *denticularia*: VIVES MORENO (1994); *extremata* (pro *extenuata*): PROUT (1935);

DIAGNOSIS: Wingspan ♂♀ 21-26 mm, second generation smaller, sometimes 18 mm only. Ground colour white, with yellow tinge of varying intensity. Postmedial fascia broad, dark brown, strongly zigzagging. Cell rings large and distinct. Elongate terminal dots (streaks) distinct. Fringe concolorous. Frons and vertex whitish, the former light brown towards vertex. Longest branches of ♂ antennae about 5 times width of flagellum. Further features under generic diagnosis.

Variability: The postmedial fascia sometimes modified to a double line, resulting from reduction of black markings. Second generation usually deeper yellow, with stronger markings, sometimes with grey suffusion. Influence of temperature and humidity on the coloration discussed in BERGMANN (1955).

MALE GENITALIA: Valva spatulate, dilated towards tip. Fibula comparatively stout, strongly curved. Base of fibula with spinulose process, fibula projection towards base short. Aedeagus short.

FEMALE GENITALIA: Ductus bursae short, straight, at ostium bursae slightly dilated.

DISTRIBUTION: European-Westasiatic. Temperate zone of Europe from southern England and Pyrenees to the Urals (e.g. AHOLA et al. 1997). In the north to southernmost

Scandinavia. Absent from the extreme south, and from most Mediterranean islands, but recently found on Sicily (GRILLO & PARENZAN 1995; PARENZAN 1998; IPPOLITO & PARENZAN 1998). – Outside Europe from northern Turkey across Caucasus, Transcaucasus to northern Iran, isolated occurrence in southern Turkey (prov. Adana: DOGANLAR 2003).

PHENOLOGY: Usually bivoltine: late April to mid-June; early July to late August, generations sometimes overlapping. Under good conditions, e.g. in the south trivoltine from late March to mid-September, with single specimens until October. In Baltic states, Scandinavia and in central Europe above ca. 1,000 m univoltine, from late May to early July. Overwintering as pupa.

BIOLOGY: Larva in nature monophagous (3) on *Acer*, or, possibly, oligophagous (2) on deciduous trees, with focus on *Acer* (Aceraceae). Repeatedly found on *A. campestre* (Germany: STEINER 2001; WEGNER pers. comm.; Austria: ERLEBACH pers. comm.; Great Britain: PORTER 1997). Recorded on *Acer* (var. cit.), preferring *A. campestre* growing in shade (BERGMANN 1955; SKOU 1986), but also *A. platanoides* (KOCH 1984), *Acer pseudoplatanus* (BERGMANN 1955; STEINER 2001). Reputedly on *Carpinus betulus*, *Betula*, *Salix*, *Prunus spinosa* and *Quercus robur* (var. cit., e.g. HOBERT 1930; LHOMME 1935; BERGMANN 1955; FORSTER & WOHLFAHRT 1981). Reared on *Acer pseudoplatanus* (ALLAN 1949; PORTER 1997; WARING & TOWNSEND 2003), and *A. campestre* (LEIPNITZ pers. comm.; KRAUS 1993). Adults nocturnal and active at dusk, attraction to light not optimal (WARING & TOWNSEND 2003). Readily attracted to bait. Nectar feeding recorded on flowers of *Berberis vulgaris* (BERGMANN 1955), and *Cornus sanguinea* (STEINER 2001). Resting on leaves during daytime (WOODFORDE 1902).

PARASITOIDS: Diptera, Tachinidae: *Ceranthia abdominalis* (HERTING 1960; 1984; ANDERSEN 1996) – Hymenoptera, Braconidae: *Apanteles vitripennis* (THOMPSON 1945; SHENEFELT 1972) – Hymenoptera, Ichneumonidae: THOMPSON (1945) mentions '*Campoplex* sp.', but this indicates no exact host-relationship, as the large genus *Campoplex* was formerly used in different generic concepts (ZWAKHALS pers. comm.).

HABITAT: Silvicolous lowland species. With the host-plant sometimes common in deciduous forests of different type, preferring hardwood (STEINER 2001), hedges and forest fringes. From 0 m up to 500 m above sea-level, exceptionally somewhat higher. Up to 1,500 m in the Alps (WOLFSBERGER 1971; FORSTER & WOHLFAHRT 1981).

SIMILAR SPECIES: No similar species in Europe.

185. Cyclophora puppillaria (HÜBNER, 1799)

Geometra puppillaria HÜBNER, 1799: Samml. Eur. Schmett. 5, Geometrae (1): pl. 13, fig. 69 (Europe). Syntype(s) lost.

Geometra nolaria HÜBNER, 1809: Samml. Eur. Schmett. 5, Geometrae (1&2): pl. 63, fig. 325 (recte 327) (Europe). Syntype(s) lost.

Geometra gyrata HÜBNER, 1813: Samml. Eur. Schmett. 5, Geometrae (1): pl. 84, fig. 434 (Europe). Syntype(s) lost.

Cabera gyraria TREITSCHKE, 1827: Schmett. Eur. 6 (1): 366. Emendation. Synonymy according to LERAUT (1997). TREITSCHKE's own material from Dalmatia possibly misidentified and referring to *C. ruficiliaria* (PROUT 1934).

Aspilates mirtalis COSTA, 1834: Annuar. zool. Napoli 1834: 74 (Italy: Foro Volcani). Syntype(s).

‡ *Cabera albipunctaria* HEYDENREICH (1851): Lepid. Eur. cat. meth. (Edn 3): 55 (Europe). Subsequently pub-

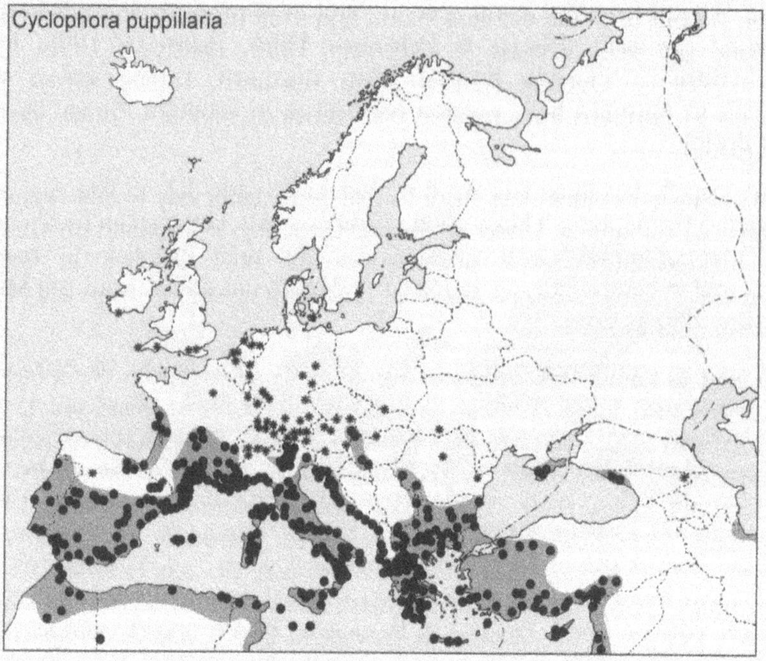

lished manuscript name of MANN, corresponding to 'f. *nolaria*'. Junior subjective homonym nec HUFNAGEL, 1767 (identity: *Cyclophora albipunctaria*).

Ephyra calaritana TURATI, 1911: Zeits. wiss. Ins. Biol. 7 (7-8): 210 (Sardinia, Cagliari Province: Bosco dei 7 Fratelli). Syntypes 8.

Cosymbia (Ephyra) pupillaria asiaeminoris AMSEL, 1935: Veröff. Dt. Kolon. u. Übersee-Mus. Bremen 1 (2): 244 (northern Israel: Waldheim). Lectotype ♂ (MNHU/PÜNGELER, examined), herewith designated to fix type locality and stabilize usage of the name; at MNHU 1♂ paralectotype from Turkey, Ankara ('Angora').

Cosymbia puppillaria granti PROUT, 1935: in SEITZ, Macrolep. 4 (Suppl.): 31, pl. 4, fig. d (Azores: San Pedro; Sta. Maria; Reguinho, Terceira). Syntypes 3♂ (BMNH). Valid at subspecific rank (MEYER 1991; SCOBLE 1999).

Cosymbia lilacinipes SCHAUS & COCKERELL, 1923: Proc. Ent. Soc. Wash. 25 (7-8): 163 (Madeira: Funchal). Syntype(s) (USNM). Valid at subspecific rank (SCOBLE 1999).

Unavailable names (misidentification): *porata*: sensu WERNEBURG (1864) and SAMMUT (2000) nec LINNAEUS, 1767. – (infrasubspecific): *agrapharia*: HOMBERG (1935); *alogaria*: SCHAWERDA (1919); *badiaria*: STAUDINGER (1871); *bucovinaria*: NEMES (1970); *decolorata*: DANNEHL (1933); *depupillata*: DANNEHL (1933); *depupillata*: DANNEHL (1933); *fasciata*: WAGNER (1927); *scorteata*: WAGNER (1924); *simplex*: THIERRY-MIEG (1916). – (incorrect subsequent spellings): *calabritana*: VIVES MORENO (1996); *calaritona*: PROUT (1914: 418); *girata*: TURATI (1911); *pupilarium*: REUTTI (1898); *pupillaria*: TREITSCHKE (1825); *puppilaria*: PROUT (1914: 418); *pupullaria*: HULST (1896).

DIAGNOSIS: Wingspan ♂♀ 23-28 mm, subsequent generations smaller, sometimes 19 mm only. Forewing apex pointed. Extremely variable in colour and pattern of wings. Ground colour usually light brown with rosy tinge, but specimens with sand coloured, brown ('f. *badiaria*') or orange ground colour not rare. Postmedial line usually dotted, but not rarely weak or absent. Medial fascia usually absent, but sometimes distinct. Cell rings often distinct and with white centre, but not rarely reduced to a dark spot or completely absent. Terminal dots absent or very weak, somewhat more distinct in the forewing apex. Fringe concolorous, but in subsp. *lilacinipes* from Madeira darker. Abdomen usually with a dorsal row of dark spots. Frons light brown or orange, paler towards proboscis and on vertex. Palpi long. Longest branches of ♂ antennae 7-10 times width of flagellum in the first generation, 5-7 times in subsequent generations. Further features under generic diagnosis.

MALE GENITALIA: Fibula comparatively stout, strongly curved, often bent twice, U-shaped. Base of fibula slightly dilated, strongly spinulose without process. Aedeagus large, with sclerite at tip.

FEMALE GENITALIA: Ductus bursae short (about 0.7-0.9 mm), at ostium bursae strongly dilated.

DISTRIBUTION: Mediterranean, migrant. Resident and usually common in southern Europe south of Alps, on all Mediterranean islands and southern Crimea. Under good conditions migratory flights of different intensity up to British Isles (SKINNER 1984), central Germany and western Ukraine. Rarely recorded also in Denmark, southern Sweden, and southern Poland (BUSZKO 2000). Possibly resident also on Isles of Scilly for short periods (SKINNER 1984), and in Hungary (FORSTER & WOHLFAHRT 1981). – Outside Europe from Morocco to Egypt (TURATI 1924; WILTSHIRE 1949), from Turkey, Cyprus and the Levant to Caucasus and northern Transcaucasus (e.g. VIIDALEPP 1996). Some isolated records from Iran (e.g. STEINER 2001) down to the south (prov. Fars, ZSM), and, doubtfully, from Afghanistan (Nuristan: WILTSHIRE 1966). Populations from Madeira ('*lilacinipes*') and Azores ('*granti*') considered synonym (HERBULOT 1968), or valid at subspecific rank (SCOBLE 1999).

PHENOLOGY: Plurivoltine: mid-March to late October, under good conditions to late November. In North Africa and the Levant in all months of the year. Migration flights north of Alps late June to mid-October, but chiefly from mid-August to early October, under good conditions with a subsequent brood at site, but not overwintering (SKINNER 1984, FORSTER & WOHLFAHRT 1981, STEINER 2001).

BIOLOGY: Larva probably polyphagous, possibly preferring evergreen oaks. Found on *Quercus* (Fagaceae) (France: ERLEBACH pers. comm.), *Cistus* (Cistaceae), *Myrtus* (Myrtaceae), *Quercus calliprinos* and *Q. ithaburensis* (Fagaceae) (Israel: BYTINSKI-SALZ & STERNLICHT 1967; HALPERIN & SAUTER 1992). Repeatedly recorded on *Quercus* (Fagaceae) (var. cit.), preferring young, fresh growth of *Quercus ilex*, *Q. coccifera* (Fagaceae), and on *Cistus salvifolius, C. monspeliensis* (Cistaceae), *Myrtus communis* (Myrtaceae), *Phillyrea angustifolia* (Oleaceae), *Erica,* and *Arbutus unedo* (Ericaceae) (Spain: SEEBOLD 1879; RIBBE 1912; France: GRASLIN 1863; MILLIÈRE 1868; LHOMME 1935; Great Britain: SKINNER 1984, PORTER 1997; Egypt: WILTSHIRE 1949a). Rearing easy on *Myrtus communis* (SCHWINGENSCHUSS 1942; COCKAYNE 1954), on young twigs of *Quercus robur* (original data; VORBRODT 1928; ERLEBACH, MÜLLER pers. comm.), and on *Ligustrum* (PORTER 1997). Adults readily attracted to light. Dispersal activities of adults high, with 'emigration flights' (EITSCHBERGER et al. 1991), r-strategy.

HABITAT: Ubiquitous in the Mediterranean region, most common in the Mediterranean maquis and in sparse oak forest (*Quercus pubescens, Q. ilex*), north of Alps preferring warm and dry localities. Usually from 0 m up to 1,000 m above sea-level in southern Europe, up to 1,500 m in Turkey. During migration flights recorded at altitudes up to 3,200 m (MAZZUCCO 1966).

SIMILAR SPECIES: *C. lennigiaria* (see no. 181), *C. ariadne* (182). Usually well characterized and easily to be distinguished from its congeners by the intense brownish, orange or rosy coloration, the dorsal abdominal spots and the falcate forewing apex.

REMARKS: Longer antennal branches have been reported erroneously for the Sardinian populations ('*calaritana*') (TURATI 1911). Extensive description of the immature stages in

HAGGETT (1958). Outside the resident distribution area hybrid mating seems to occur at higher frequency: Descendants of a female *C. puppillaria* from Switzerland revealed to be hybrids with *C. albipunctata* (THOMANN 1955).

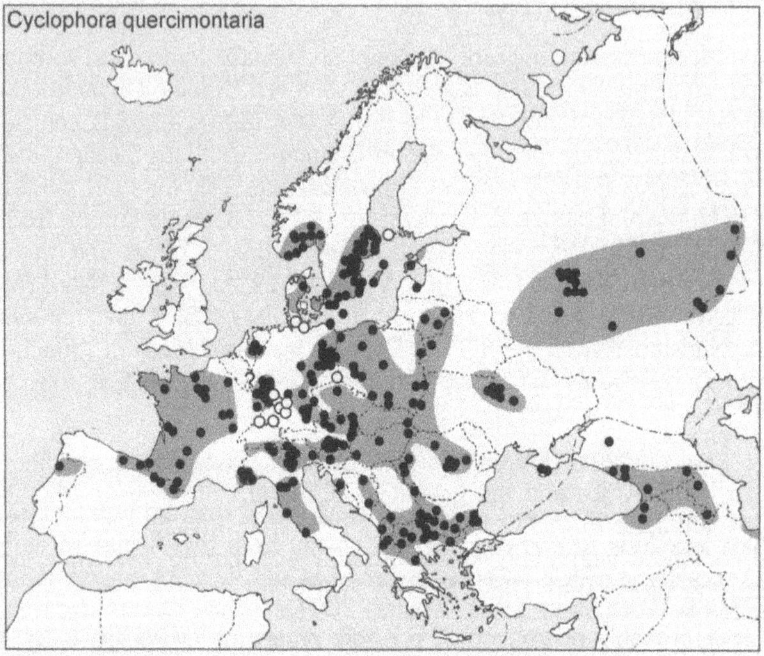

186. *Cyclophora quercimontaria* (BASTELBERGER, 1897)

Zonosoma quercimontaria BASTELBERGER, 1897: Stettin. ent. Ztg. 58 (1-3): 120 (western Germany, Rheingau: Eichberg). Syntype(s).

Cyclophora quercimontaria nigrosparsaria SCOBLE, 1999: Geometrid Moths of the World 2: 207. First available name after *nigrosparsaria* HEYDEMANN, 1938 (Ent. Z. Frankf. a.M. 51 (44): 392, pl. 1, figs 2, 3: infrasubspecific). Herewith downgraded from subspecific rank (SCOBLE 1999) to synonymy (syn.n.).

Cyclophora quercimontaria privataria SCOBLE, 1999: Geometrid Moths of the World 2: 207. First available name after *nigrosparsaria* HEYDEMANN, 1938 (Ent. Z. Frankf. a.M. 51 (44): 392, pl. 1, figs 1, 4: infrasubspecific). Herewith downgraded from subspecific rank (SCOBLE 1999) to synonymy (syn.n.).

Cosymbia quercimontaria elbursica PROUT, 1935: in Seitz, Macrolep. 4 (Suppl.): 31, pl. 4, fig. e (northern Iran, Elburs Mts.: Darekeroudbar, Sabatku, Mazanderan). Holotype ♂ (BMNH). Valid at subspecific rank.

Unavailable names (incorrect subsequent spellings): *quircimontaria*: GANEV (1983). – (infrasubspecific): *communifasciata*: LEMPKE (1949); *uniformata*: LEMPKE (1949). – (hybrid): *querciporata*: URBAHN (1939) (*quercimontaria* ♂ x *porata* ♀).

DIAGNOSIS: Wingspan ♂♀ 21-25 mm, second generation smaller, sometimes 17 mm only, in rearings up to 28 mm. Fore- and hindwings rounded. Rather constant in colour and pattern of wings. Wings sand coloured, with strong red brown suffusion. Medial line distinct, rather straight, red brown. Postmedial line dotted. Discal spots small, white, without dark ring. Terminal area concolorous with ground colour. Terminal line fine, sometimes interrupted at veins. Frons light brown, vertex sand coloured. Longest branches of ♂ antennae 5-7 times width of flagellum. Further features under generic diagnosis.

MALE GENITALIA: Valva slender, with costa concave at base. Fibula very long. Base of fibula not dilated, with spinulose process very weakly sclerotized. Aedeagus long and narrow.

FEMALE GENITALIA: Ductus bursae long (1.2-1.5 mm), dilating towards ostium bursae, near corpus bursae usually curved.

DISTRIBUTION: Euro-Caucasian. From Iberian peninsula to Urals (e.g. AHOLA et al. 1997). In the north to southern Scandinavia, in the south to central Italy (RACHELI et al. 1978; PROLA & RACHELI 1979) and Balkan peninsula. Absent from British Isles and Mediterranean islands. Reputedly in northern Portugal (SILVA CRUZ 1974; PASSOS DE CARVALHO 1984; MÜLLER 1996), data await confirmation. Usually rare and local. Abundance decreasing and strongly threatened in many German regions (ERLACHER & FRIEDRICH 1994; STEINER 2001; WOLF & HACKER 2004). Extinct in Finland (HULDEN et al. 2000). – Outside Europe in northern Turkey, Caucasus, Transcaucasus, and northern Iran (subsp. *elbursica*). Zoogeography extensively discussed in MENHOFER (1938; 1941; 1942a), HEYDEMANN (1941; 1942), HERBULOT (1947).

PHENOLOGY: Bivoltine: mid-April to early June, exceptionally until late June; early July to mid-August, single specimens until early October. North of Alps with incomplete second generation (STEINER 2001). South of Alps with second generation emerging from mid-June. Overwintering as pupa.

PARASITOIDS: Unidentified parasitoids mentioned by BOLDT (1933).

BIOLOGY: Larva monophagous (2) on *Quercus* (Fagaceae), reputedly preferring saplings (FORSTER & WOHLFAHRT 1981; BERGMANN 1955; PATOČKA 1994). Found on leaves of old *Quercus petraea* (WEGNER 1998), on shrubs of *Q. robur* (Netherlands: BOLDT 1933; Germany: HEYDEMANN 1929; Czech Republic: MICHEL 1944), and on unidentified species of *Quercus* (Germany: STEINER 2001; Austria: ERLEBACH pers. comm.). Repeatedly recorded on *Quercus* (var. cit.). Reared on young twigs of *Q. robur* (ERLEBACH, MÜLLER pers. comm.), however, refusing young leaves in rearings of WEGNER (1998), and accepting only hard leaves. Adults readily attracted to light, easily flushed from branches of *Quercus* during daytime (PROUT 1913), preferably resting on dry leaves (MICHEL 1944).

HABITAT: North of Alps xerothermophilous, restricted to the warm lowlands. Often in dry, sun-exposed places, on southern slopes, in sites with young oak saplings, preferably *Q. pubescens* (WOLFSBERGER 1966; 1971), in scattered pine forest, scrubby places, fringes of deciduous forests rich in oak, borders of heathland, sometimes also in hot humid localities (MENHOFER 1942a). Preferring sandy and silicious soils (BERGMANN 1955). Usually from 0 m up to 600 m above sea-level, up to 1,100 m in southern Alps, up to 1,400 m in southern Europe.

SIMILAR SPECIES (see Text-figs 225-229): *C. punctaria* and *C. porata* on average larger, ground colour less warm, with less reddish suffusion, antemedial and medial lines grey or lighter red, fore- and hindwings less rounded, often with dark spots in the terminal area. The small white hindwing discal spot without ring of *C. quercimontaria*, and the darker fringe often allows secure discrimination from both the above mentioned species. *C. ruficiliaria* on average larger, suffusion and medial line grey, fringe often yellowish with small terminal dots. For secure identification dissection is sometimes needed.

REMARKS: For hybrid rearings see URBAHN & URBAHN (1939). Reared descendants of a female *C. quercimontaria*, collected in nature, revealed to be natural hybrids with *C. porata* (Germany, Berlin: GELBRECHT pers. comm.).

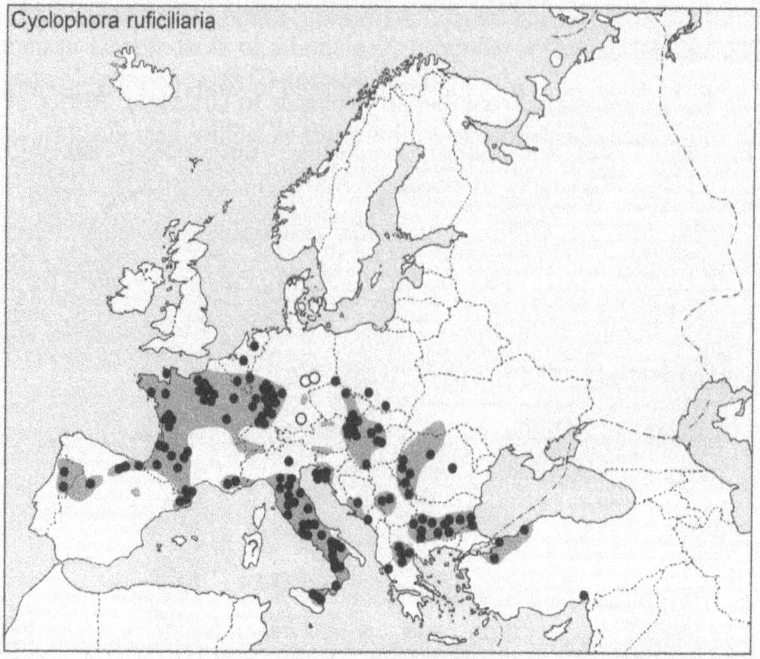

187. Cyclophora ruficiliaria (HERRICH-SCHÄFFER, 1855)

Zonosoma ruficiliaria HERRICH-SCHÄFFER, 1855: Syst. Bearb. Schmett. Eur. 6 (68): 135 (locality not stated, [Europe]?). Syntype(s).
Ephyra schaefferaria LA HARPE, 1855: Neue Denkschr. Allg. schweiz. Ges. ges. Naturw. 14 (7): 5 (non binominal), 24 (binominal), fig. 1.
unilinearia SCHARFENBERG: original reference not traced (see PROUT 1913).
Unavailable names (misidentification): *pupillaria*: sensu ZELLER (1847) nec *puppillaria* HÜBNER, 1799; ?*suppunctaria*: sensu CULOT (1918) nec ZELLER, 1847. – (infrasubspecific): *circumdataria*: BASTELBERGER (1898); *mattiacata*: BASTELBERGER (1898); *privataria*: BASTELBERGER (1898); *ruberrima*: BASTELBERGER (1898). – (incorrect subsequent spellings): *privata*: PROUT (1913); *ruciciliaria*: YLLA et al. (1990); *ruficilaria*: WOLFSBERGER (1971); *rurciliaria*: HERBULOT (1946).

DIAGNOSIS: Wingspan ♂♀ 25-30 mm, second generation smaller, 20-25 mm. Forewing with straight costa, apex not rounded. Fairly variable in colour and pattern of wings. First generation usually with orange or yellowish leather brown tinge, strongly suffused with dark scales and with strongly marked pattern. Second brood paler, with medial line often weaker. In both generations antemedial line fine and usually strongly convex. Medial fascia curved. Discal spots white, not encircled on the forewing, sometimes with a thin ring on the hindwing, sometimes completely absent. Terminal dots elongate but not fused to form a line. Frons light brown, vertex sand coloured. Longest branches of ♂ antennae 6-8 times width of flagellum. Further features under generic diagnosis.

MALE GENITALIA: Valva broad, truncate at tip. Fibula short, dilated at tip, sometimes beak-shaped. Base of fibula not dilated, with a spinulose, rounded lobe.

FEMALE GENITALIA: Ductus bursae comparatively short (0.85-1.3 mm), slightly dilating towards ostium bursae, sometimes slightly curved at junction with corpus bursae.

DISTRIBUTION: European-Westasiatic. Local, and usually rare in southern Europe, from

Iberian peninsula and France to Balkan peninsula. In the north up to the Netherlands, western Germany and southern Poland. Reputedly present on Sardinia (MÜLLER 1996). Almost all examined specimens from the Ukraine (except for the south-west; e.g. EFETOV & BUDASHKIN 1990) revealed to be misidentified (*C. suppunctaria*, *C. quercimontaria*), the records from southern European Russia and the Urals (VIIDALEPP 1996) also require re-examination. Decreasing and threatened in Germany (STEINER 2001; WOLF & HACKER 2004), extinct from Thuringia (ERLACHER & FRIEDRICH 1994). – Outside Europe in north-western Turkey, with a very isolated population in southern Turkey (prov. Adana: DOGANLAR 2003). All other records from Turkey to Caucasus, Transcaucasus, northern Iran, Iraq, and Israel (e.g. VIIDALEPP 1974; 1996; WILTSHIRE 1939; 1957; HALPERIN & SAUTER 1992) doubtful: all examined '*ruficiliaria*' from these regions revealed to belong to the very similar first generation of *C. suppunctaria*, such as the Armenian female genitalia figured in WARDIKJAN (1985). Other misidentifications (e.g. 1♀ from Georgia, ZISP) occurred with *C. linearia f. strabonaria*. Distribution and zoogeography discussed in MENHOFER (1942b).

PHENOLOGY: Bivoltine: mid-April to late June; mid-July (in the south mid-August) to early October. In southern Europe emerging from late March. Second generation often less abundant. Overwintering as pupa.

BIOLOGY: Larva monophagous (3) on *Quercus* (Fagaceae). Found (Germany: STEINER 2001) and recorded on *Quercus* (PROUT 1913; Germany: BERGMANN 1955), preferring *Q. cerris* (Czech Republic: PATOČKA 1994). Reared on *Quercus* (KRAUS 1993). Adults readily attracted to light.

PARASITOIDS: Diptera, Tachinidae: *Ceranthia abdominalis* (ANDERSEN 1996).

HABITAT: Xerothermophilous. Scattered, warm oak forest, mixed forest and scrubby places with oak, not necessarily dry. Often in sites with young oak saplings, on sunny slopes, preferably on sandy and calcareous soils. Usually from 0 m up to 600 m above sea-level, rarely higher, e.g. up to 1,000 m in northern Portugal, southern Alps, Apeninnes and Balkan peninsula, exceptionally up to 1,500 m (FLAMIGNI & BASTIA 2003b).

SIMILAR SPECIES (see Text-figs 225-229): *C. quercimontaria* (see no. 186), *C. linearia*, 'f. *strabonaria*' (192). Second generation of *C. suppunctaria* with flesh-coloured tinge and absence of dark suffusion, medial line light red, fine, straight. First generation of *C. suppunctaria* extremely similar, but transverse lines straighter, that of hindwing usually finer, and whitish discal spot usually with a fine dark ring. In ♂ identification is possible by brushing tip of abdomen to remove scales and reveal shape of fibula, in ♀ dissection is needed for secure identification (ductus bursae larger). *C. punctaria*, *C. porata*, and *C. quercimontaria* larger, with darker suffusion. *C. punctaria* and *C. porata* with slightly narrower forewings and hindwing termen more angled at M3, often with dark spots in the terminal area. Second generation of *C. porata* in southern Europe often very similar, usually to be distinguished by dark cell rings, sometimes brushing of abdomen (♂) or dissection (♀) is needed. *C. punctaria* with discal spots usually absent and antemedial line often dissolved to dots, less densely suffused with dark scales.

REMARKS: The taxon '*delaeveri* BERGER, 1949' may refer to natural hybrids between *C. punctaria* and *C. ruficiliaria*, see remarks to *C. punctaria* and LAEVER (1954).

The *porata* species-group

Male genitalia with a large, weakly setose ventral projection from base of valva. One further, dorsad, setose projection arising from this 'sacculus'. Homology to fibula and basal projections of the other two species-groups uncertain. See introduction to genus. Further features under generic diagnosis. Infraspecific variability and interspecific differences comparatively large in male genitalia.

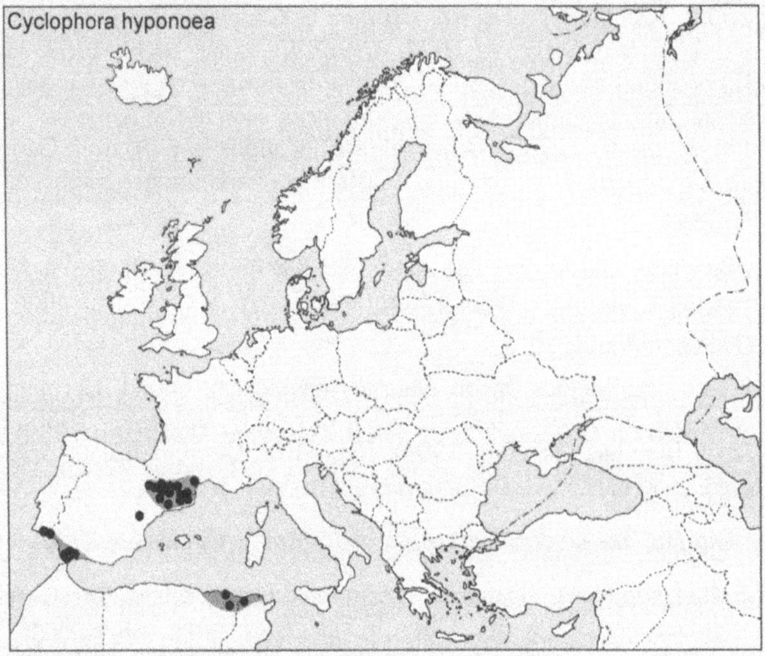

188. *Cyclophora hyponoea* (PROUT, 1935)

Cosymbia hyponoea PROUT, 1935: in SEITZ, Macrolep. 4 (Suppl.): 31, pl. 4, fig. e (eastern Algeria: Philippeville). Holotype ♂ (BMNH, examined).

Unavailable names (misidentification): *porata*: sensu PASSOS DE CARVALHO & CORLEY (1995) nec LINNAEUS, 1767. – (incorrect subsequent spelling): *hyponaea*: BOLLAND (1981).

DIAGNOSIS: Wingspan ♂♀ 24-30 mm. Forewing narrow, termen concave below the tapered apex. Wings sand coloured with orange or red brown tinge. Postmedial line strongly dentate, rarely dotted. Medial line distinct, usually dentate. Antemedial line usually well marked, zigzagging. Discal spots small, encircled by a fine dark ring. Frons light brown, vertex sand coloured. Longest branches of ♂ antennae 7-8 times width of flagellum. Further features under generic diagnosis.

MALE GENITALIA: Valva long and slender. Ventral process broad, tapering, with a short setose dorsal projection. Aedeagus comparatively short.

FEMALE GENITALIA: Ductus bursae comparatively short (1.1-1.4 mm, 0.60-0.75 times length of corpus bursae), dilating near ostium bursae, not curved at junction with corpus bursae.

DISTRIBUTION: West-Mediterranean. Iberian peninsula: Southern Spain and southern

Portugal (CORLEY et al. 2000), north-eastern Spain (prov. Huesca: BOLLAND 1981; REDONDO 1990; Catalonia: VALLHONRAT 1999b; DANTART 1991; DANTART 2000; DANTART & REQUENA, 1993), Andorra (LAEVER 1958; DANTART & ROCHE 1992). Newly recorded for southern France (Tuchan, prov. Aude; EMEM in ZSM). – Outside Europe in north-eastern Algeria and northern Tunisia.

PHENOLOGY: Bi- or trivoltine: In the north bivoltine, from early May to early September, generations overlapping in late June or early July, and with single specimens of a partial third generation until late September. In southern Iberian peninsula regularly trivoltine, from late March to mid-October.

BIOLOGY: Larval host-plants unknown, but supposedly monophagous on *Quercus* (Fagaceae). Adults attracted to light.

HABITAT: Xerothermophilous. Hot coastal lowlands and hills, usually at sites with occurrence of spinose leaved *Quercus* (Fagaceae) (SKOU pers. comm.). Usually from 0 m up to 700 m above sea-level, exceptionally up to 1,300 m (DANTART & ROCHE 1992), up to 1,400 m in Morocco.

SIMILAR SPECIES: *C. serveti* (see no. 183). *C. porata* usually smaller, medial line not dentate, cell rings larger, on average less reddish or rosy, sometimes, however, extremely similar. In the Iberian peninsula usually restricted to higher elevations, but sympatric occurrence possible at about 1,000 m above sea-level. Brushing of the ♂ abdomen to remove scales and reveal shape of fibula allows secure identification, ♀ needs dissection.

REMARKS: Specific treatments: LAEVER (1977), BOLLAND (1981) and DANTART (1991).

189. Cyclophora porata (LINNAEUS, 1767)

Phalaena Geometra porata LINNAEUS, 1767: Syst. Nat. (Edn 12) 1 (2): 866 (Italy). Syntype(s) (LSL).

‡? *Phalaena circulata* FOURCROY, 1785: Entom. Parisiensis 2: 308 (France: Paris). Retained synonym of *C. porata* by LERAUT 1997, but the size indicated by FOURCROY corresponds to that of *Geometra papilionaria*. Nomen dubium.

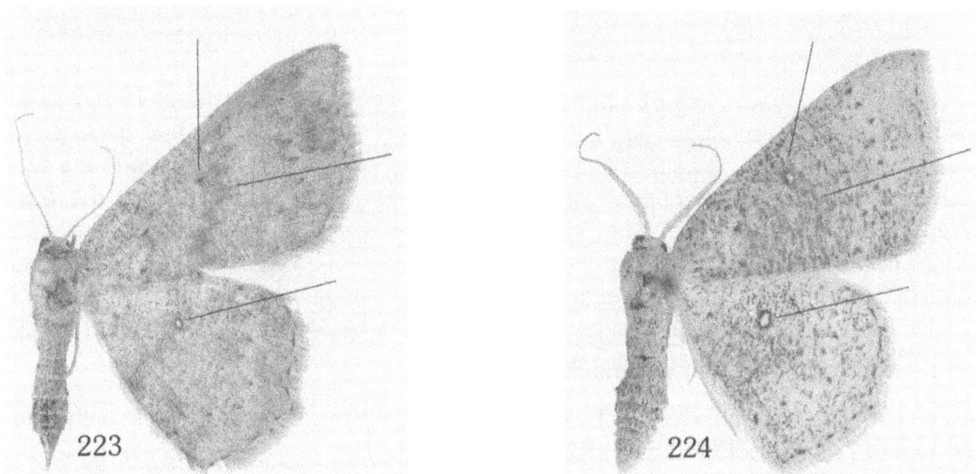

Text-figs 223-224: Diagnostic characters (indicated) of European *Cyclophora* species. Text-fig. 223: *C. hyponoea* (PROUT, 1935). Text-fig. 224: *C. porata* (LINNAEUS, 1767).

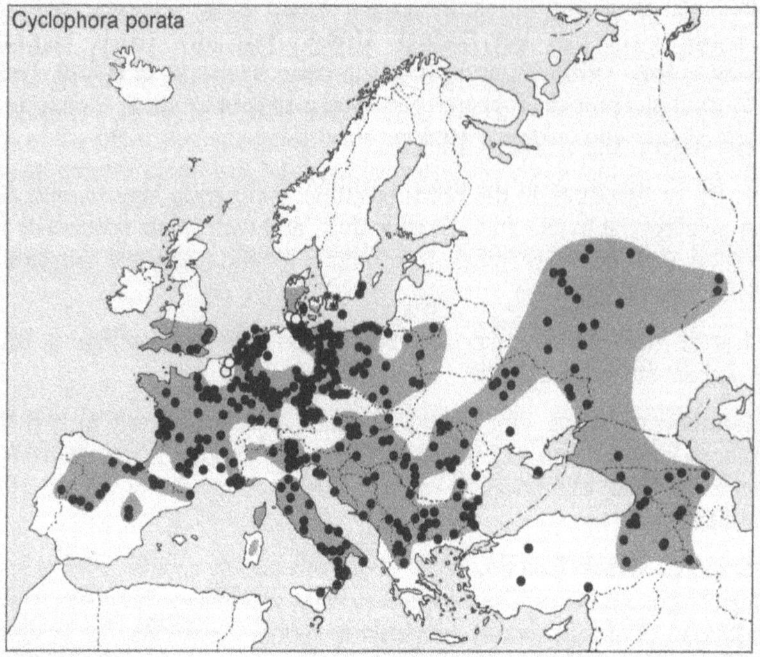

‡ *Geometra huebneraria* HAWORTH, 1809: Lepid. Brit. (2): 312. Diacritic mark in '*hübneraria*' needs emendation. Introduced in synonymy, unavailable.

‡ *Phalaena poraria* LATREILLE, 1825: Encyclopédie Méth. 10 (1): 89. Incorrect subsequent spelling (emendation according to SCOBLE 1999).

Unavailable names (misidentification): *circularia*: sensu WERNEBURG (1864) nec FABRICIUS, 1798 (identity: *Cyclophora albipunctata*); *ocellaria*: sensu HAWORTH (1809; also homonym) nec HÜBNER, 1799 (identity: *Cyclophora albiocellaria*); *punctaria*: sensu DENIS & SCHIFFERMÜLLER (1775) and HÜBNER (1799) nec LINNAEUS, 1758. – (infrasubspecific): *aestiva*: HORMUZAKI (1899); *basirubra*: COCKAYNE (1952); *delinearia*: SILBERNAGEL (1944); *depulsa*: LEMPKE (1949); *flava*: BRETSCHNEIDER (1951); *linearia*: LAMBILLON (1905); *marginaria*: LAMBILLON (1905); *nigrosparsaria*: LEMPKE (1949); *plumbomarginata*: COCKAYNE (1950); *punctularia*: LAMBILLON (1905); *rubearia*: LAMBILLON (1905); *uniformata*: LEMPKE (1949); *venata*: PROUT (1913); *vinicolor*: COCKAYNE (1952); *visperaria*: FUCHS (1884). – (nomen nudum): *hubneraria*: HAWORTH (1802). – (incorrect subsequent spelling): *prorata*: BREZ (1791).

DIAGNOSIS: Wingspan ♂♀ 22-26 mm, second generation smaller, sometimes 19 mm only. Wings sand coloured, often with strong red suffusion, mainly in the first generation. Rather variable in the contrast of wing pattern, usually much more contrasted in the second generation. Habitus of third generation sometimes approaching that of the first. Medial and antemedial lines often diffuse. Postmedial line dotted. Black cell rings with white filling distinct, conspicuous in the second generation. Grey shadows of terminal area diffuse in the first generation, much sharper in the second ('f. *visperaria*'). Frons light brown, vertex sand coloured. Longest branches of ♂ antennae 7-9 times width of flagellum. Further features under generic diagnosis.

MALE GENITALIA: Ventral process of valva very broad, tapering, 'beak-shaped', with long and narrow, slightly setose dorsal projection.

FEMALE GENITALIA: Ductus bursae comparatively long (1.45-1.7 mm, 0.7-0.9 times length of corpus bursae), strongly dilating towards ostium bursae, very narrow and curved towards corpus bursae.

DISTRIBUTION: Sub-Mediterranean. Widespread in southern and eastern Europe from Portugal to southern Urals. In western and central Europe in isolated populations, in the west to southern England, in the north to southernmost Scandinavia. Present on the larger Mediterranean islands, except Crete. Occurrence on Malta (VALLETTA 1973; MÜLLER 1996) doubtful, misidentified in SAMMUT (2000), identity: *C. puppillaria*. Deleted from the fauna of the Algarve (CORLEY et al. 2000). Abundance decreasing in England (COLLINS 1997; WARING & TOWNSEND 2003), central and southern Germany (ERLACHER & FRIEDRICH 1994; STEINER 2001; WOLF & HACKER 2004). – Outside Europe in Turkey, Caucasus, Transcaucasus and northern Iran (prov. Mazandaran, new for the fauna of Iran, coll. SOMMERER), and with isolated occurrence in Israel (HAUSMANN 1997).

PHENOLOGY: Bi- or trivoltine: late April to mid-June, mid-July to mid-September. Exceptionally single specimens of a third generation until early October. Shifts of flight periods by two weeks possible, but usually without overlap of generations. In England with incomplete second generation (SKINNER 1984). Under good conditions in southern Europe regularly trivoltine from mid-April to early October, generations overlapping, second generation emerging from June. Overwintering as pupa.

BIOLOGY: Larva probably monophagous (3) on *Quercus* (Fagaceae). Found on *Quercus robur*, and *Q. petraea* (Germany: WEGNER pers. comm.; Netherlands: BOLDT 1933; Poland: URBAHN & URBAHN 1939), on *Q. calliprinos* (Israel: BYTINSKI-SALZ & STERNLICHT 1967; HALPERIN & SAUTER 1992), preferring shrubs and small twigs budding from stems.

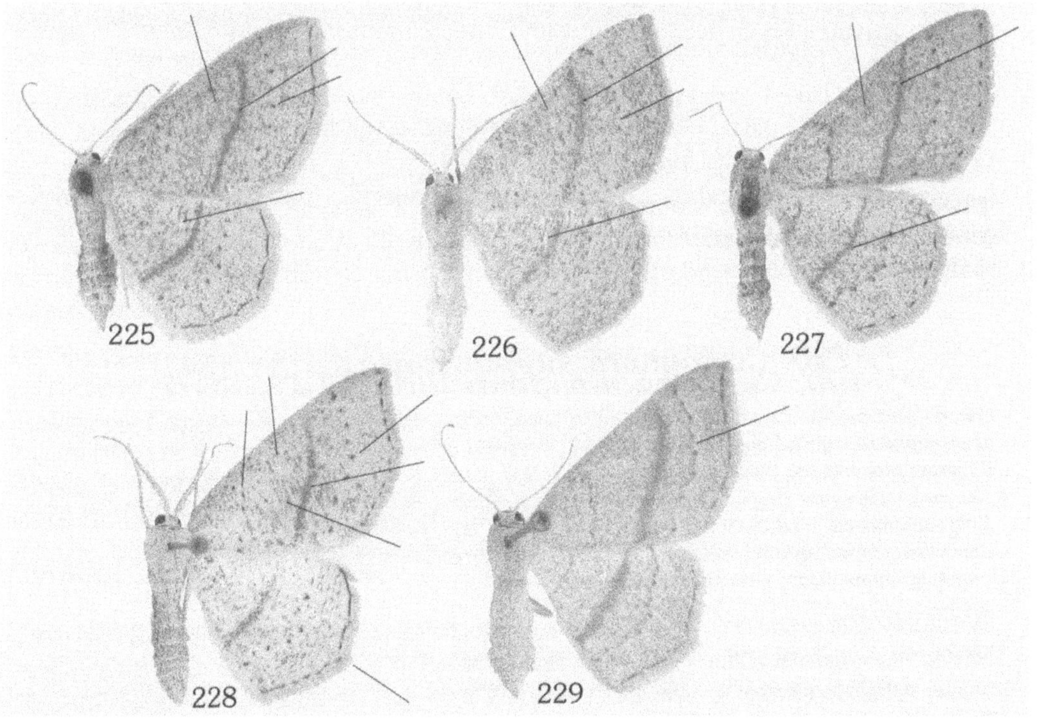

Text-figs 225-229: Diagnostic characters (indicated) of European *Cyclophora* species. Text-fig. 225: *C. quercimontaria* (BASTELBERGER, 1897). Text-fig. 226: *C. ruficiliaria* (HERRICH-SCHÄFFER, 1855). Text-fig. 227: *C. suppunctaria* (ZELLER, 1847). Text-fig. 228: *C. punctaria* (LINNAEUS, 1758). Text-fig. 229: *C. linearia* (HÜBNER, 1799), 'f. *strabonaria*'.

Recorded on *Quercus* (var. cit.), rarely on *Betula* (Betulaceae) (var. cit., e.g. ALLAN 1949; BERGMANN 1955; STEINER 2001), *Fraxinus excelsior* (Oleaceae) (BERGMANN 1955), doubtful records on *Crataegus* (PROUT 1913). Reared on young twigs of *Quercus robur* (MÜLLER pers. comm.; KRAUS 1993). Pupation on the leaf underside, falling with it to the ground (WARING & TOWNSEND 2003). Adults readily attracted to light, sometimes also to bait. During daytime resting on stems or leaves of old *Quercus ilex* trees (PROUT 1913; WOODFORDE 1902).

PARASITOIDS: Diptera, Tachinidae: *Ceranthia abdominalis* (HERTING 1960; 1984; ANDERSEN 1996).

HABITAT: Xerothermophilous. Scrubby places, sparse forest, and sunny forest fringes rich in *Quercus*, also at free standing trees. Preferring sites with young shrubs of *Quercus ilex*, often on sandy soils. In southern Europe in a wide range of different habitats, e.g. common in the Mediterranean maquis. North of Alps only at the most xerothermic sites, usually from 0 m up to 500 m, in the Alps, southern Europe and south-western Asia sometimes up to 1,700 m, in Spain and Turkey montane, never found below 1,000 m.

SIMILAR SPECIES: *C. serveti* (see no. 183), *C. quercimontaria* (186), *C. ruficiliaria* (187), *C. hyponoea* (188). From the similar species *C. punctaria, C. ruficiliaria* and *C. quercimontaria* differing in the distinct black cell ring, with white centre on all wings, males usually with grey shadows in the terminal area. Correct identification sometimes requires brushing of abdomen (♂) or dissection (♀).

REMARKS: In nature rarely hybridization with *C. quercimontaria* occurs (GELBRECHT pers. comm.; see no. 186).

The ***punctaria*** species-group

Male genitalia with large and strongly setose ventral projection from base of valva. Female genitalia with strongly sclerotized anterior margin of sterigma. Further features under generic diagnosis. Infraspecific variability and interspecific differences comparatively large in male genitalia.

190. Cyclophora suppunctaria (ZELLER, 1847)

Cabera suppunctaria ZELLER, 1847: Isis 1847 (7): 496 (Italy, Tuscany: Agnano Lake). Syntypes 3♂1♀ (BMNH: 1♂ examined, with red 'type' label).

‡ *Ephyra subpunctaria* HERRICH-SCHÄFFER, 1852: Syst. Bearb. Schmett. Eur. 6 (55): 69 (Austria). Syntype(s). Incorrect subsequent spelling, referring to ZELLER's authorship. Referring to vol. 3: 30, thus binominal, replacing *subpunctaria* HERRICH-SCHÄFFER, 1848 (pl. 67, fig. 415: non binominal).

Unavailable (infrasubspecific): *amabilis*: SCHAWERDA (1922); *grisea*: SCHAWERDA (1938). – (incorrect subsequent spelling): *suppunctava*: HERBULOT (1946).

DIAGNOSIS: Wingspan ♂♀ 23-30 mm, subsequent generations smaller, ♂♀ 21-24 mm. Strong seasonal variability: Wings of first generation sand coloured, with strong grey brown suffusion ('rough appearance'), medial, ante- and postmedial lines usually distinct. Ground colour of subsequent generations with flesh-coloured tinge, without dark suffusion ('smooth appearance'), medial line fine, ante- and postmedial lines indistinct or absent. Habitus of third generation sometimes approaching that of the first. In all generations medial line rather straight, terminal area concolorous, discal spots pale, without ring,

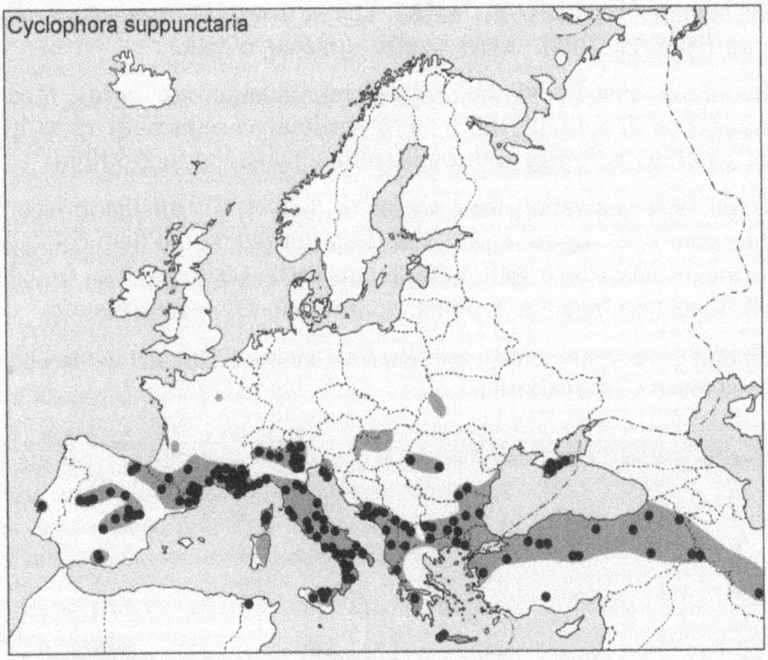

almost invisible. Frons pale brown, vertex sand coloured. Longest branches of ♂ antennae 7-9 times width of flagellum. Further features under generic diagnosis.

MALE GENITALIA: Uncus subapically dilated. Valva slender, tapering. Fibula weakly sclerotized, broad, truncate at tip. With setose process arising dorsad from fibula. Ventral projection of valva large, round, with very long sigmoid or curved setae. Aedeagus long, corresponding to the long ductus bursae of the female.

FEMALE GENITALIA: Ductus bursae very long (1.8-2.4 mm, usually 0.9-1.1 times length of corpus bursae), strongly dilating towards ostium bursae, strongly curved at junction with corpus bursae.

DISTRIBUTION: West-Palaearctic (Mediterranean). Southern Europe from Portugal (CORLEY pers. comm.) to Crimea. In the north to southern Alps, southern Slovakia and Hungary. Present on Corsica, Sardinia, Sicily, and Crete (coll. SKOU). Occurrence in southern Poland (MÜLLER 1996) not confirmed by BUSZKO (2000). Often rare and local. – Outside Europe in Tunisia and Turkey; isolated occurrences in the mountains of western and southern Iran (prov. Fars and Lorestan, new for the fauna of the Iran, leg. PÖLL, ZSM, and coll. SOMMERER) and Saudi Arabia (prov. Riyadh, new for the fauna of Saudi Arabia, coll. ZSM).

PHENOLOGY: Usually trivoltine with overlapping generations, the first from early April, the second from early June and the third until late October. In the north and on high mountains sometimes bivoltine: late April to mid-June; early July to late September. Overwintering as pupa.

BIOLOGY: Larva in nature possibly monophagous (3) on *Quercus* (Fagaceae). Found on *Quercus pubescens* (Fagaceae) (France: HEIDEMANN 1996) and *Q. pyrenaica* (CORLEY pers. comm.). Reared on young leaves of *Q. robur* (original data; HEIDEMANN 1986;

MÜLLER pers. comm.). Reputedly on 'herbs', such as *Artemisia campestris, Hippocrepis* and *Melilotus* (PATOČKA 1994). Adults readily attracted to light.

HABITAT: Xerothermophilous. Scrubby places and scattered oak forests, Mediterranean maquis. Usually from 0 m up to 700 m above sea-level, exceptionally up to 1,200 m. In Spain up to 1,600 m, in Turkey and southern Iran higher, up to 2,600 m.

SIMILAR SPECIES: First generation very similar to that of *C. ruficiliaria* (see no. 187). Second generation of *C. suppunctaria* well characterized by the flesh-coloured tinge of wing colour, and the absence of dark suffusion, medial line light red, fine, straight. Correct identification sometimes requires brushing of abdomen (♂) or dissection (♀).

REMARKS: Shape of valva and fibula, and also the habitus of the first generation, reminiscent of equivalents in *C. ruficiliaria*.

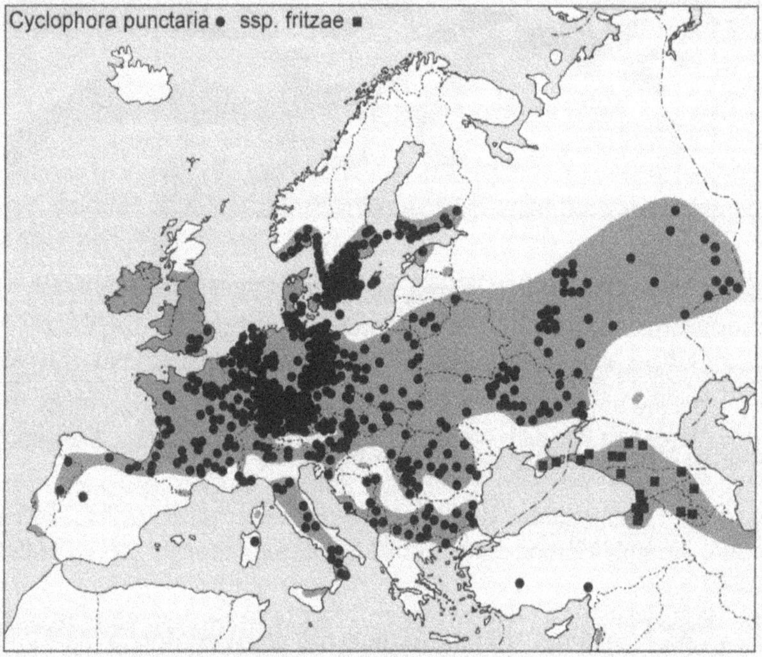

191. Cyclophora punctaria (LINNAEUS, 1758)

Phalaena Geometra punctaria LINNAEUS, 1758: Syst. Nat. (Ed. 10) 1: 522 (Finland). Syntype(s) (NHRS).
Phalaena Geometra amata LINNAEUS, 1758: Syst. Nat. (Ed. 10) 1: 524 (Europe). Syntype(s).
Phalaena Geometra teutonaria LINNAEUS, 1758: Syst. Nat. (Ed. 10) 1: Appendix: 823 (Europe). Syntype(s).
‡ *Phalaena Geometra amataria* LINNAEUS, 1761: Faun. Suec. (Edn 2): 322. Incorrect subsequent spelling of *amata*.
Phalaena Geometra fultaria VILLERS, 1789: Linn. ent. 2: 330 (France). Syntype(s).
Phalaena communifasciata DONOVAN, 1808: The nat. history of Brit. insects 13: 49, pl. 456 (England). Syntype(s).
Phalaena unifasciata DONOVAN, 1808: The nat. history of Brit. insects 13: 49 (United Kingdom). Syntype(s).
Geometra subangularia HAWORTH, 1809: Lepid. Brit. (2): 313 (England). Syntype(s).
Geometra (Ennomos) acutaria ROQUETTE, 1857: Allg. dt. naturh. Ztg. (N.S.) 3: 350 (type locality not stated). Syntype(s).
Zonosoma punctaria var. *naevata* BASTELBERGER, 1900: Dt. ent. Z. Iris 13: 84 (western Germany: Wiesbaden, Köln). Syntype(s) (SMF).

Cosymbia delaeveri BERGER, 1949: Lambillionea 49 (3-4): 34, fig. (Belgium: Membre). Holotype ♂ (DEI). Herewith downgraded from species rank (SCOBLE 1999) to synonymy (syn. n., see remarks).

Cyclophora punctaria fritzae HAUSMANN, 2003: Ent. Zeits. Stuttgart 113 (11): 326, figs 10, 26 (Russia, northern Caucasus: Pyatigorsk). Holotype ♂ (ZSM, examined). Valid at subspecific rank.

Unavailable names (misidentification): *subpunctaria*: sensu FUCHS (1889) nec *suppunctaria* ZELLER, 1847. – (infrasubspecific): *arcufera*: REUTER (1890); *cingulata*: FUCHS (1900); *demptaria*: FUCHS (1900); *erythrescens*: PREISSECKER (1923); *foliata*: FUCHS (1900); *infuscata*: REUTER (1890); ?*nigra*: MICHEL; *nigrotransversata*: HÖRHAMMER (1933); *ochreifusa*: PROUT (1913); *pulcherrimata*: FUCHS (1900); *radiomarginata*: JOANNIS (1908); *signata*: COCKAYNE (1950); *unifasciata*: DONOVAN (1808); *venata*: PROUT (1913). – (nomen nudum): *pseudopunctaria*: SVENSSON (1993). – (incorrect spellings): *fritzi*: HAUSMANN (2003a: figs 10, 26); *punctata*: HERRICH-SCHÄFFER (1840); *tentonaria*: WERNEBURG (1864).

DIAGNOSIS: Wingspan ♂♀ 25-29 mm, second generation smaller, sometimes 22 mm only. Forewing termen concave below apex. Hindwing termen angled at M3. Wings sand coloured, hindwing usually paler, not rarely with grey or red suffusion, causing a 'rough appearance' (SKOU 1986). Medial line usually distinct and convex, close to postmedial line, rarely absent. Postmedial line reduced to small dots. Terminal area usually concolorous. Terminal dots elongate, distinct. Discal spots concolorous, almost invisible. Frons light brown, vertex sand coloured. Longest branches of ♂ antennae 7-9 times width of flagellum. Further features under generic diagnosis.

Variability: Wing pattern fairly variable. Terminal area sometimes with conspicuous reddish or dark grey spots, almost restricted to the second generation and induced by high temperatures (MERRIFIELD 1893; PROUT 1913; BERGMANN 1955).

Subsp. ***fritzae*** without reddish suffusion or red markings, with with strong dark grey suffusion. Medial line dark grey or grey brown. Differences in genitalia see below.

MALE GENITALIA: Fibula curved or bent at an angle of 30-70°. Towards the south-eastern parts of the distribution fibula broader and more strongly curved. Base of fibula with small setose projection. Ventral projection of valva (sub-)rectangular, with straight setae of medium size. In subsp. *fritzae* fibula curved back towards base (gen.-fig. 191b).

FEMALE GENITALIA: Length of ductus bursae 1.3-1.6 mm (usually 0.55-0.7 times length of corpus bursae), somewhat dilating at ostium bursae, slightly curved towards corpus bursae. Subsp. *fritzae* (gen.-fig. 191b) with shorter ductus bursae, 1.05-1.2 mm (0.5-0.6 times length of corpus bursae), straight, hardly dilated at ostium bursae.

DISTRIBUTION: European-Westasiatic. Widely distributed and usually common. In the east to the Urals, in the north to the British Isles and southernmost Scandinavia, in southern Europe with more local distribution. Present on Corsica, Sardinia, and Sicily (MÜLLER 1996). Absent from large parts of the Iberian peninsula, Balearics, Malta, southern Greece, and Crete. – Outside Europe in north-eastern Turkey, Caucasus, Transcaucasus, northern Iran (subsp. *fritzae*), with very isolated populations in southern Turkey (Akshehir: ZSM; prov. Adana: DOGANLAR 2003) and in Lebanon (WILTSHIRE 1939a).

PHENOLOGY: Usually bivoltine: In central Europe from mid-April to mid-September, generations overlapping, or with a gap of 2-3 weeks between, rarely with single specimens of a third generation from late September to mid-November. Under good conditions rarely emerging from mid-March. Second generation in the north incomplete (SKOU 1986), on mountains univoltine, from June to July. Overwintering as pupa.

BIOLOGY: In nature usually monophagous (2) on *Quercus* (Fagaceae). Larva found on *Quercus robur*, *Q. petraea* (Finland: SEPPÄNEN 1954; Great Britain: PORTER 1997;

Germany: STEINER 2001; WEGNER, GELBRECHT pers. comm.; Austria: ERLEBACH pers. comm.; Netherlands: BOLDT 1933; Poland: URBAHN & URBAHN 1939), and *Tilia* (Tiliaceae) (Finland: SEPPÄNEN 1954). Repeatedly recorded in literature on *Quercus* (var. cit.), preferring shrubs (FORSTER & WOHLFAHRT 1981; BERGMANN 1955), but living also in the canopy of trees (WARING & TOWNSEND 2003; GELBRECHT pers. comm.). Reputedly on *Betula* (var. cit., e.g. Prout 1913; BERGMANN 1955; data possibly copied from each other). Adults readily attracted to light and bait, easily flushed out from oaks during daytime. Nectar feeding recorded on *Calluna vulgaris* (PROUT 1913; BERGMANN 1955). Stronger dispersal of adults in the second generation suggested by mark-recapture experiments (HAUSMANN 1990a).

PARASITOIDS: Hymenoptera, Braconidae: *Apanteles juniperatae* (SHENEFELT 1972), *Apanteles solitarius* (THOMPSON 1945; SHENEFELT 1972).

HABITAT: Mesophilous, silvicolous. Common in deciduous and mixed lowland forests with oak, pure oak forests, oak scrub, with much wider tolerance of mesophilous conditions than all its oak-feeding congeners. Preferring sandy soils (BERGMANN 1955). North of Alps from 0 m up to 700 m above sea-level, exceptionally up to 1,200 m, in southern Europe regularly up to 1,300 m.

SIMILAR SPECIES: *C. serveti* (see no. 183), *C. quercimontaria* (186), *C. ruficiliaria* (187), *C. porata* (189). Characterized by the 'rough appearance' (SKOU, 1986), hindwing often paler than forewing, discal spots often absent. Correct identification sometimes requires brushing of the abdomen (♂) or dissection (♀).

REMARKS: The taxon '*delaeveri* BERGER, 1949' probably refers to natural hybrids between *C. punctaria* and *C. ruficiliaria* (HERBULOT pers. comm.; LAEVER 1954), or to aberrant *C. punctaria*. Hybridization between these two species failed in captivity (LAEVER 1954).

192. *Cyclophora linearia* (HÜBNER, 1799)

Geometra linearia HÜBNER, 1799: Samml. Eur. Schmett. 5, Geometrae (1): pl. 13, fig. 68 (Europe). Syntype(s) lost.

‡? *Phalaena zonata* FOURCROY, 1785: Entom. Parisiensis 2: 296 (France: Paris). Synonymy to *C. linearia* (LERAUT 1997) awaits confirmation.

‡? *Phalaena Geometra luteolaria* VILLERS, 1789: Linn. ent. 2: 329 (France, Bresse near Lyon ('Bressia')). Syntype(s). Synonymy to *C. linearia* (LERAUT 1997; SCOBLE 1999) awaits confirmation. The author recommends not replacing the name *linearia* (nomen protectum), which has been in common use over the long period of two centuries, by this possible senior synonym (nomen oblitum; see Code art. 23.9. ICZN 1999).

‡ *Phalaena Geomerra* (sic!) *trilinearia* BORKHAUSEN, 1794: Naturg. eur. Schmett. 5: 502 (Germany: Gladenbach; Bergstraße, Zwingenberg). Syntypes 2♂. Junior primary homonym nec HÜBNER, 1787 (identity: *Idaea aureolaria*).

Ephyra strabonaria ZELLER, 1851: Zeits. Ent. (Lepid.) Bresl. 5 (17): 66, pl. 1, fig. 4 (Poland, Silesia: Lauban). Syntypes 2♂1♀ (MNHU: 1♀, examined, without abdomen).

Unavailable names (misidentification): ?*suppunctaria*: sensu ZELLER (1867) nec ZELLER, 1847. – (infrasubspecific): *alba*: COCKAYNE (1952); *approximans*: PROUT (1913); *arcufera*: COCKAYNE (1952); *bicolor*: OBERTHÜR (1916); *cingulata*: LEMPKE (1949); *carnearia*: LA HARPE (1855, non binominal); *demptaria*: PROUT (1913); *fasciata*: PROUT (1912); *infuscata*: PROUT (1913); *mesoorthia*: SCHULTZ (1930); *nigrosparsaria*: FUCHS (1901); *ophthalmaria*: OBERTHÜR (1916); *rufescens*: LEMPKE (1949); *semifasciata*: DERENNE-MEYERS (1929); *signata*: COCKAYNE (1950); *simplificaria*: CULOT (1918); *trilineata*: SCHAWERDA (1922). – (nomen nudum): *hybridaria*: SELYS-LONGCHAMPS (1845). – (incorrect subsequent spellings): *ophtalmaria*: OSTHELDER (1929); *strabomaria*: CAPUSE & KOVACS (1987).

DIAGNOSIS: Wingspan ♂♀ 23-29 mm, second generation 20-24 mm. Ground colour of

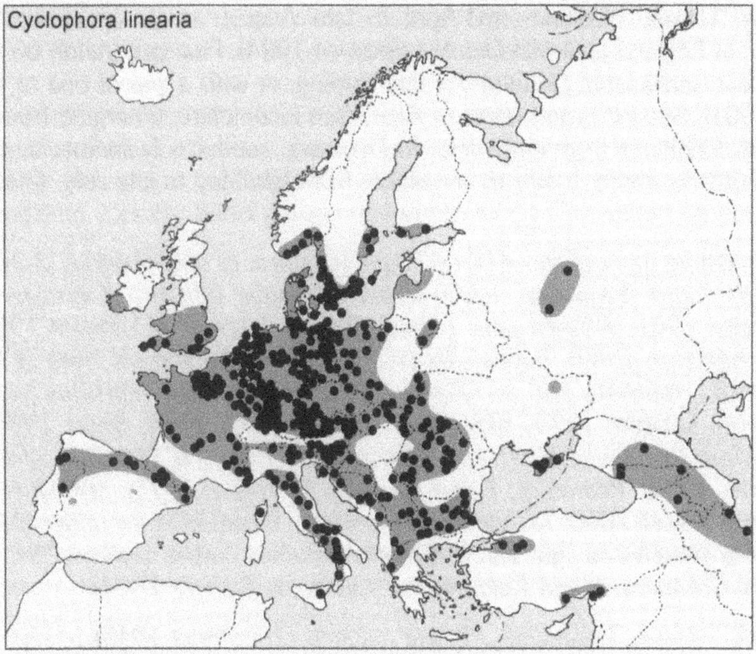

first generation yellow, rarely with brown suffusion, mainly on forewing. Wings sand coloured in the second generation ('f. *strabonaria*'), with pale brown, reddish or yellow suffusion. Medial line usually sharp and distinct in the first generation, more diffuse and weak in the second. Antemedial line usually present, postmedial line usually uninterrupted. Terminal area concolorous. Discal spots whitish, indistinct, sometimes, however, with a dark ring, mainly on the hindwing and more frequently in the second generation. Frons pale brown to reddish brown, vertex yellowish. Longest branches of ♂ antennae 7-9 times width of flagellum. Further features under generic diagnosis.

Variability: The normal form of the second brood ('f. *strabonaria*') induced by high temperatures (PROUT 1913; URBAHN & URBAHN 1939; BERGMANN, 1955), but, very rarely, occurs also in the first generation (SCHULTZ 1930). On the other hand, single specimens of the second generation may, rarely, resemble those of the first in colour and pattern.

MALE GENITALIA: Valva narrow medially, dilated in the distal half. Fibula sharply bent medially. Setose projection absent. Ventral process of valva round, with stout setae in spiral position.

FEMALE GENITALIA: Ductus bursae long, 1.35-1.75 mm (0.6-0.9 times length of corpus bursae), strongly dilating towards ostium bursae, furcate in ventral view, and curved at junction with corpus bursae (lateral view).

DISTRIBUTION: European-Westasiatic. Widely distributed in Europe, in the north up to southernmost Scandinavia. Absent from large parts of southern Europe, e.g. the lowlands and all Mediterranean islands except Corsica and Sicily. In the east to Crimea and central European Russia. In Finland only as immigrant (HULDEN et al. 2000), in Ireland very rare (SKINNER 1984).– Outside Europe local in mountainous regions of Turkey, Caucasus, Transcaucasus, and northern Iran.

PHENOLOGY: Usually bivoltine: mid-April to late August, single specimens until late September, in England until mid-October (SKINNER 1984). First generation on mountains emerging 2-3 weeks later. Generations overlapping, or with a gap of one or two weeks (STEINER 2001). Second brood north of Alps often incomplete, emerging from late July, under good conditions from early July. In Denmark, southern Scandinavia, on Corsica and on mountains above 1,300 m univoltine, from late May to late July. Overwintering as pupa.

BIOLOGY: In nature monophagous (1) on *Fagus sylvatica*, or oligophagous (2-3), but clearly preferring *Fagus* (Fagaceae). Larva found on *Fagus sylvatica* (Germany: SCHULTZ 1930; STEINER 2001; WEGNER pers. comm.; Poland: URBAHN & URBAHN 1939; Spain: GÓMEZ DE AIZPÚRUA 1989). Repeatedly recorded on shrubs and old trees of *Fagus sylvatica* (var. cit.), reputedly also on *Quercus, Betula, Vaccinium myrtillus* (var. cit., e.g. REBEL 1910b; LHOMME 1935; BERGMANN 1955; SKINNER 1984; SKOU 1986; STEINER 2001), *Sorbus torminalis* (Germany: BERGMANN 1955) and *Frangula alnus* (PROUT 1913; VOGEL 1924). Reared on *Fagus sylvatica* (FRIEDRICH 1976), and *Quercus robur* (HAUSMANN & MILLER 2000; ERLEBACH pers. comm.). Pupation at the underside of leaves. Adults readily attracted to light and bait, easily disturbed during daytime. Nectar feeding recorded on *Cotoneaster* and *Calamagrostis epigejos* (STEINER 2001).

PARASITOIDS: Diptera, Tachinidae: *Phryxe vulgaris* (THOMPSON 1945). – Hymenoptera, Braconidae: *Apanteles solitarius* (THOMPSON 1945; SHENEFELT 1972).

HABITAT: Silvicolous, mesophilous. Preferring closed *Fagus* forests, mixed deciduous woods with *Fagus*, also in cold and shady sites, e.g. on northern slopes. From 0 m up to 1,000 m above sea-level, exceptionally up to 1,900 m in the Alps. In southern Europe, Corsica and south-western Asia usually common in the *Fagus* forest belt, i.e. mainly on the mountains from about 700 up to 1,600 m, but sometimes lower, mainly along deep river valleys and shady slopes.

SIMILAR SPECIES: Discrimination from congeners usually easy by the yellow ground colour and the dotted postmedial line. The form '*strabonaria*' differs from *C. ruficiliaria* in less dark suffusion of wings, postmedial dots usually linked by a dark line. The latter feature is valuable also for discrimination from *C. quercimontaria* (compare LARSEN 1977).

REMARKS: For larval morphology and rearing see GÓMEZ DE AIZPÚRUA (1987; 1989).

Tribus **Rhodometrini** AGENJO, 1952

16 species (SCOBLE 1999) in the three genera *Rhodometra, Casilda,* and *Ochodontia*. Distributed on all continents except Australia, with distributional focus on the Afrotropical region. Accurate, worldwide revision of this pretty group is needed at generic level. Originally described at subfamily rank, subordinated again under Sterrhinae in HAUSMANN (1993a), confirmed by HOLLOWAY (1996), SCOBLE (1999) and SIHVONEN & KAILA (2004). Other authors, e.g. WILTSHIRE 1982, considered Rhodometrini a tribe of Larentiinae. For relations to Larentiinae see introduction to the subfamily, weak relations to *Lythria* (Larentiinae) seem to be supported by habitus, certain genitalic features and the rarely stalked R1 of forewing. For sister group relationship to Cosymbiini and Timandrini ('Timandrini lineage') see introduction to subfamily. For relationships to the extralimital genera *Pseudosterrha* and *Chlorerythra* see remarks to Cosymbiini and HAUSMANN

(1993a). Preliminary data sets from own molecular research (mitochondrial genes ND1, COI, and 16S rRNA) support monophyly of the 'Timandrini lineage' (MILLER et al. 2001; HAUSMANN et al. in prep.). The monophyly of the tribe is supported by seven synapomorphies according to SIHVONEN & KAILA (2004): terminal line absent, straight line dividing apex only on forewing, forewing colour yellow or reddish, fusion of hindwing veins Sc+R1 and Rs, bare uncus, medially spinose apex of sacculus, presence of transtilla, but arms not meeting dorsally. Sister group relationship to the Timandrini supported by the presence of socii in ♂ genitalia, by larval association with Polygonaceae, and by pupation in a loose silky web (HOLLOWAY 1997; SIHVONEN & KAILA 2004), pupal cremaster reminiscent of that of Cosymbiini (HOLLOWAY 1997; SIHVONEN 2003). Full number of ♂♀ hindtibial spurs previously suggested to indicate a basal position in the phylogeny of Sterrhinae, but molecular studies contradict this hypothesis, see introduction to the subfamily. Position of the genus *Ochodontia* in the Rhodometrini rather than in Larentiinae still controversial, but supported by some genitalic features, mainly the signum in ♀ genitalia. Transfer to the Timandrini may, however, be discussed too, considering habitus, larval morphology, and structure of uncus region in ♂ genitalia.

DIAGNOSIS: Wing colour usually yellow, often polychroic, with reddish forms. Hindwings rounded (exception: *Ochodontia*), usually whitish, without pattern. Forewing with straight, oblique postmedial streak. Venation (Text-figs 232, 236, 237): One single areole; exception: *Ochodontia*. Cross vein between areole and discocellular vein short, thus cell apex close to areole and areole distally exceeding cell. Hindwing with Sc+R1 and Rs fused over 1/3 to 2/3 length of cell. Frons projecting. Hindtibia of both sexes with four spurs. Antennae of ♂ bipectinate, appressed type, flagellum flattened horizontally,

Text-fig. 230: ♂ antennal structure of *Rhodometra sacraria* (LINNAEUS, 1767) (SEM). Scale bar = 0.05 mm.

Text-fig. 231: *Rhodometra sacraria* (LINNAEUS, 1767), in tectiform resting position.

branches arising close to each other from ventral part (Text-fig. 230). Antennae of ♀ ciliate-setose. Tympanal ansa medially dilated (Text-fig. 8). Sternum A2 of ♂ without additional sclerites or setose patches.

MALE GENITALIA: Uncus and socii present. Saccus round. Valva usually with setose or spinose ventral patch. Sternum A8 normal. *Ochodontia* with modified characters, e.g. divided uncus and broad saccus.

FEMALE GENITALIA: Ovipositor without ventral projections. Apophyses anteriores with appendages leading to antrum. Corpus bursae elongate. Signum developed as rhombic sclerite with longitudinal ridge, reminiscent of equivalents in Cosymbiini and Timandrini.

BIOLOGY: Larval host-plants with certain focus on Polygonaceae. Adults with wings in tectiform resting position; exception: *Ochodontia*.

IMMATURE STAGES: Egg elongate, reminiscent of eggs of other Sterrhinae. Larva usually long, with smooth skin. Pupa cuneiform, anteriorly tapering, cremaster broad, with four pairs of normal setae (PATOČKA 2003). Larva of *Ochodontia* with small head, dilated from segments T1 to A1, reminiscent of Timandrini.

Rhodometra MEYRICK 1892

(=*Sterrha*: sensu HERRICH-SCHÄFFER, 1847 nec HÜBNER, 1825)

TYPE SPECIES: *Phalaena sacraria* LINNAEUS, 1767. Type species almost cosmopolitan, see distribution of *R. sacraria*. Genus embracing ten other species, eight in the Afrotropical, two in the western Neotropical regions (SCOBLE 1999). The genus needs accurate, worldwide revision. Such study may prove, possibly, that true *Rhodometra* s.str. is monotypic.

DIAGNOSIS: Forewing triangular, hindwing elongate, rounded. Oblique postmedial streak of forewing leading into the apex. Forewing triangular, hindwing round, white. Discal spots dark. Venation (Text-fig. 232): Forewing with one large areole, R1-R4 stalked, R1-R4 and R5 connate. Hindwing with Sc+R1 and Rs fused over 1/3 to 1/2 length of cell, in the type species 1/2. Rs and M1 very shortly stalked or connate. Frons smoothly scaled, prominent. Proboscis well developed. Palpi bushy scaled, comparatively long, 1.5-2 times diameter of eye in both sexes. Antennae of ♂ bipectinate, distal fourth filiform, longest branches 15-18 times width of flagellum (Text-fig. 230). Antennae of ♀ ciliate-setose with short cilia. Legs long and slender, reminiscent of Pyralidae. Hindtibia of both sexes with two pairs of spurs of unequal length. Further features under diagnosis of tribe.

Text-fig. 232: Venation of *Rhodometra sacraria* (LINNAEUS, 1767).

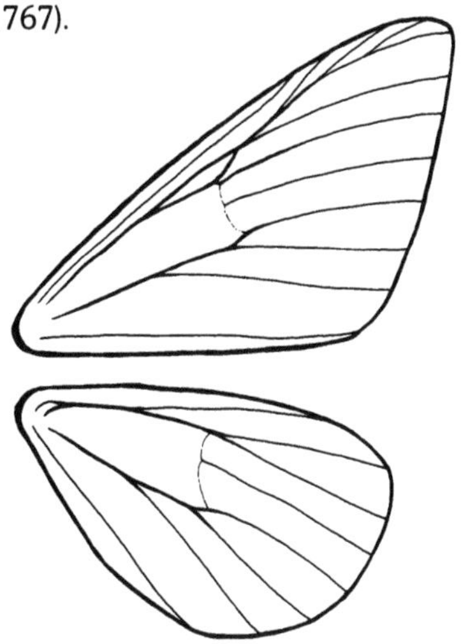

MALE GENITALIA: Uncus broad. Valva angled, with spinulose patch at tip.

FEMALE GENITALIA: Ductus bursae posteriorly with some irregularly shaped sclerites (antrum).

BIOLOGY: Adults usually with wings in tectiform resting position.

IMMATURE STAGES: Egg elongate. Larva long, with smooth skin. For morphology of immature stages see SKULE (1980).

193. Rhodometra sacraria (LINNAEUS, 1767)

Phalaena Geometra sacraria LINNAEUS, 1767: Syst. Nat. (Ed. 12) 1 (2): 863 (North Africa ('Barbaria')). Syntype(s) (LSL).

Pyralis sacralis THUNBERG, 1784: Diss. ent. sistens Ins. Suecica 1: 14 (Sweden). Syntype(s). Emendation.

Phalaena fulvaria FABRICIUS, 1794: Ent. Syst. 3 (2): 160 (Italy). Syntype(s) lost.

Phalaena Geometra labda CRAMER, 1777: Uitl. Kapell. (Pap. Exot.) 2 (16): 129, 149 (index), pl. 181, fig. D (Surinam, wrong locality according to PROUT 1914).

‡ *Phalaena Geometra sanguinaria* ESPER, 1801: Die Schmett. in Abbild. 5 (7): 173, pl. 30, figs 10, 11 (Italy: Naples distr.; Florence; southern France). Lectotype ♂ (LMW, examined by HACKER 1999). Junior objective homonym nec HÜBNER, 1787 (identity: *Idaea muricata*).

Aspilates lividaria COSTA, 1848: Fauna Regno Napoli, Lep. Geom.: [365], pl. 5, fig. 2 (Italy: Naples). Junior synonym (f. *sanguinaria*).

Aspilates minervae GISTEL, 1856: Myst. Eur. Insectenwelt: 349 (Greece: Minerva Temple by Athens). Syntype(s).

? *Rhodometra plectaria debiliaria* ROTHSCHILD, 1914: Novit. zool. 21: 345 (central Algeria: Guelt-es-Stel). Holotype ♂ (BMNH). Tentatively placed here as synonym, referring to an individual aberration with rosy longitudinal streak in the forewing cell. Listed at species rank in SCOBLE (1999).

Sterrha (Rhodometra) sacraria var. *desertorum* STAUDER, 1914: Zeitschr. f. wiss. Insektenbiol. 10: 172, fig. 13 (Algeria: Biskra, El Kantara). Syntypes 5♂6♀.

Unavailable names (infrasubspecific): *atrifasciaria*: STEFANELLI (1870); *aucta*: KRAUSSE (1913); *excaecaria*: FUCHS (1903); *fumosa*: PROUT (1935); *ochracearia*: FUCHS (1903); *rosea*: OBERTHÜR (1896); *sarothamnaria*: BROWN (1886). – (incorrect subsequent spellings): *exaecaria*: CULOT (1918); *labdaria*: GUENÉE (1858).

DIAGNOSIS: Wingspan ♂♀ 23-26 mm, subsequent generations smaller, ♂♀ 16-26 mm. Ground colour of forewing usually yellow. Straight postmedial streak leading into forewing apex, red, brown or blackish ('f. *labda*'), the last form usually restricted to ♀.

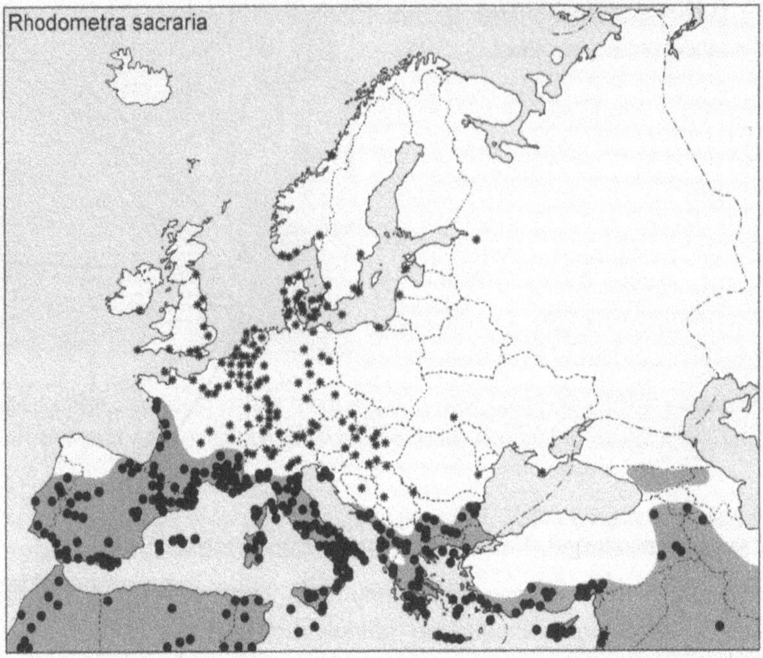

Width of postmedial streak varying. Dark discal spot small, sometimes absent. Hindwing white without any pattern. Frons and vertex concolorous with ground colour. Further features under generic and tribe diagnosis.

Variability: Forewing sometimes dirty sand coloured or rosy ('f. *sanguinaria*'), at a frequency of 1-5% in all populations. Reasons and possible genetical background for the polychroic patterns unknown, meriting special attention and rearing experiments. Coloration sensitive to temperature experiments (SKINNER 1984). Rarely with an additional rosy longitudinal streak in the forewing cell.

MALE GENITALIA: See generic diagnosis. In Afrotropical specimens socii somewhat longer.

FEMALE GENITALIA: See generic diagnosis.

DISTRIBUTION: Cosmopolitan, migrant. Breeding throughout the Mediterranean region, including all islands, in the east to Crimea. At varying frequency migration flights northwards up to Scotland and southern Scandinavia (e.g. SKULE & SVENDSEN 1984). Migration flights mainly across western European routes. North of Alps rarely with a subsequent generation, but no successful hibernation possible. Erroneously recorded for southern European Russia (EVERSMANN 1844; VIIDALEPP 1996), data referring to *C. antophilaria*. – Outside Europe on Canary islands, Madeira, Azores, African continent down to South Africa, St. Helena island, and Madagascar (PROUT 1935a). In the east across south-western Asia to Afghanistan and central Asian mountains (VIIDALEPP 1988; 1996), and eastern Mongolia (VIIDALEPP 1975). In the Indopacific region across India to south-eastern China (PROUT 1939b; HÄUSER 2001). In central America recorded from San Salvador (1♂, ZSM), awaiting confirmation.

PHENOLOGY: Plurivoltine. In the Mediterranean from late March to mid-November, in North Africa and the Levant in all months of the year. Immigration to western, central and northern Europe usually from early June, most records from mid-September to late October, descendants of immigrated specimens in central Europe from August to November (FORSTER & WOHLFAHRT 1981; HÄUSER 2001). Larva sometimes hatching from egg two days after deposition (original data), in rearings developing very quickly.

BIOLOGY: Larva oligophagous (2) with strong focus on Polygonaceae, but recorded also on other families such as Anacardiaceae, Asteraceae, Chenopodiaceae, Rosaceae (e.g. PROUT 1914; SCOBLE 1999). Found on *Polygonum aviculare* (Polygonaceae) (Great Britain: SKINNER 1984; Morocco: RUNGS 1981), and on *Emex spinosa* (Polygonaceae) (Morocco: RUNGS 1981). Repeatedly recorded on *Polygonum aviculare* (Polygonaceae) (var. cit., e.g. REBEL 1910b; WILTSHIRE 1949b; FORSTER & WOHLFAHRT 1981; SKOU 1986; PORTER 1997), on *Beta* (Chenopodiaceae) (WILTSHIRE 1949b), *Rumex* (Polygonaceae), and *Anthemis* (Asteraceae) (ALLAN 1949; DUFAY 1961; MAZEL & PESLIER 1997). Reared on *Polygonum aviculare* (WAGNER 1926; FRIEDRICH 1976; GÓMEZ DE AIZPÚRUA 1989; SKULE 1980; KRAUS 1993; HÄUSER 2001; ERLEBACH, LEIPNITZ pers. comm.), and on *Rumex* (ALLAN 1949; SKINNER 1984; FRIEDRICH 1976; PORTER 1997; CARTER & HARGREAVES 1987), strongly fixed on these host-plants and hardly accepting other plant genera (original data; FRIEDRICH 1976). Numerous other plants, however, mentioned in SCOBLE (1999) and ROBINSON et al. (2001): *Rhus, Emex, Oxygonum, Malus pumila.* Adults sometimes actively flying during daytime, easily flushed up from vegetation, but also readily attracted to light. Nectar feeding recorded on *Centaurea jacea* (HÄUSER 2001). High dispersal, with 'emigration flights' (EITSCHBERGER et al. 1991), r-strategy. Moths rest in tectiform wing position, similar to that of many Pyralidae (Text-fig. 231). Flight is reminiscent of that of Pyralidae, too.

HABITAT: In the Mediterranean ubiquitous, most abundant in the lowlands near coast, in the maquis and sand dunes. In western, central and northern Europe mainly recorded in dry localities with pioneer vegetation, fallows, or even in urban centres. From 0 m up to 2,300 m above sea-level, up to 2,700 m in Morocco.

SIMILAR SPECIES: *C. consecraria* and *C. antophilaria* differ in whitish discal spots, postmedial line ending subapically on forewing costa. *Aspitates gilvaria* (DENIS & SCHIFFERMÜLLER, 1775) (Ennominae) larger, underside of hindwing with a dark fascia, resting position planiform.

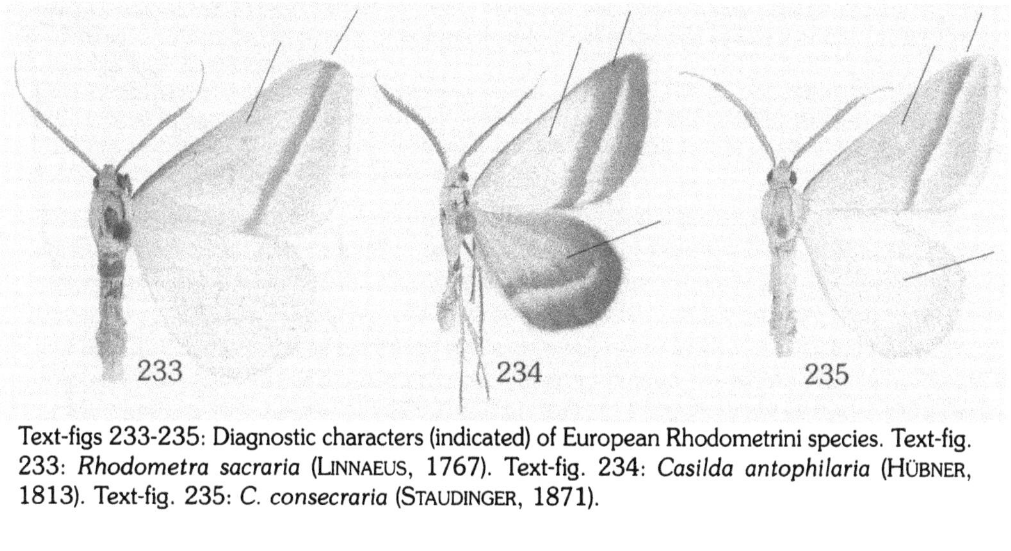

Text-figs 233-235: Diagnostic characters (indicated) of European Rhodometrini species. Text-fig. 233: *Rhodometra sacraria* (LINNAEUS, 1767). Text-fig. 234: *Casilda antophilaria* (HÜBNER, 1813). Text-fig. 235: *C. consecraria* (STAUDINGER, 1871).

REMARKS: Inheritance of rare forms with excised wing termen discussed in WILLIAMS (1950). For larval morphology and rearing see SKULE (1980), and GÓMEZ DE AIZPÚRUA (1989).

Casilda AGENJO, 1952

TYPE SPECIES: *Sterrha consecraria* STAUDINGER, 1871. Four species in the western Palaearctic and eastern Afrotropical regions. Relationships to *Rhodometra* very close. Against the background of intermediate feature patterns in some African species it should be discussed whether *Casilda* could be subordinated under a broader concept of *Rhodometra* or the latter might be split into a several new genera. Revisions: AGENJO (1952), SUTTON (1963), RAINERI (1992b).

DIAGNOSIS: See diagnosis of tribe. Oblique postmedial streak of forewing leading to the forewing costa. Forewing triangular, hindwing round, sometimes with a dark pattern. Discal spot white. Frons slightly prominent. Proboscis developed. Palpi bushy scaled, long, 1.5-2 times diameter of eye in both sexes. Antennae of ♂ bipectinate, longest branches about 15-20 times width of flagellum. Distal part of ♂ antenna filiform, varying individually from one fourth to one sixth, sometimes varying even between the antennae of the same specimen. Antennae of ♀ ciliate-setose, with short cilia. Legs long and slender, reminiscent of Pyralidae. Hindtibia of both sexes with two pairs of spurs of unequal length. Venation (Text-fig. 236) as in *Rhodometra*, but hindwing with longer fusion of Sc+R1 and Rs.

MALE GENITALIA: Uncus slender. Valva straight or curved, with a setose or spinulose patch on the ventral margin. Aedeagus curved.

FEMALE GENITALIA: Ductus bursae posteriorly sclerotized to tubular antrum. Sterigma (lamella antevaginalis) with posterior indentation.

IMMATURE STAGES: Larva long and slender.

BIOLOGY: Pupation in a loose cocoon on the ground (PROUT 1914). Adults usually with wings in tectiform resting position.

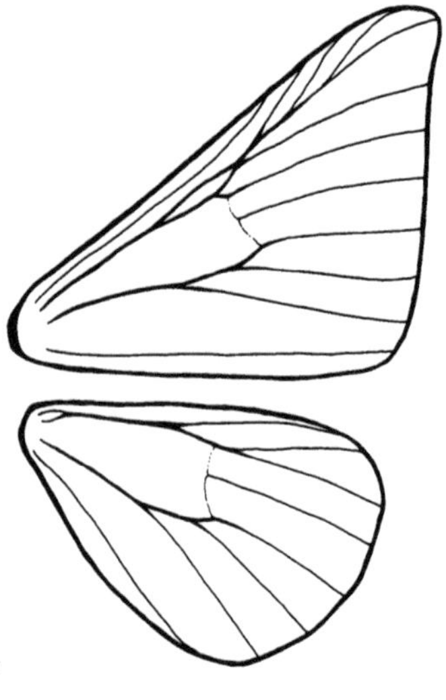

Text-fig. 236: Venation of *Casilda antophilaria* (HÜBNER, 1813).

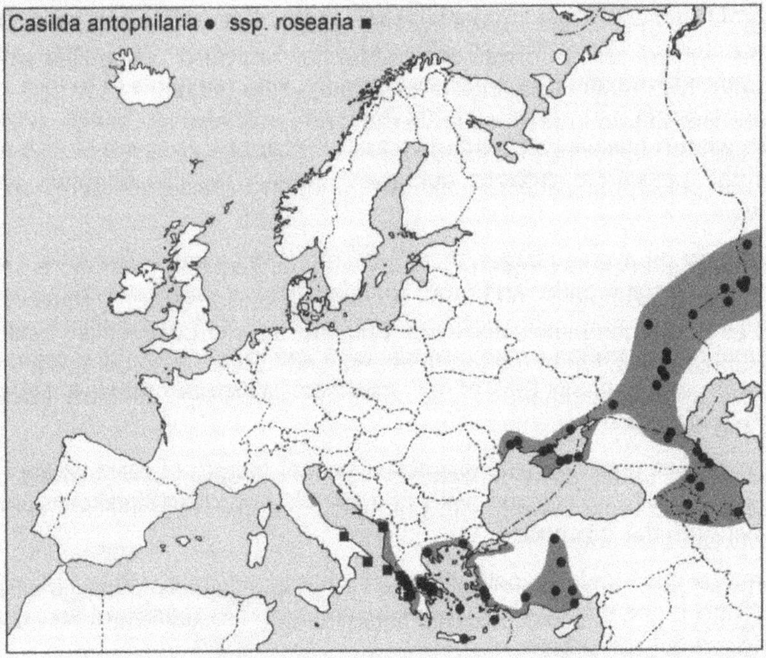

194. *Casilda antophilaria* (HÜBNER, 1813)

Geometra antophilaria HÜBNER, 1813: Samml. Eur. Schmett. 5, Geometrae (1): pl. 84, fig. 433 (south-eastern European Russia). Syntype(s) lost. For type locality see STAUDINGER (1871: 176), PROUT (1935a: 70).

‡ *Eusarca anthophilaria* HÜBNER, 1822: Syst.-Alph. Verz.: 39. (Europe). Syntype(s) lost. Incorrect subsequent spelling.

Aspilates rosearia TREITSCHKE, 1828: Schmett. Eur. 6 (2): 298 (Greece: Korfu). Holotype ♂. Herewith downgraded from species (MÜLLER 1996; SCOBLE 1999) to subspecific rank (stat. n.).

Sterrha anthophilaria var. *albidaria* ERSCHOFF, 1874: Lepidoptera. In Fedschenko, A.P.: Reise Turkest. (Mém. Soc. Amis Sci. Nat. Mosc. 11): 69 (Turkestan: Syr Darja). Syntypes 12 ♀♀. Raised to species rank in ALPHÉRAKY (1883), valid at subspecific rank (VIIDALEPP 1996). Possibly referring to *C. consecraria*.

Sterrha albidaria var. *gegenaria* ALPHÉRAKY, 1883: Horae Soc. ent. ross. 17 (3/4): 193 (China: Yining (Gulja) ('Kuldscha')). Syntypes 4♂.

Sterrha rosearia var. *elvira* THIERRY-MIEG, 1911: Annls Soc. ent. Belg. 54: 466 ('Spain: probably Albarracin'). Holotype ♂. Acoording to SUTTON (1963) and RAINERI (1992b) the type locality is erroneous, and this form belongs to *C. antophilaria*.

Unavailable names (misidentification): *sacraria*: sensu EVERSMANN (1844) nec LINNAEUS, 1767 (identity: *Rhodometra sacraria*). – (infrasubspecific): *albipunctaria*: ALPHÉRAKY (1883); *subrosearia*: STAUDINGER (1871); *subsacraria* STAUDINGER (1871) (formally raised to subspecific rank in DIDMANIDZE 2002). – (incorrect subsequent spelling): *autophilaria*: SHERBORN (1923).

DIAGNOSIS: Wingspan ♂♀ 19-24 mm, subsequent generations smaller, ♂♀ 17-22 mm, smaller specimens, exceptionally, also in springtime. Ground colour of forewing yellow, reddish or light brown. Postmedial streak leading subterminally into the forewing apex. Discal spots whitish. Hindwing usually dark grey brown, with yellowish white postmedial line and discal spot. Fringe yellowish white or rosy, often in strong contrast to the ground colour. Frons and vertex concolorous with ground colour. Venation: Hindwing fusion of Sc+R1 and Rs half length of cell or slightly longer. Further features under generic and tribe diagnosis.

Variability: Yellow forms at a frequency of 40-60% in the populations of Russia (n=42), western Kazakhstan (n=20) and Turkey (n=14), reddish forms ('f. *subrosearia*') occurring mainly in the female. Yellow forms with reddish or brownish postmedial streak, dark forms with yellowish streak. Hindwings rarely white, with remnants of the dark postmedial line ('f. *subsacraria*'). In Turkey, central Asia, and rarely also in Greece ground colour brown, with whitish hindwings ('f. *albipunctaria*'). Reasons and possible genetical background for the polychroic patterns unknown, meriting special attention and rearing experiments.

Subsp. ***rosearia*** on average larger, ♂♀ 19-27 mm, subsequent generations smaller, ♂♀ 18-22 mm. Forewings broader and more rounded. Yellow forms at a frequency of about 55% (n=28) as in the nominate subspecies, but with the red postmedial streak broader, with additional red coloration in the terminal area and, often, along the costa. Habitus in eastern Greece approaching that of the nominate subspecies. Reddish forms usually restricted to the females.

MALE GENITALIA: See under generic diagnosis. Ventral margin of valva without digitiform process. Aedeagus strongly curved, with one spinule at tip. No significant and constant difference between the subspecies.

FEMALE GENITALIA: See under generic diagnosis. Sterigma (lamella antevaginalis) strongly sclerotized, with U- or V-shaped posterior indentation. No significant and constant difference between the subspecies.

DISTRIBUTION: (East-)European-Westasiatic. Adriatic coast of eastern Italy (Marche: SPADA 1893; Puglia: RAINERI 1992b) and Greece (subsp. *rosearia*), Greek islands, and from southern Ukraine to southern Urals (nominate subspecies). Erroneously recorded for France (MÜLLER 1996). – Outside Europe in Turkey, Caucasus, Transcaucasus, southwestern Siberia, and northern parts of central Asian mountains (Kazakhstan: VIIDALEPP 1988; 1996). All records south, east and west of the indicated area refer to the next species. Populations from western Turkey and Greek islands reminiscent of subsp. *rosearia*, those from central Turkey need further study to clear up subspecific relationships.

PHENOLOGY: Bi- or trivoltine: early May to late September. Sometimes emerging two weeks earlier. Some small and dark forms reputedly collected in March (PROUT 1914).

BIOLOGY: Larva in nature possibly monophagous on *Limonium* (Plumbaginaceae). Found and reared on petals and leaves of *Limonium gmelini* (subsp. *rosearia*; Greece: LEIPNITZ pers. comm.). In literature, recorded on flowers of *L. gmelini*, too (BARTEL 1914; PROUT 1914), and on an unidentified species of *Alisma* (Alismataceae) (subsp. *rosearia*; PROUT 1914). Imagines very easily flushed up from vegetation in the sunshine during daytime. Stenotopic, K-strategy.

HABITAT: Halophilous. In Europe restricted to lowland sandy steppes, salt steppes or coastal salt marshes. From 0 m up to about 200 m above sea-level. In Turkey also around salt lakes up to 1,100 m above sea-level.

SIMILAR SPECIES: *Rhodometra sacraria* (see no. 193). Allopatric sister species *C. consecraria* with whitish hindwings, in European populations forewing termen and costa straighter. For secure identification dissection is needed.

REMARKS: Taxonomy and distribution discussed in SUTTON (1963). Egg and larva described in BARTEL (1914).

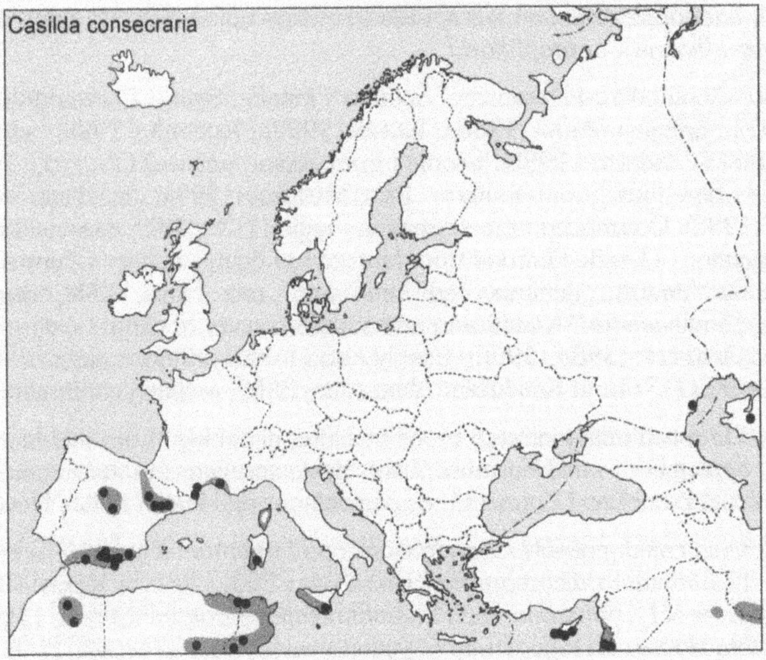

195. Casilda consecraria (STAUDINGER, 1871)

Sterrha anthophilaria var.? *consecraria* STAUDINGER, 1871: in STAUDINGER & WOCKE, Cat. Lepid. eur. Faunengeb. (Edn 2) 1: 176. First available name after *consecraria* RAMBUR, 1866 (non binominal) (Spain, Andalusia: Grenada). Lectotype ♂ (of RAMBUR: ZSM/HERBULOT, examined), selected by SUTTON (1963).

Sterrha intermediaria TURATI, 1930: Atti Soc. ital. Sci. nat. 69 (1): 62, pl. 2, fig. [11] (Libya, Cyrenaica: Barce). Syntype(s) (MSNT: 1 specimen ex coll. TURATI, examined by RAINERI (1996)). Synonymy recognized by SUTTON (1963).

Unavailable names (misidentification): *anthophilaria*: sensu AMSEL (1933) nec *antophilaria* HÜBNER, 1813; *subsacraria*: sensu TURATI (1934) nec STAUDINGER, 1871. – (infrasubspecific): *excaecaria*: FUCHS (1903); *paralellaria*: KRÜGER (1934).

DIAGNOSIS: Wingspan ♂♀ 20-25 mm, subsequent generations smaller, ♂♀ 16-22 mm, smaller specimens, exceptionally, also in springtime. Wing coloration, pattern and dichroic variability as in *C. antophilaria*, but hindwing usually white, sometimes darker at base and towards termen. Fringe usually concolorous with ground colour. Frons and vertex concolorous with ground colour. Venation: Hindwing fusion of Sc+R1 and Rs about 2/3 length of cell. Further features under generic and tribe diagnosis.

Variability: Yellow forms in the western Mediterranean at a frequency of about 35% (n=50), on Cyprus and in the Levant about 80% (n=24), in central Asia about 70% (n=31). Reasons and possible genetical background for the polychroic patterns unknown, meriting special attention and rearing experiments.

MALE GENITALIA: See under generic diagnosis. Generally smaller than genitalia of *C. antophilaria*. Ventral margin of valva with digitiform process. Aedeagus only slightly curved, usually with two spinules at tip.

FEMALE GENITALIA: See under generic diagnosis. Sterigma (lamella antevaginalis) poorly sclerotized, posterior indentation hardly visible in microscopic slides. Antrum more strongly sclerotized than in *C. antophilaria*.

DISTRIBUTION: (South-)West-Palaearctic. Southern France, Spain, Corsica (PROUT 1914; RUNGS 1991), Sardinia (PROUT 1938a; RAINERI 1992b; PARENZAN 1998), and southern Sicily (RAINERI & ZANGHERI 1995). Records from Balkan peninsula (SUTTON 1963) misidentified, as also those from southern Italy (PARENZAN 1994), the latter rectified in PARENZAN (1998). Occurrence in the Astrachan region (1 ♂, ZISP, dissected) awaits further confirmation. – Outside Europe from Morocco to Egypt, southern Cyprus, from the Levant, across Arabian peninsula and Iran (e.g. prov. Fars; ZSM, dissected) to Turkmeniya, south-western Kazakhstan and southern parts of central Asian mountains (Uzbekistan: VIIDALEPP 1988; 1996). Possibly also in the northern parts of the central Asian mountains (1 ♂ from Kazakhstan, Alma Ata; ZSM), awaiting confirmation.

PHENOLOGY: European data indicate bi- or trivoltine phenology from mid-March to late May; early September to mid-November, with single specimens in the summer months. In North Africa, Levant and Cyprus tri- or plurivoltine early March to late November.

BIOLOGY: Larva in nature possibly monophagous on *Limonium* (Plumbaginaceae). Found on flowers of *Limonium dichotomum* (Spain: KING 2002; KING & ROMERA 2004), on flowers and leaves of *L. sinuatum,* and *L. gmelinii/angustifolium* (Cyprus: LEIPNITZ pers. comm.). Recorded also on *Polygonum* (Polygonaceae) (TORMO & MUÑOZ 1995), possibly referring to data from rearing, but not accepting *Taraxacum officinale* and *Polygonum* in rearings of LEIPNITZ (pers. comm.). Adults easily flushed up from vegetation, but also actively flying during daytime, readily attracted to light. Stenotopic, never moving far from sites of occurrence (LEIPNITZ pers. comm.), K-strategy.

PARASITOIDS: Hymenoptera, Braconidae: *Homolobus truncator* (KING & ROMERA 2004). – Hymenoptera, Ichneumonidae: *Alcima orbitale* (preliminary identification), *Triclistus* spec. (KING & ROMERA 2004).

HABITAT: Halophilous. In Europe restricted to lowland salt steppes and coastal salt marshes. The latter may be temporarily flooded. From 0 m up to about 200 m above sea-level, in central Spain up to 900 m, in Morocco recorded at 1,500 m, in central Asian mountains from 500 m up to 1,000 m.

SIMILAR SPECIES: *Rhodometra sacraria* (see no. 193), *C. antophilaria* (194).

REMARKS: Taxonomy and distribution discussed in SUTTON (1963). Central Asian populations need further study on the basis of extensive material.

Ochodontia LEDERER, 1853

TYPE SPECIES: *Ennomos sareptanaria* FREYER, 1841 (junior synonym of *O. adustaria*). Monotypic. Systematic position controversial, placed by some authors in the Rhodometrini (VIIDALEPP 1996; STEKOLNIKOV 1996; SIHVONEN & KAILA 2004), but previously often subordinated under Larentiinae (PROUT 1913; STERNECK 1941: 230; MÜLLER 1996). Position in Rhodometrini supported by structure of tergum A8 and antrum sclerites in ♀ genitalia. Planiform resting position, habitus, larval morphology and some genitalic features support also relationships to the Timandrini, see introductions to the tribes Timandrini and Rhodometrini and similar suggestions in SIHVONEN & KAILA (2004). Subsequent misspelling: *Ochrodontia* (VASILENKO 1998a).

Text-fig. 237: Venation of *Ochodontia adustaria* (FISCHER DE WALDHEIM, 1840).

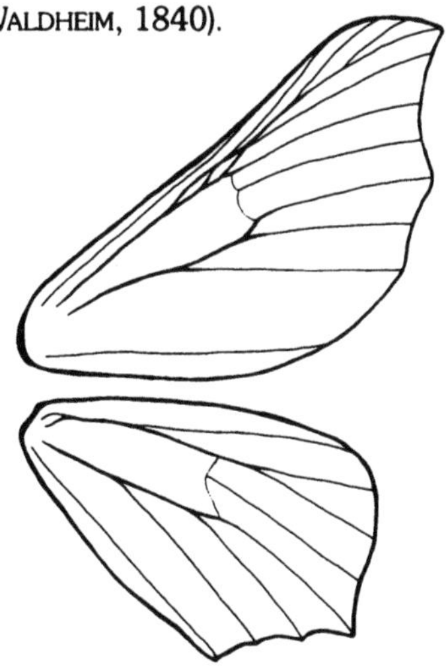

DIAGNOSIS: Forewing termen concave below apex, apex pointed. Hindwing tailed at M3. Proboscis very short. Frons slightly prominent. Palpi ♂♀ 1.2-1.5 times diameter of eye. Antennae of ♂ bipectinate, longest branches about 13-16 times width of flagellum, filiform in the distal fourth or fifth. Antennae of ♀ shortly and densely ciliate-setose. Hindtibia thin, with four spurs in both sexes. Sightly variable in venation (Text-fig. 237): Forewing with double areole, reminiscent of that of Rhodostrophiini, R1 arising subapically, R2-R4 and R5 connate. Hindwing with Sc+R1 and Rs fused over half length of cell. Rs and M1 stalked, M3 and CuA1 separate. Tympanal ansa with distinct medial dilatation, curved at tip.

MALE GENITALIA: Uncus cleft into two very narrow processes. Saccus broad, bilobed. Valva broad, membranous, with sclerotized ventral process at base. Aedeagus very short.

FEMALE GENITALIA: Tergum A7 with very short, broad apophyses anteriores and broad lateral appendages. Corpus bursae slender and very long, signum sclerite sub-rhombic, with medial ridge, rather small.

IMMATURE STAGES: Larva dilated from segments T1 to A1, reminiscent of Timandrini. With deep incisions between segments (PROUT 1914).

196. Ochodontia adustaria (FISCHER DE WALDHEIM, 1840)

Ennomos adustaria FISCHER DE WALDHEIM, 1840: Bull. Soc. imp. Nat. Moscou 1840 (1): 88 (Russia: lower Volga). Syntypes ♂♀.
Ennomos sareptanaria FREYER, 1841: Neuere Beitr. Schmett.kde. 4 (59): 113, fig. 2 (south-eastern European Russia: Krasnoarmeysk, 'Sarepta'). Syntype(s).
Ochodontia adustaria waldheimi WEHRLI: original reference not traced (compare DIDMANIDZE 2002).
Unavailable names (incorrect subsequent spelling): *sareptaria*: EVERSMANN (1844).

DIAGNOSIS: Wingspan ♂♀ 25-30 mm, second generation much smaller, ♂♀ 19-24 mm. Well characterized by the wing shape. Variability low in coloration and pattern of wings. Forewing sand coloured, hindwing paler. Antemedial line of forewing present. Postmedial fascia on both wings broad, consisting of 3-4 lines. Fringe and termen dark in the anterior half of the forewing termen and on the hindwing except for the apex. Frons light brown, vertex pale sand coloured. Further features under generic diagnosis.

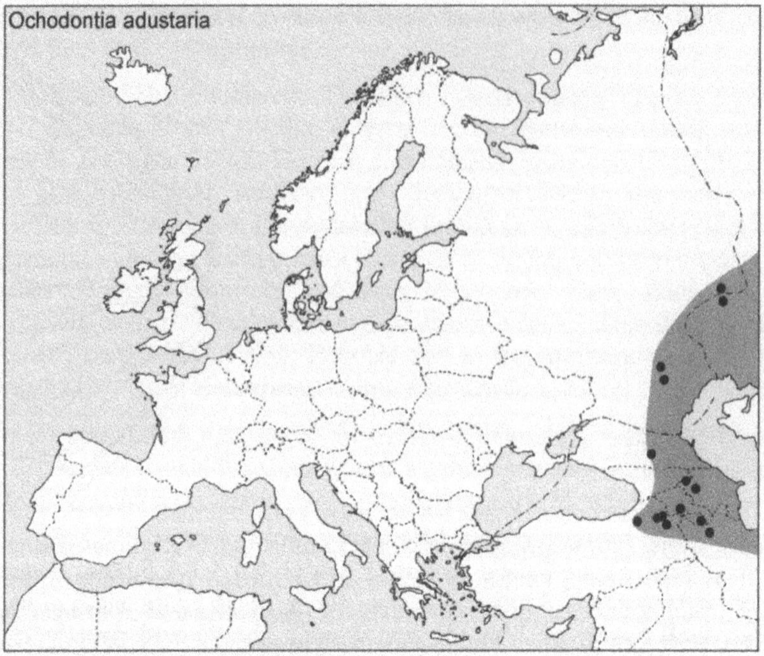

MALE AND FEMALE GENITALIA: see under generic diagnosis.

DISTRIBUTION: Turanian. Southern European Russia. – Outside Europe in north-eastern Turkey, Caucasus, Transcaucasus, northern Iran, in the east to northern Afghanistan and central Asian mountains. In ZISP one specimen from Mongolia.

PHENOLOGY: Bivoltine: late April to mid-June; late June to mid-September. First generation on mountains up to one month later, overlapping with the second generation.

BIOLOGY: Larva recorded on *Atraphaxis* (Polygonaceae) (south-eastern Kazakhstan: MIRONOV pers. comm., data from a ZISP specimen), and on *Euonymus europaeus* (Celastraceae) (PROUT 1914). Adults readily attracted to light, resting in planiform wing position (GELBRECHT, DRECHSEL, BUSSE pers. comm.).

HABITAT: Limited information available about habitat preference. In Europe restricted to lowland steppes from 0 m up to 300 m above sea-level, outside Europe also in xeromontane sites up to 2,500 m.

SIMILAR SPECIES: No similar species in Europe, except for a rather superficial similarity to some *Ennomos* species, such as *Ennomos fuscantaria* (HAWORTH, 1809) (Ennominae).

Colour plates

PLATE 1 (all specimens 1.5 x natural size)

1a. ***Anthometra plumularia* Bsd.**, ♂, St. Ildefonso, C. Spain (ZSM)
1b. *A. plumularia* Bsd., ♂, Cuenca, C. Spain (ZSM)
1c. *A. plumularia* Bsd., ♀, St. Ildefonso, C. Spain (ZSM)
1d. *A. plumularia* Bsd., ♀, Granada, S. Spain (ZSM)
1e. *A. plumularia* Bsd., ♀, Cuenca, C. Spain (ZSM)
2a. ***Emmiltis pygmaearia* Hbn.**, ♂, Wippach, Carniolia, Slowenja (ZSM)
2b. *E. pygmaearia* Hbn., ♂, Terlan, S. Tirol, N. Italy (ZSM)
2c. *E. pygmaearia* Hbn., ♂, Tivoli, Lazio, C. Italy (ZSM)
2d. *E. pygmaearia* Hbn., ♂, Picinisco, Abruzzi, C. Italy (ZSM)
2e. *E. pygmaearia* Hbn., ♀, Socerb nad Ospom, Slowenja (coll. SKOU)
2f. *E. pygmaearia* Hbn., ♀, Tivoli, Lazio, C. Italy (ZSM)
3a. ***Cleta ramosaria* Vill.**, ♂, Bilbao, N. Spain (NHMW)
3b. *C. ramosaria* Vill., ♂, Algeciras, S. Spain (ZSM)
3c. *C. ramosaria* Vill., ♂, Algeciras, S. Spain (NHMW)
3d. *C. ramosaria* Vill., ♀, Lagos, Portugal (coll. MALKIEWICZ)
3e. *C. ramosaria* Vill., ♀, Algeciras, S. Spain (ZSM)
3f. *C. ramosaria* Vill., ♀, Guadalquivir, Sevilla, S. Spain (NHMW/VARTIAN)
3g. ***Cleta transiens* Prt.**, ♂, Biskra, E. Algeria (MNHU)
3h. *C. transiens* Prt., ♀, Biskra, E. Algeria (MNHU)
4a. ***C. perpusillaria* Ev.**, ♂, Sarepta, S. European Russia (ZSM)
4b. *C. perpusillaria* Ev., ♂, Sarepta, S. European Russia (ZSM)
4c. *C. perpusillaria* Ev., ♂, Sarepta, S. European Russia (ZSM)
4d. *C. perpusillaria* Ev., ♀, Konia, C. Turkey (ZSM)
5a. ***Cleta filacearia* H.-S.**, ♂, Moriez, Ht. Provence, SE. France (coll. GELBRECHT)
5b. *C. filacearia* H.-S., ♂, Ohrid, SW. Macedonia (ZSM)
5c. *C. filacearia* H.-S., ♀, Albarracin, Teruel, E. Spain (ZSM)
5d. *C. filacearia* H.-S., ♀, Skopje, N. Macedonia (ZSM)
5e. *C. filacearia* H.-S., ♂, St. Barnabé, Alpes maritimes, SE. France (ZSM)
5f. *C. filacearia* H.-S., ♂, 'Rossia mer.' (NHMW)
6a. ***Idaea serpentata* Hufn.**, ♀, Vevais, Brandenburg, E. Germany (coll. GELBRECHT), reared
6b. *I. serpentata* Hufn., ♂, Schlaubetal, Brandenburg, E. Germany (coll. GELBRECHT)
6c. *I. serpentata* Hufn., ♂, Ilmenau, Thuringia, E. Germany (coll. MÜLLER)
6d. *I. serpentata* Hufn., ♀, Jamkov, S. Bohemia, Czech Republic (ZSM)
6e. *I. serpentata* Hufn., ♂, Pyrenees, N. Spain (NHMW)
7a. ***Idaea luteolaria* Const.**, ♂, Hoyos de Espina, Sierra de Gredos, C. Spain (NHMW)
7b. *I. luteolaria* Const., ♂, Hoyos de Espina, Sierra de Gredos, C. Spain (NHMW)
7c. *I. luteolaria* Const., ♂, Sierra del Montseny, Gerona, NE. Spain (coll. GELBRECHT), reared
7d. *I. luteolaria* Const., ♀, Sierra del Montseny, Gerona, NE. Spain (coll. GELBRECHT), reared
7e. *I. luteolaria* Const., ♂, Sierra del Montseny, Gerona, NE. Spain (coll. MÜLLER)
8a. ***Idaea aureolaria* Den. & Schiff.**, ♂, Oderberg, Brandenburg, E. Germany (coll. GELBRECHT), reared
8b. *I. aureolaria* Den. & Schiff., ♀, Oderberg, Brandenburg, E. Germany (coll. GELBRECHT), reared
8c. *I. aureolaria* Den. & Schiff., ♂, Kyffhäuser, Thuringia, E. Germany (coll. GELBRECHT), reared
8d. *I. aureolaria* Den. & Schiff., ♀, Kyffhäuser, Thuringia, E. Germany (coll. GELBRECHT), reared
8e. *I. aureolaria* Den. & Schiff., ♀, Kyffhäuser, Thuringia, E. Germany (ZSM)
8f. *I. aureolaria* Den. & Schiff., ♂, Hackelsberg, Burgenland, E. Austria (ZSM), reared
8g. *I. aureolaria* Den. & Schiff., ♂, Pra-Loup, Alpes maritimes, SE. France (coll. GELBRECHT)
8h. *I. aureolaria* Den. & Schiff., ♂, Porté, E.Pyrenees, S. France (ZSM/HERBULOT)
8i. *I. aureolaria* Den. & Schiff., ♂, Aj-Petri, Crimea, Ukraine (coll. SZEÖKE)
8k. *I. aureolaria* Den. & Schiff., ♂, Aj-Petri, Crimea, Ukraine (coll. SZEÖKE)
9a. ***Idaea flaveolaria* Hbn.**, ♂, Pra-Loup, Alpes maritimes, SE. France (coll. GELBRECHT)
9b. *I. flaveolaria* Hbn., ♂, Pra-Loup, Alpes maritimes, SE. France (coll. GELBRECHT)
9c. *I. flaveolaria* Hbn., ♂, Rhemes Notre Dame, Graische Alpen, SE. France (coll. GELBRECHT)
9d. *I. flaveolaria* Hbn., ♀, Cogne, Val d'Aosta, NW. Italy (coll. GELBRECHT)
9e. *I. flaveolaria* Hbn., ♂, St. Véran, Ht. Alpes, SE. France (ZSM/HERBULOT)
9f. *I. flaveolaria* Hbn., ♀, 'Mt. Gen.', [SE. France] (ZSM/HERBULOT)

2a
2b
2c
2d
2e
2f
3a
3b
3c
3d
3e
3f
3g
3h
4a
4b
4c
4d
5a
5b
5c
5d
5e
5f
6a
6b
6c
6d
6e
7a
7b
7c
7d
7e
8a
8b
8c
8d
8e
8f
8g
8h
8i
8k
9a
9b
9c
9d
9e
9f

PLATE 2 (all specimens 1.5 x natural size)

10a. ***Idaea muricata* Hufn.,** ♂, Dannenreich, Brandenburg, E. Germany (coll. GELBRECHT), reared
10b. *I. muricata* Hufn., ♀, Dannenreich, Brandenburg, E. Germany (coll. GELBRECHT), reared
10c. *I. muricata* Hufn., ♂, Besenhorst, Hamburg, N. Germany (ZSM), reared
10d. *I. muricata* Hufn., ♂, Oyter Moor, Niedersachsen, N. Germany (ZSM)
10e. *I. muricata* Hufn., ♂, Cleve, Niedersachsen, N. Germany (ZSM)
11a. ***Idaea determinata* Stgr.,** ♀, St. Tropez, Var, SE. France (ZSM/Herbulot), reared
11b. *I. determinata* Stgr., ♀, Aspromonte, S. Calabria, S. Italy (ZSM)
11c. *I. determinata* Stgr., ♂, Liljanovo, Pirin mts., SW. Bulgaria (coll. GELBRECHT), reared
11d. *I. determinata* Stgr., ♀, Liljanovo, Pirin mts., SW. Bulgaria (coll. GELBRECHT), reared
12a. ***Idaea litigiosaria* Bsd.,** ♂, Bize, Aude, S. France (ZSM/HERBULOT)
12b. *I. litigiosaria* Bsd., ♀, Port Bou, NE. Spain (ZSM), reared
12c. *I. litigiosaria* Bsd., ♀, Les Ambollas, E. Pyrenees (ZSM/HERBULOT)
12d. *I. litigiosaria* Bsd., ♀, Benaojan, Malaga, S. Spain (ZMUC)
12e. *I. litigiosaria* Bsd., ♂, Veletastrasse, Sierra Nevada, S. Spain (ZSM)
13a. ***Idaea mutilata* Stgr.,** ♂, S. Martino delle Scale, Palermo, Sicily (ZSM)
13b. *I. mutilata* Stgr., ♀, S. Martino delle Scale, Palermo, Sicily (ZSM)
13c. *I. mutilata* Stgr., ♂, Cesaro, NE. Sicily (ZSM)
14a. ***Idaea lusohispanica* Herb.,** ♀, Albarracin, Teruel, E. Spain (coll. SKOU)
14b. *I. lusohispanica* Herb., ♂, Albarracin, Teruel, E. Spain (coll. SKOU)
14c. *I. lusohispanica* Herb., ♀, Noguera, Teruel, E. Spain (ZSM/EMEM)
14d. *I. lusohispanica* Herb., ♀, Noguera, Teruel, E. Spain (ZSM/EMEM)
14e. *I. lusohispanica* Herb., ♂, Andalusia, S. Spain (NHMW)
14f. *I. lusohispanica* Herb., ♀, Sierra de Alfacar, Granada, S. Spain (ZSM)
14g. ***Idaea lambessata* Obth.,** ♀, Ifrane, Middle Atlas, Morocco (coll. SKOU)
14h. *I. lambessata* Obth., ♂, Ifrane, Middle Atlas, Morocco (ZSM)
15a. ***Idaea sardoniata* Homb.,** ♀, Albarracin, Teruel, E. Spain (coll. GRÜNEWALD)
15b. *I. sardoniata* Obth., ♀, Les Arcs, S. France (ZSM/HERBULOT)
15c. *I. sardoniata* Obth., ♀, Grabels, Montpeiller, S. France (ZSM)
15d. *I. sardoniata* Obth., ♀, Albarracin, Teruel, E. Spain (ZSM)
15e. *I. sardoniata* Obth., ♀, Valdevecar, Teruel, E. Spain (coll. SKOU)
15f. *I. sardoniata* Obth., ♀, Valdevecar, Teruel, E. Spain (coll. SKOU)
15g. *I. sardoniata* Obth., ♀, St. Ildefonso, C. Spain (ZSM/HERBULOT)
15h. *I. sardoniata* Obth., ♂, La Granja, Segovia, C. Spain (ZSM/HERBULOT)
16a. ***Idaea korbi* Püng.,** ♀, Cuenca, C. Spain (MNHU), 'type' (syntype)
16b. *I. korbi* Püng., ♀, Albarracin, Teruel, E. Spain (ZSM)
16c. *I. korbi* Püng., ♀, Fuente del Tajo, Teruel, E. Spain (coll. GRÜNEWALD)
16d. *I. korbi* Püng., ♀, Puerto de Navafria, Sierra de Guadarrama, C. Spain (coll. SKOU)

11a
11b
11c
11d
12a
12b
12c
12d
12e
13a
13b
13c
14a
14b
14c
14d
14e
14f
14g
14h
15a
15b
15c
15d
15e
15f
15g
15h
16a
16b
16c
16d

PLATE 3 (all specimens natural size)

17a. ***Idaea mediaria*** **Hbn.,** ♂, Albarracin, Teruel, E. Spain (ZSM)
17b. *I. mediaria* Hbn., ♀, Albarracin, Teruel, E. Spain (coll. SKOU)
17c. *I. mediaria* Hbn., ♀, Siniscola, Sardinia (coll. FRIEDRICH), reared
17d. *I. mediaria* Hbn., ♂, St. Tropez, Var, SE. France (ZSM/HERBULOT), neotype
17e. *I. mediaria* Hbn., ♀, St. Tropez, Var, SE. France (ZSM/HERBULOT)
17f. *I. mediaria* Hbn., ♀, Mt. Pacanaglia, Alpes marit., SE. France (ZSM/HERBULOT)
17g. *I. mediaria* Hbn., ♀, Sierra Espuña, Murcia, SE. Spain (ZSM)
18a. ***Idaea leipnitzi*** **Hausm. sp.n.,** ♂, Dubrovnik (Gravosa), S. Croatia (NHMW), paratype
18b. *I. leipnitzi* Hausm. sp.n., ♂, Dubrovnik (Gravosa), S. Croatia (ZSM), holotype
18c. *I. leipnitzi* Hausm. sp.n., ♀, Mercuore, Sicily (ZSM)
18d. *I. leipnitzi* Hausm. sp.n., ♂, Dubrovnik (Gravosa), S. Croatia (ZSM), paratype
18e. *I. leipnitzi* Hausm. sp.n., ♂, Dubrovnik (Gravosa), S. Croatia (ZSM), paratype
18f. *I. leipnitzi* Hausm. sp.n., ♂, Dubrovnik (Gravosa), S. Croatia (coll. SKOU), paratype
18g. *I. leipnitzi* Hausm. sp.n., ♂, Dubrovnik (Gravosa), S. Croatia (ZSM), paratype
19a. ***Idaea rufaria*** **Hbn.,** ♂, Sondershausen, Thuringia, E. Germany (coll. GELBRECHT), reared
19b. *I. rufaria* Hbn., ♀, Sondershausen, Thuringia, E. Germany (coll. GELBRECHT), reared
19c. *I. rufaria* Hbn., ♂, Sondershausen, Thuringia, E. Germany (coll. GELBRECHT), reared
19d. *I. rufaria* Hbn., ♂, Ebermannstadt, N. Bavaria, C. Germany (ZSM)
19e. *I. rufaria* Hbn., ♂, Aglasterhausen, Heidelberg, W. Germany (ZSM), reared
19f. *I. rufaria* Hbn., ♂, Guillaumes, Alpes marit., SE. France (ZSM/HERBULOT)
19g. *I. rufaria* Hbn., ♂, Mt. Paradiso, Abruzzi, C. Italy (ZSM)
19h. *I. rufaria* Hbn., ♀, Albena, Varna, E. Bulgaria (coll. GELBRECHT), reared
19i. *I. rufaria* Hbn., ♂, Albena, Varna, E. Bulgaria (coll. GELBRECHT), reared
19k. *I. rufaria* Hbn., ♀, Albena, Varna, E. Bulgaria (coll. GELBRECHT), reared
19l. *I. rufaria* Hbn., ♂, Zhalpak Saline, N. Kazakhstan (coll. SKOU)
19m. *I. rufaria* Hbn., ♂, Zhalpak Saline, N. Kazakhstan (coll. SKOU)
20a. ***Idaea consanguiberica*** **Rezb. & Exp.,** ♂, Camino del Ojen, Malaga, S. Spain (coll. SKOU)
20b. *I. consanguiberica* REZB. & EXP., ♂, Camino del Ojen, Malaga, S. Spain (coll. SKOU)
20c. *I. consanguiberica* REZB. & EXP., ♂, 10 km NW Otivar, Granada, S. Spain (coll. SKOU)
20d. *I. consanguiberica* REZB. & EXP., ♂, Tosos, Zaragoza, N. Spain (coll. SKOU)
20e. *I. consanguiberica* REZB. & EXP., ♀, Tosos, Zaragoza, N. Spain (coll. SKOU)
20f. *I. consanguiberica* REZB. & EXP., ♀, Tosos, Zaragoza, N. Spain (coll. SKOU)
21a. ***Idaea admiranda*** **Hausm. sp.n.,** ♀, Siverek, Urfa, SE. Turkey (ZSM), holotype
21b. *I. admiranda* Hausm. sp.n., ♂, Malkara Region, Tracia, European Turkey (coll. SKOU), paratype
22a. ***Idaea consanguinaria*** **Led.,** ♂, Mercuore, Mistretta, NW. Sicily (ZSM)
22b. *I. consanguinaria* Led., ♂, Mercuore, Mistretta, NW. Sicily (ZSM)
22c. *I. consanguinaria* Led., ♀, Monemvasia, S. Peloponnes, S. Greece (coll. SKOU)
22d. *I. consanguinaria* Led., ♀, Assenovgrad, Rhodopi Mts., S. Bulgaria (coll. SKOU)
22e. *I. consanguinaria* Led., ♂, Liljanovo, Pirin mts., SW. Bulgaria (ZSM), reared
22f. *I. consanguinaria* Led., ♀, Micurin, E. Bulgaria (coll. GELBRECHT), reared
22g. *I. consanguinaria* Led., ♀, Liljanovo, Pirin mts., SW. Bulgaria (ZSM), reared
22h. *I. consanguinaria* Led., ♀, Madjarovo, E. Rhodopi Mts., S. Bulgaria (coll. SKOU)
22i. *I. consanguinaria* Led., ♀, Mercuore, Mistretta, NW. Sicily (ZSM)
22k. *I. consanguinaria* Led., ♂, Liljanovo, Pirin mts., SW. Bulgaria (ZSM), reared
23a. ***Idaea ossiculata*** **Led.,** ♂, Ohrid, SW. Macedonia (ZSM)
23b. *I. ossiculata* Led., ♂, Kalavryta, Peloponnes, S. Greece (ZSM)
23c. *I. ossiculata* Led., ♀, Ohrid, SW. Macedonia (ZSM)
23d. *I. ossiculata* Led., ♀, Ohrid, SW. Macedonia (ZSM)
23e. *I. ossiculata* Led., ♀, Elton Salt Lake, S. European Russia (coll. SKOU)
23f. *I. ossiculata* Led., ♂, Garygala, Kopet Dagh, Turkmeniya (coll. SKOU)
24a. ***Idaea sericeata sericeata*** **Hbn.,** ♂, Les Vigneaux, Htes. Alpes, SE. France (ZSM/HERBULOT)
24b. *I. sericeata sericeata* Hbn., ♂, Budaörs, Budapest, N. Hungary (coll. GELBRECHT), reared
24c. *I. sericeata sericeata* Hbn., ♂, Vizzavona, Corse (ZSM/HERBULOT)
24d. *I. sericeata sericeata* Hbn., ♀, Aritzo, Sardinia (coll. SOMMERER)
24e. *I. sericeata sericeata* Hbn., ♀, Budaörs, Budapest, N. Hungary (coll. GELBRECHT), reared
24f. *I. sericeata sericeata* Hbn., ♀, Kara-Dagh, Crimea, S. Ukraine (coll. SKOU)
24g. *I. sericeata calvaria* Whli., ♂, Albarracin, Teruel, E. Spain (ZSM/EMEM)
24h. *I. sericeata calvaria* Whli., ♂, Albarracin, Teruel, E. Spain (ZSM)
24i. *I. sericeata calvaria* Whli., ♂, Diezma, Granada, S. Spain (ZSM)
24k. *I. sericeata calvaria* Whli., ♂, Pto. de la Ragua, Sierra Nevada, S. Spain (ZSM)
24l. *I. sericeata calvaria* Whli., ♀, Diezma, Granada, S. Spain (ZSM)
24m. ***Idaea allardiata*** **Mab.,** ♂, Beni Kheddache, Ksour Mts. Tunisia (coll. SKOU), reared
25a. ***Idaea macilentaria*** **H.-S.,** ♂, Las Ambollas, E. Pyrenees (ZSM/HERBULOT)
25b. *I. macilentaria* H.-S., ♀, Dompierre, Charente inf., W. France (ZSM/HERBULOT)
25c. *I. macilentaria* H.-S., ♂, Riopar, Albacete, SE. Spain (ZSM)
25d. *I. macilentaria* H.-S., ♂, Riopar, Albacete, SE. Spain (ZSM)
25e. *I. macilentaria* H.-S., ♂, Uña, Cuenca, C. Spain (coll. SKOU)
25f. *I. macilentaria* H.-S., ♀, Pont d'Arc, Vallon, Provence, S. France (coll. SKOU)
26a. ***Idaea ochrata ochrata*** **Scop.,** ♂, Oderberg, Brandenburg, E. Germany (coll. SKOU), reared
26b. *I. ochrata ochrata* Scop., ♀, Oderberg, Brandenburg, E. Germany (coll. SKOU), reared
26c. *I. ochrata ochrata* Scop., ♀, Fürstenwalde, Brandenburg, E. Germany (coll. GELBRECHT), reared
26d. *I. ochrata ochrata* Scop., ♀, Fürstenwalde, Brandenburg, E. Germany (coll. GELBRECHT), reared
26e. *I. ochrata ochrata* Scop., ♀, Gedesby Falster, Denmark (coll. SKOU)
26f. *I. ochrata ochrata* Scop., ♂, St. Boleslau, Bohemia, Czech Republik (coll. SKOU)
26g. *I. ochrata ochrata* Scop., ♀, Albena, E. Bulgaria (coll. GELBRECHT)
26h. *I. ochrata ochrata* Scop., ♂, Praia a Mare, N. Calabria, S. Italy (ZSM)
26i. *I. ochrata ochrata* Scop., ♂, Miturin, E. Bulgaria (coll. GELBRECHT)
26k. *I. ochrata albida* Zerny, ♀, Lozoya, Madrid, C. Spain (coll. SKOU)
26l. *I. ochrata albida* Zerny, ♂, Candasnos, Huesca, N. Spain (coll. SKOU)
26m. *I. ochrata albida* Zerny, ♂, Candasnos, Huesca, N. Spain (coll. SKOU)

18a 18b 18c 18d 18e 18f 18g
19a 19b 19c 19d 19e 19f
19g 19h 19i 19k 19l 19m
20a 20b 20c 20d 20e 20f
21a 21b 22a 22b 22c 22d
22e 22f 22g 22h 22i 22k
23a 23b 23c 23d 23e 23f
24a 24b 24c 24d 24e 24f
24g 24h 24i 24k 24l 24m
25a 25b 25c 25d 25e 25f
26a 26b 26c 26d 26e 26f
26g 26h 26i 26k 26l 26m

PLATE 4 (all specimens 1.5 x natural size)

27a. ***Idaea nevadata* Whli.**, ♂, Ruta del Veleta, Sierra Nevada, S. Spain (ZSM)
27b. *I. nevadata* Whli., ♂, Granada, Sierra Nevada, S. Spain (ZSM)
27c. *I. nevadata* Whli., ♀, Granada, Sierra Nevada, S. Spain (ZSM)
27d. *I. nevadata* Whli., ♂, Ruta del Veleta, Sierra Nevada, S. Spain (ZSM/HERBULOT)
28. ***Idaea figuraria* B.-H.**, ♀, La Granja, Sierra de Guadarrama, C. Spain (ZSM)
29a. ***Idaea alicantaria* Rssr.**, ♀, El Pozo del Esparto, Almeria, SE. Spain (coll. SKOU)
29b. *I. alicantaria* Rssr., ♀, Alhama de Murcia, SE. Spain (NHMW/VARTIAN)
29c. *I. alicantaria* Rssr., ♀, Cabo da Gata, Almeria, SE. Spain (coll. KRAUS)
29d. *I. alicantaria* Rssr., ♂, Sierra Alhamilla, Almeria, SE. Spain (coll. GRÜNEWALD)
29e. *I. alicantaria* Rssr., ♂, Cabo de Gata, Almeria, SE. Spain (coll. SKOU)
29f. *I. alicantaria* Rssr., ♀, Alhama de Murcia, SE. Spain (NHMW/VARTIAN)
30a. ***Idaea nigrolineata* Chrét.**, ♀, Camino de Istan, Andalusia, S. Spain (ZSM)
30b. *I. nigrolineata* Chrét., ♂, Camino de Istan, Andalusia, S. Spain (coll. SKOU)
30c. *I. nigrolineata* Chrét., ♀, Camino de Istan, Andalusia, S. Spain (ZMUC)
30d. *I. nigrolineata* Chrét., ♂, Ifrane, Middle Atlas, Morocco (ZSM/HERBULOT)
30e. *I. nigrolineata* Chrét., ♂, Gafsa, Tunisia (ZSM/HERBULOT), holotype
30f. *I. nigrolineata* Chrét., ♂, Mischliffen, Middle Atlas, Morocco (ZSM)
31a. ***Idaea completa* Stgr.**, ♂, Zahara de los Atunes, Cadiz, S. Spain (coll. MÜLLER), reared
31b. *I. completa* Stgr., ♀, Zahara de los Atunes, Cadiz, S. Spain (coll. MÜLLER), reared
31c. *I. completa* Stgr., ♂, Tabarka, Tunisia (ZSM/HERBULOT)
31d. *I. completa* Stgr., ♂, Gafsa, Tunisia (ZSM/HERBULOT), reared
31e. *I. completa* Stgr., ♂, Gafsa, Tunisia (ZSM/HERBULOT), reared
31f. *I. completa* Stgr., ♀, Gafsa, Tunisia (ZSM/HERBULOT), reared
32a. ***Idaea intermedia* Stgr.**, ♂, Athen, C. Greece (NHMW/VARTIAN)
32b. *I. intermedia* Stgr., ♂, Drepanon, E. Peloponnes, S. Greece (ZSM)
32c. *I. intermedia* Stgr., ♀, Assistes, C. Crete (NHMW/VARTIAN), reared
32d. *I. intermedia* Stgr., ♂, Fatiraki, Rhodos island (ZSM)
32e. *I. intermedia* Stgr., ♀, Agios Nikolaos, Crete (ZSM), reared
32f. *I. intermedia* Stgr., ♂, Pamukkale, Turkey (ZSM)
33a. ***Idaea rusticata* Den. & Schiff.**, ♂, Mödling, Vienna, E. Austria (ZSM)
33b. *I. rusticata* Den. & Schiff., ♀, Neustadt, Pfalz, W. Germany (ZSM/EMEM)
33c. *I. rusticata* Den. & Schiff., ♀, Palermo, Sicily (ZSM)
33d. *I. rusticata* Den. & Schiff., ♂, Liljanovo, 'Macedonia', SW. Bulgaria (ZSM)
33e. *I. rusticata* Den. & Schiff., ♀, Topolia, W. Crete (coll. SKOU)
34a. ***Idaea mustelata* Gmpp.**, ♂, Puig Ventos, Gerona, N. Spain (coll. MÜLLER)
34b. *I. mustelata* Gmpp., ♀, Albarracin, Teruel, E. Spain (ZSM)
34c. *I. mustelata* Gmpp., ♀, Albarracin, Teruel, E. Spain (ZSM), reared
34d. *I. mustelata* Gmpp., ♀, Granada, S. Spain (ZSM)
34e. *I. mustelata* Gmpp., ♀, Ifrane, Middle Atlas, Morocco (ZSM), reared
35a. ***Idaea filicata* Hbn.**, ♀, Nagy Nyir, Hungary (NHMW/VARTIAN)
35b. *I. filicata* Hbn., ♂, Budapest, Hungary (ZSM), reared
35c. *I. filicata* Hbn., ♀, Budapest, Hungary (coll. MÜLLER), reared
35d. *I. filicata* Hbn., ♀, Eslida, Sierra de Espadan, E. Spain (coll. SKOU)
35e. *I. filicata* Hbn., ♂, Agay, Var, SE. France (ZSM)
35f. *I. filicata* Hbn., ♀, Trecchina, S. Basilicata, S. Italy (ZSM)
35g. *I. filicata* Hbn., ♂, Tivoli, Lazio, C. Italy (ZSM)
35h. *I. filicata* Hbn., ♂, Liljanovo, Pirin mts., SW. Bulgaria (coll. MÜLLER)
35i. *I. filicata* Hbn., ♀, Tracino, Pantelleria island, Italy (ZSM)
35k. *I. filicata* Hbn., ♀, Batha-Harissa, Lebanon (ZSM)
36a. ***Idaea troglodytaria* H.-S.**, ♂, Mt. Smith, Rhodos island (ZSM)
36b. *I. troglodytaria* H.-S., ♀, Mt. Smith, Rhodos island (ZSM)
36c. *I. troglodytaria* H.-S., ♀, Paphos, W. Cyprus (ZSM)
36d. *I. troglodytaria* H.-S., ♀, Paphos, W. Cyprus (ZSM)
36e. *I. troglodytaria* H.-S., ♀, Limassol, S. Cyprus (NHMW)
36f. *I. troglodytaria* H.-S., ♂, Paphos, W. Cyprus (ZSM)

27a
27b
27c
27d
28
29a
29b
29c
29d
29e
29f
30a
30b
30c
30d
30e
30f
31a
31b
31c
31d
31e
31f
32a
32b
32c
32d
32e
32f
33a
33b
33c
33d
33e
34a
34b
34c
34d
34e
35a
35b
35c
35d
35e
35f
35g
35h
35i
35k
36a
36b
36c
36d
36e
36f

PLATE 5 (all specimens 1.5 x natural size)

37a. ***Idaea laevigata* Scop.,** ♂, Thüringen, E. Germany (ZSM), reared
37b. *I. laevigata* Scop., ♀, Jena, Thüringen, E. Germany (ZSM), reared
37c. *I. laevigata* Scop., ♀, Pfaffenbroich, Aachen, W. Germany (ZSM)
37d. *I. laevigata* Scop., ♀, Mercuore, Mistretta, NE. Sicily (ZSM)
37e. *I. laevigata* Scop., ♂, Cala Millor, Mallorca (coll. MÜLLER), reared
37f. *I. laevigata* Scop., ♀, Aragon, E. Spain (ZSM), reared
38a. ***Idaea efflorata* Z.,** ♂, Tivoli, Lazio, C. Italy (ZSM)
38b. *I. efflorata* Z., ♀, Mte. Sirente, Abruzzi, C. Italy (ZSM)
38c. *I. efflorata* Z., ♂, Siniscola, Sardinia (coll. FRIEDRICH), reared
38d. *I. efflorata* Z., ♀, Siniscola, Sardinia (coll. FRIEDRICH), reared
38e. *I. efflorata* Z., ♂, Mistretta, NE. Sicily (ZSM), reared
38f. *I. efflorata* Z., ♂, Philippeville, E. Algeria (ZSM), reared
39a. ***Idaea attenuaria* Rmb.,** ♂, Mistretta, NE. Sicily (ZSM), reared
39b. *I. attenuaria* Rmb., ♂, Calvi, N. Corsica (ZSM)
39c. *I. attenuaria* Rmb., ♀, Marbella, S. Spain (ZSM)
39d. *I. attenuaria* Rmb., ♀, Nerja, Malaga, S. Spain (ZSM/EMEM)
39e. *I. attenuaria* Rmb., ♀, Castellar de la Frontera, Cadiz, S. Spain (coll. SKOU)
39f. *I. attenuaria* Rmb., ♂, Philippeville, E. Algeria (ZSM), reared
40a. ***Idaea incalcarata* Chrét.,** ♂, Port Bou, N. Spain (NHMW/VARTIAN), reared
40b. *I. incalcarata* Chrét., ♂, Albarracin, Teruel, E. Spain (coll. BAISCH)
40c. *I. incalcarata* Chrét., ♀, Port Bou, N. Spain (ZSM), reared
40d. *I. incalcarata* Chrét., ♀, Albarracin, Teruel, E. Spain (ZSM), reared
40e. *I. incalcarata* Chrét., ♀, Sagas, Girona, NE. Spain (coll. SKOU)
40f. *I. incalcarata* Chrét., ♀, Digne, S. France (ZSM/HERBULOT), reared
41a. ***Idaea typicata typicata* Gn.,** ♂, Torbole, Lake Garda, N. Italy (ZSM)
41b. *I. typicata typicata* Gn., ♀, Noli, Liguria, NW. Italy (ZSM), reared
41c. *I. typicata typicata* Gn., ♂, Noli, Liguria, NW. Italy (ZSM)
41d. *I. typicata typicata* Gn., ♂, Tivoli, Lazio, C. Italy (ZSM), reared
41e. *I. typicata typicata* Gn., ♂, Mistretta, NE. Sicily (NHMW/VARTIAN), reared
41f. *I. typicata hornigaria* Stgr., ♂, Naturns, S. Tirol, N. Italy (ZSM/EMEM), reared
41g. *I. typicata hornigaria* Stgr., ♀, Bozen, S. Tirol, N. Italy (ZFMK/WAGENER)
41h. *I. typicata hornigaria* Stgr., ♀, Atzwang, S. Tirol, N. Italy (ZSM), reared
41i. *I. typicata hornigaria* Stgr., ♀, Waidbruck, S. Tirol, N. Italy (ZSM)
41k. *I. typicata hornigaria* Stgr., ♀, Atzwang, S. Tirol, N. Italy (ZSM)
42a. ***Idaea alyssumata alyssumata* Himm. & Mill.,** ♀, Pyrenees (ZSM/HERBULOT)
42b. *I. alyssumata alyssumata* Himm. & Mill., ♂, Trevelez, Granada, S. Spain (coll. BAISCH)
42c. *I. alyssumata alyssumata* Himm. & Mill., ♀, Ribas de Freser, E. Pyrenees, NE. Spain (ZSM), reared
42d. *I. alyssumata alyssumata* Himm. & Mill., ♂, Albarracin, Teruel, E. Spain (ZSM)
42e. *I. alyssumata alyssumata* Himm. & Mill., ♂, Albarracin, Teruel, E. Spain (ZSM), reared
42f. *I. alyssumata alyssumata* Himm. & Mill., ♂, Albarracin, Teruel, E. Spain (ZSM)
42g. *I. alyssumata alyssumata* Himm. & Mill., ♂, Albarracin, Teruel, E. Spain (ZSM), reared
42h. *I. alyssumata alyssumata* Himm. & Mill., ♂, Albarracin, Teruel, E. Spain (ZSM), reared
42i. *I. alyssumata alyssumata* Himm. & Mill., ♀, St. Jean du Gard, Gard, S. France (ZSM/HERBULOT)
42k. *I. alyssumata reynaldiata* Rouast, ♂, Neyron, Ain, E. France (ZSM/HERBULOT)
42l. *I. alyssumata alyssumata* Himm. & Mill., ♂, Parcent, Alicante, E. Spain (ZSM), 2nd gen.
43a. ***Idaea urcitana* Agenjo,** ♂, Almeria, SE. Spain (MNCN), holotype, 1st gen.
43b. *I. urcitana* Agenjo, ♂, Sierra Alhamilla, Almeria, SE. Spain (coll. GRÜNEWALD), 1st gen.
43c. *I. urcitana* Agenjo, ♂, Sierra Almagro, Almeria, SE. Spain (coll. GRÜNEWALD), 2nd gen.
43d. *I. urcitana* Agenjo, ♀, Tabernas, Almeria, SE. Spain (coll. PÖLL)
44a. ***Idaea moniliata* Den. & Schiff.,** ♂, Kyffhäuser, Thuringia, E. Germany (coll. GELBRECHT), reared
44b. *I. moniliata* Den. & Schiff., ♀, Kyffhäuser, Thuringia, E. Germany (coll. GELBRECHT), reared
44c. *I. moniliata* Den. & Schiff., ♂, Alpes maritimes, SE. France (ZSM/HERBULOT)
44d. *I. moniliata* Den. & Schiff., ♂, Liljanovo, Pirin mts., SW. Bulgaria (coll. GELBRECHT)
44e. *I. moniliata* Den. & Schiff., ♀, Karojba, Istria, Croatia (ZSM)

38a
38b
38c
38d
38e
38f
39a
39b
39c
39d
39e
39f
40a
40b
40c
40d
40e
40f
41a
41b
41c
41d
41e
41f
41g
41h
41i
41k
42a
42b
42c
42d
42e
42f
42g
42h
42i
42k
42l
43a
43b
43c
43d
44c
44b
44c
44d
44e

PLATE 6 (all specimens 1.5 x natural size)

45a. ***Idaea spissilimbaria* Mab.,** ♀, Cibinsgeb., C. Romania (ZSM)
45b. *I. spissilimbaria* Mab., ♂, Algeria (ZSM/HERBULOT), holotype
45c. *I. spissilimbaria* Mab., ♀, Chemin de Lozzi, Corsica (ZSM/HERBULOT)
46a. ***Idaea circuitaria* Hbn.,** ♂, St. Médard en J., Gironde, W. France (ZSM/HERBULOT)
46b. *I. circuitaria* Hbn., ♂, Noli, Liguria NW. Italy (ZSM)
46c. *I. circuitaria* Hbn., ♂, Mte. Sirente, Abruzzi, C. Italy (ZSM)
46d. *I. circuitaria* Hbn., ♂, Puig Ventos, Gerona, NE. Spain (coll. GELBRECHT), reared
46e. *I. circuitaria* Hbn., ♂, Lastva, Herzegovina (ZSM/HERBULOT)
46f. *I. circuitaria* Hbn., ♀, Asprovalta, N. Greece (ZFMK/WAGENER)
46g. ***Idaea mimosaria* Gn.,** ♀, Mersin, S. Turkey (ZSM)
47a. ***Idaea rainerii* Hausm.,** ♀, Alhama de Murcia, SE. Spain (NHMW/VARTIAN)
47b. *I. rainerii* Hausm., ♂, Palermo, Sicily (ZSM), paratype
47c. *I. rainerii* Hausm., ♀, Sardinia (MNHU)
47d. *I. rainerii* Hausm., ♂, S. Sardinia (MNHU)
47e. *I. rainerii* Hausm., ♀, Bone, E. Algeria (ZSM/HERBULOT)
48a. ***Idaea albarracina* Rssr.,** ♂, Puig Ventos, Gerona, NE. Spain (coll. GELBRECHT), reared
48b. *I. albarracina* Rssr., ♀, Puig Ventos, Gerona, NE. Spain (coll. GELBRECHT), reared
48c. *I. albarracina* Rssr., ♂, Albarracin, Teruel, E. Spain (ZSM), reared
48d. *I. albarracina* Rssr., ♀, Albarracin, Teruel, E. Spain (ZSM), reared
49a. ***Idaea incisaria incisaria* Stgr.,** ♀, Malaga, S. Spain (ZSM/HERBULOT), reared
49b. *I. incisaria incisaria* Stgr., ♂, Malaga, S. Spain (ZSM/HERBULOT), reared
49c. *I. incisaria incisaria* Stgr., ♂, Biskra, Sahara Atlas, E. Algeria (ZSM/HERBULOT)
49d. *I. incisaria incisaria* Stgr., ♂, Hammam-Rhira, N. Algeria (ZSM)
49e. *I. incisaria incisaria* Stgr., ♀, Hammam-Rhira, N. Algeria (ZSM)
49f. *I. incisaria incisaria* Stgr., ♀, Philippeville, E. Algeria (ZSM)
49g. *I. incisaria praecisa* Rssr., ♂, Xauen, Rif Mts., N. Morocco (ZSM), reared
49h. *I. incisaria praecisa* Rssr., ♀, Xauen, Rif Mts., N. Morocco (ZSM), reared
49i. *I. incisaria praecisa* Rssr., ♀, Xauen, Rif Mts., N. Morocco (ZSM), reared
49k. *I. incisaria praecisa* Rssr., ♀, Xauen, Rif Mts., N. Morocco (ZSM), reared
49l. ***Idaea textaria* Led.,** ♀, Maras, Taurus, S. Turkey (ZSM), reared
50a. ***Idaea calunetaria calunetaria* Stgr.,** ♀, Albarracin, Teruel, E. Spain (ZSM), reared
50b. *I. calunetaria calunetaria* Stgr., ♂, Vilamoura, Algarve, S. Portugal (ZSM), reared
50c. *I. calunetaria calunetaria* Stgr., ♂, Santa Coloma, Gerona, NE. Spain (coll. GELBRECHT)
50d. *I. calunetaria calunetaria* Stgr., ♀, Orjiva, Las Arpujarras S. Spain (NHMW/VARTIAN), reared
50e. *I. calunetaria calunetaria* Stgr., ♂, Almoraima, Cadiz, S. Spain (coll. SKOU)
50f. *I. calunetaria valesiaria* Püng., ♂, Zermatt, Valesia, S. Switzerland (ZSM), reared
50g. *I. calunetaria valesiaria* Püng., ♀, Zermatt, Valesia, S. Switzerland (ZSM), reared
50h. *I. calunetaria valesiaria* Püng., ♂, Le Rozier, S. France (ZSM/HERBULOT)
50i. *I. calunetaria valesiaria* Püng., ♀, Val Grisenche, Val d'Aosta, NW. Italy (ZSM), reared
51a. ***Idaea belemiata belemiata* Mill.,** ♀, Monchique, S. Portugal (NHMW)
51b. *I. belemiata belemiata* Mill., ♂, Monchique, S. Portugal (NHMW)
51c. *I. belemiata belemiata* Mill., ♀, La Plataforma, Sierra de Gredos, W. Spain (coll. SKOU)
51d. *I. belemiata belemiata* Mill., ♀, Estanya, Huesca, N. Spain (coll. SKOU)
51e. *I. belemiata belemiata* Mill., ♂, La Plataforma, Sierra de Gredos, W. Spain (coll. SKOU)
51f. *I. belemiata belemiata* Mill., ♂, Los Barrios, Cadiz, S. Spain (coll. KRAUS)
51g. *I. belemiata helianthemata* Mill., ♀, Frejus, Var, SE. France (coll. SKOU)
51h. *I. belemiata helianthemata* Mill., ♂, Frejus, Var, SE. France (ZSM)
51i. *I. belemiata helianthemata* Mill., ♂, Mt. Pacanaglia, SE. France (ZSM/HERBULOT)
51k. *I. belemiata helianthemata* Mill., ♂, Frejus, Var, SE. France (ZSM)

46c
46d
46e
46f
46g
47a
47b
47c
47d
47e
48a
48b
48c
48d
49a
49b
49c
49d
49e
49f
49g
49h
49i
49k
49l
50a
50b
50c
50d
50e
50f
50g
50h
50i
51a
51b
51c
51d
51e
51f
51g
51h
51i
51k

PLATE 7 (all specimens 1.5 x natural size)

52a. ***Idaea elongaria elongaria* Rmb.,** ♂, Saint Florent, Corsica (ZSM)
52b. *I. elongaria elongaria* Rmb., ♂, Zahara de los Atunes, Cadiz, S. Spain (ZSM), reared
52c. *I. elongaria elongaria* Rmb., ♀, Llansa, Gerona, NE. Spain (ZSM)
52d. *I. elongaria elongaria* Rmb., ♂, Colfelice, C. Italy (ZSM), reared
52e. *I. elongaria elongaria* Rmb., ♀, Colfelice, C. Italy (ZSM), reared
52f. *I. elongaria elongaria* Rmb., ♂, Noli, Liguria, NW. Italy (ZSM)
52g. *I. elongaria elongaria* Rmb., ♂, Kapitan Andreevo, Bulgaria (coll. GELBRECHT), reared
52h. *I. elongaria elongaria* Rmb., ♀, Kapitan Andreevo, Bulgaria (coll. GELBRECHT), reared
52i. *I. elongaria elongaria* Rmb., ♂, Stari Dorjan, S. Macedonia (ZSM)
52k. *I. elongaria elongaria* Rmb., ♀, Callergi Mts. W. Crete (coll. SKOU)
52l. *I. elongaria pecharia* Stgr., ♂, Budaörs, Budapest, N. Hungary (coll. GELBRECHT), reared
52m. *I. elongaria pecharia* Stgr., ♀, Budaörs, Budapest, N. Hungary (coll. GELBRECHT), reared
52n. *I. elongaria pecharia* Stgr., ♂, Budaörs, Budapest, N. Hungary (coll. MÜLLER), reared
52o. *I. elongaria pecharia* Stgr., ♀, Budaörs, Budapest, N. Hungary (coll. MÜLLER), reared
52p. *I. elongaria pecharia* Stgr., ♂, Budapest, N. Hungary (ZSM)
53a. ***Idaea palaestinensis* Stern.,** ♂, Assistes, Crete (NHMW/VARTIAN), reared
53b. *I. palaestinensis* Stern., ♀, Agios Nikolaos, Crete (coll. MÜLLER), reared
53c. *I. palaestinensis* Stern., ♂, Mt. Smith, Rhodos island (ZSM), reared
53d. *I. palaestinensis* Stern., ♀, Mt. Smith, Rhodos island (ZSM), reared
53e. *I. palaestinensis* Stern., ♀, Mt. Smith, Rhodos island (ZSM), reared
54a. *Idaea obsoletaria obsoletaria* Rmb., ♂, Schnalstal, S. Tirol, N. Italy (ZSM/EMEM)
54b. *I. obsoletaria obsoletaria* Rmb., ♀, by Trento, S. Tirol, N. Italy (ZSM)
54c. *I. obsoletaria lilaceola* Hausm., ♂, Albarracin, Teruel, E. Spain (ZSM)
54d. *I. obsoletaria lilaceola* Hausm., ♀, Banyuls-sur-mer, E. Pyrenees, S. France (ZSM/EMEM)
54e. *I. obsoletaria dionigii* Hausm., ♂, Tivoli, Lazio, C. Italy (ZSM)
54f. *I. obsoletaria dierli* Hausm., ♀, Madonie, N. Sicily (ZSM), paratype
54g. *I. obsoletaria rufularia* H.-S., ♂, Mödling by Vienna, E. Austria (ZFMK/WAGENER)
54h. *I. obsoletaria rufularia* H.-S., ♀, Budaörs, Budapest, N. Hungary (coll. MÜLLER), reared
54i. *I. obsoletaria rufularia* H.-S., ♂, Liljanovo, Pirin mts., SW. Bulgaria (coll. GELBRECHT), reared
54k. *I. obsoletaria rufularia* H.-S., ♂, Bishtiryak, S. Urals, E. European Russia (coll. NUPPONEN)
55a. ***Idaea obliquaria* Trti.,** ♂, Col de Vergio, Corsica (ZFMK/WAGENER)
55b. *I. obliquaria* Trti., ♀, Vizzavona, Corsica (ZSM)
55c. *I. obliquaria* Trti., ♂, Aritzo, Sardinia (ZSM)
55d. *I. obliquaria* Trti., ♂, Aritzo, Sardinia (ZSM)
55e. *I. obliquaria* Trti., ♀, Aritzo, Sardinia (ZSM)
56a. ***Idaea inquinata* Scop.,** ♂, Templin, Mecklenburg, E. Germany (coll. GELBRECHT), reared
56b. *I. inquinata* Scop., ♀, Templin, Mecklenburg, E. Germany (coll. GELBRECHT), reared
56c. *I. inquinata* Scop., ♂, Berlin, E. Germany (ZSM), reared
56d. *I. inquinata* Scop., ♀, Lippe, Niedersachsen, N. Germany (ZSM), reared
56e. *I. inquinata* Scop., ♀, Karlsruhe, Baden-Württemberg, W. Germany (ZSM), reared
56f. *I. inquinata* Scop., ♂, Klagenfurt, Carinthia, SE. Austria (ZSM), reared
56g. *I. inquinata* Scop., ♂, Rome, C. Italy (ZSM)
56h. *I. inquinata* Scop., ♂, Monastir, Serbia (ZSM), reared
56i. *I. inquinata* Scop., ♂, Vratnica, C. Macedonia (ZSM)
56k. *I. inquinata* Scop., ♂, Aghia Galini, S. Crete (ZSM), reared
56l. *I. inquinata* Scop., ♂, Rhodos island (ZSM), reared
56m. *I. inquinata* Scop., ♂, Rhodos island (ZSM), reared
57a. ***Idaea blaesii* Lenz & Hausm.,** ♂, Cabo de Gata, Almeria, SE. Spain (ZSM)
57b. *I. blaesii* Lenz & Hausm., ♂, Loule, Algarve, S. Portugal (ZSM), paratype
57c. *I. blaesii* Lenz & Hausm., ♂, Loule, Algarve, S. Portugal (ZSM), paratype
57d. *I. blaesii* Lenz & Hausm., ♂, Loule, Algarve, S. Portugal (ZSM), paratype
57e. *I. blaesii* Lenz & Hausm., ♀, Loule, Algarve, S. Portugal (ZSM), paratype
57f. *I. blaesii* Lenz & Hausm., ♀, Loule, Algarve, S. Portugal (ZSM), paratype

52a
52b
52c
52d
52e
52f
52g
52h
52i
52k
52l
52m
52n
52o
52p
53a
53b
53c
53d
53e
54a
54b
54c
54d
54e
54f
54g
54h
54i
54k
55a
55b
55c
55d
55e
56a
56b
56c
56d
56e
56f
56g
56h
56i
56k
56l
56m
57a
57b
57c
57d
57e
57f

PLATE 8 (all specimens 1.5 x natural size)

58a. ***Idaea dilutaria* Hbn.,** ♂, Bornich, W. Germany (ZSM)
58b. *I. dilutaria* Hbn., ♀, Dortmund, NW. Germany (ZSM)
58c. *I. dilutaria* Hbn., ♂, Wocheiner See, Carinthia, SE. Austria (ZSM)
58d. *I. dilutaria* Hbn., ♀, St. Martin Vesubie, SE. France (ZSM)
58e. *I. dilutaria* Hbn., ♀, Terlan, S. Tirol, N. Italy (ZSM)
58f. *I. dilutaria* Hbn., ♀, Drenovo, Kavadar, C. Macedonia (ZSM)
58g. *I. dilutaria* Hbn., ♂, Liljanovo, Pirin mts., SW. Bulgaria (coll. GELBRECHT), reared
58h. *I. dilutaria* Hbn., ♀, Liljanovo, Pirin mts., SW. Bulgaria (coll. GELBRECHT), reared
58i. *I. dilutaria* Hbn., ♂, Split ('Spalato'), W. Croatia (MNHU)
58k. *I. dilutaria* Hbn., ♀, Lipik, E. Croatia (ZSM)
59a. ***Idaea fuscovenosa* Gze.,** ♂, Finkenwärder, Hamburg, N. Germany (ZSM)
59b. *I. fuscovenosa* Gze., ♂, Linz, N. Austria (ZSM)
59c. *I. fuscovenosa* Gze., ♂, Liljanovo, Pirin mts., SW. Bulgaria (coll. GELBRECHT), reared
59d. *I. fuscovenosa* Gze., ♂, Liljanovo, Pirin mts., SW. Bulgaria (coll. GELBRECHT), reared
59e. *I. fuscovenosa* Gze., ♀, St. Raphael, Var, NE. France (ZSM/HERBULOT)
60a. ***Idaea robiginata* Stgr.,** ♂, Castilia, C. Spain (NHMW)
60b. *I. robiginata* Stgr., ♂, Pto. de Navacerrada, C. Spain (ZSM)
60c. *I. robiginata* Stgr., ♂, St. Ildefonso, C. Spain (ZSM)
60d. *I. robiginata* Stgr., ♂, Castilia, C. Spain (ZSM)
60e. *I. robiginata* Stgr., ♂, Pto. de Navacerrada, C. Spain (ZSM)
61a. ***Idaea lutulentaria* Stgr.,** ♂, Molinier, Sierra Segura, SE. Spain (NMHW)
61b. *I. lutulentaria* Stgr., ♀, Ronda, S. Spain (coll. GELBRECHT), reared
61c. *I. lutulentaria* Stgr., ♂, Ronda, S. Spain (coll. MÜLLER), reared
61d. *I. lutulentaria* Stgr., ♀, Ronda, S. Spain (coll. MÜLLER), reared
61e. *I. lutulentaria* Stgr., ♀, Estana, Huesca, NE. Spain (coll. SKOU)
62a. ***Idaea humiliata* Hufn.,** ♂, Eschelbronn, Baden, SW. Germany (ZSM)
62b. *I. humiliata* Hufn., ♂, Riekau, NE. Niedersachsen, N. Germany (ZSM)
62c. *I. humiliata* Hufn., ♀, by Linz, N. Austria (ZSM)
62d. *I. humiliata* Hufn., ♂, Liljanovo, Pirin mts., SW. Bulgaria (coll. GELBRECHT)
62e. *I. humiliata* Hufn., ♀, Schnalstal, S. Tirol, N. Italy (ZSM)
63a. ***Idaea bigladiata bigladiata* Herb.,** ♀, Riopar, Albacete, SE. Spain (ZSM)
63b. *I. bigladiata bigladiata* Herb., ♀, Ronda, S. Spain (ZSM)
63c. *I. bigladiata alberti* subsp.n., ♀, Tunis, Tunisia (ZSM), reared
63d. *I. bigladiata alberti* subsp.n., by Algier, Algeria (ZSM)
64a. ***Idaea politaria* Hbn.,** ♂, Puig Ventos, Gerona, NE. Spain (coll. GELBRECHT), reared
64b. *I. politaria* Hbn., ♀, Leurana, Istria, NW. Croatia (ZSM)
64c. *I. politaria* Hbn., ♀, Sozopol, E. Bulgaria (coll. GELBRECHT), reared
64d. *I. politaria* Hbn., ♀, Stary Dorjan, S. Macedonia (ZSM)
64e. *I. politaria* Hbn., ♂, Mt. Smith, Rhodos island (ZSM)
65a. ***Idaea longaria* H.-S.,** ♂, Cadiz, S. Spain (ZSM), reared
65b. *I. longaria* H.-S., ♀, Merida, Extremadura, SW. Spain (coll. SKOU)
65c. *I. longaria* H.-S., ♂, Capo Comino, Sardinia (coll. GRÜNEWALD)
65d. *I. longaria* H.-S., ♂, Gammouda, Tunisia (ZSM)
65e. *I. longaria* H.-S., ♀, Taormina, E. Sicily (NHMW/VARTIAN), reared
65f. *I. longaria* H.-S., ♂, Taormina, E. Sicily (NHMW/VARTIAN), reared
65g. *I. longaria* H.-S., ♂, Taormina, E. Sicily (NHMW/VARTIAN), reared
65h. *I. longaria* H.-S., ♂, Chersonissos, E. Crete (coll. FRIEDRICH)
65i. *I. longaria* H.-S., ♀, Chersonissos, E. Crete (coll. FRIEDRICH)
65k. *I. longaria* H.-S., ♂, Paradissi, Rhodos island (ZSM)

58e
58f
58g
58h
58i
58k
59a
59b
59c
59d
59e
60a
60b
60c
60d
60e
61a
61b
61c
61d
61e
62a
62b
62c
62d
62e
63a
63b
63c
63d
64a
64b
64c
64d
64e
65a
65b
65c
65d
65e
65f
65g
65h
65i
65k

PLATE 9 (all specimens 1.5 x natural size)

66a. **Idaea nexata Hbn.,** ♂, Chiclana, Andalusia, S. Spain (NHMW)
66b. *I. nexata* Hbn., ♂, Lissabon, Portugal (NHMW)
66c. *I. nexata* Hbn., ♂, Chiclana, Andalusia, S. Spain (ZSM)
66d. *I. nexata* Hbn., ♂, Andalusia, S. Spain (ZSM)
66e. *I. nexata* Hbn., ♂, Chiclana, Andalusia, S. Spain (ZSM)
66f. *I. nexata* Hbn., ♂, Mestigmeur, Morocco (ZSM/HERBULOT)
67a. **Idaea manicaria H.-S.,** ♂, Alhama de Murcia, SE. Spain (NHMW/VARTIAN)
67b. *I. manicaria* H.-S., ♂, Arriate, Ronda, S. Spain (ZMUC)
67c. *I. manicaria* H.-S., ♀, Arriate, Ronda, S. Spain (ZMUC)
67d. *I. manicaria* H.-S., ♀, Guadalquivir, Sevilla, S. Spain (NHMW/VARTIAN)
67e. *I. manicaria* H.-S., ♀, Rabat, Malta (ZSM)
67f. *I. manicaria* H.-S., ♀, Rabat, Malta (ZSM)
68a. **Idaea descitaria Chr.,** ♂, Uralsk, SE. European Russia (ZSM), reared
68b. *I. descitaria* Chr., ♀, Uralsk, SE. European Russia (coll. SKOU), reared
68c. *I. descitaria* Chr., ♂, Uralsk, SE. European Russia (MNHU), reared
68d. *I. descitaria* Chr., ♀, Uralsk, SE. European Russia (MNHU), reared
68e. *I. descitaria* Chr., ♀, Uralsk, SE. European Russia (ZSM), reared
69a. **Idaea vesubiata Mill.,** ♂, St. Martin Vesubie, Alpes maritimes, SE. France (ZSM)
69b. *I. vesubiata* Mill., ♀, St. Martin Vesubie, Alpes maritimes, SE. France (ZSM), reared
69c. *I. vesubiata* Mill., ♀, St. Martin Vesubie, Alpes maritimes, SE. France (ZSM), reared
69d. *I. vesubiata* Mill., ♂, St. Martin Vesubie, Alpes maritimes, SE. France (ZSM), reared
69e. *I. vesubiata* Mill., ♂, St. Martin Vesubie, Alpes maritimes, SE. France (ZSM), reared
70a. **Idaea libycata Bart.,** ♂, Toulon, S. France (NHMW/VARTIAN), reared
70b. *I. libycata* Bart., ♀, Toulon, S. France (NHMW/VARTIAN), reared
70c. *I. libycata* Bart., ♀, Toulon, S. France (ZSM), reared
70d. *I. libycata* Bart., ♂, Philippeville, E. Algeria (ZSM), reared
70e. *I. libycata* Bart., ♀, Philippeville, E. Algeria (ZSM), reared
71a. **Idaea consolidata Led.,** ♂, Tivoli, Lazio, C. Italy (ZSM), reared
71b. *I. consolidata* Led., ♂, Dubrovnik ('Gravosa'), S. Croatia (ZSM)
71c. *I. consolidata* Led., ♂, Dubrovnik ('Gravosa'), S. Croatia (ZSM)
71d. *I. consolidata* Led., ♀, Dubrovnik ('Gravosa'), S. Croatia (ZSM)
71e. *I. consolidata* Led., ♂, Stary Dorjan, SE. Macedonia (ZSM)
71f. *I. consolidata* Led., ♂, Chersonissos, E. Crete (coll. FRIEDRICH)
71g. *I. consolidata* Led., ♀, Agios Nikolaos, S. Crete (coll. MÜLLER), reared
71h. *I. consolidata* Led., ♀, Paphos, W. Cyprus (ZSM), reared
72a. **Idaea joannisiata joannisiata Homb.,** ♂, Torla, C. Pyrenees, N. Spain (NHMW)
72b. *I. joannisiata joannisiata* Homb., ♂, Vernet-les-Bains, E. Pyrenees, S. France (ZFMK)
72c. *I. joannisiata ibericata* Whli., ♀, Albarracin, Teruel, E. Spain (ZSM)
72d. *I. joannisiata ibericata* Whli., ♂, Bronchales, Teruel, E. Spain (ZSM/HERBULOT)
72e. *I. joannisiata ibericata* Whli., ♀, Bronchales, Teruel, E. Spain (ZSM/HERBULOT)
72f. *I. joannisiata ibericata* Whli., ♀, Alportel, Algarve, S. Portugal (ZSM)
72g. *I. joannisiata ibericata* Whli., ♀, Lozoya, Madrid, C. Spain (coll. SKOU)
72h. *I. joannisiata ibericata* Whli., ♂, Guils, Llerida, NE. Spain (coll. SKOU)
72i. *I. joannisiata ibericata* Whli., ♂, Pta. de la Ragua, Sierra Nevada, S. Spain (coll. GRÜNEWALD)
73. **Idaea acutipennis Hausm. & Hon.,** ♂, Dragonera, Balearics (BMNH), holotype
74a. **Idaea seriata seriata Schrank,** ♀, Tulla, Estonia (ZSM)
74b. *I. seriata seriata* Schrank, ♂, Oxford, England (coll. GRÜNEWALD)
74c. *I. seriata seriata* Schrank, ♂, Chemnitz, Saxonia, E. Germany (ZSM)
74d. *I. seriata seriata* Schrank, ♂, Kirkeby Sand, Fyn, Denmark (coll. SKOU)
74e. *I. seriata seriata* Schrank, ♂, Rheingau, W. Germany (ZSM)
74f. *I. seriata canteneraria* Bsd., ♀, Cannes, SE. France (ZSM/HERBULOT)
74g. *I. seriata canteneraria* Bsd., ♂, Tivoli, Lazio, C. Italy (ZSM), reared
74h. *I. seriata canteneraria* Bsd., ♂, Digne, Basses Alpes, SE. France (ZSM)
74i. *I. seriata canteneraria* Bsd., ♀, Makarska, Dalmatia, S. Croatia (coll. GELBRECHT), reared
74k. *I. seriata canteneraria* Bsd., ♂, Makarska, Dalmatia, S. Croatia (ZSM), reared

67a
67b
67c
67d
67e
67f
68a
68b
68c
68d
68e
69a
69b
69c
69d
69e
70a
70b
70c
70d
70e
71a
71b
71c
71d
71e
71f
71g
71h
72a
72b
72c
72d
72e
72f
72g
73
74a
74b
74c
74d
74e
74f
74g
74h
74i
74k

PLATE 10 (all specimens 1.5 x natural size)

74l. ***Idaea seriata canteneraria* Bsd.**, ♂, Kranevo, Varna, E. Bulgaria (coll. GELBRECHT), reared
74m. *I. seriata canteneraria* Bsd., ♀, Mte. Nai, S. Sardinia (coll. GRÜNEWALD)
74n. *I. seriata canteneraria* Bsd., ♀, Tracino, Pantelleria island (ZSM)
74o. *I. seriata canteneraria* Bsd., ♀, Mellieka, Malta (ZMUC)
74p. *I. seriata canteneraria* Bsd., ♂, Tabarka, Tunisia (ZSM), reared
75a. ***Idaea minuscularia* Ribbe,** ♂, Cala Millor, Mallorca (coll. GELBRECHT), reared
75b. *I. minuscularia* Ribbe, ♀, Cala Millor, Mallorca (coll. GELBRECHT), reared
75c. *I. minuscularia* Ribbe, ♂, Zahara de los Atunes, Cadiz, S. Spain (ZSM), reared
75d. *I. minuscularia* Ribbe, ♂, Camino de Istan, Marbella, S. Spain (coll. SKOU), reared
76. ***Idaea ibizaria* Mentz.**, ♀, Ibiza (coll. SKOU), paratype
77a. ***Idaea albitorquata albitorquata* Püng.**, ♀, Ag. Paraskevis, Epirus, NW. Greece (ZSM)
77b. *I. albitorquata albitorquata* Püng., ♀, Ohrid, SW. Macedonia (ZSM)
77c. *I. albitorquata albitorquata* Püng., ♀, Chersonissos, E. Crete (coll. FRIEDRICH)
77d. *I. albitorquata albitorquata* Püng., ♂, Chersonissos, E. Crete (coll. FRIEDRICH)
77e. *I. albitorquata albitorquata* Püng., ♂, Rhodos island (ZSM), reared
77f. *I. albitorquata albitorquata* Püng., ♂, Assites, Ida, C. Crete (ZSM), reared
77g. *I. albitorquata albitorquata* Püng., ♂, Faliraki, Rhodos island (ZSM), reared
77h. *I. albitorquata madoniensis* Hausm., ♀, Aetna mt., E. Sicily (MNHU/PÜNGELER)
77i. *I. albitorquata madoniensis* Hausm., ♂, Madonie, N. Sicily (ZSM), paratype
77k. *I. albitorquata madoniensis* Hausm., ♂, Madonie, N. Sicily (ZSM), paratype
78a. ***Idaea carvalhoi* Herb.**, ♂, los Monegros, Huesca, N. Spain (coll. GRÜNEWALD)
78b. *I. carvalhoi* Herb., ♂, los Monegros, Huesca, N. Spain (coll. GRÜNEWALD)
78c. *I. carvalhoi* Herb., ♂, Bolnuevo, Murcia, SE. Spain (coll. SKOU)
78d. *I. carvalhoi* Herb., ♂, Sierra d'Espuña, Murcia, SE. Spain (MNHU)
78e. *I. carvalhoi* Herb., ♀, Sierra d'Espuña, Murcia, SE. Spain (MNHU/PÜNGELER)
79a. ***Idaea camparia camparia* H.-S.**, ♂, Faliraki, Rhodos island (ZSM), reared
79b. *I. camparia camparia* H.-S., ♀, Pevkos, Crete (ZSM), reared
79c. *I. camparia camparia* H.-S., ♀, Sozopol, E. Bulgaria (coll. GELBRECHT), reared
79d. *I. camparia camparia* H.-S., ♀, Liljanovo, Pirin mts., SW. Bulgaria (coll. GELBRECHT), reared
79e. *I. camparia sodaliaria* H.-S., ♀, Dubrovnik ('Gravosa'), S. Croatia (ZSM), reared
79f. *I. camparia sodaliaria* H.-S., ♂, Dubrovnik ('Gravosa'), S. Croatia (ZSM), reared
79g. *I. camparia sodaliaria* H.-S., ♀, Kuziste, S. Croatia (ZSM/HERBULOT), reared
79h. *I. camparia sodaliaria* H.-S., ♀, Dubrovnik ('Gravosa'), S. Croatia (ZSM), reared
79i. *I. camparia camparia* H.-S., ♀, Nicolosi, Aetna Mt., E. Sicily (ZSM), reared
79k. *I. camparia camparia* H.-S., ♂, Nicolosi, Aetna Mt., E. Sicily (ZSM), reared
80a. ***Idaea subsericeata* Haw.**, ♂, Las Casillas, Jaen, S. Spain (ZSM)
80b. *I. subsericeata* Haw., ♀, Cravant, Yonne, E. France (ZSM)
80c. *I. subsericeata* Haw., ♂, Tivoli, Lazio, C. Italy (ZSM)
80d. *I. subsericeata* Haw., ♀, Trecchina, S. Basilicata, S. Italy (ZSM), reared
80e. *I. subsericeata* Haw., ♀, Trecchina, S. Basilicata, S. Italy (ZSM), reared
80f. *I. subsericeata* Haw., ♀, Aritzo, Sardinia (ZSM), reared
80g. *I. subsericeata* Haw., ♂, Mercuore, Mistretta, NE. Sicily (ZSM)
80h. *I. subsericeata* Haw., ♂, Makarska, Dalmatia, S. Croatia (ZSM)
80i. *I. subsericeata* Haw., ♂, Liljanovo, Pirin mts., SW. Bulgaria (coll. GELBRECHT), reared
80k. *I. subsericeata* Haw., ♀, Liljanovo, Pirin mts., SW. Bulgaria (coll. GELBRECHT), reared
81a. ***Idaea pallidata* Den. & Schiff.**, ♂, Hinterstoder, N. Austria (ZSM)
81b. *I. pallidata* Den. & Schiff., ♀, Seega, Artem, E. Germany (coll. GELBRECHT), reared
81c. *I. pallidata* Den. & Schiff., ♀, Amata, Lettonia (ZSM)
81d. *I. pallidata* Den. & Schiff., ♀, Hals Mose, N. Jutland, Denmark (coll. SKOU)
81e. *I. pallidata* Den. & Schiff., ♂, Begovica, Pirin Mt., SW. Bulgaria (coll. GELBRECHT), reared

75a
75b
75c
75d
76
77a
77b
77c
77d
77e
77f
77g
77h
77i
77k
78a
78b
78c
78d
78e
79a
79b
79c
79d
79e
79f
79g
79h
79i
79k
80a
80b
80c
80d
80e
80f
80g
80h
80i
80k
81a
81b
81c
81d
81e

PLATE 11 (all specimens 1.5 x natural size)

82a. ***Idaea mancipiata mancipiata* Stgr.,** ♂, Candasnos, Huesca, N. Spain (coll. SKOU)
82b. *I. mancipiata mancipiata* Stgr., ♀, Candasnos, Huesca, N. Spain (coll. SKOU)
82c. *I. mancipiata repagulata* Prt., ♂, Uralsk, SE. European Russia (ZSM)
82d. *I. mancipiata repagulata* Prt., ♂, Boroshtshilovgr. Obl., Streljetskaya stel., NE. Ukraine (ZMKU)
82e. *I. mancipiata repagulata* Prt., ♀, Baku, Aserbaidjan (ZSM)
83a. ***Idaea sylvestraria* Hbn.,** ♂, Skaby, Fürstenwalde, E. Germany (ZSM), reared
83b. *I. sylvestraria* Hbn., ♂, Storkow, Brandenburg, E. Germany (coll. GELBRECHT), reared
83c. *I. sylvestraria* Hbn., ♀, Skaby, Fürstenwalde, E. Germany (coll. GELBRECHT), reared
83d. *I. sylvestraria* Hbn., ♂, Selzthal, Styria, SE. Austria (ZSM)
83e. *I. sylvestraria* Hbn., ♀, Lancashire, England (MNHU)
84a. ***Idaea dimidiata* Hufn.,** ♂, Kreuzhorst, Magdeburg, E. Germany (coll. GELBRECHT), reared
84b. *I. dimidiata* Hufn., ♂, Puisieux-en-Clanlieu, Aisne, N. France (ZSM/HERBULOT)
84c. *I. dimidiata* Hufn., ♀, Tivoli, Lazio, C. Italy (ZSM)
84d. *I. dimidiata* Hufn., ♀, Liljanovo, Pirin mts., SW. Bulgaria (coll. GELBRECHT), reared
84e. *I. dimidiata* Hufn., ♀, Litochoron, N. Greece (ZFMK/WAGENER)
85a. ***Idaea subsaturata* Gn.,** ♂, Torla; C. Pyrenees, N. Spain (NHMW)
85b. *I. subsaturata* Gn., ♂, Denia Vic., Alicante (ZSM), reared
85c. *I. subsaturata* Gn., ♂, Douelle, Lot, SW. France (ZSM/HERBULOT)
85d. *I. subsaturata* Gn., ♂, Philippeville, E. Algeria (ZSM), reared
85e. *I. subsaturata* Gn., ♂, Algier, N. Algeria (ZSM)
86a. ***Idaea trigeminata* Haw.,** ♂, Bruck, Leitha, E. Austria (ZSM)
86b. *I. trigeminata* Haw., ♀, Bruck, Leitha, E. Austria (ZSM)
86c. *I. trigeminata* Haw., ♀, Cserkut, Peks, Hungary (ZSM)
86d. *I. trigeminata* Haw., ♀, Mercuore, Mistretta, NE. Sicily (ZSM)
86e. *I. trigeminata* Haw., ♀, Kiten, E. Bulgaria (coll. GELBRECHT), reared
86f. *I. trigeminata* Haw., ♂, Liljanovo, Pirin mts., SW. Bulgaria (coll. GELBRECHT), reared
86g. *I. trigeminata* Haw., ♂, Litochoron, N. Greece (ZSM)
87a. ***Idaea biselata* Hufn.,** ♂, Triebsch, Fürstenwalde, Brandenburg, E. Germany (coll. GELBRECHT)
87b. *I. biselata* Hufn., ♀, Königs Wusterhausen, Brandenburg, E. Germany (coll. GELBRECHT)
87c. *I. biselata* Hufn., ♀, Königs Wusterhausen, Brandenburg, E. Germany (coll. GELBRECHT)
87d. *I. biselata* Hufn., ♂, Linz, N. Austria (ZSM)
87e. *I. biselata* Hufn., ♂, Cibinsgebirge, C. Romania (ZSM)
87f. *I. biselata* Hufn., ♂, Clenay, Cote d'Or, NE. France (ZSM/HERBULOT)
87g. *I. biselata* Hufn., ♀, Cervione, Corsica (ZSM)
87h. *I. biselata* Hufn., ♀, Ribes, Girona, NE. Spain (coll. SKOU)
88a. ***Idaea fractilineata fractilineata* Z.,** ♂, Madonie, N. Sicily (ZSM)
88b. *I. fractilineata fractilineata* Z., ♀, Madonie, N. Sicily (ZSM)
88c. *I. fractilineata fractilineata* Z., ♂, Madonie, N. Sicily (ZSM)
88d. *I. fractilineata subrufaria* Stgr., ♀, Tracino, Pantelleria island (ZSM)
88e. *I. fractilineata subrufaria* Stgr., ♀, Tracino, Pantelleria island (ZSM)
88f. *I. fractilineata subrufaria* Stgr., ♂, Chiclana, S. Spain (ZSM)
88g. *I. fractilineata subrufaria* Stgr., ♀, Tunis, Tunisia (ZSM)
88h. *I. fractilineata subrufaria* Stgr., ♂, Beni Kheddache, Tunisia (ZSM), reared
88i. *I. fractilineata subrufaria* Stgr., ♀, Cuenca, C. Spain (ZSM), reared
88k. *I. fractilineata subrufaria* Stgr., ♀, Algier, Algeria (ZSM/HERBULOT)
88l. ***Idaea inclinata* Led.,** ♀, Palaestina (ZSM/HERBULOT)
88m. *I. inclinata* Led., ♀, Gobustan, 60 km S Baku, Azerbajdzhan (ZMKU)
89a. ***Idaea exilaria* Gn.,** ♂, Granada, S. Spain (ZSM)
89b. *I. exilaria* Gn., ♀, Granada, S. Spain (ZSM)
89c. *I. exilaria* Gn., ♂, Molinico, Sierra Segura, SE. Spain (NHMW)
89d. *I. exilaria* Gn., ♂, Hospitalet, Tarragona, NE. Spain (NHMW/VARTIAN)
89e. *I. exilaria* Gn., ♂, Lusitania [Portugal] (NHMW)
89f. *I. exilaria* Gn., ♀, Candasnos, Huesca, NE. Spain (coll. SOMMERER)

82a
82b
82c
82d
82e
83a
83b
83c
83d
83e
84a
84b
84c
84d
84e
85a
85b
85c
85d
85e
86a
86b
86c
86d
86e
86f
86g
87a
87b
87c
87d
87e
87f
87g
87h
88a
88b
88c
88d
88e
88f
88g
88h
88i
88k
88l
88m
89a
89b
89c
89d
89e
89f

PLATE 12 (all specimens 1.5 x natural size)

90a. ***Idaea cervantaria* Mill.**, ♂, Port Bou, NE. Spain (ZSM), reared
90b. *Idaea cervantaria* Mill., ♀, Port Bou, NE. Spain (ZSM), reared
90c. *Idaea cervantaria* Mill., ♀, Llanca, Gerona, NE. Spain (ZSM)
90d. *Idaea cervantaria* Mill., ♂, Albarracin, Teruel, E. Spain (ZSM)
90e. *Idaea cervantaria* Mill., ♂, Albarracin, Teruel, E. Spain (NHMW)
90f. *Idaea cervantaria* Mill., ♂, Orgiva, Granada, S. Spain (ZSM)
90g. *Idaea cervantaria* Mill., ♀, Granada, S. Spain (ZSM), reared
90h. *Idaea cervantaria* Mill., ♂, Ronda, S. Spain (ZSM)
90i. *Idaea cervantaria* Mill., ♂, Orjiva, Las Arpujarras, S. Spain (ZSM), reared
90k. *Idaea cervantaria* Mill., ♂, Loule, Algarve, S. Portugal (ZSM), reared
91a. ***Idaea contiguaria* Hbn.**, ♂, Dürnstein, NE. Austria (ZSM), reared
91b. *I. contiguaria* Hbn., ♀, melanistic, Dürnstein, NE. Austria (ZSM), reared
91c. *I. contiguaria* Hbn., ♀, Bornich, W. Germany (ZSM)
91d. *I. contiguaria* Hbn., ♀, Val Grisenche, Val d'Aosta, NW. Italy (coll. GELBRECHT), reared
91e. *I. contiguaria* Hbn., ♀, Sierra del Montseny, Gerona, NE. Spain (coll. GELBRECHT), reared
92a. ***Idaea rupicolaria* Rssr.**, ♂, Pt. de Lobo, Sierra Nevada, S. Spain (NHMW/VARTIAN)
92b. *I. rupicolaria* Rssr., ♀, Pt. de Lobo, Sierra Nevada, S. Spain (NHMW/VARTIAN)
92c. *I. rupicolaria* Rssr., ♀, M. Mediodia, Sierra Nevada, S. Spain (NHMW), syntype
92d. *I. rupicolaria* Rssr., ♀, Cam. de Veleta, Sierra Nevada, S. Spain (coll. SKOU)
93a. ***Idaea deitanaria* Rssr. & Weis.**, ♀, Penalba, Huesca, N. Spain (coll. KRAUS)
93b. *I. deitanaria* Rssr. & Weis., ♀, Motril, Malaga, S. Spain (NHMW/VARTIAN)
93c. *I. deitanaria* Rssr. & Weis., ♂, Sierra Alhamilla, Almeria, SE. Spain (coll. GRÜNEWALD)
93d. *I. deitanaria* Rssr. & Weis., ♂, Alfajarin, Zaragoza, N. Spain (coll. GELBRECHT)
93e. *I. deitanaria* Rssr. & Weis., ♀, Los Monegros, Huesca, N. Spain (coll. GRÜNEWALD)
94a. ***Idaea saleri* Dom. & Baix.**, ♂, L'Hospitalet, Tarragona, NE. Spain (coll. HALL)
94b. *I. saleri* Dom. & Baix., ♂, L'Hospitalet, Tarragona, NE. Spain (coll. HALL)
94c. *I. saleri* Dom. & Baix., ♀, L'Hospitalet, Tarragona, NE. Spain (coll. HALL)
94d. *I. saleri* Dom. & Baix., ♀, L'Hospitalet, Tarragona, NE. Spain (coll. HALL)
94e. *I. saleri* Dom. & Baix., ♀, Huercal Overa, Almeria, SE. Spain (coll. GRÜNEWALD)
94f. *I. saleri* Dom. & Baix., ♀, El Pozo del Esparto Almeria, SE. Spain (coll. SKOU)
95a. ***Idaea tineata* Th.-Mieg.**, ♂, Faliraki, Rhodos island (ZSM)
95b. *I. tineata* Th.-Mieg., ♀, Samos island (NHMW)
95c. *I. tineata* Th.-Mieg., ♂, Nazilli, Nascöy, W. Turkey (ZSM)
95d. *I. tineata* Th.-Mieg., ♀, 50 km W Konya, S. Turkey (ZSM)
95e. *I. tineata* Th.-Mieg., ♀, 50 km W Konya, S. Turkey (ZSM)
95f. *I. tineata* Th.-Mieg., ♀, 50 km W Konya, S. Turkey (ZSM)
96a. ***Idaea infirmaria* Rmb.**, ♀, Mercuore, Mistretta, NE Sicily (ZSM)
96b. *I. infirmaria* Rmb., ♂, St.-Étienne, Vallée Fr., S. France (ZSM/HERBULOT)
96c. *I. infirmaria* Rmb., ♀, Noli, Liguria, NW. Italy (ZSM)
96d. *I. infirmaria* Rmb., ♂, Torre la Higuera, Spain (coll. SKOU)
96e. *I. infirmaria* Rmb., ♀, Alpes maritimes, SE. France (ZSM), reared
96f. *I. infirmaria* Rmb., ♂, Cuenca, C. Spain (ZSM)
96g. *I. infirmaria* Rmb., ♀, Albarracin, Teruel, E. Spain (coll. KRAUS)
96h. *I. infirmaria* Rmb., ♂, Veleta, Sierra Nevada, S. Spain (coll. BAISCH)
96i. *I. infirmaria* Rmb., ♂, Agia Paraskevi, Smolikas, N. Greece (coll. KRAUS)
96k. *I. infirmaria* Rmb., ♂, Agia Paraskevi, Smolikas, N. Greece (coll. KRAUS)
96l. *I. infirmaria* Rmb., ♀, Ta Pigadia, Karpathos island, Greece (ZSM)
96m. *I. infirmaria* Rmb., ♂, Cuenca, C. Spain (NHMW/VARTIAN)
97a. ***Idaea rhodogrammaria* Püng.**, ♂, Noguera, Teruel, E. Spain (NHMW)
97b. *I. rhodogrammaria* Püng., ♂, Sierra del Tremedal, Teruel, E. Spain (ZFMK)
97c. *I. rhodogrammaria* Püng., ♀, Covarrubias, Burgos, N. Spain (ZSM/HERBULOT)
97d. *I. rhodogrammaria* Püng., ♀, Sta. Elena, Jaen, S. Spain (ZSM/HERBULOT)
97e. *I. rhodogrammaria* Püng., ♂, Siniscola, Sardinia (coll. FRIEDRICH)
97f *I. rhodogrammaria* Püng., ♂, Sierra d'Espuña, SE. Spain (MNHU/PÜNGELER), 'cotype'

90a
90b
90c
90d
90e
90f
90g
90h
90i
90k
91a
91b
91c
91d
91e
92a
92b
92c
92d
93a
93b
93c
93d
93e
94a
94b
94c
94d
94e
94f
95a
95b
95c
95d
95e
95f
96a
96b
96c
96d
96e
96f
96g
96h
96i
96k
96l
96m
97a
97b
97c
97d
97e
97f

PLATE 13 (all specimens 1.5 x natural size)

98a. ***Idaea ostrinaria* Hbn.,** ♂, Dubrovnik ('Gravosa'), Dalmatia, S. Croatia (ZSM), reared
98b. *I. ostrinaria* Hbn., ♂, Albena, E. Bulgaria (coll. Gelbrecht), reared
98c. *I. ostrinaria* Hbn., ♀, Liljanovo, Pirin mts., SW. Bulgaria (coll. Gelbrecht), reared
98d. *I. ostrinaria* Hbn., ♂, Faliraki, Rhodos island (ZSM), reared
98e. *I. ostrinaria* Hbn., ♂, Domusnovas, Sardinia (ZSM/EMEM)
99a. ***Idaea metohiensis* Reb.,** ♂, Trebevic, Bosnia (ZFMK/Wagener)
99b. *I. metohiensis* Reb., ♂, Kataphygion, Olympos, N. Greece (ZSM)
99c. *I. metohiensis* Reb., ♀, Kataphygion, Olympos, N. Greece (ZSM)
99d. *I. metohiensis* Reb., ♀, Punat Hrusta, Krk island, W. Croatia (ZSM)
100a. ***Idaea eugeniata* Mill.,** ♀, Puig Ventos, Gerona, NE. Spain (ZSM), reared
100b. *I. eugeniata* Mill., ♂, Serra de Estrela, C. Portugal (ZSM), reared
100c. *I. eugeniata* Mill., ♀, St. Pons, Hérault, S. France (ZSM/Herbulot), reared
100d. *I. eugeniata* Mill., ♂, Philippeville, E. Algeria (ZSM), reared
100e. *I. eugeniata* Mill., ♂, Philippeville, E. Algeria (ZSM), reared
101a. ***Idaea distinctaria* Bsd.,** ♂, Nice, S. France (ZSM/Herbulot)
101b. *I. distinctaria* Bsd., ♂, Orebic, Dalmatia, S. Croatia (ZSM), reared
101c. *I. distinctaria* Bsd., ♀, Dubrovnik ('Gravosa'), Dalmatia, S. Croatia (ZSM)
101d. *I. distinctaria* Bsd., ♂, Dubrovnik ('Gravosa'), Dalmatia, S. Croatia (ZSM), reared
101e. *I. distinctaria* Bsd., ♂, Vuria, Lamia, C. Greece (ZSM)
101f. *I. distinctaria* Bsd., ♀, Malia, Stalis, N. Crete (coll. Sommerer)
101g. *I. distinctaria* Bsd., ♀, Agios Nikolaos, E. Crete (ZSM), reared
101h. *I. distinctaria* Bsd., ♀, Mt. Smith, Rhodos island (ZSM), reared
101i. *I. distinctaria* Bsd., ♂, Faliraki, Rhodos island (ZSM)
101k. *I. distinctaria* Bsd., ♀, melanistic, Tuz Gölü, C. Turkey (ZSM), reared
102a. ***Idaea predotaria* Hartig,** ♂, St. Raphael, Var, S. France (ZSM/Herbulot)
102b. *I. predotaria* Hartig, ♂, Cala Millor, Mallorca, Balearics (ZSM), reared
102c. *I. predotaria* Hartig, ♀, Paguera, Mallorca, Balearics (ZSM)
102d. *I. predotaria* Hartig, ♂, St. Carles de la Rapita, Tarragona, NE. Spain (coll. Skou)
102e. *I. predotaria* Hartig, ♀, Philippeville, E. Algeria (ZSM)
103a. ***Idaea nitidata* H.-S.,** ♂, Unterwart, Styria, SE. Austria (ZSM)
103b. *I. nitidata* H.-S., ♂, Unterwart, Styria, SE. Austria (ZSM)
103c. *I. nitidata* H.-S., ♀, Unterwart, Styria, SE. Austria (ZSM)
103d. *I. nitidata* H.-S., ♂, Sejmonowsk, C. Urals, E. European Russia (ZSM)
104a. ***Idaea emarginata* L.,** ♂, Mörfelden, Hessen, W. Germany (ZSM/EMEM)
104b. *I. emarginata* L., ♂, Wallersberg, N. Bavaria, C. Germany (ZSM)
104c. *I. emarginata* L., ♂, Linz, N. Austria (ZSM)
104d. *I. emarginata* L., ♀, Linz, N. Austria (ZSM)
104e. *I. emarginata* L., ♀, Lobau, Vienna, E. Austria (ZSM)
104f. *I. emarginata* L., ♀, Vienna, E. Austria (ZSM)
104g. *I. emarginata* L., ♀, Puisieux-en-Clanlieu, Aisne, N. France (ZSM/Herbulot)
105. ***Idaea dromikos* Hausm. sp.n.,** ♂, Vernet les Bains, S. France (ZSM/EMEM), paratype
106a. ***Idaea simplicior* Prt.,** ♂, St. Ildefonso, C. Spain (ZSM)
106b. *I. simplicior* Prt., ♂, Hoyos del Espino, Sierra de Gredos, W. Spain (coll. Skou)
106c. *I. simplicior* Prt., ♀, Hoyos del Espino, Sierra de Gredos, W. Spain (coll. Skou)
106d. *I. simplicior* Prt., ♂, Ifrane, Middle Atlas, C. Morocco (ZSM)

98d
98e
99a
99b
99c
99d
100a
100b
100c
100d
100e
101a
101b
101c
101d
101e
101f
101g
101h
101i
101k
102a
102b
102c
102d
102e
103a
103b
103c
103d
104a
104b
104c
104d
104e
104f
104g
105
106a
106b
106c
106d

PLATE 14 (all specimens natural size)

107a. ***Idaea rubraria* Stgr.**, ♂, Ürzig, Mosel, W. Germany (coll. GELBRECHT), reared
107b. *I. rubraria* Stgr., ♀, Ürzig, Mosel, W. Germany (coll. GELBRECHT), reared
107c. *I. rubraria* Stgr., ♀, Dürnstein, NE. Austria (ZSM), reared
107d. *I. rubraria* Stgr., ♂, Dürnstein, NE. Austria (ZSM), reared
107e. *I. rubraria* Stgr., ♂, Kranevo, Varna, E. Bulgaria (coll. GELBRECHT), reared
107f. *I. rubraria* Stgr., ♂, Cuenca, C. Spain (coll. MÜLLER)
108a. ***Idaea aversata* L.**, ♂, Dürnstein, NE. Austria (ZSM), reared
108b. *I. aversata* L., ♀, Prebichl, Styria, SE. Austria (ZSM)
108c. *I. aversata* L., ♀, Mörnsheim, C. Bavaria, S. Germany (coll. KELLER)
108d. *I. aversata* L., ♀, Dürnstein, NE. Austria (ZSM), reared
108e. *I. aversata* L., ♂, Zuneggen, Valesia, S. Switzerland (ZSM)
108f. *I. aversata* L., ♀, Amlach, E. Tyrolia, S. Austria (ZSM), reared
108g. *I. aversata* L., ♀, Ebernburg, Pfalz, W. Germany (ZSM/EMEM)
108h. *I. aversata* L., ♀, Sattendorf, Carinthia, S. Austria (ZSM), reared
108i. *I. aversata* L., ♀, Cibins Mts., C. Romania (ZSM)
108k. *I. aversata* L., ♀, melanistic, Vienna, E. Austria (NHMW)
109a. ***Idaea degeneraria* Hbn.**, ♂, Agay, Esterel Mts., S. France (ZSM)
109b. *I. degeneraria* Hbn., ♀, Wippach, Carniolia, N. Slovenja (ZSM), reared
109c. *I. degeneraria* Hbn., ♂, Garda, N. Italy (ZSM)
109d. *I. degeneraria* Hbn., ♀, Terlan, S. Tirol, N. Italy (ZSM)
109e. *I. degeneraria* Hbn., ♂, Paguera, Mallorca, Balearics (ZSM), reared
109f. *I. degeneraria* Hbn., ♀, Cala Millor, Mallorca, Balearics (coll. GELBRECHT), reared
109g. *I. degeneraria* Hbn., ♀, Sta. Teresa, Sardinia (ZSM), reared
109h. *I. degeneraria* Hbn., ♀, Bastia, Corsica (ZSM), reared, syntype
109i. *I. degeneraria* Hbn., ♀, Evisa, Corsica (ZSM/EMEM)
109k. *I. degeneraria* Hbn., ♀, Asco valley, Corsica (ZSM)
109l. *I. degeneraria* Hbn., ♀, Assistes, Crete (ZSM), reared
109m. *I. degeneraria erschoffi* Chr., ♀, Pyatigorsk, N. Caucasus, SW. Russia (ZSM)
110a. ***Idaea straminata* Bkh.**, ♀, Göhrde, Lüneburg, N. Germany (ZSM), reared
110b. *I. straminata* Bkh., ♂, Linz, N. Austria (ZSM)
110c. *I. straminata* Bkh., ♀, Spital, N. Austria (ZSM)
110d. *I. straminata* Bkh., ♂, Dürnstein, NE. Austria (ZSM)
110e. *I. straminata* Bkh., ♂, Amfreville, Eure, NW. France (ZSM/HERBULOT)
110f. *I. straminata* Bkh., ♀, Mte. Sirente, Abruzzi, C. Italy (ZSM)
110g. *I. straminata* Bkh., ♀, Aritzo, Sardinia (ZSM)
110h. *I. straminata* Bkh., ♀, Rovinj, Istria, W. Croatia (ZSM)
110i. *I. straminata* Bkh., ♀, Cibins Mts., C. Romania (ZSM)
110k. *I. straminata* Bkh., ♀, Algez, Malaga, S. Spain (NHMW/VARTIAN)
111a. ***Idaea deversaria deversaria* H.-S.**, ♂, Göhrde, Lüneburg, N. Germany (ZSM)
111b. *I. deversaria deversaria* H.-S., ♂, Peuestorf, Niedersachsen, N. Germany (ZSM)
111c. *I. deversaria deversaria* H.-S., ♂, Sigmundskron, S. Tirol, N. Italy (ZSM)
111d. *I. deversaria deversaria* H.-S., ♀, Tivoli, Lazio, C. Italy (ZSM), reared
111e. *I. deversaria deversaria* H.-S., ♂, Suh[aja] Konka; Zaporozhskaja Obl., Ukraine (ZMKU)
111f. *I. deversaria deversaria* H.-S., ♂, Stari Dorjan, SE. Macedonia (ZSM)
111g. *I. deversaria deversaria* H.-S., ♀, Liljanovo, Pirin mts., SW. Bulgaria (ZSM/EMEM)
111h. *I. deversaria deversaria* H.-S., ♀, Zuneggen, Valesia, S. Switzerland (ZSM)
111i. *I. deversaria deversaria* H.-S., ♂, Gèdre, C. Pyrenees, S. France (ZSM/HERBULOT)
111k. *I. deversaria fallax* Hausm., ♂, Bronchales, Teruel, E. Spain (ZSM/HERBULOT), holotype
111l. *I. deversaria fallax* Hausm., ♀, St. Tropez, Var, SE. France (ZSM/HERBULOT)
112a. ***Idaea squalidaria* Stgr.**, ♂, Pyrenees, N. Spain (NHMW)
112b. *I. squalidaria* Stgr., ♀, V. de Ordesa, Pyrenees, NE. Spain (MNHU)
112c. *I. squalidaria* Stgr., ♂, Font de Cerdana, Pyrenees, NE. Spain (coll. SKOU)
112d. *I. squalidaria* Stgr., ♀, E. Pyrenees, NE. Spain (MNHU), syntype
112e. *I. squalidaria* Stgr., ♀, Font de Cerdana, Pyrenees, NE. Spain (coll. SKOU)
112f. *I. squalidaria* Stgr., ♀, Lago S. Mauricio, Pyrenees, NE. Spain (coll. SKOU)
113a. ***Brachyglossina hispanaria* Püng.**, ♂, Cam. d. Veleta, Sierra Nevada, S. Spain (coll. SKOU)
113b. *B. hispanaria* Püng., ♂, Estepona, Malaga, S. Spain (coll. SKOU)
113c. *B. hispanaria* Püng., ♂, Cam. d. Veleta, Sierra Nevada, S. Spain (coll. SKOU)
113d. *B. hispanaria* Püng., ♀, El Mirandor, Marbella, Malaga, S. Spain (ZMUC)
113e. *B. hispanaria* Püng., ♀, Camino de Ojen, Marbella, Malaga, S. Spain (coll. SKOU)
113f. *B. hispanaria* Püng., ♀, Sierra d'Espuña, Murcia, SE. Spain (MNHU), syntype
114a. ***Limeria macraria* Stgr.**, ♂, locality not stated [S. Urals] (MNHU/PÜNGELER)
114b. *L. macraria* Stgr., ♂, Karaganda, Kazakhstan (ZSM)
114c. *L. macraria* Stgr., ♂, Amurskii vill., Cheljabinsk Obl., S. Urals, SE. European Russia (coll. NUPPONEN)
114d. *L. macraria* Stgr., ♀, Amurskii vill., Cheljabinsk Obl., S. Urals, SE. European Russia (coll. NUPPONEN)
114e. *L. macraria* Stgr., ♀, locality not stated (ZISP)
114f. *L. macraria* Stgr., ♀, Aschchabad, Turkmeniya (NHMW)

107a
107b
107c
107d
107e
107f
108a
108b
108c
108d
108e
108f
108g
108h
108i
108k
109a
109b
109c
109d
109e
109f
09g
109h
109i
109k
109l
109m
110a
110b
110c
110d
110e
110f
110g
110h
110i
110k
111a
111b
111c
111d
111e
111f
111g
111h
111i
111k
111l
12a
112b
112c
112d
112e
112f
13a
113b
113c
113d
113e
113f
a
114b
114c
114d
114e
114f

PLATE 15 (all specimens natural size)

115a. ***Oar reaumuraria* Mill.**, ♂, Alicante, SE. Spain (ZSM)
115b. *O. reaumuraria* Mill., ♂, Tabernas, Almeria, SE. Spain (coll. SKOU)
115c. *O. reaumuraria* Mill., ♀, Baza-Benamaurel, Granada, S. Spain (coll. SKOU)
115d. ***Oar pratana* F.**, ♂, Biskra, E. Algeria (ZSM)
115e. *O. pratana* F., ♂, Tunis, Tunisia (coll. SKOU)
115f. *O. pratana* F., ♀, Biskra, E. Algeria (ZSM/Herbulot)
115g. *O. pratana* F., ♂, Ed Damer, Hudeiba, Sudan (ZSM)
115h. *O. pratana baezi* Hausm. subsp.n., ♂, Agadir, SW. Morocco (ZSM)
116a. ***Cinglis humifusaria* Ev.**, ♂, locality not stated [European Russia] (ZISP)
116b. *C. humifusaria* Ev., ♀, Sarepta, S. European Russia (MNHU)
116c. *C. humifusaria* Ev., ♂, Ai Deze, Turkmeniya (coll. SKOU)
117a. ***Cinglis andalusiaria* Wagn.**, ♂, Los Monegros, Huesca, N. Spain (ZSM)
117b. *C. andalusiaria* Wagn., ♀, Fayon, Ebro, Aragon, NE. Spain (ZSM)
117c. *C. andalusiaria* Wagn., ♂, Tosos, Zaragoza, N. Spain (coll. SKOU)
117d. *C. andalusiaria* Wagn., ♀, Candasnos, Huesca, N. Spain (coll. SKOU)
118a. ***Scopula immorata immorata* L.**, ♂, Linz, N. Austria (ZSM)
118b. *S. immorata immorata* L., ♀, Reval, Estonia (ZSM)
118. *S. immorata immorata* L., ♂, La Bessée, Ht. Alpes, SE. France (ZSM/HERBULOT)
118d. *S. immorata immorata* L., ♂, Dnepropetrovskaja Andreevka, Ukraine (ZSM)
118e. *S. immorata immorata* L., ♂, Sojmonowsk, C. Urals, E. European Russia (MNHU)
118f. *S. immorata duercki* Shel., ♂, Hoyos d'Espino, Sierra Gredos, C. Spain (ZSM/HERBULOT)
118g. *S. immorata riloensis* Züll., ♀, Begovica, Pirin, SW. Bulgaria (coll. GELBRECHT), reared
118h. *S. immorata riloensis* Züll., ♂, Begovica, Pirin, SW. Bulgaria (coll. GELBRECHT)
118i. *S. immorata riloensis* Züll., ♂, Ilgaz Dagl. Pontus, N. Turkey (coll. GELBRECHT)
118k. *S. immorata riloensis* Züll., ♀, Pinarlar Tamdere, Giresun, N. Turkey (coll. BAISCH)
119a. ***Scopula tessellaria* Bsd.**, ♂, Velosne, Meuse, N. France (ZSM/HERBULOT)
119b. *S. tessellaria* Bsd., ♂, la Palud sur Verdon, Provence, SE. France (coll. GELBRECHT)
119c. *S. tessellaria* Bsd., ♂, Forêt de Benon, Charente inf., W. France (ZSM/HERBULOT)
119d. *S. tessellaria* Bsd., ♀, Forêt de Benon, Charente inf., W. France (ZSM/HERBULOT)
119e. *S. tessellaria* Bsd., ♂, Salso Maggiore, Parma, NW. Italy (ZSM)
119f. *S. tessellaria* Bsd., ♀, Salso Maggiore, Parma, NW. Italy (ZSM)
119g. *S. tessellaria* Bsd., ♀, Josefsthal, Croatia (ZSM)
119h. *S. tessellaria* Bsd., ♂, Guberla, S. Urals, SE. European Russia (ZSM)
119i. *S. tessellaria* Bsd., ♀, Guberla, S. Urals, SE. European Russia (ZSM)
119k. *S. tessellaria* Bsd., ♂, Berezovka, Pavlodar, NE. Kazakhstan (ZSM)
120a. ***Scopula corrivalaria* Kretsch.**, ♂, by Berlin, E. Germany (ZSM), reared
120b. *S. corrivalaria* Kretsch., ♂, Bindow, Königs Wusterhausen, E. Germany (coll. GELBRECHT), reared
120c. *S. corrivalaria* Kretsch., ♂, Berlin, E. Germany (ZSM), reared
120d. *S. corrivalaria* Kretsch., ♀, Spandau, Berlin, E. Germany (ZSM), reared
120e. *S. corrivalaria* Kretsch., ♂, Amuré, Deux Sèvres, W. France (ZSM/HERBULOT)
120f. *S. corrivalaria* Kretsch., ♂, Amuré, Deux Sèvres, W. France (ZSM/HERBULOT)
121a. ***Scopula caricaria* Reut.**, ♂, Graben-Ndf., Baden, SW. Germany (ZSM)
121b. *S. caricaria* Reut., ♂, Inzenhof, Burgenland, SE. Austria (ZSM)
121c. *S. caricaria* Reut., ♂, Inzenhof, Burgenland, SE. Austria (ZSM)
121d. *S. caricaria* Reut., ♂, Montbonnot, Isère, E. France (ZSM/HERBULOT)
121e. *S. caricaria* Reut., ♀, Bozen, S. Tirol, N. Italy (ZSM)
121f. *S. caricaria* Reut., ♂, Lipik, Slavonia, E. Croatia (ZSM)
122a. ***Scopula nemoraria* Hbn.**, ♂, Anklamer Stadtluch, NE. Germany (coll. GELBRECHT), reared
122b. *S. nemoraria* Hbn., ♀, Anklamer Stadtbruch, NE. Germany (coll. GELBRECHT), reared
122c. *S. nemoraria* Hbn., ♂, Liegnitz, Silesia, S. Poland (ZSM)
122d. *S. nemoraria* Hbn., ♂, Mooswald, Freiburg i.Br., SW. Germany (ZSM/EMEM), reared
122e. *S. nemoraria* Hbn., ♀, Sojmonowsk, C. Urals, E. European Russia (MNHU)
123a. ***Scopula umbelaria* Hbn.**, ♂, Gleistal, Thuringia, C. Germany (coll. GELBRECHT), reared
123b. *S. umbelaria* Hbn., ♀, Gleistal, Thuringia, C. Germany (coll. GELBRECHT), reared
123c. *S. umbelaria* Hbn., ♂, Hinterstoder, Oberösterreich, C. Austria (ZSM)
123d. *S. umbelaria* Hbn., ♀, Hinterstoder, Oberösterreich, C. Austria (ZSM)
123e. *S. umbelaria* Hbn., ♀, Stainz, Styria, SE. Austria (ZSM)
124a. ***Scopula nigropunctata* Hufn.**, ♂, Göhrde, Lüneburg, N. Germany (ZSM)
124b. *S. nigropunctata* Hufn., ♀, Dannenreich, Königs Wusterhausen, E. Germany (coll. GELBRECHT)
124c. *S. nigropunctata* Hufn., ♂, Rehfelde, Strausberg, E. Germany (coll. GELBRECHT)
124d. *S. nigropunctata* Hufn., ♀, Göhrde, Lüneburg, N. Germany (ZSM)
124e. *S. nigropunctata* Hufn., ♂, Schopfheim, S. Baden, SW. Germany (ZSM)
124f. *S. nigropunctata* Hufn., ♂, Sigmundskron, S. Tirol, N. Italy (ZSM)
124g. *S. nigropunctata* Hufn., ♀, Laon, N. France (ZSM)
124h. *S. nigropunctata* Hufn., ♂, Kiten, E. Bulgaria (coll. GELBRECHT)
124i. *S. nigropunctata* Hufn., ♂, Picos de Europa, C. Pyrenees, N. Spain (ZSM)
124k. *S. nigropunctata* Hufn., ♀, Picos de Europa, C. Pyrenees, N. Spain (ZFMK)

116a
116b
116c
117a
117b
117c
117d
118a
118b
118c
118d
118e
118f
118g
118h
118i
118k
119a
119b
119c
119d
119e
119f
119g
119h
119i
119k
120a
120b
120c
120d
120e
120f
121a
121b
121c
121d
121e
121f
122a
122b
122c
122d
122e
123a
123b
123c
123d
123e
124a
124b
124c
124d
124e
124f
124g
124h
124i
124k

PLATE 16 (all specimens natural size)

125a. ***Scopula virgulata virgulata* Den. & Schiff.**, ♂, Töpchin, Königs Wusterhausen, E. Germany (coll. GELBRECHT)
125b. *S. virgulata virgulata* Den. & Schiff., ♀, Kleinwall, Grünheide, E. Germany (coll. GELBRECHT), reared
125c. *S. virgulata virgulata* Den. & Schiff., ♂, Wallgau, S. Bavaria, S. Germany (ZSM)
125d. *S. virgulata virgulata* Den. & Schiff., ♂, Bialowieza, E. Poland (coll. SKOU)
125e. *S. virgulata virgulata* Den. & Schiff., ♀, Budaörs, Budapest, N. Hungary (coll. GELBRECHT), reared
125f. *S. virgulata virgulata* Den. & Schiff., ♂, Leuk, Valesia, S. Switzerland (ZSM)
125g. *S. virgulata virgulata* Den. & Schiff., ♂, Sojmonowsk, C. Urals, E. European Russia (MNHU)
125h. *S. virgulata virgulata* Den. & Schiff., ♀, Sojmonowsk, C. Urals, E. European Russia (MNHU)
125i. *S. virgulata virgulata* Den. & Schiff., ♂, Guberla, S. Urals, SE. European Russia (ZSM)
125k. *S. virgulata virgulata* Den. & Schiff., ♀, Guberla, S. Urals, SE. European Russia (ZSM)
125l. *S. virgulata rossica* Djak., ♂, by St. Petersburg, NW. European Russia (ZISP)
125m. *S. virgulata rossica* Djak., ♀, by St. Petersburg, NW. European Russia (ZISP)
126a. ***Scopula ornata* Scop.**, ♂, Hörschlag, N. Austria (ZSM)
126b. *S. ornata* Scop., ♂, Seega, Thuringia, C. Germany (ZSM)
126c. *S. ornata* Scop., ♂, Wallersberg, N. Bavaria, C. Germany (ZSM)
126d. *S. ornata* Scop., ♂, Digne, Provence, SE. France (coll. GELBRECHT)
126e. *S. ornata* Scop., ♂, Stilfser Joch, S. Tirol, N. Italy (ZSM)
126f. *S. ornata* Scop., ♀, Praia da Rocha, Algarve, S. Portugal (ZMUC)
126g. *S. ornata* Scop., ♀, Capo Comino, Sardinia (coll. GRÜNEWALD)
126h. *S. ornata* Scop., ♂, Dubrovnik ('Gravosa'), S. Croatia (ZSM)
126i. *S. ornata* Scop., ♂, Tolbuchin, Batovo, E. Bulgaria (coll. GELBRECHT)
126k. *S. ornata* Scop., ♀, Malia, Neapolis, Crete (NHMW/VARTIAN)
126l. *S. ornata* Scop., ♂, Walouiki, SW. European Russia (NHMW)
126m. *S. ornata* Scop., ♂, Sarepta, S. European Russia (MNHU)
127a. ***Scopula concinnaria concinnaria* Dup.**, ♀, Cuenca, C. Spain (ZSM)
127b. *S. concinnaria concinnaria* Dup., ♂, Rt. Veleta, Sierra Nevada, S. Spain (ZSM/HERBULOT)
127c. *S. concinnaria universaria* Zerny, ♀, Albarracin, Teruel, E. Spain (ZSM)
127d. *S. concinnaria universaria* Zerny, ♂, Albarracin, Teruel, E. Spain (ZSM)
127e. *S. concinnaria universaria* Zerny, ♂, Albarracin, Teruel, E. Spain (ZSM/EMEM)
128a. ***Scopula orientalis* Alph.**, ♂, Drenovo, Kavedar, C. Macedonia (ZSM)
128b. *S. orientalis* Alph., ♂, Uchisar, Nevshehir, C. Turkey (coll. GELBRECHT)
128c. *S. orientalis* Alph., ♂, Göreme, Nevshehir, C. Turkey (coll. GELBRECHT)
128d. *S. orientalis* Alph., ♂, Khosrov reserve, S. Armenia (ZSM)
128e. *S. orientalis* Alph., ♀, Sevastopol, Crimea, S. Ukraine (ZISP)
129a. ***Scopula decorata decorata* Den. & Schiff.**, ♀, S. Tirol, N. Italy (ZSM)
129b. *S. decorata decorata* Den. & Schiff., ♂, Naturns, S. Tirol, N. Italy (ZSM)
129c. *S. decorata decorata* Den. & Schiff., ♂, Digne, Provence, SE. France (ZSM)
128d. *S. decorata decorata* Den. & Schiff., ♂, Albiotusch Mt., Bulgaria (NHMW)
129e. *S. decorata decorata* Den. & Schiff., ♂, Liljanovo, Pirin mts., SW. Bulgaria (coll. GELBRECHT)
129f. *S. decorata decorata* Den. & Schiff., ♀, Matka, Treska gorge, N. Macedonia (ZSM)
129g. *S. decorata decorata* Den. & Schiff., ♂, Uralsk, SE. European Russia (MNHU/PÜNGELER)
129h. *S. decorata decorata* Den. & Schiff., ♂, Sem. Kolodesej, Crimea, S. Ukraine (ZSM)
129i. *S. decorata decorata* Den. & Schiff., ♀, Campagna Romana, Lazio, C. Italy (ZSM)
129k. *S. decorata decorata* Den. & Schiff., ♀, Mtgna. Grande, Abruzzi, C. Italy (ZSM)
129l. *S. decorata decorata* Den. & Schiff. (trans. *violata*), ♀, Deutsch-Altenburg, E. Austria (ZSM)
129m. *S. decorata decorata* Den. & Schiff. (trans. *violata*), ♀, Burg, Hoyerswerda, E. Germany (coll. GELBRECHT), reared
129n. *S. decorata decorata* Den. & Schiff. (trans. *violata*), ♀, Pätz, Königs Wusterhausen, E. Germany (coll. GELBRECHT)
129o. *S. decorata decorata* Den. & Schiff. (trans. *violata*), ♂, Pätz, Königs Wusterhausen, E. Germany (coll. GELBRECHT)
129p. *S. decorata violata* Thunb., ♂, Estonia (ZSM)
129q. *S. decorata armeniaca* Th.-Mieg, ♀, Tacht i Suleiman, C. Elburs, N. Iran (ZSM)
129r. *S. decorata congruata* Z., ♂, Albarracin, Teruel, E. Spain (ZSM)
129s. *S. decorata congruata* Z., ♀, Diezma, Granada, S. Spain (ZSM)
129t. *S. decorata congruata* Z., ♂, Sierra de Alfacar, Granada, S. Spain (ZSM)
129u. *S. decorata congruata* Z., ♀, Tizi'n'Test, High Atlas, C. Morocco (ZSM/HERBULOT)
129v. *S. decorata congruata* Z., ♀, Madonie, N. Sicily (ZSM)
129w. *S. decorata congruata* Z., ♀, Mt. Cuccio, Palermo, NW. Sicily (NHMW)
129x. *S. decorata congruata* Z., ♂, Mt. Cuccio, Palermo, NW. Sicily (NHMW)
130a. ***Scopula honestata* Mab.**, ♂, Pt. Balistreri, Mt. Limbara, Sardinia (coll. SKOU)
130b. *S. honestata* Mab., ♂, Col de Vizzavona, C. Corsica (ZSM/HERBULOT)
130c. *S. honestata* Mab., ♂, Col de Vizzavona, C. Corsica (ZSM/HERBULOT)
130d. *S. honestata* Mab., ♀, Col de Vizzavona, C. Corsica (ZSM/HERBULOT)
130e. *S. honestata* Mab., ♀, Col de Vizzavona, C. Corsica (ZSM/HERBULOT)
130f. *S. honestata* Mab., ♀, Col de Vizzavona, C. Corsica (ZSM/HERBULOT)
131a. ***Scopula subtilata* Chr.**, ♂, Guberli, S. Urals, SE. European Russia (ZISP)
131b. *S. subtilata* Chr., ♂, Sarepta, S. European Russia (ZSM)
131c. *S. subtilata* Chr., ♂, Sarepta, S. European Russia (ZSM)
131d. *S. subtilata* Chr., ♂, Sarepta, S. European Russia (ZSM)
131e. *S. subtilata* Chr., ♀, Guberli, S. Urals, SE. European Russia (ZISP)

125g 125h 125i 125k 125l 125m
126a 126b 126c 126d 126e 126f
126g 126h 126i 126k 126l 126m
127a 127b 127c 127d 127e
128a 128b 128c 128d 128e
129a 129b 129c 129d 129e
129f 129g 129h 129i 129k 129l
29m 129n 129o 129p 129q 129r
29s 129t 129u 129v 129w 129x
30a 130b 130c 130d 130e 130f
131a 131b 131c 131d 131e

PLATE 17 (all specimens natural size)

132a. ***Scopula submutata submutata* Tr.**, ♂, Dubrovnik ('Gravosa'), S. Croatia (ZSM)
132b. *S. submutata submutata* Tr., ♀, Matka, Treska gorge, N. Macedonia (ZSM)
132c. *S. submutata submutata* Tr., ♂, Drenovo, Kavadar, C. Macedonia (NHMW/VARTIAN)
132d. *S. submutata submutata* Tr., ♀, Drenovo, Kavadar, C. Macedonia (ZSM)
132e. *S. submutata submutata* Tr., ♂, Drenovo, Kavadar, C. Macedonia (ZSM)
132f. *S. submutata submutata* Tr., ♂, Nauplia, Peloponnes, S. Greece (coll. FRIEDRICH)
132g. *S. submutata submutata* Tr., ♀, Latsch, S. Tirol, N. Italy (ZSM)
132h. *S. submutata submutata* Tr., ♂, Torbole, Lake Garda, N. Italy (ZSM)
132i. *S. submutata submutata* Tr., ♀, Trecchina, S. Basilicata, S. Italy (ZSM)
132k. *S. submutata submutata* Tr., ♀, Quinson, Alpes de Ht. Provence, SE. France (ZSM)
132l. *S. submutata roseonitens* Wgn., ♀, Digne, Provence, SE. France (ZSM)
132m. *S. submutata roseonitens* Wgn., ♂, Digne, Provence, SE. France (ZSM)
132n. *S. submutata gedrensis* Hausm., ♂, Gèdre, C. Pyrenees, S. France (ZSM), holotype
132o. *S. submutata nivellearia* Obth., ♂, Albarracin, Teruel, E. Spain (ZSM)
132p. *S. submutata nivellearia* Obth., ♂, Villamanrique, S. Spain (ZSM)
132q. *S. submutata nivellearia* Obth., ♂, Badajos, Extremadura, SW. Spain (coll. SKOU)
132r. *S. submutata nivellearia* Obth., ♂, Orgiva, Las Alpujarras, S. Spain (ZSM)
132s. *S. submutata nivellearia* Obth., ♀, Vilamoura, Algarve, S. Portugal (ZSM)
133a. ***Scopula vigilata* Wgn.**, ♂, Tivoli, Lazio, C. Italy (ZSM)
133b. *S. vigilata* Wgn., ♂, Tivoli, Lazio, C. Italy (ZSM)
133c. *S. vigilata* Wgn., ♂, Tivoli, Lazio, C. Italy (ZSM)
133d. *S. vigilata* Wgn., ♀, Trecchina, S. Basilicata, S. Italy (ZSM)
133e. *S. vigilata* Wgn., ♂, Mellieka, Malta (ZMUC)
133f. *S. vigilata* Wgn., ♂, Kalavrita, N. Peloponnes, S. Greece (coll. FRIEDRICH)
133g. *S. vigilata* Wgn., ♀, Messinia, Ag. Dimitrios, Peloponnes, S. Greece (SMNS)
134a. ***Scopula rubiginata* Hufn.**, ♂, Schneuerdingen, Lüneburg, N. Germany (ZSM)
134b. *S. rubiginata* Hufn., ♀, Niederlehme, Mark Brandenburg, E. Germany (coll. GELBRECHT), reared
134c. *S. rubiginata* Hufn., ♂, Echiré, Deux-Sèvres, W. France (ZSM/HERBULOT)
134d. *S. rubiginata* Hufn., ♂, Terlan, S. Tirol, N. Italy (ZSM)
134e. *S. rubiginata* Hufn., ♂, Albarracin, Teruel, E. Spain (ZSM/EMEM)
134f. *S. rubiginata* Hufn., ♂, Albarracin, Teruel, E. Spain (coll. SKOU)
134g. *S. rubiginata* Hufn., ♀, Tivoli, Lazio, C. Italy (ZSM), reared
134h. *S. rubiginata* Hufn., ♂, Cibins Mt., C. Romania (ZSM)
134i. *S. rubiginata* Hufn., ♂, Ohrid, SW. Macedonia (ZSM)
134k. *S. rubiginata* Hufn., ♂, Ohrid, SW. Macedonia (ZSM)
134l. *S. rubiginata* Hufn., ♂, Ohrid, SW. Macedonia (ZSM)
134m. *S. rubiginata* Hufn., ♂, Liljanovo, Pirin mts., SW. Bulgaria (coll. GELBRECHT)
134n. *S. rubiginata* Hufn., ♀, Kranevo, Varna, E. Bulgaria (coll. GELBRECHT)
134o. *S. rubiginata* Hufn., ♂, Tbillissi, Grusinia, Georgia (coll. GELBRECHT), reared
135a. ***Scopula ochraceata* Stgr.**, ♂, Sarepta, S. European Russia (ZFMK)
135b. *S. ochraceata* Stgr., ♀, Sarepta, S. European Russia (ZFMK)
135c. *S. ochraceata* Stgr., ♀, Sarepta, S. European Russia (MNHU)
135d. *S. ochraceata* Stgr., ♂, Matka, Treska gorge, N. Macedonia (ZSM)
135e. *S. ochraceata* Stgr., ♂, Ohrid, SW. Macedonia (ZSM)
135f. *S. ochraceata* Stgr., ♀, Drenovo, Kavadar, C. Macedonia (ZSM)
136a. ***Scopula turbidaria* Hbn.**, ♂, Granada, S. Spain (NHMW)
136b. *S. turbidaria* Hbn., ♂, Granada, S. Spain (NHMW)
136c. *S. turbidaria* Hbn., ♂, Selgua, Aragon, NE. Spain (NHMW)
136d. *S. turbidaria* Hbn., ♂, Selgua, Aragon, NE. Spain (NHMW)
136e. *S. turbidaria* Hbn., ♂, Albarracin, Teruel, E. Spain (ZSM)
136f. *S. turbidaria* Hbn., ♂, Algarve, S. Portugal (ZSM)
137a. ***Scopula turbulentaria* Stgr.**, ♂, Mercuore, Mistretta, NE. Sicily (ZSM)
137b. *S. turbulentaria* Stgr., ♀, Tivoli, Lazio, C. Italy (ZSM)
137c. *S. turbulentaria* Stgr., ♂, Kalavrita, N. Peloponnes, S. Greece (coll. GELBRECHT), reared
137d. *S. turbulentaria* Stgr., ♀, Kalavrita, N. Peloponnes, S. Greece (coll. GELBRECHT), reared
137e. *S. turbulentaria* Stgr., ♂, Faliraki, Rhodos island (ZSM)
137f. *S. turbulentaria* Stgr., ♂, Faliraki, Rhodos island (ZSM)
138a. ***Scopula decolor* Stgr.**, ♂, Mellieka, Malta (ZMUC)
138b. *S. decolor* Stgr., ♂, Barranca de Valcuerna, Candasnos, Huesca, N. Spain (coll. SKOU)
138c. *S. decolor* Stgr., ♀, Sierra Alhamilla, Almeria, S. Spain (coll. GRÜNEWALD)
138d. *S. decolor* Stgr., ♂, Barranca de Valcuerna, Candasnos, Huesca, N. Spain (coll. SKOU)
138e. *S. decolor* Stgr., ♂, Nefta, Tunisia (ZMUC)
139a. ***Scopula rubellata* Stgr.**, ♂, Albarracin, Teruel, E. Spain (ZSM)
139b. *S. rubellata* Stgr., ♀, Albarracin, Teruel, E. Spain (ZSM)
139c. *S. rubellata* Stgr., ♂, Albarracin, Teruel, E. Spain (ZSM)
139d. *S. rubellata* Stgr., ♂, Albarracin, Teruel, E. Spain (ZSM)
139e. *S. rubellata* Stgr., ♂, Albarracin, Teruel, E. Spain (ZSM)
139f. *S. rubellata* Stgr., ♀, Escorial, S. Spain (ZSM)
140a. ***Scopula scalercii* Hausm.**, ♂, Sulmona, C. Italy (MNHU), paratype
140b. *S. scalercii* Hausm., ♀, Gravina, Puglia, S. Italy (ZSM), holotype
141a. ***Scopula beckeraria beckeraria* Led.**, ♂, Sarepta, S. European Russia (ZSM)
141b. *S. beckeraria rebeli* Led., ♂, Ohrid, SW. Macedonia (ZSM)
141c. *S. beckeraria rebeli* Led., ♂, Drenovo, Kavadar, C. Macedonia (ZSM)
141d. *S. beckeraria rebeli* Led., ♀, Kirikkale, C. Turkey (ZSM)

132a 132b 132c 132d 132e 132f
132g 132h 132i 132k 132l 132m
132n 132o 132p 132q 132r 132s
133a 133b 133c 133d 133e 133f 133g
134a 134b 134c 134d 134e 134f 134g
134h 134i 134k 134l 134m 134n 134o
135a 135b 135c 135d 135e 135f
136a 136b 136c 136d 136e 136f
137a 137b 137c 137d 137e 137f
138a 138b 138c 138d 138e
139a 139b 139c 139d 139e 139f
140a 140b 141a 141b 141c 141d

PLATE 18 (all specimens natural size)

142a. **Scopula incanata incanata L.**, ♂, Reval, Estonia (ZSM)
142b. *S. incanata incanata* L., ♂, Strandhof, Estonia (ZSM)
142c. *S. incanata incanata* L., ♀, Pößneck, Thuringia, C. Germany (coll. GELBRECHT), reared
142d. *S. incanata incanata* L., ♂, Wallersberg, N. Bavaria, C. Germany (ZSM)
142e. *S. incanata incanata* L., ♀, Gaistal, N. Tirol, Austria (ZSM)
142f. *S. incanata incanata* L., ♂, Hinterstoder, C. Austria (ZSM)
142g. *S. incanata incanata* L., ♂, Dürnstein, NE. Austria (ZSM)
142h. *S. incanata incanata* L., ♂, Albula, SE. Switzerland (ZSM/EMEM)
142i. *S. incanata incanata* L., ♀, Naturns, S. Tirol, N. Italy (ZSM/EMEM)
142k. *S. incanata incanata* L., ♂, Ohrid, SW. Macedonia (ZSM)
142l. *S. incanata incanata* L., ♂, Kataphygion, Olympos, N. Greece (ZSM)
142m. *S. incanata incanata* L., ♀, Ovindoli, Abruzzi, C. Italy (ZSM/EMEM)
142n. *S. incanata incanata* L., ♂, Itkol, Caucasus, SW. Russia (ZSM)
142o. *S. incanata incanata* L., ♂, Val d'Ossoue, C. Pyrenees, S. France (ZSM)
142p. *S. incanata ibericata* Rssr., ♂, Garg. Pozas, Sierra Gredos, C. Spain (ZSM), 'cotype'
143a. **Scopula marginepunctata Gze.**, ♂, Kaiserstuhl, Baden, SW. Germany (ZSM)
143b. *S. marginepunctata* Gze., ♀, Berlin, E. Germany (ZSM), reared
143c. *S. marginepunctata* Gze., ♀, Wallersberg, N. Bavaria, C. Germany (ZSM)
143d. *S. marginepunctata* Gze., ♂, Linz, N. Austria (ZSM)
143e. *S. marginepunctata* Gze., ♀, Linz, N. Austria (ZSM)
143f. *S. marginepunctata* Gze., ♀, Schnalstal, S. Tirol, N. Italy (ZSM/EMEM)
143g. *S. marginepunctata* Gze., ♂, Ofen, W. Hungary (ZSM)
143h. *S. marginepunctata* Gze., ♂, Dentelles de Montmirsil, Vaucluse, S. France (ZFMK/WAGENER)
143i. *S. marginepunctata* Gze., ♂, Digne, Provence, SE. France (ZSM)
143k. *S. marginepunctata* Gze., ♀, Gèdre, C. Pyrenees, S. France (ZSM/EMEM)
143l. *S. marginepunctata* Gze., ♂, La Rabida, Huelva, S. Spain (ZSM/HERBULOT)
143m. *S. marginepunctata* Gze., ♂, Santo Tirso, S. Portugal (ZSM)
143n. *S. marginepunctata* Gze., ♀, Estanya, Huesca, N. Spain (coll. SKOU)
143o. *S. marginepunctata* Gze., ♀, Raec gorge, Macedonia (coll. KRAUS)
143p. *S. marginepunctata* Gze., ♂, Ohrid, SW. Macedonia (ZSM)
143q. *S. marginepunctata* Gze., ♀, Itea, C. Greece (coll. GRÜNEWALD)
143r. *S. marginepunctata* Gze., ♀, El Ksiba, Middle Atlas, C. Morocco (ZSM), reared
143s. *S. marginepunctata terrigena* Prt., ♂, Tacht i Suleiman, C. Elburs, N. Iran (ZSM)
144a. **Scopula luridata Z.**, ♂, Chersonissos, E. Crete (coll. FRIEDRICH)
144b. *S. luridata* Z., ♀, Monemvasia, S. Peloponnes, S. Greece (coll. FRIEDRICH)
144c. *S. luridata* Z., ♂, Moni Thari, Rhodos island (ZSM)
144d. *S. luridata* Z., ♂, Faliraki, Rhodos island (ZSM)
144e. *S. luridata* Z., ♂, by Antalya, S. Turkey (ZSM), reared
144f. *S. luridata* Z., ♀, by Antalya, S. Turkey (ZSM), reared
145a. **Scopula imitaria imitaria Hbn.**, ♂, Montemaderno, Lake Garda, N. Italy (ZSM), reared
145b. *S. imitaria imitaria* Hbn., ♀, Tivoli, Lazio, C. Italy (ZSM), reared
145c. *S. imitaria imitaria* Hbn., ♂, Tivoli, Lazio, C. Italy (ZSM)
145d. *S. imitaria imitaria* Hbn., ♂, La Faute, Vendée, W. France (ZSM/HERBULOT)
145e. *S. imitaria imitaria* Hbn., ♀, Tivoli, Lazio, C. Italy (ZSM), reared
145f. *S. imitaria imitaria* Hbn., ♂, Bastia, Corsica (ZSM), reared
145g. *S. imitaria imitaria* Hbn., ♂, Liljanovo, Pirin mts., SW. Bulgaria (coll. GELBRECHT), reared
145h. *S. imitaria imitaria* Hbn., ♀, Liljanovo, Pirin mts., SW. Bulgaria (coll. GELBRECHT), reared
145i. *S. imitaria imitaria* Hbn., ♀, Algeciras, S. Spain (ZSM)
145k. *S. imitaria syriacaria* Culot, ♂, Lasithi, E. Crete (NHMW/VARTIAN)
145l. *S. imitaria syriacaria* Culot, ♂, Spili, C. Crete (ZSM)
145m. *S. imitaria syriacaria* Culot, ♀, Tabarja Beach, Beirut, Lebanon (ZSM/EMEM)
146a. **Scopula immutata L.**, ♂, Göhrde, Lüneburg, N. Germany (ZSM)
146b. *S. immutata* L., ♂, Skabyer Heide, Mark Brandenburg, E. Germany (coll. GELBRECHT)
146c. *S. immutata* L., ♂, Dannenreich, Königs Wusterhausen, E. Germany (coll. GELBRECHT)
146d. *S. immutata* L., ♀, Anklamer Stadtluch, Mecklenburg, NE. Germany (coll. GELBRECHT)
146e. *S. immutata* L., ♂, by Linz, N. Austria (ZSM)
146f. *S. immutata* L., ♂, by Linz, N. Austria (ZSM)
146g. *S. immutata* L., ♂, by Linz, N. Austria (ZSM)
146h. *S. immutata* L., ♂, Vienna, E. Austria (ZSM)
146i. *S. immutata* L., ♂, Epannes, Deux-Sèvres, W. France (ZSM/HERBULOT)
146k. *S. immutata* L., ♀, Trigny, Marne, N. France (ZSM/HERBULOT)
146l. *S. immutata* L., ♀, Jagodzin, SW Poland (ZSM)
146m. *S. immutata* L., ♂, Nasice, E. Croatia (ZSM/EMEM)
147a. **Scopula frigidaria Möschl.**, ♂, Lapponia (NHMW)
147b. *S. frigidaria* Möschl., ♂, Lampela, Kuolajärvi, Lapponia, Finland (ZSM)
147c. *S. frigidaria* Möschl., ♂, Lampela, Kuolajärvi, Lapponia, Finland (ZSM)
147d. *S. frigidaria* Möschl., ♂, Kuusamo, Finland (ZSM)
147e. *S. frigidaria* Möschl., ♂, Kuusamo, Finland (ZSM)
147f. *S. frigidaria* Möschl., ♀, Lampela, Kuolajärvi, Lapponia, Finland (ZSM)

142f
142g
142h
142i
142k
142l
142m
142n
142o
142p
143a
143b
143c
143d
143e
143f
143g
143h
143i
143k
143l
143m
143n
143o
143p
143q
143r
143s
44a
144b
144c
144d
144e
144f
145a
145b
145c
145d
145e
145f
45g
145h
145i
145k
145l
145m
146a
146b
146c
146d
146e
146f
146g
146h
146i
146k
146l
146m
147a
147b
147c
147d
147e
147f

PLATE 19 (all specimens natural size)

148a. ***Scopula ternata* Schrank,** ♂, Göhrde, Lüneburg, N. Germany (ZSM), reared
148b. *S. ternata* Schrank, ♀, Berlin, E. Germany (ZSM), reared
148c. *S. ternata* Schrank, ♂, Breitenhülser Wald, C. Austria (ZSM)
148d. *S. ternata* Schrank, ♀, Hinterstoder, C. Austria (ZSM)
148e. *S. ternata* Schrank, ♂, Witoscha, Sofia, NW. Bulgaria (coll. GELBRECHT)
149a. ***Scopula floslactata floslactata* Haw.,** ♂, Grünheide, Erkner, E. Germany (coll. GELBRECHT)
149b. *S. floslactata floslactata* Haw., ♀, Stocketwald, E. Austria (ZSM)
149c. *S. floslactata floslactata* Haw., ♀, Breitfurt, W. Germany (ZSM)
149d. *S. floslactata floslactata* Haw., ♀, Hörsching, N. Austria (ZSM)
149e. *S. floslactata scotica* Cock., ♂, N. Scotland, Great Britain (ZFMK)
150a. ***Scopula subpunctaria* H.-S.,** ♂, Hinterstoder, C. Austria (ZSM)
150b. *S. subpunctaria* H.-S., ♀, Passeiertal, S. Tirol, N. Italy (coll. GELBRECHT), reared
150c. *S. subpunctaria* H.-S., ♀, Rodltal, Gramast, N. Austria (ZSM)
150d. *S. subpunctaria* H.-S., ♀, Gorizia ('Görz'), NE. Italy (ZSM)
150e. *S. subpunctaria* H.-S., ♀, Amfreville, Eure, NW. France (ZSM/HERBULOT)
151a. ***Scopula flaccidaria* Z.,** ♂, Donauauen, Vienna, E. Austria (NHMW/VARTIAN), reared
151b. *S. flaccidaria* Z., ♂, Inzenhof, Burgenland, SE. Austria (ZSM), reared
151c. *S. flaccidaria* Z., ♀, Harkany, Pecs, S. Hungary (ZSM), reared
151d. *S. flaccidaria* Z., ♀, Harkany, Pecs, S. Hungary (ZSM), reared
151e. *S. flaccidaria* Z., ♀, Harkany, Pecs, S. Hungary (coll. GELBRECHT), reared
151f. *S. flaccidaria* Z., ♂, Bursa, NW. Turkey (NHMW)
152a. ***Scopula emutaria* Hbn.,** ♂, Sylt, Schleswig-Holstein, N. Germany (ZSM)
152b. *S. emutaria* Hbn., ♂, Riopar, Albacete, SE. Spain (ZSM)
152c. *S. emutaria* Hbn., ♂, Puig Ventos, Gerona, NE. Spain (coll. GELBRECHT), reared
152d. *S. emutaria* Hbn., ♀, Tour du Vallat, Camargue, S. France (ZSM/EMEM), reared
152e. *S. emutaria* Hbn., ♂, Calvi, N. Corsica (ZSM)
152f. *S. emutaria* Hbn., ♀, Dalmatia, S. Croatia (NHMW)
153a. ***Scopula minorata* Bsd.,** ♂, Valencia, E. Spain (ZSM), reared
153b. *S. minorata* Bsd., ♂, Valencia, E. Spain (ZSM), reared
153c. *S. minorata* Bsd., ♂, Paguera, Mallorca, Balearics (ZSM), reared
153d. *S. minorata* Bsd., ♀, Paguera, Mallorca, Balearics (ZSM), reared
153e. *S. minorata* Bsd., ♂, Granada, S. Spain (ZSM)
153f. *S. minorata* Bsd., ♂, Granada, S. Spain (ZSM)
153g. *S. minorata* Bsd., ♀, by Huelva, S. Spain (ZSM/EMEM)
153h. *S. minorata* Bsd., ♀, Vilamoura, Algarve, S. Portugal (ZSM)
153i. *S. minorata* Bsd., ♂, Rome, C. Italy (ZSM/HERBULOT)
153k. *S. minorata* Bsd., ♂, Tortora, N. Calabria, S. Italy (ZSM), reared
153l. *S. minorata* Bsd., ♀, Tivoli, Lazio, C. Italy (ZSM)
153m. *S. minorata* Bsd., ♀, Tivoli, Lazio, C. Italy (ZSM)
153n. *S. minorata* Bsd., ♀, Mt. Smith, Rhodos island (ZSM)
153o. *S. minorata* Bsd., ♂, Silifke, S. Turkey (ZSM)
154a. ***Scopula cajanderi* Herz,** ♂, Krasnyi Kamen, Polar Urals, NE. European Russia (ZMH)
154b. *S. cajanderi* Herz, ♀, Krasnyi Kamen, Polar Urals, NE. European Russia (ZMH)
154c. *S. cajanderi* Herz, ♂, Vilui-Delta, Lena-region, Yakutia, N. Russia (ZISP), paralectotype
154d. *S. cajanderi* Herz, ♂, Vilui-Delta, Lena-region, Yakutia, N. Russia (ZISP), paralectotype
154e. *S. cajanderi* Herz, ♂, Irkutsk, N. Russia (ZISP)
154f. *S. cajanderi* Herz, ♀, Iktirchaja-Tshorur, N. Russia (ZISP)
155a. ***Scopula divisaria* Chr.,** ♂, Guberli, S. Urals, SE. European Russia (ZISP), syntype
155b. *S. divisaria* Chr., ♂, Guberli, S. Urals, SE. European Russia (ZSM), syntype
155c. *S. divisaria* Chr., ♂, Guberli, S. Urals, SE. European Russia (ZISP), syntype
155d. *S. divisaria* Chr., ♀, Guberli, S. Urals, SE. European Russia (MNHU)
155e. *S. divisaria* Chr. (?), ♂, Mt. Toraygir, Almaty, E. Kazakhstan (coll. SOMMERER)
155f. ***Scopula latelineata* Grs.,** ♂, Alexander Mts., Kirghizstan (MNHU), syntype
156a. ***Scopula arenosaria* Stgr.,** ♀, [Astrachan, S. European Russia], erroneously labelled 'Narün' (MNHU), syntype
156b. *S. arenosaria* Stgr., ♂, [Astrachan, S. European Russia], erroneously labelled 'Narün' (MNHU), syntype
156c. *S. arenosaria* Stgr., ♂, Urda, Ryn-Kum desert, W. Kazakhstan (coll. SKOU)
156d. *S. arenosaria* Stgr., ♀, Uralsk, S. European Russia (ZSM)
156e. *S. arenosaria* Stgr., ♂, Urda, Ryn-Kum desert, W. Kazakhstan (coll. SKOU)
157a. ***Scopula albiceraria albiceraria* H.-S.,** ♂, Sarepta, S. European Russia (MNHU)
157b. *S. albiceraria albiceraria* H.-S., ♂, Uralsk, S. European Russia (MNHU)
157c. *S. albiceraria albiceraria* H.-S., ♂, W. Kazakhstan (ZISP)
157d. *S. albiceraria vitellinaria* Ev., ♀, Amur, E. Russia (MNHU)
157e. *S. albiceraria vitellinaria* Ev., ♂, Amur, E. Russia (MNHU)
157f. *S. albiceraria vitellinaria* Ev., ♂, Sosnovka, Burgatia, Transbaikalia, C. Russia (ZSM/EMEM)
158a. ***Scopula immistaria immistaria* H.-S.,** ♂, Maras, Taurus, S. Turkey (ZSM)
158b. *S. immistaria immistaria* H.-S., ♂, Maras, Taurus, S. Turkey (ZSM)
158c. *S. immistaria immistaria* H.-S., ♂, Mardin, E. Taurus, S. Turkey (ZSM)
158d. *S. immistaria immistaria* H.-S., ♂, Sarepta, S. European Russia (MNHU)
158e. *S. immistaria beshkovi* Gelbr. & Hausm., ♂, Pobiti Kameni, Varna, E. Bulgaria (ZSM), holotype
158f. *S. immistaria beshkovi* Gelbr. & Hausm., ♀, Pobiti Kameni, Varna, E. Bulgaria (ZSM), paratype

148a 148b 148c 148d 148e
149a 149b 149c 149d 149e
150a 150b 150c 150d 150e
151a 151b 151c 151d 151e 151f
152a 152b 152c 152d 152e 152f
3a 153b 153c 153d 153e 153f 153g
53h 153i 153k 153l 153m 153n 153o
54a 154b 154c 154d 154e 154f
155a 155b 155c 155d 155e
155f 156a 156b 156c 156d 156e
157a 157b 157c 157d 157e 157f
158a 158b 158c 158d 158e 158f

PLATE 20 (all specimens natural size)

159a. ***Glossotrophia confinaria confinaria* H.-S.**, ♀, St. Marcellin de Vars, Ht. Alpes, SE. France (ZSM/HERBULOT)
159b. *G. confinaria confinaria* H.-S., ♂, Dubrovnik ('Gravosa'), S. Croatia (ZSM)
159c. *G. confinaria confinaria* H.-S., ♀, Dubrovnik ('Gravosa'), S. Croatia (ZSM)
159d. *G. confinaria confinaria* H.-S., ♂, Liljanovo, Pirin mts., SW. Bulgaria (coll. GELBRECHT), reared
159e. *G. confinaria confinaria* H.-S., ♀, Liljanovo, Pirin mts., SW. Bulgaria (coll. GELBRECHT), reared
159f. *G. confinaria confinaria* H.-S., ♀, Liljanovo, Pirin mts., SW. Bulgaria (coll. GELBRECHT), reared
159g. *G. confinaria confinaria* H.-S., ♀, Arahova, Parnass, C. Greece (coll. FRIEDRICH)
159h. *G. confinaria confinaria* H.-S., ♂, Ohrid, SW. Macedonia (ZSM)
159i. *G. confinaria confinaria* H.-S., ♀, Ohrid, SW. Macedonia (ZSM)
159k. *G. confinaria prouti* Hausm., ♂, Terlan, S. Tirol, N. Italy (ZSM), paratype
159l. *G. confinaria prouti* Hausm., ♀, Meran, S. Tirol, N. Italy (ZSM), paratype
159m. *G. confinaria scoblei* Hausm., ♂, Mercuore, Mistretta, NE. Sicily (ZSM)
159n. *G. confinaria aetnaea* Prout., ♂, Nicolosi, S. Etna, E. Sicily (MNHU)
159o. *G. confinaria aetnaea* Prout., ♀, Nicolosi, S. Etna, E. Sicily (MNHU)
159p. ***Glossotrophia diffinaria* Prt.**, ♂, Hasan Dagi, Aksaray, C. Turkey (coll. GELBRECHT)
159q. *G. diffinaria* Prt., ♂, Kackar Dagh, Artvin, NE. Turkey (coll. GELBRECHT), reared
159r. *G. diffinaria* Prt., ♂, Göreme, Nevshehir, C. Turkey (ZSM)
159s. *G. diffinaria* Prt., ♀, Nevshehir, C. Turkey (ZSM/EMEM)
160a. ***Glossotrophia alba alba* Hausm.**, ♂, Tivoli, Lazio, C. Italy (ZSM)
160b. *G. alba alba* Hausm., ♀, Mtgna. Grande, Abruzzi, C. Italy (ZSM), paralectotype
160c. *G. alba alba* Hausm., ♀, Campagna Romana, Lazio, C. Italy (ZSM)
160d. *G. alba brunellii* Hausm., ♂, Maratea, SW. Basilicata, S. Italy (ZSM)
160e. *G. alba brunellii* Hausm., ♀, Maratea, SW. Basilicata, S. Italy (ZSM)
160f. *G. alba milleri* subsp.n., ♂, Mt. Renoso, Ghisoni, Corsica (ZSM), paratype
161a. ***Glossotrophia mentzeri* Hausm.**, ♂, Assites, Psiloriti, C. Crete (ZSM), paratype
161b. *G. mentzeri* Hausm., ♀, Silva Rouva, Ida, C. Crete (ZSM), paratype, reared
161c. *G. mentzeri* Hausm., ♀, Assites, Psiloriti, C. Crete (ZSM), paratype
161d. *G. mentzeri* Hausm., ♀, Psychro, E. Crete (ZSM), paratype
161e. *G. mentzeri* Hausm., ♂, Males, S. Dikti Mts., E. Crete (coll. SKOU), reared
161f. *G. mentzeri* Hausm., ♂, Males, S. Dikti Mts., E. Crete (coll. SKOU), reared
162a. ***Glossotrophia rufomixtaria* Grasl.**, ♂, Vernet-les-Bains, E. Pyrenees, S. France (ZSM/HERBULOT)
162b. *G. rufomixtaria* Grasl., ♀, Casteil, E. Pyrenees, S. France (ZSM/HERBULOT)
162c. *G. rufomixtaria* Grasl., ♀, St. Étienne, Vallée Francaise, S. France (ZSM/HERBULOT)
162d. *G. rufomixtaria* Grasl., ♀, St. Ildefonso, Madrid, C. Spain (ZSM)
162e. *G. rufomixtaria* Grasl., ♂, Hoyos del Espino, Sierra de Gredos, W. Spain (coll. SKOU)
162f. *G. rufomixtaria* Grasl., ♀, by Huelva, S. Spain (ZSM/EMEM)
162g. *G. rufomixtaria* Grasl., ♂, Granada, S. Spain (ZSM)
162h. *G. rufomixtaria* Grasl., ♀, Orgiva, Sierra Nevada, S. Spain (NHMW/VARTIAN)
162i. *G. rufomixtaria* Grasl., ♂, Los Gallardos, Almeria, S. Spain (coll. SKOU)
162k. *G. rufomixtaria* Grasl., 'f. *annae*', ♂, Alfajarin, Zaragoza, NE. Spain (coll. GELBRECHT)
162l. *G. rufomixtaria* Grasl., 'f. *annae*', ♀, Beterrita, Zaragoza, NE. Spain (coll. SKOU), paratype (*annae*)
162m. *G. rufomixtaria* Grasl., trans. 'f. *annae*', ♀, Valdovecar, Teruel, E. Spain (ZSM/HERBULOT)
163a. ***Glossotrophia sacraria taurica* Whli.**, ♀, Recadiye Pass, Malatya, SE. Turkey (ZSM/EMEM)
163b. *G. sacraria sacraria* B.-H., ♂, Ushtagan, 120 km E Aktau, W. Kazakhstan (coll. SKOU)
163c. *G. sacraria sacraria* B.-H., ♂, Tushchibek, Mangistau Mts., W. Kazakhstan (coll. SKOU)
164a. ***Glossotrophia asellaria asellaria* H.-S.**, ♂, Col Larone, Corsica (ZSM/EMEM)
164b. *G. asellaria asellaria* H.-S., ♂, Aritzo, Sardinia (ZSM)
164c. *G. asellaria romanaria* Mill., ♀, Tivoli, Lazio, C. Italy (ZSM)
164d. *G. asellaria romanaria* Mill., ♀, Tivoli, Lazio, C. Italy (ZSM)
164e. *G. asellaria philipparia* Prt., ♀, Constantine, E. Algeria (ZSM/HERBULOT)
164f. *G. asellaria tripolitana* Trti., ♀, Gafsa, C. Tunisia (ZSM/HERBULOT), reared
164g. *G. asellaria isabellaria* Mill., ♂, St.-Étienne, Vallée-Francaise, S. France (ZSM/HERBULOT)
164h. *G. asellaria isabellaria* Mill., ♂, Collioure, E. Pyrenees, S. France (ZSM/HERBULOT)
164i. *G. asellaria isabellaria* Mill., ♀, Banyuls-sur-Mer, E. Pyrenees, S. France (ZSM/EMEM)
164k. *G. asellaria isabellaria* Mill., ♀, Collioure, E. Pyrenees, S. France (ZSM/HERBULOT)
164l. *G. asellaria isabellaria* Mill., ♂, Albarracin, Teruel, E. Spain (ZSM)
164m. *G. asellaria isabellaria* Mill., ♂, Albarracin, Teruel, E. Spain (ZSM)
164n. *G. asellaria dentatolineata* Whli., ♂, Alhama de Murcia, SE. Spain (ZSM/HERBULOT)
164o. *G. asellaria dentatolineata* Whli., ♀, La Rahige, Malaga, S. Spain (coll. SKOU)
164p. *G. asellaria dentatolineata* Whli., ♀, Granada, S. Spain (ZSM)
165a. ***Holarctias rufinaria* Stgr.**, ♂, Krasnyi Kamen, Polar Urals, NE. European Russia (ZMH)
165b. *H. rufinaria* Stgr., ♀, Krasnyi Kamen, Polar Urals, NE. European Russia (ZMH)
165c. ***Holarctias sentinaria* Geyer**, ♀, Labrador, North America (ZSM)
165d. ***Holarctias rufinularia* Stgr.**, ♂, Altai Mts., C. Russia (ZSM)
165e. *H. rufinularia* Stgr., ♂, Terelj, Mongolia (coll. GELBRECHT)
165f. *H. rufinularia* Stgr., ♀, Terelj, Mongolia (coll. GELBRECHT)
166a. ***Problepsis ocellata* Friv.**, ♂, Assites, C. Crete (NHMW/VARTIAN), reared
166b. *P. ocellata* Friv., ♀, Assites, C. Crete (NHMW/VARTIAN), reared
166c. *P. ocellata* Friv., ♀, Zachlorou, N. Peloponnes, S. Greece (ZSM/EMEM), reared
166d. *P. ocellata* Friv., ♂, Camliyayta, Tarsus, S. Turkey (coll. GELBRECHT)

159g
159h
159i
159k
159l
159m
159n
159o
159p
159q
159r
159s
60a
160b
160c
160d
160e
160f
161a
161b
161c
161d
161e
161f
162a
162b
162c
162d
162e
162f
62g
162h
162i
162k
162l
162m
3a
163b
163c
164a
164b
164c
64d
164e
164f
164g
164h
164i
64k
164l
164m
164n
164o
164p
65a
165b
165c
165d
165e
165f
166a
166b
166c
166d

PLATE 21 (all specimens natural size)

167a. ***Rhodostrophia jacularia* Hbn.,** ♂, Sarepta, S. European Russia (NHMW)
167b. *R. jacularia* Hbn., ♂, Sarepta, S. European Russia (ZSM)
167c. *R. jacularia* Hbn., ♂, Rostov-Don, SW. European Russia (ZSM/EMEM)
167d. *R. jacularia* Hbn., ♀, Sarepta, S. European Russia (MNHU)
167e. *R. jacularia* Hbn., ♀, Sarepta, S. European Russia (ZSM)
168a. ***Rhodostrophia vibicaria* Cl.,** ♂, Blåvand, W. Jütland, Denmark (ZSM), reared
168b. *R. vibicaria* Cl., ♂, Sylt, Schleswig-Holstein, N. Germany (ZSM)
168c. *R. vibicaria* Cl., ♀, Blåvand, W. Jütland, Denmark (coll. GELBRECHT), reared
168d. *R. vibicaria* Cl., ♂, Velburg, C. Bavaria, S. Germany (ZSM)
168e. *R. vibicaria* Cl., ♀, Regensburg, C. Bavaria, S. Germany (ZSM), reared
168f. *R. vibicaria* Cl., ♂, Dürnstein, NE. Austria (ZSM)
168g. *R. vibicaria* Cl., ♂, Naturns, S. Tirol, N. Italy (ZSM)
168h. *R. vibicaria* Cl., ♂, Petrina Plan, SW. Macedonia (ZSM)
168i. *R. vibicaria strigata* Stgr., ♂, Pyrenees, N. Spain (NHMW)
169a. ***Rhodostrophia pudorata pudorata* F.,** ♂, Lambèse, E. Algeria (ZSM)
169b. *R. pudorata pudorata* F., ♂, Chiclana, Andalusia, S. Spain (ZSM)
169c. *R. pudorata pudorata* F., ♀, Algezares, Andalusia, S. Spain (ZSM)
169d. *R. pudorata perezaria* Obth., ♂, Juneda, Catalonia, NE. Spain (ZSM/HERBULOT)
169e. *R. pudorata perezaria* Obth., ♂, Albarracin, Teruel, E. Spain (ZSM)
169f. *R. pudorata perezaria* Obth., ♂, Albarracin, Teruel, E. Spain (ZSM)
169g. *R. pudorata sicanaria* Z., ♂, Mt. Cuccio, N. Sicily (ZSM)
169h. *R. pudorata sicanaria* Z., ♀, Mt. Cuccio, N. Sicily (ZSM)
169i. ***Rhodostrophia sieversi* Hbn.,** ♂, Eriwan, Armenia (ZSM)
170a. ***Rhodostrophia cretacaria* Reb.,** ♂, Psychro, E. Crete (ZSM)
170b. *R. cretacaria* Reb., ♀, Lakkoi, W. Crete (ZSM)
170c. *R. cretacaria* Reb., ♂, Psychro, E. Crete (NHMW/VARTIAN)
171a. ***Rhodostrophia calabra calabra* Petagna,** ♀, Mt. Simbruini, Abruzzi, C. Italy (ZSM)
171b. *R. calabra calabra* Petagna, ♀, Ovindoli, Abruzzi, C. Italy (ZSM/EMEM)
171c. *R. calabra calabra* Petagna, ♀, Dubrovnik ('Gravosa'), S. Croatia (ZSM)
171d. *R. calabra calabra* Petagna, ♂, Dubrovnik ('Gravosa'), S. Croatia (ZSM)
171e. *R. calabra calabra* Petagna, ♂, Mentone, Alpes maritimes, SE. France (ZSM)
171f. *R. calabra calabra* Petagna, ♂, St. Paul de Fen., E. Pyrenees, S. France (ZSM)
171g. *R. calabra separata* Prt., ♂, Sierra de Espuña, Murcia, SE. Spain (ZSM)
171h. *R. calabra separata* Prt., ♀, Albarracin, Teruel, E. Spain (ZSM)
171i. *R. calabra separata* Prt., ♂, Faro, S. Portugal (ZSM)
172a. ***Rhodostrophia tabidaria tabidaria* Z.,** ♂, Faliraki, Rhodos island (ZSM)
172b. *R. tabidaria tabidaria* Z., ♀, Kusadasi, 100 km S Izmir, W. Turkey (ZSM/EMEM)
172c. *R. tabidaria cypriaria* Reb., ♂, Agia Napa, Cyprus (coll. SKOU)
173a. ***Rhodostrophia discopunctata* Ams.,** ♀, Drenovo, Kavadar, C. Macedonia (ZSM)
173b. *R. discopunctata* Ams., ♀, Drenovo, Kavadar, C. Macedonia (ZSM)
173c. *R. discopunctata* Ams., ♀, Ohrid, SW. Macedonia (ZSM)
173d. *R. discopunctata* Ams., ♀, Ohrid, SW. Macedonia (ZSM)

168a
168b
168c
168d
168e
168f
168g
168h
168i
169a
169b
169c
169d
169e
169f
169g
169h
169i
170a
170b
170c
171a
171b
171c
171d
171e
171f
171g
171h
171i
172a
172b
172c
173a
173b
173c
173d

PLATE 22 (all specimens natural size)

173e. ***Rhodostrophia terrestraria* Led.,** ♂, Shakuh, NE. Iran (ZSM)
173f. *R. terrestraria* Led., ♀, Alexander Mts., Kirghizstan (ZSM)
174a. ***Rhodostrophia badiaria* Frr.,** ♂, Sarepta, S. European Russia (ZISP)
174b. *R. badiaria* Frr., ♂, Chan Urlu Dagh, Kurdistan, E. Turkey (ZSM)
174c. *R. badiaria* Frr., ♂, Kicilcahamam, NW. Turkey (ZSM)
174d. *R. badiaria* Frr., ♀, Kaman, Ankara, Turkey (ZSM)
174e. *R. badiaria* Frr., ♀, Kirikkale, Ankara, Turkey (ZSM)
174f. ***Rhodostrophia vastaria* Chr.,** ♂, Tushchibek, Mangistau Mts., W. Kazakhstan (coll. SKOU)
175a. ***Timandra comae* Schmidt,** ♂, Dannenreich, Königs Wusterhausen, E. Germany (coll. GELBRECHT)
175b. *T. comae* Schmidt, ♀, Lage, Lippe, N. Gerrmany (ZSM), reared
175c. *T. comae* Schmidt, ♀, Karlsruhe, Baden, SW. Germany (ZSM)
175d. *T. comae* Schmidt, ♂, Regensburg, C. Bavaria, S. Germany (ZSM)
175e. *T. comae* Schmidt, ♀, Freising, S. Bavaria, S. Germany (ZSM/EMEM)
175f. *T. comae* Schmidt, ♂, Linz, N. Austria (ZSM), reared
175g. *T. comae* Schmidt, ♂, Windischgr., Austria (ZSM)
175h. *T. comae* Schmidt, ♀, Tivoli, Lazio, C. Italy (ZSM)
175i. *T. comae* Schmidt, ♀, Rovinjsko Selo, Istria, W. Croatia (ZSM)
175k. *T. comae* Schmidt, ♂, Dnepropetrovskaja Andreevka, Ukraine (ZSM)
176a. ***Timandra griseata* Peters.,** ♂, Nauvo-Haverö, Finland (coll. SKOU)
176b. *T. griseata* Peters., ♂, Nurmo, Finland (ZMH)
176c. *T. griseata* Peters., ♀, St. Reposaari, Isokatava, Finland (ZMH)
176d. *T. griseata* Peters., ♀, Nauvo-Haverö, Finland (coll. SKOU)
176e. *T. griseata* Peters., ♂, Sojmonowsk, C. Urals, E. European Russia (MNHU)
177a. ***Timandra rectistrigaria* Ev.,** ♂, Krasnyi Kamen, Polar Urals, NE. European Russia (ZMH)
177b. *T. rectistrigaria* Ev., ♀, Krasnyi Kamen, Polar Urals, NE. European Russia (ZMH)
177c. *T. rectistrigaria* Ev., ♂, Krasnyi Kamen, Polar Urals, NE. European Russia (ZMH)
177d. *T. rectistrigaria* Ev., ♀, Krasnyi Kamen, Polar Urals, NE. European Russia (ZMH)
177e. *T. rectistrigaria* Ev., ♂, Tunkinsk Weissbg., SW Irkutsk, C. Russia (ZSM)
178a. ***Cyclophora pendularia* Cl.,** ♀, Püggen, NE. Niedersachsen, N. Germany (ZSM), reared
178b. *C. pendularia* Cl., ♀, Dannenreich, Königs Wusterhausen, E. Germany (coll. GELBRECHT), reared
178c. *C. pendularia* Cl., ♂, Dannenreich, Königs Wusterhausen, E. Germany (coll. GELBRECHT), reared
178d. *C. pendularia* Cl., ♂, Oberweiden, Marchfeld, E. Austria (ZSM), reared
178e. *C. pendularia* Cl., ♂, Linz, N. Austria (ZSM), reared
178f. *C. pendularia* Cl., ♀, Lipik, E. Croatia (ZSM)
179a. C. hybr. *albipunctata* Hufn. (♂) x *pendularia* Cl. (♀), ♂, Stettin, Pommern, NW. Poland (ZSM), reared
179b. ***Cyclophora albipunctata* Hufn.,** ♀, Stafford, Great Britain (ZSM)
179c. *C. albipunctata* Hufn., ♂, 'Anglia', Great Britain (MNHU)
179d. *C. albipunctata* Hufn., ♂, Skabyer Heide, Spreenhagen, E. Germany (coll. GELBRECHT)
179e. *C. albipunctata* Hufn., ♂, Niederlehme, Mark Brandenburg, E. Germany (coll. GELBRECHT)
179f. *C. albipunctata* Hufn., ♀, Berlin, E. Germany (ZSM)
179g. *C. albipunctata* Hufn., ♀, Dannenreich, Königs Wusterhausen, E. Germany (coll. GELBRECHT), reared
179h. *C. albipunctata* Hufn., ♂, Habichtswald, Kassel, C. Germany (ZSM)
179i. *C. albipunctata* Hufn., ♀, Dettendorf, S. Bavaria, S. Germany (ZSM/EMEM)
179k. *C. albipunctata* Hufn., ♂, Gotteszell, N. Bavaria, SE. Germany (ZSM)
179l. *C. albipunctata* Hufn., ♂, Teberda, NW. Caucasus, SW. Russia (coll. GELBRECHT), reared
179m. *C. albipunctata* Hufn., ♀, Teberda, NW. Caucasus, SW. Russia (coll. GELBRECHT), reared
180a. ***Cyclophora albiocellaria* Hbn.,** ♂, Krems, N. Austria (ZSM), reared
180b. *C. albiocellaria* Hbn., ♀, Dürnstein, N. Austria (ZSM)
180c. *C. albiocellaria* Hbn., ♂, Vienna, E. Austria (ZSM), reared
180d. *C. albiocellaria* Hbn., ♂, Dürnstein, N. Austria (ZSM), reared
180e. *C. albiocellaria* Hbn., ♂, Ohrid, SW. Macedonia (ZSM)
180f. *C. albiocellaria* Hbn., ♀, Rostov, SW. European Russia (NHMW/VARTIAN)
181a. ***Cyclophora lennigiaria* Fuchs,** ♂, Oberhausen, Pfalz, W. Germany (ZSM), reared
181b. *C. lennigiaria* Fuchs, ♀, Oberhausen, Pfalz, W. Germany (ZSM), reared
181c. *C. lennigiaria* Fuchs, ♂, Schloßböckelheim, Pfalz, W. Germany (ZSM), reared
181d. *C. lennigiaria* Fuchs, ♀, Oberhausen, Pfalz, W. Germany (ZSM), reared
181e. *C. lennigiaria* Fuchs, ♂, Schloßböckelheim, Pfalz, W. Germany (ZSM), reared
181f. *C. lennigiaria* Fuchs, ♂, Digne, Provence, SE. France (ZSM)

173f
174a
174c
174d
174e
174f
175a
175b
175c
175d
175e
175f
175g
175h
175i
175k
176a
176b
176c
176d
176e
177a
177b
177c
177d
177e
178a
178b
178c
178d
178e
178f
179a
179b
179c
179d
179e
179f
179g
179h
179i
179k
179l
179m
180a
180b
180c
180d
180e
180f
181a
181b
181c
181d
181e
181f

PLATE 23 (all specimens natural size)

182a. ***Cyclophora ariadne* Rssr.,** ♂, Nida, C. Crete (NHMW/VARTIAN), reared
182b. *C. ariadne* Rssr., ♀, Psychro, E. Crete (NHMW/VARTIAN), reared
182c. *C. ariadne* Rssr., ♀, Psychro, E. Crete (SMNK), reared
182d. *C. ariadne* Rssr., ♂, Nida, C. Crete (SMNK), reared
182e. *C. ariadne* Rssr., ♂, Silva Rouva, Ida, C. Crete (SMNK), reared
182f. *C. ariadne* Rssr., ♀, Nida, C. Crete (SMNK), reared
183a. ***Cyclophora serveti* Red. & Gast.,** ♂, Cuenca, C. Spain (MNHU/PÜNGELER)
183b. *C. serveti* Red. & Gast., ♂, Cuenca, C. Spain (MNHU/PÜNGELER)
183c. *C. serveti* Red. & Gast., ♂, Alfacar, Granada, S. Spain (ZSM/EMEM)
183d. *C. serveti* Red. & Gast., ♀, Alfacar, Granada, S. Spain (ZSM/EMEM)
183e. *C. serveti* Red. & Gast., ♂, Estepar, Burgos, N. Spain (MNCN), paratype
183f. *C. serveti* Red. & Gast., ♀, Estepar, Burgos, N. Spain (MNCN), paratype
184a. ***Cyclophora annularia* F.,** ♂, Nebbeskov, Langeland, Denmark (coll. SKOU)
184b. *C. annularia* F., ♂, Schkeuditz, Saxonia, E. Germany (coll. GELBRECHT)
184c. *C. annularia* F., ♂, Wildau, Königs Wusterhausen, E. Germany (coll. GELBRECHT), reared
184d. *C. annularia* F., ♀, Linz, N. Austria (ZSM), reared
184e. *C. annularia* F., ♂, Forêt de Chizé, Deux-Sèvres, W. France (ZSM/HERBULOT)
184f. *C. annularia* F., ♂, Vulture, Basilicata, S. Italy (ZSM)
185a. ***Cyclophora puppillaria* Hbn.,** ♂, Garda, N. Italy (ZSM)
185b. *C. puppillaria* Hbn., ♀, Vienna, E. Austria (ZSM), reared
185c. *C. puppillaria* Hbn., ♂, Cannes, Alpes maritimes, SE. France (ZSM/HERBULOT), reared
185d. *C. puppillaria* Hbn., ♂, Talairan, Corbières, S. France (ZSM)
185e. *C. puppillaria* Hbn., ♀, Talairan, Corbières, S. France (ZSM)
185f. *C. puppillaria* Hbn., ♂, St. Tropez, Var, SE. France (ZSM/HERBULOT)
185g. *C. puppillaria* Hbn., ♀, Port Bou, NE. Spain (ZSM)
185h. *C. puppillaria* Hbn., ♂, Planas del Rey, Tarragona, NE. Spain (ZSM/HERBULOT)
185i. *C. puppillaria* Hbn., ♂, Barcelona, NE. Spain (MNHU)
185k. *C. puppillaria* Hbn., ♀, Philippeville, E. Algeria (ZSM), reared
185l. *C. puppillaria* Hbn., ♂, Tivoli, Lazio, C. Italy (ZSM), reared
185m. *C. puppillaria* Hbn., ♂, Tivoli, Lazio, C. Italy (ZSM)
185n. *C. puppillaria* Hbn., ♂, Gadoni, Sardinia (coll. SKOU), reared
185o. *C. puppillaria* Hbn., ♀, Tivoli, Lazio, C. Italy (ZSM)
185p. *C. puppillaria* Hbn., ♀, Dubrovnik ('Gravosa'), S. Croatia (ZSM), reared
185q. *C. puppillaria* Hbn., ♀, Dubrovnik ('Gravosa'), S. Croatia (ZSM), reared
185r. *C. puppillaria* Hbn., ♂, Dubrovnik ('Gravosa'), S. Croatia (ZSM), reared
185s. *C. puppillaria* Hbn., ♂, Nauplion, Greece (ZSM/HERBULOT)
186a. ***Cyclophora quercimontaria* Bast.,** ♂, Grünheide, Erkner, E. Germany (coll. GELBRECHT), reared
186b. *C. quercimontaria* Bast., ♀, Nassau, Taunus, W. Germany (ZSM)
186c. *C. quercimontaria* Bast., ♀, Eichberg, Rheingau, W. Germany (ZSM), reared
186d. *C. quercimontaria* Bast., ♀, Karlsruhe, Baden, SW. Germany (ZSM), reared
186e. *C. quercimontaria* Bast., ♂, Giesshübl, Vienna, E. Austria (ZSM), reared
186f. *C. quercimontaria* Bast., ♂, Linz, N. Austria (ZSM)
186g. *C. quercimontaria* Bast., ♂, Schleinbach, E. Austria (ZSM), reared
186h. *C. quercimontaria* Bast., ♂, Sattnitz, Carinthia, SE. Austria (ZSM)
186i. *C. quercimontaria* Bast., ♂, Amfreville, Eure, NW. France (ZSM/HERBULOT)
186k. *C. quercimontaria* Bast., ♀, Matka, Treska gorge, N. Macedonia (ZSM)
186l. *C. quercimontaria* Bast., ♀, Naturns, S. Tirol, N. Italy (ZSM)
186m. *C. quercimontaria* Bast., ♀, Liljanovo, Pirin mts., SW. Bulgaria (coll. GELBRECHT), reared
187a. ***Cyclophora ruficiliaria* H.-S.,** ♀, St. Goarshausen, Nassau, W. Germany (ZSM), 1st gen.
187b. *C. ruficiliaria* H.-S., ♀, Rotenfels, Pfalz, W. Germany (ZSM/EMEM), 1st gen., reared
187c. *C. ruficiliaria* H.-S., ♀, Breitenfurt, Vienna, E. Austria (ZSM), 1st gen.
187d. *C. ruficiliaria* H.-S., ♀, Neuwaldegg, Vienna, E. Austria (ZSM), 1st gen., reared
187e. *C. ruficiliaria* H.-S., ♀, Maisons Lafitte, Paris, France (ZSM/HERBULOT), 1st gen.
187f. *C. ruficiliaria* H.-S., ♂, Vulture, Basilicata, S. Italy (ZSM), 1st gen.
187g. *C. ruficiliaria* H.-S., ♂, Eichberg, Rheingau, W. Germany (ZSM/EMEM), 2nd gen., reared
187h. *C. ruficiliaria* H.-S., ♂, Pizzola-Aquila, Abruzzi, C. Italy (ZSM), 2nd gen.
187i. *C. ruficiliaria* H.-S., ♂, Kiten, E. Bulgaria (coll. GELBRECHT), 2nd gen., reared
187k. *C. ruficiliaria* H.-S., ♀, Kiten, E. Bulgaria (coll. GELBRECHT), 2nd gen., reared
188a. ***Cyclophora hyponoea* Prt.,** ♀, Le Tarf, NE. Algeria (ZSM/HERBULOT)
188b. *C. hyponoea* Prt., ♂, Le Tarf, NE. Algeria (ZSM/HERBULOT)
188c. *C. hyponoea* Prt., ♀, Algeciras, Andalusia, S. Spain (NHMW)
188d. *C. hyponoea* Prt., ♂, Aransis, Llerida, NE. Spain (coll. SKOU)
188e. *C. hyponoea* Prt., ♂, Estanya, Huesca, N. Spain (coll. SKOU)
189a. ***Cyclophora porata* L.,** ♂, Dannenreich, Königs Wusterhausen, E. Germany (coll. GELBRECHT), 1st gen., reared
189b. *C. porata* L., ♀, Dannenreich, Königs Wusterhausen, E. Germany (coll. GELBRECHT), 1st gen., reared
189c. *C. porata* L., ♀, Durlacher Wald, Karlsruhe, SW. Germany (ZSM), 1st gen.
189d. *C. porata* L., ♀, Ravone, Bologna, N. Italy (ZSM), 1st gen.
189e. *C. porata* L., ♂, Sierra Alta, Aragon, E. Spain (ZSM), 1st gen.
189f. *C. porata* L., ♂, Puerto de Orihuela, Teruel, E. Spain (coll. SKOU), 2nd gen.

183a
183b
183c
183d
183e
183f
184a
184b
184c
184d
184e
184f
185a
185b
185c
185d
185e
185f
185g
185h
185i
185k
185l
185m
185n
185o
185p
185q
185r
185s
86a
186b
186c
186d
186e
186f
186g
186h
186i
186k
186l
186m
187a
187b
187c
187d
187e
187f
187g
187h
187i
187k
188a
188b
188c
188d
188e
189a
189b
189c
189d
189e
189f

PLATE 24 (all specimens natural size)

189g. **_Cyclophora porata_ L.,** ♂, Wallersberg, N. Bavaria, C. Germany (ZSM), 2nd gen.
189h. *C. porata* L., ♂, Lorch, Rheinland, W. Germany (ZSM/EMEM), 2nd gen.
189i. *C. porata* L., ♀, Dürnstein, NE. Austria (ZSM), 2nd gen., reared
189k. *C. porata* L., ♀, Tivoli, Lazio, C. Italy (ZSM), 2nd gen.
189l. *C. porata* L., ♀, Dubrovnik ('Gravosa'), S. Croatia (ZSM), 2nd gen., reared
189m. *C. porata* L., ♀, Douelle, Lot, SW. France (ZSM/Herbulot), 3rd gen.
190a. **_Cyclophora suppunctaria_ Z.,** ♀, Digne, Provence, SE. France (ZSM/EMEM), 1st gen.
190b. *C. suppunctaria* Z., ♂, Tivoli, Lazio, C. Italy (ZSM), 1st gen.
190c. *C. suppunctaria* Z., ♀, Dubrovnik ('Gravosa'), S. Croatia (ZSM), 1st gen.
190d. *C. suppunctaria* Z., ♀, Puerto de Orihuela, Teruel, E. Spain (coll. Skou), 1st gen.
190e. *C. suppunctaria* Z., ♂, Alfacar, Granada, S. Spain (ZSM/EMEM), 1st gen.
190f. *C. suppunctaria* Z., ♀, Mercuore, Mistretta, NW. Sicily (ZSM), 2nd gen.
190g. *C. suppunctaria* Z., ♂, Dubrovnik ('Gravosa'), S. Croatia (ZSM), 2nd gen., reared
190h. *C. suppunctaria* Z., ♀, Dubrovnik ('Gravosa'), S. Croatia (ZSM), 2nd gen.
190i. *C. suppunctaria* Z., ♀, Liljanovo, Pirin mts., SW. Bulgaria (coll. Gelbrecht), 2nd gen., reared
190k. *C. suppunctaria* Z., ♂, Liljanovo, Pirin mts., SW. Bulgaria (coll. Gelbrecht), 2nd gen., reared
190l. *C. suppunctaria* Z., ♀, Noli, Liguria, NW. Italy (ZSM), 3rd gen.
191a. **_Cyclophora punctaria punctaria_ L.,** ♀, Neugraben, Hamburg, N. Germany (ZSM)
191b. *C. punctaria punctaria* L., ♀, by Königs Wusterhausen, E. Germany (coll. Gelbrecht)
191c. *C. punctaria punctaria* L., ♂, by Linz, N. Austria (ZSM), reared
191d. *C. punctaria punctaria* L., ♀, Karlsruhe, Baden, SW. Germany (ZSM), reared
191e. *C. punctaria punctaria* L., ♀, Kottenforst, Bonn, W. Germany (ZSM/EMEM)
191f. *C. punctaria punctaria* L., ♀, Donnerskirchen, Burgenland, SE. Austria (ZSM)
191g. *C. punctaria punctaria* L., ♀, by Linz, N. Austria (ZSM), reared
191h. *C. punctaria punctaria* L., ♀, Kharkov, SW. European Russia (NHMW/Vartian)
191i. *C. punctaria fritzae* Hausm., ♀, Pyatigorsk, N. Caucasus, SW. Russia (NHMW/Vartian), paratype
191k. *C. punctaria fritzae* Hausm., ♂, Pyatigorsk, N. Caucasus, SW. Russia (ZSM), holotype
192a. **_Cyclophora linearia_ Hbn.,** ♂, Rauensche Brg., Fürstenwalde, E. Germany (coll. Gelbrecht), 1st gen.
192b. *C. linearia* Hbn., ♀, Hinterstoder, C. Austria (ZSM), 1st gen., reared
192c. *C. linearia* Hbn., ♀, Hinterstoder, C. Austria (ZSM), 1st gen., reared
192d. *C. linearia* Hbn., ♀, Spital, C. Austria (ZSM), 1st gen., reared
192e. *C. linearia* Hbn., ♀, Drenovo, Kavadar, C. Macedonia (ZSM), 1st gen.
192f. *C. linearia* Hbn., ♂, Rodltal, Gramast, C. Austria (ZSM), 2nd gen., reared
192g. *C. linearia* Hbn., ♀, Linz, N. Austria (ZSM), 2nd gen., reared
192h. *C. linearia* Hbn., ♂, Dürnstein, NE. Austria (ZSM), 2nd gen., reared
192i. *C. linearia* Hbn., ♀, Wallersberg, N. Bavaria, C. Germany (ZSM), 2nd gen.
192k. *C. linearia* Hbn., ♀, Kl. Goell, Salzburg, C. Austria (ZSM), 2nd gen.
192l. *C. linearia* Hbn., ♀, Eichberg, Rheingau, W. Germany (ZSM), 2nd gen., reared
193a. **_Rhodometra sacraria_ L.,** ♂, Quiliano, Liguria, NW. Italy (ZSM)
193b. *R. sacraria* L., ♀, Busambra, NW. Sicily (ZSM)
193c. *R. sacraria* L., ♂, S. Medard en J., Gironde, W. France (ZSM/Herbulot)
193d. *R. sacraria* L., ♂, Dubrovnik ('Gravosa'), S. Croatia (ZSM)
193e. *R. sacraria* L., ♂, Dubrovnik ('Gravosa'), S. Croatia (NHMW)
193f. *R. sacraria* L., ♀, Olympia, Peloponnes, S. Greece (ZSM)
194a. **_Casilda antophilaria antophilaria_ Hbn.,** ♂, Urda, Run-Kum steppe, W. Kazakhstan (coll. Skou)
194b. *C. antophilaria antophilaria* Hbn., ♀, Bisen, Run-Kum steppe, W. Kazakhstan (coll. Skou)
194c. *C. antophilaria antophilaria* Hbn., ♂, Churjuk, Sivash Lake, Khersonskaja obl., S. Ukraine (coll. Müller)
194d. *C. antophilaria antophilaria* Hbn., ♂, Tuz Gölü, C. Turkey (ZSM)
194e. *C. antophilaria antophilaria* Hbn., ♂, Tuz Gölü, C. Turkey (ZSM)
194f. *C. antophilaria rosearia* Tr., ♂, Messolonghi, Etolia, W. Greece (coll. Grünewald)
194g. *C. antophilaria rosearia* Tr., ♂, Salaora, Arta, W. Greece (coll. Gelbrecht)
194h. *C. antophilaria rosearia* Tr., ♀, [Korfu], W. Greece (ZSM)
194i. *C. antophilaria rosearia* Tr., ♀, Kalogria-Patra, NW. Peloponnes, S. Greece (coll. Gelbrecht)
194k. *C. antophilaria rosearia* Tr., ♀, Koronissia, Arta, W. Greece (coll. Friedrich)
194l. *C. antophilaria rosearia* Tr., ♂, Kerkyra, Korfu, W. Greece (coll. Grünewald)
195a. **_Casilda consecraria_ Stgr.,** ♂, Sierra Espuña, Murcia, SE. Spain (NHMW/Vartian)
195b. *C. consecraria* Stgr., ♂, S. Maries de la mer, Camargue, S. France (coll. Grünewald)
195c. *C. consecraria* Stgr., ♂, Aranjuez, Madrid, C. Spain (coll. Kraus)
195d. *C. consecraria* Stgr., ♂, Baza, 110 km NE. Granada, S. Spain (NHMW/Vartian)
195e. *C. consecraria* Stgr., ♀, Tabernas, Almeria, SE. Spain (coll. Skou)
195f. *C. consecraria* Stgr., ♀, Aranjuez, Madrid, C. Spain (coll. Kraus)
196a. **_Ochodontia adustaria_ Wald.,** ♂, Sarepta, S. European Russia (ZISP)
196b. *O. adustaria* Wald., ♀, Sarepta, S. European Russia (ZISP)
196c. *O. adustaria* Wald., ♀, Sarepta, S. European Russia (ZSM)
196d. *O. adustaria* Wald., ♂, Sarepta, S. European Russia (NHMW)
196e. *O. adustaria* Wald., ♀, Khosov Reserve, S. Armenia (ZSM)

190a
190b
190c
190d
190e
190f
190g
190h
190i
190k
190l
191a
191b
191c
191d
191e
191f
191g
191h
191i
191k
192a
192b
192c
192d
192e
192f
192g
192h
192i
192k
192l
193a
193b
193c
193d
193e
193f
194a
194b
194c
194d
194e
194f
194g
194h
194i
194k
194l
195a
195b
195c
195d
195e
195f
196a
196b
196c
196d
196e

Genitalia plates

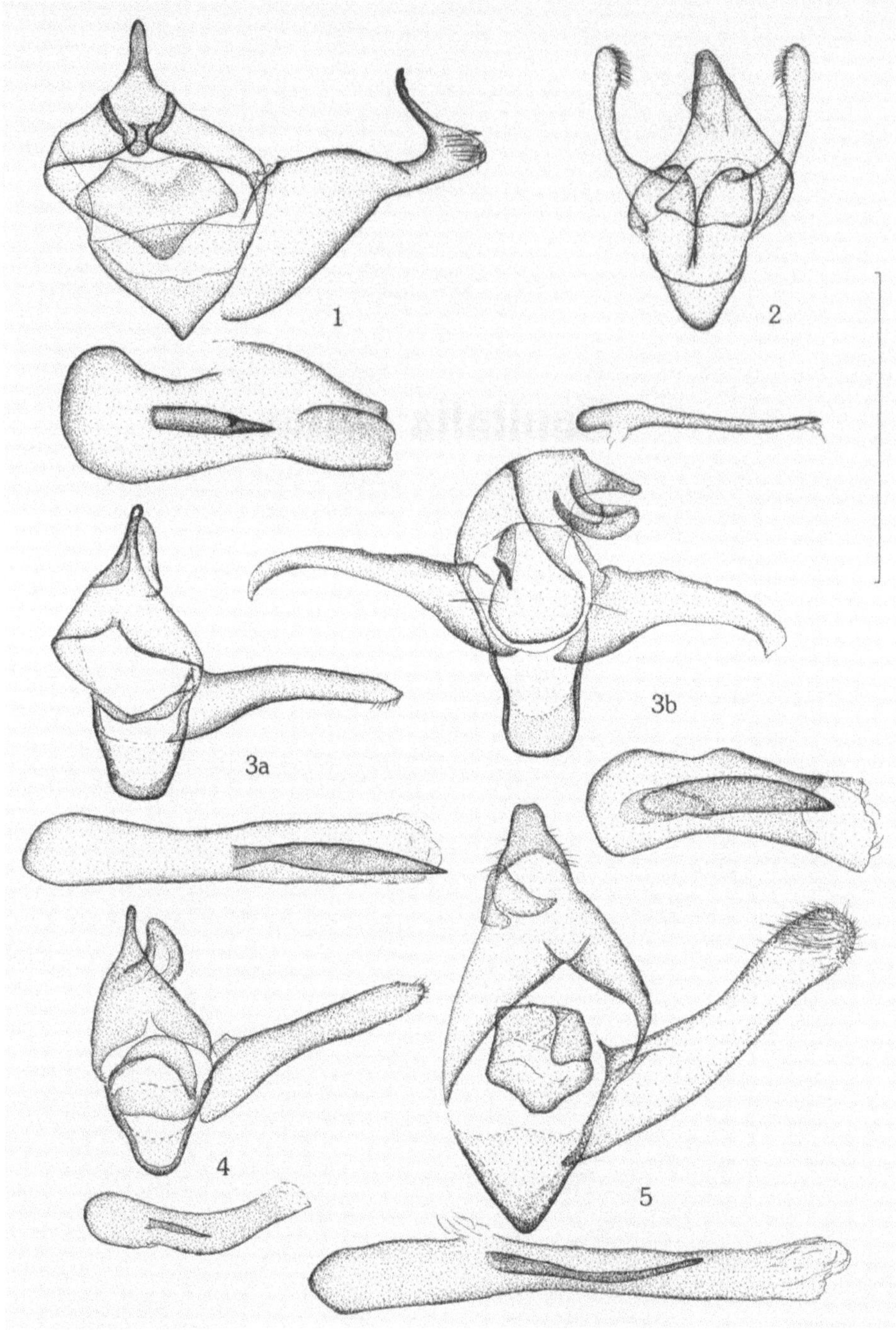

MALE GENITALIA (scale bar = 1 mm)

1. *Anthometra plumularia* Bsd., C. Spain
2. *Emmiltis pygmaearia* Hbn., N. Italy
3a. *Cleta ramosaria* Vill., S. Spain
3b. *Cleta transiens* Prt., E. Algeria
4. *Cleta perpusillaria* Ev., S. European Russia
5. *Cleta filacearia* H.-S., C. Italy

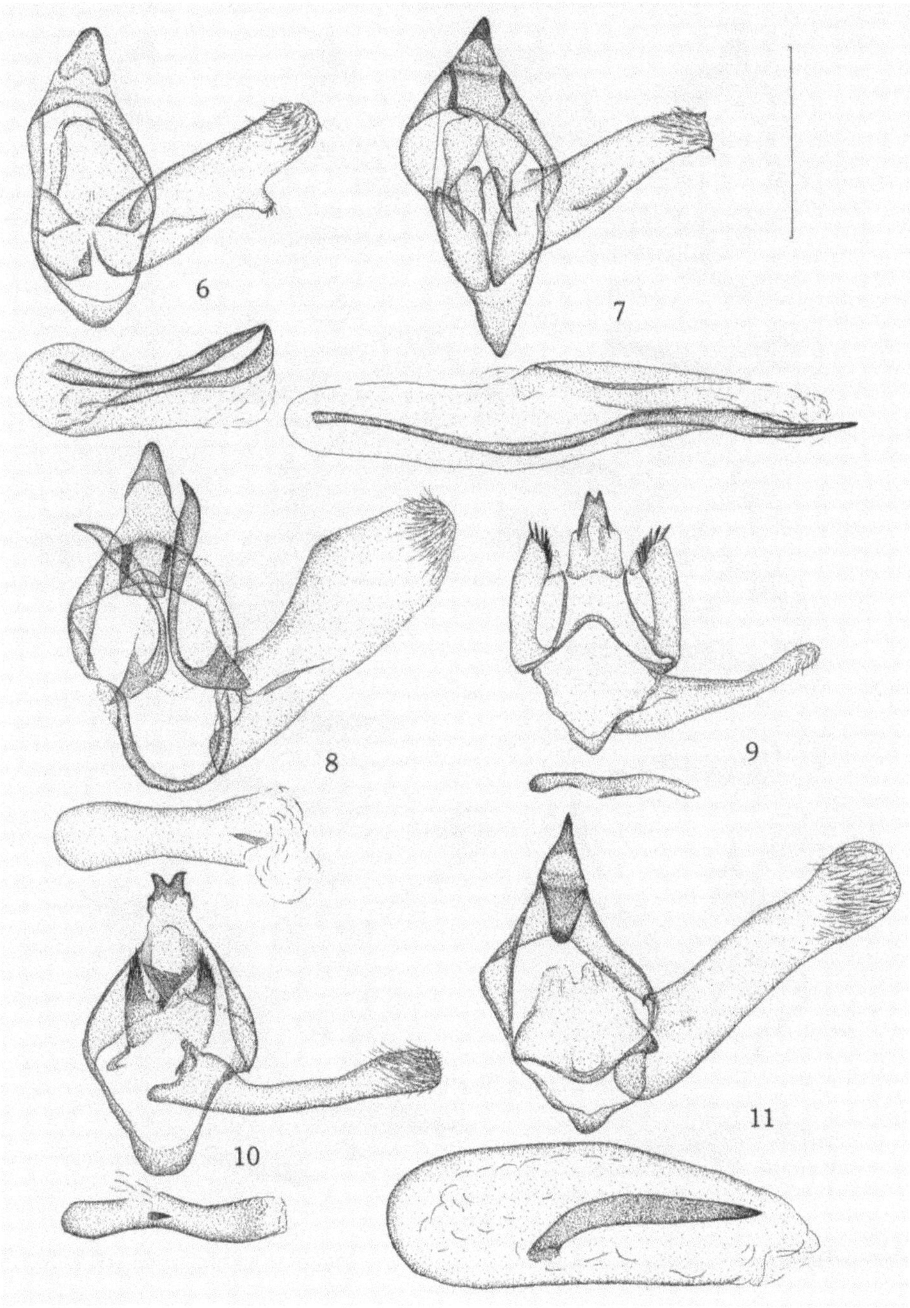

MALE GENITALIA (scale bar = 1 mm)

6. *Idaea serpentata* Hufn., S. Germany
7. *Idaea luteolaria* Const., C. Spain
8. *Idaea aureolaria* Den. & Schiff., S. Germany
9. *Idaea flaveolaria* Hbn., N. Italy
10. *Idaea muricata* Hufn., Hungary
11. *Idaea determinata* Stgr., Turkey

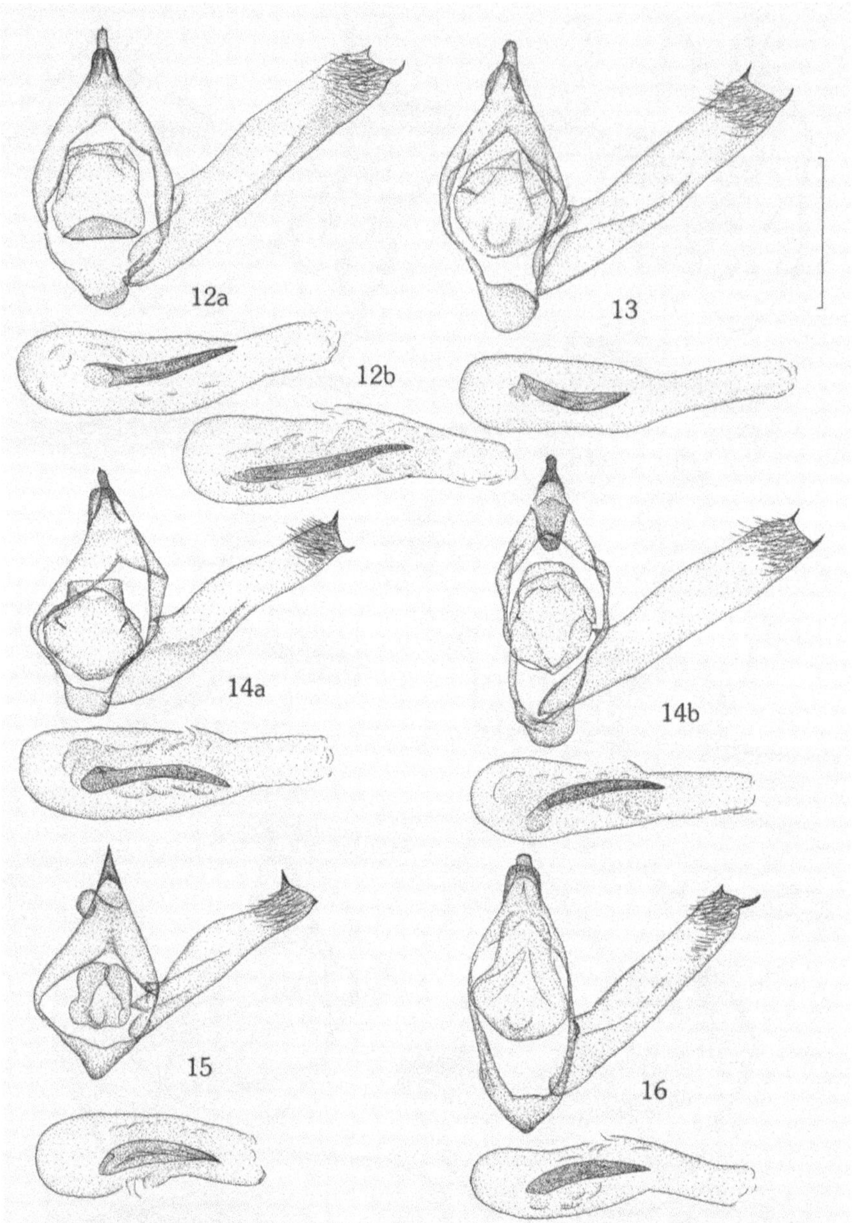

MALE GENITALIA (scale bar = 1 mm)

12a. *Idaea litigiosaria* Bsd., S. Spain
12b. *Idaea litigiosaria* Bsd., Tunisia, aedeagus
13. *Idaea mutilata* Stgr., Sicily
14a. *Idaea lusohispanica* Herb., S. Spain
14b. *Idaea lambessata* Obth., Morocco
15. *Idaea sardoniata* Homb., N. Spain
16. *Idaea korbi* Püng., C. Spain

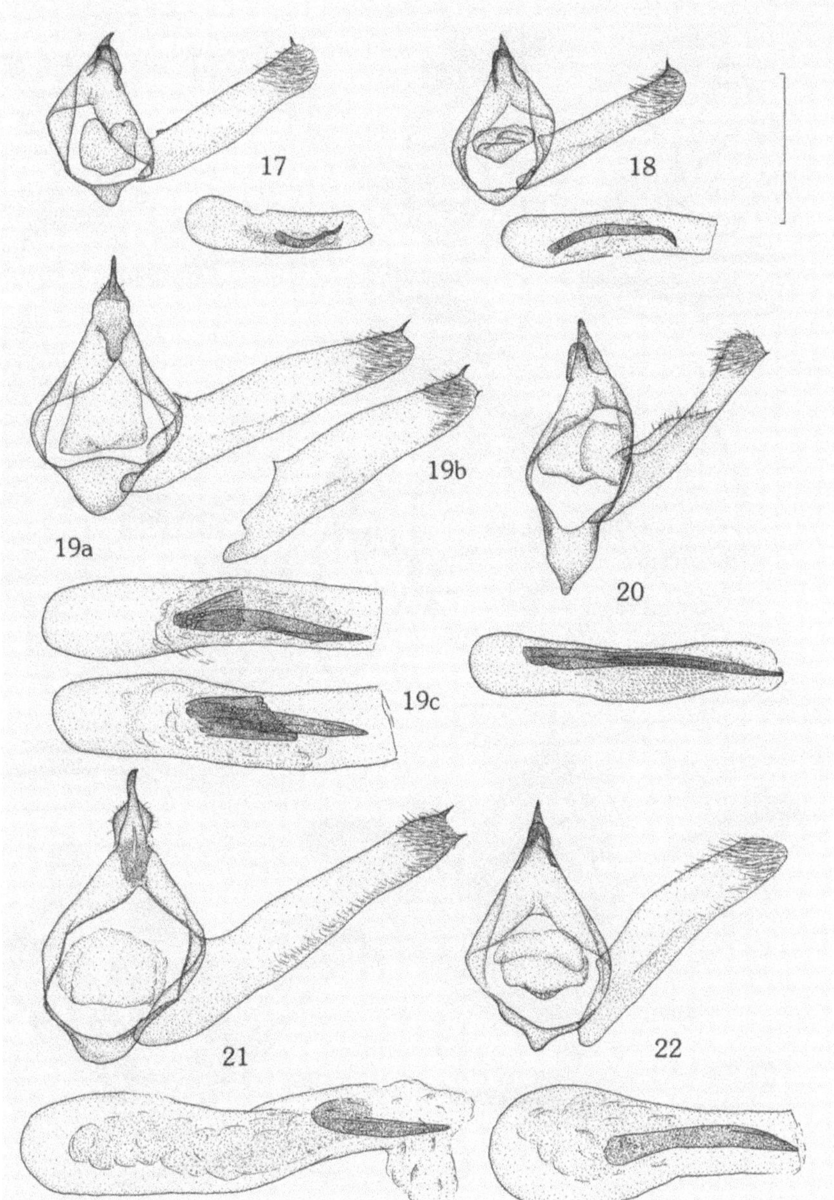

MALE GENITALIA (scale bar = 1 mm)

17. *Idaea mediaria* Hbn., S. France
18. *Idaea leipnitzi* Hausm. sp.n., paratype, SW. Croatia
19a. *Idaea rufaria* Hbn., S. France
19b. *Idaea rufaria* Hbn., E. Germany (valva)
19c. *Idaea rufaria* Hbn., Turkmeniya (aedeagus)
20. *Idaea consanguiberica* Rezb. & Exp., W. Spain
21. *Idaea admiranda* Hausm. sp.n., paratype, European Turkey
22. *Idaea consanguinaria* Led., Rhodos island

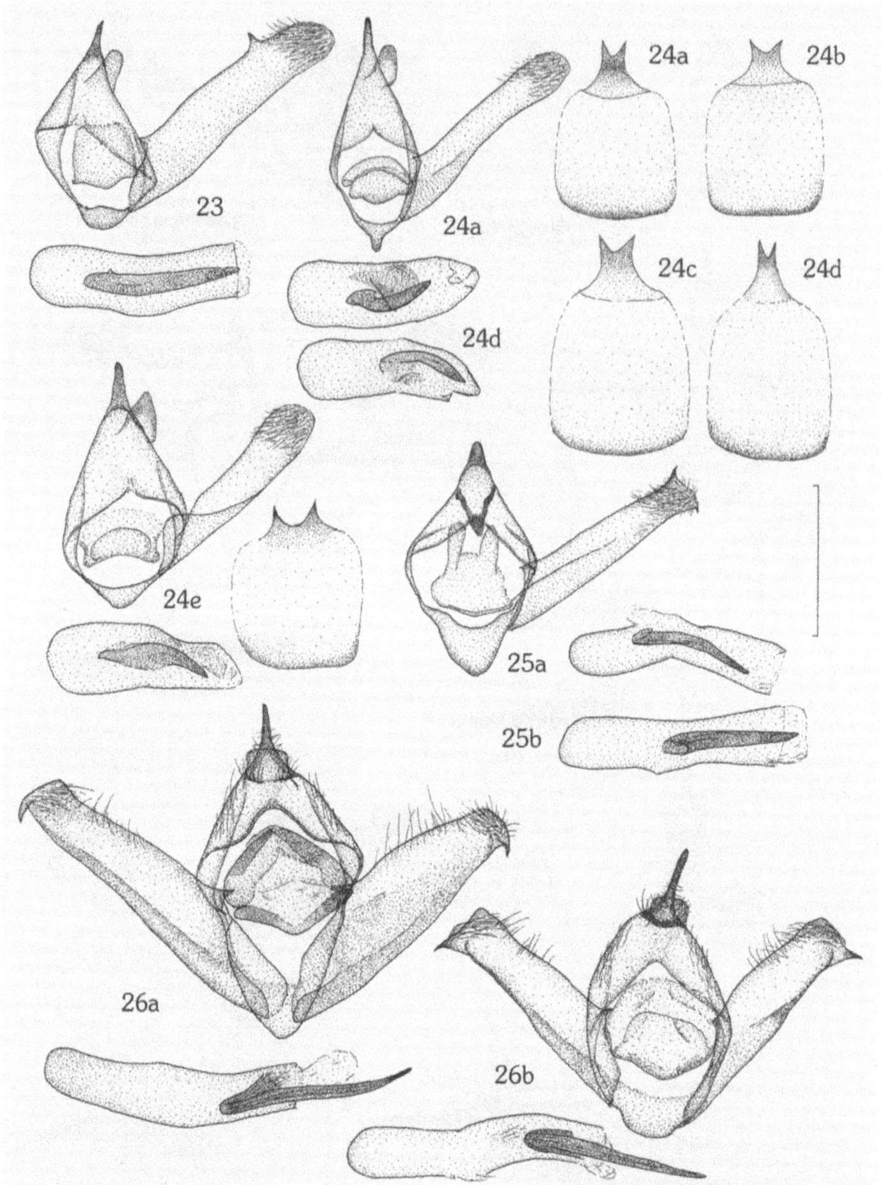

MALE GENITALIA (scale bar = 1 mm)

23. *Idaea ossiculata* Led., Macedonia
24a. *Idaea sericeata sericeata* Hbn., E. Austria, with sternum A8
24b *Idaea sericeata sericeata* Hbn., Corsica, sternum A8
24c. *Idaea sericeata calvaria* Whli., C. Spain, sternum A8
24d. *Idaea sericeata calvaria* Whli., Morocco, aedeagus and sternum A8
24e. *Idaea allardiata* Mab., Tunisia, with sternum A8
25a. *Idaea macilentaria* H.-S., S. France
25b. *Idaea macilentaria* H.-S., W. Germany, aedeagus
26a. *Idaea ochrata ochrata* Scop., E. Germany
26b. *Idaea ochrata albida* Zerny, C. Spain

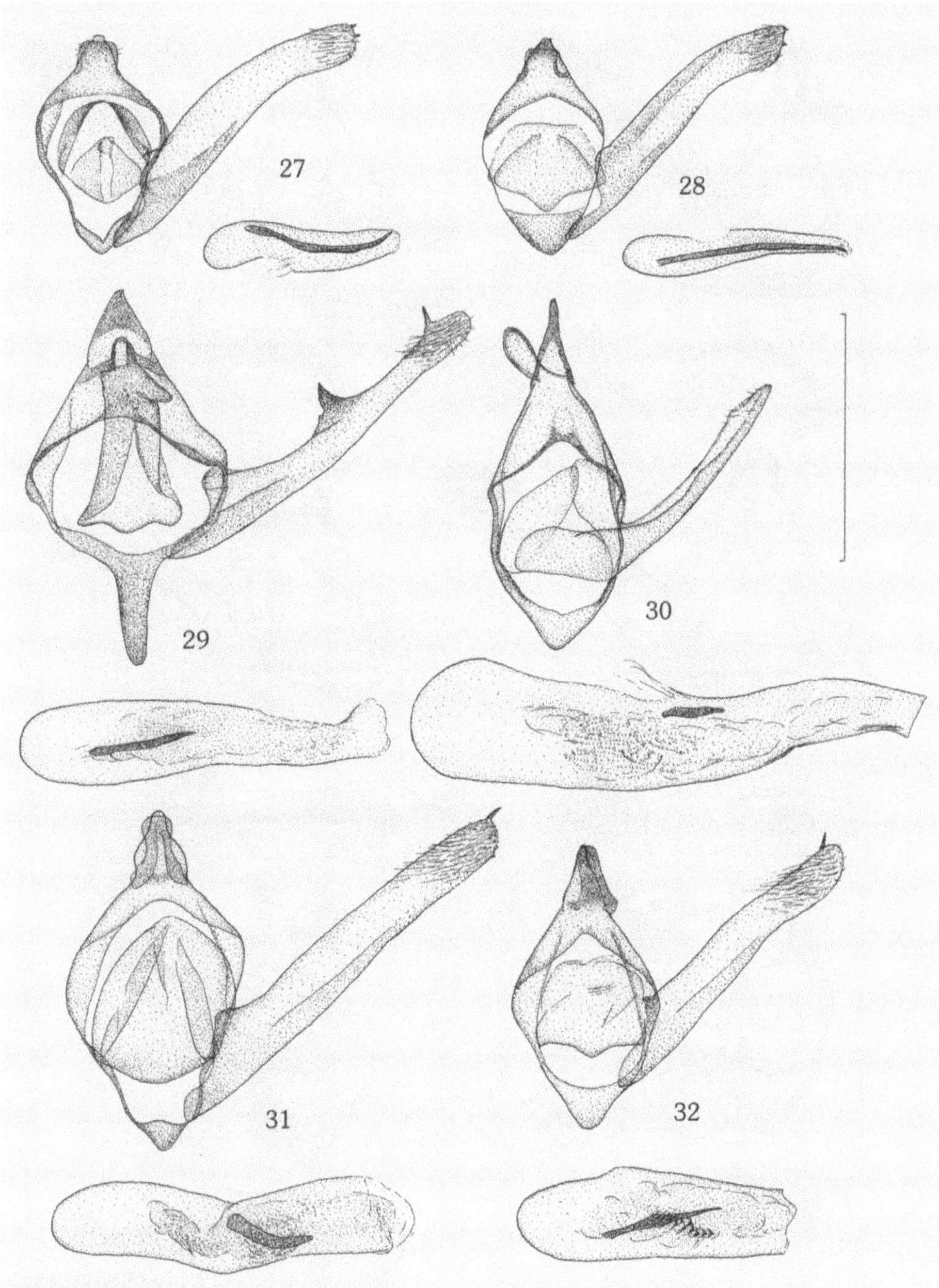

MALE GENITALIA (scale bar = 1 mm)

27. *Idaea nevadata* Whli., S. Spain
28. *Idaea figuraria* B.-H., Morocco
29. *Idaea alicantaria* Rssr., SE. Spain
30. *Idaea nigrolineata* Chrét., holotype, Tunisia
31. *Idaea completa* Stgr., S. Spain
32. *Idaea intermedia* Stgr., Rhodos island

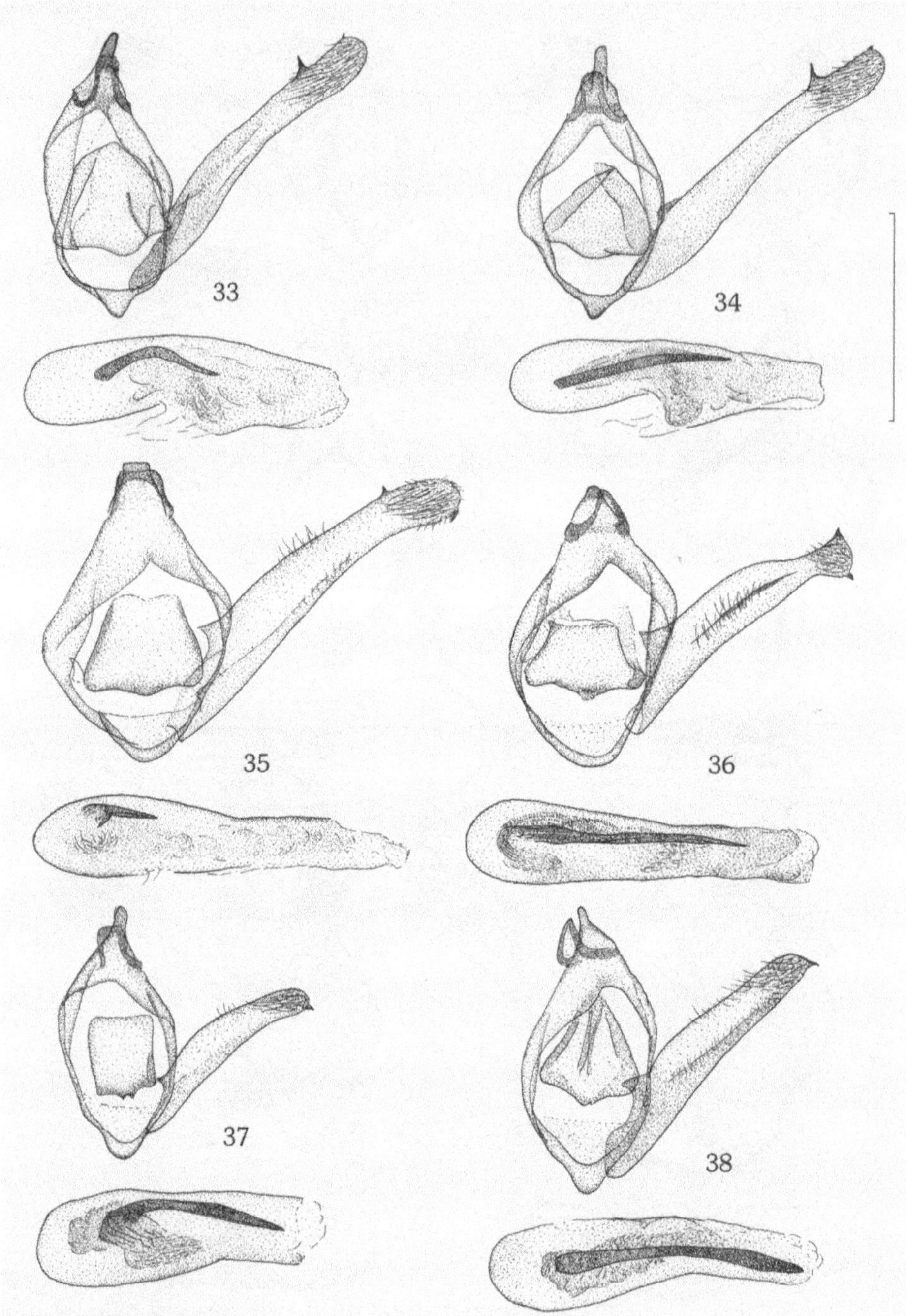

MALE GENITALIA (scale bar = 1 mm)

33. *Idaea rusticata* Den. & Schiff., W. Germany
34. *Idaea mustelata* Gmpp., C. Spain
35. *Idaea filicata* Hbn., N. Italy
36. *Idaea troglodytaria* H.-S., Rhodos island
37. *Idaea laevigata* Scop., E. Spain
38. *Idaea efflorata* Z., Sicily

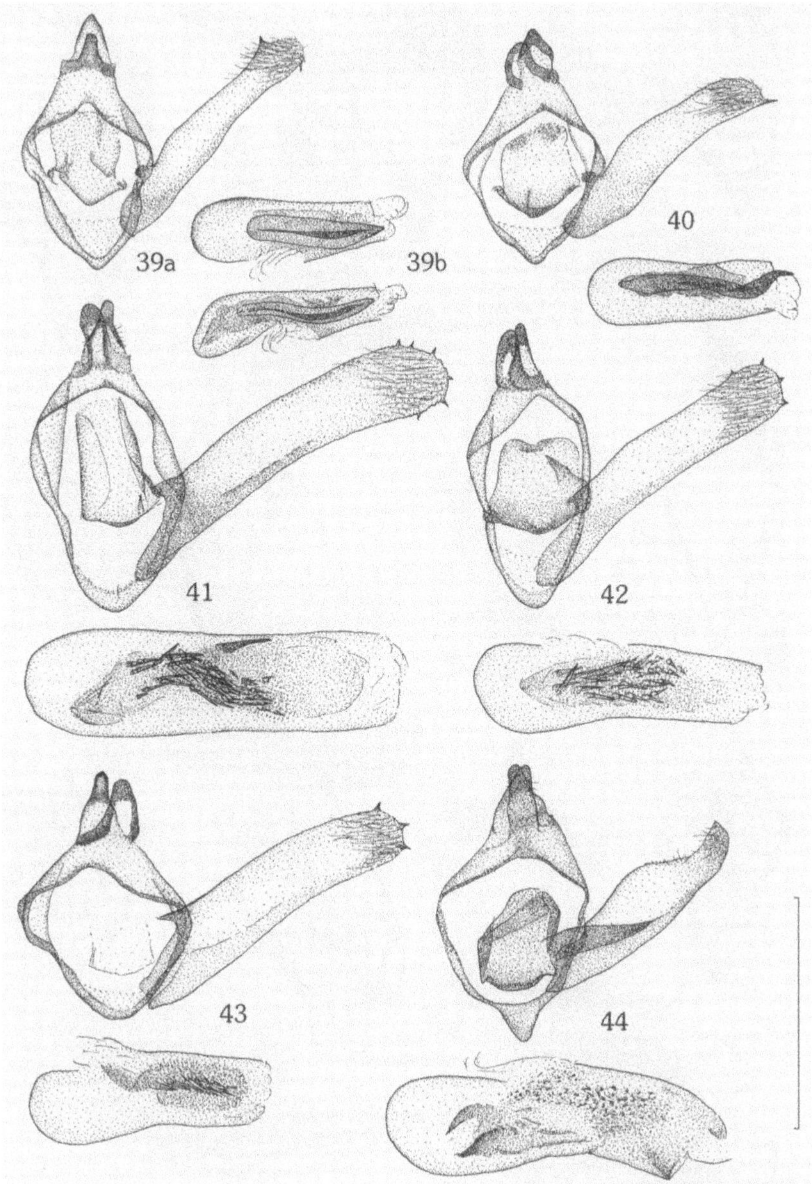

MALE GENITALIA (scale bar = 1 mm)

39a. *Idaea attenuaria* Rmb., Corsica
39b. *Idaea attenuaria* Rmb., S. Spain, aedeagus (lateral view)
40. *Idaea incalcarata* Chrét., NE. Spain
41. *Idaea typicata* Gn., NW. Italy
42. *Idaea alyssumata* Himm. & Mill., NE. Spain
43. *Idaea urcitana* Agenjo, holotype, SE. Spain
44. *Idaea moniliata* Den. & Schiff., Greece

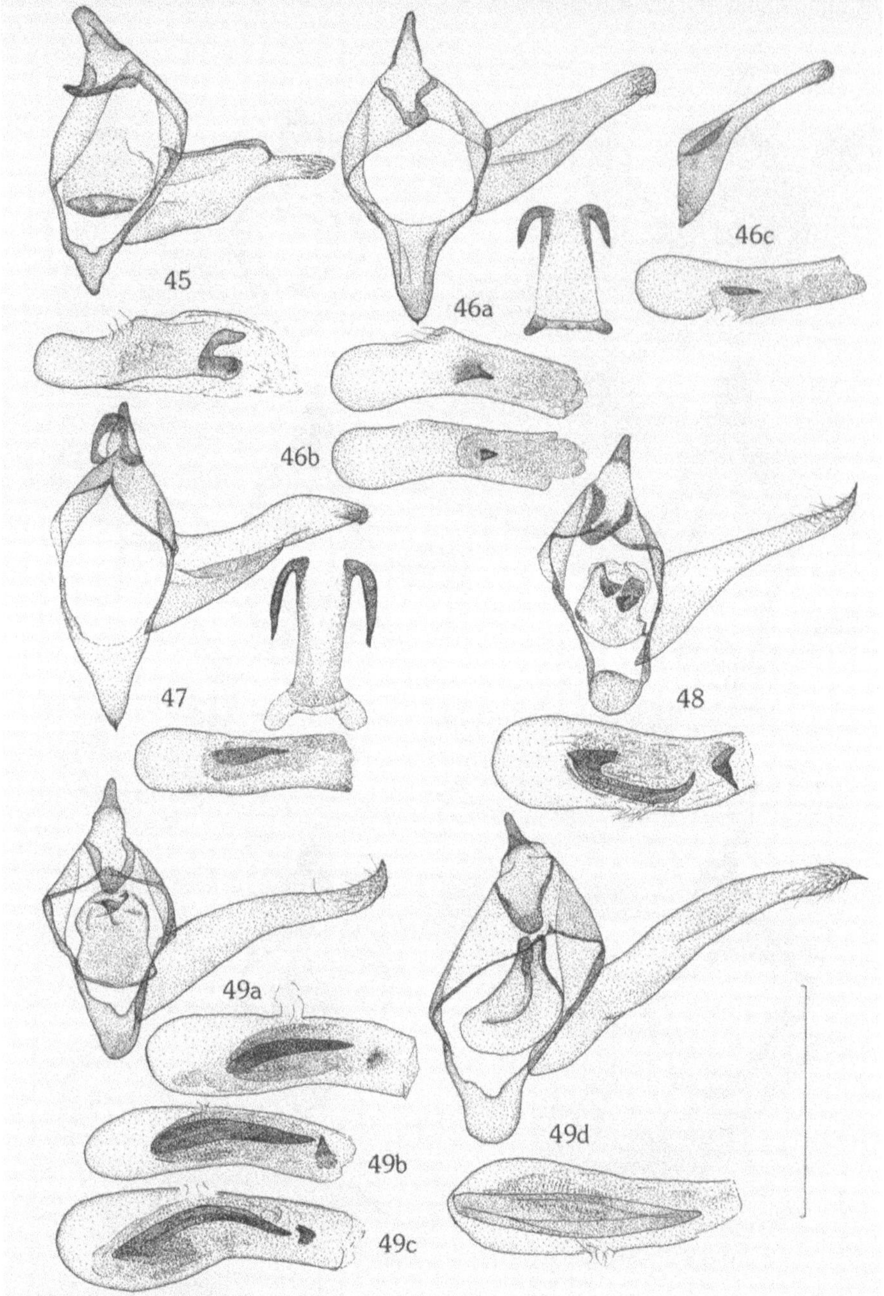

MALE GENITALIA (scale bar = 1 mm)

45. *Idaea spissilimbaria* Mab., holotype, 'Algeria'
46a. *Idaea circuitaria circuitaria* Hbn., C. Italy, juxta separate
46b. *Idaea circuitaria expositoi* Hausm., holotype, S. Spain, aedeagus
46c. *Idaea mimosaria wimmeri* Hausm., paratype, Cyprus, valva and aedeagus
47. *Idaea rainerii* Hausm., SE. Spain, juxta separate
48. *Idaea albarracina* Rssr., NE. Spain
49a. *Idaea incisaria incisaria* Stgr., S. Spain
49b. *Idaea incisaria praecisa* Rssr., N. Morocco, aedeagus
49c. *Idaea incisaria pulverulenta* Rssr., E. Algeria, aedeagus
49d. *Idaea textaria* Led., Cyprus

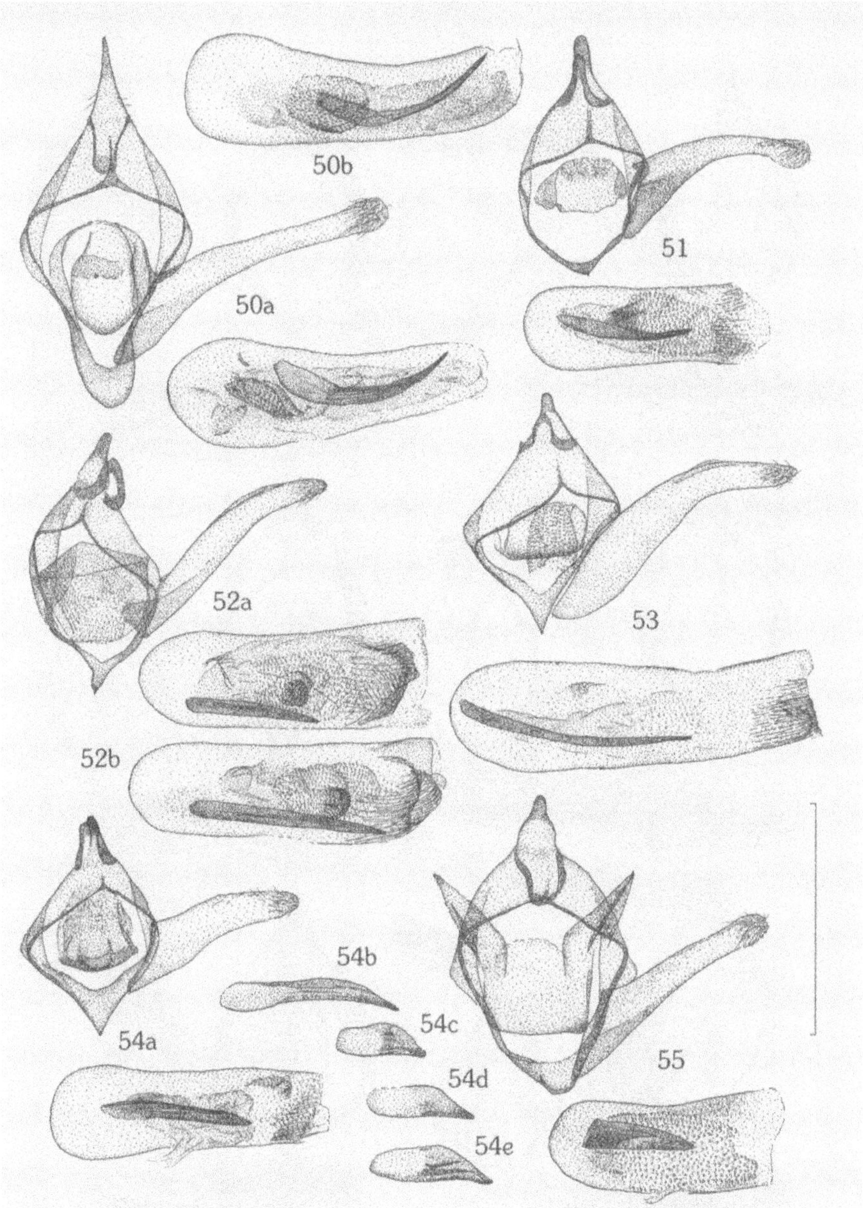

MALE GENITALIA (scale bar = 1 mm)

50a. *Idaea calunetaria calunetaria* Stgr., S. Spain
50b. *Idaea calunetaria valesiaria* Püng., S. Switzerland, aedeagus
51. *Idaea belemiata* Mill., E. Spain
52a. *Idaea elongaria elongaria* Rmb., S. Spain
52b. *Idaea elongaria pecharia* Stgr., N. Hungary, aedeagus
53. *Idaea palaestinensis* Stern., S. Greece
54a. *Idaea obsoletaria obsoletaria* Rmb., Corsica
54b. *Idaea obsoletaria dionigii* Hausm., paratype, S. Italy, cornutus
54c. *Idaea obsoletaria dierli* Hausm., paratype, N. Sicily, cornutus
54d. *Idaea obsoletaria lilaceola* Hausm., E. Spain, cornutus
54e. *Idaea obsoletaria rufularia* H.-S., E. Austria, cornutus
55. *Idaea obliquaria* Trti., Corsica

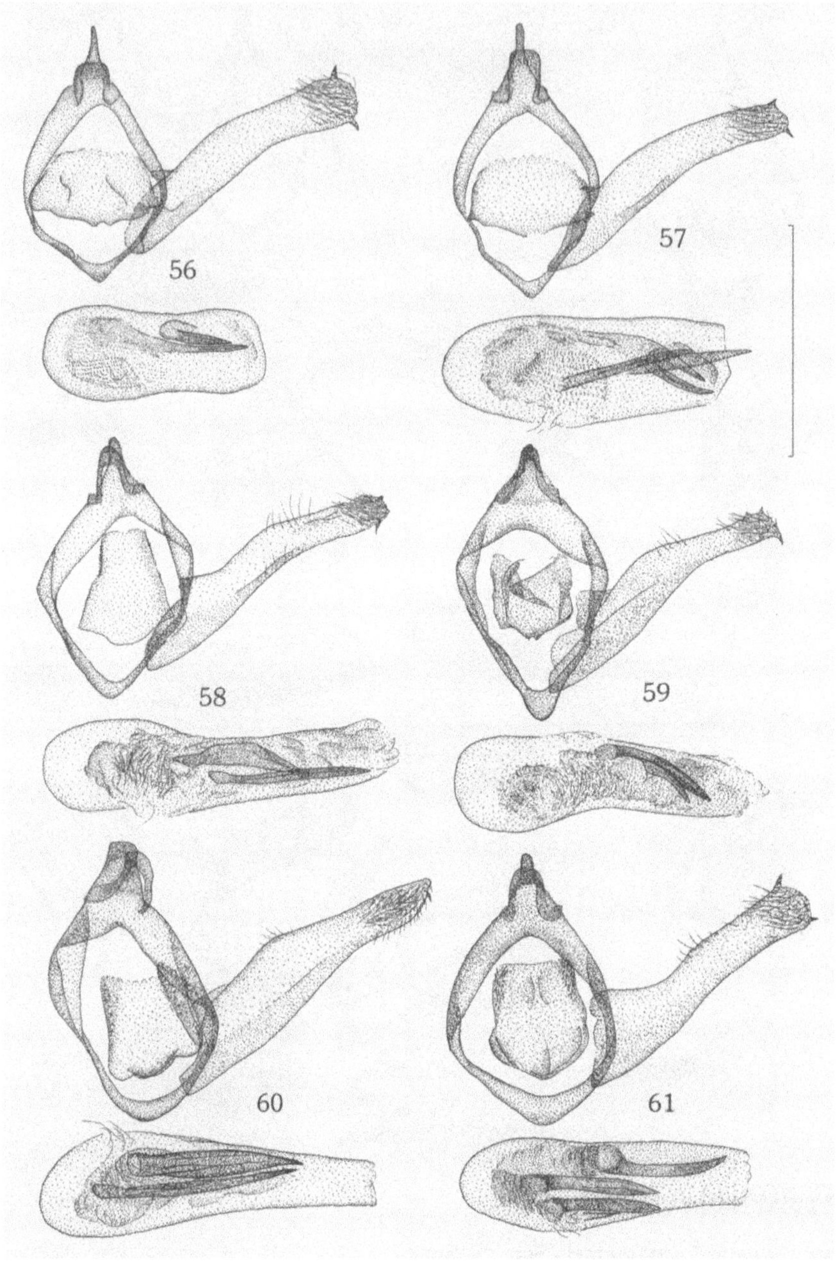

MALE GENITALIA (scale bar = 1 mm)

56. *Idaea inquinata* Scop., Macedonia
57. *Idaea blaesii* Lenz & Hausm., NE. Spain
58. *Idaea dilutaria* Hbn., NW. Serbia
59. *Idaea fuscovenosa* Gze., N. Germany
60. *Idaea robiginata* Stgr., C. Spain
61. *Idaea lutulentaria* Stgr., SE. Spain

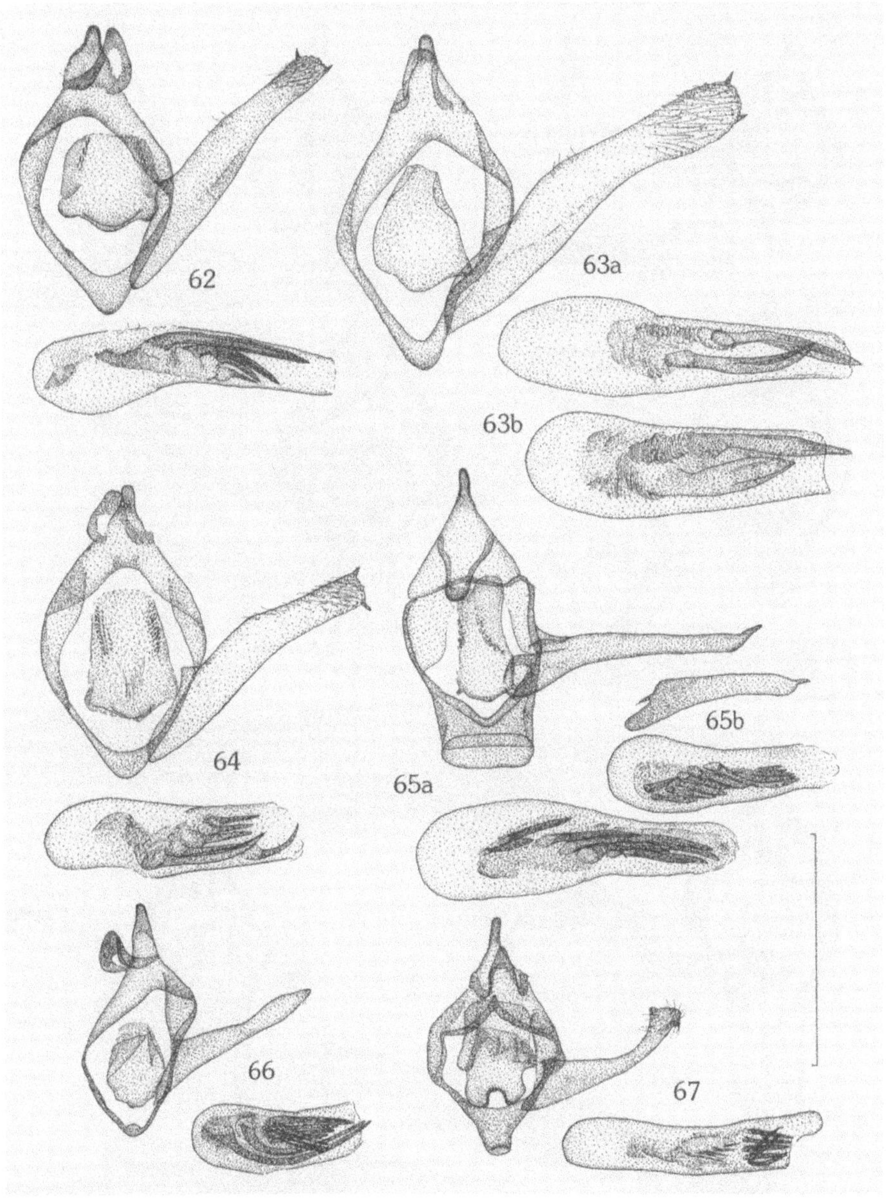

MALE GENITALIA (scale bar = 1 mm)

62. *Idaea humiliata* Hufn., C. Spain
63a. *Idaea bigladiata bigladiata* Herb., S. Spain
63b. *Idaea bigladiata alberti* subsp. n., Tunisia, aedeagus
64. *Idaea politaria* Hbn., Macedonia
65a. *Idaea longaria* H.-S., Sicily
65b. *Idaea longaria* H.-S., S. Spain, aedeagus and valva
66. *Idaea nexata* Hbn., S. Spain
67. *Idaea manicaria* H.-S., Malta

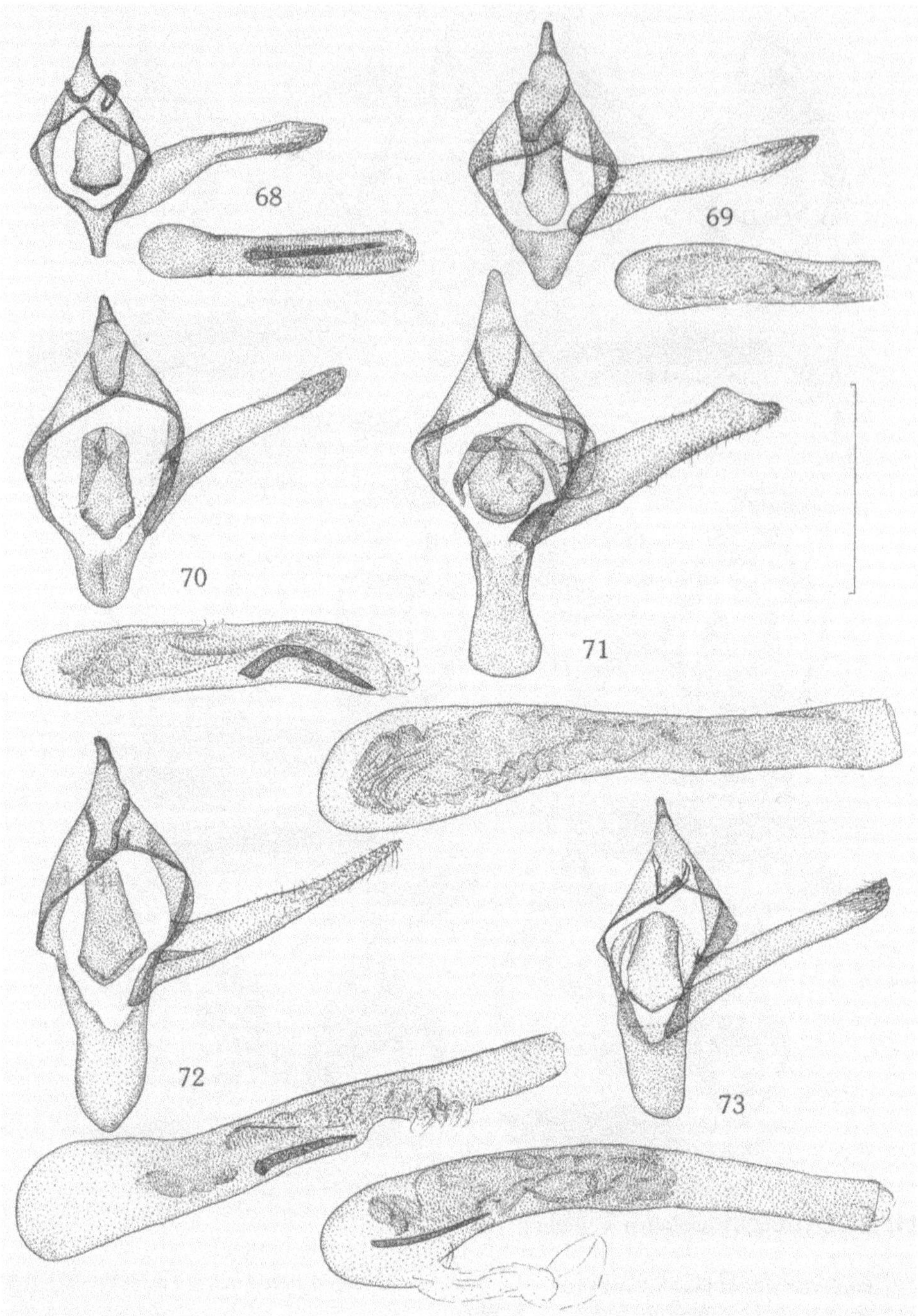

MALE GENITALIA (scale bar = 1 mm)

68. *Idaea descitaria* Chr., SE. European Russia
69. *Idaea vesubiata* Mill., SE. France
70. *Idaea libycata* Bart., S. France
71. *Idaea consolidata* Led., C. Italy
72. *Idaea joannisiata* Homb., N. Spain
73. *Idaea acutipennis* Hausm. & Honey, holotype, Balearics

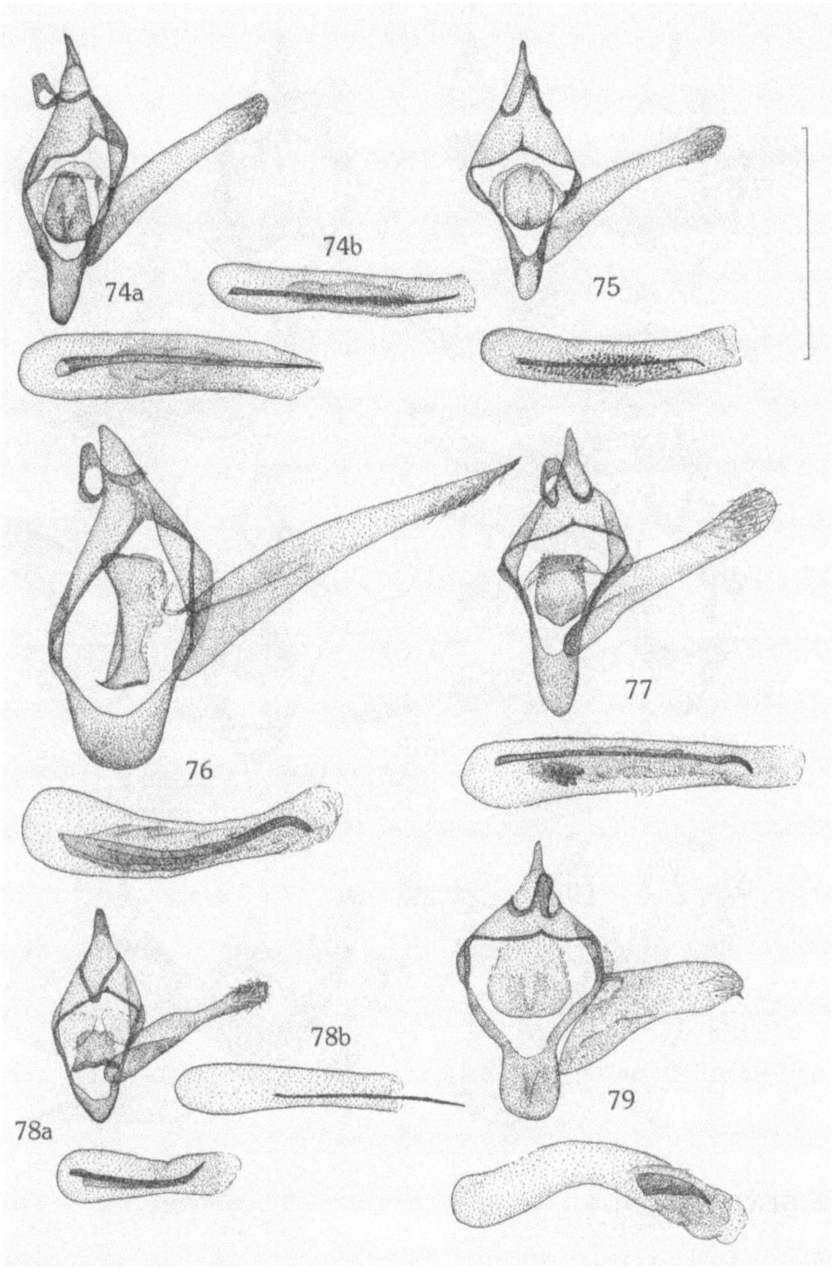

MALE GENITALIA (scale bar = 1 mm)

74a. *Idaea seriata seriata* Schrank, Hungary
74b. *Idaea seriata canteneraria* Bsd., C. Algeria, aedeagus
75. *Idaea minuscularia* Ribbe, S. Spain
76. *Idaea ibizaria* Mentz., paratype, Balearics
77. *Idaea albitorquata* Püng., paralectotype, S. Greece
78a. *Idaea carvalhoi* Herb., SE. Spain
78b. *Idaea fathmaria* Obth., Algeria, aedeagus
79. *Idaea camparia* H.-S., Israel

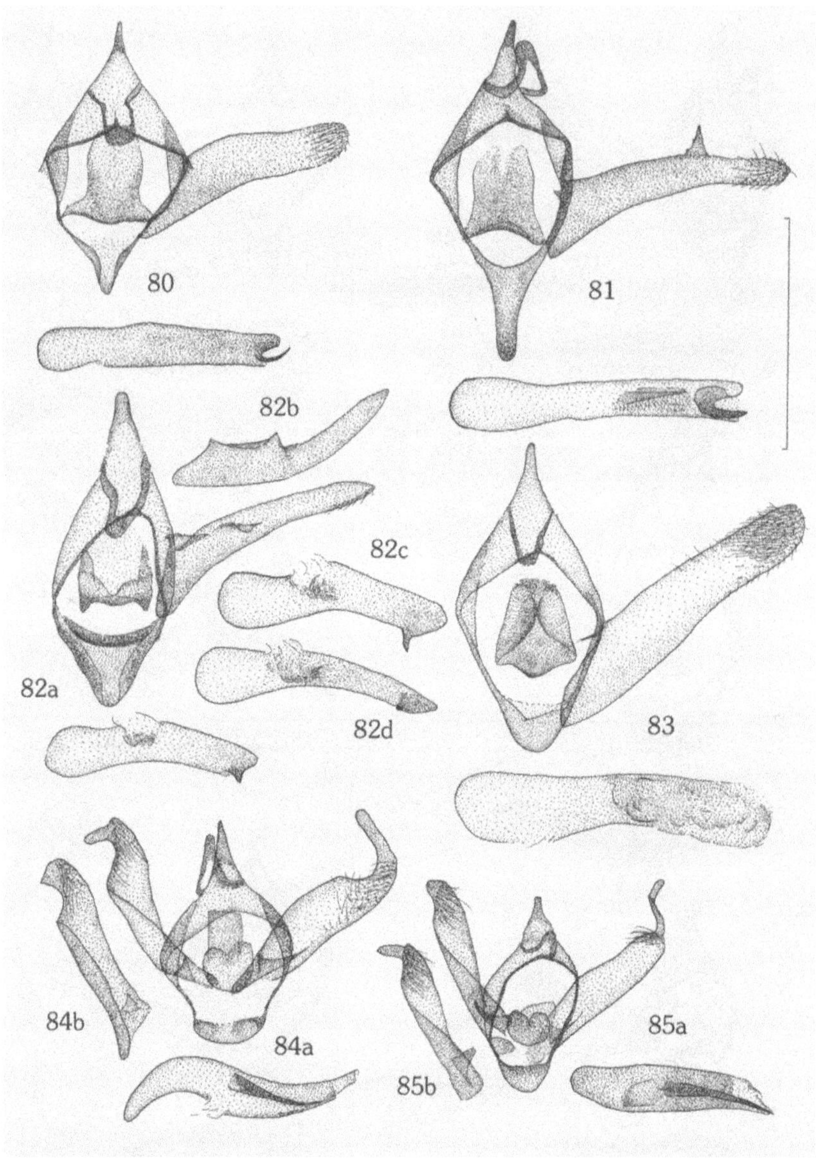

MALE GENITALIA (scale bar = 1 mm)

80. *Idaea subsericeata* Haw., C. Italy
81. *Idaea pallidata* Den. & Schiff., N. Italy
82a. *Idaea mancipiata mancipiata* Stgr., N. Spain
82b. *Idaea mancipiata mancipiata* Stgr., N. Spain, valva in lateral view
82c. *Idaea mancipiata repagulata* Prt., SE. European Russia, aedeagus
82d. *Idaea mancipiata repagulata* Prt., Azerbaijan, aedeagus
83. *Idaea sylvestraria* Hbn., N. Italy
84a. *Idaea dimidiata dimidiata* Hufn., S. France
84b. *Idaea dimidiata antitaurica* Whli., Cyprus, left valva
85a. *Idaea subsaturata subsaturata* Gn., N. Spain
85b. *Idaea subsaturata lecerfiata* Homb., Algeria, left valva

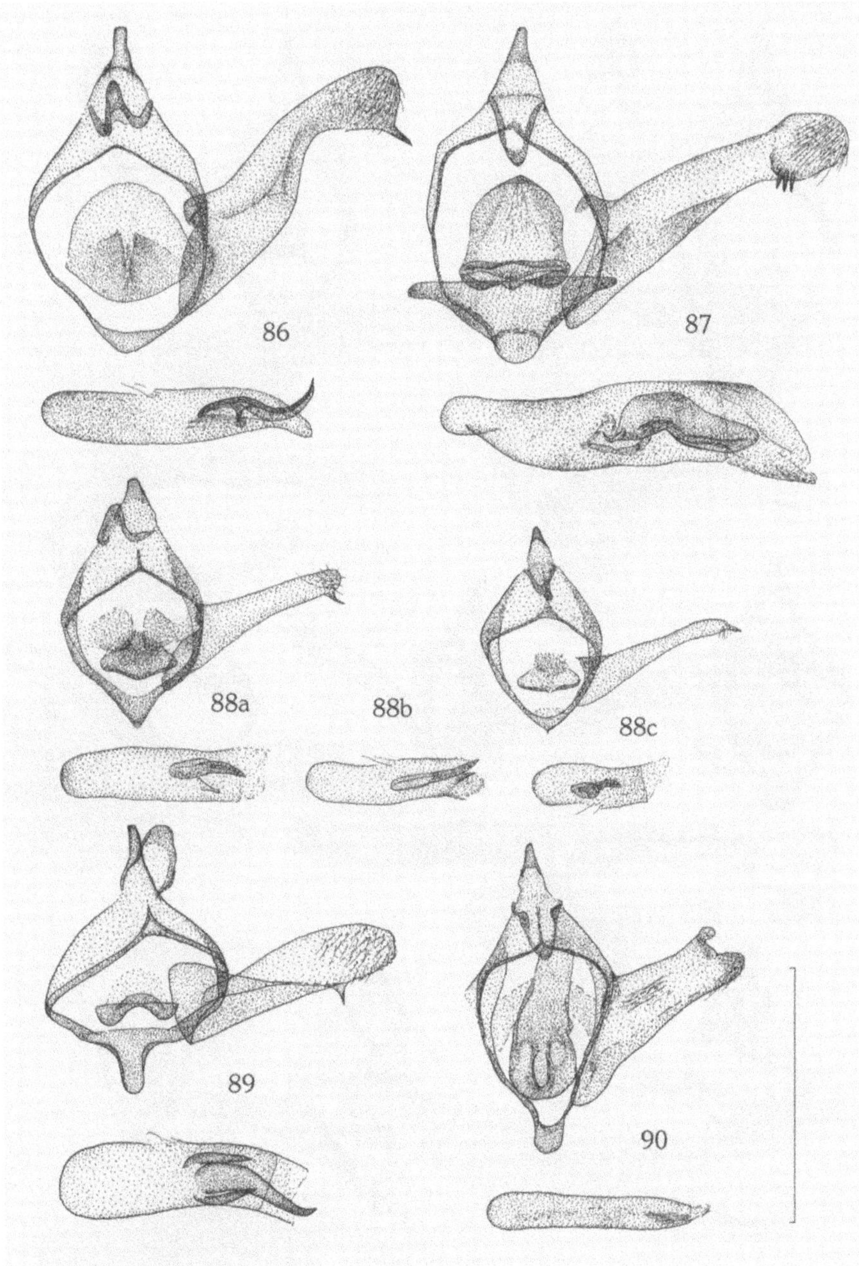

MALE GENITALIA (scale bar = 1 mm)

86. *Idaea trigeminata* Haw., W. Croatia
87. *Idaea biselata* Hufn., Romania
88a. *Idaea fractilineata fractilineata* Z., Sicily
88b. *Idaea fractilineata subrufaria* Stgr., S. Spain
88c. *Idaea inclinata* Led., Cyprus
89. *Idaea exilaria* Gn., N. Spain
90. *Idaea cervantaria* Mill., NE. Spain

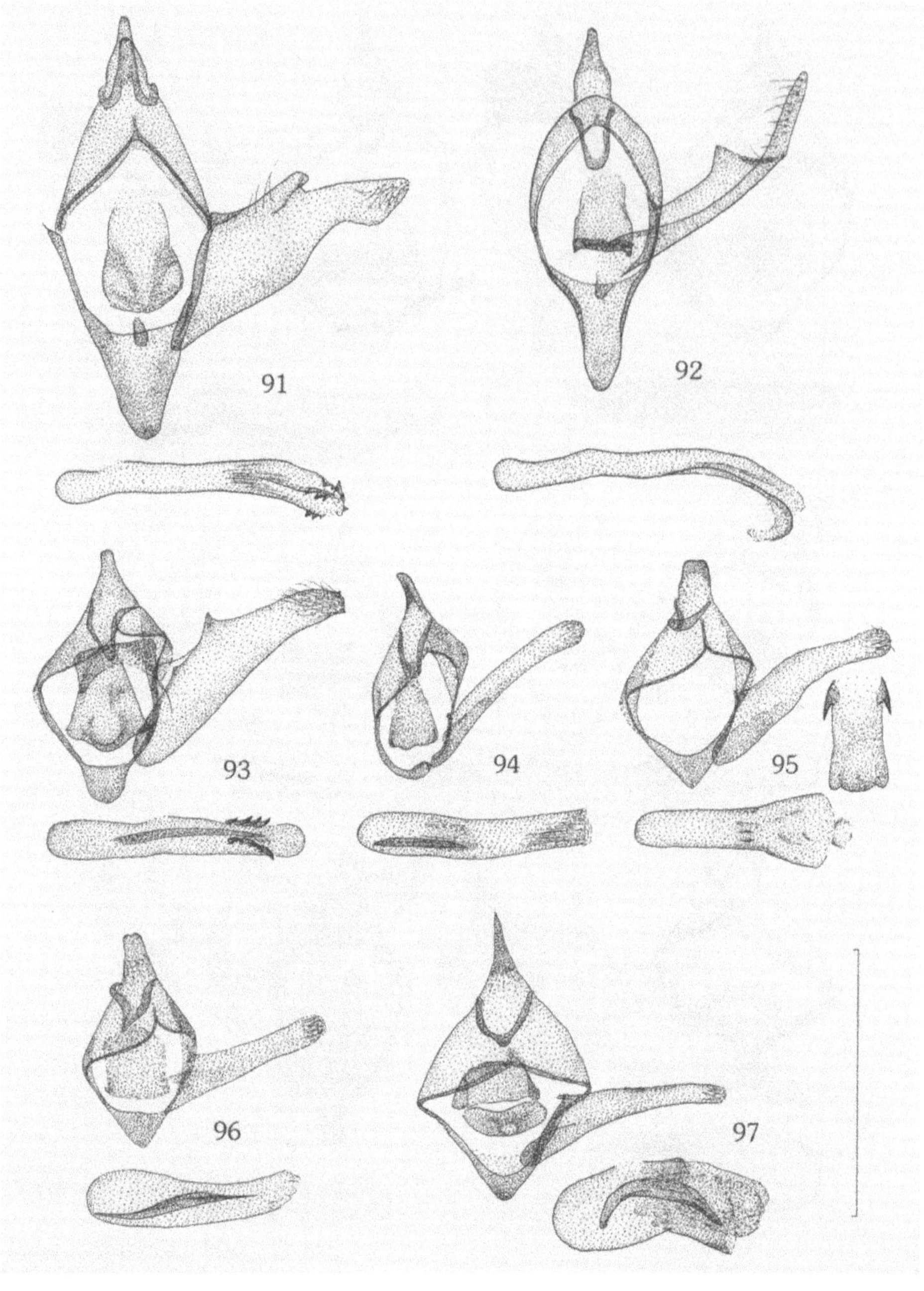

MALE GENITALIA (scale bar = 1 mm)

91. *Idaea contiguaria* Hbn., W. Germany
92. *Idaea rupicolaria* Rssr., S. Spain
93. *Idaea deitanaria* Rssr. & Weis., SE. Spain
94. *Idaea saleri* Dom. & Baix., SE. Spain
95. *Idaea tineata* Th.-Mieg., Cyprus, juxta separate
96. *Idaea infirmaria* Rmb., Corsica
97. *Idaea rhodogrammaria* Püng., E. Spain

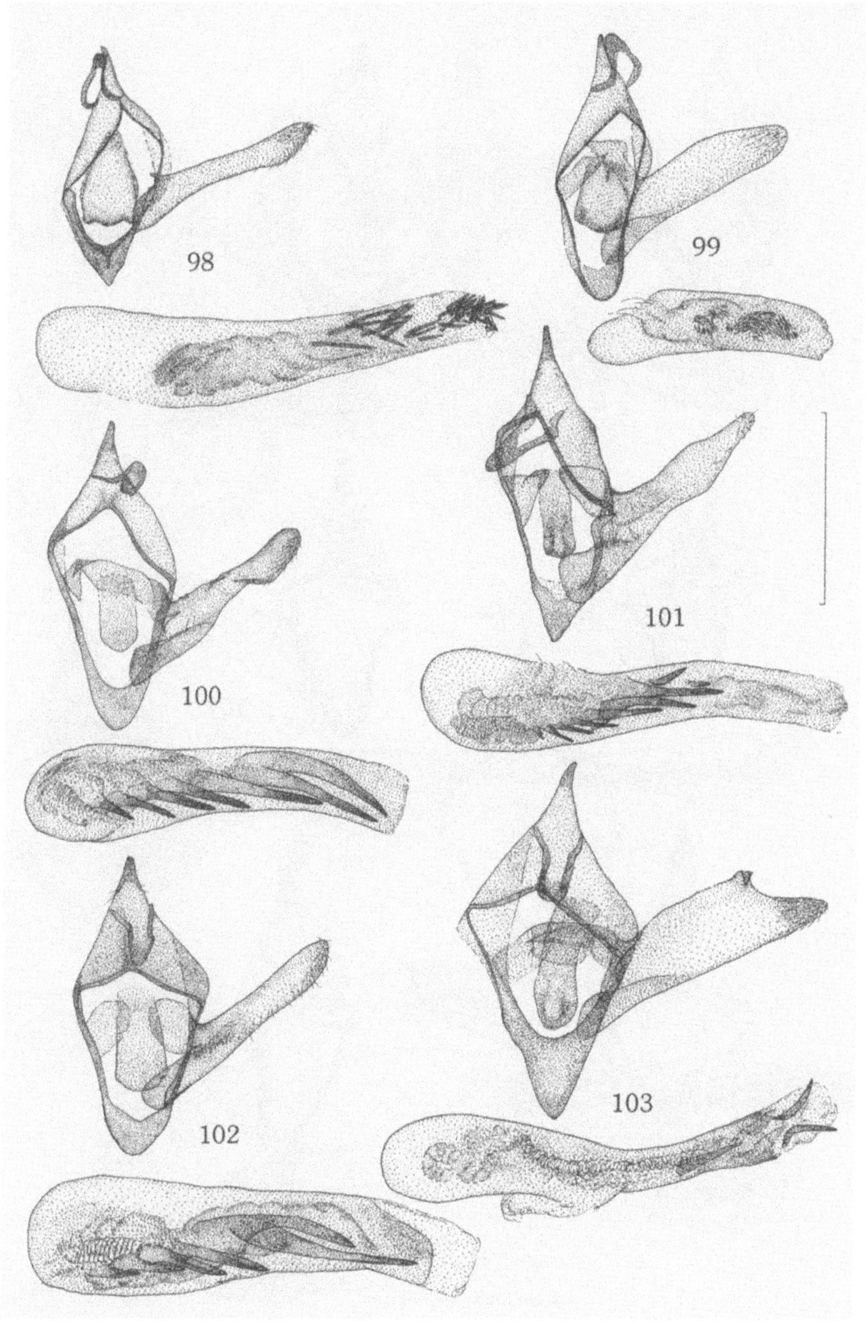

MALE GENITALIA (scale bar = 1 mm)

98. *Idaea ostrinaria* Hbn., Macedonia
99. *Idaea metohiensis* Reb., N. Greece
100. *Idaea eugeniata* Mill., NE. Spain
101. *Idaea distinctaria* Bsd., Liguria
102. *Idaea predotaria* Hartig, S. France
103. *Idaea nitidata* H.-S., SE. Austria

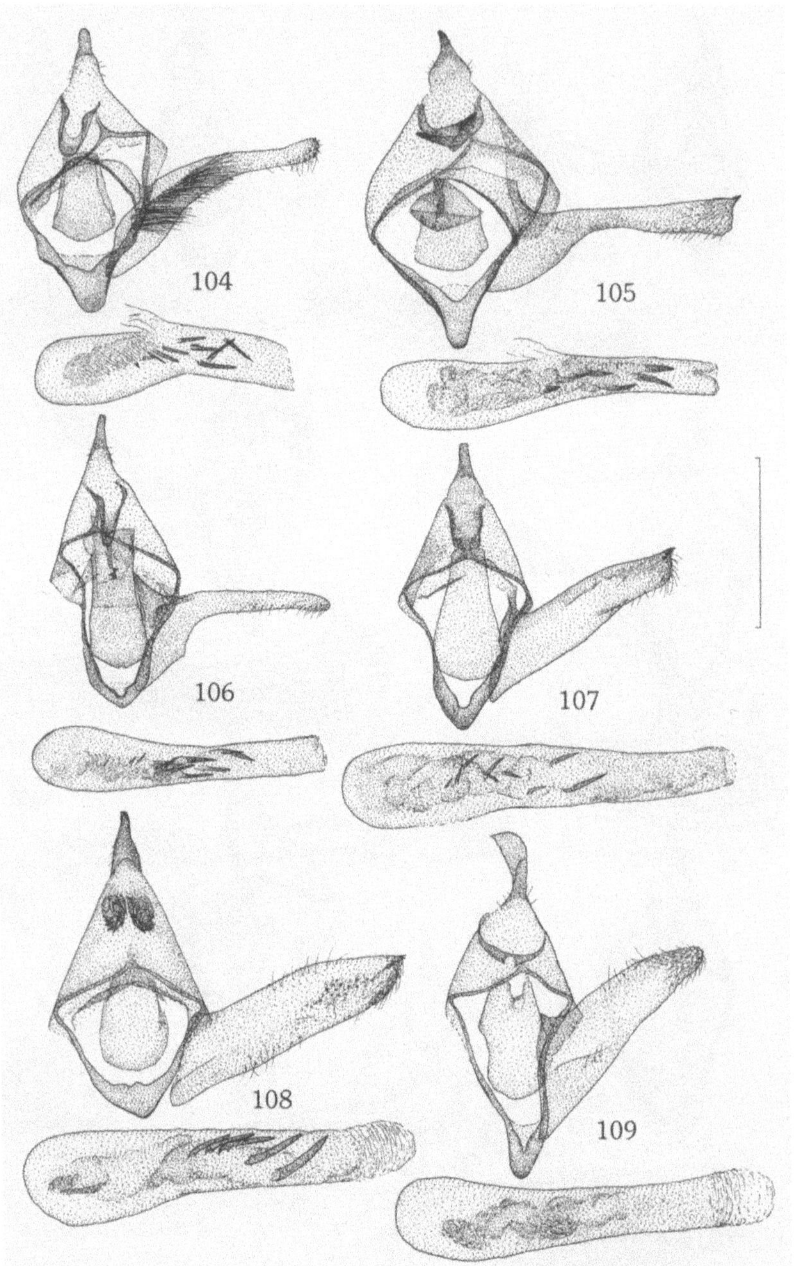

MALE GENITALIA (scale bar = 1 mm)

104. *Idaea emarginata* L., Sweden
105. *Idaea dromikos* Hausm. sp.n., paratype, S. France
106. *Idaea simplicior* Prt., C. Spain
107. *Idaea rubraria* Stgr., Turkey
108. *Idaea aversata* L., C. Germany
109. *Idaea degeneraria* Hbn., S. Italy

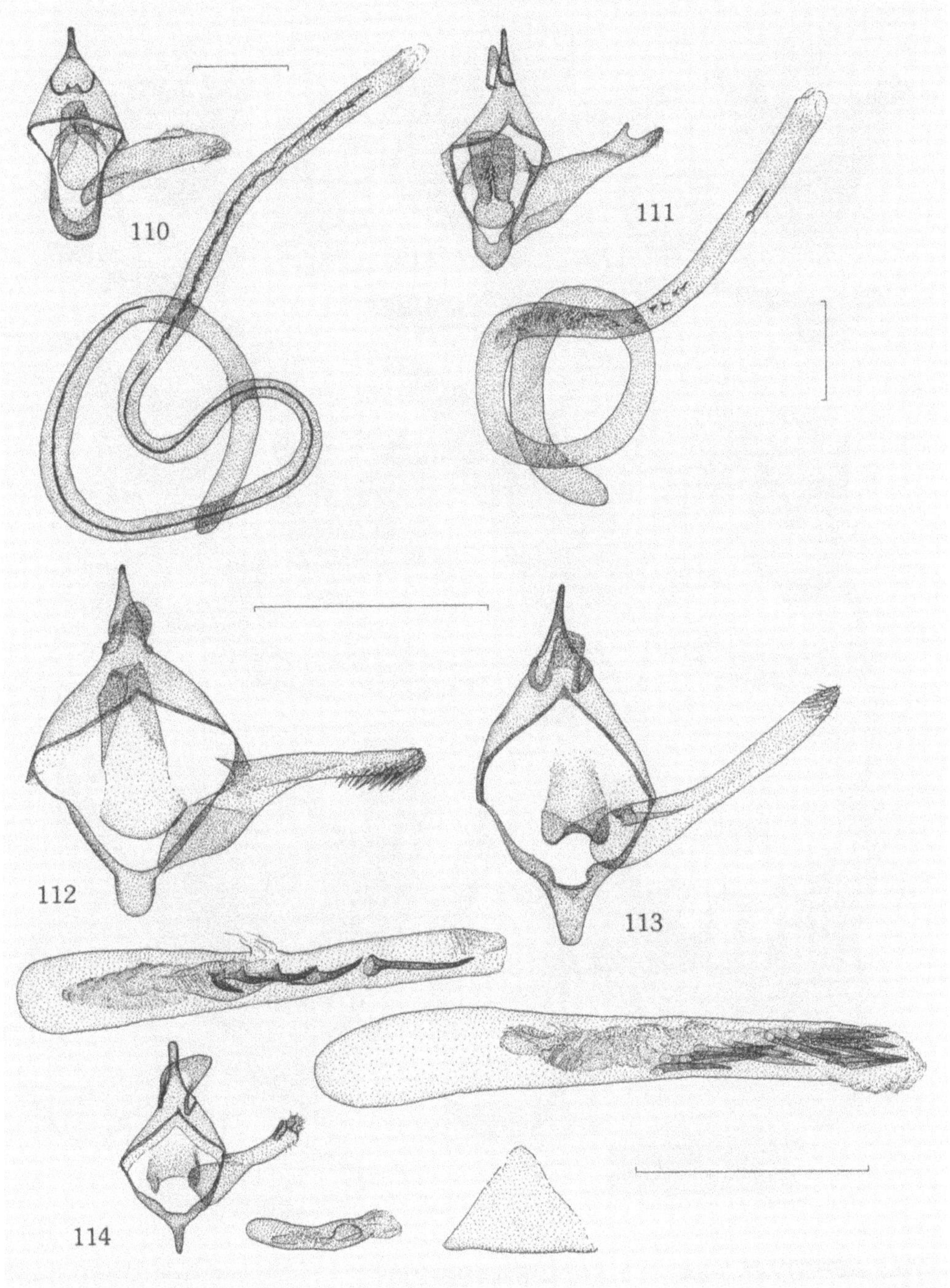

MALE GENITALIA (scale bars = 1 mm)

110. *Idaea straminata* Bkh., S. Italy
111. *Idaea deversaria* H.-S., W. Croatia
112. *Idaea squalidaria* Stgr., N. Spain
113. *Brachyglossina hispanaria* Püng., S. Spain
114. *Limeria macraria* Stgr., SE. European Russia, with sternum A8

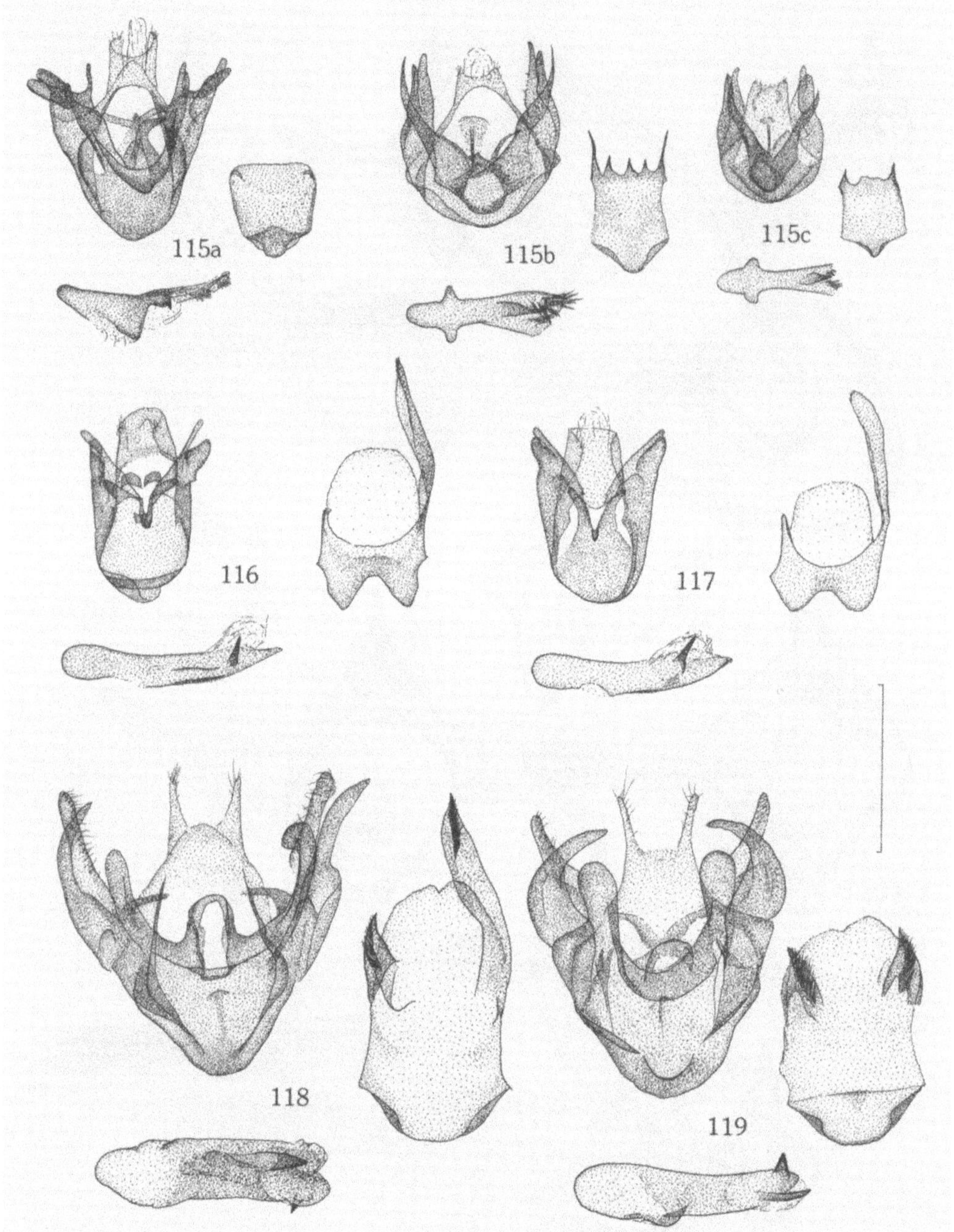

MALE GENITALIA (scale bar = 1 mm, with sternum A8)

115a. *Oar reaumuraria* Mill., SE. Spain
115b. *Oar pratana* F., Israel
115c. *Oar pratana baezi* subsp. n., paratype, Canary islands
116. *Cinglis humifusaria* Ev., Turkey
117. *Cinglis andalusiaria* Wagn., S. Spain
118. *Scopula immorata* L., S. France
119. *Scopula tessellaria* Bsd., S. France

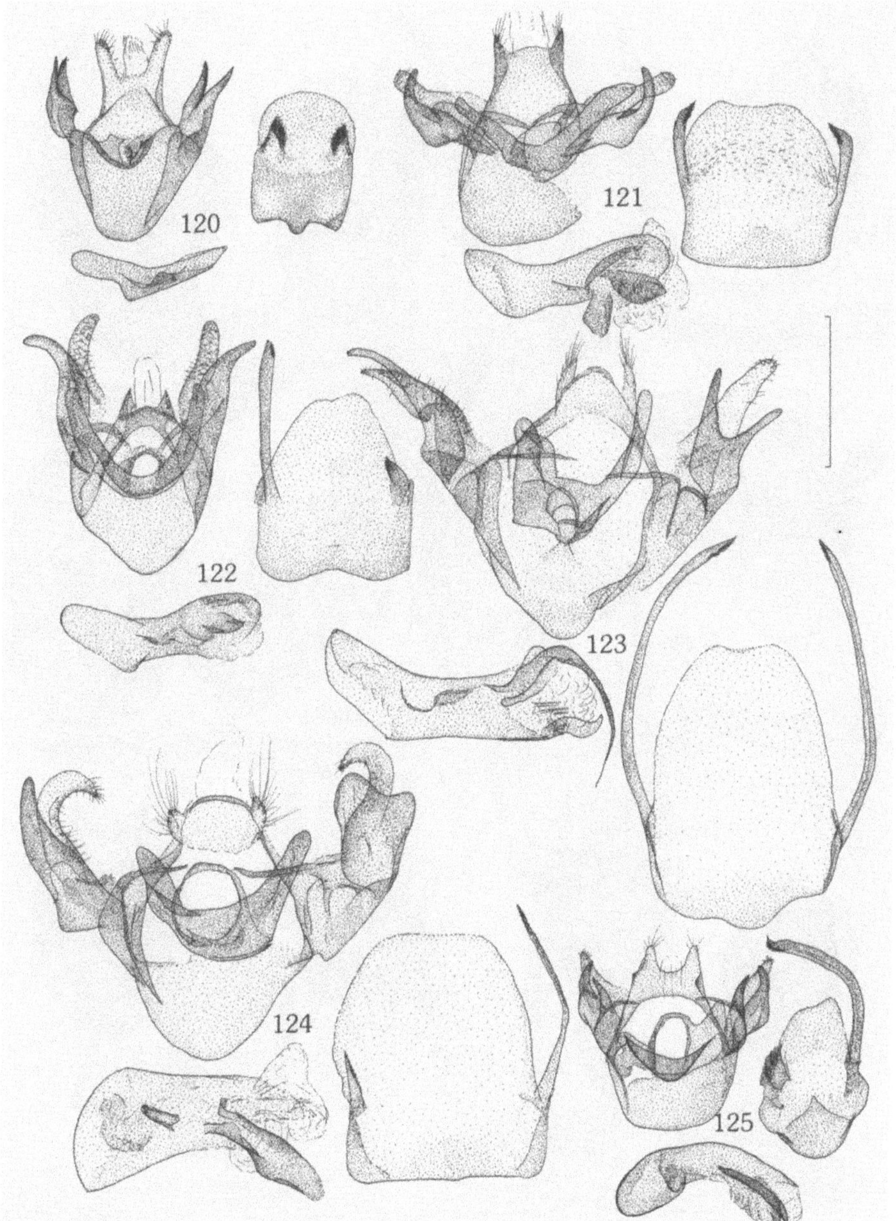

MALE GENITALIA (scale bar = 1 mm, with sternum A8)

120. *Scopula corrivalaria* Kretsch., Hungary
121. *Scopula caricaria* Reut., N. Serbia
122. *Scopula nemoraria* Hbn., S. Germany
123. *Scopula umbelaria* Hbn., NE. Austria
124. *Scopula nigropunctata* Hufn., S. Germany
125. *Scopula virgulata* Den. & Schiff., SW. Germany

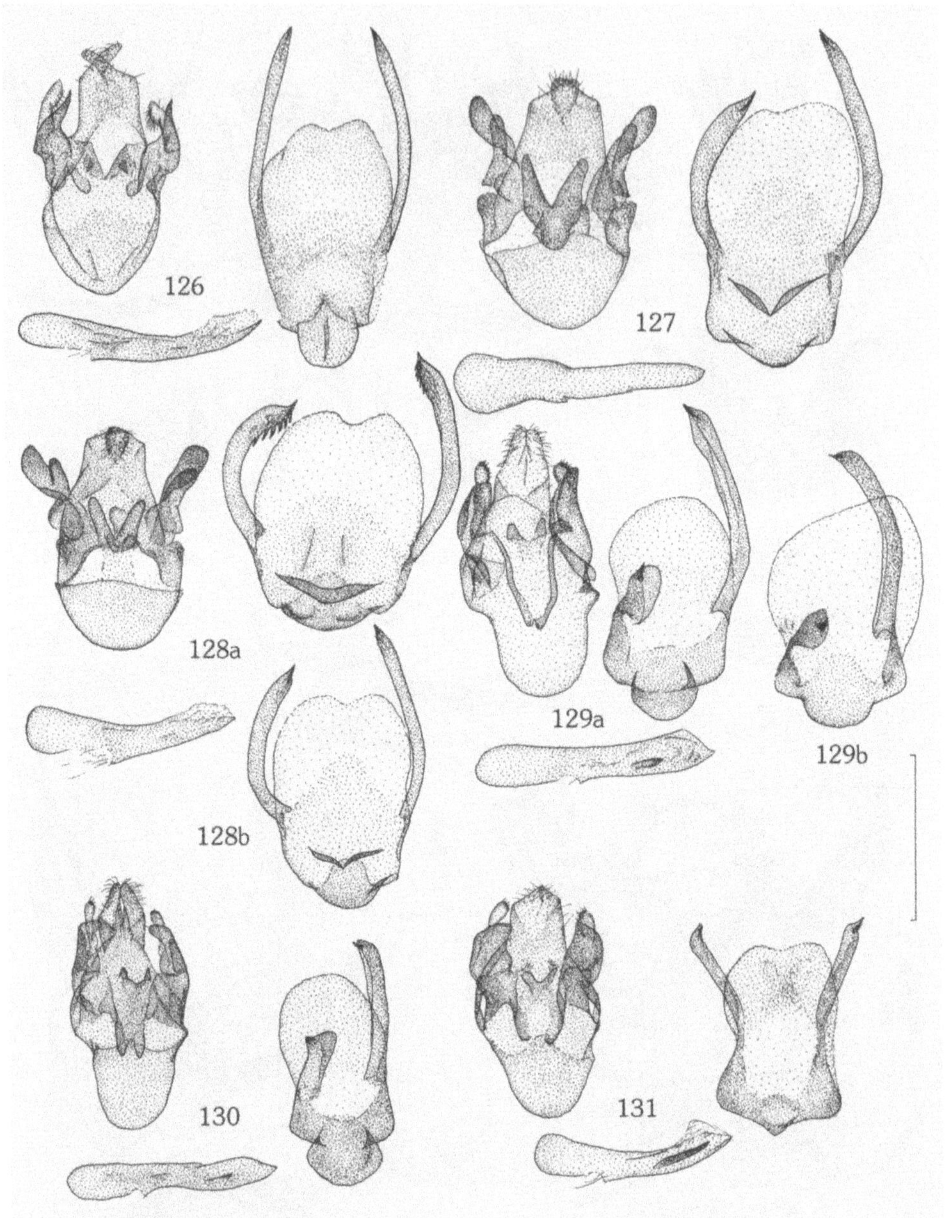

MALE GENITALIA (scale bar = 1 mm, with sternum A8)

126. *Scopula ornata* Scop., C. Romania
127. *Scopula concinnaria* Dup., S. Spain
128a. *Scopula orientalis* Alph., SE. European Russia
128b. *Scopula orientalis* Alph., SE. European Russia, sternum A8
129a. *Scopula decorata decorata* Den. & Schiff., Greece
129b. *Scopula decorata congruata* Z., S. Spain, sternum A8
130. *Scopula honestata* Mab., Corsica
131. *Scopula subtilata* Chr., S. European Russia

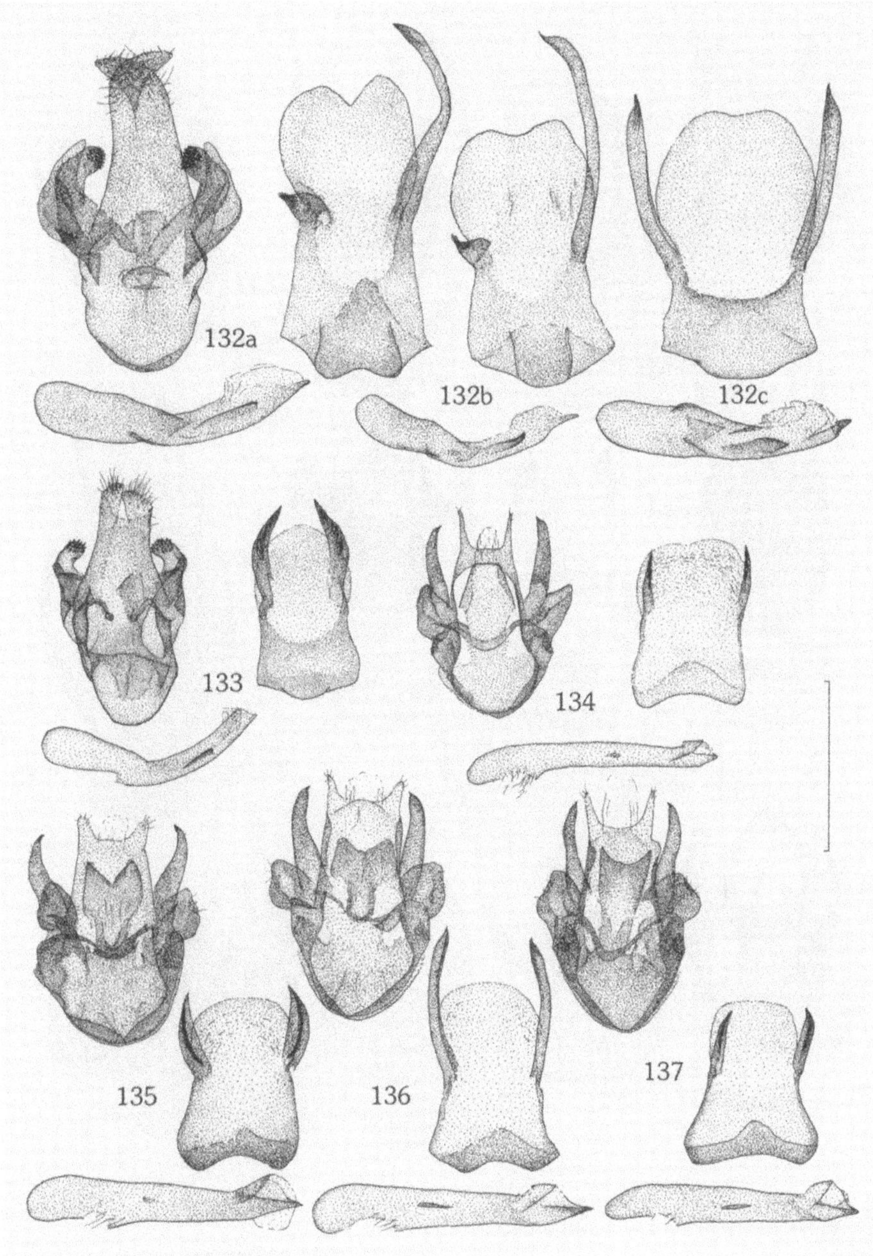

MALE GENITALIA (scale bar = 1 mm, with sternum A8)

132a. *Scopula submutata submutata* Tr., SE. Austria
132b. *Scopula submutata nivellearia* Obth., SE. Spain, sternum A8 and aedeagus
132c. *Scopula transcaspica* Prt., Turkmeniya, sternum A8 and aedeagus
133. *Scopula vigilata* Wgn., S. Italy
134. *Scopula rubiginata* Hufn., Macedonia
135. *Scopula ochraceata* Stgr., Macedonia
136. *Scopula turbidaria* Hbn., S. Spain
137. *Scopula turbulentaria* Stgr., C. Italy

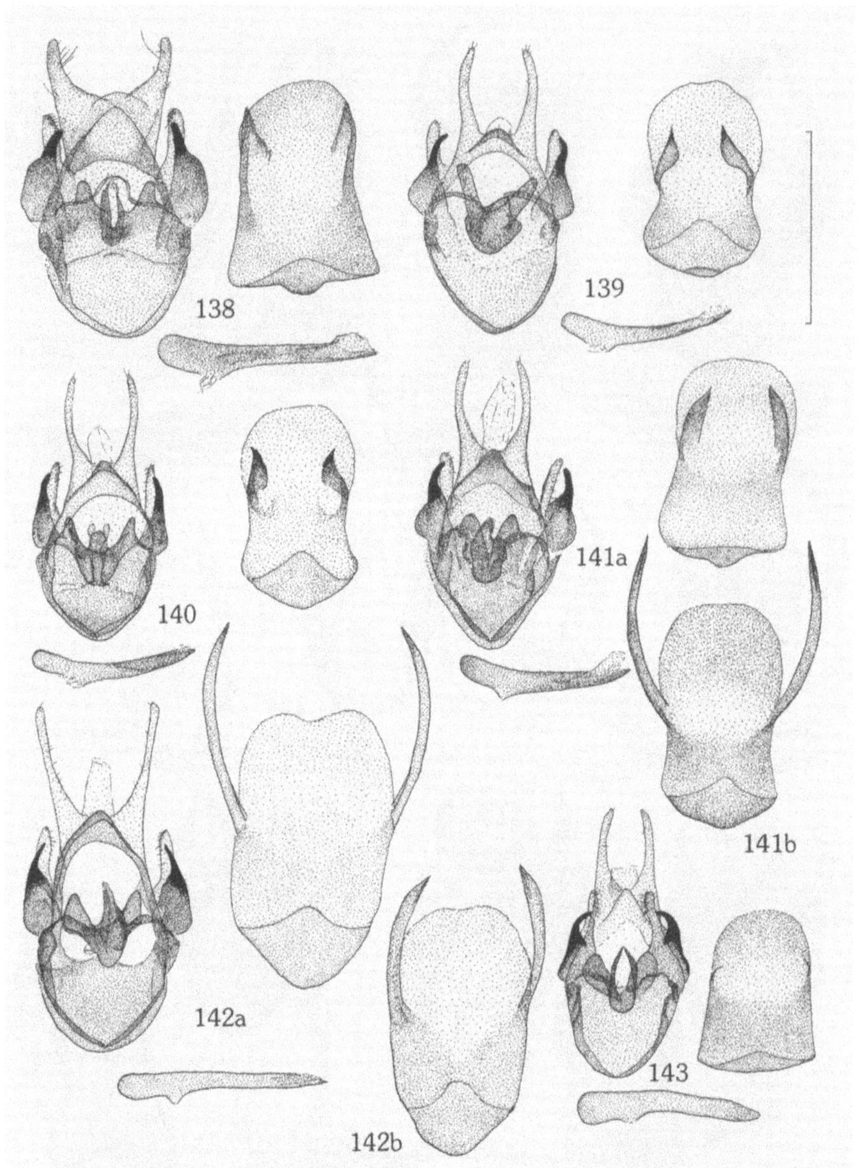

MALE GENITALIA (scale bar = 1 mm, with sternum A8)

138. *Scopula decolor* Stgr., NE. Spain
139. *Scopula rubellata* Stgr., E. Spain
140. *Scopula scalercii* Hausm., C. Italy
141a. *Scopula beckeraria beckeraria* Led., S. Urals
141b. *Scopula beckeraria rebeli* Led., Macedonia
142a. *Scopula incanata* L., S. Germany
142b. *Scopula incanata* L., Romania, polymorphic variant, sternum A8
143. *Scopula marginepunctata* Gze., S. Italy

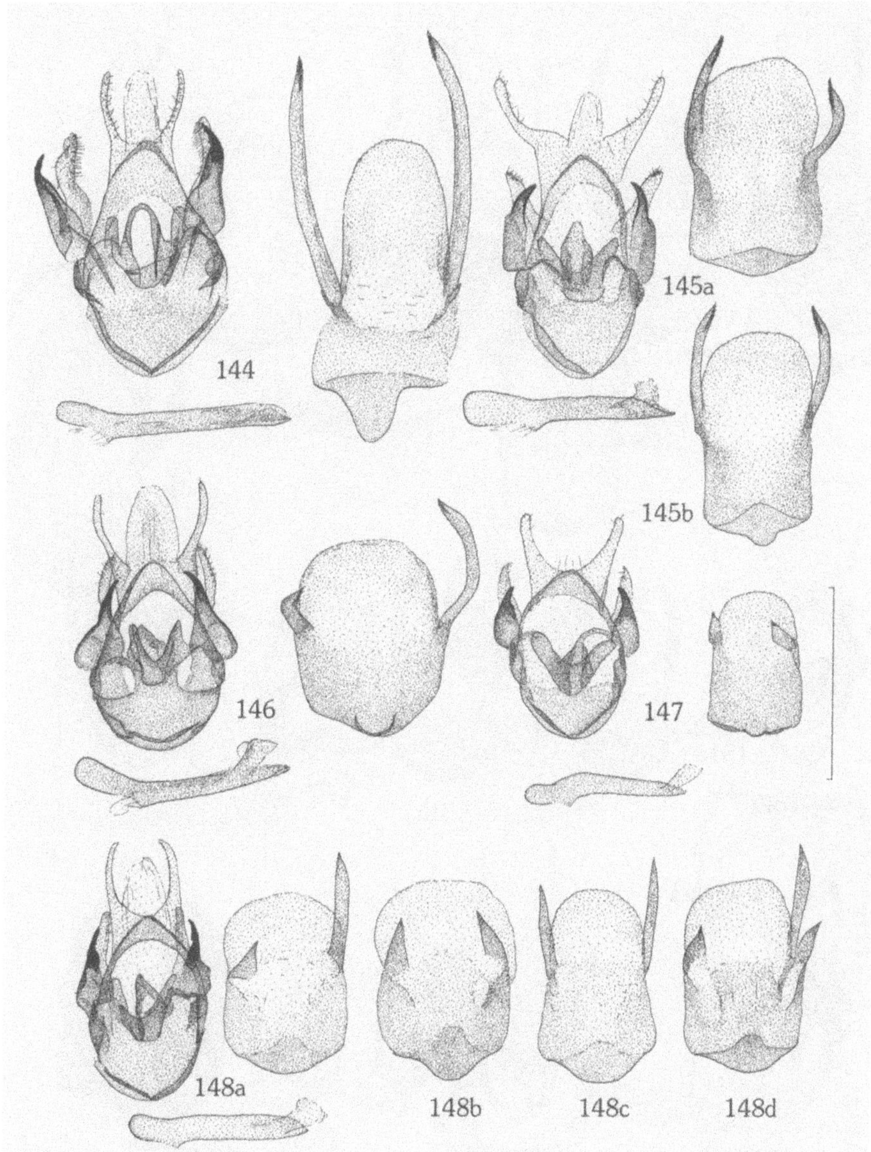

MALE GENITALIA (scale bar = 1 mm, with sternum A8)

144. *Scopula luridata* Z., Cyprus
145a. *Scopula imitaria imitaria* Hbn., C. Italy
145b. *Scopula imitaria syriacaria* Culot, Cyprus, sternum A8
146. *Scopula immutata* L., Denmark
147. *Scopula frigidaria* Möschl., Finland
148a. *Scopula ternata* Schrank, Norway
148b. *Scopula ternata* Schrank, Norway, polymorphic variant, sternum A8
148c. *Scopula ternata* Schrank, Norway, polymorphic variant, sternum A8
148d. *Scopula ternata* Schrank, Norway, aberration, sternum A8

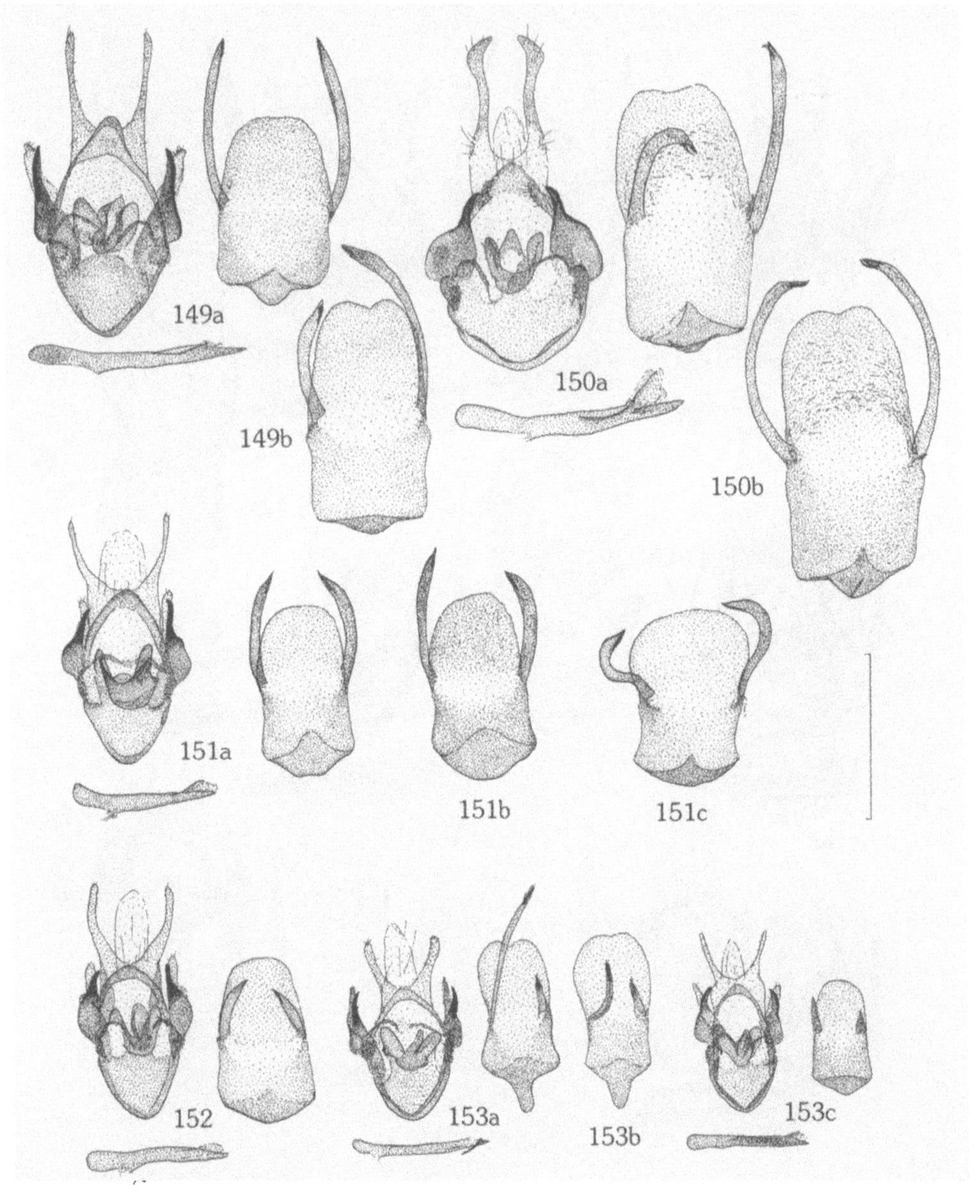

MALE GENITALIA (scale bar = 1 mm, with sternum A8)

149a. *Scopula floslactata* Haw., SW. Germany
149b. *Scopula floslactata* Haw., C. Austria, polymorphic variant, sternum A8
150a. *Scopula subpunctaria* H.-S., SE. Austria
150b. *Scopula subpunctaria* H.-S., SE. Austria, polymorphic variant, sternum A8
151a. *Scopula flaccidaria* Z., E. Austria
151b. *Scopula flaccidaria* Z., E. Austria, variant, sternum A8
151c. *Scopula albidaria* Stgr., C. Asia, sternum A8
152. *Scopula emutaria* Hbn., Portugal
153a. *Scopula minorata* Bsd., Portugal
153b. *Scopula minorata* Bsd., Israel, polymorphic variant, sternum A8
153c. *Scopula* spec. n.?, SE. France

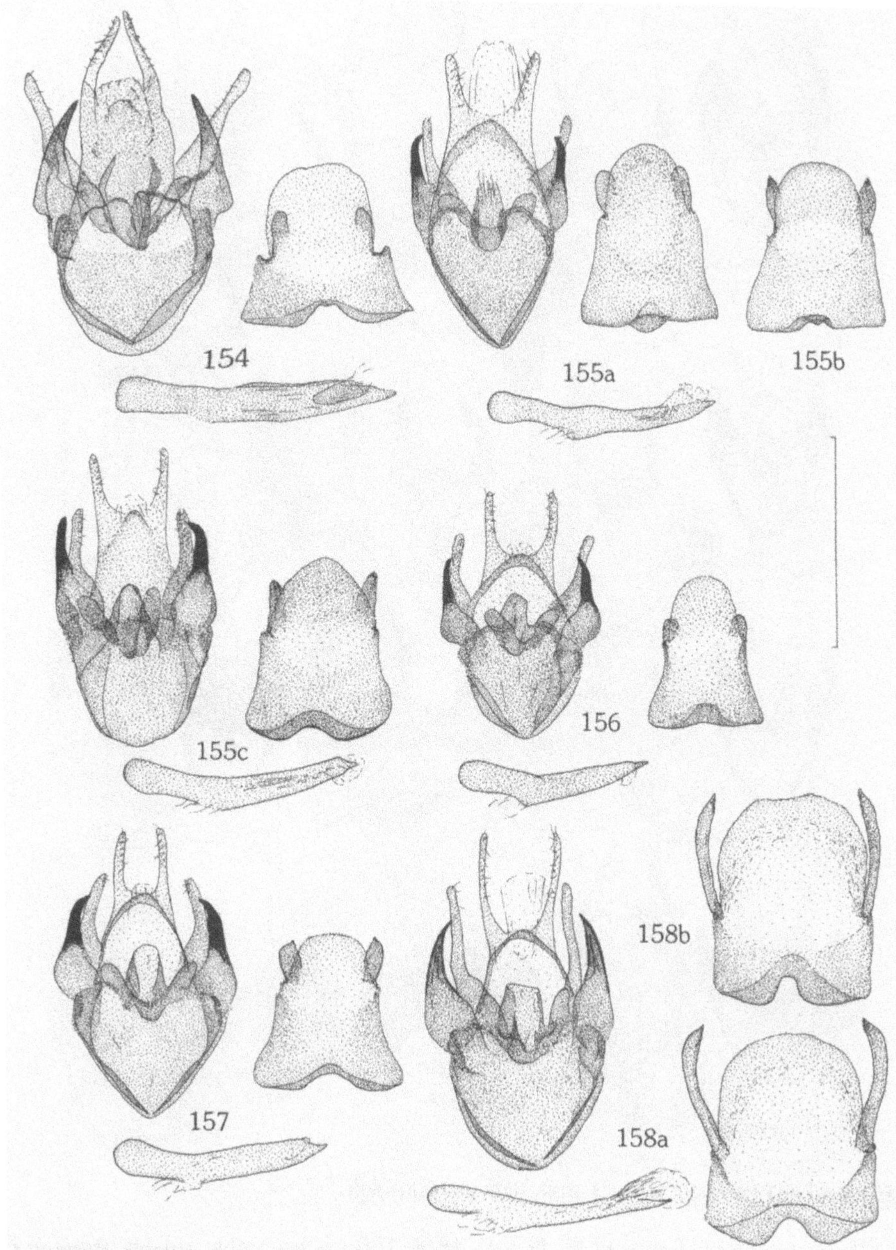

MALE GENITALIA (scale bar = 1 mm, with sternum A8)

154. *Scopula cajanderi* Herz, NE. European Russia
155a. *Scopula divisaria* Chr., SE. European Russia
155b. *Scopula divisaria* Chr. (?), E. Kazakhstan, sternum A8
155c. *Scopula latelineata* Grs., Kirghizstan, syntype
156. *Scopula arenosaria* Stgr., SE. European Russia, syntype
157. *Scopula albiceraria* H.-S., SE. European Russia
158a. *Scopula immistaria immistaria* H.-S., Caucasus
158b. *Scopula immistaria beshkovi* Gelbr. & Hausm., E. Bulgaria, paratype, sternum A8

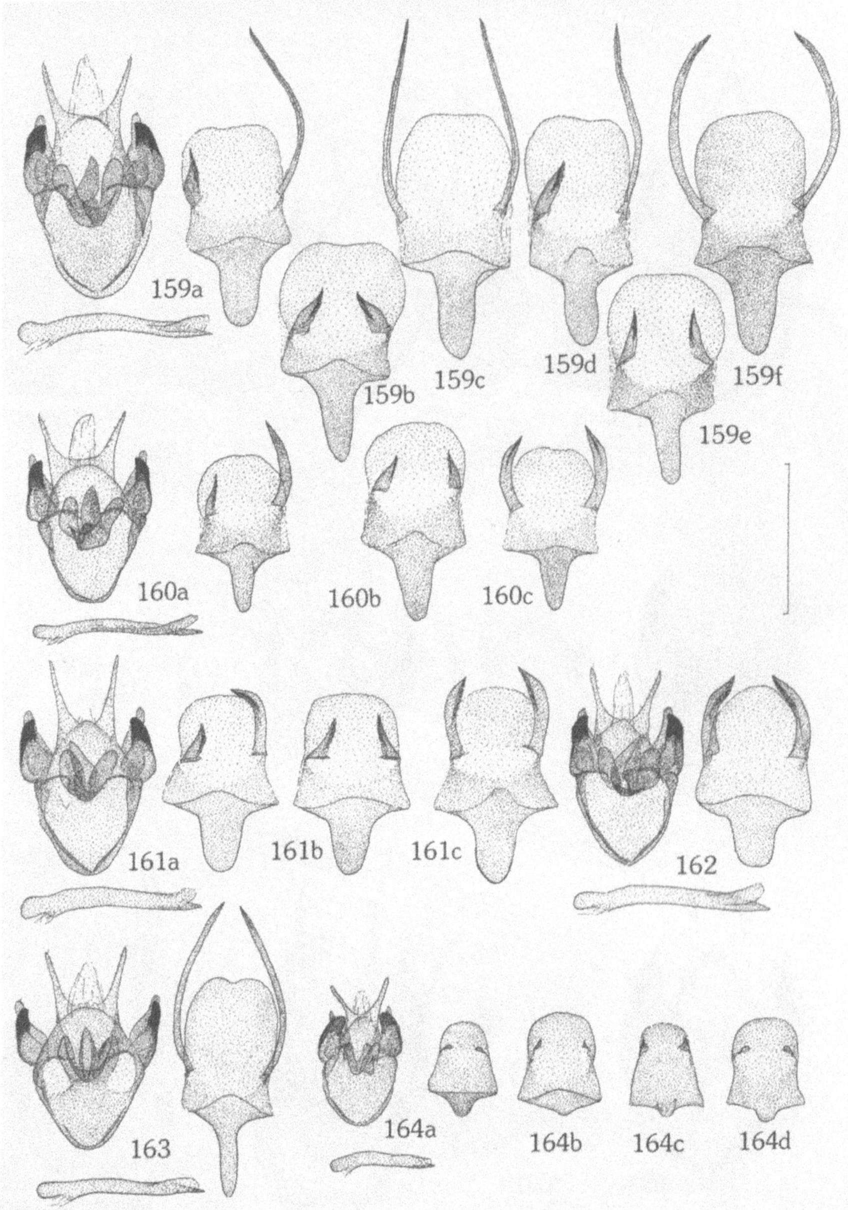

MALE GENITALIA (scale bar = 1 mm, with sternum A8)

159a. *Glossotrophia confinaria* H.-S., N. Italy; 159b, 159c: polymorphic variants, sternum A8
159d. *Glossotrophia diffinaria* Prt., C. Turkey, sternum A8 only; 159e, 159f: polymorphic variants, sternum A8
160a. *Glossotrophia alba* Hausm., C. Italy; 160b, 160c: polymorphic variants, sternum A8
161a. *Glossotrophia mentzeri* Hausm., Crete; 161b (paratype), 160c (holotype): polymorphic variants, sternum A8
162. *Glossotrophia rufomixtaria* Grasl., S. Spain
163. *Glossotrophia sacraria* B.-H., SE. European Russia
164a. *Glossotrophia asellaria asellaria* H.-S., Sardinia
164b. *Glossotrophia asellaria romanaria* Mill., C. Italy, sternum A8
164c. *Glossotrophia asellaria isabellaria* Mill., N. Spain, sternum A8
164d. *Glossotrophia asellaria dentatolineata* Whli., SE. Spain, sternum A8

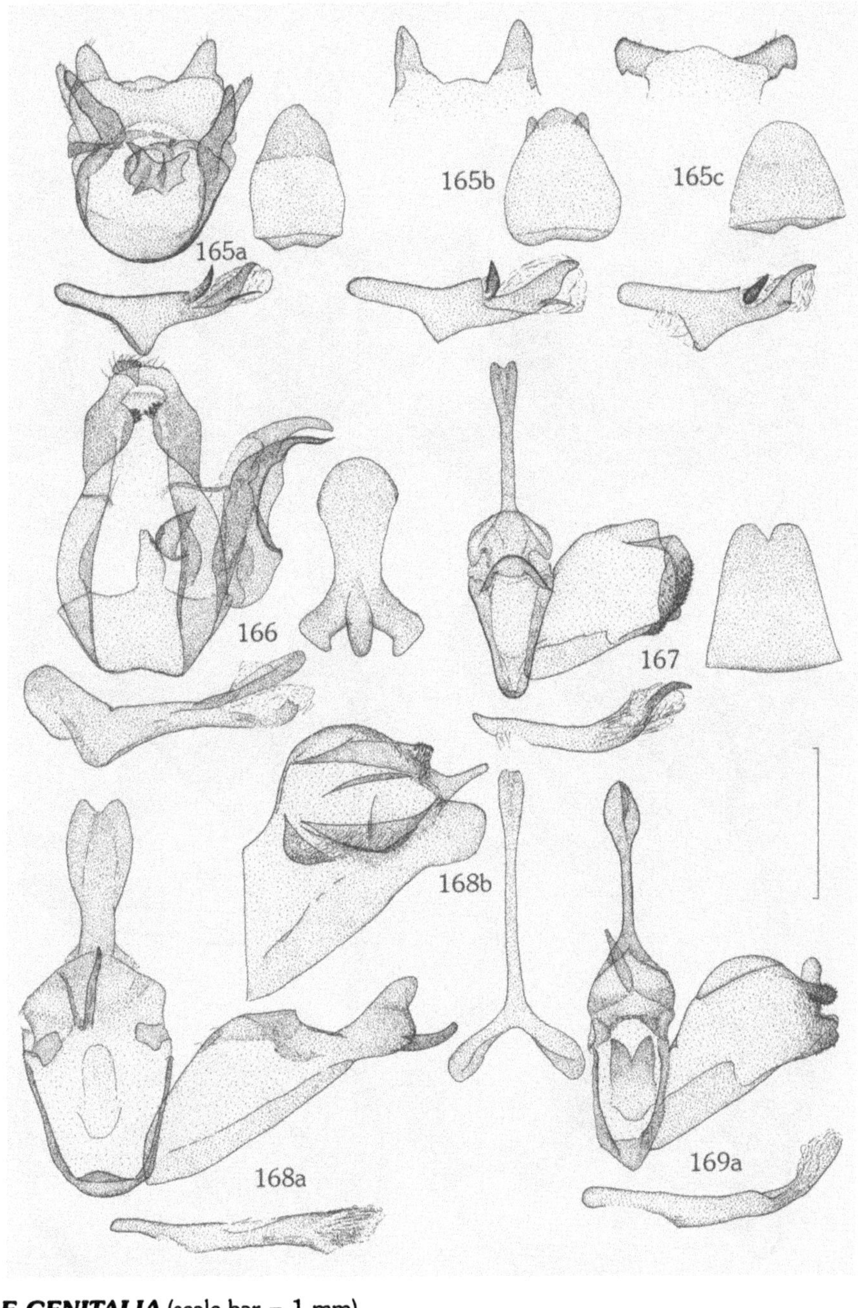

MALE GENITALIA (scale bar = 1 mm)

165a. *Holarctias rufinaria* Stgr., C. Russia (Jakutia), with sternum A8
165b. *Holarctias sentinaria* Geyer, E. Canada, socii, aedeagus and sternum A8
165c. *Holarctias rufinularia* Stgr., C. Russia (Sajan), socii, aedeagus and sternum A8
166. *Problepsis ocellata* Friv., Cyprus, with sternum A8
167. *Rhodostrophia jacularia* Hbn., S. European Russia, with sternum A8
168a. *Rhodostrophia vibicaria* Cl., Sicily
168b. *Rhodostrophia auctata* Stgr., Turkey, valva and uncus
169a. *Rhodostrophia pudorata* F., S. Spain

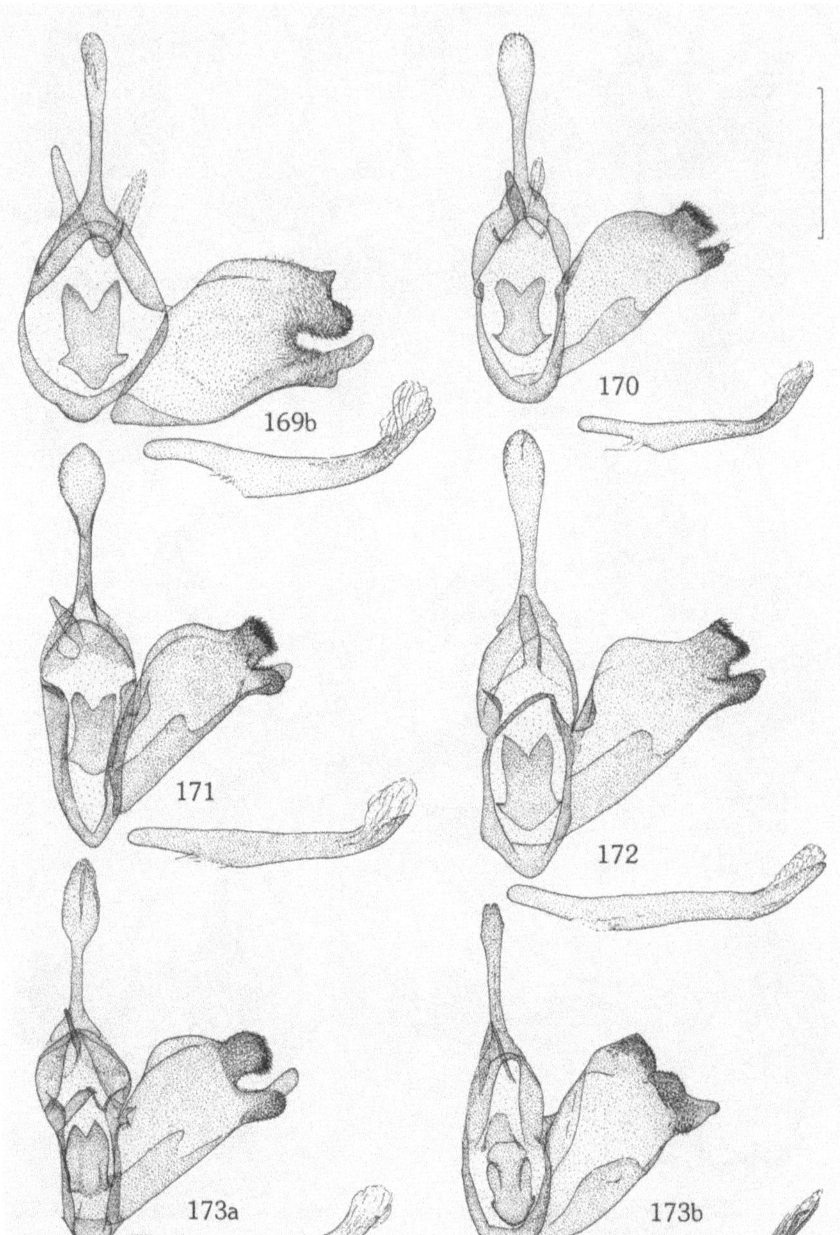

MALE GENITALIA (scale bar = 1 mm)

169b. *Rhodostrophia sieversi* Hbn., Armenia
170. *Rhodostrophia cretacaria* Reb., Crete
171. *Rhodostrophia calabra* Petagna, N. Italy
172. *Rhodostrophia tabidaria* Z., Rhodos island
173a. *Rhodostrophia discopunctata* Ams., Bulgaria
173b. *Rhodostrophia terrestraria* Led., 'S. European Russia' (locality uncertain)

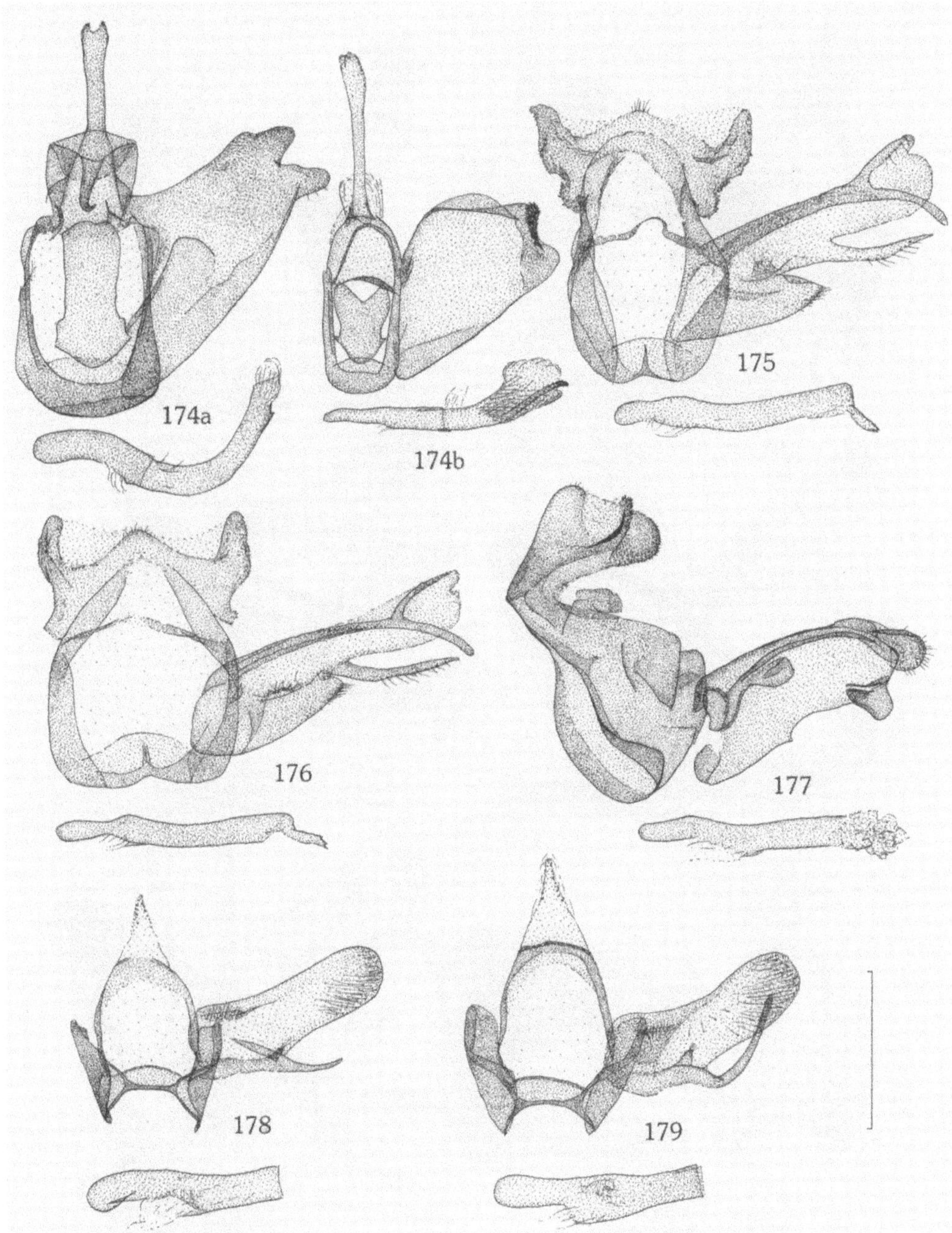

MALE GENITALIA (scale bar = 1 mm)

174a. *Rhodostrophia badiaria* Frr., S. European Russia
174b. *Rhodostrophia vastaria* Chr., W. Kazakhstan
175. *Timandra comae* Schmidt, NW. Poland
176. *Timandra griseata* Peters., Finland
177. *Timandra rectistrigaria* Ev., Polar Urals, NE. European Russia, lateral view
178. *Cyclophora pendularia* Cl., SW. Germany
179. *Cyclophora albipunctata* Hufn., N. Italy

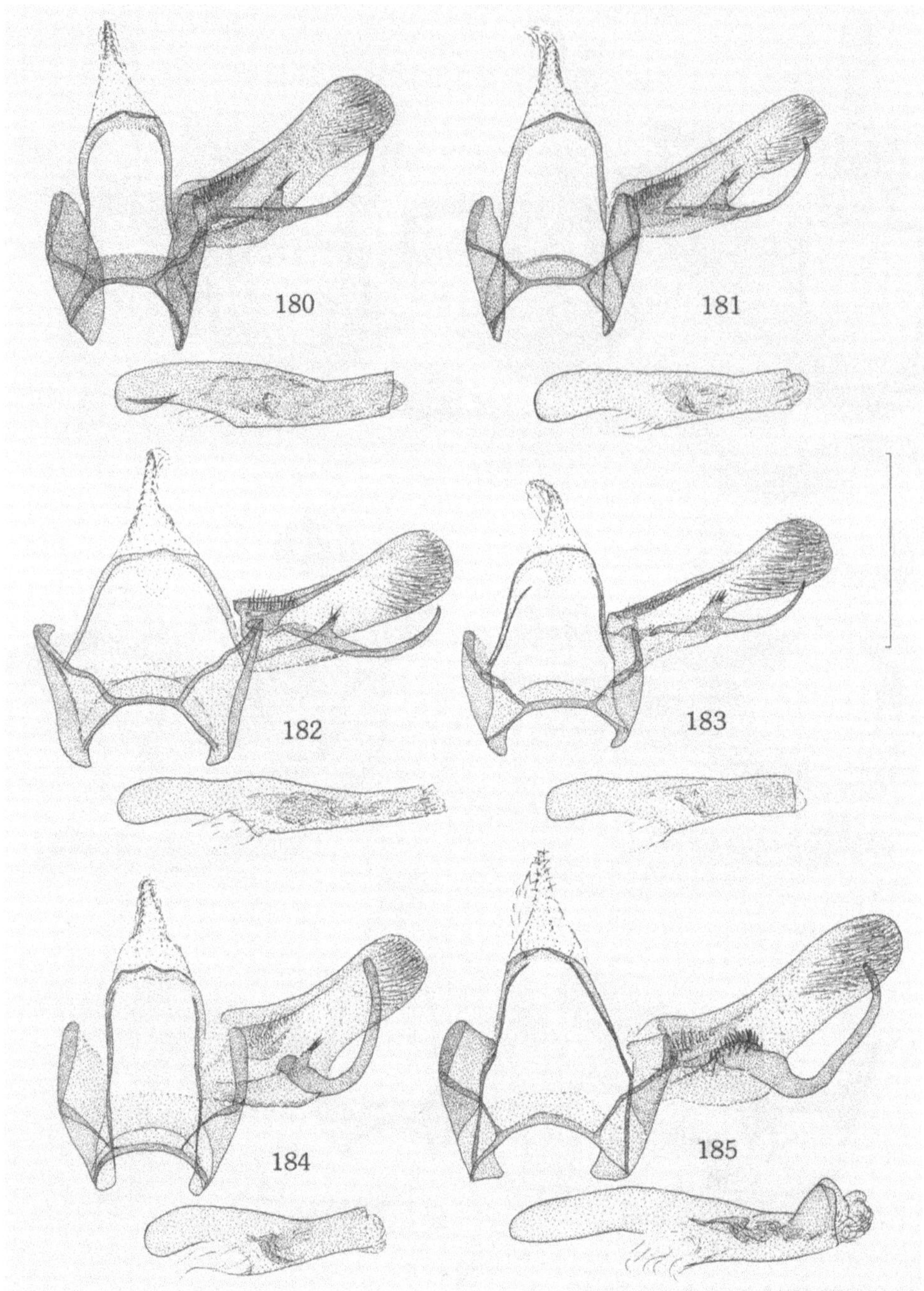

MALE GENITALIA (scale bar = 1 mm)

180. *Cyclophora albiocellaria* Hbn., S. Serbia
181. *Cyclophora lennigiaria* Fuchs, S. France
182. *Cyclophora ariadne* Rssr., C. Crete
183. *Cyclophora serveti* Red. & Gast., paratype, N. Spain
184. *Cyclophora annularia* F., Great Britain
185. *Cyclophora puppillaria* Hbn., Corsica

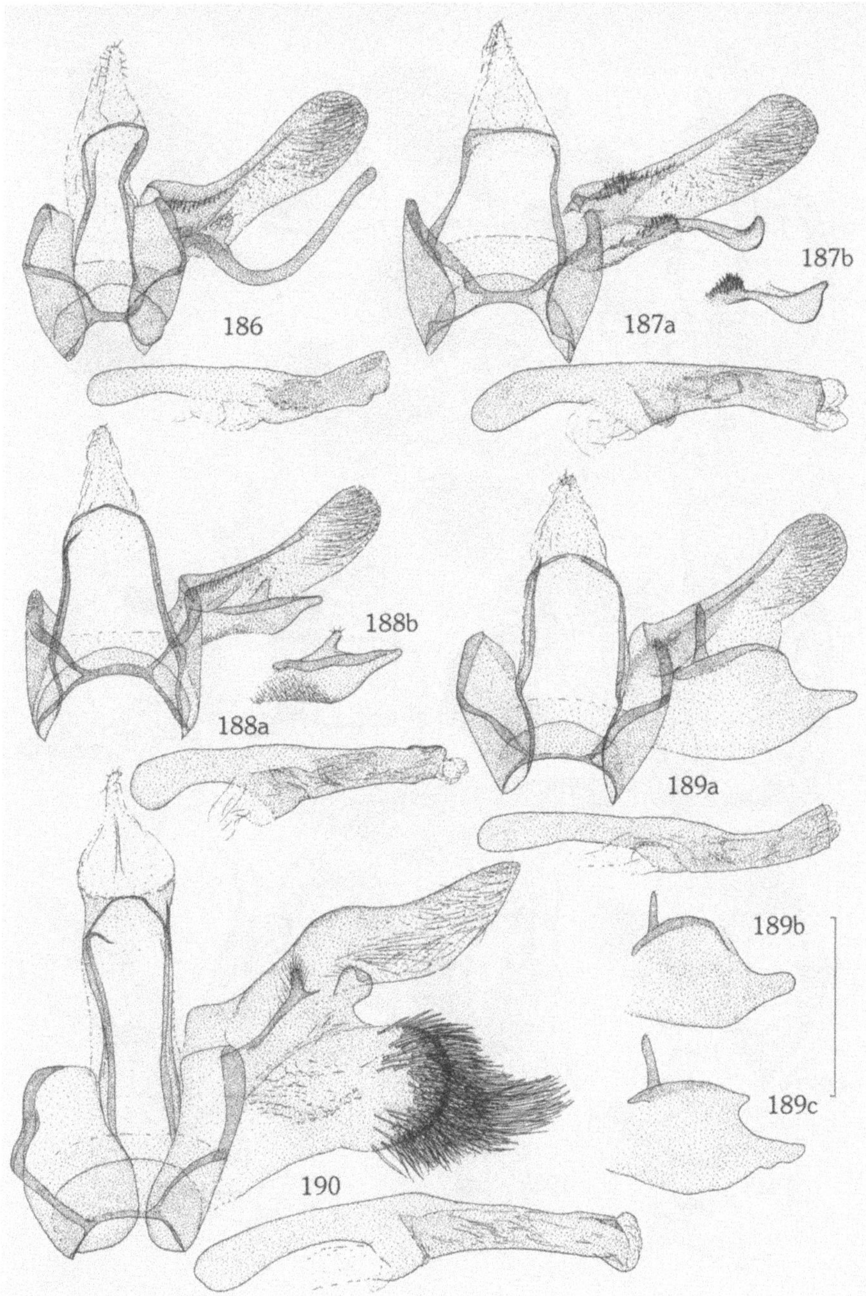

MALE GENITALIA (scale bar = 1 mm)

186. *Cyclophora quercimontaria* Bast., W. Germany
187a. *Cyclophora ruficiliaria* H.-S., NW. Turkey
187b. *Cyclophora ruficiliaria* H.-S., W. Germany, fibula (variant)
188a. *Cyclophora hyponoea* Prt., NE. Spain
188b. *Cyclophora hyponoea* Prt., Algeria, ventral projection of valva (variant)
189a. *Cyclophora porata* L., NE. Poland
189b. *Cyclophora porata* L., E. Germany, ventral projection of valva (variant)
189c. *Cyclophora porata* L., C. Italy, ventral projection of valva (variant)
190. *Cyclophora suppunctaria* Z., W. Croatia

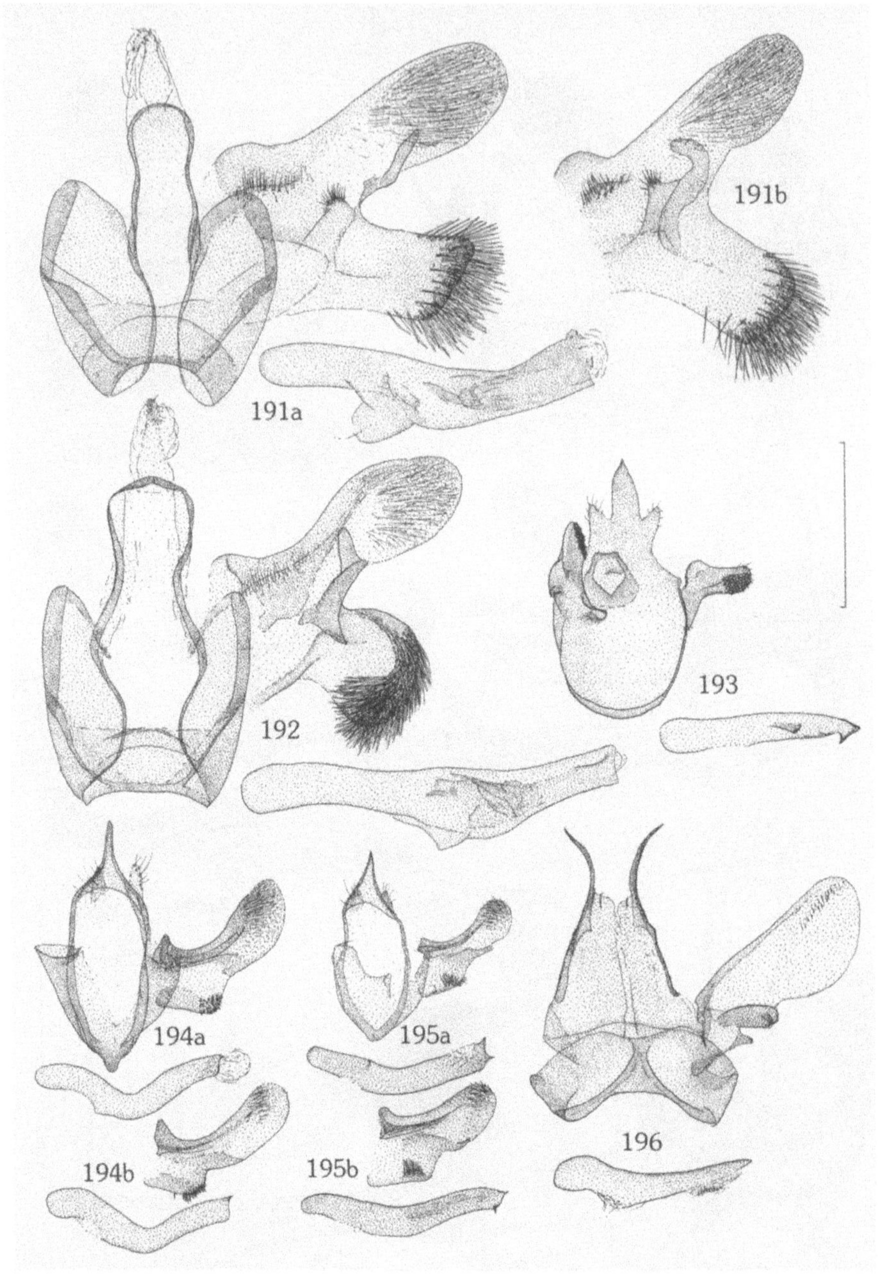

MALE GENITALIA (scale bar = 1 mm)

191a. *Cyclophora punctaria punctaria* L., W. Germany
191b. *Cyclophora punctaria fritzae* Hausm., paratype, N. Caucasus, SW. Russia
192. *Cyclophora linearia* Hbn., Macedonia
193. *Rhodometra sacraria* L., S. France, left valva in natural position, right valva spread
194a. *Casilda antophilaria antophilaria* Hbn., S. European Russia
194b. *Casilda antophilaria rosearia* Tr., C. Greece
195a. *Casilda consecraria* Stgr., SE. Spain
195b. *Casilda consecraria* Stgr., Israel
196. *Ochodontia adustaria* Wald., S. European Russia

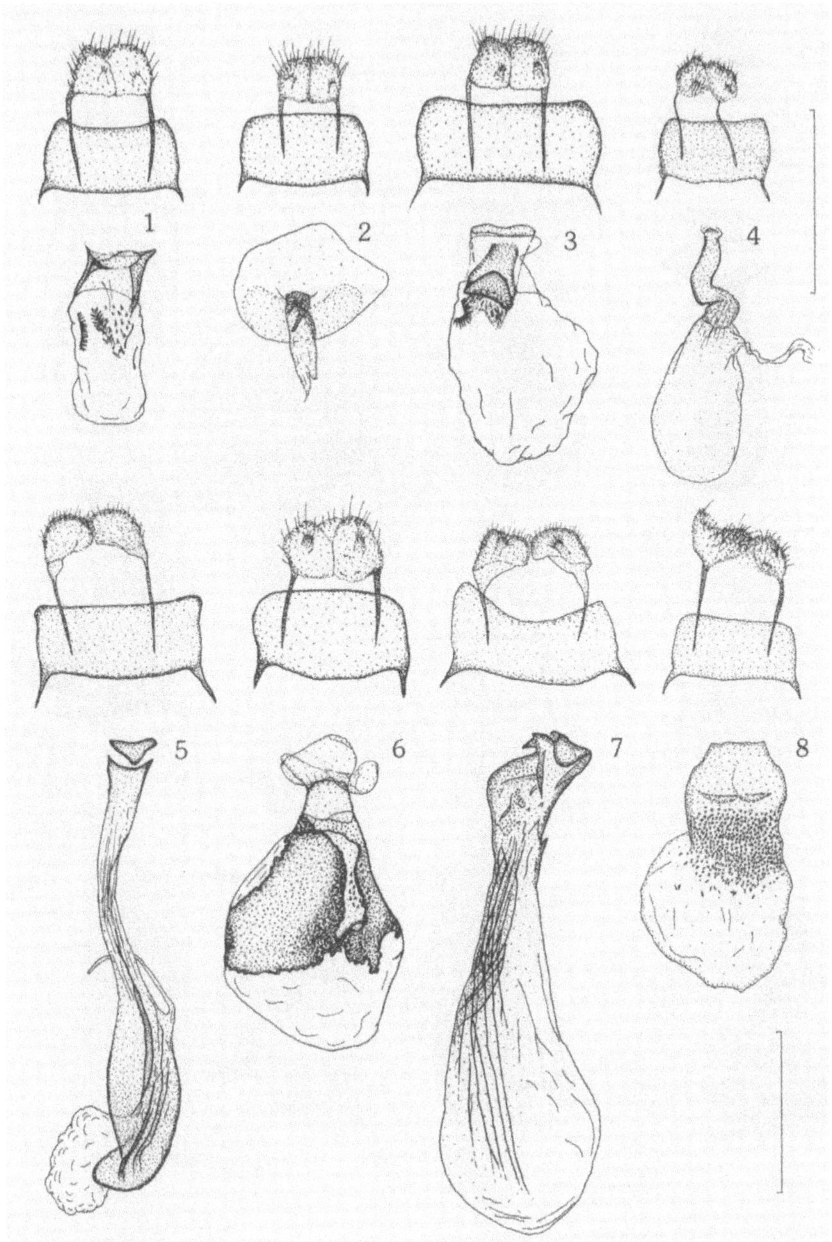

FEMALE GENITALIA (scale bars = 1 mm)

1. *Anthometra plumularia* Bsd., C. Spain
2. *Emmiltis pygmaearia* Hbn., C. Italy
3. *Cleta ramosaria* Vill., S. Spain
4. *Cleta perpusillaria* Ev., S. European Russia
5. *Cleta filacearia* H.-S., E. Spain
6. *Idaea serpentata* Hufn., Romania
7. *Idaea luteolaria* Const., C. Spain
8. *Idaea aureolaria* Den. & Schiff., E. Germany

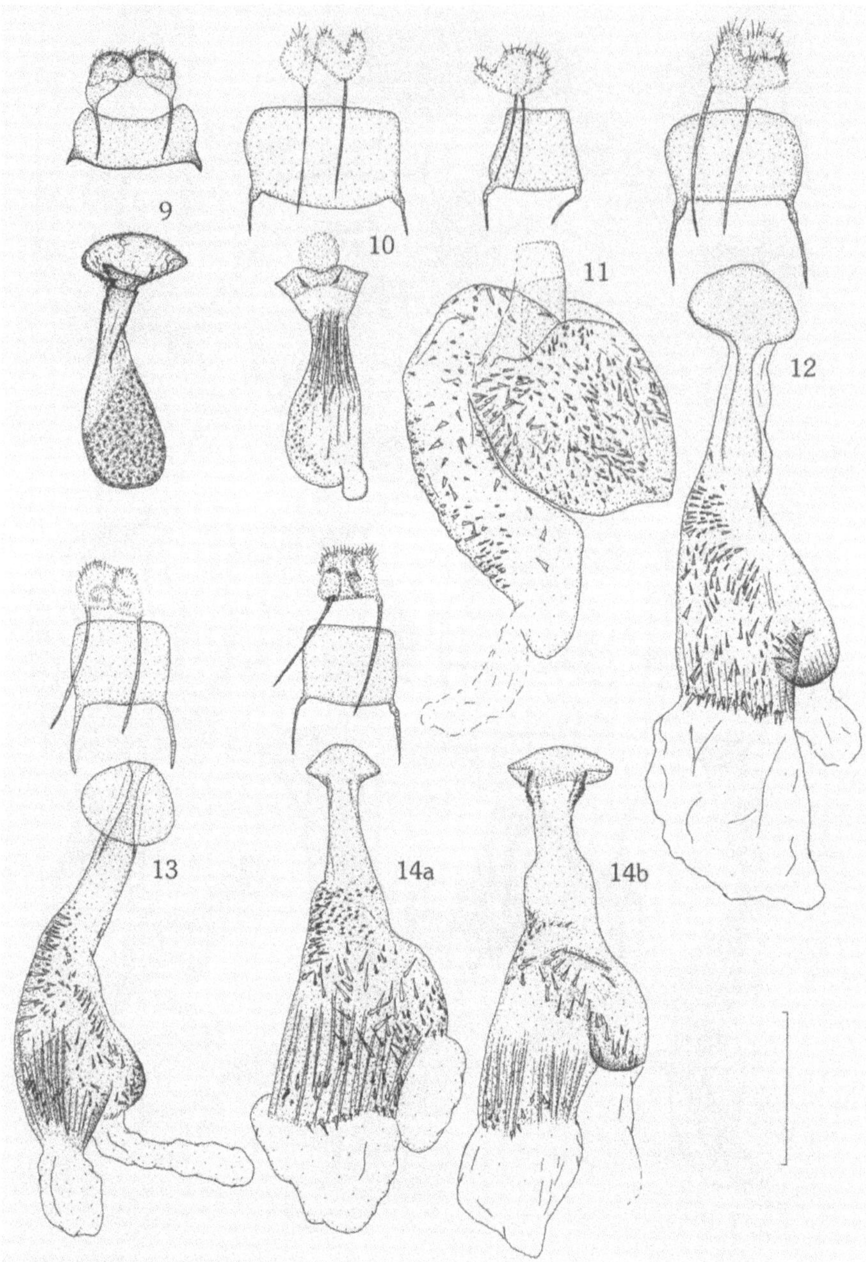

FEMALE GENITALIA (scale bar = 1 mm)

9. *Idaea flaveolaria* Hbn., S. Switzerland
10. *Idaea muricata* Hufn., Hungary
11. *Idaea determinata* Stgr., S. Italy
12. *Idaea litigiosaria* Bsd., S. Spain
13. *Idaea mutilata* Stgr., S. Italy
14a. *Idaea lusohispanica* Herb., S. Spain
14b. *Idaea lambessata* Obth., Morocco, without ovipositor and tergum A8

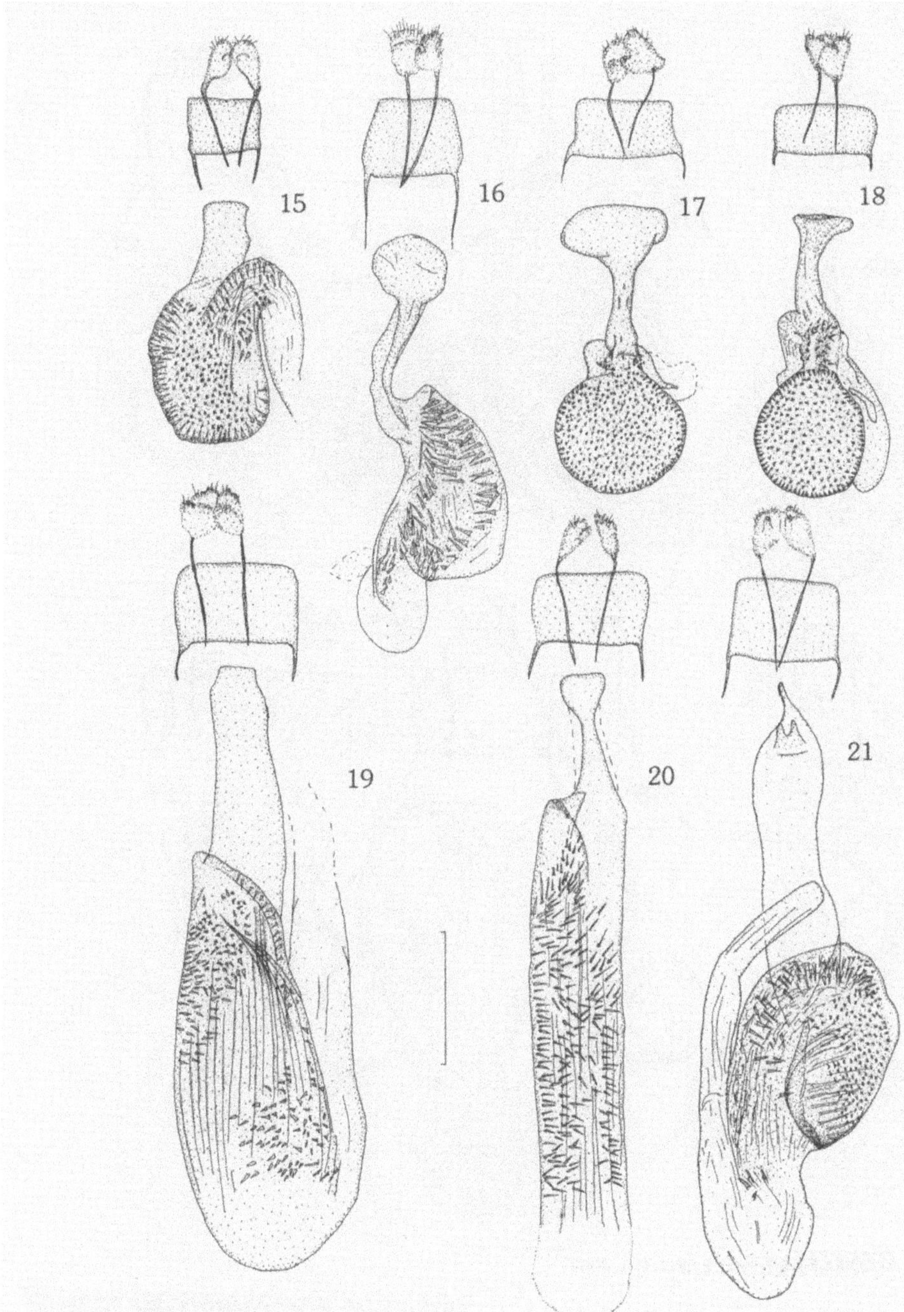

FEMALE GENITALIA (scale bar = 1 mm)

15. *Idaea sardoniata* Homb., S. France
16. *Idaea korbi* Püng., C. Spain
17. *Idaea mediaria* Hbn., NW. Italy
18. *Idaea leipnitzi* Hausm. sp.n., paratype, SW. Croatia
19. *Idaea rufaria* Hbn., Caucasus
20. *Idaea consanguiberica* Rezb. & Exp., W. Spain
21. *Idaea admiranda* Hausm. sp.n., paratype, E. Turkey

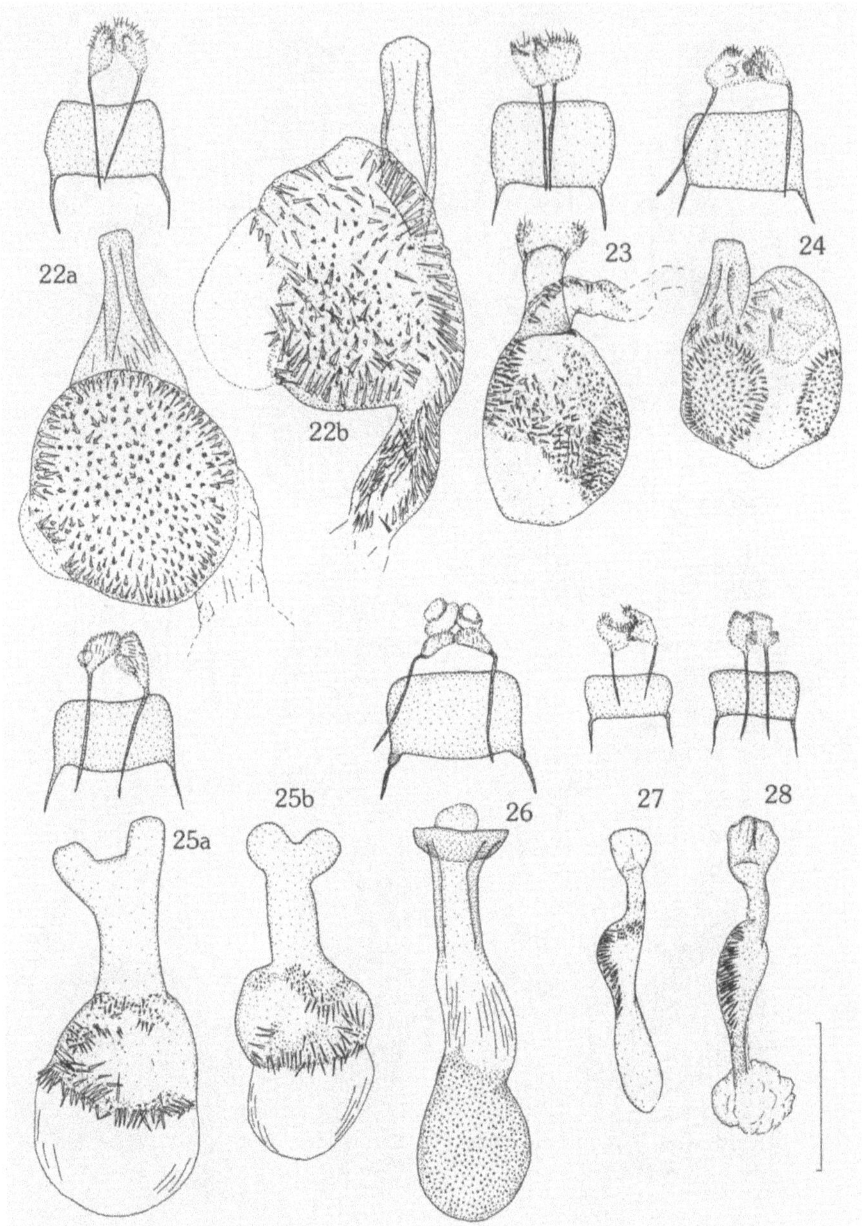

FEMALE GENITALIA (scale bar = 1 mm)

22a. *Idaea consanguinaria turatii* B.-H., Sicily
22b. *Idaea consanguinaria consanguinaria* Led., SW. Croatia
23. *Idaea ossiculata* Led., N. Greece
24. *Idaea sericeata* Hbn., S. Urals
25a. *Idaea macilentaria* H.-S., S. Spain
25b. *Idaea macilentaria* H.-S., W. Germany
26. *Idaea ochrata* Scop., Hungary
27. *Idaea nevadata* Whli., S. Spain
28. *Idaea figuraria* B.-H., C. Spain

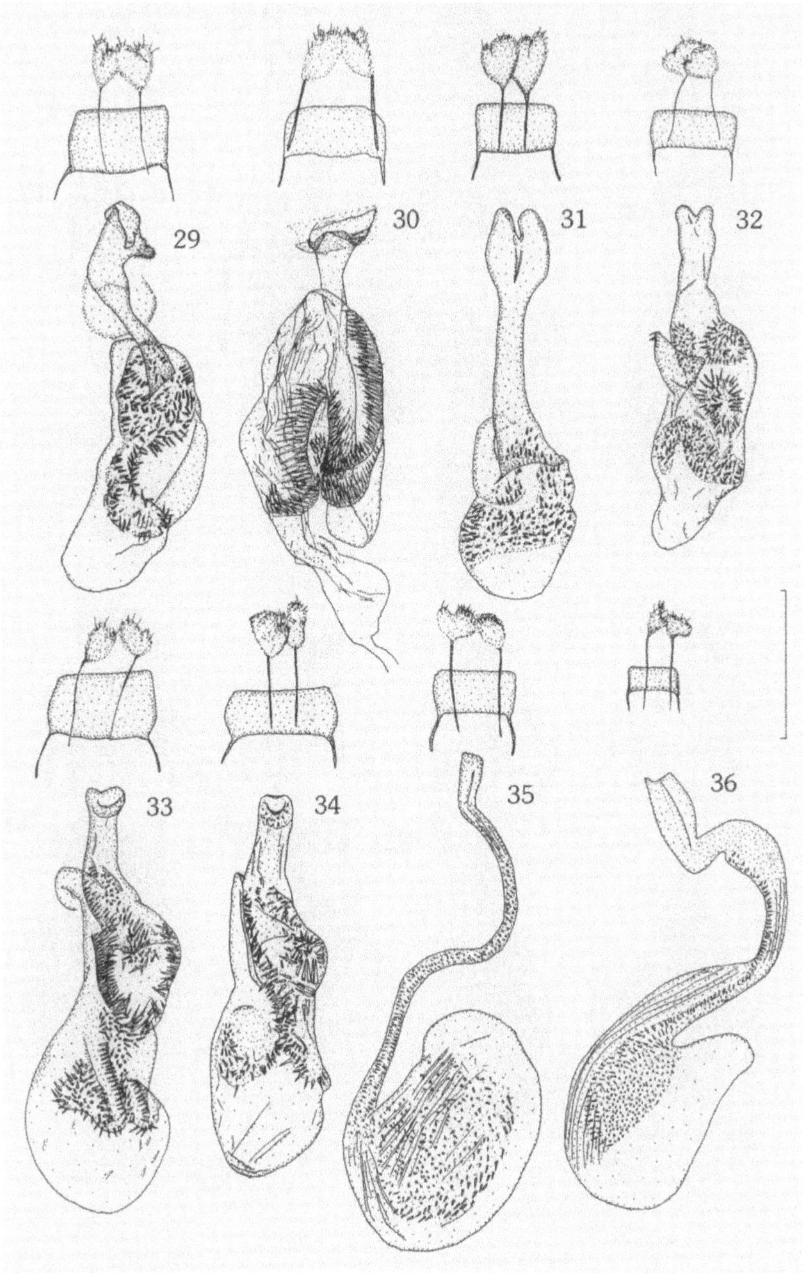

FEMALE GENITALIA (scale bar = 1 mm)

29. *Idaea alicantaria* Rssr., SE. Spain
30. *Idaea nigrolineata* Chrét., Morocco
31. *Idaea completa* Stgr., Tunisia
32. *Idaea intermedia* Stgr., Crete
33. *Idaea rusticata* Den. & Schiff., W. Germany
34. *Idaea mustelata* Gmpp., C. Spain
35. *Idaea filicata* Hbn., N. Italy
36. *Idaea troglodytaria* H.-S., Cyprus

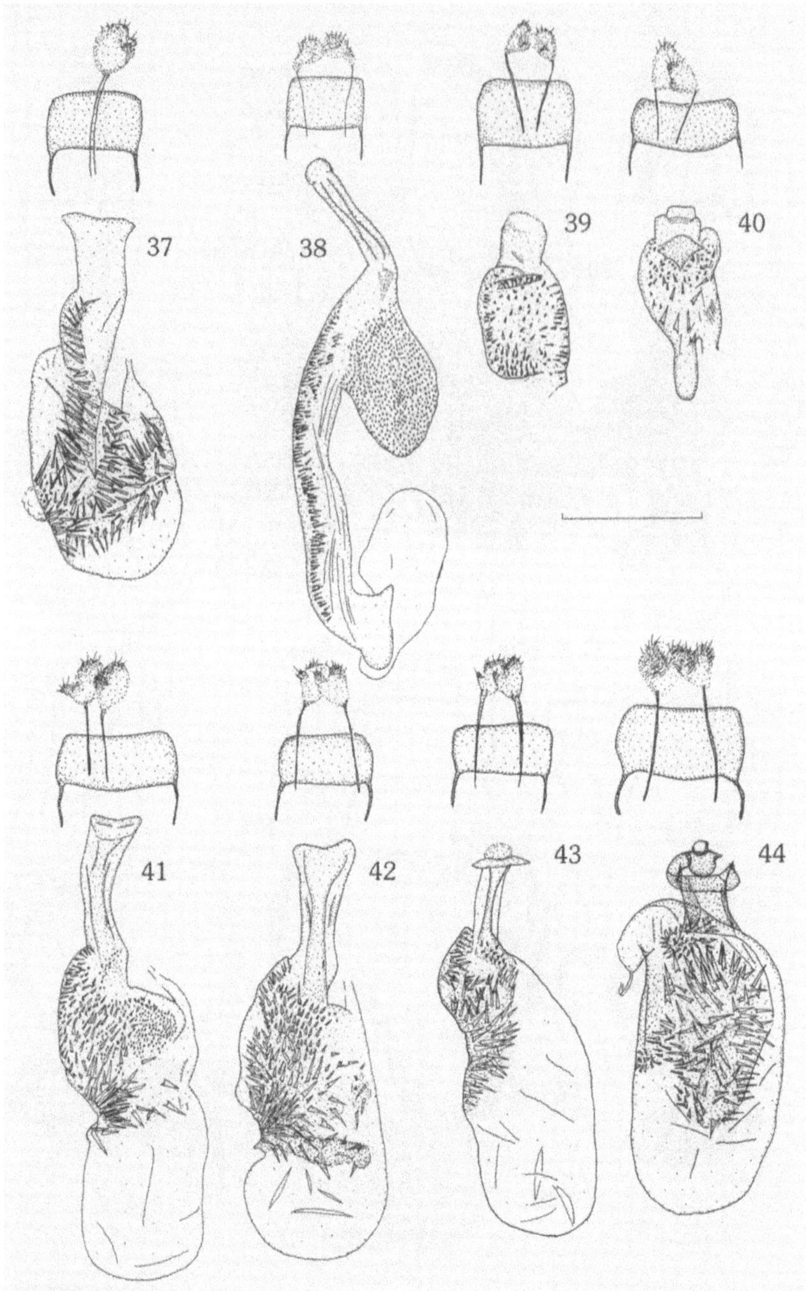

FEMALE GENITALIA (scale bar = 1 mm)

37. *Idaea laevigata* Scop., E. Germany
38. *Idaea efflorata* Z., C. Italy
39. *Idaea attenuaria* Rmb., Algeria
40. *Idaea incalcarata* Chrét., NE. Spain
41. *Idaea typicata* Gn., NW. Italy
42. *Idaea alyssumata* Himm. & Mill., NE. Spain
43. *Idaea urcitana* Agenjo, holotype, SE. Spain
44. *Idaea moniliata* Den. & Schiff., Slovenja

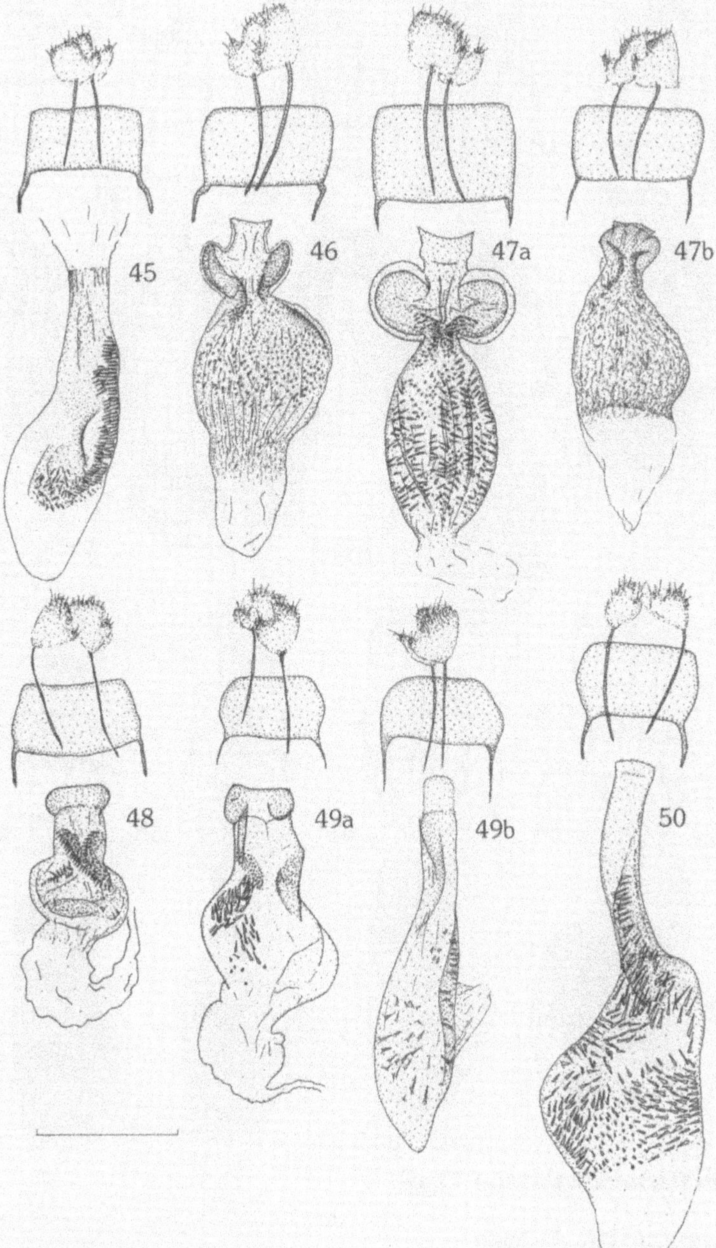

FEMALE GENITALIA (scale bar = 1 mm)

45. *Idaea spissilimbaria* Mab., Corsica
46. *Idaea circuitaria* Hbn., C. Italy
47a. *Idaea rainerii* Hausm., paratype, Sicily
47b. *Idaea mimosaria wimmeri* Hausm., C. Turkey
48. *Idaea albarracina* Rssr., NE. Spain
49a. *Idaea incisaria* Stgr., S. Spain
49b. *Idaea textaria* Led., Cyprus
50. *Idaea calunetaria* Stgr., S. Spain

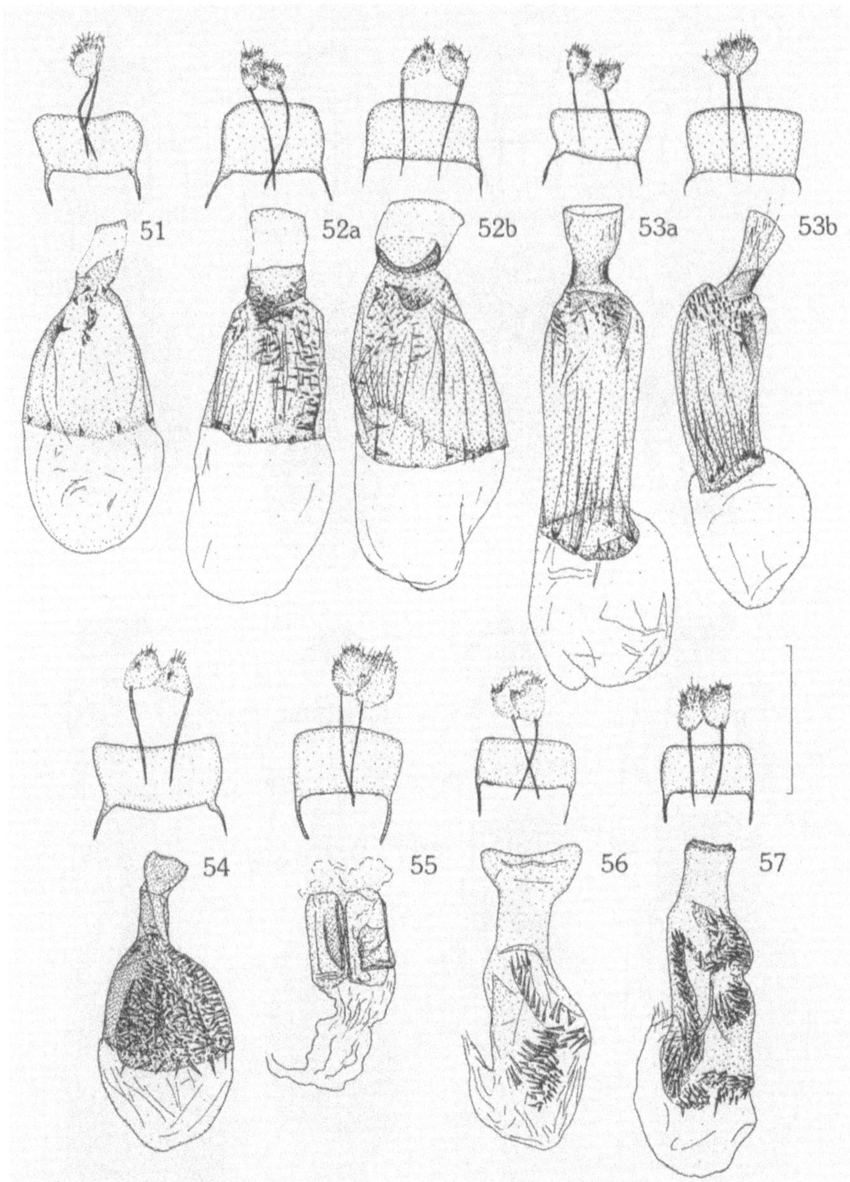

FEMALE GENITALIA (scale bar = 1 mm)

51. *Idaea belemiata* Mill., S. Spain
52a. *Idaea elongaria elongaria* Rmb., Sicily
52b. *Idaea elongaria pecharia* Stgr., N. Hungary
53a. *Idaea palaestinensis* Stern., Crete
53b. *Idaea palaestinensis* Stern., Samos, variant
54. *Idaea obsoletaria* Rmb., Hungary
55. *Idaea obliquaria* Trti., Corsica
56. *Idaea inquinata* Scop., Macedonia
57. *Idaea blaesii* Lenz & Hausm., Portugal

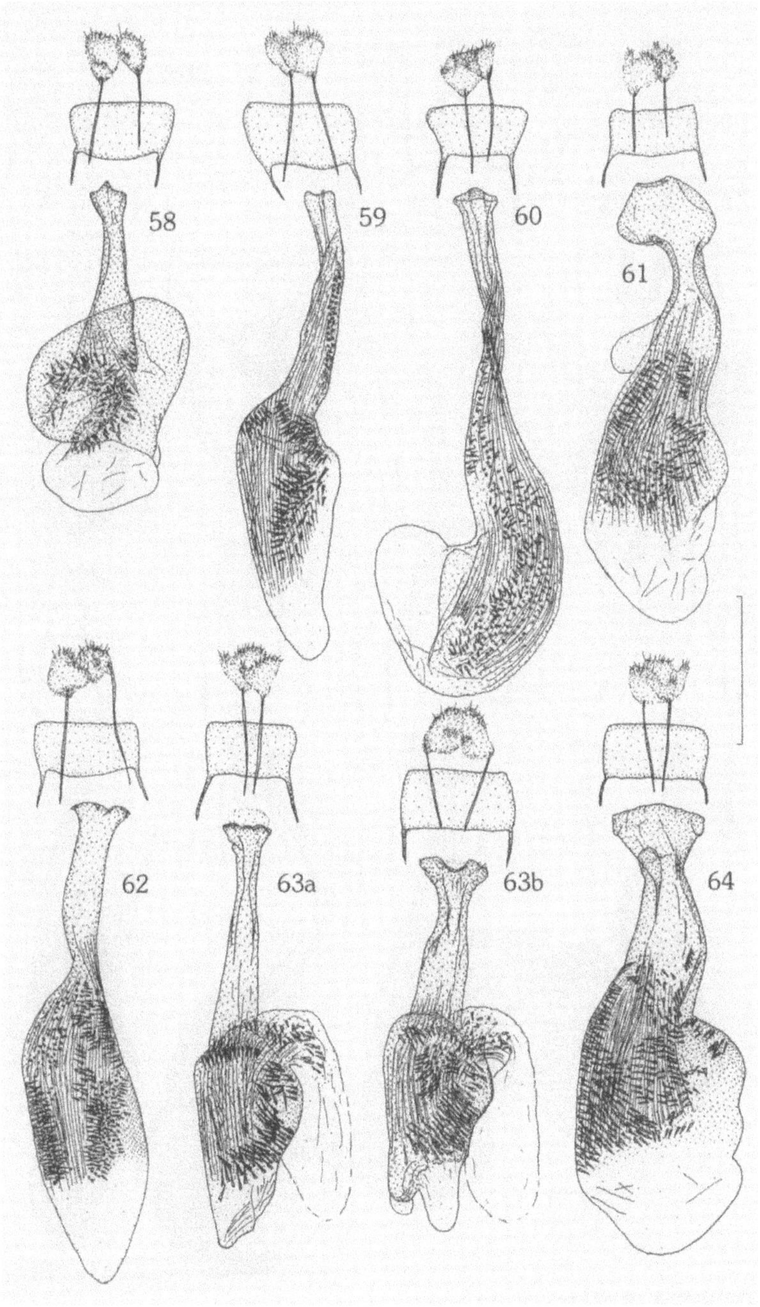

FEMALE GENITALIA (scale bar = 1 mm)

58. *Idaea dilutaria* Hbn., Hungary
59. *Idaea fuscovenosa* Gze., S. Germany
60. *Idaea robiginata* Stgr., C. Spain
61. *Idaea lutulentaria* Stgr., S. Spain
62. *Idaea humiliata* Hufn., S. Germany
63a. *Idaea bigladiata bigladiata* Herb., Morocco
63b. *Idaea bigladiata alberti* subsp. n., E. Algeria
64. *Idaea politaria* Hbn., NE. Spain

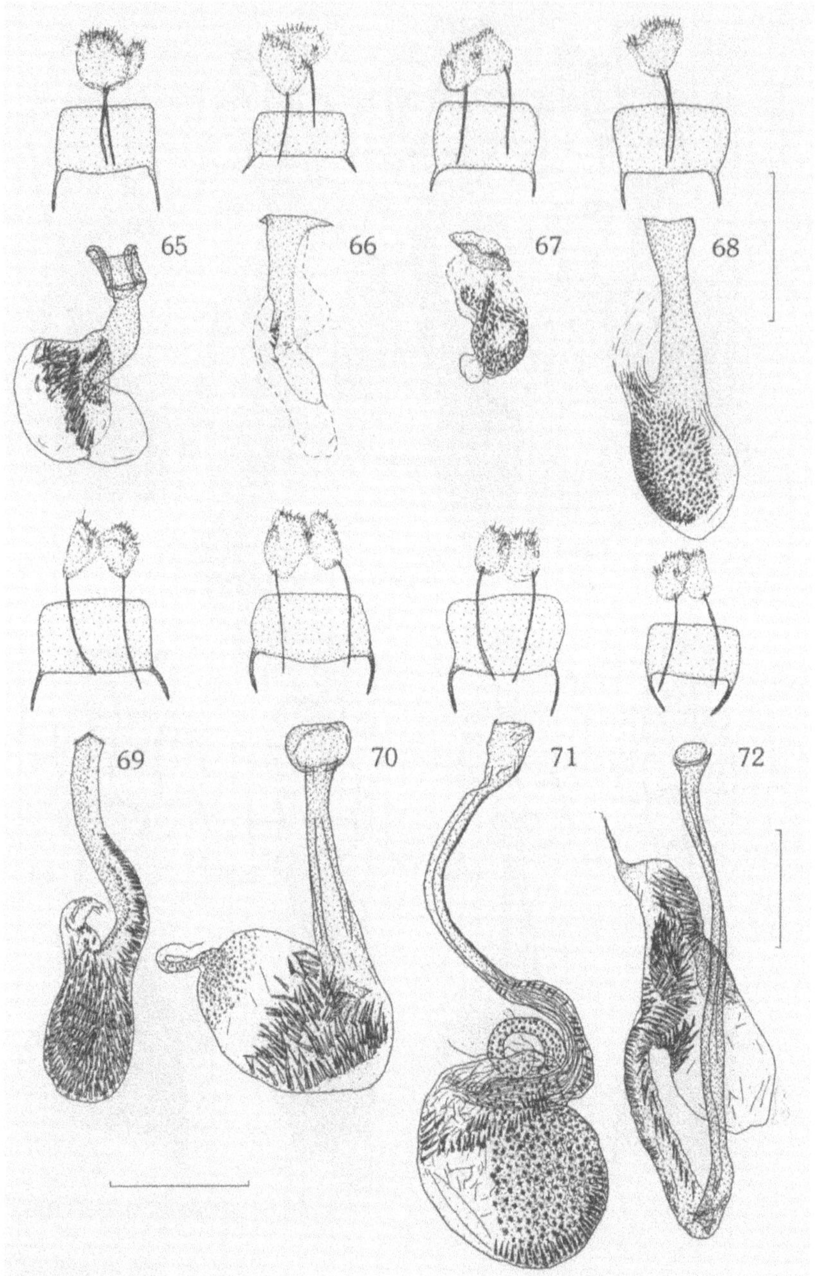

FEMALE GENITALIA (scale bar = 1 mm)

65. *Idaea longaria* H.-S., Sicily
66. *Idaea nexata* Hbn., S. Spain
67. *Idaea manicaria* H.-S., Malta
68. *Idaea descitaria* Chr., SE. European Russia
69. *Idaea vesubiata* Mill., SE. France
70. *Idaea libycata* Bart., S. France
71. *Idaea consolidata* Led., C. Italy
72. *Idaea joannisiata* Homb., N. Spain

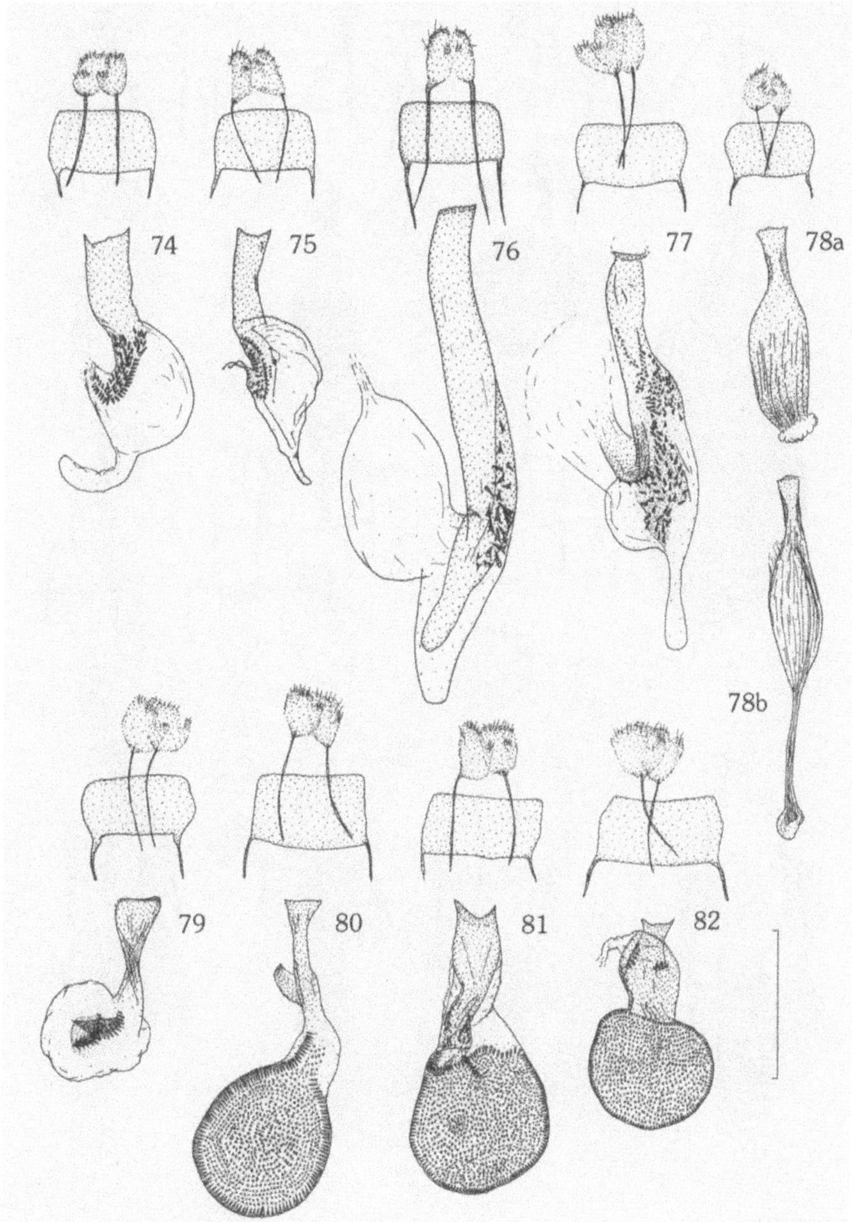

FEMALE GENITALIA (scale bar = 1 mm)

74. *Idaea seriata* Schrank, C. Greece
75. *Idaea minuscularia* Ribbe, Portugal
76. *Idaea ibizaria* Mentz., paratype, Balearic islands
77. *Idaea albitorquata* Püng., S. Greece
78a. *Idaea carvalhoi* Herb., S. Spain
78b. *Idaea fathmaria* Obth., Morocco, without ovipositor and tergum A8
79. *Idaea camparia* H.-S., Israel
80. *Idaea subsericeata* Haw., E. Spain
81. *Idaea pallidata* Den. & Schiff., Hungary
82. *Idaea mancipiata* Stgr., N. Spain

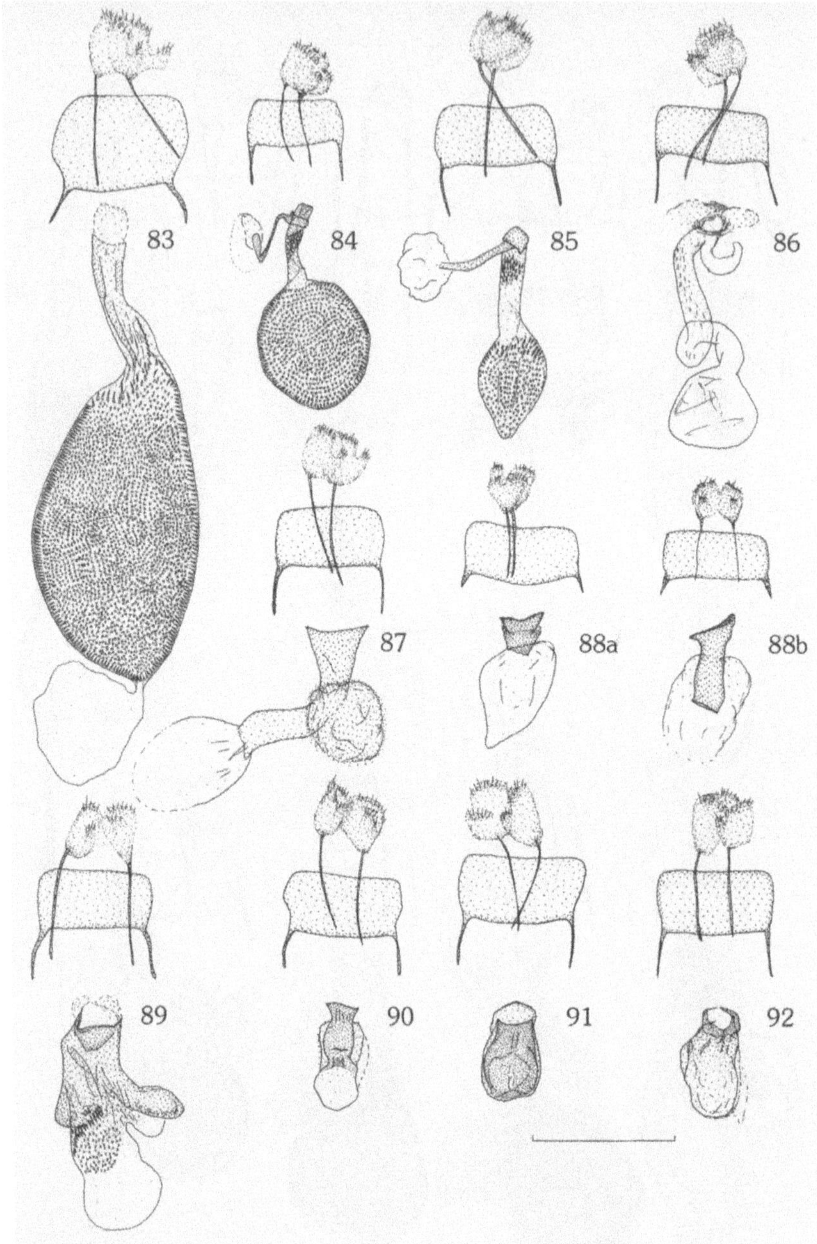

FEMALE GENITALIA (scale bar = 1 mm)

83. *Idaea sylvestraria* Hbn., C. Germany
84. *Idaea dimidiata* Hufn., NE. Spain
85. *Idaea subsaturata* Gn., Morocco
86. *Idaea trigeminata* Haw., N. Greece
87. *Idaea biselata* Hufn., Romania
88a. *Idaea fractilineata* Z., Sicily
88b. *Idaea inclinata* Led., Cyprus
89. *Idaea exilaria* Gn., S. Spain
90. *Idaea cervantaria* Mill., NE. Spain
91. *Idaea contiguaria* Hbn., C. Germany
92. *Idaea rupicolaria* Rssr., S. Spain

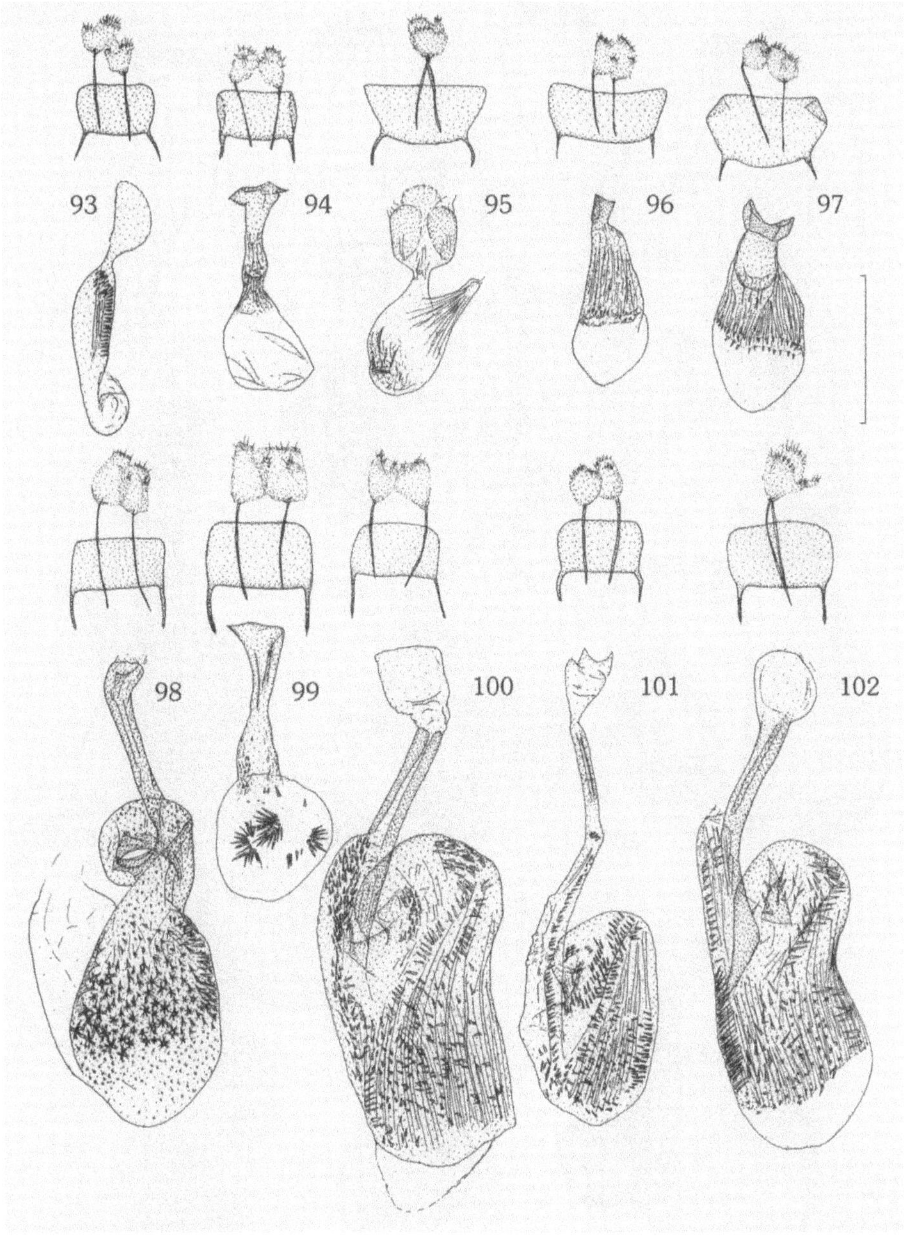

FEMALE GENITALIA (scale bar = 1 mm)

93. *Idaea deitanaria* Rssr. & Weis., E. Spain
94. *Idaea saleri* Dom. & Baix., SE. Spain
95. *Idaea tineata* Th.-Mieg., S. Turkey
96. *Idaea infirmaria* Rmb., S. Spain
97. *Idaea rhodogrammaria* Püng., E. Spain
98. *Idaea ostrinaria* Hbn., E. Spain
99. *Idaea metohiensis* Reb., Macedonia
100. *Idaea eugeniata* Mill., Tunisia
101. *Idaea distinctaria* Bsd., Crete
102. *Idaea predotaria* Hartig, NE. Spain

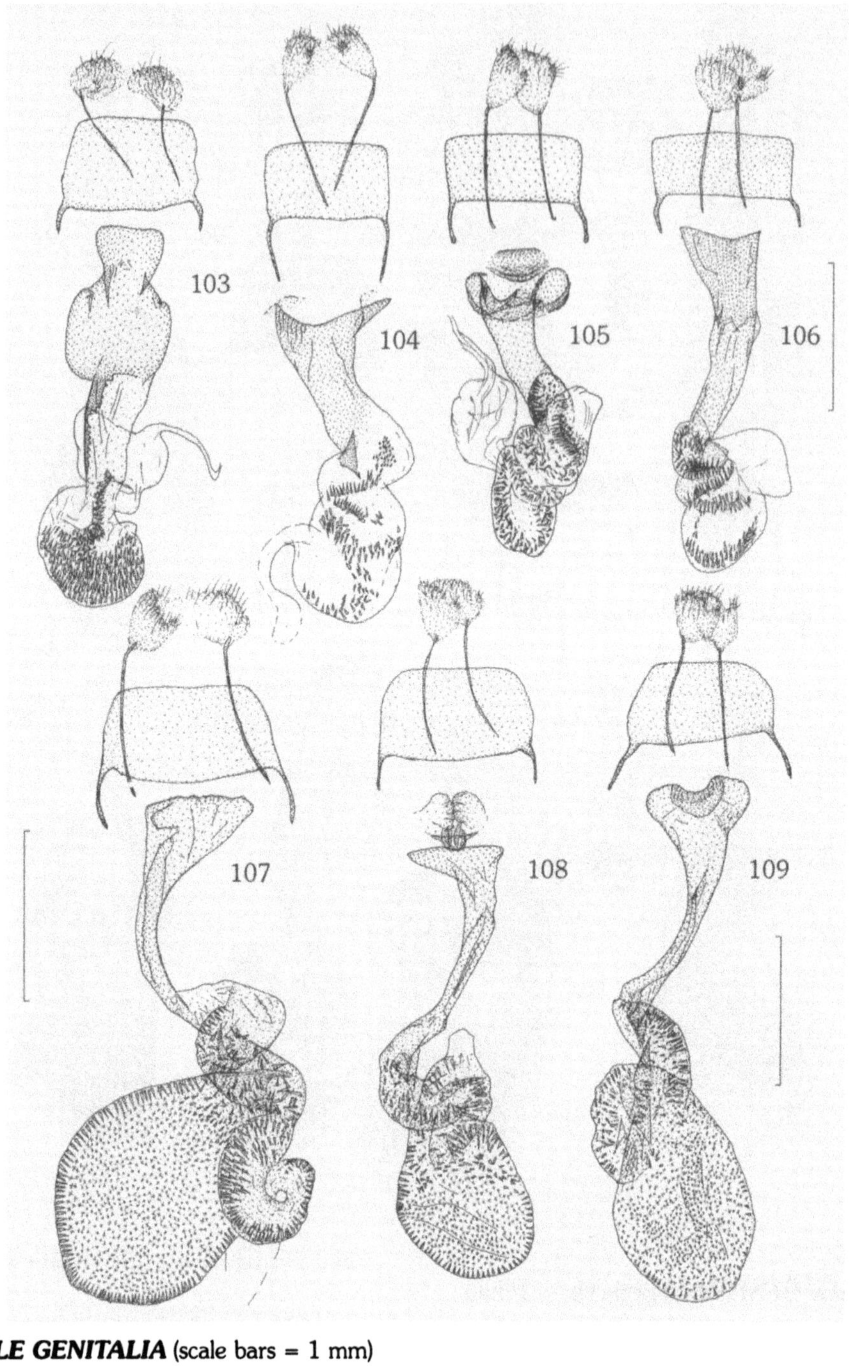

FEMALE GENITALIA (scale bars = 1 mm)

103. *Idaea nitidata* H.-S., SE. Austria
104. *Idaea emarginata* L., Sweden
105. *Idaea dromikos* Hausm. sp.n., paratype, S. France
106. *Idaea simplicior* Prt., C. Spain
107. *Idaea rubraria* Stgr., Turkey
108. *Idaea aversata* L., C. Germany
109. *Idaea degeneraria* Hbn., S. Italy

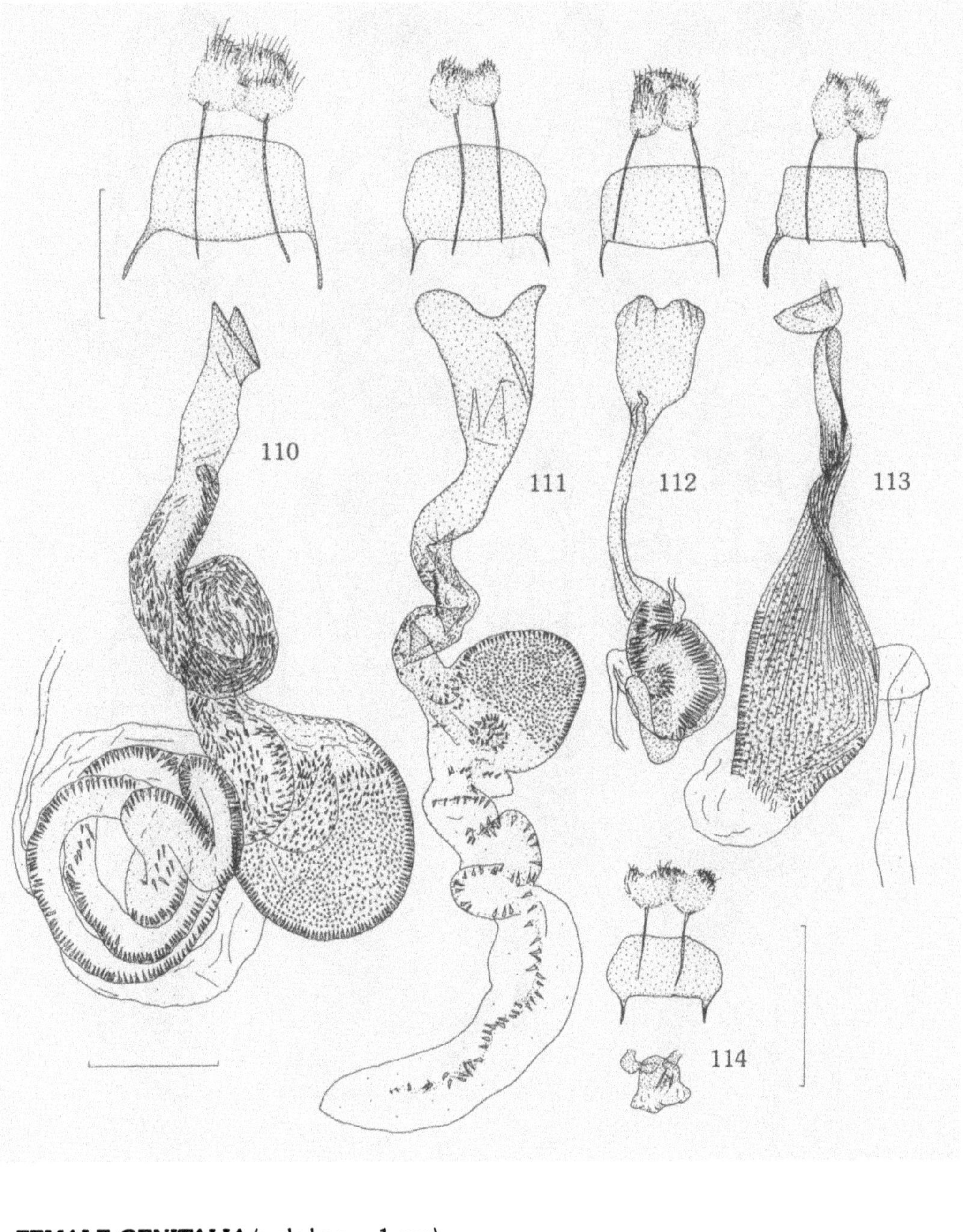

FEMALE GENITALIA (scale bars = 1 mm)

110. *Idaea straminata* Bkh., S. Italy
111. *Idaea deversaria* H.-S., W. Croatia
112. *Idaea squalidaria* Stgr., N. Spain
113. *Brachyglossina hispanaria* Püng., S. Spain
114. *Limeria macraria* Stgr., SE. European Russia, with sternum A8

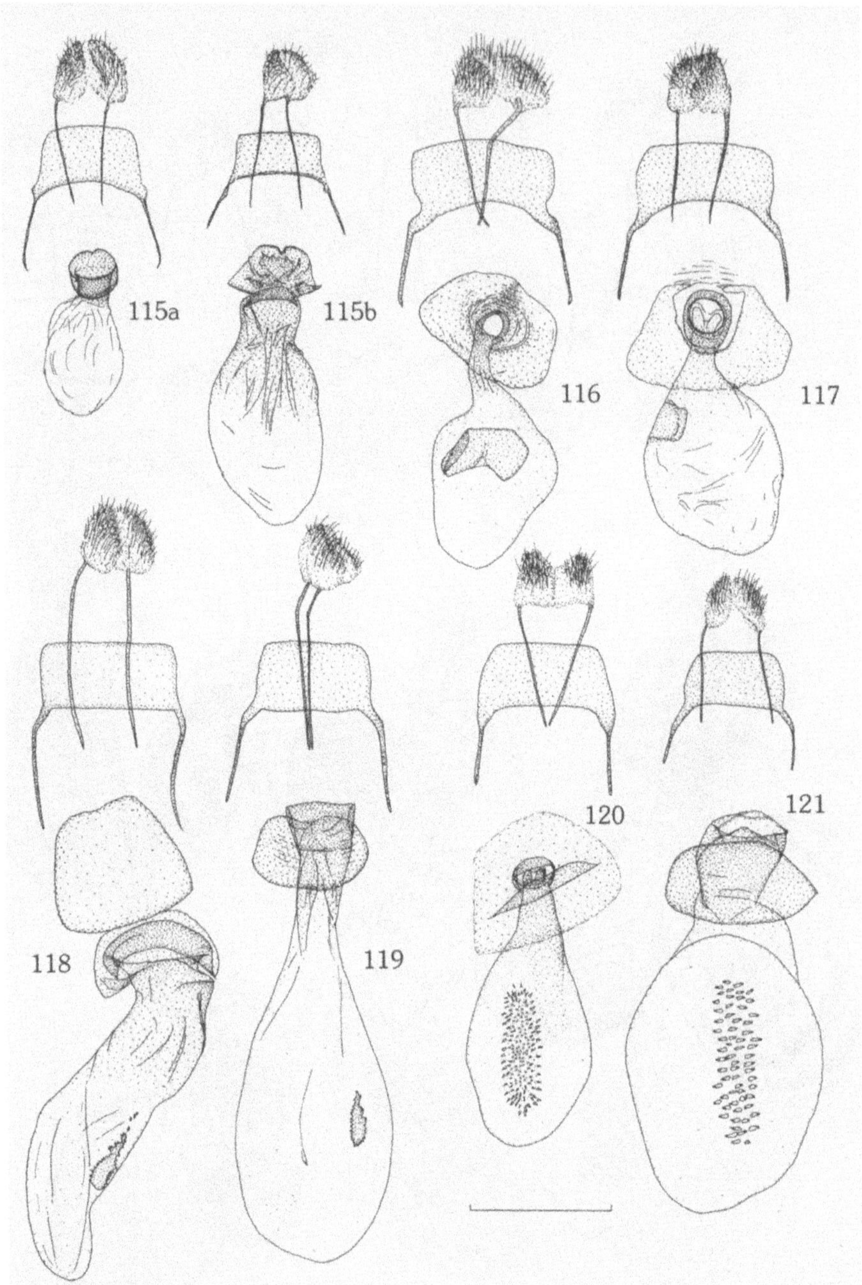

FEMALE GENITALIA (scale bar = 1 mm)

115a. *Oar reaumuraria* Mill., SE. Spain
115b. *Oar pratana* F., Tunisia
116. *Cinglis humifusaria* Ev., SE. European Russia
117. *Cinglis andalusiaria* Wagn., E. Spain
118. *Scopula immorata* L., Bulgaria
119. *Scopula tessellaria* Bsd., SE. European Russia
120. *Scopula corrivalaria* Kretsch., E. Germany
121. *Scopula caricaria* Reut., SW. Germany

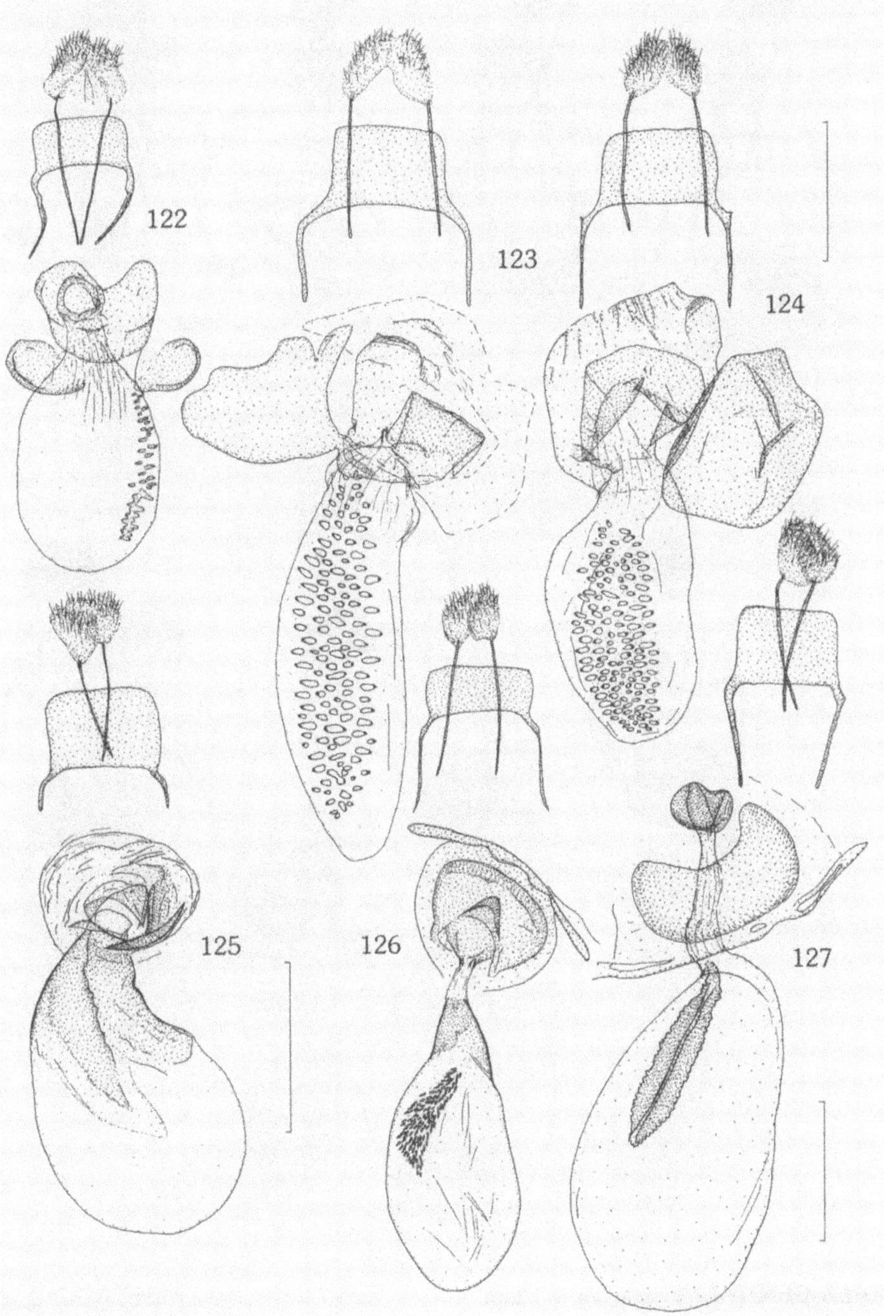

FEMALE GENITALIA (scale bars = 1 mm)

122. *Scopula nemoraria* Hbn., S. Germany
123. *Scopula umbelaria* Hbn., NE. Austria
124. *Scopula nigropunctata* Hufn., S. Germany
125. *Scopula virgulata* Den. & Schiff., Hungary
126. *Scopula ornata* Scop., N. Greece, sterigma folded forward
127. *Scopula concinnaria* Dup., S. Spain

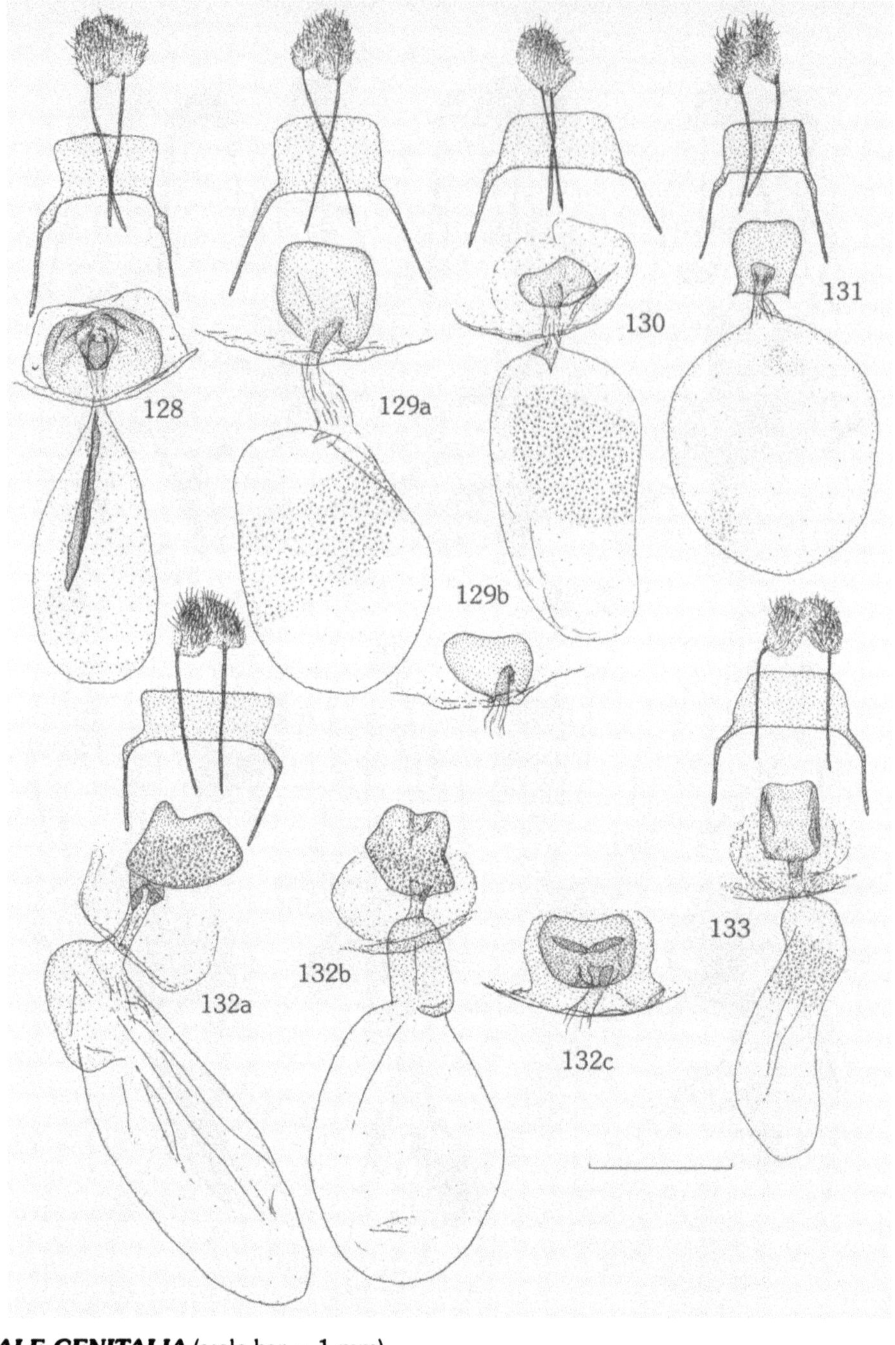

FEMALE GENITALIA (scale bar = 1 mm)

128. *Scopula orientalis* Alph., C. Turkey
129a. *Scopula decorata decorata* Den. & Schiff., N. Italy
129b. *Scopula decorata congruata* Z., Sicily, sterigma and antrum
130. *Scopula honestata* Mab., Corsica
131. *Scopula subtilata* Chr., S. European Russia
132a. *Scopula submutata submutata* Tr., SE. Austria
132b. *Scopula submutata roseonitens* Wagn., S. France, without ovipositor and tergum A8
132c. *Scopula transcaspica* Prt., Turkmeniya, sterigma and antrum
133. *Scopula vigilata* Wgn., Cyprus

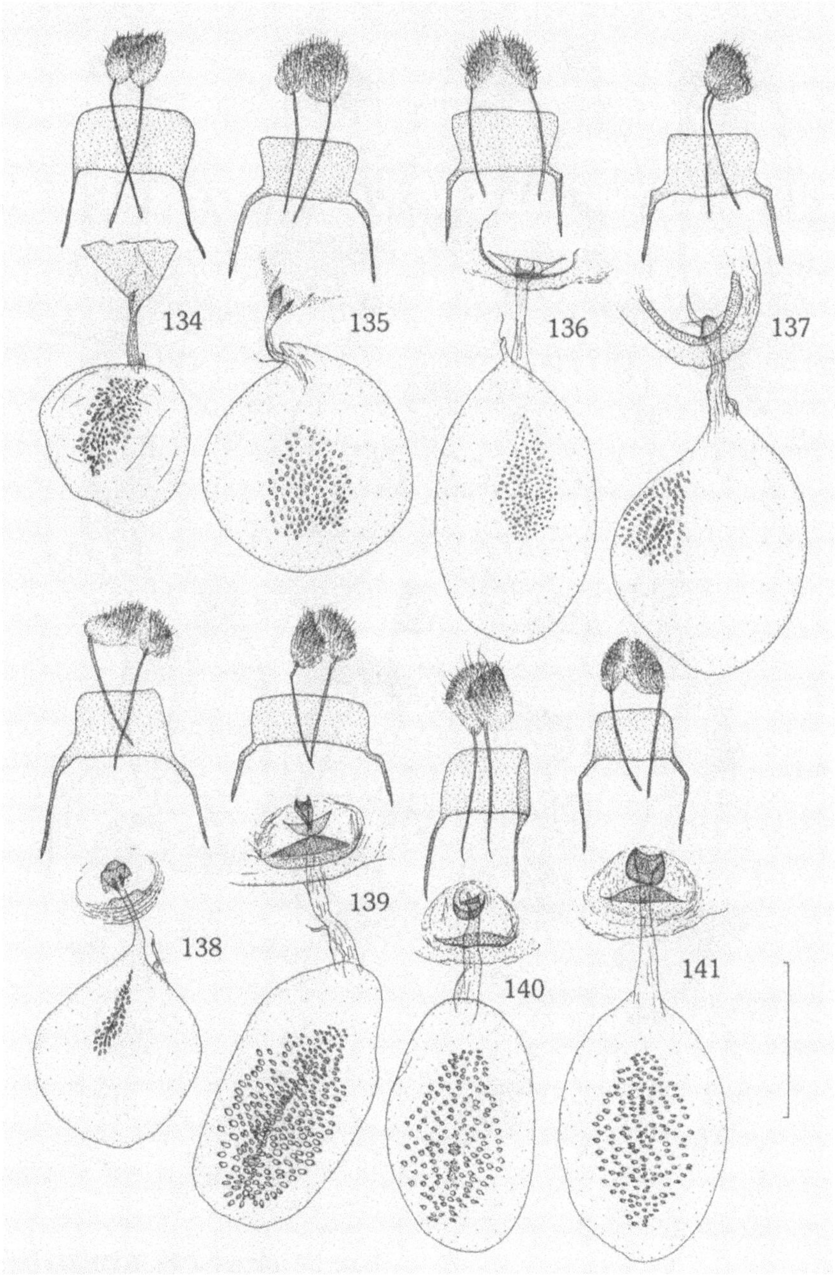

FEMALE GENITALIA (scale bar = 1 mm)

134. *Scopula rubiginata* Hufn., SE. Spain
135. *Scopula ochraceata* Stgr., Macedonia
136. *Scopula turbidaria* Hbn., E. Spain
137. *Scopula turbulentaria* Stgr., S. Italy
138. *Scopula decolor* Stgr., Algeria
139. *Scopula rubellata* Stgr., Portugal
140. *Scopula scalercii* Hausm., S. Italy
141. *Scopula beckeraria* Led., Macedonia

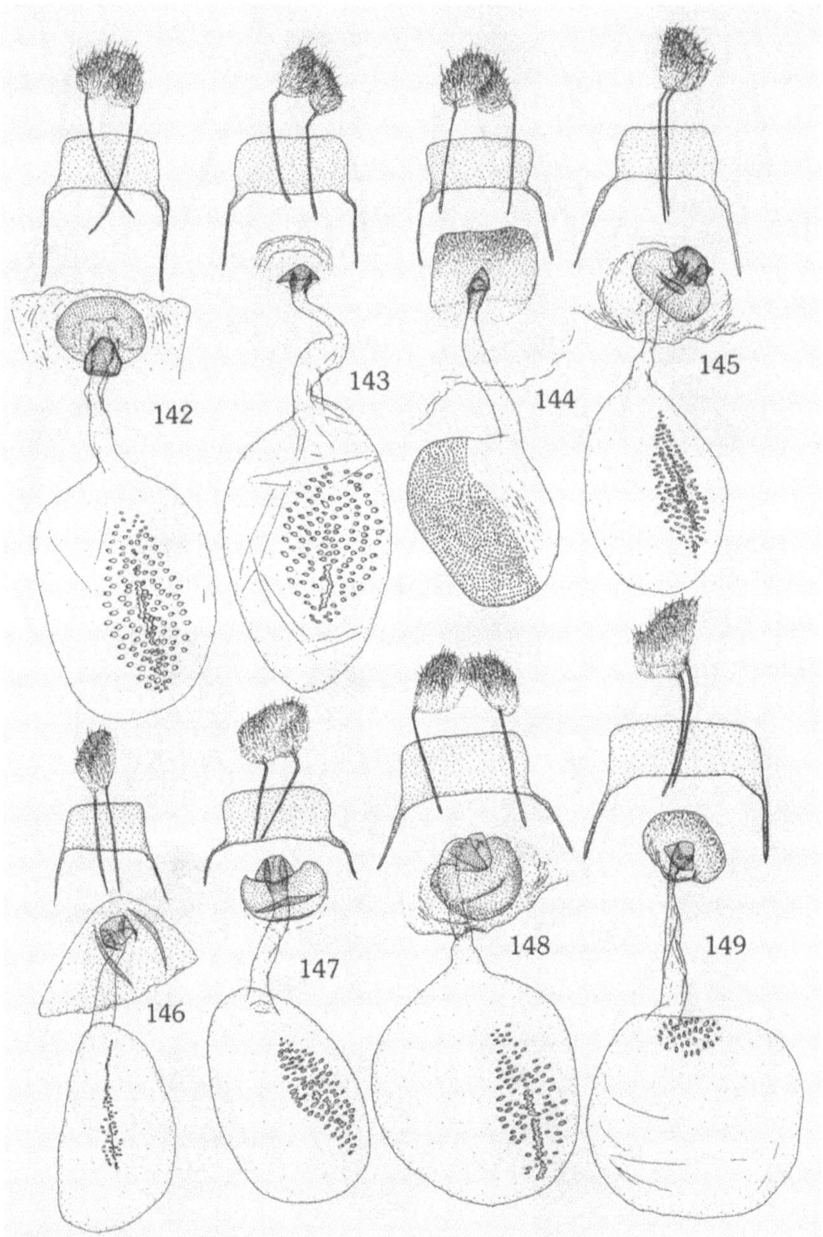

FEMALE GENITALIA (scale bar = 1 mm)

142. *Scopula incanata* L., Estonia, sterigma folded forward
143. *Scopula marginepunctata* Gze., Turkey
144. *Scopula luridata* Z., Israel
145. *Scopula imitaria* Hbn., S. Italy
146. *Scopula immutata* L., Estonia
147. *Scopula frigidaria* Möschl., Alaska
148. *Scopula ternata* Schrank, Pyrenees, antrum bent down (natural position)
149. *Scopula floslactata* Haw, N. France

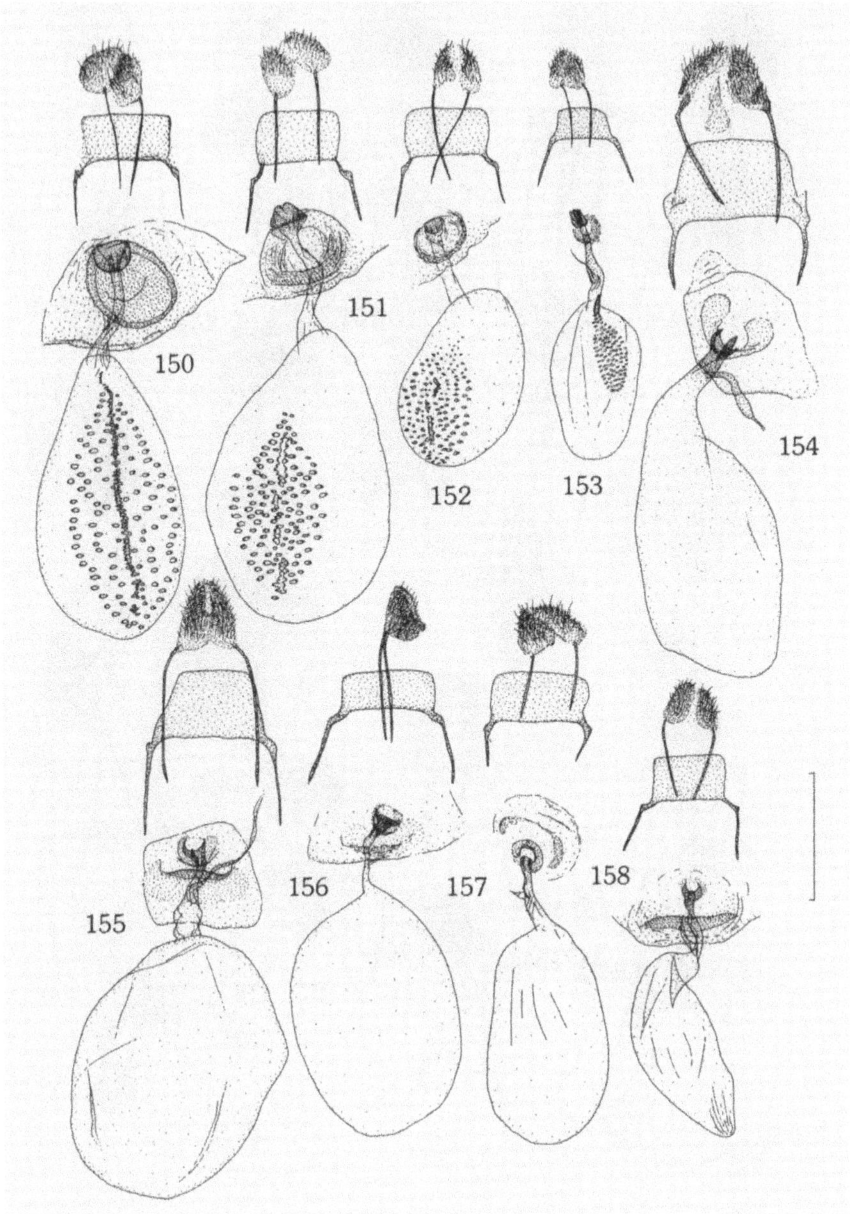

FEMALE GENITALIA (scale bar = 1 mm)

150. *Scopula subpunctaria* H.-S., N. Italy
151. *Scopula flaccidaria* Z., Turkey
152. *Scopula emutaria* Hbn., S. France
153. *Scopula minorata* Bsd., Sardinia
154. *Scopula cajanderi* Herz, Polar Urals, Russia
155. *Scopula divisaria* Chr., S. Urals, Russia
156. *Scopula arenosaria* Stgr., S. European Russia
157. *Scopula albiceraria* H.-S., S. European Russia, sterigma folded forward
158. *Scopula immistaria* H.-S., Bulgaria

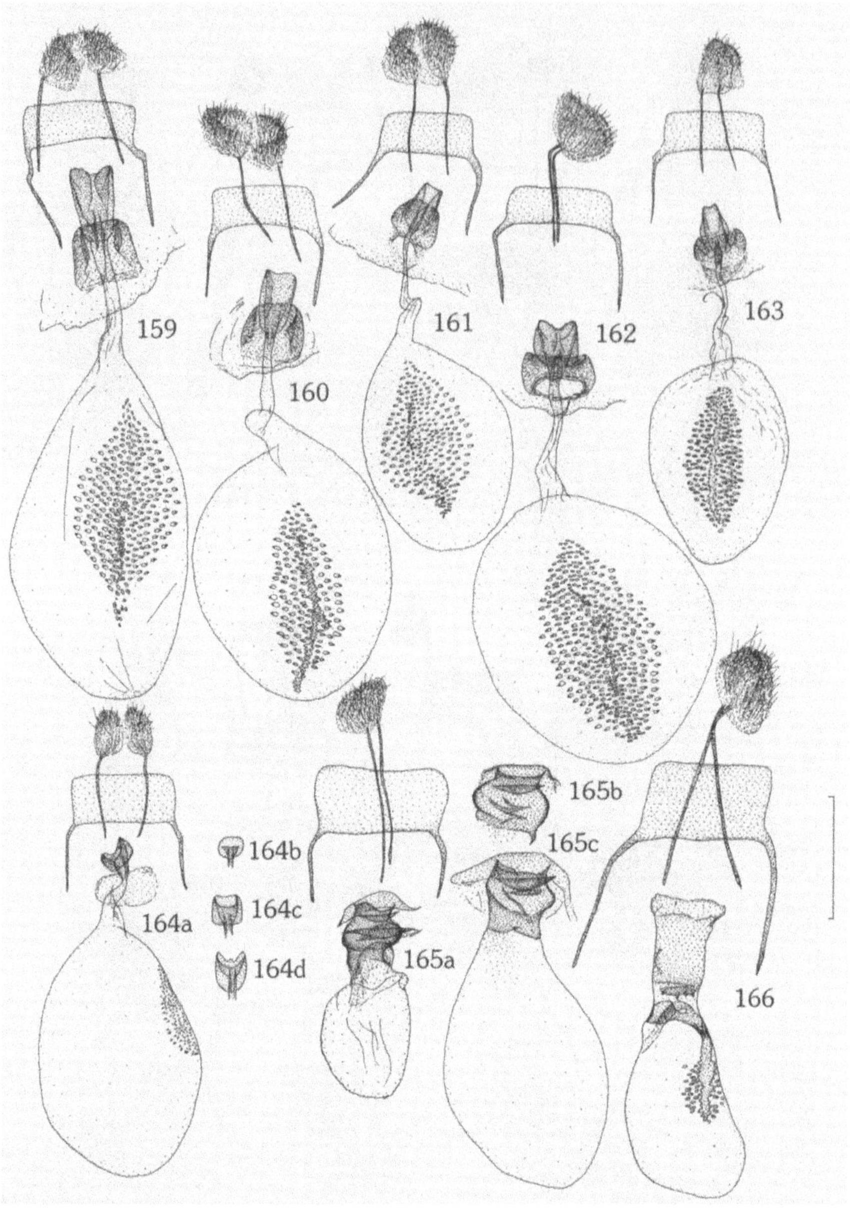

FEMALE GENITALIA (scale bar = 1 mm)

159. *Glossotrophia confinaria* H.-S., N. Italy
160. *Glossotrophia alba* Hausm., C. Italy
161. *Glossotrophia mentzeri* Hausm., Crete
162. *Glossotrophia rufomixtaria* Grasl., NE. Spain
163. *Glossotrophia sacraria* B.-H., Cyprus
164a. *Glossotrophia asellaria asellaria* H.-S., Corsica; fig. 164b.: subsp. *isabellaria* Mill., NE. Spain; fig. 164c: subsp. *dentatolineata* Whli., S. Spain; fig. 164d: subsp. *romanaria* Mill., C. Italy
165a. *Holarctias rufinaria* Stgr., Polar Urals, Russia; fig. 165b: *Holarctias rufinularia* Stgr., C. Russia; fig. 165c: *Holarctias sentinaria* Geyer, Labrador (antrum and sterigma)
166. *Problepsis ocellata* Friv., Crete

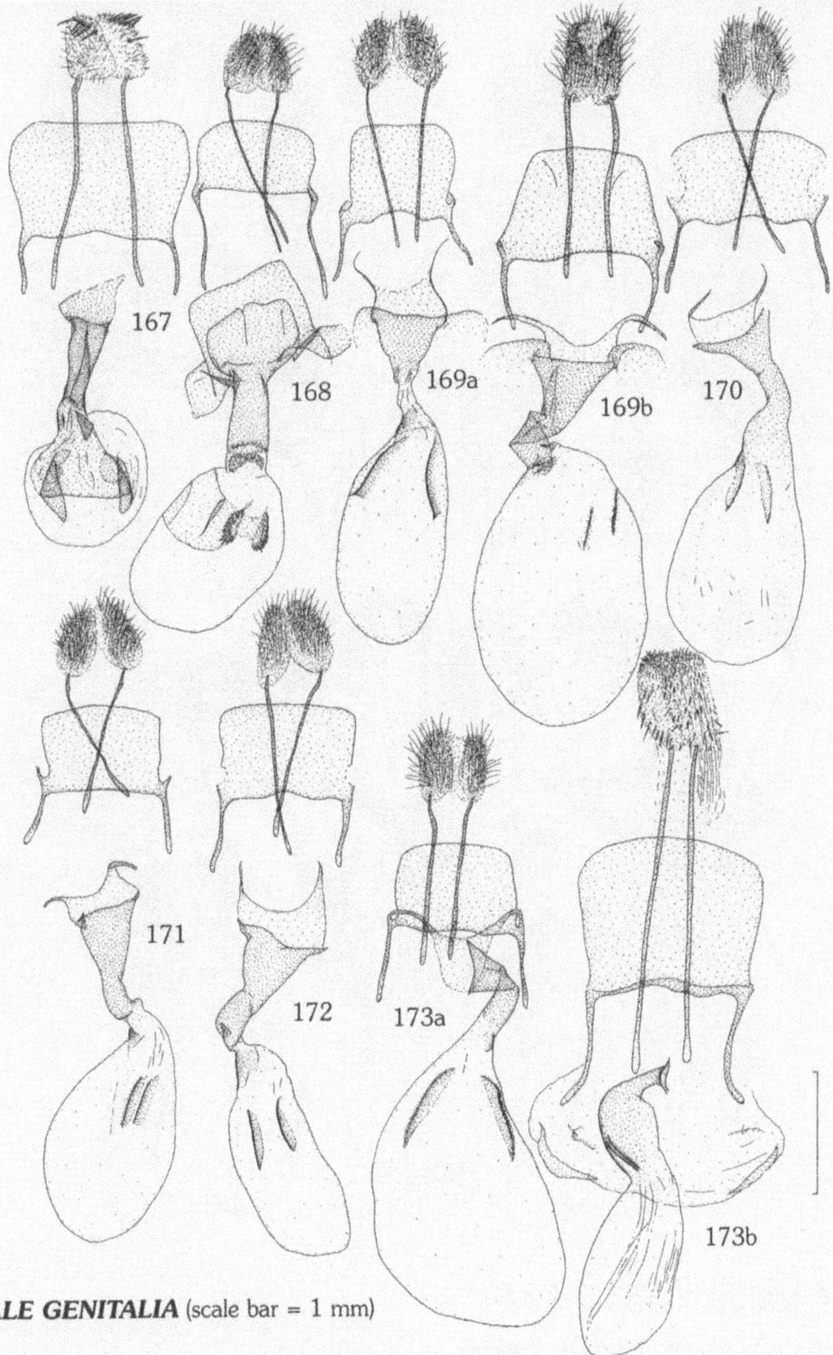

FEMALE GENITALIA (scale bar = 1 mm)

167. *Rhodostrophia jacularia* Hbn., S. European Russia
168. *Rhodostrophia vibicaria* Cl., NE. Spain
169a. *Rhodostrophia pudorata* F., S. Spain
169b. *Rhodostrophia sieversi* Hbn., N. Turkey
170. *Rhodostrophia cretacaria* Reb., Crete
171. *Rhodostrophia calabra* Petagna, S. Spain
172. *Rhodostrophia tabidaria* Z., Rhodos island
173a. *Rhodostrophia discopunctata* Ams., Macedonia
173b. *Rhodostrophia terrestraria* Led., Uzbekistan

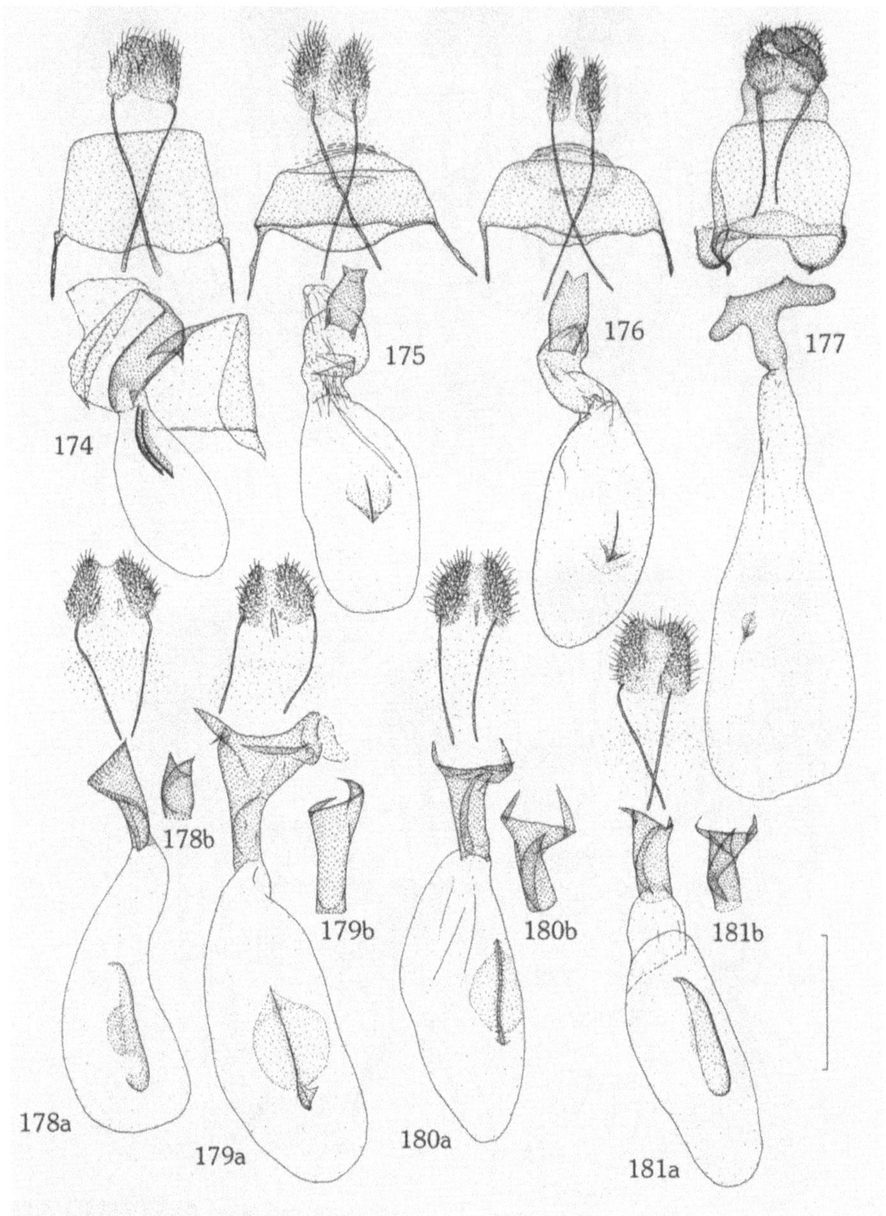

FEMALE GENITALIA (scale bar = 1 mm)

174. *Rhodostrophia badiaria* Frr., Turkey
175. *Timandra comae* Schmidt, Macedonia
176. *Timandra griseata* Peters., Finland
177. *Timandra rectistrigaria* Ev., Polar Urals, NE. European Russia
178a. *Cyclophora pendularia* Cl., S. France, signum lateral; 178b: England, ductus bursae
179a. *Cyclophora albipunctata* Hufn., England; 179b: Hungary, ductus bursae
180a. *Cyclophora albiocellaria* Hbn., E. Austria; 180b: S. European Russia, ductus bursae
181a. *Cyclophora lennigiaria* Fuchs, W. Germany, signum lateral; 181b: S. France, ductus bursae

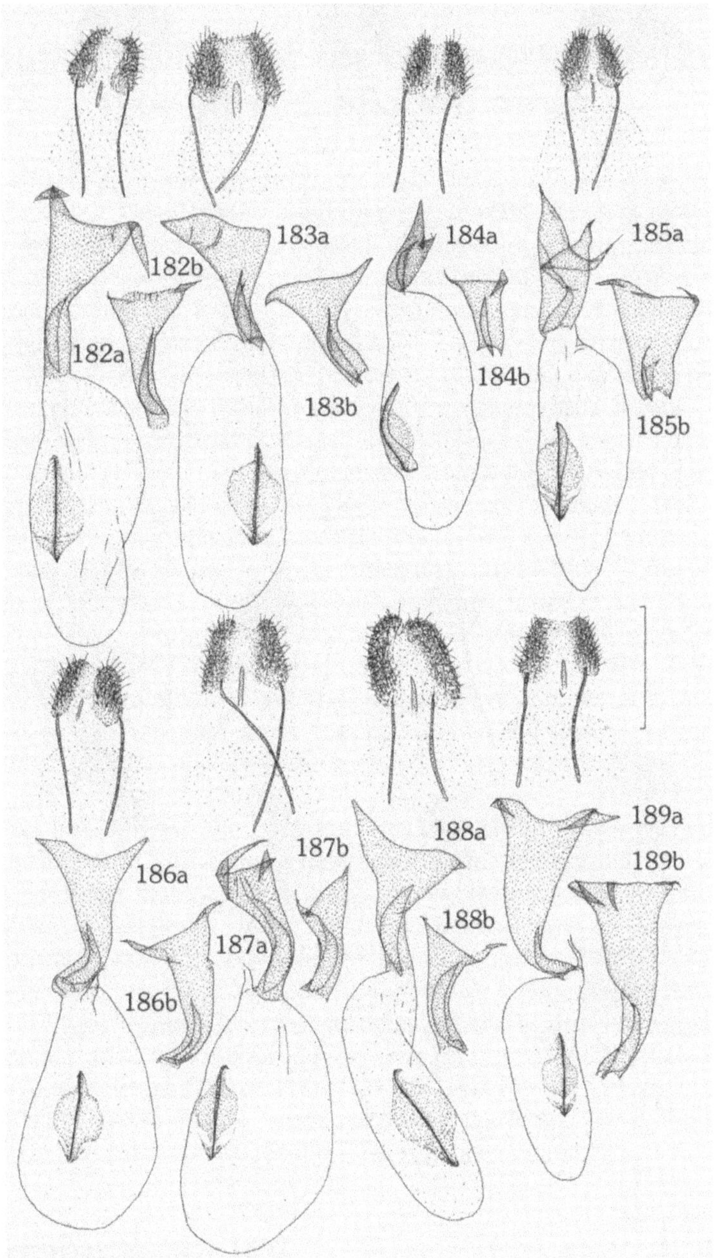

FEMALE GENITALIA (scale bar = 1 mm)

182a. *Cyclophora ariadne* Rssr., Crete; 182b: Crete, ductus bursae
183a. *Cyclophora serveti* Red. & Gast., N. Spain; 183b: N. Spain, ductus bursae
184a. *Cyclophora annularia* F., C. Italy; 184b: England, ductus bursae
185a. *Cyclophora puppillaria* Hbn., S. Spain; 185b: Tunisia, ductus bursae
186a. *Cyclophora quercimontaria* Bast., N. Italy; 186b: SE. Austria, ductus bursae
187a. *Cyclophora ruficiliaria* H.-S., NW. Turkey; 187b: S. Italy, ductus bursae
188a. *Cyclophora hyponoea* Prt., S. Spain; 188b: NE. Spain, ductus bursae
189a. *Cyclophora porata* L., NW. Italy; 189b: England, ductus bursae

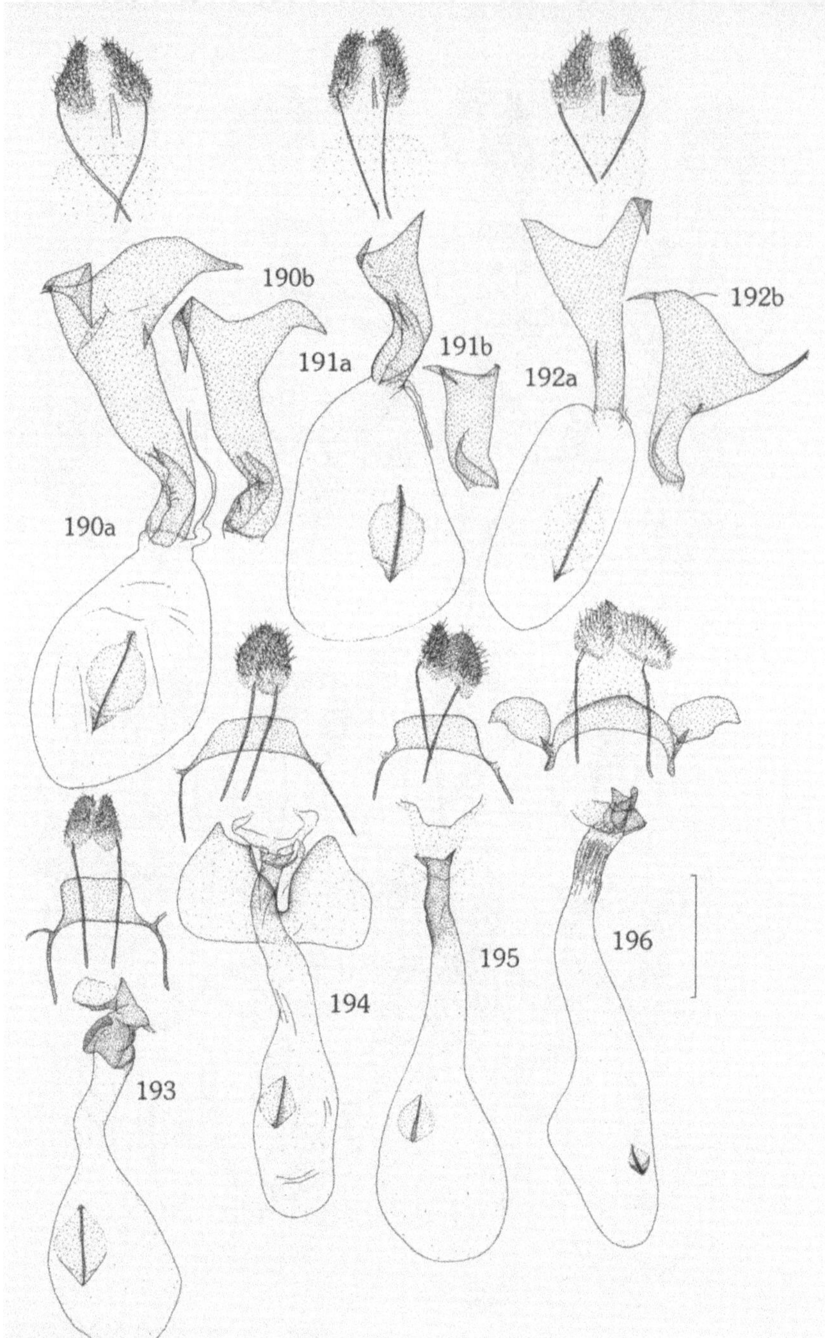

FEMALE GENITALIA (scale bar = 1 mm)

190a. *Cyclophora suppunctaria* Z., C. Italy; 190b: N. Turkey, ductus bursae
191a. *Cyclophora punctaria punctaria* L., C. Italy; 191b: N. Turkey, ductus bursae
191b. *Cyclophora punctaria fritzae* Hausm., Georgia, paratype, ductus bursae
192a. *Cyclophora linearia* Hbn., Germany; 192b: N. Turkey, ductus bursae
193. *Rhodometra sacraria* L., Cyprus
194. *Casilda antophilaria* Hbn., S. European Russia
195. *Casilda consecraria* Stgr., Cyprus
196. *Ochodontia adustaria* Wald., S. Urals, Russia

Systematic checklist of the Geometridae of Europe (Vol. 2) and adjacent areas

The checklist shows, in systematic order, the taxa (from subfamily level to subspecific level) of the European Geometridae treated in Vol. 2. The numbers after the species name refer to the species numbers in the text. 128 species occurring in areas immediately adjacent to Europe (as defined here) are inserted (marked with an asterisk) in the list with an indication of their distribution in these adjacent countries. It will be interesting to see whether, and when, some of these species will be found in Europe. Some of these species are known to occur very close to the borders of Europe, e.g. *Idaea textaria* from the western Turkish coast, less than 20 km from the Greek island Samos. Three other possible elements of the European fauna are treated in separate species chapters in this book: *Idaea lambessata* (with one, possibly mislabelled, specimen from central Spain), *Rhodostrophia sieversi* (with one, possibly mislabelled, specimen from Dalmatia), *Rhodostrophia terrestraria* (with one, possibly mislabelled, specimen from Volga plain). Many other species are distributed in North Africa and may be found some day on Sicily, Sardinia, Malta or Pantelleria, others approach southern Spain to just about 50 km in their northernmost occurrences in Morocco, e.g. *Idaea renataria* (OBERTHÜR, 1878), or *I. fathmaria* (OBERTHÜR, 1876). *Rhodostrophia vastaria* (CHRISTOPH, 1876) was collected about 200-300 km south of the south-easternmost steppes of European Russia and may belong to the European fauna, too, considering the absence of zoogeographical barriers. *Rhodostrophia auctata* (STAUDINGER, 1879) is recorded from north-western Turkey, about 100 km east of Istanbul.

For some taxa deleted from the European checklist due to nomenclatural synonymy see abstract. In literature further taxa have been listed, erroneously, as valid species names for the European fauna, such as *Idaea effeminata* (misidentification in ANIKIN et al. 2000, identity: *I. descitaria*), *Idaea nitidulata* (species rank in SCOBLE 1999, see under *Idaea mutilata*, no. 13), *'Scopula' arcuaria* (species rank in SCOBLE 1999, see under *Idaea belemiata*, no. 51), *Cyclophora maderensis lundbladi* (see BRYK 1940 and SCOBLE 1999; but types mislabelled, from Madeira rather than from Portugal, KARSHOLT 2001 pers. comm.). *Idaea allongata sublongaria* STAUDINGER, 1900 was recorded as doubtful from Crete by REBEL (1916), PROUT (1934d) and REISSER (1974), but no reliable verification for that record could be traced, neither in the REBEL nor in the REISSER collection (NHMW, LÖDL pers. comm.; SMNK, TRUSCH pers. comm.).

Subfamilia **Sterrhinae** MEYRICK, 1892

Tribus **Sterrhini** MEYRICK, 1892

Anthometra BOISDUVAL, 1840
A. plumularia BOISDUVAL, 1840 1

Emmiltis HÜBNER, 1825
E. pygmaearia (HÜBNER, 1809) 2

Cleta DUPONCHEL, 1845
C. ramosaria (VILLERS, 1789) 3
C. transiens PROUT, 1935 [Algeria, Tunisia]
C. perpusillaria (EVERSMANN, 1847) 4
C. cinneretharia AMSEL, 1935 [Levant]
C. filacearia (HERRICH-SCHÄFFER, 1847) 5

Idaea TREITSCHKE, 1825

The ***aureolaria*** species-group
I. serpentata (HUFNAGEL, 1767) 6
I. luteolaria (CONSTANT, 1863) 7
I. aureolaria (DENIS & SCHIFFERMÜLLER, 1775) 8
I. flaveolaria (HÜBNER, 1809) 9
I. muricata (HUFNAGEL, 1767) 10

The ***rufaria*** species-group
I. determinata (STAUDINGER, 1876) 11
I. litigiosaria (BOISDUVAL, 1840) 12

I. mutilata (Staudinger, 1876) 13
**I. hausmanni* (Trusch & Müller, 2001) [Tunisia, Libya]
I. lusohispanica Herbulot, 1991 14
**I. lambessata* (Oberthür, 1887) [Morocco to Tunisia, Spain?]
I. sardoniata (Homberg, 1912) 15
I. korbi (Püngeler, 1916) 16
I. mediaria (Hübner, 1819) 17
  **I. m. flavidior* (Rothschild, 1914) [Algeria]
I. leipnitzi Hausmann, sp. n. 18
**I. bengasiaria* (Turati, 1926) [Libya]
**I. renataria* (Oberthür, 1878) [Morocco to Libya]
I. rufaria (Hübner, 1799) 19
I. consanguiberica Rezbanyai-Reser & Exposito, 1992 20
**I. debiliata* (Sterneck, 1934) [Morocco]
I. admiranda Hausmann, sp.n. 21
I. consanguinaria (Lederer, 1853) 22
  I. c. turatii (Sohn-Rethel, 1929)
  **I. c. consecrata* (Staudinger, 1897) [Levant]
I. ossiculata (Lederer, 1871) 23
I. sericeata (Hübner, 1813) 24
  I. s. calvaria Wehrli, 1927
**I. allardiata* (Mabille, 1869) [Algeria to Libya]
I. macilentaria (Herrich-Schäffer, 1847) 25
  (=*antiquaria* Herrich-Schäffer, 1847)
**I. numidaria* (Lucas, 1849) [Algeria]
I. ochrata (Scopoli, 1763) 26
  I. o. albida (Zerny, 1936)

The ***rusticata*** species-group
I. nevadata (Wehrli, 1934) 27
I. figuraria (Bang-Haas, 1907) 28
  (=*subcompleta* Fernández, 1931)
I. alicantaria (Reisser, 1963) 29
**I. oberthueri* (Lucas, 1917) [Tunisia]
I. nigrolineata (Chrétien, 1910) 30
I. completa (Staudinger, 1892) 31
  **I. c. apostolica* Hausmann, 1994 [Cyprus]
**I. erlacheri* Hausmann, 2003 [Tunisia]
I. intermedia (Staudinger, 1879) 32
I. rusticata (Denis & Schiffermüller, 1775) 33
  (=*vulpinaria* Herrich-Schäffer, 1852)
I. mustelata (Gumppenberg, 1892) 34
I. filicata (Hübner, 1799) 35
I. troglodytaria (Heydenreich, 1851) 36
I. laevigata (Scopoli, 1763) 37
  (=*couloniata* Balestre, 1907)
**I. vilaflorensis* (Rebel, 1911) [Canary islands]
**I. charitata* (Rebel, 1914) [Canary islands]
I. efflorata Zeller, 1849 38
  (=*extarsaria* Herrich-Schäffer, 1852)

I. attenuaria (Rambur, 1833) 39
I. incalcarata (Chrétien, 1913) 40

The ***typicata*** species-group
I. typicata (Guenée, 1858) 41
  I. t. hornigaria (Staudinger, 1901)
**I. dyraria* (Zerny, 1934) [Morocco]
I. alyssumata (Himmighoffen & Millière, 1871) 42
  I. a. reynaldiata (Rouast, 1876)
I. urcitana (Agenjo, 1952) 43
I. moniliata (Denis & Schiffermüller, 1775) 44
**I. ludovicaria* (Culot, 1918) [Morocco to Tunisia]
**I. unicalcarata* (Prout, 1922) [Morocco, Algeria]

The ***spissilimbaria*** species-group
I. spissilimbaria (Mabille, 1888) 45

*The ***microptera*** species-group
**I. microptera* (Warren & Rothschild, 1905) [Egypt, Levant]
**I. sordida* (Rothschild, 1913) [Morocco to Libya]
  **I. s. nili* (Wiltshire, 1949) [Egypt, Levant]
**I. sordidior* (Wiltshire, 1949) [Egypt]
**I. granulosa* (Warren & Rothschild, 1905) [Egypt, Levant]
**I. miserrima* (Turati, 1930) [Libya]
**I. scabraria* (Chrétien, 1909) [Algeria, Tunisia]
**I. sanctaria* (Staudinger, 1900) [Levant]
  **I. s. outayana* (Wehrli, 1931) [Morocco to Tunisia]
  (=*bucephalaria* Chrétien, 1909)
  **I. s. transcatenulata* (Rothschild, 1915) [Algeria]
**I. ortali* Hausmann, 1991 [Levant]
**I. romeii* (Turati, 1930) [Libya]

*The ***volloni*** species-group
**I. volloni* (Lucas & Joannis, 1907) [Canaries to Levant]

The ***circuitaria*** species-group
I. circuitaria (Hübner, 1819) 46
  I. c. expositoi Hausmann, 1994
I. rainerii Hausmann, 1994 47
**I. mimosaria* (Guenée, 1858) [Turkey, Levant]
  **I. m. wimmeri* Hausmann, 1994 [Cyprus]

The ***incisaria*** species-group
I. albarracina (Reisser, 1933) 48
I. incisaria (Staudinger, 1892) 49
  **I. i. praecisa* (Reisser, 1933) [Morocco]
  **I. i. pulverulenta* (Reisser, 1933) [Morocco to Libya]

*I. *textaria* (LEDERER, 1861)
[Turkey, Levant, Transcaucasus]
I. calunetaria (STAUDINGER, 1859) 50
(=*dorycniata* BELLIER, 1862)
I. c. valesiaria (PÜNGELER, 1888)
**I. c. episticta* (WEHRLI, 1930)
[Morocco, Algeria]
**I. mareotica* (DRAUDT, 1912)
[Morocco to Levant]
(=*fuscularia* TURATI, 1930)
**I. allongata* (STAUDINGER, 1897) [Levant]
**I. a. sublongaria* (STAUDINGER, 1899) [Levant]
**I. pectinata* (STERNECK, 1933) [Levant]

The ***obsoletaria*** species-group
**I. hathor* (WILTSHIRE, 1949) [Egypt, Levant]
I. belemiata (MILLIÈRE, 1868) 51
(=*substraminata* PROUT, 1913)
I. b. helianthemata (MILLIÈRE, 1870)
(=*euphorbiata* BALESTRE, 1906)
(=*oberthuriata* BALESTRE, 1907)
(=*cavenacata* CHRÉTIEN, 1929)
I. elongaria (RAMBUR, 1833) 52
I. e. pecharia (STAUDINGER, 1863)
**I. maxima* (WEHRLI, 1934) [Morocco]
**I. palmata* (STAUDINGER, 1901) [Canary islands]
**I. substriata* (TURATI, 1930) [Libya]
**I. peluraria* (REISSER, 1939) [Turkey, Cyprus]
**I. persidis* (WILTSHIRE, 1966) [Turkey]
I. palaestinensis (STERNECK, 1933) 53
I. obsoletaria (RAMBUR, 1833) 54
I. o. lilaceola HAUSMANN, 2003
I. o. dierli HAUSMANN, 1991
I. o. dionigii HAUSMANN, 1991
I. o. rufularia (HERRICH-SCHÄFFER, 1847)
**I. o. algeriensis* (BETHUNE-BAKER, 1888)
[Algeria]
**I. epaphrodita* (WEHRLI, 1934) [Levant]
**I. e. eremita* (WILTSHIRE, 1949) [Egypt]
**I. wittmeri* (WILTSHIRE, 1982)
**I. w. mienisi* HAUSMANN, 1991 [Levant]
I. obliquaria (TURATI, 1913) 55

The ***inquinata*** species-group
I. inquinata (SCOPOLI, 1763) 56
(=*herbariata* FABRICIUS, 1798)
**I. i. adherbariata* (STAUDINGER, 1898) [Levant]
**I. saida* (WILTSHIRE, 1968) [Levant]
**I. holliata* (HOMBERG, 1909) [Levant]
**I. banghaasi* (PROUT, 1934) [Levant]
**I. affinitata* (BANG-HAAS, 1907) [Levant]
I. blaesii LENZ & HAUSMANN, 1992 57
I. dilutaria (HÜBNER, 1799) 58
I. fuscovenosa (GOEZE, 1781) 59
I. f. corsula (SCHAWERDA, 1929)
I. robiginata (STAUDINGER, 1863) 60
I. lutulentaria (STAUDINGER, 1892) 61
**I. terminolineata* (ROTHSCHILD, 1933)
[Morocco, Algeria]
I. humiliata (HUFNAGEL, 1767) 62
I. bigladiata HERBULOT, 1975 63
**I. b. alberti* HAUSMANN, subsp. n.
[Algeria, Tunisia]
I. politaria (HÜBNER, 1799) 64
**I. roseofasciata* (CHRISTOPH, 1882)
[Turkey, Transcaucasus]
**I. capnaria* (PÜNGELER, 1909) [Levant]
**I. maronitaria* (ZERNY, 1933) [Levant]

The ***longaria*** species-group
I. longaria (HERRICH-SCHÄFFER, 1852) 65
**I. abnorma* (PINKER, 1960) [Canary islands]
**I. atlantica* (STAINTON, 1859) [Madeira]
(=*I. illuminata* PROUT, 1939)
**I. margaritacea* (TURATI & KRÜGER, 1936) [Libya]

The ***nexata*** species-group
I. nexata (HÜBNER, 1813) 66
I. manicaria (HERRICH-SCHÄFFER, 1852) 67
(=*fatimata* STAUDINGER, 1895)

The ***consolidata*** species-group
I. descitaria (CHRISTOPH, 1893) 68
I. vesubiata (MILLIÈRE, 1873) 69
I. libycata (BARTEL, 1906) 70
I. consolidata (LEDERER, 1853) 71
I. joannisiata (HOMBERG, 1911) 72
I. j. ibericata (WEHRLI, 1927)
I. acutipennis HAUSMANN & HONEY, 2004 73
**I. jacobi* HAUSMANN, 1991 [Levant]
**I. striolata* (STAUDINGER, 1892) [Levant]
**I. detritaria* (STAUDINGER, 1897) [Levant]

The ***seriata*** species-group
I. seriata (SCHRANK, 1802) 74
(=*virgularia* sensu HÜBNER, 1799)
I. s. canteneraria (BOISDUVAL, 1840)
I. minuscularia (RIBBE, 1912) 75
(=*herbuloti* AGENJO, 1952)
I. ibizaria MENTZER, 1980 76
I. albitorquata (PÜNGELER, 1909) 77
I. a. madoniensis HAUSMANN, 1993
**I. a. maculifera* HAUSMANN, 1994 [Cyprus]
I. carvalhoi HERBULOT, 1979 78
**I. fathmaria* (OBERTHÜR, 1876)
[Morocco, Algeria]
**I. muelleri* HAUSMANN, 2003 [Caucasus]
**I. maderae* (BETHUNE-BAKER, 1891) [Madeira]
(=*zargi* BETHUNE-BAKER, 1891)

The ***camparia*** species-group
I. camparia (HERRICH-SCHÄFFER, 1852) 79
I. c. sodaliaria (HERRICH-SCHÄFFER, 1852)

The ***subsericeata*** species-group
I. subsericeata (HAWORTH, 1809) 80
(=*zernyi* LUCAS, 1934)
**I. maurusia* (TURATI, 1924) [Libya]
I. pallidata (DENIS & SCHIFFERMÜLLER, 1775) 81
I. mancipiata (STAUDINGER, 1871) 82
(=*bustilloi* AGENJO, 1967)
I. m. repagulata (PROUT, 1934)
**I. proclivata* (FUCHS, 1902) [Turkey, Transcaucasus]
(=*osthelderi* WEHRLI, 1932)
I. sylvestraria (HÜBNER, 1799) 83
I. dimidiata (HUFNAGEL, 1767) 84
**I. d. antitaurica* (WEHRLI, 1931) [Turkey, Levant]
I. subsaturata (GUENÉE, 1858) 85
**I. s. lecerfiata* (HOMBERG, 1910) [Morocco to Libya; Levant?; Cyprus?]
I. trigeminata (HAWORTH, 1809) 86
**I. t. tenuirussata* (ZERNY, 1933) [Cyprus, Levant]
I. biselata (HUFNAGEL, 1767) 87

The ***fractilineata*** species-group
I. fractilineata (ZELLER, 1847) 88
I. f. subrufaria (STAUDINGER, 1899)
**I. inclinata* (LEDERER, 1855) [Libya to Turkey, Cyprus]
I. exilaria (GUENÉE, 1858) 89
**I. erubescens* HAUSMANN, 2003 [Algeria, Tunisia]
**I. truschi* HAUSMANN, 2003 [Morocco]
**I. merklaria* (OBERTHÜR, 1885) [Algeria]

The ***cervantaria*** species-group
I. cervantaria (MILLIÈRE, 1869) 90
**I. c. carneotincta* (ZERNY, 1936) [Morocco]
I. contiguaria (HÜBNER, 1799) 91
I. c. britanniae (MÜLLER, 1936)
I. rupicolaria (REISSER, 1927) 92

The ***deitanaria*** species-group
I. deitanaria (REISSER & WEISERT, 1977) 93

The ***infirmaria*** species-group
I. saleri DOMÍNGUEZ & BAIXERAS, 1992 94
I. tineata (THIERRY-MIEG, 1911) 95
(=*antennata* WEHRLI, 1931)
**I. t. praetineata* HAUSMANN, 1994 [Levant]
I. infirmaria (RAMBUR, 1833) 96
**I. i. mitescens* SCOBLE, 1999 [Morocco to Tunisia]
I. rhodogrammaria (PÜNGELER, 1913) 97
**I. lobaria* (CHRÉTIEN, 1909) [Algeria, Tunisia, Levant]
(=*balestraria* LUCAS, 1909)
**I. okbaria* (CHRÉTIEN, 1911) [Algeria?, Tunisia]
**I. bacalladoi* (PINKER, 1974) [Canary islands]
**I. alutaceata* (RUNGS, 1945) [Morocco]
**I. purpurariata* (PINKER, 1974) [Canary islands]
**I. fuerteventurensis* (PINKER, 1974) [Canary islands]

*The ***gracilipennis*** species-group
**I. gracilipennis* (WARREN, 1901) [Turkey, Levant]
**I. purpureomarginata* (BOHATSCH, 1880) [Turkey, Levant]
**I. p. stefani* HAUSMANN, 1991 [Levant]
**I. subpurpurata* (STAUDINGER, 1900) [Libya, Levant, Turkey]

The ***ostrinaria*** species-group
I. ostrinaria (HÜBNER, 1813) 98
**I. o. demarginata* (SCHWINGENSCHUSS, 1938) [Turkey]
I. metohiensis (REBEL, 1900) 99
**I. m. anatolica* (SCHWINGENSCHUSS, 1938) [Turkey]

The ***eugeniata*** species-group
I. eugeniata (MILLIÈRE, 1870) 100
**I. e. algeriaca* (CULOT, 1917) [Morocco to Tunisia]
I. distinctaria (BOISDUVAL 1840) 101
(=*ruficostata* ZELLER, 1849)
I. predotaria (HARTIG, 1951) 102
(=*griseanova* REZBANYAI-RESER, 1987)

The ***nitidata*** species-group
I. nitidata (HERRICH-SCHÄFFER, 1861) 103

The ***aversata*** species-group
I. emarginata (LINNAEUS, 1758) 104
I. dromikos HAUSMANN, sp. n. 105
I. simplicior (PROUT, 1934) 106
I. rubraria (STAUDINGER, 1901) 107
(=*bilinearia* FUCHS, 1889)
I. aversata (LINNAEUS, 1758) 108
**I. a. indeviata* (PROUT, 1929) [Algeria, Tunisia]
**I. gelbrechti* HAUSMANN, 2003 [Morocco]
I. degeneraria (HÜBNER, 1799) 109
I. d. erschoffi (CHRISTOPH, 1872)
I. straminata (BORKHAUSEN, 1794) 110
(=*inornata* HAWORTH, 1809)
I. deversaria (HERRICH-SCHÄFFER, 1847) 111
(?=*maritimaria* BRUAND, 1846)
I. d. fallax HAUSMANN, 2003
I. squalidaria (STAUDINGER, 1882) 112
**I. neglecta* HAUSMANN & WERNO, 2003 [Canaries]

incertae sedis
**I. protrusa* (TURATI, 1926) [Libya]
**I. bicertaria* (STAUDINGER, 1892) [Tunisia]

Brachyglossina WAGNER, 1914
B. hispanaria (PÜNGELER, 1913) 113
**B. brevitarsata* (HAUSMANN, 1991) [Levant]
**B. consociata* (STAUDINGER, 1900) [Turkey, Levant]
**B. improbata* (STAUDINGER, 1898) [Levant]
**B. hassan* HERBULOT, 1974 [Algeria]
**B. tantalidis* (TURATI, 1924) [Algeria?, Tunisia, Libya]
**B. mauritanica* (BETHUNE-BAKER, 1885) [Algeria]
**B. mzabensis* PROUT, 1938 [Algeria]
**B. ochrolutearia* (TURATI, 1934) [Libya]
**B. acidalaria* WAGNER, 1914 [Tunisia]
**B. macracantha* PROUT, 1938 [Algeria]
**B. maroccana* (WEHRLI, 1930) [Morocco]
**B. paroranaria* WEHRLI, 1938 [Algeria]
  **B. p. suboranaria* WEHRLI, 1938 [Algeria]
**B. seitzi* PROUT, 1938 [Algeria]
**B. vindicata* PROUT, 1938 [Algeria]
**B. fulta* PROUT, 1938 [Algeria]
**B. culoti* (WEHRLI, 1930) [Morocco, Algeria] (=*pseudoranaria* ZERNY, 1936)
**B. oranaria* (BANG-HAAS, 1907) [Algeria, Morocco]
**B. staudingeri* (PROUT, 1932) [Levant]
**B. sonyae* (WILTSHIRE, 1990) [Levant]
**B. williamsi* WILTSHIRE, 1949 [Egypt, Levant]

***Tineigidia** STERNECK, 1934
**T. eremica* AMSEL & STERNECK, 1934 [Levant]
**T. aegyptiaca* WILTSHIRE, 1949 [Egypt]

Limeria STAUDINGER, 1892
L. macraria STAUDINGER, 1892 114

Tribus **Scopulini** DUPONCHEL, 1845

Oar PROUT, 1913
O. reaumuraria (MILLIÈRE, 1864) 115
**O. pratana* (FABRICIUS, 1794) [Algeria to Libya]
  **O. p. oppressa* (WALKER, 1870) [Levant, Egypt, Arabia]
  **O. p. baezi* HAUSMANN, subsp. n. [Canary islands, Morocco]

Cinglis GUENÉE, 1858
C. humifusaria (EVERSMANN, 1837) 116
C. andalusiaria WAGNER, 1935 117

Pseudocinglis HAUSMANN, 1994
**P. moralesi* (RUNGS, 1945) [Morocco]

Scopula SCHRANK, 1802

subgenus ***Scopula*** SCHRANK, 1802

The ***immorata*** species-group
S. immorata (LINNAEUS, 1758) 118
  S. i. duercki SHELJUZHKO, 1955
  S. i. riloensis (ZÜLLICH, 1936)
S. tessellaria (BOISDUVAL, 1840) 119
S. corrivalaria (KRETSCHMAR, 1862) 120
S. caricaria (REUTTI, 1853) 121
  (=*virginalis* FOURCROY, 1785)
S. nemoraria (HÜBNER, 1799) 122
S. umbelaria (HÜBNER, 1813) 123
S. nigropunctata (HUFNAGEL, 1767) 124
**S. nepheloperas* (PROUT, 1916) [Egypt, Sinai]

The ***virgulata*** species-group
S. virgulata (DENIS & SCHIFFERMÜLLER, 1775) 125
  S. v. rossica (DJAKONOV, 1926)

The ***ornata*** species-group
S. ornata (SCOPOLI, 1763) 126
S. concinnaria (DUPONCHEL, 1842) 127
  S. c. universaria (ZERNY, 1927)
S. orientalis (ALPHÉRAKY, 1876) 128
S. decorata (DENIS & SCHIFFERMÜLLER, 1775) 129
  S. d. violata (THUNBERG, 1784)
  S. d. congruata (ZELLER, 1847)
S. honestata (MABILLE, 1869) 130
S. subtilata (CHRISTOPH, 1867) 131
S. submutata (TREITSCHKE, 1828) 132
  S. s. roseonitens (WAGNER, 1926)
  S. s. gedrensis HAUSMANN, 2003
  S. s. nivellearia (OBERTHÜR, 1922)
  **S. s. taurilibanotica* (WEHRLI, 1932) [Turkey, Cyprus, Levant]
**S. transcaspica* PROUT, 1935 [Turkey, Transcaucasus]
S. vigilata (WAGNER, 1926) 133

The ***rubiginata*** species-group
S. rubiginata (HUFNAGEL, 1767) 134
S. ochraceata (STAUDINGER, 1901) 135
S. turbidaria (HÜBNER, 1819) 136
S. turbulentaria (STAUDINGER, 1870) 137

subgenus ***Calothysanis*** HÜBNER, 1823
The ***incanata*** species-group
S. decolor (STAUDINGER, 1897) 138
  **S. d. flaccata* (STAUDINGER, 1897) [Levant]
  S. rubellata (STAUDINGER, 1871) 139
S. scalercii HAUSMANN, 2003 140
S. beckeraria (LEDERER, 1853) 141
  S. b. rebeli (PROUT, 1913)
  **S. b. hermonicola* HAUSMANN, 1997 [Levant]
S. incanata (LINNAEUS, 1758) 142
  S. i. ibericata (REISSER, 1935)
S. marginepunctata (GOEZE, 1781) 143
  **S. m. argillacea* REISSER, 1993 [Morocco to Tunisia]

S. *luridata* (ZELLER, 1847) 144
(=*fulminataria* TURATI, 1927)
*S. *bimaculata* (TURATI & KRÜGER, 1936) [Libya]
(?=*barcearia* TURATI & KRÜGER, 1936)
S. *imitaria* (HÜBNER, 1799) 145
S. *i. syriacaria* (CULOT, 1918)
S. *immutata* (LINNAEUS, 1758) 146
S. *frigidaria* (MÖSCHLER, 1860) 147

The ***ternata*** species-group
S. *ternata* SCHRANK, 1802 148
S. *floslactata* (HAWORTH, 1809) 149
S. *f. scotica* (COCKAYNE, 1951)
S. *subpunctaria* (HERRICH-SCHÄFFER, 1847) 150
S. *flaccidaria* (ZELLER, 1852) 151
S. *emutaria* (HÜBNER, 1809) 152
S. *minorata* (BOISDUVAL, 1833) 153
(=*ochroleucaria* HERRICH-SCHÄFFER, 1847)
(=*inustaria* HERRICH-SCHÄFFER, 1847)
(=*medioumbraria* TURATI, 1930)
*S. *andresi* (DRAUDT, 1912) [Egypt]
*S. *adelpharia* (PÜNGELER, 1894) [Levant]
*S. *a. pharaonis* STERNECK, 1933 [Egypt]
*S. *donovani* (DISTANT, 1892)
[Morocco, Egypt, Levant]
*S. *irrorata* (BETHUNE-BAKER, 1891) [Madeira]
*S. *guancharia* (ALPHÉRAKY, 1889)
[Canary islands]
*S. *g. illustris* PINKER, 1968 [Canary islands]
*S. *g. mus* PINKER, 1968 [Canary islands]
*S. *g. uniformis* PINKER, 1968
[Canary islands]
*S. *g. punctabilineatella* (LUCAS, 1937)
[Morocco]

The ***immistaria*** species-group
S. *cajanderi* (HERZ, 1903) 154
S. *divisaria* (CHRISTOPH, 1893) 155
(=*latelineata* sensu STAUDINGER, 1901)
S. *arenosaria* (STAUDINGER, 1879) 156
S. *albiceraria* (HERRICH-SCHÄFFER, 1847) 157
*S. *lehmanni* HAUSMANN, 1991 [Levant]
S. *immistaria* (HERRICH-SCHÄFFER, 1852) 158
S. *i. beshkovi* GELBRECHT & HAUSMANN, 1997

incertae sedis
*S. *gastonaria* (OBERTHÜR, 1876) [Algeria]
*S. *g. luteofasciata* (ROTHSCHILD, 1913)
[Algeria]

Glossotrophia PROUT, 1913
subgenus ***Glossotrophia*** PROUT, 1913
G. *confinaria* (HERRICH-SCHÄFFER, 1847) 159
G. *c. prouti* HAUSMANN, 1993
G. *c. aetnaea* PROUT, 1935
G. *c. scoblei* HAUSMANN, 1993
*G. *diffinaria* PROUT, 1913
[Turkey, Transcaucasus]
*G. *uberaria* ZERNY, 1933 [Cyprus, Levant]
G. *alba* HAUSMANN, 1993 160
G. *a. milleri* HAUSMANN, subsp. n.
G. *a. brunellii* HAUSMANN, 1993
*G. *a. africana* HAUSMANN, 1993 [Tunisia]
G. *mentzeri* HAUSMANN, 1993 161
G. *rufomixtaria* (GRASLIN, 1863) 162
(=*annae* MENTZER, 1990)
*G. *r. sahariensis* HAUSMANN, 1993
[Algeria, Tunisia]
*G. *chalcographata* BRANDT, 1938
*G. *c. sinaica* REBEL, 1948 [Levant]

subgenus ***Libanonia*** HAUSMANN, 1993
G. *sacraria* (BANG-HAAS, 1910) 163
*G. *s. semitata* PROUT, 1913 [Levant]
*G. *s. fumata* HAUSMANN, 1993 [Cyprus]
*G. *s. taurica* WEHRLI, 1930
[Turkey, Transcaucasus]

subgenus ***Parenzanella*** HAUSMANN, 1993
G. *asellaria* (HERRICH-SCHÄFFER, 1847) 164
G. *a. romanaria* (MILLIÈRE, 1869)
G. *a. isabellaria* (MILLIÈRE, 1868)
G. *a. dentatolineata* WEHRLI, 1926
*G. *a. gerstbergeri* HAUSMANN, 1993
[Canary islands]
*G. *a. lenzi* HAUSMANN, 1993 [Morocco]
*G. *a. philipparia* PROUT, 1913
[Algeria, Tunisia]
*G. *a. tripolitana* TURATI, 1929
[Algeria to Libya]

subgenus ***Microglossotrophia*** HAUSMANN, 1993
*G. *romanarioides* (ROTHSCHILD, 1913) [Algeria]
*G. *alfierii* WILTSHIRE, 1949 [Egypt, Levant]

Holarctias PROUT, 1913
H. *rufinaria* (STAUDINGER, 1861) 165

Problepsis LEDERER, 1853
P. *ocellata* (FRIVALDSZKY, 1845) 166

Tribus **Rhodostrophiini** PROUT, 1935

Rhodostrophia HÜBNER, 1823
The ***jacularia*** species-group
R. *jacularia* (HÜBNER, 1813) 167

The ***vibicaria*** species-group
R. *vibicaria* (CLERCK, 1759) 168
R. *v. strigata* (STAUDINGER, 1871)

The ***calabra*** species-group
*R. *auctata* (STAUDINGER, 1879)
[Turkey, Transcaucasus]
R. *pudorata* (FABRICIUS, 1794) 169

R. p. perezaria (OBERTHÜR, 1875)
R. p. sicanaria (ZELLER, 1852)
**R. sieversi* (CHRISTOPH, 1882)
R. cretacaria REBEL, 1916 170
R. calabra (PETAGNA, 1787) 171
R. c. separata PROUT, 1935
**R. c. transcaucasica* PROUT, 1920 [Caucasus, Transcaucasus]
R. tabidaria (ZELLER, 1847) 172
**R. t. cypriaria* REBEL, 1916 [Cyprus]
R. discopunctata AMSEL, 1935 173
(=*tabidaria* sensu GUENÉE, 1858)

The ***terrestraria*** species-group
**R. terrestraria* (LEDERER, 1869) [European Russia?; Turkey, Transcaucasus]

The ***badiaria*** species-group
R. badiaria (FREYER, 1841) 174
**R. iranica* SCHWINGENSCHUSS, 1939 [Transcaucasus]
**R. bahara* BRANDT, 1938 [Turkey, Levant]
**R. vastaria* (CHRISTOPH, 1876) [W. Kazakhstan]

The ***cuprinaria*** species-group
**R. cuprinaria* (CHRISTOPH, 1876) [Transcaucasus; Turkey]
**R. peripheres* (PROUT, 1938) [Transcaucasus]

Tribus **Timandrini** STEPHENS, 1850

Timandra DUPONCHEL, 1829
T. comae SCHMIDT, 1931 175
(=*amata* sensu STAUDINGER, 1871)
**T. c. palmonii* HAUSMANN, 1997 [Levant]
T. griseata PETERSEN, 1902 176
T. rectistrigaria (EVERSMANN, 1851) 177

Tribus **Cosymbiini** PROUT, 1911

Cyclophora HÜBNER, 1822
The ***pendularia*** species-group
C. pendularia (CLERCK, 1759) 178
C. albipunctata (HUFNAGEL, 1767) 179
C. albiocellaria (HÜBNER, 1789) 180
C. lennigiaria (FUCHS, 1883) 181
**C. l. mauretanica* REISSER, 1933 [Morocco]
C. ariadne (REISSER, 1939) 182
C. serveti REDONDO & GASTON, 1999 183
C. annularia (FABRICIUS, 1775) 184
C. puppillaria (HÜBNER, 1799) 185
**C. p. granti* (PROUT, 1935) [Azores]
**C. p. lilacinipes* (SCHAUS & COCKERELL, 1923) [Madeira]
**C. maderensis* (BETHUNE-BAKER, 1891) [Madeira]
(=*lundbladi* BRYK, 1940)
C. quercimontaria (BASTELBERGER, 1897) 186
C. ruficiliaria (HERRICH-SCHÄFFER, 1855) 187

The ***porata*** species-group
C. hyponoea (PROUT, 1935) 188
C. porata (LINNAEUS, 1767) 189

The ***punctaria*** species-group
C. suppunctaria (ZELLER, 1847) 190
C. punctaria (LINNAEUS, 1758) 191
C. p. fritzae HAUSMANN, 2003
C. linearia (HÜBNER, 1799) 192

****Pseudosterrha*** WARREN, 1888
**P. rufistrigata* (HAMPSON, 1896) [Egypt, Levant]
**P. paulula* (SWINHOE, 1887) [Egypt, Levant]

****Chlorerythra*** WARREN, 1895
**C. rubriplaga* WARREN, 1895
**C. r. sinaica* WILTSHIRE, 1949 [Egypt: Sinai]

Tribus **Rhodometrini** AGENJO, 1952

Rhodometra MEYRICK 1892
R. sacraria (LINNAEUS, 1767) 193
(=*debiliaria* ROTHSCHILD, 1914)

Casilda AGENJO, 1952
C. antophilaria (HÜBNER, 1813) 194
C. a. rosearia (TREITSCHKE, 1828)
C. consecraria (STAUDINGER, 1871) 195
(=*intermediaria* TURATI, 1930)

Ochodontia LEDERER, 1853
O. adustaria (FISCHER DE WALDHEIM, 1840) 196

References

AARVIK, L., BERGGREN, K. & L.O. HANSEN (2000): Catalogus Lepidopterorum Norvegiae. Lepid. Arbeitsgruppe Zool. Mus., Univ. Oslo, Norsk Inst. Skogforskn., Ås. Oslo, 2000, 192 pp.

ABBOTT, P.W. (1898): A few notes on *Acidalia humiliata* Hufn.. – The Entomologist 31 (427): 274-276.

ABOS CASTEL, F. (1980): Lepidópteros de la provincia de Huesca (I). – Shilap Revta lepid. 8 (29): 41-43.

ABOS CASTEL, F.P. (1988): Lepidópteros de la provincia de Huesca. Addenda a los capítulos publicados con anterioridad sobre el tema en Shilap Revta. lepid. – Shilap Revta. lepid. 16 (64): 311-330.

ABOS CASTEL, F.P. (1990): Lepidópteros de la provincia de Huesca. Addenda segunda a los capítulos publicados con anterioridad sobre el tema en Shilap Revta. lepid. – Shilap Revta. lepid. 18 (70): 159-168.

ABOS CASTEL, F.P. (1994): Lepidópteros del Parque Natural de la Sierra y Canone de Guara. – Zapateri 4: 61-74.

ABOS CASTEL, F.P. (1995): Lepidópteros de la provincia de Huesca (España). Addenda tercera a los capítulos publicados sobre el tema en Shilap Revta. lepid. (Insecta: Lepidoptera). – Shilap Revta. lepid. 23 (89): 5-21.

ABOS, L., DANTART, J. & J. PIBERNAT (1999): Resultats de l'excursió collectiva a Sant Joan de Fàbregues (Osona). – Bull. Soc. Cat. Lep. 83: 23-24.

ABRAHAM, D. (2002): Molecular aspects of pheromone evolution in moths. – Dept. Ecology, University of Lund, ISBN 91-7105-174-0: 1-41.

ABRAHAM, D., LÖFSTEDT, C. & J.-F. PICIMBON (2002d): Pheromone-binding proteins of Geometrid moths with different pheromone types. – Dept. Ecology, University of Lund, ISBN 91-7105-174-0 V: 1-14.

ABRAHAM, D., PICIMBON, J.-F. & C. LÖFSTEDT (2002b): A phylogenetic reconstruction of the evolution of different pheromone types within Geometridae. – Dept. Ecology, University of Lund, ISBN 91-7105-174-0 II: 1-18.

ABRAHAM, D., RYRHOLM, N., WITTZELL, H., HOLLOWAY, J.D., SCOBLE, M.J. & C. LÖFSTEDT (2001): Molecular Phylogeny of the Subfamilies in Geometridae (Geometroidea: Lepidoptera). – Mol. Phylo. Evol. 20 (1): 65-77.

ADKIN, R. (1903): *Acidalia aversata*. – Proc. South London ent. nat. Hist. Soc. 1903: 42-43.

AGENJO, R. (1946-1947): Catálogo ordenador de los Lepidopteros de España. – Graellsia IV (3-6); V (1-3).

AGENJO, R. (1952): Faunula Lepidopterologica Almeriense. Consejo superior de investigaciones cientificas, Madrid. 370 pp..

AGENJO, R. (1953): Geometridae espanoles de la coleccion D.Hospital, de Barcelona. – EOS 29: 289-328, lams. X-XI.

AGENJO, R. (1954): R. P. Ambrosio Fernandez, O. S. A. 1882-1953. – Graellsia 12: 1-19.

AGENJO, R. (1967). *Sterrha bustilloi* geométrido *nova species* de la provincia de Madrid. – EOS 42 (1966): 299-304.

AGENJO, R. (1971): Ocho generos y veinte especies de Geometridae nuevos para España (Lep.). – Graellsia 27: 3-21, pls 1-4.

AGENJO, R. (1976). Las muy poco conocidas *Idaea korbi* (Püngeler, 1916) e *Idaea hispanaria* (Püngeler, 1913) descritas de España y acceptaciòn de la presencia en el país de la *Scopula (Eucidalia) submutata* (Treitschke, 1828) (Lep. Geometridae). – EOS 51 (1975): 39-51.

AHOLA, M., KAITILA, J.-P., NUPPONEN, K., JUNNILAINEN, J., OLSCHWANG, V.N. & YU.E. MIKHAILOV (1997): Materialy k poznaniju babochek Urala. Nauchnye rezultaty russko-finskoj lepidopterologicheskoj ekspedicii na Juzhnom Urale v 1996 g. Macrolepidoptera. [Materials to knowledge of Macrolepidoptera of the Urals. The scientific results of Russian-Finnish lepidopterological expedition to the Southern Urals in 1996]. Uspehi entomologii na Urale. Ekaterinburg (1997): 98-104.

AICHELE, F. (1927): Die Biologie von *Acid. nemoraria* (Geometr.). – Mitt. Münchn. ent. Ges. 17: 13.

AISTLEITNER, E. & U. AISTLEITNER (2001): Weitere Notizen zur Lepidopterenfauna von Pantelleria, den Pelagischen und Ägadischen Inseln (Italien, Sizilien). – Z. Arb.Gem. Öst. Ent. 53: 93-100.

AISTLEITNER, U. & E. AISTLEITNER (2000a): Bemerkenswerte Nachweise und Erstfunde von Schmetterlingen für Vorarlberg (Austria occ.) und das Fürstentum Liechtenstein (Lepidoptera). – Opusc. zool. flumin. 183: 1-24.

AISTLEITNER, U. & E. AISTLEITNER (2000b): Biodiversitätsforschung im Natura 2000-Gebiet „Bregenzerach-Schlucht": Die Großschmetterlinge (Macrolepidoptera) im Bereich des Öko-Lehrpfades der Vorarlberger Kraftwerke (Bregenzerwald, Vorarlberg, Österreich). – Vorarlberger Naturschau 11. 165-202.

ALBERTI, B. & J. SOFFNER (1962): Zur Kenntnis der Lepidopteren-Fauna Süd- und Südostrusslands. – Mitt. Münchn. ent. Ges. 52: 146-198.

ALLAN, P.B.M. (1949): Larval foodplants. – The Garden City Press, Letchworth, 126 pp..

ALPHÉRAKY, S. (1876): Cheshuekrylye okrestnostej Taganroga. [Lépidoptères des environs de Taganrog]. –

Horae Soc. ent. Ross. 10: 35-53.
ANDREAS, K. (1910): Beschreibung der Raupe von *Acid. subsaturata* und einer Abart dieses Spanners. – Ent. Z. Frankf. a.M. 24 (31): 170.
ANDRES, A. & A. SEITZ (1924): Die Lepidopteren-Fauna Ägyptens. – Senckenbergiana, Frankfurt a.M. 6 (1/2): 13-83.
ANE (ARBEITSGEMEINSCHAFT NORDBAYERISCHER ENTOMOLOGEN Ed.) (1988): Prodromus der Lepidopterenfauna Nordbayerns. – Neue ent. Nachr. 23: 1-159.
ANIKIN, V.V., SACHKOV, S.A., ZOLOTUHIN, V.V. & E.M. ANTONOVA (2000): "Fauna lepidopterologica Volgo-Uralensis" 150 years later: changes and additions. Part 3. Geometridae (Insecta, Lepidoptera). – Atalanta 31 (1/2): 293-326.
ANTONOVA E.M. (1978): [Contribution to the fauna of Geometridae (Lepidoptera) of the reserve "Galichya gora" (Lipetsk region)]. in: [New and little known species of the insects of the European part of the USSR (1977)]: 41-46 (in Russian).
ANTONOVA, E.M. & A.M. TIKHOMIROV (1992): Regionalnyj kadastr pjadenitz Ivanovskoj oblasti (*Archiearinae, Hemitheinae, Sterrhinae* i *Larentiinae*). Voprosy inventarizatzii fauny, 1992, Ivanovo, pp. 17-26 (in Russian).
ANTONOVA, E.M. & A.N. SMETANIN (1978): Vidovoj sostav, trophicheskie svjazi i zoogeographicheskie osobennosti pjadenitz zapovednika "Trostjanetz" (Geometridae, Lepidoptera). [Species composition, throphic relationships and specific zoogeographical features of loopermoths (Geometridae, Lepidoptera) in the Nature Reserve "Trostyanets"]. – Bjul. Mosk. obsch. ispytatelej prirody (otd. biol.) 83 (1): 52-61 (in Russian).
ANTONOVA, E.M. & L.V. BOLSHAKOV (1995): Pjadenitzy (Lepidoptera, Geometridae) Tul'skoj oblasti. [Geometrid moths (Lepidoptera) of Tula Region. – Actias 2 (1-2): 13-32 (in Russian).
ANTONOVA, E.M. & L.V. BOLSHAKOV (2001): Pjadenitzy (Lepidoptera, Geometridae) Tul'skoj oblasti. Dopolnenie (s vkljucheniem dannyh po sopredel'nym rajonam Kaluzhskoj, Moskovskoj i Orlovskoj oblastej). [The Geometrid moths (Lepidoptera) of Tula Area. Addition (with including the data on the neighbouring districts of Kaluga, Moscow and Orel Areas]. – In: Biologicheskoe raznoobrazie Tul'skogo kraja na rubezhe vekov, Tula, 1: 53-63 (in Russian).
ANTONOVA, E.M. & M.A. KLEPIKOV (2000): Pjadenitzy (Lepidoptera: Geometridae) Jaroslavskoj oblasti [The Geometrid Moths (Lepidoptera) of Yaroslavl Area]. – Russ. ent. J. 9 (1): 79-90 (in Russian).
ANTONOVA, E.M. (1972a): K sostavu fauny babochek pjadenitz (Lepidoptera, Geometridae) Berezinskogo Gosudarstvennogo zapovednika i ih ekologicheskomu raspredeleniju [To the fauna of geometrid moths (Lepidoptera, Geometridae) of the Berezinski state reserve and their ecological distribution]. – Berezinskij zapovednik 2: 140-159 (in Russian).
ANTONOVA, E.M. (1972b): K poznaniju fauny i ekologii babochek pjadenitz (Geometridae, Lepidoptera) Mordovskogo zapovednika. – Trudy Mordovsk. Gos. Zapovedn. im. P.G. Smidovitcha 6: 121-133 (in Russian).
ANTONOVA, E.M. (1977): K poznaniju fauny pjadenitz (Geometridae, Lepidoptera) zapovednika "Galich'ja Gora" (Lipetzkaja obl.). In: Novye i maloizvestnye vidy nasekomyh Evropejskoj chasti SSSR, Leningrad, pp. 41-46 (in Russian).
ANTONOVA, E.M. (1979): Die Geometriden-Fauna der osteuropäischen Waldsteppe und ihre Besonderheiten (Pensaer Gebiet). – Verhandlungen VII. Internationales Symposium über Entomofaunistik in Mitteleuropa,: 244-246.
ANTONOVA, E.M. (1985): Pjadenitzy Bashkirii (Lepidoptera, Geometridae) i ih zoogeograficheskie osobennosti [Geometridae of Bashkiria (Lepidoptera, Geometridae) and their zoogeographical peculiarity]. – Sborn. trudov Zool. Mus. MGU 23: 183-197 (in Russian).
ANTONOVA, E.M. (1989): Fauna i zoogeographia pjadenitz (Geometridae, Lepidoptera) Il'menskogo Gosudarstvennogo zapovednika [Geometrids of Il'mensky Nature Reserve: Fauna and biogeography]. In: Nasekomye v ekosistemah Sibiri i Dal'nego Vostoka. Issledovanija po faune Sovetskogo Sojuza. Moskva (MGU press), pp. 180-191 (in Russian).
ANTONOVA, E.M. (1994): Pjadenitzy (Geometridae). In: Cheshuekrylye Prioksko-Terrasnogo zapovednika (Lepidoptera). [Butterflies and moths of the Prioksko-terrasnogo Natural Reserve. Flora and fauna of natural reserves]. – Flora i fauna zapovednikov Moscow, (1994) 55: 23-31 (in Russian).
ANTONOVA, E.M., MIRONOV, V.G., SVIRIDOV, A.V. & I.V. SHMYTOVA (1999): Novye i redkie nahodki raznousyh cheshuekrylyh (Lepidoptera: Heterocera) dlja bassejna verhnej Oki [New and rare records of moths (Lepidoptera: Heterocera) for the upper Oka River Basin]. – Russ. ent. J. 8 (1): 53-56.
ANTONOVA, E.M., SVIRIDOV, A.V. & V.I. ROSTSHINENKO (1989): Fauna pjadenitz i sovok (Geometridae, Noctuidae) v Udmurtii [Fauna of geometrid and noctuid moths (Geometridae, Noctuidae) of Udmurtia]. In: Fauna i oekologia zhivotnyh UASSR i prilezhastshih rajonov [Fauna and ecology of animals of UASSR and

adjacent territories]. Izhevsk, pp. 108-121 (in Russian).
ARNSCHEID, W.R. (2000): Die Macrolepidopteren-Fauna Westliguriens (Riviera dei Fiori und Ligurische Alpen in Oberitalien (Insecta, Lepidoptera). – Neue ent. Nachr. 47: 1-310.
BALSAC, H.H. de & M. CHOUI (1984): Les Lépidoptères de la Gaume franco-belge (Esquisse zoogeographique et liste des espèces), Addenda. – Alexanor 13 (6): 253-273.
BANG-HAAS, A. (1907): Neue oder wenig bekannte palaearktische Makrolepidopteren. – Dt. ent. Z. Iris 20: 69-88.
BÄNZIGER, H. & D.S. FLETCHER (1985): Three new zoophilous moths of the genus *Scopula* (Lepidoptera, Geometridae) from South-east Asia. – Journal Nat. Hist. 19: 851-860.
BARANIAK, E. & J. SOSINSKI (2000): Geometrid Moths (Lepidoptera, Geometridae) of the Wielkopolski National Park. – Badania Fiz. Nad Polska Zach. C 47: 29-44.
BARTEL, M. (1914): Ueber einige Lepidopteren-Arten der Uralsteppen. – Mitt. Münchn. ent. Ges. 5 (1-4): 5-22.
BASTELBERGER (1900): Über *Zonosoma lenigiaria* Fuchs und ihre Beziehung zu *albiocellaria* Hb.. – Jahrb. Nassau. Ver. Naturk. 53: 203-215.
BAYNES, E.S.A. (1964): A revised catalogue of Irish Macrolepidoptera (Butterflies and Moths). – E.W. Classey Ltd., Hampton, Middlesex.
BAYNES, E.S.A. (1970): Supplement to a revised catalogue of Irish Macrolepidoptera (Butterflies and Moths). E.W. Classey Ltd., Hampton, 28 pp..
BAYNES, E.S.A. (1973): A revised catalogue of Irish Macrolepidoptera. – Hampton, 109 pp..
BELJAEV, E.A. & R.B. KURANISHI (2000): Geometrid moths (Lepidoptera: Geometridae) collected from the Kamchatka Peninsula and the North Kuril Islands in 1996-1997. – Nat. Hist. Res. Special Issue 7: 235-242.
BELJAEV, E.A. & S.V. VASILENKO (2002): An annotated checklist of geometrid moths (Lepidoptera: Geometridae) from the Kamchatka Peninsula and adjacent islands. – Ent. Fennica 13: 195-235.
BELJAEV, E.A. (2001): New data on Geometridae (Lepidoptera) from Sakhalin island. – Far Eastern Entomologist 106: 1-5.
BELLA, S., RUSSO, P. & P. PARENZAN (1999): Contributi alla conoscenza dela Lepidotterofauna siciliana VII. I Lepidotteri Eteroceri del Pantano Longarini (Siracusa, Sicilia Sud-orientale). – Phytophaga 9: 15-37.
BELLIER, M. (1861): Observations sur la Faune Entomologique de la Sicile. – Annls. Soc. ent. Fr. 8 (4): 667-713.
BENDER, R. (1963): Beiträge zur Lepidopterenfauna der Insel Rhodos. – Z. Wien. ent. Ges. 48 (74, 1): 11-20.
BERGER, L.-A. (1949a): Un nouveau *Cosymbia*. – Lambillionea 49 (3-4): 33-34.
BERGER, L.-A. (1949b): Espèces nouvelles pour la faune belge. – Lambillionea 49 (3-4): 23.
BERGMANN, A. (1938): Vorläufige Mitteilungen über Versuche zur experimentellen Erzeugung von melanistischen und albinistischen Formen. – Ent. Z. Frankf. a.M. 52 (31): 245-248.
BERGMANN, A. (1955): Die Grossschmetterlinge Mitteldeutschlands. Band 5 (1/2), Spanner. Urania-Verlag, Leipzig/Jena, 1267 pp..
BERTACCINI, E. & V. CAMPRI (2003): Alcuni Lepidotteri inediti ed altri poco conosciuti per la Liguria occidentale (Insecta Lepidoptera Heterocera). – Quad. Studi Not. Stor. Nat. Romagna 17 suppl.: 119-126.
BESHKOV, S. & B. GOATER (2000): Macrolepidoptera and Microlepidoptera (Alucitidae and Pyralidae) recorded in Bulgaria, 12-24 September 1995 (Lepidoptera). – Hist. nat. Bulg. 12: 41-58.
BESHKOV, S. & K. MISJA (1995): A Contribution to the knowledge of the Lepidoptera fauna of Albania. 1. Some materials from the collection of K.Misja in the Natural History Museum Tirana and some results of the collecting trip of Beshkov during 1992 (Lepidoptera, Macrolepidoptera). – Atalanta 26 (1/2): 345-363.
BESHKOV, S. & S. ABADJIEV (2000): The butterfly and moth fauna of the areas around the lakes of Durankulak, Ezerets, Shabla and Shablenska Tuzla in NE Bulgaria (Insecta: Lepidoptera). – Atalanta 31 (3/4): 543-573.
BESHKOV, S. (1992): Faunistic advances on Bulgarian Lepidoptera. – Boll. Ass. Rom. Ent. 46: 37-56.
BESHKOV, S., MISJA, K. & S. ABADJIEV (1996): A Contribution to the knowledge of the Lepidoptera fauna of Albania. 3. Some results of the collecting in April 1994 and some materials from the collection of the Natural History Museum of Tirana (Lepidoptera, Macrolepidoptera, Heterocera). – Atalanta 27 (3/4): 623-648.
BESHKOV, S., NOWACKI, J. & K. PAŁKA (1999): Contribution to the knowledge on the distribution of Macrolepidoptera in Bulgaria. – Wiad. ent., Poznan, 18 (3): 169-185.
BIDZILYA, O.V., BUDASHKIN, YU.I., KLYUCHKO, Z.F. & I.YU. KUSTJUK (2002): A contribution to the knowledge of the Lepidoptera fauna of the Ukok plateau in south-eastern Altai, Russia. – Entomofauna 23 (17): 201-220.
BIERMANN, H. (1990): Beitrag zur Schmetterlingsfauna der Insel Montecristo im Toskanischen Archipel (Lepidoptera). – Nachr. ent. Ver. Apollo, Frankf. a.M., N.F. 11 (3): 179-184.
BINDER, A. (1994): Beitrag zur Schmetterlingsfauna Böhmens (Insecta, Lepidoptera). – Neue ent. Nachr. 32: 1-150.
BLASCHKE, P. (1955): Raupenkalender für das mitteleuropäische Faunengebiet. – Alfred Kernen Verlag

Stuttgart, 149 pp..
BLATTNER, S. (1975): Die Geometridenfauna von Basel und Umgebung. – Tät.ber. naturf. Ges. Basel 1971-1974 [1975]: 371-437.
BLÁZQUEZ, A, HERNÁNDEZ-ROLDÁN, J., NIETO, M.A. & A. GARCÍA-SANTANO (2000): Nuevos datos sobre la fauna de macroheteróceros de la provincia de Cáceres (España) III (Insecta: Lepidoptera). – Shilap Revta. lepid. 28 (110): 173-186.
BLESZYNSKI, S. (1960): Klucze do Oznaczania Owadów Polski Cz. XXVII, Zeszyt 46a, Miernikowce - Geometridae: 1-149a. – Polski Zwiazek Ent., Warszawa.
BODENHEIMER, F.S. (1930): Die Schädlingsfauna Palästinas. – P. Parey, Berlin 438 pp..
BOHATSCH, O. (1885): Lepidopterologische Mitteilungen. – Wien. ent. Z. 4 (5): 143-146.
BOISDUVAL, J.A. (1940): Genera et index methodicus europaeorum lepidopterorum. – Paris, 238 pp..
BOLDT, R. (1929): Geometriden-Raupenfang. – Int. ent. Zeits. 23 (20): 243-248; (23): 273-277; (24): 285-293.
BOLDT, R. (1933): *Ephyra quercimontaria* Bast. aus der Fauna von Nijmwegen (Holland) (Lep. Geom.). – Ent. Zeits. 47 (11): 89.
BOLLAND, F. (1975): Essai de compréhension du sol et de sa flore par l'etude du peuplement d'un biotope en hétérocères. – Shilap Revta. lepid. III (10): 112-115.
BOLTE, K.B. (1990): Guide to the Geometridae of Canada (Lepidoptera), VI Subfamily Larentiinae, 1. Revision of the Genus *Eupithecia*. – Mem. ent. Soc. Canada 151, Ottawa, 253 pp..
BONNES, M. & W. SCHMITZ (1993): Ergänzungen zur Schmetterlingsfauna der Balearen (Lepidoptera) nebst einigen Netzflüglerdaten (Neuropteroidea). – Ent. Z. 103 (8): 129-152.
BORKHAUSEN, M.B. (1794): Naturgeschichte der Europäischen Schmetterlinge nach systematischer Ordnung. 5. Theil: der Phalaenen dritte Horde: Spanner. – Frankfurt, 1- 572.
BRANDT, W. (1938): Beitrag zur Lepidopteren-Fauna von Iran. Neue Gattungen, Arten und Formen (Macrolepidoptera). – Ent. Rdsch. 55 (50): 584-588; 597-602; (60): 701-702.
BRANDT, W. (1941): Beitrag zur Lepidopteren-Fauna von Iran (4). Einige neue Geometriden. – Mitt. Münchn. ent. Ges. 31 (3): 864-886.
BREHM, G. (2002): Diversity of geometrid moths in a montane rainforest in Ecuador. – Doctoral thesis, University of Bayreuth, 196 pp..
BRETSCHNEIDER, R. (1951): Neue Geometriden-Formen in Sachsen. – Z. Wien. ent. Ges. 36 (62) (1/3): 21-22.
BROOKS, M. (1991): A complete Guide to British Moths. – Jonathan Cape, London, 248pp..
BROS, E. de & W. SCHMIDT-KOEHL (1979): Beitrag zur Kenntnis der Lepidopteren-Fauna des Rif-Atlas in Nord-Marokko. – Mitt. ent. Ges. Basel 29 (1): 1-26.
BRUSSEAUX, G. & M. JACQUIN (1996): Contribution à la connaissance de la faune de l'Île-de-France: Inventaire des Lépidoptères des coteaux d'Avron (Seine-Saint-Denis) (Lepidoptera, Rhopalocera et Heterocera). – Alexanor 19 (5): 299-315.
BRUSSEAUX, G. & N. LAVENU. (1991): Contribution à la connaissance de la faune de l'Île-de-France: Inventaire des Lépidoptères de la forêt de Notre-Dame (Val-de-Marne) (Lepidoptera, Rhopalocera et Heterocera). – Alexanor 17 (4): 217-228, 2 fig.
BRUSSEAUX, G. (1994): Cinquième contribution á l'inventaire des Lépidoptères de Corse. Onze Lépidoptères nouveaux pour l'île (Lep., Geometridae, Noctuidae). – Alexanor 18 (6): 369-374.
BRUSSEAUX, G. (1995): Septième contribution a la connaissance des Lépidoptères de Corse. Six nouvelles espèces inédites pour l'île (Lepidoptera, Geometridae, Sterrhinae et Larentiinae). – Alexanor 19 (3): 165-166.
BRUSSEAUX, G. (1997): Inventaire des Lépidoptères de la forêt de Notre-Dame (Val-de-Marne), II (Lepidoptera, Rhopalocera et Heterocera). – Alexanor 20 (3): 131-140.
BRUSSEAUX, G. (1998a): Huitième contribution à la connaissance des Lépidoptères de Corse. Sept espèces nouvelles pour l'île, dont deux nouvelles pour la France (Lepidoptera, Pyralidae, Geometridae et Noctuidae). – Alexanor 20 (4): 215-216.
BRUSSEAUX, G. (1998b): Inventaire des Lépidoptères de l'ancienne carrière de Belle-Assise, en forêt de Ferrières (Seine-et-Marne) (Lepidoptera, Rhopalocera et Heterocera). – Alexanor 20 (7-8): 393-411.
BRYK, F. (1940): Die von Prof. Dr. Lundblad gesammelten Grossschmetterlinge der Iberischen Halbinsel. – Arkiv för Zool. 32 (22). 36 pp, 7 pls.
BUBACEK, O. & H. REISSER (1926): Neue Makrolepidopteren-Formen aus den andalusischen Gebirgen. – Z. Österr. Ent.-Ver. 11 (12): 115-120.
BUCKLER, W. (1897): The Larvae of the British Butterflies and Moths, 7: 1-132. London.
BUDASHKIN, YU.I. & I.YU. KOSTJUK (1987): Tsheshuyekrylye, soobstshenie 2. Pyadenitsy (Lepidoptera, Geometridae). – Tsheshuyekrylye Karadaghskogo zapovednika [Butterflies and Moths of the Karadagh

Reserve]. – Moscow, p. 22-31. (in Russian).

BURESCH, J. (1919): Beitrag zur Lepidopterenfauna des Piringebirges (Pirin-Planina) in Mazedonien. – Zeitschr. wiss. Insektenbiol. 14: (97-107, 137-144, 224-231), 271-280.

BURGERMEISTER, F. (1964): Makrolepidopteren aus dem Raume Dubrovnik (Süddalmatien, FVR Jugoslavien). – Z. Wien. ent. Ges. 49 (75, 9/10): 137-152.

BURGERMEISTER, F. (1981): Neue Sammelreisen in das Gebiet der Ligurischen Voralpen (Hinterland von Alassio-italienische Riviera). – Jahresb. Steyrer Ent. 1981: 68-88.

BUSSE, R. & F. OCKRUCK (1991): Ein Beitrag zur Kenntnis der Schmetterlingsfauna des Pirin-Gebirges im Süd-Westen Bulgariens (Lepidoptera). – Phegea 19 (1): 5-23.

BUSTILLO, G.M.R. & M.A. VARELA (1981): Catalogo sistematico de los lepidópteros Ibericos, I, Macrolepidoptera. Madrid. 499 pp..

BUSZKO, J. & J. NOWACKI (2000): The Lepidoptera of Poland. A Distributional Checklist. – Polish Ent. Monographs, Poznan & Torun, 1: 1-178.

BUSZKO, J. (1992): Motyle (Lepidoptera) reserwatu las Piwnicki. – Parki Narodowe i Rezerwaty Przyrody 10 (1-2): 5-60.

BUSZKO, J. (2000): Atlas motyli Polski. Czes? III. Falice, wycinki, miernikowce (Thyatiridae, Drepanidae, Geometridae). Warszawa. 518 pp..

BUSZKO, J., KOKOT, A., PALIK, E. & Z. SLIWINSKI (1996): Motyle wieksze (Macrolepidoptera) Puszy Bialowieskiej [Macrolepidoptera of Bialowieza Forest]. – Parki nar. Rez. przyr. Bial. 15 (4): 3-46 (in Polish).

BYTINSKI-SALZ, H. & W. BRANDT (1937): New Lepidoptera from Iran. – Entom. Record 1937: 1-15.

BYTINSKI-SALZ, H. (1934): Ein Beitrag zur Kenntnis der Lepidopterenfauna Sardiniens. – Int. ent. Zeits. Guben 28: 1-41.

BYTINSKI-SALZ, H. (1937): Secondo contributo alla conoscenza della lepidotterofauna della Sardegna. – Mem. Soc. ent. Ital. Genova 15: 194-212.

CALBERLA, E. (1889): Elenco dei Lepidotteri raccolti in Sicilia da sig. Enrico Calberla nel giugno e luglio 1889. – Il Naturalista Siciliano, IX (2): 42-49.

CALBERLA, H. (1890): Die Macrolepidopterenfauna der Römischen Campagna und der angrenzenden Provinzen Mittelitaliens. – Dt. ent. Z. Iris 3: 47-94.

CALLE, J.A. DE LA, LENCINA, F., GONZALEZ, F. & A.S. ORTIZ (2000): Las mariposas de la Región de Murcia. Macrolepidópteros diurnos y nocturnos. – Nausícaä, Murcia. 396 pp..

CAPUŞE, I. & A. KOVÁCS (1987): Catalogul colectiei de Lepidoptere "László Diószeghy" de la Muzeul Judetean covasna, Sfîntu Gheorghe. [Catalogue de la collection de Lépidoptères "László Díoszeghy" du Musée départamental covasna Sfîntu Gheorghe]. – Inst. de Spéologie "Émile Racovitza", Bucarest, 397 pp..

CARADJA, A. (1931): Beiträge zur Lepidopterenfauna der südlichen Dobrogea, insbesondere der sogenannten "Coasta del Argint" für das Jahr 1930 (Dritte biogeographische Skizze: Europa). – Acad. Rom. Mem., Sect. St. 7 (8): 293-344.

CARNEIRO MENDES F. (1950): Lepidopteros (macro) da regiao de Sintra.- Bol. Soc. Porto Sci. nat., 2a ser. 3 (1): 47-65.

CARNELUTTI, J. & S. MICHIELI (1958): Beitrag zur Lepidopterenfauna der Crna Gora. – Fragm. Balcan., Mus. Maced. Sci. nat. 2 (10): 67-81.

CARNELUTTI, J. (1992a): Rdeči seznam ogroženih metuljev (Macrolepidoptera) v Sloveniji. [The Red List of Endangered Macrolepidoptera in Slovenia]. – Varstvo Narave 17: 61-104.

CARNELUTTI, J. (1992b): Popravki/errata.- Varstvo narave, Ljubljana, 18, p. 189-190.

CARTER, D.J. & B. HARGREAVES (1987): Raupen und Schmetterlinge Europas und ihre Futterpflanzen, Verlag Paul Parey, Hamburg-Berlin.

CARTER, D.J. (1984): Pest Lepidoptera of Europe with Special Reference to the British Isles. Dr. W. Junk Publishers, Dordrecht, Boston, Lancaster, 431 pp..

CASSULO, L.A. & V. RAINERI (1996): Primi dati sui Lepidotteri della Garzaia di Valenza Po (Piemonte). – Boll. Soc. ent. Ital. Genova 127 (3): 257-266.

CHALMERS-HUNT, J.M. (1961): Aberrations of British Lepidoptera. – The Entomologist 94: 281-284.

CHAPELON, P. (1992): Contribution à l'inventaire des Géométrides de Saône-et-Loire (Lepidoptera Geometridae). – Alexanor 17 (8): 451-466.

CHRÉTIEN, P. (1899a): Les premiers états de l'*Acidalia litigiosaria* B. (Lép.). – Bull. Soc. ent. Fr. 1899: 181-183.

CHRÉTIEN, P. (1899b): Les premiers états de l'*Acidalia sericeata* Hb. (Lép.). – Bull. Soc. ent. Fr. 1899 (7): 127-129.

CHRÉTIEN, P. (1905): Les chenilles des santolines. – Le Naturaliste, (2 série) 27 (435): 89-91; (437): 113-114; (438): 129-131; (439): 143-144.

CHRÉTIEN, P. (1917): Contribution à la connaissance des Lépidoptères du Nord de l'Afrique. Notes biologicues et critiques. – Annls Soc. ent. Fr. 85 (3): 369-400.
CHRÉTIEN, P. (1926): Premiers états de l'*Acidalia subsaturata* Gn.. – Amat. Pap. 3: 17-19.
CHRÉTIEN, P. (1928): Remarques sur quelques Acidalies de la faune Française. – Amat. Pap. 4: 33-37; 49-51; 65-69; 81-83; 97-99; 113-118; 129-133; 145-153.
CHRÉTIEN, P. (1929): Remarques sur quelques Acidalies de la faune Française, IV-V. – Amat. Pap. 4: 161-169.
CHRISTOPH, H. (1867): Beschreibung einiger neuer Schmetterlinge aus der Umgegend von Sarepta. – Stettin. ent. Ztg. 28 (4-6): 233-240.
CIFUENTES, J. (1989): Algunos macroheteróceros interesantes para el Norte de España. – Shilap Revta. lepid. 17 (67): 277-280.
CIFUENTES, J. (1990): Nueva contribución al conocimiento de la distribución de los heteróceros de Navarra. – Shilap Revta. lepid. 18 (70): 181-185.
CLANCY, S.P (2003): The Bright Wave Moth *Idaea ochrata cantiata* (Prout, 1913), Survey 2003. – Unpublished confidential report commissioned and funded by Butterfly Conservation, London.
CLERCK, C.A. (1759): Icones insectorum rariorum cum nominibus erorum, locisque e C. Linnaei ... Systema Naturae allegatus. – Stockholm, 21 pp., 55 pls.
COCKAYNE, E.A. (1950): Aberrations of British Lepidoptera (Geometridae). – The Entomologist 83: 49-55.
COCKAYNE, E.A. (1951): Two new subspecies of British Geometridae (Lep.). – The Entomologist 84 (1058): 154-155.
COCKAYNE, E.A. (1952): Aberrations of British Macrolepidoptera (Geometridae). – The Entomologist's Rec. J. Var. 64: 65-68.
COCKAYNE, E.A. (1953): Lectotypes of one subspecies and two aberrations of British Geometridae. – The Entomologist's Rec. J. Var. 65: 193.
COCKAYNE, E.A. (1954): Notes on five rare species of British Lepidoptera illustrated in colour in Plate A. – Entomologist's Gazette 5: 189-192.
COLLINS, G. A. (1997): Larger moths of Surrey. – Surrey Wildlife Trust, 333 pp..
COMMON, I.F.B. (1990): Moths of Australia. – E.J. Brill, and Melbourne University Press, 535 pp., 32 pls..
COOK, M.A. & M.J. SCOBLE (1992): Tympanal organs of geometrid moths: a review of their morphology, function, and systematic importance. – Systematic Entomology 17: 219-232.
CORLEY, M.F.V. (2002): Albert Zerkowitz's Portuguese Lepidoptera. – Entomologist's Gazette 53: 249-258.
CORLEY, M.F.V., GARDINER, A.J., CLEERE, N. & P.D. WALLIS (2000): Further additions to the Lepidoptera of Algarve, Portugal (Insecta: Lepidoptera). – Shilap Revta. lepid. 28 (111): 245-319.
COSTA, O.G. (1848-50): Fauna del Regno di Napoli. Geometre. – Napols, 104 pp..
COVELL, C. (1983): The state of our knowledge of the Neotropical Sterrhinae (Geometridae). – Second Symposium on Neotropical Lepidoptera Arequipa, Peru, 1983. Supplement 1: 17-23.
CUELLO i SUBIRANA, J. (1981): Els Lepidòpters de les Illes Balears. Cens provisional. – Treb. Soc. Cat. Lep. 4: 33-53.
CULOT, J. (1917-1919): Noctuelles et Géomètres d'Europe. Vol. 3, Géomètres I (1917; 269 pp.). Vol. 4, Géomètres II (1919; 269 pp.). – Genève.
DANIEL, F. (1968): Die Makrolepidopteren-Fauna des Sausalgebirges in der Südsteiermark. – Mitt. Abt. Zool. Bot. Landesmus. Joanneum 30: 87-176.
DANIEL, F., FORSTER, W. & L. OSTHELDER (1951): Beiträge zur Lepidopterenfauna Mazedoniens. – Veröff. Zool. Staatssammlung München 2: 1-78.
DANNEHL, F. (1927): Beiträge zur Lepidopteren-Fauna Südtirols. – Ent. Z. Frankf. a.M., 40: 403-408; 453-468; 41 (1): 17-31; (13): 272-284; (15): 309-319.
DANNEHL, F. (1933): Neues aus meiner Sammlung (Macrolepidoptera). – Ent. Z. Frankf. a.M. 47 (4): 32-33.
DANTART, J. & E. REQUENA (1993): Aproximació a la Fauna de Geomètrids de la Comarca d'Anoia (Lepidoptera: Geometridae). – Treb. Soc. Cat. Lep. 12 (1992): 61-94.
DANTART, J. & P. ROCHE. (1992): Aproximación a un catálogo de los macroheteróceros de Andorra (III) (Lepidoptera: Geometridae). – Shilap Revta. lepid. 20 (78): 125-139.
DANTART, J. (1988): Datos para el conocimiento de los Geométridos Ibéricos (II): Siete especies nuevas para la fauna Ibérica (Lep. Geometridae). – Treb. Soc. Cat. Lep. 8: 59-67.
DANTART, J. (1990): Los Macroheteróceros estivales de Sant Jordi Desvalls (Gerona) (Lepidoptera: Macroheterocera). – Treb. Soc. Cat. Lep. 10 (1989): 65-92.
DANTART, J. (1991): Datos para el conocimiento de los Geométridos Ibéricos (III): Sobre la distribución de algunas especies en el NE Ibérico (Lepidoptera: Geometridae). – Treb. Soc. Cat. Lep. 11 (1990-1991): 57-82.
DANTART, J. (1997): Dades per al coneixement dels Geomètrids Ibèrics (IV): *Eupithecia undata* (Freyer, 1840), espècie nova per a la fauna Ibèrica i altres citacions interessants (Lepidoptera: Geometridae). – Treb. Soc.

Cat. 14: 115-135.

DANTART, J. (2000): Lista sistemàtica dels geomètrids de Catalunya (Lepidoptera: Geometridae). – Treb. Soc. Cat. Lep. 15: 121-182.

DANTART, J., PEREZ DE-GREGORIO, J.J. & F. VALLHONRAT (1993): Contribució a la Fauna de Macroheteròcers de les illes Balears (Lepidoptera: Macroheterocera). – Treb. Soc. Cat. Lep. 12 (1992): 95-114.

DANTART, J., VALLHONRAT, F. & A. MIQUEL (1999): Aproximació a la fauna de geomètrids de les muntanyes de Prades (Lepidoptera: Geometridae). – Bull. Soc. Cat. Lep. 84: 45-60.

DAPPORTO, L. & F. FABIANO (2000): Notes on some interesting Geometridae collected in Tuscany (Italy). – Nota lepidopterologica 23 (2): 185-190.

DAPPORTO, L., CECCHI, B., LO CASCIO, P. & A. SFORZI (1999): Contributi alla conoscenza dell'artropodofauna dell'Isola di Pianosa (Arcipelago Toscano). II. Prima nota sui macrolepidotteri (Insecta Lepidoptera). – Boll. Soc. ent. Ital. 131 (3): 245-252.

DARGE, P. (2002): Contribution à l'étude de la faune entomologique de Bourgogne (troisième note): les insectes des sables éoliens de la réserve naturelle de la Truchère (Saône et Loire). – Bull. sci. Bourg. 50 (1): 17-24.

DAVEY, P. (2001): The Dingy Mocha moth *Cyclophora pendularia* (Clerck, 1759), survey 2000, Unpublished confidential report commissioned by Butterfly Conservation, London.

DENIS, M. & I. SCHIFFERMÜLLER (1775): Ankündung eines systematischen Werkes von den Schmetterlingen der Wienergegend. – Vienna, 324 pp., 3 pls.

DERZHAVETS, YU.A., IVANOV, A.I., MIRONOV, V.G., MISHCHENKO, O.A., PRASOLOV, V.N. & S.YU.SINYOV (1986): [Macrolepidoptera of the Leningrad oblast; in Russian]. – (Akad. Nauk SSSR) Trudy vsesoyuznogo entom. obsh. 67: 186-270.

DESCHKA, G. & J. WIMMER (1996): Ökologische Valenzanalyse mit Grossschmetterlingen als Indikatoren in der Gemeinde Waldhausen in Oberösterreich. – Jb. Oö. Mus.-Ver. 141/I: 341-404.

DESCHKA, G. & J. WIMMER (2000): Die Schmetterlingsfauna der Kreuzmauer (Insecta: Lepidoptera). – Beitr. Naturk. Oberösterreichs 9: 65-186.

DIDMANIDZE, E.A. (1978): Butterflies of arid regions of Georgia (Lepidoptera, Heterocera). – Verlag "Metzniereba", Tbilisi (Tiflis), 319 pp..

DIDMANIDZE, E.A. (2001): [Faunistic list of Geometridae (Lepidoptera: Geometridae) of Georgia]. Tbilisi: 42-47.

DIDMANIDZE, E.A. (2002): Faunistic list of Geometras (Lepidoptera: Geometridae) of Georgia. – Proceedings of the Institute of Zoology (Tbilisi) 21: 176-206.

DIÓSZEGHY, L. (1935): Die Lepidopterenfauna des Retyezatgebirges, Nachtrag I. – Verh. Mitt. Siebenb. Ver. Naturw. Hermannstadt 83/84: 107-126.

DJAKONOV, A.M. (1968): Cheshuekrylye (Macrolepidoptera) Leningradskoj oblasti [The Macrolepidoptera of the Leningrad oblast']. – Trudy Leningr. obshch. estestvoisp. 74 (4): 1-116.

DOGANLAR, F. & A. HAUSMANN (2004): *Rhodostrophia bahara* Brandt, 1938, new for the fauna of Turkey and Syria (Lepidoptera: Geometridae). – Israel Journal of Zoology 49: 317-318.

DOGANLAR, F. (2003): Dogu akdeniz bölgesi Geometridae (Lepidoptera) familyasi üzerinde faunistik ve sistematik arastimalyar [Systematic and faunistic studies of Geometridae (Lepidoptera) in the East Mediterranean region of Turkey]. – Cukurova Üniversitesi fen bilimleri enstitüsü, Adana, 275 pp. (in Turkish).

DOMINGUEZ, M. & J. BAIXERAS (1989): Aportació al millor coneixement dels Drepanidae Comstock, 1893 i Geometridae Leach, 1815, Valencians (Lepidoptera). – Treb. Soc. Cat. Lep. 9 (1988): 59-67.

DOMINGUEZ, M., YELA, J.L. & J. BAIXERAS (1994): Les Géomètres de la Région de la Alcarria (Guadalajara, Espagne). – Alexanor 18 (5): 265-274.

DÖRING, E. (1955): Zur Morphologie der Schmetterlingseier. – Akademie-Verlag, Berlin, 154 pp., 61 pl..

DUFAY, C. & R. MAZEL (1981): Les Lépidoptères des Pyrénées-Orientales. Suppl. a la faune de 1961. – Vie et Milieu 31 (2): 183-191.

DUFAY, C. (1961): Faune terrestre et d'eau douce des Pyrénées-Orientales, Fasc. 6, Lépidoptères I, Macrolépidoptères. – Vie et Milieu 12 (1) Suppl., 153 pp..

DUFAY, C. (1965-1966): Contribution a la connaissance du peuplement en Lépidoptères de la Haute-Provence. – Bull. mens. Soc. linn. Lyon 34: 197-212; 35 (8): 377-392. 34: 145—376, 35: 17-488 (interrupted pagination).

DUFAY, C. (1972): Sur la géonemie de divers Lépidoptères rares ou nouveaux pour certaines regions (Pyralidae, Geometridae, Notodontidae). – Alexanor 7 (5): 219-223.

DUFAY, C. (1981): *Idaea incisaria* (Staudinger), espèce nouvelle pour la France. – Alexanor 12 (1): 19-21.

DUFAY, C. (1983): *Idaea spissilimbaria* (Mabille), espèce nouvelle pour la France continentale et la Bulgarie. – Alexanor 13 (1): 13-15.

DUFRANE, A. (1931): Quelques Sterrhinae. – Lambillionea 11: 184-188.

DUKE, N.J. & A.J. DUKE (1998): An annotated list of larval host-plants utilized by Southern African Geometridae

(Lepidoptera). – Metamorphosis 9 (1): 5-22.
DUPONCHEL, P.A.J. (1842): Geometridae. In: GODART & DUPONCHEL: Histoire Naturelle des Lépidoptères ou Papillons de France. Paris, Suppl. 4.
DÜRCK, H. & H. REISSER (1934): Beitrag zur Lepidopterenfauna des Rifgebirges von Spanisch-Marokko. – EOS 9 (3-4): 211-300.
DUTREIX, C. & D. MOREL (2000): Inventaire détaillé des Insectes „Macro-Lépidoptères" de Bourgogne. – IDEA-Infos 16: [1-9].
EBERT, G. & A. STEINER (2001): Sterrhinae. In: EBERT, G. (ed.) (2001): Die Schmetterlinge Baden-Württembergs, Band 8: Nachtfalter VI: 74.
EBERT, G. (1965): Afghanische Geometriden (Lep.) I. – Stuttg. Beitr. Naturk. 142: 1-32.
EBERT, G. (2001): Sterrhinae (p.p.). In: EBERT, G. (ed.) (2001): Die Schmetterlinge Baden-Württembergs, Band 8: Nachtfalter VI: 101-209.
EBERT, G. (Hrsg.) (2001): Die Schmetterlinge Baden-Württembergs, Band 8: Nachtfalter VI, Geometridae Geometridae. – Verlag E. Ulmer, Stuttgart, 541 pp..
EBERT, K. (1993): Die Grossschmetterlinge des Vogtlandes (Insecta, Lepidoptera). – Neue ent. Nachr. 31: 172 pp..
EFETOV, K.A. & BUDASHKIN, Yu.I. (1990): [Moths of Crimea].– Simferopol. 112 pp. (in Russian).
EITSCHBERGER, U., REINHARDT, R. & H. STEINIGER (1991): Wanderfalter in Europa (Lepidoptera). – Atalanta 22 (1): 1-67.
ELLISON, R.E., E.P. WILTSHIRE (1939): The Lepidoptera of the Lebanon with notes on their season and distribution. – Trans. Royal Ent. Soc. London 88 (1), 1-56.
EL-SAWAF, S.K., HAMMAD, S.M., DONIA, A.R. & A.H. SAAD (1968): Biological studies on *Scopula coenosaria luridata* Zett. (Lepidoptera: Geometridae). – Bull. Soc. ent. Egypte 52: 35-40.
EMBACHER, G. (1990a): Prodromus der Schmetterlingsfauna Salzburgs. – Naturwissenschaftliche Arbeitsgemeinschaft Haus der Natur, Salzburg, 151 pp..
EMBACHER, G. (1990b): Kritische Bemerkungen zu zweifelhaften Lepidopterenfunden inklusive Nachtrag zur Bibliographie der Schmetterlingsfauna des Landes Salzburg (Insecta, Lepidoptera). – Entomofauna 11 (11): 177-213.
EMBACHER, G. (2000a): Kleiner Beitrag zur Lepidopterenfauna Griechenlands (Insecta: Lepidoptera). – Z. Arb.Gem. Öst. Ent. 52 (3-4): 65-70.
EMBACHER, G. (2000b): Prodromus 2000. Die Großschmetterlinge Salzburgs. – Naturschutz Beiträge 25/00, Salzburg, 85 pp..
ERLACHER S.-I. & E. FRIEDRICH (1994): Verzeichnis der Spanner (Lepidoptera: Geometridae) Thüringen. – Check-Listen Thüringer Insekten 2, Jena: 55-64.
ERLACHER, S., FRIEDRICH, E. & C. SCHÖNBORN (1993): Kommentiertes Verzeichnis der Spanner Thüringens (Lepidoptera: Geometridae). – Mitt. Int. Ent. Ver. Frankf. a.M. 18 (3-4): 121-142.
ERLACHER, S.-I. (1998): Wiederfunde verschollen geglaubter Spannerarten für die Thüringer Fauna (Lep., Geometridae). – Ent. Nachr. Ber. 42 (1/2): 45-49.
ERLEBACH, S., FRANZ, R. & P. HUEMER (1997): Rezente Erstnachweise und erwähnenswerte Beobachtungen von Schmetterlingen in Nordtirol (Insecta, Lepidoptera). – Veröff. tirol. Landesmus. Ferdinandeum 77: 223-232.
ESCHE, T. (1996): Konkurrieren Nachtschmetterlinge um Blüten? Untersuchungen zu Nischentrennung und Bestäubungseffektivität (Insecta, Lepidoptera). – Neue ent. Nachr. 35: 1-194.
ESPER, E.J.C. (1776-1830): Die Schmetterlinge in Abbildungen nach der Natur mit Beschreibungen. – Erlangen, 5 vol..
EVERSMANN, E. (1844): Fauna lepidopterologica Volga-Uraliensis exhibiens lepidopterorum species. – Casani, Typis universitatis, 633 pp..
EXPÓSITO HERMOSA, A. (1978): Catálogo Provisional de la Familia Geometridae Leach, 1815 (Fauna Ibérica). I. – Shilap Revta. lepid. 6 (21): 37-44.
EXPÓSITO HERMOSA, A. (1980): Geometridae de la Sierra de Cazorla. In: VIEDMA, M.G. de et al.: Fauna de Cazorla. Invertebrados. Monografia no. 23. Madrid: 85-95.
FABRICIUS, J.C. (1775): Systema Entomologicae, sistens Insectorum classes, ordines, genera, species, adjectis, synonymis, locis, descriptionibus, observationibus. – Flensburg, Lipsia.
FABRICIUS, J.C. (1777): Genera Insectorum eorumque characteres naturales ... adjecta mantissa specierum nuper detectarum. – Cologne, 324 pp..
FABRICIUS, J.C. (1781-[1782]): Species insectorum exhibentes eorum differentias specificas, synonyma auctorum, loca natalia, metamorphosis adjectis observationibus, descriptionibus. – Hamburg, Cologne, 2 vols.
FABRICIUS, J.C. (1787): Mantissa Insectorum sistens eorum species nuper detectas adjectis characteribus gener-

icis, differentiis specificis emendationibus, observationibus. – Hafniae 2: 1-382.
FABRICIUS, J.C. (1794): Entomologica systematica emendata et aucta secundum classes, ordines, genera, species adjectis synonymis, locis, observationibus, descriptionibus. – Hafniae 3 (2): 1-349.
FABRICIUS, J.C. (1798): Supplementum Entomologicae systematicae. – Hafniae 3: 1-572.
FAILLIE, L. & R. PASSIN (1983): Les Lépidoptères de la Sarthe (suite). – Alexanor 13 (3): 98-117.
FAJČIK, J. & F. SLAMKA (1996): Motýle strednej Európy. I. zväzok. Určovanie-Rozšírenie a Stanovište Motýl'a-Bionómia. *Drepanidae*, Geometridae, Lasiocampidae, Endromidae, Lemoniidae, Saturniidae, Sphingidae, Notodontidae, Lymantriidae, Arctiidae. (Die Schmetterlinge Mitteleuropas, I. Band. Bestimmung-Verbreitung-Flugstandort-Bionomie). Bratislava, 1996. pp. 1-113, tab. I-XX.
FAJČIK, J. (2003): Motýle strednej a severnej Európy. Urćovanie-Rozšírenie a Stanovište Motýl'a-Bionómia. Drepanidae, Geometridae, Lasiocampidae, Endromidae, Lemoniidae, Saturniidae, Sphingidae, Notodontidae, Lymantriidae, Arctiidae. (Die Schmetterlinge Mittel- und Nordeuropas. Bestimmung-Verbreitung-Flugstandort-Bionomie). Bratislava, 2003, 172 pp. 38 pl..
FALCK, P., JEPPESEN, G. & K. LARSEN (1985): *Scopula imitaria* (Hübner 1799) fundet i Danmark (Lepidoptera, Geometridae). – Lepidoptera, København 4 (9): 329-331.
FÄNGER, H. (1999): Comparative morphology of tergal phragmata occurring in the dorsal thoracoabdominal junction of ditrysian Lepidoptera (Insecta). – Zoomorphology 119: 163-183.
FAZEKAS, I. (1979): Faunára új Geometrida faj, a *Sterrha vulpinaria* H.-Sch. Magyarországon (Lepidoptera). – Fol. Ent. Hung. 32 (1): 219-223.
FEICHTENBERGER, E.W. (1962): Die Macrolepidopteren-Fauna des Stubachtales (Salzburg, Hohe Tauern). – Z. Wien. ent. Ges. 47 (9): 142-149.
FEICHTENBERGER, E.W. (1965): Die norwegische Lepidopterenfauna am Polarkreis, mit Bezug auf die der Ostalpen. – Z. Wien. ent. Ges. 50 (76, 6-8): 80-118.
FIBIGER, M. & P. SVENDSEN (1981): Danske natsommerfugle. Ændringer i den danske natsommerfuglefauna i perioden 1966-1980. Dansk faunistisk bibliotek. Bind 1. Scandinavian Science Press Ltd. Klampenborg, 272 pp., 6 pls.
FIBIGER, M. (1976): Ny dansk måler – *Scopula emutaria* Hbn.. – Lepidoptera, København 3 (1): 1-4.
FIUMI, G. & S. CAMPORESI (1988): I Macrolepidotteri. – Amministr. Prov. di Forlí, 262 pp., 10 pls.
FLAMIGNI, C. & G. BASTIA (1998): Osservazioni sulle specie emiliane del genere *Idaea* Treitschke, 1825 (Insecta, Lepidoptera, Geometridae). – Quad. Studi Not. Stor. Nat. Romagna 9: 13-42.
FLAMIGNI, C. & G. BASTIA (2003a): Nuovi dati faunistici su alcune specie di Geometridi, Nottuidi e Nolidi dell'Italia centrale, meridionale e insulare (Insecta, Lepidoptera, Geometridae, Noctuidae, Nolidae). – Quad. Studi Not. Stor. Nat. Romagna 17 suppl.: 79-86.
FLAMIGNI, C. & G. BASTIA (2003b): I Geometridi del Bolognese. I parte (Insecta, Lepidoptera, Geometridae: Ennominae, Alsophilinae, Geometrinae, Sterrhinae). – Quad. Studi Not. Stor. Nat. Romagna 17 suppl.: 87-118.
FLAMIGNI, C. & G. BASTIA (2004): Contributo alla conoscenza di Geometridi delle Alpi italiane (Lepidoptera, Geometridae). – G. it. Ent. Cremona (in press).
FLAMIGNI, C. (1999): *Idaea emarginata*. – Boll. Soc. ent. Ital. 131 (1): 268.
FLAMIGNI, C., BASTIA, G. & L. DAPPORTO (2001): Nuove segnalazioni e note critiche sui Geometridi di Emiglia, Romagna e Toscana, I parte. – Quad. Studi Not. Stor. Nat. Romagna 14: 89-122.
FLETCHER, D.S. (1961): Lepidoptera der Deutschen Nepal-Expedition 1955, I, Geometridae. – Veröff. Zool. Staatssammlung München 6: 163-178.
FLETCHER, D.S. (1979): Geometroidea in: NYE, I.W.B. (ed.): The generic names of moths of the world, vol. 3. – British Museum (N.H.). London, 243 pp..
FÖHST, P. (1991): Ein Beitrag zur Makrolepidopterenfauna des südlichen Teils des Lago d'Iseo in der Lombardei (prov. Bergamo). – Atalanta 22 (2/4): 271-351.
FOLTIN, H. (1954): Die Macrolepidopterenfauna der Hochmoore Oberösterreichs. – Z. Wien. ent. Ges. 39 (65, 3): 98-115.
FORSTER, W. & T.A. WOHLFAHRT (1981): Die Schmetterlinge Mitteleuropas, 5. Band, Spanner. Franckh'sche Verlagsbuchhandlung, Stuttgart, 312 pp., 26 pls.
FORSTER, W. (1981): Änderung eines Namens. – Nachr.bl. Bayer. Ent. 22 (1): 15.
FOURCROY, DE, A.F. (ed.) (1785): Entomologia parisiensis, sive Catalogus Insectorum quae in Agro Parisiensi reperiuntur. Pars I. – Paris, 544 pp..
FRIEDEMANN, H. (1937): *Acidalia ochrata* und *similata*. – Ent. Zeitschr. 51 (28): 266.
FRIEDRICH, E. (1984): Handbuch der Schmetterlingszucht, Europäische Arten. – Franck'sche Verlagshandlung, Kosmos-Verlag, Stuttgart, 186 pp..
FRIEDRICH, E. (1986): Breeding Butterflies and Moths - a practical Handbook for British and European Species,

English Edition. – B.H. & A. Harley Ltd., Colchester, Essex, England, 176 pp.
FRILLI, F. & W. PIZZAGHI (1975): Contributo alla conoscenza dell'entomofauna dell'Apennino e della pianura attorno a Piacenza. – Entomologica Bari 11: 29-80.
FUCHS, A. (1875): Lepidopterologische Betrachtungen. – Stettin. ent. Ztg. 36 (4-6): 225-235.
FUCHS, A. (1879): Lepidopterologische Mittheilungen aus dem nassauischen Rheinthale. – Stettin. ent. Ztg. 40 (1-3): 40-43.
FUCHS, A. (1883): Macrolepidopteren des unteren Rheingaues. – Stettin. ent. Ztg. 44 (4-6): 248-275.
FUCHS, A. (1889): Lepidopterologische Beobachtungen aus dem unteren Rheingau, VIII-X. – Jb. Nassau. Ver. Naturk. 42: 210-219.
FUCHS, F. (1904): Neue Schmetterlingsformen. – Soc. Ent. 19 (3): 17-18..
FUESSLY, I.C. (1781-1786): Archiv der Insektengeschichte. – Zürich.
GAEDIKE, R. & W. HEINICKE (1999): Verzeichnis der Schmetterlinge Deutschlands. Entomofauna Germanica, Bd 3, Dresden, Ent. Nachr. Ber. 5: 1-216.
GALVAGNI, E. (1909): Beiträge zur Kenntnis der Lepidopterenfauna der adriatischen Inseln. – Mitt. naturw. Ver. Univ. Wien 7: 5-10, 154-254.
GANEV, J. & S. BESHKOV (1987): Records of Macrolepidoptera from Bulgaria. – Phegea 15 (2): 113-117.
GANEV, J. (1982): Record of new and local species of Heterocera from Bulgaria. – Nota lepidopterologica 5 (4): 157-168.
GANEV, J. (1983a): Die Schmetterlingsfauna der Hochgebirge Bulgariens. I Ossogovo Gebirge (Lepidoptera, *Noctuidae* und Geometridae). – Atalanta 14 (1): 60-79.
GANEV, J. (1983b): Macrolepidoptera of Zemen Gorge in SW-Bulgaria. – Phegea 11 (4): 81-96.
GANEV, J. (1983c): Systematic and Synonymic List of Bulgarian Geometridae. – Phegea 11: 31-42.
GANEV, J. (1984a): Die Schmetterlinge in den Hochgebirgen Bulgariens II. Die Noctuidae und Geometridae des Rhodopen-Gebirge. – Ent. Nachr. Ber. 28: 123-135.
GANEV, J. (1984b): Die Schmetterlingsfauna des Vulkanhügels Kozuch in Südwest-Bulgarien (Lepidoptera: Macrolepidoptera). – Phegea 12 (4): 121-136.
GAUCKLER, H. (1924): Etwas über die Zucht von *Acidalia muricata* Hufn. – Ent. Z. Frankf. a.M. 37 (23/24): 53.
GELBRECHT J. & B. MÜLLER (1987): Kommentiertes Verzeichnis der Spanner der DDR nach dem Stande von 1986 (Lep., Geometridae). – Ent. Nachr. Ber. 31 (3): 97-106.
GELBRECHT, J. & A. HAUSMANN (1997): *Scopula immistaria beshkovi* ssp. n. – neu für Bulgarien und den Balkan (Lepidoptera, Geometridae). – Linzer biol. Beitr. 29/2: 983-990.
GELBRECHT, J. (1999): Die Geometriden Deutschlands - eine Übersicht über die Bundesländer (Lep.). – Ent. Nachr. Ber. 43 (1): 9-26.
GELBRECHT, J., A. RICHERT & H. WEGNER (1995): Biotopansprüche ausgewählter vom Aussterben bedrohter oder verschollener Schmetterlingsarten der Mark Brandenburg. – Ent. Nachr. Ber. 39: 183-201.
GERSTBERGER, M. & L. STIESY (1987): Schmetterlinge in Berlin-West, Teil II, Fördererkreis der Naturwissenschaftlichen Museen Berlins e.V. (Hrsg.), Berlin.
GERSTBERGER, M. (1990): Zur Verbreitung von *Scopula vigilata* (PROUT) in Europa (Geometridae). – Nota lepidopterologica 13 (4): 213-214.
GERSTBERGER, M., STIESY, L., THEIMER, F. & M. WOELKY (1991): Standardliste und Rote Liste der Schmetterlinge von Berlin (West): Großschmetterlinge und Zünsler.
GIBEAUX, C. (1993): *Idaea nexata* (Hübner) en France (Lep. Geometridae). – Ent. Gall. 4 (4): 165.
GIULIANO, F. & P. PARENZAN (1994): Contributi alla conoscenza della Lepidotterofauna siciliana II. Macrolepidotteri dei Monti Nebrodi. – Phytophaga 5: 85-111.
GLASER, W. (1958): Beschreibung der ersten Stände von *Acidalia longaria* HS. (Lep. Geometridae). – Ent. Nachr.bl. Österr. und Schweiz. Ent. 10 (2): 29-32.
GOEZE, I.A.E. (1781): Entomologische Beyträge zu des Ritter Linné zwölften Ausgabe des Natursystems. – Leipzig, III. Theil, Bd. 3.
GOMBOC, S. (1996): Vier weitere Neufunde für die Großschmetterlingsfauna Sloweniens. – Acta Ent. Slovenica 4 (2): 101-105.
GOMEZ BUSTILLO, M. & A. EXPÓSITO HERMOSA (1979): Revisión de la superfamília Geometroidea en la Península Ibérica. – Shilap Revta. lepid. 28: 287-299.
GÓMEZ DE AIZPÚRUA, C. (1987): Biologia y morfologia de las orugas, Lepidoptera, tomo III, Geometridae. – Boletin de Sanidad Vegetal, Madrid, nr. 8, 230 pp..
GÓMEZ DE AIZPÚRUA, C. (1988a): Atlas Provisional de los Lepidópteros (Heterocera) de Alava, Bizkaia y Guipúzcoa, tomo VI. – Ed. Servicio Central de Publicaciones de Gobierno Vasco, 366 pp..
GÓMEZ DE AIZPÚRUA, C. (1988b): Catálogo de los Lepidópteros de Actividad Nocturna (Heterocera) de Alava,

Bizkaia y Guipúzcoa, tomo III. – Ed. Servicio Central de Publicaciones de Gobierno Vasco.

GÓMEZ DE AIZPÚRUA, C. (1989): Biologia y morfologia de las orugas. Lepidoptera, tomo VII, Geometridae. – Boll. de Sanidad Vegetal 7 (15): 1-224.

GRANVILLE, J.J. DE (1968): Voyage entomologique en Corse. – Alexanor 5: 247-249.

GRASLIN, M.A. DE (1863): Notice sur deux explorations entomologiques faites dans les Pyrenées orientales en 1847 et en 1857. – Annls Soc. ent. Fr. (4e série) 3: 297-372.

GRELIER, Y. (1989): Dix ans de chasses de nuit a Marsaneix (Dordogne) (Lepidoptera). – Alexanor 16 (4): 201-211.

GREMMINGER, A. (1944): Zur Gattung *Cosymbia* Hb. (*Codonia* Hb., *Ephyra* Dup., *Zonosoma* Ld.) aus dem Oberrheingebiet. – Z. Wien. ent. Ges. 29: 219-224.

GRILLO, N. & P. PARENZAN (1995): Contributi alla conoscenza della Lepidotterofauna siciliana IV. Geometridae. – Phytophaga 6: 111-129.

GRILLO, N., PARENZAN, P. & F. PORCELLI (1996): Segnalazioni faunistiche di Geometridi e Nottuidi per l'Italia settentrionale e centrale (Lepidoptera). – Entomologica, Bari, 30: 167-184.

GRÖNBLOM, T., JALAS, I., KAISILA, J., KROGERUS, H. & E. SUOMALAINEN (1962): Catalogus Lepidopterorum Fenniae et regionum adiacentium. I. Macrolepidoptera. Helsingfors, 28 pp..

GROSS, H. (1887): Zur Biologie der *Acidalia punctata* Tr.. – Stettin. ent. Ztg. 1887: 48.

GROSSER, N. (1982): Ein Beitrag zur Makrolepidopterenfauna des Pirin-Gebirges (Südwestbulgarien) und seines südlichen Vorlandes. – Entomol. Nachr. Ber. 26 (5): 217-225.

GUENÉE, A. (1858): In BOISDUVAL, M.M. & M.A. GUENÉE. Histoire naturelle des insectes (Lepidoptera), Species Général des Lépidoptères. Tom. IX. X. Uranides et Phalenites I. II. – Paris.

GUMPPENBERG, C.F.v. (1888-1890): Systema Geometrarum zonae temperatioris septentrionalis. Systematische Bearbeitung der Spanner der nördlichen gemässigten Zone, II, III. – Nova Acta Ksl. Leop.-Carol. Dt. Akad. Naturforsch. 54 (4): 125-255, 269-431.

HABELER, H. (1976): Beitrag zur Lepidopterenfauna Dalmatiens. – Acta ent. Jugosl. 12 (1-2): 67-87.

HABELER, H. (1998): Neue und bemerkenswerte Arten für die Schmetterlingsfauna Kroatiens von der Insel Krk (Lepidoptera). – Ent. Croat. 3 (1-2): 33-44.

HABICH, O. (1884): Lepidopterologisches. – Wien. ent. Ztg. 3: 245-247.

HACKER, H. & G. DERRA (1985): Zweiter Beitrag zur Heterocerenfauna (Lepidoptera) Griechenlands. – Atalanta 16 (1/2): 114-157.

HACKER, H. & H. KOLBECK (1996): Die Schmetterlingsfauna der Naturwaldreservate Dianensruhe, Wolfsee, Seeben, und Fasanerie. – Schriftenreihe Naturwaldreservate in Bayern 3: 77-120.

HACKER, H. (1981): Beitrag zur Lepidopterenfauna des nördlichen Fränkischen Jura, Teil 2: Geometridae. – Atalanta 12 (4): 260-284.

HACKER, H. (1999): Die Typen der von E.J.Ch. Esper (1742-1810) in seinem „Die Schmetterlinge in Abbildungen nach der Natur" beschriebenen Bombycoidea, Drepanoidea, Geometroidea, Hepialoidea, Lasiocampoidea, Noctuoidea, Pyraloidea, Tineoidea (Lepidoptera) II. – Esperiana 7: 443-461.

HAGGETT, G. (1958): Descriptions of the early stages of *Cosymbia puppillaria* (Hübner) Lep: Geometridae. Entomologist's Gazette 9: 103-106.

HALPERIN, J. & W. SAUTER (1992): An annotated list with new records of Lepidoptera associated with forest and ornamental trees and shrubs in Israel. – Israel J. Entom. 25/26: 105-147.

HANDFIELD, L., LANDRY, J.-F., LANDRY, B. & D. LAFONTAINE (1997): Liste des Lépidoptères du Québec et du Labrador. – Fabreries, Suppl. 7, 155 pp..

HANSON, B. (1962): Bidrag till kännedomen om Gotska Sandöns fjärilsfauna I [contribution to the knowledge of the lepidopterous fauna of Gotland]. – Entomol. Tids. Årg. 83 (3/4): 123-131 (in Swedish).

HARKORT, W. & H.-J. WEIGT (1969): Beobachtungen zur Schmetterlingsfauna im Raum Dortmund-Hagen-Iserlohn. Teil 3: Spanner. – Dortm. Beitr. Landeskd. 3: 19-67.

HARPE, DE LA J.C. (1853): Faune Suisse. Lépidoptères. – IV Partie. Phalénides. – Neue Denkschrift. allgem. Schweiz. Ges. ges. Naturwiss. (Nouveaux Mém. Soc. Helv. Sci. Natur.) 13: 1-160, 1 pl.

HARRISON, H. (1927): *Acidalia (Sterrha) laevigata* Sc. in Durham. – The Entomologist 60: 222.

HARTIG, F. (1930): Note di Lepidotterologia. Nuove aggiunte alla Fauna Lepidotterologica della Venezia Tridentina. – Trento, A. Scotoni, 56 pp., 1 pl..

HARTIG, F. (1938): I Macrolepidotteri di Madonna di Campiglio. – Mem. Soc. ent. Ital. 16: 232-270.

HARTIG, F. (1939) Nuovi contributi alla conoscenza della Fauna delle Isole Italiane dell'Egeo. XIII. Conoscenza attuale della fauna lepidotterologica dell'Isola di Rodi. – Boll. Lab. Ent. Agr. Portici, 3: 221-246.

HARTIG, F. (1951): In HARTIG, F. & H.G. AMSEL. Lepidoptera Sardinica, Parte I. Macrolepidoptera. – Fragm. Ent. 1 (1): 1-159.

HÄUSER, C. (2001): *Rhodometra sacraria*. In: EBERT, G. (ed.) (2001): Die Schmetterlinge Baden-Württembergs,

Band 8: Nachtfalter VI: 209-210.
HAUSMANN A. & S. DÖTTERL (2004): Nectar-plants and larval food-plants of the genus *Glossotrophia* (Lepidoptera: Geometridae, Sterrhinae): studies on pollen grains attached to museum specimens. – Nota lepidopterologica 26 (3/4): 127-136.
HAUSMANN, A (2003b) (ed.): The Forum Herbulot world list of family group names in Geometridae. – (http://www.herbulot.de), 9 pp..
HAUSMANN, A. & A. WERNO (2003): A new *Idaea* species from the Canary islands (Lepidoptera, Geometridae, Sterrhinae). – Ent. Zeits. Stuttgart 113 (12): 329-330.
HAUSMANN, A. & GY.M. LÁSZLÓ (1999): Taxonomic and faunistic studies on Turkmenian Sterrhinae (Lepidoptera: Geometridae). – Folia ent. Hung. 60: 317-324.
HAUSMANN, A. & M. HONEY (2004): A new *Idaea* species from the Balearic islands (Lepidoptera, Geometridae, Sterrhinae). – Ent. Zeits. Stuttgart 114 (1): 12-13.
HAUSMANN, A. & M.A. MILLER (ed.) (2000): Atlas der Raupen europäischer und kleinasiatischer Schmetterlinge fotographiert von Burkhard Nippe. Pfeil Verlag, München, 95 pp..
HAUSMANN, A. & P. PARENZAN (1990): Neue und interessante Geometridenarten für die Süditalienfauna (Lepidoptera, Geometridae). – Entomofauna 11 (29), 497-503.
HAUSMANN, A. & U. AISTLEITNER (1998): Beitrag zur Kenntnis der Spanner-Fauna Spaniens (Lepidoptera, Geometridae). – Nachr.bl. Bayer. Ent. 47 (3/4): 101-105.
HAUSMANN, A. (1988): Großschmetterlinge im Münchner Norden. – Schriftenreihe Bayer. Landesamt für Umweltschutz, 83: 61-95.
HAUSMANN, A. (1990a): Zur Dynamik von Nachtfalter-Artenspektren: Turnover und Dispersionsverhalten als Elemente von Verbreitungsstrategien. Spixiana Suppl. 16, 1-222.
HAUSMANN, A. (1990b): Die Bedeutung des genauen Lichtfallen-Standortes für die Aussagekraft des Fangergebnisses (Lepidoptera, Macroheterocera). Atalanta 21 (3/4), 301-312.
HAUSMANN, A. (1991a): Ergänzungen zur Großschmetterlingsfauna des Münchner Nordens (Lepidoptera). – Entomofauna 12 (6), 81-96.
HAUSMANN, A. (1991b): Zur Abhängigkeit des apparenten Artenaustausches von der Stichprobengröße (Lepidoptera, Macroheterocera). – Spixiana 14 (2), 237-242.
HAUSMANN, A. (1991c): Taxonomische Untersuchungen an der *Idaea obsoletaria* Rmb.-Gruppe (Lep., Geometridae). – Nachr.bl. Bayer. Ent. 40 (2), 33-41.
HAUSMANN, A. (1991d): Beitrag zur Geometridenfauna Palästinas: Die Spanner der Klapperich-Ausbeute aus Jordanien (Lepidoptera, Geometridae). – Mitt. Münchn. ent. Ges. 81, 111-163.
HAUSMANN, A. (1992): *Idaea fittkaui* sp.n. aus Afghanistan mit weiteren Bemerkungen zur Systematik der 13. Gruppe der Gattung *Idaea* (Lepidoptera, Geometridae). – Nachr.bl. Bayer. Ent. 41 (2): 46-55.
HAUSMANN, A. (1993a): *Heterolocha xerophilaria* Püngeler, 1902, - ein Synonym von *Pseudosterrha rufistrigata* (Hampson, 1896), comb. n., mit weiteren Anmerkungen zur Systematik der Sterrhinae (Lepidoptera, Geometridae). – Nota. lep. 16 (1): 23-33.
HAUSMANN, A. (1993b): Zweiter Beitrag zur Taxonomie und Systematik der Gattung *Glossotrophia* Prout, 1913 (Lepidoptera: Geometridae, Sterrhinae). – Mitt. Münchn. ent. Ges. 83: 77-107.
HAUSMANN, A. (1993c): Der Aussagewert struktureller Unterschiede des 8. Sternits. Beitrag zur Systematik der italienischen Vertreter der Gattung *Glossotrophia* Prout, 1913 (Lepidoptera, Geometridae). – Atalanta 24: 265-297.
HAUSMANN, A. (1993d): *Idaea eugeniata* Millière, 1870, neu für die Fauna Mitteleuropas (Lepidoptera, Geometridae). – Nachr.bl. Bayer. Ent. 42 (1): 24-25.
HAUSMANN, A. (1993e): Contributo alla conoscenza della fauna siciliana: I rappresentanti della famiglia Geometridae presso la collezione dello Zoologische Staatssammlung di Monaco (Insecta: Lepidoptera). – Naturalista siciliana S. IV, XVII (1-2) 1993: 83-101.
HAUSMANN, A. (1994a): "*Idaea circuitaria* (Hübner, [1819])" - ein Komplex von mehreren verschiedenen Arten aus dem Mittelmeerraum (Lepidoptera: Geometridae, Sterrhinae). – Zeits. Arb.gem. Österr. Ent. 46 (1/2): 45-55.
HAUSMANN, A. (1994b): Notes on the Morphology of Some Little Known Species of the Genus *Idaea* Treitschke, 1825, in the Middle East (Lepidoptera: Geometridae, Sterrhinae. – Bonner Zool. Nachr. 45 (1): 67-78.
HAUSMANN, A. (1994c): Beitrag zur Geometridenfauna Zyperns. – Zeits. Arb.gem. Österr. Ent. 46 (3/4): 81-98.
HAUSMANN, A. (1994d): Dritter Beitrag zur Revision der Gattung *Glossotrophia* Prout, 1913 nebst Beschreibung zweier neuer Gattungen (Lepidoptera: Geometridae, Sterrhinae). – Nota lepidopterologica 17 (1): 195-211.

HAUSMANN, A. (1994e): *Idaea blaesii* Lenz & Hausmann, 1992, neu für die Faunen Spaniens und Marokkos (Lepidoptera: Geometridae, Sterrhinae). – Nachr.bl. Bayer. Ent. 43 (1/2): 28.
HAUSMANN, A. (1995): Neue Geometriden-Funde aus Zypern und Gesamtübersicht über die Fauna. – Mitt. Münchn. ent. Ges. 85: 79-111.
HAUSMANN, A. (1997): The Geometrid Moths of various entomological collections in Israel (Lepidoptera, Geometridae). – Entomofauna 18, 1-19.
HAUSMANN, A. (1998): Interessante nordafrikanische Geometridenarten aus der Sammlung Herbulot, Paris (Lepidoptera, Geometridae). – Facetta 14 (1/2): 39-41.
HAUSMANN, A. (1999): Falsification of an entomological rule: Polymorphic genitalia in Geometrid moths. – Spixiana 22 (1): 83-90.
HAUSMANN, A. (2001): Introduction. Archiearinae, Orthostixinae, Desmobathrinae, Alsophilinae, Geometrinae. – In: A. HAUSMANN (ed.): The Geometrid Moths of Europe 1. – Apollo Books, Stenstrup, 282 pp. & 8 colour pls.
HAUSMANN, A. (2003a): New Sterrhinae from Europe, North Africa, and the Caucasus (Lepidoptera, Geometridae). – Ent. Zeits. Stuttgart 113 (11): 319-328.
HAUSMANN, A. (2005): Zur Morphologie und systematischen Stellung von *Pseudocinglis moralesi* (RUNGS, 1945), comb. n. aus Marokko (Lepidoptera, Geometridae). – Ent. Zeits. Stuttgart 115 (in prep.).
HAUSMANN, A., ERLACHER, S. & M.A. MILLER (in prep.): Molecular analysis of the mitochondrial COX I gene: Evidence for infra- and suprageneric relationships within the geometrid moth subfamily Sterrhinae.
HAWORTH, A.H. (1803-1828): Lepidoptera Britannica, sistens digestimen novam lepidopterorum quae in Magna Britannica repertiuntur ... adjunguntur dissertationes variae ad historiam naturalam spectantes. – London, 4 parts.
HEIDEMANN, H. (1986): Raupe und Puppe des graugesprenkelten Ringfleckspanners (*Cyclophora suppunctaria* Z. sensu Forster & Wohlfahrt) (Lepidoptera, Geometridae). – Neue Ent. Nachr. 19 (1/2): 19-22.
HEIN, S. (1936): Ein neuer Geometriden-Hybride (Lepid.). – Z. Österr. Ent.-Ver. 21 (2/3): 9-11.
HELLMANN, F. (1987): Die Macrolepidopteren der Brenta-Gruppe (Trentino - Oberitalien) (Lepidoptera). – Studi Trentini Sci. nat., Acta Biologica 63: 3-166.
HELLMANN, F., BROCKMANN, E. & P.M. KRISTAL (1999): I Macrolepidotteri della Valle d'Aosta. Monografie-2. Saint-Pierre-Valle d'Aosta, Aosta, 284 pp..
HELLRIGL, K. (1996): Die Tierwelt Südtirols. – Veröffentlichungen des Naturmuseums Südtirol, Bozen, 831 pp..
HERBULOT, C. (1945): Une Géométride nouvelle pour la faune française: *Sterrha nitidata* H.-S.. – Rev. Fr. Lép. 10: 69-70.
HERBULOT, C. (1947a): Matériaux pour une faunule des Lépidoptères des environs de Saint-Tropez. – Rev. Fr. Lép. 10: 305-321.
HERBULOT, C. (1947b): Captures de Geometridae dans la région parisienne. – Rev. Fr. Lép. 11: 15-18.
HERBULOT, C. (1947c): Une géométride nouvelle pour la faune de France: *Sterrha fathmaria* Obth.. – Rev. Fr. Lép. 11: 168-170.
HERBULOT, C. (1950a): On the presence of middle spurs in the hind tibiae of a male *Scopula* (Lep. Geometridae). – Entomologist 83: 255.
HERBULOT, C. (1950b): Troisième note sur Saint-Tropez. – Lambillionea 50: 48-52.
HERBULOT, C. (1953): Note synonymique. – Bull. Soc. ent. Fr. 58:48.
HERBULOT, C. (1954): Une interessante acquisition pour la faune Française: *Scopula rubellata* Stgr. (Lep. Geometridae). – Bull. mens. Soc. linn. Lyon 23 (7): 181-182.
HERBULOT, C. (1955): Remarques sur le catalogue des Lépidoptères du Pont-de-l'Arche. – L'Entomologiste 11 (2-3): 42-47.
HERBULOT, C. (1957): Nouvelles formes de Geometridae d'Europe occidentale et d'Afrique du Nord. – Bull. mens. Soc. linn. Lyon 26: 190-192.
HERBULOT, C. (1959): Lépidoptères Geometridae nouveaux pour le département des Ardennes. – Bull. Soc. Hist. nat. Ardennes 48: 25-27.
HERBULOT, C. (1961a): Troisième note sur les Lépidoptères des environs de Pont-de-l'Arche (Eure). – Alexanor 1: 234-236.
HERBULOT, C. (1961b): Captures de Geometridae faites par P.Dardenne dans le département de l'Yonne. – Alexanor 2 (1): 19-21.
HERBULOT, C. (1961c-1963): Mise a jour de la liste des Geometridae de France. – Alexanor (1961) 2 (4): 117-124; (1962) 2 (5): 147-154; (1963) 3 (1): 17-24; 3 (2): 85-93.
HERBULOT, C. (1962a): Les noms de genres de Geometridae employés par Hübner dans son Systematisch-alphabetisches Verzeichniss. – Lambillionea 62: 53-57.
HERBULOT, C. (1962b): Sur quelques Geometridae des Llanos de Urgel (Aragon). – Alexanor 2 (6): 229-230.

HERBULOT, C. (1965a): Lépidoptères Geometridae du Tibesti. – Lambillionea 63 (5-8): 25-40.
HERBULOT, C. (1965b): Note sur les Lépidoptères Geometridae du département des Ardennes. – Bull. Soc. Hist. nat. Ardennes 54: 22-23.
HERBULOT, C. (1966): Note sur quelques Geometridae des Monts de Vaucluse. – Lambillionea 65 (5-6): 35-41.
HERBULOT, C. (1968a): Sur Cinq Geometridae de Corse. – Alexanor 5: 244-246.
HERBULOT, C. (1968b): Sur quelques Geometridae de Sardaigne. – Alexanor 5 (5): 231-232.
HERBULOT, C. (1975): Un nouvel *Idaea* Ibero-Mauritien. – Alexanor 9 (4): 192.
HERBULOT, C. (1978): Liste des Geometridae de Corse qui ne se trouvent pas en France continentale. – Alexanor 10 (6): 245-248.
HERBULOT, C. (1979): Un nouvel *Idaea* Franco-Ibérique (Lep. Geometridae, Sterrhinae). – Alexanor 11 (4): 177-178.
HERBULOT, C. (1982): Les types de Geometridae décrits par Pierre Chrétien (Lepidoptera). – Alexanor 12 (5): 207-208.
HERBULOT, C. (1983): Contribution a l'inventaire des Lépidoptères du departement de l'Aisne. Geometridae pris à Puisieux-et-Clanlieu. – Lambillionea 82 (5-6): 34-40.
HERBULOT, C. (1984a): Les types de Geometridae decrits par Daniel Lucas (Lepidoptera). – Misc. Ent. 50: 47-52.
HERBULOT, C. (1984b): Complément à la liste des Geometridae pris à Puisieux-et-Clanlieu. – Lambillionea 83 (9-19): 66.
HERBULOT, C. (1985): Sur la présence en France de *Scopula nemoraria* (Hübner) (Lep. Geometridae Sterrhinae). – Ent. gall. 1 (4): 271.
HERBULOT, C. (1986): Sur quelques Lépidoptères géométrides de Catalogne. – Bull. Sciences Nat. 50: 25.
HERBULOT, C. (1987): Les types de Lépidoptères Géométrides decrits ou figurés par Pierre Rambur. – Miscellanea Ent. 51 (2): 33-40.
HERBULOT, C. (1991): Un nouvel *Idaea* ibérique. – Alexanor 16 (7): 387-388.
HERBULOT, C. (1992): Combien d'espèces de Lépidoptères dans le monde? – Alexanor 17 (7): 447-448.
HERBULOT, C. (1994): Un nouvel *Idaea* d'Asie Centrale. – Bull. Soc. ent. Mulhouse 1994: 66.
HERNANDEZ-ROLDAN, J., BLAZQUEZ, A., MARTIN-DIAZ, J. & M.A. NIETO (1999): Nuevos datos sobre la fauna de macroheteróceros de la provincia de Caceres (España) II (Insecta: Lepidoptera). – Shilap Revta. lepid. 27 (107): 327-348.
HERRENSCHMIDT, F. (1986): Contribution à la connaissance de la faune des Macrolépidoptères de la Drôme meridionale (Lepidoptera). – Alexanor 14 (8): 339-362.
HERRICH-SCHÄFFER, G.A.W. (1829-1844): Deutschlands Insekten, Fortsetzung von Panzers Fauna insectorum Germanica, 10 vol., H. 111-190.
HERRICH-SCHÄFFER, G.A.W. (1843-1856): Systematische Bearbeitung der Schmetterlinge von Europa, zugleich als Text, Revision und Supplement zu Jakob Hübner's Sammlung europäischer Schmetterlinge. – Regensburg, 6 vol..
HERRICH-SCHÄFFER, G.A.W. (1861): Neue Schmetterlinge aus Europa und den angrenzenden Ländern. Bd. 3. – Regensburg., 32 pp..
HEYDEMANN, F. (1929): *Cosymbia quercimontaria* auch in Schleswig-Holstein gefunden. – Int. ent. Zeits. 22: 434.
HEYDEMANN, F. (1941): Zu *Cosymbia quercimontaria* Bastlb.. – Z. Wien. Ent.-Ver. 26 (10): 238-240.
HEYDEMANN, F. (1942): Zu *Cosymbia quercimontaria* Bast.. – Z. Wien. Ent.-Ver. 27 (11): 277.
HEYDENREICH, G.H. (1851): Lepidopterorum Europaeorum Catalogus Methodicus. Systematisches Verzeichniss der Europaeischen Schmetterlinge. 3 Ausgabe. – Leipzig, 131 pp..
HICKS, M.E. & J.W. HALLE (1998): Lepidoptera of St. Agnes Isles of Scilly. A Systematic list and analysis of the species recorded on St. Agnes 1992-1997. A brochure, 74 pp..
HOBERT, G. (1930): Die Schmetterlingsfauna (Lepidoptera) von Mühlhausen in Thür. und Umgebung. – Ent. Z. Frankf. a.M. 44: 50, 93, 109, 126, 141, 169, 186, 218, 243, 308.
HODGES, R.W. (ed.) (1983): Check List of the Lepidoptera of America North of Mexico, including Greenland. London, XXIV + 284 pp.
HOFFMANN, F. & R. KLOS (1917): Die Schmetterlinge Steiermarks IV. – Mitt. naturw. Ver. Steiermark 53: 66-70.
HOFFMEYER, S. (1966): De danske malere (2nd Edn). – Unversitetsforlaget i Aarhus, 361 pp..
HOLLOWAY, J.D. (1984): The larger moths of the Gunung Mulu National Park; a preliminary asessment of their distribution, ecology, and potential as environmental indicators. – The Sarawak Mus. Journ. 30 (2): 149-190, pl. 12.
HOLLOWAY, J.D. (1993): The Moths of Borneo: Family Geometridae, Subfamily Ennominae. - London, 309

pp..
HOLLOWAY, J.D. (1997): The Moths of Borneo, pt. 10, Geometridae: Sterrhinae, Larentiinae. – Malayan Nature Journal 51: 1-242.
HOLLOWAY, J.D., KIBBY, G. & D. PEGGIE (2001): The families of Malesian moths and butterflies. – Leiden, Fauna Malesian Handbooks, Boston, Köln London, 455 pp..
HOMBERG, R. (1910a): Déscription d'une nouvelle variété de *Acidalia subsaturata* Guen. (Lep. Geometridae). – Bull. Soc. ent. Fr. 1910: 18-20.
HOMBERG, R. (1910b): Déscription de la chenille d'*Acidalia libycata* Bartel (Lep. Geometridae). – Bull. Soc. ent. Fr. 1910: 61.
HOMBERG, R. (1911): Déscription des premiers états de *Acidalia completa* Stgr. (Lep. Geometridae). – Bull. Soc. ent. Fr. 16: 144-146.
HORN W., KAHLE, I., FRIESE, G. & R. GAEDIKE (1990): Collectiones entomologicae. – Akad. Landw. DDR, Berlin, 2 vols, 573 pp..
HREBLAY, M. & A. LOBMAYER (1992): Die Schmetterlingsfauna des Nord-Tarna Gebietes, Ungarn (Lepidoptera). – Folia ent. Hung. 52: 35-49.
HRUBY, K. (1964): Prodromus Lepidopter Slovenska. – SAV, Bratislava, 962 pp..
HÜBNER, J. (1786-1790): Beiträge zur Geschichte der Schmetterlinge. – Augsburg, 2 vols.
HÜBNER, J. (1796-1838): Sammlung Europäischer Schmetterlinge. – Augsburg, 7 vols.
HÜBNER, J. (1816-1826): Verzeichniss bekannter Schmettlinge. – Augsburg, 431+72 pp..
HUEMER, P. & G. TARMANN (1993): Die Schmetterlinge Österreichs (Lepidoptera). Systematisches Verzeichnis mit Verbreitungsangaben für die einzelnen Bundesländer. – Tiroler Landesmuseum Ferdinandeum, Innsbruck, 224 pp..
HUEMER, P. (1994): Schmetterlinge (Lepidoptera) im Naturschutzgebiet Rheindelta (Vorarlberg, Österreich): Artenbestand, Ökologie, Gefährdung. – Linzer biol. Beitr. 26 (1): 3-132.
HUEMER, P. (1996): Schmetterlinge (Lepidoptera) im Bereich der Naturschutzgebiete Bangser Ried und Matschels (Vorarlberg): Diversität – Ökologie – Gefährdung. – Vorarlberger Naturschau 2: 141-202.
HUFNAGEL (1767): Tabellen von den Tag-, Abend- und Nachtvögeln der Umgebung von Berlin (Geometridae: Vierte Tabelle ... der Nachtvögel hiesiger Gegend). – Berlinisches Magazin oder gesammelte Schriften für die Liebhaber der Arzneywissenschaft, Naturgeschichte und der angenehmen Wissenschaften überhaupt, Bd. 2-4.
HUGGINS, H.S. (1950): *Sterrha muricata* Hufnagel in Ireland. – The Entomologist 83: 234-235.
HULDEN, L., ALBRECHT, A., ITÄMIES, J., MALINEN, P. & J. WETTENHOVI (2000): Suomen Suurperhosatlas Finlands Storfjärilsatlas [Atlas of Finnish Macrolepidoptera], Geometridae, Sterrhinae. – Helsinki, 328 pp..
ICZN (1999): see INTERNATIONAL COMMISSION ON ZOOLOGICAL NOMENCLATURE
INOUE, H. (1958): Descriptions and records of some Japanese Geometridae (II). – Tinea 4 (2): 241-256.
INOUE, H. (1961): Geometridae, Insecta Japonica, Series 1, Part 4: 1-106. – Hokuryukan, Tokyo.
INOUE, H. (1971): Icones Heterocerorum Japonicorum in Coloribus Naturalibus (Geometridae in Part). – Osaka Hoikusha Publishing Co. Ltd.
INOUE, H. (1977): Catalogue of the Geometridae of Japan (Lepidoptera). – Bull. Fac. Dom. Sci., Otsuma Woman's University, 13: 227-346.
INTACHAT, J. & J.D. HOLLOWAY (2000): Is there stratification in diversity or preferred flight height of geometrid moths in Malaysian tropical rainforest? – Biodiv. and Conservation 9: 1417-1439.
INTERNATIONAL COMMISSION ON ZOOLOGICAL NOMENCLATURE (1999): International Code of Zoological Nomenclature, Fourth Edition, adopted by the International Union of Biological Sciences. – The International Trust for Zoological Nomenclature, c/o The Natural History Museum, London, + 306 pp..
IPPOLITO, F. & P. PARENZAN (1998): I Macrolepidotteri del Demanio Forestale di Santo Stefano Quisquina (Agrigento). Contributi alla Conoscenza della Lepidotterofauna della Sicilia VI. – Phytophaga 8: 57-84.
ITÄMIES, J., PYÖRNILÄ, M., KUUSELA, K. & A. PYÖRNILÄ (1980): Flight periods of geometrid moths in central and northern Finland (Lepidoptera). – Notulae Ent. 60: 77-87.
IVANOV, A.I., MATOV, A.Yu., KHRAMOV, B.A., GRIGOREV, G.A., MIRONOV, V.G. & O.A. MISHCHENKO (1999): Artenverzeichnis der Macrolepidoptera von Sankt-Petersburg und des Sankt-Petersburger Gebietes nach Aufsammlungen in den Jahren 1960-1998 (Insecta, Lepidoptera). – Atalanta 30 (1/4): 293-356.
IVINSKIS, P. (1993): Geometridae. In: Check list of Lithuanian Lepidoptera. – Ecol. Instit. Vilnius, 115-132.
IVINSKIS, P., KARALIUS, S., MANIKAS, A., PACEVIÄIUS, V. & G. ·VITRA (1988): 3 novyh i 62 redkih dlja fauny Litovskoj SSR vidov pjadenitz, obnaruzhennyh v 1968-1985 gg. [3 new and 62 Geometridae species rare to the Lithuanian SSR, found in 1968-1985]. Novye i redkie dlja Litovskoj SSR vidy nasekomyh. Soobstsh. i opisanija 1988 goda, Vilnjus, pp. 33-44.
JALAVA, J. & J. KULLBERG (in press): A contribution to the knowledge of the fauna of Lepidoptera of Polar Urals

with some new additional distributional records of the species from Siberia and North America. – Ent. Fenn.
JANSE, A.J.T. (1933-1935): The Moths of South Africa, Vol. II Geometridae: 1-448 (Sterrhinae: 1934). – E.P. & Commercial Printing Co. Ltd., Durban.
JANSSEN, A. (1985): Katalogus van de Antwerpse Lepidoptera, Geometridae. – Vlaamse Verein. Ent., Phegea 13-15, Suppl., 159-217.
JOSEPH, C. & P. JOSEPH (1991): Contribution à l'établissement de la liste des Lépidoptères du departement du Jura, II (Lep., Zygaenoidea et Geometroidea). – Alexanor 17 (2): 79-87.
JUGAN, D. & Ch. JOSEPH (1988): Contribution à la connaissance des Macrolépidoptères de Haute-Saône (Lepidoptera). – Alexanor 15 (6): 323-381.
JÜRIVETE, U., KAITILA, J., KESKÜLA, T., NUPPONEN, K., VIIDALEPP, J. & E. ÕUNAP (2000): Eesti Liblikad, Katalog (Estonian Lepidoptera, Catalogue). – Tallinn, 151 pp..
KAILA, L. & A. ALBRECHT (1994): The classification of the *Timandra griseata* group (Lepidoptera: Geometridae, Sterrhinae). – Ent. Scandinavica 25: 461-479.
KAILA, L. & A. ALBRECHT (1995): Suolaheinämittarit *Timandra* (Lepidoptera, Geometridae) – kaksi lajila Suomessa [The Geometrid genus *Timandra* (Lepidoptera, Geometridae) contains two species in Finland] [in Finnish]. – Baptria 20 (3): 149-156.
KALCHBERG, A. VON (1895): [*Acidalia longaria* H.S.]. – Jahresb. Wien. ent. Ver. 5: 5.
KARISCH, T. (1991): Heteroceren-Nachweise aus der Permer Bezirk, UdSSR (Insecta, Lepidoptera). 2. Beitrag zur Fauna der Ural-Vorlandes. – Entomofauna 12 (21): 337-352.
KARSHOLT, O. & E.S. NIELSEN (1976): Systematisk fortegnelse over Danmarks sommerfugle. (Catalogue of the Lepidoptera of Denmark). – Scand. Sci. Press Ltd., Klampenborg, Denmark, 128 pp.
KARSHOLT, O. & E.S. NIELSEN (1985): The Lepidoptera described by C.P. Thunberg. – Ent. scand. 16 (4): 433-463.
KARSHOLT, O. & J. RAZOWSKI (1996): The Lepidoptera of Europe, a Distributional Checklist. – Apollo Books, Stenstrup, 380 pp..
KARSHOLT, O. & P. SKOU (1987): Sommerfugle (Lepidoptera) fra Skallingen. – Ent. Meddr., København, 54 (1): 67-92.
KARSHOLT, O. & P. STADEL NIELSEN (1998): Revideret katalog over de danske Sommerfugle. – København, 144 pp..
KARSHOLT, O., KRISTENSEN, N.P., KAABER, S., LARSEN, K., SCHMIDT NIELSEN, E., SCHNACK, K., SKOU, P. & B. SKULE (1985): Katalog over de danske Sommerfugle (Catalogue of the Lepidoptera of Denmark). – Ent. Meddr., København, 52 (2-3): 1-163.
KASY, F. (1965): Zur Kenntnis der Schmetterlingsfauna des östlichen Neusiedlersee-Gebietes. – Wiss. Arbeiten Burgenl. 34: 75-211.
KASY, F. (1978): Die Schmetterlingsfauna des Naturschutzgebietes Hackelsberg, Nordburgenland. – Z. Arb.Gem. Österr. Ent. 30 (Suppl.): 1-44.
KASY, F. (1987): Die Schmetterlingsfauna des Naturschutzgebietes "Glaslauterriegel-Heferlberg" südlich von Wien. – Z. Arb.Gem. Österr. Ent. 38 (Suppl.): 1-35.
KELLNER, J. (1995): Die Grossschmetterlinge von Dessau und Umgebung. – Naturw. Beitr. Mus. Dessau. Sonderheft 1. 204 pp..
KHOMENKO, V.N., GOLOVUSHKIN, M.I. & I.YU. KOSTJUK (1994): [Ecological-Faunistic Structure and Species Composition Dynamics of the Geometrid Moths (Lepidoptera, Geometridae) of Dahursky Nature Reserve] (in Russian). – Trudy sapow. ‚Dahurskyi' 2: 59-72.
KIEFER, H. (1941): III. Nachtrag zur Makrolepidopteren-Fauna des steirischen Ennstales. – Ent. Zeits. 55 (20): 157-160; 181-184.
KING, G. & L. ROMERA (2004): Contribución a la biologia de los geométridos en Madrid, con datos nuevos sobre sus plantas huéspedes (Lepidoptera: Geometridae). – Shilap Revta. lepid. (in print).
KING, G. (2002): Notas sobre las plantas huéspedes de *Phaiogramma etruscaria* (Zeller, 1849), *Microloxia herbaria* (Hübner, 1813), *Casilda consecraria* (Staudinger, 1871) y *Lithostege castiliaria* (Staudinger, 1877) en las estepas de Madrid, España (Lepidoptera: Geometridae). – Shilap Revta. lepid. 30 (119): 251-256.
KITSCHELT, R. (1925): Zusammenstellung der bisher in dem ehemaligen Gebiete von Südtirol beobachteten Grossschmetterlinge. – Wien. 421 pp..
KLIMASZEWSKI, S.M. (1960): Studia nad srodkowoeuropejskimi gatunkami z rodzaju *Cosymbia* Hbn. (Lepidoptera, Geometridae) [studies on the central European species of the genus *Cosymbia* Hbn. (Lepidoptera, Geometridae)]. – Fragm. Faun. 8 (30): 447-467 (in Polish).
KLIMESCH, J. (1933): Legère contribution a l'étude de la vaiabilité de *Ptychopoda (Sterrha) eburnata* Wocke. – Amat. Pap. 6 (19/20): 319-322.
KNOCH, A.W. (1781-1783): Die Schmetterlinge. In: Beiträge zur Insecten-Geschichte. – Leipzig, pts 1-3.

KNÖLKE, S. (2002): Elektronenmikroskopische Untersuchungen der Antennenmerkmale einiger Vertreter der Gattung *Idaea*. – Master thesis at the Ludwig-Maximilians-University Munich, 100 pp., 13 pls.
KOCAK, A.Ö. (1982-86): On the Validity of the Species Group Names proposed by Denis & Schiffermüller, 1775 in Ankündigung (sic) eines Systematischen Werkes von den Schmetterlingen der Wiener Gegend. – Priamus 2 (1): 5-42 (1982); 3 (3) 98-130; 3 (4): 133-154 (1984); 4 (1/2): 22-36 (1986).
KOCAK, A.Ö. (1990a): Ecological notes on the Turkish Lepidoptera. – Miscell. papers, Centre Ent. Stud. Ankara 4: 1.
KOCAK, A.Ö. (1990b): Additional faunistical and ecological notes to the Lepidoptera of the Ayas district (central Turkey). – Miscell. papers, Centre Ent. Stud. Ankara 4: 3-7.
KOCAK, A.Ö. (1990c): An annotated List of the Lepidoptera of Karadere and Bolu District (Prov. Bolu, N. Turkey). – Misc. Papers Centre Ent. Stud. Ankara 6: 1-11.
KOCAK, A.Ö. (1991): Faunistisch-ökologische Notizen über die Tagfalterfauna von Kizilcahamam (Prov. Ankara, Türkei) (Lepidoptera). – Priamus 5 (4): 1-22.
KOCAK, A.Ö. (1993): Über die Tagfalterfauna des Gebirges Hodulca bei Kizilcahamam (Prov. Ankara, Türkei). – Priamus 6 (3/4): 97-113.
KOCAK, A.Ö. (1994): „Türkiye kod listesi" ile ilgili taksonomik notlar – II (Lepidoptera). – Miscell. papers, Centre Ent. Stud. Ankara 19: 5-8.
KOCH, M. (1984): Wir bestimmen Schmetterlinge. Teil IV Spanner (Geometridae). Verlag J. Neumann-Neudamm, Leipzig, pp. 531-760.
KÖPPEL, C. (1997): Die Großschmetterlinge (Makrolepidoptera) der Rastatter Rheinaue: Habitatwahl sowie Überflutungstoleranz und Überlebensstrategien bei Hochwasser. – Neue Ent. Nachr. 39: 624 pp..
KOSCHWITZ, U., KRAUS, W. & E. BLUM (1985): Contribution to the Butterfly-fauna of Albarracin (Prov. Teruel) and adjacent areas (I). – Shilap Revta. lepid. 13 (50): 125-130.
KOSTJUK, I.YU. (1985): Novye dannye k faune pjadenitz (Lepidoptera, Geometridae) Kanevskogo zapovednika i ego okrestnostej. [New data to the fauna of geometrid Moths (Lepidoptera, Geometridae] of Kanevskij Nature Reserve and adjacent territories]. – In: "Problemy obschej i molekuljarnoj biologii", Kiev 4: 78-80. (in Russian).
KOSTJUK, I.YU. (1986): Materialy k izucheniyu pyadenits (Geometridae) doliny Srednego Dnepra. [Materials for a study of the Geometrid-Moths of the middle Dnieper valley]. – Problemy obshchei i molekulyarnoi biologii, Kiev, vyp. 5: 55-58. (in Russian).
KOSTJUK, I.YU. (1990): Geometridae. In: EFETOV, K.A. & YU.I. BUDASHKIN: Babotshki Kryma. – Sinferopol, (111 pp.): 86-92.
KOSTJUK, I.YU. (1992): [The Geometrid Moth fauna (Lepidoptera, Geometridae) of south-eastern Transbaikalia] (in Russian). – In: Insects of Dauria and adjacent territories, Central res. lab., Moscow:52-64.
KOSTJUK, I.YU. (2004): Modern state of study of Geometridae (Lepidoptera) in Ukraine: preliminary list and new records (in print).
KOSTJUK, I.YU., & M.I. GOLOVUSHKIN (2003): [A Contribution to the Knowledge of Geometrid Moths (Lepidoptera, Geometridae) of Chita Oblast] (in Russian). – Proc. zool. Mus. Kiev Taras Schev. Nat. Univ. 1 (1): 16-42.
KOSTJUK, I.YU., BUDASHKIN, Yu.I. & M.I. GOLOVUSHKIN (1994): [The Lepidoptera of the Dahursky Nature Reserve], in Russian. – Acad. Sci. Ukraine, Kiev, 36 pp..
KOSTJUK, I.YU., SHESHURAK, P.N., PLYUSHCH, I.G. & E.M. GALKINA (1998): [Geometridae of the Chernigov Region, northern Ukraine] Raznousye cheshuekrylye (Lepidoptera: Heterocera) Chernigovskoj oblasti Ukrainy. Chast 3. Pjadenitzy (Geometridae). – Nezhin, "Nauka-servis", 158 pp. (in Russian).
KOUTSAFTIKIS, A. (1973): Vergleichend-zoogeographische Untersuchung einiger Heterocerenfamilien der nordägäischen Inseln Thasos, Samotraki und Limnos. – Ann. Mus. Goulandris 1: 185-238.
KOVACS, L. (1953): A magyarorszagi nagylepkek es elterjedesuk (Die Gross-Schmetterlinge Ungarns und ihre Verbreitung). – Folia ent. Hung. (series nova) 6 (2): 76-164, 1 map.
KOVACS, L. (1956): A magyarorszagi nagylepkek es elterjedesuk. II [The Macrolepidoptera of Hungary and their distribution]. – Folia ent. Hung. (series nova) 9 (4): 89-140.
KOVACS, L. (1958): Valtozasok a magyarorszagi nagylepkek adataiban a Fauna Regni Hungariae, illetoled Abafi-Aigner lepkekonyvenek megjelenese ota (Die Veränderungen in der Grossschmetterlingsfauna von Ungarn seit dem Erscheinen der Fauna Regni Hungariae bzw. des Schmetterlingsbuches von Abafi-Aigner). – Folia ent. Hung. (series nova) 11 (19): 309-364.
KOVACS, L. (1965): Araszolólepkék I., Geometridae I. – Fauna Hungariae 74 (8), Budapest, 55 pp..
KOZLOV, M.V. & J. JALAVA (1994): Lepidoptera of the Kola Peninsula, northwestern Russia. – Entom. Fennica 5 (2): 65-85.
KRAMPL, F. & J. MAREK (1980): Beitrag zur Lepidopterenfaunistik der Tschechoslowakei. – Acta faun. Ent. Mus.

Nat. Pragae 16 (188): 89-95.
KRAMPL, F. (1973): Píd'alky južních čech. [Die Spanner Südböhmens (Geometridae, Lepidoptera)]. – Prírodoved. čas. jihočes., česke Budejovice, 13 (Suppl. 1): 1-74.
KRAMPL, F. (1984). Three Geometrid species new to Czechoslovakia, with remarks on their identification (Lepidoptera, Geometridae). – Acta ent. Bohemosl. 81: 280-291.
KRAMPL, F. (1992): Boreal macro-moths in central Europe (Czechoslovakia) and their eco-geographical characteristics (Lepidoptera: Geometridae, Noctuidae, Notodontidae). – Acta Entom. Bohemoslov 89 (4): 237-262.
KRAUS, W. (1993): Verzeichnis der Großschmetterlinge (Insecta: Lepidoptera) der Pfalz. – Pollichia 27, Bad Dürkheim, 618 pp..
KRAUS, W. (1997): A contribution to the knowledge of the Lepidoptera of the „Parque natural Cabo de Gata", Nuja, Almeria, Spain. – Shilap Revta. lepid. 25 (97): 63-64.
KRAUS, W. (1999): Beobachtungen zur Macrolepidopterenfauna der Iberischen Halbinsel. Teil 2: Artenliste Drepanidae bis Notodontidae. – Nachr. ent. Ver. Apollo, N.F. 20 (2): 231-263.
KRISTENSEN, N.P. (1966): Notes on *Sterrha ochrata*, a moth new to the Danish fauna (Lep., Geometridae). – Ent. Meddr. 34: 214-220.
KRISTENSEN, N.P. (ed.) (1999): Handbook of Zoology, vol. IV (35), Lepidoptera, Moths & Butterflies, 1: Evolution, Systematics, and Biogeography. – W. de Gruyter, Berlin & New York, (491 pp.): 423-462.
KROGERUS, H. (1954): Investigations on the Lepidoptera of Neufundland, I. Macrolepidoptera. – Acta Zool. Fennica 82: 1-80.
KRÜGER, G.C. (1913): Beiträge zur Entwicklungsgeschichte der italienischen Lepidopteren. – Ent. Mitt. Ent. Mus. 2: 106-120.
KRÜGER, G.C. (1939): Notizie sulla fauna della Sirtica occidentale: Lepidotteri. – Annali del Mus. Libico Stor. Nat. 1, 317-357.
KUCHLEIN, J.H. & R. DE VOS (1999): Annotated checklist of the dutch Lepidoptera. – Backhuys Publishers, Leiden, 302 pp..
KUDLER, J. (1978): Geometroidea. In: SCHWENKE, W. (ed.): Die Forstschädlinge Europas, Vol. 3 Schmetterlinge. – Verlag Paul Parey, Hamburg und Berlin: 218-264.
KÜHN, A. & M. VON ENGELHARDT (1940): Ein das Flügelmuster beeinflussender Letalfaktor bei *Ptychopoda seriata*. – Biol. Zentralblatt 60 (11-12): 561-566.
KULAK, A.V. (2001): [New and interesting geometrid Moths from Byelorus (Lepidoptera, Geometridae)] (in Russian). – Ves. N. Akad. Nabuk Byel. 2001 (1): 111-115.
KULAK, A.V. (2002): [Faunistic Materials to Geometrid Moth subfamily Sterrhinae in Byelorus (Lepidoptera, Geometridae)]. – In print (in Russian).
KUMAKOV, A.P. & YU.P. KORSHUNOV (1979): Cheshuekrylye Saratovskoj oblasti. [The Lepidoptera of Saratov Province]. – Izdat. Saratov. Univ., Saratov, 240 pp.
KUSDAS, K. & J. THURNER (1955): Beitrag zur Insektenfauna der Provinz Udine (Oberitalien). – Atti del 1 convegno Friulano di Sci. nat. Udine: 273-334.
KUTENKOVA, N.N. (1989): Cheshuekrylye zapovednika "Kivach". [The Lepidoptera of "Kivach" Nature Reserve]. – Flora i Fauna zapovednikov SSSR. Moskva, 61 pp..
KUZNETZOV, V.I. (1960): [On the fauna and biology of Lepidoptera of the western Kopet-Dagh]. – Trudy zool. Inst. Akad. Nauk SSR 27: 65-93 (in Russian).
KVALHEIM, O. (1944): Bidrag til kunnskapen om *Sterrha pallidata* Schiff. (Lep. Geom.). – Norsk Ent. Tids. 7 (1-2): 81-84 (in Norwegian, English summary).
LA HARPE, DE (1853): Faune Suisse, Lépidoptères.
LAEVER, DE, E. (1954): Notes sur le genre *Cosymbia* Hübner. – Lambillionea 54 (5-8): 35-41.
LAEVER, DE, E. (1960): Note sur *Sterrha incisaria* Staudinger. – Lambillionea 60 (1-2): 18-19.
LAEVER, DE, E. (1977): *Cyclophora* Hübner (=*Cosymbia* Hb) *hyponoea* Prout en Espagne (Lep. Geometridae). – Shilap Revta. lepid. 5 (4): 316.
LAINÉ, M. (1977): Macrolépidoptères de Normandie. II - Hétérocères (Hepialoidea, Cossoidea, Zygaenoidea, Geometroidea). – Ann. Mus. Du Havre 9: 1-58.
LAJONQUIÈRE, DE, E. (1963): *Scopula decolor* Stgr., espèce nouvelle pour la faune Européenne. – Alexanor 2 (3): 40.
LARSEN, K. (1977): Om 2. kuld af *Cyclophora linearia* Hb. og *C. quercimontaria* B.. – Lepidoptera, København 3 (4): 89-90.
LARSEN, K. (1995): *Timandra griseata* Petersen, 1902 ny for den dansken fauna (Lepidoptera, Geometridae). – Ent. Meddr. 63 (3): 53-55.
LASAN, M. (1997): Nekaj novih vrst v favni metuljev (Lepidoptera) Slovenije. – Acta ent. Slovenica 5 (1): 51-58.
LAŠTUVKA, Z. (1993): Katalog von Faltern der Mährisch-Schlesischen Region (Lepidoptera). – Brno, 130 pp..

LAŠTUVKA, Z. (1994): Motýli rozšíreného území CHKO Pálava. [Lepidoptera of the Protected Landscape Area Pálava]. – Univ. Agr. Brno, 118 pp..
LAŠTUVKA, Z. (1998): Seznam motýlu české a Slovenské republiky. [Checklist of Lepidoptera of the Czech and Slovak Republics (Insecta, Lepidoptera)]. – Konvoj, Brno, 117 pp..
LAŠTUVKA, Z. & H. ŠEVROVÁ (2004): Communities of the families Geometridae and Noctuidae in the Moravian Karst (Czech Republic) and their value of bioindication (Lepidoptera). – Biologia, Bratislava (in print).
LÁSZLÓ, G.M. & B. HERCZIG (1995): A general survey of the Geometridae (Lepidoptera) fauna of the Checheno-Ingoush Republic (Russia, NE Caucasus). – Folia ent. Hung. 56: 75-84.
LATTIN, DE, G. (1951): Türkiye Kelebekleri Hakkinda II. [Türkische Lepidopteren. II]. – Rev. Fac. Sci. l'Univ. d'Istanbul, Ser. B, 16: 45-73.
LATTIN, DE, G. (1967): Grundriss der Zoogeographie. – VEB Gustav Fischer Verlag, Jena, Hochschullehrbücher für Biologie 12: 1-602.
LEDERER, J. (1853): Versuch, die europäischen Lepidopteren in möglichst natürliche Reihenfolge zu stellen, nebst Bemerkungen zu einigen Familien und Arten. – Verh. zool.-bot. Ver. Wien 3: 165-270.
LEHMANN, L. (1990): Beitrag zur Makrolepidopterenfauna Südwestbulgariens mit Bemerkungen zur subspezifischen Gliederung (Lepidoptera). – Atalanta 21 (1/2): 121-138.
LEMPKE, B.J. (1936) Catalogus der Nederlandse Macrolepidoptera. – Tijdschr. Ent. 79: 238ff..
LEMPKE, B.J. (1949-1950): Catalogus der Nederlandse Macrolepidoptera, Geometridae. – Tijdschr. Ent. 90: 146-196; 92: 113-218.
LEMPKE, B.J. (1967): Catalogus der Nederlandse Macrolepidoptera (veertiende supplement). – Tijdschr. Ent. 110 (10): 223-342, 7 pls.
LENZ, J. & A. HAUSMANN (1992): *Idaea blaesii* sp. n., eine neue Sterrhine aus Portugal (Lepidoptera, Geometridae). – Mitt. Münchn. ent. Ges. 82: 47-53.
LERAUT, P. (1994): *Scopula marginepunctata britonaria* (Oberthür, 1917) stat. rev., dans l'ouest de la France (Lep. Geometridae). Ent. gall. 4 (4): 211-212.
LERAUT, P.J.A. (1980): Liste systématique et synonymique des Lépidoptères de France, Belgique et Corse. – Alexanor (Suppl.), Paris, 334 pp..
LERAUT, P.J.A. (1997): Liste systématique et synonymique des Lépidoptères de France, Belgique et Corse (deuxième édition). – Alexanor (Suppl.), Paris, 526 pp..
LERAUT, P.J.A. (2003): Le guide entomologique. – Delachaux et Niestlé, Paris, 527 pp..
LHOMME, L. (1935) (‚1923-1935'): Catalogue des Lépidoptères de France et de Belgique; fam. xviii Geometridae. Tome I: Macrolépidoptères (2ème partie): 378-622. Le Carriol, Douelle (Lot).
LINNAEUS, C. (1758): Systema Naturae, I, 10th Edn. – Stockholm, 823 pp..
LINNAEUS, C. (1761): Fauna Svecica, 2nd Edn. – Stockholm, 578 pp..
LINNAEUS, C. (1767): Systema Naturae, 12th Edn. – Stockholm, 533-1327.
LUIG J. & T. KESKÜLA (1995): Catalogus Lepidopterorum Estoniae. – Tartu, 130 pp..
LUQUET, G.C. (1982): Macrolépidoptères recoltés dans les environs de La Ferte-Mace (Orne): Quelques complements aux "Macrolépidoptères de Normandie" du Dr Marcel Lainé. I. Rhopalocera, Bombycoidea et Geometroidea (2-e partie). – Linneana Belgica 8 (9): 399-416.
LUQUET, G.C. (2000): Biocenotique des Lépidoptères du Mont Ventoux (Vaucluse). – Alexanor Suppl. 2000: 400 pp..
LUTRAN, G., MAZEL, R., PESLIER, S. & C. TAVOILLOT (1998): Mise à jour au 31 décembre 1997 de la liste des Lépidoptères des Pyrénées-Orientales (Insecta, Lepidoptera). – Alexanor 20 (6): 325-332.
LVOVSKY, A.L. (1971): Materialy po faune cheshuekrylyh (Macrolepidoptera) Astrakhanskoj oblasti. – Ent. obozr. 50 (4): 800-810.
MABILLE, P. (1867): Notices sur les Lépidoptères de la Corse. – Annls Soc. ent. Fr. 6 (4 sér.): 545-564.
MABILLE, P. (1872): Recherches et observations lépidoptèrologiques. – Annls Soc. ent. Fr. 2 (5 sér.): 489-502.
MACK, W. (1985): Lepidoptera. In: Die Nordost-Alpen im Spiegel ihrer Landtierwelt, vol. 5: 1-484.
MAIDORN, G. (1914): Zucht und Beschreibung der Raupe von *Acidalia litigiosaria* B.. – Z. wiss. Insektenbiol. 10 (3): 85-86.
MALICKY, H. (1966): Lepidopterologische Ergebnisse eines Schweizer Aufenthaltes. – Mitt. Schweiz. ent. Ges. 38 (3/4): 173-182.
MALKIEWICZ, A. & J. SOSINSKI (1999): Systematic annotated check-list of Polish Geometridae (Lepidoptera). – Polskie Pismo Ent. 68 (2) 197-208.
MANN, J. (1855): Die Lepidopteren, gesammelt auf einer entomologischen Reise in Corsika im Jahre 1855. – Verh. zool.-bot. Ver. Wien 5: 528-572.
MANN, J. (1864): Nachtrag zur Schmetterlingsfauna von Brussa. – Wien. ent. Monatschr. 8: 173-190.
MARIANI, M. (1939): Fauna Lepidopterorum Siciliae (Catalogo ragionato). – Mem. Soc. ent. Ital. 17 (2): 129-

187.
MARIANI, M. (1940-1943): Fauna Lepidopterorum Italiae. Parte I. Catalogo ragionato dei Lepidotteri d'Italia. Fasc. II e III. - Giorn. Sci. nat. Econ., 42 (3): 81-227.
MARINI, M. & I. RUSSO (1980): Interessanti reperti di Lepidotteri in Calabria. - Boll. Ist. Ent. Univ. Bologna, 35: 249-265.
MARINI, M. & M. TRENTINI (1985): Osservazioni sui Lepidotteri dell'Appennino Lucchese. IV. Geometridae. - Boll. Ist. ent. Univ. Bologna 39: 17-36.
MATHESON, C.D., MÜLLER, G., VERNON, K., JUNNILA, A., HAUSMANN, A., GREENBLATT, C. & Y. SCHLEIN (in press, 2004): Species identification of plant residues found in the gut of insects by a PCR method. - Molecular Ecology.
MAY, H. (1892): Ueber die ersten Stände einiger Geometriden. - Jahresb. Wien. ent. Ver. 2: 23-29.
MAY, H. (1893): Ueber die ersten Stände einiger Geometriden. - Jahresb. Wien. ent. Ver. 3: 39-43.
MAZEL, R. & S. PESLIER (1997): Cartographie des Lépidoptères des Pyrénées-Orientales, 1, Geometridae. - Revue Assoc. Roussill. Ent., Suppl., 115 pp..
MAZEL, R. (1995): Sur quelques *Idaea* du Roussillon, dont *Idaea bustilloi* Agenjo, 1967, espèce nouvelle pour la France (Lepidoptera, Geometridae). - Alexanor 18 (7): 401-408.
MAZZUCCO, K. (1966): *Cosymbia pupillaria* Hbn. bei 3200 m. - Z. Wien. ent. Ges. 51: 154.
MCGUFFIN, W.C. (1967): Guide to the Geometridae of Canada (Lepidoptera), I, Subfamily Sterrhinae. - Mem. ent. Soc. Canada 50, Ottawa, 67 pp..
MENDES, C. DE AZEVEDO (1910): Lagartas ineditas de Lepidopteros da fam. Geometridae. - Broteria 9: 105-109.
MENDES, C. S.J. (1912): Lepidopteros de S. Fiel (Beira Baixa - Portugal). - Broteria 10 (3): 161-182.
MÉNDEZ GARNICA, J.M. (1983): Contribución al conocimiento de los Lepidópteros del norte de Burgos (Cuadricula UTM VN-65). Heteroceros - I. - Shilap Revta. lepid. 11 (42): 133-140.
MÉNDEZ GARNICA, J.M. (2001): Resultado de 20 años de muestreo y seguimiento de lepidópteros en la Cerca (Burgos). - Est. Mus. Cienc. Nat. de Álava 16: 179-201.
MENHOFER, H. (1938): *Ephyra quercimontaria* Bast. neu für Thüringen. - Ent. Z. Frankf. a.M. 13 (52): 101-102.
MENHOFER, H. (1941): *Cosymbia quercimontaria* Bast. - ein Falter ozeanischer Heiden und seine Verbreitung (Lep.). - Z. Wien. Ent.-Ver. 26 (5): 137-144.
MENHOFER, H. (1942a): Zur Verbreitung der *Cosymbia quercimontaria* Bast.. - Z. Wien. Ent.-Ver. 27 (7): 161-166.
MENHOFER, H. (1942b): Gehört *Cosymbia ruficiliaria* H.-S. zum mediterranen oder orientalischen Faunenelement? - Z. Wien. Ent.-Ver. 27: 283-287.
MENTZER, VON, E. (1980): *Idaea ibizaria* n. sp. from Spain (Lepidoptera, Geometridae). - Ent. scand. 11: 206-208.
MENTZER, VON, E. (1984): Die Genera bei Denis & Schiffermüller als Nomenklaturfrage (Lepidoptera). - Nota lepidopterologica 7: 59-70.
MENTZER, VON, E. (1990a): *Idaea boeklini* sp. n. from South Europe, with notes on the nomenclature and systematics of the *Idaea virgularia* (Hübner) species group (Lepidoptera: Geometridae). - Nota lepidopterologica 12 (4): 290-302.
MENTZER, VON, E. (1990b): *Glossotrophia annae* sp. n. from Spain (Lepidoptera: Geometridae). - Nota lepidopterologica 13 (1): 43-49.
MERRIFIELD, F. (1893): On effects of temperature on rearings of *Ephyra punctaria*. - Trans. ent. Soc. London 1893: 65-67.
MERZHEEVSAYA, O.I., LITVINOVA, A.N. & R.V. MOLCHANOVA (1976): Tsheshuyekrylye (Lepidoptera) Byelorussii. - Minsk, 131 pp..
MÉSZÁROS, Z., RONKAY, L., HERCZIG, B., SZEOKE, K. & C. SZABOKY (1985): Datos para el conocimiento de la fauna de los Lepidopteros de Bulgaria (3a Parte). - Shilap Revta lepid. 12 (45): 67-70; (47): 99-201; 13: 297-299; 14: 67-72.
MEYER, M. & M. HELLERS (1990): Les Lépidoptères de Madère. Note preliminare: Les Macrolépidoptères observés en mai 1989 (Lepidoptera, Geometridae, Sphingidae, Noctuidae, Papilionidae). - Nota lepidopterologica 13 (2-3): 147-159.
MEYER, M. (1991): Les Lépidoptères de la région macaronésienne, II, Liste des Macrolépidoptères observés en jouillet-aout 1990 aux Acores (Lepidoptera, Geometridae, Sphingidae, Noctuidae). - Linneana Belgica XIII (3): 117-134.
MICHAEL, F. (1963): Beitrag zur Kenntnis des Formenkreises der Geometride *Sterrha muricata* Hufn. (Lep.). - Ent. Berichte 1: 44-49.

MICHEL, J. (1944): Ein Beitrag zum Vorkommen und zur Lebensweise von *Codonia quercimontaria* Bastelb. – Ent. Zeits. 57 (20): 154-160.

MIKAT, M. & L. MARSIK (1997): Prispevek k poznani fauny motylu navrhovaneho chraneneho uzemi "Na Plachte" v Hradci Kralove (Lepidoptera). [Contribution to the knowledge of fauna of butterflies (Lepidoptera) of protected area "Na Plachte" (Hradec Kralove, Czech Republic)]. – Acta Mus. Reginaehrad., Ser. A, 25: 163-192.

MIKKOLA, K. & K. SPITZER (1983): Lepidoptera associated with peatlands in central and northern Europe: a synthesis. – Nota lepidopterologica 6 (4): 216-229.

MIKKOLA, K. (1979): Vanishing and declining species of Finnish Lepidoptera. – Notulae ent. 59: 1-9.

MIKKOLA, K. (1985): The Geometroidea and Noctuoidea described by Carl Clerck (Lepidoptera). – Ent. scand. 16: 121-129.

MILLER, E. (1915): [Notes on the character of the lepidopterous fauna of the Murmansk coast]. – Bull. Soc. ent. Moscow 1915: 124-135.

MILLER, E., ZUBOWSKI, N. & A. RUSCINSKI (1930): Materiale pentru fauna entomologica din Bessarabia. Macrolepidoptera. Suppl. 3. – Bul. Muz. Nat. Ist. Nat. Chişinau, 2 (3): 97-130.

MILLER, M.A., ERLACHER, I. & A. HAUSMANN (in prep.): Species rank for both *Timandra comae* Schmidt, 1931 and *T. griseata* Petersen, 1902? Results from the molecular analysis of the mitochondrial COX I gene (Lepidoptera, Geometridae, Sterrhinae).

MILLER, M.A., HAUSMANN, A. & R. TRUSCH (2001). The phylogenetic relationships in Geometrid moths. An approach using mitochondrial DNA (mtDNA) sequences. – Spixiana 24 (3): 201-202.

MILLIÈRE, P. (1864-1868): Iconographie et description de chenilles de Lépidoptères inédits, vol. II, Paris, 1-506.

MILLIÈRE, P. (1869-1874): Iconographie et description de Chenilles et Lépidoptères inédits. – Annls Soc. Linn. Lyon 3 (23-35): 1-488, 101-154 pls.

MILLIÈRE, P. (1873): Catalogue Raisonné des Lépidoptères des Alpes-Maritimes (2). – Mem. Soc. Sci. nat. 1873: 141-455.

MIRONOV, V. (1986): [New Geometrid moths from Tian Shan] (Geometridae). – Trudy vses. ent. Obshch. 67: 94-100 (in Russian).

MIRONOV, V. (2003): Larentiinae II (Perizomini and Eupitheciini). – In: A. HAUSMANN (ed.): The Geometrid Moths of Europe 4. – Apollo Books, Stenstrup, 463 pp. & 16 colour pls.

MLADINOV, L. (1980): Lepidoptera iz dolinegornjeg toka rijeke Kupe IV. Geometridae. – Acta ent. Jugosl. 16 (1-2): 69-91.

MONTEIRO, T. & J. PASSOS DE CARVALHO (1984): Lepidópteros do Algarve. – Anais Fac. Ciênc. Porto 64 (1-4): 95-219.

MONTEIRO, T. (1969): Lépidoptères de l'Algarve. – Publ. Inst. zool. "Dr. Augusto Nobre" 105. 39 pp..

MOTHIRON, P. (2001): Inventaire commenté des Lépidoptères de l'Île-de-France II. Géomètres (Lepidoptera, Geometridae). – Alexanor, Suppl. 21: 164 pp..

MOUCHA, J. & I. NOVAK (1962): Ergebnisse der Erforschung der Lepidopteren-Fauna der Slowakei (1946-1960). II. Teil. – Acta faun. ent. Mus. Nat. Pragae 8: 95-137.

MÜLLER, B. & J. GELBRECHT (1992): Veränderungen in der Spannerfauna der DDR seit 1945 (Lep., Geometridae). – Nota lepidopterologica, Suppl. 3: 70-81.

MÜLLER, B. (1996): Geometridae. In: KARSHOLT, O. & J. RAZOWSKI (ed.): The Lepidoptera of Europe, a Distributional Checklist. (pp. 218-249). – Apollo Books, Stenstrup, 380 pp..

MÜLLER, L. (1936): *Acidalia contiguaria* Hb.. – Mitt. Münchn. ent. Ges. 26: 1-32, pl..

NAKAJIMA, H. & R. SATO (1979): A list of food plants of the Japanese Geometridae II, Archiearinae, Oenochrominae and Geometrinae. – Japan. Heteroc. J. 100: 663-680.

NAKAMURA, M. (1987): Pupae of Japanese Geometridae I (Lepidoptera). – Tinea 12 (Suppl.): 213-219.

NAKAMURA, M. (1994): Pupae of Japanese Geometridae (II). – Transactions of the Shikoku Ent. Soc. 20: 247-256.

NEMES, I. (1970): Contributii la Cunoasterea Geometridelor din Republica Socialista Romania II (Lepidoptera). – Studii si Comunicari Stiinte Naturale Suceava 1970: 73-86 (in Romanian).

NEMTZEV, V.V., ANTONOVA, E.M. & A.V. SVIRIDOV (1991): [Butterflies and moths of the Darwin Nature Reserve] Cheshuekrylye Darvinskogo zapovednika. – Flora i Fauna zapovednikov SSSR. Moscow, 1991, 49 pp. (in Russian).

NESTOROVA, E. (1998): Catalogus faunae Bulgaricae, 2. Lepidoptera: Geometridae. – Acad. Sci. Bulg. Ist. Zool. Pensoft Publ. Sofia – Moscow, 194 pp..

NIELSEN, E.S., EDWARDS, E.D. & T.V. RANGSI (eds) (1996): Checklist of the Lepidoptera of Australia. Monographs on Australian Lepidoptera 4. – Csiro Publishing, Collingwood, Australia 529 pp..

NIELSEN, F.J. & B. BAUNGAARD (1999): Fangst og klaekning af *Idaea ochrata*. – Lepidoptera, København 7 (8):

250.
NOLCKEN, F. H. W. (1870): Lepidopterologische Fauna von Estland, Livland und Kurland, 2. - Riga, 849 pp..
NORDSTRÖM, F. (1941): Om tydandet av några av Linnés och Clercks fjärilarter. - Ent. Tidskr. 62: 127-130.
NORDSTRÖM, F. (1943): Forteckning over Sveriges storfjarilar [Catalogus Insectorum Sueciae. III. Macrolepidoptera]. - Opusc. Ent. 8 (3-4): 59-120.
NORDSTRÖM, F. (1953): Catalogus Insectorum Sueciae. III. Macrolepidoptera. - Opusc. Ent. 18 (1): 75-87.
NORDSTRÖM, F., WAHLGREN, E. & A. TULLGREN (1941): Svenska Fjarilar (Macrolepidoptera). - Stockholm, IV & 353 pp. & 49 Tafn.
NOVOA PEREZ, J.M., NIETO MANZANO, M.A., GARCIA-VILLANUEVA, V. & J.A. MORENO TAMUREJO (2002): Proyecto de muestreo y catalogación de los Macroheteróceros de Extremadura, España (Insecta: Lepidoptera). - Bol. S.E.A. 30: 121-142.
NYST, R.H. (1987): Contribution à l'établissement de la liste des Macrolépidoptères de la Côte-d'Or et du Var. - Alexanor 15 (1): 3-19.
OBERTHÜR, C. (1922): Les Lépidoptères du Maroc. - Ét. Lepid. Comp. 19: 1-402.
OBERTHÜR, C. (1923): Catalogue des Lépidoptères des Pyrénées-Orientales. - Ét. Lepid. Comp. 21 (2): 9-75.
OBRAZTSOV, N.S. (1932): *Acidalia orientalis* Alph., bona species. - Int. ent. Zeits. 25: 235-240.
OBRAZTSOV, N.S. (1935): Untersuchungsergebnisse der nächtlichen Lepidopterenfauna in der Stadt Nikolaew. - Trav. Mus. Zool. Kiev 15: 141-161.
OBRAZTSOV, N.S. (1936): Materialien zur Lepidopterenfauna des Parkes von Vessjolaja Bokovenjka (Ukraine). - Folia Zool. et Hydrobiol. Riga 9 (1): 29-57.
OKYAR, Z. & N. AKTAÇ (1999): Trakya Bölgesi Geometridae Türlerinin Taksonomik ve Faunistik Yönden Ara?tirilmasi (Faunistic and Taxonomic Studies on the Geometridae Species of Turkish Thrace). - Tr. J. Zool. 23 (1): 99-132.
OLIVIER, A. (2000): CHRISTIAN FRIEDRICH FREYER's "Neuere Beiträge zur Schmetterlingskunde mit Abbildungen nach der Natur": an analysis, with new data on its publication dates (Insecta, Lepidoptera). - Beitr. Ent. 50 (2): 407-486.
OPHEIM, M. (1972): Geometridae. In: Catalogue of the Lepidoptera of Norway, pt. 3. - Zool. Mus. Oslo: 1-36.
ORHANT, G. (1996): Répartition de *Scopula tessellaria* (Boisduval, 1840) en France et chez nos voisins (Lepidoptera: Geometridae). - Linneana Belgica 15 (8): 329-331.
OSTHELDER, L. (1929): Die Schmetterlinge Südbayerns und der angrenzenden nördlichen Kalkalpen, 1. Teil Großschmetterlinge, 3. Heft, Spanner. - Mitt. Münchn. ent. Ges. 19, Supplement.
OZORSKI, E. (1948): Tableau de détermination des espèces françaises des genres *Scopula* Schr. et *Sterrha* Hb.. - Bull. Soc. ent. Mulhouse 1948: 1-22.
PALM, E., JENSEN, K., NIELSEN, U.S. & B.H. THOMSEN (1982): Lepidoptera VI, Nordvestsjaellands sommerfugle. - København, 83 pp..
PARENZAN, P. & A. HAUSMANN (1992): Nuovi interessanti reperti di Geometridi (Lepidoptera) in Italia meridionale. - Entomofauna 13 (8), 157-172.
PARENZAN, P. (1976): Contributi alla conoscenza della lepidotterofauna dell'Italia meridionale. II. Nuovi reperti di Noctuidae e Geometridae. - Entomologica, XII: 153-169.
PARENZAN, P. (1988): Nuove segnalazioni di Geometridae (Lepidoptera) per l'Italia meridionale. - Entomologica, XXIII: 139-160.
PARENZAN, P. (1994): Contributi alla conoscenza della Lepidotterofauna dell'Italia meridionale: XVII. Geometridae. - Entomologica Bari 28: 99-246.
PARENZAN, P. (1998): Osservazioni preliminari sulla Macrolepidotterofauna della riserva naturale di Vendicari (Siracusa) e prima segnalazione di *Leucania palestinae* Staudinger, 1897 (Lepidoptera, Noctuidae) per la fauna italiana. - Naturalista sicil. S. IV, 22 (3-4): 355-364.
PARENZAN, P., BELLA, S. & P. RUSSO (1998a): *Idaea completa* (Staudinger, 1892) (Lepidoptera: Geometridae) specie nuova per la fauna di Sicilia e italiana. - Entomologica Bari 32: 195-199.
PARENZAN, P., HAUSMANN, A. & S. SCALERCIO (1998b): Addenda e corrigenda ai Geometridae dell'Italia meridionale. - Entomologica (Bari) 32: 51-79.
PARSONS, M., GREEN, D. & N. BOURN (2001): Lepidoptera Conservation Bulletin incorporating the National Moth Conservation Project, Number 2, April 2000 - March 2001, 50 pp..
PASSOS DE CARVALHO, J. & M.F.V. CORLEY (1995): Additions to the Lepidoptera of Algarve, Portugal (Insecta: Lepidoptera). - Shilap Revta. lepid. 23 (91): 191-230.
PASSOS DE CARVALHO, J. (1984): Contribuição para o conhecimento da Lepidopterofauna do Parque Nacional da Peneda-Gerês. Família Geometridae. Parque Nacional da Peneda-Gerês, 1-40 pp..
PATOČKA, J. (1994a): Die Puppen der Tribus Cyclophorini Mitteleuropas (Lepidoptera: Geometridae). - Nota lepidopterologica 17 (1/2): 73-86.

PATOČKA, J. (1994b): Die Puppen der Spanner Mitteleuropas (Lepidoptera, Geometridae). Charakteristik. Bestimmungstabelle der Gattungen. – Tijdschr. Ent. 137: 27-56.

PATOČKA, J. (2003): Die Puppen der Spanner (Lepidoptera, Geometridae) Mitteleuropas: Unterfamilie Sterrhinae. – Bonner zool. Beitr. 51 (4): 269-296.

PEARSON, D.L. (1994): Selecting indicator taxa for the quantitative assessment of biodiversity. – Phil. Transact. Roy. Soc. London. Series B, Biol. Sci. 345: 75-79.

PEREGOVITS, L. (1995): Data to the Macrolepidoptera fauna of Transylvania (Romania). I. Faunistical data. – Folia ent. Hung. 56: 163-178.

PETERSEN, W. (1937): Zur Frage der Artberechtigung von *Rhodostrophia tabidaria* Zell. (Lepidoptera Geometridae). – Ann. Mus. zool. Polon. Warsaw 11 (27): 443-458.

PFEIFFER, E. (1927): Ein Beitrag zur Insektenfauna von Kleinasien (Anatolien). – Mitt. Münchn. ent. Ges. 17 (1-6): 35-55.

PIERCE, F.N. (1914): The Genitalia of The Group Geometridae of the Lepidoptera of the British Islands. – E. W. Classey Ltd. Faringdon, U.K. XXIX & 88 pp.+ 48 pl.

PINKER, R. (1957): Beitrag zur Lepidopterenfauna des südlichen Burgenlandes. – Z. Wien. ent. Ges. 43 (Bd. 69, 8): 147-151.

PINKER, R. (1968): Die Lepidopterenfauna Mazedoniens, III Geometridae. – Prirod. Muz. Skopje, Pos. Izdanie 4, 71 pp..

PINKER, R. (1969): Neue und interessante Lepidopteren aus Madeira und den Azoren mit faunistischen Hinweisen auf die Kanaren. – Z. Wien. ent. Ges. 54 (Bd 80, 9-12): 101-131, taf. 12-14.

PIZZETTI, L. & V. RAINERI (1998): Segnalazioni di alcuni Geometridi della collezione Barajon interessanti per la fauna italiana (Lepidoptera Geometridae). – Boll. Soc. ent. Ital. 130 (2): 159-166.

PLEISCH, E. (1981): Beobachtungen über die Grossschmetterlingsfauna (Macrolepidoptera) von Mergoscia/TI und der näheren Umgebung (1966-1979). – Mitt. Ent. Ges. Basel 30: 140-160.

POHIER, F. & P. LEFORT (1997): Lépidoptères recueillis sur l'Hautil (Yvelines) par Gustave Trachier entre 1941 et 1974 (Insecta, Lepidoptera). – Alexanor 20 (4): 195-213.

POPE, F. (1908): Note on the larva of *Acidalia osseata*. – The Entomologist 41 (538): 62-63.

POPESCU-GORJ, A. & I. DRAGHIA (1967): L'entomofauna des forets du Sud de la Dobrudja. – Trav. Mus. d'hist. nat. "Gr. Antipa" 7: 181-202.

POPESCU-GORJ, A. & I. DRAGHIA (1968): Lepidoptera (Entomofaune de Letea). – Trav. Mus. d'hist. nat. 'Gr. Antipa' 9: 227-277.

POPESCU-GORJ, A. (1959): Date noi Asupra Lepidopterelor din Dobrogea. – Studii si cercetari de Biologie, ser. Biol. Anim. 11 (1): 15-20.

POPESCU-GORJ, A. (1964): Catalogue de la Collection de Lépidoptères ‚Prof. A. Ostrogovich'. – Du Mus. d'Hist. nat. 'Gr. Antipa', Bucarest, 293 pp..

POPESCU-GORJ, A. (1976): Entomofauna du Nord de la Dobrogea. Lepidoptera. – Trav. Mus. d'hist. nat. "Gr. Antipa" 17: 159-172.

POPESCU-GORJ, A. (1987): La liste systematique revisée des éspèces de Macrolépidoptères mentionnées dans la faune de Roumanie. Mise à jour de leur classification et nomenclature. – Trav. Mus. hist. nat. 'Gr. Antipa' 29: 69-123.

POPESCU-GORJ, A. (1995): Lepidopterans from the surroundings of the town Sinaia and from Bucegi Mountains (Romania). – Trav. Mus. hist. nat. 'Gr. Antipa' 35: 161-220.

POPESCU-GORJ, A., OLARU, V. & I. DRAGHIA (1972): L'entomofaune du "grind" de Caraorman (Delta du Danube). Ord. Lepidoptera. – Trav. Mus. hist. nat. 'Gr. Antipa' 12: 181-206.

PORTER, J. (1997): The Colour Identification Guide to Caterpillars of the British Isles (Macrolep.). – Viking (Penguin Books) London, 275 pp..

PRINS, DE, W. (1983): Systematische naamlijst van de Belgische Lepidoptera. – Entomobrochure (Antwerpen), 4: 1-57.

PRINS, DE, W. (1984): Some faunistic remarks on the Spanish Lepidoptera-fauna (Part III). – Shilap Revta lepid. 12 (46): 131-134.

PRINS, DE, W. (1998): Catalogue of the Lepidoptera of Belgium. – Dokum. trav. Inst. royal Sci. nat. Belg., Brussels, 236 pp..

PROLA, C. & T. RACHELI (1979): I Geometridi dell'Italia centrale. Parte I. – Boll. Ist. Ent. Un. Bologna, 34: 191-246.

PROUT, L.B. (1905): Notes on the wave moths (genus *Acidalia* auct.). – Entomologist 38: 43-48.

PROUT, L.B. (1912-1916): Die spannerartigen Nachtfalter. In: SEITZ, A. (ed.): Die Groß-Schmetterlinge der Erde, vol. 4. – Verlag A. Kernen, Stuttgart.

PROUT, L.B. (1920-1941b): Die indoaustralischen Spanner. In: SEITZ, A. (ed.): Die Groß-Schmetterlinge der

Erde, vol. 12. – Verlag A. Kernen, Stuttgart.
PROUT, L.B. (1930-1938c): Die spannerartigen Nachtfalter (Fauna Africana). In: SEITZ, A. (ed.): Die Groß-Schmetterlinge der Erde, vol. 16. – Verlag A. Kernen, Stuttgart.
PROUT, L.B. (1934-1935a, additions 1938a): Brephinae, Oenochrominae, Hemitheinae, Sterrhinae, Larentiinae. In: SEITZ, A. (ed.): Die Groß-Schmetterlinge der Erde, Suppl. 4. – Verlag A. Kernen, Stuttgart.
PROUT, L.B. (1934d): Lepidopterorum Catalogus, Pars 61: Geometridae, Subfamilia Sterrhinae I. – Verlag W. Junk, Berlin, 486 pp..
PUCCI, C. & B. PAPARATTI (1989). Contributo alla conoscenza della lepidotterofauna Umbra: catture di adulti effettuate cona una trappola luminosa tipo 'Pennsylvania'. – Frustula Entomol. 10: 1-40.
PUECHREDON, E.B. DE (1954): Lépidoptères de la Tremezzina (Lac de Come). – Revue franç. Lepid. 14 (15-17): 211-217.
PÜNGELER, R. (1889): Lepidopterologische Mittheilungen aus der Schweiz. – Stettin. ent. Ztg. 50 (4-6): 143-151.
PÜNGELER, R. (1892): Zur Kenntnis einiger Acidalien. – Stettin. ent. Ztg. 53 (10-12): 363-368.
PÜNGELER, R. (1896): Lepidopterologische Mitteilungen aus der Schweiz. – Stettin. ent. Ztg. 57: 217-241.
PÜNGELER, R. (1906): Neue paläarktische Makrolepidopteren. – Dt. ent. Z. Iris 19: 78-98.
PÜNGELER, R. (1909): Neue palaearctische Macrolepidopteren. – Dt. ent. Z. Iris 21: 286-303.
QUINETTE, J.-P. & N. LEPERTEL (1993): Inventaire préliminaire des Macrolépidoptères du departement de la Manche (Lepidoptera, Rhopalocera et Heterocera). – Alexanor 18 (3): 157-177.
RACHELI, T., PROLA, C. & P. PROVERA (1978): Note su alcuni Lepidotteri dell'Italia centrale. – Boll. Assoc. rom. ent. 33: 19-39.
RAGUSA, E. (1889-1893): Note Lepidotterologiche. – Il Naturalista Sicil. 8-13, 40 pp..
RAINERI, V. & A. SCIARETTA (1996): Contributo alla conoscenza dei Lepidotteri Geometridi del Molise (Lepidoptera: Geometridae). – Doriana 6 (295): 1-11.
RAINERI, V. & S. ZANGHERI (1995): Lepidoptera Drepanoidea, Axioidea, Geometroidea. In: MINELLI, A., RUFFO, S. & S. LA POSTA (eds.), Checklist delle specie della fauna Italiana, 90. – Ed. Calderini, Bologna, 23 pp..
RAINERI, V. (1985): Ricerche sui Geometridi delle Alpi Liguri (Lepidoptera). – Boll. Soc. ent. Ital., Genova, 117 (4-7): 102-112.
RAINERI, V. (1986a). Materiali per una fauna dell'arcipelago Toscano. 27. I lepidotteri dell'isola di Montecristo. – Ann. Mus. Civ. Storia Nat. "G. Doria" 6 (nr. 257). 6 pp..
RAINERI, V. (1986b): Ricerche sui Geometridi delle Alpi Liguri. Parte seconda (Lepidoptera). – Boll. Soc. ent. Ital., Genova, 118 (1-3): 25-30.
RAINERI, V. (1989): *Idaea predotaria* (Hartig, 1951) stat. n., comb. n., the valid senior synonym of *I. griseanova* Rezbanyai-Reser, 1987 (Lepidoptera, Geometridae). – Nota lepidopterologica 12 (3): 187-191.
RAINERI, V. (1991): Some Geometridae new to Italy (Lepidoptera). – Nota lepidopterologica 14 (3): 234-240.
RAINERI, V. (1992a): *Idaea boeklini* von Mentzer, 1990, syn. n. of *Idaea camparia* (Herrich-Schäffer, 1852) (Lepidoptera: Geometridae). – Nota lepidopterologica 15 (2): 152-157.
RAINERI, V. (1992b): Some considerations about *Casilda antophilaria* (Hübner, 1813) and *C. consecraria* (Staudinger, 1871) (Lepidoptera: Geometridae). – Atalanta 23 (3/4): 619-622.
RAINERI, V. (1992c): I Lepidotteri Geometridi dell'Isola di Capraia. – Riv. Mus. civ. Sci. nat. E. Caffi Bergamo 16: 143-148.
RAINERI, V. (1992d): I Geometridi raccolti in Spagna durante le campagne di ricerca dell'Istituto di Zoologia dell'Universita di Genova (Lepidoptera). – Boll. Soc. ent. Ital., Genova, 124 (1): 3-11.
RAINERI, V. (1992e): Aggiornamento sui Geometridi raccolti nella Penisola Iberica durante una campagna di ricerca dell'Istituto di Zoologia dell'Universita di Genova (Lepidoptera). – Mem. Soc. ent. Ital. 71 (2): 653-658.
RAINERI, V. (1995): Breve nota su alcuni interessanti Geometridi raccolti in Grecia (Lepidoptera, Geometridae). – Boll. Soc. ent. Ital., Genova, 126 (3): 277-284.
RAINERI, V. (1996): La collezione di Geometridi del conte Turati conservata al Museo Regionale di Scienze Naturali di Torino (Lepidoptera Geometridae). – Boll. Mus. reg. Sci. nat. Torino 14 (1): 179-247.
RÁKOSY, L. & C. WIESER (2000): Das Macin Gebirge (Rumänien, Nord-Dobrudscha). – Carinthia 190 (110): 7-116.
RÁKOSY, L. & W. WEBER (1986): Die Gross-Schmetterlinge von Sighisoara (Schassburg) und Umgebung, Siebenburgen, Rumanien (Lepidoptera). – Atalanta 16 (3/4): 315-392.
RAMBUR, M.P. (1833): Catalogue des Lépidoptères de l'île de Corse avec la désription et la figure des espèces inédites. – Annls Soc. ent. Fr. 2: 1-59, 2 pls.
RAPPAZ, R. (1979): Les Papillons du Valais. Macrolépidoptères. – Martigny, 377 pp..
RAUH, W. & K. SENGHAS (1976): Schmeil-Fitschen, Flora von Deutschland, 86th ed.. – Quelle & Meyer,

Heidelberg, 516 pp..
REBEL, H. & H. ZERNY (1934): Die Lepidopterenfauna Albaniens (mit Berücksichtigung der Nachbargebiete). Wissenschaftliche Ergebnisse der im Auftrage und Kosten der Akademie der Wissenschaften in Wien im Jahre 1918 entsendeten Expedition nach Nordalbanien. – Denkschr. Akad. Wiss. Wien, math.-nat. Kl. 103: 37-161.
REBEL, H. (1893): Beitrag zur Lepidopteren-Fauna Südtirols. – Verh. zool.-bot. Ges. Wien 42: 509-520.
REBEL, H. (1902): Lepidopteren aus Morea gesammelt von Herrn Martin Holtz im Jahre 1901. – Berl. ent. Z. 47 (1-2): 83-110.
REBEL, H. (1903a): Lepidopteren aus Morea. – Berl. ent. Z. 48: 243-249.
REBEL, H. (1903b): Studien über die Lepidopterenfauna der Balkanländer. I. Teil. Bulgarien und Ostrumelien. – Ann. k. k. Naturhist. Hofmus., Wien 18: 123-347 + 1 Taf.
REBEL, H. (1904): Studien über die Lepidopterenfauna der Balkanlander. II. Teil. Bosnien und Hercegovina. – Ann. k. k. Naturhist. Hofmus., Wien 19 (2-3): 97-377.
REBEL, H. (1905a): Lepidopteren aus Morea. II. Nachtrag. – Berl. ent. Z. 50: 291-314.
REBEL, H. (1905b): In: PENTHER, A. & E. ZEDERBAUER. Ergebnisse einer naturwissenschaftlichen Reise zum Erdschias-Dagh (Kleinasien). Lepidopteren. – Ann. k. k. Naturhist. Hofmus., Wien 20: 189-219.
REBEL, H. (1905c): Lepidopteren aus dem Gebiete des Triglav und der Crna Prst in Krain. – Jahresb. Wien. ent. Ver. 1905: 123-144.
REBEL, H. (1909-1912): Lepidopteren aus dem Gebiete des Monte Maggiore in Istrien. – Jahresb. Wien. ent. Ver. 21: 97-110, 22: 227-240, 23: 177-205.
REBEL, H. (1910a): Beitrag zur Lepidopterenfauna der Ionischen Inseln. – Verh. zool.-bot. Ver. Wien 60: 418-431.
REBEL, H. (1910b): Fr. Berge's Schmetterlingsbuch nach dem gegenwärtigen Stande der Lepidopterologie, 9. Auflage. – E. Schweizerbart'sche Verlagsbuchhandlung, Nägele & Dr. Sproesser, Stuttgart.
REBEL, H. (1911a): Versammlung am 3. Februar 1911. – Vehr. zool.-bot. Ges. 61: (79)-(90).
REBEL, H. (1911b): Versammlung am 3. November 1911. – Verh. zool.-bot. Ges. 61: (174)-(181).
REBEL, H. (1911c): Lepidopteren aus dem Gebiete des Triglav und der Crna Prst in Krain. III. Nachtrag. – Jahresb. Wien. ent. Ver. 21: 111-144.
REBEL, H. (1911d): Die Lepidopterenfauna von Herkulesbad und Orsova. – Ann. k. k. Naturhist. Hofmus., Wien 25: 253-430.
REBEL, H. (1911e): Die Lepidopterenfauna von Herkulesbad und Orsova. – Ann. k. k. Naturhist. Hofmus., Wien 25: 253-430.
REBEL, H. (1912): Nachtrag zur Lepidopterenfauna von Korfu. – Verh. zool.-bot. Ges. Wien 62: (12) - (15).
REBEL, H. (1913): Studien über die Lepidopterenfauna der Balkanländer. III. Teil. Sammelergebnisse aus Montenegro, Albanien, Mazedonien und Thrazien. – Ann. k. k. Naturhist. Hofmus., Wien 27: 281-334.
REBEL, H. (1914): Lepidopteren aus dem nordalbanisch-montenegrischen Grenzgebiet. – Sitzungsber. Akad. Wiss. Wien, math-nat. Kl., Abt. 1, 123: 1111-1128.
REBEL, H. (1916a): Zur Lepidopteren-Fauna der Insel Rhodus. – Jahresb. Wien. ent. Ver. 1916: 113-115.
REBEL, H. (1916b): Die Lepidopterenfauna Kretas. – Ann. k. k. Naturhist. Hofmus., Wien 30: 66-172.
REBEL, H. (1916c): Über die Lepidopterenfauna Cyperns. – Jahresb. Wien. ent. Ver. 1915: 93-109.
REBEL, H. (1917): Lepidopteren aus Neumontenegro. – Sitzungsber. Akad. Wiss. Wien, math.-nat. Kl., Abt. 1, 126: 765-813.
REBEL, H. (1918): Lepidopteren aus Mittelalbanien. – Z. Österr. Ent.-Ver. 3: 75-77, 85-88.
REBEL, H. (1924): [Geometriden aus Triest]. – Verh. zool.-bot. Ges. Wien (1923) 73: (6)-(7).
REBEL, H. (1926): Lepidopteren von den Balearen. – Dt. ent. Z. Iris 40 (2/3): 135-146.
REBEL, H. (1927): Beitrag zur Lepidopteren-Fauna der Insel Cypern. – Verh. zool.-bot. Ver. Wien 77: (58)-(63).
REBEL, H. (1928): Zur Lepidopteren-Fauna Cyperns V. – Verh. zool.-bot. Ver. Wien 78: (29)-(32).
REBEL, H. (1930): Über eine Lepidopterenausbeute von den Pityusen. – Verh. zool.-bot. Ges. Wien 80 (1-2): (1)-(5).
REBEL, H. (1934a): Lepidopteren von den Balearen und Pityusen. – Dt. ent. Z. Iris 48: 122-133.
REBEL, H. (1934b): Griechische Lepidopteren III. – Z. Österr. Ent.-Ver. 19: 63-66..
REBEL, H. (1935): Lepidopteren von den Ägäischen Inseln. – Sitzungsber. Akad. Wiss. Wien, Math.-naturw. Kl. 144: 243-262.
REBEL, H. (1938): Zur Lepidopterenfauna Kretas. – Dt. ent. Z. Iris 52: 30-36.
REBEL, H. (1939): Zur Lepidopterenfauna Cyperns. – Mitt. Münchn. ent. Ges. 29: 487-564.
REDONDO, V.M. & F.J. GASTÓN (1999): Los Geometridae (Lepidoptera) de Aragón (España). – Soc. ent. Aragon. (Monografias S.E.A.), Zaragoza, 3: 1-130.
REDONDO, V.M. & F.J. GASTÓN (2003): Designación de neotipo para *Idaea subcompleta* (Fernández, 1931),

especie válida de la Península Ibérica y su diferenciación anatómica con *Idaea nevadata* (Wehrli, 1926) (Lepidoptera, Geometridae, Sterrhinae). – Bol. S.E.A. 33: 245-252.

REDONDO, V.M. (1986): Introduccion al conocimiento de los Macrolepidópteros del Valle Medio del Ebro en Aragon (2a Parte). – Shilap Revta. lepid. 14 (54): 17-31.

REDONDO, V.M., BLASSCO-ZUMETA, J. & G.E. KING (2001): Macrolepidópteros de un sabinar de *Juniperus thurifera* L. en Los Monegros, Zaragoza, España (Insecta: Lepidoptera). – Shilap Revta. lepid. 29 (116): 323-370.

REDONDO, V.M., GASTÓN, F.J. & R. JIMENO (in prep.): Los Geometridae de la Peninsula Iberica e Islas Baleares (introduccion a su estudio).

REID, J. (1986): Geometridae. In: FRIEDRICH, E.: Breeding Butterflies and Moths, English Edition. - Harley Books, Colchester: 83-101.

REINHARDT, R., FRIEDEMANN, P. & U. EITSCHBERGER (1999): Fragmentarisches Verzeichnis der Schmetterlinge Europas und angrenzender Regionen mit einem vorläufigen Vorschlag zur Festlegung von Identifikationsnummern. Alphabetischer Gesamtindex. – Neue Ent. Nachr. 44 (1) und (2): 1.000 pp..

REIPRICH, A. & I. OKÁLI (1989): Dodatky k Prodromu Lepidopter Slovenska. 3. zväzok. – Biol. Práce, Bratislava, 141 pp..

REIPRICH, A. (1972): Verzeichnis der Falterarten aus dem Naturschutzgebiet Slovensky raj und der Umgebung von Spisska Nova Ves (Ergebnisse der Erforschung aus den Jahren 1945-1969). – Ent. problemy, Bratislava, 10: 5-104.

REIPRICH, A. (1977): Doplnky k Prodromu lepidopter Slovenska. – Ent. problemy, Bratislava, 14: 13-69.

REISSER, H. & F. WEISERT (1977): Eine neue *Sterrha* aus Spanien: *Sterrha deitanaria* n.sp. (Lepidoptera, Geometridae). – Zeits. Arb.Gem. österr. Ent. 28 (4): 106-108.

REISSER, H. (1926): Bericht über eine Sammelreise nach Corsica und Beitrag zur dortigen Fauna. – Verh. zool.-bot. Ver. Wien. 76: 1-25.

REISSER, H. (1927): Sammelergebnisse aus Andalusien mit spezieller Berücksichtigung der Sierra Nevada. – Verh. zool.-bot. Ges. Wien. 77 (2): (64)-(68).

REISSER, H. (1928): Lichtfang in der Sierra Nevada. – Verh. zool.-bot. Ges. Wien. 78: (8)-(13).

REISSER, H. (1929): Beschreibung der unbekannten ersten Stände einiger corsischer Geometriden. – Z. Österr. Ent.-Ver. 14: 3-6; 79-81.

REISSER, H. (1930): Herbstlicher Lichtfang von Lepidopteren in Corsica. – Verh. zool.-bot. Ges. Wien. 80 (1-2): 32-44.

REISSER, H. (1931): Acidaliinen (Lep., Geom.) aus der Sierra Nevada. – Int. ent. Zeits. 24 (44): 453-457.

REISSER, H. (1933a): Beitrag zur Lepidopterenfauna des Rifgebirges von Spanisch-Marokko, Geometridae. – EOS 9: 239-272.

REISSER, H. (1933b): Neue Heteroceren aus dem Grossen Atlas. – Z. Österr. Ent.-Ver. 18: 39-44, 49-52, 3 pls.

REISSER, H. (1933c): Sammelreise nach Korsika. – Verh. zool.-bot. Ges. Wien 83: (10)-(13).

REISSER, H. (1935): Neue Heteroceren aus der Sierra de Gredos. – Ent. Rundschau 53 (5): 77-80.

REISSER, H. (1936a): Über *Ptychopoda incisaria* Stgr.. – Z. Österr. Ent.-Ver. 19: 15-16.

REISSER, H. (1936b): Reise nach Sierra de Gredos, Hispania. – Verh. zool.-bot. Ges. Wien 85: 121-125.

REISSER, H. (1937). Über die ersten Stände einiger südeuropäischer Heteroceren. – Z. Österr. Ent.-Ver. 22: 72-78, 90-92.

REISSER, H. (1939a): Beiträge zur Kenntnis der Sterrhinae (Lep. Geom.) I. *Sterrha palaestinensis* (Stck.) *pelu-raria* subsp. nov.. – Z. Österr. Ent.-Ver. 24: 26-27.

REISSER, H. (1939b): Beiträge zur Kenntnis der Sterrhinae (Lep. Geom.) II. *Codonia (Cosymbia) ariadne* spec. nov.. – Z. Wien. Ent.-Ver. 24: 169-170.

REISSER, H. (1946): Lepidopteren von den Ägäischen Inseln. – Z. Wien. ent. Ges. 31: 44-49.

REISSER, H. (1949): Neue oder bemerkenswerte Falterfunde aus dem Burgenland und aus Niederösterreich. – Z. Wien. ent. Ges. 34 (10/12): 141-146.

REISSER, H. (1957): Beiträge zur Kenntnis der Sterrhinae (Lep. Geom.) III. Über *Problepsis ocellata* Friv. und ihre ersten Stände. – Z. Wien. ent. Ges. 42 (11): 161-172.

REISSER, H. (1958): Neue Heteroceren aus Kreta. – Z. Wien. ent. Ges. 43 (69, 7): 105-128.

REISSER, H. (1961): Beiträge zur Kenntnis der Sterrhinae (Lep. Geom.) IV. Über gleichgerichtete Formenbildungen bei vielen Arten. – Z. Wien. ent. Ges. 46: 173-189.

REISSER, H. (1962a): Beiträge zur Kenntnis der Sterrhinae (Lep. Geom.) V. Was ist *Sterrha dilutaria praeustaria* Lah. (Mann i.l.)? – Z. Wien. ent. Ges. 47: 21-24.

REISSER, H. (1962b): Weitere neue Heteroceren aus Kreta. – Z. Wien. ent. Ges. 47 (73, 12): 193-216.

REISSER, H. (1963): Beiträge zur Kenntnis der Sterrhinae (Lep. Geom.) VI. *Sterrha alicantaria* sp.n., eine neue Art der südspanischen Küstengebiete. – Z. Wien. ent. Ges. 48: 45-50.

REISSER, H. (1974): Zur Lepidopterenfauna Kretas: neue Zugänge zum Artenbestand. – Ann. Mus. Goulandris 2: 137-140.
REMM, H. & J. VIIDALEPP (1986): Catalogus Macrolepidopterorum Estoniae. [Eesti Suurliblikate Nimestik]. – Tartu. 34 pp..
RESSLER, W. (1928): Ein Beitrag zur Biologie von *Ac. nemoraria*. – Int. ent. Zeits. 22 (9): 97-100.
RETZLAFF, H., DUDLER, H., FINKE, C., PÄHLER, R., SCHNELL, K. & W. SCHULZE (1993): Zur Schmetterlingsfauna von Westfalen, Ergänzungen, Neu- und Wiederfunde. – Mitt. Arbgem. ostwestf.-lippischer Ent. 9 (2): 37-53.
REUTER, E. (1890): Länsi suoni Västra Finland 5.
REZBANYAI, L. (1974): Die Grossschmetterlingsfauna des Köszeger-Gebirges. – Folia ent. Hung. 27: 139-182.
REZBANYAI-RESER, L. & A. EXPÓSITO HERMÓSA (1992): *Idaea consanguiberica* n.sp. aus Spanien (Lepidoptera: Geometridae). – Ent. Z. 102 (15): 273-288.
REZBANYAI-RESER, L. (1985): Bestätigung des Vorkommens von *Idaea (Sterrha) obsoletaria* Rmbr. und *I. politata* Hbn. in der Südschweiz (Lep. Geometridae). – Mitt. ent. Ges. Basel 35 (3): 92-95.
REZBANYAI-RESER, L. (1987): *Idaea griseanova* sp.n., eine bisher verkannte Zwillingsart von *ruficostata* Zeller, 1849, aus dem Westmediterraneum (Lepidoptera, Geometridae). – Mitt. ent. Ges. Basel 37 (4): 141-182.
REZBANYAI-RESER, L. (1988a): Weitere Angaben zum Vorkommen von *Idaea (Sterrha) obsoletaria* Rmbr. in der Südschweiz (Lep., Geometridae). – Ent. Berichte Luzern 20: 131-132.
REZBANYAI-RESER, L. (1988b): *Idaea pallidata* D.Sch. und *subsericeata* Haw.: Berichtigung zu den Genitalabbildungen in Bleszynski 1960 und Forster & Wohlfahrt 1981 (Lep., Geometridae). – Ent. Berichte Luzern 20: 133-135.
REZBANYAI-RESER, L. (1989): *Idaea straminata* Bkh. (=*inornata* Haw.) und *deversaria* H.-Sch.: eine Berichtigung zu Bleszynski 1960 und Koch 1984 (Lepidoptera, Geometridae). – Ent. Zeits. 99 (3): 27-30.
REZBANYAI-RESER, L. (1990a): *Idaea (Sterrha) rubraria* Staudinger, 1871, neu für den Tessin und die Schweiz (Lepidoptera, Geometridae). – Mitt. ent. Ges. Basel 40 (1/2): 72-75.
REZBANYAI-RESER, L. (1990b): Zur Insektenfauna von Obergütsch (500-600 m), Stadt Luzern. – Ent. Berichte Luzern 24: 1-94.
REZBANYAI-RESER, L. (1990c): Zur Macrolepidopterenfauna der Insel Brissago, Kanton Tessin (Lepidoptera: "Macroheterocera" - "Nachtgrossfalter"). – Entomol. Ber. Luzern 23: 37-130.
REZBANYAI-RESER, L. (1991a): Zwolf Tage Lichtfallenfang in Calas de Mallorca (Malorca: Balearen, Spanien), Ende September 1980 (Macrolepidoptera). – Ent. Z. 101 (10): 173-192.
REZBANYAI-RESER, L. (1991b): Zur Insektenfauna des Kantons Schaffhausen (Osterfingen, Hakkau-Egg und Löhningen. II. Lepidoptera 1: 'Macroheterocera' ('Nachtgrossfalter'). – Entomol. Ber. Luzern 26: 21-124.
REZBANYAI-RESER, L. (1992): Zur Insektenfauna vom Rüss-Spitz (Kanton Zug) bei Maschwanden ZH. II. Lepidoptera 1: 'Macrolepidoptera' ('Grossschmetterlinge'). – Entomol. Ber. Luzern 27: 25-114.
REZBANYAI-RESER, L. (1993a): Elenco critico aggiornato dei Macrolepidotteri del cantone Ticino, Svizzera meridionale (Insecta, Lepidoptera). – Boll. Soc. Tic. Sci. Natur. (Lugano) 81 (1): 39-96.
REZBANYAI-RESER, L. (1993b): Anmerkungen zu ‚Aktuelle, kritische Liste der Makrolepidoptera des Tessin, Südschweiz' (Lepidoptera). – Ent. Berichte Luzern 30: 31-48.
REZBANYAI-RESER, L. (1993c): Zur Insektenfauna der Umgebung von Lauerz, Kanton Schwyz. 1. Sägel (455 m) und Schuttwald (480 m). II. Lepidoptera 1: 'Macrolepidoptera' ('Grossschmetterlinge') (Fortsetzung). – Ent. Ber. Luzern 29: 1-28.
REZBANYAI-RESER, L. (1997): Herbstliche Lichtfallen- und Tagfänge von Gross- Schmetterlingen an drei Orten in Mittelitalien (Marche, Toscana, Lazio) (Lepidoptera: Macrolepidoptera). – Ent. Z. 107 (9): 362-390.
REZBANYAI-RESER, L. (1998): Bemerkungen und Berichtigungen zur Macrolepidopterenfauna des schweizerischen Nationalparks und seiner Umgebung von Pictet, 1942 (Lepidoptera). – Mitt. Schweiz. Ent. Ges. 71 (1-2): 177-186.
RIBBE, C. (1912): Beiträge zu einer Lepidopteren-Fauna von Andalusien. Geometridae. – Dt. ent. Z. Iris 23: 299-343.
RIEMIS, A. (1994): Geometridae of Turkey 3. A provisional list of the Geometridae of Turkey (Lepidoptera). – Phegea 22 (1): 15-22.
RIEMIS, A. (1996): Geometridae of Turkey 5. Data on 38 new Turkish Geometridae species (Lepidoptera). – Phegea 24 (2): 83-87.
RIEMIS, A. (1998): Geometridae of Turkey 7. Additions and corrections to the provisional list of Geometridae of Turkey (Lepidoptera: Geometridae). – Phegea 26 (2): 76-80.
ROBINSON, G.S., ACKERY, P.R, KITCHING, I.J. BECCALONI, G.W. & L.M. HERNÁNDEZ (2001): Hostplants of the moth and butterfly caterpillars of the Oriental Region. – Southdene Sdn Bhd, Kuala Lumpur, 744 pp..
ROMANISZYN J. & F. SCHILLE (1929): Fauna motyli Polski. I. – Prace monogr. kom. fizjogr. 6. 516 pp..
ROMANO, F.P. & M. ROMANO (1995): Lepidoptera. In: MASSA, B.: Arthropoda di Lampedusa, Linosa e

Pantelleria (Canale di Sicilia, mar Mediterraneo). – Il Naturalista Siciliano 19 Suppl. (Ser. 4): 714-716.
ROMERA, L., CIFUENTES, J., EXPÓSITO, A. & E. CASCALES (2004): Nuevos datos sobre la fauna de Geometridae de Madrid (España) (Insecta: Lepidoptera). – Shilap Revta. lepid. 32 (125): 39-43.
ROMERA, L., CIFUENTES, J., FERNÁNDEZ, J. & J.L. VIEJO (2001): Contribución al conocimiento de los Geometridae de Madrid (España) II: Subfamilias Oenochrominae Guenée, 1757, Geometrinae Leach, 1815 y Sterrhinae Meyrick, 1892 (Insecta: Lepidoptera). – Shilap Revta. lepid. 29 (114): 93-114.
RONDOU, J.P. (1934): Catalogue des Lépidoptères des Pyrenées. Geometridae. – Ann. Soc. ent. France 103 (3-4): 257-320.
RONKAY, L. & C. SZABOKY (1981): Investigations on the Lepidoptera fauna of the Zemplen Mts. (NE Hungary). I. The valley of Kemence stream. – Folia ent. Hung. 42 (2): 167-184.
ROTHSCHILD, W. (1914): A preliminary account of the Lepidopterous fauna of Guelt-es-Stel, Central Algeria. – Novit. Zool. 21: 299-357.
ROTUNDO, G., PARENZAN, P. & A. DE CRISTOFARO (1999): Contributo alla conoscenza della fauna della Riserva Naturale di „Collemeluccio" (Molise): Lepidoptera, Heterocera. – Phytophaga 9: 39-62.
ROUGEMONT, DE, F. (1903): Catalogue des Lépidoptères du Jura neuchâtelois. – Bull. Soc. neuchat. Sci. nat. 29-31: 1-366.
RÜCKER, E. (1921): Das Ei von *Acidalia corrivalaria* Kretschmar. – Dt. ent. Zeits. (1920): 175-176.
RUHLAND (1912): Hat *Acidalia trigeminata* eine 2. Generation? – Int. ent. Zeits. (Guben) 6 (34): 238-239.
RUNGS, C.E.E. & H. POWEL[L] (1941): Notes de Lépidoptérologie marocaine (9e note). Additions a la faune des Lépidoptères du Maroc. – Bull. Soc. Sci. nat. Maroc 20: 106-112.
RUNGS, C.E.E. (1946): Notes de Lépidoptérologie marocaine (XIII). Nouvelles additions a la faune des Lépidoptères du Maroc. – Bull. Soc. Sci. nat. Maroc 24: 111-119.
RUNGS, C.E.E. (1981): Catalogue raisonné des Lépidoptères du Maroc. Inventaire faunistique et observations écologiques. Tome II. – Trav. Inst. Sci. Série Zool., Rabat 40: 1-365.
RUNGS, C.E.E. (1988): Liste-inventaire systématique et synonymique des Lépidoptères de Corse. Contribution à la connaissance de la faune entomologique de la Corse: Lepidoptera. Liste-inventaire des espèces de l'île. – Alexanor 15 (Suppl.): [1]-[86].
RUNGS, C.E.E. (1991): Additions à la liste des Lépidoptères de Corse. – Alexanor 17 (2): 120.
RYRHOLM, N. (1989): The influence of the climatic energy balance on living conditions and distribution patterns of *Idaea* spp. (Lepidoptera: Geometridae): An expansion of the species energy theory. – Acta universitatis Upsaliensis 208: 109 pp.
SALKELD, E.H. (1983): A catalogue of the eggs of some Canadian Geometridae (Lepidoptera), with comments. – Mem. Ent. Soc. Canada, Nr. 126, Ottawa, 271 pp..
SÄLZL, M. (1935): Spanner (XIX. Geometridae). In: METSCHL, C. & M. SÄLZL. Die Schmetterlinge der Regensburger Umgebung. I. Teil Grossschmetterlinge. – Dt. ent. Z. Iris 49: 58-132.
SAMMUT, P. (1983): Die Geometriden der Maltesischen Inseln (Lepidoptera, Geometridae). – Neue ent. Nachr. 6: 61-64.
SAMMUT, P.M. (1984): A Systematic and Synonymic List of the Lepidoptera of the Maltese Islands. – Neue ent. Nachr. 13: 1-124.
SAMMUT, P.M. (2000): Il-Lepidoptera. – Pubblikazzjonijiet Indipendenza, Malta, 245 pp..
SANCHEZ EGUIALDE, D. & J. CIFUENTES (1990): Contribution al conocimiento de los Geometroidea de Tudela y la Ribera Sur de Navarra. Otras citas de Geometroidea en la Peninsula Iberica. – Shilap Revta. lepid. 18 (69): 73-82.
SANNINO, L. & A. BALBIANI (1984): Su un attacco di *Scopula ochroleucaria* H.S. (Lepidoptera, Geometridae) al tabacco in Campania. – La difesa delle piante 5-6: 289-300.
SANNINO, L. & B. ESPINOSA (1999): Aspetti morfologici ed etologici di *Scopula turbidaria* (Lepidoptera, Geometridae). – Fragmenta Entom. 31 (2): 377-395.
SATO, R. (2001): Type specimens of the Geometridae (Lepidoptera) described by Matsumura in the Hokkaido University collection, Japan. – Insecta Matsumurana N.S. 58: 115-138.
SATPATHI, C.R. (1995): Insects infesting groundnut (*Arachis hypogaea*) in West Bengal. – Annals of Entomology (Dehra Dun) 13: 97-98.
SATTLER, K. & W.G. TREMEWAN (1973): The entomological Publications of Pierre Millière (1811-1887). – Bull. Brit. Mus. (Nat. Hist.), Hist. ser. 4 (4): 221-280; 3 pls.
SATTLER, K. & W.G. TREMEWAN (1984): The Lepidoptera names of Denis & Schiffermüller - a case for stability. – Nota lepidopterologica 7: 282-285.
SATTLER, K. (1969): Das „Wiener Verzeichnis" von 1775. – Z. Wien. ent. Ges. 54: 2-7, 3 pl..
SAVENKOV, N., ŠULCS, I., KERPPOLA, S. & L. HULDÉN (1996): Checklist of Latvian Lepidoptera - Latvijas Taurizu Katalogs. – Baptria 21 (3a): 1-71.

SAVENKOVS, N.V. (1984): Nekotorye dannye po faune i ekologii cheshuekrylych Latvii. [Some data on fauna and ecology of Lepidoptera of the Latvia]. – Latv. Ent. 27: 46-63.
SCALERCIO, S. (1999): Macrolepidotteri notturni catturati nel Vincese (Toscana-Italia) (Lepidoptera). – Mem. Soc. ent. Ital. 77: 311-316.
SCALERCIO, S. & M. INFUSINO (2003): I Macrolepidotteri di fosso Scuotrapiti, lago dell'Angitola (Calabria, Italia meridionale) (Lepidoptera). – Phytophaga 13: 25-52.
SCHAWERDA, K. (1913): Siebenter Nachtrag zur Lepidopterenfauna Bosniens und der Herzegowina. – Verh. zool.-bot. Ges. Wien. 63 (3-4): 141-178.
SCHAWERDA, K. (1914): Achter Nachtrag zur Lepidopterenfauna Bosniens und der Herzegowina. – Verh. zool.-bot. Ges. Wien. 64: 349-378.
SCHAWERDA, K. (1921): Beitrag zur Lepidopteren-Fauna der kroatischen Küste und Neubeschreibungen. – Dt. ent. Z. Iris 1921: 111-129.
SCHAWERDA, K. (1941): Sammelergebnis aus Warasdin Teplitz. – Ent. Zeitschr. 54: 261-264.
SCHEURINGER, E. (1972): Die Macrolepidopteren-Fauna des Schnalstales (Vintschgau-Südtirol). – Studi Trent. Sc. Nat. B 49: 229-448. Trento.
SCHEURINGER, E. (1983): Die Macrolepidopteren-Fauna des Schnalstales (Vintschgau-Südtirol), 1. Nachtrag. – Nachr.bl. Bayer. Ent. 32 (3): 65-74.
SCHMIDLIN, A. (1964): Übersicht über die europäischen Arten der Familie Geometridae (Lep.). – Mitt. ent. Ges. Basel, N. F. 14 (4/5): 77-137.
SCHMIDT-KOEHL, W. (1966): Neue Beiträge zur Macrolepidopteren-Fauna Oberandalusiens aus den Provinzen Granada, Almeria und Malaga (Südspanien). – Ent. Z. 76 (1/2): 1-15.
SCHMIDT-KOEHL, W. (1979): Die Grossschmetterlinge des Saarlandes (Insecta, Lepidoptera). Noctuidae, Geometridae. Monographischer Katalog. – Abh. Arbeitsgem. tier- und pflanzengeogr. Heimatforschung im Saarland 9. 242 pp..
SCHMITT, T, SUMMKELLER, R., WERNO, A. & N. ZAHM (1991): Neue und wiederentdeckte Arten für die Saarländische Lepidopterenfauna. – Faun.-flor. Notizen Saarland 22 (3-4): 100-137.
SCHNACK, K. (1985): Katalog over de danske Sommerfugle. – Ent. Meddr. 52 (2-3): 1-163.
SCHNEIDER, C. ([1936]-1939): Beiträge zur Biologie einiger Geometriden. – Int. ent. Zeits..
SCHNEIDER, C. (1929): Beiträge zur Biologie einiger Geometriden. V. *Rhodostrophia vibicaria* Cl.. – Ent. Anzeiger 9: 443-444.
SCHNEIDER, C. (1932): *Acidalia immorata* L. (Lep. Geometr.). – Ent. Z. Frankf. a.M. 46: 149-150.
SCHNEIDER, C. (1934): Die in Württemberg vorkommenden Arten der Geometridengattungen *Brephos* Zink. – *Timandra* Dup. – Ent. Z. Frankf. a.M. 48 (4): 29-32.
SCHNEIDER, C. (1938): Beitrag zur Biologie der Geometriden. – Int. ent. Zeits. 17 (20): 159-160.
SCHRANK, F. (1801-1802): Fauna Boica. – Nürnberg, 2 vols.
SCHROTH, M. (1984): Die Makrolepidopteren aus der Umgebung von Hanau am Mai (Hessen). – Nachr. ent. Ver. Apollo, Frankf. a.M., Suppl. 3: 1-83.
SCHROTH, M. (1985): 1.Nachtrag zu den "Makrolepidopteren aus der Umgebung von Hanau am Main (Hessen). – Nachr. ent. Ver. Apollo, Frankf. a.M., N.F., 6 (1): 21-33.
SCHULTZ, V.G.M. (1930a): Lepidopterologische Studien. 4. Einiges über Lebensgeschichte und Zucht von *Acidalia muricata* Hufn. – Int. ent. Zeits. 23 (45): 513-518.
SCHULTZ, V.G.M. (1930b): Lepidopterologische Studien. 5. Ein Einwand gegen die auf „Relikt"arten beruhende Annahme einer postglazialen Steppenzeit. – Int. ent. Zeits. 23 (41): 465-468.
SCHULTZ, V.G.M. (1930c): Beitrag zur Kenntnis der Lippischen Großschmetterlingsfauna. – Lage (Lippe).
SCHULTZ, V.G.M. (1930d): Lepidopterologische Studien. 3. *Ephyra linearia* Hb. gen. aest. *strabonaria* Z. auch in der Frühjahresgeneration! – Int. ent. Zeits. 23 (39): 445-446.
SCHULTZ, V.G.M. (1953): Neue Beiträge zur Schmetterlingskunde. Über die Raupe, die Puppe und eine anormale Falterform von *Scopula ternata* Schrk. (*Acidalia fumata* Steph.) (Lep. Geometridae). – Z. Wien. ent. Ges. 38: 95-99.
SCHWINGENSCHUSS, L. & F. WAGNER (1926): Beitrag zur Macro-Lepidopteren-Fauna Süddalmatiens insbesondere der Umgebung Gravosas, Geometridae. – Z. Österr. Ent.-Ver. 11: 35-53.
SCHWINGENSCHUSS, L. & F. WAGNER. (1927): Beitrag zur Macro-Lepidopteren-Fauna Süddalmatiens insbesondere der Umgebung Gravosa's. – Z. Österr. Ent.-Ver. 12 (5): 45-50; (6): 62-64; (7): 68-72; (8): 73-75.
SCHWINGENSCHUSS, L. (1930): Lepidopterologische Ergebnisse einer Herbstreise nach Andalusien. – Verh. zool.-bot. Ges. Wien. 80 (1-2): 1-31.
SCHWINGENSCHUSS, L. (1938): Beitrag zur Lepidopterenfauna Bithyniens, besonders von Boli (jetzt Bolu). – Ent. Rdsch. 55: 141-147; 158-164; 173-177; 181-184; 199-202; 223-226; 299-300; 337-340; 411-412; 454-457.

SCHWINGENSCHUSS, L. (1939): Beitrag zur Lepidopterenfauna von Iran (Persien). – Ent. Zeitschr. 53 (25) (Ent. Rdsch. 38): 201-210.

SCHWINGENSCHUSS, L. (1940): Beitrag zur Fauna von Bithynien, besonders von Boli (jetzt Bolu). – Ent. Zeits. 53 (32): 281-283.

SCHWINGENSCHUSS, L. (1942): Eine Falterausbeute aus Sizilien. I Teil. Macrolepidoptera. – Z. Wiener Ent.-Ver. 27: 177-184, 222-224, 244-249.

SCHWINGENSCHUSS, L. (1953): Beitrag zur Lepidopterenfauna von Niederösterreich: St. Peter i.d.Au, Seitenstetten und Umgebung, Geometridae. – Z. Wien. ent. Ges. 38: 217-223, 251-255, 282-285.

SCHWINGENSCHUSS, L. (1962): Nachträge, Ergänzungen und Berichtungen zur "Lepidopterenfauna von Albarracin in Aragonien" von H. Zerny. – Z. Wien. ent. Ges. 47 (Bd 73, 1): 4-11.

SCIARRETTA, A & V. RAINERI (2001): Nuove conoscenze faunistiche sui Geometridi del Molise (Italia centrale) (Lepidoptera Geometridae). – Boll. Zool. Agr. Bachic. Ser. II, 33 (1): 19-41.

SCOBLE, M.J. (1999): Geometrid Moths of the World: a catalogue (Lepidoptera, Geometridae). Vol. 1 and 2. CSIRO Publishing and Apollo Books, Stenstrup. 1016 pp. (+ 129 pp. of Index).

SCOBLE, M.J., GASTON, K.J. & A. CROOK (1995): Using taxonomic data to estimate species richness in Geometridae. – Journal Lep. Soc. 49 (2): 136-147.

SCOPOLI, G.A. (1763): Entomologia Carniolica exhibens insecta Carnioliae indigena et distributa in ordines, genera, species, varietates. Methodo Linneana. – Vindobonae, 422 pp., 43 pls.

SEEBOLD, T. (1879): Catàlogo de los Lepidópteros observados en los alrededores de Bilbao. – Anales de Hist. Nat. 8, Geometridae: 107-113.

SEGERER, A.H. (1997): Verifikation älterer und fraglicher Regensburger Lepidopterenmeldungen (Insecta: Lepidoptera). – Beitr. bayer. Entomofaunistik 2: 177-265.

SEPPÄNEN, E.J. (1954): Suomen Suurperhostoukkien Ravintokasvit (The food-plants of the larvae of the Macrolepidoptera of Finland). – Suomen Elaimet (Animalia Fennica) 8.

SEPPÄNEN, E.J. (1970): Suurperhostoukkien Ravintokasvit (The food-plants of the larvae of the Macrolepidoptera of Finland). – Suomen Elaimet (Animalia Fennica) 14.

SEVEN, S. (1991): Trakya Lepidoptera faunasi üzerine bibliyo?rafik ara?tirmalar [Bibliographical Studies on the Lepidoptera Fauna of Turkey-in-Europe]. – Priamus 6 (1/2): 1-95. (in Turkish).

SHELJUZHKO, L. (1955): Neue und wenig bekannte Noctuiden und Geometriden der Zoologischen Staatssammlung in München. – Mitt. Münchn. ent. Ges. 44/45: 277-292.

SHENEFELT, R.D. (1972): Hymenopterorum Catalogus. Pars 7. Braconidae 4 (Microgasterinae, Apanteles). Uitgeverij Dr. W.Junk N.V. 's-Gravenhage, pp. 429-668.

SHIN, Y.-H. (1996): Synonymic List and Distribution of the Geometridae of Korea (Lepidoptera). – Center for Insect Systematics, Korea, Insecta Koreana, 153 pp..

SICH, A. (1904): Notes on the ova and early stage of the larva of *Acidalia emutaria*. – The Entomologist 37: 108-109.

SIHVONEN, P. & L. KAILA (2004): Phylogeny and tribal classification of Sterrhinae with emphasis on delimiting Scopulini (Lepidoptera: Geometridae). – Syst. Ent. R. Ent. Soc. London 29: 324-358.

SIHVONEN, P. & M. SILJANDER (2003): Diversity of Scopulini moths on a worldwide scale: spatial and temporal patterns (Lepidoptera: Geometridae, Sterrhinae). – in SIHVONEN, P.: Diversity and classification of the Scopulini (Lepidoptera: Geometridae, Sterrhinae). – Helsinki, Edta Prima Oy, 158 pp. (academic dissertation, University of Helsinki, ISBN 952-91-6385-1).

SIHVONEN, P. (2001a): Everted vesicae of the *Timandra griseata* group: methodology and differential features (Geometridae, Sterrhinae). – Nota lepidopterologica 24 (3): 57-64.

SIHVONEN, P. (2001b): Revision of the *Scopula cajanderi* (Lepidoptera: Geometridae, Sterrhinae) species group with description of a new species. – Canadian Entomologist 133: 467-486.

SIHVONEN, P. (2003a): Diversity and classification of the Scopulini (Lepidoptera: Geometridae, Sterrhinae). – Helsinki, Edta Prima Oy, 158 pp. (academic dissertation, University of Helsinki, ISBN 952-91-6385-1).

SIHVONEN, P. (2003b): Phylogeny and classification of the Scopulini moths (Lepidoptera: Geometridae, Sterrhinae). – in SIHVONEN, P.: Diversity and classification of the Scopulini (Lepidoptera: Geometridae, Sterrhinae). – Helsinki, Edta Prima Oy, 158 pp. (academic dissertation, University of Helsinki, ISBN 952-91-6385-1).

SILBERNAGEL, A. (1944): Die Schmetterlinge der Ochrid-Gegend in Mazedonien. II Nachtrag und Berichtigungen. – Z. Wien. ent. Ges. 29: 29-187 (interrupted pagination).

SILVA CRUZ, DA, M.A. & J.T. WATTISON (1934-1935). Heteroceros de Portugal. – Mem. e Est. Mus. Zool. Univ. Coimbra, Ser. 1, 78 (1934), 39 pp.; 87 (1935), 104 pp..

SILVA CRUZ, DA, M.A. & T. GONÇALVES (1966): Notas Lepidopterologicas. V. Contribuicao sistematica para o inventario dos Heteroceros portugueses. – Mem. Est. Mus. Univ. de Coimbra 298: 1-8.

Silva Cruz, da, M.A. & T. Gonçalves (1974): Macrolepidópteros da Região do Alto Douro. – Inst. zool. "Dr. Augusto Nobre", Facul. Ciênc. do Porto 127: 9-50, 1 map.

Silva Cruz, da, M.A. & T. Gonçalves (1977): Catálogo sistemático dos Macrolepidópteros de Portugal. – Inst. zool. "Dr. Augusto Nobre", Fac. Cienc. do Porto 133: 1-56.

Silva Cruz, da, M.A. (1967): Lepidópteros da Região de Abrantes. – Inst. zool. "Dr. Augusto Nobre", Facul. Ciênc. do Porto 99: 9-35.

Singh, B. (1951): Immature stages of Indian Lepidoptera, No. 8. Geometridae. – Ind. For. Rec. (n.s.) 8 (7): 61-158.

Skala, H. (1910a): Diverses. – Ent. Jahrbuch 19: 132-134.

Skala, H. (1910b): Lepidopterologische Beiträge. – Int. ent. Zeits. (Guben) 3 (41): 223-224.

Skinner, B. (1984): Colour Identification Guide to Moths of the British Isles. (Macrolepidoptera). 267 pp., 42 pls. Viking, London.

Skou, P. (1984): Nordens målere. – Fauna bøger & Apollo bøger. København & Svenborg. 332 pp..

Skou, P. (1986): The Geometroid Moths of North Europe (Lepidoptera: Drepanidae and Geometridae). – Entomonograph 6: 1-348, pls. 1-24.

Skule, B. & P. Svendsen (1980): Indflyvningen af *Rhodometra sacraria* (Linnaeus, 1767) i Nordeuropa i 1983. – Lepidoptera, København 4 (8): 285-295.

Skule, B. (1980): *Rhodometra sacraria* L. Lidt om artens biologi og udbredelse, samt om den anden europaeiske *Rhodometra*-art: *Rhodometra anthophilaria* Hb. – Lepidoptera, København 3 (9): 261-268.

Sliwinski, Z., Wiackowski, S. & B. Marciniak. (1991): Motyle (Lepidoptera) Swietokrzyskiego Parku Narodowego. – Fragm. Faun., Warszawa 35 (8): 123-145.

Sobczyk, T. & J. Gelbrecht (2004): Zur Arealregression der an Thymian (*Thymus*) gebundenen Arten *Scopula decorata* ([Denis & Schiffermüller], 1775) und *Pseudophilotes vicrama* (Moore, 1864) in Deutschland (Lepidoptera, Geometridae et Lycaenidae). – Märkische Ent. Nachr. 6 (1): 1-16.

Söderman, G., Lundsten, K.-E. & R. Leinonen (1998): Luoteis-Venäjän yöperhosseurannan tulokset 1995-1997. [Results from the Moth Monitoring Scheme in Northwestern Russia 1995-97]. – Baptria 23 (4): 219-230.

Soffner, J. (1955): *Sterrha (Ptychopoda) eugeniata* Mill.. – Z. Wien. ent. Ges. 40 (Bd 66) (10): 291-293.

Sohn-Rethel, O. (1929): Beiträge zur Heteroceren-Fauna Italiens. – Dt. ent. Z. Iris 43: 1-23.

Sommerer, M. (2002): Opinion. To agree or not to agree – the question of gender agreement in the International Code of Zoological Nomenclature. – Nota lepidopterologica 25 (2/3): 191-204.

Sorhagen, L. (1881): Aus meinem entomologischen Tagebuche. – Berl. Ent. Zeits. 25 (1-2): 17-21.

Sotavalta, O. (1995): Itä-Fennoskandian mittareiden (Geometridae) levinneisyys. [Distribution atlas of Macrolepidoptera: Geometridae of eastern Fennoscandia]. – Baptria 20 (1a): 1-64.

South, R. (1961): The Moths of the British Isles, Vol. 2. – Warne, London, 379 pp..

Spada, L. (1893): Contribuzione alla Fauna marchegiana. I Lepidotteri finora trovati nel territorio di Osimo. – Il Naturalista Siciliano 12 (3/10): 53 pp..

Speiser, P. (1989): Die Schmetterlingsfauna der Provinzen Ost- und Westpreussen. – Beitr. Naturk. Preussens, Königsberg 9: 1-148.

Spuler, A. (1903-1910): Die Schmetterlinge Europas, 4 volumes. – Schweizerbart'sche Verlagsbuchhandlung, Stuttgart, vol. 2: 523 pp, vol. 3: 91 pl..

Stamm, K. (1981): Prodromus der Lepidopteren-Fauna der Rheinlande und Westfalens, 229 pp..

Stanescu, M. & A. Hausmann (2002): On the presence of *Idaea spissilimbaria* (Mabille, 1888) (Lepidoptera: Geometridae: Sterrhinae) in Romania. – Trav. Mus. Hist. nat. Grigore Antipa 44: 249-251.

Stanescu, M. & D. Rusti (1997): New data concerning the Lepidoptera (Insecta) of Maramures (Romania). – Trav. Mus. Hist. nat. 'Gr. Antipa' 37: 89-111.

Stanescu, M. (1995): The catalogue of "Ludovic Beregszaszy" Lepidopteran collection (Insecta: Lepidoptera). – Trav. Mus. Hist. nat. 'Gr. Antipa' 35: 221-346.

Stanescu, M. (1997a): New data concerning the Lepidoptera fauna of southern Dobrogea (Romania). – Trav. Mus. Natl. Hist. Nat. 'Gr. Antipa' 39: 91-107.

Stanescu, M. (1997b): The updated systematic and synonymic list of the Geometridae (Insecta: Lepidoptera) of Romania. – Trav. Mus. Hist. nat. 'Gr. Antipa' 39: 109-160.

Stange, G. (1881): Lepidopterologische Bemerkungen. – Stettin. ent. Ztg. 42: 113-114.

Starnecker, G. (2001): Die Gürtelpuppe von *Cyclophora linearia*. In: Ebert, G. (ed.) (2001): Die Schmetterlinge Baden-Württembergs, Band 8: Nachtfalter VI: 78-79.

Staude, H.S. & A.I. Curle (1997): A classification of visual-signals emanating from the wings of African Lepidoptera. – Metamorphosis, Supplement 3: 156-188.

Stauder, H. (1924): Edelwildjagden, Beiträge zur Sammeltechnik und Biologie begehrter Arten und Formen

von Lepidopteren; Beschreibung berühmter Fundplätze. - Ent. Anz. 4 (14): 127-128.
STAUDINGER, O. & H. REBEL (1901): Catalog der Lepidopteren des Palaearktischen Faunengebietes. Teil I. Verlag Friedländer & Sohn, Berlin, 411 pp..
STAUDINGER, O. (1861): Catalog der Lepidopteren Europas und der angrenzenden Länder. - Dresden.
STAUDINGER, O. (1870): Beitrag zur Lepidopterenfauna Griechenlands. - Horae Soc. ent. Ross. 7: 3-304.
STAUDINGER, O. (1871): Macrolepidoptera. In: STAUDINGER, O. & M. WOCKE: Katalog der Lepidopteren des europäischen Faunengebietes. Dresden.
STAUDINGER, O. (1879): Lepidopteren-Fauna Kleinasiens. - Horae Soc. ent. Ross. 14: 176-482.
STAUDINGER, O. (1892a) Neue Arten und Varietäten von Lepidopteren des palaearctischen Faunengebietes. - Dt. ent. Z. Iris 4: 224-339.
STAUDINGER, O. (1892b) Neue Arten und Varietäten von paläarktischen Geometriden aus meiner Sammlung. - Dt. ent. Z. Iris 5: 141-260.
STAUDINGER, O. (1894): Neue Lepidopterenarten und Varietäten aus dem paläarktischen Faunengebiet. - Dt. ent. Z. Iris 7: 241-296.
STAUDINGER, O. (1895): Neue paläarktische Lepidopteren. - Dt. ent. Z. Iris 8: 288-299.
STAUDINGER, O. (1897-1898): Neue Lepidopteren aus Palästina. - Dt. ent. Z. Iris 10: 271-319.
STAUDINGER, O. (1899): Neue paläarktische Lepidopteren. - Dt. ent. Z. Iris 12: 156-163.
STAUDINGER, O. (1900): Neue Lepidopteren des paläarktischen Faunengebietes. - Dt. ent. Z. Iris 12: 352-403.
STEINER, A. (2001): Die Gattung *Cyclophora*. In: EBERT, G. (ed.) (2001): Die Schmetterlinge Baden-Württembergs, Band 8: Nachtfalter VI: 74-101.
STEKOLNIKOV, A.A. (1993): Pjadenitzy (Lepidoptera, Geometridae) zapovednoj lesostepnoj dubravy "Les na Vorskle". [The Geometroid Moths (Lepidoptera) of the reserve oak-grove "Forest on Worskla"]. - Vestn. St.-Petersburg Gos. Univ., Serie 3 (Biologia) 1 (3): 22-29.
STEKOLNIKOV, A.A. (1996): [Keys to species of some genera in subfamily Sterrhinae (Lepidoptera, Geometridae) from European parts of Russia] (in Russian). - Vestn. St. Petersb. Univ. 3 (4): 3-40.
STEPHENS, J.F. (1927-1835): Illustrations of British Entomology (Haustellata), vol. 1-4.
STERNECK, J. (1932/33): Studien über Acidaliinae (Geometr.). I. - Z. Österr. Ent.-Ver. 17 (9/10): 67-71; 17 (11): 77-80; 17 (12): 82-84; 18 (1): 1-6; 18 (2): 9-15.
STERNECK, J. (1933): Studien über Acidaliinae (Geometr.). II. - Z. Österr. Ent.-Ver. 18: 63-68, pl..
STERNECK, J. (1940): Versuch einer Darstellung der systematischen Beziehungen bei den palaearktischen Sterrhinae (Acidaliinae), Studien über Acidaliinae (Sterrhinae) VII, I. Teil: Die Gattung *Sterrha* und deren nächste Verwandte. - Z. Wien. Ent.-Ver. 25: 6-17; 25-36; 56-59; 77-79; 98-107; 126-128; 136-142; 152-159; 161-176, pl. 1-10.
STERNECK, J. (1940/1941): Versuch einer Darstellung der systematischen Beziehungen bei den palaearktischen Sterrhinae (Acidaliinae), Studien über Acidaliinae (Sterrhinae) VIII, II. Teil: Die Gattung *Scopula* und deren nahe Verwandte. - Z. Wien. Ent.-Ver. 25 (1940): 200-208, 214-218, pl. 23-28; - 26 (1941): 17-31, 41-55, 88-96, 105-116.
STERNECK, J. (1941): Versuch einer Darstellung der systematischen Beziehungen bei den palaearktischen Sterrhinae (Acidaliinae). Studien über Acidaliinae (Sterrhinae) IX, III.-VI. Teil. - Z. Wien. Ent.-Ver. 26: 150-159; 176-183; 191-198; 211-216; 217-222; 222-230; 248-262, pl. 12-15.
STERZL, O. (1967): Prodromus der Schmetterlingsfauna von Niederösterreich (I. Teil der 2. Auflage). - Verh. zool.-bot. Ver. Wien 107: 75-193.
SUDHAUS W. & K. REHFELD (1992): Einführung in die Phylogenetik und Systematik. - Gustav Fischer Verlag.
ŠULCS, A. & I. ŠULCS (1978): Neue und wenig bekannte Arten der Lepidopteren-Fauna Lettlands. 7. Mitteilung. - Notulae Ent. 58 (4): 141-150.
ŠULCS, A. & I. ŠULCS (1981): Neue und wenig bekannte Arten der Lepidopteren-Fauna Lettlands. 8. Mitteilung. - Notulae Ent. 61 (2): 91-102.
ŠULCS, A. & J. VIIDALEPP (1972): Verbreitung der Gross-Schmetterlinge im Balticum. IV. Spanner (Geometridae). - Dt. Ent. Z., N.F. 19: 151-209.
ŠULCS, A. (1973): Neue und wenig bekannte Arten der Lepidopteren-Fauna Lettlands. 5. Mitteilung. - Ann. Ent. Fenn. 39 (1): 1-16.
ŠULCS, A., VIIDALEPP, J. & P. IVINSKIS (1981): 1. Nachtrag zur Verbreitung der Gross-Schmetterlinge im Balticum (Lepidoptera). - Dt. ent. Z., N.F. 28 (1-3): 123-146.
SUOMALAINEN, E., KAISILA, J. & K. MIKKOLA (1980): Noteworthy records of Finnish Lepidoptera 1955-1974. I. Hesperioidea, Papilionoidea, Bombycoidea and Geometroidea. - Notulae Ent. 60 (1): 49-61.
SUTTON, S.L. & H.E. BEAUMONT (1989): Butterflies and Moths of Yorkshire. - Yorkshire Naturalist's Union, Doncaster.
SUTTON, S.L. (1963): South Caspian insect fauna 1961. I. Systematic List of Lepidoptera with notes. - Ann.

& Mag. Nat. Hist., London, Ser. 13a 6 (66): 353-374.
SVENDSEN, P. & E. HAURITZ (1991): En ny aberration af *Scopula immorata* L. – Lepidoptera, København 6 (2): 54-55.
SVENDSEN, P. (ed.) & M. FIBIGER (1992): The Distribution of European Macrolepidoptera. – European Faunistical Press, Copenhagen, 1992.
SVENSSON, I. (1967): *Scopula aequicerata* Träff not a separate species (Lep.). – Opusc. Ent. 32 (3): 252-254.
SVENSSON, I. (1993): Fjärilkalender, Lepidoptera-calendar. – Författere Ingvar Svensson, Kristianstad (Sweden), 110 pp..
SVENSSON, I., ELMQUIST, H., GUSTAFSSON, B., HELLBERG, H., IMBY, L. & G. PALMQVIST (1994): Catalogus Lepidopterorum Sueciae. Code list L1. – Stockholm (Naturhistoriska Riksmuseet & Entomologiska Föreningen), 302 pp..
SVIRIDOV, A.V., ANTONOVA, E.M., BLINUSHOV, A.E. & O.M. BUTENKO (1998): Vysshie cheshuekrylye Okskogo zapovednika (annotirovannyj spisok vidov). – Flora i Fauna zapovednikov 70, Moscow, 1998, pp. 1-40.
SZÉKELY, L. (1989): [Data to the knowledge of Macrolepidoptera from Marosvasarhely (Central Transylvania) and its surroundings] Adatok Marosvasarhely (Kozep-Erdely) videke nagylepkefaunajanak ismeretehez (Lepidoptera: Macrolepidoptera). – Folia ent. Hung. 50: 137-145 (in Hungarian).
SZEOKE, K. (1987): Results of the investigations on the Lepidoptera-fauna of the eastern Bakony Mts., Hungary. – Folia Mus. Hist.-Nat. Bakonyiensis 6: 119-122.
TAUTEL, C. (1993): Aspect de la fauna des papillons en Auvergne: la région du Livradois (Lepidoptera). – Ent. gall. 4 (2-3): 90-96.
TEOBALDELLI, A. (1976): I Macrolepidotteri del Maceratese e dei Monti Sibillini (Apennino Umbro-Marchigiano) (Primo contributo alla conoscenza dei Lepidotteri delle Marche). – Note ed app. sperim. ent. agr. Perugia 16: 81-346.
TEOBALDELLI, A. (1979): Lépidoptères capturés en Val d'Aoste. – Alexanor 11 (4): 145-152.
TEOBALDELLI, A. (1994): Nuovi rinvenimenti di Macrolepidotteri nel territorio Marchigiano (Lepidoptera). – Boll. Soc. ent. Ital., Genova 126 (1): 69-74.
THOMANN, H. (1955): *Codonia* hybr. *kessleri* m. hybr. nov. (Lep., Geom.) = *C. pendularia* Cl. ♂ x *C. pupillaria* Hb. ♀. – Mitt. ent. Ges. Basel 5: 33-35.
THOMPSON, W.R. (1945): A Catalogue of the Parasites and Predators of Insect Pests. Sect. 1 (Parasite Host Catalogue), Part 6 (Parasites of the Lepidoptera). The Imperial Parasite Service. Belleville, Ontario, Canada, 131-258 pp..
THOMPSON, W.R. (1947): A Catalogue of the Parasites and Predators of Insect Pests. Sect. 1 (Parasite Host Catalogue), Part 8 (Parasites of the Lepidoptera). The Imperial Parasite Service. Belleville, Ontario, Canada, 386-523 pp..
THOMSON, E. (1967): Die Großschmetterlinge Estlands. – Helmut Rauschenbusch Verlag, Stollkamm, 203 pp..
THUNBERG, C.P. (1784-1794): Dissertatio Entomologica sistens Insecta Suecica, quorum partem orimam ... publico examini subjicit Joh. Borgstrom. – Uppsala.
THURNER, J. (1938): Die Schmetterlinge der Ochrid-Gegend in Macedonien. – Izvest. tzarsk. prirodonauchni inst., Sofia [Mitt. Königl. naturwiss. Inst. Sofia] 11: 121-179.
THURNER, J. (1967): Lepidopteren aus Morea. Ein weiterer Beitrag zur Fauna der Peloponnes (Griechenland). – Z. Wien. ent. Ges. 52 (Bd 78, 6-8): 50-58.
TIKHONOV, V.V. (1993): Dendrofileous Geometridae of the Northern Caucasus. – Vestnik St. Petersb. Univ. 3 (1): 29-36.
TOMIĆ, D., MIHAJLOVIĆ, L. & M. RISTIĆ (1994): Fauna arborikolnih sovica (Lepidoptera, Noctuidae) i zemljomerki (Lepidoptera, Geometridae) Deliblatskog Peska. – Bull. Nat. Hist. Mus., Belgrade B 48: 147-164.
TONGE, A.E. (1932): The Ova of British Lepidoptera. Pt. III. Geometridae etc. – Trans. & Proc. Ent. & Nat. Hist. Soc. 1932-1933: 11-25.
TORMO, J.E. & B. MUÑOZ (1995): Guía ilustrada de los lepidópteros nocturnos de la provincia de Alicante. – Collección técnica núm. 14. Inst. Cult. Juan Gil-Albert. Diputación Prov. de Alicante, Alicante, 223 pp..
TRÄFF, G. (1965): *Scopula aequicerata* n. sp. (Lep.) nebst lepidopterologischen Beobachtungen im nördlichen Lappland. – Opusc. Ent. 30 (1-2): 131-134.
TREITSCHKE, F. (1827/1828): Die Schmetterlinge von Europa. (Fortsetzung des Ochsenheimer'schen Werkes). – Leipzig, vol. 6.
TRUSCH, R. & B. MÜLLER (2001): Zur Geometridenfauna des Dahar-Berglandes in Süd-Tunesien (Lepidoptera: Geometridae). – Ent. Zeits. Stuttgart 111 (5): 133-140.
TRUSCH, R., MILLER, M., HILLE, A. & A. HAUSMANN (2002): Unraveling the Gordian knot: molecular approach to a new and better understanding of sibling species complexes in geometrid moths. – Proceedings of the 13 European Congress of Lepidopterology, p. 59-60; København (Zoologisk Museum).

TUGWELL (1880): Description of the larva of *Acidalia ochrata*. – The Entomologist 13 (211): 306-312.
TULESHKOV, K. (1951): Faunistichni prouchvanija na pl. Olimp v Grcija (Insecta-Lepidoptera). [Faunistic studies on the Olymp mountain, Greece (Insecta, Lepidoptera)]. – Izv. Zool. Inst. 1: 305-328. (in Bulgarian)
TURATI, C.E. & A. FIORI (1930): Lepidotteri di Rodi. – Mem. Soc. ent. Ital., Genova 9: 196-214.
TURATI, E. & G. KRÜGER (1936): Contributi alla fauna Cirenaica. Novità di Lepidotterologia. – Mem. Soc. ent. Ital. 15: 55-77, pl. 11.
TURATI, E. & V. ZANON (1922): Materiali per una faunula Lepidotterologica di Cirenaica. – Atti Soc. It. Sci. nat. Milano 61: 1-47, 1 pl..
TURATI, E. (1911): Lepidoptera aus Sardinien. – Zeits. f. Insektenbiol. 7: 205-213.
TURATI, E. (1913): Un Record Entomologico. Materiali per una faunula dei Lepidotteri della Sardegna. – Atti Soc. Ital. Sci. nat., Milano, 51 (3-4): 265-365.
TURATI, E. (1914): Contribuzioni alla Fauna d'Italia e descrizione di specie e forme nuove di Lepidotteri. – Atti Soc. It. Sci. nat. Milano 53: 468-487; 533-619, 2 pl..
TURATI, E. (1924): Spedizione lepidotterologica in Cirenaica 1921 - 1922. – Atti Soc. It. Sci. nat. Milano 63: 21-191, 6 pl..
TURATI, E. (1925): Missione Zoologica del. Dr. E. Festa in Cirenaica. – Boll. Mus. Zool. Anat. comp. Univ. Torino 39 (27): 1-9.
TURATI, E. (1926): Novità di Lepidotterologia in Cirenaica. – Atti Soc. It. Sci. nat. Milano 65: 25-83.
TURATI, E. (1927): Novità di Lepidotterologia in Cirenaica. – Atti Soc. It. Sci. nat. Milano 66: 313-344, 1 pl..
TURATI, E. (1930a): Eteroceri di Tripolitania. – Boll. Lab. Zool. gen. agr. Portici, XXIII: 98-128.
TURATI, E. (1930b): Novità di Lepidotterologia in Cirenaica III. – Atti Soc. It. Sci. nat. Milano 69: 1-47, 1 pl..
TURATI, E. (1934): Novità di Lepidotterologia in Cirenaica IV. – Atti Soc. It. Sci. nat. Milano 73: 159-212.
UFFELN, K. (1908): Die Grossschmetterlinge Westfalens, mit besonderer Berücksichtigung der Gegenden von Warburg, Rietberg und Hagen. – Regensbergsche Buchdruckerei, Münster.
UFFELN, K. (1913): Zur Kenntnis von *Acidalia muricata* Hufn.. – Z. wiss. Insektenbiol. 9 (3): 95-96.
UHERKOVICH, A. (1976): Adatok a Del-Dunantul nagylepkefaunajahoz (Makrolepidoptera) [Contribution to the knowledge of the Macrolepidoptera of southern Transdanubia]. – Folia ent. Hung. 29 (1): 119-137.
UHERKOVICH, A. (1981): Data to the Macrolepidoptera Fauna of South Transdanubia (Lepidoptera) II. – Folia ent. Hung. 34 (2): 239-252.
URBAHN, E. & H. URBAHN (1939): Die Schmetterlinge Pommerns mit einem vergleichenden Überblick über den Ostseeraum. – Stettin. ent. Ztg. 100: 185-826.
URBAHN, E. & H. URBAHN (1979): In der DDR verschollene Geometriden (Lep.). – Ent. Ber. Berlin 1979 (2): 57-61.
URBAHN, E. (1935): Zur Kenntnis von *Acidalia (Ptychopoda) inornata* Haw. und *deversaria* H.-S. (Lep. Geom.). – Int. ent. Zeits. 29 (9): 101-105.
URBAHN, E. (1966): Ergebnisse der Albanien-Expedition 1961 des Deutschen Entomologischen Institutes, 51 Beitrag, Lepidoptera: Geometridae. – Beitr. Entom. 16 (3/4): 407-446.
URBAHN, E. (1969): Verzeichnis der Spanner (Geometridae) der DDR. – Ent. Nachr. (Dresden) 13 (7/8): 78-86.
URBAHN, E. (1971): Zur Unterscheidung der europäischen *Cosymbia*-Arten (Lep. Geom.). – Dt. ent. Z., N.F. 18 (1-3): 271-281.
URBAHN, E. (1972): Zunahme von Melanismus-Beobachtungen bei Makrolepidopteren Europas in neuerer Zeit. – Mitt. Münchn. ent. Ges. 61: 1-15.
URBAHN, E. (1973): Bestimmungshilfen für leicht verwechselbare Spanner-Arten. – Entomol. Nachr. 17 (1): 1-10.
VALLE, K.J. (1946): Suurperhoset (Macrolepidoptera). IV Mittarit (Geometrae). – Animalia Fennica 5: 1-370, 22 pls.
VALLETTA, A. (1973): The Moths of the Maltese Islands. – Progress Press, Malta, 118 pp..
VALLHONRAT, F. & E. PETITPIERRE (1985): Contribuciò a la Fauna de Geomètrids de les Belears. – Treb. Soc. Cat. Lep. 7: 39-42.
VALLHONRAT, F. (1980): Geometridae nous o interressants per a la fauna Catalana. – Treb. Soc. Cat. Lepid. 3: 41-50.
VALLHONRAT, F. (1989): Aproximació a la Fauna de Geomètrids de Susqueda i Les Guilleries (Lepidoptera: Geometridae). – Treb. Soc. Cat. Lep. 9 (1988): 23-32.
VALLHONRAT, F. (1999a): Els geomètrids dels Ports de Tortosa-Beseit: noves citacions (Lepidoptera: Geometridae). – Bull. Soc. Cat. Lep. 84: 27-30.
VALLHONRAT, F. (1999b): Alguns geomètrids del massís del Montsant (Lepidoptera: Geometridae). – Bull. Soc. Cat. Lep. 84: 23-26.
VALLHONRAT, F., CERVELLÓ, A., XAUS, A., PÉREZ DE-GREGORIO, J.J., PARÉS, J.M., REQUENA, E., DANTART, J.,

MARTÍ, J., OROZCO, A. & F. PALOU (2002): Introducció a la fauna de lepidòpters de la vall d'Alinyà (Catalunya). – In print.
VARIS, V. (ed.), AHOLA, M., ALBRECHT, A., JALAVA, J., KAILA, L., KERPPOLA, S., & J. KULLBERG (1995): Checklist of the Finnish Lepidoptera - Suomen perhosten luettelo. – Sahlbergia 2 (1): 1-80.
VASILENKO, S.V. (1984): [Some features of the life cycle of *Idaea biselata* (Lepidoptera, Geometridae) in the southern part of the west Siberian plain] (in Russian). – Sibir. Biol. Journ., SO RAI, Novosibirsk 1994: 125-128.
VASILENKO, S.V. (1998a): Die seltenen Spannerarten (Lepidoptera: Geometridae) aus Turkmenistan. Teil II. – Russian Entomol. J. 7 (3-4): 221-224.
VASILENKO, S.V. (1998b): Pyadenitzy (Lepidoptera, Geometridae) Kurganskoj oblasti. In: Bespozvonochnye zhivotnye Juzhnogo Zaural'ja i sopredel'nyh territorij. Mater. Vseross. konf., 14-16 April 1998. – Kurgan: Kurgan State Univ. publ. house: 82-86.
VASILENKO, S.V. (1999): [Addition to the faunistic list of Geometrid moths (Lepidoptera, Geometridae) of the Daurskii State Reserve] (in Russian). – Nasek. Daurii i Sopr. Terr. II, Novosibirsk: 224-227.
VASILENKO, S.V. (2001): [*Scopula subtilata* (Lepidoptera, Geometridae) –new species for the Fauna of Siberia] (in Russian). – Vestnik zool. 35 (5): 89-91.
VIEIRA, V. (1997): Lepidoptera of the Azores Islands. – Bol. Mus. Mun. Funchal, 49 (273): 5-76.
VIGNAL, P. & M. VINTÉJOUX (1986): Liste des Macrolépidoptères de Corrèze. – Misc. ent. 51: 1-30.
VIGNAL, P. & M. VINTÉJOUX (1991): Additions à la liste des Macrolépidoptères de Corrèze (Lepidoptera). – Alexanor 17 (4): 229-231.
VIIDALEPP, J. & H. REMM (1996): Eesti liblikate määraja [A key-book of Estonian Macrolepidoptera]. Tallinn, "Valgus". 444 pp., 40 pls. (in Estonian)
VIIDALEPP, J. & V.P. SOLJANIKOV (1977): [On the geometrid moths (Lepidoptera, Geometridae) of the northern part of the Mongolian People's Republic]. – Nasek. Mongolyi. 5: 620-641 (in Russian).
VIIDALEPP, J. (1974): [Geometrid Moth fauna of Armenia SSR]. – Trudy Zool TGY 8 (1974): 65-85 (in Russian).
VIIDALEPP, J. (1976): Spisok pyadenitz (Lepidoptera, Geometridae) fauny SSSR, I [A list of Geometridae (Lepidoptera) of the USSR, I]. – Entom. Obosr. 55 (4): 842-852 (in Russian).
VIIDALEPP, J. (1977): Spisok pyadenitz (Lepidoptera, Geometridae) fauny SSSR, II [A list of Geometridae (Lepidoptera) of the USSR, II]. – Ent. obozr. 56 (3): 564-576 (in Russian).
VIIDALEPP, J. (1987): [Geometridae fauna Praeamur]. Nasekom. son. BAM 1987: 74-82 (in Russian).
VIIDALEPP, J. (1988): [Geometrid moths of mountainous Middle Asia]. – Moscow, Nauka. 240 pp. (in Russian)
VIIDALEPP, J. (1995a): Catalogus Macrolepidopterorum Estoniae. [Eesti suurliblikate nimestik levikutabelina]. Tallinn-Tartu, ELUS, Abiks Loodusevaatlejale 95. 65 pp..
VIIDALEPP, J. (1995b): Distribution patterns, conservation and bioindicators of Macrolepidoptera in Estonia. – Proc. 9th Internat. Colloquium EIS, Helsinki. WWF Finland Report 7: 60-64.
VIIDALEPP, J. (1996) Checklist of the Geometridae (Lepidoptera) of the former U.S.S.R.. – Apollo Books, Stenstrup, 111 pp..
VIIDALEPP, J.R., KRASILNIKOVA, G.A. & M.A. DARITSCHEVA (1992): [Ecological-faunistical treatment of the Geometrid Moths of Turkmeniya] (in Russian). – Aschabad, 1992: 89-152.
VILLERS DE, C. (1789): Caroli Linnaei Entomologica, Faunae Sueciae descriptionibus aucta ... generum specierumque rariorum iconibus ornata. – Lyon, 4 vols.
VINTÉJOUX, M. (1976): Contribution a l'etude des Lépidoptères de Corrèze (suite et fin). – Alexanor 9 (6): 279-287.
VIVES MORENO, A. (1994): Catálogo sistemático y sinonímico de los Lepidópteros de la Península Ibérica y Baleares (Insecta: Lepidoptera, II). – Minist. Agr., Pesca & Alim., Madrid, 775 pp..
VIVES MORENO, A. (1995): Primera addenda et corrigenda al 'Catálogo sistemático y sinonímico de los lepidópteros de la Península Ibérica y Baleares (Segunda Parte)' (Insecta: Lepidoptera). – Shilap Revta. lepid. 23 (91): 307-341.
VIVES MORENO, A. (1996): Segunda addenda et corrigenda al 'Catálogo sistemático y sinonímico de los lepidópteros de la Península Ibérica y Baleares (Segunda Parte)' (Insecta: Lepidoptera). – Shilap Revta. lepid. 24 (95): 275-315.
VOGEL, G. (1924): Nachtrag zur Schmetterlingsfauna der Provinzen Ost- und Westpreußen. – Beitr. Naturk. Preussens, Königsberg 30: 42-82.
VOJNITS, A. & L. RONKAY (1981): The Geometridae of the Hortobagy National Park (Lepidoptera). – Natur. Hist. Nat. Parks of Hungary, Budapest 1: 197-201.
VOJNITS, A. (1976): Geometrinae and Sterrhinae from Mongolia (Lepidoptera, Geometrinae). – Annls Hist.-Nat. Mus. Nat. Hung. 68: 169-174.
VOJNITS, A. (1977): Archiearinae, Rhodometrinae, Geometrinae II, Sterrhinae II and Ennominae III

(Lepidoptera, Geometrinae) from Mongolia. – Annls Hist.-Nat. Mus. Nat. Hung. 69: 165-175.
VOJNITS, A. (1984): Berichtigungen und Ergänzungen zur Geometridae-Fauna Ungarns. – Articulata II (5): 110-118.
VOLDRICH, M. (1963): La faune des papillons des montagnes de Sumava Centrale. – Acta faun. ent. Mus. Nat. Pragae 9 (75): 5-55.
VOLLRATH, G., BAUER, H. & E. FISCHER (1981): Die Grossschmetterlinge des Fichtelgebirges. Eine Bestandsaufnahme. – Naturwiss. Ges. Bayreuth 13: 163-186.
VORBRODT, C. (1928): Die Schmetterlinge von Zermatt. – Dt. ent. Z. Iris 42 (1/2): 7-80; 81-130.
VORBRODT, C. (1930): Tessiner und Misoxer Schmetterlinge. – Mitt. Schweiz. ent. Ges. 14 (6): 199-396.
VORBRODT, K. & J. MÜLLER-RUTZ (1914): Die Schmetterlinge der Schweiz. Bd 2, 726 pp., Bern.
VORBRODT, K. (1921): Die Schmetterlinge der Schweiz (4. Nachtrag). – Mitt. Schweiz. ent. Ges. 13 (3-4): 175-206.
VORBRODT, K. (1925): Die Schmetterlinge der Schweiz (5. Nachtrag). – Mitt. Schweiz. ent. Ges. 13 (8): 434-466.
WAGNER, F. (1929): Weiterer Beitrag zur Lepidopteren-Fauna Inner-Anatoliens. – Mitt. Münchn. ent. Ges. 19 (10-12): 175-206 & Taf. II.
WAGNER, F. (1930): Zweiter (III). Beitrag zur Lepidopteren-Fauna Inner-Anatoliens. – Int. ent. Zeits. 23: 545-559.
WAGNER, F. (1935): *Cinglis andalusiaria* sp. nov.. – Z. Öst. Ent.-Ver. 20 (3): 17-19, 36.
WARDIKJAN, S.A. (1955): [Genitalia of Transcaucasian geometrid moths (Lepidoptera, Geometridae)]. – Entomol. Obosr. 34: 252-274 (in Russian).
WARDIKJAN, S.A. (1970a): [Geometrid moths of the genus *Sterrha* Hb. in Armenia SSR (Lepidoptera, Geometridae) I]. – Akad. Nauk Arm. SSR 23 (10): 5-20 (in Russian).
WARDIKJAN, S.A. (1970b): [Geometrid moths of the genus *Scopula* Schrank in Armenia SSR (Lepidoptera, Geometridae)]. – Akad. Nauk Arm. SSR: 5-20 (in Russian).
WARDIKJAN, S.A. (1985): [Atlas of the genitalia of geometrid moths] Atlas genital'nogo apparata pyadenitz (Lepidoptera, Geometridae). – Armjanskoj SSR, Erevan, AN Arm. SSR: 5-135 (in Russian).
WARING, P. & M. TOWNSEND (2003): Field Guide to the Moths of Great Britain and Ireland. – British Wildlife Publishing, Hampshire, 432 pp..
WARLET, J.M. (1968): Une seconde génération de *Scopula tessellaria* Boisd. – Lambillionea 67 (1-2): 1-4.
WARNECKE, G. (1916): Die Geometriden-Fauna Schleswig-Holsteins, IIA. – Int. ent. Zeits. 10 (17): 94-96.
WARNECKE, G. (1933): Über südwesteuropäische Faunenelemente in der Macrolepidopteren-Fauna des Mittel-Rheingebietes. I. *Ephyra (Cosymbia) lennigiaria* Fuchs. – Ent. Zeits. 47 (7): 53-67.
WARNECKE, G. (1936a): *Acidalia (Ptychopoda) ochrata* Hb. (Lepid. Geom.) auf der ostfriesischen Insel Borkum. – Mitt. Dt. ent. Ges.6 (5-10): 61-63.
WARNECKE, G. (1936b): *Acidalia emutaria* Hb. (Lep. Geom.) an den Küsten der Nordsee. – Mitt. Dt. ent. Ges.7 (4/5): 61-63.
WARNECKE, G. (1936c): Die Grossschmetterlinge der nordfriesischen Insel Sylt, 2. – Ent. Rdsch. 53 (25): 345-350.
WEGNER, H. (1979): Faunistisch bemerkenswerte Beobachtungen 342. (Lep. Geom.). – Bombus 2 (65): 257-260.
WEGNER, H. (1984): Neue und bemerkenswerte Funde aus unserem Faunengebiet. – Bombus 2 (72): 285-286.
WEGNER, H. (1986): Zur Demotopfrage von *Scopula emutaria* 399. (Lep. Geom.). – Bombus 2 (74): 295-298.
WEGNER, H. (1996): Neue und besonders bemerkenswerte Großschmetterlingsbeobachtungen der letzten 15 Jahre in Schleswig-Holstein. – Bombus 3 (17-20): 71-75.
WEGNER, H. (1998a): Die Geometriden-Arten Nordwestdeutschlands. – Eine Auflistung des aktuellen und historischen Artenbestandes in den Bundesländern Schleswig-Holstein einschließlich Hamburg (SH/HH) und Niedersachsen einschließlich Bremen (NS/HB). – Bombus 3 (35-38): 137-149.
WEGNER, H. (1998b): Ein weiterer Beitrag zur Aktualisierung der Geometridenfauna im Faunengebiet und in Niedersachsen. – Bombus 3 (35-38): 149-150.
WEGNER, H. (2000): *Idaea ochrata* (Scopoli, 1763) hat sich im südöstlichen Faunengebiet angesiedelt. – Bombus 3 (45-46): 184.
WEHRLI, E. (1924): Nice - St.Martin Vésubie - Digne. Ein Beitrag zur Lepidopterenfauna der Alpes Maritimes und der Basses Alpes. – Dt. ent. Z. Iris 38: 50-57; 59-98.
WEHRLI, E. (1926): Ein Streifzug in die andalusischen Gebirge. Beitrag zur Geometriden-Fauna Andalusiens. – Dt. ent. Z. Iris 40 (2/3): 113-129.
WEHRLI, E. (1927a): Einige interessante Acidalien und Eupithecien aus den Sierra Nevada (Esp.). – Int. ent. Zeits. (Guben) 37: 340-346.

WEHRLI, E. (1927b): Ein weiterer Streifzug in die andalusischen Gebirge. 22.VI. bis 13.VII.1926. Beitrag zur Geometridenfauna Andalusiens. (Lepid.). – Dt. ent. Z. Iris 41 (1): 49-80.

WEHRLI, E. (1931a): Über einige nordafrikanische Geometriden (Lepid. Heterocera). – Int. ent. Zeits. 24 (41): 425-444.

WEHRLI, E. (1931b): Einige neue paläarktische Geometriden aus Syrien, Algerien und Sicilien (Lepid. Het.). – Mitt. Münchn. ent. Ges. 21: 41-46.

WEHRLI, E. (1934a): In: OSTHELDER, L. & E. PFEIFFER. Lepidopterenfauna von Marasch in türkisch Nordsyrien. – Mitt. Münchn. ent. Ges. 24 (2): 25-55.

WEHRLI, E. (1934b): Ueber unbekannte alte und neue Sterrhinae (Acidaliinae). (Lepid.Het.). – Ent. Z. Frankf. a.M. 47 (19): 156; (20): 159-161; (21): 166-167.

WEIDNER, H. (1993): Bestimmungstabellen der Vorratsschädlinge und des Hausungeziefers Mitteleuropas. – Gustav Fischer Verlag Stuttgart - Jena - New York, 328 pp..

WEIGT, H.-J. (1982): Lepidoptera Westfalica. – Abh. Westf. Mus. Naturkunde 44: 3-111.

WEST, B.K. (1974): Discovery of Larvae and Natural Foodplant of Least Carpet, *Idaea vulpinaria* Herrich-Schäffer (*rusticata* sensu auct.) (Lep.: Geometridae). – Ent. Rec. 86: 258-259.

WILLIAMS, H.B. (1950): On a hereditary abnormality in the shape of the wings in Lepidoptera. – Ent. Gazette 1 (2): 109-113.

WILTSHIRE, E.P. (1939a): In: ELLISON, R.E. & E.P. WILTSHIRE. The Lepidoptera of the Lebanon: with notes on their season and distribution. – Trans. Royal ent. Soc. London 88 (1): 1-56.

WILTSHIRE, E.P. (1939b): More notes on Kurdish Lepidoptera. The Entomologist's Rec. and J. Var. 51: 97-138.

WILTSHIRE, E.P. (1945): 70 new records of Lepidoptera from Iran. – The Entomologist's Rec. and J. Var. 57: 77-85.

WILTSHIRE, E.P. (1948): Middle East Lepidoptera, IX: Two new forms or species and thirty-five new records from Cyprus. – The Entomologist's Rec. and J. Var. 60: 79-87.

WILTSHIRE, E.P. (1949a): The Lepidoptera of the Kingdom of Egypt, Pt. 2. – Bull. Soc. Fouad Ier d'Ent. 33: 381-457.

WILTSHIRE, E.P. (1949b): Some more new records of Lepidoptera from Cyprus, Iraq and Iran. – The Entomologist's Rec. and J. Var. 61: 73-76.

WILTSHIRE, E.P. (1957): The Lepidoptera of Iraq. Nicholas Kaye Limited. Government of Iraq (Ministry of Agriculture), 162 pp. & 17 pls.

WILTSHIRE, E.P. (1966): Österreichische entomologische Iran-Afghanistan-Expeditionen, Beiträge zur Lepidopterenfauna, Teil 9, Subfamilie Sterrhinae (Lepidoptera, Geometridae). Middle East Lepidoptera XXII. – Z. Wien. ent. Ges. 51 (77) (9-11): 113-138.

WILTSHIRE, E.P. (1967): Middle East Lepidoptera XX. – Beitr. naturk. Forsch. SW.Deitschl. 26 (3): 137-169.

WILTSHIRE, E.P. (1970): Österreichische Entomologische Expeditionen nach Persien und Afghanistan. – Ann. Naturhist. Mus. Wien 74: 371-386.

WILTSHIRE, E.P. (1990): An Illustrated, Annotated Catalogue of the Macro-Heterocera of Saudi Arabia. – Fauna of Saudi Arabia 11: 91-250.

WOJTUSIAK, J. (1966): Fauna makrolepidopterologiczna Wzniesienia Gubalowskiego i jej charakterystyka zoogeograficzna [Macrolepidopterous Fauna of the Gubalowka Elevation and its Zoogeographical Charakteristic]. – Acta zool. Cracov. 11 (4): 75-109 + 1 tab..

WOLF, W. & H. HACKER (2004): Rote Liste gefährdeter Nachtfalter (Lepidoptera: Sphinges, Bombyces, Noctuidae, Geometridae) Bayerns. – Schriftenreihe Bayer. LfU 186: 217-233.

WOLF, W. (1981): Die sogenannten Grossschmetterlinge des nordöstlichen Bayreuther Umlandes (Insecta, Lepidoptera). Ein Beitrag zur Kenntnis des Obermain-Hügellandes. – Ber. naturwiss. Ges. Bayreuth 17: 145-254.

WOLF, W. (1988a): Systematische und synonymische Liste der Spanner Deutschlands unter besonderer Berücksichtigung der Denis & Schiffermüller'schen Taxa (Lep., Geometridae). – Neue ent. Nachr. 22, 1-78.

WOLF, W. (1988b): Geometridae. Prodromus der Lepidopterenfauna Nordbayerns, ANE e.V. Neue ent. Nachr. 22/23: 95-107.

WOLFF, N.L. (1964): The Lepidoptera of Greenland. – Meddr. Grønland (København) 159 (11): 1-74, 21 pls.

WOLFSBERGER, J. (1945-1949): Neue und interessante Macrolepidopterenfunde aus Südbayern und den angrenzenden nordlichen Kalkalpen (1st contribution). – Mitt. Münchn. ent. Ges. 35-39: 308-329.

WOLFSBERGER, J. (1950): Neue und interessante Macrolepidopterenfunde aus Südbayern und den angrenzenden nordlichen Kalkalpen (2nd contribution). – Mitt. Münchn. ent. Ges. 40: 207-236.

WOLFSBERGER, J. (1954-1955): Neue und interessante Macrolepidopterenfunde aus Südbayern und den angrenzenden nordlichen Kalkalpen (4th contribution). – Mitt. Münchn. ent. Ges. 44-45: 300-347.

WOLFSBERGER, J. (1955): Neue und interessante Macrolepidopterenfunde aus Südbayern und den angrenzenden nordlichen Kalkalpen (3rd contribution). – NachrBl. bayer. Ent. 3: 5-7; 13-21.

WOLFSBERGER, J. (1958): Neue und interessante Macrolepidopterenfunde aus Südbayern und den angrenzenden nordlichen Kalkalpen (5th contribution). – NachrBl. bayer. Ent. 7: 49-62; 65-72.

WOLFSBERGER, J. (1960): Neue und interessante Macrolepidopterenfunde aus Südbayern und den angrenzenden nordlichen Kalkalpen (6th contribution). – Mitt. Münchn. ent. Ges. 50: 35-54.

WOLFSBERGER, J. (1966): Die Macrolepidopteren-Fauna des Gardaseegebietes. – Estratto dalle Mem. Mus. Civ. Stor. Nat. Verona Vol. XIII, 1-385.

WOLFSBERGER, J. (1971): Die Macrolepidopteren-Fauna des Monte Baldo in Oberitalien. – Museo Civico di Storia Naturale di Verona, Memorie fuori serie N. 4, 1-335.

WOLFSBERGER, J. (1974): Neue und interessante Macrolepidopterenfunde aus Südbayern und den angrenzenden nordlichen Kalkalpen (7th contribution). – NachrBl. bayer. Ent. 23 (3): 33-56.

WOLFSBERGER, J., (1975): Die Macrolepidopteren-Fauna des Gardaseegebietes, 1. Nachtrag. – Boll. Mus. Civ. Stor. Nat. Verona I, 1974: 167-193.

WOLLRATH, G., BAUER, H. & E. FISCHER (1966): Die Grossschmetterlinge des Fichtelgebirges. – Naturw. Ges. Bayreuth, Bericht 12 /1964-1966/: 45-164.

WOODBRIDGE, F.C. (1906): *Acidalia immorata*. – Entomologist 39: 231-232.

WOODFORDE, F.C. (1902): The north Staffordshire form of *Zonosoma (Ephyra) pendularia*). – Entomologist 35: 275-276.

YLLA, J., TURET, J. & J. GARCIAMORENO (1990): La família Geometridae al massís de Cabrera (Osona). – Treb. Soc. Cat. 10: 27-43.

YOUNG, C. (2003): Characterisation of the Australian Nacophorini and a Phylogeny for the Geometridae from Molecular and Morphological Data. – Doctoral thesis, University of Tasmania, Hobart, 773 pp..

ZAHM, N. (1989): Lepidopteren aus der Oasi di Serranella (Mittelitalien, Abruzzen). – Boll. Ass. Romana Entomol. 43 (1988): 1-9.

ZANGHERI, P. (1969): Repertorio della Flora e Fauna della Romagna, 3 (Regno Animale). – Mem. Mus. Civ. Storia nat. Verona, Fuori ser. 1: 855-1414.

ZANGHERI, S. (1960): Richerche faunistiche e zoogeografiche sui Lepidotteri delle Puglie e della Lucania. – Mem. Soc. Ent. Ital. 39: 5-35.

ZANGHERI, S. (1963): Considerazioni sulla fauna Lepidotterologica dei massicci montani della Calabria. – Arch. Bot. Biogeogr. Ital. 39: 1-23.

ZANGHERI, S. (1975): La Lepidotterofauna delle Prealpi Venete sulla base della Collezione A.Ancilotto. – Boll. Ist. ent. Univ., Bologna 31: 215-250.

ZECEVIC M. & S. RADOVANOVIC (1974): Leptiri Timocke Krajine (makrolepidoptera), prilog poznavanju faune leptirova Srbije. – Stamparija Bor, Zajecar, 185 pp..

ZELLER, P.C. (1847a): Verzeichniß der vom Professor Dr. Loew in der Türkei und Asien gesammelten Lepidoptera. – Isis (Oken) 1847 (1): 3-39.

ZELLER, P.C. (1847b): Bemerkungen über die auf einer Reise nach Italien und Sicilien beobachteten Schmetterlingsarten. – Isis (Oken) 1847 (7): 481-522.

ZELLER, P.C. (1855): Die Lepidopteren in Scopoli's Entomologica Carniolica. – Stett. ent. Ztg. 16: 233-257.

ZELLER, P.C. (1869): Beiträge zur Naturgeschichte der Lepidopteren. – Stettin. ent. Ztg. 1869: 401-429.

ZERNY, H. (1927a): Die Lepidopterenfauna von Albarracin in Aragonien. – EOS 3 (3-4): 299-488.

ZERNY, H. (1927b): Die Lepidopteren-Fauna von Algeciras und Gibraltar in Süd-Andalusien. – Dt. ent. Z. Iris 41 (2): 83-146.

ZERNY, H. (1933): Lepidopteren aus dem nördlichen Libanon. – Dt. ent. Z. Iris 47: 60-109.

ZERNY, H. (1934): Neue Heteroceren aus dem Grossen Atlas in Marokko. – Z. Österr. Ent.-Ver. 19 (5): 33-39; (6): 43-48; (7/8): 52-53, Taf. 5-7.

ZERNY, H. (1936): Die Lepidopterenfauna des Grossen Atlas in Marokko und seiner Randgebiete. – Mém. Soc. Sci. nat. Maroc 42: 1-163 & 2 pls.

ZHANG, B.-C. (1994): Index of Economically Important Lepidoptera. – CAB International, Wallingford, 599 pp..

ZHU, J.W., RYRHOLM, N., LJUNGBERG, H., HANSSON, B.S., HALL, D., REED, D. & C. LÖFSTEDT (1996): Olefinic acetates, DELTA-9,11-14: OAc and Δ7,9-12:OAc used as sex-pheromone components in three geometrid moths, *Idaea aversata*, *I. straminata* and *I. biselata* (Geometridae, Lepidoptera). – J. Chem. Ecol. 22: 1505-1526.

ZHURAVLEV, S.M. (1910): Contribution sur la faune des Lépidoptères des environs d'Ouralsk et d'autres de la province de l'Oural. – Horae Soc. ent. Ross. 39: 415-463.

ZILLI, A., MALTZEFF, P., PINZARI, M. & V. RAINERI (2001): I Lepidotteri della tenuta presidenziale di Castelporziano (Lepidoptera). – Boll. Ass. Romana Ent. 56 (1-4): 13-48.

Index to scientific names
of the Systematic Account, Vol. 2

The index gives page reference to the main treatment. Numbers in semibold refer to species numbers and figures.